International Congress on Energy 2017

Topical Conference at the 2017 AIChE Annual Meeting

Minneapolis, Minnesota, USA
29 October – 3 November 2017

Volume 1 of 2

ISBN: 978-1-5108-5776-6

Printed from e-media with permission by:

Curran Associates, Inc.
57 Morehouse Lane
Red Hook, NY 12571

Some format issues inherent in the e-media version may also appear in this print version.

Printed by Curran Associates, Inc. (2018)

For permission requests, please contact AIChE
at the address below.

AIChE
120 Wall Street, FL 23
New York, NY 10005-4020

Phone: (800) 242-4363
Fax: (203) 775-5177

www.aiche.org

Additional copies of this publication are available from:

Curran Associates, Inc.
57 Morehouse Lane
Red Hook, NY 12571 USA
Phone: 845-758-0400
Fax: 845-758-2633
Email: curran@proceedings.com
Web: www.proceedings.com

TABLE OF CONTENTS

VOLUME 1

VOLUME 2

Some abstracts in this book refer to images that were not made available by the publisher

505003 Introductory Remarks by Nada Anid

Abstract:

PAIC aims to identify, respond to, and communicate public-policy issues at local, national, and international levels. It also aids AIChE members in learning how to communicate better with government and the public. In light of this mission, the goal of the Town Hall meeting is 1) to share the findings of three PAIC Task Forces and get the membersâ€™ input on each and on other matters of importance to the AIChE membership and 2) to consider the best means of communication to enhance the standing of the AIChE community and express its stand on policy matters.

504000 Food-Energy-Water Issues

Abstract:

The overlapping needs and technologies for food, energy, and water call for contributions from many people around the world, but chemical engineers bring especially valuable expertise. This briefing will summarize the issues and preview the topical conference that is taking place this week.

504002 Advanced Manufacturing

Abstract:

New â€œAdvanced Manufacturingâ€ approaches are expected to impact the production of chemical products significantly. Recent advances include process intensification, manufacturing for sustainability, new sensors and actuators, data science, and high-level simulations. We will discuss what advances and further developments need to be communicated, why they should be adopted, and how they can be communicated effectively.

504001 Climate-Change Review and Adaptation

Abstract:

The PAIC is leading an AIChE review of climate change and our policy on it. AIChE members are participating in a series of online Engage discussions, each introduced by an article posted on ChEnected. By drawing from the knowledge, dedication, and passion of AIChE members, AIChE can shape a Climate-Change Policy that captures membersâ€™ concerns while helping our profession to grapple with this issue.

505002 PAIC Town Hall

Abstract:

In this Town-Hall-style, moderated discussion, the emphasis will be on identifying other critical public-policy issues where AIChE and chemical engineers can have impact. Attendees are encouraged to submit questions beforehand to the organizers by email or at the session. The goal is to shape action paths and find participants who are interested in engaging with each other and the public on these issues.

499972 Bioprocess Model and Economic Analysis of Microalgae Production in Flat-Panel Photobioreactors Taking into Account Geospatial Factors

Abstract:

Microalgae are a promising feedstock for production of fuels and value-added chemicals. In order to assess the economic, energetic and environmental performance of algae based system in an industrial scale biorefinery, it is important to model algae productivity accurately. In this work, a first principle based growth model of microalgae is developed for flat panel photobioreactors taking into account the transport and biological processes. The primary factors that affect microalgae growth, such as solar irradiance, temperature, specific characteristics of species, uptake rates of nutrients and carbon dioxide, are incorporated in the bioreactor process model. Year-round algae productivity potential is computed for different locations to elucidate the implications of climate variations for the locations. Based on the yearly microalgal productivity for different regions, economic analysis is conducted to quantify the microalgae production cost for the different locations. As there are significant uncertainties associated with the different parameters of the growth model and economic analyses, detailed sensitivity analyses are performed. Results show that microalgae production cost is heavily influenced by climatic factors like solar irradiance and air temperature as well as by different equipment and process design factors. Results from the dynamic model can be used to optimize and scaling up of algae systems for efficient production of fuels and chemicals for variety of algae species as well as geospatial locations.

497873 Continuous Photobioreactor Cultivation of *Nannochloropsis oculata* to Isolate Cosmetical Grade Phospholipids

Abstract:

Microalgae are a promising source of natural ingredients as their cell mass contains polysaccharides, proteins, lipids, pigments, and vitamins. Phospholipids, in particular, are of interest to the cosmetics industry for various formulations, such as emulsifiers, liposome formers, solubilizers, wetting agents, and bioactive compounds. They are currently extracted from food sources, but this practice raises sustainability questions because of the competition with food. Microalgae can serve as a source of phospholipids on a more sustainable basis. To advance the status of microalgae bioproducts for cosmetics applications, the marine microalga *Nannochloropsis oculata* was cultivated in a vertical flat panel photobioreactor (VFPPBR) in batch and continuous mode. Batch cultivation was employed to optimize the growth parameters of *N. oculata*, calculate its maximum specific growth rate (μ_{max}), and determine its phospholipid productivity. The measured μ_{max} of 0.015 hr^{-1} was then used to select the dilution rate for designing and operating a VFPPBR-based continuous cultivation that is deemed more scalable and productive from a commercialization standpoint. Cell harvesting was followed by organic solvent extraction to solubilize and recover the algal phospholipids. The experimental findings in continuous cultivation of *N. oculata*, the robustness of the strain, and its polar lipid productivity, bioreactor designs, performance of the strain in batch and continuous reactors, and green pathways for downstream processing to isolate, concentrate and fractionate algal phospholipids were analyzed and discussed.

498781 Effects of pH on Cell Growth, Lipid Production and CO_2 Demand of Microalgae *Chlorella sorokiniana*

Abstract:

Microalgae have emerged as one of the most promising alternative energy feedstocks. Some advantages include the simple cellular structure, short production cycle, high lipid content, and fast growth. However, high production costs due to high CO_2 usage and low lipid productivity have been one of the major challenges impeding the commercial production of algal biofuels, Here, cell growth and lipid contents of *Chlorella sorokiniana* DOE 1412 were tested at different pH in flask cultivation. Culture pH was manipulated by CO_2 addition. The optimal pH for DOE1412 is 5.8-6.0 when only taking into account the cell growth and lipid production. A flat panel airlift photobioreactor (PBR) was used for scale-up cultivation at five different pH levels (6.5, 7.0, 7.5, 8.0 and 8.5). Data of pH value and the CO_2 addition was collected by a data logger. Biomass productivity increased with decreasing pH. By taking into account not only the cell growth and lipid production but also CO_2 demand, the lowest value of g CO_2/g dry biomass was achieved at pH 8. Also, the fatty acid profiles and biodiesel properties in different culture pH were investigated. Cetane number (CN) of biodiesel in the pH 6.5, 7 and 7.5 treatments satisfied the US standard ASTM D6751; among them, the pH 6.5 treatment met the Europe standard EN 14214. Finally, protein content in algae biomass increased with increasing pH, while C/N ratio in cells decreased.

484990 Design of Marine Macroalgae Photobioreactor Integrated into Building with Natural Sun Illumination

Abstract:

Seagriculture, which can provide offshore grown macroalgae biomass for food, platform chemicals and biofuels production will play a significant role in bioeconomy and biorefineries. Nevertheless, seagriculture development has been hindered by the lack of laboratory photobioreactors that enable fundamental macroalgae research. Here we report on the indoor macroalgae photobioreactor (MPBR) system integrated into the LEED certified building that enables maintenance and cultivation of marine macroalgae at controlled environment under natural illumination. Depending on specific reactor location, illumination varied between reactors in 238-348 μmole photons m^{-2} s^{-1} . The total direct energy consumption of the system was 8.15-11.45kWh during the 24 hours of operationThe total operation volume of the system was 6,400 L of seawater (salinity 3.9%, pH 8.2). Initial water stock was provided from the Tel Aviv Reading marine area and was partially (1/4-1/3) exchanged every 3 months. The compensation for evaporation was done by adding deionized water to the system to maintain the salinity at 3.9%.

The MPBR operation was been demonstrated for 6 months on continuous maintenance and cultivation of three green macroalgae species *Cladophora sp.*, *Ulva compressa* and *Ulva rigida* isolated form the Eastern Mediterranean coast. The growth rates, protein, rhamnose, xylose, arabinose, glucose, galactose and glucuronic acid content of the cultivated species were quantified. *Cladophora sp.*, isolated from Rosh-ha-Nikra sampling site was cultivated for 122 days. The growth rates changed from -3.66 %day^{-1} to 9.78 %day^{-1}, with an average 3.47 %day^{-1}. The maximum growth was for 12 continuous days. *Cladophora sp.* isolated from Mikhmoret sampling site was cultivated for 122 days. The growth rates changed from 9.32%day^{-1} to 95.5 %day^{-1}, average 36.63 %day^{-1}. The maximum growth was for 12 continuous days. *Cladophora sp.* isolated from Reading sampling site was cultivated for 97 days. The growth rates were stable at 95.47 %day^{-1} during the entire period of 38 cultivation days.

Carbohydrate profiling was done for *Cladophora sp.* harvested at Mikhmoret site and cultivated at the system at the growing rate of 63.9%$^{-1}$ during harvesting. Rhamnose content was 3.89 μg mg_{DW}^{-1}, arabinose 30.46 μg mg_{DW}^{-1}, galactose 26.59 μg mg_{DW}^{-1}, glucose 61.72 μg mg_{DW}^{-1}, xylose 17.79 μg mg_{DW}^{-1} and glucuronic acid 0.98 μg mg_{DW}^{-1}. The total carbohydrate yield was 14.10%. The protein content in the biomass was 11%.

U. compressa isolated from Rosh-ha-Nikra sampling site was cultivated for 187 days. The growth rates changed from -0.06% day^{-1} to 3.5 %day^{-1}, average 1.01 %day^{-1}. The maximum growth was for 9 continuous days. *U. compressa* isolated from Mikhmoret sampling site was cultivated for 172 days. The growth rates changed from 1.46 %day^{-1} to 5.51 %day^{-1}, average 3.65 %day^{-1}.The maximum growth was for 27 continuous days. *U.compressa* isolated from Reading sampling site was cultivated for 143 days. The growth rates were mostly negative, with the biomass loss at -82.34 %day^{-1}, growth rate of only 0.83 %day^{-1} was observed 40 continuous days.

Carbohydrate and protein profiling was done for *U. compressa* harvested at Rosh-ha-Nikra site and cultivated at the system at the growing rate of 3.5% day^{-1} at the time of a harvest. Rhamnose content was 48.58 μg mg_{DW}^{-1}, arabinose 0.29 μg mg_{DW}^{-1}, galactose 9.03 μg mg_{DW}^{-1}, glucose 75.13 μg mg_{DW}^{-1}, xylose 23.83 μg mg_{DW}^{-1} and glucuronic acid 18.13 μg mg_{DW}^{-1}. The total carbohydrate yield was 17.50%. The total protein content of this biomass was 25.5%.

U. rigida isolated from Rosh-ha-Nikra sampling site was cultivated for 89 days. The growth rates changed from -3.13 %day^{-1} to 19.51 %day^{-1}, average 4.61 %day^{-1}. The maximum growth was for 16 continuous days. *U.* rigida isolated from Haifa sampling site was cultivated for 155 days. The growth rates changed from -0.09 %day^{-1} to 9.4 %day^{-1}, average 2.01 %day^{-1}. The maximum growth was for 64 continuous days.

Carbohydrate and protein profiling was done for *U. rigida* harvested at Haifa site and cultivated at the system at the growing rate of 1.4%/day at the time of a harvest. Rhamnose content was 37.29 μg mg_{DW}^{-1}, arabinose 0.46 μg mg_{DW}^{-1}, galactose 3.60 μg mg_{DW}^{-1}, glucose 54.84 μg mg_{DW}^{-1}, xylose 9.11μg mg_{DW}^{-1} and glucuronic acid 16.62 μg mg_{DW}^{-1}. The total carbohydrate yield was 12.20%. The protein content from the biomass harvested was 29%.

This work provides detailed insights on the design of practical indoors photobioreactors for isolation, selection, cultivation maintenance and analysis of macroalgae species required for biorefinery feedstock development.

488865 An Integrated Approach for Bioenergy Production from Microalgae Using Solar Energy, CO_2 and Wastewater

Abstract:

New biomass sources for alternative fuels has grown to be a subject of increasing importance as the world struggles to resolve the economic and strategic impacts of limited fossil fuel resources on environment and global climate. Microalgae are among the most promising non-food-crop-based biomass feedstocks. Microalgae are photosynthetic microorganisms with simple growing requirements (light, CO_2 and nutrients) that can produce lipids, proteins and carbohydrates in large amounts over short periods of time. These products can be processed into both bioenergy and valuable co-products. On the other hand, carbon dioxide (CO_2) is one of the most important contributors for the increase of the greenhouse effect. CO_2 concentrations are increasing in the last decades mainly due to the increase of emissions due to human activities. A promising technology is the biological capture of CO_2 using microalgae. These microorganisms can fix CO_2 using light with efficiency ten times greater than terrestrial plants.

To be economically competitive with carbon capture/sequestration methodologies, an intensive research is needed to integrate all microalgal culture benefits: flue gas and wastewater treatment and biomass production. In this research, an integrated process is introduced to produce bioenergy from microalgal biomass combined with simultaneous CO_2 capture/sequestration and bioremediation of wastewater nutrients. Also, this research investigates different processes and operations to optimize microalgal biomass production using open ponds or photobioreactors. Finally, this research focuses on the potential of producing biogas from microalgal biomass through anaerobic digestion and further upgrading of biogas into higher alkenes through catalytic processes

The main objective of this research is to investigate the growth kinetics of the appropriate microalgae species at different feed composition of CO_2 with air to optimize the growth rate of microalgae at maximum CO_2 consumption and in various wastewater media to reduce wastewater nutrients concentrations (COD, ammonia, nitrate and phosphate) to the potable clean water level. Also, this research aims to study the growth kinetics of the microalgae at other important parameters such as the effect of concentrated solar energy, pH, temperature, wastewater dilution and algae circulation rate. The anaerobic digestion of the harvested algae is investigated at different operating conditions such as liquid/solid ratio, temperature and pH. Methane concentration in the outlet stream will be optimized as function of these operating conditions.

There is a number of challenging issues in the realization of the proposed technology. One of the major challenges is to find appropriate microalgae species tolerant to high concentrations of CO_2. The flue gas of a power plant contains up to 500 times higher CO_2 content than that in the atmosphere. Therefore, different microalgae species will be studied under similar composition of CO_2 in the flue gas from a fossil fuel-based power generating station to find the proper type of microalgae adapting to higher CO_2 concentrations. Another major challenge is to find the proper technology for microalgal biomass production. There are two main technologies; open ponds and photobioreactors. Even that the open ponds are relatively cheap and easy to clean and maintain, they have poor biomass productivity, mixing, light and CO_2 utilization and they require large area of land. On the other hand, photobioreactors have different designs such as tubular, flat plate and column. Usually, the photobioreactors produce high microalgal biomass. Therefore, this research will explore different designs of open ponds and photobioreactors to optimize the growth of desired microalgae species. Finally, integrated processes propose different challenges that need to be faced. One of the powerful tools to meet these challenges is modelling and simulation software. For example, an essential part of any integrated process is heat and energy integration. Aspen simulation packages proved their capabilities to design and simulate such integration. In this research, modelling and simulation packages will be used to optimise the overall proposed integrated process along with analysing and modelling collected data from the experimental work.

It is expected that the innovation and implementation of the integrated approach for microalgal bioenergy production CO_2 will offer a solution of providing sustainable energy demand and addressing the global commitment to reduce greenhouse gases emissions by CO_2 capture/sequestration.

490777 Identification and Reconstruction of Pathways for Lignin Catabolism

Abstract:

Cellulosic biofuel production yields a substantial lignin byproduct stream that currently has few applications. Biological conversion of lignin compounds into chemicals and fuels has the potential to improve the economics of cellulosic biofuels. Microbial strains have been isolated with the ability to specifically degrade challenging lignin dimers, and the associated pathways could substantially improve the conversion of lignin into valuable biochemicals. However, no dimer degradation pathway has been characterized to the level necessary for reconstruction in a heterologous host, largely due to the combined challenges of fully identifying the relevant enzymes and then optimizing the resulting pathways for heterologous function. We have used a barcoded transposon library to rapidly identify entire catabolic pathways in genetically-intractable microbes such as the lignin-degrading strain *Novosphingobium aromaticivorans*. Simultaneously, we are reconstructing pathways for lignin catabolism in *Escherichia coli* to enable pathway characterization in a tractable and well-defined host. Evolutionary optimization of these heterologous pathways identified factors limiting pathway activity and will allow the design of characterized, optimized synthetic operons for lignin dimer degradation.

499050 Investigating New Pathways and a 'Funneling' Strategy to Enhance Production of Chemicals Derived from Shikimic Acid Biosynthesis

Abstract:

The engineering of non-natural pathways is a promising to approach to developing renewable and sustainable alternatives to many conventional petrochemicals. This is true of a multitude of useful chemicals that can be synthesized from precursors of native shikimic acid biosynthesis, including various aromatics and other molecules. For example, several recent studies have demonstrated how *Escherichia coli* can be engineered for the efficient production of muconic acid which, in turn, can be used as a precursor to adipic acid – a platform chemical used for the synthesis of many different polymers (e.g., nylon 6,6). To date, no fewer than five distinct pathways have been reported for muconic acid production from glucose, each involving different precursors of the shikimic acid pathway and enzyme chemistries. This unique flexibility (i.e., the ability to synthesize the same product via multiple different pathways) arises in part due to the fact that muconic acid is also a naturally-occurring metabolite that arises as a key intermediate in the beta-ketoadipate pathway, used by various microbes to degrade a range of aromatic compounds. Further exploiting this principle, here we demonstrate how three additional routes to muconic acid can be constructed by linking a series of recently-engineering pathways for phenol biosynthesis with its subsequent, partial degradation. The resultant phenol-dependent pathways support muconic acid production at up to 0.5 g/L in shake flasks. We also report a novel, four-step pathway for muconic acid production from endogenous chorismate via p-hydroxybenzoate which supports muconic acid titers of up to about 1 g/L while circumventing auxotrophic limitations experienced via the original, 3-hydroshikimate-derived route. Moreover, as inspired by the mechanisms of 'biological funneling' employed by soil microbes for the simultaneous degradation of diverse aromatic mixtures, we investigate the potential of a synthetic 'funneling' strategy for enhancing achievable muconic acid titers and yields. More specifically, by co-expressing multiple pathways that stem from different endogenous precursors before then converging on a common intermediate, muconic acid yields have been enhanced by as much as 40% relative to the best performing, single pathway control. In addition to maximizing the availability of precursors, this novel strategy offers a more holistic approach to balancing pathway performance and host fitness. Analogous 'funneling' approaches have similarly been extended to investigate the general utility of this strategy for improving the production of aromatic products, including phenol, catechol, and others.

498494 Engineering of ADP1 for Production of Vanillin from Lignin

Abstract:

Vanillin is one of the most widely used natural flavoring agents in the world. Its popularity exceeds the supply from traditional plant extraction. Therefore, chemical synthesis has been used to meet global demand. Although this has successfully improved vanillin availability and pricing, growing consumer preferences for “natural” ingredients have steered food and agriculture firms to instead rely on biologically derived vanillin in their products. This forces the need for a creative solution for vanillin products to be both natural and affordable. Biotechnological approaches have been used to meet this need, with glucose and ferulic acid used as starting materials for a biological conversion to vanillin to date. Here, we look to harness the bacteria ADP1's natural ability to degrade lignin monomers to produce vanillin from biomass. Our approach involves engineering ADP1 by deleting ADP1's natural vanillic acid and protocatechuate degradation pathways and by introducing a catechol o-methyltransferase to convert protocatechuate to vanillic acid. In addition, we look to introduce a decarboxylase enzyme to convert vanillic acid to vanillin, to potentially provide an affordable and renewable source of natural vanillin.

Dow AgroSciences

C4-C9 Alcohols Production By '+1' Pathway Engineered E. Coli: Evaluation of Vector Expression Systems and Fermentation Process Conditions

Authors: Mamatha Devarapalli, Paresh Sanghani, Chris Stowers, Sarah Delaplane, Ryan Hill and Devon Rosenfeld

Solutions for the Growing World

Renewable Approaches for C4-C9 Alcohol Production

- C4-C9 linear alcohols are commonly used as solvents and perfumes
- C4-C9 alcohols can also be reduced into alkanes and alkenes to serve as precursors for other molecules
- Petrochemical processes for medium-chain alcohol (C4-C9) production are costly due to high catalyst cost and poor selectivity
  - Odd chains are even more challenging
- Three main pathway options
  - Fatty acid synthesis
  - Reverse beta oxidation (pioneered by R. Gonzalez lab at Rice)
  - **'+1 Pathway'** (pioneered by J. Liao at UCLA)
    - Branch chain alcohols
    - **Linear alcohols**

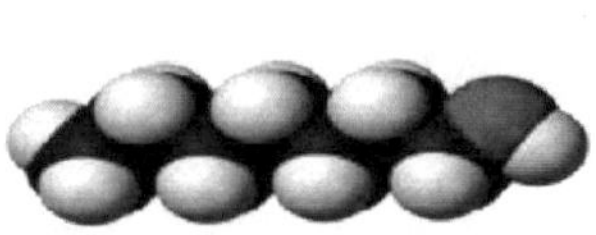

Dow AgroSciences | 2

Background on the LeuABCD "+1 Pathway"

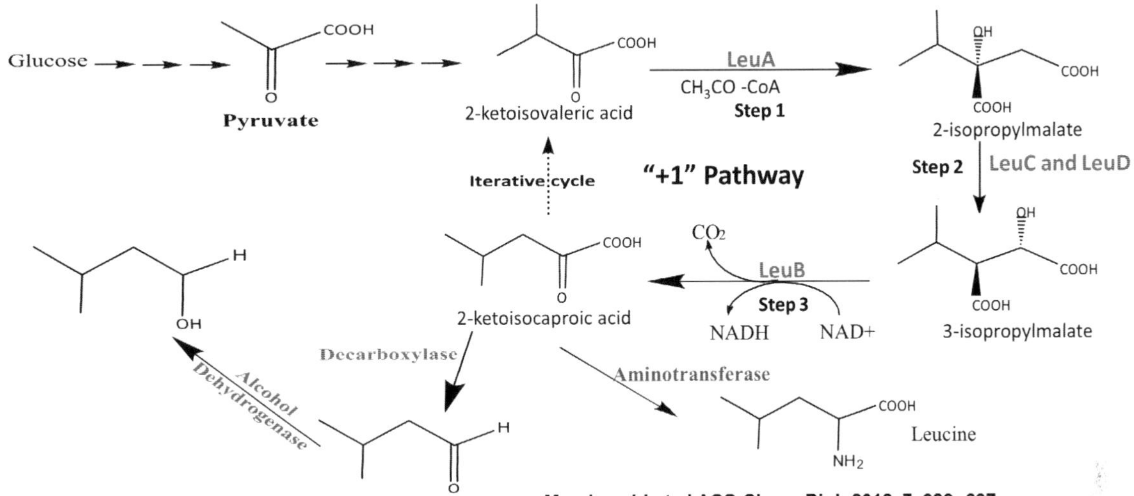

Marcheschi et al ACS Chem. Biol. 2012, 7, 689–697

Dow AgroSciences

| 3

"+1 Pathway' Pathway Development Strategy

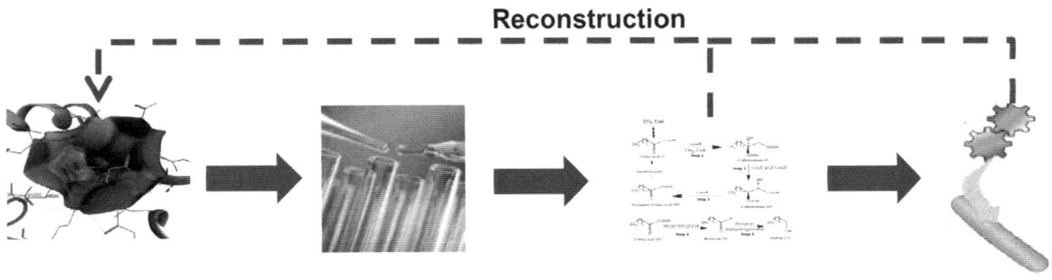

In silico Enzyme Design
- **Protein crystallography**
- **Active site analysis**

In vitro Enzyme Characterization
- **Kinetics k_{cat}, k_M**
- **Substrate specificity**
- **pH dependence**

In vitro Pathway Evaluation
- **Substrate inhibition**
- **Pathway kinetics**
- **Carbon balance**

In vivo Pathway Characterization
- **Titer**
- **Overall yield**
- **Byproduct analysis**
- **Fermentation Optimization**

Dow AgroSciences

| 4

Strain Engineering to Drive Flux Towards '+1 Pathway' LCAs

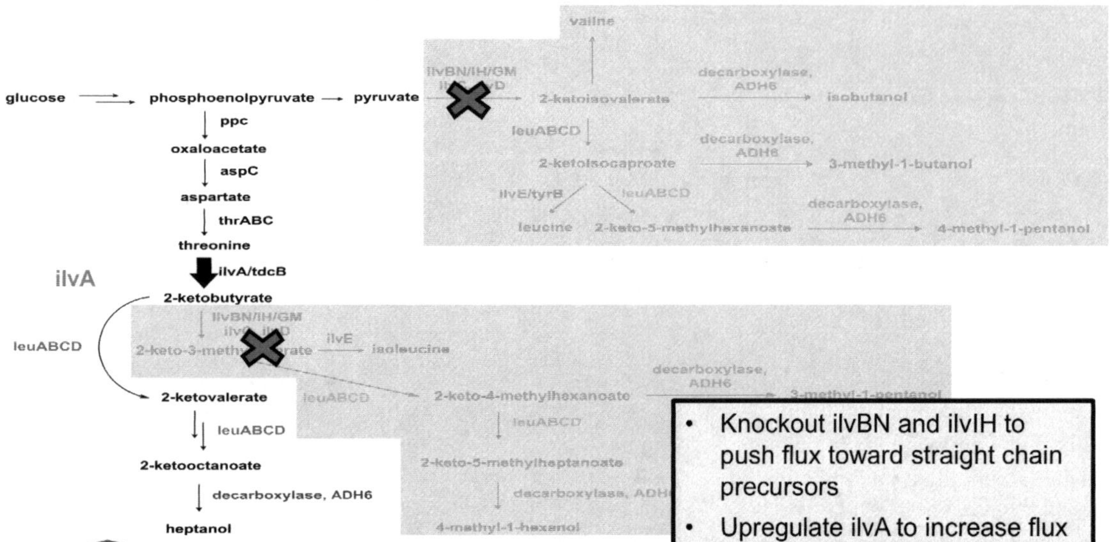

- Knockout ilvBN and ilvIH to push flux toward straight chain precursors
- Upregulate ilvA to increase flux toward 2-ketobutyrate

+1 Pathway Expression in Duet Vectors

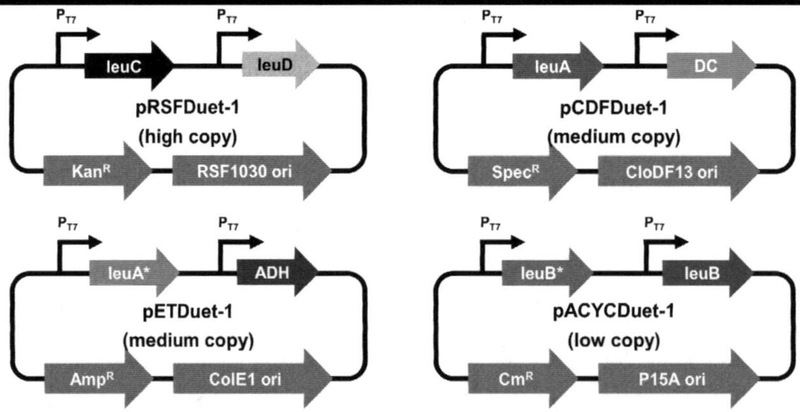

- Engineered leuA (leuA*), leuA, engineered leuB (leuB*), leuB, leuC, leuD, DC, and ADH6 in *E. coli* using the duet vector system
- Plasmid copy number can be varied to tune pathway performance

Modified LeuB Shifts Alcohol Distribution to Longer Chain Length

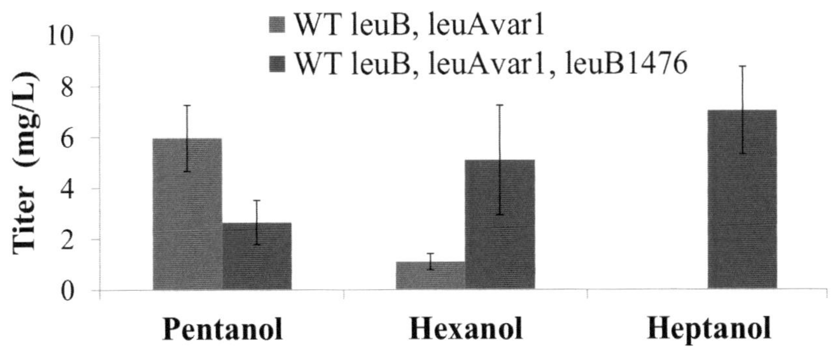

- Initial shake flasks experiments accumulated only 15 mg/L of long-chain alcohols
- Modified LeuB seemed to shift selectivity to longer chain lengths
- **Titers were far lower than compared to published results**

Dow AgroSciences | 7

Liao Construct Design

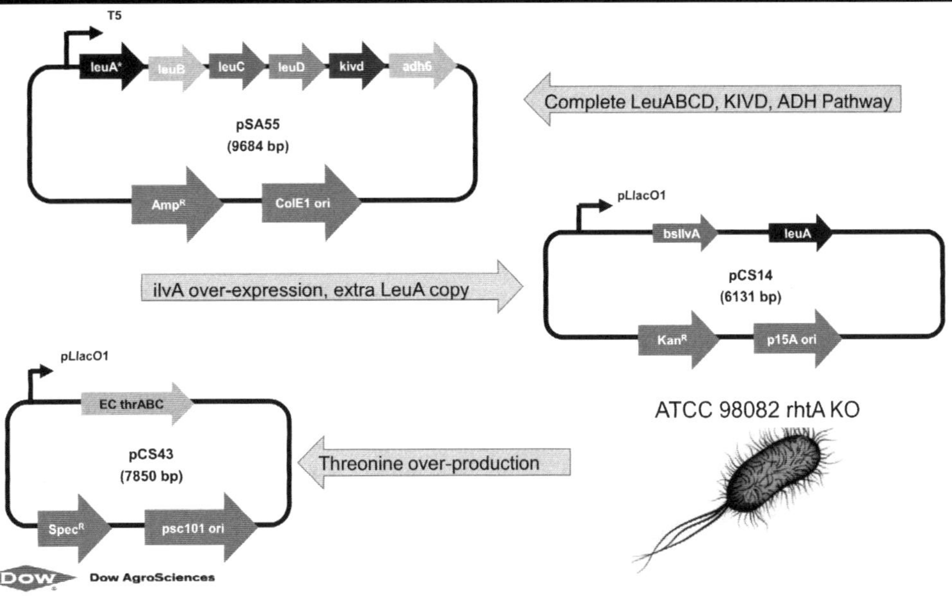

Dow AgroSciences | 8

Total Alcohol Titer

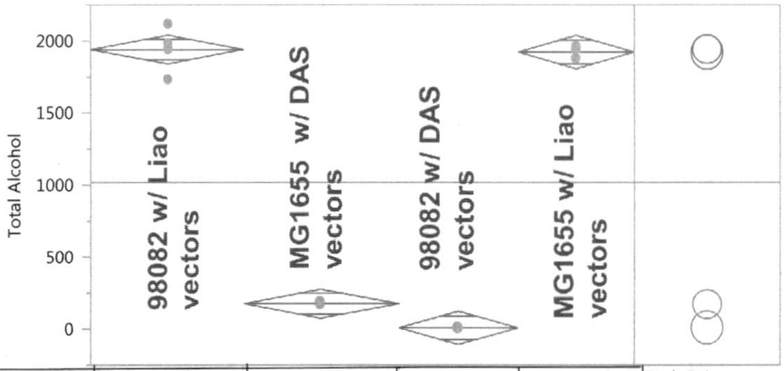

Strain	Liao	DAS	Liao	DAS
Vectors	Liao	DAS	DAS	Liao

Lower performance in DAS strains seem to be due to the expression strategy (vectors).

Proteomics Evaluation of Liao vs. DAS Vector System

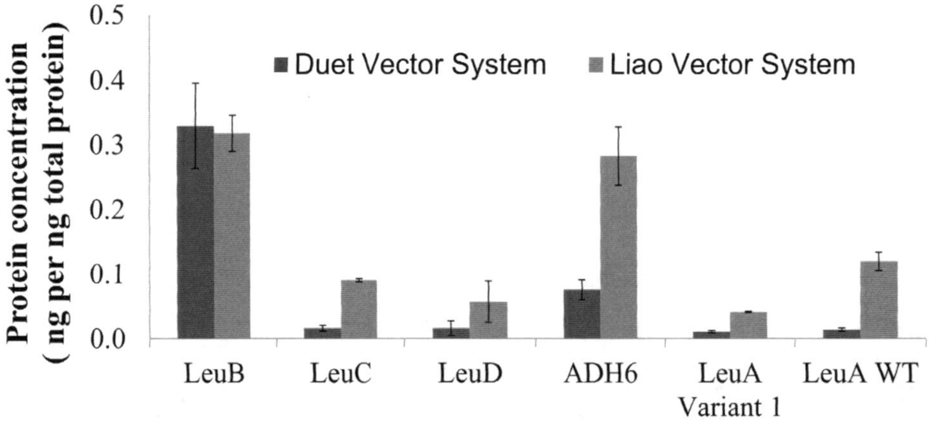

Surprisingly the Liao vector system had higher expression levels of several key enzymes in the pathway

MG1655 3-vector vs. 2-vector Expression System

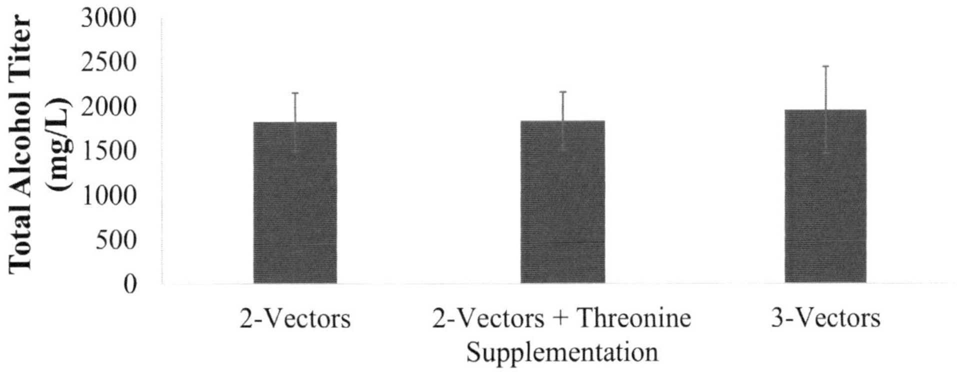

- Elimination of the ThrABC over-expression vector seemed to have little impact on performance.
- Threonine supplementation seemed to have no effect.
- No effect on alcohol distribution was observed.

Is both LeuA and ilvA over-expression needed?

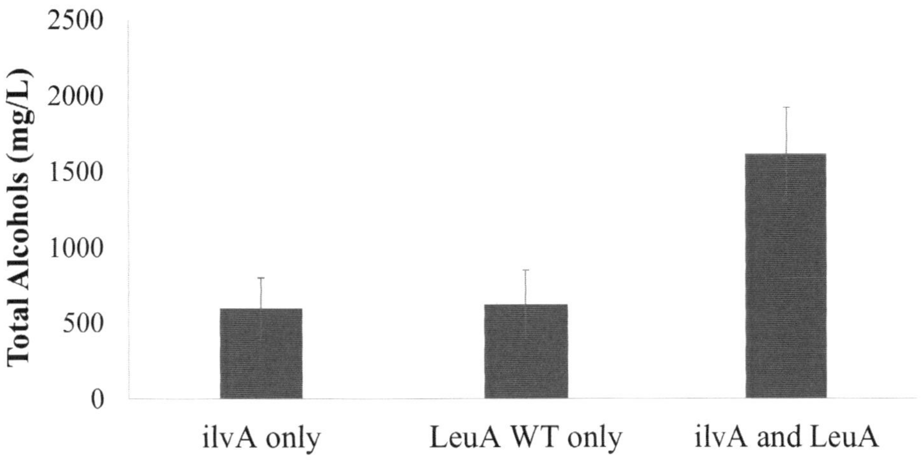

Both IlvA and LeuA are needed in order to achieve high alcohol titers

Metabolomic Characterization Reveals Enzymatic Limitations

- Large amount of 2-KB and short-chain 2-alkyl malates detected with threonine feed
- Accumulation of 2-alkymalates is consistent with a LeuCD limitation
- Drop off in metabolite concentration at increased chain length could indicate LeuA limitation

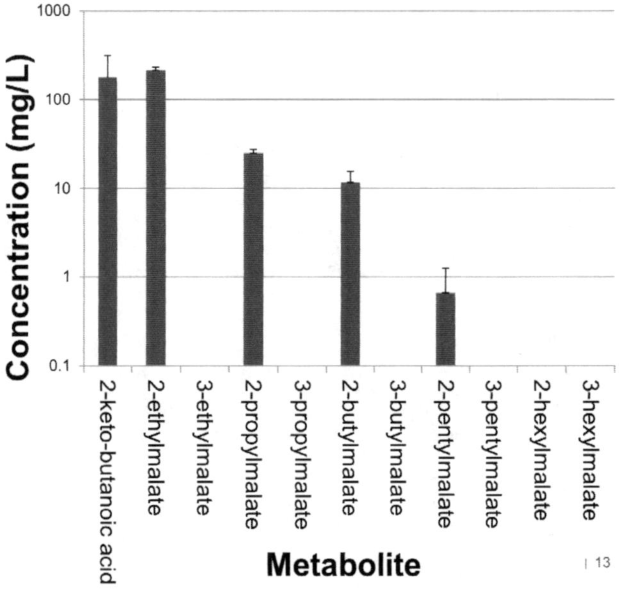

2-HM Feeding

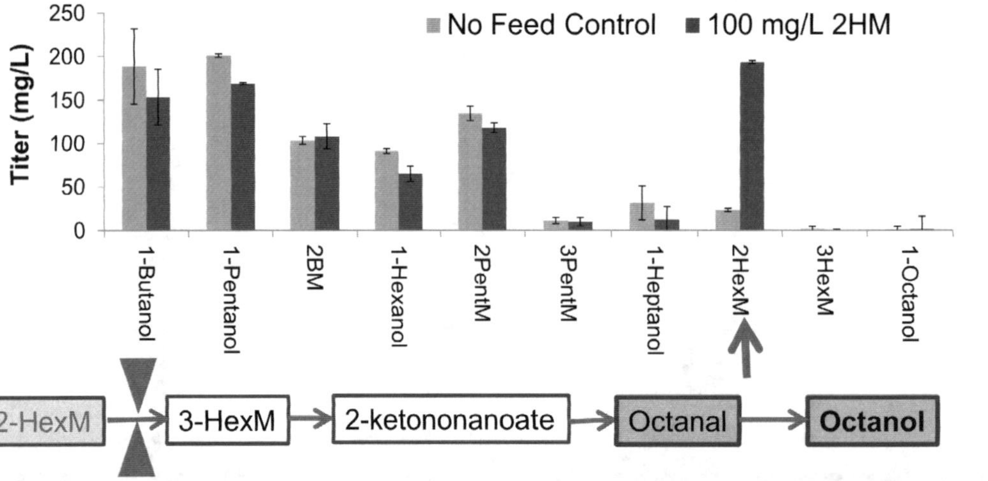

No conversion is observed when 2-HexM is fed indicating LeuCD bottleneck

3-HM Feeding

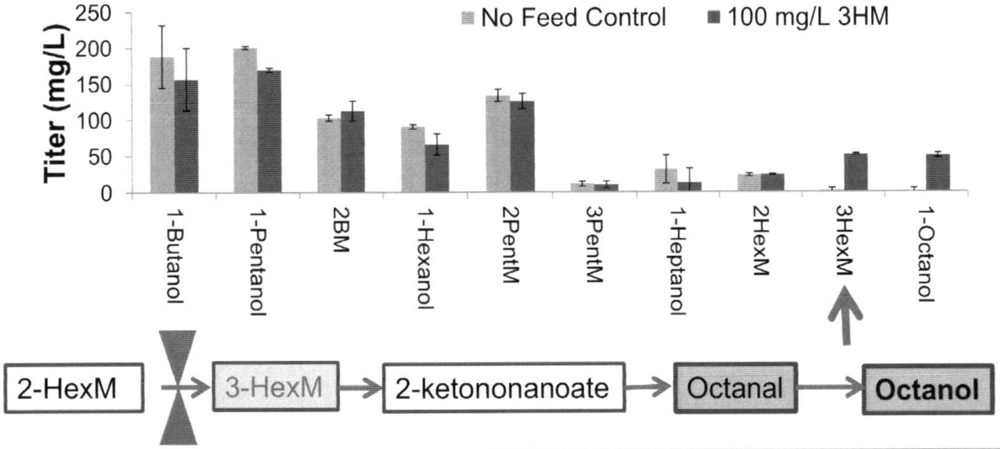

Half of the fed 3-HexM converted into octanol. Near perfect mass balance occurred between residual 3-HexM and octanol.

C8 Keto acid Feeding

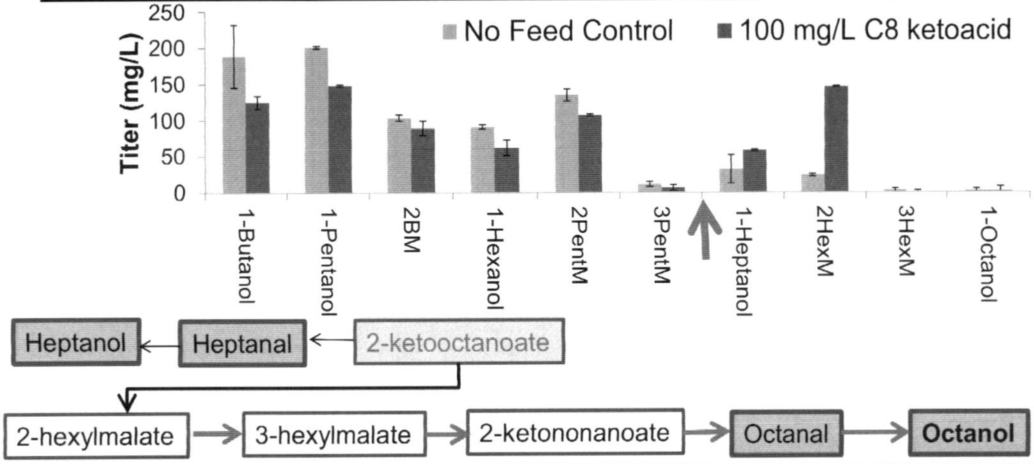

An increase in heptanol and 2-HM was observed when C8 keto acid was fed. Most of the carbon is trapped in 2-HM indicating LeuCD limitation

pH effect on alcohol production- Bioreactor studies

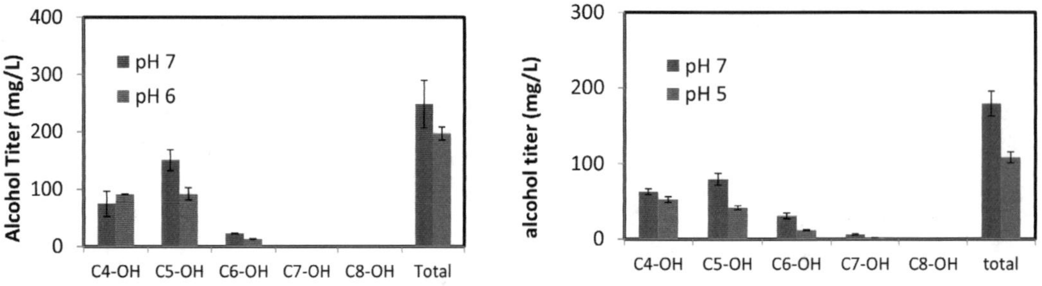

Alcohol production was higher at pH 7 followed by pH 6 and then pH 5

Effect of induction time and post induction temperature

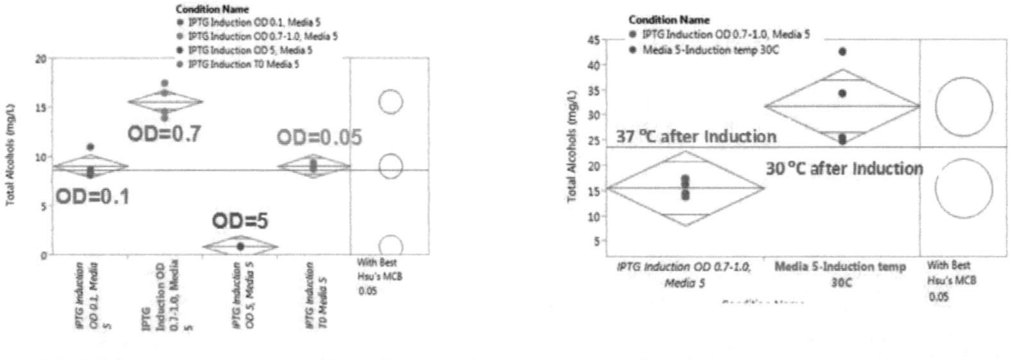

Lowering temperature to 30 °C post-induction and inducing the E.coli cultures at 0.7- 1.0 OD(600 nm) showed higher production of total alcohols

Conclusions

- '+1 Pathway' is functional for long-chain alcohols production
- Improved selectivity to longer chain lengths can be achieved by engineering LeuB, PPDC and LeuCD, but work remains to shift the alcohol distribution more towards C6-C9
- 'Omics' is a powerful tool to guide strain development and enzyme engineering.
  - ➢ Proteomics quickly demonstrated the underlying cause for performance differences in two expression systems
  - ➢ Metabolomics illuminated pathway bottlenecks and guided future enzyme engineering
- Fermentation conditions such as pH, induction timing and post-induction temperature has significant impact on the total alcohol production

Dow AgroSciences | 19

Acknowledgements

- Mike Harris
- Scott Greenwalt
- Bryan Ward
- Matt Roach
- Matt Robinson
- Sowji Garapati
- Pam Torkelson
- Mark Mikola
- Allison Lutocka
- Samantha Hall

Dow AgroSciences | 20

HTP Enzyme Characterization of LeuB

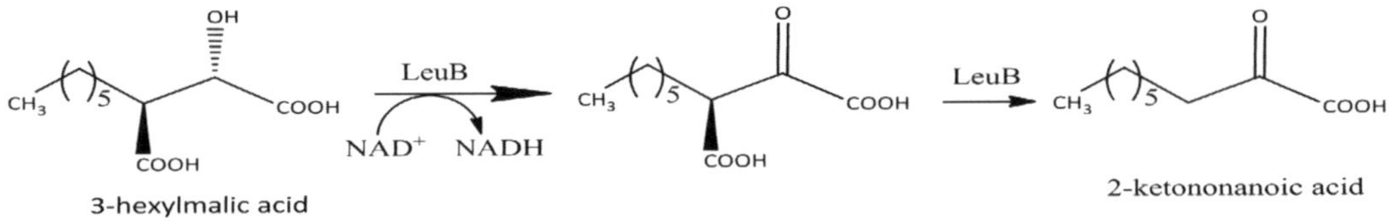

Substrates used for evaluating LeuB variants:

Substrate	Cycle Number	Product	Terminal Product
3-isopropylmalic acid (3-IPM)	1	2-ketoisocaproate (C6)	Isopentanol
3-butylmalic acid (3-BM)	3	2-ketoheptanoate (C7)	Hexanol
3-hexylmalic acid (3-HM)	5	2-ketononanoate (C9)	Octanol

Activity of Wild Type LeuB Against Native and Non-native Substrates

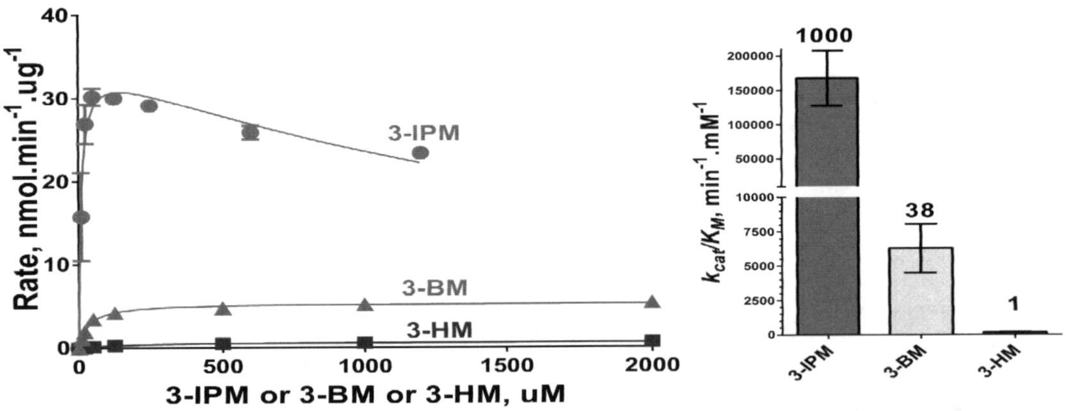

LeuB requires modification to improve activity towards longer substrates

Active Site Model of LeuB (Homology Modeling)

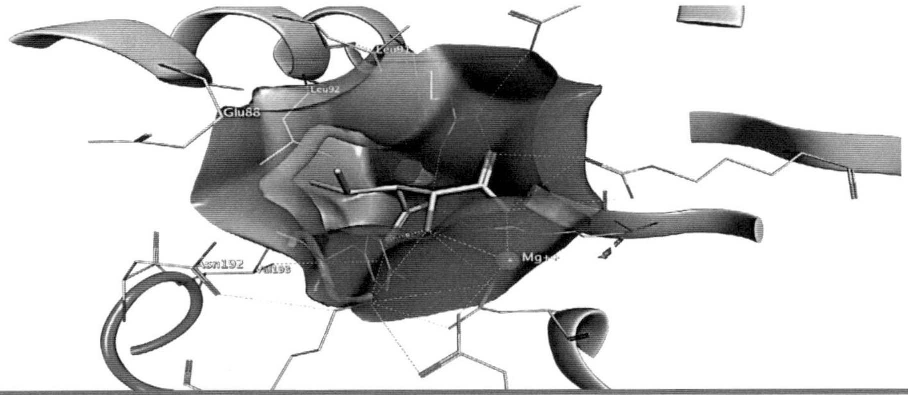

29 LeuB variants designed by rational engineering (P. Sanghani)

LeuB variants with improved activity against 3-HM

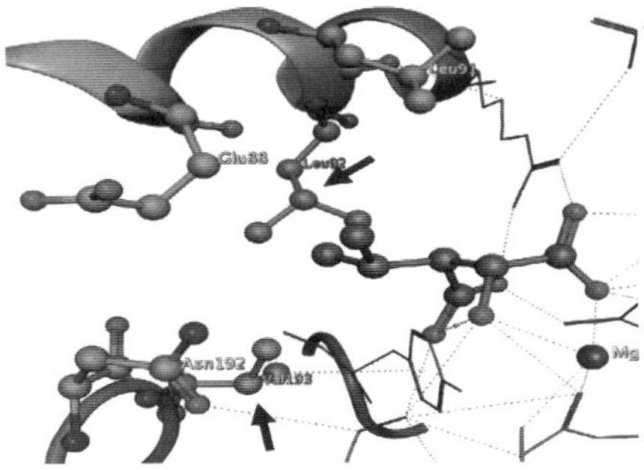

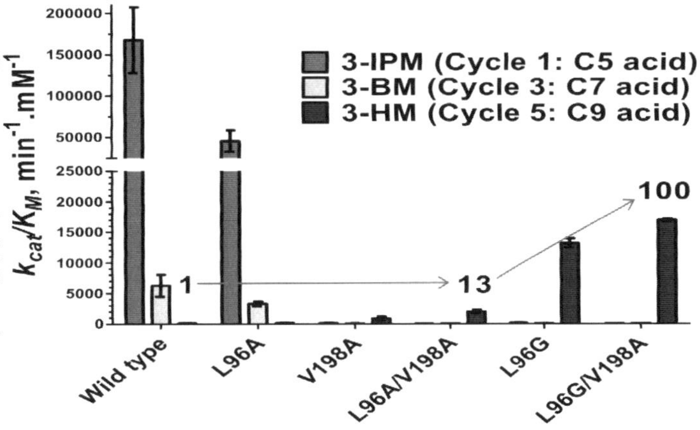

Pathway Validation in vitro Using Engineered Enzymes

- LeuB and decarboxylase enzymes engineered to improve selectivity to long chain length alcohols
- In vitro pathway evaluation methods developed
- Pathway performance evaluated at varying enzyme concentrations

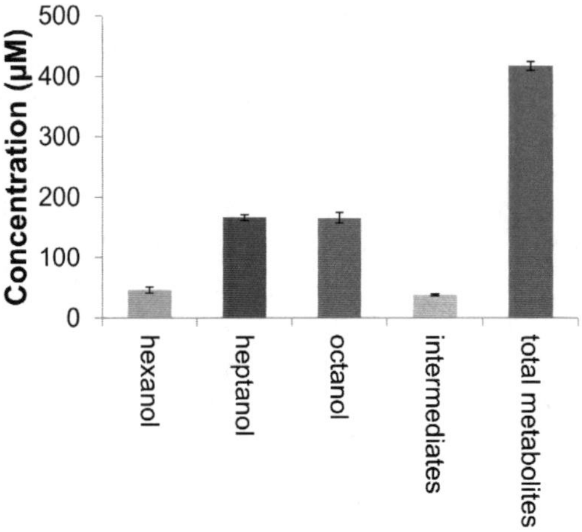

2-KB + Acetyl CoA + NADPH → Alcohols + Intermediates

Liao Results w/ the LeuACBD "+1 Pathway"

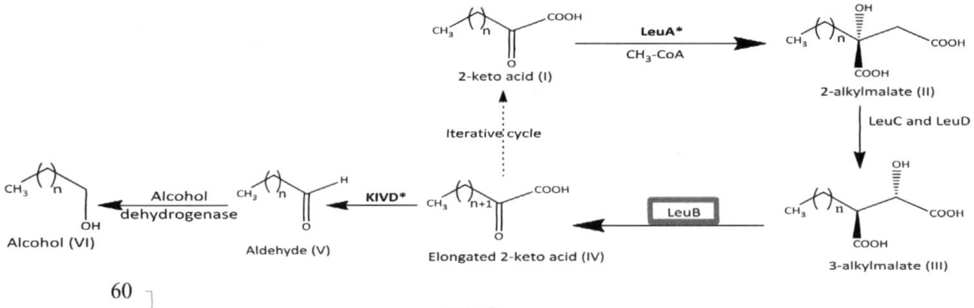

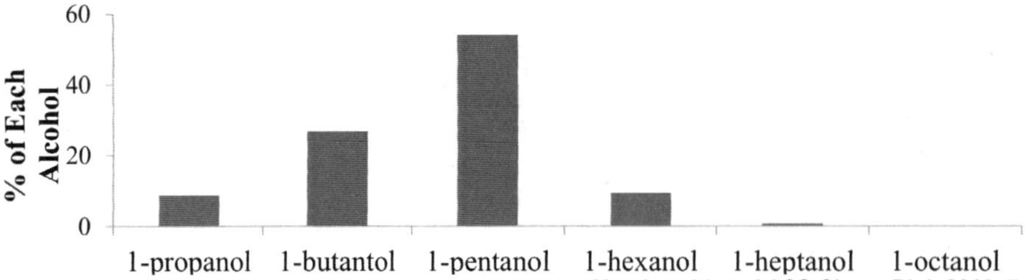

Marcheschi et al ACS Chem. Biol. 2012, 7, 689–697

Metabolomic Characterization of Pathway to Identify Pathway Bottleneck

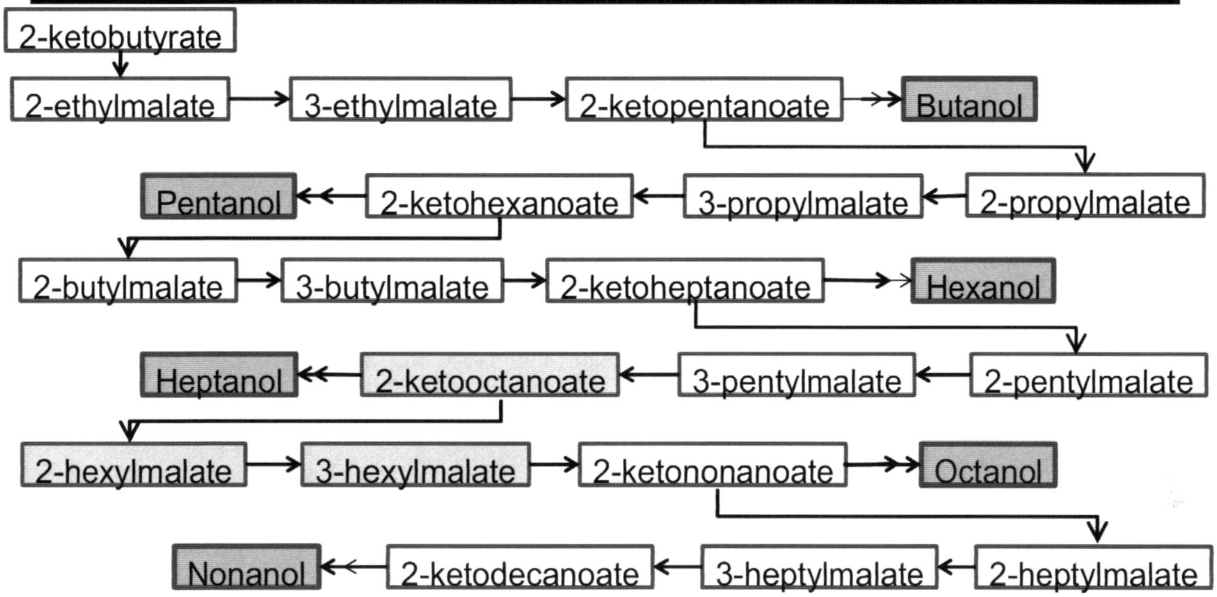

493164 Establishing a Platform *Escherichia coli* Strain to Generate Xylose-Derived Value-Added Products

Abstract:

Xylose is the most abundant C5 sugar in lignocellulosic biomass and also represents a source of carbon from non-edible feedstocks. Here, we report the construction of a platform *Escherichia coli* strain for the production of high value compounds via the nonphosphorylative xylose metabolism. Firstly, a 3,4-dihydroxybutanal over-producing pathway was constructed by efficient enzymes screening and host strain engineering. Then two 3,4-dihydroxybutanal dehydrogenases were identified to efficiently convert 3,4-dihydroxybutanal into 3,4-dihydroxybutyric acid (3,4-DHBA), which is the hydrolyzed form of 3-Hydroxy-Î³-butyrolactone (3HBL). 3HBL is one of the top value-added building block for synthesis of various drugs and nutraceuticals. This novel 3,4-DHBA biosynthetic pathway produced 1.27 g/L of 3,4-DHBA in shake flasks, which is the highest titer reported so far. The application of this platform was further demonstrated by building an artificial pathway for biosynthesis of 1,4-butanediol (1,4-BDO). The 3,4-dihydroxybutanal over-producing strain was transformed into a 1,2,4-butanetriol (1,2,4-BTO) over-producing strain by over-expression of endogenous alcohol dehydrogenase, which enabled 1.5 g/L 1,2,4-BTO produced from xylose. Then the *Klebsiella oxytoca* diol dehydratase was engineered to achieve non-native catalysis of 1,2,4-BTO into 1,4-BDO by a series of rational protein engineering strategies. Those efforts enabled 209 mg/L 1,4-BDO produced via a novel metabolic route from xylose. This work demonstrates great potential for large-scale production of 3,4-DHBA and sets an example to build novel biosynthetic pathways via rational protein engineering.

500133 Understanding and Enhancing Bioconversion of Acetic Acid to Biodiesel Using *Yarrowia lipolytica*

Abstract:

Acetate is an inexpensive and renewable carbon source that can be generated from various sources such as pyrolysis of lignocellulosic biomass and anaerobic digestion of fermentable wastes. The oleaginous yeast *Yarrowia lipolytica* is a promising biocatalyst that can utilize acetate and convert it into triacylglycerides (TAGs), a precursor for green diesel. However, currently there is limited knowledge on the metabolism of *Y. lipolytica* when cultured on acetate. To develop a better understanding, we used sodium acetate as the sole carbon source and performed ^{13}C-MFA on two different strains as well as two fermentation phasesâ€"growth (nitrogen present) and lipid production (nitrogen depleted and cells no longer synthesize biomass). The resolved flux distributions demonstrated that the glyoxylate shunt pathway is constantly active in both strains and phases. Furthermore, the cells were able to divert an adequate amount of glyoxylate shunt flux to gluconeogenesis through the PEP-pyruvate node. In this way, the oxidative pentose phosphate pathway remained active even when the cells no longer divide to generate sufficient lipogenic NADPH for TAG accumulation. Based on these conclusions, we also demonstrated that the enhancement of acetate utilization and conversion into TAGs can be achieved by providing the cells with another substrate to increase flux through upper glycolysis.

504091 Metabolic Engineering of Bacteria for Production of Oleochemicals

Abstract:

Finding a sustainable alternative for todayâ€™s petrochemical industry is a major challenge facing chemical engineers and society at large. To be sustainable, routes for converting carbon dioxide and light into organic compounds for use as both fuels and chemical building blocks must be identified, understood, and engineered. Advances in metabolic engineering, synthetic biology, and other bioengineering disciplines have expanded the scope of what can be produced in a living organism. As in other engineering disciplines, synthetic biologists want to apply a general understanding of science (e.g. microbiology and biochemistry) to construct complex systems from well-characterized parts (e.g. DNA and protein). Once novel synthetic biological systems (e.g. enzymes for biofuel synthesis) are constructed, they must be engineered to function inside evolving cells without negatively impacting the hostâ€™s physiology.

In this talk, I will describe pathways for producing high-value commodity chemicals derived from fatty-acids and how my group and others have combined synthetic biology and systems biology to improve oleochemical production in bacteria using sustainable feedstocks. The talk will describe the critical regulatory points in native fatty acid metabolism, strategies for deregulating the pathway, and alternatives that by-pass it altogether. I will highlight the use of heterologous plant and bacterial enzymes to alter the chain length distribution of products from common long-chain molecules to higher-value medium-chain analogs. I will also describe our collaborative efforts with the Maranas lab at Penn State to use computational tools to redesign the substrate specificity of enzymes used to produce medium chain oleochemicals. Here, we have used the IPRO software to identify a small (~60) library of thioesterase mutants that resulted in a new variant that produces 2-fold more fatty acid with over 50% C_8 chain lengths, molecules not produced by the native enzyme.

486988 Impacts of CO_2 Supply Systems for Algal-Based Biorefineries on Biofuel Life Cycle Assessments

Abstract:

Biofuels have great potential as low-carbon alternatives to fossil-based transportation fuels. This talk will focus on utilization of CO_2 in an advanced biofuel system and the integration of that system with CO_2 generation sources. The biofuel system is the Algenol Direct to Ethanol® technology, which provides an efficient, cyanobacteria-based system for producing ethanol and bio-crude. The Algenol process consumes CO_2 in a process that yields a biofuel with much lower greenhouse gas emissions than gasoline.[1-3] In this talk, different options for CO_2 capture and utilization (CCU) are considered and their impact on the overall system operation evaluated in terms of carbon footprint reduction. The boundary of the life cycle analysis includes the CO_2 source. Overall 13 scenarios are described including an artificial reference case with CO_2 arriving at the boundary limits of a 2000 acre, model biorefinery with zero carbon footprint burden and zero CO_2 cost at a 10% IRR. Most scenarios involve combined heat and power (CHP) systems with no power supplied from the grid. Both natural gas and cellulosic fuel sources are considered for the CHP system. Five of the scenarios are "stand-alone" in that they do not require co-location with an anthropogenic CO_2 source. This is achieved by generating enough CO_2 on site to supply the needs of the biorefinery and selling excess power back to the grid. Four of the five stand-alone systems have carbon footprints and economics that are fully competitive with the commonly considered coal flue gas case. Overall carbon footprint reductions range from 23% to 116%. Equivalent CO_2 costs (based on comparisons to the reference case) range from \$35/-\$96/tonne with coal flue gas being \$45/tonne.

[1]D. Luo et al, "Lifecycle Energy and Greenhouse Gas Emissions from Ethanol Produced by Algae", *Environmental Science and Technology* **44**, 8670 (2010).

[2]Notice of EPA pathway approval: http://epa.gov/otaq/fuels/renewablefuels/new-pathways/approved-pathways.htm

[3]R. P. Lively et al, "Anthropogenic CO_2 as a feedstock for the production of algal-based biofuels", *Biofuels, Bioproducts, and Biorefining* **9**, 72-81 (2015)

498698 Carbon Fixation By Rubisco-Nanostructure Complex to Produce 3-Phosphoglyceric Acid: A Life Cycle Assessment

Abstract:

As an effort to eliminate greenhouse gases (GHGs), biochemists, microbiologists, and engineers are working toward systems that mimic the plant cellular environment's ability to convert GHGs into higher value products [1]. In the Calvin cycle of photosynthesis, ribulose-1,5-bisphosphate carboxylase oxygenase (RubisCO), which is an enzyme present in chloroplasts, catalyzes the conversion of ribulose 1,5-bisphosphate (RuBP) to 3-phosphoglyceric acid (3-PGA) while incorporating atmospheric CO_2 into the organic molecule [1,2]. It has been said that RubisCO is the most abundant protein in the world and accounts for most of the biological CO_2 fixed on earth [3]. Satagopan et al. have developed multienzyme-nanostructure complexes that catalyze cascade reactions in the Calvin cycle to enhance the catalytic properties of enzymes [1]. The complexes are made of three enzymes that include Rubisco and a nanostructure to support the enzymes.

In this work, a carbon footprint and other environmental impacts of 3-PGA produced through carbon fixation by the multienzyme-nanostructure complexes are evaluated using the Life Cycle Assessment (LCA) approach [4]. Two types of nanostructures for the complex are investigated as follows: Camptothecin (CPT)-dipeptide nanotube [1,5] and fluorenylmethyloxycarbonyl (Fmoc) tetrapeptide nanofiber. To examine the effectiveness of carbon fixation by those emerging technologies, another route that doesn't fix CO_2 to prepare 3-PGA is investigated [6] and the carbon footprint of each route is compared. Also, the potential use of 3-PGA and its impacts are examined to account for a cradle-to-grave life cycle boundary. From the results of life cycle impact assessment, processes that show the highest contribution to the footprint are identified.

Many challenges exist in applying the LCA to emerging technologies due to the nascent nature of these technologies [7]. If any life cycle inventory (LCI) data are not available from existing database [8,9], the data are estimated from laboratory experiments or obtained from literatures and patents. To account for the uncertainty of those data, a sensitivity analysis is performed. Three cases are considered based on the amount of fugitive emissions and the recyclability of unreacted reactants and solvents. The reusability of the complexes is also taken into consideration through the sensitivity analysis because they function as catalysts for the carbon fixation. Also, most of energy flows are excluded from the inventory analysis because most of biological processes take place around room temperature. If product yields are unknown, they are determined by a stoichiometry calculation.

CPT is extracted from the stem of a specific tree species, *Camptotheca acuminate* [10]. Since its extraction yield from the tree and the subsequent product yield of the CPT-dipeptide nanotube are low, the total carbon footprint is affected significantly depending on whether biogenic CO_2 emissions are included in the analysis or not. In this work, therefore, the effects of the biogenic CO_2 emissions on the footprint are examined and discussed.

Even though this work accounts for some data uncertainty via sensitivity analysis, other sources of uncertainty are still left for a more robust analysis. Some data, such as labor, equipment, and other unavailable data, are excluded from the analysis boundary due to the lack of information. Also, inventories from different LCI database [8,9] often base on different location. Moreover, since many unavailable data from the LCI database are obtained from literatures and patents, there may be errors in the life cycle network as well as the inventory data. Those potential uncertainty and errors are mostly attributed to the nature of emerging technologies. If a more complete analysis that accounts for those issues is developed, it would present significant benefits for evaluating the impacts of emerging technologies.

References

[1] Sriram Satagopan, Yuan Sun, Jon R. Parquette, and F. Robert Tabita. Self-assembled nanostructures for capturing atmospheric CO_2. Unpublished manuscript.

[2] Spencer M. Whitney, Robert L. Houtz, and Hernan Alonso. Advancing our understanding and capacity to engineer nature's CO_2-sequestering enzyme, Rubisco. Plant Physiology. 155(1): 27-35, 2011.

[3] John A. Raven. Rubisco: still the most abundant protein of Earth? New Phytologist. 198(1): 1-3, 2013.

[4] ISO. Environmental management—Life cycle assessment—Principles and framework (ISO 14040:2006). ISO. Geneva, Switzerland, 2016.

[5] Se Hye Kim et al. The self-assembly of anticancer camptothecin–dipeptide nanotubes: A minimalistic and high drug loading approach to increased efficacy. Chemistry - A European Journal. 21(1): 101-105, 2015.

[6] Ines Mandl and Carl Neuberg. Preparation of D(-)-3-phosphoglyceric acid. Methods in Enzymology. 3:208-214, 1957.

[7] Marcelle C. McManus et al. Challenge clusters facing LCA in environmental decision-making–what we can learn from biofuels. The International Journal of Life Cycle Assessment. 20(10):1399-1414, 2015.

[8] U.S. Life Cycle Inventory Database. National Renewable Energy Laboratory, 2012.

[9] Rolf Frischknecht et al. The ecoinvent database: Overview and methodological framework. International Journal of Life Cycle Assessment, 10(1):3–9, 2005.

[10] J. George Buta and Michael J. Novak, Isolation of camptothecin and 10-methoxycamptothecin from camptotheca acuminate by gel permeation chromatography. Industrial and Engineering Chemistry Product Research and Development. 17(2): 160-161, 1978.

496672 Life Cycle Assessment of Biofuel and Bioenergy Production from Brown Algae through Biochemical Pathways

Abstract:

Brown algae *Laminaria Japonica* is the most cultivated type of seaweed with a large potential for sustainable biofuel and bioenergy production. This study aims to evaluate the environmental impacts of seaweed cultivation and utilization for biofuel production through biochemical conversion routes including the sugar platform (SP), methane platform (MP), and volatile fatty acids platform (VFAP). In the SP, bioethanol is produced through saccharification of carbohydrates and fermentation of produced sugars. In the MP and VFAP, methane and volatile fatty acids (VFA) are produced through anaerobic digestion of seaweed. In the MP, methane is utilized as fuel source in a boiler/turbogenerator for heat and power production. In VFAP, VFAs are recovered through extraction/distillation and then hydrogenated to mixed alcohols. All three processes are simulated in Aspen Plus v.8.8 to obtain heat and material balances. Simulation results along with data from literature and Ecoinvent V3 databases for seaweed cultivation, chemicals, nutrients, and waste treatment operation are used to obtain life cycle inventory. Then, life cycle models are developed in SimaPro V8.2 software for each process. The environmental impact categories considered include global warming potential, acidification, eutrophication, fresh water consumption, and non-renewable energy use. Results are used to identify bottlenecks, limitations, and main contributors to the environmental profile of each route and assess the potentials to reduce the emissions and enhance the sustainability of the production process.

497439 Integrated Design and Analysis of Chemical Production from Biomass Feedstocks

Abstract:

Integrated Design and Analysis of Chemical Production from Biomass Feedstocks

Abhay Athaley[1], Praneeth Annam[1], Basudeb Saha[2], Marianthi Ierapetritou[1]

[1] Department of Chemical and Biochemical Engineering, Rutgers - The State University of New Jersey

[2.] Catalysis Center for Energy Innovation & Department of Chemical & Biomolecular Engineering, University of Delaware

This work principally concentrates on the integrated design, techno-economic and the life-cycle analysis of the production of chemicals from the lignocellulose biomass. The priority for the development of variate fine chemicals from biomass feedstock stepped-up over the changing time. The idea of bio-refinery has been proposed that uses different conversion technologies to produce multiple products. The success of bio-refinery is highly dependent on its mature technology and its applicability to produce not only high volume low cost fuels but also to produce low volume high cost chemicals.[1]

Bio-based products acceptance in the market depends on the competitiveness of economics and sustainability when compared to oil-based chemicals and products. In our previous work, [2]we investigated a novel hydrolysis process[3] for the productions of sugars and the subsequent production of para-Xylene and compared it with two other hydrolysis process (Concentrated acid hydrolysis and Dilute acid hydrolysis) using techno-economic and life cycle assessment. In our present work, we integrate the novel hydrolysis process to production of other high value chemicals such as surfactants,[4] butadiene, 2-methyl styrene etc. and high value jet fuels and combine it with the production process of para-Xylene and phthalic anhydride.[5] The aforementioned chemicals are lucrative and also it furthers our goal of designing an integrated bio-refinery.

This work uses techno-economic analysis and life cycle assessment to design and evaluate the production alternatives. The detailed process flowsheet is developed and simulated using Aspen Plus. The steps for the production of different chemicals and fuels are integrated with the novel hydrolysis process into a single flowsheet. Next, heat integration and economic analysis is carried out to calculate the minimum selling price of each of the product. Furthermore, sensitivity analysis is performed to analyze the bottlenecks of entire process and to compare the cost of the chemicals with fluctuating cost of raw materials and capacity of the plant. It is also seen that the overall cost of raw materials increases considerably for the production of surfactants as compared to other chemicals such as butadiene or phthalic anhydride. To check the competitiveness of the above process in terms of sustainability Life Cycle Analysis is carried out on SimaPro®. The results with respect to carbon production and water consumption are shown and compared to the

current oil based processes.

References:

1. Lin, Z., V. Nikolakis, and M. Ierapetritou, *Alternative Approaches for p-Xylene Production from Starch: Techno-Economic Analysis.* Industrial & Engineering Chemistry Research, 2014. **53**(26): p. 10688-10699.

2. Athaley A., S.B., Ierapetritou M.,, *TECHNO-ECONOMIC ANALYSIS OF DIFFERENT TYPES OF HYDROLYSIS PROCESS FOR THE PRODUCTION OF P-XYLENE.* 2017.

3. Sadula, S., Athaley, A., Zheng, W., Ierapetritou, M., Saha, B., *Process Intensification for Cellulosic Biorefineries.* ChemSusChem, 2017. **10**.

4. Park D., J.K., Koehle M., Krumm C., Ren L., Damen J., Shete M., Lee H., Zuo X., Lee B., Fan W., Vlachos., Lobo R., Tsapatsis M., Dauenhauer P.,, *Tunable Oleo-Furan Surfactants by Acylation of Renewable Furans.* ACS Catalysis, 2016. **2**: p. 820-824.

5. Lin, Z., Ierapetritou, M., Nikolakis, V.,, *Phthalic anhydride production from hemicellulose solutions: Technoeconomic analysis and life cycle assessment.* AIChE Journal, 2015. **61**: p. 3708-3718.

499590 Life-Cycle Analysis (LCA) of Bio-Derived Terephtalic Acid (TPA) and Bio-Derived-Hexamethylenediamine (HMDA)

Abstract:

Increasing efforts to reduce environmental impacts while boosting U.S chemical production have been an important driver in the growing bioeconomy. These efforts not only stimulate economic development but also promote competitiveness in the bioeconomy. Under the analysis led by the Clean Energy Manufacturing Analysis Center (CEMAC), screening of the impact of initial 170 chemicals derived from biomass were conducted, resulting in three relevant bioproducts: 3- hydroxypropionic acid (HP), terephtalic acid (TPA), and hexamethylenediamine (HMDA). These bio-derived chemicals can be used in the production of superabsorbent polymers, coatings, adhesives, plastics, fibers, paints, etc. In this work, we use the Greenhouse gases, Regulated Emissions and Energy use in Transportation (GREET®) model to conduct a life-cycle analysis (LCA) on the last two bioproducts: TPA and HMDA. We evaluate the key drivers of the GHG emissions and water consumption. We present thermochemical and biochemical pathways as alternative production pathways for these chemicals. These biomass-derived chemicals can displace fossil-derived products potentially and reduce GHG emissions, water and fossil fuel consumption as compared to their fossil-derived counterparts.

495175 Sustainability Analysis of Condensed-Phase n-Butanol Production from Ethanol

Abstract:

Bioethanol is one of the most valuable replacements for or additives to fossil fuels. However, ethanol is not without its limitations that include low energy density, corrosivity, and contamination issues when mixed with water. In contrast, n-butanol has a notable advantage in these aspects, as it is more miscible in gasoline, has higher energy density, and has an air to fuel ratio comparable to that of gasoline. As a result, there has been growing interest in the production of biobutanol and higher alcohols as biofuels as well as for use in the chemical industry.

The condensation of ethanol to butanol and higher alcohols is known as Guerbet chemistry [1]. The reactions are well known and have been substantially investigated. The path to Guerbet chemistry commercialization, however, has been hindered by substandard yields, catalyst stability concerns, and high capital and energy costs.

Our research group has studied the condensed-phase Guerbet reaction of ethanol to higher alcohols [2], [3]. Recently, we have optimized a condensed phase continuous flow reactor using multi-metallic catalysts on γ-Al_2O_3 support modified with La_2O_3. We have achieved higher alcohols selectivity of 80% at ethanol conversions of 40%. These encouraging results have prompted us to explore the economic feasibility and environmental impact of industrial-scale higher alcohol production using Guerbet chemistry.

An economic and environmental assessment of a first-generation facility producing 25 million gallons of n-butanol per year has been performed for a range of butanol and higher alcohol selectivities. Simulation of the catalytic reactor and the recovery and purification processes have been conducted using AspenPlus. Economic parameters such as capital costs, operating costs, annual profit, and return on investment have been determined for several scenarios. The results obtained show n-butanol required selling prices are reasonably competitive for conversion and selectivity combinations achieved in laboratory reactions. Interestingly, analysis of the separation train indicate significant economic benefit by carrying out distillations at elevated pressures. Environmental impacts such as climate change, resource depletion, water use, and eutrophication are also assessed.

References

[1] D. Gabriëls, W. Y. Hernández, B. Sels, P. Van Der Voort, and A. Verberckmoes, "Review of catalytic systems and thermodynamics for the Guerbet condensation reaction and challenges for biomass valorization," *Catal. Sci. Technol.*, vol. 5, pp. 3876–3902, 2015.

[2] T. L. Jordison, L. Peereboom, and D. J. Miller, "Impact of Water on Condensed Phase Ethanol Guerbet Reactions," *Ind. Eng. Chem. Res.*, vol. 55, no. 23, pp. 6579–6585, 2016.

[3] T. L. Jordison, C. T. Lira, and D. J. Miller, "Condensed-Phase Ethanol Conversion to Higher Alcohols," *Ind. Eng. Chem. Res.*, vol. 54, no. 44, pp. 10991–11000, 2015.

490780 Hydrothermal Upgrading of Algal Bio-Oil By Supercritical Water

Abstract:

Global warming is a significant problem the world is currently facing as result of greenhouse gasses being emitted from the continual usage of fossil fuels. One particular solution that has been examined to reduce the rate of global warming is using bio-crude as a fuel replacement. Bio-crude is a more desired renewable fuel because liquid fuels are easier to transport and have higher energy densities than solid and gaseous fuels. Bio-crude is produced by the thermochemical or biochemical conversion of biomass through methods such as liquefaction, pyrolysis, solvent extraction, etc. Just like crude oil, bio-crude needs to be refined before it can be used; this is to remove oxygen and other heteroatoms that lower its fuel quality. The most common method of upgrading crude oil is hydro-deoxygenation which involves treating crude with hydrogen at high temperatures. Hydrogen is predominantly produced by natural gas reforming, thus it would be counter intuitive to use a fossil fuel in order to improve a renewable energy technology.

Supercritical water (SCW) treatment is examined as an alternative method to upgrade crude in a more sustainable manner. Water at supercritical conditions becomes a single phase and exhibits compressibility and density similar to that of gasses and liquids, respectively. Furthermore, its dielectric constant significantly decreases with increasing temperature, causing it to behave like a nonpolar solvent. SCW is also considered an environmentally-friendly solvent as it turns to regular water at ambient temperatures. These ideal solvent properties are proposed to be useful in crude upgrading as it is expected that SCW breaks down longer chain hydrocarbons while promoting oxidation reactions to remove heteroatoms from the crude.

This goal of this study was to upgrade bio-crudes produced from hydrothermal liquefaction (HTL) of microalgae (*N. salina*) with SCW in order to remove impurities, decrease oxygen content, reduce viscosity, and produce desired low carbon (C_6-C_{14}) alkane products. The bio-crude was treated with and without Pd/C catalyst at temperatures of 380, 400, and 420 °C at a fixed SCW density of 278 kg m^{-3}, thus allowing pressure to be monitored than controlled. The upgraded bio-crude quality was assessed by evaluating energy content, ultimate analysis, GCMS, FTIR, NMR, viscosity, and density. The energy content of the treated bio-oil increased with reaction temperature and catalyst loading as a result of the decrease in oxygen and sulfur content. Viscosity and density decreased due to the decomposition of large molecules, creating shorter straight chain hydrocarbons.

500381 Model Aldehyde Chemistry Using Cezrox Catalyst for Hydrothermal Liquefaction Upgrading

Abstract:

Green processes that upgrade biomass feedstocks and recycled bio-waste can both help meet our growing energy demands as well as reduce our overall carbon footprint. One promising technology is hydrothermal liquefaction (HTL), which breaks down wet organic solids at moderate temperatures and high pressures to form a carbon rich oil phase. While the bio-oil formed has a high heating value, the process often produces other byproducts such as solid biochar, carbon dioxide and wide distribution of organics that are water soluble. To design an economical HTL process, the water soluble organic byproducts must be minimized to increase oil product yields and reduce wastewater generation. Metal oxide catalysts are a promising option for HTL upgrading chemistry due to their catalytic activity as well as stability in hot liquid water conditions. $CeZrO_x$, in particular, is active for acid coupling reactions which simultaneously increase the carbon number of the product and decrease its oxygen to carbon ratio, both of which act to increase the oil yield and decrease the carbon loss to the aqueous phase. In our study, we sought to understand the key reactions that take place in the HTL process when using $CeZrO_x$ as a catalyst.

In preliminary runs, use of $CeZrO_x$ increased oil yields relative to other commonly used catalysts in the HTL conversion of food waste at 300 Â°C and 20.7 MPa. Subsequently, tests were performed to expose $CeZrO_x$ to hot liquid water conditions at 300 Â°C and 20.7 MPa for up to 165 hours. Post-run characterization was performed using both XRD and XPS to verify the hydrothermal stability of $CeZrO_x$, Model HTL reactions using $CeZrO_x$ were also performed with water soluble alcohols, aldehydes, ketones and carboxylic acids as reactants. $CeZrO_x$ had minimal activity for catalyzing reactions involving alcohol-alcohol, carboxylic acid-carboxylic acid or ketone-ketone coupling. In contrast, $CeZrO_x$ was highly active â€“ and selective â€“ for aldehyde coupling reactions, with the primary product being alkenes. For example, pentanal self-condensation reactions under HTL conditions produced a decene product that phase separated into an oil phase at room conditions. On-going work is focused on optimizing the catalyst properties and catalytic reactor conditions.

495412 Hydrothermal Liquefaction of Wild-Harvested Cyanobacteria from a Hypereutrophic Lake in Wisconsin

Abstract:

Processing by hydrothermal liquefaction (HTL) of wild-harvested cyanobacteria could have the potential to provide a sustainable, domestic fuel supply and nutrients for use in agriculture while removing a source of hypoxia from over-burdened waterways. Lab-scale cultured strains of microalgae has shown freshwater cyanobacteria as a promising feedstock with low ash content, leading to high biocrude yields. However, HTL of wild-harvested microalgal strains has yet to be fully investigated. Additionally, the intermittent nature of cyanobacteria blooms raises questions about the need to store harvested microalgae and the effect of storage time on the products.

Wild-harvested microalgae sampled from Lake Tainter, a hypereutrophic lake in Northwest Wisconsin, was collected and prepared in two ways: dried while fresh and stored long-term, and stored wet long term then dried before HTL processing. Fractions of the feedstock, mostly consisting of Microcystis Aeruginosa, were treated via HTL at a range of temperatures (280-300°C), solid loading ratios (10-25% w/w), and reaction holding times (30-60 min) in 30 mL batch reactors. The percent yield of the products and chemical composition of the biocrude and aqueous fraction (PHWW) was characterized by GC-MS, ICP-MS, and elemental analyzer.

496834 Multi-Scale Modeling of Heavy Oil Upgrading in Near-Critical/Supercritical Water

Abstract:

The upgrading of heavy oil using near-critical or supercritical water can be pushed towards better outcomes (higher light liquid product yield and lower coke formation) by suppressing the condensation reactions of light hydrocarbon species to heavier hydrocarbons and coke precursors. A potential method to achieve this is to ensure a much lower residence time for the light hydrocarbon species in the high-temperature reactor zone once they are formed by thermal cracking reactions of the heavy hydrocarbons.

This difference in residence times between the light and heavy species in the oil can in turn be realized in practice in a reactor configuration like the extractive upflow reactor which exploits gravity effects and the higher solubility of light hydrocarbons in the water phase. The oil phase being 2-8 times heavier than the water phase (depending on the operating pressure and temperature) circulates at the bottom of the reactor, while the water phase injected from the bottom flows through the bed of oil and exits the reactor at the top, in the process, extracting the light hydrocarbons soluble in the water phase. While experiments demonstrating the efficacy of the extractive upgrading process of heavy oil can be found in the literature, there remains a need to model and understand the complex coupling between the chemical reactions of interest and the interfacial mass transfer of hydrocarbon species and evaluate their effect on the final product yields.

With a view to address this need, we employ a two-phase well-stirred reactor (WSR) model with lumped chemical kinetics to study the overall coupling of the chemical reactions and the oil-water phase equilibrium in an extractive upflow reactor at P = 30MPa and T = 663,683 and 703K. Using this low-fidelity model, we demonstrate the improved yields of the upgrading process at higher inlet flow rates of water directly corresponding to higher rates of extraction of the light species from the oil to the water phase. This is in qualitative agreement with experimental demonstrations of the extractive upgrading configuration in literature. The WSR model also shows that the total interfacial species mass transfer rates in the reactor must be higher than the rates of retrograde reactions (condensation and coking reactions) for the best outcomes. This necessitates the need to develop methods to obtain realistic estimates for the interfacial species mass transfer rates in a realistic extractive upgrading reactor configuration.

In order to obtain a realistic range of values for the interfacial species mass transfer rates for different operating conditions of interest we use a multi-scale modeling approach, wherein the total interfacial species mass transfer rates in the reactor are estimated as the product of an average total interfacial area within the reactor and an averaged interfacial mass flux per unit area of the light hydrocarbon species.

The total average oil-water interfacial area in the reactor is calculated by simulating the macro-scale mixing process in a realistic lab-scale reactor geometry (diameter=30mm, height=100mm and 20g of oil feed) using two-phase flow simulations in 2-D with the Volume of Fluid (VoF) method with fluid properties like density, viscosity and interfacial tension estimated at the different given P, T conditions. The total average interfacial area in the reactor scales as the square root of the Weber number for a range of different operating conditions (pressure, temperature and inlet velocities). This is a useful intermediate result which can help engineers design extractive upgrading reactors for specified target light species extraction rates.

The average interfacial mass flux of light hydrocarbon species is estimated using 1-D droplet mixing simulations in spherically symmetric co-ordinates which account for the detailed oil-water phase equilibrium and diffusive fluxes at the interface. The heavy oil is modeled as a mixture of two pseudo-components, one corresponding to the light fraction and another corresponding to the heavy fraction in the mass ratio 1:1. The average interfacial mass flux of the light pseudo-component at P=30MPa increases from T=623K to 703K by more than an order of magnitude which is consistent with the rapidly increasing solubility of the light oil fraction in water with increasing temperature.

The total interfacial mass transfer rates estimated at P=30MPa and T=663-703K for inlet water mass flow rates in the range of 2.5-10 g/min and 20g of initial oil feed are an order of magnitude higher than the rates of the lumped cracking and coking reactions at these temperatures. This implies that the extractive upgrading reactor has the potential to effectively remove the light hydrocarbons from the oil phase as soon as they are formed and transport them safely out of the high temperature reaction zone. Although, this medium-fidelity approach to estimating the interfacial species mass transfer rates does not account for complex effects like the enhancement of the species fluxes due to sharpening of gradients in a flow field or the reduction in interfacial area due to interphase mass transfer, it nevertheless yields a reasonable first-order estimate which can be used to aid the design process of extractive upgrading reactors.

493158 Effects of Process Conditions on Metal Content in Biocrude from Hydrothermal Liquefaction of Microalgae

Abstract:

Hydrothermal liquefaction (HTL) converts wet biomass into a crude bio-oil that could be further treated to be acceptable for processing in petroleum refineries. HTL takes advantage of the properties of hot, compressed water near its critical point to break down biomacromolecules in the feedstock into smaller molecules. Several issues presently prevent the substitution of HTL biocrude for petroleum and one is the presence of metals in the crude bio-oil. Iron is found to plug the head of the catalyst beds that are used to hydrotreat algal biocrude oil [1]. Metal impurities in crude bio-oil can also cause catalyst deactivation, coke formation, corrosion, and deposition. Thus, further studies are required to understand how the HTL conditions influence metal content and to develop methods to reduce the metal content in algal biocrude for downstream processing operations.

We hydrothermally treated microalgae using different processing conditions (isothermal reaction vs. rapid heating, different temperatures and reaction times, different reactor loadings) and determined the amounts of thirteen different metals in the biocrude oil as well as in the other product phases. The metal content in biocrude oil undergoes dramatic change within the first 5 minutes of HTL reaction. Generally, lighter elements are transferred into bio-oil faster than heavier metals. Iron in bio-oil behaves differently than many of the other metals, and is the target metal to be removed due to its abundance. As the reaction conditions become more severe, less iron resides in bio-oil phase. Optimum reaction conditions could be identified in terms of a high bio-oil yield and low metal concentration in the bio-oil. The results from this study provide the first glimpse into how and whether HTL process conditions can be used to manipulate metal content in algae biocrude.

Reference

[1] Jarvis, Jacqueline M., et al. "Impact of iron porphyrin complexes when hydroprocessing algal HTL biocrude." *Fuel* 182 (2016): 411-418.

498900 Bio-Crude Quality and Composition from Algal Solids Cultivated in Varying Media, Growth Stage, and Solids Loading

Abstract:

Hydrothermal liquefaction (HTL) has become a promising technique for converting algal biomass to high-value products because it utilizes algaeâ€™s natural, wet environment for the conversion making it both energy and cost-effective. Furthermore, the necessary water and nutrients to cultivate algal biomass can be provided in a sustainable and inexpensive manner with the use of wastewater (WW) effluent. Algal biomass harvested from WW can also contain large amounts of inorganics, primarily calcium, depending on the treatment operation at the plant. During the HTL reaction these inorganics produce an abundant, solid co-product. X-ray Diffraction (XRD) has confirmed these solids to be one of three calcium materials: hydroxyapatite (HAp), $Ca_5(PO_4)_3OH$, tricalcium phosphate (TCP), $Ca3(PO4)_2$, or calcium carbonate ($CaCO_3$). The purpose of this presentation is two-fold. First, to demonstrate the impact media and growth stage have on biocrude composition. Second, to understand the role these calcium solids have on biocrude quality.

Conventional HTL reactions at 350°C for 60 minutes were performed on various WW and raceway-cultivated algal solids. To simulate seasonal and regional differences of WW, algae were grown in two separate raceway ponds: one raceway that had N-limited media and the other P-limited. Algal solids were then harvested at different growth stages thus altering their biomolecular content and ultimately varying the biocrude composition. To regulate inorganic content of the algal solids, the algae were harvested via auto-induced flocculation that occurred from a manual increase in media pH. At each growth stage the total Ca:P molar ratio of the algal solids were manipulated to range between 1 and 3 with the addition of lime ($Ca(OH)_2$), a common chemical used to balance the alkalinity and pH in WW treatment.

GC-MS results displayed dramatic change in oil composition dependent upon growth media, growth stage, and solids formed. Elemental CHN/O analysis indicated little change in N/C or O/C ratio between growth stages and media. However, the H/C ratio did change with the HTL solids produced from the reaction. Thus, further characterization of HTL solids were measured for their catalytic potential. In addition, fundamental reactions of model inorganic and algal compounds were also performed to better understand the mechanism of solid synthesis and biocrude upgrading.

Thermochemistry of Alkylaromatics Reconsidered

Presentation at the AIChE National Meeting

Lawrence Lai[1]

William H. Green[1]

1 – Massachusetts Institute of Technology, Cambridge, MA

1

Increase in Oil Demand

OPEC World Oil Outlook (2016)

Global oil demand growth in the long-term

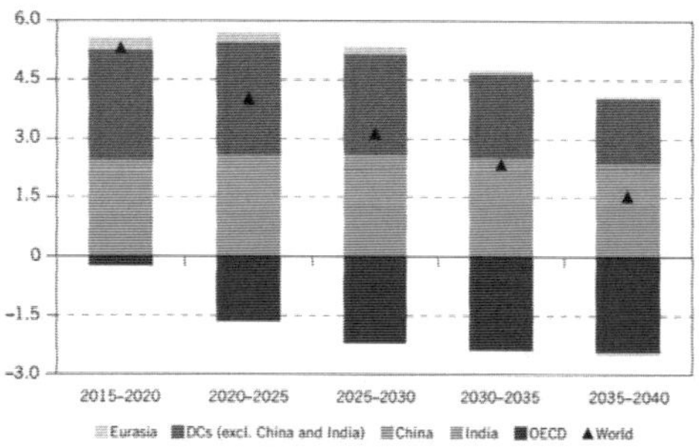

- Increase in global oil demand
- Demand to be met by heavy oil & bitumen
  - Difficult to vaporize - pyrolysis takes place
  - Not as clean combustion
  - High viscosity
  - Higher sulfur content

Caniaz and Erkey, *Process Intensification for Heavy Oil Upgrading Using Supercritical Water*

2

Supercritical Water Upgrading

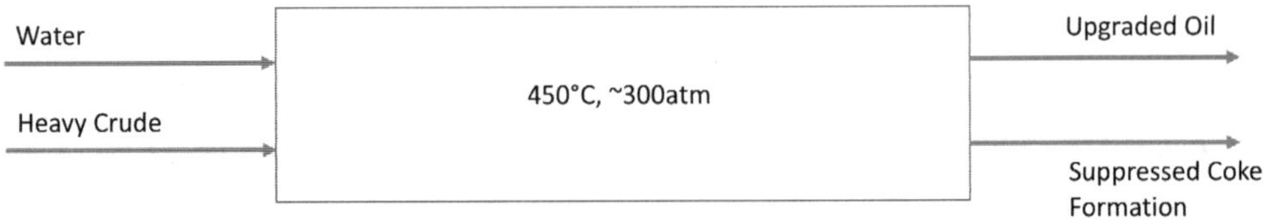

- Shown to suppress coke formation

With Water

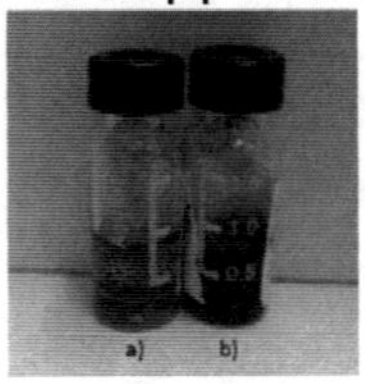

Without Water

aramco

Choi, *Supercritical Water Process to Upgrade Petroleum, WO2013066852 A1, 2013*

3

Hexylbenzene

- Focus of study to represent alkylbenzenes in crude oil
- Moderate size allows computational analysis using Reaction Mechanism Generator (RMG)
- Long alkyl chain to study cleavage chemistry
- Falls within GC-detection range

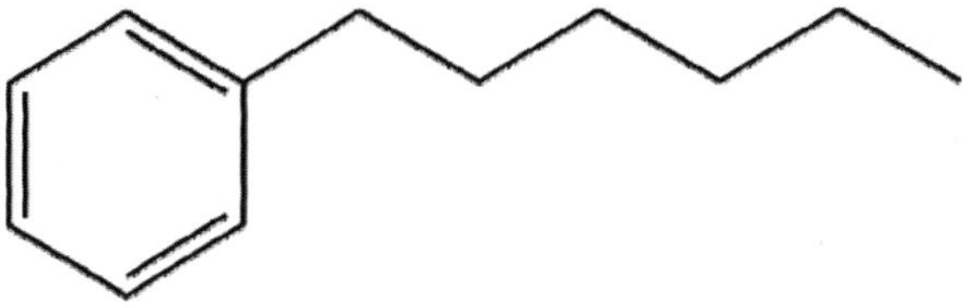

4

Hexylbenzene

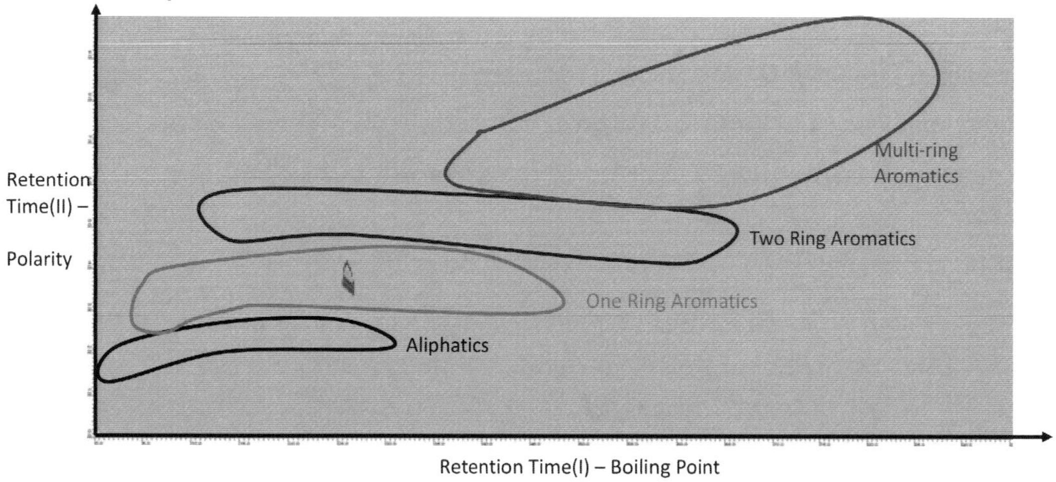

5

Supercritical Water Treated Hexylbenzene, 450°C, 45 minutes

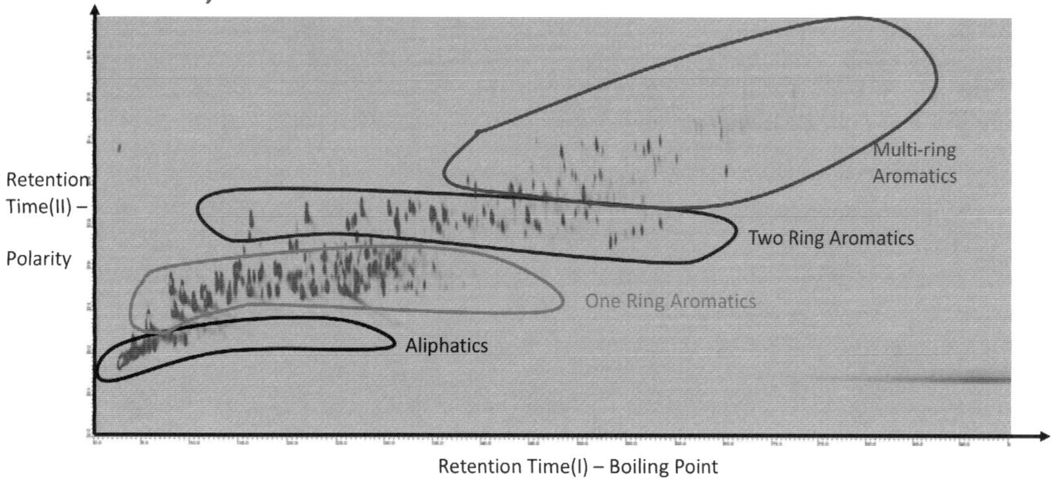

6

Intractable Experiment

- One reactant + water led to >500 Species
- Each species could be formed through multiple different reactions
- Radicals and other unstable intermediates are not observed on GCxGC
- Cannot use experimental means to uncover chemistry

Use the Green Group Reaction Mechanism Generator (RMG) to better understand the system

RMG does not show differences in presence of water

7

Reaction Mechanism Generator (RMG)

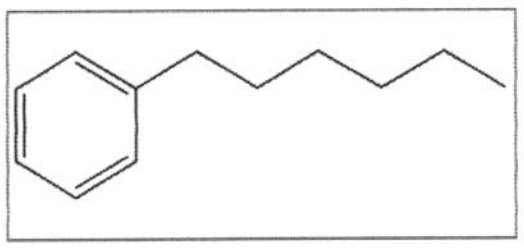

Core

8

Reaction Mechanism Generator (RMG)

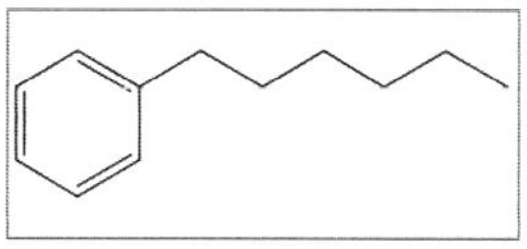

Core

Data Base of homolytic scission

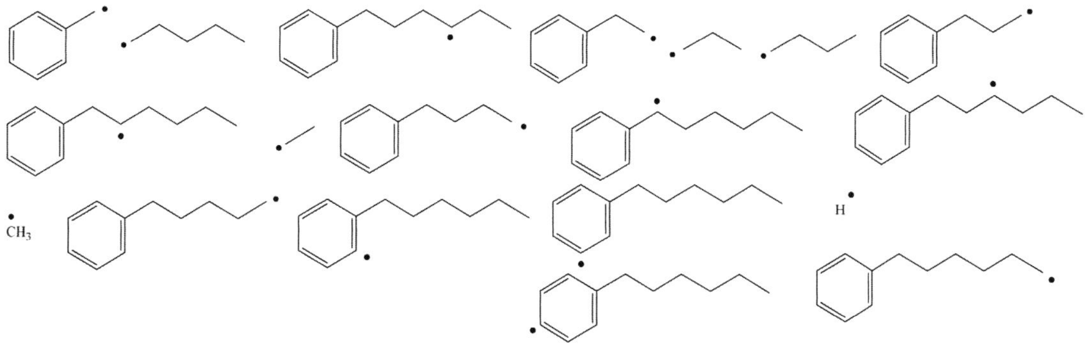

9

Reaction Mechanism Generator (RMG)

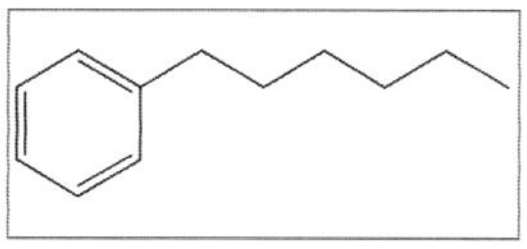

Core

Highest Flux Products

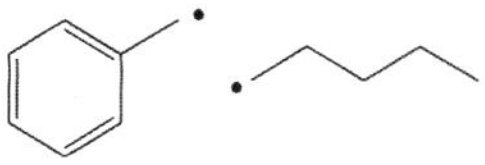

10

Reaction Mechanism Generator (RMG)

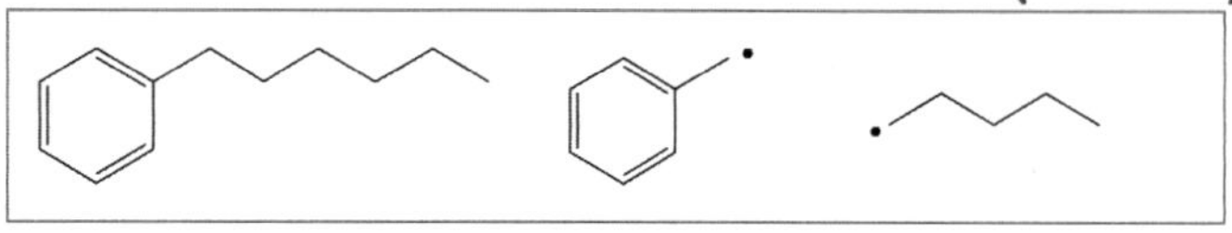

Core

11

Reaction Mechanism Generator (RMG)

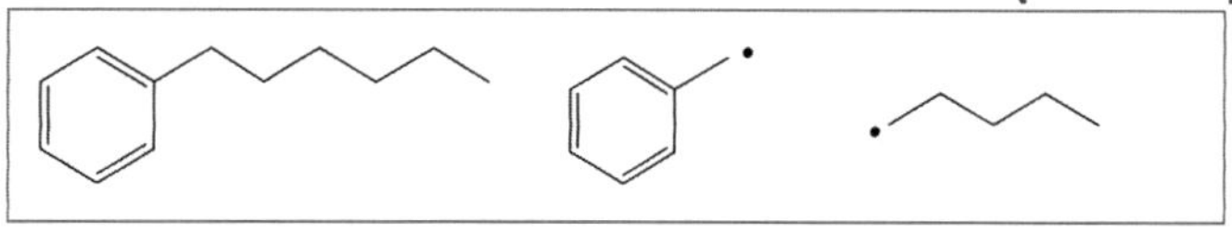

Core

New types of reactions can now take place:

Hydrogen Abstraction

Beta Scission

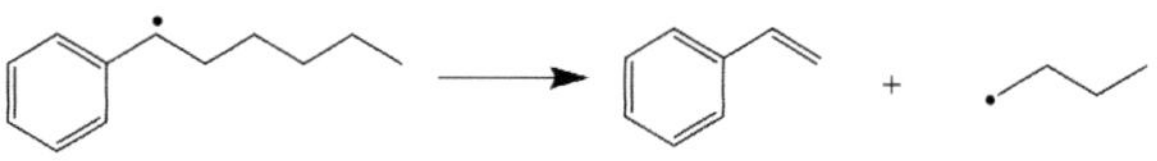

12

Overview of RMG algorithm

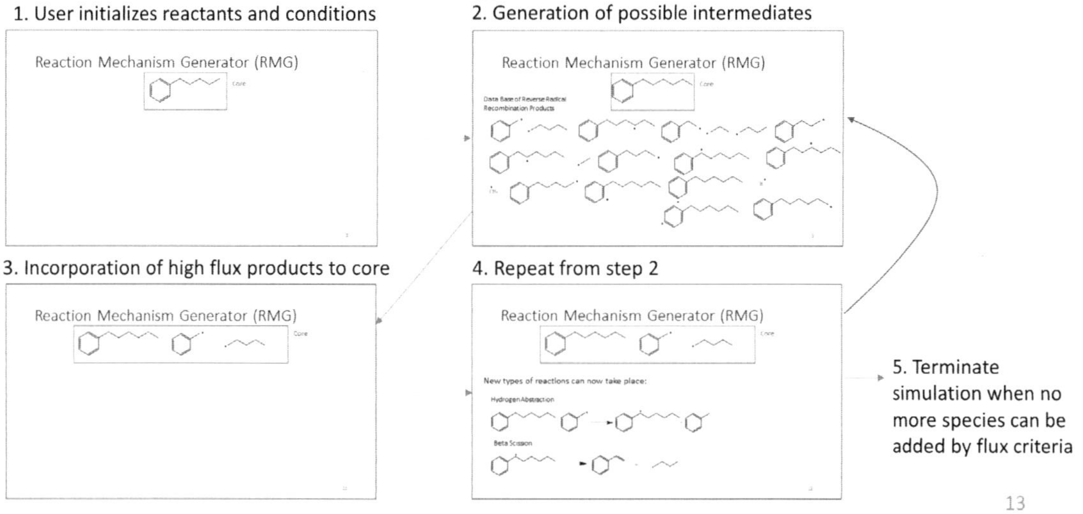

13

Model Conversion – Carr et al.[1]

- Model conversion factor of 2 off
- Inaccurate conversion leads to inaccurate product yields

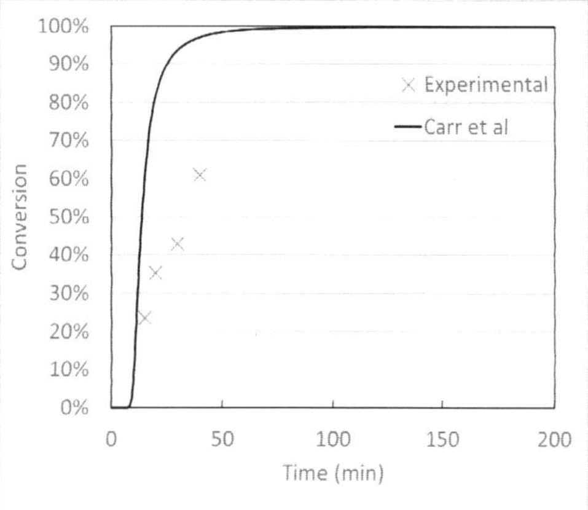

[1] Carr et al., *Energy and Fuels,* 2015.

14

"Retro-ene" Reaction

- "Retro-ene" Reaction thought to be important

(1)

	A (s^{-1})	n	E_a (kcal/mol)	Rate Constant at (723 K)
Burklé-Vitzthum[1] (Used in model)	6.8 x 10^{12} s^{-1}	0	54.4	2.44 x 10^{-4}
CBS-QB3	1.63 x 10^{7} s^{-1}	1.69	73.9	5.07 x 10^{-11}

- Rate used in Carr et al. was 7 orders of magnitude faster than reality
  - Highly accelerates hexylbenzene conversion

[1] V. Burklé-Vitzthum and R. Michels, *Ind. Eng. Chem. Res.*, 2005.

15

Model Conversion – Removed Retroene

- Conversion too slow
- Inspected model details

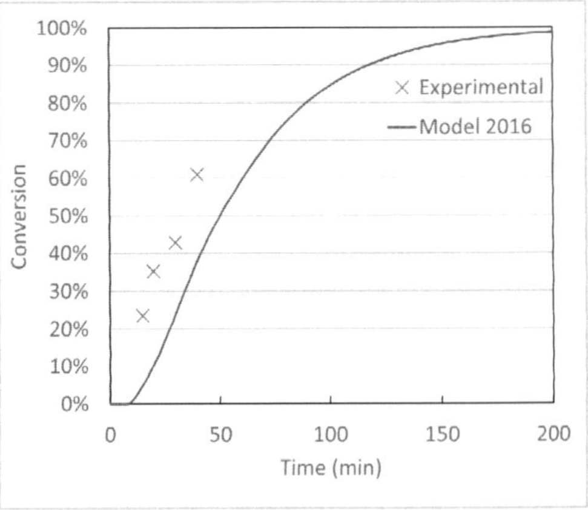

16

Hexylbenzene Model Summary

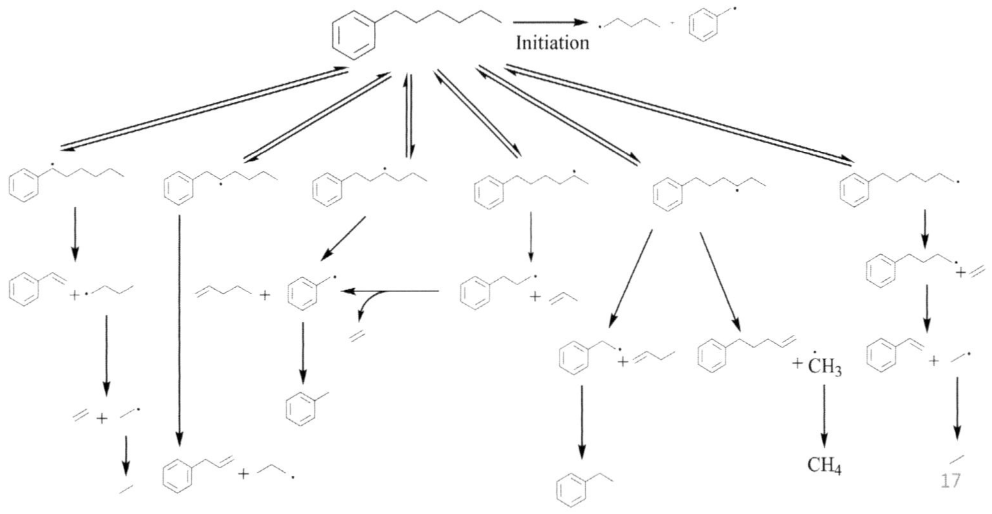

17

Initiation Step

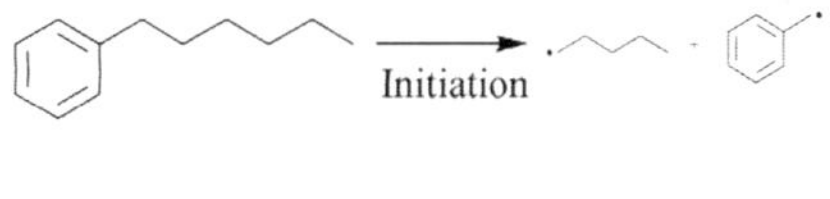

18

Hexylbenzene Radical Hydrogen Abstraction

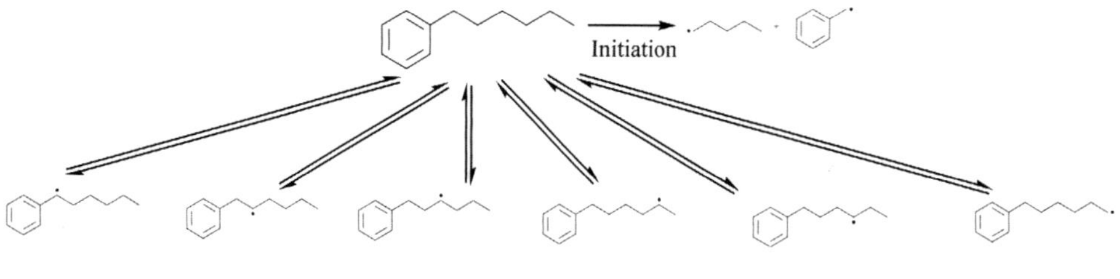

19

Main Beta Scission Pathway

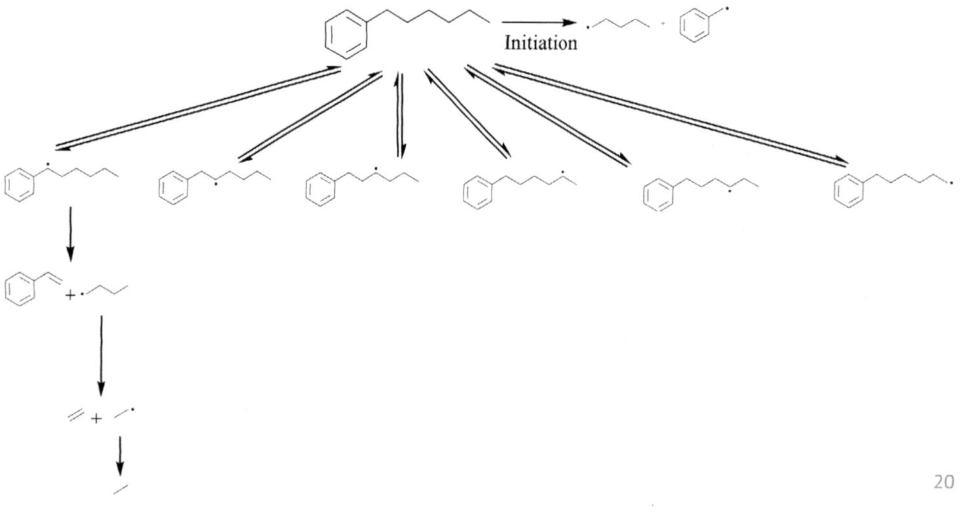

20

Main Product Flux

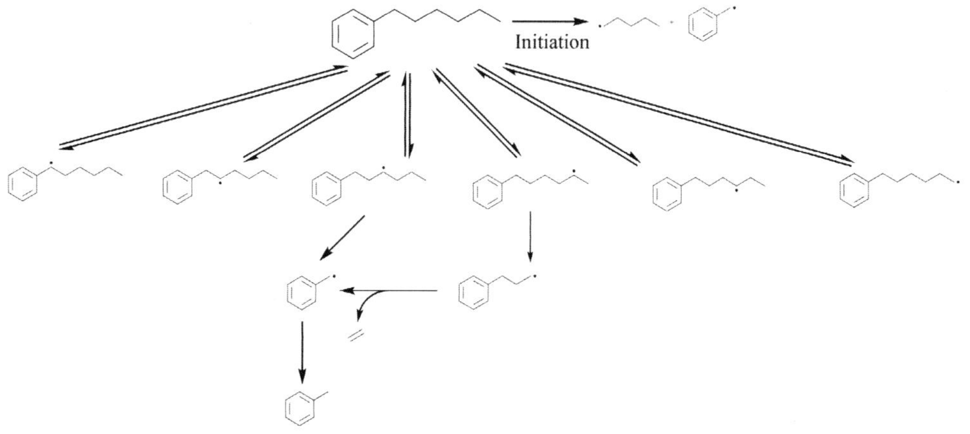

21

Hexylbenzene Model Summary

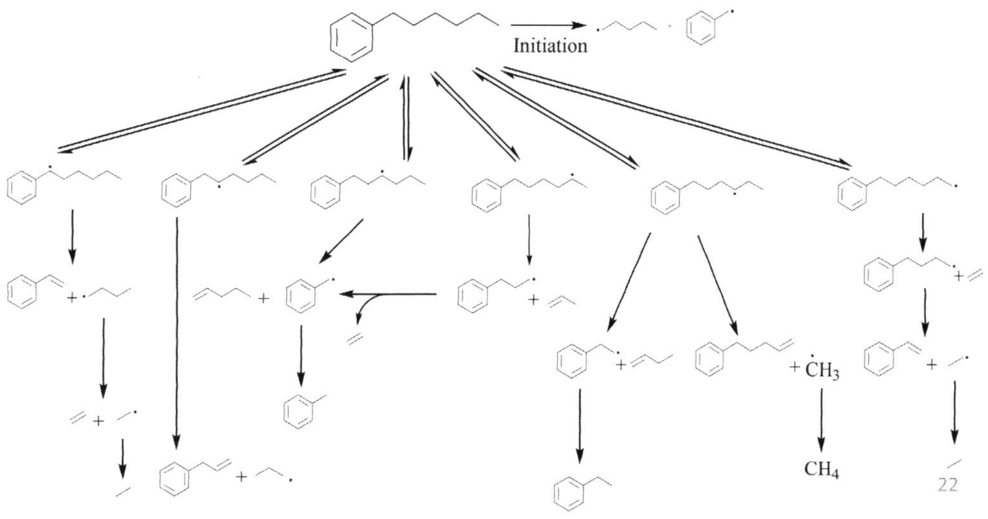

22

Sensitivity Analysis

$$s_i = \frac{\mathrm{d} \ln c_i}{d \ln k_j}$$

Where s is sensitivity

c is concentration

K is the rate constant

Subscript i indicates a species

Subscript j indicates a reaction

- Reaction (2)

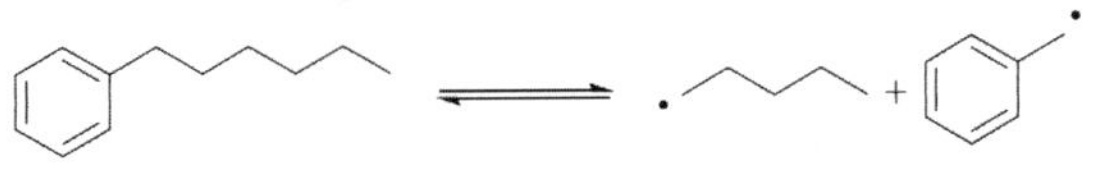

- Reaction (3)

23

Discrepancy too large to be due to uncertainty in rate coefficients

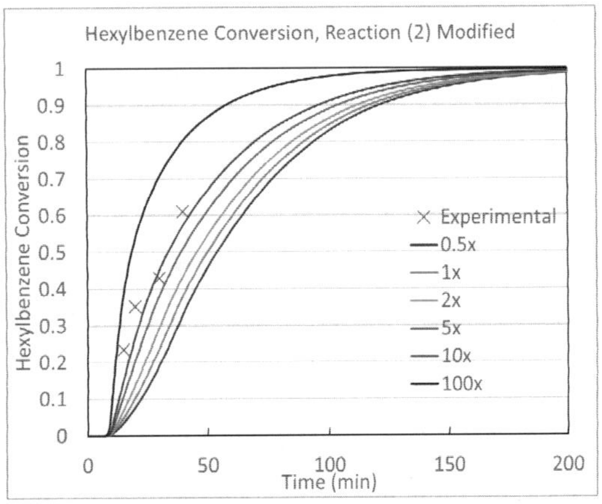

$k(T) = 2.1 \times 10^{13} \exp(-0.13 \text{ kcal mol}^{-1} / RT) \text{ cm}^3 \text{ mol}^{-1} \text{ s}^{-1}$

New Rate Calculated

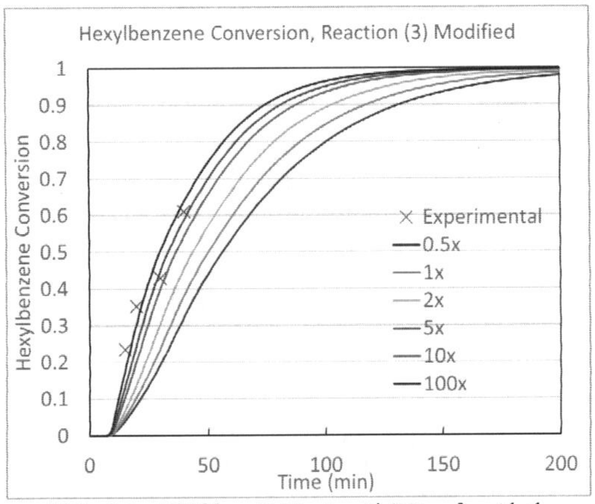

$k(T) = 2.4 \times 10^2\, T^{2.8} \exp(3.7 \text{ kcal mol}^{-1} / RT) \text{ cm}^3 \text{ mol}^{-1} \text{ s}^{-1}$

24

High Sensitivity to Thermochemistry

- Thermochemistry (Gibbs free energy) changes equilibria of reactions
- Small change in G leads to huge change in conversion

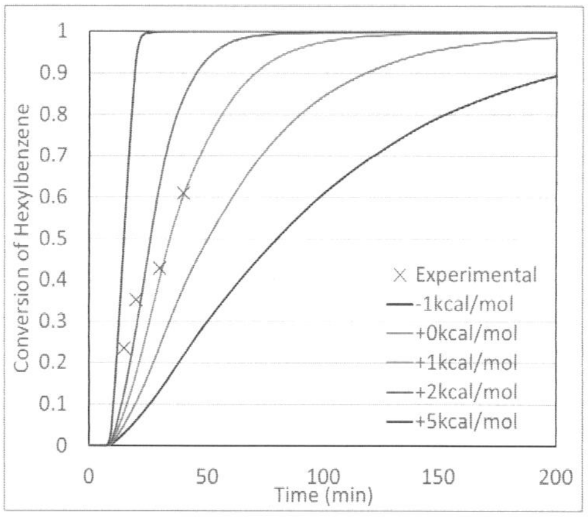

(Hexylbenzene Gibbs Free Energy = 118kcal/mol)

25

Thermochemistry Update

- Calculations done in CBS-QB3 level of theory; all hindered rotors included
- Species of interest:
  - Hexylbenzene and all aliphatic hexylbenzene radicals
  - All alkylbenzenes from toluene to pentylbenzene
  - All alkylbenzyl radicals adjacent to aromatic ring from benzyl to pentylbenzyl
  - Species involved in reactions (2) and (3): butyl, pentyl radical, and styrene

26

Updating Thermochemistry Resolves Discrepancy in Conversion

- Model matches experiments within 30% difference
- Updated thermochemistry available in RMG's thermochemistry database
  - Library named Lai_Hexylbenzene

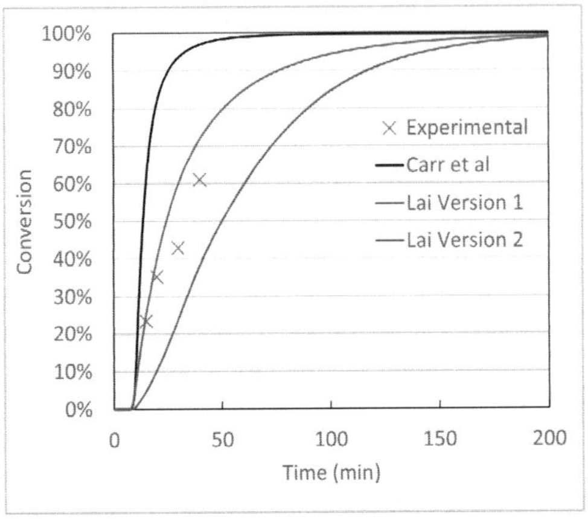

27

Individual Species Yield

- Species yield match at early times
- Model does not correctly depict chemistry at high conversion

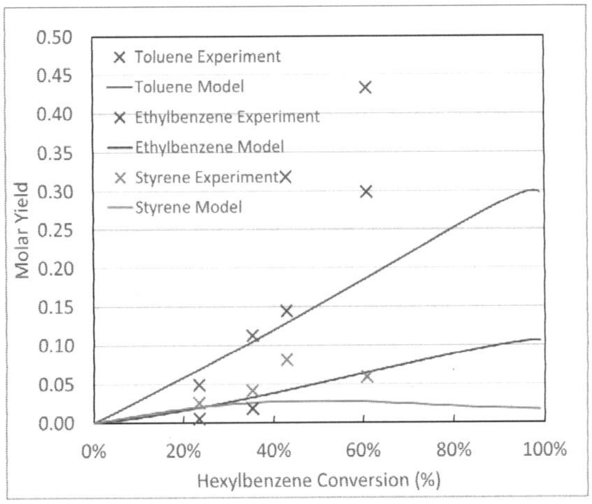

28

Error Analysis

- CBS-QB3 has an error of ±2.4kcal/mol [1]
- With error bounds introduced, conversion of model is at an appropriate match with experiments

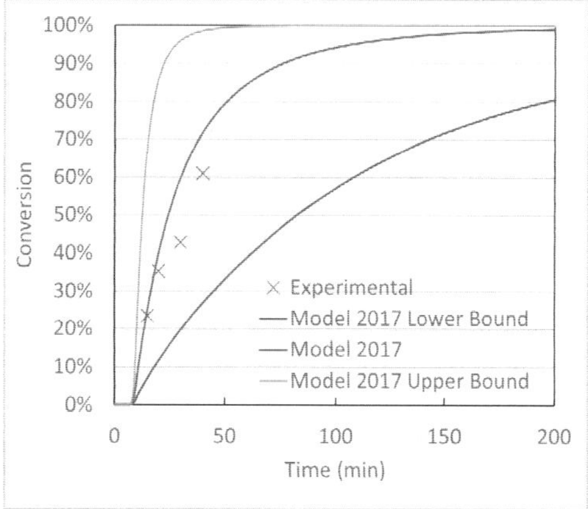

[1] K. P. Somers and J. M. Simmie, *The Journal of Physical Chemistry*, 2015.

29

Conclusion

- RMG was used to model pyrolysis of hexylbenzene
  - System is intractable through experimentation – use of RMG neccesary
- Thermochemistry affects conversion of reactant much more than kinetics
  - Pyrolysis at 723K
  - 2kcal/mol change in Gibbs Free Energy can affect conversion dramatically
  - 10x change in rate constant for initiation step required for similar change
- CBS-QB3 level of theory not precise
  - Model can be further improved using methods which compute/determine thermochemistry with tighter uncertainty ranges.

30

Acknowledgments

- Green Group
  - W. H. Green
  - Soumya Gudiyella
  - Max Liu
  - Zachary Buras
  - Kehang Han
- WPI
  - Mike Timko
- Funding

aramco

Founding member of the MIT Energy Initiative

31

Questions

- Lawrence Lai
- laitcl@mit.edu

32

References

- A. G. Carr, C. A. Class, L. Lai, Y. Kida, T. Monrose and W. H. Green, "Supercritical Water Treatment of Crude Oil and Hexylbenzene: An Experimental and Mechanistic Study on Alkylbenzene Decomposition," *Energy & Fuels,* vol. 29, pp. 5290-5302, 2015.

- V. Burklé-Vitzthum and R. Michels, "Experimental Study and Modeling of the ROle of Hydronaphthalenics on the Thermal Stability of Hydrocarbons under Laboratory and Geological Conditions," *Ind. Eng. Chem. Res.,* vol. 44, pp. 8972-8987, 2005.

- K. P. Somers and J. M. Simmie, "Benchmarking Compound Methods (CBS-QB3, CBS-APNO, G3, G4, W1BD) against the Active Thermochemical Tables: Formation Enthalpies of Radicals," *J. Phys. Chem. A,* vol. 119, pp. 8922-8933, 2015.

33

Backup Slides

34

Hexylbenzene Experimentation

RMG

Modeling
- Reaction Mechanism Generator (RMG)
- Species susceptible of formation included in initial conditions

Loading
- Hexylbenzene
- Naphthalene Standard
- SCW or Pyrolysis

Reaction
- Heating by sand bath
- Heat up time of 10 minutes - 450°C
- Pressure monitored
- 300 or 75 bar (SCW or Pyro)

Quenching
- Cooling by water
- Cool down time of 2-3 minutes

Analysis
- GC-MS Liquids
- GCxGC-FID for diagnosis
- GC-FID for gases

35

505271 Session Keynote - Beneficial Uses of Stranded Assets -- Emerging Opportunities

Abstract:

Stranded Assets: Industrialized economies have very high productivity of manufactured goods and agricultural products. This high productivity is associated with high level of wasted resources: for example waste heat from process plants, municipal solid waste going to landfills, agricultural biomass waste, sewage sludge, waste water and flare gas associated with oil and gas production. This wasted resources represent lost economic value, but also pose risk to environmental sustainability. These **Stranded Assets,** when converted to higher value co-products, represent multi-billion dollars market potential. States like California are leading the efforts and technology development to create Win-Win solution for these stranded assets as discussed next.

Emerging Opportunities: The wasted resources described above can be converted to higher value co-products such as hydrogen, methane, syn-gas, bio-oil, bio-char and chemicals. The conversion technologies include anaerobic digestion, gasification, pyrolysis, etc. among others. Efforts are underway to demonstrate these conversion technologies at commercial scale and validate them for economic viability and environmental sustainability. California is the first state to recognize the solid waste as a renewable energy source, extending it with policy support for financial incentives. One such example is: Hydrogen or Methane produced from human waste now qualifies for renewable fuel and low-carbon fuel credits. A Multi-Purpose Energy Station being developed by T2M Global Team offers a promising pathway for increased revenue from sale of higher value co-products, and thus a new opportunity of beneficial uses of the stranded assets. The innovative energy station provides H2 for fuel cell electric cars, charging for battery electric vehicles, methane for CNG vehicles, and grid support services to eliminate intermittency problems associated with solar and wind energy.

Policy Support Opportunities: Waste heat and flare gases deserve wasted resource recognition for their potential of economic value and environmental benefits. It may be appropriate to consider them as Tier 2 Renewables, similar to the precedence in CT, NY and NJ for clean power from natural gas fuel cells.

505134 Session Keynote: Low-Carbon Power Generation with Methane Fueled SOFCs

Abstract:

LOW-CARBON POWER GENERATION WITH METHANE FUELED SOFCs

Introduction

Natural gas fueled power generation systems, such as large scale thermal power plants and solid oxide fuel cells (SOFCs), are promising technologies with high electrical efficiencies of higher than 50% (LHV), and give high temperature exhaust heats[1]. While the CO_2 emission coefficient of the electrical power with a natural gas fuel will decrease with increasing the electrical efficiency, it remains to have non-negligible value even at the considerably high efficiency. Methane is a main component of natural gas. Recently thermal decomposition and catalytic thermal decomposition of methane have attracted attentions as advanced technologies producing COx-free hydrogen, which are simple ways to reduce the CO_2 emission caused by methane fuel[2].

Abstract

As a fuel of the SOFCs, methane is typically mixed with steam and reformed to hydrogen and carbon monoxide by steam-reforming reaction before supplied to SOFC stacks. The thermal energy needed for the endothermic steam-reforming reaction is supplied by the heat generated as a co-product accompanied by power generation from SOFCs. Methane fueled SOFCs have high electrical efficiencies, however, they accompany CO_2 emission of 0.198 η^{-1} kg-CO_2 / kWh, where η is electrical efficiency. While the CO_2 emission coefficient decreases with increasing the electrical efficiency, the reduction of the coefficient has a lower limit of 0.198 kg-CO_2 / kWh by means of merely increasing the electrical efficiency.

We have conducted feasibility study of the innovative low-carbon methane-fueled power generation, in which methane was partly decomposed to form hydrogen and solid carbon before power generation, and a mixture of the generated hydrogen and undecomposed methane was used as a fuel. Figure 1 shows the schematic illustration of the system designed for the low-carbon power generation with the methane fueled SOFC. In the SOFC system, partially replacing the steam-reforming methane fuel by the decomposed methane fuel was found to result in a lower CO_2 emission coefficient of the electrical power generated in the system without a serious increase of the thermal energy needed for the fuel processing. The coefficient had strong dependence on the ratio of the decomposition to the steam-reforming.

The solid carbon deposited by the decomposition will be able to be utilized as carbon materials such as carbon black and carbon fiber, or source material of syngas which leads to liquid chemicals. Thus by using this system, carbon fixation in methane fueled power generation will be materialized through forming the co-products such as solid carbon, syngas, and/or liquid organic compounds.

Fig. 1. Schematics of the system used in this study,

Acknowledgement

This research is supported by The Japan Science and Technology Agency (JST) through its "Center of Innovation Science and Technology-based Radical Innovation and Entrepreneurship Program (COI Program).

References

1. Yoshio Matsuzaki, Yuya Tachikawa, Takaaki Somekawa, Toru Hatae, Hiroshige Matsumoto, Shunsuke Taniguchi, Kazunari Sasaki (2015)., "Effect of proton-conduction in electrolyte on electric efficiency of multi-stage solid oxide fuel cells", *Sci. Rep.* 5, 12640, pp 1-10

2. Hazzim F. Abbas, W.M.A. Wan Daud (2010), "Hydrogen production by methane decomposition: A review", *Int. J. Hydrog. Energy* 35 (3) pp. 1160-1190

PNNL-SA-130024

Pacific Northwest NATIONAL LABORATORY

Proudly Operated by **Battelle** *Since 1965*

Performance Modeling of Materials-Based Hydrogen Storage Systems for Automotive Applications

Kriston Brooks, Pacific Northwest National Laboratory
David Tamburello, Savannah River National Laboratory
Sam Sprik, National Renewable Energy Laboratory
Matthew Thornton, National Renewable Energy Laboratory

October 30, 2017

SRNL

NREL National Renewable Energy Laboratory

Pacific Northwest NATIONAL LABORATORY

U.S. DEPARTMENT OF ENERGY

Hydrogen Storage Engineering CENTER OF EXCELLENCE

Pacific Northwest NATIONAL LABORATORY

Proudly Operated by **Battelle** *Since 1965*

Materials-Based Hydrogen Storage

- Conventional H_2 Storage
  - Gaseous storage: 350 – 700 bar, ambient temperature
  - Liquid storage: 20K, near atmospheric pressure
  - Low volumetric and gravimetric density
- Materials-Based Hydrogen Storage
  - Adsorbents
    - Physisorption on high surface materials at 77K and ~ 100 bar
    - Activated carbons, MOF materials
  - Metal Hydrides
    - Reversible chemisorption: refuel with gaseous hydrogen
    - $NaAlH_4$, $TiCrMnH_2$, FeTi
  - Chemical Hydrides
    - Non-reversible chemisorption: return to central facility to regenerate
    - NH_3BH_3, AlH_3

Pacific Northwest NATIONAL LABORATORY
Proudly Operated by Battelle Since 1965

Problem in Developing New Materials

- Challenge for Materials Developers
  - DOE Technical Targets are based on system properties not material properties
  - Material developers cannot quickly evaluate materials to determine if promising
    - Address engineering problems
    - Estimate storage system mass and volume
    - Determine system performance relative to drive cycles

DOE Technical Targets
- System Volume and Mass
- Vehicle Range
- Onboard Efficiency
- Vehicle Performance
  - Transient response
  - Refueling
  - Ambient temperature

November 21, 2017 | 3

Pacific Northwest NATIONAL LABORATORY
Proudly Operated by Battelle Since 1965

DOE Hydrogen Storage Engineering Center of Excellence

- Multi-Lab/Industry project Under the Department of Energy (DOE) Fuel Cell Technologies Office
- Address the significant **engineering** challenges associated with developing materials-based hydrogen storage systems
- Enable fuel cell vehicles to meet customer expectations for driving range and performance for light duty vehicle applications
  - Design, build and evaluate **subscale prototype** systems.
  - Develop, validate, and disseminate new system, **engineering and design models**
  - Material developers use models to evaluate new materials.

NREL
Los Alamos NATIONAL LABORATORY
OSU Oregon State UNIVERSITY
HRI UQTR
GM
BASF The Chemical Company
SRNL
United Technologies Research Center
MICHIGAN
Ford
JPL Jet Propulsion Laboratory California Institute of Technology
HEXAGON LINCOLN
Hydrogen Storage Engineering CENTER OF EXCELLENCE

November 21, 2017 | 4

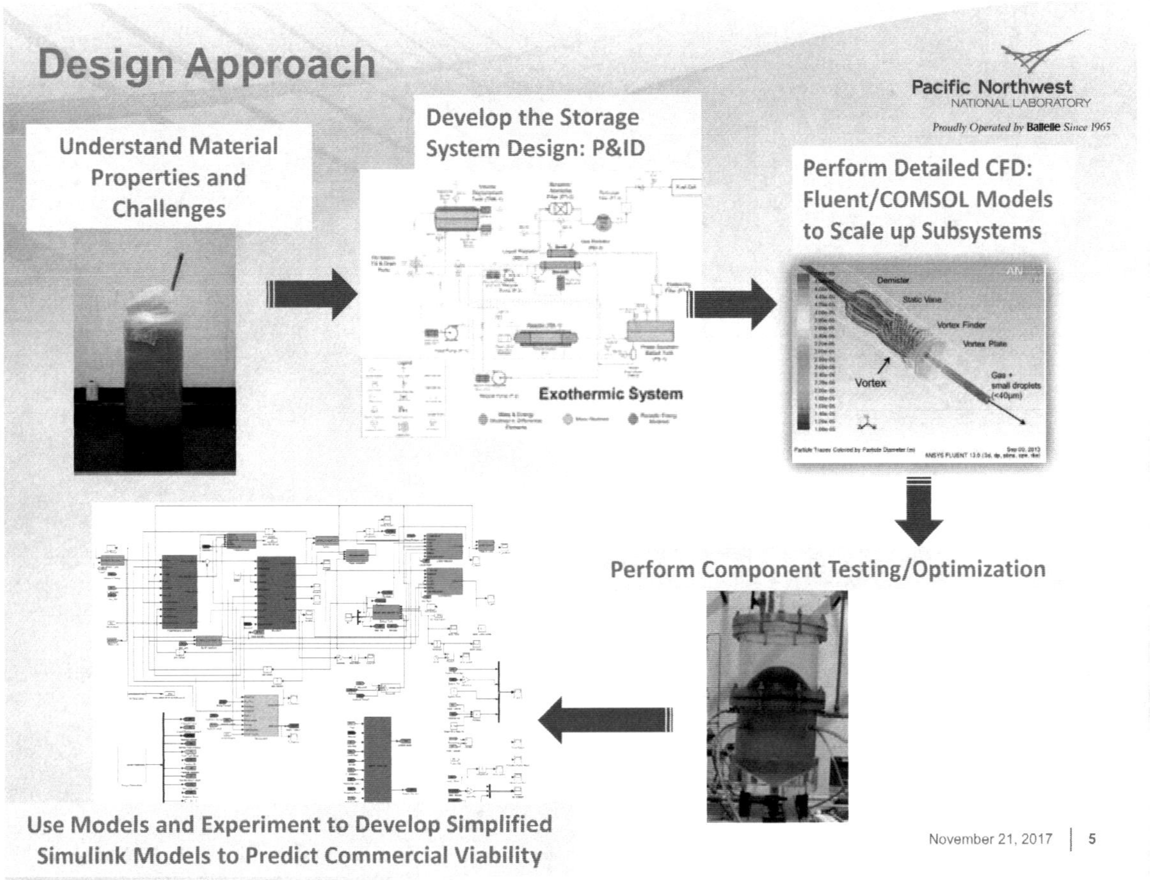
Design Approach
Pacific Northwest
NATIONAL LABORATORY
Proudly Operated by Battelle Since 1965
Understand Material Properties and Challenges
Develop the Storage System Design: P&ID
Exothermic System
Perform Detailed CFD: Fluent/COMSOL Models to Scale up Subsystems
Demister
Static Vane
Vortex Finder
Vortex Plate
Vortex
Gas + small droplets (<40μm)
Perform Component Testing/Optimization
Use Models and Experiment to Develop Simplified Simulink Models to Predict Commercial Viability
November 21, 2017 | 5

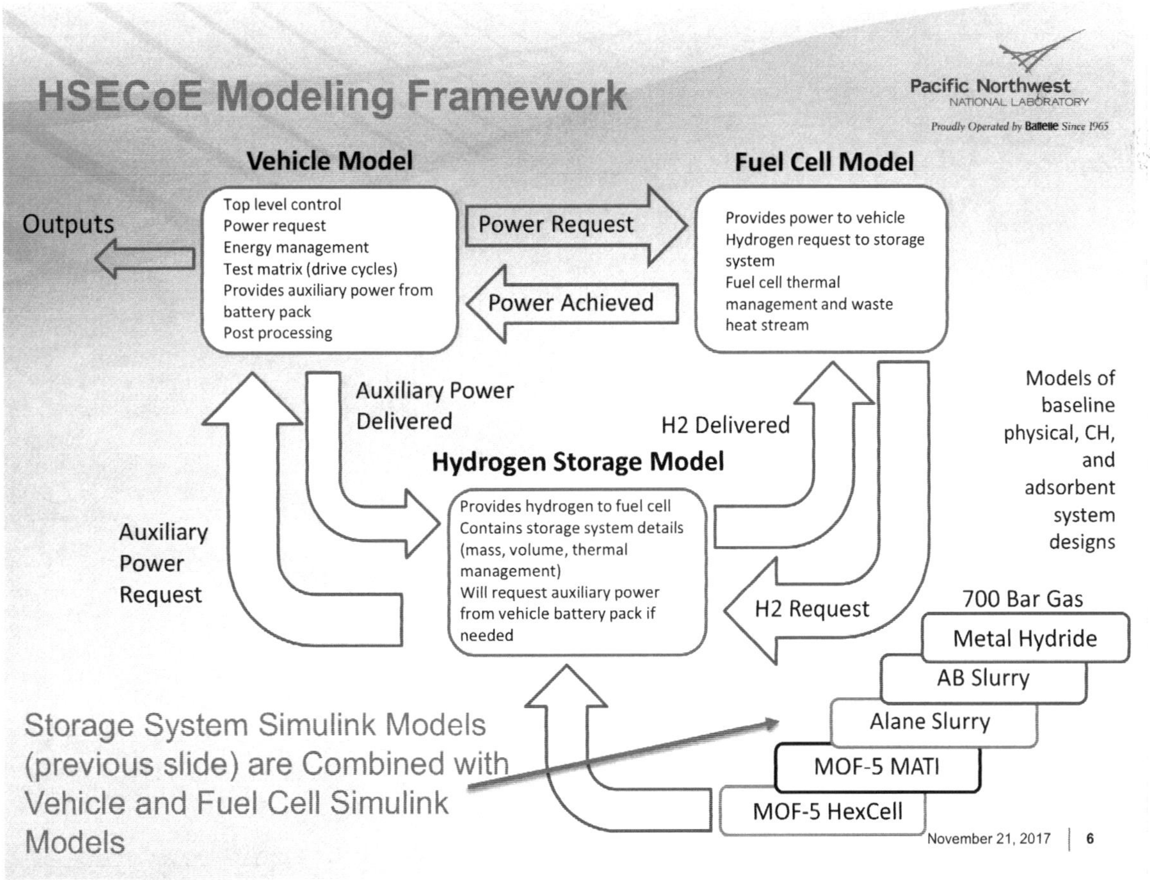
HSECoE Modeling Framework
Pacific Northwest
NATIONAL LABORATORY
Proudly Operated by Battelle Since 1965
Vehicle Model
Top level control
Power request
Energy management
Test matrix (drive cycles)
Provides auxiliary power from battery pack
Post processing
Outputs
Power Request
Power Achieved
Fuel Cell Model
Provides power to vehicle
Hydrogen request to storage system
Fuel cell thermal management and waste heat stream
Auxiliary Power Delivered
H2 Delivered
Hydrogen Storage Model
Provides hydrogen to fuel cell
Contains storage system details (mass, volume, thermal management)
Will request auxiliary power from vehicle battery pack if needed
Auxiliary Power Request
H2 Request
Models of baseline physical, CH, and adsorbent system designs
700 Bar Gas
Metal Hydride
AB Slurry
Alane Slurry
MOF-5 MATI
MOF-5 HexCell
Storage System Simulink Models (previous slide) are Combined with Vehicle and Fuel Cell Simulink Models
November 21, 2017 | 6

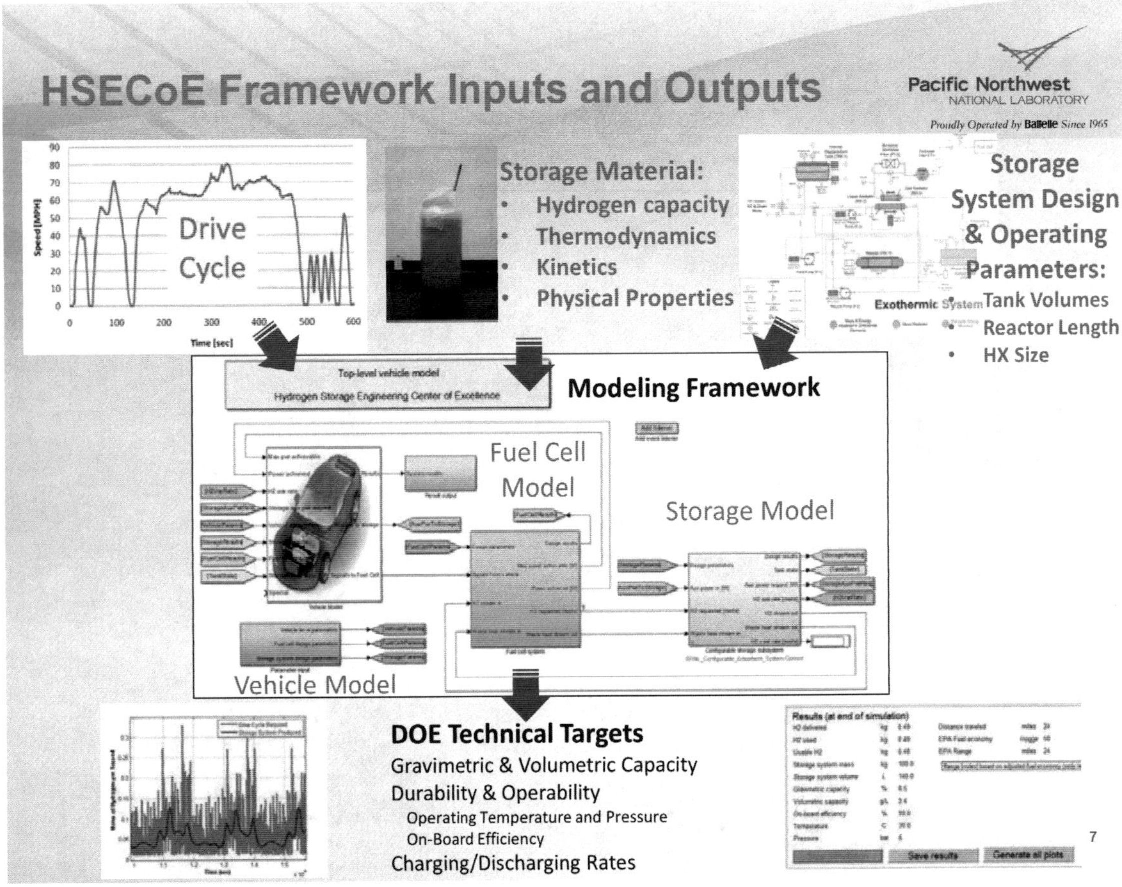
HSECoE Framework Inputs and Outputs
Pacific Northwest
NATIONAL LABORATORY
Proudly Operated by Battelle Since 1965
Drive Cycle
Speed [MPH]
Time [sec]
Storage Material:
• Hydrogen capacity
• Thermodynamics
• Kinetics
• Physical Properties
Storage System Design & Operating Parameters:
• Tank Volumes
• Reactor Length
• HX Size
Exothermic System
Top-level vehicle model
Hydrogen Storage Engineering Center of Excellence
Modeling Framework
Fuel Cell Model
Storage Model
Vehicle Model
DOE Technical Targets
Gravimetric & Volumetric Capacity
Durability & Operability
Operating Temperature and Pressure
On-Board Efficiency
Charging/Discharging Rates
Results (at end of simulation)
Save results
Generate all plots
7

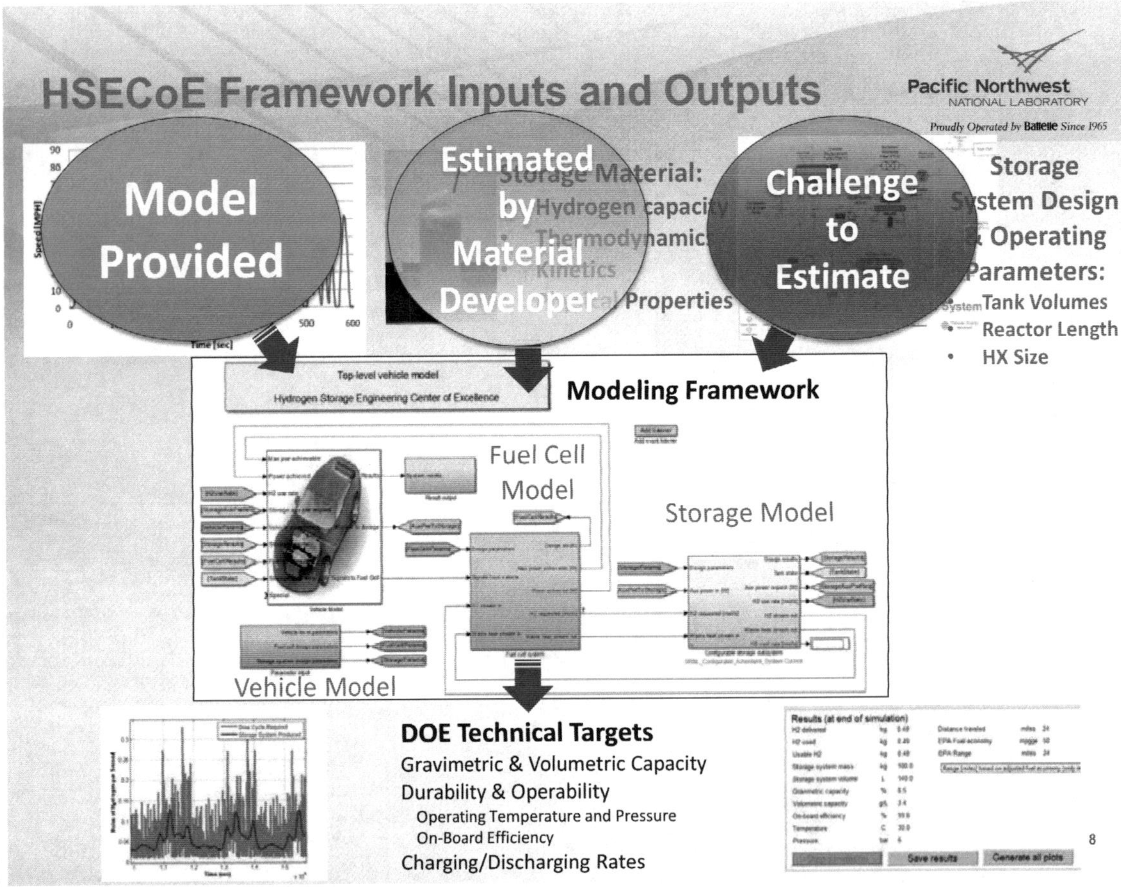
HSECoE Framework Inputs and Outputs
Pacific Northwest
NATIONAL LABORATORY
Proudly Operated by Battelle Since 1965
Model Provided
Estimated by Material Developer
Challenge to Estimate
Storage Material:
Properties
Storage System Design & Operating Parameters:
Tank Volumes
Reactor Length
• HX Size
Top-level vehicle model
Hydrogen Storage Engineering Center of Excellence
Modeling Framework
Fuel Cell Model
Storage Model
Vehicle Model
DOE Technical Targets
Gravimetric & Volumetric Capacity
Durability & Operability
Operating Temperature and Pressure
On-Board Efficiency
Charging/Discharging Rates
Results (at end of simulation)
Save results
Generate all plots
8

HSECoE Framework Inputs and Outputs

Pacific Northwest NATIONAL LABORATORY
Proudly Operated by Battelle Since 1965

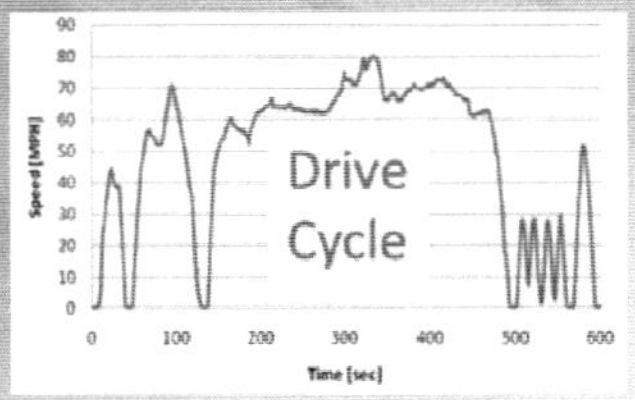

Storage Material:
- Hydrogen capacity
- Thermodynamics
- Kinetics
- Physical Properties

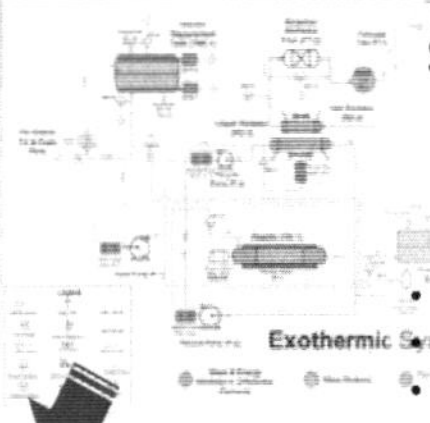

Storage System Design & Operating Parameters:
- Tank Volumes
- Reactor Length
- HX Size

New System Estimator Tool

Stand-Alone Values

Estimated Gravimetric & Volumetric Capacity

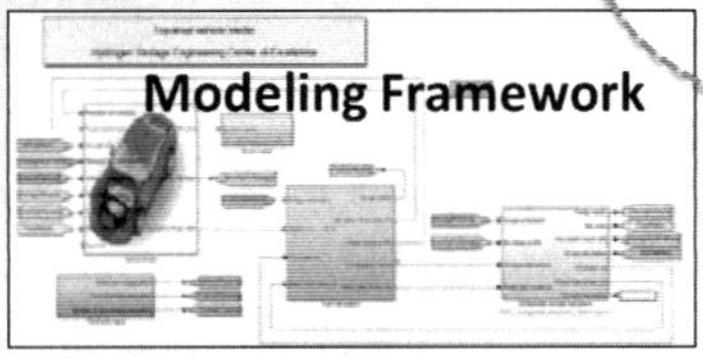

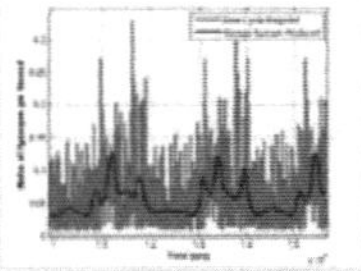

DOE Technical Targets

Gravimetric & Volumetric Capacity
Durability & Operability
  Operating Temperature and Pressure
  On-Board Efficiency
Charging/Discharging Rates

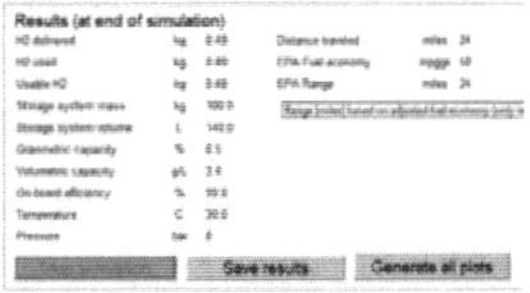

System Estimator Tool

Pacific Northwest NATIONAL LABORATORY
Proudly Operated by Battelle Since 1965

- Developed for Adsorbents, Chemical Hydrides
  - Under development now for Metal Hydrides and Cryo-Compressed Systems
- Written in MatLAB
  - Can be included in the Framework
  - Can be run as a stand-alone executable file (no license required)
- User Inputs/Outputs
  - Microsoft Excel Table

Adsorbent System Design

Pacific Northwest
NATIONAL LABORATORY

- Full tank pressure
  - 10 bar ≤ P_{tank} ≤ 100 bar
- Full tank temperature
  - 77 K ≤ T_{tank} ≤ 300 K
- Model was tuned toward a temperature increase of 80 K
  - Target usable hydrogen
  - 0.5 kg ≤ H2,usable ≤ 12 kg
- Type of Adsorbent / System
  - Adsorbent/System # 1 – Powder (0.18 g/cc)
    - HexCell heat exchanger
  - Adsorbent/System #2 – Compacted pucks (0.4 g/cc)
    - MATI isolated-fluid heat exchanger

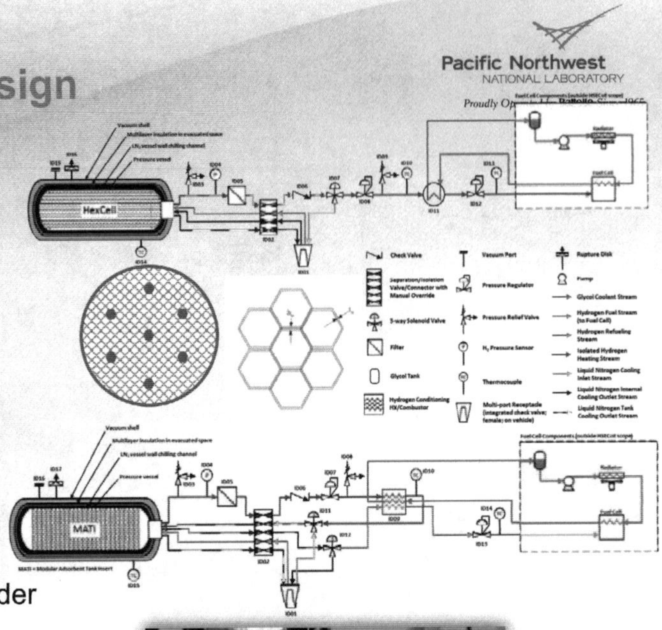

System Estimator Inputs for Adsorbents

Pacific Northwest
NATIONAL LABORATORY
Proudly Operated by Battelle Since 1965

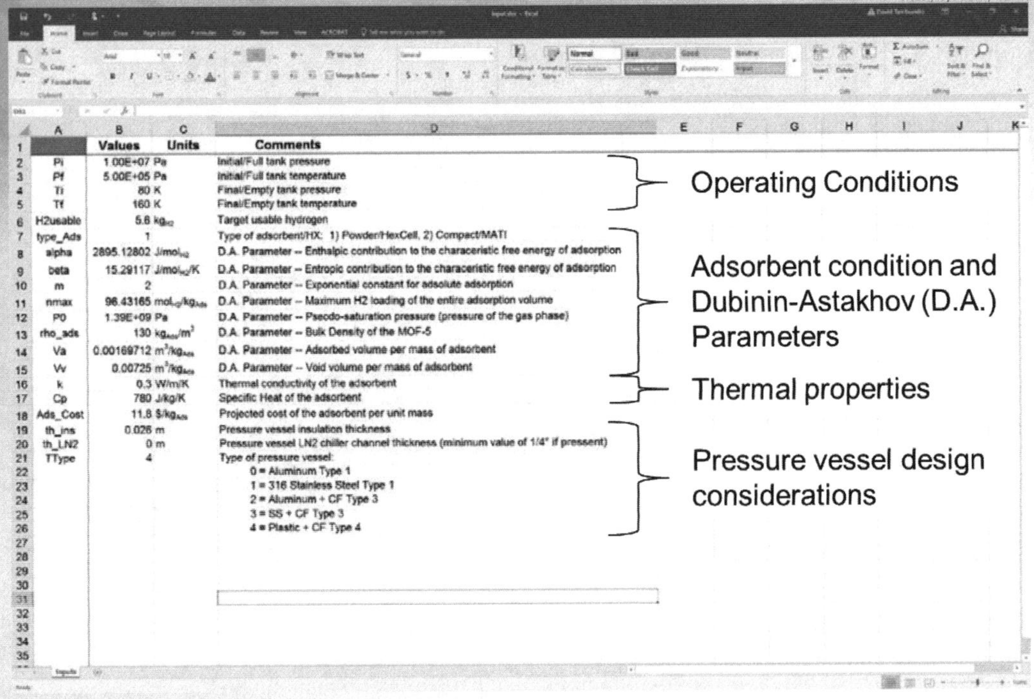

	Values	Units	Comments
Pi	1.00E+07	Pa	Initial/Full tank pressure
Pf	5.00E+05	Pa	Initial/Full tank temperature
Ti	80	K	Final/Empty tank pressure
Tf	160	K	Final/Empty tank temperature
H2usable	5.6	kg_{H2}	Target usable hydrogen
type_Ads	1		Type of adsorbent/HX: 1) Powder/HexCell, 2) Compact/MATI
alpha	2895.12802	J/mol_{H2}	D.A. Parameter -- Enthalpic contribution to the characeristic free energy of adsorption
beta	15.29117	$J/mol_{H2}/K$	D.A. Parameter -- Entropic contribution to the characeristic free energy of adsorption
m	2		D.A. Parameter -- Exponential constant for adsolute adsorption
nmax	96.43165	mol_{H2}/kg_{Ads}	D.A. Parameter -- Maximum H2 loading of the entire adsorption volume
P0	1.39E+09	Pa	D.A. Parameter -- Pseodo-saturation pressure (pressure of the gas phase)
rho_ads	130	kg_{Ads}/m^3	D.A. Parameter -- Bulk Density of the MOF-5
Va	0.00169712	m^3/kg_{Ads}	D.A. Parameter -- Adsorbed volume per mass of adsorbent
Vv	0.00725	m^3/kg_{Ads}	D.A. Parameter -- Void volume per mass of adsorbent
k	0.3	W/m/K	Thermal conductivity of the adsorbent
Cp	780	J/kg/K	Specific Heat of the adsorbent
Ads_Cost	11.8	$\$/kg_{Ads}$	Projected cost of the adsorbent per unit mass
th_ins	0.026	m	Pressure vessel insulation thickness
th_LN2	0	m	Pressure vessel LN2 chiller channel thickness (minimum value of 1/4" if pressent)
TType	4		Type of pressure vessel:
			0 = Aluminum Type 1
			1 = 316 Stainless Steel Type 1
			2 = Aluminum + CF Type 3
			3 = SS + CF Type 3
			4 = Plastic + CF Type 4

Operating Conditions

Adsorbent condition and Dubinin-Astakhov (D.A.) Parameters

Thermal properties

Pressure vessel design considerations

System Estimator Tool Results for Adsorbents

Pacific Northwest NATIONAL LABORATORY
Proudly Operated by **Battelle** *Since 1965*

	Gravimetric Capacity $[g_{H2}/g_{sys}]$	Volumetric Capacity $[g_{H2}/L_{sys}]$	Source
MOF-5 Powder [130 kg/m^3]	0.0338 g_{H2}/g_{sys}	18.6 g_{H2}/L_{sys}	HSECoE
MOF-5 Compact [406 kg/m^3]	0.0314 g_{H2}/g_{sys}	21.3 g_{H2}/L_{sys}	HSECoE
DUT-23 (Co) Powder* [200 kg/m^3]	0.0348 g_{H2}/g_{sys}	20.7 g_{H2}/L_{sys}	Ford / University of Michigan
IRMOF-20 Powder* [200 kg/m^3]	0.0341 g_{H2}/g_{sys}	20.3 g_{H2}/L_{sys}	Ford / University of Michigan
MOF-5 Powder [200 kg/m^3]	0.0332 g_{H2}/g_{sys}	19.6 g_{H2}/L_{sys}	HSECoE

***Special thanks to Ford and the University of Michigan for sharing their data**

System Design Assumptions:

- Operating Conditions: 80 K, 100 bar to 160 K, 5 bar
- 5.6 kg of usable hydrogen (~5.714 kg of actual hydrogen storage)
- Type 1 aluminum pressure vessel
- LN_2 pressure vessel chiller channel thickness of 9.525 mm
- Uniform insulation thickness of 23 mm, with a 2 mm outer aluminum shell

Chemical Hydride System Design

Pacific Northwest NATIONAL LABORATORY
Proudly Operated by **Battelle** *Since 1965*

- System pressure
  - 5 bar ≤ P_{system} ≤ 25 bar
- System temperature
  - 20°C ≤ T_{system} ≤ 300°C
- Model developed
  - exothermic
  - endothermic
  - liquid/slurry phase materials
- Target usable hydrogen:
  - 0.5 kg ≤ H2,usable ≤ 12 kg
- Volume Displacement Tank
- Thermolysis of reactants
- Phase Separation/Purification

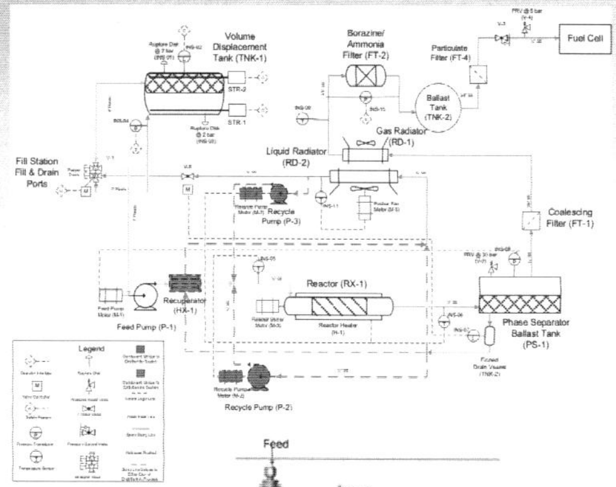

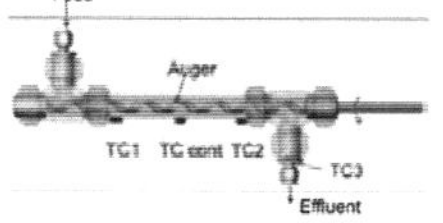

System Estimator Tool Inputs for Chemical Hydrides

Pacific Northwest NATIONAL LABORATORY

Proudly Operated by **Battelle** *Since 1965*

	Values	Units	Comments
ExoEndo	1	--	Exothermic/Endotheric Flag (Exo = 1, Endo = 0)
Kinetic_Model	1	--	Kinetic Model Flag (Avrami Kinetics = 1, nth Order Kinetics = 2)
MW_CH	30.8	g/mol	molecular weight Chemical Hydrogen Material
slurry	1	--	Fluid Properties Flag (Slurry (1) or Liquid (0))
x_H2	0.152	--	Wt Fraction H2 in the CHS Material
n_rxn	1	--	Number of Reactions to Model (1 or 2)
DH_rxn_1	-17981	J/mol H2	Reaction Enthalpy Rxn 1 (negative=exothermic)
Beta1	2.355	mol H2/mol CH	Molar Ratio H2 maximum for CH material Rxn 1
A1	244	sec-1	Pre-exponential factor for Rxn 1
E1	29900	J/mol H2	Activation Energy for Rxn 1
n1	3.1	--	Exponent for Avrami or Reaction Order for Rxn 1
DH_rxn_2	0	J/mol H2	Reaction Enthalpy Rxn 2 (negative=exothermic)
Beta2	0	mol H2/mol CH	Molar Ratio H2 maximum for CH material Rxn 2
A2	0	sec-1	Pre-exponential factor for Rxn 2
E2	0	J/mol H2	Activation Energy for Rxn 2
n2	1	--	Exponent for Avrami or Reaction Order for Rxn 2
x_inert	0.5	--	Weight fraction inert with CHS Material to Slurry
Cp_CH	2694	J/kg/K	Heat Capacity CHS Material
Cp_i	1846	J/kg/K	Heat Capacity inert slurrying agent
Cp_p	774	J/kg/K	Heat Capacity CHS Material Product
rho_CH	780	kg/m3	Density CHS Material
rho_i	1000	kg/m3	Density inert slurrying agent
rho_P	1640	kg/m3	Density CHS Material Product
ppm_imp	500	ppm	Concentration of impurity 1
A_imp	0.1	g impurity/g adsorbent	Adsorbent maximum loading impurity 1
MW_imp	17	g/mol	molecular weight impurity 1
ppm_imp2	2000	ppm	Concentration of impurity 2
A_imp2	0.35	g impurity/g adsorbent	Adsorbent maximum loading impurity 2
MW_imp2	80.5	g/mol	molecular weight impurity 2
Useable H2	5.6	kg	Mass of usable hydrogen required
Power	40	kW	Average Hydrogen Storage H2 Production Required
Pset	25		Ballast Tank Pressure Initial Condition and Setpoint
Q_heater	8000		Reactor heater per length
Tmax	400	°C	Maximum acceptable reactor temperature

Reaction Parameters
Weight Fraction H_2

Heat Capacity & Density

Impurities

Operating Conditions

System Estimator Results for Chemical Hydrides

Pacific Northwest NATIONAL LABORATORY

...dly Operated by **Battelle** *Since 1965*

Calculated System Parameter	Ammonia Borane	Alane	CBN*
Total System Mass (kg)	133	188	117
Total System Volume (L)	146	161	135
System Gravimetric Capacity (kg H_2/kg system)	0.042	0.029	0.048
System Volumetric Capacity (kg H_2/L system)	0.038	0.035	0.041
Reactor Length (m)	0.64	1.28	2.3
Ballast Tank Volume (L)	14	22	32
Mass Chemical Hydride (kg)	77	128	70
Fraction Chemical Hydride	0.5	0.5	1
Liquid Radiator Length (m) (3 tubes)	2.4	1.6	0.9
Gas Radiator Length (m) (1 tube)	1.2	1.0	1.4
Recuperator Length (m) (3 tubes)	0	3.22	0.8
Startup Temperature (°C)	178	202	279
Ballast Time (s)	75	117	176

*CBN:

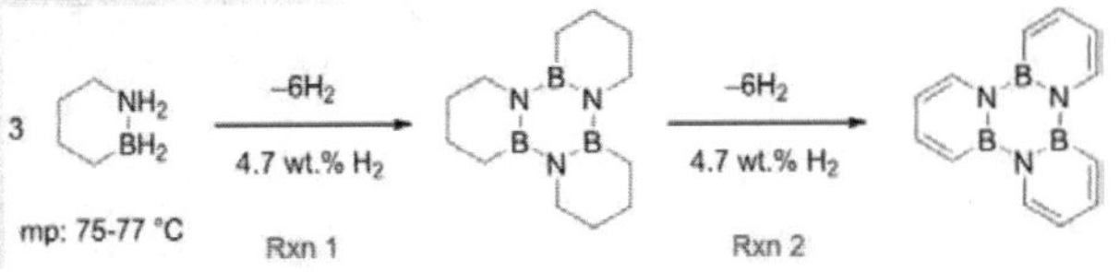

Storage System Design Tool Incorporated into Framework

Pacific Northwest NATIONAL LABORATORY
Proudly Operated by **Battelle** *Since 1965*

- Input user specified parameters for new material/design
- Design tool calculates framework inputs for new material/design
- New material/design can then added to future vehicle simulation

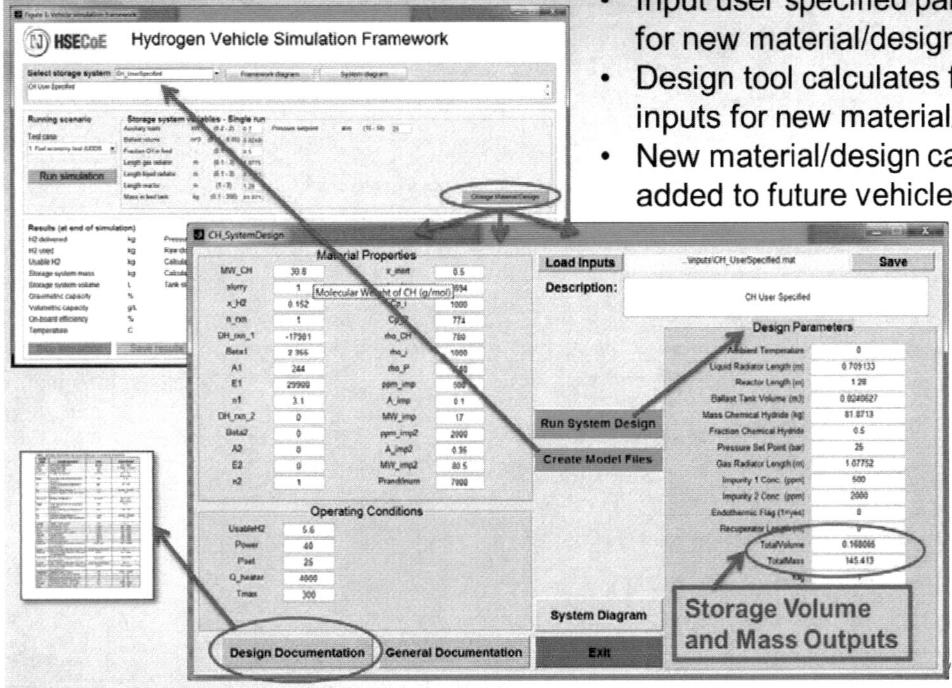

Model Access Website and Support

Pacific Northwest NATIONAL LABORATORY
Proudly Operated by **Battelle** *Since 1965*

HSECoE website:
http://hsecoe.org/

Model support:
HSECoE@nrel.gov

Model access/description sub-page

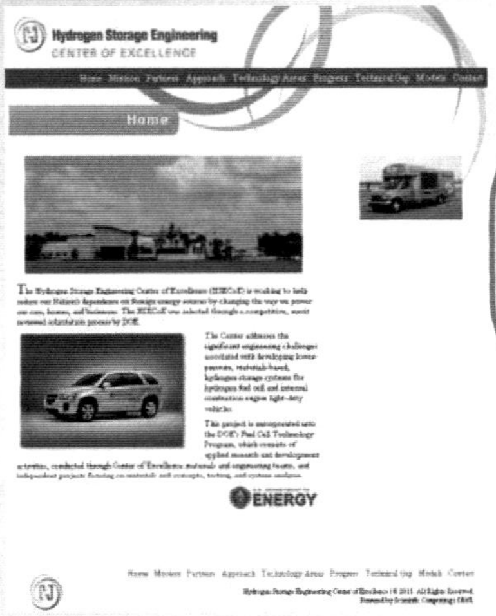

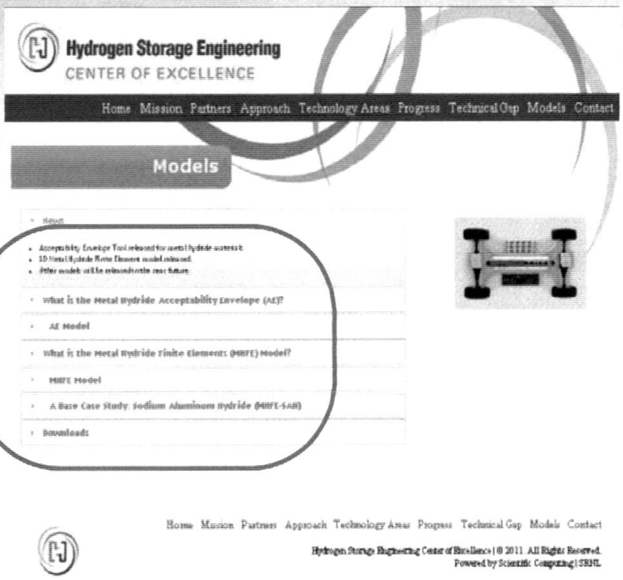

Storage System Models Available for Download

- Adsorbent Storage System Design Tool
- Chemical Hydride Storage System Design Tool
- Hydrogen Vehicle Simulation Framework
  - Chemical Hydride
    - Exothermic (AB)
    - Chemical Hydride Endothermic (Alane)
  - Adsorbent
    - Powder (HexCell)
    - Compacted Powder (MATI)
  - Metal Hydride (Generic hydride 30 kJ/mol)
  - Compressed Hydrogen (350 and 700 bar)

Conclusions

Pacific Northwest
NATIONAL LABORATORY
Proudly Operated by Battelle Since 1965

- Hydrogen storage using materials-based systems provide an alternative to conventional high pressure or liquid H_2 storage
- HSECoE developed the vehicle framework to evaluate hydrogen storage materials relative to the DOE technical targets
- Design tool allows material researchers to estimate
  - Parameters needed to run the framework
  - System mass and volume
- The framework and design tool have been developed for adsorbents and chemical hydrides.

Future Work

Pacific Northwest
NATIONAL LABORATORY
Proudly Operated by Battelle Since 1965

- Develop design tools for cryo-compressed and metal hydrides
- Develop volume-based design to target specific vehicle volume limitations/designs
- Expand model to other vehicle classes (beyond light duty)
- Update Adsorbent model with Unilan (or the 2-state Langmuir) models in addition to the D-A model.
- Work with Material Based H_2 Storage Developers to Apply model to their materials

Questions?

Pacific Northwest
NATIONAL LABORATORY
Proudly Operated by Battelle Since 1965

HSECoE Models on the WEB Team:

Matthew Thornton
David Tamburello
Kriston Brooks
Sam Sprik

SRNL
NREL National Renewable Energy Laboratory
Pacific Northwest NATIONAL LABORATORY

With support from Bob Bowman and Mike Veenstra

Adsorption data provided by Ford, University of Michigan, and University of California Berkeley

Special thanks to the rest of the HSECoE, Jesse Adams, and Ned Stetson

501938 Immobilized Ferrite Nanoparticles for H_2 Generation *Via* Low Temperature Thermochemical Water-Splitting Process

Abstract:

H_2 in its cleaner form can be efficiently produced from the thermochemical water-splitting process. As the process is typically performed at higher temperatures of 900°-1600°C, significant energy input is needed for continuous H_2 production during multiple thermochemical cycles. At higher temperatures though grain growth and spalling of redox materials in thermal reactors often lead to decrease in the H_2 volume during multiple thermochemical cycles. In order to reduce the grain growth and achieve thermal stabilization, we attempted to synthesize novel immobilized ferrite nanoparticles using the sol-gel technique. In particular, highly porous ceramic supports were utilized to immobilize the ferrite nanoparticles. The morphologies of these immobilized materials were investigated by the scanning electron microscopy and X-ray diffraction. A number of ferrite nanoparticles were immobilized and their H_2 generation ability was compared at lower temperatures of 600°-900°
and H_2 volume generation results will be presented.

492691 Application of Liquid Organic Hydrogen Carrier in Mobility

Abstract:

A new, dense and safe form of transport of hydrogen is a Liquid Organic Hydrogen Carrier (LOHC). Meanwhile, the company Hydrogenious Technologies is selling loading and unloading equipment throughout the world.

A new challenge is to run the mobility on a CO2-free basis. One option is the battery car, but what still suffers from a very limited range and certain special drawbacks. An application to battery powered trucks for long distances is out of imagination.

So, the German government asked the author, to set up a plan, how these trucks can be operated with hydrogen in the form of LOHC. The lecture will describe the existing heavy duty hydrogen vehicles and the limits and advantages of this option.

501373 Super-Saturation of Nitrogen and Hydrogen in Transition Metals and Metal Hydrides/Nitrides

Abstract:

Super-saturation of metals with small gas molecules may occur when the amount of dissolved gas exceeds the theoretical maximum solubility at a given temperature and pressure. Since supersaturation effectively increases the amount of gas uptake, it could be useful in material storage and clean energy applications, such as in the case of hydrogen storage or carbon capture. One of the additional consequences of super-saturation is to induce a change in the physical properties of the bulk solid. Super-saturation may possibly enhance small gas diffusivity in metals, a factor that could improve separations in metallic membrane applications. A fundamental understanding of how super-saturation influences material properties can help guide engineers in the design of more effective materials.

The potential for super-saturation-induced property enhancement depends on two factors: (i) in what context is the desired property change considered desirable, and (ii) can super-saturation effectively induce this desired change. Other sub-related concerns involve the stability of this super-saturated state, the reversibility of the induced change, and the energy cost associated with establishing and maintaining the super-saturated state. In order to gain insight into the behavior of super-saturation in metals and related materials, density functional theory is used to investigate the fundamental mechanisms of super-saturation in both pure, alloy and defected (with vacancies) materials.

492233 Process Synthesis and Economical Evaluation of Torrefied Wood Pellet Production Processes

Abstract:

Torrefied wood pellet is a solid biofuel derived from forestry, agriculture, or industrial woody residues. The torrefied wood pellets, with properties close to coal [1], can be potentially used to replace coal for power generation. The design and optimization of the wood pellet production process can help save energy consumption, reduce CO_2 emissions, and increase economic benefits.

A torrefied wood pellet production process involves several unit operations including drying, grinding, torrefaction, pelletizing and cooling, with major heat consumption in drying and torrefaction, and power mainly consumed for grinding and pelletization. During torrefaction combustible organic volatiles are also generated from biomass decomposition, which can be used to provide heat for the drying and torrefaction to make the overall process auto-thermal, thus improve process energy and material efficiency [2]. The heating value of organic volatiles and their mass fraction can be adjusted by controlling torrefaction reactor conditions, such as reactor temperature and mean solid residence time. There are several potential configurations of the above biomass torrefaction and densification processes based on desired pellet product quality and energy efficiency, such as 1) drying " torrefaction " grinding " pelletization; 2) drying " grinding " torrefaction " pelletization, 3) drying " grinding " pelletization " torrefaction. Previous studies of the torrefied wood pellet production process have focused on either single energy balances or single reactor models. Our study combined both considering detail reactor equipment design with particle mechanics, thermal dynamics, and kinetics etc., as well as different heat integration strategies.

In this study, we conducted a systematic investigation of the above three configurations for the production of torrefied pellets using biomass feedstocks of different physical properties. Different biomass feedstocks of different properties require the use of different equipment for drying and torrefaction. For wood chips, rotary dryers and torrefiers are considered with combined direct and indirect heat transfer from combustion flue gases. For sawdust powders, fluidized bed dryer and torrefier with or without build-in heat exchangers are applied. The equipment types and heat transfer mode are selected based on the constraints of operating conditions e.g. inlet gas temperature, gas velocity and availability of gas blowers for high temperature flue gases etc., as well as the overall flowsheet heat integration strategies. Sensitivity analysis was also performed to search for the optimum operating conditions. After the flowsheet synthesis, we carried out economical evaluation to identify the most promising configurations for making desired products. Different strategies were proposed and discussed for different purpose e.g. high energy qualified product, low energy demand, or low production costs. All the synthesized processes were modeled and simulated using Aspen Plus under steady state.The developed flowsheet simulation packages could be applied to assist the development and scale-up of commercial biomass torrefaction and densification process.

References

[1] WH Chen, J Peng, XT Bi., 2015. A state-of-the-art review of biomass torrefaction, densification, and applications. Renewable and Sustainable Energy Reviews, 44, 847-866.

[2] Prins, M.J., 2005. Thermodynamic analysis of biomass gasification and torrefaction. Ph.D. Thesis. Eindhoven University of Technology.

492924 Implementation of Six Sigma in a Methane Generation Process

Abstract:

Distillery process waste water contains organic and inorganic compounds with high COD strength, which causes negative impact on environment. Instead of discharging this waste water to sewer system, it can be pre-treated and used for energy generation before discharging it to the environment. The pre-treatment that possibly used is the anaerobic digestion process, several groups of bacteria will feed on this waste water to generate methane gas. An expanded granular sludge bed reactor has been utilized as the main reactor for this process. The biogas produced will be mainly consists of methane and carbon dioxide. In this paper we studied, how to improve the biogas production and improving methane composition in the generated gas using Six Sigma methodology. DMAIC (Define Measure Analyze Improve Control) method has been implemented in this biogas production process so, process failure could be identified, improve the gauges used in measurement and finally, enhance the yield and composition of methane.

Six Sigma DMAIC methodology behaves as a roadmap to understand various unexplored areas in this process that could help in organizing and updating the standard operating procedure for the experiments. The improvement was observed after introducing six sigma tools and concepts in our experiment, the yield was increased from 11 gallons to 28 gallons in 60 hours which is an impressive 154.5% increase in yield and increase in the percentage composition of methane in the yield from 50% to 76%.

495569 Performance of Fed-Batch Relative to Batch Processes in Unsteady State Fermentation and *in situ* Gas Stripping Simulations Under Different Conditions

Abstract:

The production of biofuels and chemicals from lignocellulosic biomass on the fermentation platform is considered a sustainable energy alternative to fossil fuels mainly because of the mitigation of environmental impacts and energy independence. Butanol and ethanol have properties that make them excellent liquid transportation fuel alternatives. Fermentation is frequently operated as a batch or semi-batch process. The batch fermentation using microorganisms is characterized by low reactor productivities, yield and total solvent produced because of product toxicity and substrate inhibition to the microorganisms. The integrated fed-batch fermentation and *in situ* gas stripping circumvents substrate inhibition by using a substrate feeding strategy that utilizes a lower initial substrate concentration at the beginning of fermentation. As fermentation proceeds and the substrate is used up, fresh substrate is added at concentrations that sustains microbial growth and product formation, and below substrate concentrations inhibitory to the microorganisms. Furthermore, product inhibition is alleviated by removing solvents as they are produced to prevent the build-up of solvents to inhibitory concentrations.

In this study, the batch reactor, RBatch block, in Aspen Plus is linked to a Fortran user kinetics subroutine to simulate an unsteady state fed-batch fermentation process. A fed-batch and *in situ* gas stripping process will be simulated by continuously feeding nitrogen gas to an RBatch block. With this simulation approach, the advantages of the integrated fed-batch and *in situ* gas stripping to improve the productivity, yield and total solvent production by reducing substrate and product inhibitions relative to a batch fermentation will be highlighted under different conditions. An important feature of the procedure developed in this study will be to show that the trends predicted from the model are consistent with general trends and microbial growth kinetics observed in fermentation systems that use microorganisms in the face of high substrate and product inhibitions. This study provides a platform that can be used to as guide to the fermentation experimentalist and assess the accuracy of literature data. Additionally, a universally accepted process simulator, Aspen Plus, can be used in the simulation of fermentation processes linked with separation unit operations to provide strategies for the design and optimization of bioprocesses.

496926 Technoeconomic Assessment of Biofuel and Biochemical Production Using New Pretreatment Processes for Lignocellulosic Biomass and Lignin Valorization

Abstract:

Recently developed extractive ammonia (EA) and copper–bipyridine catalyzed alkaline hydrogen peroxide (AHP) pretreatments are capable of solubilizing a large portion of lignin in lignocellulosic biomass. Liquefied lignin can be valorized through electrochemical processes to produce value-added biochemicals such as cyclohexanol which is a Nylon 6,6 precursor. In addition, lignin removal improves saccharification by increasing the carbohydrate accessibility to the enzymes. This study evaluates the potential of two pretreatment processes for bioethanol and biochemical production from lignocellulosic biomass, namely: 1) extractive ammonia (EA) pretreatment and 2) copper–bipyridine catalyzed alkaline hydrogen peroxide (AHP) pretreatment. Both processes are simulated in Aspen Plus v8.8 to obtain heat and material balances required for energy and economic assessments. Technoeconomic models are developed to calculate the minimum ethanol selling price considering a 10% internal rate of return and 30 years of plant life. Results showed that high ammonia-to-biomass ratio in EA makes the ammonia recovery highly energy intensive. Similarly, high hydrogen peroxide and bipyridine demand in AHP makes it a costly process. Therefore, reducing the amount of pretreatment chemicals is required to improve the economic viability of both processes. Finally, sensitivity analyses on different process parameters were performed to identify the pitfalls and potential of each process for further development and economic profitability.

498923 Technoeconomic and Strategic Insights on Cellulosic Biofuel Production with Ionic Liquid Pretreatment

Abstract:

Biomass pretreatment is a key step in the biochemical conversion of cellulosic biomass to biofuels. Some ionic liquids (ILs) are very efficient solvents for biomass pretreatment (1-4). Given the possibility of numerous anions and cations, there is a significant opportunity to *design* ILs and *tune*their properties to improve sugar yields. Due to the highly-integrated nature of the cellulosic biorefineries, however, the pretreatment process needs to be developed not only to improve sugar yields but also to minimize any negative impacts on the other parts of the biorefinery (including enzymatic hydrolysis, microbial fermentation wastewater treatment, and on-site cogeneration facility to produce steam and electricity). Design of such processes requires understanding of the underlying material & energy balances, mass & heat transfer effects, thermodynamics, and kinetics. To this end, detailed model based technoeconomic analysis (TEA) offers valuable guidance. Our recent work (1-4) highlighted the importance of biorefinery processes with reduced water and energy footprint, and high IL recovery to ensure economic viability. With further advancements, some of the recently developed processes could be cost-competitive, with a minimum selling price of <$4/gal of advanced biofuel. This presentation will discuss TEA driven insights and challenges that must be addressed to realize an economical biomass conversion process at industrial scale .

References:

1. Konda et al. (2014). Understanding cost drivers and economic potential of two variants of ionic liquid pretreatment for cellulosic biofuel production. Biotechnology for Biofuels 7: 86.
2. Konda et al. (2015). Towards a cost-competitive ionic liquid based pretreatment: Insights from technoeconomic analysis. AIChE Annual Meeting, November 2015, Salt Lake City, UT.
3. Sun et al. (2016) CO_2 enabled process integration for the production of cellulosic ethanol using bionic liquids." Energy & Environmental Science 9.9: 2822-2834.
4. Xu et al. (2016) Transforming biomass conversion with ionic liquids: process intensification and the development of a high-gravity, one-pot process for the production of cellulosic ethanol." Energy & Environmental Science 9.3: 1042-1049.

500238 Process Simulation of Integrated Biomass Torrefaction and Pelleting Plant

Abstract:

Process simulation is a useful technique to perform a detailed mass and energy balance of a system using different operating conditions. design parameters and equipment specifications. Simulation is beneficial to scale-up the process and estimation of the investment costs. The process of torrefied pellet production including different unit operations such as combustion, drying, torrefaction, grinding, and pelletizing was simulated with Aspen Plus software platform. In this study, a process simulation of converting wood chips into torrefied pelleting plant was developed. Drying section was simulated through a solid conventional dryer applying the experimental data for drying kinetics. In the torrefaction section, a two-step kinetic model consists of intermediate products was employed and the solid product, as well as volatile products, were produced at various torrefaction temperature. A volatile stream was considered as a combination of ten major components with fixed composition deriving from the torrefaction experiments of forest biomass. At last, a pinch analysis was implemented, and several scenarios were proposed based on different capacities. The size reduction section and pelletizing were also modeled based on experimental particle size distribution and industrial energy consumption data. This simulation model might be applied to a wide range of feedstock types through woody biomass to energy crops, and the effects of various process parameters on the production cost and environmental impacts can be evaluated.

Keywords: process simulation, torrefaction, drying, pelletizing, Aspen Plus

499952 Influence of Solvent Molecular Structure on Energy Consumption of Post-Combustion CO_2 Capture Processes

Abstract:

Influence of Solvent Molecular Structure on Energy Consumption of Post-Combustion CO_2 Capture Processes

Kevin G. Joback[1] J.R. Heberle[2] and Abhoyjit S. Bhown[2]

[1]Molecular Knowledge Systems, Bedford, NH 03110-0755 USA

[2]Electric Power Research Institute, Palo Alto, CA 94304, USA

We present an updated physical property-based model for calculating the energy requirement for the post-combustion capture of carbon dioxide by absorption in aqueous amine solutions for any arbitrary amine. The only input the model requires is the amine's molecular structure.

Equations 1 through 6 show the chemical reactions between carbon dioxide, water and an amine:

$$CO2g \rightarrow CO2aq \quad (1)$$

$$2H2Ol \rightarrow H3O{+}{+}OH{-} \quad (2)$$

$$CO2aq{+}2H2Ol{-}{>}H3O{+}{+}HCO3{-} \quad (3)$$

$$H2Ol{+}HCO3{-}{=}H3O{+}{+}CO3{-}2 \quad (4)$$

$$RR'R''Naq{+}H3O{+}{=}RR'R''NH{+}{+}H2Ol \quad (5)$$

$$HCO3{-}{+}RR'NH(aq){=}H2Ol{+}RR'NCOO{-} \quad (6)$$

Primary and secondary amines undergo reactions 5 and 6 while tertiary amines undergo only reaction 6, i.e., tertiary amines do not form carbamates.

Our updated model uses:

· An equation for the Henry's law constant of carbon dioxide in aqueous solvent mixtures (Equation 1). This equation was developed from experimental data on carbon dioxide solubility in water and nitrous oxide solubility in pure amines.

· Equilibrium constants regressed from data on the water dissociation reaction (Equation 2) and the carbon dioxide ionization reactions (Equation 3 and Equation 4).

· A new group contribution technique for estimating a solvent's pKa as a function of temperature (Equation 5).

· A new, general equation for the equilibrium constant of the carbamate formation reaction for primary and secondary amines (Equation 6). This equation requires only knowledge of the amine's molecular structure; no additional experimental data is needed.

· A group contribution methods to estimate the liquid-phase activity coefficients.

· A group contribution method to estimate pure component heat capacities.

· A simple model used to estimate the heat capacity of aqueous amine solutions loaded with carbon dioxide.

Combined, these equations are used to calculate the concentration of chemical species, carbon dioxide partial pressure and the enthalpies of reaction over a range of solvent concentrations, loadings, and system temperatures. Finally, the effective heat capacity is used to calculate the equivalent electrical load (reduction in net electrical output from a power plant) using a simplified process-independent model.

The development of the model was guided by a simplified sensitivity analysis. We varied values for parameters and inputs—e.g., Henry's law constants, pKa as a function of temperature, equilibrium constants, activity coefficients, and heat capacities—to identify those parameters that had the greatest affect on overall model accuracy. Those parameters with the greatest significance became the focus of improvement. The final model developed has good accuracy when compared with experimental data for species concentrations, partial pressures, and enthalpies of reaction for several common aqueous amine solvents. We also show that our new model has accuracy comparable to several current models that have been regressed directly from experimental data.

While the model uses multiple equations and regressed parameters, we reemphasize that the only input needed to evaluate the performance of any amine solvent is the amine's molecular structure. We used our model to computationally evaluate hundreds of candidate solvent structures. For the most promising candidates we calculated the species concentrations, partial pressures, enthalpies of reactions and imposed loads. From these calculations, we have drawn some general conclusions on the influence basic physical properties on the energy consumption of post-combustion carbon dioxide capture processes.

503008 Highly Efficient Warm Gas Carbon Capture System for IGCC Power Plants

Abstract:

TDA is developing a novel sorbent that removes CO_2 via physical adsorption from synthesis gas. The relatively strong affinity of the sorbent to CO_2 enables effective operation at temperatures up to 300°C (well above the dew point of synthesis gas stream generated by commercial gasifiers). However, because the sorbent and the CO_2 do not form a true covalent bond, the energy needed to regenerate TDAâ€™s sorbent (5.4 kcal per mol of CO_2) is much lower than that observed for either chemical absorbents (e.g., 29.9 kcal/mol CO_2 for sodium carbonate) or amine-based solvents (e.g., 14.2 kcal/mol CO_2 for monoethanolamine). TDAâ€™s sorbent can be regenerated isothermally and CO_2 can be recovered at pressure (as high as 150 psia). Thus, the energy needed to regenerate the sorbent and compress the CO_2 for sequestration is significantly lower than that for any other technology reported to date. The high surface area and favorable porosity of the sorbent also provides a unique platform to introduce additional functionality, such as active groups to catalyze the water-gas-shift (WGS) reaction or to remove trace metals (e.g., Hg, As). Hence, TDAâ€™s CO_2 sorbent could be used to do a warm gas capture of CO_2 above the dew point of the synthesis gas in an Integrated Gasification Combined Cycle (IGCC) power plant. TDA's pre-combustion CO_2 capture system achieves a net plant efficiency of 34% on a coal high heating value basis when operated on bituminous coal, which is significantly higher than that can be achieved with the Selexol™ scrubber at 31.6% (state-of-the-art pre-combustion carbon capture technology). The levelized cost of electricity including the transport, storage and monitoring costs for CO_2 is calculated as \$92.9/MWh for TDA's system, in comparison to \$105.2/MWh for Selexol™. In a current DOE sponsored project, TDA is carrying out field tests with the warm gas CO_2 capture unit at 0.1 MWe scale at the National Carbon Capture Center (NCCC) in Wilsonville, AL and has so far completed over 500 hours of field tests achieving greater than 90% carbon capture. In this paper we present the results from the system analyses and the field test with the 0.1 MW_e pilot unit at NCCC

486301 the Development of Data Mining-Molecular Modeling Approach to Screen Physical Solvents for Gas Separation

Abstract:

Integrating the design of both process and solvent for gas separation involves simultaneously optimizing the process structures, operating conditions and solvent molecules. This task is challenging partly due to the large number of non-linear equations for process models, the large number of variables (such as the operating conditions) and the large solvent chemical space to be explored. One way to alleviate this complicated problem is through sub-optimizations by restricting the search space in each step of process design and solvent development. In this work, we focus on solvent design to physically and selectively absorb one gas from the other in the gas mixture. Specifically, a hybrid computational approach was developed to systematically screen solvents to separate CO_2 from a fuel gas mixture which contains CO_2, H_2 and H_2O. The objective of this work is to search for solvents which have desired properties such as hydrophobicity, high CO_2 loading (volume based), high CO_2/H_2 solubility selectivity, non-foaming, low viscosity, low heat capacity, good thermal stability, no environmental and safety issues, and cheap price. The computational approach involves searching the NIST database for pure organic compounds, the in-house computational database development, and atomistic simulations. By using this computational approach, several promising solvents have been identified. For example, we have identified one super-hydrophobic solvent, which has a high boiling point of ~275 °C, large CO_2 loading of 1.06 mol/MPa/L at 25 °C, CO_2 heat of absorption of 10.6 kJ/mol (not high), large CO_2/H_2 solubility selectivity of 50 at 25 °C, low viscous (~4 cP) at ambient condition, heat capacity of 1.64-1.87 J/g°C at 25-100 °C (not high and much less than the heat capacity value for water), non-foaming, and reasonably cheap. The predictions are being confirmed by the experiments at NETL. In addition, we will present the largest CO_2 solubility and CO_2/H_2 solubility selectivity in any organic compound, which would be useful to implement the techno-economic analysis in process modeling to estimate the minimum operating and capital costs.

498520 Analysis of Post-Combustion Inertial CO_2 Extraction System

Abstract:

Analysis of Post-Combustion Inertial CO_2 Extraction System

Adam H. Berger[1] Yuqi Wang[1*] Abhoyjit S. Bhown[1]

Robert Kielb,[2] Anthony Castrogiovanni[2] Vladimir Balepin[3]

[1]Electric Power Research Institute (EPRI), Palo Alto, CA, USA

[2]ACENT Laboratories, Bohemia, NY, USA

[3]Orbital ATK, Ronkonkoma, NY, USA

Abstract

One method for removing CO_2 from flue gas that has been proposed for post-combustion CO_2 capture is cryogenic separation. To cool the flue gas to the necessary temperature to remove CO_2 as a solid, Orbital ATK and ACENT Labs have proposed the Inertial CO_2 Extraction System (ICES) as shown in Figure 1. ICES uses the cooling effect from accelerating flue gas to supersonic speeds in order to promote the formation of solid CO_2 particles. The solids can then agglomerate to form larger particles that grow sufficiently to be inertially separated from the remaining gas flow by an appropriately-curved gas flow duct. At the duct curve, CO_2 solids can be removed using a simple knife edge separator, sent to a cyclone, and pressurized in dense-phase, minimizing the energy of compression. The remaining CO_2-free gas stream that is still at high speed in the duct but at low temperature and pressure can recover its pressure in a diffusor for emission to the atmosphere. This work provides a thermodynamic analysis of this process and the effects of condensing material in supersonic, two phase flow in the duct.

When gas is isentropically accelerated to supersonic speeds through a nozzle, the temperature and pressure of the gas are significantly lowered to a point where CO_2 forms a solid as described by isentropic supersonic flow equations. However, these simple relations are complicated by the introduction of multi-phase flow, heat of sublimation released during phase change, friction losses, and other non-idealities. In addition, to ensure that the solid particles that are formed can grow to the appropriate mass to inertially migrate towards the solid-capture duct, nucleation sites can be introduced in the form of recycled solid CO_2 particles. In order to analyze the performance of the ICES capture system for post-combustion CO_2 capture, the accelerating flow through a converging-diverging nozzle, solid CO_2 capture, and subsequent pressure recovery in a decelerating diffusor were computationally simulated at a range of initial conditions and with different levels of solid CO_2 recycle.

Figure 1. Inertial CO_2 Extraction System.

To simulate the ICES process, an equation of state for CO_2 valid in the solid-gas region was implemented. This was combined with real-gas property data to calculate the phase behaviour and thermodynamic state of the system at the temperatures and pressures required for solid CO_2 formation. The equation of state with phase equilibrium calculations was used to calculate the thermodynamic state of the process at each velocity during acceleration and deceleration. The solids separation and gas deceleration in the diffusor were subsequently calculated as well using similar techniques. The simulation results were verified through forwards and backwards comparison to isentropic flow, single-phase supersonic flow, and Rayleigh flow. This integrated simulation provides an understanding of the ideal flow case and allows analysis of the ICES process.

This paper will present the results of these calculations, including the effect of inlet temperature, inlet pressure, and the presence of a recycled medium to increase nucleation. Several lessons learned from the simulations and parametric tests of the process will be described. Thermodynamic limits to supersonic cryogenic CO_2 capture processes will also be discussed, for post combustion CO_2 capture from coal-derived flue gas as well as for other gas streams of interest.

*Present Address: Department of Mechanical Engineering, University of California, Los Angeles

499031 Process Modeling and Optimization of a Novel Membrane-Assisted Chilled Ammonia Process for CO_2 Capture

Abstract:

Recent focus on climate change and clean energy has led to a large research focus into the area of post-combustion carbon capture. According to the Intergovernmental Panel on Climate Change, fossil fuels are a leading cause on climate change[1]. Monoethanolamine (MEA) has been the standard of solvents used for carbon capture but there has been recent interest in ammonia. Ammonia is an attractive substitute due to its low energy requirements for regeneration, high CO_2 loading capabilities, and its resistance to oxidative and thermal degradation[2]. One drawback of ammonia is the solvent loss due to the high vapor pressure of ammonia. Two methods, implemented separately or concurrently, are currently being employed to reduce absorber slip for the ammonia system; operating the absorber at low temperatures as done in what is known as the chilled ammonia process (CAP) to lower ammonia vapor pressure, and adding a wash system to remove ammonia from the flue gas before it is released to the atmosphere. This work focuses on developing a full process model of the CAP coupled with a water wash section to reduce the ammonia loss. To reduce the energy penalty in the stripper for the water-wash section as well as CO_2 regenerator, a novel membrane-based water separation technique is investigated.

The electrolyte NRTL model is used to calculate liquid properties and PC-SAFT equation of state to calculate vapor properties. A rate-based model of the towers with two-film characterization and true components is developed. Model validation for the CAP process has been done in the literature but is relatively sparse. Due to the extreme non-ideality of these systems, parameter regression can greatly improve the performance of these models. Simultaneous regression of mass transfer coefficients, interfacial area, diffusivity, and reaction kinetics are carried out by using wetted wall column (WWC) and pilot plant data together. This results in a model that is valid over a large range of operating conditions and equipment sizes. Furthermore, a model of a reverse osmosis membrane is developed to be used in the water wash and CO_2 desorber sections. The membrane model is developed for multi-component separation following the solution diffusion mechanism. Finally, the capture system is optimized to minimize the annuitized cost by considering tradeoffs between the capital cost of the membranes and the energy benefits that are achieved by using them.

REFERENCES:

1 IPCC, 2014: Climate Change 2014: Synthesis Report. *Contribution of Working Groups I, II and III to the Fifth Assessment Report of the Intergovernmental Panel on Climate Change [Core Writing Team, R.K. Pachauri and L.A. Meyer (eds.)]*. IPCC, Geneva, Switzerland, 151 pp.

2 Puxty,G., Rowland, R., Attalla, M., Comparison of the rate of CO2 absorption into aqueous ammonia and monoethanolamine. *Chemical Engineering Science*. 2010; 65, 915-922

495282 Design and Operations Optimization of Membrane Separation for Flexible Carbon Capture from Natural Gas Combined Cycle Systems

Abstract:

We explore the concept of flexible carbon capture using membrane separation. Flexible carbon capture has been studied in recent years as a measure to decarbonize fossil-fired power generation in response to electricity market conditions, including varying electricity prices and/or fluctuating electricity supply requirements. The importance of flexible operation of carbon capture systems is highlighted by the increasing penetration of intermittent energy generation from renewable sources such as wind and solar PV.

To accommodate the integration of renewable energy into the grid, fossil-fired power plants would need to generate varying electricity load, producing flue gas with varying characteristics and at varying volumes. Carbon capture systems in a high-renewables grid will be required to operate flexibly to respond to these changes.

Flexible carbon capture has been studied for amine absorption, [1, 2] a leading CCS technology. This work explores the possibility of flexible carbon capture using membrane separation, a promising alternative to amine absorption. In particular, membranes are considered to be very responsive to system changes and require very short start-up times. [3] In addition, membrane separation has advantages such as having a smaller footprint, being more environmentally benign (no corrosive chemicals involved in the separation process), and potentially incurring lower separation energy.

In this work, we perform optimization to determine the optimal process design and time-varying operations of a polymeric membrane system separating CO_2 from a natural gas combined cycle (NGCC) with wind energy integration. The total net present value (NPV) of the gas turbines and membrane system is maximized. Both design and operations of the capture plant are optimized. Design decision variables include membrane process configuration, membrane size, CO_2/N_2 selectivity and CO_2 permeance (two key membrane properties relevant to separation performance), and compressor and vacuum pump sizes. These parameters are determined before a capture unit is built. After the plant is built, operational decision variables include gas flowrates, the pressure ratio across the membrane and permeate-side (low-pressure side) pressure. These are parameters that can be adjusted, within limits, given electricity market conditions for a given time period.

Time-varying electricity output from gas turbines as well as the associated flue gas flowrate and composition will be determined by HyPPO, an in-house software developed at Stanford University for modeling and optimization of flexible and renewable-integrated power systems. HyPPO models beneficial operating strategies for a set of statistically representative days. HyPPO results will be used with membrane separation models for various process configurations as modeled in MATLAB.

[1] Cohen, S.M., G.T. Rochelle, and M.E. Webber. "Optimizing post-combustion CO_2 capture in response to volatile electricity prices." *Int. J. Greenhouse Gas Control* 8 (2012): 180â€"195.

[2] Mac Dowell, N., and N. Shah. "The multi-period optimisation of an amine-based CO_2 capture process integrated with a super-critical coal-fired power station for flexible operation." *Comput. Chem. Eng.* 74 (2015): 169â€"183.

[3] Brunetti, A., F. Scura, G. Barbieri, and E. Drioli. "Membrane technologies for CO_2 separation." *J. Membr. Sci.* 359 (2010): 115â€"125.

494893 Technical and Economic Feasibility Study of Membranes for Carbon Capture from Low CO_2 Concentration Sources

Abstract:

In the past decade, membranes and membrane processes have been studied extensively for carbon capture from various gas sources in the energy sector, such as syngas (~35% CO_2) from an integrated gasification combined cycle, flue gas (11 â€“ 15% CO_2) in coal-fired power plants, and gas turbine flue gas (~4% CO_2) of a natural gas combined cycle. However, there are other more challenging candidates for carbon capture where the CO_2 concentration is low but the total CO_2 emission is large. In this presentation, a first-of-a-kind membrane process is proposed for carbon capture from low CO_2 concentration (~1%) sources. The low CO_2 concentration sources have included the treated flue gas in coal-fired power plants in which 90% CO_2 has been removed by a primary carbon capture unit, e.g., membrane or aqueous amine absorption, as well as coal mine gas streams. The low CO_2 concentration feed gas is passed to a 2-stage enriching membrane cascade where the CO_2 is sequentially enriched to 95% purity. Vacuum is pulled on the permeate sides of both membrane stages to provide the transmembrane driving force. The techno-economic analysis shows that a high CO_2/N_2 selectivity is crucial to achieve the 95% CO_2 purity. If the CO_2 permeance is higher than 1000 GPU (1 GPU = 10^{-6} cm^3 (STP)·cm^{-2}·s^{-1}·$cmHg^{-1}$), a minimum selectivity of 80 is required. A high selectivity of about 200 is preferred from a cost and energy standpoint. Further increasing the selectivity shows little benefit. By using the state-of-the-art amine-containing facilitated transport membrane with a CO_2 permeance of 1100 GPU and a CO_2/N_2 selectivity of 200, a capture cost of $294.7/tonne CO_2 (in the 2011 dollar basis) can be achieved for 90% CO_2 recovery. This corresponds to an increase in the cost of electricity of 21.2%. The membrane process is feasible for CO_2 mitigation with a feed CO_2 concentration spectrum of 0.6 â€“ 2.0%. If the feed CO_2 concentration falls within this range, 90% CO_2 recovery and 95% purity can be achieved with similar total capital and operational expenditures. It is also found that the 90% CO_2 removal may not be the most economic option for low CO_2 concentration sources. Reducing the CO_2 recovery leads to a smaller energy penalty, and a lowest capture cost is reached at 70% recovery. These cost calculations in the 2011 dollar basis have included both membrane module installation cost and 20% process contingency.

492531 Doping Polybenzimidazole (PBI) with Phosphoric Acid to Enhance Membrane H_2/CO_2 Separation Properties

Abstract:

Membranes that permeate H_2 and reject CO_2 at temperatures above 150 °C are of great interest for low cost CO_2 capture in pre-combustion processes. The current leading polymeric material for this application is poly[2,2'-(m-phenylene)-5,5'-bisbenzimidazole] (PBI), which exhibits good thermal stability, strong size-sieving ability and good H_2/CO_2 selectivity. In this study, we demonstrate that H_2/CO_2 selectivity in PBI can be significantly enhanced by doping with phosphoric acid (H_3PO_4). H_3PO_4 can bond strongly with imidazole rings in PBI, and the complexes are thermally stable up to 200 °C under vacuum. We prepare a series of H_3PO_4 doped PBI samples by immersing PBI thin films (12 μm) in solutions containing H_3PO_4 and methanol. The H_3PO_4 concentration can be varied from 0.05 wt.% to 1.0 wt.% to achieve different doping levels in the PBI films (0.3 - 1.0, defined as the molar ratio of H_3PO_4 to PBI repeating unit). Increasing the doping level increases H_2/CO_2 selectivity and decreases H_2 permeability. For example, pure PBI exhibits pure-gas H_2 permeability of 27 Barrers and H_2/CO_2 selectivity of 14 at 150 °C, while the PBI with a doping level of 0.44 exhibits H_2 permeability of 6.1 Barrers and H_2/CO_2 selectivity of 59 at 150 °C. As the doping level increases to 1.0, the PBI shows an impressive H_2/CO_2 selectivity of 136, though the H_2 permeability decreases to 1.3 Barrers at 150 °C. This performance is above the upper bound in the Robeson's plot for H_2/CO_2 separation. This presentation will also discuss the effect of H_3PO_4 doping on structural changes such as free volume, as well as CO_2 sorption and diffusion, and the structure/property relationships in these H_3PO_4 doped PBI thin films.

487347 Hydrodeoxygenation of Anisole As Bio-Oil Model Compound over Supported Non-Sulphided CoMo Catalysts: Effect of Co/Mo Ratio and Support

Abstract:

Hydrodeoxygenation of anisole as bio-oil model compound over supported non-sulphided CoMo catalysts: effect of Co/Mo ratio and support

Chanakya Ranga[a], Rune Lødeng[b], Vaios I. Alexiadis[a], Joris W. Thybaut[a*]

[a]Laboratory for Chemical Technology, Ghent University, Technologiepark 914, B-9052 Ghent, Belgium.

[b]SINTEF Materials & Chemistry, Department of Kinetics and Catalysis, N-7465 Trondheim, Norway.

*Corresponding Author: Joris W. Thybaut, E-mail: Joris.Thybaut@UGent.be

Introduction

Fast pyrolysis is considered as one of the prominent approaches to obtain liquid organic products with high yield from lignocellulosic biomass. However, pyrolysis oils (bio-oil) cannot be directly used as a fuel or fuel additive due to the high oxygen content [1], which results in undesirable properties such as low thermal and chemical stability, low heating value, immiscibility with conventional, fossil fuels and high acidity. These oils, hence, require upgrading via oxygen removal and increase in hydrogen content. Catalytic hydrodeoxygenation (HDO) is one of the most promising routes to upgrade pyrolysis oils for producing liquid transportation fuels [2].

The complexity of fast pyrolysis oil has prompted the use of model compounds such as phenolics, furans, ethers, acids etc. to study the intricacies of hydrodeoxygenation [1, 3]. Anisole, because of its methoxy group, has already been widely investigated as a model compound for lignin derived bio-oil [4]. The low sulfur content in bio-oil renders the use of traditional sulfided catalysts (NiMoS and CoMoS) for hydroprocessing less suited, as they require the presence of sulfur to retain their activity and, hence, result in contaminated end products. To mitigate this issue, non sulfided transition metal (Ni, Co, Mo) catalysts on various supports have been investigated for hydrotreatment of bio-oil model compounds [5-8], yet significant challenges remain in improving the catalyst activity, selectivity and stability. Our previous work on MoO_3/ZrO_2 catalysts provided insight in the effect of catalyst structural composition with varying preparation conditions on catalyst performance [7-8]. The present work explores the intricacies of anisole HDO reaction pathways upon addition of Co to the Mo catalysts by assessing their intrinsic kinetics performance in non-sulphided form. The catalyst performance is correlated with properties such as reducibility, active site dispersion, crystallite size, metal-support interaction.

Results

A series of supported CoMo catalysts were prepared with varying Co/Mo ratio (0.25, 0.58, and 1.07) while keeping the Mo loading between 8.3 – 10.2 wt%. A sequential incipient wetness impregnation (with Mo being introduced first) method using aqueous solutions of the corresponding precursor salts was employed during the synthesis procedure. Two different supports, i.e., Al_2O_3 and ZrO_2, were used. Material physicochemical characteristics were evaluated through ICP-OES, BET, (in-situ) XRD, H_2-TPR, NH_3-TPD, CO chemisorption, (S)TEM, and XPS techniques. The performance of these CoMo catalysts was tested for anisole HDO at gas phase conditions in a fixed bed tubular reactor in plug flow regime. A high-throughput kinetics screening set-up (HTK-S) [9] comprising 16 plug flow reactors with an internal diameter of 0.00211 m and a length of 0.85 m was used for the acquisition of intrinsic kinetic data.

CoMo/Al_2O_3 catalysts promote the transalkylation and isomerization reactions whereas, ZrO_2 supported ones promote mainly hydrodeoxygenation and transalkylation reactions. Catalyst stability as well as anisole conversion increased with Co/Mo ratio. Figure 1 displays the anisole conversion and product selectivity for CoMo/ZrO_2 and CoMo/Al_2O_3 catalysts with Co/Mo = 1.07 during a stability test (TOS of 50 h). Al_2O_3 supported catalysts displayed higher anisole conversions compared to ZrO_2 ones at the tested operating conditions. Further reaction pathway elucidation is done with the use of intrinsic kinetic measurements varying space time (50 – 250 kg_{cat} s mol^{-1}), temperature (523 - 623 K) and H_2/anisole inlet molar ratio (50-150 mol mol^{-1}) with best performing catalysts. Catalyst deactivation causes will be explored through characterization (O_2-TPO, XRD, XPS, (S)TEM) of spent catalyst samples.

Addition of Co improved the catalyst stability compared to the MoO_3 catalysts in our previous study [7-8]. Catalyst activity-structure correlations drawn from present study can serve significantly in guiding the synthesis of next generation catalysts with better catalyst stability and activity during anisole HDO.

(a)

(b)

Figure 1 Time on stream (TOS) experiment with (a) CoMo/ZrO_2 and (b) CoMo/Al_2O_3 at T = 573 K, P_T = 0.5 MPa, H_2/anisole = 50 mol mol^{-1}, space-time = 220 kg_{cat} s $mol^{-1}{}_{anisole}$. Co/Mo = 1.07 for both catalysts.

Conclusions

Supported CoMo catalysts observed to be relatively stable than the MoO_3 catalysts at the tested operating conditions during anisole HDO over 50 h TOS. Al_2O_3 supported catalysts exhibited higher anisole conversions and relatively lesser stability compared to ZrO_2 ones at the tested operating conditions. Catalysts stability as well as HDO activity observed during an extensive experimental campaign are correlated to the key properties such as active site dispersion, reducibility, crystallite size, and metal-support interaction. This will further enhance our understanding of the catalyst activity, stability and selectivity towards anisole HDO and provide valuable insights into the corresponding reaction pathways.

Acknowledgments

This research work is supported by the Innovative Catalyst Design for large-scale sustainable processes (i-CaD) project, which is a European Research Council (ERC) consolidator grant, by the European Commission in the 7th Framework Programme (GA n° 615456). It also fits in the framework of the FAST industrialization by Catalyst Research and Development (FASTCARD) project, which is a large scale collaborative project supported by European Commission in the 7th Framework Programme (GA n° 604277).

Keywords:

hydrodeoxygenation, reaction pathways, catalyst stability

References

1. Ruddy, D. A.; Schaidle, J. A.; Ferrell III, J. R.; Wang, J.; Moens, L.; Hensley, J. E., *Green Chemistry* **2014,** *16* (2), 454-490.

2. Li, C.; Zhao, X.; Wang, A.; Huber, G. W.; Zhang, T., *Chem. Rev* **2015,** *115* (21), 11559-11624.

3. Furimsky, E., *Applied Catalysis A: General* **2000,** *199*, 147–190.

4. He, Z.; Wang, X., *Catalysis for Sustainable Energy* **2013,** *1* (1), 28–52.

5. Sankaranarayanan, T. M.; Berenguer, A.; Ochoa-Hernandez, C.; Moreno, I.; Jana, P.; Coronado, J. M.; Serrano, D. P.; Pizarro, P., *Catalysis Today* **2015,** *243*, 163-172.

6. Zuo, H.; Liu, Q.; Wang, T.; Ma, L.; Zhang, Q.; Zhang, Q., *Energy & Fuels* **2012,** *26* (6), 3747-3755.

7. Lødeng, R.; Ranga, C.; Rajkhowa, T.; Alexiadis, V. I.; Bjørkan, H.; Chytil, S.; Svenum, I. H.; Walmsley, J.; Thybaut, J. W., *Biomass Conversion and Biorefinery* **2017**, 1-17.

8. Ranga, C.; Lødeng, R.; Alexiadis, V. I.; Rajkhowa, T.; Bjørkan, H.; Chytil, S.; Svenum, I. H.; Walmsley, J.; Detavernier, C.; Poelman, H.; Van Der Voort, P.; Thybaut, J. W., *Chemical Engineering Journal* **2017,** (submitted).

9. Van der Borght, K.; Toch, K.; Galvita, V. V.; Thybaut, J. W.; Marin, G. B., *Catalysts* **2015,** *5*, 1948-1968.

497051 Sulfur-Tolerant Molybdenum Carbide Catalysts Enabling Low-Temperature Stabilization of Fast Pyrolysis Bio-Oil

Abstract:

Fast pyrolysis is a cost effective lignocellulosic biomass liquefaction technology. However, the liquid product, bio-oil, is not suitable for direct use as transportation fuels or as a crude oil substitute due to undesirable qualities such as low heating value, corrosiveness, and instability directly related to its high oxygen and water content. The quality of bio-oils can be improved by catalytic upgrading (e.g., hydroprocessing, cracking) to eliminate oxygenated functionalities.

Developing catalytic bio-oil upgrading technology has proven challenging due in large part to short catalyst lifetime. Bio-oil contains a large quantity of carbonyl compounds which, due to high tendency toward polymerization reactions, lead to rapid catalyst fouling and bed plugging. Stabilization of bio-oil by converting reactive carbonyls prior to higher temperature upgrading is therefore critical. Low-temperature hydrogenation of carbonyl fractions has been recognized as a promising approach to bio-oil stabilization. While reduced precious metal catalysts such as ruthenium are effective in low-temperature hydrogenation reactions, they are highly sensitive to sulfur species present in biomass-derived feedstocks. Therefore, to address rapid performance loss due to S-poisoning of active sites, sulfur management strategies need to be implemented (e.g., sulfur guard, catalyst regeneration) which entails increased process cost and complexity.

Molybdenum carbides have shown potential in various biomass conversion reactions especially with high selectivity towards C-O bond cleavage. In this work, we have assessed Mo carbides in low-temperature hydrogenation of bio-oil with particular emphasis on understanding their sulfur tolerance. Our study revealed that molybdenum carbides are effective in low-temperature conversion of carbonyl compounds. Furthermore, carbides showed excellent sulfur tolerance, and structural robustness with respect to morphology, crystallography, and bulk composition. These properties enabled carbides to maintain catalytic performance in both aqueous-phase furfural (380 ppm S, 150 °C, 120 bar, batch reactor) and real bio-oil (52 ppm S, 140 °C, 124 bar, flow reactor, 100 h TOS) reactor feeds. Due to surface bifunctionality (metal-acid), carbides were able to catalyze both hydrogenation and C-C coupling reactions retaining most of carbon atoms in liquid products as more stable and higher molecular weight compounds. These findings (sulfur tolerance, durability, H and C atom economy) indicate that molybdenum carbides can provide a catalyst solution for the development of commercially viable bio-oil stabilization technology.

492815 Controlling Selectivity of Bio-Oil Model Compound Upgrading with Metal Promoters on Molybdenum Carbide

Abstract:

The upgrading of thermochemical biofuels could lead to cleaner, more homogeneous fuels[1], however, the processes and reactions are not very selective. For research described in this paper, we examine the use of various promoters on molybdenum carbide to improve the reaction selectivities. Crotonaldehyde is used as a model compound because it possesses multiple functionalities commonly found in pyrolysis bio-oils, while also allowing us to probe how changes to the catalyst affect selectivity to several different potential products[2].

In previous work, we have shown that potassium addition to Mo_2C catalyst increased the base site density and decreased the acid site density, and consequently shifted the crotonaldehyde and acetic acid conversion selectivities in the presence of H_2. The results suggested that the increase in base sites was responsible for the increase in selectivities to 3-butenal and acetone, respectively. In this paper, we report on how other metal promoters, including Fe, shift the crotonaldehyde upgrading selectivity relative to Mo_2C.

The Mo_2C catalysts were synthesized by a temperature-programmed reaction method as described previously[3]. Metal salt was added to the native (unpassivated) Mo_2C catalyst by incipient wetness in an oxygen-free, inert atmosphere. The catalysts were dried in H_2 for 2 hrs at 110‹C, reduced in H_2 for 4 hrs at 450‹C, and then passivated in 1% O_2/He for at least 6 hrs at room temperature. All of the catalysts were characterized using x-ray powder diffraction, N_2 physisorption, and inductively coupled plasma optical emission spectrometry. Base and acid site concentrations were measured via temperature programmed desorption of CO_2 and NH_3. The activities and selectivities were measured at atmospheric pressure and 275-350‹C in a flow reactor with a H_2/crotonaldehyde ratio of 6. The ratio was selected because it is twice the stoichiometric amount of H_2 required for complete saturation and deoxygenation of crotonaldehyde to butane.

While K shifted the selectivity to 3-butenal, Fe on Mo_2C favored the formation of butene and suppressed formation of butyraldehyde at high temperatures. The model compound results presented in this paper will demonstrate the ability to tune the product selectivity of molybdenum carbide catalysts for specific desired reaction pathways by adding various metal promoters. Combining these results to promote Mo_2C with a mixture of these promoters could offer a powerful way to tune the selectivity of Mo_2C for bio-oil upgrading, depending on the desired character of the upgraded product.

---

[1] (a) Czernik, S.; Bridgwater, A.V.; *Energy & Fuels*, **2004**, *18*, 590-598; (b) Zhang, Q.; Chang, J.; Tiejun, W.; Yung, X. *Energy Conversion and Management*, **2007**, *48*, 87-92.

[2] Mullen, C.A.; Strahan, G.D.; Boateng, A.A. *Energy & Fuels*, **2009**, *23*, 2707-2718.

[3] Schaidle, J.A.; Schweitzer, N.M.; Ajenifujah, O.T.; Thompson, L.T. *J. Catalysis*, **2012**, *289*, 210-217.

490490 Reductive Conversion of Lignin with Copper-Doped Catalysts

Abstract:

The lignin component of biomass has potential as a renewable source for industrially useful aromatic chemicals. While technologies for the selective conversion of the carbohydrate components have been successful, lignin is generally treated as waste and burned for low grade heat. We have developed a catalytic system, based on a copper-doped porous oxide, which can reductively disassemble lignin in supercritical methanol with little to no char formation. While the copper-doped porous oxide catalyzes C-O hydrogenolysis of aryl-ether bonds linking lignin monomers, it also catalyzed ring methylation and hydrogenation, leading to lignin disassembly product proliferation. We have found that adding dimethyl carbonate can significantly suppress the hydrogenation of the phenolic intermediates responsible for much of the undesirable product diversity by methylation of phenolic hydroxyl groups to form a more stable aryl ether species. Using the methanol and dimethyl carbonate solvent system with the copper-doped porous oxide catalyst, O-methylation of phenolic hydroxyl groups was effective at increasing aromatic yields from a number of lignin model polymer systems and poplar organosolv lignin. These results demonstrate the promise of our copper-doped porous oxide catalyst when used in the presence of methanol and dimethyl carbonate to facilitate aromatic compound production from lignin.

498094 Hierarchical Nickel Incorporated USY Zeolites for Hydrodeoxygenation of Lignin-Derived Pyrolysis Oil Model Compounds

Abstract:

Pyrolysis of biomass and agricultural wastes is a readily achieved method for production of sustainable fuels and platform chemicals. While pyrolysis oil yields of up to 75 wt.% of the original biomass have been demonstrated,[1] raw pyrolysis oil is not suitable for combustion fuel applications due to poor stability and low heating value. Hydrodeoxygenation (HDO) of the oil fraction is a common method for removal of pyrolysis oil oxygen, which is eliminated in the form of water, thus preserving atomic carbon in the pyrolysis oil. Traditional HDO catalysts include NiMo, CoMo and precious metals, however the former require sulfidation and the latter are costly. Here, we propose Ultra-stable Y (USY) zeolite supported Ni catalysts for liquid-phase HDO of anisole, a common product formed from lignin pyrolysis.

Traditional zeolite catalysts are highly microporous, and as a result may limit diffusion of reactants and products to or from catalyst acid sites. Moreover, the low external surface area may result in increased micropore blockage after Ni impregnation. A variety of mesoporous USY zeolites were created using top-down alkaline desilication (DS), desilication with a pore-directing agent (DS-PDA)[2] and surfactant-mediated desilication (SA)[3]. As shown in Figure 1(A), DS formed a broad pore-size distribution with very little micropore volume, whereas DS-PDA preserved some microporosity and produced a narrower range of mesopores. The SA method preserved about 90% of the micropore volume and produced a narrow range of mesopore size at about 30 Å. Material acidity and crystallinity decreased after all treatments, and a clear correlation between material crystallinity, acidity and micropore volume was observed. Figure 1(B) shows a visualization of the types of pores formed from each treatment. Alkaline treatment strength and PDA concentration were varied in order to determine the effect of each variable on zeolite properties.

Ni was impregnated by the incipient wetness method, and NiO crystal size was approximately 30 nm, determined by application of the Scherrer equation to X-ray diffraction (XRD) patterns. After reduction, the Ni crystal size decreased significantly on the more microporous materials, but remained high on the larger-pore materials. Larger crystal sizes and greater interaction with the DS and DS-PDA zeolites was observed from temperature programmed reduction (TPR).

Each material was evaluated as a catalyst for the liquid-phase HDO of anisole at 200 °C and 750 psi. Overall reaction rate increased in the order Ni-SA-USY<Ni-USY<Ni-DS-USY<Ni-DS-USY-PDA, shown in Figure 1(C). A simplified kinetic model of the anisole reaction pathways was proposed, and kinetic parameters were fit to the data and normalized to the number of Ni and acid sites. Analysis of the fitted kinetic parameters showed that the larger mesoporous zeolites enhanced access of reactant anisole to the Ni sites for more efficient hydrogenation of the aromatic ring, whereas acid-catalyzed deoxygenation of the hydrogenated intermediate was about an order of magnitude faster. The Ni-DS-USY-PDA was the most effective catalyst overall, as hydrogenation of the aromatic ring was shown to be the rate determining step in this reaction system.

Figure 1. (A) BJH pore size distribution of mesoporous USY materials. (B) Visualization of each mesoporous USY material, white space indicates pore area. (C) Reaction rates for anisole hydrodeoxygenation (bars correspond to left axis), in addition to final anisole conversion and mass balance (corresponding to right axis).

References

[1] O. Onay, Fuel 86 (2007) 1452–1460.

[2] D. Verboekend, G. Vile, J. Perez-Ramirez, Cryst. Growth Des. 12 (2012) 3123–3132.

[3] J. García-Martínez, M. Johnson, J. Valla, K. Li, J.Y. Ying, Catal. Sci. Technol. 2 (2012) 987.

500416 *Operando* Solid-State NMR Study of Hydrogenolysis of Lignin Ether Linkages

Abstract:

Lignin is the most abundant and readily available feedstocks for bio-derived aromatics. Under mostly reductive conditions, transition metals like Ni and Pd catalyze the hydrogenolysis of lignin ether linkages in the liquid phase. However, the conversion chemistry is lack of molecular understanding especially at the solid-liquid interface, mostly owing to the demanding reaction conditions including high temperature/pressure. Customized high temperature/pressure solid-state NMR rotors were adapted, allowing *operando* solid-state NMR measurements of conversion of lignin catalyzed with supported metal catalysts. For example, benzylphenylether (BPE) as the a-O-4 model compound was studied with supported Ni catalysts at 120-225 °C. Kinetic and mechanistic study was performed with selective isotope labeling (^{13}C, ^{2}H) of both reactants and solvent. The pressure of 50 bar H_2 allows the simultaneous study of the further hydrogenation of phenol to cyclohexanol while toluene showed much higher stability with almost no conversion to methylcyclohexane. When no extra hydrogen gas is charged in the rotor, solvent 2-proponal can also serve as the hydrogen source and Ni can catalyze the transfer hydrogenation reaction with acetone formed and detected *in situ*.

490499 in-Situ Spectroscopy of Lignin Disassembly

Abstract:

Lignin (an aromatic macromolecule in biomass) is considered as waste and under-utilized. A more appealing and value-added use of lignin is its disassembly into aromatic platform chemicals, which can be achieved by appropriate combinations of catalyst, solvent, and reaction conditions. Mechanistic reaction network studies of lignin disassembly are challenging, because the reactions require relatively high temperatures (above 200 °C) and pressures (above 80 bar), as well as the co-existence of solid, liquid, and gas phases. We have used *in-situ* magic-angle-spinning (MAS) solid-state nuclear magnetic resonance (NMR) and small angle neutron scattering (SANS) investigations to study catalytic disassembly of lignin at multiple length scales and in real-time. Our project leverages unique high-temperature and -pressure MAS NMR technology at the Pacific Northwest National Laboratory (PNNL) and a pressure reaction cell for collecting time-resolved neutron scattering data during thermochemical treatment of biomass at Oak Ridge National Laboratory.

501759 UnderstandingÂ Catalyst Selectivity forÂ CO_2 Electro-Reduction

Abstract:

Developing new processes that can employ renewable electricity to produce fuels and chemicals using water and CO_2 as inputs could play an important role in a sustainable future. New catalysts are needed for these processes, particularly those that are selective to desired products, one of the principal challenges in the field. In this talk, we will discuss some of our latest efforts to understand catalyst selectivity for CO_2 electro-reduction, including in-depth studies of catalyst surface structure. We then leverage that knowledge to develop catalyst materials and interfaces with improved performance.

491315 Ag-Sn Bimetallic Catalyst with a Core-Shell Structure for CO_2 Reduction

Abstract:

Converting greenhouse gas carbon dioxide (CO_2) to value-added chemicals is an appealing approach to tackle CO_2 emission challenges. The chemical transformation of CO_2 requires suitable catalysts that can lower the activation energy barrier to minimize the energy penalty associated with the CO_2 reduction reaction. First-row transition metals are potential candidates as catalysts for electrochemical CO_2 reduction; however, their high oxygen affinity makes them easy to be oxidized, which could strongly affect the catalytic properties of metal-based catalysts.[1] Recently, we proposed a strategy to synthesize ultra-thin electrocatalysts using a core-shell nanostructure that contains a bimetallic core responsible for high electronic conductivity and an ultra-thin partially oxidized shell for catalytic CO_2 conversion.

In this presentation, we will show a model core-shell electrocatalyst that has a silver-tin bimetallic core and an ultra-thin partially oxidized tin oxide shell (denoted as AgSn/SnO_x) for CO_2 reduction to formate with high Faradaic efficiencies and high current densities.[2] A structure-activity correlation between the AgSn/SnO_x core-shell structure and CO_2 electrocatalytic activity was established by studying a series of Ag-Sn catalysts with different Ag/Sn compositions. More importantly, a volcano-like correlation between the bulk composition and electrocatalytic performance was observed, in which the optimal thickness of the partially oxidized SnO_x shell was ~1.7 nm.

References:

1. Lu, Q., & Jiao, F.* Electrochemical CO_2 reduction: electrocatalyst, reaction mechanism, and process engineering. *Nano Energy* 29, 439-456 (2016).

2. Luc, W., Collins, C., Wang, S. W., Xin, H. L., He, K., Kang, Y. J. & Jiao, F.* Ag-Sn bimetallic catalyst with a core-shell structure for CO_2 reduction. *Journal of the American Chemical Society* 139, 1885-1893 (2017).

501199 Electrochemical CO_2 Reduction over Cuag Bimetallic Electrodes and Well-Defined Surface Alloys with Enhanced Oxygenate Selectivity

Abstract:

Atmospheric CO_2 is a potential source of renewable carbon for the production of fuels and chemicals. However, for this process to be sustainable the hydrogen required must be derived from water and the necessary energy must be supplied by the sun. One approach to this goal is the utilization of renewable electricity to drive the electrochemical CO_2 reduction reaction (CO_2RR). Previous research has shown that the overall efficiency of CO_2RR and the distribution of generated products depend primarily on the electrocatalyst used as the cathode. Cu is the only monometallic electrocatalyst capable of reducing CO_2 into potential fuels or multi-carbon chemicals with a total Faradaic efficiency (FE) in excess of 1%. The principal products formed over polycrystalline Cu are H_2, CH_4, C_2H_4, and C_2H_5OH. However, at potentials cathodic of -1 V vs RHE excessive H_2 evolution leads almost exclusively to the methanation of CO_2. Since multi-carbon products are more valuable precursors to fuels and chemicals than CH_4, recent studies have focused on identifying factors governing the multi-carbon product selectivity. These studies have revealed that the multi-carbon product selectivity is sensitive to the surface morphology, electrolyte composition, and the applied potential. However, these studies have not managed to substantially reduce the fraction of current lost to the parasitic evolution of H_2.

Herein we report our investigations of electrochemical CO_2 reduction over CuAg bimetallic electrodes and well-defined surface alloys, which we have found to exhibit an enhanced activity for the evolution of products derived from CO relative to H_2 compared to pure Cu. Furthermore, we have found that these electrocatalysts display an unusually high selectivity for the formation of carbonyl-containing compounds, such as acetaldehyde. While this bimetallic system was investigated on the hypothesis that CO spillover from phase-segregated domains of Ag to Cu would increase the rate of C-C coupling, further characterization by photoelectron spectroscopy and electrochemical cycling led to a very different conclusion. The revised hypothesis was confirmed by synthesizing well-defined Cu+Ag surface alloys, which reproduced the activity and selectivity trends observed over the bimetallic electrodes. The results are explained in terms of an observable shift in the electronic properties of Cu and the resulting impact that these modifications have on the adsorption energies of key reaction intermediates.

493375 A Carbon Nanotube Supported Gold Catalyst for the Electroreduction of Carbon Dioxide (CO_2)

Abstract:

The electrochemical reduction of carbon dioxide (CO_2) to value added chemicals (such as formic acid, carbon monoxide (CO), ethylene, and ethanol) has been proposed as a potential strategy for the utilization of excess anthropogenic CO_2 emissions.[1, 2] In particular, the electroreduction of CO_2 to CO seems interesting, as CO (either alone or in combination with H_2 *i.e.,* syngas) can be used as a platform chemical to produce a variety of hydrocarbons. However, even after a few decades of research, electrochemical systems that can produce CO *via* the electroreduction of CO_2 at industrially relevant rates (partial current density for CO >150 mA cm^{-2}) and low overall cell overpotentials (absolute value <1V),[3] remain difficult to develop.

In this presentation, we will report a supported gold (Au) catalyst that exhibits remarkably high activity (partial current density for CO ~158 mA cm^{-2}) at low cell overpotentials (~-0.94 V), for the electroreduction of CO_2 to CO. The catalyst was synthesized *via* an *in situ* chemical reduction of the Au precursor (chloroauric acid) on a polybenzimidazole wrapped carbon nanotube support. The electrochemical characterization was performed in a flow electrolyzer to avoid mass transfer limitations associated with the low solubility of CO_2 in aqueous solutions.[4] Furthermore, we will also discuss the effect of electrolyte composition (concentration and pH) with respect to increasing the activity and lowering the overpotential required for the electroreduction of CO_2 to CO on the supported Au catalyst. The mechanistic implications of such a behavior will also be discussed.

References:

[1] A.M. Appel, J.E. Bercaw, A.B. Bocarsly, H. Dobbek, D.L. DuBois, M. Dupuis, J.G. Ferry, E. Fujita, R. Hille, P.J.A. Kenis, C.A. Kerfeld, R.H. Morris, C.H.F. Peden, A.R. Portis, S.W. Ragsdale, T.B. Rauchfuss, J.N.H. Reek, L.C. Seefeldt, R.K. Thauer, G.L. Waldrop, *Chem. Rev.* 2013, 113, 6621-6658.

[2] H.R.M. Jhong, S. Ma, P.J.A. Kenis, *Curr. Opin. Chem. Eng.* 2013, 2, 191-199.

[3] S. Verma, B. Kim, H.R.M. Jhong, S. Ma, P.J.A. Kenis, *ChemSusChem* 2016, 9, 1972-1979.

[4] D.T. Whipple, E.C. Finke, P.J.A. Kenis, *Electrochem. Solid-State Lett.* 2010, 13, B109-B111.

487255 Interplay of Mass Transfer and Local pH Effects in CO_2 Reduction Electrocatalysis on Cu Nanowires

Abstract:

Besides the catalyst structures, the electrocatalysis for CO_2 reduction is also dependent on the electrolyte and CO_2 partial pressure, which can affect the catalytic activity and selectivity via mass transfer and/or local pH effects. While protons are readily available in aqueous solutions, the supply of reactants to the electrode surface is limited by the concentration and diffusivity of dissolved CO_2 molecules, which places a hurdle on the current density and efficiency of CO_2 reduction. The reduction of CO_2, as well as the evolution of H_2, produces hydroxide anions ($OH^{â€“}$) and causes rise of the local pH on the electrode surface.

Here we report on the study of mass transfer and local pH effects on CO_2 reduction by using Cu nanowires as the electrocatalysts. Highly dense Cu nanowires were synthesized by first thermal oxidation of Cu mesh in air, and then reduction by applying a cathodic electrochemical potential or thermal annealing in a reducing atmosphere. Our previous report has shown that the electrochemically reduced (ECR) Cu nanowires are highly active for CO_2 reduction, with favorable production of CO and formate at potentials more positive than -0.5 V (versus reversible hydrogen electrode) and C_2 species (ethanol, ethylene, and ethane) at more negative potentials. We have further performed more systematic studies in the high-overpotential region, i.e., potentials more negative than -0.5 V, where the mass transport of CO_2 and high local pH near the electrode surface are believed to play a significant role in determining the electrocatalytic performance. Simulations have been developed to understand the interplay between the mass transfer and local pH effects and elucidate their roles in the structure-performance relationships of the high-surface-area nanowire catalysts. Based on these results, discussions are provided for optimizing the electrochemical conditions for CO_2 reduction to produce valuable C_2 products.

497824 Measurement of Intrinsic Activity of Electrocatalytic Reduction of CO_2 over Cu

Abstract:

Electrocatalytic reduction of CO_2 offers a sustainable route to mitigate global warming and store intermittent solar and wind energy. Various electrocatalysts have been studied over the past three decades for the conversion of CO_2 to oxygenates such as CO, HCOOH, $C_2H_2O_4$, CH_3OH, and C_2H_5OH; and hydrocarbons such as CH_4, C_2H_4, and C_2H_6. Among these electrocatalysts, Cu is the only catalyst that can produce hydrocarbons. Several electrochemical cells have been proposed in the literature to measure activity and selectivity of the CO_2 reduction reaction. However, all of these measurements were affected by the poor mass transfer of CO_2 to the cathode at potentials more negative than -1 V vs RHE. The limiting factor in the electrochemical reduction of CO_2 is the depletion of dissolved CO_2 coupled with the increase in pH near the cathode. Here we propose a compact design of an electrochemical cell with the cathode being a Rotating Disk Electrode (RDE). The RDE sparged with gaseous CO_2 minimizes the cathode polarization and hence enable the measurement of intrinsic activity. The effect pH and applied potential on the intrinsic activity and selectivity of electrocatalytic CO_2 reduction reaction over Cu will be presented in this talk.

495014 Electrocatalytic Carbon Fixation on Molecular-Functionalized Surfaces

Abstract:

Global warming and related environmental issues have been exacerbated by the accelerated consumption of fossil fuels as energy resources. Utilizing renewable energy input to transform the abundant chemical feedstock carbon dioxide (CO_2) back into fuels and other value-added chemicals and materials can complete a potential carbon-neutral cycle to help the sustainability of our planet. To this end, we are developing hybrid homogeneous-heterogeneous catalysts for electrochemical carbon fixation based on the combination of inorganic molecules and surfaces. Our latest work on the development of systems capable of CO_2 and CO reduction to C-C coupled products will be discussed in this presentation.

494963 Electrochemical Reduction of CO_2 on Metal Doped Graphene

Abstract:

Single atom transition metals embedded at single vacancies of graphene provide a unique paradigm for catalytic reactions.[1,2] We present a density functional theory study of such systems for the electrochemical reduction of CO_2. Using an explicit description of the solvent at the electrochemical interface, the activation energies for key proton-electron transfer steps were determined. On transition metals, the first proton transfer ($*CO + H^+ + e^- \rightarrow *CHO$) has been shown to be the critical rate determining step in C_1 product formation. On these single atom sites, the corresponding barrier scales more favorably with the CO binding energy than for 211 and 111 transition metal surfaces, in the direction of improved activity.[3] Hydrogen evolution intermediates, on the other hand, are found to less stable than those on transition metals, which suggests a lower hydrogen evolution activity. Using a simple microkinetic model based on the calculated electrochemical barriers, we present a rate volcano for the production of methane from CO. All the considered transition metal dopants (Cr, Mo, W, Mn, Fe, Ru, Os, Co, Rh, Ir, Ni, Pd, Pt, and Cu) form stable doped-graphene structures, and a subset of these dopants (Ru, Fe, Ni, Pd, Pt, Rh, Co, and Ni) are not expected to be poisoned by oxide species. The presence of nitrogen near the active site was found to affect intermediate and transition state energies, but variations in these energetics followed the same scaling determined for these materials without nitrogen. We identify promising candidates with high activity, stability, and selectivity for the reduction of CO_2. This work highlights the potential of these systems as improved electrocatalysts over pure transition metals for CO_2 reduction.

1. Krasheninnikov, A.V; Lehtinen, P.O.; Foster, A.S.; Pyykkö P.; Nieminen, R.M. *Phys. Rev. Lett.* **102**, 126807 (2009).
2. Stolbov, S.; Ortigoza, M.A. *J. Chem. Phys.* **142**, 154703 (2015).
3. Liu, X.; Xiao, J.; Peng, H.; Hong, X.; Chan, K.; Nørskov, J.K. *Nat. Comm.* (2017).

506141 Into Hot Water: Utilizing Thermal Distributed Energy Resources to Improve Grid Reliability

Abstract:

Combined heat and power (CHP) technology along with grid-integrated water heaters (GIWH) are underutilized thermal distributed energy resources. They can be implemented on a large scale to provide ancillary services for the grid, including load shifting, frequency regulation, and improved grid reliability. The market is ripe for CHP and GIWH to be rapidly implemented on a large scale. This paper discusses the legislative and technical background of CHP and GIWH, the way these technologies provide benefits to a rapidly modernizing smart grid, and concludes with potential policies on the state and federal level that would support greater use of thermal distributed resources. Three policies are discussed in detail. The first of these is to expand the Department of Energyâ€™s CHP Technical Assistance Partnership program to facilitate knowledge and implementation of a shared ownership model for CHP systems between utilities and an institution that will utilize the energy produced. Next, states should encourage utilities to create incentives for consumers to install GIWH to be used in utility programs; these incentives should focus in particular on assisting low-income households to purchase GIWHs. Last, states should set goals to have all utilities that function on retail markets to adopt time-based pricing structures for electricity.

506142 Turning the Tide: Policies to Advance Saltwater Desalination in the United States

Abstract:

Although water remains one of the worldâ€™s most abundant resources, most of the water resides in the ocean. With an already limited supply of freshwater, the increasing global population, changing climate, and worsening droughts are placing additional strains on the worldâ€™s water supply. Currently, several regions of the U.S. are experiencing some level of water shortage, and while the majority of the affected states reside in the West, the effects are felt nationwide. Although both water conservation and reuse remain key components of the national water portfolio, these practices alone will not sufficiently accommodate future water demands. Desalination, which removes salts and other suspended solids from seawater and brackish water, is a technique widely used around the world to convert high-salinity water into water suitable for human use. The unique benefits of seawater desalination remain to be recognized in the U.S., and since the options for augmenting the current water supply are limited, seawater desalination should hold a distinct place in the U.S. water portfolio.

506143 Microgrids for the Macrogrid: Advancing Community Microgrids for Grid Modernization

Abstract:

Global trends - the digital age, prospect of cyberwarfare, climate change, consumer preferences â€“ are irreversibly changing the supply and demand for electricity as well as the risks that threaten access to electricity. The existing grid was designed to serve a different kind of society and is ill-suited for meeting new challenges. A modern electricity system will not only have more actors, but will need to be fundamentally different in how it dispatches resources, communicates information, responds to grid disturbances, and plans for the future. A modern grid is decentralized, flexible, adaptable, and smart (i.e. automated, digitized, self-healing). By these characteristics, electricity access becomes reliable, efficient, and resilient. Microgrids add both resilience and flexibility to the system and policy initiatives can encourage them through technology-agnostic market rules and economic forces.

510175 Small Scale Shock-Proof Biogas Digesters

Abstract:

Affordable, easily maintained biogas digesters have the potential to greatly increase the standard of living in several key communities around the world. From food scraps and animal wastes, these reactors can output high-quality fertilizer in addition to natural gas that can be used for heating, cooking, etc. Designing an assembly guide with the key features of a safe biogas reactor is key to developing this technology in the communities in question; particularly communities with limited access to education or manufacturing capabilities. Development of this assembly guide requires implementation and development of a small â€œtestâ€ reactor, including various mechanical systems, embedded electronics, and other features that would allow a more comprehensive and thorough understanding on the feasibility and overall maximum output possible by this kind of reactor. Secondary research would strengthen this knowledge base and further increase the ability for these reactors to be beneficial to their intended users, this research includes lab-scale experimentation on the growth rates of critical bacteria, determining decomposition rates of certain safety features, and detailing how other physical stimuli may result in differing output rates of the reactor. The presence of a simple, customizable biogas reactor in an impoverished community could potentially raise the standard of living for the surrounding populace by eliminating excess food waste, producing high-quality fertilizer for crops, and supplying a steady stream of clean natural gas for heating or cooking.

491918 Rapid Wood Fractionation ≤ 80°C for Sustainable and Economic Biorefinery

Abstract:

Here we report a rapid and low temperature (≤ 80°C) wood fractionation process for producing wood pulp, cellulose and lignin nanomaterials, chemicals, and sugar/biofuels. Approximately 90% of wood lignin can be dissolved at 80°C for 20 min. Equivalent delignification using alkaline pulping can be achieved only at 170°C for 2 h. The novel process fractionated wood into two fractions: (1) a primarily carbohydrate-rich water-insoluble solid fraction that can be used for sustainable production of fibers such as dissolving pulp, lignocellulosic nanomaterials, and/or sugars through subsequent enzymatic hydrolysis, and (2) a spent acid liquor stream containing mainly dissolved lignin that can be easily precipitated as lignin nanoparticles simply by diluting the spent acid liquor. The dissolved hemicellulose sugars can be used to produce furfural after lignin precipitation. The chemical used has a low water solubility, which facilitates efficient recovery simply using commercially proven crystallization technology by cooling concentrated spent acid solution to ambient temperatures to achieve environmental sustainability.

502673 Insights into Biomass Recalcitrance

Abstract:

The utilization of biomass for biofuels and bio-derived chemicals and materials based on the biological platform is contingent on efficient separation, deconstruction and conversion. Todayâ€™s cellulosic ethanol plants, especially those based on the cellulase deconstruction, have demonstrated that the conversion of plant polysaccharides to ethanol can be accomplished on an industrial scale but current process challenges are hindering wide-spread adaption. One of the key issues is the recalcitrance of biomass and the need to utilize aggressive pretreatment technologies. Recent advances in our understanding of the plant cell wall structure and our ability to reduce biomass recalcitrance by altering these constituents by controlling genetic features of the plant and innovations in pretreatment chemistries provides innovative pathways to facilitate the biorefining of plant polysaccharides to biofuels and yield a lignin rich process stream that can be valorized to chemicals and/or materials. This seminar review our contributions in this field and highlight future opportunities.

492489 Low Temperature and High Efficiency Biomass Fuel Cell and Bio-Hydrogen Production

Abstract:

Title: Low temperature and high efficiency biomass fuel cell and bio-hydrogen production

Name: Yulin Deng

School of Chemical & Biomolecular Engineering, Georgia Institute of Technology, Atlanta, GA 30302-0620, yulin.deng@rbi.gatech.edu

A novel fuel cell which can directly use native polymeric biomasses, such as starch, cellulose, lignin, and even switchgrass and wood powders will be discussed. This fuel cell combines some features of solar cells, fuel cells, and redox flow batteries. Specifically, an oxidation/reduction ion pair, such as Fe^{2+}/Fe^{3+}, or polyoxomatelate, are used as catalyst to oxidize biomass by either absorbing solar light or heat energy. The power density of the solar-induced hybrid fuel cell powered with raw biomass could reach ~50 mW/cm^2. Unlike most cell technologies that are sensitive to impurities, the cell reported in this study is inert to most organic and inorganic contaminants present in the fuels. The fuel cell is completely noble metal free. The similar fuel cell system can also be converted to electrolysis cell for hydrogen production at low temperature using native biomass directly.

504077 Circular Economy: A Path Towards Innovation and Commercialization of Biocomposites for Sustainable Manufacturing

Abstract:

"Circular economy" is an emerging concept that targets waste minimizations through a closed loop economy system, unlike the traditional linear model ("make and dispose" type economy), and therefore supports sustainable development. The underlying principle of this emerging economy is "Nothing is waste; waste is a co-product waiting for another opportunity". In this system, the byproducts and co-products of one industry can find commercially viable applications in another industry, with appropriate design and engineering. There is a significant potential for utilization of biobased coproducts from agro and forest industries, food and beverage industry as well as biofuel and biomass-based pyrolysis industries. Our focus is on biocomposite applications that utilize these materials as reinforcing and filler materials for plastics, including biobased and biodegradable plastics. The developed eco-friendly biocomposites are suitable for use in consumer products, light-weight auto-parts and 100% compostable packaging.

Acknowledgements: Agriculture and Agri-Food Canada and Competitive Green Technologies (CGtech) through AgriInnovation Program (Project #052882 and 051910); the Ontario Ministry of Agriculture, Food, and Rural Affairs - University of Guelph Bioeconomy Industrial uses Research Program (Project # 200425); Natural Sciences and Engineering Research Council, Canada for the Discovery Grants (Project #401111); and the Ontario Research Fund, Research Excellence Program; Round-7 (ORF-RE07) from the Ontario Ministry of Research and Innovation (MRI), currently known as the Ontario Ministry of Research, Innovation and Science (MRIS) (Project # 052665 and #052644).

506138 Computing at the Nexus of Food, Energy, and Water

Abstract:

In coming decades, the world population is projected to grow significantly increasing the demand for food, water, energy, and other resources. Furthermore, these resource challenges may be amplified due to climate change and urbanization. In addition, Food, energy and water (FEW) systems were traditionally analyzed and planned independently to address the challenges of population growth, climate change and urbanization. However, such piece-meal approaches (e.g., bio-fuel subsidy, fertilizers in agriculture) to solving problems in one system (e.g., energy, food) led to unanticipated harms to other systems (e.g., food price increase, water resource depletion and degradation). Thus, understanding the interdependent and interconnected nature of food, energy, and water systems (FEW nexus) is a societal priority.

Computing is crucial for understanding the problem, the interconnections, and the impacts withing FEW nexus. It is also needed for monitoring a variety of Earth resources (e.g., agriculture fields, fresh water lakes, energy needs for cooling or heating, etc.), and trends (e.g., deforestation, pollution, etc.) for timely detection and management of risks, such as impending crop failures and crop-stress anywhere in the world. It is also needed to reduce waste and to improve efficiency, e.g., amount of water and energy needed to produce food.

Computing success stories go beyond the cyber-infrastructure for simulations (e.g., GCMs, AgMIP) to include precision agriculture and GEOGLAM. Precision agriculture uses cyber-physical systems and data science to increase yield, and reduce fertilizer and pesticide runoffs. The Global Agricultural Monitoring (GEOGLAM) , an international system, uses remotely sensed satellite imagery to monitor major crops for yield forecasts to enable timely interventions to reduce disruptions in global food supply.

However, the FEW nexus presents new challenges and opportunities for computing. For example, data science methods need to not only re-examine assumptions such as non-stationarity (e.g., climate change) but also address nexus challenges such as high cost of false positives, (social) feedback loops, and multiple spatio-temporal scale.

505270 Circling the Nexus

Abstract:

Growing support for circular economy approaches in the public and private sectors is drawing attention to the food-energy-water nexus. In the United States, the Materials Marketplace is an award-winning regional and national platform to facilitate company-to-company industrial reuse of all kinds. Through the cloud-based platform, traditional and non-traditional industrial water, energy, food and waste streams are uploaded and proactively matched with new product and revenue opportunities, ultimately enabling the culture shift to a circular, closed-loop economy. In addition to diverting waste from landfills, these recovery activities generate significant cost savings, energy savings, and create new jobs and business opportunities.

492541 Limits to Growth and Global Sustainability of Food-Energy-Water Nexus

Abstract:

A mathematical representation of the major dynamics of the World, inspired by the "Limits to Growth" model[1] constructed in the 1970s, has been under construction for approximately the past ten years[2]. The mathematical representation includes an ecosystem with food-web model, a human population, industry, energy generation, a macro-economy, and rudiments of a legal system with private property, taxation, and regulation. The work was started in 2004 at the U.S. Environmental Protection Agency under the supervision of Dr. Heriberto Cabezas. Details of the model can be found in the scientific literature[3]. Around 2010, the project was transferred to the Vishwamitra Research Institute led by Dr. Urmila Diwekar in Chicago, Illinois, USA where the work has proceeded under the joint supervision of Dr. Urmila Diwekar and Professor Heriberto Cabezas from the Peter Pazmany Catholic University. Several publications have resulted from this research effort[4],[5], two of which are cited here. The mathematical model now includes water also making it a system where food-energy-water nexus can be studied and is developed sufficiently so that scenario analysis shows that consumption increase predicted for the future makes the compartments, i.e. species, of this model unsustainable. However, complex network analysis offers a pathway to analyze various versions of this model. The value of the model is it provides the conceptual machinery "steering" the system away from

---

[1] Meadows, D. H., Meadows, D. L., Randers, J., & Behrens, W. W. (1972). *The limits to growth*. New York, 102.

[2] Cabezas, H., Pawlowski, C.W., Whitmore, H.Wm., & A.L. Mayer, "On the Sustainability of Integrated Model Systems with Industrial, Ecological, & Macroeconomic Components," Res., Cons. & Recycl., 50, 122-129 (2007).

[3] Kotecha, P., Diwekar, U.M. and H. Cabezas, "Model Based Approach to Study the Impact of Biofuels on the Sustainability of an Integrated System," Clean Tech. Env. Pol.," 15 (1), 21-33 (2013).

[4] Benavides, P.T., Diwekar, U. and H. Cabezas, "Controllability of Complex Networks for Sustainable System Dynamics," J. Complex Networks (2015), published on-line. DOI: 10.1093/comnet/cnu051.

[5] Doshi, R., Diwekar, U., Benavides, P.T., Yenkie, K.M. & H. Cabezas, "Maximizing Sustainability of Ecosystem Model through Socio-Economic Policies Derived from Multivariable Optimal Control Theory," Clean Tech. Env. Policy, 17 (6), 1573-1583 (2015). DOI: 10.1007/s10098-014-0889-2.

504195 Sustaining Innovation / Innovating Sustainably

Abstract:

Optimism is at the heart of both innovation and sustainability. Research creates discoveries that lead to inventions, creating innovations that make the world better. R&D starts with the optimistic view that you can create something better. It is an investment in the future, where resources remain for future generations. At Dow, weâ€™ve earned the ability to continue innovating based on our past performance. R&D isnâ€™t left to chance, we take steps to stack the deck by:

- Rigorously manage our R&D portfolio based on quantifiable metrics
- Focus on R&D productivity and timely delivery of technology
- Collaborating to extend access to new inventions and to create customer intimacy
- Application of technical excellence in both product and process design

Portfolio management, built on a well-established foundation, guides our efforts and incorporates sustainability and other metrics to influence decisions. We use the life-cycle impact of our products to guide our efforts. Far too often, hype has surpassed reality, boundary conditions have been neglected, and actions taken that do not truly lead to a sustainable future. Looking across the life-cycle compels decisions to reduce raw material use, energy use and to limit toxics use. Innovation will create a sustainable future.

504196 Creating Chemistry for a Sustainable Future

Abstract:

Imagine a future where we can use our creativity and innovation power to help solve some of the largest global sustainability challenges of our times. Are we doing enough to enable this?

Investing in businesses that seek positive economic, ecological and social impact is becoming imperative. We will be judged by our communities, by our work-force, by the talent pool of Millennials, by investors. We will be judged in the courts of social media and public opinion. We have to act now; authentically, and for the right reasons

Chemical companies are positioned uniquely to solve the global sustainability challenges. But we have to take a holistic look at these challenges; going beyond our typical value chains and including the larger eco-system of stakeholders. That is, going beyond current business models to co-creating solutions.

For example, Chemistry plays an active role in improving energy efficiency of homes. There are many opportunities to co-create solutions to improve this situation. BASF therefore, is partnering with other stakeholders to change state laws and make it easier to update residential building codes. This challenged us to work with new stakeholders: industry partners, architects, city planners, builders, citizens, governments and many more. This is what we mean by co-creation.

BASF is convinced of the importance of the UN Sustainability Development goals to our long-term success. This is not solely about recycling or bio-based raw materials; those are of course important. Moving forward, however, requires us to build upon existing platforms and address critical issues such as transparency, education, diversity and inclusion â€“ to just name a few.

No single company can address all the seventeen UN sustainability development goals on their own. Co-creation is the way forward to address both the overlaps and the gaps.

504198 Continued Delivery on Impactful Sustainable and Innovative Business and R&D Strategies

Abstract:

Expanding our impact through sustainability is a core value at Owens Corning. From the products we make to the way that we make them, sustainability is an essential element of our business. We design innovative solutions that meet expectations of customers/markets and contribute toward addressing significant global issues such as energy conservation, environmental impact and overall quality of life. We consistently leverage our technological expertise to deliver sustainability-driven innovations to the Insulation, Roofing and Composites markets. Our insulation products enhance energy efficiency; cool roof shingles help reduce air-conditioning and power consumption; reinforced lightweight glass-composites deliver superior mechanical properties with lower environmental impact than traditional materials. Our business strategies promote innovative ways to achieve high-performing, greener and healthier products via high-efficiency manufacturing processes with reduced energy and water consumption. The discussion in this panel will highlight Owens Corningâ€™s innovative strides throughout its core businesses, its continued commitment to robust sustainability progress in its handprint and footprint, and the companyâ€™s significant safety improvement.

491786 Experimental Study of Hydrocarbon Trapping over Passive NO_x Adsorber

Abstract:

Experimental Study of Hydrocarbon Trapping over Passive NOx Adsorbers

Sam Malamis, Kyle Karinshak, Michael P. Harold*

Dept. of Chemical & Biomolecular Engineering, University of Houston, Houston, Texas

* corresponding author e-mail: mpharold@central.uh.edu

Introduction

Automotive catalysts including TWC, LNT, and SCR catalysts are all highly effective in reducing NO_x and HC emissions above 200°C [1]. Most engines, however, require a warm-up period during which the catalyst undergoes a "cold start" environment. During this period of up to 200s, large amounts of pollutants such as NOx, N_2O, CO and Hydrocarbons slip through the catalyst into the atmosphere [2]. To mitigate this occurrence, it is beneficial to develop a passive NOx adsorber, or PNA, that can store NOx and hydrocarbons at these low temperatures, and then to release them at higher temperatures to improve the effectiveness of the aftertreatment system [3].

Few comprehensive studies have been performed on this topic, and only recently has there been a movement toward understanding the chemistry behind these low temperature traps. To date, a wide variety of materials has been explored, including palladium on ceria-zirconia (Pd-CZO) [4], CeO_2-M_2O_3 with supported Pt or Pd [5], and Palladium on various zeolites including BEA, MFI, CHA [3]. While it has been reported that Pd-CZO is a potentially effective NO_x trap at low temperatures, it also strongly absorbs SO_2, which inhibits NO_x storage [3,4]. This study focuses on the use of zeolites as PNA materials in order to trap both NOx and HC during the cold start period, with the potential to act as a partial oxidation catalyst upstream of the SCR or LNT. The study mainly explores the effect of temperature and various feed gases on NOx and HC trapping.

Materials and methods

Experiments were performed using 2wt% Pd-BEA zeolite supported on a cordierite monolith. A feed gas was flowed over the catalyst in a vertical quartz-tube reactor, and the effluent species concentrations were measured using FT-IR. The catalyst was first pre-treated in 5% O_2 (bal. Ar) for 30 min at 550°C. To determine the amount of NOx stored on the catalyst during an experiment, a feed gas of 2% O_2 and 400ppm NO was flowed over the catalyst for 5 min at a fixed feed temperature, either 50°C, 80°C or 150°C. After 5 min., the NO was turned off and the reactor was maintained at the adsorption temperature until the concentration of NO dropped to less than 2ppm. To measure the species desorbing after adsorbing on the catalyst surface, a temperature programmed desorption (TPD) was performed at 20ºC/min from the adsorption temperature to 550ºC. The surface was then pretreated to re-oxidize the Pd to PdO to ensure proper behavior during subsequent experiments [6]. To measure the effect of hydrocarbons on catalyst performance, experiments similar to those described above were performed, using a feed with either 800ppm C_3H_6 and 2% O_2 or a mixture of 400ppm NO, 800ppm C_3H_6 and 2% O_2.

Results

It was found that using a feed of only NO and O_2, Pd-BEA stored between 0.4 and 1.0 mol NOx/gcat*10^4, but released it too early during the TPD (i.e. at a temperature <200ºC) (Figure 1). It was also found that as the adsorption temperature increased, the amount of NOx stored and desorbed decreased (Figure 2). Experiments using only propylene in the feed indicate a very strong propensity for the catalyst to trap hydrocarbons at low temperatures (1.5-6.0 mol C_3H_6/gcat*10^4), more so than in trapping NOx (Figures 3). Adding C_3H_6 to the NO feed revealed a competitive adsorption effect of both NOx and C_3H_6, as the stored quantities of these species was significantly lower than the pure feeds, as shown in Figures 2 and 3. While the total amount of stored NOx decreased with the addition of C_3H_6, our studies show that the majority of NOx was released above 200ºC, which is a promising result (Figure 1).

Figure 1. Percentage of NOx that desorbs at a catalyst temperature higher than 200ºC for feeds with only NO and with added C_3H_6 on 2wt% Pd-BEA. Values are reported for each adsorption temperature. Note that the addition of propylene in the feed significantly delays NOx desorption. Feed: 400ppm NO with or without 800ppm C_3H_6, 2% O_2, bal. Ar.

Figure 2. NOx adsorption and desorption quantities at various adsorption temperatures and for feeds with or without C_3H_6. Values are reported as moles of NOx/g cat*10^4. Note that as the adsorption temperature increases, adsorption/desorption of NOx decreases. Feed: 400ppm NO with or without 800ppm C_3H_6, 2% O_2, bal. Ar.

Figure 3. C_3H_6 adsorption and desorption quantities at various adsorption temperatures and for feeds with or without NO. Adsorption values are reported as moles of C_3H_6/g cat*10^4, while desorption values are reported as moles of C/g cat*10^4. Note that as the adsorption temperature increases, adsorption/desorption of C_3H_6 decreases. Feed: 800ppm C_3H_6 with or without 400ppm NO, 2% O_2, bal. Ar.

References

1. Twigg M: **Progress and future challenges in controlling automotive exhaust gas emissions.** *Applied Catalyst B* 2007, **70**:2-15.

2. Wilenmann M, Favez J, Alvarez R: **Cold-start emissions of modern passenger cars at different low ambient temperatures and their evolution over vehicle legislation categories.** *Atmospheric environment* 2009, **43**:241-2429.

3. Chen HY *et al.*: **Low temperature NO storage of zeolite supported Pd or low temperature diesel engine emission control.** *Catal. Lett.* 2016.

4. Theis JR, Lambert C: **The effects of CO, C_2H_4, and H_2O on the NO_x storage performance of low temperature NO_x adsorbers for diesel applications.** *SAE International* 2017, **10.**

5. Jones S *et al.*: **CeO_2**
2016 **CLEERS Issue.**

498784 CuO/Co_3O_4: A Highly Active and Selective Low Temperature NO Decomposition Catalyst

Abstract:

Abstract:

The direct decomposition of NO is thermodynamically favorable within the temperatures encountered by an automotive catalytic converter; however, activity remains too low for practical application [1]. Platinum group metals (PGM), such as Pd, are active for NO decomposition at temperatures above 600 °C, and attempts to improve their activity below 600 °C have been unsuccessful due oxidation of the metal and subsequent deactivation [2]. Regarding non-PGM catalysts, Cu-ZSM5 is perhaps the most widely known material displaying low temperature (≤ 500 °C) direct NO decomposition activity, however it has propensity to produce an undesired side product: the greenhouse gas N_2O [3]. The spinel oxide Co_3O_4 does not exhibit direct NOx activity as high as Cu-ZSM5, but even at 350 °C it is able to decompose 50% of N_2O to N_2 [4,5]. To address the performance gap below 500 °C, we therefore propose combining the benefits of both CuOx and Co_3O_4 into a single NO decomposition catalyst. Supported copper oxide on cobalt oxide CuOx/Co_3O_4 was synthesized by incipient wetness impregnation. We hypothesized that this material can utilize the N_2O decomposition capabilities of Co_3O_4 to reduce the N_2O generated by NO decomposition over the CuOx species.

The CuOx/Co_3O_4 were synthesized with Cu surface densities ranging from 0.7 to 10.3 Cu/nm^2 and characterized by XRD, XPS, and TEM. The Cu surface density corresponding to the monolayer loading for CuOx/Co_3O_4 was found to be 6.7 Cu/nm^2, therefore, the catalysts set included sub-monolayer, monolayer, or 3D-microcrystalline CuOx species. Thus, this catalyst set formed a strong basis for identifying a direct NOx decomposition structure-activity relationship. The catalysts were evaluated for direct NOx decomposition activity using a packed bed reactor. The maximum activity for NO decomposition by CuOx/Co_3O_4 was found at the CuOx surface density corresponding to monolayer. There is a subsequent decrease in the activity at higher surface densities due to the presence of CuOx microcrystals. When compared to Cu-ZSM5, the specific activity from 300-400 °C is similar, however, the surface area normalized (areal) activity exceeds that of Cu-ZSM5 as the CuOx/Co_3O_4 has a specific surface area that is an order of magnitude lower. Additionally, at or below 400 °C, the N_2 selectivity is higher over CuOx/Co_3O_4 compared to Cu-ZSM5, as a result of decreased N_2O formation. The CuOx/Co_3O_4 catalysts maintain relatively good activity in the presence of O_2 and CO_2. Nitric oxide adsorption/desorption reveals an optimized adsorption strength over CuOx/Co_3O_4. Additional mechanistic insight from *in situ* FTIR will also be presented.

References:

1. H. Falsig, T. Bligaard, C.H. Christensen, J. K. Norskov, *Pure Appl. Chem.*, Vol. 79, No. 11, pp. 1895–1903, 2007.
2. G.K. Reddy, C. Ling, T. Peck, H. Jia, *RSC Adv.*, 2017, 7, 19645
3. B. Moden, P.D. Costa, D. K. Lee, E. Iglesia, *J. Phys. Chem. B* 2002**,** *106,* 9633-9641
4. P.W. Park, J.K. Kil, H.H. Kung, M. C. Kung, *Catal. Today*, Volume 42, Issues 1-2, June 1998, 51-60.
5. L. Xue, H. He, C. Liu, C. Zhang, and B. Zhang. *Environ. Sci. Technol.*, **2009**, 43 (3), 890-895.

488866 Effect of Different Structure Directing Agents on Deactivation of Cu/SAPO-34 during Low-TemperatureÂ NH_3-SCR

Abstract:

Cu/SAPO-34 and Cu/SSZ-13 having CHA structure have attracted significant attention because of their high activity and N_2 selectivity during SCR reaction as well as superior hydro carbon resistance. The Cu/SAPO-34 has shown better hydrothermal durability than Cu/SSZ-13 at high temperature [1]. However, we have earlier observed that Cu/SAPO-34 prepared by the morpholine route deteriorated under water exposure at low temperature, where the activity decreased from 87% conversion to 6% conversion after 9 h of low temperature experiment [2]. Nevertheless, there is still a lack of understanding of the deactivation mechanism of Cu/SAPO-34 during low temperature NH_3-SCR reaction.

In this study, Cu/SAPO-34 using three different templates, i.e., morpholine(MO), trimethylamine (TEA), and tetraethylammonium hydroxide (TEAOH), as structure directing agent were prepared by incipient wetness impregnation (IWI) method. Characterizations of those catalysts suggest that they have different physical/chemical properties as well as various species of Cu. For example, analysis of H_2-TPR profiles of Cu/SAPO-34 catalysts proposed that they contain different amount of Cu species depending on the structure directing agents. Those catalysts were systematically investigated for low-temperature NH_3-SCR system under water exposure in order to find the correlation of their chemical/physical structures and deactivation. It was observed that the activity of Cu/SAPO-34 declined when they were exposed to water at low temperature as observed in the previous study [2]. Interestingly, deactivation of Cu/SAPO-34 catalysts behave differently depending on the choice of template. Additionally we found that activity of Cu/SAPO-34 using MO as template after water exposure at low temperature declined regardless of a variety of regeneration methods including a dry atmosphere at 600 °C. To conclude, the choice of template over Cu/SAPO-34 catalysts significantly influence their activities and deactivation during low temperature NH_3-SCR.

[1] D.W. Fickel, E. Dâ€™Addio, J.A. Lauterbach, R.F. Lobo, Appl. Catal. B: Environ. 102 (2011) 441-448

[2] K. Leistner, L. Olsson, Appl. Catal. B: Environ. 165 (2015) 192-199

497426 A Perovskite-like Catalyst for the Simultaneous Removal of Soot and NO: Effect of the Synthesis Method

Abstract:

Due to the environmental problems associated with the use of diesel engines, and the strengthening of the regulations concerning pollutant emissions in the transport field. The study of the catalytic simultaneous removal of soot and NO at relatively low temperatures (within the range of diesel engine exhaust: 150-450°C) has taken a paramount importance in pollution control area.

Mixed oxides coming from perovskite-like materials, seems to be a promising catalyst for such purpose. In this way, the present work tested the $La_{0.7}Ag_{0.3}MnO_3$ perovskite-like material, prepared for different synthetic routes, for the simultaneous removal of soot and NO. Three different synthetic routes are compared: microwave-assisted hydrothermal (HMW), microwave-assisted (MW) and solid state (SS).

The catalytic activity of the prepared catalysts was tested in a temperature-controlled furnace inside which a soot-catalyst-sand fixed bed was placed. The mixture of carbonaceous material and catalysts was carried out with spatula (i.e. loose contact), keeping always a 1:9 weight ratio. The soot model used in this study consisted of a carbon black, FW200, from Orion Engineering. The soot-catalyst-sand mixture was exposed at a constant heating rate of 10°C/min up to 600°C, under a 2200 ppm NO/ 10% vol.O2/ He atmosphere. The evolution of CO_2, CO, NO, NOx, N_2,O_2 gases with temperature were followed by mass spectrometry and a KIGAZ 310 gas analyzer. The catalyst materials were also characterized by SEM, TEM, XRD, H_2-TPR and O_2-TPD.

The results show a good catalytic behavior of the materials tested for the oxidation of soot. Figure 1 and Table 1 present the temperature at which soot ignition begins (T_{ig}), and the temperatures at which the conversion of soot to CO_2 was 50%, 90% and 100%, named as T50, T90 and T100, respectively. When it is compared the catalytic activity results with the non-catalytic one; it was found an ignition temperature shift of 278°C when the HMW-synthesized catalyst was used, and a shift of 241°C for the MW-synthesized catalyst. Also, the maximum soot oxidation peak, (T50), for all the catalytic systems evaluated in this study was shifted to lower temperatures according to the following decreasing order: non-catalytic > SS > HMW > MW. Another important result regarding to catalytic soot oxidation, it is that no CO emission was found in the gas stream, indicating 100% selectivity toward CO_2. In contrast, for the non-catalytic one, the selectivity CO_2/CO ratio was near one.

Figure 1. CO_2 Yield as a function of soot oxidation temperature over $La_{0.7}Ag_{0.3}MnO_3$ perovskite-like material synthesized by different methodologies (MW, HMW, SS).

Table 1. Ignition temperature (Tig); temperatures at which the soot conversion was 50%, 90% and 100%, respectively; and CO_2 selectivity for catalytic and the non-catalytic test.

Test	Tig (°C)	T50 (°C)	T90 (°C)	T100 (°C)	S_{CO2} (%)
$NO+O_2$	428	634	676	728	N/A
HMW	150	409	465	577	100
MW	187	367	487	665	100
SS	170	439	502	679	100

Regarding NO removal, Figure 2 shows the NO evolution with temperature, for the experiments carried out with $La_{0.7}Ag_{0.3}MnO_3$ synthesized by methodologies. Figure 2 shows that during soot oxidation, the catalysts tested presented some NO adsorption/desorption events followed by a NO reduction toward N_2 around the soot oxidation peak. The catalyst synthesized through MW route shows a prominent desorption event between 210°C and 350°C compared with the other two materials. Nevertheless, its adsorption or reduction capacity was not as good as the one obtained from the HMW-synthesized catalyst.

Figure 2. NO evolution as a function of temperature for each of the synthesized catalysts tested.

Figure 3 shows the CO_2 and NO/NOx profiles as function of temperature when the HMW-synthesized catalyst was used. From figure, it is possible to see that this catalyst was able to remove 45% of the initial NO concentration as N2. Also, it is possible to see that the temperature for the NO/NOx peak reduction was found around 400 °C matching the temperature at which the evolution of CO_2 was maximum. This result shows the ability for the simultaneous removal of soot and NO of the $La_{0.7}Ag_{0.3}MnO_3$ perovskite-like material synthesized by the HMW method.

Figure 3. CO_2, NO/NOx and N_2 profiles as a function of temperature when the HMW-synthesized solid was used as catalyst.

The results presented above demonstrate that the selection of the synthetic route has a significant effect on the catalytic performance of a material. Also, it was evidenced that the perovskite-like material ($La_{0.7}Ag_{0.3}MnO_3$) synthesized by microwave-assisted hydrothermal (HMW) method has an excellent catalytic activity for the simultaneous removal of soot and NO by shifting the soot oxidation peak toward lower temperature and removing up to 45% of the initial NO as N2.

497655 Fast Cycling to Achieve High NO_x Conversion in Lean Exhaust: Role of Ceria

Abstract:

Fast Cycling to Achieve High NOx Conversion in Lean Exhaust: Role of Ceria

Zhiyu Zhou, Michael P. Harold*, Dan Luss**

Department of Chemical and Biomolecular Engineering, University of Houston, Houston, TX 77204

* *mharold@uh.edu,* ** *dluss@central.uh.edu*

The abatement of NOx ($NO+NO_2$) remains a challenge under the continuously more stringent emission standards. Commercialized deNOx systems, NOx storage and reduction (NSR) and selective catalytic reduction (SCR), function well from ~250 °C to ~400 °C, but have unsatisfactory performance outside that temperature range. Toyota researchers recently invented a new deNOx system, Di-Air (diesel NOx aftertreatment by adsorbed intermediate reductants), which involves fast injection of fuel into the exhaust feed to the NSR converter [1]. Previous studies show that ceria increases the deNOx performance of the Di-Air system in both low [2] and high [3] temperature regions. Data suggest that NOx conversion occurs via the reaction of an intermediate produced through the reaction with O_2 and NO [1, 2]. However, a recent study [3] suggests that catalysts containing ceria may involve an alternative pathway of NOx abatement via NO decomposition at high temperatures. We studied in a bench-scale flow reactor system the NO decomposition and reduction on ceria under different frequencies with CO as sole reductant.

In our study, a lean to rich feed time ratio 6 to 1 of was controlled to keep the same fuel penalty (fuel to oxidant ratio). The lean/rich cycling conditions included lean/rich times (in seconds) of 90/15, 60/10, 30/5 and 6/1. Different reaction conditions were applied to study NO conversion behavior under favorable and adverse conditions. 500 ppm NO was added to both rich phase and lean phase. 2.5% CO was only added to rich phase and O_2 with varied concentration was only added to lean phase. In some cases, CO_2 and H_2O were introduced continuously along with NO, O_2 and CO to simulate lean exhaust.

When CO is used as the sole reductant and no H_2O is introduced in the feed, NO is mainly converted to N_2 with only trace amounts of N_2O and NO_2 detected. Figure 1 shows the cycle-averaged NO conversion with four cycle times from 450 °C to 625 °C in two feed conditions. In the first case, when 1400 ppm O_2 is introduced in lean feed, Figure 1(a) shows that increase of lean/rich switching frequency enhances NO conversion. The optimized lean/rich switching frequency is always 6/1 above 500 ^{o}C. In the second case, when 2100ppm O_2 is introduced in the lean feed, Figure 1(b) shows that a different optimal lean/rich switching frequency exists for different temperatures. The optimal lean/rich switching frequency is 30/5 at intermediate temperatures (500 ^{o}C and 550 ^{o}C) and 6/1 at high temperatures (600 ^{o}C and 625 ^{o}C) respectively. Figure 1 shows that the optimal lean/rich switching frequency changes with reaction conditions (e.g., temperature and feed condition).

Figure 1. (a) NO conversion with different lean/rich switching frequency; (a) rich: 2.5%CO, 500ppm NO; lean: 500ppm NO, 1400ppm NO; (b) rich: 2.5%CO, 500ppm NO, lean: 500ppm NO, 2100ppm O_2

Figure 2 shows the cycle-averaged NO conversion with the same four cycle times at 600 ^{o}C with fixed ratio of lean and rich feeds. The short cycle limit represents NO conversion obtained with total mixed lean and rich feeds. The long cycle limit represents weight averaged NO conversion in lean phase and rich phase. The optimal lean/rich switching frequency of 30/5 exceeds those of both short and long cycle time limit. This shows that the lean/rich cycling operation is beneficial to NO conversion.

Figure 2. NO conversion at 600C, with a fixed switching frequency as 90s lean/15s rich; rich: 2.5%CO, 500ppm NO, lean: 500ppm NO, 3600ppm O_2

Figure 3 shows that NO conversion changes when CO_2 and H_2O are fed. By comparing the blue line (reference) with either red line (with H_2O only) or purple line (with CO_2 only), it is obvious that both H_2O and CO_2 inhibit NO decomposition on ceria but H_2O has a smaller detrimental effect than CO_2. The green line (with both CO_2 and H_2O) lies between the red line (with H_2O only) and the purple line (with CO_2 only), which indicates that H_2O can mitigate the detrimental effect of CO_2.

Figure 3. NO conversion at 550C, with a fixed switching frequency as 90s lean/15s rich; (rich: 2.5%CO, 500ppm NO, lean: 500ppm NO, varied O_2 concentration, CO_2 and H_2O in both lean and rich if introduced)

This study on NO decomposition on ceria during lean and rich cycling helps explain the beneficial function of ceria on NOx reduction at high temperatures, which also provide guidance for optimization of catalyst formulation and operation strategies.

Reference

1. Y. Bisaiji et al., *SAE Int. J. Fuels Lubr.* 5 (2012) 380-388
2. Y. Zheng et al., *Catal. Today*, 276 (2016), 192-201.
3. Y. Wang et al., *Top. Catal.*, 59 (2016), 854-860

495352 NO Reduction By CO over CeO_2 Supported Co_xO_y Catalysts

Abstract:

Nitrogen oxides (NO_x) are a main contributorâ€"in addition to hydrocarbons (HC), carbon monoxide (CO), sulfur oxides (SO_x), and particulate matter (PM)â€"to global air pollution [1]. Significant investigations have been performed for catalytic NO_x reduction resulting in technologies such as NH_3-SCR and NO_x storage and reduction [2]. In the last several decades, platinum group metals (PGMs) and zeolites have been extensively applied to NO_x abatement [3]. Although these catalysts exhibit a high level of catalytic activity, NO_x decomposition over PGMs and zeolites has some practical issues such as high cost and high light-off temperature [4]. To resolve these issues, transition metal oxide catalysts have been investigated for low temperature NO_x decomposition reaction by CO with varied O_2 concentrations, which is a primary reaction in three-way catalysis (TWC).

In this study, the series of CeO_2 supported Co_xO_y catalysts were prepared by incipient wetness impregnation method with different metal oxide loadings (0.5 wt%-30 wt%) for NO reduction by CO. Catalytic activity, catalyst physicochemical property, and molecular/electronic structure were investigated by FTIR, BET, Raman spectroscopy, X-ray diffraction (XRD), and X-ray absorption spectroscopy (XAS). No diffraction peaks due to crystalline CoO or Co_3O_4 were observed in XRD patterns under to 10 wt%, while nanocrystaline Co_3O_4 peaks at ~610cm^{-1} was observed in the Raman spectra with 6wt% sample. Both Raman and XRD spectroscopy results provided that 5~6wt% Co_xO_y /CeO_2 contains monolayer coverage. It was observed that both Co_xO_y coverage and oxidation state of Co are closely related to the NO reduction by CO. Monolayer and above monolayer coverage catalysts showed 80~90% CO and NO conversion at 300~400°C with high N_2 and CO_2 selectivity (>90%). Based on the time on stream results, it was also observed that the crystalline Co_3O_4 nanoparticles above monolayer coverage improve the stability of catalyst.

ACKNOWLEDGMENT: We gratefully acknowledge the financial support for this study from the Department of Materials Science & Chemical Engineering at Stony Brook University through start-up research funding.

[1]. K. Skalska, et al. Sci. Total Env., 408 (2010) 3976-3989, [2]. J.H. Kwak, et al., J. Catal., 287 (2012) 203-209, [3]. S. Roy, et al., Applied Energy, 86 (2009) 2283-2297, [4]. J. Liu, et al., Applied Catalysis B: Env., 84 (2008) 185-195

498251 Nb_2O_5 Promoted Base Metal Oxide Catalysts Used for Air Pollution Abatement in Stationary Plants

Abstract:

Catalytic oxidation is a very efficient way for the abatement of low concentrations of Volatile Organic Compounds (VOCs). The most active catalysts used for oxidation are supported noble metals. However the main drawback with noble metals is their high price. Our goal is to consider transition metal oxides as possible replacements for expensive precious metals for low-temperature stationary abatement applications. Current research in our group has found that niobium oxide (Nb_2O_5) is worthy of additional studies as a promoter in altering the oxidation states of base metal oxides. Consequently, bimetallic catalysts supported on Nb_2O_5 were synthesized to achieve the development of stable oxidation. Both Nb_2O_5 and Al_2O_3 carriers were pre-calcined at 800ËšC for 2hr. Then, the calcination temperature for metal-supported catalysts was experimentally determined using temperature programmed decomposition in air using thermogravimetric analysis (TGA). The C_3H_8 oxidation reactions were carried out in a fixed-bed quartz flow reactor at atmospheric pressure. The reaction conditions were as follows: C_3H_8: 0.1 vol. %, O_2: 18.9 vol. % and N_2 balance. Under ambient conditions, the gas hourly space velocity (GHSV) was 65 000 h^{-1} and catalysis volume is 0.1 ml. Effluent gases of were continuously analyzed by micro GC. Isothermal aging tests were conducted to assess the stability of the catalysts at 400 ËšC in the air. Results suggest a strong interaction between niobium oxide and the active components.

499320 N Doped TiO_2 nanoparticles Synthesized By One Step Liquid Flame Aerosol Method (LFSP): Understanding the Effect of Synthesis Parameters and Photocatalytic Degradation of VOCs

Abstract:

Environmental photocatalysis Nitrogen doped titanium dioxide is attracting a continuously increasing. Conventional wet synthesis methods need multiple processing steps, high energy input and long processing times, In contrast, flame synthesis, especially liquid flame spray pyrolysis (LFSP), is a versatile one-step process for synthesizing a variety of anion doped nanoparticles. Nitrogen-doped TiO
synthesis. A combination of various physico-chemical characterizations were performed. The photocatalytic activity and stability of the N-doped TiO_2 samples were evaluated through using the decomposition of phenol as model reaction under visible light irradiation. Characterization results show that the nitrogen dopants have a significant effect on the crystallite size and optical absorption of TiO_2. It is found that the N-doped TiO_2 catalysts have enhanced absorption in the visible light region, and exhibit higher activity for photocatalytic degradation of VOCs. We have produced The UV-vis spectroscopy results of N incorporated TiO_2 showed enhancement of light absorption in the visible range by selected composite materials in the UV range (250-400 nm). We have found the Ti/N ratio of 11% for the materials synthesized from our XPS data. To the best of our knowledge, this is the first report which demonstrates single-step, direct flame synthesis of N-doped TiO_2 and their application as a photocatalyst better performance. The method is extremely simple, low cost, energy efficient, very effective, and can be easily scaled up for large scale production. These interesting results will be discussed in the presentation.

503980 Bridging the Gap between Feedstock Growers and Users: A Study of Poplar Coppice Based Biorefinery

Abstract:

In the advanced biofuel industry, the land productivity is important to feedstock growers and the product yield of conversion process is important to biorefinery facilities. The crop productivities are often not positively correlated to their bioconversion yields. It is therefore critical to examine the sugar yield from a given feedstock and relate that to its productivity. In current study, the 2-year old poplar coppices, including one low-productive hybrid and one high-productive hybrid, were collected from two plantations. Through steam pretreatment and enzymatic hydrolysis, the bioconversion yields of low and high productive poplar hybrids were compared for both sites. Although they have the similar chemical composition, the low-productive hybrids had 9% to 19% higher sugar yields than the high-productive hybrids. This suggests that to evaluate the feedstocks and estimate the product yield, knowing the feedstock chemical composition is not enough. Biorefineries should conduct conversion trials to better understand the processing performance of feedstocks. Based on the findings, we investigated the economics from both perspectives of feedstock growers and users by examining a biorefinery with 700,000 dry tonne feedstock processing capacity.

Study on Liquefaction of Straw Biomass catalyzed by Sulfonate Ionic Liquid [HSO_3-BMIM][HSO_4] in Ethanol

Reporter: Qian Guan
Work Unit: 1.Energy Research Institute Co., Ltd, Henan Academy of Sciences
2.Henan Key Lab of Biomass Energy

CONTENTS

一 Background and significance

一、Background and significance

Oil price

SOS

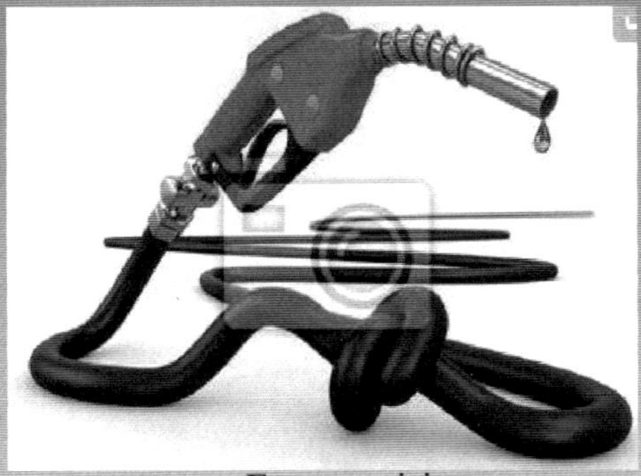

Energy crisis

Environmental issues

一、Background and significance

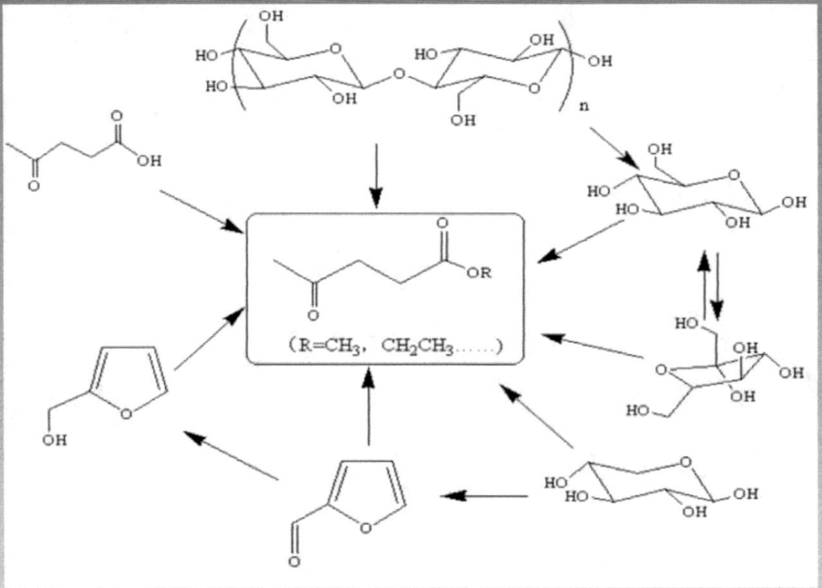

Raw material of levulinates

A Démolis, N Essayem, F Rataboul. Synthesis and Applications of Alkyl Levulinates. ACS Sustainable Chemistry & Engineering, 2014, 2 (6): 1338-1352.

一、Background and significance

Major issues

- Energy crisis
- Environmental issues from traditional energy
- The utilization of biomass is low

significance

Solve the existing problems
achieve efficient and efficient use of biomass

二
Research approach

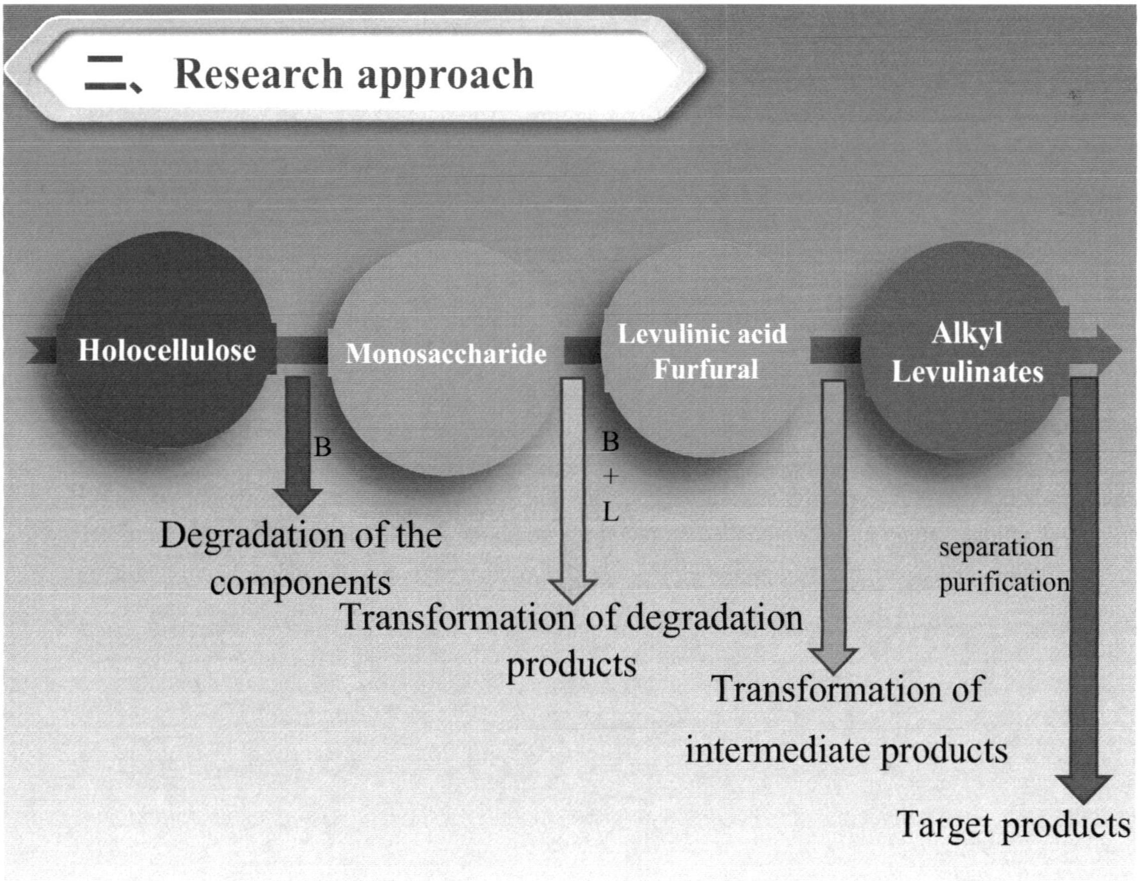
二、Research approach
Holocellulose
Monosaccharide
Levulinic acid
Furfural
Alkyl
Levulinates
B
Degradation of the
components
B
+
L
Transformation of degradation
products
Transformation of
intermediate products
separation
purification
Target products

二、Research approach

Ionic liquid as catalyst

Non-toxic, non-corrosive, low vapor pressure

Thermal stability and other performance stability
Safety for the equipment

With the characteristics of Lewis acid and Brφnsted acid catalytic performance is better

三 Contents and procedure

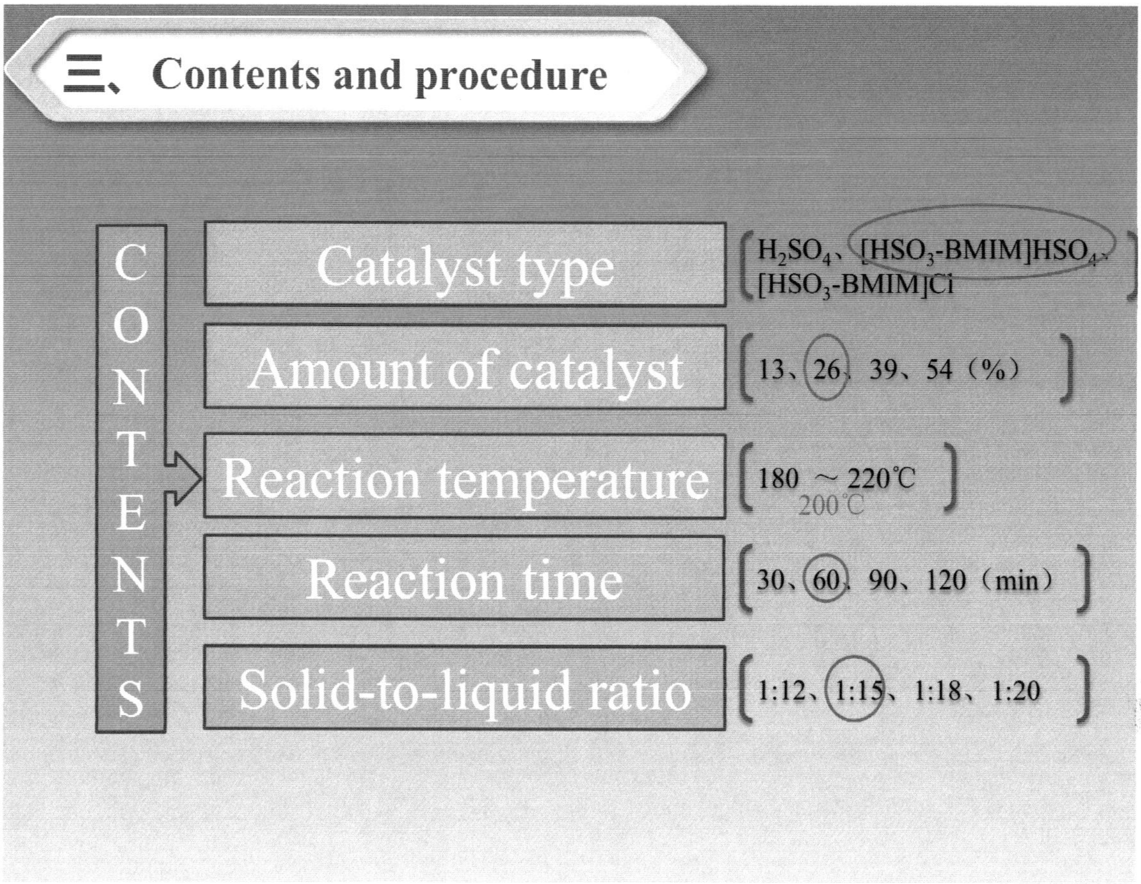
三、Contents and procedure
CONTENTS
Catalyst type
H2SO4、[HSO3-BMIM]HSO4、[HSO3-BMIM]Cl
Amount of catalyst
13、26、39、54（%）
Reaction temperature
180 ～ 220℃
200℃
Reaction time
30、60、90、120（min）
Solid-to-liquid ratio
1:12、1:15、1:18、1:20

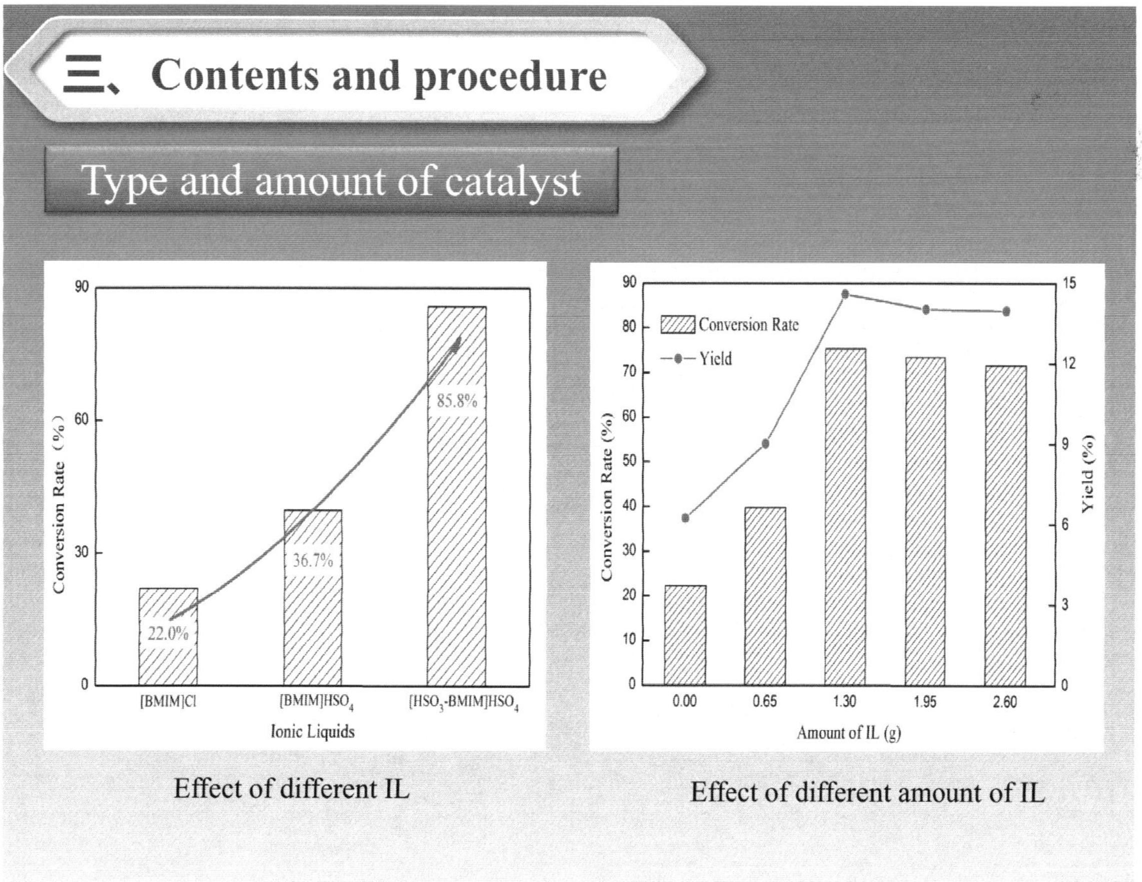
三、Contents and procedure
Type and amount of catalyst
Conversion Rate（%）
Ionic Liquids
[BMIM]Cl
[BMIM]HSO4
[HSO3-BMIM]HSO4
22.0%
36.7%
85.8%
Effect of different IL
Conversion Rate
Yield
Conversion Rate (%)
Yield (%)
Amount of IL (g)
Effect of different amount of IL

三、Contents and procedure

Reaction temperature and time

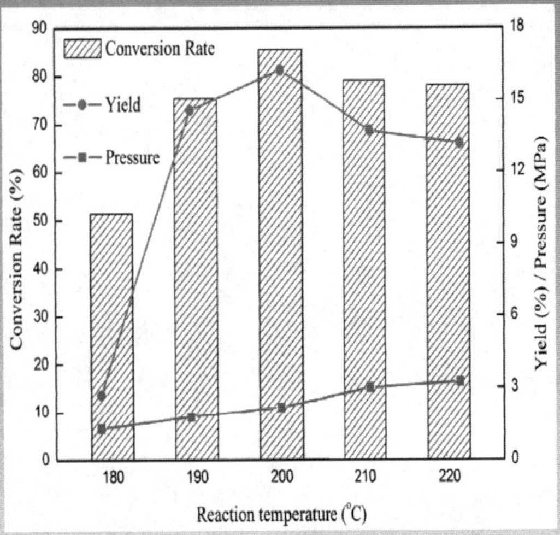

Effect of different reaction temperature

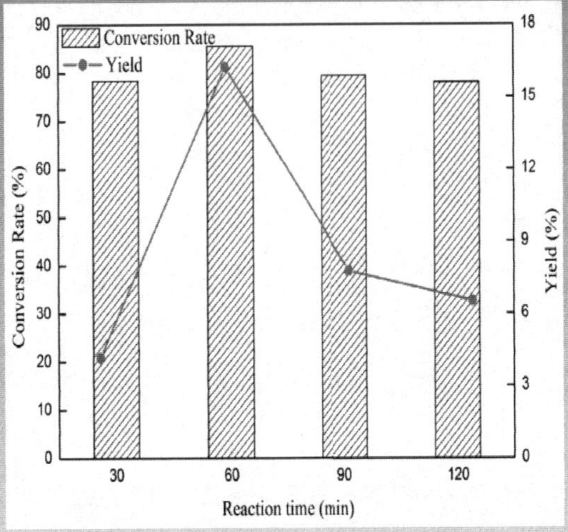

Effect of different reaction time

三、Contents and procedure

Residue analysis

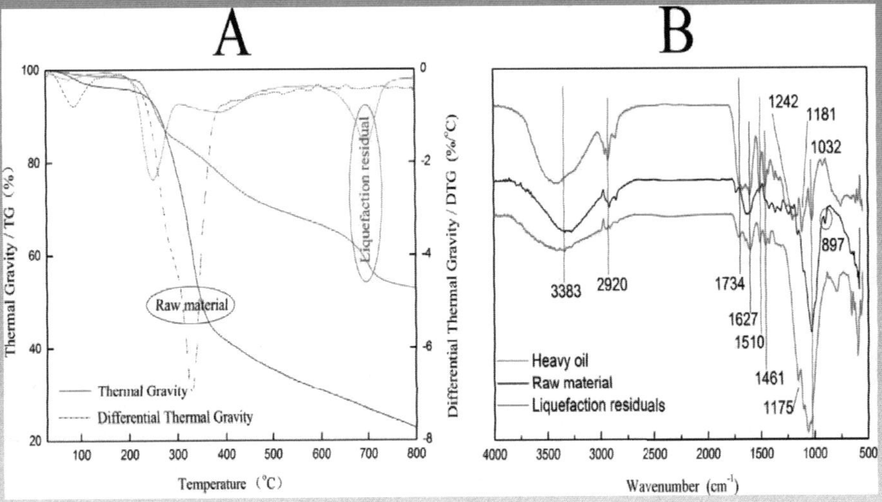

Results of liquefaction residue characterization （optimum condition）

三、 Contents and procedure

Product analysis

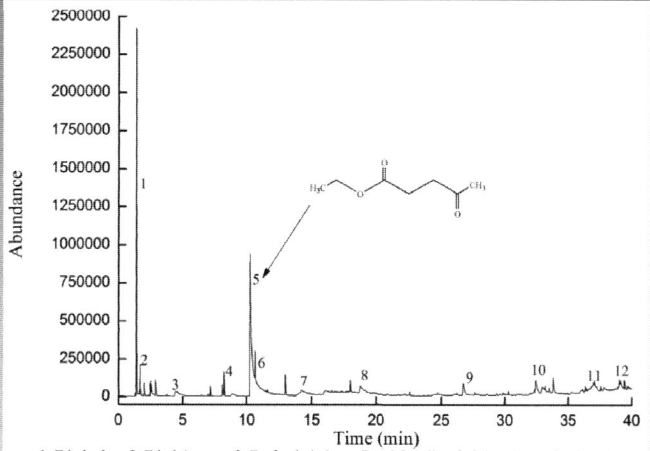

1. Ethyl ether; 2. Ethyl Acetate; 3. Furfural; 4. 6-tert-Butyl-2,4-dimethylphenol; 5. Ethyl levulinate; 6. 2-Furaldehyde diethyl acetal; 7. 3-Methyl-2-oxo-2H-pyrane-6-carboxylic acid; 8. 3,3-dimethyl-4-oxo-Cyclopentanecarboxylic acid methyl ester; 9. Ethyl citrate; 10. Ethyl (2E)-3-(4-hydroxy-3-methoxyphenyl)-2-propenoate; 11. Hexadecanoic acid, ethyl ester; 12. (Z)-6-Octadecenoic acid

Total ion chromatogram of liquid product （optimum condition）

三、 Contents and procedure

Composition and content

serial number	Retention time/min	Products	Relative percentage content/%	matching degree
1	1.398	Ethyl ether	6.3	86
2	1.696	Ethyl Acetate	0.78	64
3	2.021	Diethoxy-methane	0.59	53
4	2.910	4-Ethylbenzoic acid-2-ethoxyethyl ester	4.69	50
5	4.236	Furfural	3.34	86
6	7.167	6-tert-Butyl-2,4-dimethylphenol	0.47	30
7	8.121	Ethoxyacetaldehyde diethylacetal	0.92	50
8	10.072	2-Furancarboxylic acid, ethyl ester	0.74	91
9	10.288	Ethyl levulinate	28.11	95
10	10.682	2-Furaldehyde diethyl acetal	9.87	95
11	11.920	2-Furanmethanol	1.53	46
12	13.385	2,2-dimethyl-1,3-cyclopentanedione	1.52	42
13	14.353	3-Methyl-2-oxo-2H-pyrane-6-carboxylic acid	11.19	27
14	15.537	Cyclopentanecarboxylic acid-2-pentyl ester	4.48	22
15	15.943	Butanedioic acid, dimethyl ester	1.35	35
16	16.526	Butanedioic acid, 4ydroxyl-, diethyl ester	1.96	58
17	18.632	3,3-dimethyl-4-oxo-cyclopentanecarboxylic acid methyl ester	10.46	50
18	19.559	Succinic acid, 2-decyl ethyl ester	1.38	59
19	26.749	Ethyl citrate	2.89	91
20	33.933	Ethyl (2E)-3-(4-hydroxy-3-methoxyphenyl)- 2-propenoate	4.45	99
21	36.790	Hexadecanoic acid, ethyl ester	1.35	95
22	38.132	(Z)-6-Octadecenoic acid	0.29	91
23	38.542	Ethyl oleate	0.37	99
24	39.467	Octadecanoic acid, ethyl ester	0.22	95

三、Contents and procedure

mechanism

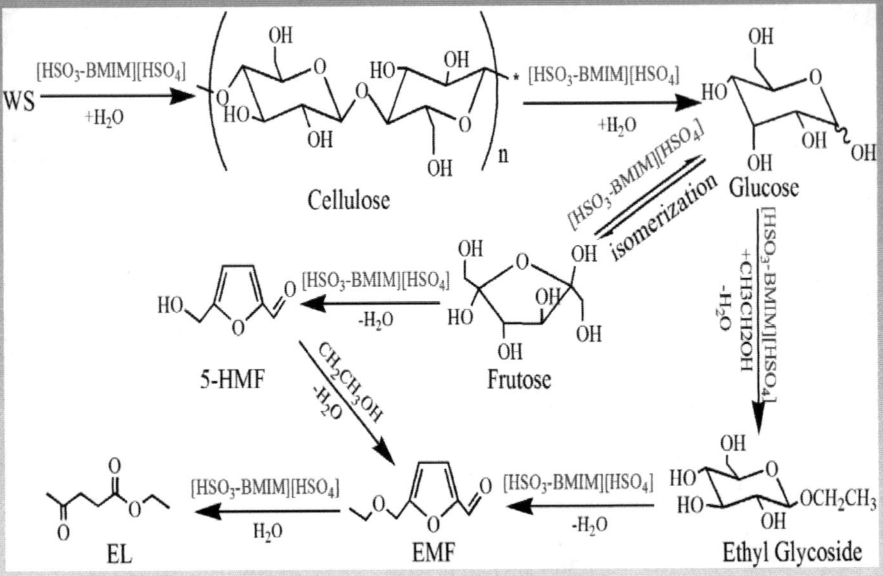

Reaction Scheme of EL Formation catalyzed by $[HSO_3\text{-}BMIM][HSO_4]$

四 Conclusion and prospect

四、Conclusion and prospect

Conclusion

- Acidic ionic liquid 1-methyl-3-(4-sulfobutyl) imidazole bisulphate was used as a catalyst under relatively mild conditions for the liquefaction of biomass, resulting in the formation of ethyl levulinate.
- Lignin was more resistant to liquefaction than were cellulose and hemicellulose.
- The surface of the liquefaction residuals was covered with un-degraded lignin.
- The liquid product of the reaction contained primarily oxygenates such as esters, acids, ketones, alcohols, and phenols.
- GC-MS data show that the light oil fraction of the liquid product was composed primarily of the degradation products of cellulose and hemicellulose, whereas the main constituents of the heavy oil fraction were derived from the degradation products of lignin.

四、Conclusion and prospect

Prospect

- Recycling of ionic liquids
- Separation and purification of alkyl levulinates
- Application of alkyl levulinates in the energy field

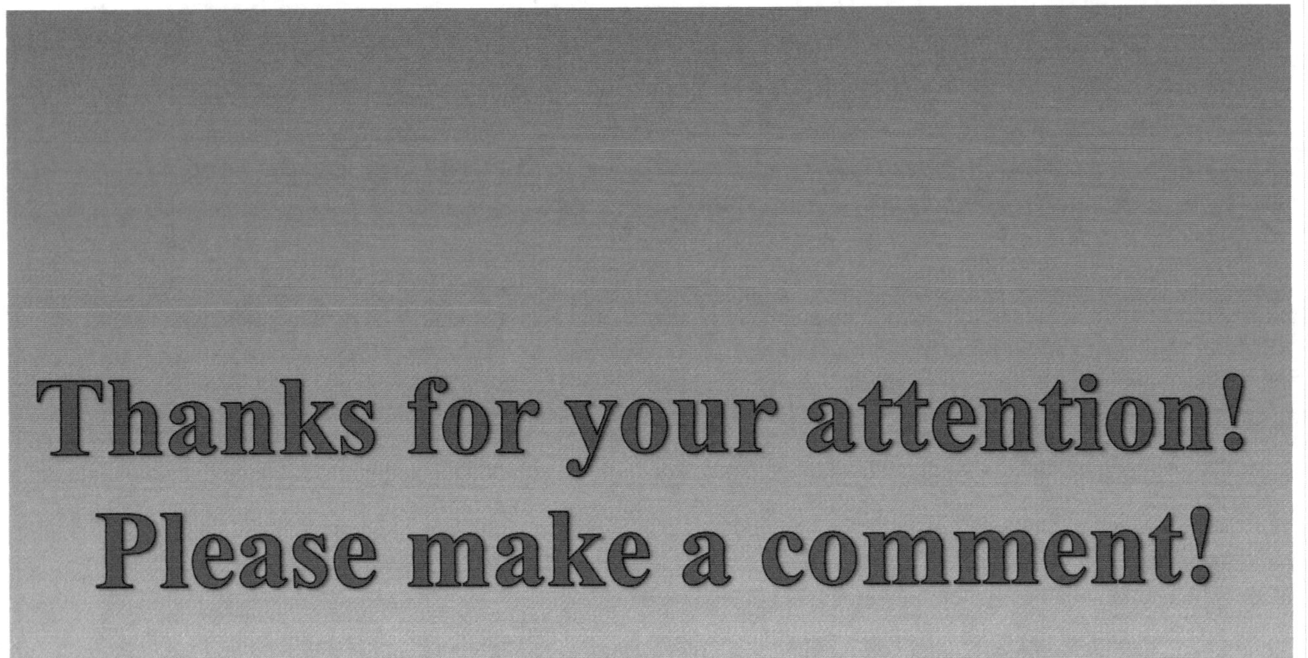
Thanks for your attention!
Please make a comment!

493016 Douglas-fir Tannin Inhibition of *Trichoderma Reesei* Cellulase

Abstract:

Forestry residues from timber harvest and forest thinnings are an abundant potential source of biomass for biofuels. This material is comprised of braches, broken or defective tree parts, tops, and trees not meeting grade specification. Biofuel production from forestry residues is currently being investigated at the pilot plant production level for processes based on pretreatment, saccharification, and fermentation. Finding ways to reduce the price of every step in the process will help make this technology reach commercial production.

Trees have evolved to contain a variety of different molecules that defend the tree from degradation. This broad class of molecules are often called â€œextractivesâ€, and they can make up 5 - 25% of the dry mass of forestry residues[1]. The tannins in Douglas-fir bark are the most potent extractive inhibitor and can be abundant in bark rich feedstocks. Better understanding of how tannins cause inhibition can help us adapt biofuel processing to mitigate this inhibition.

In this study we will show how Douglas-fir tannins inhibit the cellulase Cellobiohydrolase I (Cel7a) from *Trichoderma reesei.* This cellulase can make up 55% of the enzyme mass in a cellulase cocktail[2] and is noted for its ability to hydrolyze crystalline cellulose into cellobiose. Experimental saccharification results show that inhibition from Douglas-fir bark tannins can reduce cellobiohydrolase activity 20% for high concentration bark feedstocks. The binding of tannins onto Cel7a are being characterized through isothermal titration calorimetry and this information is paired with molecular dynamics simulations to identify how a Douglas-fir tannin binds to Cel7a. Tannin binding to both the carbohydrate binding module and the catalytic core is being investigated via molecular simulations.

As biofuel processing considers forest residues as a renewable feedstock, it is necessary to devise process strategies to remove or neutralize tannin inhibition, or create cellulases resistant to their inhibitory effects. Additional process steps or chemical additives could increase saccharification yields with the downside of additional processing costs, whereas inhibition resistant cellulases would make process modification unnecessary. This work provides molecular level information that can be used identify possible residues for mutagenesis to make cellulases tannin-resistant.

1. Oleson, K. R.; Schwartz, D. T., Extractives in Douglas-fir forestry residue and considerations for biofuel production. *Phytochemistry Reviews* **2016,** 15, (5), 985-1008.

2. Evans, B. R.; Margalit, R.; Woodward, J., VERATRYL ALCOHOL OXIDASE ACTIVITY OF A CHEMICALLY-MODIFIED CELLULASE PROTEIN. *Archives of Biochemistry and Biophysics* **1994,** 312, (2), 459-466.

494965 Inhibition Effect of Aromatic Aldehydes on Butanol Fermentation By *Clostridium Acetobutylicum*

Abstract:

A large number of degradation compounds are formed during biomass pretreatment and they significantly inhibit the efficiency of biomass conversion to biofuels. Of those identified potential inhibitors, aromatic aldehydes play an important role in inhibition activity. Hence the effect of 13 aromatic aldehydes on acetoneâ€"butanolâ€"ethanol (ABE) fermentation was assessed at four concentrations in the present work. It was found that the inhibition severity was affected by the ortho substituents ($OH > OCH_3 > CHO$) and strongly related to the position of hydroxyl group instead of the number of hydroxyl groups. The ortho-hydroxyl group significantly contributed to the aromatic aldehyde inhibition. The ortho-substituted 2-hydroxybenzaldehyde caused at least 20-fold stronger inhibition than meta- and para-substituted analogues of 3- and 4-hydroxybenzaldehydes. The presence of ortho-hydroxyl group can form an intramolecular hydrogen bond with carbonyl hydrogen and potentially increase the cell membrane permeability and electrophilicity. Quantitative structureâ€"activity relationship (QSAR) analysis was used to establish a correlation between inhibition activity (IC_{50}) and physicochemical descriptors. A strong correlation was observed between IC_{50} and the energy of the highest occupied molecular orbital E_{HOMO}.

486017 Renewable Isoprene By Sequential Hydrogenation of Itaconic Acid and Dehydra-Decyclization of 3-Methyl-Tetrahydrofuran

Abstract:

The catalytic conversion of biomass-derived feedstocks to value added chemicals is an important challenge to alleviate the dependence on petroleum-based resources. To accomplish this, the inherently high oxygen content of biomass compounds, such as that of lignocellulosic biomass, requires significant reduction via hydrodeoxygenation strategies. The unsaturated carboxylic acid itaconic acid (IA) can be produced from biomass via fermentation pathways, for example. A pathway of interest is the conversion of IA to isoprene, facilitating the renewable production of an industrially relevant diolefin. IA is successively hydrogenated to yield 3-methyl tetrahydrofuran (3-MTHF), in a one-pot cascade reaction (Figure 1A), where a Pd-Re/SiO_2 bimetallic catalyst results in the highest 3-MTHF yield (80%, Figure 1B). The 3-MTHF is then converted to isoprene, and other pentadienes, through an acid catalyzed vapor-phase dehydra-decyclization. Multiple solid acid catalysts, including aluminosilicates, metal oxides and phosporous modified zeolites, were screened for the dehydra-decyclization step. All silicon phosphorous containing zeolites were found to be the most selective (70% isoprene and 20% pentadienes), where the major side reaction involved a retro-prins condensation of 3-MTHF to butane and formaldehyde. Through kinetic studies, an investigation into the effect of Brønsted acid strength, pore size and operating conditions on the selectivity to isoprene are discussed.

Figure 1. A. Itaconic acid, produced by glucose fermentation, is converted by one-pot hydrogenation to 3-methyl-tetrahydrofuran (3-MTHF). Gas-phase dehydra-decyclization of 3-MTHF with solid acid catalysts yields isoprene. **B.** Three conditions of experimental conversion of IA (5 wt% in H_2O) to 3-MTHF (red) with supported metal catalysts and hydrogen pressure after 24 hours. Products include methyl-butanol (¦, MBO), resulting from the over-hydrogenation of 3-MTHF.

[1] Abdelrahman, O. A.; Park, D. S.; Vinter, K. P.; Spanjers, C. S.; Ren, L.; Cho, H. J.; Zhang, K.; Fan, W.; Tsapatsis, M.; Dauenhauer, P. J. *ACS Catal.* **2017**, *7*, 1428-1431.

498237 Using Microkinetic Analysis to Predict Product Selectivity during Propionic Acid Hydrodeoxygenation over Supported Pt and Ru Catalysis

Abstract:

Recent industrial expansion in the biotechnology sector has raised interest in using succinic acid as an oxygenated platform for commodity chemicals. Most notably, succinic acid serves as a biogenic alternative to maleic anhydride, which is the typical industrial precursor to gamma-butyrolactone, butanediol, and tetrahydrofuran. Each of these can alternatively be prepared by partial hydrodeoxygenation (HDO) of succinic acid; however, each target is susceptible to deep hydrogenation/hydrogenolysis, and selectivity control is nontrivial. The central challenge in partial hydrogenation of succinic acid is that carboxylic acid reduction is difficult relative to the array of secondary and tertiary reactions that occur under similar conditions. These include aldehyde decarbonylation, alcohol hydrogenolysis, alkane hydrogenolysis, and methanation, which broadly result in the formation of low-value carbon oxides and light alkanes.

Although noble metals, such as Pt and Ru, show good activity in hydrodeoxygenation, they generally also activate secondary and tertiary reaction pathways, resulting in the formation of undesired fragmentation and deep hydrogenation products. A popular method for tuning their selectivity is to pair these noble metals with oxophilic "promoter" metals, such as Sn. Unfortunately, our present understanding of how secondary metals impact reaction kinetics to modulate selectivity is murky, and the design of such materials remains largely alchemical. A more rational framework for material synthesis requires an elementary understanding of reaction kinetics governing both carboxylic acid HDO and its myriad competing pathways. Furthermore, we must understand how the addition of promoter metals will perturb this free energy landscape to enhance (or detract from) monometallic selectivity. The goal of the present study is to establish kinetic expectations for carboxylic acid HDO over monometallic Pt and Ru, which provides the necessary foundation for subsequent discussion of how the addition of secondary metals impacts observed selectivity.

As a model system, we focus on propionic acid HDO and attempt to reconcile mechanistic descriptions of HDO chemistry with experimentally observed product formation rates through microkinetic analysis. Because the carboxylic acid HDO occurs alongside multiple side reactions—all of which are generally expected to occur under typical reaction conditions—it is exceedingly difficult to infer rates of primary reactions that consume carboxylic acids. In general, we are only able to directly quantify reaction rates for terminal pathways—such as methanation and alkane hydrogenolysis. Accordingly, we first consider the elementary kinetics of these terminal pathways and subsequently apply these insights as we build toward a quantitative description of carboxylic acid HDO.

490362 Renewable Adipic Acid Production Via Metal-Free Cleavage of C-O Bonds in the Presence of Molecular H_2 in Organic Acid Solvents

Abstract:

A central challenge in biomass valorization is the selective removal of oxygen-containing functional groups from platform chemicals, e.g., 5-hydroxymethylfurfural, to valuable chemicals without overhydrogenating the substrate to undesired products. Supported metal or metal oxide catalysts have shown promise in removing C-O bonds; however, many catalytically active metals/metal oxides are economically costly and can lead to major selectivity challenges, such as overhydrogenation or C-C cracking. Recently, we have developed a system for the selective removal of C-O single bonds, through employment of homogeneous or solid acids with molecular H_2 in organic acid solvents. In this system, cleavage of C-OH and ether bonds occurs readily without reducing higher order oxygenated functional groups such as carboxylic acids. This system is particularly relevant in forming adipic acid (AA), a key polymer precursor for producing Nylon 66, from biomass-derived tetrahydrofuran-2,5-dicarboxylic acid (THFDCA), where we have achieved an AA yield of 89% in the absence of any metal catalyst. In this work, we demonstrate key mechanistic insights in THFDCA hydrogenolysis to AA via computational input, kinetic measurements, and reactivity data, where we shed light onto the interplay between acids, iodides, molecular H_2, and the organic acid solvent.

490678 Influence of Sn Promoter on Pd and Pt Catalysts for Conversion of Heptanoic Acid and Propane

Abstract:

Silica-supported Pd catalysts can be used to selectively remove oxygen from biomass-derived carboxylic acids through decarbonylation. This work aims to improve Pd catalyst stability and selectivity towards desirable products by varying metal particle size as well as by introducing Sn to the catalyst. Catalysts were prepared by ion exchange or incipient wetness impregnation of metal precursors on Davisil 636 silica gel supports, followed by calcination in air and reduction in dihydrogen. Monometallic Pd and Sn catalysts, in addition to chemically-mixed and physically-mixed PdSn catalysts, were tested for activity in heptanoic acid conversion. Two-stage reactor beds were used to probe the effect of Sn proximity on Pd. X-ray diffraction, hydrogen chemisorption, TEM, and XANES-TPR were used to characterize the catalysts. The stability of the monometallic Pd catalyst during decarbonylation of heptanoic acid was influenced by the metal particle size, with smaller particles deactivating slower than larger particles. Whereas catalyst deactivation was reduced by adding Sn to Pd by impregnation, improved stability was also observed when Sn/SiO_2 was physically mixed with Pd/SiO_2. The physically-mixed PdSn initially produced products characteristic of monometallic Pd, however, the two bimetallic catalysts showed similar product selectivities after 10 h on stream. The PdSn catalysts also expanded the reaction network to include dehydration and decarboxylative ketonization, with the primary product shifting from hexene to heptanal. STEM-EDS of physically-mixed PdSn catalyst indicated that Sn migrated to the silica support particles containing Pd. In situ XANES analysis during butyric acid deoxygenation indicated that SnO_x is the primary Sn phase associated with Pd in both the physically-mixed and chemically-mixed catalysts. Ongoing work is exploring Sn promotion of Pt for propane dehydrogenation. The high temperatures and reducing conditions of the alkane dehydrogenation reaction are substantially different from the carboxylic acid reactions and likely result in a more reduced state of the Sn promoter. Bimetallic PtSn catalysts showed improved stability compared to monometallic Pt catalysts which deactivated rapidly with time on stream. X-ray diffraction indicated the presence of Pt_xSn_y in the bimetallic catalysts, suggesting that the alloy phase plays a critical role in improving catalyst stability.

491400 Selectivity Control during the One-Pot Conversion of Aliphatic Carboxylic Acids to Linear Olefins through Tandem Hydrogenation/Dehydration

Abstract:

Olefins are important building blocks for the synthesis of higher value chemicals in the petrochemical industry and they are typically produced from fossil fuels. As an alternative to fossil fuel-based chemicals, biomass-derived carboxylic acids could serve as a viable replacement feed for olefin production. However, the production of olefins from carboxylic acids in one step remains a challenge due to the stability of carboxylic acids and the tendency of olefins to over-hydrogenate during hydrogenation, leading to alkanes.

In the present study, we have studied the one-pot catalytic conversion of linear aliphatic carboxylic acids to linear olefins via tandem hydrogenation/dehydration reactions. Hexanoic acid was converted to a mixture of hexenes over Cu nanoparticles supported on commercial silica-alumina (Siral 70) in a continuous flow reactor. The combination of the hydrogenation properties of earth-abundant Cu and the acid sites on silica-alumina allows hydrogenation and dehydration to occur simultaneously, a process that normally calls for different reaction conditions. At 483 K and 5 bar H_2 gauge pressure, we obtained 98.5% conversion of hexanoic acid, with a molar product distribution of 88.9% hexenes and 11.1% hexane. Remarkably, we observed a brusque selectivity switch to 99.9% hexane at 100% conversion. We hypothesize that the presence of a small amount of hexanoic acid on the surface of the catalyst prevents overhydrogenation of the hexenes. Current efforts are targeted at reaction kinetic studies as well as surface studies using FT-IR spectroscopy to ascertain the nature of this selectivity switch. Process runs using carboxylic acids produced from biomass fermentation will also be carried out to ensure process compatibility.

Catalyst for Conversion of Methyl Ethyl Ketone to Butenes

Zahraa Alauda[a,b], Hayder Alatabi[a,b], Quanxing Zheng[a], Xu Li[a], Keith Hohn[a]

a: Department of Chemical Engineering, Kansas State University, 1005 Durland Hall, Manhattan, KS 66506, United States

b: Department of Chemical Engineering , The University of Technology, Baghdad, Iraq

Abstract

The aim of this work is to figure out a suitable catalyst to convert methyl ethyl ketone (MEK), to butene directly in one step. The catalytic gas phase reaction of MEK was investigated in a fixed bed reactor over Cu-alumina, Cu-zeolite Y sodium (Cu-ZYNa) and Cu-zeolite Y hydrogen (Cu-ZYH) as bifunctional heterogeneous catalysts, which prodvide both hydrogenation and dehydration functionalities. In the reaction, MEK is hydrogenated to 2-butanol which is further dehydrated to produce butene. Experimental results showed that selectivity of butene was the highest over Cu-ZYNa, and it was improved by finding the optimum reaction temperature and the percentage of copper loaded on ZYNa. The highest selectivity of butene (97.9%) was obtained at 270 °C and 20 wt% Cu-ZYNa. Over Cu-alumina, the selectivity of butenes was less than Cu-ZYNa since subsequent hydrogenation of butene occurred to produce butane. Also, it was observed that with increasing the H_2/MEK molar ratio, butane selectivity increased. However, when this ratio was decreased, hydrogenation of butene was reduced but dimerization to C_8 alkenes begane to be favored. The main products over 20% Cu-alumina were butene and butane, and the maximum selectivity of butene (87%) was achieved at an H_2/MEK molar ratio of five. The lowest selectivity of butene was obtained using Cu-ZYH, reaching ~40%. All catalysts were characterized by (TPD-NH_3), (TPD-CO_2), TPR and XPS to probe catalyst acidity, basicity and the reducibility of Cu loaded on supports. It was found that the amount of acidity in Cu-ZYH is much than in Cu-ZYNa. This could have caused the selectivity of butene to decrease as a result of the motivation of dimerization, oligomerization and cracking reactions.

496585 The U.S. Department of Energy's R&D Program for Carbon Use and Reuse

Abstract:

The U.S. Department of Energy Office of Fossil Energyâ€™s National Energy Technology Laboratory has an active carbon capture and storage (CCS) program. A focus area within the carbon storage program is Carbon Use and Reuse. The objective of this area is to identify and develop a suite of technologies to beneficially use CO_2 to produce useful products excluding enhanced oil recovery technologies. Identification of candidate chemistries and economically feasible approaches that utilize large amounts of CO_2 as a feedstock for high-demand product is very challenging while also ensuring that the overall process has a smaller greenhouse gas footprint that traditional pathways. Current projects in the portfolio include algae, cements/concretes and novel concepts.

490858 Design of New Bimetallic Catalysts for More Selective CO_2 Hydrogenation to Olefins and Methanol

Abstract:

CO_2 recovered from industrial flue gas can be hydrogenated using H_2 produced from H_2O using renewable energy, for manufacturing chemicals (lower olefins such as ethylene and propylene, methanol, and carbonates), and fuels (such as liquid transportation fuels including gasoline, diesel and jet fuels, methanol/DME, or synthetic natural gas). The challenges are to develop highly selective and active catalysts and energy-efficient catalytic processing, apart from the need to lower the cost of H_2 production using renewable energy. This presentation will discuss some new bimetallic catalysts (M1-M2, M1=Fe, Pd, Co, M2=Co, Cu, Ni, Pd) based on our recent research at Penn State and based on our collaboration with researchers in China, Thailand, and Japan. The hydrogenation of CO_2 over novel bimetallic catalysts can selectively produce either C_2-$C_4^{=}$ olefins (and C5+ liquid hydrocarbons) using Fe-Co catalysts, or methanol over Pd-Cu catalysts. Nano-structured catalysts based on hollow zeolite-encapsulated Ni nanoparticles can create a highly stable and carbon-resistant catalysts for co-conversion of CO_2 and CH_4 in bio-gas and natural gas to syngas for chemicals and ultra-clean fuels. Some recent advance in industrial CO2 conversion will also be briefly discussed.

497138 Direct Carbonation of $Ca(OH)_2$ Using Liquid and Supercritical CO_2

Abstract:

Direct carbonation of Ca(OH)2 using liquid and supercritical CO2

Daniel Klingenberg, Ali Zolhadr and Joseph J. Biernacki

Abstract:

The production of portland cement is responsible for about 7% of the global carbon dioxide (CO2) emissions. Roughly one ton of CO2 is produced for every ton of cement. The direct carbonation of calcium hydroxide (Ca(OH)2) has recently been suggested as a possible CO2-neutral replacement for portland cement in some applications, e.g. prefabricated cement-based construction components. The art of burning limestone ($CaCO_3$), the formation of slaked lime (calcium hydroxide) and subsequent carbonation has been practiced since at least Roman times as a form of cement. The process, however, is generally too slow to be practical for most modern applications. This study examines the use of liquid and supercritical carbon dioxide to accelerate carbonation. Discs of Ca(OH)2 were compacted at various pressures and thicknesses (different mass) to produce samples of different density and size. The primary goal of this study was to determine the practical limits of carbonation Ca(OH)2 and to glean relevant kinetic and morphological (microstructural) information. X-ray diffraction and electron microscopy was used to characterize carbonated samples.

UCLA Engineering

Civil and Environmental Engineering

CARBON UPCYCLING

CO_2 upcycling by mineralization of a carbonate-based cementation material: Processing-property relationships of CO_2NCRETE™

Prepared By: Gaurav N. Sant, Ph.D.
Associate Professor and Henry Samueli Fellow
University of California, Los Angeles (UCLA), CA 90095
Phone: (310) 206-3084, Email: gsant@ucla.edu

Collaborators: G. Falzone, H. Kweon, I. Mehdipour, B. Wang, et al.
Team Carbon Upcycling UCLA: http://www.co2upcycling.com

Developed for: Session 138: CO_2 use and reuse, AIChE Annual Meeting 2017, Minneapolis, MN, October 30, 2017

UCLA Engineering

Civil and Environmental Engineering

CARBON UPCYCLING

Presentation outline

- Making a case for CO_2 utilization
- Critical requirements for ensuring scalable CO_2 utilization and why construction offers a target of opportunity
- CO_2 utilization by the direct carbonation of $Ca(OH)_2$
- Demonstration of CO_2NCRETE™ production by direct CO_2 conversion from flue gas (i.e., dilute CO_2 streams)
- Summary and conclusions

Background: Why we should utilize CO_2

- Capture costs \$ 60-90 per ton *(2016)*
- There are additional costs associated with disposing CO_2 in reservoirs (e.g., saline aquifers, O&G reservoirs ...)
- Questions regarding permanence of sequestration solution, and suitable sites are not always readily accessible
- Unless "sovereign guarantees" are provided CO_2 sequestration is difficult
- There is a need to mitigate industrial CO_2 emissions, and create a waste-to-value or carbon-to-value economy

Why I count on concrete (and you should too)

- Concrete is truly the foundation of modern infrastructure (bridges, ...)
- The production of OPC (ordinary portland cement) at industrial scale in the late 1800s set off our ability to construct like never before
- Why: cheap, abundant, forgiving, and offers an "open time" for transport and placement (casting)
- We produce > 20 billion tons of concrete annually (global) from a production base of > 4 B of OPC

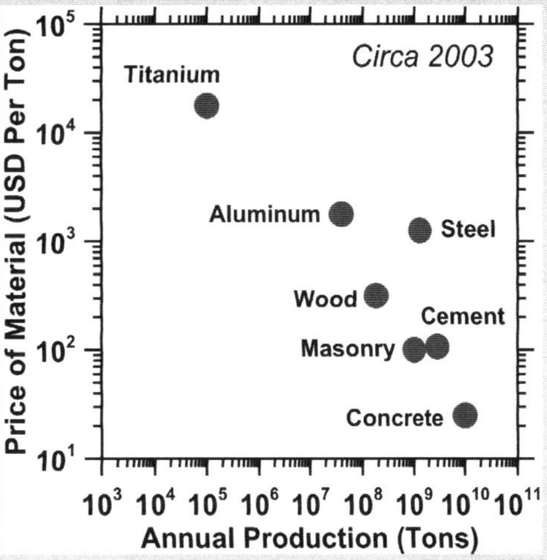

Count on Concrete

But CO_2 emissions from OPC production ...

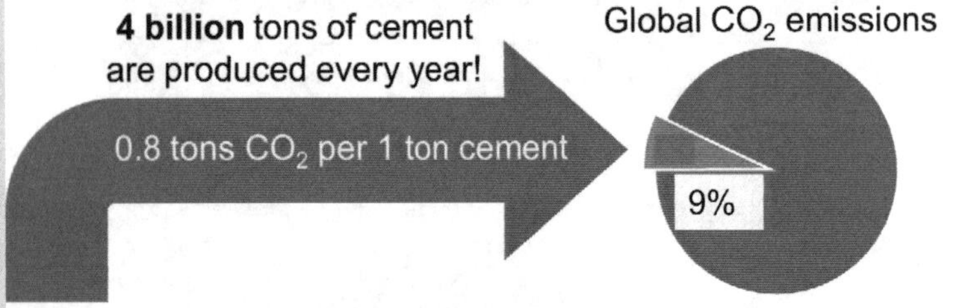

Calcination

$$CaCO_3 \rightarrow CaO + CO_2$$

- **Industry risk**
- CO_2 penalty
- Impact on OPC pricing
- Impact on construction
- In time, 9% will turn to 25% (2050)

What do we do about CO_2 (from cement) ?

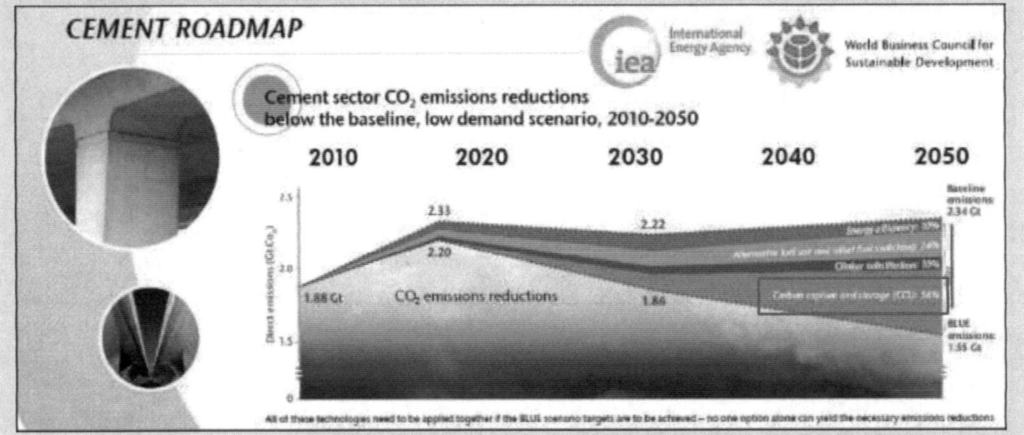

56 % of CO_2 emissions reductions is projected to come from carbon capture and storage (CCS). Clearly, this is unaffordable for a commodity (chemical) which costs $ 100 per ton. This would impede our ability to construct buildings, infrastructure and beyond.

UCLA Engineering

Civil and Environmental Engineering

CARBON UPCYCLING

Why CO_2 mineralization ?

UCLA Engineering

Civil and Environmental Engineering

CARBON UPCYCLING

Mineralizing CO_2 into [Ca, Mg] carbonates

@ Brucite is expected to offer similar carbonation behavior and potential

Carbonated mortar building element** | **Quarried, crushed and homogenized

***** Maintain mineralized CO_2 within closed-loop***

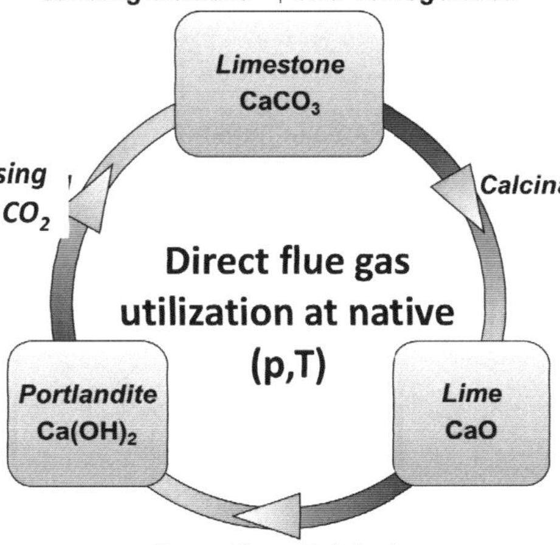

 # Portlandite can take up around 50 mass % CO_2 cheaply ($ 100/t)

** Portlandite can be produced by low-temperature pathways*

UCLA Engineering
Civil and Environmental Engineering
CO2 CARBON UPCYCLING

Why focus on cement and concrete?

- Gigaton scale CO_2 utilization, e.g., by replacing OPC and concrete aggregates; estimate of 1-to-5 billion tons by 2030
- Low(er) temperature routes for cement production (8% global industry energy)
- Allows use of stack-tap CO_2, i.e., with SO_x and NO_x and no need for capture
- The construction industry offers ready markets and it could accommodate the largest quantity of CO_2 for lowest cost
- Offers a multi-gigaton market without "parasitic energy or financial" losses

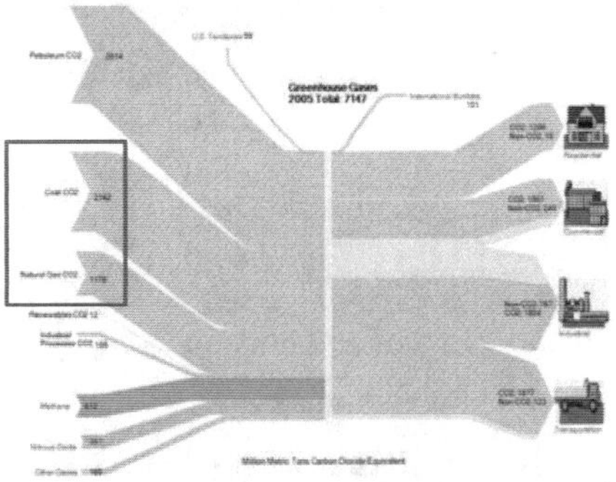

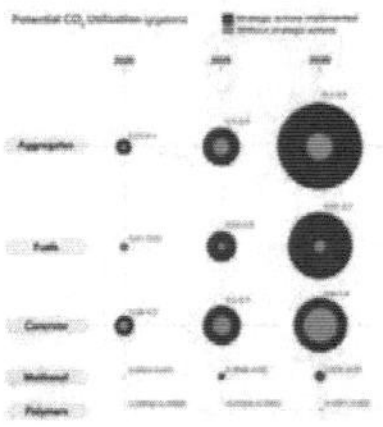

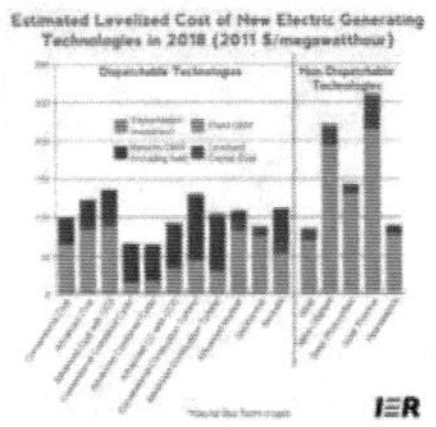

UCLA Engineering
Civil and Environmental Engineering
CO2 CARBON UPCYCLING

CO_2 mineralization reactions

- The reaction of gas phase $CO_2$with mineral to form products that embed CO_2within a stable solid
- $Ca(OH)_2$ (and $Mg(OH)_2$) is a very attractive carbonation agent, i.e., far more so than CaO for example
- This approach has been historically used to produce lime mortars, but the rate or carbonation has been far too slow to be practical
- The formation of limestone from through solution reactions is known to offer cementation (sea-shells)

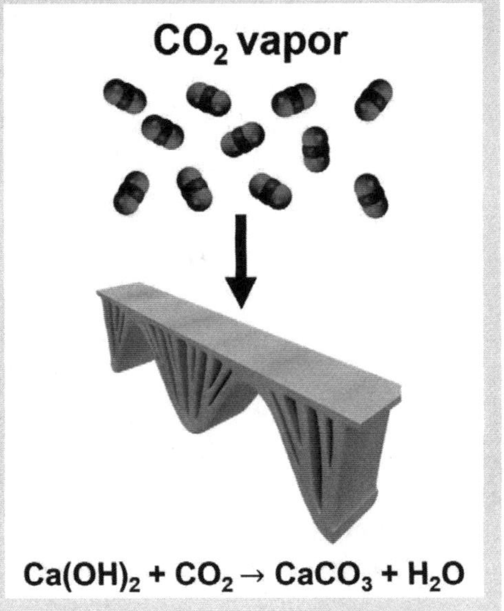

The carbonation product(s)

- When $Ca(OH)_2$ is reacted with CO_2 in gas, liquid, or supercritical states calcite forms as the stable polymorph (of $CaCO_3$)
- Thermogravimetric analysis (TGA) readily quantifies the extent of carbonation reaction (i.e., the quantity of CO_2 that is taken up into the products)
- Water that is released during carbonation is also detected – an important feature in catalyzing portlandite carbonation

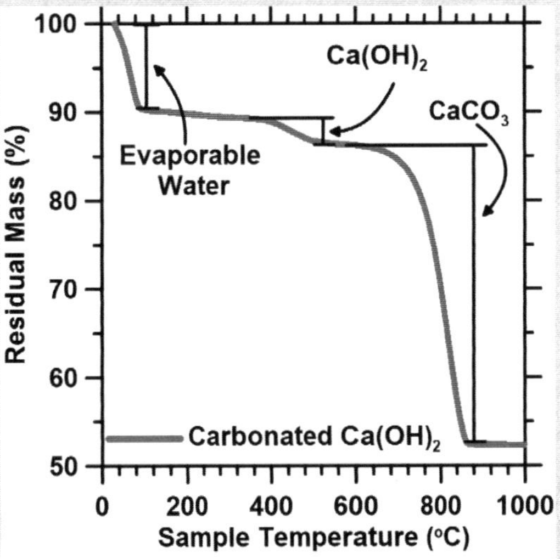

$$Ca(OH)_2 + CO_2 \rightarrow CaCO_3 + H_2O$$

Insights into the carbonation reaction

- Instantaneous surface nucleation
- Solid reactant separated from fluid by boundary layer; process switches to diffusion-control
- Reaction proceeds by OH^--for-CO_3^{2-} exchange (carbonate entry appears to be rate limiting)
- OH^- recombines with H^+ resulting in formation of water – which in turn "catalyzes" the process
- Also confirmed by DFT calculation – 32% reduction in energy (ΔG)

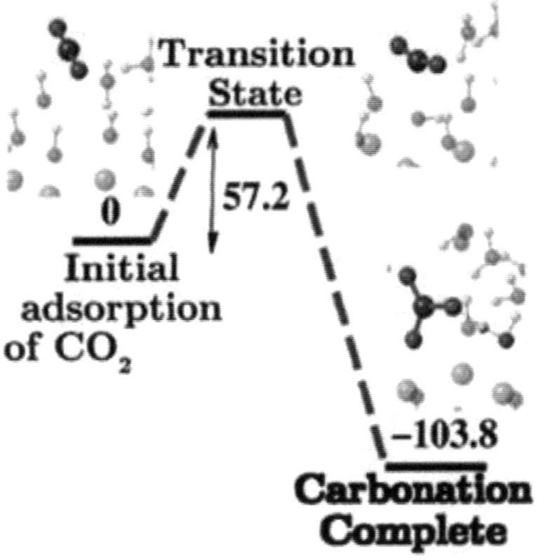

Rapid carbonation kinetics

- For $Ca(OH)_2$ particulates that react with pressurized CO_2, carbonation is rapid; reaction achieves 80% completion in 2 h
- Kinetics suggest initial region of surface control (surface area is influential); and a secondary region of diffusion control
- In the secondary step, reaction proceeds by CO_2-for-H_2O exchange (intermediate C and O species; not really "CO_2")

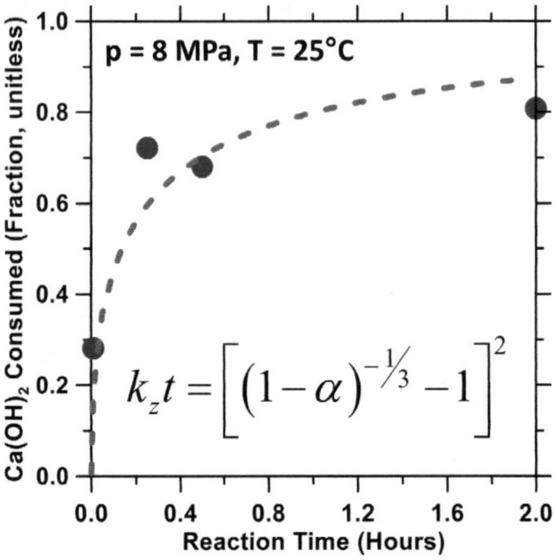

UCLA Engineering
Civil and Environmental Engineering
CO2 CARBON UPCYCLING

Role of (p,T) on portlandite carbonation

- $Ca(OH)_2$ powders, d_{50} = 10 µm
- Little dependence on 'p', slightly more so on 'T'. This is a critical reason for use for $Ca(OH)_2$
- Arrhenius analysis reveals the activation volume (ΔV_R = -27.3 cm^3/mole) and activation energy to be: (E_a = 7.5 kJ/mole)
- Carbonation proceeds rapidly for RH > 20%; so some water (even adsorbed) is required for progress of reaction

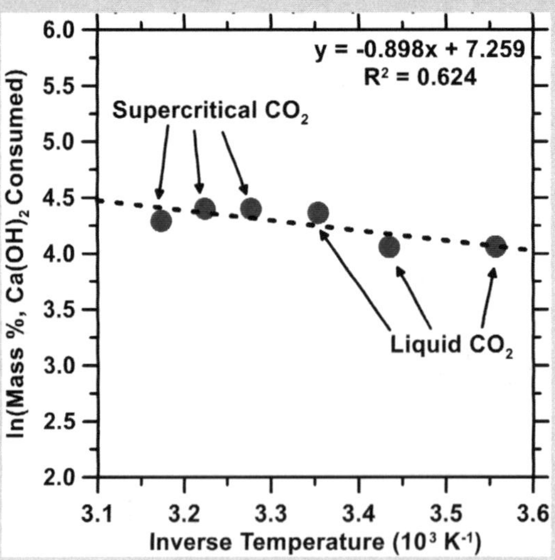

Progress of carbonation at p = 0.1 kPa, T ≥ 45°C

- Examined carbonation progress of bicomponent $Ca(OH)_2$ + OPC slurries formed into cylindrical samples: d x l [100 mm x 150 mm] in 17% CO_2 (v/v) over 24 h
- Previously, slurries were dried at 75°C for 12 hours to react OPC, and create open pore pathways
- In general, progress of reactions is rapid although reaction rate is maximized around 60°C (related to exothermic nature of reaction and Le Chatelier's principle)

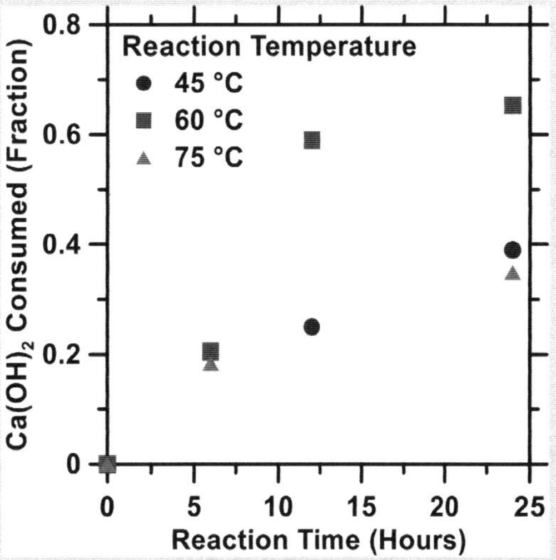

Cementation due to CO_2 mineralization?

- $CaCO_3$ products are expected to bridge and 'cement' sand grains in the style OPC does in concrete
- This is seen in SEM observations
- Based on our data, the gradient in carbonation is small, so limited impact on section properties
- Moreover, carbonation results in a 10% increase in solid volume
- This increase in volume should in-fill the porosity, ensure strength gain with increasing carbonation

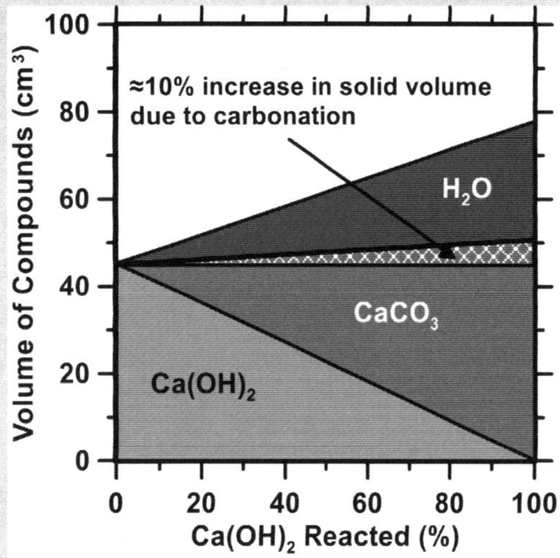

UCLA Engineering
Civil and Environmental Engineering

CARBON UPCYCLING

Mechanical strength gain with carbonation

- Created a range of formulations which contain portlandite, fly ash (filler), OPC (to offer green strength), sand, stone and water *a la* concrete
- In general, carbonation results in around 35-to-50% strength increase based on formulation in binders that may contain ≤ 30 mass % OPC, or are OPC-free
- Expect to achieve 35 MPa after 7 days of aging and processing

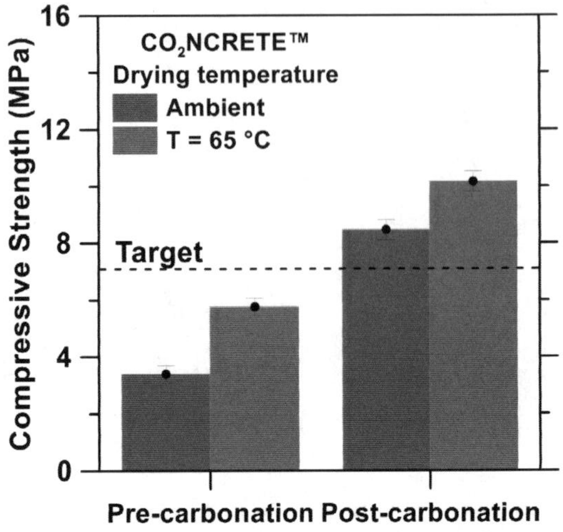

Gaurav Sant AIChE 2017 Annual Meeting, Minneapolis, MN Slide 17 of 21

UCLA Engineering
Civil and Environmental Engineering

CARBON UPCYCLING

CO_2NCRETE™ pilot-scale demonstration

- Designed, developed and validated system of 6 reactors to process 200 kg of CO_2 in 24 hours
- Input as a N_2-CO_2 mixture for CO_2 concentration range of 5-17% (v/v)
- Seeks to replicate conditions of fossil fuel power plant with heat secured from flue gas; T < 100°C
- System has achieved > 65% CO_2 capture (mineralization) efficiency per 24 hours carbonation period
- Embeds 0.2 g of CO_2 / g of reactant

Gaurav Sant AIChE 2017 Annual Meeting, Minneapolis, MN Slide 18 of 21

$CO_2NCRETE^{TM}$ workflow

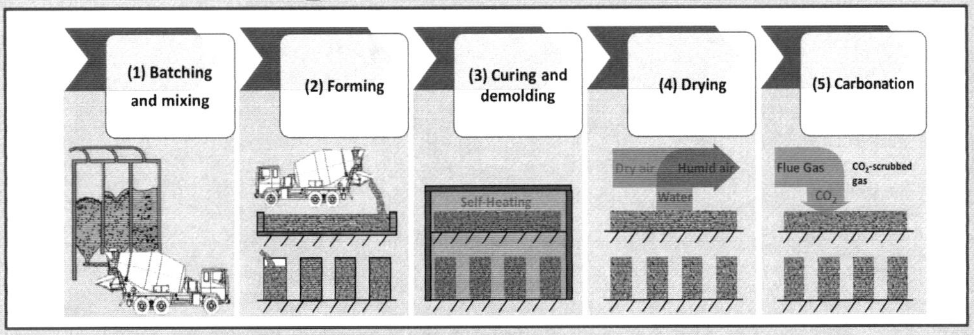

- CO_2NCRETE is batched/mixed using traditional concrete production equipment, and then formed into the desired geometry (e.g., slabs)
- The forms are removed after CO_2NCRETE gains green strength (i.e., ≤ 3 MPa) to maximize surface area that is available for CO_2 contact
- Batch production in parallel reactors allows continuous CO_2 utilization

UCLA Engineering
Civil and Environmental Engineering
CO2 CARBON UPCYCLING

Ensuring CO_2-neutral cementation ...

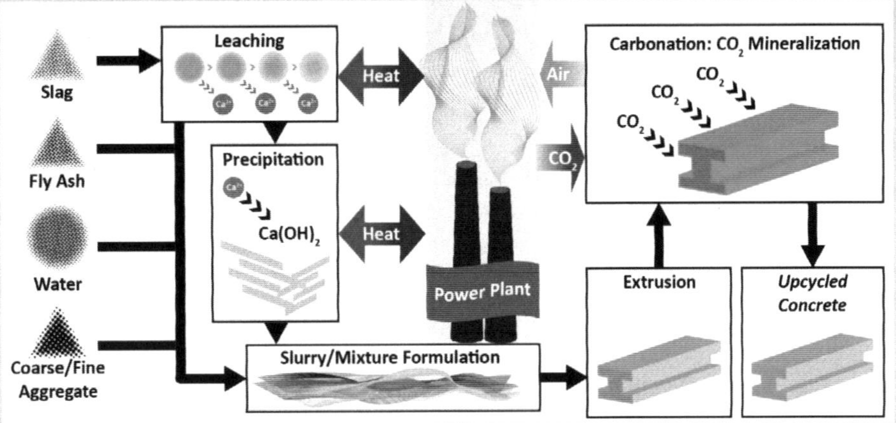

- *Upcycling* low-value industrial wastes into a CO_2-negative concrete
- The process utilizes Ca (and Mg) -extracted from wastes as precursors to produce $Ca(OH)_2$ at low temperatures for CO_2 mineralization

UCLA Engineering

Civil and Environmental Engineering

CARBON UPCYCLING

Summary and conclusions

- Direct carbonation of $Ca(OH)_2$ using dilute CO_2 streams is a viable route towards large-scale CO_2 utilization
- CO_2 mineralization occurs rapidly, and results in the rapid development of strength; indicative of cementation
- The process has been scaled up to demonstrate a pilot-scale process 200 kg of CO_2 in a day; while producing a range of element geometries including cylinders and slabs
- The success of this approach provokes new pathways for CO_2 mineralization that could achieve CO_2-neutral (and perhaps CO_2-negative) actions

Gaurav Sant AIChE 2017 Annual Meeting, Minneapolis, MN Slide 21 of 21

498101 CO_2 Conversion to Fuels and Chemicals through Microbial Electrolysis Cells

Abstract:

With a fully oxidized carbon atom, CO_2 is notoriously stable under most environmental conditions. Thus, the conversion of CO_2 to fuels and chemicals has been pursued through chemical, photochemical, electrochemical, and biological means, in addition to reforming. Through these different routes, CO_2 can be converted to chemicals such as formic acid, acetic acid, and urea; alcohols such as ethanol and methanol; and gases such as methane, syngas (CO), and acetylene (C_2H_2). The majority of these reactions require a reducing agent, most commonly hydrogen (H_2). However, hydrogen is quite expensive, so the resulting products would not be cost-competitive; thus, the need for alternative approaches is obvious.

To reduce CO_2 to fuels and chemicals, our laboratory has enriched a microbial community that has demonstrated repeatable high rate production of methane and organic acids from CO_2. Two unique features make this process extremely interesting for further exploration, and profoundly attractive for potential technological development. First, CO_2 conversion to value-added products takes place in a one-chamber membrane-less Microbial Electrolysis Cell (MEC). The methane production rate is 10-56x higher than typically reported for methane release from CO_2 in two-chamber MEC units. Second, the enriched microbial community thrives on a defined medium without the presence of any organic carbon and nitrogen sources.

In this talk, we will present our recent data on this project. Results, such as the microbial community composition, the enrichment process, electrode modification and fundamentals related to the single chamber MECs will be discussed. All of these will assist in future research and development in making MEC a promising platform for converting CO_2 to useful commodities.

489037 High-Selectivity, Gas Fermentation of CO_2 to Ethanol

Abstract:

LanzaTech has developed a fully integrated, platform for commercial-scale production of fuels and chemicals from sustainable biomass and waste feedstocks. The platform comprises of a highly-optimized, gas-utilizing host organism, an advanced strain engineering and process optimization pipeline, validated metabolic and process models, and scalable reactor designs with demonstrated performance.

LanzaTechâ€™s innovative gas fermentation platform can capture and utilize CO_2, H_2 and CO for the production of fuels and chemicals. The ability of the fermentation platform to utilize gas mixtures of varying compositions enables the use of a diverse range of feedstocks, including syngas generated from agricultural residues or municipal solid waste and waste gases from industrial sources (e.g. refineries and steel mills).

Refinery waste gases which are primarily composed of H_2 and COÂ

products directly from low value gas streams while reducing CO_2 and air emissions at the refinery site. Furthermore, the ethanol gas fermentation pathway exhibits a high selectivity which significantly exceeds the selectivity of catalytic pathways for conversion of CO_2, H_2 and CO. This ability of the LanzaTech gas fermentation platform to selectively produce ethanol from dilute refinery waste gases provides the next step in the drive to increase efficiency and reduce waste in the refinery setting.

Production of over 30 molecules has been demonstrated, adding value to carbon emissions and highlighting the opportunities that exist through carbon capture and utilization technologies.

Proprietary, scalable reactor designs and optimized process chemistry, ensure efficient, continuous, single-pass gas conversion with a high selectivity to the product of interest. The process has been successfully scaled up from the laboratory bench through in-lab and in-field pilot plants to fully integrated 100,000-gallon/year pre-commercial demonstration plants with a total of over 55,000 hours on stream. First full commercial units are in construction in US, Europe and Asia.

Carbon utilization technologies such as LanzaTechâ€™s offer both feedstock holders a path to add value, and chemical users a path to off take large volumes of recycled carbon products.

493824 Role of Nitrogen Addition to Reduced Graphene Oxide-TiO_2 Nanocompostites in Enhancing CO_2 Photoreduction

Abstract:

Carbon dioxide (CO_2) capture and utilization is a promising potential technology to address global climate change. Photocatalytic conversion of CO_2 into hydrocarbon fuels has attracted abundant research attention in recent years. However, there have been few reports about CO_2 capture and conversion on the same material so far. Addition of nitrogen has been reported to be a feasible approach to enhance materialsâ€™ CO_2 capture performance. In this work, we demonstrated two ways to fabricate the nitrogen contained reduced graphene oxide (NrGO)-TiO_2 nanocomposites and discussed the function of nitrogen surface modification, as well as the nitrogen structure doping. We found that both nitrogen surface modification and structure doping could enhance the CO_2 photoreduction of rGO-TiO_2 materials significantly.

It was demonstrated that NrGO (N surface modified rGO or N structure doped rGO) with an appropriate N quantity and N-bonding configuration acted as a dual-functional promoter, simultaneously enhancing CO_2 adsorption on the catalyst surface and facilitating electron-hole separation, while eventually boosted the photocatalytic performance. Experimental results in this work provide a better understanding of the critical roles of nitrogen addition in the synthesized composites and also inspire the ongoing interest in a better design of other N-contained graphene based materials for photoreduction of CO_2.

Reference:

Lin, L. Y., Nie, Y., Kavadiya, S., Soundappan, T., & Biswas, P. (2017). N-doped reduced graphene oxide promoted nano TiO_2 as a bifunctional adsorbent/photocatalyst for CO_2 photoreduction: Effect of N species. *Chemical Engineering Journal*, *316*, 449-460.

487709 Design of Advanced Nanomaterials for Electrocatalysis

Abstract:

Hydrogen is a promising energy carrier for the replacement of fossil fuels. The hydrogen cycle in the sustainable energy system involves four important reactions, namely, oxygen reduction reaction (ORR), hydrogen oxidation reaction (HOR), oxygen evolution reaction (OER) and hydrogen evolution reaction (HER). Our research emphasizes the catalysts for these key reactions. Here, we introduce a "model surface - nano verification - catalyst design" research system for discovery and production of advanced nanomaterials for sustainable energy and fuel.

491078 Layered Hybrid Structure of Cobalt/N-Doped Carbon Derived from Metal-Organic Frameworks for Electrocatalytic Hydrogen Evolution

Abstract:

This study reports in situ synthesis of Co/N-Carbon hybrid structures via the pyrolysis of layered cobalt based metal–organic frameworks. As excellent hydrogen evolution reaction (HER) catalysts, the 2D hybrid layered structure composed by ultrafine metallic cobalt nanoparticles and N-carbon, showed a small overpotential of 103 mV (vs RHE) at 10mA/cm2, high cycle durability and considerable long-term stability for HER. The superior electrochemical performances of Co/N-carbon for HER can be attributed to such 2D configuration which includes the low H atom binding energy of metallic cobalt, the high conductivity of carbon and the stability of carbon-encapsulated Co nanoparticles. Therefore, we have worked on the synthesis and characterization of the 2D hybrid structure of Co/N-Carbon derived from metal organic frameworks. The as-prepared 2-D sheet structures retained the well-defined morphology of the initial Co-based MOFs, and cobalt nanoparticles (ca. 7.2 nm) were well embedded on the N-carbon. Due to the integration of the superior features of ultrafine cobalt nanoparticles and N-carbon, the resultant Co/N-Carbon showed enhanced electrocatalytic activity and durability when compared to the commercial Pt/C electrocatalyst at the alkaline media. More importantly, Co/N-carbon's electrocatalytic activities in terms of overpotential at 10mA/cm2 for hydrogen evolution have been more advantageous over other nanocarbon-based electrocatalysts.

Acknowledgements: Financial support from Interdisciplinary Graduate School, Nanyang Technological University Singapore is gratefully acknowledged.

References

[1] Walter, M. G.; Warren, E. L.; McKone, J. R.; Boettcher, S. W.; Mi, Q.; Santori, E. A.; Lewis, N. S. Chem. Rev 2010, 110 (11), 6446-6473.

[2] Hou, Y.; Wen, Z.; Cui, S.; Ci, S.; Mao, S.; Chen, J. Adv. Funct. Mater 2015, 25 (6), 872-882.

492501 Comparison of Hydroxide-Mediated and Direct Mechanisms for Alkaline Hydrogen Electrocatalysis

Abstract:

It has long been recognized that the reaction rates of the hydrogen oxidation and hydrogen evolution reactions (HOR and HER) are slower in basic than acidic electrolytes, even though the surface intermediate of adsorbed hydrogen is independent of solution pH. Understanding the root of this observation is critical to designing catalysts for a multitude of electrochemical reactions with relevance to energy conversion and storage. In this work, we undertake a fundamental investigation to determine if adsorbed hydroxide plays an active or spectator role in HER/HOR kinetics.

One proposed explanation for the pH dependence of HER/HOR kinetics, suggested by the research groups of both Gasteiger and Yan, is that in base, hydroxide ions stabilize the Pt-H bond for stronger binding and slower catalysis [1,2]. Central to this hypothesis is an experimentally measured shift with pH in the peak potential for hydrogen underpotential deposition (H-UPD) on the (100) and (110) steps on the Pt surface. However, the Janik and Koper groups have found that the shift in H-UPD potential is caused not by changes to hydrogen adsorption, but instead by competitive adsorption of hydroxide in the presence of alkalai cations [3,4]. Markovic et al have proposed that such hydroxide adsorption is actually necessary to facilitate water recombination/dissociation in base, and that oxophilic sites such as Ni(OH)2 generate a bifunctional mechanism for efficient alkaline hydrogen electrocatalysis [5].

In this work, we investigate specifically the hypothesis that adsorbed hydroxide is an active participant in the alkaline HER and HOR by combining electroanalytical techniques with microkinetic modeling. We develop expressions for time-dependent current and voltage as a function of hydrogen and hydroxide binding strength for the indirect (hydroxide-mediated) and direct (hydroxide-as-spectator) Volmer steps in base. These expressions are compared to experimental cyclic voltammograms of H-UPD on polycrystalline platinum. The free energy of hydroxide adsorption is controlled experimentally by the electrolyte cation via changes in cation solvation strength [6]. Varying the sweep rate results in greater peak separation that can be used to extract rate constants for the adsorption reaction via traditional electroanalysis [7].

Our computations show that at both slow and fast scan rates, either the direct or indirect mechanism can describe the peak potential splitting and peak current. However, hydroxide binding strength affects the mechanisms differently. In the indirect mechanism, stronger hydroxide binding reduces overpotential and, therefore, peak splitting. In the direct mechanism, stronger hydroxide binding decreases available surface sites, leading to lower exchange current densities and greater overpotential. Comparison with experiment shows that on the (110) and (100) facets of polycrystalline platinum, moving from 0.1M KOH to 0.1M LiOH decreases the peak potential location, but increases peak potential splitting. This trend shows that stronger hydroxide binding has detrimental effects on H-UPD kinetics, and that the direct mechanism is therefore more likely to dominate observed behavior.

Our results strongly suggest that adsorbed hydroxide serves as a competitive spectator in the alkaline Volmer step, and that the bifunctional HOR/HER mechanism plays only a minor role at best. This study contributes to resolving a long-standing paradox in electrocatalysis and surface science by determining that oxophilicity is not an accurate descriptor for alkaline hydrogen electrocatalysts. Other parameters, such as water orientation and non-covalent interactions, must play a greater role in overall activity. Future work will discuss these and other factors in the relation between H-UPD kinetics and HER/HOR catalyst activity .

(1) Durst et al. Energy Environ. Sci. 2014, 2255.

(2) Sheng et al. Nat. Commun. 2015, 5848.

(3) McCrum et. al. J. Phys. Chem. C, 2015, 457.

(4) Van Der Niet et al. Catal. Today 2013, 105.

(5) Strmcnik et al, Nat. Chem. 2013, 1.

(6) Danilovic, N. et al. Electrocatalysis 3, 2012, 221.

(7) Angerstein-Kozlowska et al. J. Electroanal. Chem. Interfacial Electrochem. 1979, 1.

497033 Alkaline Hydrogen Evolution from Ni–Mo Intermetallics with High Mo Content

Abstract:

The transition from acidic to alkaline conditions promises considerable cost savings for water electrolysis systems based on polymer ion-exchange membrane electrolytes. Binary transition metal alloys containing Ni and Mo have been known for decades to exhibit excellent performance as alkaline water electrolyzer cathodes,[1] but the specific reason that these alloys are more active toward the hydrogen evolution reaction (HER) than either Ni or Mo alone remains the subject of some considerable debate.[2] Moreover, previously reported mass-specific activities for Ni–Mo alloys are still at least 100-fold lower than that of Pt, which makes them difficult to deploy in a practical electrolysis system that operates at suitably high energy efficiency and power density.

Prior work on Ni–Mo electrocatalysts has suggested that the alloy phase exhibits monotonically increasing HER activity as the bulk Mo mole fraction is increased.[3] However, the accessible composition range for binary Ni–Mo alloys is constrained by the solid solubility of Mo in the Ni fcc lattice, which is only ~25 mol% at 1000 °C. Surface-science experiments on bulk $Ni_{0.8}Mo_{0.2}$ alloys have indicated that the topmost atomic layer becomes considerably enriched in Mo, up to approximately equimolar Mo and Ni.[4] Thus, the question remains as to how much Mo is present in the most active Ni–Mo binary HER electrocatalyst surface, let alone what is the mechanism for enhanced activity.

In this presentation, I will discuss recent and ongoing work aimed at understanding the influence of bulk composition on the activity of nanostructured Ni–Mo alkaline HER catalysts.[5] We synthesized unsupported powders of Ni, $Ni_{0.92}Mo_{0.08}$ alloy, and the intermetallic compound Mo_7Ni_7. Mass-specific activity was found to be highest for the 8 mol% Mo alloy, but these data were clearly convoluted by catalyst morphologies, as the 50 mol% Mo intermetallic exhibited very low surface roughness. Attempts to account for electrochemically active surface area using a straightforward method based on electrochemical impedance spectroscopy (EIS) yielded unexpected results that conflicted with morphological analysis from electron microscopy. We have therefore postulated that Mo undergoes a pseudocapacitive redox transition near the onset potential for the HER, leading to erroneous results from EIS measurements. These results further implicate a plausible bifunctional catalytic mechanism that explicitly involves Mo redox transitions.

(1) Brown, D. E.; Mahmood, M. N.; Man, M. C. M.; Turner, A. K. *Electrochimica Acta* **1984**, *29* (11), 1551–1556.

(2) McKone, J. R.; Marinescu, S. C.; Brunschwig, B. S.; Winkler, J. R.; Gray, H. B. *Chem. Sci.* **2014**, *5* (3), 865–878.

(3) McKone, J. R.; Sadtler, B. F.; Werlang, C. A.; Lewis, N. S.; Gray, H. B. *ACS Catal.* **2013**, *3* (2), 166–

169.

(4) Soriaga, M. P.; Baricuatro, J. H.; Cummins, K. D.; Kim, Y.-G.; Saadi, F. H.; Sun, G.; McCrory, C. C.; McKone, J. R.; Velazquez, J. M.; Ferrer, I. M.; I, C., Azhar. *Surf. Sci.* **2015**, *631*, 285–294.

(5) Csernica, P. M.; McKone, J. R.; Mulzer, C. R.; Dichtel, W. R.; Abruña, H. D.; DiSalvo, F. J. *ACS Catal.* **2017**, 3375–3383.

485062 Increased Photocatalytic Activity of TiO_2 Nanoparticles with Defects for Sustainable Hydrogen Production

Abstract:

Hydrogen gas is touted as a clean-burning energy carrier that can be used in fuel cells for stationary and transportation power at higher efficiency than conventional internal combustion engines. The primary source of H_2today is from nonrenewable natural gas-derived methane steam reforming (MSR). MSR suffers from being a highly endothermic and endergonic reaction (though less so than water splitting), requiring significant energy input as heat generated from burning half of the feed methane, limiting atom efficiency and emitting greenhouse gases. Photocatalytic MSR circumvents the high heat requirement and lowers the reaction temperature, thereby avoiding coke buildup problems as well.

Titanium dioxide (TiO_2) is of particular interest in photocatalysis due to its high photoactivity under ultraviolet illumination and its stability under catalytic reaction conditions. Together with supported metal nanoparticles, TiO_2has been used to catalyze a number of important chemical and photochemical reactions.

We have synthesized defect-rich anatase TiO_2 that shows unique selectivity for hydrogen formation from methane. We have also demonstrated that ultra-violet light illumination boosts the activity and selectivity of the catalyst for hydrogen formation. We have demonstrated hydrogen yield of 27% under steady state conditions using a nickel cocatalyst. Through a combination of experimental and computational techniques, we investigate the mechanisms of defect formation and how these defects activate methane and selectively catalyze hydrogen formation.

486781 Transition Metal Nitride Core-Noble Metal Shell Nanoparticles As Highly CO Tolerant Catalysts

Abstract:

Introduction

Core-shell nanoparticles (NPs) have become increasingly attractive as a catalytic design platform due to their unique ability to simultaneously enhance the activity, increase the durability, and reduce the loading of noble metal (NM) catalysts. However, these materials have been limited by the lack of independent control over the core and shell compositions, shell thickness, and particle size [1]. To address these issues, our group recently developed a novel method for the synthesis of transition metal carbide (TMC) core-NM shell (denoted NM/TMC) NPs with precise control over the core-shell architecture [2]. Group 4-6 TMCs are earth-abundant compounds with high sinter-resistance, electrical conductivity, and chemical stability, making them ideal candidates for core materials [3].

Transition metal nitrides (TMNs) share all of these favorable material properties of TMCs but have rarely been investigated in core-shell materials. Here, we present the ability to synthesize NM/TMN core-shell NPs with varying core and shell compositions and controllable shell thickness via nitridation of the corresponding NM/TMC. The impacts of titanium tungsten nitride (TiWN) and titanium tungsten carbide (TiWC) cores on the electronic properties of a Pt shell are compared, and the effect of these differences on catalytic properties was investigated by measuring the activity for the hydrogen oxidation reaction (HOR) under the presence of CO to assess the CO tolerance. The results illustrate how the use of nitride versus carbide cores offers an effective way to tune the reactivity of NM catalysts.

Materials and Methods

NM/TMC NPs were nitrided under NH_3 at 800°C for 3 h to form NM/TMN NPs. Synthetic details for NM/TMC are provided elsewhere [2]. All materials were supported on carbon. Electrochemical measurements were performed at 30°C on a rotating disk electrode with a Ag/AgCl reference and platinum counter electrode. All potentials are reported versus a reversible hydrogen electrode (RHE). For comparison, catalyst loadings were targeted to have roughly equal Pt surface areas. HOR scans were performed in H_2-sat 0.1 M $HClO_4$ at 2 mV/s and 1600 rpm with iR compensation. To test CO tolerance, the working electrode was first held at 0.025 V until the electrolyte equilibrated with 1000 ppm CO before starting the HOR scan.

Results and Discussion

Pt/TiWC NPs with varying Pt coverage were synthesized using the method previously reported by our group and transformed into Pt/TiWN NPs via nitridation at 800°C. The high sinter-resistance and strong interfacial bonds of Pt/TiWC prevent particle agglomeration and preserve the core-shell structure while the core is converted into a nitride, which is confirmed by XRD, STEM-EDX, TEM, XPS, and CHNS analysis (Figure 1A).

XPS spectra of the core-shell materials reveal shifts in the core electron binding energies, called core level shifts (CLSs), for both the W 4f and Pt 4f signals. These CLSs indicate a significant change in the Pt electronic structure, as shifts in the core level energies of metal overlayers are reflected in similar shifts in the valence d-band center, which is known to correlate strongly with adsorbate binding energies and thus the reactivity [4-5]. The XPS results suggest downward shifts in the Pt d-band center for the core-shell materials compared to a pure Pt surface and an even lower d-band center and weaker binding for Pt/TiWN compared to Pt/TiWC.

To probe the difference in adsorbate binding and reactivity among our core-shell and commercial Pt (denoted Pt_{comm}) catalysts, we used CO as a model adsorbate and measured the HOR activity in the presence of CO (Figure 1B). Strong CO binding is a common problem for Pt catalysts, poisoning the surface and impeding reactivity. In the presence of 1000 ppm CO, the HOR activity of Pt_{comm} is completely suppressed by CO adsorption and is not recovered until above 0.5 V when CO can be oxidized off the surface. In contrast, both core-shell materials retain HOR activity at potentials as low as 0.1 V, thus demonstrating dramatically improved CO tolerance and reduced CO bond strength. Furthermore, 2-3 monolayer (ML) Pt/TiWN displays even greater CO tolerance than 2-3 ML Pt/TiWC, indicative of weaker CO binding as predicted from the XPS CLSs.

Figure 1. A) STEM-EDX map of 2-3 ML Pt/TiWN. B) HOR polarization curves in H_2-sat 0.1 M $HClO_4$ at 1600 rpm with (solid line) and without (dotted line) 1000 ppm CO.

Significance

In addition to Pt/TiWC, the synthetic method can be applied to core-shell NPs of varying NM/TMC compositions, demonstrating a promising way of synthesizing new core-shell architectures with tunable electronic properties that can serve as next generation catalysts.

References

1. Yang, H. *Angew. Chem. Int. Ed.* 50, 2674-2676 (2011).

2. Hunt, S.T., Milina, M., Alba-Rubio, A.C., Hendon, C.H., Dumesic, J.A., and Román-Leshkov, Y. *Science* 352, 974-978 (2016).

3. Zhong, Y., Xia, X., Shi, F., Zhan, J., Tu, J., and Fan, H.J. *Adv. Sci.* 3, 1500286 (2016).

4. Weinert, M., and Watson, R.E. *Phys. Rev. B* 51, 17168-17180 (1995).

5. Kitchin, J.R., Nørskov, J.K., Barteau, M.A., and Chen, J.G. *J. Chem. Phys.* 120, 10240-10246 (2004).

502612 Understanding the Electrochemical Behavior of a Thin, Flexible Micro Fuel Cell

Abstract:

Portable fuel cell systems in the power range of few milliwatts to a few hundred watts have become the focus of vigorous research and development activities. Fuel cells promise to provide more reliable, longer portable power than batteries. Micro fuel cell (mFC) technology can take advantage of fuels with energy densities of an order of magnitude higher than the energy stored in batteries. The air breathing cathode reduces the size and weight of the cathodic plate, and simplifies system requirements since oxygen inlet is fully passive. On the other hand, working conditions for the air breathing cell are more demanding due to the absence of an air flow behind the cathode gas diffusion layer to assist the removal of water.

The importance of micro-structures on polymers for micro fuel cell fabrication is enormous, particularly when considered as a low-cost alternative to the silicon- or glass-based MEMS technologies, for disposable small electronic gadgets. Many polymer-based microfabrication techniques via microinjection molding, casting, and micro-hot embossing have been developed.

Hot embossing provides advantages such as a relatively low-cost for embossing tools, simple operation and higher accuracy in the replication of small features. The embossing master stamp can be a silicon wafer, glass, electroplated nickel mold or other stamp with microfeatures. The hot embossing process introduces less residual stress in the polymer because the polymer stretches for a very short distance from the substrate into microstructure during hot embossing. In addition, the temperature variation range for the polymer is smaller than that required in injection molding, thus can reduce shrinkage during cooling and the friction forces acting on the microfeatures during de-molding. However, still micro-hot embossing is facing challenge in terms of process feasibility, since it is difficult to make the polymer to fill completely into microfeatured geometry of high aspect ratio and it is also delicate to separate the embossed structures from the mold without breakage.

An air-breathing micro fuel cell with direct hydrogen flow through porous anode electrode is realized and reported in this study. The performance of the microdevice with embedded flow channels and electrodes is characterized in ambient conditions. To achieve this goal, we implement a SU-8 microchannel stamp to transfer a pattern into a Nafion 1110 membrane by hot embossing. Nafion 212 thermally seals the whole device as a blanket top layer. The fabrication process of microchannels and micro fuel cell construction is evaluated with optical and scanning electron microscopy (SEM) step by step. Variations of hydrogen feed rate on performance were investigated. The shared-anode characteristic design of the double-sided cell is further studied by separate performance and electrochemical impedance spectroscopy (EIS) measurements taken from each side and compared with data for the complete cell. The maximum power density per superficial (footprint) area is 68.4 mWcm-2 for the shared anode stack. In the present micro fuel cell architecture, the external package is constructed of Nafion polymer, which offers a thin, light and flexible energy source with low cost of manufacturing. The device performance offers a high volumetric and

gravimetric energy density for portable applications.

492950 Meeting Challenges in a Land Constrained Solar Economy: Co-Production of Energy and Food Using Photovoltaic Systems over Farmland

Abstract:

The main purpose of this presentation is to demonstrate that contrary to the popular belief that ample land area is available to install photovoltaic modules, in an economy mainly supported by solar energy, there will be severe land availability constraints in many regions of the world—including many states in the USA. This will inevitably lead to competition for land between food and energy [1]. *We therefore propose several innovative PV systems that will allow for the co-production of food and energy from the same unit area of farmland thereby allowing local energy needs to be met by local supply of photons*. This allows us to overcome feedstock (land) and transport (long-distance transmission) limitations for electrical generation that would otherwise be difficult to overcome in a solar economy.

A recent study by MacKay showed that the energy consumption per unit area in densely populated countries such as the UK, Germany, Republic of Korea, and Japan is approaching the energy generation per unit area of photovoltaic farms [2]. This suggests that a transition to a solar powered economy in these nations would require deployment of solar on a large fraction of the nation's total land area, which will directly compete with land requirements for other needs (e.g. agriculture). Many other countries, including the United States are also trending toward this scenario [2]. A recent study by National Renewable Energy Laboratories confirms the large land area requirements of solar farms [3].

Using the available data on current U.S. energy consumption and land availability [4]-[5], we have estimated that as many as 31 states may be unable to meet their needs without importing energy long distances from other states or offshore. This scenario arises because much of the open land in the U.S. is devoted to cropland (21.5%), pasture land (32.3%), and forest land (30.4%). Only 3.6% of the land falls in the category of miscellaneous/other use land [5].

As it should remain a priority to preserve our native forest and wildlife areas, one of the greatest opportunities for increasing the availability of land for solar energy generation is over farmland. These land areas are often easily accessed and surround population centers. However, if PV is to be successful deployed over farmland, novel systems will need to be designed that do not negatively impact crop production. As most photosynthetic activity requires only the visible portion of the spectrum (400-750 nm), these wavelengths can be allowed to past through to plants while the IR portion of the spectrum is harvested as electricity. It is also clear that plants can go relatively extended periods of time without solar irradiance (night) and survive extended periods of reduced lighting (cloudy days) without significant impact to crop production so it may be possible to harvest a portion of the visible portion as well.

Figure 1 shows several possible configurations for harvesting solar energy over farmland. Fig. 1 (a)-(b) are systems that could operate optimally using current industrially-proven solar absorbers (e.g. silicon). Fig. 1 (c)-(f) shows systems for which light splitting would result in even higher performance by optimizing the band gap of the absorber for the portion of the spectrum it is absorbing. Fig. 1(a) is the simplest case in which south-facing PV modules are altered such that every other cell is left open or replaced with a light diffuser. In this system crops never experience shading for very long as the shadow will be constantly moving underneath the panel. Fig. 1(b) is the case in which single- or dual-axis tracking panels are used. Again, the shadow will move throughout the day. In the event that crops cannot tolerate extended periods of shade, these panels can also be patterned and be made twice the size to decrease the time intervals of shading, though the total time each plant would spend in the shade over the course of the day would be the same.

Fig. 1(c)-(f) are systems based on industrially available dichroic mirrors that split the spectrum [6]-[7]. These offer the advantage of concentrating the light for higher efficiencies. As a downside, diffuse portions of the spectrum may not be harvested efficiently. In Fig. 1(c), a tracking parabolic trough system is used in which the IR portion of the spectrum is reflected to a PV cell of appropriate band gap while the visible light passes through. As an option, a thin bifacial PV cells may be placed at the back (or a pattern thereof) to harvest a portion of the visible light and light scattered back from the crops. In Fig. 1(d), the system is non-tracking for mechanical simplicity and directs much of the IR spectrum to bifacial PV cells of appropriate band gap. During non-growing seasons, PV panels could be inserted in place of the transparent support. Fig. 1(e) depicts a tracking, long-pass system in which the visible light is concentrated and collected in fiber optic cables and then redistributed to plants underneath (and potentially other uses such as lighting. Fig. 1(f) depicts a non-tracking, long-pass system in which visible light is again collected in fiber optic cables and then redistributed to plants underneath, while the IR passes through to be collected on PV of appropriate band gap underneath. Promising preliminary calculations have been completed and are on-going to determine how much power can be produced and the optimal band gaps for each system.

The choice of which system to use will depend on how the crops respond to different shading characteristics and how much light deprivation and which portions of the spectrum they can tolerate being deprived of. It will also depend on the relative percentage of diffuse to direct light in a geographic location as the systems in Fig. 1(a)-(b) are better suited for areas with more diffuse light. We explore these options and make some general conclusions on system selection.

References

[1] Solar Projects Sow Tension, Wall Street Journal, Wall Str. J. (2017) A3.

[2] D.J.C. MacKay, Solar energy in the context of energy use, energy transportation and energy storage., Philos. Trans. A. Math. Phys. Eng. Sci. 371 (2013) 20110431. doi:10.1098/rsta.2011.0431.

[3] S. Ong, C. Campbell, P. Denholm, R. Margolis, G. Heath, Land-Use Requirements for Solar Power Plants in the United States - Technical Report NREL/TP-6A20-56290, 2013.

[4] State Profiles and Energy Estimates, (n.d.). http://www.eia.gov/state/.

[5] Major Land Uses, (n.d.). https://www.ers.usda.gov/data-products/major-land-uses/.

[6] T. Ulavi, T. Hebrink, J.H. Davidson, Analysis of a hybrid solar window for building integration, Sol. Energy. 105 (2014) 290–302. doi:10.1016/j.solener.2014.03.006.

[7] E. Gençer, C. Miskin, X. Sun, M.R. Khan, P. Bermel, M.A. Alam, R. Agrawal, Directing solar photons to sustainably meet food, energy, and water needs, Sci. Rep. (2017).

498203 Embodied Phosphorus in Interstate U.S. Food Transfers: Sustainability Implications for Food-Energy-Water Nexus

Abstract:

Modern agriculture relies heavily on nitrogen, phosphorus, and potassium based fertilizers to reduce the yield gap and sustain current crop yields. Recently inorganic phosphorus fertilizers, an essential input to agriculture have garnered attention as they are primarily obtained from non-renewable and geographically concentrated rock phosphate sources. Phosphorus content in the soil is heavily skewed across the world with many regions facing water quality issues due to nutrient overloading in surface waters, while some face nutrient deficiency and soil degradation issues. Food trade is essential for supplying diverse and large quantities of food items to feed the growing population at global, national, and regional scale. Food trade also presents a pathway for transferring large volume of virtual nutrients and associated environmental impacts. The spatial heterogeneity of food production systems can increase or decrease the consumption based environmental impact of food trade in a particular region. At domestic level, vast quantities of food is traded between states resulting in a complex interstate food trade network. Although, estimates of United States (U.S.) centric domestic production and international trade of embodied phosphorus are available, we lack an understanding of interstate flux of embodied phosphorus for the U.S. To avoid unintended consequences and ensure optimal outcomes for a resilient food supply system, a holistic systems approach to studying interconnected resources is required. We model the origin and destination of food and embodied phosphorus flows to quantify trends in regional phosphorus use, identify hotspots, and propose interventions for enhancing the fertilizer use efficiency and overall environmental sustainability of the U.S. food system.

Using publicly available disparate datasets, we develop a food trade and associated virtual phosphorus model for the U.S. To this end, we evaluate over 60 crop items covering commodity groups of cereal grains, and fruits, nuts and vegetables. The interstate food trade network is translated to virtual phosphorus network by using crop yields and phosphate fertilizer application rates from estimates provided by the U.S. Department of Agriculture. Preliminary results indicate that over 655 million tons of food commodities were transferred across the U.S. in 2012 along with 61 million tons of embodied phosphorus fertilizer. Additionally, fruits and vegetables category contributed only 30% by mass for food trade but accounted for half of embodied phosphorus trade compared to cereal grains. We also evaluate virtual phosphorus savings for each individual item to identify phosphorus use efficiency in the US food production system. Additionally, we quantify the embodied energy flows associated with embodied phosphorus trade and discuss the inorganic phosphorus impact on water quality. Furthermore, we apply a network theory framework to understand the structure, robustness, and environmental sustainability of the considered trade networks. From a network perspective, our previous work has found that interstate food trade forms a well-connected, robust network with a majority of states participating in a high volume of trade. We characterize interstate virtual phosphorus network through series of higher order network measures to understand robustness and resilience patterns in virtual phosphorus trade. This will provide a necessary framework to evaluate series of interventions in a highly interconnected food trade system.

495017 Comparing Beccs and DAC for Climate Change Mitigation: The Water-Land-Energy Nexus

Abstract:

The Paris Agreement signaled global commitment to limit average global temperature rise to 2°C and to pursue efforts to achieve 1.5°C. The IPCC AR5 cites negative emission technologies, particularly bio-energy with carbon capture and storage (BECCS), as critical to achieving this[1]. BECCS' ability to deliver negative emissions has been questioned with many uncertainties around land-use change emissions. Furthermore, large-scale biomass use poses several biophysical and socioeconomic risks such as exacerbating scarce water resources, biodiversity loss, and large opportunity costs of land-use for biomass. Capturing carbon dioxide (CO_2) directly from the atmosphere, *direct air capture* (DAC), presents an alternative source of negative emissions. Established capture technologies can be used to extract it from the atmosphere, seemingly without the significant natural resource requirements of BECCS. However, the low concentration of CO_2 in air necessitates a high energy cost of capture.

In this work, a comparison of the strengths and weaknesses of each technology based on key performance metrics (land, water and energy requirements) has been done. A negative emissions target of 12 Gt_{CO2}/y was used; this is the central estimate of what is required from RCP2.6[1] and has been cited as the target beyond which BECCS is no longer economically feasible. A *'farm-to-NET plant'* model of BECCS accounting extensively for the biomass supply chain and power plant specifications (coal co-firing proportions, capture rates) has been built to determine the overall GHG emission balance of the technology. Data from DAC processes proposed in the literature and existing pilot projects have been used to quantify the potential land, water and energy requirements of a large-scale system.

For BECCS, high variability in carbon balance outcomes was observed depending on the feedstock type and source, and assumptions on indirect land-use changes. Depending on decisions taken throughout the supply chain, a BECCS facility could take over 50 years to begin removing CO_2 from the atmosphere, or it could start removing CO_2 from day one[2]. Stark contrasts in resource requirements to meet the mitigation target were found for BECCS and DAC. Depending on the conditions of its deployment, a BECCS plant operating at 100% load factor could need 363-2100 Mha of land and 4-16 Tm^3/yr of water dedicated.

DAC's needs were 3-5 orders of magnitude lower at 0.04-3.3 Mha and 180 km^3/yr[3], thus posing little to these resources. It is important to note that DAC's water requirements are very sensitive to relative humidity and hence location of the plant[3]. Furthermore, the land area associated with the electricity used for DAC process was not accounted for, and could be significant when considering low carbon power, such as solar and wind.

BECCS' status as an energy-positive mitigation option was found to be untrue in some scenarios. Its annual energy output in the case of US-sourced Switchgrass could be as high as 22 EJ/yr, however up to 37 EJ/yr could be required if Brazil-sourced Miscanthus is the choice biomass. DAC energy costs, however, are still significantly higher at 81-274 EJ/y, even when the lower-bound from literature is assumed.

Surplus resource availability, particularly land and water, are highly-dependent on geography. BECCS and DAC have been shown to require contrasting amounts of these. Therefore, the preferred technology must be discussed within the context of a specific location, and in the case of BECCS, biomass source if not locally-grown.

References

1. IPCC. *Climate Change 2014: Mitigation of Climate Change. Working Group III Contribution to the Fifth Assessment Report of the Intergovernmental Panel on Climate Change* (2014). doi:10.1017/CBO9781107415416

2. Fajardy, M. & Mac Dowell, N. Can BECCS deliver sustainable and resource efficient negative emissions? *Energy Environ. Sci.* 0–42 (2017). doi:10.1039/C7EE00465F

3. Socolow, R. *et al.* Direct Air Capture of CO 2 with Chemicals Panel on Public Affairs. *Am. Phys. Soc. - Panel Public Aff.* 100 (2011).

499908 Food Manufacturing: The Shift from Centralised to DistributedÂ Production

Abstract:

Projections for global food demand for a growing world population indicates an increase of 70% by 2050 (United Nations, 2017). The competitiveness among the food producers to satisfy production demand and consumer expectations has led to the search of new alternatives and innovations to make food products more attractive for the consumers, thus increasing company profits (Ingram et al., 2013).

With the growing concern about climate change, the consumer expectations are not limited to the relation quality-price, but also they include a low environmental impact of the product (Kemp et al., 2010).

As in other industries, the food manufacturing has benefited from the economy of scale, where the mass production allows the reduction of the manufacturing cost. However, a centralised production leads to a complex distribution chain, where the distances travelled for the product distribution increase and with them the cost and the CO_2emissions related to transportation.

Another production alternative is distributed manufacturing with shorter distances among the facilities (i.e. farm-factory, factory-distribution centre), thus the transportation cost and hence environmental impact are reduced. Nevertheless, other factors can affect the environmental impact of a production-distribution system, e.g. the carbon footprint of the raw materials.

The aim of this work is to analyse from an economic and environmental perspective several manufacturing-distribution scenarios, and identify the factors with greater influence in the choice between a centralised manufacturing and a more distributed one. For this purpose, a novel framework we call the *honeycomb model* is presented for the techno-economic and sustainability assessment of different manufacturing scenarios. This framework can be applied for the estimation of the optimum configuration that minimises the cost and CO_2emissions of the manufacturing-distribution chain, as well as perform studies of sensitivity of the optimum number of processing plants to certain parameters/characteristics of the system. Here, the tomato paste production is used as case study.

The results indicate that increments in the area of the system analysed and the product demand favour a more distributed manufacturing, while the origin of the raw materials and their carbon footprint could lead to the centralisation of the production.

Sensitivity analysis of the optimum manufacturing configuration provides important information about when a more distributed manufacturing is preferred over a centralised one. This information can be useful for further design of strategies to shift a centralised manufacturing-distribution configuration to a more distributed one when the purpose is to boost local economies.

This project is part of the Centre for Sustainable Energy Use in Food Chains (CSEF) from RCUK.

References

K. Kemp, A. Insch, D.K. Holdsworth, J.G. Knight. 2010. Food miles: Do UK consumers actually care? Food Policy 35:504â€'513.

J.S.I. Ingram, H.L. Wright, L. Foster, T. Aldred, D. Barling, T.G. Benton, et al. 2013.Priority research questions for the UK food system. Food Sec. 5:617â€"636.

United Nations. Available in: http://www.un.org/en/development/desa/news/population/2015-report.html. Date accessed: February 18th, 2017.

494133 Understanding Biomass Value Chains and the Environment-Food-Energy-Water Nexus through Whole-Systems Analysis and Optimisation

Abstract:

Biomass has the potential to provide access to a sustainable supply of resources such as energy, chemicals and materials. Since the production of biomass, i.e. plant matter such as forestry, agricultural crops and residues, requires a large amount of land and water, competes with food production and can affect the ecosystem (e.g. deforestation, leaching of fertilisers and pesticides into water bodies), there is a delicate balance between all of these elements. Together, these are known as the environment-food-energy-water nexus and all of the benefits associated with the use of biomass to meet our requirements for energy and other products must be considered along with the potential impact to the nexus as a whole. Therefore, it is important not only to understand how best to exploit the great potential of biomass but also to understand the detailed interactions among all of the elements in the nexus.

The BEFEW (Biomass and the Energy-Food-Environment-Water Nexus) project funded by the Newton Fund and EPSRC (Grant No. EP/P018165/1) is examining the many different possibilities for producing energy and high-value products from biomass that also maintain the balance in the nexus. Detailed maps of existing land use and soil quality, water availability, quality and tolerance to contamination, and climate projections in terms of temperature and rainfall are used in order to understand the potential for growing biomass. Different processing facilities that convert the biomass to energy or other chemicals are researched and a comprehensive database of such technologies, including a site suitability analysis is developed. Thus, a detailed and rich representation of the influence, on the whole of the nexus, of the types of biomass grown, where they are grown, the types of technologies used (and therefore the products that will be made from biomass) and where they are located can be built. A mixed-integer linear programming (MILP) model is developed in order to determine the combinations of crops grown and technologies used to convert biomass to energy and high-value products (thus forming a biomass value chain) that provide the greatest benefit with the lowest impact on the nexus. Using the model, the efficient and robust value chains that are synergistic with the nexus and have the potential for deployment over a wide range of scenarios with different combinations of key socio-economic and environmental performance indicators and constraints are identified. In this conference, the latest progress in the BEFEW project will be presented, with a focus on the MILP model and case studies.

505039 Innovative Approaches to Achieving FEW Nexus Goals that can Influence Near-Term & Long-Term Decision Making

Abstract:

The world population is estimated to reach to 9.7B by 2050, resulting in increased demand for resources and increased production of waste. As a result, today's waste disposal systems and approach to waste management will not be adequate for future applications. Therefore, a systematic approach to sustainable food, energy and water management is needed to meet increased demand for resources and efficient waste utilization systems. Currently, the food, energy and water (FEW) sectors use technologies developed to address the needs of a given sector (*i.e.* energy, water, food or agriculture), and wastes from each sector are usually managed separately. Production systems underlying these sectors have traditionally treated pollution and waste as externalities that are diluted into the ambient environment, and separate infrastructure systems have consequently been developed with unnecessary inefficiencies that are increasingly problematic. Because neither landfilling nor industrial composting provide ultimate reutilization of organic waste, and neither can achieve both energy and nutrient recovery in one process from organic waste, we need to transition from a "Traditional Linear Waste Disposal" approach to a sustainable "Closed Loop Resource Recovery" approach. In addition, finding efficient and affordable solutions to avoid GHG emission leakages from landfills and composting facilities through the use of Anaerobic Digestion infrastructure will be a key element of a sustainable system for future waste management. Thus, exploring how FEW systems can efficiently be optimized to (1) repurpose or recycle organic waste products, (2) internalize traditional externalities, and (3) integrate wastes with resource inputs across systems, by deploying integrative management strategies is essential. In addition, integrating other relevant disciplines that bring the necessary economic, social and educational perspectives to addressing this problem, will achieve broader impacts across the system.

504671 Rapid and near-complete dissolution of wood lignin at â‰¤ 80Â°C using a recyclable acid hydrotrope for sustainable production of high-value building blocks

Abstract:

Here we report the discovery of the hydrotropic properties of a recyclable aromatic acid, *p*-toluenesulfonic acid (*p*-TsOH), for low-cost and efficient fractionation of wood through rapid and near-complete dissolution of lignin. Approximately 90% of poplar wood (NE222) lignin can be dissolved at 80°C for 20 min. Equivalent delignification using known hydrotropes such as aromatic salts can be achieved only at 150°C or higher for over 10 h, or at 170°C for 2 h with alkaline pulping. *p*-TsOH fractionated wood into two fractions: (1) a primarily carbohydrate-rich water-insoluble solid fraction that can be used for sustainable production of high-value building blocks, such as dissolving pulp fibers and lignocellulosic nanomaterials, and/or sugars through subsequent enzymatic hydrolysis, and (2) a spent acid liquor stream containing mainly dissolved lignin that can be easily precipitated as lignin nanoparticles simply by diluting the spent acid liquor to below the minimal hydrotrope concentration. *p*-TsOH has a low water solubility, which facilitates efficient recovery simply using commercially proven crystallization technology by cooling concentrated spent acid solution to ambient temperatures to achieve environmental sustainability.

490632 Tuning the Physicochemical Properties of Biochar Derived from Ashe Juniper By Vacuum Pressure and Temperature

Abstract:

Ashe juniper is a problematic plant in central Texas due to rapid expansion causing the alteration of rangeland plant communities and hydrological landscapes. The thermochemical conversion of Ashe juniper into biochar with specific properties is a promising way for valorization of Ashe juniper. This study presents the effects of temperature (350, 450 and 520°C) and vacuum pressure (0.09, 0.7 and 3 kPa) on physical and chemical properties of biochar produced by vacuum pyrolysis from Ashe juniper. We found that temperature had a major effect on the degree of carbonization, surface functionality, as well as the O/C and H/C ratios regardless of vacuum pressure. The severity of vacuum pressure effects on product properties was found to depend on pyrolysis temperature. It was also confirmed that the vacuum pressure was a crucial factor for controlling the surface area and porosity at each constant pyrolysis temperature. This work also demonstrated that the biochar from Ashe juniper under vacuum pyrolysis showed improved aromaticity and surface area over the reported biochar derived from other raw biomass. Biochar produced at 450 and 520°C showed potential application as an adsorbent, a catalyst support and a carbon sequestration agent. Methylene blue (MB) adsorption test confirmed that biochar produced at 450°C and 0.7 kPa showed potential as an adsorbent.

495552 Renewable Transportation Biofuel Converted from Wet Biowaste Via Hydrothermal Liquefaction

Abstract:

This study demonstrates a proof-of-concept in the production of high quality renewable biofuel from wet biowaste via hydrothermal liquefaction (HTL). Distillation was employed to effectively separate the biocrude oil converted from swine manure (SW), food processing waste (FPW), and *Spirulina platensis* (SP) via HTL into different fractions. Distillation curves of different types of HTL biocrude oil were reported. Physicochemical characterizations, including viscosity, elemental test, chemical compositions, and acidity, were conducted on distillates separated from different feedstocks. SW-, FPW-, and SP-derived biocrude respectively contains 15 wt.% , 56 wt.%, and 15 wt.% distillates with heating values of 43-46 MJ/kg and alkanes with carbon numbers ranging from C8 to C18. Compared to the distillates from SW- and SP-derived biocrude oil, the distillates from FPW-derived biocrude demonstrate the closest energy content to petroleum diesel, though this type of distillates contained an excessively high acidity that need to be reduced from 35.3 mg KOH/g to â‰¤ 3 mg/g (the requirements suggested by the ASTM standard). An orthogonal array design of esterification experiment was performed to optimize the reaction temperature (50-70Â°C), reaction time (0.5h-6h), catalysts concentration (0.5 wt.%-2 wt.%), and the molar ratio of FPW-distillates to methanol (1:5-1:15), for achieving the lowest acidity. Compared to other available methods to upgrade HTL biocrude oil, the integrative upgrading approach proposed by this study (distillation plus esterification) demonstrates a competitive energy consumption ratio (0.03-0.06) to zeolite cracking (0.07), supercritical fluid (SCF) treatment (0.17), and hydrotreating (0.24). Moreover, the reaction severity (R_0) of the upgrading approach used in this study (with log R_0 of 5.9-9.5) is much lower than those of zeolite cracking (with log R_0 of 11.0), SCF treatment (with log R_0 of 10.6), and hydrotreating (with log R_0 of 11.3). Finally, the fuel specification analysis and engine test were conducted with the drop-in biodiesel, which was prepared with 10 vol.% (HTL10) and 20 vol.% (HTL20) upgraded distillates and 80-90 vol.% petroleum diesel. HTL10 and HTL20 exhibited a qualified Cetane number (>40 min), lubricity (<520 μm), and oxidation stability (>6 hr), as well as a comparable viscosity (0.2%-19% lower) and net heat of combustion (3%-4% lower) to those of petroleum diesel. Further, diesel engine tests demonstrated that HTL10 can lead to a superior power output (8% higher) and lower emissions of NO_x (3-7%), CO (1-44%), CO_2 (1-4%), and unburned hydrocarbons (10-21%). The present study showcases an energy-efficient and technically cohesive approach to produce renewable high-quality drop-in biofuels for demanding transport applications.

502978 Co-Gasification of Woody Biomass and Chicken Manure

Abstract:

In this work, the potential of gasification as a means of disposal of chicken manure, by converting this animal waste at hen layer farms into clean energy, was studied. Gasification experiments were conducted in a downdraft gasifier and the quality of producer gas generated was analysed. While gasification of pure wood waste resulted in producer gas with a lower heating value (LHV) of about 4.68 MJ/Nm^3, co-gasification of 30 wt% chicken manure and 70 wt% wood waste also generated producer gas of comparable quality, with LHV of 5.23 MJ/Nm^3. From this observation, chicken manure could be a suitable feedstock for co-gasification with woody biomass. In the second part of the study, the solid residue generated â€" biochar â€" was used to study its potential in water treatment, in particular the removal of emerging contaminant, as biochar is known for its superb adsorptive properties. Our results demonstrated that artificial sweeteners were successfully removed by the biochar. By studying the relation of pH, zeta potential, and adsorption efficiency, it was also predicted that the mechanism governing the adsorptive removal was likely electrostatic interaction and specific interaction. Last but not least, a cost-benefit analysis was conducted using a Monte Carlo simulation model to evaluate the economic feasibility of deploying a gasification system in a hen layer farm from business point of view.

Simulation Study of Single-channel Closed Cold Compression Molding for Straw Biomass using ANSYS Workbench

Xiaofei Xin[1,2], Zhiwei Wang[1,2,], Tingzhou Lei[2,*], Miao Yang[1,2], Xiaofeng He[1,2], Qian Guan[1,2], Gaofeng Chen[1,2], Tian Qi[1,3], Xueqin Li[1,2], Yifeng Wu[1,2], Yi Gao[1,2]*

(1Energy Research Institute Co. Ltd., Henan Academy of Sciences, Zhengzhou 450008, China

2Henan Key Lab. of Biomass Energy, Zhengzhou 450008, China

3Tianjin University, Tianjin 300072, China)

ABSTRACT: To simulate and analyze the creep process during the cold compression of agricultural and forestry waste and other biomass, we combined the biomass chemistry principles of elastic-plastic theory and contact analysis to establish a three-dimensional simulation model using SOLIDWORKS software. In addition, a single-channel closed compression process was simulated using ANSYS Workbench software. The displacement, equivalent stress, shear stress, X-direction strain, and elastic-plastic strain change laws of straw biomass during cold compression and the equivalent stress of the mold were determined. Furthermore, the creep law of the cold compression densification process was theoretically revealed. These results provide a basis for the formation mechanism and mold design scheme of the tested process.

KEYWORDS: ANSYS Workbench; biomass; cold compression; simulation

1. Introduction

China produces a large amount of agricultural and forest waste each year, which can be converted into about 500 million tons of standard coal[1]. Biomass energy is the only renewable energy that can be easily stored and transported. Given its uneven distribution and relatively low energy density, many physical conversion technologies, such as biomass densification molding, have been adopted domestically and internationally[2]. Molding technology transforms loose biomass into solid compression-molded fuel, which can improve the density of the raw material up to 1 t/m^3. Molding can improve the combustion characteristics and can be used as a pretreatment for biomass gas fuels, liquid fuels, and other industrial products[3]. It is an important technique for the comprehensive realization of large-scale biomass resource use. Cold compression molding has the advantages of high adaptability to raw materials and water contents of materials, and low power consumption, and has been widely promoted in the industry[4].

A number of studies have examined the cold compression molding process. Dong et al. [5] and Wang et al. [6] determined the change laws of the load during the material densification process using sawdust as the research material and a finite element simulation of the extrusion process, which was performed using ANSYS software. Combined with information on the typical structure of the biological properties, Sun et al. [7] simulated and analyzed the relationship between pressure deformation and pressure compression density during the extrusion stage. Based on a theoretical study of nonlinear problems, Hu and colleagues [8–10] studied the influence of the structural parameters of different molds on the compression process. Sun et al. [11] determined the cohesion and internal friction angle of straw powder using a triaxial compression test and analyzed the compression process of straw powder with thermal coupling. The ANSYS Workbench platform has been developed rapidly in recent years, and features a user-friendly interface and good three-dimensional (3D) effects. In this study, the biomass cold compression molding process was simulated using ANSYS Workbench based on the theories of elastic-plastic mechanics and contact mechanics. This analysis of the displacement, stress, strain, shear stress, and elastic-plastic strain laws of biological material under compression and the stress of the mold provide a basis for the formation mechanism and mold design scheme of the tested process.

2. Constitutive Relationship of the Biomass Elasto-plastic Model

2.1 Yield Criterion

The biomass compression process is divided into the loose stage, transition stage, compression stage, and push stage. A simple elastic model cannot satisfy the simulation demand of analyzing the complexity of the compression process, and the plastic constitutive equation (stress-strain relationship) of the material is nonlinear. The critical value of the material entering the plastic state is the yield limit under uniaxial tension and compression[3]. The stress-strain relationship of elastic phase materials obeys the generalized Hooke's law $\sigma_{i,j} \sim \varepsilon_{i,j}$,

$$
\begin{aligned}
\varepsilon_x &= \frac{1}{E}\left[\sigma_x - \upsilon\left(\sigma_y + \sigma_x\right)\right], \gamma_{yz} = \frac{\tau_{yz}}{G} \\
\varepsilon_y &= \frac{1}{E}\left[\sigma_y - \upsilon\left(\sigma_z + \sigma_x\right)\right], \gamma_{zx} = \frac{\tau_{zx}}{G} \\
\varepsilon_z &= \frac{1}{E}\left[\sigma_z - \upsilon\left(\sigma_x + \sigma_y\right)\right], \gamma_{xy} = \frac{\tau_{xy}}{G}
\end{aligned} \tag{1}
$$

The yield condition formula can be written as follows:

$$F(\sigma_{i,j}) = 0 \tag{2}$$

Equation (2) represents the yield surface around the origin. When $F(\sigma_{i,j}) < 0$, the stress point should be within the surface and the material is in the elastic state. When the material has satisfied $F(\sigma_{i,j}) = 0$, it begins to yield and enters the plastic state. Assuming that the material is initially isotropic, the yield condition is independent of the coordinate axis and can be written as the principal stress function or stress invariant function:

$$F(\sigma_1, \sigma_2, \sigma_3) = 0 \text{ or } f\left(J_1, J_2, J_3\right) = 0 \tag{3}$$

When the Mises yield condition considers the influence of the intermediate principal stress, when Equation (4) is satisfied, the material will enter the yield state, and C is constant in the formula.

$$J_2' = C \tag{4}$$

2.2 Hardening Criterion

Because the isotropic model is considered to be similar to the expansion of the loading surface within a space, the material is considered to maintain its isotropic properties after entering the plastic deformation state only under the conditions of minimal deformation and an unchanged stress ratio. Because of the internal structure of plant cells, biomass material is predominantly anisotropic. In this study, straw biomass was considered to follow the kinematic hardening rule, which is applied to large plastic deformation where the stress changes repeatedly. Under the uniaxial stress state, the Bauschinger effect decreases the yield stress in the opposite direction, and the total elastic range remains constant. Assuming that the size of the yield surface remains constant and only moves in the yield direction, the initial material will no longer remain isotropic after yielding. At the inner center of the yield surface, the coordinates are zero during the initial yield, and the loading surface is expressed as follows:

$$f\left(\sigma_{ij},H_a\right)=f\left(\sigma_{ij}-\alpha_{ij}\right)=f^*\left(\sigma_{ij}-\alpha_{ij}\right)-k=0 \tag{5}$$

In the formula, $f^*\left(\sigma_{ij}\right)-k=0$ is the initial yield surface and moves with α_{ij}, which is the moving tensor, and depends on the amount of plastic deformation. When $\alpha_{ij}=c\,\dot{\varepsilon}_{ij}^{p}$, for linear hardening, C is considered as a constant.

3. ANSYS Workbench Finite Element Analysis

Figure 1 presents a flow chart of the basic analysis of the biomass compression molding simulation process.

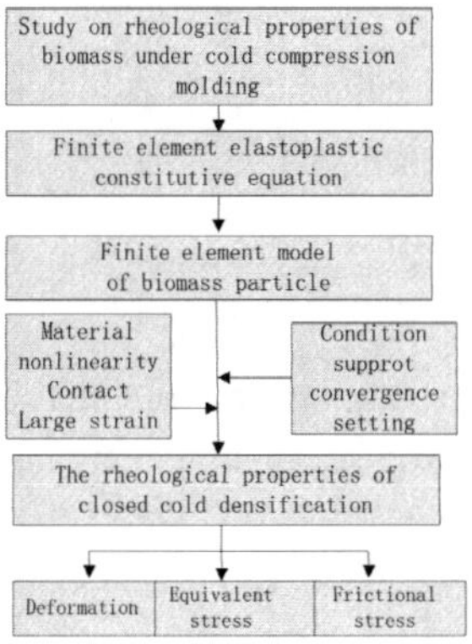

Fig. 1 Analysis flow chart of biomass compression

3.1 Self-Made Mold Design

A 3D simulation model was established using SOLIDWORKS software, where the cone angle of the mold was 60°. The biomass materials were compressed from the upper entrance of the single channel and compressed in the biomass formation chamber. The 3D model was directly imported into ANSYS Workbench and checked for data loss.

3.2 Meshing and Material Settings of the 3D Model

Since the mold was cylindrical, one-quarter of the mold was used as a simplified analysis model. The model was divided into three parts, the mold, biomass, and pressure bar, which was constructed of structural steel, so that the small section could be intercepted. Considering the nonlinearity of the material, the contact nonlinear problem, and the large strain problem of the compression process, the high-order SOLID186 element type and the structure form of meshing method were adopted. The meshing is shown in Figure 2; the model was divided into 1,478 units and 8,235 nodes.

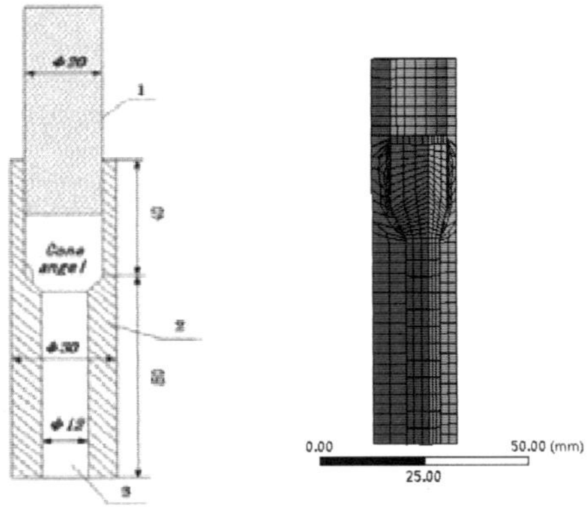

Fig. 2 Meshing of the 3D Models

The biomass, mold, and pressure bar parameters were each set in Engineering Data, and the material data were set as listed in Table 1.

Table 1 Setting parameters of the materials

Material type	Parameter	Setting
	Density (ρ /kg • m^{-3})	330
Straw/	Poisson's ratio (υ)	0.35
Biomass	Elasticity modulus (E/GPa)	160
	Yield strength (MPa)	90
	Density (ρ /kg • m^{-3})	7750
Stainless steel/Mold	Poisson's ratio (υ)	0.28
	Elasticity modulus (E/GPa)	210
Structural	Density (ρ /kg • m^{-3})	7850
steel/	Poisson's ratio (υ)	0.3

Pressure bar	Elasticity modulus (E/GPa)	200

3.3 Contact Settings

During the cold compression densification process, a highly nonlinear problem existed at the point of contact between the bar and biomass as well as the biomass and mold. SOLIDWORKS formed the contact pairs automatically. The contact setting parameters were defined in the Connections section. Considering that the gap of the elastic modulus between the three components was large, the entire biomass body was defined as the contact body, and surface contact was defined as rigid-soft contact. The contact types were defined as Bonded and Frictional, the friction coefficient was defined as 0.2, and the behavior was defined as Asymmetric. The augmented Lagrange formulation, which is applicable to large strain problems and nonlinear problems, was used and the normal stiffness factor was defined as 0.1[12].

3.4 Constraints and Loads Settings and Solve Settings

The bottoms of the mold and biomass were provided with Fixed Support, and the dividing surfaces were provided with Displacement in the corresponding direction set to zero. Pressure bar displacement was applied to restrict the Displacement. Considering the complexity of the compression process, the full Newton-Raphson option, line search in the nonlinear option, and large deflection option were applied to promote the convergence of the simulation results[13].

4 Results and Analysis

4.1 Stress and Strain Distribution of Biomass

Figure 3 shows a flow diagram of the four stages in the biomass compression phase. As can be seen in Figure 3, as compression displacement increased gradually, the biomass compression state gradually moved downward, and the compressive deformation of the material maintained a degree of hierarchy from top to bottom. The initial stage was uniform and the flow state changed after entering the variable cross-section. The flow state of the middle layer of biomass was markedly faster than that of the outer layer, and the compression path was funnel-shaped and intensified inward. Because of the pressure, relatively large shear stress and friction force formed at the cross-section, and the material of the biomass was compacted under the various forces, resulting in a degree of non-uniformity of the density.

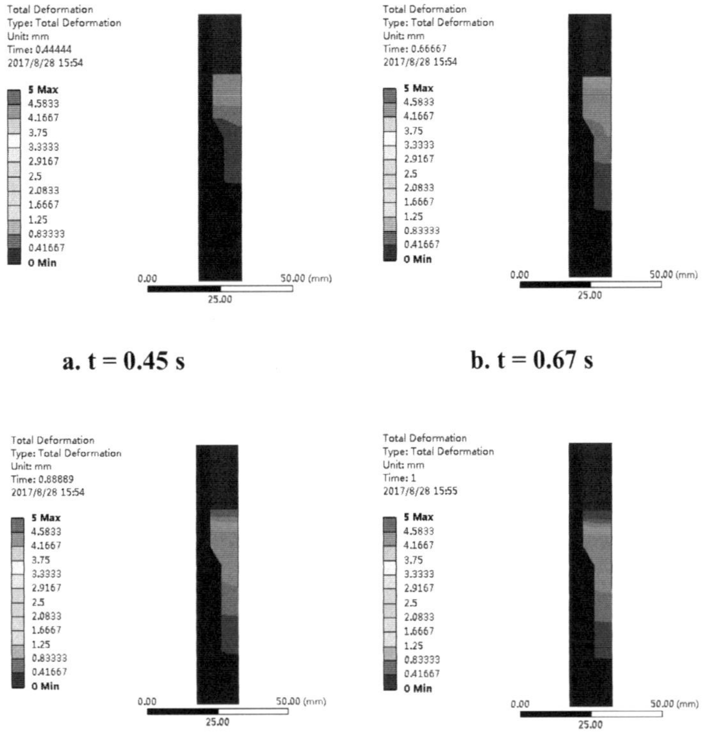

a. t = 0.45 s **b. t = 0.67 s**

c. t = 0.89 s **d. t = 1 s**

Fig. 3 Flow diagram of biomass at different stages of compression

The path of direction 1-2 on the variable cross-section was selected for further analysis (Fig. 4). As the biomass passed through the variable cross-section, the displacement distribution curve changed markedly [Fig. 4(b)]. As the variable cross-section moved downward, the biomass deformation increased rapidly. Deformation remained roughly stable when the length reached 3 mm but declined around 5 mm. The variation of the equivalent stress curve was relatively stable, and the range of stress was 75–90 MPa.

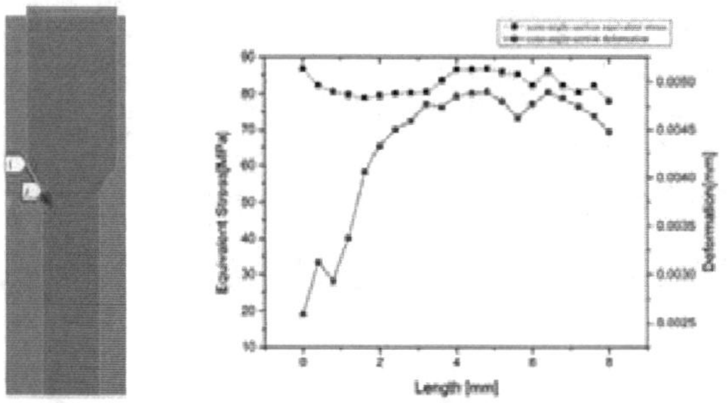

a. Path of the cross-section (1-2) b. Displacement and stress distribution

Fig. 4 Variations in the cross-section

To further assess the stress and strain of the biomass, three paths were selected as biomass cross-sections, inlet section 1, middle section 2, and outlet section 3. The deformation and equivalent stress distribution curves are shown in Figures 5 and. 6.

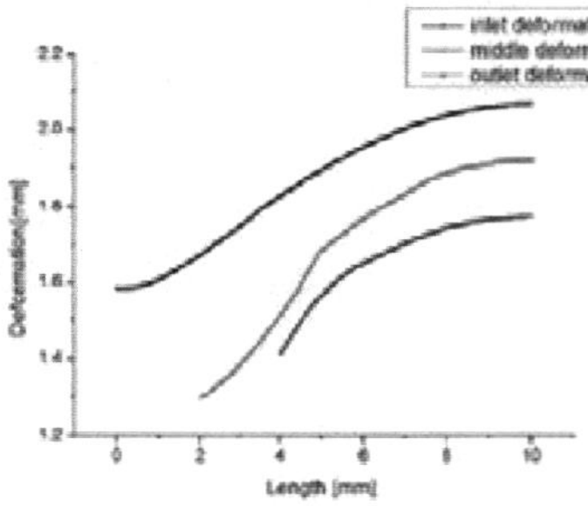

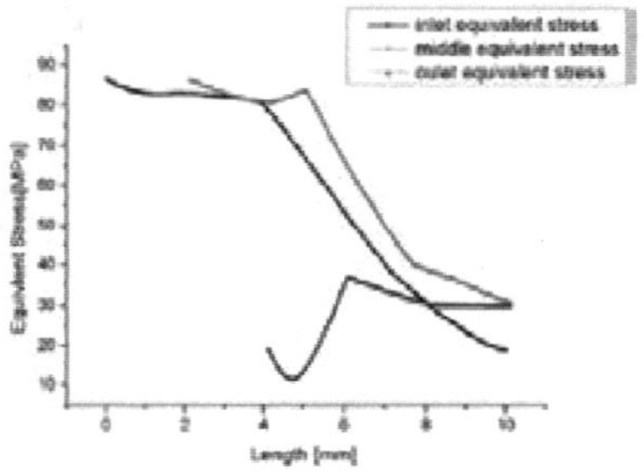

Fig. 5 Path deformation distribution **Fig. 6 Path equivalent stress distribution**

The biomass was compressed into the variable cross-section, where the middle area had the largest displacement and the complete section flow state was parabolic (Fig. 5). At middle section 2, raw material near the inner wall of the mold was in a state of stagnation under bar pressure, frictional force, and cone force, and elastic deformation occurred. The minimum equivalent stress area was in the middle of the biomass material, which gradually increased toward the mold wall (Fig. 6). The equivalent stress reached a maximum force of 80–90 MPa near the wall under the action of bar pressure and friction, which was consistent with the minimal displacement. Plastic deformation occurred at the near inner wall of outlet section 3, resulting in reverse shear force that led to a marked decrease in equivalent stress. With the downward compression of the material, the displacement gap between the intermediate and both sides gradually narrowed, and the flow state tended to stabilize. In addition, the equivalent stress distribution became more stable.

4.2 XY Plane Shear Stress and X-direction Strain of Biomass

Due to the pressure, the shear stress at the variable cross-section was relatively large, and the XY plane shear stress of the biomass changed over time (Fig. 7). Shear stress first occurred at the inflection points of the entrance and exit at the variable cross-section. As the compression process progressed, the shear stress gradually increased to 49 MPa, the two regions gradually expanded, and a layered force field formed. Near outlet section 3, a reverse shear stress of 20 MPa appeared in the middle of the material by the mold wall. The raw biomass material was further broken down under the joint action of shear stress and compressive stress, and the section of the material near the inner area of the mold was gradually broken down and gathered toward the center for bonding.

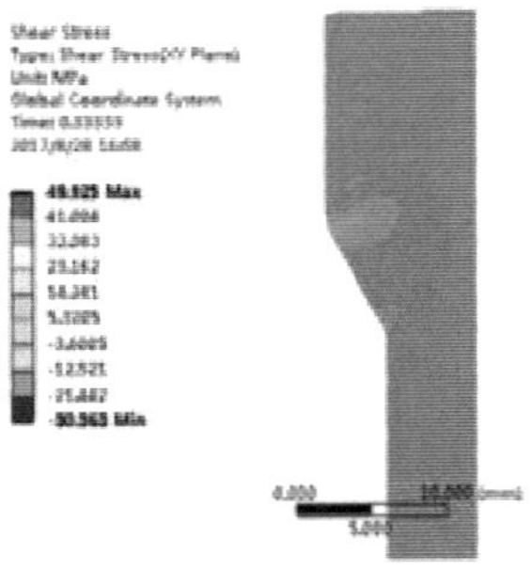

a. t = 0.33 s

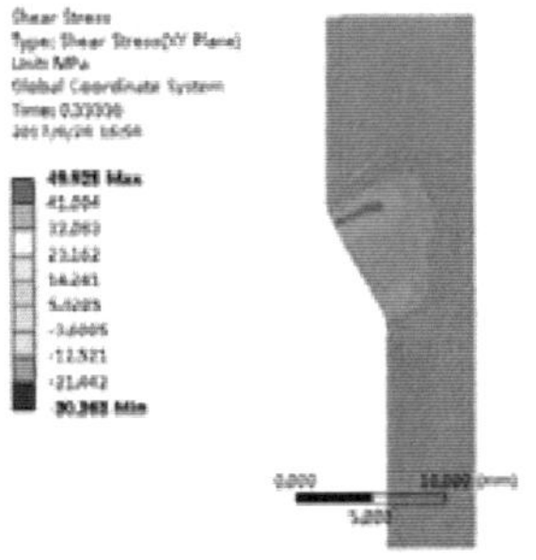

b. t = 0.56 s

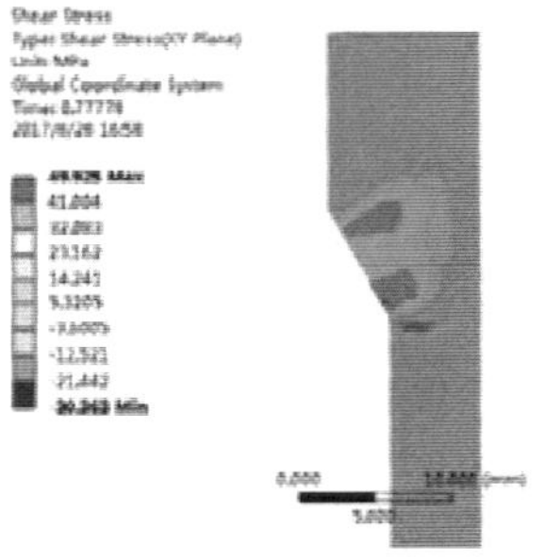

c. t = 0.78 s

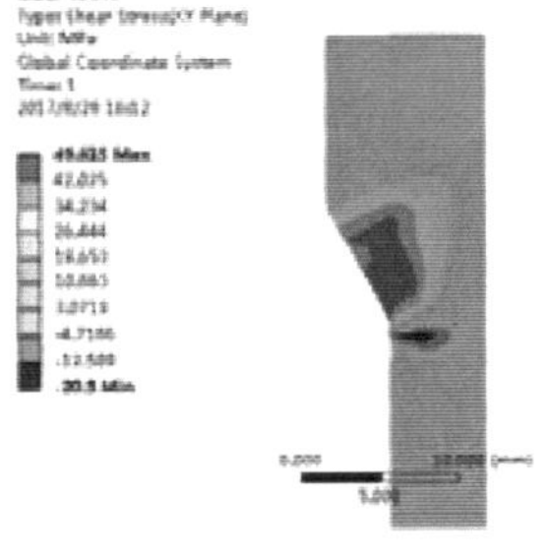

d. t = 1 s

Fig. 7 XY plane shear stress distribution

The X-direction strain began at the inflection point of the cross-section entrance, and gradually expanded downward and toward the middle region (Fig. 8). The X-direction strain was closely related to the shear stress of the biomass. As shear stress gradually increased, the material near the inner wall of the mold was constantly squeezed toward the middle region.

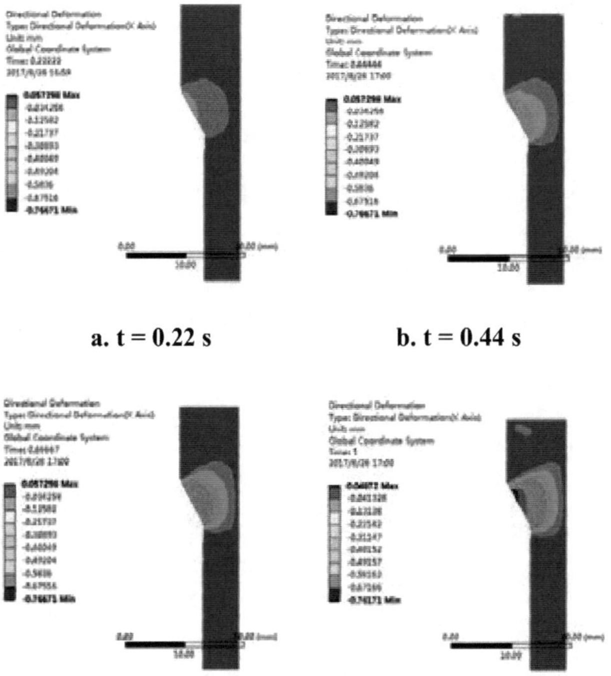

a. t = 0.22 s **b. t = 0.44 s**

c. t = 0.67 s **d. t = 1 s**

Fig. 8 X-direction strain distribution

4.3 Elastic Strain and Plastic Strain

Figure 9 shows the equivalent elastic strain and equivalent plastic strain of biomass. Compared with the above analysis, the biomass inlet material was mainly in the elastic stage, and elastic changes mainly occurred in the middle area of the material. Plastic strain mainly occurred at the contact point of the inlet and outlet corner of the variable cross-section, as well as in other contact areas.

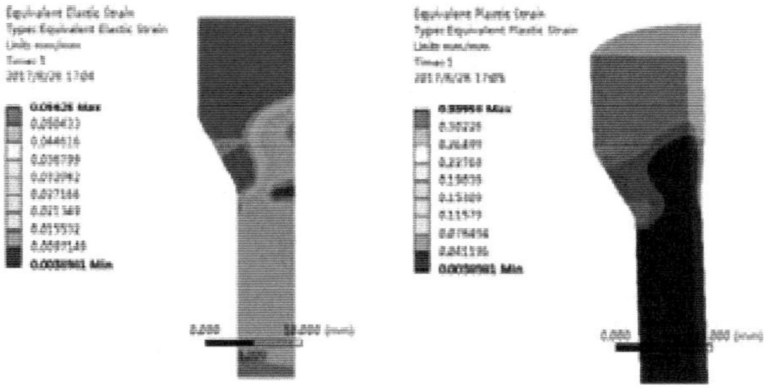

a. Equivalent elastic strain b. Equivalent plastic strain

Fig. 9 Elastic and plastic strain distribution

4.4 Equivalent Stress of the Mold

Figure 10 presents the equivalent stress distribution of the 60° mold. The equivalent stress was mainly concentrated on the inside of the variable cross-section, where the stress concentration was much larger than that of the other parts of the mold. If the angle of the cone changed, the equivalent stress of the mold would change substantially.

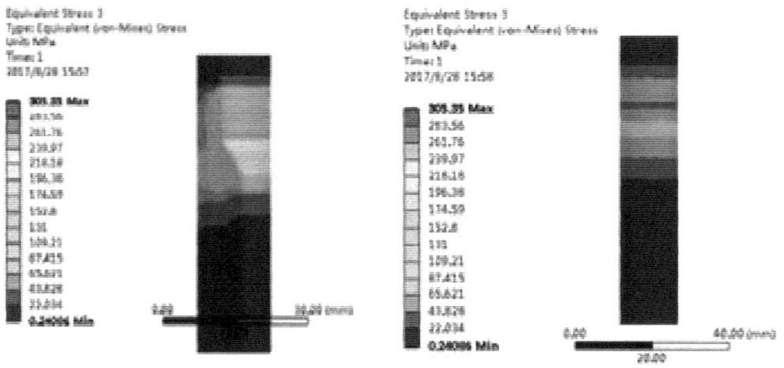

a. Rear view of the mold b. Front view of the mold

Fig. 10 Equivalent stress distribution of the mold

4.5 Contact Analysis

In the initial stage of compression, the frictional stress between the biomass material and the mold was in an elastic relationship, indicating that the compressive force of the bar to the biomass obeyed Hooke's law over time, mainly in the elastic stage. As the pressure increased and the compression process progressed into the compression stage, the frictional stress increased accordingly. As a result of elastic and plastic deformation, the friction appeared to fluctuate slightly, but still exhibited an overall increasing trend.

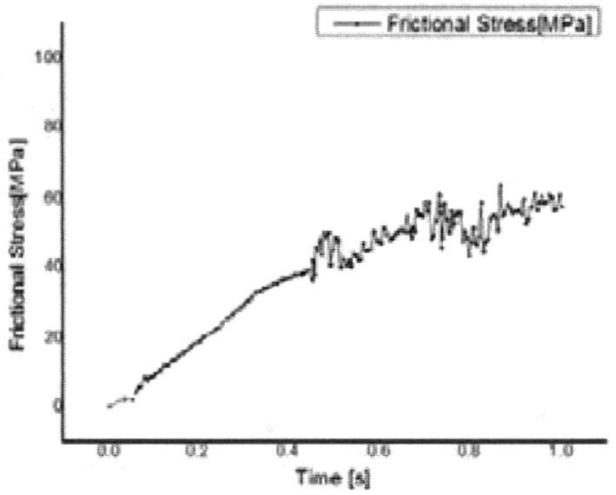

Fig. 11 Frictional stress distribution between the biomass and the mold

5 Conclusions

We established an accurate 3D model of the characteristics of biomass undergoing compression and the compression process. The closed cold compression process was simulated with ANSYS Workbench, and the flow characteristics of the creep process were analyzed. The major results are summarized as follows.

(1) In the single-channel closed cold compression molding process of biomass, the pressure was transferred from top to bottom, and the flow of material showed a degree of change. Under the joint action of various forces, the flow in the middle region was faster than that in the inner wall of the mold, and assumed a funnel-shaped flow. The displacement and equivalent stress distribution curves of different sections of the mold were compared, and the displacement deviation of the middle region and the area on both sides gradually decreased and stabilized, and the equivalent stress distribution tended to become uniform.

(2) Shear stress first occurred at the inlet and outlet inflection points of the variable cross-section, which gradually expanded from the center of the two regions. Meanwhile, the center material exhibited reverse shear stress near outlet section 3. The strain in the X-direction was closely related to the shear stress of the biomass. As the shear stress gradually increased, the material near the inner wall of the mold was squeezed toward the middle region. Elastic strain mainly occurred at the entrance of the raw material and in the middle area, and plastic strain mainly occurred in the contact region of the inlet and outlet inflection point.

(3) The stress concentration at the inner cross-section of the mold was much larger than that of other areas, and the lateral area experienced transition stress. The frictional stress between the mold

and the biomass predominantly obeyed Hooke's law, and entered the plastic state and fluctuated under the action of pressure force and variable section force.

Acknowledgments: This study was supported by the National Natural Science Foundation of China (grant no. 51476050) and the Henan Science and Technology Tackling Project (grant nos. 172102410030, 162102210390, and 152102310392).

References:

1. Zh. Yuan, Ch. Wu, and L. Ma, Principles and Technologies of Bioenergy Utilization (Chemical Industry Press, 2016).

2. S.J. Liu, L.P. Abrahamson, and G.M. Scott, Biomass Bioenergy 39, 1 (2012).

3. Z. Wang, T. Lei, M. Yang, Z. Li, T. Qi, X. Xin, X. He, A. Ajayebi, and X. Yan, Appl. Energy 192, 83 (2017).

4. R. Wang, Plastic Mechanics Foundation (Peking University Press, 1982).

5. Y.P. Dong, M.W. Gao, and Q.X. Sun, J. Shandong Univ. Sci. 35, 9 (2005).

6. W. Wang, Y. Dong, S. Zhou, H. Sun, and W. Ding, J. Agric. Mech. Res. (2011).

7. Q.X. Sun, R.J. Zhang, and Y.P. Dong, Trans. Chin. Soc. Agric. Mach. 130 (2009).

8. J.J. Hu, G.Y. Xu, J.W. Liu, T.Z. Lei, and S.Q. Shen, Bioresources 5, 2447 (2010).

9. J.J. Hu, Straw Pellet Fuel Cold Molding by Compression: Experimental Study and Numerical Simulation, Dalian University of Technology, 2008.

10. J. Hu, T. Lei, X. He, Z. Li, J. Liu, Z. Wang, and J. Zhu, Acta Energiae Solaris Sin. 29, 241 (2008).

11 Y. Sun, X. Bai, and Y. Li, J. Agric. Mech. Res. 237 (2016).

12 Y. Dong and J. Gao, Acta Energiae Solaris Sin. 31, 661 (2010).

13 R.X. Zhao, Study on Optimization of Biomass Fuel Pressure Formation Process Parameter, North China University of Science and Technology, 2016.

The English in this document has been checked by at least two professional editors, both native speakers of English. For a certificate, please see:

http://www.textcheck.com/certificate/M7L8iq

489620 High-Performance Magnetic Activated Carbon from Solid Waste from Lignin Conversion Processes. 1. Their Use As Adsorbents for CO_2

Abstract:

Wood is a major source of raw materials for renewable fuels, chemicals and materials. Depending on its type, lignin can account for roughly one-third of the mass. It is considered a waste product in most of the processes that convert wood into valuable products. In the paper and pulp industry, where most of the lignin is available, 98 % is combusted to supply the processes with energy and low-grade heat[1]. This combustion is wasting potential value of lignin, both in an economical aspect and in the potential replacement of fuels, materials and chemicals based on fossil resources. Lignin can be used to produce value-added and renewable products in lignin-to-liquid processes. In these processes, lignin is depolymerized to form bio-oil containing various useful organic molecules for use in chemicals, fuels and materials[2]. Integrating such processes can be important in future biorefineries, where renewable resources are converted in to useful products. However, these processes produce large amounts solid by-products which, so far, has not found any constructive use. In the biorefinery concept, it is imperative to maximize the use of the renewable raw materials, both for the sustainability of the process and to generate enough added values to make the sustainable processes economically viable. In this work, we used the solid by-products of lignin-to-liquid processes to produce highly porous and magnetic activated carbons for use in carbon dioxide separation processes. Activated carbons have many commercial applications within separation, deodorization, purification and catalysis[3]. However, the majority of activated carbons are produced from fossil resources[4] and there is a need to replace these fossil-based materials with renewable alternatives. In carbon dioxide separation processes, activated carbons can potentially lower the carbon dioxide capture cost, which is a problem for CCS (carbon capture and storage) systems[5]. The general advantages of activated carbons compared to other solid sorbents are high adsorption capacities, stable adsorption/desorption cycling and low production costs. To produce the magnetic activated carbons in this work, Spruce and Eucalyptus lignin was heated at 380 ^{o}C with water, formic acid and a catalyst, the solid by-product of this bio-oil producing process was mixed with an aqueous solution of potassium hydroxide, heated at 200 ^{o}C for 5 hours and activated under nitrogen gas flow for 4 hours at 700-800 ^{o}C. The carbons activated at 700 ^{o}C exhibited carbon dioxide adsorption capacity of up to 6.0 mmol/g at 273K and 1 atm and very high ultramicropore volumes. The carbons activated at 800 ^{o}C exhibited BET specific surface areas, measured by nitrogen adsorption at 77K, of up to 2875 m^2/g. The activated carbon based on Eucalyptus lignin and activated at 700°C showed an isosteric heat of CO_2 adsorption was 27-30 kJ/mol, a simplified description indicated a CO_2-over-N_2selectivity of 14, while a full IAST prediction indicated a selectivity of between 3 and 4, and very good cyclic performance. All the activated carbons exhibited soft magnetic behaviour due to embedded nanoparticles of iron and magnetite. We show that the activation temperature does not only alter the porosity of the carbons but also the textual properties. The combination of the highly porous nature, the magnetic properties and the option to optimize the porosity through the

activation temperature makes these materials highly interesting for many other potential applications outside CCS as well. This work has been published elsewhere[6].

[1] R.J.A. Gosselink, E. De Jong, B. Guran, A. Abächerli, Co-ordination network for lignin - Standardisation, production and applications adapted to market requirements (EUROLIGNIN), Ind. Crops Prod. 20 (2004) 121–129. doi:10.1016/j.indcrop.2004.04.015.

[2] M. Kleinert, T. Barth, Towards a lignincellulosic biorefinery: Direct one-step conversion of lignin to hydrogen-enriched biofuel, Energy and Fuels. 22 (2008) 1371–1379. doi:10.1021/ef700631w.

[3] R.C. Bansal, M. Goyal, Activated carbon adsorption, CRC Press, 2005.

[4] A. Ahmadpour, D.D. Do, The preparation of active carbons from coal by chemical and physical activation, Carbon N. Y. 34 (1996) 471–479. doi:10.1016/0008-6223(95)00204-9.

[5] V. Presser, J. McDonough, S.-H. Yeon, Y. Gogotsi, Effect of pore size on carbon dioxide sorption by carbide derived carbon, Energy Environ. Sci. 4 (2011) 3059. doi:10.1039/c1ee01176f.

[6] W. Hao, F. Björnerbäck, Y. Trushkina, Mikel O. Bengoechea, G. Salazar-Alvarez, T. Tarth, N. Hedin, High-Performance Magnetic Activated Carbon from Solid Waste from Lignin Conversion processes. 1. Their Use As Adsorbents for CO2, ACS Sustainable Chem. Eng., 2017, 5 (4), pp 3087–3095, doi: 10.1021/acssuschemeng.6b02795

493008 Assessing the Thermodynamic Viability of Mixed Metal Oxides for Solar Thermochemical Water Splitting

Abstract:

While solar energy is the most abundant renewable energy resource, the capture, storage, and distribution of it remains a challenge. Solar thermochemical water splitting (STWS) provides a promising route for efficient utilization of this disperse resource since it allows for use of the entire solar spectrum to convert water to an energy dense fuel, H_2. However, despite a significant number of materials having been examined, an optimal redox material to drive this process has yet to be developed. In order to be viable for economic hydrogen production, materials must for have high hydrogen productivity, fast reduction and oxidation kinetics, low thermal reduction temperatures, and long term stability and reactivity. In this work we will utilize *ab initio* methods to assess the thermodynamic viability of a large number of spinel and perovskite metal oxides for STWS. Materials will be screened for stability, oxygen vacancy formation energy, and extent of reduction. The effect of temperature on the parameters including crystal structure, cation disorder, and magnetic order and their relationship to predicted STWS ability will be discussed. Compositional (doping) control will be utilized to optimize materialsâ€™ thermodynamic properties. Experimental results from redox cycling in a stagnation flow reactor will be presented and compared to DFT models.

487359 Thermodynamic Characterization of Charge Compensating Double Doped Ceria for Improved Redox Performance of Solar Thermochemical H_2O/CO_2 Splitting Cycles

Abstract:

Solar-driven thermochemical redox cycles utilizing nonstoichiometric metal oxides are promising for splitting H_2O and CO_2 to produce H_2 and CO (syngas), the precursor of synthetic liquid hydrocarbon fuels. The state of the art redox material is pure ceria but it has been suggested that charge compensating double doped ceria might be a novel doping strategy to expand the range of viable dopants. In this work, charge compensating double doped ceria was fabricated using 3+ and 5+ dopants, namely $Ce_{0.9}A_{0.05}Nb_{0.05}O_2$ (A=Y, La, Sc). These materials, along with $Ce_xLa_{(1-x)/2}Nb_{(1-x)/2}O_2$ (x=0.75,0.95), were investigated by thermodynamic characterization and compared to pure and single doped ceria ($Ce_{0.9}B_{0.1}O_2$ where B=Y, La, Nb, Hf). Oxygen nonstoichiometry was measured by thermogravimetric analysis in the temperature range T = 1173-1773 K and oxygen partial pressure range $p_{O2} = 10^{-15}$-10^{-1} atm. At a given T and p_{O2}, the reduction extent of the double doped materials is between that of $Ce_{0.9}Hf_{0.1}O_2$ and pure ceria as predicted by DFT calculations, with increasing values for decreasing dopant concentrations. We found that the experimental data is accurately described by a defect model based on a combined point and cluster defect. Thermodynamic properties, namely the partial molar enthalpy and entropy were extracted from the defect model. The calculated partial molar enthalpy of the charge compensating double doped materials is within the range of 360-410 kJ/mol and varies based on composition and dopant concentration. These novel redox materials exhibit promising behavior and their properties can be tuned for improved redox performance of solar thermochemical H_2O/CO_2 splitting cycles.

495695 Solar Thermochemical Methane Reforming Using Metal Oxide Oxygen Exchange Materials

Abstract:

Methane reforming is a mature industrial process that is still heavily leveraged in modern energy infrastructure. For instance, 95% of the hydrogen produced in the United States is made via methane reforming processes. These processes are highly endothermic and typically take place at 700â€“1100Â°C in the presence of a catalyst. Natural gas reforming processes, coupled with the high-energy/high-temperature process conditions, will continue to be a major pathway for fuels and commodities for the near future, which provides motivation to investigate solar-driven alternative process routes. Methane steam reforming, methane dry reforming, and methane partial oxidation are all processes that can leverage high-temperature solar energy to simultaneously alleviate economic and environmental impacts. These processes also achieve increased energy efficiency in the conversion of methane, the primary component of natural gas, to value-added commodities such as syngas, fuels, or ammonia. The catalytic materials used in methane reforming are typically Ni-based metals on a support structure. Although these active materials can be effective, they often suffer from deactivation, coking, sulfur poisoning, and oxidation. For this reason, this work investigates the use of metal oxide oxygen exchange materials that function as two-step redox cycle engines for the methane reforming application. Additionally, the viability of using Ni and Co-doped iron oxide materials for continuous, solar-driven methane reforming processes is explored in detail.

496715 Oxide-Independent Thermodynamics of Solar Chemical-Looping Reforming for Producing Synthesis Gas and Hydrogen

Abstract:

The solar upgrading of natural gas to produce synthesis gas is an attractive option for producing more sustainable transportation fuels. Synthesis gas is a valuable feedstock for producing high octane gasoline, kerosene, and high cetane diesel. Synthesis gas produced from natural gas, water, and process heat from concentrated solar energy contains up to 28% more energy than the chemical feedstock, offering a corresponding 22% reduction in carbon dioxide emissions on an energetic basis. Solar chemical-looping reforming (SCLR) with a metal oxide oxygen carrier has attracted attention for the application because it offers an opportunity to produce separate product streams of pure hydrogen and synthesis gas with a two to one ratio of hydrogen to carbon monoxide. Furthermore, the solar to chemical energy conversion via SCLR has potential for high efficiency. With realistic accounting energy requirements including parasitic work, the projected efficiency for ideal chemical conversion with SCLR is 54%.

In this work, oxide-independent chemical and process thermodynamics analyses are applied to frame the requirements for attaining high efficiency with SCLR in practice. Chemical equilibrium influences on process efficiency are presented for both the syngas-producing methane partial oxidation reaction and the hydrogen-producing water splitting reaction. Key drivers of efficiency include the ratio of methane to oxygen in the methane partial oxidation reaction and the conversion of oxidizer in the splitting reaction. The equilibrium oxygen partial pressure provides a link between the performance of the two reactions. A process thermodynamics analysis quantifies the impact of methane conversion, hydrogen and carbon monoxide selectivity, and oxidizer conversion on efficiency. The parasitic work for separating synthesis gas from the products of the methane partial oxidation reaction is also considered. Methane conversion and oxidizer conversion are shown to have the largest impact on efficiency. With complete conversion of methane to hydrogen and carbon monoxide in the methane partial oxidation reaction, the projected efficiency increases from 18% to 54% as the conversion of water to hydrogen in the splitting reaction is increased from 10% to 100%.

494447 Experimental Investigation of Carbon Dioxide Capture and High-Temperature Thermal Energy Storage Via Metal Oxide Chemical Looping in a Prototype Reactor for Solar Thermochemical Metal Oxide Looping

Abstract:

Two-step metal oxide carbonationâ€“calcination chemical looping is a promising approach for solar-driven CO_2 capture from flue gases and as a high-temperature thermochemical energy storage system for solar-thermal power plants. The reversible two-step process involves the endothermic calcination of a metal carbonate (e.g. $CaCO_3$, $SrCO_3$) to the metal oxide driven by concentrated solar energy, followed by the exothermic carbonation of the metal oxide to capture CO_2 or provide a source of high-temperature heat.

A 1 kW solar-driven packed-bed reactor prototype has been designed and built to experimentally evaluate the process using CaCO3 as the sorbent. The packed bed of spherical sorbent particles is contained in an annular reaction zone surrounding a solar cavity and heated with concentrated solar radiation via the diathermal cavity wall. Gases are supplied to the reactor via inlet and outlet manifolds and distributed inside the reaction zone with gas distributors.

Reaction steps in experiments are conducted with alternating inert and CO_2 gas flows at varying flow rates and thermal radiation input from the ANU High flux solar simulator. Reaction rates and extents are determined by measuring the changes between inlet and outlet gas compositions under controlled operation. Sorbent particles are analysed before and after experiments for composition, structure, size, and morphology. The pressure drop across the particle bed and the temperature at selected points of the reaction bed, inlet gas temperature, and outlet gas temperature are measured and controlled.

We will present and discuss the reactor design, experimental set-up, and most recent experimental results obtained with the prototype reactor.

492840 The Role of Oxygen Vacancies of CeO_2 Supported Catalyst on Catalytic Performance of the CO_2 to CO

Abstract:

The performance of a catalyst depends on many factors. It is usually hard to discuss these effects separately due to their interwoven relationship. Here, we focus on the catalytic activity of nanosized Cu supported on cerium oxide by the space-confined method and mesoporous silica, whereby Cu nanocatalysts are deliberately controlled under the same loading, particle size and coordination number to highlight the interactions caused by different supports. It is interesting to note that the redox temperature of cerium oxide is dramatically decreased from >600°C to <300°C by CuCeOx solid solution so that its conversion is 18% with 100% selectivity approaching the equilibrium conversion 23% of the CO_2 to CO at 1 bar and 300°C the by flow reactor.

Oxygen vacancies generated by the reducible cerium oxide were found to be able to contribute more carbonate with bidentate type absorption. The space confined method was capable of generating high Cu dispersion (~50%) from nano-size CuCeOx solid solution. The Cu maintains the CeO_2 support to keep the oxygen vacancies recycling by hydrogen spill over behavior at 300°C when the reaction of CO_2 to CO oxidized the catalyst. However, similar metal-support was not seen in the mesopeorous silica system.

We have demonstrated that the active sites locate in the proximity between support and reduced nano-sized metal by synchrotron analysis, TEM mapping, in-situ FTIR and catalytic reaction. Furthermore, the work presented illustrates the significant role of oxygen vacancies in the metal-support interaction and how the potential of oxide supported nanocatalysts can be further explored and tuned by for the space confined method and thermal treatment for CO_2 to CO.

Advanced Ni-CeO_2 Nanocatalysts for Chemical CO_2 Recycling

*T. Smith, T. Stroud, E. Le Saché, H. Arellano-Garcia, T.R. Reina**

Department of Chemical and Process Engineering, University of Surrey, Guildford, GU2 7XH, UK

*Corresponding author: t.ramirezreina@surrey.ac.uk

Introduction

Carbon dioxide is one of the major issues facing the world currently, the adverse effects on the environment that CO_2 emissions cause are well documented. Major steps are being taken by industry to mitigate these effects, with a proportion of the CO_2 produced being captured each year. This scheme to reduce carbon emissions into the atmosphere is part of a worldwide initiative, whereby the carbon captured can be stored, but there is still a lot of progress to make within this field before this is a truly viable option. There is however, another exciting alternative which is the chemical upgrading of CO_2 to obtain fuels and chemicals [1] [2]. This way, the depletion of CO_2 emissions is accompanied by an extra motivation in terms of the generation of added-value products.

Among the different alternatives for CO_2 utilisation, dry reforming of methane is a promising route leading to the production of syngas (H_2 and CO) [3]. This is a highly useful and valuable intermediary, which can be used in processes like the Fischer-Tropsch Synthesis to create long chain hydrocarbons, such as diesel [4].

The issue that arises in the dry reforming process, is the deactivation of the catalysts used in the reaction of the feed chemicals. This deactivation is due to sintering of the active phase and/or carbon deposition [3] [5]. In order to overcome these problems noble metal based catalysts have been employed but their cost make them unsuitable for a real application.

The aim of this work is to develop a highly active dry reforming catalyst based on Ni. Considering the disadvantages of Ni in terms of deactivation, we have developed multicomponent materials based on monometallic and bimetallic Ni/CeO_2/Al_2O_3 catalysts with excellent features for reforming reactions. In this study, all the catalysts contain 10 %wt. Ni and 20 %wt. CeO_2. As for the bimetallic formulations, Pt and Sn have been selected as promoters with a nominal loading of 0.3 and 0.8 % wt. respectively.

Experimental

The catalysts were prepared by wet sequential impregnation using alumina (Sasol) and the relevant precursors (Sigma Aldrich) employing a Rotovap, with a session of drying and calcining between each step of the impregnation. The drying was performed at 80°C and was left overnight, the calcination was performed at 700°C for 4 h.

The dry reforming of methane was carried out at atmospheric pressure in a quartz tube reactor. 0.1 g of catalyst was used per reaction supported upon a reactor bed of quartz wool. The catalyst was reduced in situ at 800°C for 1 h, in a 100 mL/min flow containing 20% H_2 and 80% N_2. The reaction was then conducted at 700°C for over 20 h, in a 100 ml/min flow containing 12.5% CH_4, 12.5% CO_2 and 75% N_2 (WHSV = 60,000 mL g^{-1} h^{-1}). The gas products of the reaction where analysed using an ABB AO2208 gas analyser. The spent catalysts were recovered and characterized.

All the fresh catalyst samples have been characterized fully, by XRD, H_2-TPR and N_2 adsorption. The spent samples have also been characterized fully, post reaction, by XRD, Raman and SEM.

Results and Discussion

The textural properties of the sample catalysts are shown below in **Table 1**. What can be observed from the data is that the specific surface area as well as the pore volume of the catalyst is reduced by the addition of ceria. However, the addition of Sn and Pt does not affect the textural properties of the catalysts. The decrease in the surface area can be attributed to ceria species covering the surface of Al_2O_3, this is corroborated by the decrease in pore volume.

Table 1: *Textural properties of the catalyst samples.*

Catalyst	BET (m^2 g^{-1})	Pore Volume (cm^3 g^{-1})
Ni/Al	164	0.42
Ni/CeAl	116	0.28
SnNi/CeAl	120	0.29
PtNi/CeAl	118	0.28

The XRD data shown below in **Figure 1** provides insight into the chemical structure of the prepared catalyst samples. The absence of any crystalline peak attributed to Ni species indicates a good metallic dispersion and small particle sizes. However, the presence of $NiAl_2O_4$ cannot be fully discarded since the spinel peaks overlap with those of the bare Al_2O_3. In addition, the presence of ceria fluorite type structures is evidenced in the XRD patterns. As for the bimetallic samples, the XRD depicted for PtNi/CeAl is comparable to that of the Ni/CeAl base catalyst except for the small PtO peak at $2\Theta = 40°$, while the SnNi/CeAl catalyst shows a very similar XRD pattern to Ni/CeAl.

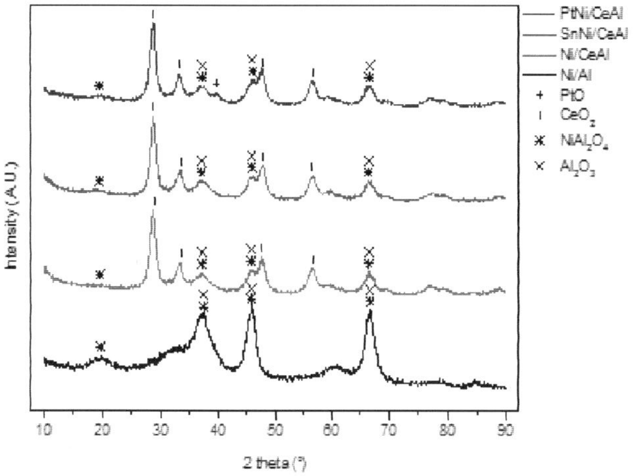

Figure 1: *XRD patterns of the catalysts.*

The redox properties of the catalysts are shown in the H_2-TPR profiles displayed in **Figure 2**. All the profiles present, contain a peak at high temperatures (650-850°C), this is due to the bulk reduction of the $NiAl_2O_4$ spinels, which due to their structure require a high temperature to be reduced. The later agrees with the XRD data. The three catalysts containing ceria exhibit a shoulder at 850°C as well as much larger H_2 consumption peak than the base Ni/Al catalyst. This is attributable to the bulk ceria reduction occurring in the temperature range 700-900°C. Also, a reduction process at 250°C on the PtNi/CeAl profile was identified and attributed to the reduction of PtO.

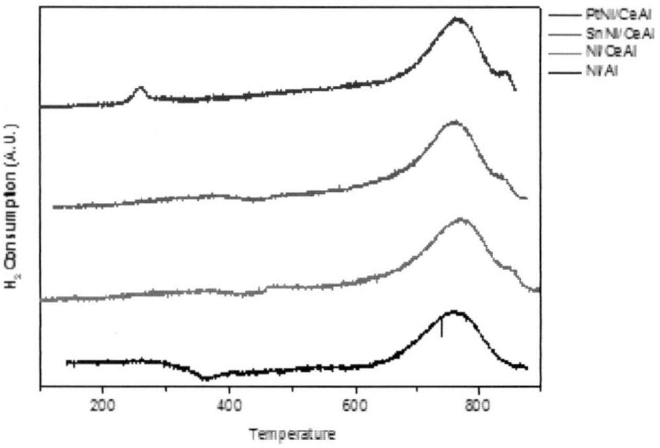

***Figure 2:** H_2-TPR profiles of the catalysts.*

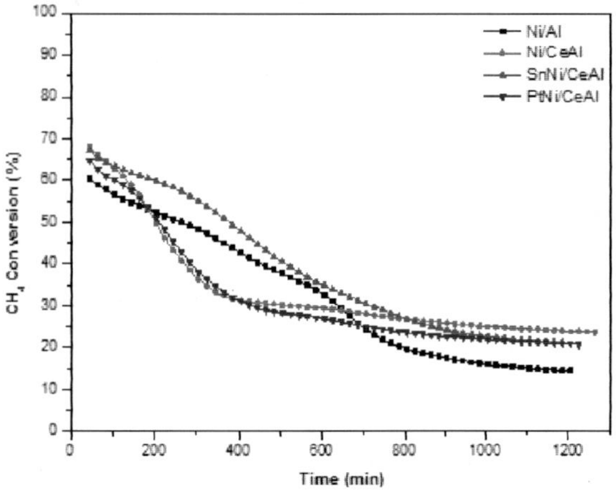

***Figure 3:** CH_4 conversions of the catalysts.*

Figure 3 and **Figure 4** below show the CH_4 and CO_2 conversions for all the catalyst after 20+ h on the reaction stream. It can be seen that the addition of ceria helps to significantly stabilise the catalyst due to its high oxygen storage capacity (OSC), in comparison the Ni/Al conversions are observed to drop especially after 10 hours. This is attributed to the formation of hard carbon on the surface of the catalyst blocking the accessibility of the molecules to the active sites. The effect the promoters have on the conversions differs, the Sn promoted catalyst has higher conversions to begin with before stabilising at similar conversions to the other ceria based catalyst. This has been attributed to the smaller particle size and higher particle dispersion the Sn promotion causes. By contrast the addition of Pt as a promoter hardly affects the catalytic performance of the Ni/Ce/Al sample. The Sn promoted catalyst also had the highest H_2/CO ratio to begin with, of 0.97, before stabilising at a similar value to the other catalysts of around 0.75.

The characterisation performed post reaction on the spent catalysts (not shown for sake of briefness) concur with the conversion hypothesis. Large concentration of hard carbon were deposited on the Ni/Al while smaller concentrations and softer carbon haven been found on the ceria and ceria-tin promoted catalysts.

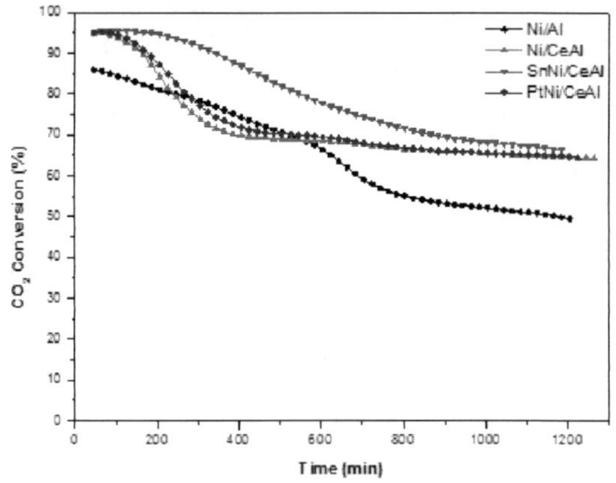

***Figure 4:** CO_2 conversions of the catalysts.*

Conclusions

This work evidences that multicomponent Ni based catalysts can be efficiently utilised for the chemical CO_2 recycling via dry reforming. High levels of CO_2 conversion and good H_2/CO ratios have been achieved in general. The bimetallic Ni-Sn system supported on CeO_2-Al_2O_3 was the most promising material among the studied catalysts. The excellent behaviour of this catalyst is due to its improved redox properties and the Ni-Sn synergy which prevents active phase sintering and carbon poisoning.

Further studies are currently being undertaken by our group to discern the effect of the operating conditions as well as the long-term stability of these advanced materials, in order to explore their potential application in real CO_2 utilisation processes.

References

1. Pires JCM, Martins FG, Alvim-Ferraz MCM, Simoes M. Recent Developements on Carbon Capture and Storage: An Overview. Chemical Engineering Research and Design. 2011; 89(9): p. 1446-1460.
2. Leung DYC, Caramanna G, Maroto-Vler MM. An Overview of Current Status of Carbon Dioxide Capture and Storage Technologies. Renewable and Sustainable Energy Reviews. 2014; 39: p. 426-443.
3. Pakhare D, Spivey J. A Review of Dry (CO2) Reforming of Methane over Noble Metal Catalysts. Chemical Society Reviews. 2014; 43: p. 7813-7837.
4. de Araujo GC, de Lima SM, Assaf JM, Pena MA, Fierro JLG, Rangel MdC. Catalystic Evaluation of Perovskite-Type Oxide LaNi1-xRuxO3 in Methane Dry Reforming. Catalysis Today. 2008; 133: p. 129-135.
5. Charisiou ND, Siakavelas G, Papageridis KN, Baklavaridis A, Tzounis L, Avraam DG, et al. Syngas Production via the Biogas Dry Reforming Reaction Over Nickel Supported on Modified with CeO2 and/or La2O3 Alumina Catalysts. Journal of Natural Gas Science and Engineering. 2016; 31: p. 164-183.

497365 One-Step Synthesis of Transition Metal Doped Titanium Dioxide Catalysts for Oxygen Removal

Abstract:

Carbon dioxide (CO_2) capture and sequestration have been considered as promising techniques to alleviate the emission of greenhouse gas. However, it is difficult to cost effectively deploy them in conventional coal-fired combustion plants due to the diluted CO_2 concentration in the flue gas, requiring additional methodologies for concentrating the stream. An oxy-fuel combustion technique which uses pure oxygen (O_2) to obtain a concentrated CO_2 stream has been developed to solve this problem, and has obtained a sequestration-ready gas stream. However, this gas stream contains O_2 as well as CO_2, which can cause damage to the gas infrastructure through corrosion. To overcome this shortcoming, catalytic O_2 removal system with hydrocarbons has been recently applied. In this study, we develop 1) a one-step synthesis of transition metal doped catalysts and 2) an efficient catalytic O_2 removal system with methane (CH_4). To achieve this, we first synthesize different transition metal (Cu, Fe, Ir, Ni, Mn, Rh, Ru) doped titanium oxide (TiO_2) catalysts in a flame aerosol reactor. Compared to liquid phase synthesis methods such as wet impregnation and sol-gel, flame synthesis is a gas phase technique that can produce homogeneous nanoparticles and can be scaled up for high throughput production. The different catalysts are characterized by BET, X-ray diffraction, scanning electron microscopy and transmission electron microscopy to relate nanoparticle characteristics to performance. Second, we test the synthesized catalysts for O_2 removal with CH_4 to optimize the catalyst. Third, we optimize the synthesis conditions (dopant concentration (0-5 wt.%), fuel flow rate (0.8-1.8 LPM)) to get the highest O_2 removal efficiency.

494289 Carboxylation of Propylene Oxide to Propylene Carbonate

Abstract:

Polycarbonates, a class thermoplastic polymer, have wide ranging applications in automotive & transportation, construction, packaging, consumer goods, optical media, electrical & electronics and medical devices. Conventionally, manufacturing of polycarbonates involves phosgenation of bis-phenol-A, which has drawbacks due to toxic phosgene and corrosive products like HCl in stoichiometric amounts. Alternative non-phosgene route for polycarbonates involves carboxylation of propylene oxide as a key step. This is also an example of CO_2 utilization for value added chemicals. Cyclic carbonates are also widely used as aprotic solvents, anti-foaming agents, antifreeze, plasticizers, and monomer for various commodity polymers. Several studies reported carboxylation of epoxides employing either homogeneous or heterogeneous catalysts. Reported heterogeneous catalysts include metal oxides, zeolites, polymer supported quaternary onium salts, ion exchange resins, and polymer supported ionic liquids. However, the main challenges in this reaction are low catalyst activity, and high CO_2 pressure. In this study, we systematically investigated the influence of solvents on ion exchange resin catalyzed carboxylation of propylene oxide with the aim of enhancing the activity and selectivity. Intrinsic kinetics of carboxylation of propylene oxide has also been studied in a batch slurry reactor to understand the effect of reaction conditions and develop a suitable rate equation.

The experiments for solvent effects and kinetic studies were carried out in a stirred pressure reactor with 100 cm^3 capacity with provisions for control of agitation speed, temperature and sampling of liquids. In these experiments, the CO_2 pressure in the reactor was kept constant by continuous supply through a CO_2 reservoir using a constant pressure regulator such that the temporal reaction progress was followed from the pressure decrease in the reservoir. At the end of each experiment, liquid products were analyzed for PO and PC to assess the material balance. The effect of solvents and the order of activity is: Methanol > Ethanol > Isopropyl alcohol > Dimethyl formamide > N-methyl-2-pyrrolidone > Propylene carbonate > Dimethyl carbonate and it was found that using methanol as a solvent rate of carboxylation were enhanced by a factor of 3. The polar protic solvents (such as linear alcohols) are strong nucleophiles, strong hydrogen-bond donors and acceptors and interact strongly with the electron-deficient electrophiles, epoxides and CO_2. In contrast, polar aprotic solvents are weak nucleophiles compared to the polar protic solvents resulting in low activity. These conclusions, based on CO_2 solubility measurements in methanol and PC solvents along with kinetic studies of PO carboxylation, will be presented and discussed.

Improving the Performance of Sn Promoted Ni/Al_2O_3 Catalysts for the Dry Reforming of Methane

*T. Stroud, T. Smith, E. Le Saché, H. Arellano-Garcia, T.R. Reina**

Department of Chemical and Process Engineering, Faculty of Engineering and Physical Sciences, University of Surrey, Guildford, UK

Corresponding author: t.ramirezreina@surrey.ac.uk

Introduction

Prevention of CO_2 reaching the upper atmosphere is a topic of great interest in the world today. Co-currently, green energy production and chemical feedstock sources are being highly researched, where combining these issues represents the most efficient solution. Dry reforming of methane (DRM) represents one such solution:

$$CH_4 + CO_2 \rightleftharpoons 2CO + 2H_2$$

$$\Delta H_{298K} = +247 \text{ kJ/mol} \qquad \Delta G_{298K} = +170 \text{ kJ/mol}$$

The products of this reaction (known colloquially as syngas) are a highly useful precursor for a number of chemical processes, such as the production of methanol and synthesis of liquid fuels via Fischer-Tropsch (FTS).

Given the endothermic nature of the reaction, high reaction temperatures are required to achieve favourable conversions of the reactants to syngas. Unfortunately, elevated temperatures also promote the formation of solid carbon via a number of further side reactions [1]. This carbon then deposits on the surface of the catalyst which (alongside metal sintering) causes the catalyst to deactivate.

Hence an effective catalyst must resist sintering and be resilient towards the formation of carbon upon the surface; whilst also being relatively inexpensive, such that it can be realistically employed within the wider industry. Nickel has previously been identified as a viable active metal to use, whilst an alumina (Al_2O_3) support helps disperse the Ni over a wider surface area, improving the activity over that of just Ni alone [1].

However, this still suffers from severe carbon deactivation and hence there remains potential for improvement. The aim of this work is to develop advanced catalysts for chemical CO_2 recycling via DRM using tin and ceria as promoters of Ni/Al_2O_3

Experimental

Catalysts were prepared by the method of sequential impregnation, where a base of 10 wt.% Ni was first prepared by adding the appropriate amount of $Ni(NO_3)\bullet 6H_2O$ (Sigma-Aldrich) to an alumina (SASOL) base, using acetone (Sigma-Aldrich) as the solvent. This was then thoroughly mixed for 30 minutes, before evaporating the acetone off with the aid of a vacuum pump. The sample was then further air dried in an oven overnight at 80 °C, after which calcination was performed at 700 °C for 4 hours. The resultant Ni/Al catalyst was employed as a reference material in this paper. Following the same preparation method, 3 promoted catalysts were prepared: Sn 0.02 mol.% Ni/Al, Sn 0.04 mol.% Ni/Al, Sn 0.02 mol.% Ni/Ce-Al. The tin was added using a $SnCl_2$ (Sigma-Aldrich) precursor and the ceria by $Ce(NO_3)_2\bullet H_2O$ (Sigma-Aldrich) aiming for a nominal CeO_2 loading of 20 wt.%.

Experiments were run in a quartz tube reactor at atmospheric pressure, utilising quartz wool as a reactor bed. An electric fired oven was used to ramp up (15 °C/min) and maintain the temperatures used for reduction (800 °C) and reaction (700 °C). Reduction was performed in a total flow of 100 ml/min for 1 hour in an atmosphere consisting 20 % H_2 (BOC) and 80 % N_2 (BOC).

Each reaction was run for over 20 hours, using 0.1 g of catalyst and a total inlet flow of 100 ml/min, corresponding to a weight hourly space velocity (WHSV) of 60, 000 ml/(g-cat h). Of the inlet flow, 12.5 % was CO_2 (BOC), 12.5 % CH_4 (BOC) and the remainder being that of N_2.

All fresh catalysts were characterised by means of XRD, H_2-TPR and N_2 physisorption (BET) analysis. Selected spent catalysts also underwent SEM and Raman.

Results and Discussion

H_2-TPR analysis (not shown here) confirms that a reduction temperature of 800 °C is sufficient to reduce the metal oxides to their elemental forms, whilst the XRD data (**Figure 1**) shows the Ni particles have been well dispersed. The increase of Sn loading as a promoter appears to have little effect on the crystalline structure of the samples, whereas cerium shows typical diffraction peaks of CeO_2 fluorite phases.

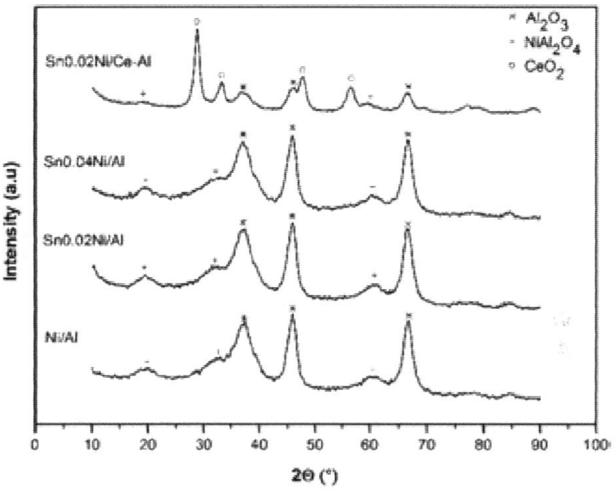

Figure 1 XRD profiles of the sample catalysts

N_2 physisorption results (**Table 1**) show the addition of Sn does not affect the textural properties of the reference Ni/Al material. However CeO_2 reduces the surface area and volume of the pores, which is down to the CeO_2 nanoparticles covering the surface of the primary support (Al_2O_3).

Table 1 Results of N_2 physisorption analysis on fresh catalysts

Sample	S_{Bet} (m^2/g)	V_{Pore} (cm^3/g)	D_{Pore} (nm)
Ni/Al	164	0.42	10.1
Sn0.02 Ni/Al	173	0.42	9.8
Sn0.04 Ni/Al	172	0.42	9.8
Sn0.02 Ni/Ce-Al	120	0.29	9.6

In terms of catalytic performance, both methane and CO_2 conversions (**Figure 2** and **Figure 3** respectively) are improved by the addition of Sn, showing less deactivation with time when compared with the un-promoted Ni/Al catalyst. Initially, both the un-promoted and 0.04 Sn sample appear to perform better than the 0.02 Sn catalyst, but this quickly changes around the 500 minute mark for CH_4 conversion and 450 minute mark for CO_2. Increasing the quantity of Sn from 0.02 mol.% to 0.04 mol.% exhibits a decrease in performance with time. This is theorised to be down to the extra Sn blocking the active sites [2], therefore reducing performance; hence a ratio of 0.02 mol.% of Sn was selected as the optimum to be further improved upon.

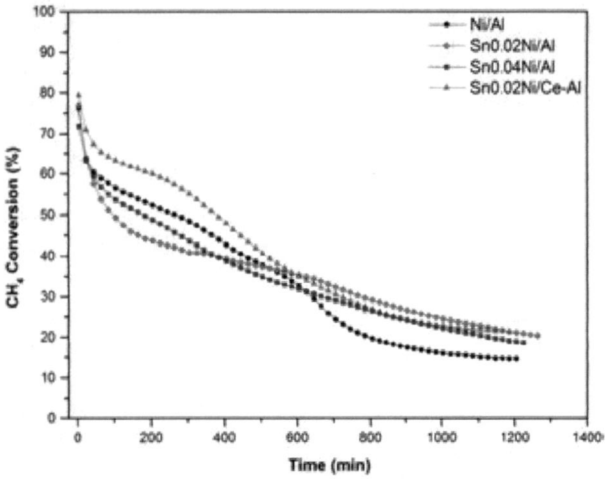

Figure 2 Methane conversions for the catalysts over a period of approximately 20 hours

Ceria is an excellent redox promoter due to its high oxygen storage capacity [3]. Therefore by adding CeO_2, the amount of surface oxygen available increases [1]. This helps with the removal of carbon species from the catalyst surface, increasing the lifespan and improving activity. This has been observed in the case of the multicomponent catalyst (Ni-Sn/Ce-Al), with conversions being greater for longer periods of time in the case of methane. An impressively high peak of 95 % CO_2 conversion can be seen, with a seemingly stable 66 % conversion after approximately 20 hours of on-stream operation. Lower conversions of methane (around 20 % after 20 hours) in comparison to CO_2 are to be expected, due to decomposition of CH_4 being the rate determining step and also the primary depositor of carbon [4].

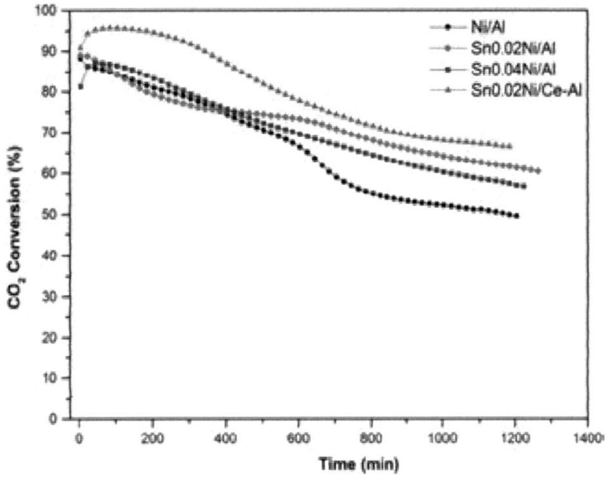

Figure 3 Carbon dioxide conversions for the catalysts over a period of approximately 20 hours

For the H_2/CO ratio (**Figure 4**), the Ce doped sample appears the best candidate early on, peaking at a ratio of 0.97 after *ca.* one hour of operation. However this eventually decreases down to the same level (a ratio of around 0.75) of the un-promoted and 0.04 Sn samples after 20 hours. The Sn0.02 Ni/Al catalyst shows a slightly more favourable ratio of 0.81 after 20 hours, showing the addition of tin is also effective at improving the H_2/CO ratio; with the optimum ratio again being 0.02 mol.%.

With the addition of ceria, a slight drop in performance is observed when compared with the 0.02Sn promoted catalyst. At the end of the experiment, a higher conversion of CO_2 is present for Ce, with a similar methane conversion. This therefore results in a comparatively greater abundance of CO being present in the product stream (as the H_2 is only formed by methane decomposition), giving a lower ratio. The observed trend could be due to the ability of ceria to catalyse the reverse water shift reaction, RWGS ($CO_2 + H_2 \rightleftharpoons H_2O + CO$) which explains greater CO_2 conversion and the smaller H_2/CO ratio for the ceria promoted sample.

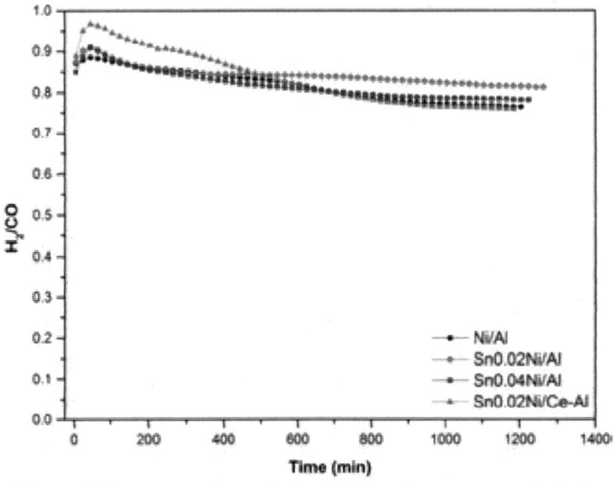

Figure 4 Syngas product ratio for each catalyst over a period of approximately 20 hours

Conclusions

The addition of tin as a promoter to a nickel alumina catalyst shows a marked improvement in conversions of both carbon dioxide and methane to syngas. However, adding too large amounts of tin can have a detrimental effect on catalytic performance by restricting access of reactants to the active nickel sites. It has been found that 0.02 mol.% is an optimum tin loading.

Addition of ceria to the alumina support increases resistance to carbon formation on the surface, exhibiting higher conversions for a longer period of time. In parallel, ceria aids the RWGS reaction boosting the CO_2 conversion. When included together, tin and ceria work synergistically, resulting in a very promising catalyst for dry reforming reactions.

References

1. Charisiou N.D., Baklavaridis A., Papadakis V.G., Goula M.A. *Syn Gas Production via the Biogas Reforming Reaction Over Ni/MgO–Al_2O_3 and Ni/CaO-Al_2O_3 Catalysts.* Waste and Biomass Valorization. 2016; 7: p. 725-736.

2. Hou Z., Yokota O., Tanaka T., Yashima T. *Surface Properties of a Coke-Free Sn Doped Nickel Catalyst for the CO_2 Reforming of Methane.* Applied Surface Science. 2004; 233: p. 58-68.

3. Reina T.R., Ivanova S., Centeno M.A., Odriolzola J.A. *Boosting the Activity of a Au/CeO_2/Al_2O_3 Catalyst for the WGS Reaction.* Catalysis Today. 2015; 253: p. 149-154.

4. Wei J., Iglesia E. *Isotopic and Kinetic Assessment of the Mechanism of Reactions of CH_4 With CO_2 or H_2O to Form Synthesis Gas and Carbon on Nickel Catalysts.* Journal of Catalysis. 2004; 224(2): p. 370-383.

498418 The Effects of Aging on Steady State Ammonia Oxidation over Pt/Al_2O_3

Abstract:

Three-way catalyst (TWC) light-off (LO) performance is critical in cost-effectively meeting emission standards. Improvements in catalyst LO should include catalyst modifications that minimize inhibition effects of exhaust species such as CO and hydrocarbons (HCs), lower precious group metal (PGM) loading, or use a lower cost PGM such as Pd. Alumina (Al_2O_3) and zirconia (ZrO_2) supports provide high surface area and stability. Furthermore, ceria (CeO_2) provides the key benefit of oxygen storage, compensating for deviations from stoichiometric TWC operating conditions by supplying oxygen for CO and HC oxidation under rich transient conditions. Inclusion of ceria in PGM catalysts enhances CO and HC conversion and minimizes self-inhibition; this improved activity is mainly attributed to ceria containing more efficient oxygen activation sites at the PGM-support interface relative to other support materials such as Al_2O_3 [1-4].

Monolith catalysts having washcoats with 1 wt% Pd, 3 g/in^3 loading, 400 cpsi and formulations Pd/Al_2O_3, Pd/CZO, and Pd/CZO/Al_2O_3 (17 wt% CZO, balance Al_2O_3) were synthesized and compared. Catalyst performance experiments were conducted to study the role of Pd and support compositions and loadings on inhibition during oxidation of CO, acetylene (C_2H_2), and propylene (C_3H_6) under near-stoichiometric conditions ($\lambda = 1.01$). Transient and steady-state bench scale reactor studies with simulated exhaust gas mixtures were conducted to extract oxidation kinetics, compare LO behavior, and develop a predictive model for understanding and optimizing catalyst performance.

Catalyst activities were compared via temperature-programmed oxidation experiments. Figure 1 compares CO LO curves using feed concentrations of 0.5% CO and 1% CO. The Pd/CZO catalyst was tested when "fresh" (initial tests after synthesis) and 10+ months later ("deactivated"). Using all three catalysts, LO temperatures increase with feed CO concentration, showing that CO oxidation is self-inhibiting. With increasing CZO content, LO temperatures decrease, demonstrating the promotional effects of using ceria.

Figure 2 shows C_2H_2 LO (feed concentrations of 375 ppm C_2H_2 and 750 ppm C_2H_2) and C_3H_6 LO (feed concentrations of 250 ppm C_3H_6 and 500 ppm C_3H_6) using the deactivated Pd/CZO catalyst. Like CO, C_2H_2 and C_3H_6 are self-inhibiting. For the same concentrations of carbon in the feed mixture (750 ppm C_1 or 1500 ppm C_1), C_2H_2 oxidation occurs at higher temperatures than C_3H_6 oxidation.

Exhaust hydrocarbons propylene and acetylene reveals some interesting differences during co-oxidation with CO. The individual LO features of 1% CO and 750 ppm C_2H_2 in Figures 1 and 2 are compared to their co-oxidation (1% CO + 750 ppm C_2H_2) in Figure 3. CO lights off at higher temperatures in the mixture, i.e. C_2H_2 inhibits CO oxidation. During co-oxidation, C_2H_2 oxidation is inhibited by the presence of CO below

~15% conversion, but is otherwise improved compared to oxidation of C_2H_2 alone. Whereas 1% CO lights off before 750 ppm of C_2H_2 during individual oxidation, the reverse is true during co-oxidation. This is in contrast to earlier findings of CO + C_3H_6 trends in which was found that CO inhibits C_3H_6; i.e. propylene conversion follows CO light off (see below).It is instructive to compare the behavior of the two hydrocarbons at the same C1 concentration. Though the same feed concentration of 1% CO + 1500 ppm C_1 (from HCs) is used, C_3H_6 and C_2H_2 co-oxidation with CO exhibits different trends in terms of LO temperatures and order of species oxidation. As shown in Figure 4, using the deactivated Pd/CZO catalyst, 1% CO lights off before 500 ppm C_3H_6 both during individual oxidation and mixture LO. During co-oxidation, 500 ppm C_3H_6 is 50% converted at ~ 246 °C, compared to ~ 266 °C for 750 ppm C_2H_2.

Differential kinetics studies were conducted to quantify reaction orders and activation energies. The reaction order with respect to CO was found to be approximately −1 using Pd/Al_2O_3 and Pd/CZO/Al_2O_3 and 0 to slightly negative using Pd/CZO, consistent with literature values [5-7]. The activation energies for CO oxidation were found to be 134-213 kJ/mol using Pd/Al_2O_3 and Pd/CZO/Al_2O_3, and 25-70 kJ/mol using Pd/CZO. Using all three catalysts, the reaction order with respect to C_3H_6 was found to be approximately −0.89 to −0.75 and the activation energy in the range of 72-105 kJ/mol, approximately half of that observed for CO oxidation and similar to literature values [5]. Using the Pd/CZO catalyst, the reaction order with respect to C_2H_2 was found to be approximately −0.52 to −0.96 and the activation energy of C_2H_2 oxidation approximately 73 kJ/mol.

Steady-state and transient CO, C_2H_2, and C_3H_6 oxidation experiments using fresh Pd/Al_2O_3, Pd/CZO, and Pd/CZO/Al_2O_3 monolith catalysts were conducted to compare LO behavior and kinetic parameters (i.e. reaction orders and activation energies). CO LO temperatures decreased with increasing proportion of ceria used in the TWC, and C_2H_2 and C_3H_6 LO studies using Pd/CZO demonstrated self- and mutual-inhibition during species oxidation. Kinetics studies demonstrated further the self-inhibiting oxidation behavior.

Ongoing work involves the use of a mechanistic-based kinetic model incorporated into a low-dimensional monolith model [8-9]. The global reactor models are used with CO, C_2H_2, and C_3H_6 oxidation on Pd with steps involving ceria. The 1+1 dimensional model formulation comprises coupled species balances with transverse-average equations having external and internal mass transfer coefficients to verify experimental LO behavior during individual and co-oxidation using catalysts and feed mixtures of different formulations. Finally, experiments involving C_2H_2 oxidation on Pd/Al_2O_3 and Pd/CZO/Al_2O_3 catalysts and other hydrocarbons including toluene are being conducted to investigate inhibition during LO, extract kinetic parameters for oxidation reactions, and provide a basis for validating modeling simulations.

Figure 1: CO conversion (%) versus monolith temperature using Pd/Al_2O_3, Pd/CZO/Al_2O_3, and Pd/CZO catalysts for feed concentrations of 0.5% and 1% CO.

Figure 2: C_2H_2, C_3H_6 conversion (%) versus monolith temperature using deactivated Pd/CZO for feed concentrations of 375 ppm C_2H_2, 750 ppm C_2H_2, 250 ppm C_3H_6, and 500 ppm C_3H_6.

Figure 3: CO, C_2H_2 conversion (%) versus monolith temperature using deactivated Pd/CZO for feed concentrations of 750 ppm C_2H_2 and/or 1% CO.

Figure 4: CO, C_3H_6 conversion (%) versus monolith temperature using deactivated Pd/CZO for feed concentrations of 500 ppm C_2H_2 and/or 1% CO.

References

[1] Fernandez-Garcia, M.; Martinez-Arias, A.; Iglesias-Juez, A.; Hungria, A.B.; Anderson, J.A.; Conesa, J.C.; and Soria, J. *App. Cat. B* 2001, **31**, 39-50.

[2] Martinez-Arias, A., Fernandez-Garcia, M.; Iglesias-Juez, A.; Hungria, A.B.; Anderson, J.A.; Conesa, J.C.; and Soria, J.. *App. Cat. B* 2001, **31**, 51-60.

[3] Sharma, S.; Hegde, M.S.; Das, R.N.; and Pandey, M. *App. Cat. A* 2008, **337**, 130-137.

[4] Harmsen, J.M.A.; Hoebink, H.B.J.; and Shouten, J.C. *Chem. Eng. Sci.* 2001, **56**, 2019-2035.

[5] Yao, Y-.F. Y. *J. Catal.*, 1984, **87**, 152-162.

[6] Cant, N.W.; Hicks, P.C.; and Lennon, B.S. *J. Catal.* 1978, **54**, 372-383.

[7] Rainer, D.R.; Koranne, M.; Vesecky, M.; and Goodman, D.W. *J. Phys. Chem. B* 1997, **101**, 10769-10774.

[8] Joshi, S.; Harold, M.P.; Balakotaiah, V. *AIChE J.*, 2009, **55,** 1771-1783.

[9] Raj, R.; Harold, M.P.; Balakotaiah, V. *Chem. Eng. J.*, 2015, **281**, 322-333.

494315 Two-Step Selective Bromination of Thiophenic Compounds for Gasoline Desulfurization Under Mild Conditions

Abstract:

The removal of trace amount of refractory thiophenic compounds from fuels has raised intense concern globally. The development of less energy-intensified desulfurization process is under great demand. In this work, we proposed a new approach of two-step selective bromination of thiophenic compounds for gasoline desulfurization under mild conditions. The Ag modified MoO_3 nanorods (AMO) were synthesized and characterized by XRD, HRTEM and N_2 adsorption, and evaluated in selective bromination of thiophenic compounds. Results showed that (a) thiophenic compounds in model gasoline can be selectively brominated over AMO catalyst with the addition of Br_2, and their bromination production can be removed selectively by followed-up extraction. (b) At the optimized Ag loading of 10%, the apparent rate constants for T and BT reached as high as $95.8h^{-1}$ and $142.7h^{-1}$, respectively under ambient conditions, much higher than those cases in oxidative desulfurization, adsorptive desulfurization, etc. Product analyses showed that the portion of dibromo-thiophenes increased dramatically with increasing loading of Ag, due likely to the formation of $MoAg_2O_4$. (c) The bromination selectivity of thiophenic compounds reached four orders of magnitude than those of aromatic compounds in fuel. Meanwhile, the competition from olefins can be strategically suppressed by HBr pretreatment. (d) With the integration of the 2nd-step of extraction, desulfurization capacity of thiophenes reached as high as 30 mg-S/g-cat, while the integrated system showed excellent bromination desulfurization capacity of thiophene at ~260 mg-S/g-cat under high fuel-to-catalyst mass ratio of 5000. The adsorption of various bromides over $MoAg_2O_4$ was simulated to elucidate the desulfurization mechanism. The fast kinetics, high desulfurization capacity and selectivity of the proposed two-step approach under mild conditions make it promising for gasoline desulfurization.

Keywords: desulfurization, gasoline, selectivity, Ag modified MoO_3

494394 Improvement of Low Temperature Activity for CO Oxidation on Ceria Catalyst By Transition Metal Substitution

Abstract:

Recently, reduction of exhaust gas has been an important issue with increasing the sales of automobiles. CO reduction in various exhaust gases is particularly critical because it damages people and environment. A ceria-based catalyst that can reduce automotive emissions instead of three-way catalysts have been developed. When the temperature is lowered, however, the efficiency is rapidly decreased. In order to improve the activity of ceria catalyst, we synthesized transition metal (TM)-doped ceria. In this study, we examined four transition metal (TM=Cu, Co, Mn, and Fe) doped and their co-doped ceria. TM-doped ceria ($Ce_{0.95}TM_{0.05}O_{2-î'}$) were synthesized by EDTA-citrate complexing method. By means of this method, we obtained highly pure and homogeneous nano-powder. The nanoparticles were then characterized by X-ray diffraction (XRD), scanning electron microscopy (SEM), and energy dispersive spectroscopy (SEM-EDS). To clarify the positive effects of doping, the surface area was evaluated by Brunauer-Emmett-Teller (BET) measurement. It turns out that the TM-doped ceria powders had higher specific surface area than that of undoped ceria, and correspondingly, the larger specific surface area enhanced the catalytic activity. Our density functional theory results demonstrated that this is attributed to the different interactions between dopant and support. Our results will be helpful for effectively exploring the TM dopant candidates of ceria and determining their optimal concentration for the enhanced catalytic activity.

487852 Characterization of the Active Site and Mechanism for 1,6-Hexanediol Production from Tetrahydrofuran-Dimethanol over Pt-Based Catalysts

Abstract:

Characterization of the Active Site and Mechanism for 1,6-Hexanediol Production from Tetrahydrofuran-Dimethanol over Pt-Based Catalysts

Terminal diols such as 1,4-butanediol (1,4-BDO) and 1,6-hexanediol (1,6-HDO) are important building-block chemicals and find application in the synthesis of specialty chemicals and a variety of polymers, primarily in polyesters and polyurethanes, and polyamides such as nylon-6,6.[1,2] Currently, 1,6-HDO is produced industrially from cylohexanone/cyclohexanol (KA oil) by oxidation with nitric acid to form adipic acid, followed by hydrogenation of the dimethoxy ester to yield 1,6-HDO. The attractiveness of the overall process is tempered by low conversions (4-8% for the oxidation of KA oil), difficult separations, the use of non-renewable fossil feedstocks, and emission of greenhouse gases, namely N_2O.[3] Various attempts to synthesize 1,6-HDO, as well as other alpha,omega-diols, from biomass resources have been reported. Using supported catalysts that contain precious metals with a reducible metal oxide such as Rh-ReO_x,[4-9] and Ir-ReO_x,[10,11] tetrahydropyran-2-methanol (THP2M), a biomass-derived feedstock,[12] can be converted into 1,6-HDO with high selectivity. However, the high cost of the catalyst, in combination with a low productivity, limits the industrial applicability of those systems.[13] More recently, it has been shown that 1,6-HDO can be produced from THP2M without the use of precious metals. However, 1,6-HDO selectivity only reached ~40%.[14] Therefore, this route is not likely to be industrially applicable soon either.

In an effort to make the route to 1,6-HDO from biomass more industrially viable, our group has developed a complete reaction network producing 1,6-HDO from woody biomass.[15] The early steps in this process convert biomass to tetrahydrofuran-dimethanol (THFDM) with high selectivity, and will likely be discussed elsewhere at this conference. In this study, we focus on the final steps to 1,6-HDO from THFDM, shown in Scheme 1.

Scheme 1. Reaction sequence to 1,6-hexanediol (1,6-HDO) from tetrahydrofuran-dimethanol (THFDM)

We have developed a Pt-based catalyst that performs each step of the above reaction sequence with a total 1,6-HDO selectivity >90%. In comparison to the state-of-the-art, a previously described Rh-ReO_x/C catalyst is not active for this reaction. Moreover, while Rh-ReO_x/C is active for 1,6-HDO production from THP2M with selectivity >90%, the reaction rate is an order of magnitude lower than the total production rate of 1,6-HDO from THFDM over our Pt-based catalysts. Our catalyst, therefore, produces 1,6-HDO from a further upstream product, and at a much higher rate than the state-of-the-art Rh-ReO_x/C system.

In situ characterization of this catalyst has revealed important information with regards to the active site and mechanism for 1,6-HDO production over our Pt-based catalysts. Our primary findings are that the oxidation state and acidity of the oxophilic additive (Mo, V, W, Re, for example) are crucial for this reaction. All catalysts are reduced with H_2 prior to reaction, so the same conditions were used for characterization. We studied these catalysts with pyridine FTIR spectroscopy to obtain the Bronsted/Lewis acid ratio in each of our Pt-based catalysts, and compare that to reactivity data we had already compiled for the reaction of interest. As we observed that Bronsted acidity is correlated to the turnover frequency. Moreover, we found that this ratio increases in all catalysts during pre-reduction, a necessary step to achieve any reactivity. This data seems to indicate that the Bronsted acid site necessary for this reaction to take place, which are also confirmed by XPS and Raman characterizations. Furthermore, we are able to alter and characterize the distance between Pt nanoparticles and the oxophilic additive by changing the catalyst synthesis technique. Based on these studies we proposed a different mechanism to those which have been proposed previously, which will be discussed in more detail during the conference.

1 P. Werle, M. Morawietz, S. Lundmark, K. Ssrensen, E. Karivinen and J. Lehtonen, Alcohols, Polyhydric, in *UllmannÕs Encyclopedia of Industrial Chemistry*, 2012.

2 W. Tianfu, M. S. Ide, M. R. Nolan, R. J. Davis and B. H. Shanks, *Energy Environ. Focus*, 2016, **5**, 13Ð17.

3 S. Van de Vyver and Y. Roman-Leshkov, *Catal. Sci. Technol.*, 2013, **3**, 1465Ð1479.

4 M. Chia, Y. J. Pag‡n-Torres, D. Hibbitts, Q. Tan, H. N. Pham, A. K. Datye, M. Neurock, R. J. Davis and J. A. Dumesic, *J. Am. Chem. Soc.*, 2011, **133**, 12675Ð12689.

5 M. Chia, B. J. OÕNeill, R. Alamillo, P. J. Dietrich, F. H. Ribeiro, J. T. Miller and J. A. Dumesic, *J. Catal.*, 2013, **308**, 226Ð236.

6 P. U. Karanjkar, S. P. Burt, X. Chen, K. J. Barnett, M. R. Ball, M. D. Kumbhalkar, J. B. Miller, I. Hermans, J. A. Dumesic and G. W. Huber, *Catal. Sci. Technol.*, 2016, **6**, 7841Ð7851.

7 Y. Kojima, S. Kotani, M. Sano, T. Suzuki and T. Miyake, *J. Jpn. Pet. Inst.*, 2013, **56**, 133Ð141.

8 K. Chen, S. Koso, T. Kubota, Y. Nakagawa and K. Tomishige, *ChemCatChem*, 2010, **2**, 547Ð555.

9 Y. Nakagawa, M. Tamura and K. Tomishige, *ACS Catal.*, 2013, **3**, 2655Ð2668.

10 S. Koso, I. Furikado, A. Shimao, T. Miyazawa, K. Kunimori and K. Tomishige, *Chem. Commun.*, 2009, **15**, 2035Ð2037.

11 Y. Nakagawa, Y. Shinmi, S. Koso and K. Tomishige, *J. Catal.*, 2010, **272**, 191Ð194.

12 B. Xiao, M. Zheng, X. Li, J. Pang, R. Sun, H. Wang, X. Pang, A. Wang, X. Wang and T. Zhang, *Green Chem.*, 2016, **18**, 2175Ð2184.

13 Z. J. Brentzel, K. J. Barnett, K. Huang, C. T. Maravelias, J. A. Dumesic, and G. W. Huber, *ChemSusChem.*, 2017, **10**, 1-6.

14 S. P. Burt, K. J. Barnett, D. J. McClelland, P. Wolf, J. A. Dumesic, G. W. Huber, and I. Hermans, *Green Chem.*, 2017, **19**, 1390-1398.

15 J. He, K. Huang, K. J. Barnett, S. Krishna, D. M. Alonso, Z. Brentzal, S. P. Burt, T. W. Walker, W. Banholzer, C. T. Maravelias, I. Hermans, J. A. Dumesic, and G. W. Huber, *Faraday Discussions*, 2017.

492877 Mechanism and Kinetics of 1-Dodecanol Etherification and Dehydration over Tungstated Zirconia

Abstract:

As atmospheric carbon dioxide levels continue to rise as a result of burning fossil fuels, there is a growing interest in developing renewable fuels derived from biomass to meet the worldâ€™s increasing energy demands. Ethers have attracted recent interest as diesel additives and specialty chemicals due to their high cetane numbers and excellent lubricant properties, and can be produced via direct etherification of biomass-derived alcohols in the liquid phase. The competing reaction for alcohol dehydration over an acid catalyst is unimolecular dehydration to form alkenes, which is thermodynamically favored above approximately 350 K. In order to optimize the direct etherification of long chain alcohols in the liquid phase, it is necessary to develop an understanding of the mechanism and kinetics of etherification and dehydration reactions. In this study, tungstated zirconia was identified as a selective solid acid catalyst for the liquid phase etherification of 1-dodecanol. Through kinetic modeling and mechanistic probing, this study suggests that a cooperative effect between Brønsted and Lewis acid sites on tungstated zirconia enhances the selectivity to ether by increasing the surface concentration of adsorbed alcohol molecules, promoting bi-molecular etherification over unimolecular dehydration. Kinetic isotope effects for linear alcohol dehydration were measured in order to elucidate the rate limiting steps in the mechanism. Effects of alcohol concentration and product inhibition were measured and fit to kinetic models consistent with the proposed mechanisms. In addition, acid site characterization and selective poisoning experiments were used to probe the role of the acid sites and support the proposed mechanism.

WISCONSIN
UNIVERSITY OF WISCONSIN–MADISON

A Three-Step Catalytic Pathway for the Scalable Production of 1,5-Pentanediol from Biomass-Derived Tetrahydrofurfuryl Alcohol

Kevin J. Barnett[1], Zachary J. Brentzel[1], Kefeng Huang[1], Ling Li[1,2], Guozhu Liu[2], Christos T. Maravelias[1], James A. Dumesic[1] and George W. Huber[1]

(1)Department of Chemical and Biological Engineering, University of Wisconsin-Madison, Madison, WI,
(2)School of Chemical Engineering and Technology, Tianjin University, Tianjin, China

AIChE National Conference: October 30, 2017

C4-C6 α,ω-Diols Production and Market

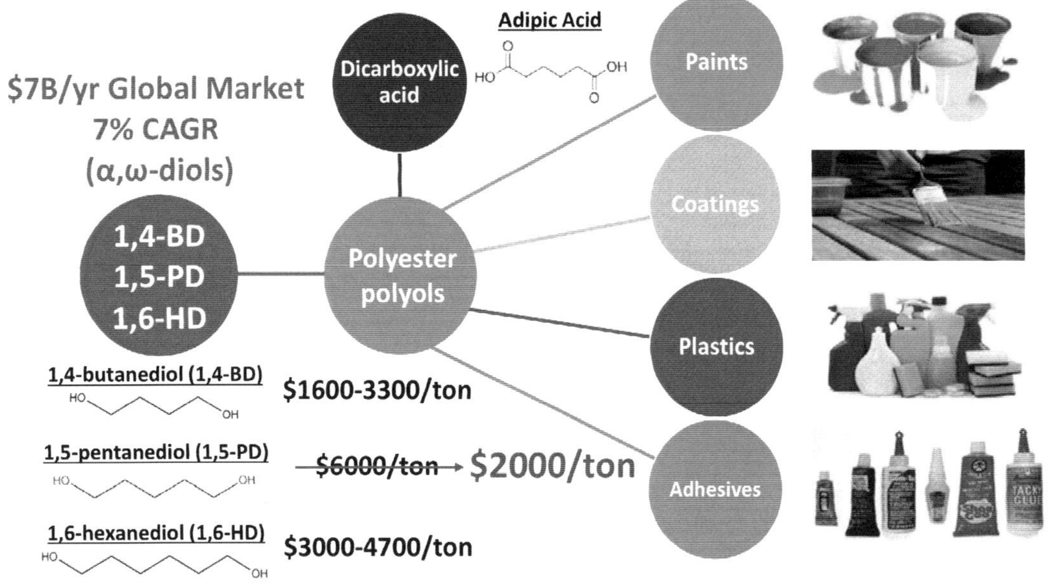

2

1,5-PD Production from Biomass

- Direct Hydrogenolysis Approach:
  Hydrogenolysis over bimetallic, noble-metal catalysts (i.e. RhRe)[1,2]
  - Uneconomical due to high catalyst costs

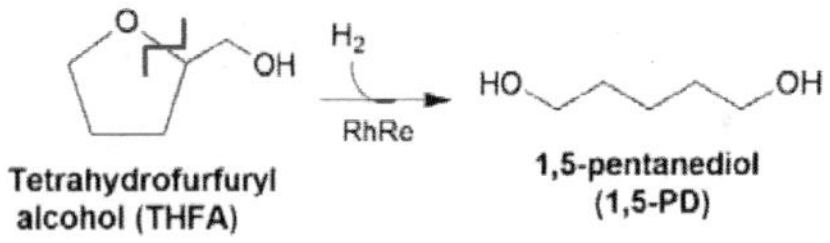

[1] M. Chia et al. *J. Am. Chem. Soc.* **2011**, *133*, 12675–89.
[2] Y. Nakagawa et al. *Catal. Today* **2012**, *195*, 136–143.

- New Approach: 3-step **D**ehydration-**H**ydration-**H**ydrogenation (**DHH**) pathway

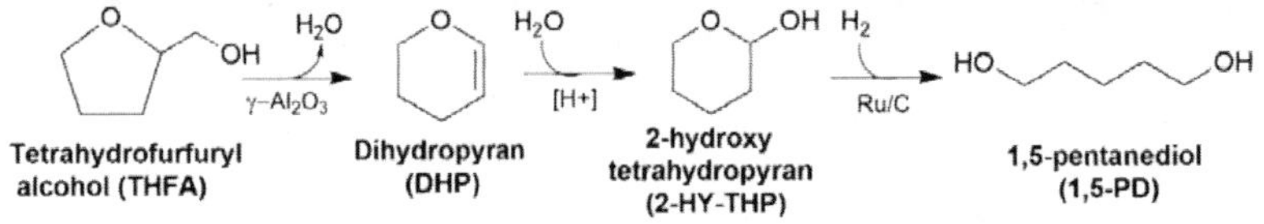

- Previously attempted by Geller et. al in 1946[3]

[3] *L. E. Schniepp, H. H. Geller, J. Am. Chem. Soc.* ***1946****, 68, 1646–1648.*

3

DHH pathway for 1,5-PD production

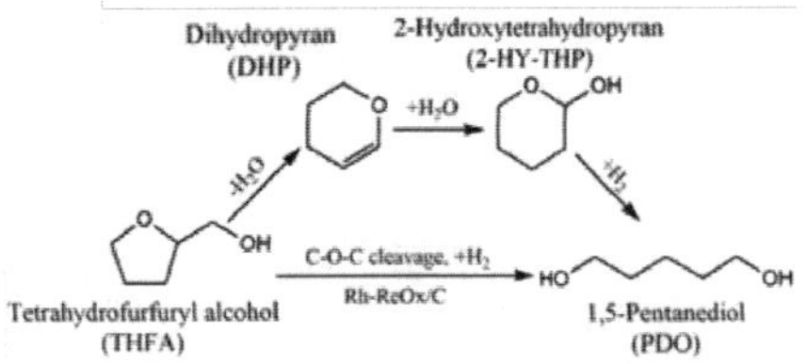

- Brentzel et. al. Chemicals from Biomass: Combining Ring-Opening Tautomerization and Hydrogenation Reactions to Produce 1,5-Pentanediol from Furfural. *ChemSusChem* **2017**.*10*, 1351-1355.
- Huang et al. Conversion of Furfural to 1,5-Pentanediol: Process Synthesis and Analysis. *ACS Sustainable Chemistry and Engineering*. **2017**. *5* (6), 4699-4706.
- Huang et al. Improving Economics of Lignocellulosic Biofuels: An Integrated Strategy for Coproducing 1,5-Pentanediol and Ethanol. **2017**. *In revision.*

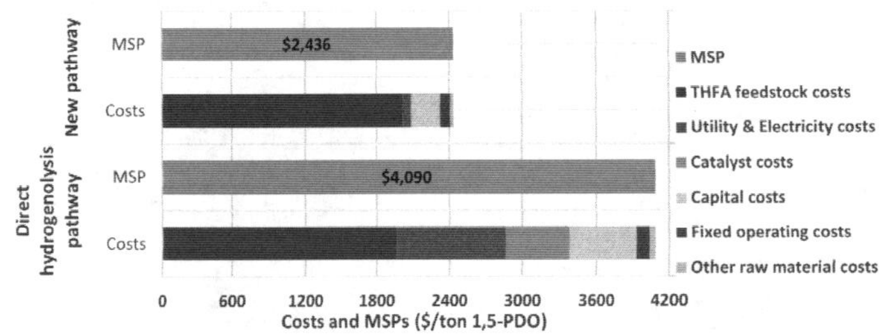

- Production Cost of DHH pathway is 6.6 times lower (excluding feedstock)
- Catalyst cost of DHH pathway is 51 times lower

4

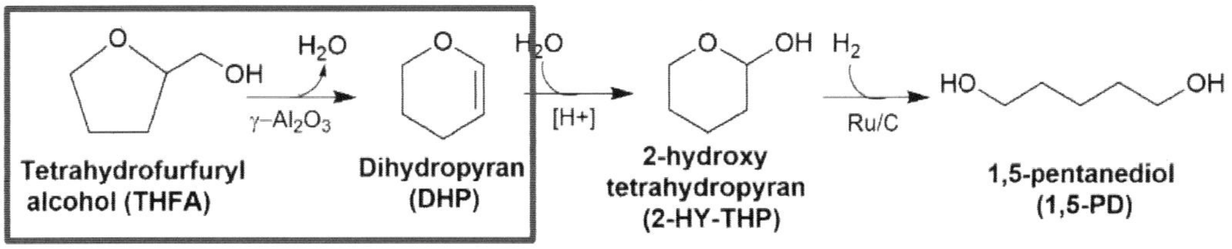

THFA DEHYDRATION

THFA Dehydration to Dihydropyran

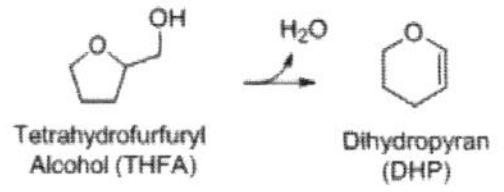

- THFA dehydrated over γ-Al_2O_3 in vapor phase
- Catalyst regenerated with 500˚C calcination step

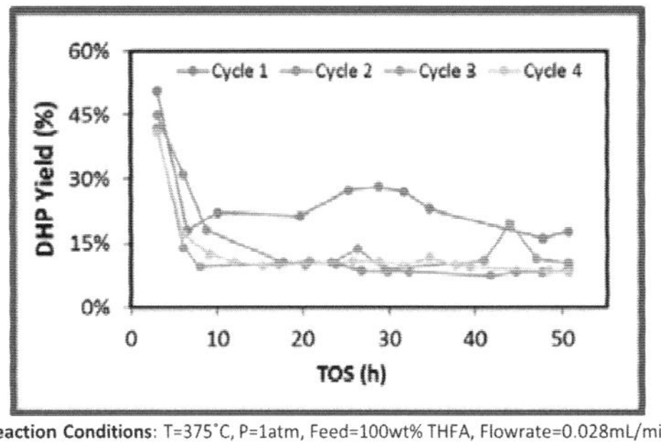

Reaction Conditions: T=375˚C, P=1atm, Feed=100wt% THFA, Flowrate=0.028mL/min

Regeneration recovers surface area lost to coking

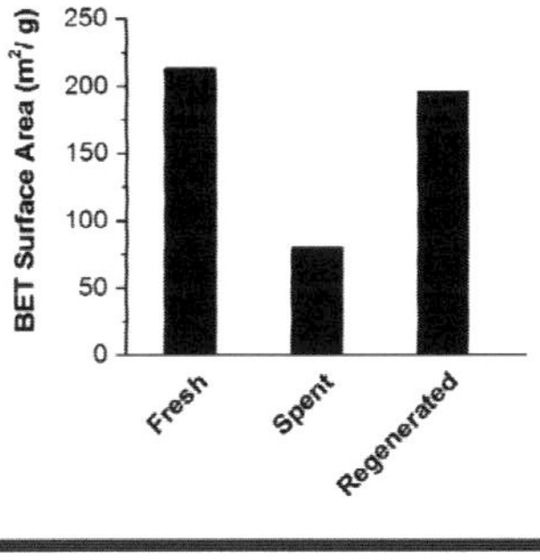

- ***Catalyst nearly completely regenerable***
  - Only 2.5% loss in activity from Cycle 3 to Cycle 4
- ***90% DHP Yields achieved at high conversion***

DHH Pathway Advantages

- **Step 1:** THFA Dehydration
  - Low-cost γ-Al_2O_3 catalyst
  - Catalyst regenerable

Reactant Concentration	Product Yield
100%	**90%**

7

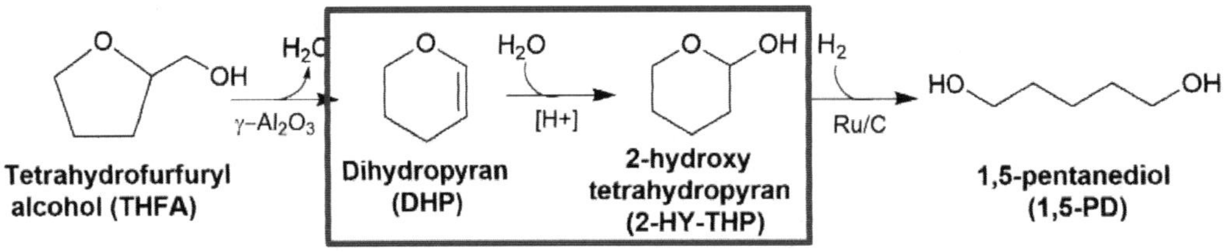

DIHYDROPYRAN HYDRATION

8

DHP Hydrates to 1,5-PD precursors in >99% Yields

- DHP hydrated in water with <u>no catalyst added</u>

Temperature (°C)	Reaction Time (h)	2-HY-THP Yield	5-tetrahydropyranyloxy-pentanal Yield	2-tetrahydropyranyl ether Yield	Overall Yield
70	4	84.5%	5.3%	1.9%	91.7%
100	4	92.0%	6.6%	1.2%	99.8%
130	4	89.5%	6.4%	1.0%	96.9%
160	4	65.4%	4.2%	1.3%	70.9%

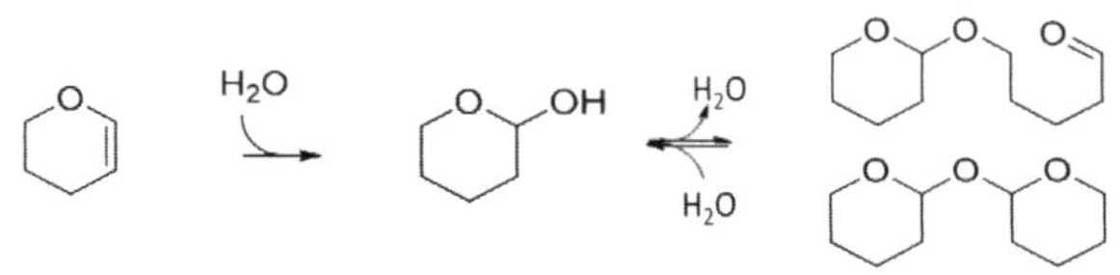

2-phase Batch Reaction

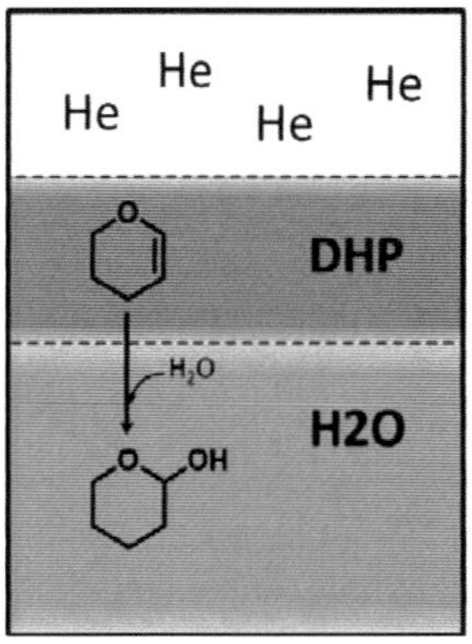

Reaction Conditions: P=250 psi He, Feed=20wt% DHP/H_2O

DHP Hydration is Autocatalytic

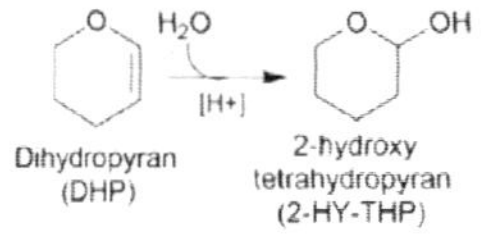

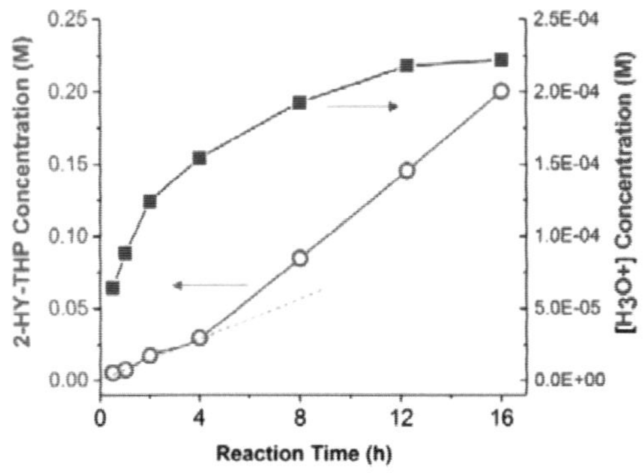

Reaction Conditions: T=25°C, P=500 psi Ar, Feed=20wt% DHP/H_2O

- Performed DHP hydration at 25°C to study initiation period of reaction
- Reaction rate doubles after 4h due to increase in solution acidity
- Small amount of acidic species (likely 5-hydroxyvaleric acid) form
- Initiation period very short! Occurs at <1% conversions

Acidic solid coke autocatalyzes DHP hydration in flow reactor

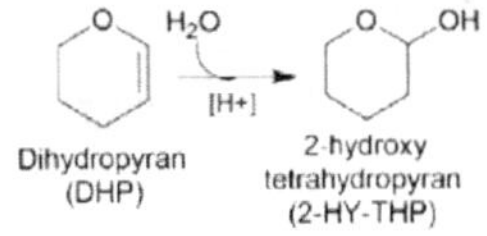

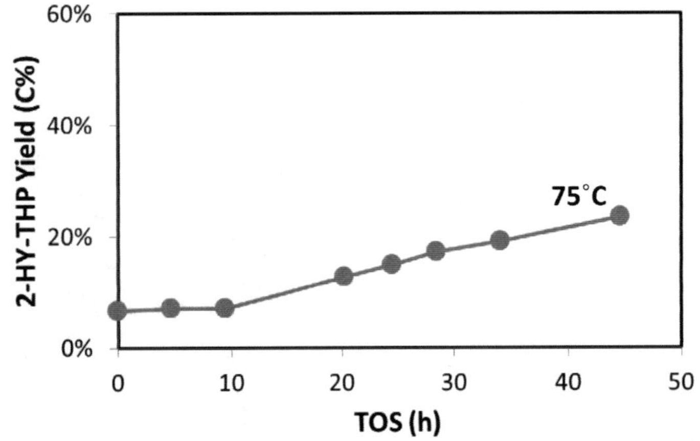

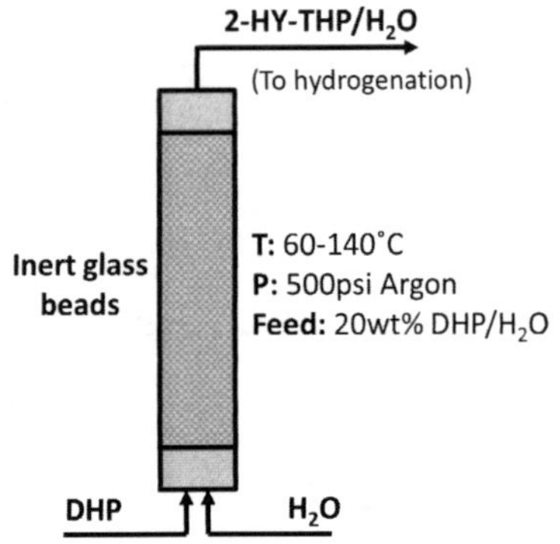

- DHP hydration over inert glass beads performed in continuous flow reactor

11

Acidic solid coke autocatalyzes DHP hydration in flow reactor

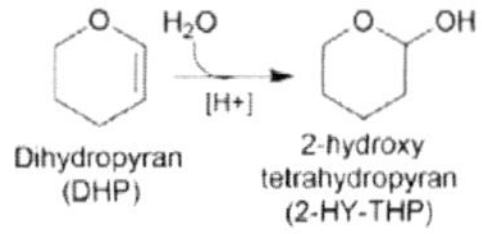

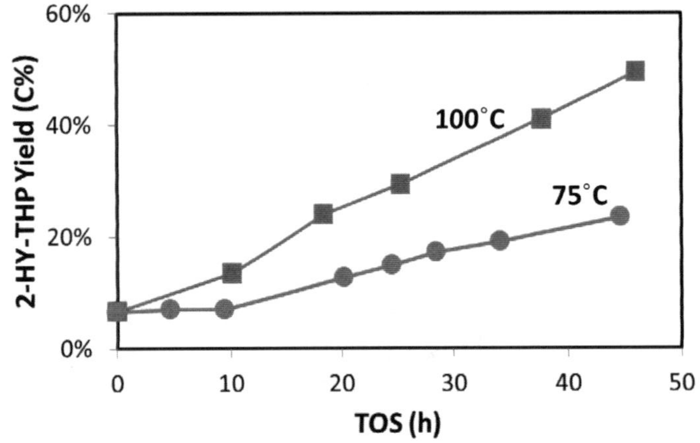

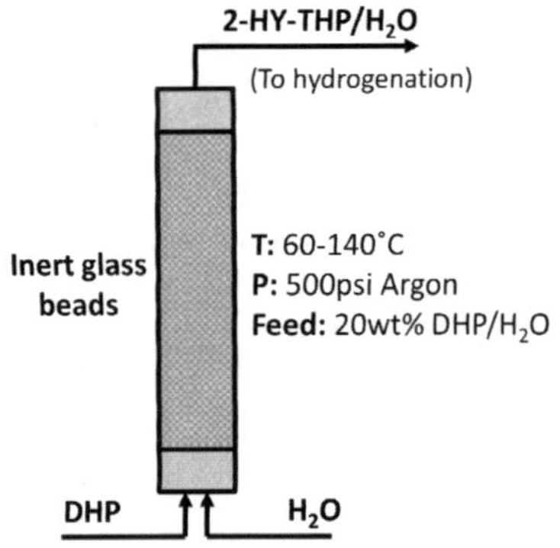

- Increased acidic coke formation at higher temperature increases hydration rate

12

Acidic solid coke autocatalyzes DHP hydration in flow reactor

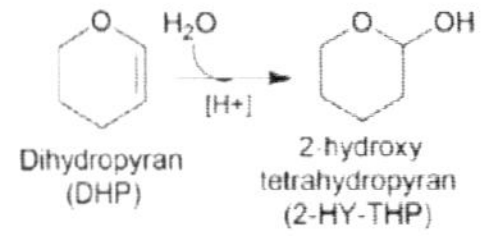

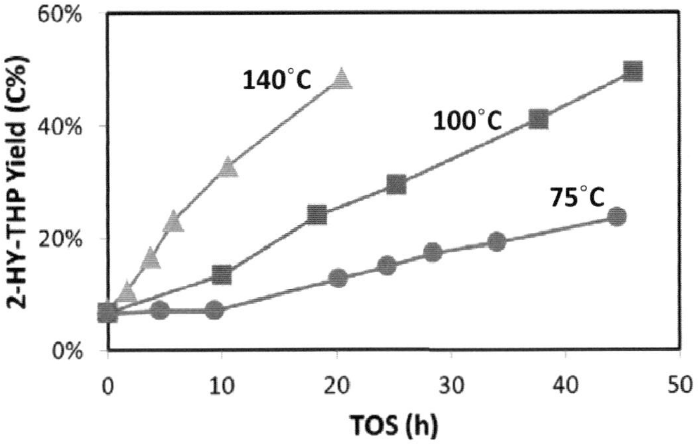

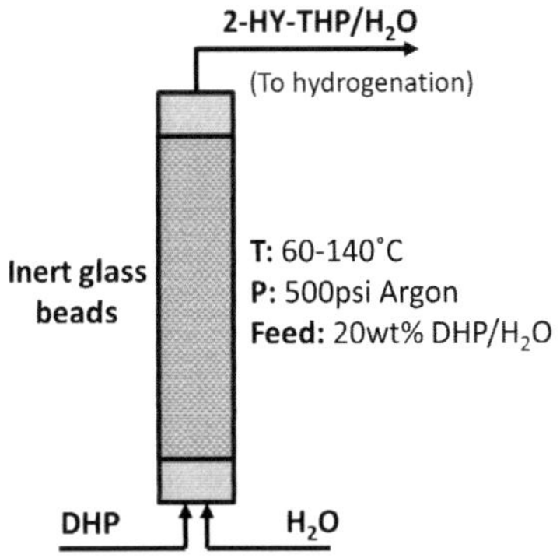

- Increased acidic coke formation at higher temperature increases hydration rate

13

Acidic solid coke autocatalyzes DHP hydration in flow reactor

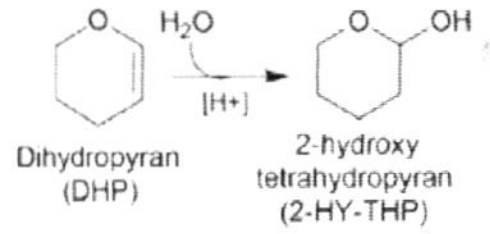

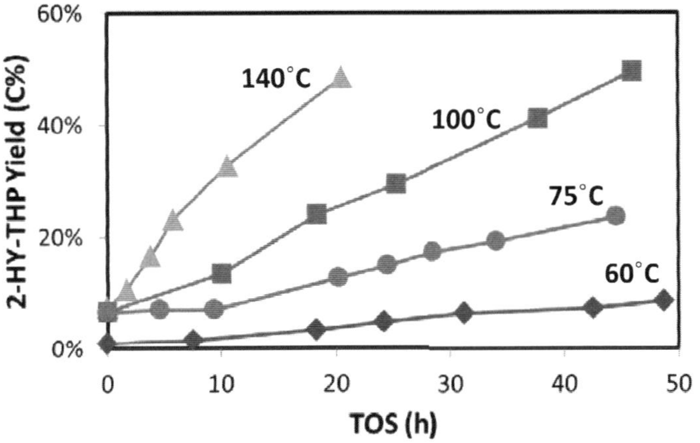

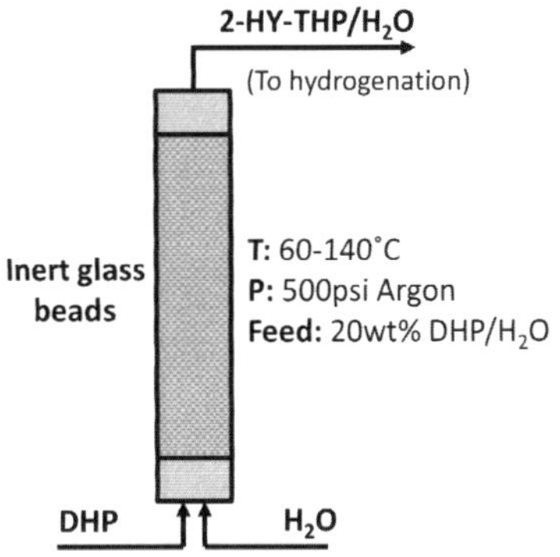

- Increased acidic coke formation at higher temperature increases hydration rate

14

Solid-acid Catalyzed DHP Hydration

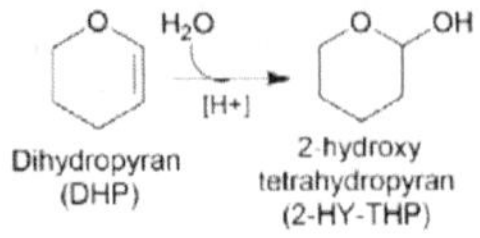

- **Goal:** Improve hydration rates by using solid acid catalysts

- Solid acid catalysts screened for DHP hydration in batch reactors:

Catalyst	Reaction Rate (μmol/gcat-min)	Bronsted site TOF (s-1)
No Catalyst	0	-
γ-Al2O3	4	0.00038
HZSM5	2822	0.115
Amberlyst-70	26017	0.152

← **No activity contribution from autocatalytic hydration**

Reaction Conditions: T=50°C, P=500psi Ar, Feed=0.5wt% DHP/H2O, Reaction Time=1h

- Solid acids show high activity at condition where autocatalytic hydration does not occur
- Hydration rates correlate with Brønsted site density

15

Why does DHP so readily hydrate?

- DHP hydration rates >> cyclohexene hydration rates
- **Hypothesis:** due to formation of stable oxocarbenium transition state

Cyclohexene Mechanism

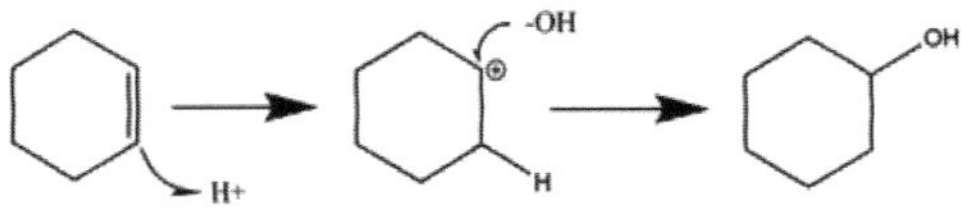

2° Carbocation

Dihydropyran Mechanism

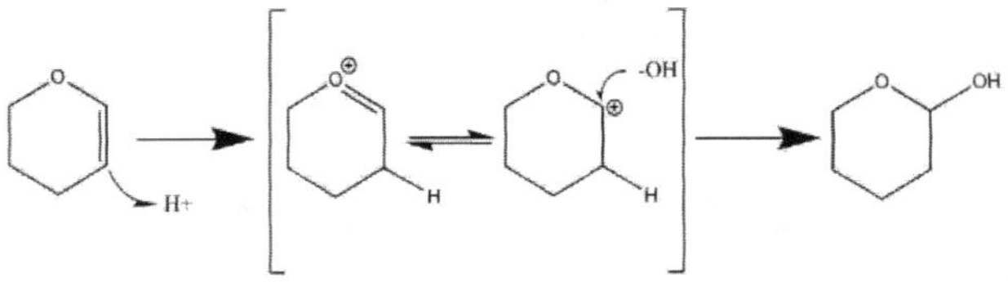

Oxocarbenium

16

Why does DHP so readily hydrate?

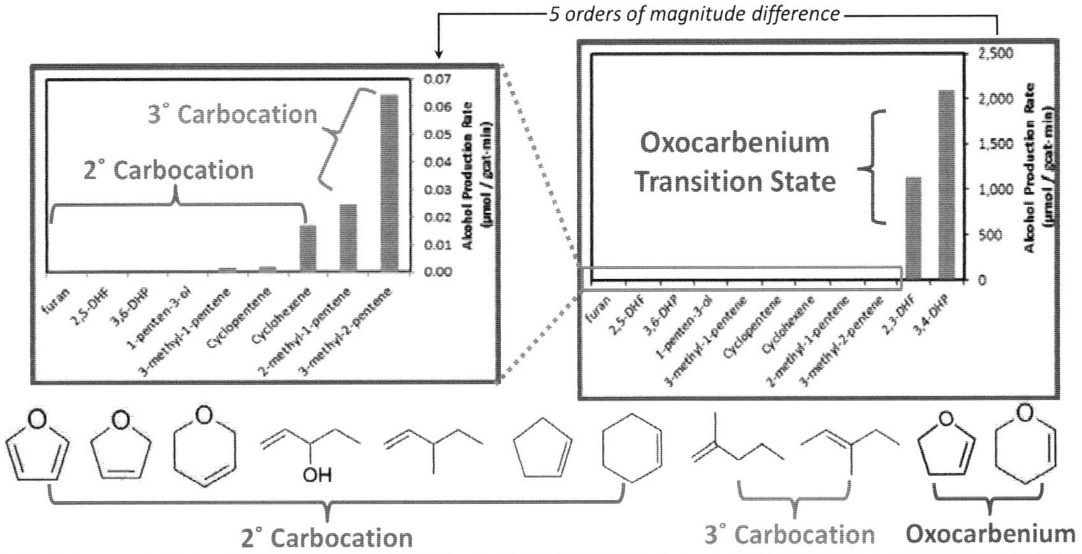

17

DHH Pathway Advantages

- **Step 1:** THFA Dehydration
  - Low-cost γ-Al_2O_3 catalyst
  - Catalyst regenerable
- **Step 2:** DHP Hydration
  - DHP hydrates with no catalyst
  - Rate increased by *in situ* formation of acids
  - Low-cost solid acid catalysts
  - DHP readily hydrates due to stable transition state

	Reactant Concentration	Product Yield
Step 1	**100%**	90%
Step 2	**50%**	98%

18

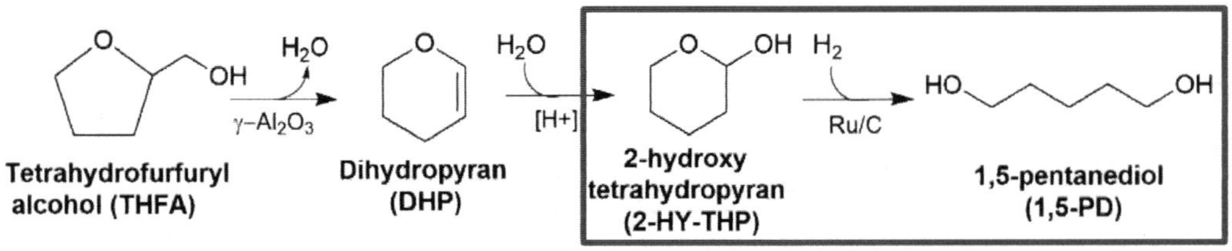

2-HY-THP HYDROGENATION

19

2-HY-THP and Dimers convert to 1,5-pentanediol at ~100% Yields

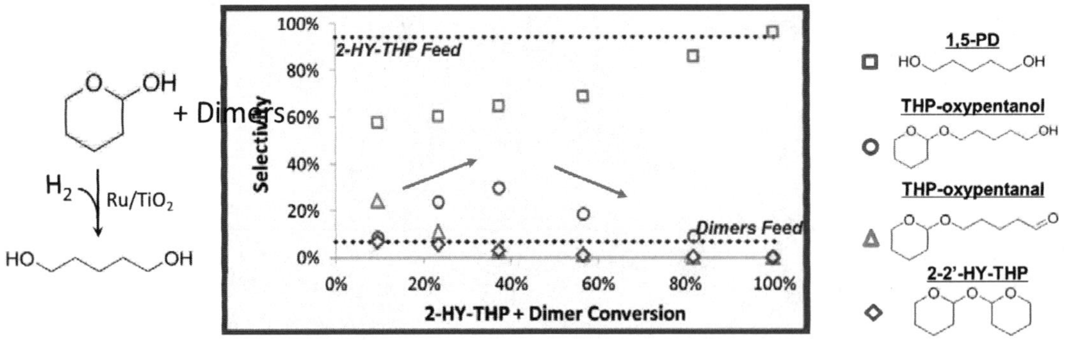

Reaction Conditions: Feed=20wt% 2-HY-THP/H_2O, Catalyst=1% Ru/TiO2, T=120°C, P=650psi H_2

- Dimers increase at low conversion due to shift in equilibrium
- 97% 1,5-PD yields achieved at complete conversion

20

Ring-opening Tautomerization of 2-HY-THP into 5-hydroxyvaleraldehyde: NMR Studies

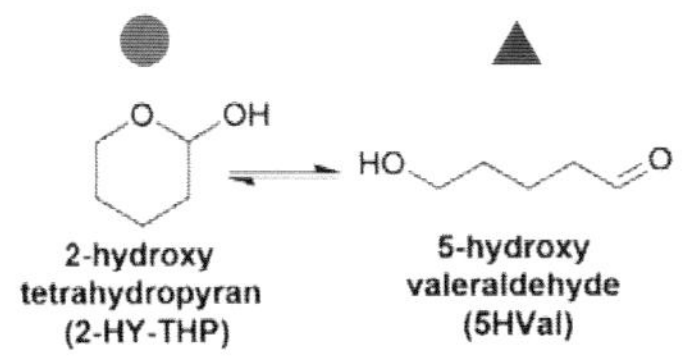

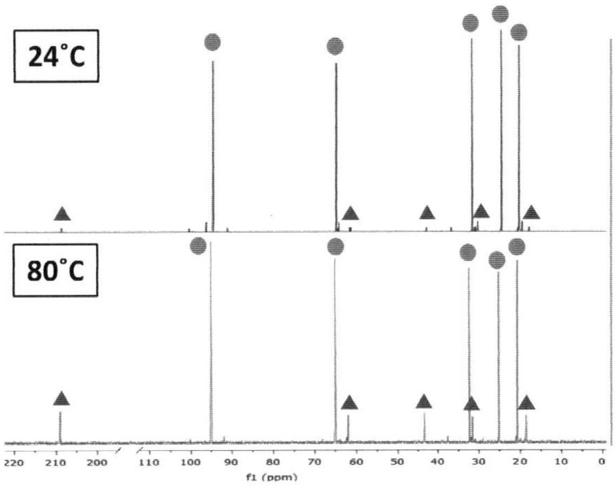

- Variable temperature quantitative ^{13}C NMR studies confirm the existence of 5-hydroxyvaleraldehyde (5HVal)
- Higher concentrations of 5HVal are observed at elevated temperatures
  - Tautomerization reaction is endothermic

21

Ring-opening Tautomerization of 2-HY-THP into 5-hydroxyvaleraldehyde: NMR Studies

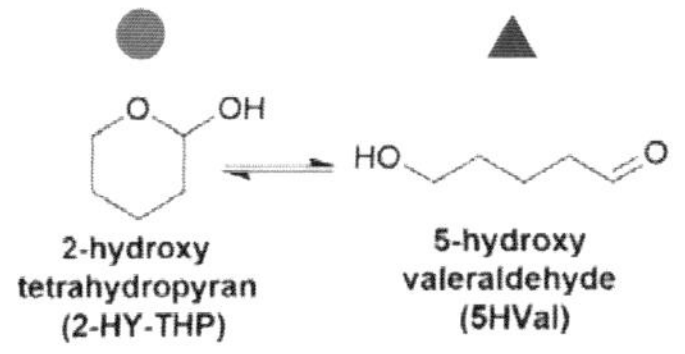

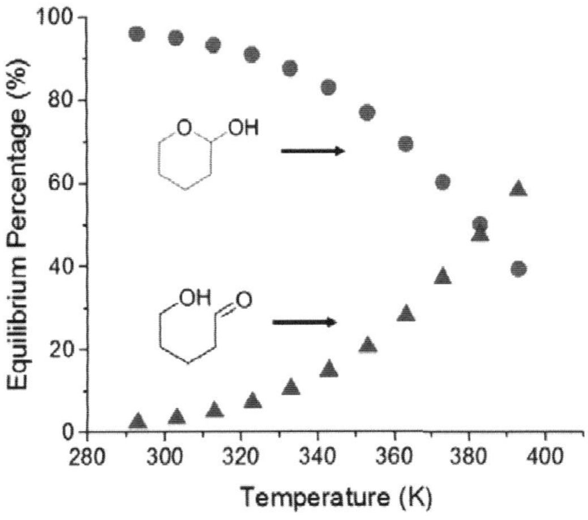

- Variable temperature quantitative ^{13}C NMR studies confirm the existence of 5-hydroxyvaleraldehyde (5HVal)
- Higher concentrations of 5HVal are observed at elevated temperatures
  - Tautomerization reaction is endothermic

22

Tautomerization vs. Hydrogenolysis

- Direct Hydrogenolysis Approach:
  Hydrogenolysis over bimetallic, noble-metal catalysts (i.e. RhRe)

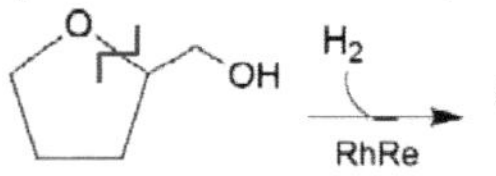

- New Approach: Ring-opening tautomerization and hydrogenation

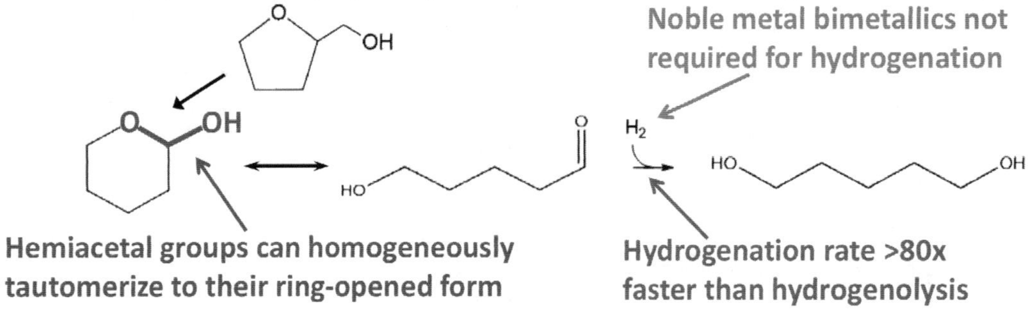

DHH chemistry applicable to C4 and C6 streams

- Gamma-butyrolactone to 1,4-butanediol

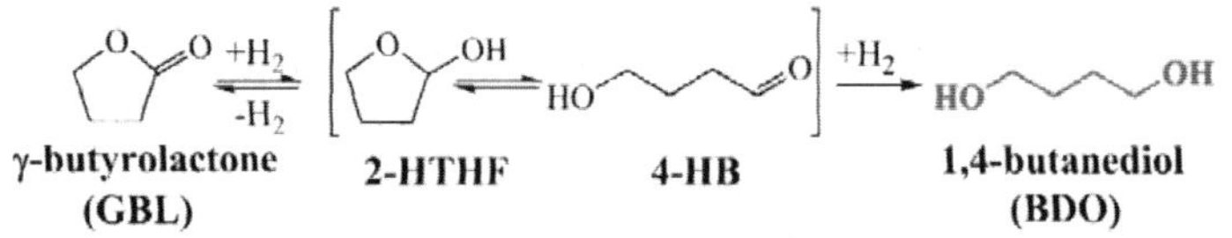

Z. Huang et al. Hydrogenation of γ-butryolactone to 1,4-butanediol over bimetallic $CuCo/TiO_2$ catalysts. *Submitted.*

- Tetrahydropyran-2-methanol to 1,6-hexanediol

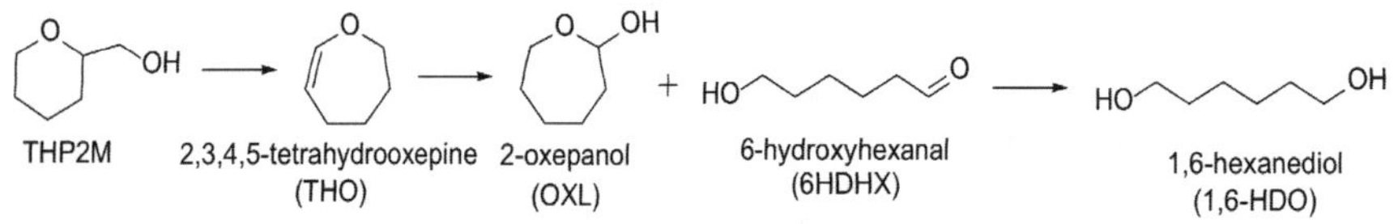

S. P. Burt et al. Production of 1,6-hexanediol from tetrahydropyran-2-methanol by dehydration-hydration and hydrogenation. *Green Chemistry.* ***2017**, 19, 1390.*

DHH Pathway Advantages

	Reactant Concentration	Product Yield
Step 1: THFA Dehydration	100%	90%
Step 2: DHP Hydration: Autocatalytic	50%	98%
Step 3: 2-HY-THP Hydrogenation	50%	99%

87% Overall Yield

- **Step 1:** THFA Dehydration
  - Low-cost γ-Al_2O_3 catalyst
  - Catalyst regenerable
- **Step 2:** DHP Hydration: Autocatalytic
  - DHP hydrates with no catalyst
  - Rate increased by *in situ* formation of acids
  - Low-cost solid acid catalysts
  - DHP readily hydrates due to stable transition state
- **Step 3:** 2-HY-THP Hydrogenation
  - Tautomerization of 2-HY-THP to aldehyde
  - Lower-cost, monometallic catalysts

25

Conclusions

- DHH pathway: 51x lower catalyst cost than direct hydrogenolysis
- High DHP hydration rates over solid acids or with no added catalyst
- Ring-opening tautomerization allows for high hydrogenation rates over lower-cost monometallic catalysts
- 1,5-PD minimum selling price <$2000/ton from furfural
  - Current 1,5-PD price = $6000
  - Historic 1,6-HD price = $3000-4700
- In process of commercializing 1,5-PD process

Pyran

26

Acknowledgements

Advisor: Professor George Huber

Collaborators: Zachary Brentzel, Kefeng Huang, Professor James Dumesic, and Professor Christos Maravelias

Huber Group Members

Funding: Department of Energy - Bioenergy Technologies Office (DOE BETO)

495407 Continuous Condensed Phase Conversion of Ethanol to Higher Alcohols over Bimetallic Catalysts

Abstract:

The production of bio-derived fuels and additives is one of the major avenues for reducing consumption of fossil fuels. Ethanol plays a central role in the bio-fuel arena as a gasoline substitute and oxygenate additive. It can also be used in the production of n-butanol, which has advantages over ethanol in terms of its applications in fuels. Such benefits include higher miscibility with gasoline, more resistance to water contamination, and chemical properties such as energy density, stoichiometric air-fuel ratio, heat of vaporization, research octane number, and motor octane number that are closer to gasoline [1][2]. The general pathway for the production of bio-butanol from bio-ethanol is known as the Guerbet reaction. There have been several studies in recent years for conducting Guerbet reaction using numerous catalysts and reaction conditions [3]. Most of these experiments were done in batch systems or gas phase continuous packed bed reactors. In this study, the previous results obtained from our research group in a batch reactor [4][5] have been extended to a condensed-phase continuous reactor using similar catalysts (8% Ni/γ-Al_2O_3 modified with La_2O_3) and varying reaction parameters such as temperature, WHSV, and feed composition. Also, the catalyst has been modified using bi-metallic (Ni-Cu, Ni-Pt, Ni-Pd, and Ni-Co), and tri-metallic catalysts. Results show that adding Cu to Ni catalyst improves reaction selectivity to higher alcohols because of reduced byproduct gas formation. i. However, a decreases in catalyst activity is observed because Cu has a lower activity for alcohol dehydrogenation that is the initial step in the Guerbet reaction. Ethanol conversion of 40% and higher alcohol selectivity of 80% were observed at optimal condition for the Ni-Cu bimetallic catalysts.

References

[1] R. L. Wingad, P. J. Gates, S. T. G. Street, and D. F. Wass, "Catalytic Conversion of Ethanol to n-Butanol Using Ruthenium P-N Ligand Complexes," *ACS Catal.*, vol. 5, no. 10, pp. 5822–5826, 2015.

[2] H. S. Ghaziaskar and C. (Charles) Xu, "One-step continuous process for the production of 1-butanol and 1-hexanol by catalytic conversion of bio-ethanol at its sub-/supercritical state," *RSC Adv.*, vol. 3, no. 13, pp. 4271–4280, 2013.

[3] D. Gabriëls, W. Y. Hernández, B. Sels, P. Van Der Voort, and A. Verberckmoes, "Review of catalytic systems and thermodynamics for the Guerbet condensation reaction and challenges for biomass valorization," *Catal. Sci. Technol.*, vol. 5, pp. 3876–3902, 2015.

[4] T. L. Jordison, L. Peereboom, and D. J. Miller, "Impact of Water on Condensed Phase Ethanol Guerbet Reactions," *Ind. Eng. Chem. Res.*, vol. 55, no. 23, pp. 6579–6585, 2016.

[5] T. L. Jordison, C. T. Lira, and D. J. Miller, "Condensed-Phase Ethanol Conversion to Higher Alcohols," *Ind. Eng. Chem. Res.*, vol. 54, no. 44, pp. 10991–11000, 2015.

501011 The Effects of Metal-Acid Site Proximity on Bifunctional Isomerization of Alkanes and Deoxygenation of Alcohols

Abstract:

Hydroisomerization of alkanes occurs through kinetic cascades mediated by alkene intermediates that require a (de)hydrogenation metal function within diffusion distances of the kinetically-relevant acid function. Hydrodeoxygenation of alkanols similarly requires a hydrogenating metal function proximal to kinetically relevant acid sites to scavenge dehydrated product alkenes and prevent catalyst deactivation with no influence on rates. In practice, the location and distribution of metal clusters with respect to acid sites—and thus their proximity—is seldom well-characterized because metal precursors are deposited onto bound solid acids that also contain inert metal-oxide binders during synthesis protocols. The accepted intimacy criterion for bifunctional isomerization [1] requires that the metal and acid functions reside at distances (~100 um for mesoporous solid acids) that prevent concentration gradients of alkenes—derived from reactant alkanes in isomerization and formed from oxygenates via dehydration for hydrodeoxygenation—within acid domains. Indeed, for zeolite crystals with an extracrystalline metal function present in the form of Pt/SiO_2, distances of ~150 nm are sufficient to maintain constant alkene reactant concentration within acid domains, as shown by isomerization rates that are unaffected by the size or intracrystalline site density of zeolite crystals. We find, however, that as metal-acid distances approach the nanometer scale via introduction of the metal function as intracrystalline clusters using exchange methods, isomerization rates are markedly higher. These remarkable yet previously unexplained effects of site proximity and their mechanistic underpinnings are addressed in this study. Turnover rates (per H^+) (i) for n-heptane isomerization on Pt-exchanged MFI and on physical mixtures of Pt/SiO_2 with MFI (of varying site density ($[H^+]$) and crystal size) and (ii) for 4-methylcyclohexanol, 2-butanol, and 2-propanol hydrodeoxygenation on physical mixtures of Pt/SiO_2 with FAU and BEA and on physical mixtures of Pt/SiO_2 with Pt-exchanged FAU and Pt-exchanged BEA. We also report turnover rates for probe reactions selective to each of the two functions to assess their monofunctional reactivities as the distances separating them decrease. Acid sites are characterized by turnover rates for methanol dehydration [2], a reaction that sensitively probes acid reactivity and any consequences of changes in acid strength or confining environment as a result of proximity; the metal function is assessed using CO oxidation rates [3], which are able to report on changes in electronic properties of Pt clusters as a result of their proximity to zeolitic protons.

n-Heptane isomerization turnover rates (548 K, 60-101 kPa H_2, 5 kPa n-heptane: ~70 molec. $(H^+ ks)^{-1}$) on MFI (of different crystallite size and site density; mixed with Pt/SiO_2) are invariant with the Thiele modulus, the dimensionless quantity that rigorously describes reactant concentration gradients, until large values of the Thiele parameter (corresponding to metal-acid separation of ~250 nm) at which point rates decrease by about twofold. Thus, smaller metal-acid distances lead to negligible intracrystalline heptene concentrations, the required intimacy criterion [1] has been satisfied for these bifunctional catalysts with metal function present only as extracrystalline Pt/SiO_2. Alkanol dehydration turnover rates (393 K, 2 MPa H_2, 2.5 kPa 4-methylcyclohexanol: ~25 molec. $(H^+ ks)^{-1}$) were measured on BEA and FAU (0.60 and 0.74 nm apertures, respectively, compared to 0.47 nm in MFI [4]). Dehydration turnover rates are much lower than those for n-heptane isomerization, which, taken together with the larger molecular diffusivities afforded by large pore frameworks, precludes the presence of concentration gradients of methylcyclohexene products. Concentrations of products measured are much lower than those predicated by equilibrated reactant dehydration, indicating that further decreases in metal-acid site proximity should be inconsequential. In contrast to these expectations, we measure unexpectedly significant increases (by factors of 2-9) in reaction rate constants (Figure 1a for 2-butanol and 4-methylcyclohexanol on BEA, BEA-Pt/SiO_2, and PtBEA-Pt/SiO_2; Figure 1b for n-heptane isomerization over MFI-Pt/SiO_2 and PtMFI-Pt/SiO_2 for MFI of different site density) for both isomerization and hydrodeoxygenation when Pt clusters and H^+ reside within the intracrystalline framework at distances of ~5-20 nm—as estimated assuming uniform distribution of each of the two functions during synthesis procedures.

Figure 1. **(a)** Measured dehydration turnover rate constants (per H^+) for 2-propanol, 2-butanol, and 4-methylcyclohexanol on BEA, BEA-Pt/SiO_2, and PtBEA-Pt/SiO_2 (393 K; 0.5-2.5 MPa H_2; 0.1-10 kPa alkanol). Inset equations are rate expressions. *2-Propanol dehydration is independent of P(H_2) for PtBEA-Pt/SiO_2. **(b)** Measured n-heptane isomerization turnover rate constants (per H+) on MFI-Pt/SiO_2 and PtMFI-Pt/SiO_2 for MFI of varying proton density (548 K; 60-101 kPa H_2; 0.1-25 kPa n-heptane).

The nanoscale proximity between metal and acid function could potentially result in cooperative behavior changing the inherent monofunctional reactivity. Alkene skeletal isomerization and alkanol dehydration steps limit the respective rates of alkane isomerization and alkanol hydrodeoxygenation. Any potential changes in acid reactivity, caused by modification of acid strength and local environment for acid sites as a result of proximate Pt clusters, were assessed using methanol dehydration turnover rates (per H^+; 433 K) were measured on both MFI and on PtMFI (Figure 2a). Turnover rates and rate constants (inset in Fig. 2a) are the same, consistent with unperturbed reactivity and properties of the acid sites. CO oxidation turnover rates and rate constants (per Pt_s; 443 K) on Pt/SiO_2 and Pt-MFI are also the same (Figure 2b and inset), indicating that the reactivity and properties of Pt clusters are unaffected by their confinement or by their proximity to acid sites within intracrystalline regions. Thus, we conclude that the monofunctional reactivity of the individual metal and acid functions are unaffected by their close proximity.

Figure 2. **(a)** Measured methanol dehydration turnover rate (per H^+) on MFI and PtMFI (433 K). Dashed curve represents average regressed fit to the inset rate expression; regressed values of first- and zero-order rate constants inset. **(b)** Parity plot for CO oxidation turnover rates (per Pt_s) on Pt/SiO_2 and PtMFI (443 K). Rate expression and values of effective rate constants inset. Dashed line represents parity.

These significant rate enhancements for isomerization and dehydration reactions must occur via highly reactive intermediate species. We surmise that these short-lived intermediates are present only at low concentrations and thus require metal clusters within nanometer distance from acid sites. Rate enhancements for 2-butanol and 4-methylcyclohexanol but not 2-propanol dehydration (Figure 1a) shed light on the identity of such reactive intermediates. Additionally, alkanol dehydration turnover rates are independent of $P(H_2)$ on BEA and on BEA-Pt/SiO_2; however, turnover rates for 2-butanol and 4-methylcyclohexanol become inversely dependent on $P(H_2)$ when Pt is present in nanometer proximity from acid sites (Fig. 1a inset). These kinetic data, together with insights from density functional theory, implicate unsaturated alkenone/alkenol analogs of reactants as plausible intermediates for dehydration reactions. Such a pathway is unavailable for 2-propanol because the analogous alkenol cannot undergo dehydration and consequently does not impact its measured rates. The identity of highly reactive intermediates mediating isomerization remains under investigation. Unlike for alkanol dehydration, isomerization rates show no change in $P(H_2)$ dependence with encapsulation of Pt clusters, thus excluding diene-derived intermediates. We continue to apply density functional theory treatments to assess the plausibility of species as alkenederived intermediates for isomerization.

We have addressed previously uninterpreted enhancements in reaction rates as they arise from nanoscale metal-acid site proximity. Such insights can be applied to optimize industrial catalysts for these and other analogous reactions.

References

[1] P.B. Weisz, Adv. Catal. 13 (1962) 137–190. doi:10.1016/S0360-0564(08)60287-4.

[2] A.J. Jones, R.T. Carr, S.I. Zones, E. Iglesia, J. Catal. 312 (2014) 58–68. doi:10.1016/j.jcat.2014.01.007.

[3] A.D. Allian, et al., J. Am. Chem. Soc. 133 (2011) 4498–4517. doi:10.1021/ja110073u.

[4] C. Baerlocher, L. McCusker, Database of Zeolite Structures. http://www.iza-structure.org/databases/.

The authors thank Zhichen (Jane) Shi for measurement of diffusion timescales of zeolite samples. Additionally, the authors acknowledge financial support from Chevron Energy Technology Company and computational resources from the National Science Foundation's Extreme Science and Engineering Discovery Environment (ACI-1053575).

501210 The Influence of Solvent on Acid Catalyzed Dehydration of Model Polyols

Abstract:

The catalytic conversion of biomass and biomass-derived intermediates to energy-dense fuels and chemicals is often carried out in solvents or complex liquid mixtures. The reaction media can play an important role in influencing the catalytic molecular transformations. Herein we examine the catalytic dehydration of different model alcohols in solvent mixtures comprised of water and a polar organic solvent such as DMSO and GVL. Experiments along with theoretical simulations show that the acid-catalyzed dehydration reactions can be significantly promoted without product degradation by the appropriate choice of the solvent system. The net effect yields a significant improvement in both selectivity and reactivity over the reactions in these specific aqueous based systems. Molecular dynamics simulations were carried out to establish the faithful models of the co-solvent systems. These resulting structures provided input for more rigorous ab initio molecular dynamics simulations to follow the free energies changes for the detailed reaction paths involving proton addition, water elimination and proton abstraction to form the corresponding olefin products. The results show that the extent of solvation for different reactant alcohols can be rather different. There are non-uniform changes that occur with increasing the composition of the co-solvent or changing the alcohol reactant. In some cases there is a significantly lowering of activation barriers for the different alcohols while in others the barrier increases. These solvent effects were subsequently utilized to help promote the production of hydroxymethylfurfural (HMF) from fructose, while limiting the subsequent degradation of HMF to levulinic acid. The results provide important insights that may help to provide guidance into solvent selection for the selective production of targeted chemicals, along with the development of new acidic catalytic systems.

496817 Generalizing the Design Principles of Pt-Based Alloy Catalysts with Improved ORR Performance and Durability

Abstract:

Since the discovery of Pt skin type Pt_3Ni catalysts with superior performance towards oxygen reduction reaction,[1] there has been a blossoming of Pt-based alloy catalysts for ORR and various other applications.[2,3] In last couple years, the focus has slowly shifted from catalysts with monolayer skins to those with thicker Pt overlayers for the sake of improving the stability.[4-6] But the decrease in the activity has been observed with increasing overlayer thickness. Thus, how to increase the stability of catalysts, with no harm on the activity, or even simultaneously increase activity, have been the focus of the current catalyst development, but few systematic design principles have been established to guide these efforts. It is worsened by a lacking of atomic-scale understanding of water/electrode interface and consequently the bonding nature of ORR intermediates. For example, it is still in the debate regarding whether hydrated OH or non-hydrated OH is the plausible ORR intermediate.

In the present talk, we will start by determining the atomic-scale structure of oxygenated species at water/electrode interfaces under electrochemical ORR conditions, by combining a careful calibration of Density Functional Theory (DFT)-determined energetics, the liquid water/Pt interfaces generated from *ab-initio* molecular dynamics (AIMD) simulations and a detailed simulation of X-ray Photoelectron Spectroscopy (XPS) signatures.[7] In the second part, we will investigate the structure and the stability of Pt overlayers with various thickness. We will further analyze their activity based on a scheme established through above calibration. Then, general design principles will be proposed for searching for ORR catalysts with improved stability and activity. We will close by comparing the predicted stability and activity of various catalysts with those available in the literature and from our collaborators.

[1] V.R. Stamenkovic, B. Fowler, B.S. Mun, G.F. Wang, P.N. Ross, C.A. Lucas, N.M. Markovic, Science 315 (2007) 493.

[2] J. Greeley, I.E.L. Stephens, A.S. Bondarenko, T.P. Johansson, H.A. Hansen, T.F. Jaramillo, J. Rossmeisl, I. Chorkendorff, J.K. Nørskov, Nat. Chem. 1 (2009) 552.

[3] P. Strasser, S. Koh, T. Anniyev, J. Greeley, K. More, C. Yu, Z. Liu, S. Kaya, D. Nordlund, H. Ogasawara, M.F. Toney, A. Nilsson, Nat Chem 2 (2010) 454.

[4] M. Escudero-Escribano, P. Malacrida, M.H. Hansen, U.G. Vej-Hansen, A. Velázquez-Palenzuela, V. Tripkovic, J. Schiøtz, J. Rossmeisl, I.E.L. Stephens, I. Chorkendorff, Science 352 (2016) 73.

[5] N. Todoroki, H. Watanabe, T. Kondo, S. Kaneko, T. Wadayama, Electrochim. Acta 222 (2016) 1616.

[6] M. Asano, R. Kawamura, R. Sasakawa, N. Todoroki, T. Wadayama, ACS Catal. 6 (2016) 5285.

[7] Z. Zeng, J. Greeley, Nano Energy 29 (2016) 369.

490188 Computational Insights into Thermal and Photoresponse of Pd/C and Pd/N Co-Doped TiO_{2}Â using DFT

Abstract:

TiO_2 has attracted increasing attention during the past few decades as a promising catalytic and photocatalytic material due to its unique physicochemical properties. Specifically, in heterogeneous catalysis, it has been used as a catalytic material for treatment of environmental pollutants (greenhouse gases), photocatalytic reduction of CO_2 with water vapour and photodegradation of dyes. In recent years, doping of transition metal and nonmetal ions in TiO_2 has been reported to enhance the catalytic and photocatalytic performance of TiO_2. The main objective of this work is to study the effect of co-doping on the structure-property relationships of anatase TiO_2 using density functional theory (DFT+U) approach. Palladium ion and carbon (C)/Nitrogen (N) were co-doped in TiO_2 bulk and surface (001) and (100) planes in order to investigate the thermal and photoresponse of TiO_2. Differential bond length formations and associated net oxygen activations (NOA) were determined for the bulk and surface exposed Pd/C and Pd/N doped TiO_2. Effects of anion vacancies on the bond length distributions and net oxygen activations were found for the co-doped structures. NOA was found to be higher for Pd/N co-doped catalysts with 18.8 % and 28% in case of bulk and surface exposed codoped TiO_2 when compared to 17 % and 25 % in case of bulk and surface exposed Pd/C co-doped TiO_2. Band structure analyses were performed in order to examine the photoresponse of co-doped TiO_2. Band gaps obtained from the band structure analyses were in a range of 1.9-3 eV for Pd/C-doped TiO_2 whereas 2.7-3.1 eV for Pd/N-doped TiO_2, respectively, thereby establishing the importance of codoping for thermal and photocatalytic behaviour of TiO_2-based materials.

491746 Transition Metal-Doped C_2N as Active Catalysts for the Oxygen Reduction Reaction

Abstract:

Meeting ever-increasing worldwide energy demands in a sustainable way is one of the most pressing challenges of our times. The oxygen reduction reaction (ORR) plays a critical role in pollution-free energy conversion technologies, such as hydrogen fuel cells and metal-air batteries. Platinum (Pt) is one of the best known catalysts for ORR, but the high cost of this precious metal and the sluggish ORR kinetics inhibit the widespread growth and implementation of fuel cell technologies. [1,2]

Two-dimensional (2D) transition metal-doped carbon nitride catalysts have emerged as abundant and low-cost alternatives to Pt for ORR. [3] A more recent member of this family, C_2N, [4] exhibits several promising features for ORR, including a conductive framework and the potential to create and tune active sites through controlled transition metal doping. [5]

In this work, we use density functional theory (DFT) to investigate the ORR mechanism for C_2N catalysts doped with various transition metals. We find that the choice of metal dopant significantly affects the coverages of the ORR intermediates (*OOH, *O, *OH) on this material. These coverages directly influence the predicted ORR mechanism and overpotentials, which determine catalytic activity. By including explicit solvent layers in our predictions, we also investigate solvation effects on the binding energies. By studying the ORR mechanism and activity of various metal-doped C_2N catalysts, we identify critical material properties that influence the performance of single site 2-D catalysts for electrochemical reactions.

References:

[1] Xia, W., Mahmood, A., Liang, Z., Zou, R., & Guo, S. (2016). Earthâ€Abundant Nanomaterials for Oxygen Reduction.Â *Angewandte Chemie International Edition*,Â *55*(8), 2650-2676.

[2] United States Department of Energy. (2016). Comparison of Fuel Cell Technologies. https://energy.gov/eere/fuelcells/comparison-fuel-cell-technologies.

[3] Zheng, Y. et al. (2017). Molecule-Level g-C_3N_4 Coordinated Transition Metals as a New Class of Electrocatalysts for Oxygen Electrode Reactions. *Journal of the American Chemical Society*, 139(9), 3336-3339.

[4] Mahmood, J., et. al. (2015). Nitrogenated holey two-dimensional structures. *Nature communications*, *6*, 6486.

[5] Li, X., Zhong, W., Cui, P., Li, J., & Jiang, J. (2016). Design of Efficient Catalysts with Double Transition Metal Atoms on C_2N Layer. *The Journal of Physical Chemistry Letters*, *7*(9), 1750-1755.

496687 Electrochemical Synthesis of Hydrogen Peroxide Via Water Electrolysis: Challenges and Opportunities

Abstract:

Electrochemical production of hydrogen peroxide (H_2O_2) from water oxidation could provide a way to produce a chemically valuable product from an earth abundant material. Today there are only a few experimental studies that produce measurable amount of hydrogen peroxide through electrochemical oxidation of water.[1,2] A major challenge is that the electrocatalyst must suppress the thermodynamically favored four-electron pathway (O_2 evolution) and direct selectivity towards two-electron path (H_2O_2 evolution).[3] We have recently developed a complete understanding of water oxidation incorporating the formation of hydroxyl radical, a one-electron water oxidation reaction.[4] Based on thermodynamic analysis, we demonstrated leads to OH radical, H_2O_2 and O_2 evolution and rationalized product selectivity in several different experimental reports.

In this work, by combining the above knowledge and density functional theory (DFT) calculations we search for other materials to identify suitable electrocatalyst with high selectivity towards two-electron path (H_2O_2 evolution).

References

(1) Fuku, K.; Sayama, K.; Fujishima, A.; Honda, K.; Maeda, K. *Chem. Commun.* **2016**, *52* (31), 5406–5409.

(2) Izgorodin, A.; Izgorodina, E.; MacFarlane, D. R. *Energy Environ. Sci.* **2012**, *5* (11), 9496.

(3) Viswanathan, V.; Hansen, H. A.; Nørskov, J. K. *J. Phys. Chem. Lett.* **2015**, *6* (21), 4224–4228.

(4) Siahrostami, S.; Li, G.-L.; Viswanathan, V.; Nørskov, J. K. *J. Phys. Chem. Lett.* **2017**, 1157–1160.

497976 Modeling Electrochemical Reactions:Â DFT-Based Models Including Explicit Solvation, Electrolyte, and Electrochemical Potential

Abstract:

Modeling Electrochemical Reactions: DFT-Based Models Including

Explicit Solvation, Electrolyte, and Electrochemical Potential

Jason D. Goodpaster*

Department of Chemistry, University of Minnesota Twin Cities, Minneapolis, Minnesota 55455, USA

**Corresponding author: jgoodpas@umn.edu*

I report an electrochemical model based on periodic Kohn-Sham Density Functional Theory (DFT) that includes the effects of the electrochemical potential, solvent, and electrolyte. In this model, the electrode surface and all adsorbates are treated using DFT. In the region close to the surface, the solvent is treated explicitly, whereas in the region far from the surface the solvent is treated using a continuum dielectric. The electrolyte ions are implicitly represented by a linearized Poisson-Boltzmann model.[1,2] The electrode potential was calculated from the Fermi energy. The Fermi energy can be varied by changing the number of electrons in the simulation cell; therefore, the electrochemical potential was set by calculating the number of electrons the simulation cell required for a specific Fermi energy.[3] Using this model, free energy profiles for a reaction can be calculated as a function of the applied potential. The free energy of activation for each elementary step can then used to determine the rate coefficient for that step and combined with microkinetic models to determine the rate of product formation.

I applied this model to investigate the mechanistic pathways by which CO_2 on copper and silver surfaces are electrochemically reduced.[3,4] I compare this model to vacuum DFT, implicit solvation DFT, explicit solvation DFT, and electrochemical potential DFT models. Our results demonstrated that including electrochemical potential is critical to accurate calculations of free energy barriers and kinetics for electrochemical reactions. Additionally, we show that for some reactions explicit solvation plays an important role in solvating adsorbed intermediates. Finally, I discuss progress in using these free energies for microkinetic modeling.

References:

1. K. Letchworth-Weaver and T. A. Arias Phys. Rev. B 86, 075140 (2012).
2. K. Mathew, R. Sundararaman, K. Letchworth-Weaver, T. A. Arias, and R. G. Hennig. J. Chem. Phys. 140, 084106 (2014).
3. J. D. Goodpaster, Alexis T. Bell, and M. Head-Gordon J. Phys. Chem. Lett. 7, 1471 (2016).

4. J. D. Goodpaster, Alexis T. Bell, and M. Head-Gordon *in prep.*

491234 Efficiently Determining Solvent Effects at Metal-Water Interfaces

Abstract:

Accurately modeling interfaces between liquid and solid phase is one important frontier in computational catalysis. Explicitly modeling these interfaces can be challenging due to the large number of molecules needed, as well as the large number of possible configurations. Implicit or continuum solvent models can be much quicker, but have only recently been included in plane wave density functional theory (DFT) codes. In contrast, implicit solvation methods are much more mature in molecular DFT codes. How implicit solvation methods perform in modeling periodic catalytic systems is still unclear. In this work we systematically evaluated how interfaces between water can metals can be treated using implicit solvation methods.

In total we considered 41 different adsorbates and used three different solvation methods implemented in the NWChem, VASP, and JDFTx codes. Agreement between the three programs was strong. We classified the adsorbates into five different groups, and found that trends within each group were consistent. For instance, adsorbed atomic or weakly polar species showed almost negligible solvation effects, while adsorbed aromatic compounds were strongly stabilized by the presence of water solvent (up to 0.44 eV). We also modeled four prototypical reactions in vacuum and water: oxygen reduction reaction, formic acid oxidation, C-C cleavage, and water-gas shift. Reaction energies changed up to 0.23 eV when solvent was present. We analyzed the contributing energy terms to the solvation energy of adsorbed species, and found that relevant desciptors (e.g. Bader charge of adsorbates, dipole moments) are correlated to these solvation energy terms. We furthermore created several artificial neural networks, where we evaluated potential descriptors for predicting solvation energies of adsorbates. Our best neural network indicated that Bader charge of adsorbed species, solvation energies of gas species, gas dipole moments, and molecular surface area of adsorbates are all simple descriptors that can be used to predict solvation energies of adsorbates. Our work provides guidelines on when solvation effects may be important for surface chemistry, and also provides valuable insights on modeling such effects in an efficient manner using implicit solvation.

498550 Machine-Learning Model Development for Electrocatalyst Discovery

Abstract:

Machine-Learning Model Development for Electrocatalyst Discovery

Zheng Li, and Hongliang Xin*

*hxin@vt.edu

Session: Catalysis and Reaction Engineering Division- Electrocatalysis and Photoelectrocatalysis

Transition metal alloys, metal oxides and molecular complexes have shown great promise for catalyzing many chemical reactions. For a given type of catalyst, its surface reactivity can be tailored by varying geometric characteristics, metal ligands, organic functional groups and extrinsic factors (e.g., solvation, support) in the vicinity of active sites [1]. With hierarchical complexities in catalyst design, a priori estimation of chemical reactivity of surface atoms is attractive. However, it is very time-consuming and costly, if at all possible, to search for highly optimized catalyst structures by the empirical testing and/or quantum-chemical calculations. Therefore, we tackle this problem by developing the predictive, data-driven machine-learning-augmented models (e.g., artificial neural network, kernel ridge regression) that can unravel the complex, non-linear interactions between the electronic structure properties with reactivity descriptors, thus enabling fast and accurate predictions of surface reactivity of a broad range of material space without resorting to the expensive quantum-chemical calculations.

Since the efficiency and accuracy of machine-learning models are strongly dependent on the input features, we develop various types of data representation of catalyst surfaces. For metal alloys and metal oxides, we engineer numerical representation of surface metal atoms by using geometry and electronic structure based primary features [2-4] including the local electronegativity and the d-band features that are dependent on the surroundings of an adsorption site, together with the intrinsic properties of active metal atoms including the electronegativity, ionic potential, electron affinity, etc. For the description of molecular complexes, we apply the Coulomb matrix [5], which is a simple matrix form crystal structure representation based on the molecular complex structure coordination and atoms physical properties. Another type of representation method for molecular complexes is based on the similarity indexing of molecular complexes with the associating functional groups and structure components. The machine-learning-augmented models exhibit outstanding prediction power in screening for the bimetallic, metal oxide and molecular complexes surfaces with desired adsorption properties. Compared with the traditional high-throughput computational and experimental trial-and-error approach, the machine-learning models have great potential in accelerating the discovery of new catalytic materials with desired properties. Sensitivity analysis of the machine-learning models can shed light on the nature of chemical bonding and its governing factors, providing chemical insights into catalytic processes and fundamental knowledge base of molecule-catalyst interactions.

[1] Vojvodic, Aleksandra, and Jens K. Nørskov. 2015. "New Design Paradigm for Heterogeneous Catalysts." *National Science Review* 2 (2): 140–49.

[2] Z. Li, X. Ma, and H. Xin, Feature engineering of machine-learning chemisorption models for catalyst design, Catalysis Today. 280, Part 2 (2017) 232–238.

[3] X. Ma, Z. Li, L.E.K. Achenie, and H. Xin, Machine-Learning-Augmented Chemisorption Model for CO2 Electroreduction Catalyst Screening, J. Phys. Chem. Lett. 6 (2015) 3528–3533.

[4] Z. Li, S. Wang, W. Chin, and H. Xin, Rapid Screening of Bifunctional Alloy Catalysts Enabled by Machine Learning, Submitted to Journal of Materials Chemistry A. (2017).

[5] M. Rupp, A. Tkatchenko, K.-R. Müller, and O.A. von Lilienfeld, Fast and Accurate Modeling of Molecular Atomization Energies with Machine Learning, Phys. Rev. Lett. 108 (2012) 58301.

494685 Evaluating Combined Heat and Power Deionization Systems for Efficient Water Reuse at Thermoelectric Power Plants

Abstract:

Efforts to increase both the United States and global water supply has primarily taken place through seawater and brackish water desalination. According to the International Desalination Association, the United States capacity for desalination increased in 2013 by 50% from 4 million to 6 million m^3/day. Unlike most areas of the world, nearly 77% of all desalination in the United States comes from the treatment of brackish water. Brackish water has lower salt concentrations (~<10 g/l) than seawater (35 g/l) and thus the energy consumed to remove salt ions in theory is much less than seawater desalination. Current technologies utilized for seawater desalination, reverse osmosis (RO) and Membrane distillation (MD), require a significant amount of energy regardless of the type of water (brackish or seawater). Furthermore, these technologies have approached their thermodynamic energy limits (1-3 kWh m^â€“3^), indicating that further energy savings cannot be achieved. RO and MD are energy intensive technologies because they operated through extracting the main component water (H_2O) from the salt (NaCl). Alternatively, capacitive deionization (CDI), an emerging separation-based technology, operates by adsorbing the minority component (NaCl) from the feeding brackish water. Traditional CDI cells operate in a cyclic process where the capacitive electrodes are brought in contact with a concentrated salt solution (e.g. brackish water) and charged. Once the charged electrodes are fully saturated with salt ions, the electrodes are regenerated through discharging the electrodes in a brine solution. Analogous to a P-V diagram for traditional heat engines, the enclosed area of the charge versus voltage represents the energy that is consumed during this charging-discharging process. Until recently the CDI three step intermittent process has been relatively untouched in terms of literature and system operation. In this study, we explore the effect of temperature on salt removal with CDI. Experimentally, we use traditional electrochemical characterization methods, cyclic voltammetry (CV) and electrochemical impedance spectroscopy (EIS), to evaluate the capacitance and resistance changes with temperature. Here it was found that through increasing the operating temperature from 25°C to 45°C, the specific capacitance of carbon electrodes increased from 34.60±2.15 to 42.56±3.57 F/g, and the cell resistance decreased from 20.20±2.97 to 15.57±2.52 ohm. Additionally, the salt adsorption capacity decreased from 13.82±1.13, to 6.58±1.01 milligram NaCl/mg carbon. This indicates that although the electrochemical performance of CDI can be increased with temperature, the ion storage within the electric double layer is hindered at higher temperatures.

493557 Biochar Amendments for Increased Crop Yields: How Can Biochars Improve Crop Nutrient Availability?

Abstract:

Background:

Maintaining a sustainable food supply for the foreseeable future without excessive environmental degradation will require novel engineering approaches for remediating the human disruption of the nitrogen cycle. Over the past several decades, food production has been able to keep pace with human population growth only thanks to the development of new high-yielding crop varieties grown with the help of fertilizers. Thus, vast amounts of energy are consumed every year to produce the approximately 200 Mt of ammonia needed to make the nitrogen (N) fertilizers used by farmers around the globe.

Unfortunately, more than half of the N fertilizer applied to a field is not available for plant growth due to losses caused by surface runoff, leaching into surface and ground water, or volatilization. Because of these losses, the increased use of nitrogen fertilizers has been linked to a variety of water pollution problems ranging from the expansion of the hypoxic zone in the Gulf of Mexico caused by eutrophication to the contamination of wells and groundwater with nitrate N. Moreover, denitrification reactions convert nitrates to nitrous oxide, a major greenhouse gas and nitrogen oxide, a major atmospheric pollutant and significant contributor to the depletion of atmospheric ozone.

The amendment of soils with biochar has been heralded as a sustainable method for increasing crop yield and preventing pollution problems caused by fertilizer runoff. However, the available experimental results are highly variable [1-3]. While some studies reported significant beneficial effects with up to 60% higher yield after biochar addition, other studies reported that biochar amendments either had no effect or even resulted in up to 30% lower agricultural productivity [3]. One of the most important reasons for this variability is that biochar is not a single entity. Dozens of biomass feedstocks are used to produce biochars in multiple types of reactors under varying temperature and oxygen conditions, leading to thousands of biochars with widely varying properties. To complicate things even more, soil properties, climate conditions (like rainfall and temperature), plant requirements and many other parameters vary from application to application. As a result, such studies and meta-analyses have not provided much insight into the fundamental mechanisms that modulate the environmental performance of biochar amendments.

Our Previous Work on this Problem:

To begin bridging this knowledge gap, we focused our recent research on the mechanisms that control the ability of biochars to improve crop nutrient availability, one of the main mechanisms proposed to explain crop yield increases in biochar-amended soils [2].

As a first step in this direction, we have recently developed and tested a mathematical model that uses first principles to describe nutrient transport in soils amended with biochar [4]. Using a simple Langmuir isotherm to describe nutrient adsorption, our simulation results revealed that nutrient transport in biochar-amended soils is modulated by a complex interplay between the physical mechanisms governing flow in porous media and nutrient transport, and the chemical mechanisms modulating nutrient adsorption of the pore surface of biochar solids. Nutrient transport rates through the topsoil decreased when biochars with the proper adsorption capacity and affinity were mixed with the soil. This was due to the differential dynamics of the adsorption and desorption processes. Nutrient adsorption was significantly faster than desorption, with biochar essentially acting as a slow (or controlled) release medium for the adsorbed nutrient [4].

Current Work and Summary of Results:

The present study extends our previous work with a systematic study of the adsorption and desorption dynamics on a much broader class of biochars. We report the development of a mathematical model that describes the response of biochar particles to a pulse of nutrient that simulates a single fertilizer application. Nutrient adsorbs on the biochar particle when the fertilizer solution flows past the particle and starts desorbing when the fertilizer solution is replaced by pure water. Two nutrient transport mechanisms are considered: (a) diffusion through liquid-filled pores and (b) surface diffusion. We assume that nutrient adsorption follows a Sips isotherm whose parameters can be varied to describe adsorption on both homogeneous and heterogeneous solids. The Freundlich and Langmuir isotherms commonly used to describe adsorption of ammonium or nitrates on biochars are special cases of the Sips isotherm. The highly nonlinear partial differential equations describing this transient adsorption/desorption problem are discretized using high-order finite differences and the resulting system of first order differential equations are integrated with a variable-step, variable-order integrator based on numerical differentiation formulas.

As noted previously, a "good" biochar should be able to quickly adsorb the nutrient when it is present in the liquid flowing around it, but it should desorb it much more slowly in order to act as slow (or controlled) release medium. This is the mechanism through which a "good" biochar will improve crop nutrient availability [2, 4].

Our simulations clearly show that the ability of a biochar to slowly release the adsorbed nutrient depends on a complex interaction of (a) external mass transfer, (b) intraparticle diffusion (both pore and surface diffusion), and (c) adsorption dynamics (affinity constant and adsorption capacity).

The interaction between external mass transfer and intraparticle pore diffusion is a key factor for modulating the release of nutrient that has adsorbed on the biochar. Our system is characterized by low values of the Biot number, the dimensionless number that represents the ratio of external mass transport rate over the intraparticle diffusion rate. This is because of the low interstitial water velocities (10-90 cm/day) in biochar-amended soils. Low Biot numbers are one of the key factors responsible for the long desorption tails that characterize the slow release of nutrients from "good" biochars.

Our simulations reveal that all the processes mentioned above can significantly affect the nutrient release rates. For example, surface diffusion accelerates nutrient desorption from small (2 mm in diameter) and highly porous biochar particles (0.6 total porosity). But, surface diffusion will slow down desorption from larger (4-8 mm in diameter) and less porous particles, even though the latter biochar particles may have exactly the same adsorption isotherm parameters. Adsorption affinity and site heterogeneity can also interact with porosity and particle size to modulate nutrient desorption. Finally, surface diffusion again improves the ability of a biochar to act a slow release medium at high interstitial velocities.

By revealing that the response of biochar particle to a pulse of nutrient is governed by complex interactions among external mass transfer, intraparticle diffusion and adsorption, this study provides new insights into the fundamental mechanisms that influence the environmental performance of biochar amendments. It may also explain why biochars that share some key properties (e.g. carbon content) or produced at the same pyrolysis temperature can potentially have widely different impacts on crop yields, as has been repeatedly reported in the literature [2,3]. Much better characterization of biochar properties (porosity, adsorption isotherm, intraparticle transport mechanism) and more precise control of amendment or soil parameters (like particle size, biochar aging, soil permeability) are necessary before we can fully elucidate the mechanisms of biochar action.

References:

[1] Spokas, K. A.; Cantrell, K. B.; Novak, J. M.; et al. "Biochar: A Synthesis of Its Agronomic Impact beyond Carbon Sequestration." *J. Environmental Quality* **41**(2012) 973-989.

[2] Jeffery, S; F.G.A. Verheijena; M. van der Velde; A.C. Bastos. "A quantitative review of the effects of biochar application to soils on crop productivity using meta-analysis." *Agriculture, Ecosystems and Environment* **144**(2011) 175–187.

[3] Crane-Droesch, A; S. Abiven; S. Jeffery; M.S. Torn. "Heterogeneous global crop yield response to biochar: a meta-regression analysis." *Environmental Research Letters* **8**(2013) 044049.

[4] Sun, S; C. E. Brewer; C. A. Masiello; K. Zygourakis. "Nutrient Transport in Soils Amended with Biochar: A Transient Model with Two Stationary Phases and Intraparticle Diffusion" *Industrial and Engineering Chemistry Research*, **54** (2015) 4123–4135.

501354 Moisture Retention in Emulated Soil Micromodels: Development and Performance of Sustainable Agriculture Biotechnology

Abstract:

Soil bacteria living near plant roots secrete exopolysaccharides (EPS), a polymeric substance which can help modulate soil moisture and promotes plant growth. EPS functions in several different ways: (i) EPS behave as hydrogel, swelling during wet soil conditions and remaining hydrated during dry soil conditions; (ii) EPS promotes aggregation of soil particles and thereby promotes water retention in dry soil conditions through stronger capillary forces; and (iii) EPS alters surface properties of soil, creating more water repellent surfaces to inhibit rapid re-wetting in wet soil conditions. EPS functions in concert with the physical microstructure of real soil; therefore, it is necessary to study EPS performance in physically realistic micro-structures. Here we show pore-scale changes in moisture distribution within emulated soil micromodels that have systematically varied physical microstructures and surface water repellencies, and can be loaded with well-defined pore solutions. Experimental results showed dilute EPS solutions took twice as long to dry compared with pure water in identical micromodels. Microstructure geometry was found to influence drying rate over a range of pore saturation values as well as control the spatial distribution of liquid and gas phases within unsaturated soil micromodels. Deionized water took four times longer to completely dry from micromodels with higher water repellency surfaces compared with micromodels with lower water repellency surfaces. These results illustrate the various mechanisms whereby EPS impacts the spatial distribution and evaporation resistance of soil moisture. We provide a systematic method to evaluate performance of natural EPS solutions in modulating soil water and promoting resiliency of terrestrial ecosystem; we anticipate our approach may speed the development of sustainable agriculture biotechnology to enable food production using less water.

498287 Molecular Modeling and Simulation Studies of the Structural and Energetic Evolution during Dehydration of Food Systems

Abstract:

From a molecular point of view, foodstuffs are mostly comprised of biopolymeric molecules that form highly porous structures in which solvents, solutes, additives, and other natural or artificial ingredients are accommodated. Because of the complexity of the food structures and compositions and because of the sensitivity of the pore structures and properties of food to physical and chemical changes arising externally or internally, foodstuffs are one of the most complex types of soft materials and are inclined to exhibit significant structure-property relationships in the physicochemical, functional, processing, and nutritional aspects of food science and engineering. In this respect, water (moisture) is a major factor in determining the properties of foods in various stages as it is not only a major component of food materials but also acts as space filler for stabilizing food structures, solvent for dissolving additive molecules, medium for heat and mass transfer, and participant in the chemical and/or biological reactions in food systems. .

One of the important ways for preparing and preserving foods is dehydration (water removal), which is a process that requires energy to be supplied to enable water molecules to break loose from their interactions and escape from the porous structures of the foods. The process of food dehydration and the preservation of dehydrated foods can thus be directly related to the interactions of water molecules with food macromolecules, solute ions, and additive molecules, because these interactions dictate the energy requirement for water removal during food dehydration as well as the structures of foods before, during, and after dehydration. Moreover, the magnitudes of these interactions depend on the densities and intermolecular distances of mutually interacting molecules, rendering the water molecules in foods to exist in different energetic states and making single system-wide thermodynamic parameters (e.g, water activity) inadequate in representing the diverse states of water molecules in food porous structures and in elucidating the microscopic and macroscopic structural evolution of food materials caused by dehydration. Capable of tackling these challenges is the molecular dynamics (MD) method which employs molecular-based interaction potentials to construct a model system that provides a satisfactory physicochemical representation of the real system and allows various structural and dynamic properties to be studied at a molecular level. Recently, we have employed MD modeling and simulations to construct an amylose-based food material and study the dehydration process in a continuous dynamic manner by periodically examining the energetics of all water molecules at the water-vapor interface and removing the one with the highest energy from the model food system. It was found that the strong interactions of water molecules with the food polymer stabilize the porous structure of the food material and makes the dehydration energy requirement to be greater than that involved in pure water vaporization. They also cause the water interaction energetics to be highly nonuniform in all three spatial dimensions. As dehydration proceeds, the surface tension of the descending water interface coupled with strong water-polymer interactions exert downward stresses on the food polymer to induce polymer conformational compaction and food structural reduction.

We further added two functional food ingredients to the model food systems in our MD simulation studies. The first one is potassium sorbate, which is a commonly used food preservative against the growth of bacteria, fungi, molds, and yeast in foods. The second one is fructose-1,6-diphosphate (FDP), which is a bioprotective agent and a promptly available high energy substrate and can promote growth and improve health. The stability of the food porous structures was found to be enhanced by the presence of the food additive molecules which function as additional fillers in the food pore space and interact strongly with the polymer chains and water molecules of the food material.Â During dehydration, the descending water level tends to accumulate the additive molecules, in particular sorbate with a nonpolar hydrocarbon moiety, at the vapor-liquid interface, which helps to form dense clusters of food polymer chains at the interface, reduces the level of structural reduction, but increases the energy requirement for water removal.Â The additive molecules in the interior regions of the food structures have strong tendencies to stay close to food polymer and hence exhibit nonuniform density distributions as well. Extensively dehydrated food porous structures were found to have substantially improved stability due to the water molecules remaining in the pore structure that effectively â €œbindâ€ the food polymer and additive molecules together. Which makes it very difficult to recover the original pore structures of the food systems that existed before the commencement of dehydration. The findings of these MD studies suggest that the original pore structures could be reduced less during dehydration and could be recovered more effectively by rehydration, if (i) the food material has a higher polymer density, (ii) solute or additive molecules have elongated shapes to act as cross-linkers, and (iii) the effects of water surface tension can be minimized or eliminated using special dehydration methods such as freeze drying. The MD modeling and simulation approach developed in this work could form a basis for systematic evaluation of food porous structures and their structural and energetic evolution and mechanisms in molecular detail and for providing scientific explanations to the macroscopic trends and empirically observed phenomena during food dehydration.

493109 Metallic Membranes for N_2 Separation and NH_3 Production

Abstract:

The availability of nitrogen, phosphorus, and water are the three main factors that limit our ability to produce enough food to feed the growing population of the planet. The nitrogen cycle is one of the most significant biogeochemical cycles on Earth, as nitrogen is an essential nutrient for all forms of life. Although freely available in the atmosphere as dinitrogen (N_2), access to fixed forms of nitrogen constitutes, in many cases, the most limiting factor for plant growth. The industrial production of ammonia for fertilizers via the current Haber-Bosch process is an energy-intensive process that consumes 1-2% of the worldâ€™s annual energy supply. For these reasons, the need for advanced catalytic methods for the reduction of N_2 to ammonia remains a requirement for sustainability in the food, energy and water systems cycle.

The aim of this work is to explore the potential of metallic membranes for N_2 separation with the final intent to produce NH_3. Based on a preliminary theoretical investigation using density functional theory, the Group V transition metals (e.g., vanadium (V), niobium (Nb) and tantalum (Ta)) show strong affinity toward N_2. Moreover, from solubility and diffusivity values taken from the literature, iron (Fe) is a suitable fit for this application. Therefore, V, Nb, Ta and Fe foils with a thickness of 40 Î¼m have been chosen as membrane materials in this study.

Permeation tests with pure gases (He, N_2, CO_2 and CH_4) have been performed to characterize the membranes in terms of the N_2 permeating flux and ideal selectivity at different temperature and pressure conditions, varying from 400 to 600 Â°C and from 2.0 to 60 bar, respectively. Moreover, scanning electron microscopy (SEM) and an electron microprobe analyzer (EMPA) have been used to investigate the effect of the different operating conditions on the membrane surface. From the experimental results, all of the membranes show infinite selectivity towards N_2 permeation at each operating condition investigated and V showed better performance with respect to Nb, Ta, and Fe.

The V membrane is housed in the membrane reactor where a H_2 stream is used as the sweep gas to promote the NH_3 reaction. Lower pressure than the conventional Haber-Bosch process is used. Specifically, an operating pressure of 60 bar, instead of 200 bar (Haber-Bosch process), and a temperature of 500 °C are used as operating conditions to produce NH_3. The performance of the membrane reactor in terms of NH_3 conversion is, hence, investigated.

485442 Capacitive Deionization of Brackish Water for Irrigation and Energy Storage

Abstract:

Irrigated agriculture water represents the bulk of demand for water in most of the countries suffering from acute water scarcity in the Near East and North Africa, as well as Mexico, Pakistan, South Africa, and large parts of China and India. Greenhouses are emerging solution. Current testing in Mozambique and Sierra Leone has demonstrated significantly reduced water use. Yet even a few gallons per day per greenhouse is challenging because many local water sources are brackish. Capacitive deionization (CDI) is a sustainable, energy efficient, and cost effective technology for removal of low or moderate salinity typical of brackish water as compared to technologies that remove the water instead. Indigenous production of the carbon for the electrodes holds the key to such technology being adopted as a source of irrigation water, simultaneously facilitating the use of such using renewable energy sources. Biochars have recently been explored as electrode materials in CD, and can be easily produced from local plant-based agricultural waste materials.

The goal of this study is thus to identify biomass process conditions to enable the best performance as electrode material in CDI. Studies on CDI so far have been targeted towards achieving the best adsorption capacity, which directs towards the use of highly processed and tailored carbons. This is mostly achieved through chemical activation processes using toxic or corrosive and expensive materials which require deionized water for subsequent washing of the carbon, for achieving better CDI performance. Hence, with the effort directed towards making the process viable for use locally in regions with acute water scarcity, a change in the approach to tailor the electrode material for use in CDI is proposed. This approach is to determine the best possible material utilization by ensuring maximum possible contact between the liquid and the biochar surface (wettability) and by generating the least Ohmic loss thus resulting in highest electrical efficiency.

498868 High Performance CO_2 Electrolyzers

Abstract:

This paper will consider the combination of CO_2 electrolysis and GTL processing as a way to convert waste CO_2 into useful fuels and chemicals. First electrolysis is used to convert the CO_2 to CO and O_2

The main emphasis of the talk will be on the CO_2 electrolysis. A cell with optimized cathode and anion exchange membrane has run for over 3500 hrs at 200 mA/cm^2 and 3V at room temperature with stable high selectivity of ~95%. Currents rise to 600 mA/cm^2 at 50°C. A new flow-field design has been created that allows large cells to be easily manufactured Solid. Initial results on coupling the devices to fuel production systems will also be discussed. These results pave the way for the commercialization of CO_2 conversion to useful fuels.

500835 CO Conversion to Selected Chemicals Using Electrochemically Generated CO from CO_2

Abstract:

Dioxide Materials has developed new electrochemical cell technology capable of generating CO from the electrochemical reduction of CO_2 at high current density and Faradic efficiency. The electrochemical cell operates in a current density range of 200 to 600 mA/cm^2 at voltages in theÂ 2.9 â€“ 3.2 V range. The conversion selectivity of CO_2 to CO is in the range of 95% to 98%.

The electrochemical cell utilizes a two-compartment design, consisting of an anode and cathode compartment that are separated by a special anion exchange membrane. The anode consists of a GDE utilizing a nanoparticle IrO_2 catalyst deposited onto a carbon fiber paper support and using a potassium bicarbonate anolyte solution that is recirculated in anode flow field. The cathode utilizes a nanoparticle Ag-based catalyst in combination with a Sustainion^{TM} ionomer that is deposited on a GDE carbon paper cathode to efficiently reduce the CO_2feed to CO at a low cathodic overpotential.

The special membrane utilized in the cell design is a highly conductive Sustainionâ„¢ X-37 anion exchange membrane, developed by Dioxide Materials, which allows the transport of OH^- and HCO_3^-ions from the cathode to the anode.

The surprisingly high conversion performance and current density of this electrochemical CO cell at low voltages allows us to consider the use of this technology, when scaled-up, in potentially producing selected renewable chemicals when operating on renewable energy sources such as solar, wind, and nuclear.

In this presentation, we will discuss the details of the new electrochemical cell technology in addition to identifying the production of potential renewable chemicals such as methanol and acetic acid using the cell CO product.

482789 Nano-Engineered Catalyst for the Utilization of CO_2 in Dry Reforming to Produce Syngas

Abstract:

The objective of this study was to develop a novel catalytic reactor containing nano-engineered catalyst for the utilization of CO_2 (captured from coal-fired power plants) in dry reforming of methane (DRM) ($CO_2 + CH_4$ â†' 2 H_2 + 2 CO) to produce synthesis gas (syngas). The technology aims to reduce CO_2 emissions by developing beneficial uses for CO_2 from coal-fired power plants. It also offers an alternative to mitigate CO_2 emissions in areas where geologic storage may not be an optimal solution and/or utilization could significantly offset the costs of carbon capture and sequestration.Â

The nano-engineered Ni-based catalyst was prepared by atomic layer deposition (ALD). The Ni particles were as small as ~2-4 nm. The nano-engineered catalyst supported on porous Î³-Al_2O_3 particles showed extremely high methane reforming rates at different temperatures (1840 $Lh^{-1}g_{Ni}^{-1}$ at 850 °C, 1740 $Lh^{-1}g_{Ni}^{-1}$ at 800 °C, 1320 $Lh^{-1}g_{Ni}^{-1}$ at 750 °C). These are the highest DRM reaction rates reported to date as compared to data in the literature. The Ni-based ALD catalyst also showed good stability in DRM reaction during a 300-hour continuous operation at temperatures that ranged from 700°C to 850°C. The good stability is due to a strong bonding between the nanoparticles and substrates since the particles were chemically bonded to the substrate during the ALD process. The high thermal stability maintains high dispersion of Ni nanoparticles, which can inhibit coke formation, because their step edges are small enough to limit carbon nucleation and growth.

498262 Ethylene Production from Shale and Coal Derived Flue Gases

Abstract:

As an inexpensive and abundant source of energy, coal, provides ~41%[1] of the worldâ€™s electricity needs. Yet, coal based power plants are under serious scrutiny due to their damaging impact on the environment. In 2015, although coal contributed to 33%[2] of the total electricity production in the U.S., it was responsible for 71%[3] of the total CO_2 emission from the U.S. electric power sector for the same year. One of the most attractive approaches to combat high GHG emission from coal-fired plants is CO_2 capture and conversion to higher value products.

Contemporaneously, thanks to the shale gas growth in the U.S., an excess supply for pure (extracted) ethane and other light hydrocarbons is available currently at low cost. As a result, an increasing number of steam cracker plants are using ethane as raw material to produce ethylene, the highest volume petrochemical intermediate produced in the world with a current global production capacity of 334 billion lb/year. Ethylene production from steam cracking is also an energy intensive process. A typical ethylene steam cracker plant consumes ~17-21 GJ/ton ethylene energy and emits ~1.0-1.2 ton CO_2/ton of ethylene[4].

Southern Research (SR), in partnership with DOE/NETL is developing a low temperature ethylene production process using CO_2, notably from coal-fired flue gas streams, and readily available low cost ethane as raw materials. The proposed process proceeds through a less energy intensive oxidative dehydrogenation pathway by using SRâ€™s recently developed low-cost mixed metal oxide catalyst which balances the functionalities between ethane activation and ethylene selectivity with a goal to ensure yields (>60%) higher than reported in literature. The process is also anticipated to benefit from the presence of several components of the flue gas from coal-fired plants and ensure lower downstream separation costs.Â IT is anticipated that this new process would (i) reduce the reaction temperature for ethylene production by at least 200°C from the conventional steam cracking and (ii) reduce overall GHG emission by ~35% for the combined coal-fired power plant and ethylene production process. With an existing SR catalyst formulation, high selectivity (99%) and high turn-over rates have already been achieved. This paper will discuss the catalyst screening and key performance results for further optimizing catalysts. The studied parameters include reaction temperature, ratio of ethane to oxidant and oxidizing agents (CO_2/O_2), catalyst deactivation, and impact on separations as a result of product selectivity and yield. References

[1] http://data.worldbank.org/indicator/EG.ELC.COAL.ZS, [2] Electric power monthly, Jan 2017, U.S. Energy Information Administration, [3] Monthly energy review. August 2016. U.S. Energy Information Administration. www.eia.gov/mer. [4] Ren, Tao, Martin Patel, and Kornelis Blok. *Energy* 31.4 (2006): 425-451.

503439 CO_2 Utilization in the Production of Ethylene Oxide

Abstract:

A conventional hydrocarbon feedstock that is of particular interest for CO_2 utilization is ethylene, which is used to make ethylene oxide. Ethylene oxide is an important feedstock for the chemical industry and is used to make many useful products such as polyurethanes, polyols, glycols, nitriles, alcoholamines, and ethers. Ethylene oxide is manufactured by several closely related industrial processes by major chemical makers worldwide (e.g., DOW Chemical, Japan Catalytic Company, Shell International Chemicals, Sumitomo Chemical, BASF, Scientific Design).

The chemical methodology by which ethylene oxide is currently produced is the partial oxidation of ethylene using oxygen (this process is called epoxidation). A silver-based catalyst is used, and either air or preferably oxygen is the oxidant. Ethylene epoxidation is performed at lower temperatures (220 to 280°C) compared with other selective oxidation processes and at high pressure (10 to 30 bar).[1]

The Captured CO_2 Catalyst for the Production of Ethylene oxide (C3-PEO), a technology being developed at RTI International, aims at producing ethylene oxideâ€"a high-value chemicalâ€"while consuming CO_2â€"a greenhouse gas. This technology is based on novel catalysts utilizing the following key discoveries:

1. Abstraction of oxygen from CO_2 using the reduced mixed-metal oxide catalysts
2. Transfer of the abstracted oxygen from CO_2 to react with hydrocarbons to form desired products
3. Operation at temperatures which make these catalysts commercially practical to produce ethylene oxide

   Our proposed process uses a cofeed reactor as shown in Figure 1. The proposed ethylene oxide process will have a concentrated CO byproduct stream, which could be used for the manufacture of many products such as methanol, dimethyl ether, acetic acid, acetic anhydride, vinyl acetate, styrene, terephthalic acid, formic acid, *n*-butanol, 2-methylpropanal, acrylic acids, neopentylacids, propanoic acid, dimethyl formamide, and Fischer-Tropsch hydrocarbons.[2] Therefore, the two marketable product streams from the proposed process are ethylene oxide and CO, which are both valuable intermediates for the petrochemical industry.

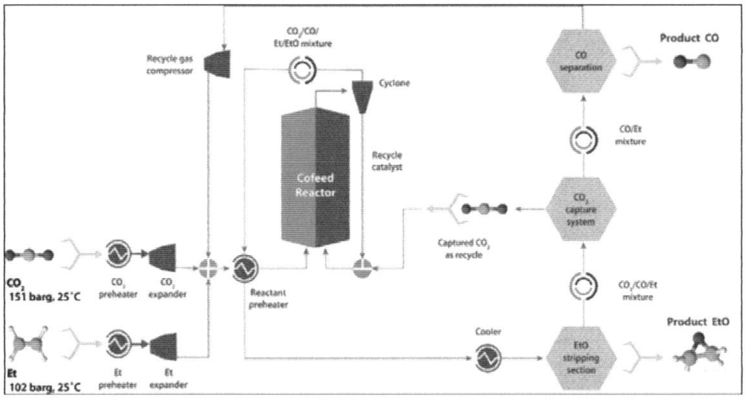

Figure 1. Conceptual schematic of the transport reactor process used in this process.

RTI has improved on its previous catalyst formulations for CO_2 utilization[3] and developed families of catalysts which can both remove oxygen from CO_2 and transfer the oxygen to ethylene to make ethylene oxide. The catalyst families are based on metal oxide phases which were found to be similar to iron in terms of reacting with CO_2, but are more selective than iron for ethylene epoxidation. Improvements on the production of ethylene oxide have been made by the use of promoters, probing the catalyst support to identify a correlation with support acidity, and observing the impact of surface area on dispersion. Ethylene oxide has been produced at small scale in the fixed-bed catalyst test reactor in both a continuous cofeed and a transport mode of operation. However, a cofeed system was found to be optimal and studied for scale-up of the technology.

Life cycle analysis has shown that the process avoids more than 2.8 kg of CO_2 per kg of ethylene oxide produced. Techno-economic analysis results show that the process can produce an IRR above 15%, and sensitivity analyses reveal the most important factors regarding the economic viability of the technology. We have compared the RTI CO_2 oxidation process with the state of art partial oxidation process and found that the new process can be economically favorable depending on market conditions.

(1) Chongterdtoonskul, A.; Schwank, J. W.; Chavadej, S. *Journal of Molecular Catalysis A: Chemical* **2013**, *372.*

(2) Kolb, K. E.; Kolb, D. *Journal of Chemical Education* **1983**, *60*, 57.

(3) Shen, J. P.; Mobley, P. D.; Douglas, L. M.; Peters, J. E.; Lail, M.; Norman, J. S.; Turk, B. *RSC Advances* **2014**, *4*, 45198.

Keywords: CO2 Utilization, Ethylene Epoxidation

FARADAY TECHNOLOGY, INC.

Pulsed Electrodeposition of High-Activity Gas-Diffusion Electrocatalysts for CO_2 Reduction

2017 AIChE Annual Meeting

Minneapolis, MN

30 Oct 2017

Brian Skinn[a], Sujat Sen[b], Rajeswaran Radhakrishnan[a], McLain Leonard[b], Dan Wang[a], Tim Hall[a], Stephen Snyder[a], Fikile R. Brushett[b], EJ Taylor[a]

[a] Faraday Technology, Inc., Englewood, OH 45315

[b] Dept. of Chemical Engineering, Massachusetts Institute of Technology, Cambridge, MA 02139

BrianSkinn@FaradayTechnology.com

17AIChE Annual Meeting, Minneapolis, MN

MIT

Outline

- Electrochemical Processing
  - Overview
  - Advantages of Pulse-Processing
- Pulsed FARADAYIC® ElectroCatalyzation of GDEs
  - Rationale and Electrodeposition Approach
  - Catalyst Material Selection
  - Prepared GDE Electrocatalysts
  - Electrocatalyst Performance Testing

FARADAY
TECHNOLOGY, INC.

Electrochemical Processing

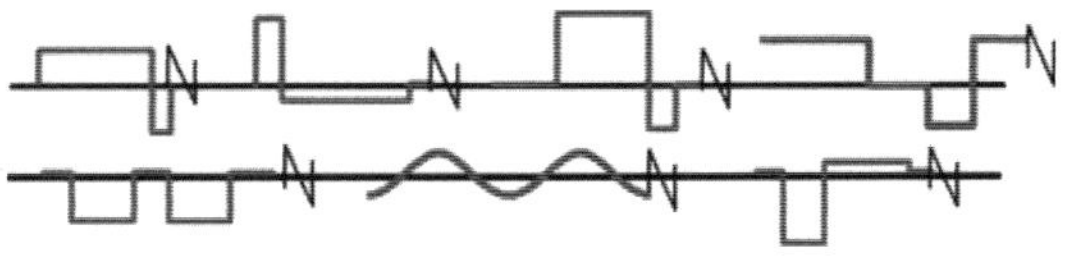

Conventional DC vs Pulse/Pulse-Reverse

Applied Electrochemistry

Red ⇋ n *e* ⁻ + *Ox*

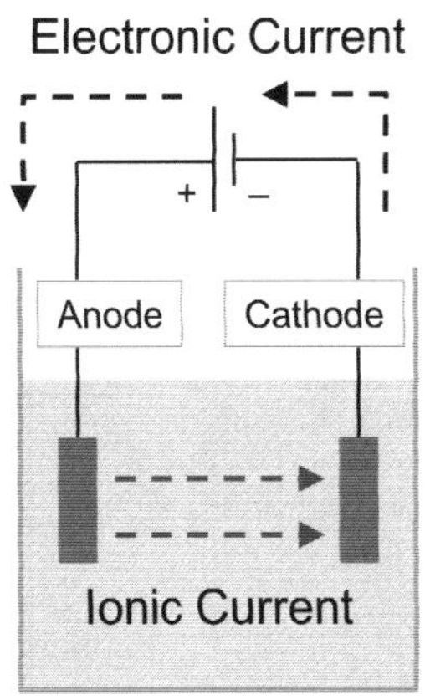

Current is net flow of
POSITIVE CHARGE

Applied Electrochemistry

$Red \leftrightharpoons n\, e^- + Ox$

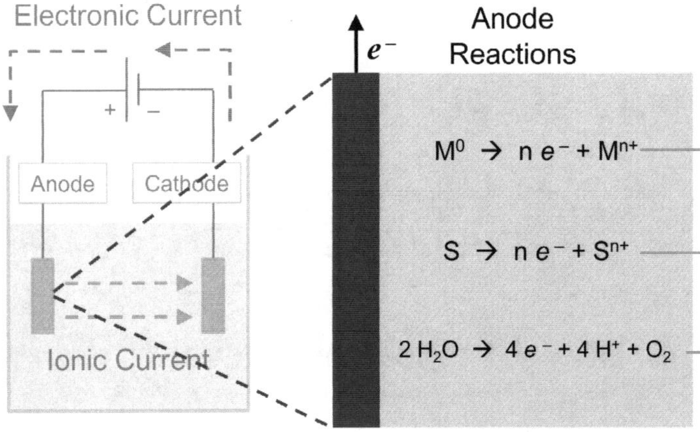

Material Processing
Anodic (subtractive)
Electropolishing

Cathodic (additive)
Electroplating

Chemical Modification
Anodic / Cathodic
Chlor-alkali
CO_2 conversion

Water Electrolysis
Anodic / Cathodic
H_2 / O_2 production
Bubble generation

Current is net flow of
POSITIVE CHARGE

Pulsed Waveforms vs DC

Know: *Linearized Concentration Profile*

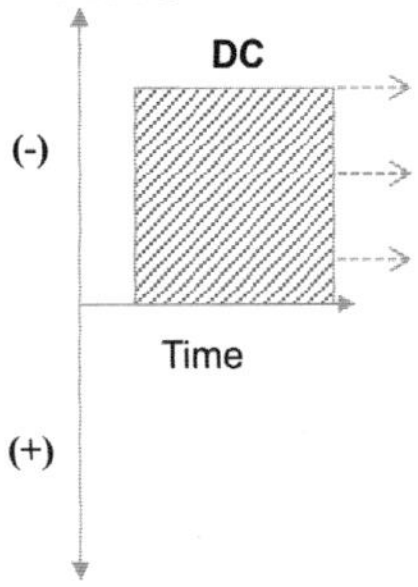

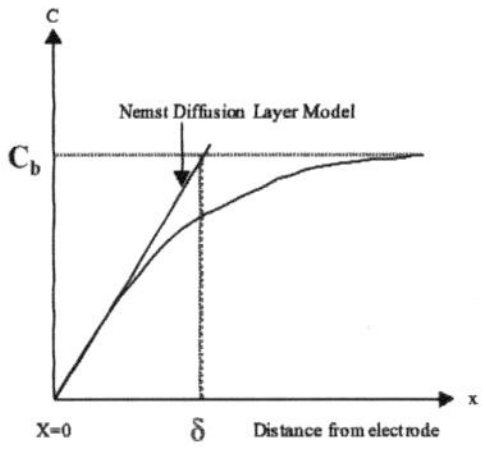

Pulsed Waveforms vs DC

Know: *Linearized Concentration Profile*

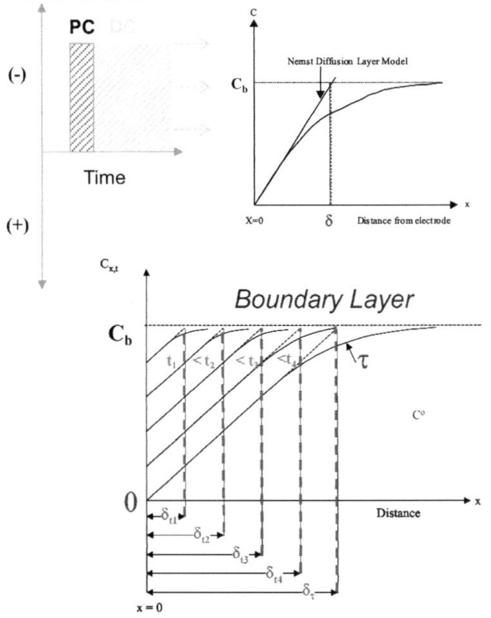

Pulsed Waveforms vs DC

Know: *Linearized Concentration Profile*

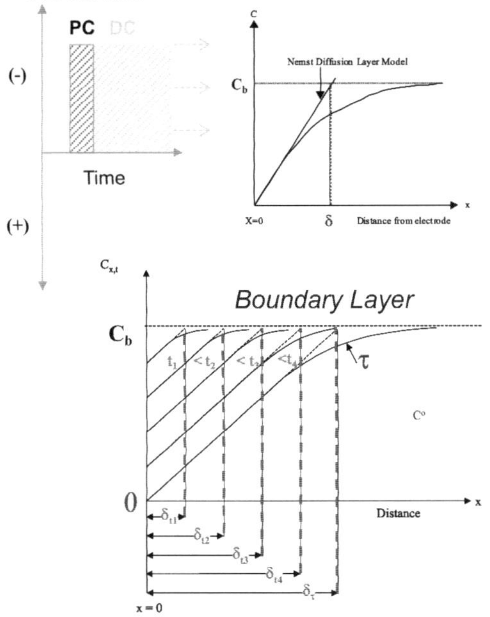

Assume:

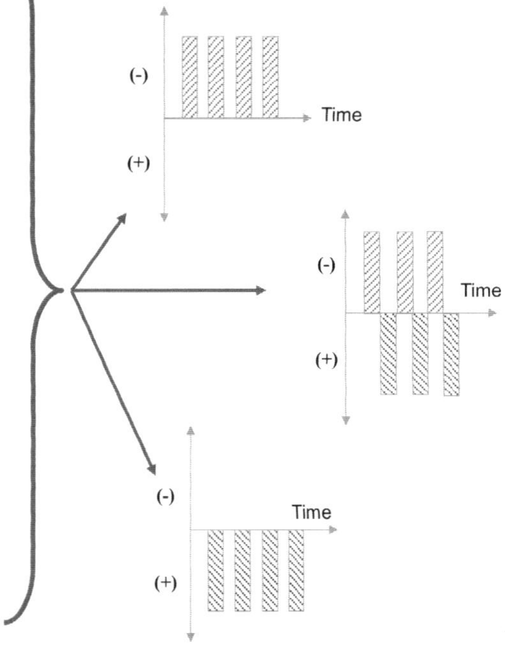

Prior Work: Copper Electrodeposition for PCBs

Standard DC Recipe

$CuSO_4$
H_2SO_4
"Accelerator"
"Suppressor"
"Leveler"
Other Additives

Pulsed Waveform Recipe

$CuSO_4$
H_2SO_4
Trace Cl^-
Trace PEG

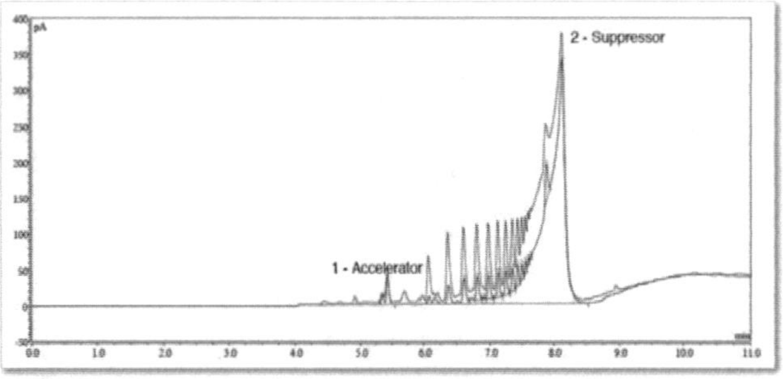

Plante, et al. "Quantization and Characterization of Copper Plating Bath Additives by Liquid Chromatography with Charged Aerosol Detection." Thermo Scientific. http://www.dionex.com/en-us/webdocs/113469-Pittcon12_1820-6_MPlante_PN70008_PlatingBaths.pdf

Z-Interconnects Cu-Plated by Pulse-Reverse FARADAYIC® ElectroDeposition

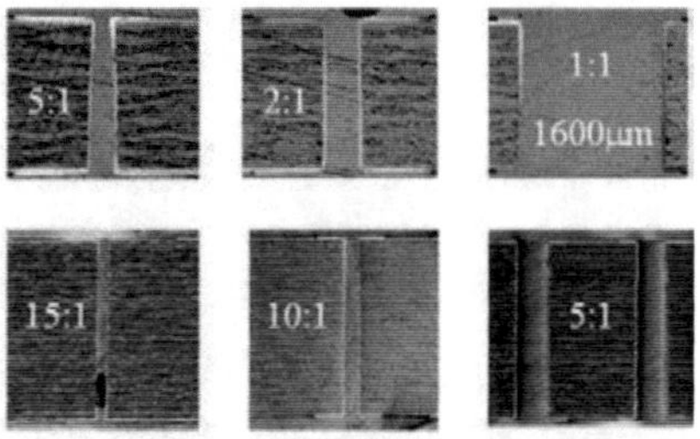

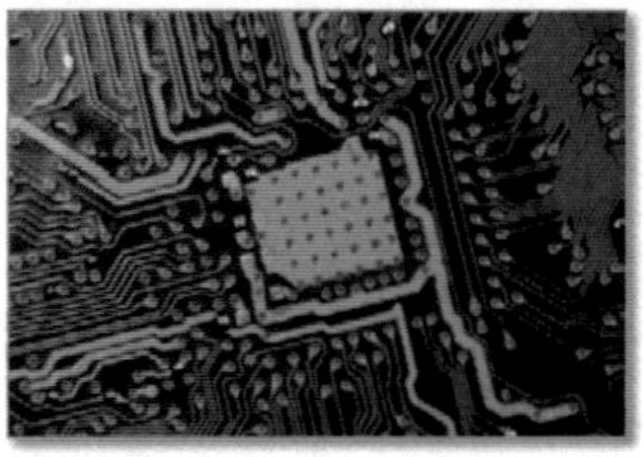

Prior Work: Lead-Free Tin

Rotating Disk Plating Tool

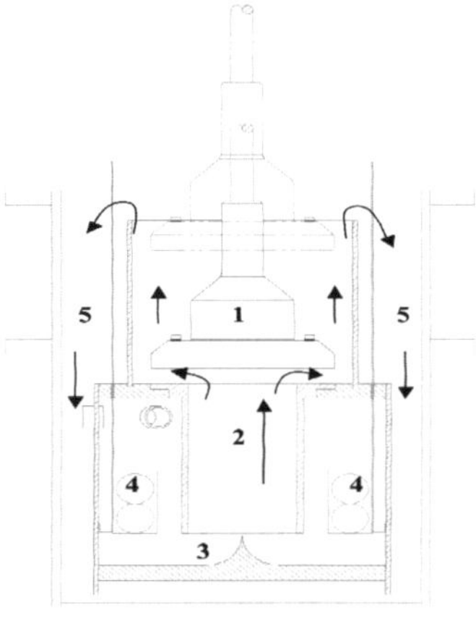

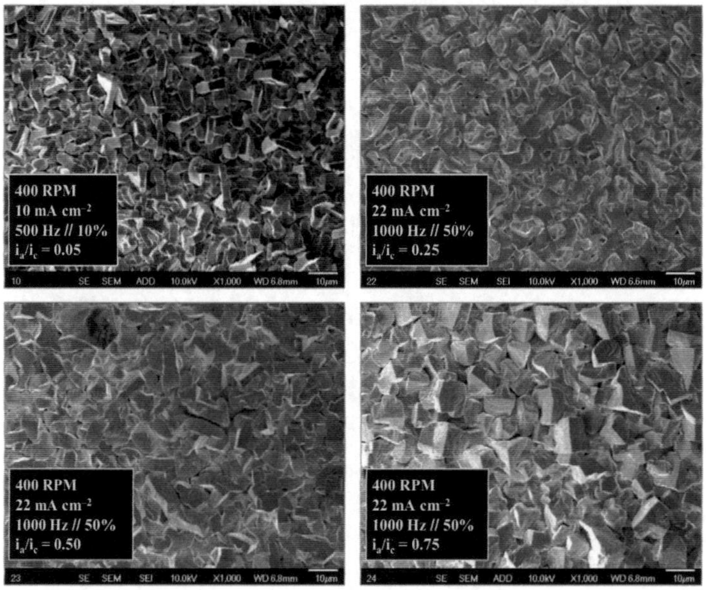

Sn Deposits on Cu Substrates — 1000× Magnification

U.S. Patent 8,603,315 Granted

FARADAYIC® ElectroCatalyzation

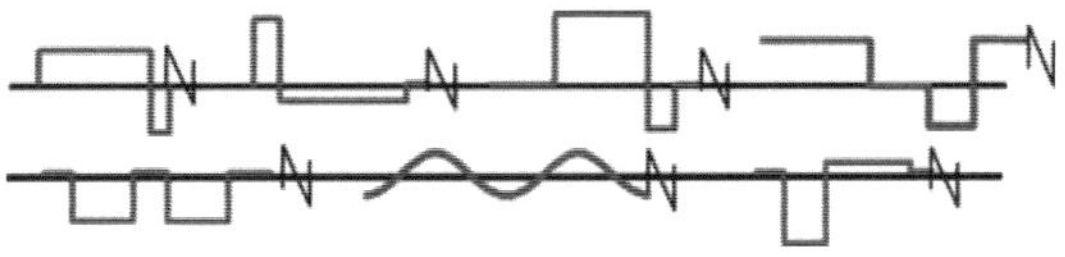

Gas Diffusion Electrode (GDE) Electrocatalysts

FARADAYIC® GDE ElectroCatalyzation

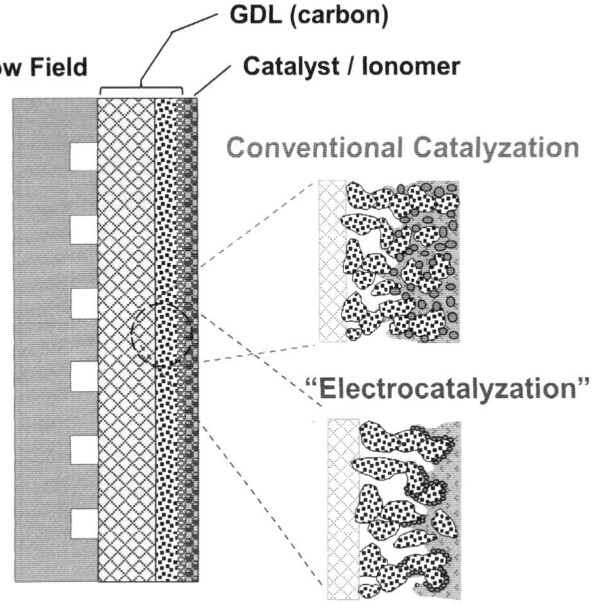

TEM of Pt on Carbon

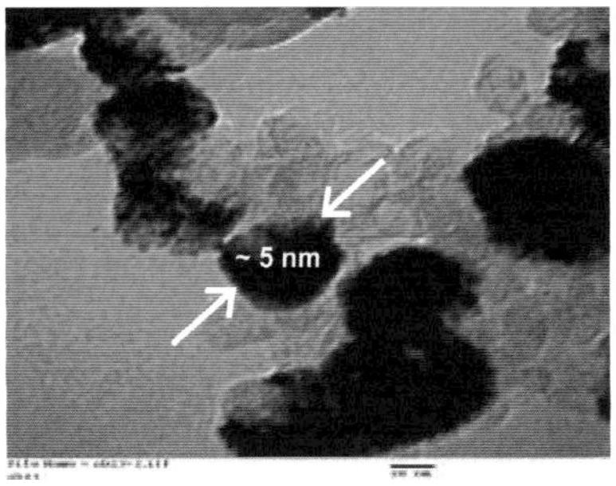

Performance Comparison

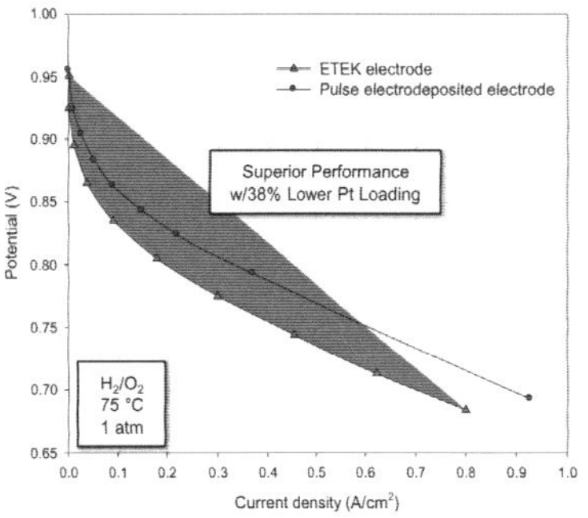

FARADAYIC® GDE ElectroCatalyzation

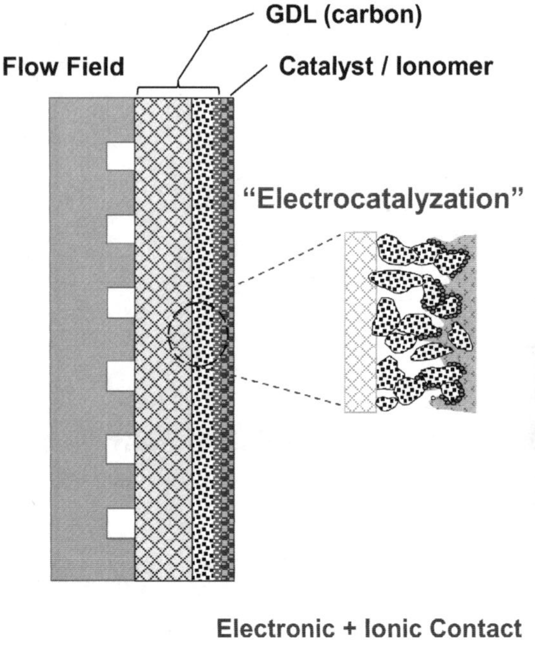

FARADAY
TECHNOLOGY, INC.

FARADAY
TECHNOLOGY, INC.

FARADAYIC® GDE ElectroCatalyzation

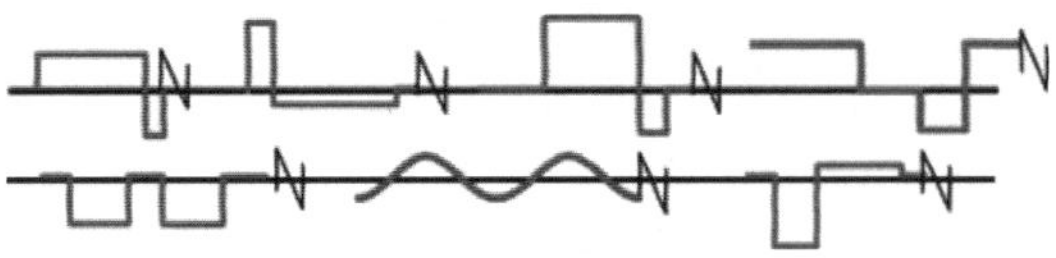

Electrocatalyst Materials Selection

CO_2 Electrocatalyst Selection

Product distribution over single-metal electrocatalysts as determined by Hori et al.
Electrochimica Acta 39(11-12): 1833 (1994).

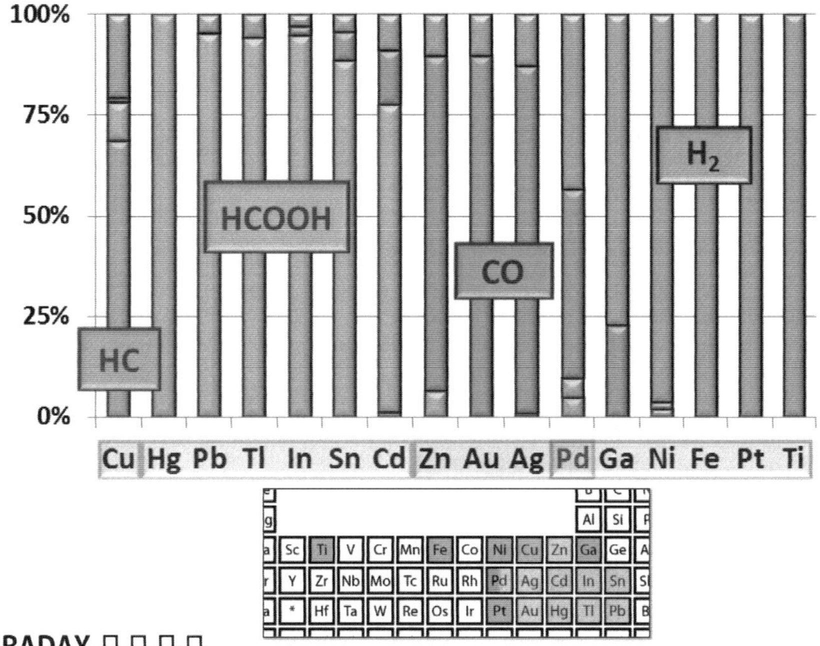

FARADAY TECHNOLOGY, INC.

CO_2 Electrocatalyst Selection

Product distribution over single-metal electrocatalysts as determined by Hori et al.
Electrochimica Acta 39(11-12): 1833 (1994).

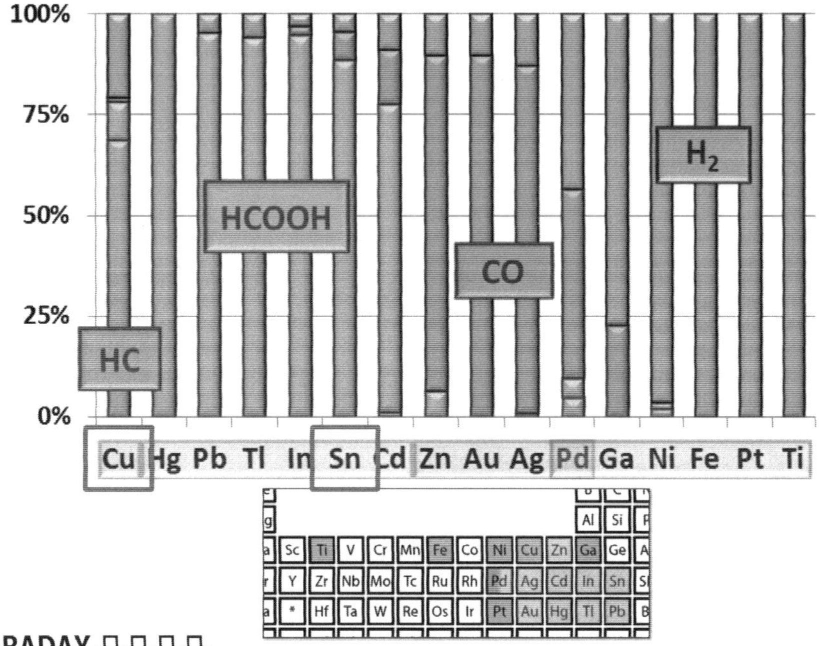

FARADAY TECHNOLOGY, INC.

FARADAY
TECHNOLOGY, INC.

FARADAYIC® GDE ElectroCatalyzation

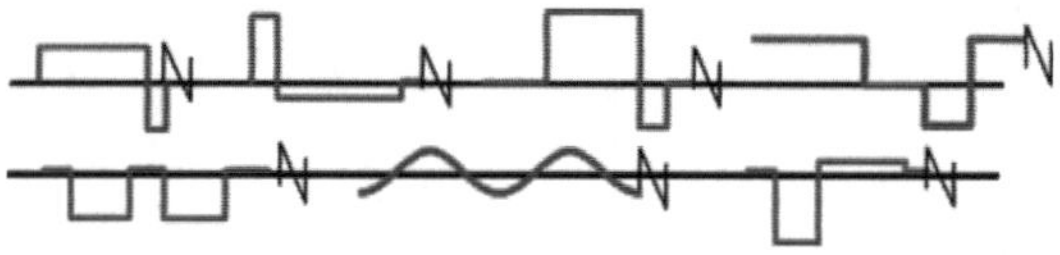

Prepared Electrocatalysts

GDE Catalysts – Cu

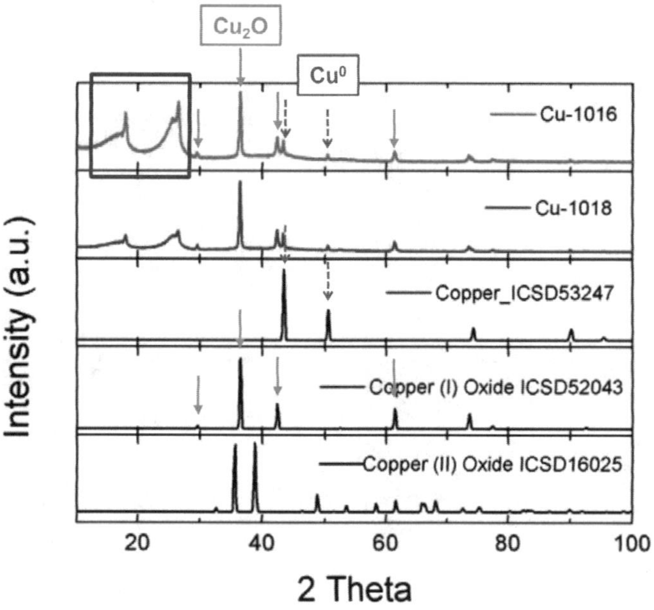

GDE Catalysts – Sn

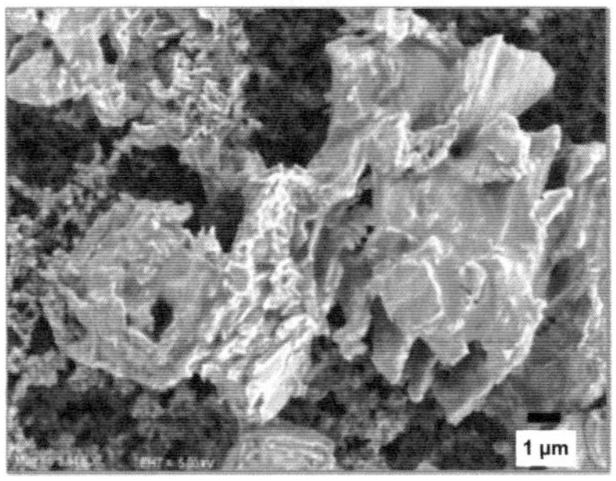

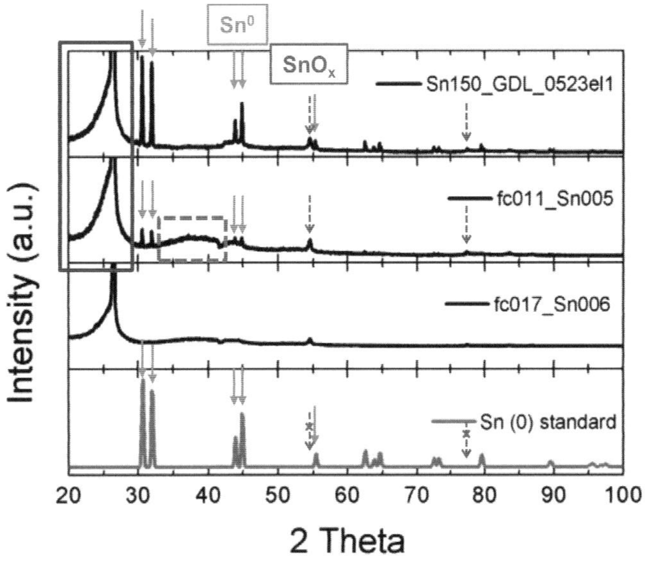

FARADAYIC® ElectroCatalyzation

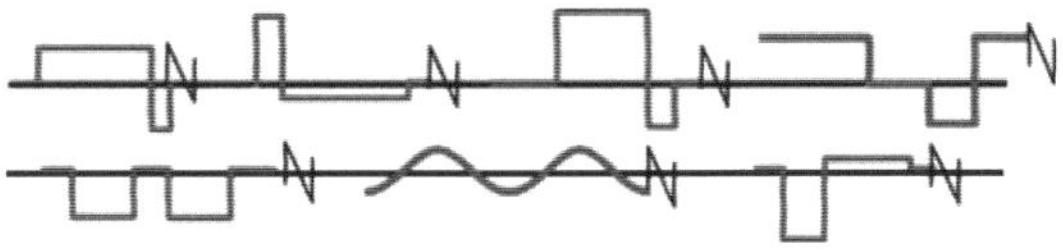

Electrocatalyst Testing

W-Cell Apparatus

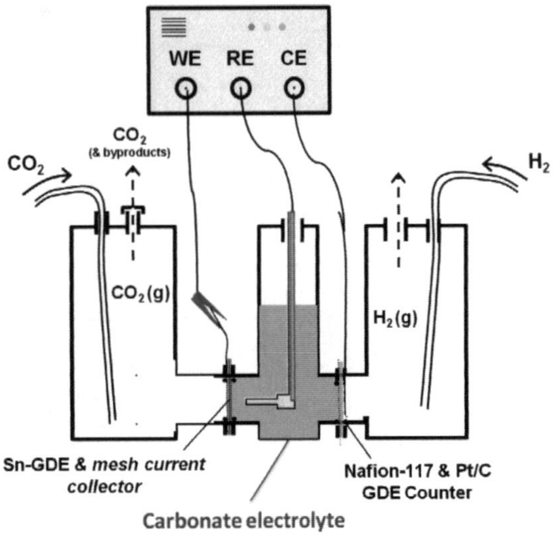

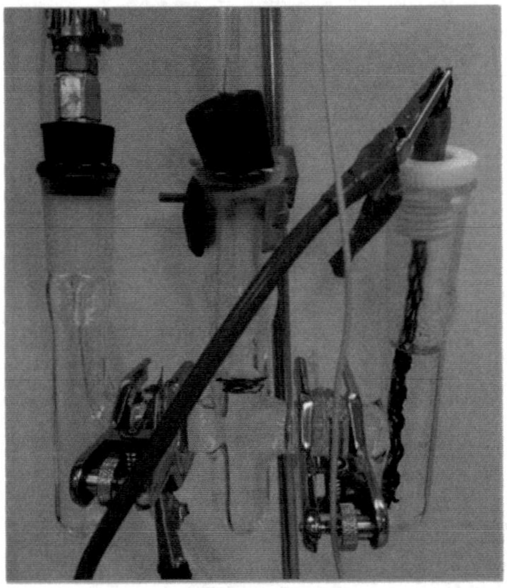

FARADAY TECHNOLOGY, INC.

Flow Cell Apparatus

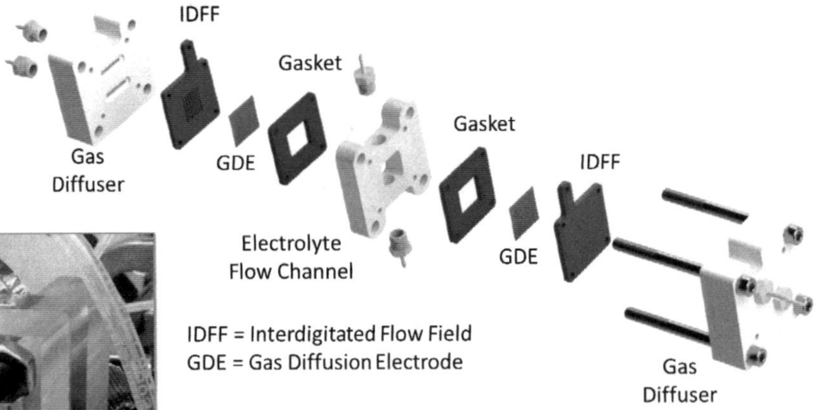

IDFF = Interdigitated Flow Field
GDE = Gas Diffusion Electrode

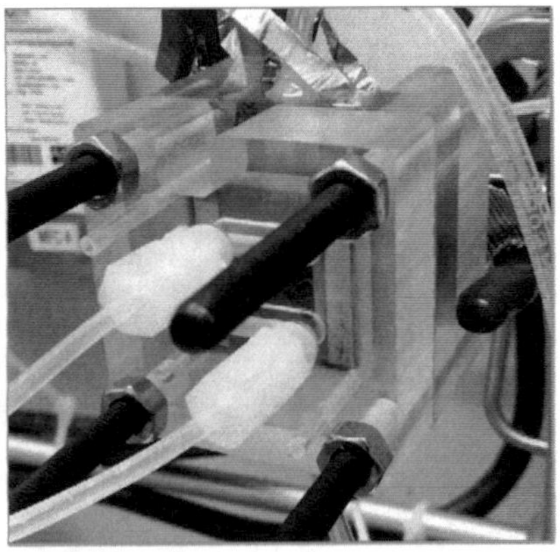

FARADAY TECHNOLOGY, INC.

Sn GDE – W-Cell Electrocatalysis

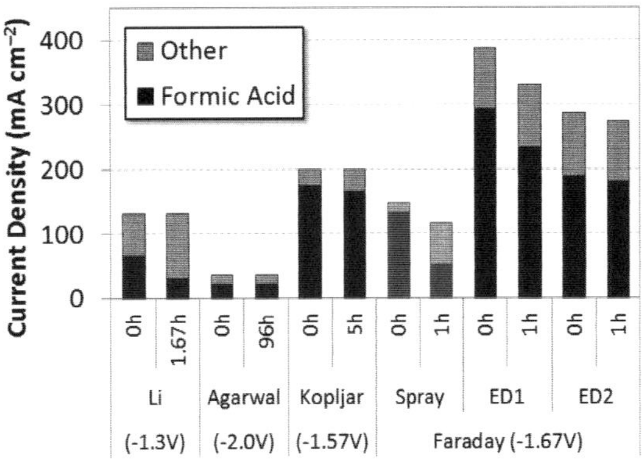

(Half-cell potentials vs SHE)

WE: $CO_2 + 2H^+ + 2e^- \rightarrow HCOOH$

CE: $H_2 \rightarrow 2H^+ + 2e^-$

Cell: $CO_2 + H_2 \rightarrow HCOOH$

Li and Oloman. *J Appl Electrochem* 35: 955, 2005.

Agarwal et al. *ChemSusChem* 4: 1301, 2011.

Kopljgar et al. *J Appl Electrochem* 44: 1107, 2014.

Cu GDE – Flow Cell Electrocatalysis

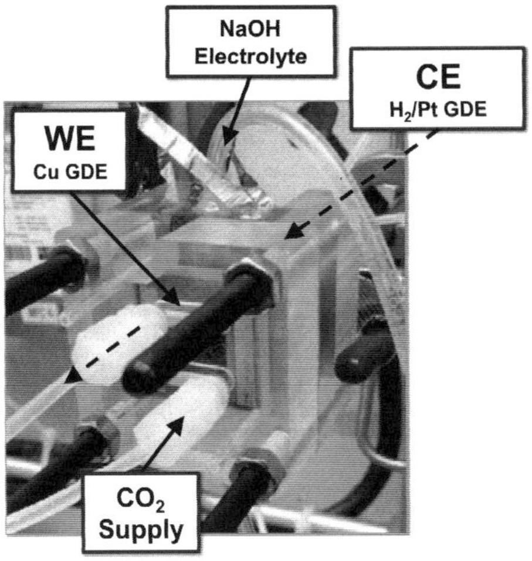

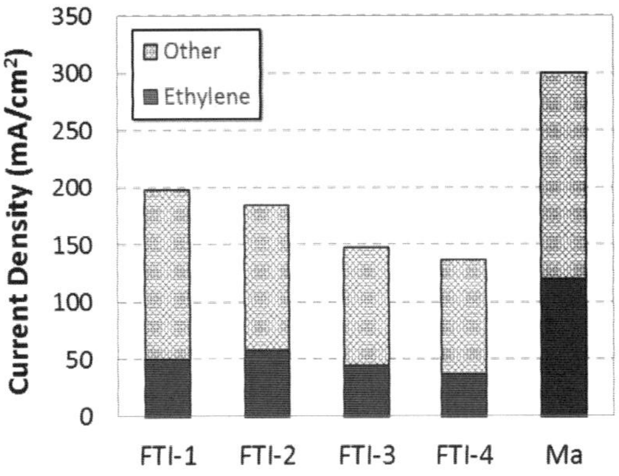

Ma et al. *J Power Sources* 301: 219, 2016.

WE: $2\ CO_2 + 8\ H_2O + 12\ e^- \rightarrow C_2H_4 + 12\ OH^-$

CE: $12\ OH^- \rightarrow 6\ H_2O + 3\ O_2 + 12\ e^-$

Cell: $2\ CO_2 + 2\ H_2O \rightarrow C_2H_4 + 3\ O_2$

Summary

- Wide variety of active GDE catalyst materials can be deposited via pulsed FARADAYIC® ElectroCatalyzation
  - Readily scalable process
- High-activity CO_2 conversion catalysts
  - Demonstrated in industrially-relevant, scalable form factor
  - Performance competitive with or superior to literature
    - Total activities (mA/cm^2)
    - Formate/ethylene selectivity
  - Durability appears promising; further testing is pending

Acknowledgment

- This presentation is based upon work supported by the United States Department of Energy (contracts #DE-SC0015173 and #DE-SC0017199). Any opinions, findings, conclusions and recommendations expressed in this material are those of the authors and do not necessarily reflect the views of the US DOE.

- Contact Information:
  Brian T. Skinn
  Ph: +1 937 836 7749
  Email: BrianSkinn@FaradayTechnology.com

Thank You!

490489 Process Intensification for Electrochemical Utilization of CO_2

Abstract:

Conversion of anthropogenic CO_2 emissions to useful chemical fuels or feedstocks is a desirable strategy for reducing atmospheric CO_2 concentration. Formic acid is a diverse product currently used in the textile and agricultural sectors, as well as a component in cleaning products and preservatives. Formic acid is also valuable as chemical fuel, a potential liquid H_2 carrier in the "centralized de-carbon economy". Production of formic acid with intermittent renewable energy sources would allow for excess energy produced during low demand load to be stored and used during peak load times, while also reducing CO_2 emissions. Progress toward the development of an electrochemical "one-step" CO_2 utilization process to produce formic acid will be discussed. The proposed CAER-ECUS process combines three previously developed components, a protective redox hydrogel, a CO_2 reducing catalyst, and the CAER proprietary Xerogel® electrode, to synthesize formic acid directly from CO_2 in a flexible process that can be modified based on changing economics and market conditions. These three key components work in concert to make the CAER-ECUS process a potentially disruptive technology by: 1) directly producing a viable and flexible chemical feedstock that can be sold or used for energy storage, 2) protecting valuable catalyst against deactivation by overpotentials and faradaic inefficiencies from the electrode or oxidation from flue gas contaminants, and 3) preforming the reaction on a large surface area (> 200 m^2/g) electrode with large pore structures reduce mass transport issues. We show that the redox-active hydrogel increases electrochemically-active surface area, and that the effective charge carrier has a redox potential within the desirable range for CO_2 reduction. Hydrogel properties, catalyst selection, process design and integration, as well as future directions for the CAER-ECUS process will we discussed.

494879 A DEM Modeling of Biomass Fast Pyrolysis in a Double Screw Reactor

Abstract:

Biomass is an abundant renewable resource that has been used for thousands of years to provide heat and energy in rural areas. More efficient thermochemical conversion ways of utilizing biomass such as fast pyrolysis are introduced in recent studies [1, 2, 3]. Screw reactors are one of the commonly used reactors for biomass fast pyrolysis [3, 4, 5]. In a screw reactor, mechanical forces are provided to the bed to enhance particle mixing and heat transfer, and achieve particulate matter transport for continuous operation. The pyrolysis process in the screw reactor configuration is a reactive granular flow system which involves particle flow, heat transfer and biomass devolatilization reactions at the same time. However, reactor-scale evaluation of heat and mass transfer effects on the biomass fast pyrolysis is very limited. Efforts are needed to discover the underlying physical phenomena and the interplay between the physical transport phenomena and chemical reactions for reactor design and optimization. Numerical simulation is capable of predicting the granular flow and heat transfer behavior in the reactor and providing useful information of particle flow such as particle residence time, particle mixing degree, and heat transfer coefficient [6].

In this research, we proposed an extended DEM method for modeling the reactive granular flow in a double screw reactor. A semi-detailed biomass devolatilization kinetics is adopted in the simulation and the decomposition dynamics of the major biomass components and pyrolysis products are analyzed at different operating conditions. The heat transfer coefficient information is extracted from the DEM simulation and provided as heat transfer boundary conditions for a single particle model. The intra-particle heat and mass transport phenomena are modeled and the effects on the biomass pyrolysis process are evaluated. We report that the heat transfer coefficient in the double screw reactor varies within a wide range of 20 to 90 $W/(m^2K)$ for different operating conditions. Results also indicate the particle-fluid-particle conductive heat transfer pathways are the dominant contributors to the total heat flux, which accounts for 70%-80% in the total heat flux. Radiation heat transfer contributes 14%-26% to the total heat flux. The reported result is helpful for reactor performance evaluation and reactor design optimization.

Reference

[1] C. Stevens, R. C. Brown, Thermochemical processing of biomass: conversion into fuels, chemicals and power, John Wiley & Sons, 2011.

[2] H. Goyal, D. Seal, R. Saxena, Bio-fuels from thermochemical conversion of renewable resources: a review, Renewable and sustainable energy reviews 12 (2) (2008) 504–517.

[3] R.Venderbosch,W.Prins,Fastpyrolysistechnologydevelopment,Biofuels,bioproductsandbiorefining 4 (2) (2010) 178–208.

[4] E. Butler, G. Devlin, D. Meier, K. McDonnell, A review of recent laboratory research and commercial developments in fast pyrolysis and upgrading, Renewable and Sustainable Energy Reviews 15 (8) (2011) 4171–4186.

[5] A. V. Bridgwater, Review of fast pyrolysis of biomass and product upgrading, Biomass and bioenergy 38 (2012) 68–94.

[6] F. Qi, T. J. Heindel, M. M. Wright, Numerical study of particle mixing in a lab-scale screw mixer using the discrete element method, Powder Technology 308 (2017) 334–345.

484227 Hydrocarbon Produced from Upgrading Rich Phenolic Compound Bio-Oil with Low Catalyst Coking

Abstract:

Catalytic upgrading of raw bio-oil and liquidâ€"liquid extracted bio-oil (high concentrated phenolic with trace acid and acetaldehyde) with methanol over ZSM-5 catalyst had been studied in this work. Temperature played vital function and leaded to increasing gas yield but less catalyst coking. It also changed both chemical distribution and selectivity on both gas and liquid products, with aromatic concentration increasing by 34.42%. Temperature of 400 _C was selected as the optimized reaction conditions with liquid yield of 10.47 wt.% and 75.00% aromatic hydrocarbon in liquid product from 100 g biomass, only with coke yield of 1.42 wt.%. Phenolic-rich extracted bio-oil obtained higher aromatic hydrocarbon yield (7.3 wt.% increased from 1.1 wt.%) and lower coke yield (1.42 wt.% decrease from 15.79 wt.%) than raw bio-oil. Catalyst regenerated from this type of feedstock also achieved higher activity and longer useable running times on aromatic yield compared to fresh catalyst. This result suggested that lignin derived phenolic and guaiacol compounds were not the only reason caused catalyst coking on ZSM-5 catalyst. Small high active molecules of acetic acid and acetaldehyde also acted as important precursors of catalyst coke formation.

501928 Renewable Hydrocarbon Production from Catalytic Upgrading of Fast Pyrolysis Bio-Oil from a Feedstock Blend

Abstract:

Production of renewable fuel blendstock from catalytic hydroprocessing of fast pyrolysis oil has been considered as one of the promising thermochemical conversion pathways of biomass to renewable fuel. However, this technology requires overcoming technical and economic barriers for it to be considered competitive with petroleum-derived fuels. The use of low-cost, blended feedstocks in place of a single source of biomass is one way to reduce the overall process costs. In this presentation, we report the catalytic upgrading of a fast pyrolysis bio-oil derived from a mix of clean pine, forest residue and demolition waste. Due to the thermal instability of the bio-oil, it was pretreated in the presence of Ru/C catalyst using a continuous flow reactor (400 ml) with nominal bio-oil space velocity of 0.25-0.5 h^{-1} and 500-1000 L H_2/L bio-oil and operated at 140°C and 1200 psig. Further catalytic hydrodeoxygenation and hydrocracking reactions were accomplished in a two-bed continuous flow reactor consisting of a sulfided Ru/C catalyst held at 170°C, followed by a sulfided commercial hydrotreating/hydrocracking catalyst held nominally at 400°C. The operating pressure was set at 2000 psig, with 0.1 h^{-1} bio-oil space velocity and 2500 L H_2/L bio-oil. The two-phase product was subsequently separated, with the organic phase (less than 3 wt% oxygen) distilled in a batch set-up into gasoline (20-184°C), diesel (184°C-338°C), jet (150°-250°C) and heavy (>338°C) fractions. Characterization of these fractions will be presented.

496589 CFD-DEM Model for Predicting Producer Gas Contaminants during Biomass Gasification

Abstract:

Biomass gasification is valuable method for extracting energy in the form of producer gas from biomass. The presence of contaminants in producer gas however creates minor and severe downstream problems. We developed a quasi-3D CFD simulation of biomass gasification using Eulerian-Lagrangian concept in an open source C++ code (OpenFOAM). The particle-particle collisions were described using Hertz-Mindlin particle contact model. Navier-Stokes equation was utilized to resolved the gas flow field variables and turbulence was included through a standard Î°-Îµ turbulence model. The pyrolysis, homogeneous and heterogeneous reactions in the gasifier were considered. Six case evaluating the effects of gasification temperature (790, 934, and 1078 Â°C) and equivalence ratio (0.15, 0.25, and 0.35) were simulated. The gas and particle flow field variables such as velocity, pressure, and void fractions inside the gasifier were obtained. Also, the concentration of permanent gases (CO, CO_2, H_2O, CH_4, and H_2O) and contaminant gases (NH_3, HCN, and H_2S) inside the gasifier and across the gasifier outlet were obtained. The simulation results were then compared with experimental data from a bench-scale bubbling fluidized bed gasifier.

501946 Insights into the Relationship between Structure and Reactivity Descriptors in Realistic Nanoparticle Models: A DFT Approach

Abstract:

Fischerâ€“Tropsch (FT) synthesis is a promising road for the conversion of syngas into valuable hydrocarbons such as gasoline, diesel, chemicals and so forth. The chemical reactions are catalyzed more commonly by transition metals including Fe, Co and Ru. Albeit, only Fe and Co catalysts are commonly employed in industry. Due to the various steps involved during the conversion and the variety of products that can be generated, significant variation in catalyst composition and shape affects the nature and content of the products. That is, the catalytic activity of metal nanoparticles is greatly size-dependent and even just an additional atom can entirely change the reactivity and stability. Due to encouraging results with previous studies on 13-atom cluster models of Co, Fe, Ni, Pd, Pt, and Ru, we are now exploring 55-atom cluster models of the same metals using Density Functional Theory (DFT) calculations. These cluster models have shown to retain the accuracy of the periodic slab models at a lower computational cost. Since experimental information at the atomic level is difficult to obtain, we carried out systematic DFT calculations for evaluating reactivity descriptors of larger models. These cluster-based models offer a platform for the study of the adsorption and desorption of CO and H2 at the initial stages of the FT synthesis.

498860 Transportation Fuels from Renewable Hydrogen and Green Carbon Dioxide – a Technical, Economic, and Environmental Evaluation

Abstract:

The desire to reduce greenhouse gas emissions from the transportation sector along with the increasing penetration of fluctuating renewable power sources in several countries have led to an growing interest in the production of synthetic fuels from electricity [1]. Most of the corresponding processes employ water electrolysis in order to produce hydrogen. Apart from direct use in fuel cell electric vehicles, hydrogen can also be reacted with CO_2 to form various gaseous or liquid organic fuels that are more compatible with existing infrastructure and engines. Several pathways are being suggested in literature that use CO_2 from different sources [2] and lead to different fuels [3]. However, comprehensive comparisons to assess the benefits and drawbacks of these options including both fuel production and use in transportation are still scarce.

In this contribution, we present a detailed technical, environmental, and economic evaluation of several pathways that are considered promising. This includes direct application of hydrogen in a fuel cell vehicle, as well as conversion of hydrogen and CO_2 to combustion engine fuels like methane, methanol, and different methyl ethers, some of which can also be used as blend components with conventional fuels [4]. The CO_2 used is assumed to be obtained from biogas upgrading, where it occurs as a currently unused byproduct in high concentrations. Biogas upgrading is already being conducted at a number of operating biogas plants in order to obtain biomethane of higher purity [3]. The pathways are evaluated with respect to their well-to-wheel efficiency, greenhouse gas emissions, and fuel cost. A technical analysis is conducted based on detailed process simulations in order to highlight what causes the differences in the performance of the different pathways, and an outlook is given on the potential for improved production processes.

References:

[1] M. Bailera, P. Lisbona, L.M. Romeo, S. Espatolero. Renew. Sust. Energ. Rev. 69: 292–312, 2017.

[2] G. Reiter, J. Lindorfer. J. CO_2 Util. 10: 40–49, 2015.

[3] W. Wang, S. Wang, X. Ma, J. Gong. Chem. Soc. Rev. 40: 3703–3727, 2011.

[4] L. Lautenschütz, D. Oestreich, P. Seidenspinner, U. Arnold, E. Dinjus, J. Sauer. Fuel 173: 129–137, 2016.

492889 Hydraulic Retention Time and Temperature Impacts on Biogas Production in Expanded Granular Sludge Bed Reactor

Abstract:

Renewable sources play a significant role in providing sustainable energy with reduced Greenhouse Gas emissions considered to be a significant cause of global warming. Renewable energy resources combined in a hybrid energy system can also provide reduced energy costs. Anaerobic Digestion is one of these renewable energy sources, which involves the degradation of biomass in the absence of air to produce biogas that can be used in many applications. This technology also helps reduce the effluent waste discharged to a waste-water treatment system that further improves the eco-friendly nature of the hybrid system by lowering water loss to the environment.

An expanded granular sludge bed reactor, updated from the up-flow anaerobic sludge reactor, has been used to investigate the effect of hydraulic retention time and temperature on biogas production rate and methane composition of the biogas. The process consists of a two-stage reactor with a pre-acidification stage where hydrolysis and acidogenesis steps occur followed by the main reactor where the acetogenesis and methanogenesis steps occur.

Equivalent specific values of organic loading rates for industrial waste water with different COD levels (20, 30 and 40 g COD/l) have been used. The biogas production increases with decreasing hydraulic retention time (i.e. 24 gallons/ day) and decreases when biogas production is inhibited by lower residence times below that required to complete degradation of the wastes and eventually affect the activity of the overall biodigestion process.

Methane composition and COD removal efficiency are (65-78%) and (90-96.5 %) respectively. A detailed analysis has been conducted for the process effluent using a Benchtop Spectrophotometer to monitor the process stability.

500041 Reverse Water Gas Shift Reaction over Cu-Fe/Al_2O_3 Catalyst in Solid Oxide Electrolysis Cells

Abstract:

Reverse water gas shift (RWGS) reaction is a practical method to convert CO_2 to CO utilizing hydrogen supplied from clean energy sources. Solid oxide electrolysis cells (SOECs) are new energy conversion devices with high efficiency and environmental compatibility. They can split steam (H_2O) into hydrogen and oxygen by using electricity generated from clean energy sources (such as solar energy and wind energy). Thus, applying the SOECs in the RWGS reaction could be a feasible way for efficient conversion of CO_2 using renewable energy sources. RWGS reaction was carried out over Fe-Cu/Al_2O_3 catalyst by a fixed-bed quartz tube reactor using externally supplied hydrogen at temperature of 700 °C, while fixing total flow rates of H_2 and CO_2 at 100 ml/min (CO_2:H_2 is 1:1). The catalyst showed good catalytic activity with CO
results confirmed that the sintering of active sites was prevented during the reaction. Cu-Fe/Al_2O_3 catalyst was then utilized as a functional reforming layer for the cathode of a conventional SOEC single cell. The single cell consisted of Ni-yttria stabilized zirconia (YSZ) cathode support/YSZ/$Ce_{0.8}Sm_{0.2}O_{1.9}$ (SDC) bi-layer electrolyte/$La_{0.6}Sr_{0.4}Co_{0.2}Fe_{0.8}O_{3-\text{î}'}$ anode. Using this SOEC single cell with the catalysts layer, the RWGS reaction was investigated under the applied electric load condition. Our results indicated that by electrolysing H_2O over the cathode of SOEC cell, a reduced amount of external hydrogen is needed to obtain a similar CO_2 conversion as the fixed-bed quartz tube test.

494882 Electrochemical Hydrogenation of Bio-Oil Compounds Using a Polymer Electrolyte Membrane Reactor

Abstract:

Electrochemical hydrogenation (ECH) is an option for stabilizing or upgrading bio-oil that employs mild conditions (less than 80ºC and ambient pressure). In this study, acetone, furfural and phenol are chosen as model substrates because of their abundance in bio-oil. The ECH method in a Polymer Electrolyte Membrane Fuel Cell (PEMFC) reactor has been considered to be one of the feasible ways of hydrogenating chemical compounds. In this study, acetone, furfural, and phenol which are present in pyrolysis bio-oil were studied for ECH. The main products from the ECH of acetone were isopropanol with selectivity approximately 90%. Diisopropyl ether was a byproduct. 16% propanol and 1% diisopropyl was identified in the final product using a new membrane, a holed bag collection method, the reaction temperature 65ºC, and the reaction relative humidity of 90%. Propanol selectivity could be further improved by reducing the input power supply voltage.

The main challenge in ECH using a proton exchange membrane reactor is that a large fraction of current goes towards generating hydrogen gas that is not used in the hydrogenation reaction. High current efficiency (59.7%) could be achieved by using a PEMFC reactor. The mild operation condition is the greatest advantage of the proposed method, the optimized reaction condition is at 65 ºC and 1 atmosphere. Relative humidity positively effects H_2 Utilization Efficiency and Current Efficiency, while temperature did not show linear influence during acetone ECH. The challenging part was the membrane contamination. Suitable reaction pressure, low reaction temperature, and low reaction relative humidity could prolong membrane life.

Parametric optimization study of downdraft biomass gasification using a comprehensive transport and kinetic model

Tapas Kumar Patra, Pratik N Sheth
Department of Chemical Engineering
BITS Pilani, Pilani Campus
INDIA

AIChE Annual Meeting 2017
October 29- November 3
Minneapolis Convention Center, Minneapolis, USA

17AIChE
Annual Meeting, Minneapolis, MN

Plan of Presentation

- Introduction
- Details of the developed model
  - Model Equations
  - Boundary conditions
  - Modeling Approach
- Operating Parameters
- Results and Discussion
- Conclusion
- References

Introduction

Why Biomass Energy ??

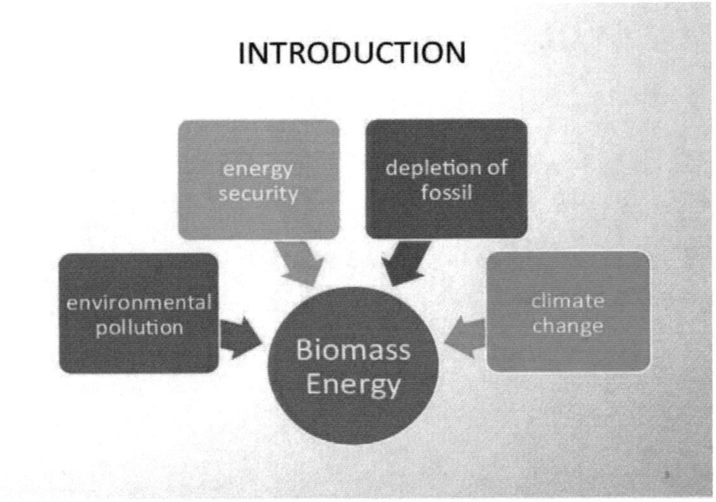

Downdraft Biomass Gasification

- Biomass gasification is the process converting biomass into gaseous fue
- Sub stoichiometric amount of air/oxygen is supplied
- In a downdraft gasifier both biomass and gasifying agent move in the downward direction
- Biomass passes through four zones
- Each zone performs different reactions and are distinguished based on temperature
- Boundaries of the zones are not rigid but transitional

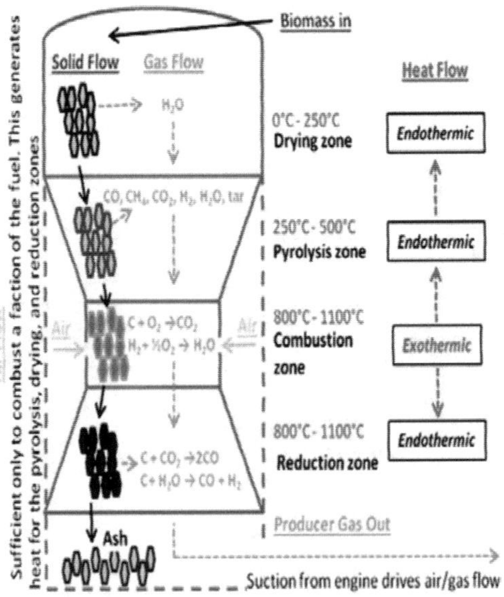

Gasification Reactions and Thermodynamics

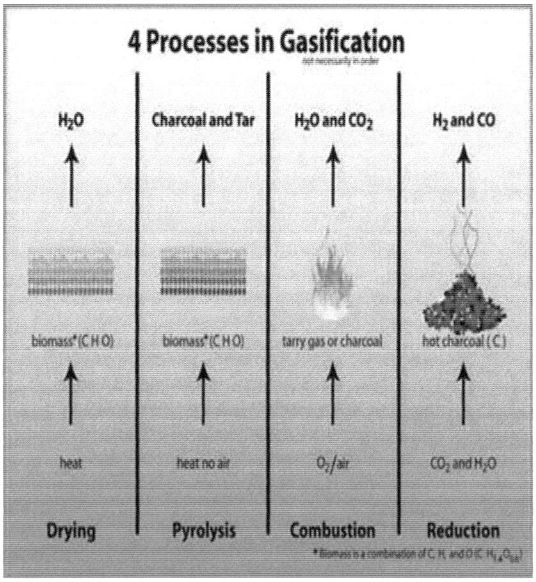

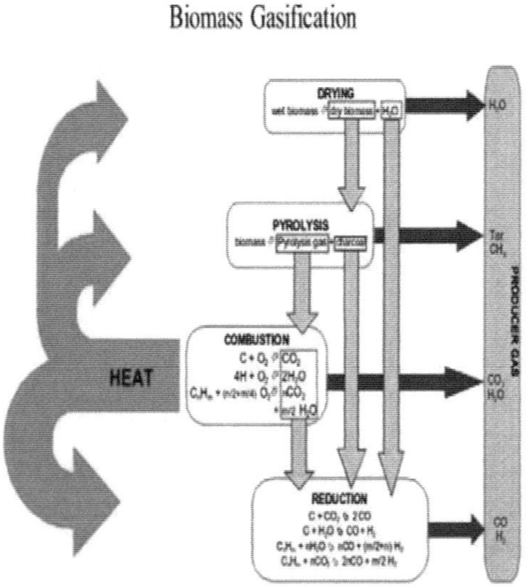

Overall Gasification Process

Preprocessing of biomass ⇒ Gasification ⇒ Gas clean-up & reforming ⇒ Gas utilization

Preprocessing of biomass	Gasification	Gas clean-up & reforming	Gas utilization
Size reduction, Drying, Preparation of gasifying agents	Heating, Chemical reactions, Catalysis	Cleaning of tar from syngas, Reforming of the syngas, Catalysis	Gas turbine, Gas burner, Fuel cell, Combined heat and power
Upstream processing	Gasification	Downstream processing	

Application of syngas

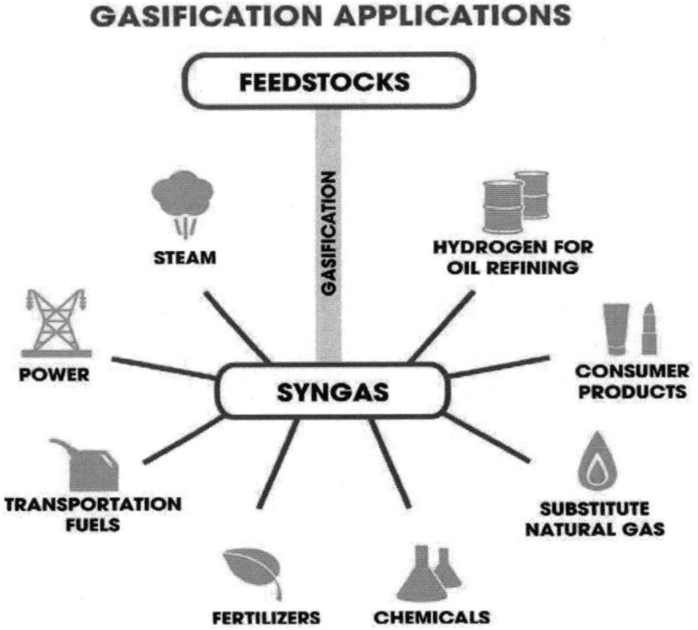

6

Gasification Process Simulation

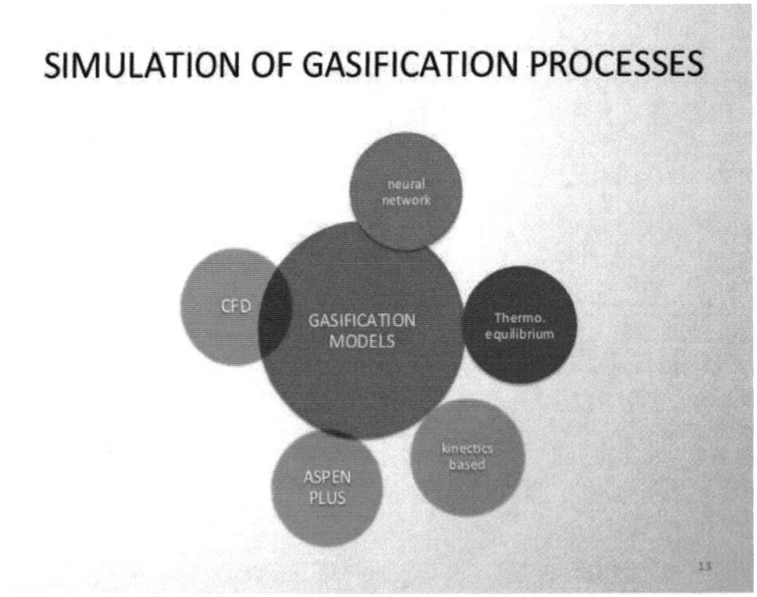

7

Schematic diagram of developed Model

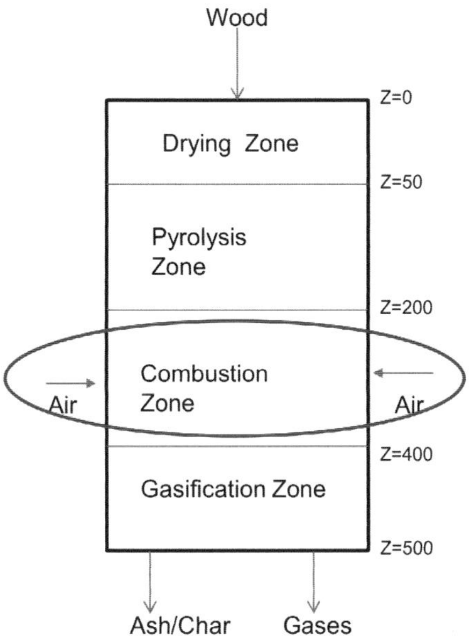

8

Model Equations

- **Wood**

$$\frac{\partial \rho_w}{\partial t} + \frac{\partial U_s \rho_w}{\partial z} = -r_{p1}$$

p1—primary pyrolysis

- **Moisture**

$$\frac{\partial \rho_M}{\partial t} + \frac{\partial U_s \rho_M}{\partial z} = -m_M$$

m_M-moisture evaporation/condensation

- **Oxygen**

$$\varepsilon \frac{\partial \rho_O}{\partial t} + \frac{\partial U_g \rho_O}{\partial z} = \frac{\partial}{\partial z}\left(\varepsilon D_O \rho_g \frac{\partial Y_O}{\partial z} \right) + M_O \sum_j v_{Oj} r_j + U_{gOi} \rho_{gO} Y_{OO} \delta(z - z_{Oi})$$

j=c1-c6

- **Gas Species**

$$\varepsilon \frac{\partial \rho_i}{\partial t} + \frac{\partial U_g \rho_i}{\partial z} = \frac{\partial}{\partial z}\left(\varepsilon D_i \rho_g \frac{\partial Y_i}{\partial z} \right) + M_i \sum_j v_{ij} r_j + v_i r_{p1} + v_i^* r_{p2}$$

i=H_2, CO,CO_2,CH_4, j=c1-c6,g1-g3,wg,sr1,sr2

- **Steam**

$$\varepsilon \frac{\partial \rho_{H_2O}}{\partial t} + \frac{\partial U_g \rho_{H_2O}}{\partial z} = \frac{\partial}{\partial z}\left(\varepsilon D_{H_2O} \rho_g \frac{\partial Y_{H_2O}}{\partial z} \right) + M_{H_2O} \sum_j v_{H_2Oj} r_j + U_{gOi} \rho_{gO} Y_{H_2OO} \delta(z - z_{Oi}) + m_M + v_{H_2O} r_{p1} + v_{H_2O} r_{p2}$$

j=c1-c6,g1-g3,wg,sr1,sr2

- **Primary Tar**

$$\varepsilon \frac{\partial \rho_{T1}}{\partial t} + \frac{\partial U_g \rho_{T1}}{\partial z} = \frac{\partial}{\partial z}\left(\varepsilon D_{T1} \rho_g \frac{\partial Y_{T1}}{\partial z} \right) - M_{T1} r_{c1} + v_{T1} r_{p1} - r_{p2}$$

- **Nitrogen**

$$\rho_{N_2} = \frac{P M_{N2}}{RT} \qquad \rho_g = \Sigma \rho_i$$

- **Total Gas Continuity**

$$\varepsilon \frac{\partial \rho_g}{\partial t} + \frac{\partial (U_g \rho_g)}{\partial z} = +\sum_i \sum_j v_{ij} M_i r_j + m_M + (1 - v_C) r_{p1} + U_{gOi} \rho_{gO} \delta(z - z_{Oi})$$

i=O_2,H_2,CO,CO_2,H_2O,CH_4; j=c6,g1-g3

9

Model Equations

- **Solid Phase Enthalpy**

$$\frac{\partial(\sum_i \rho_i H_{si})}{\partial t} = \frac{\partial}{\partial z}\left(\lambda_s^* \frac{\partial T_s}{\partial z}\right) + \frac{\partial(U_s \sum_i \rho_i H_{si})}{\partial z} - \sum_j r_j \Delta H_j - Q_{sg} + Q_{sw} - m_M \Delta$$

$$H_{si} = C_{si}(T_s - T_o), \quad i = W, M, C$$

- **Gas Phase Enthalpy**

$$\varepsilon \frac{\partial(\sum_i \rho_i H_{gi})}{\partial t} = \frac{\partial}{\partial z}\left(\lambda_g^* \frac{\partial T_g}{\partial z}\right) + \frac{\partial(U_g \sum_i \rho_i H_{gi})}{\partial z} - \sum_j r_j \Delta H_j + Q_{sg} + Q_{gw} + U_{g0i}\rho_{g0}C_{g0}T_{g0}\delta(z - z_0)$$

$$H_{gi} = C_{gi}(T_g - T_o), \quad i = N_2, CO_2, CO, H_2, O_2, CH_4, H_2O, T$$

Solid-wall heat transfer

$$Q_{sw} = \frac{4h_{sw}}{D}(T_w - T_s)$$

Gas-wall heat transfer

$$Q_{gw} = \frac{4h_{gw}}{D}(T_w - T_g)$$

Solid-gas heat transfer

$$Q_{sg} = h_{sg} A_p \upsilon_p (T_s - T_g)$$

$$A_p \upsilon_p = \frac{6(1 - \varepsilon)}{d_p}$$

Boundary Conditions

- At the top of the reactor (z=0), temperature, velocity and density of the inlet stream is assumed constant and the solid is assumed to be at the ambient temperature T_e

ρ_{wood}= 668.0 kg/m^3

$\rho_{moisture}$= 66.8 kg/m^3 *(Considering 10% Moisture content)*

$\rho_{H_2} = \rho_{CO} = \rho_{CO_2} = \rho_{CH_4} = \rho_{H_2O} = \rho_{T_1} = 0.0$

$T_s = T_e$= 300.0 K

At the air inlet,

ρ_{O_2} = 0.248619kg/m^3*(0.21 * air density at 25 C temperature (1.1839))*

ρ_{N_2}= 0.935281kg/m^3 *(0.79 * air density at 25 C temperature (1.1839))*

ρ_g= 1.1839 kg/m^3

$Ug = U_{g0}$=0.2m/s

- At the bottom of the reactor(Z=L)

$$\frac{\partial T_s}{\partial z} = \frac{\partial T_g}{\partial z} = \frac{\partial \rho_i}{\partial z} = 0, (i = CO_2, CO, H_2, CH_4, H_2O, T, N_2, O_2)$$

Modeling Approach

innovate achieve lead

- The reactor is divided into a set of elementary cylindrical cells
- Finite Difference Method (FDM) was applied for discretization
- After discretization, the equations were solved with Semi – implicit procedure i.e. each conservation equation is implicit in the corresponding variable being conserved whereas the other variables are taken as the last available values.
- Time step = 10^{-5}s
- Space step=0.01m
- Grid length = length/space step
- All the constant values were taken as stated in the literature

Model Validation and Parametric Study

innovate achieve lead

- The model is validated with the experimental data of Sheth *et al.* (2010)
- The model has been validated for experimental parameters such as
  - Producer gas flow rate
  - Biomass consumption rate
  - Char production rate
- Parameters chosen for the study are
  - Air flow rate
  - Biomass moisture content
  - Reduction zone length

Results: Air Flow Rate

- Air flow rate is one of the vital parameters for gasification process
- Sub-stoichiometric amount of air is supplied to avoid complete combustion
- Increase in air flow rate dilutes the product gas since nitrogen dominates in the composition
- From figure it is seen that the composition of CO and H_2 decreases continuously and for N_2 and CO_2 it increases.
- It clearly shows that combustion dominates with the increase in air flow rate

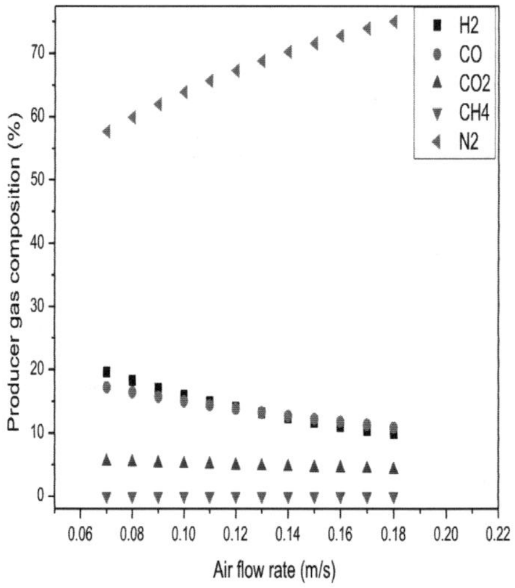

Model predicted product gas composition after 30 minutes for different air flow rates

Results: Air Flow Rate

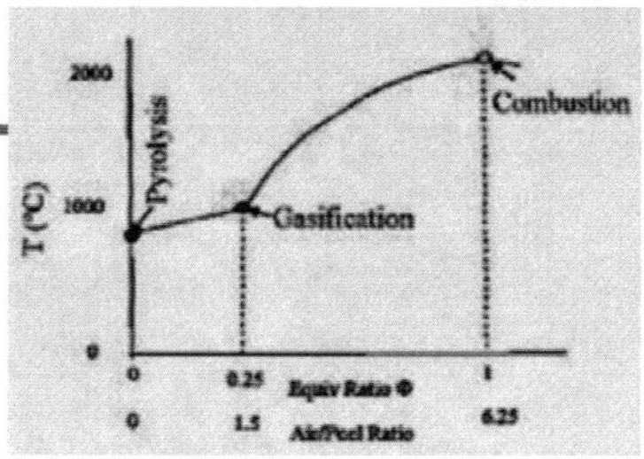

$$Equivence\ ratio = \frac{\left(Air/_{Fuel}\right)_{actual}}{\left(Air/_{Fuel}\right)_{Stoichiomeric}}$$

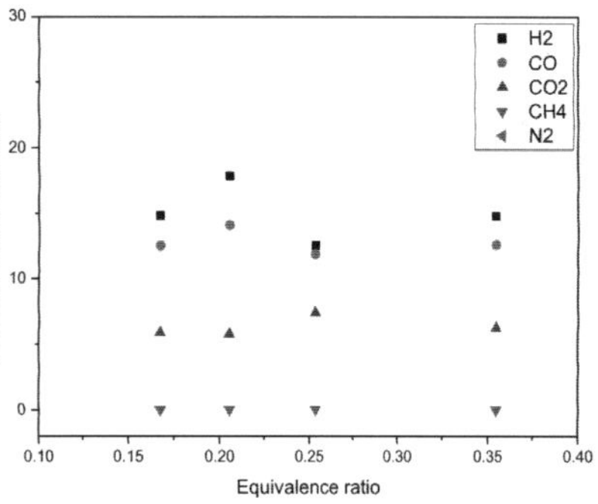

Model predicted product gas composition for different ER

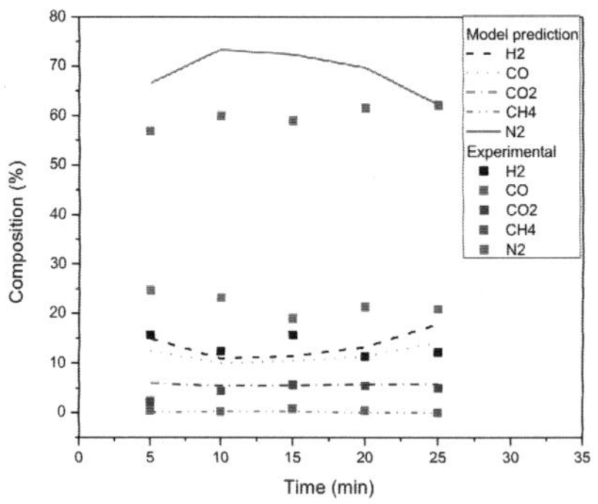

Comparison of Model predicted product gas composition with experimental data for ER 0.2054

Results: Moisture Content

- Moisture content of biomass decides the composition of syngas
- Suitability of a biomass depends on the gasification process
- For downdraft biomass gasification the moisture content range varies between 5 % to 18 % wet basis
- Higher moisture content leads to deterioration of product gas quality

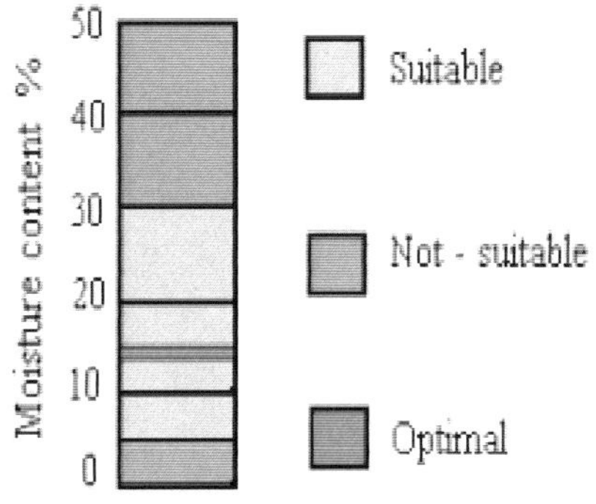

Moisture Scale for Gasification

Results: Moisture Content

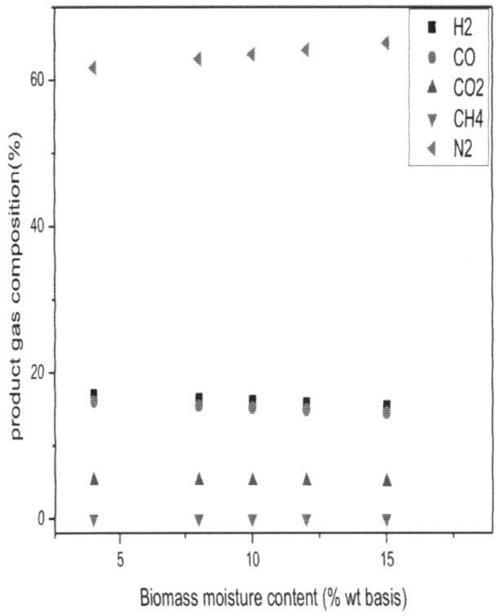

Model predicted product gas composition for different moisture content

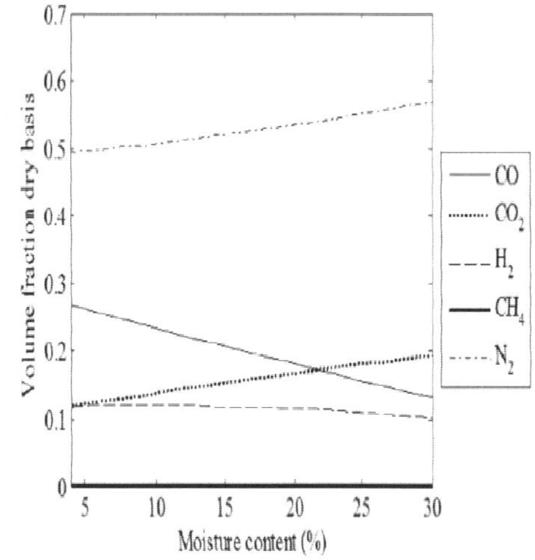

product gas composition for different moisture content reported by Atnaw *et al.* (2017)

ATNAW, S., SULAIMAN, S., SINGH, L., WAHID, Z., KU MOHAMMAD FAIZAL BIN CHE KU YAHYA, C.. Modeling and Parametric Study for Maximizing Heating Value of Gasification Syngas. **BioResources**, North America, 12, feb. 2017

Results: Reduction zone length

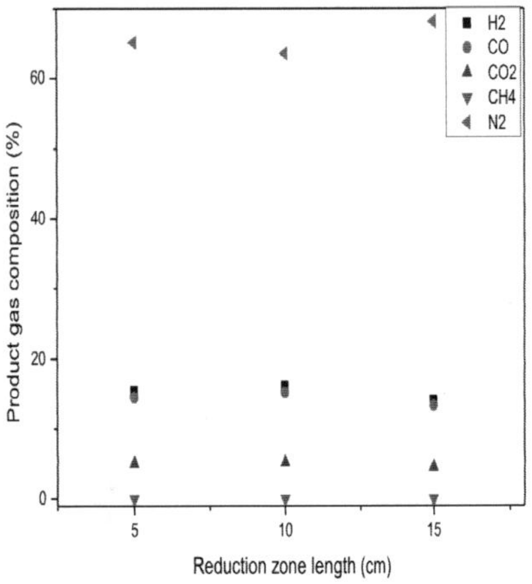

Model predicted product gas composition for different reduction zone lengths

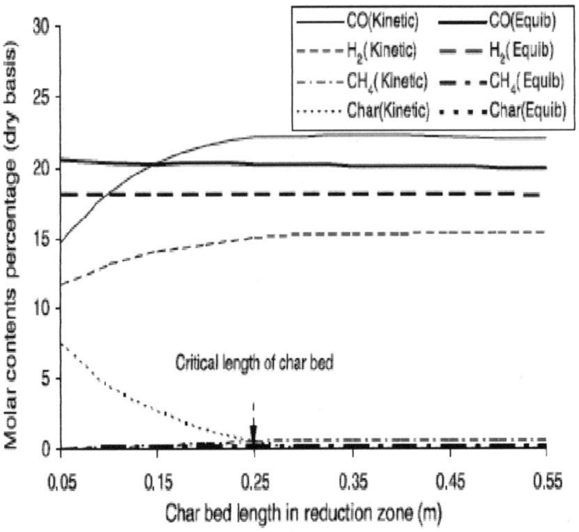

Model predicted product gas composition for different char bed lengths reported by Sharma (2008)

Sharma AK. Equilibrium and kinetic modeling of char reduction reactions in a downdraft biomass gasifier: A comparison. Solar Energy 2008;82(10):918-28.

Conclusions

- A comprehensive transport and kinetic model for downdraft biomass gasification has been developed
- The model has been validated with experimental data reported by Sheth *et al.* (2010).
- A parametric study using Air flow rate, biomass moisture content and reduction zone length has been carried out.
- The increase in air flow rate from 0.08 to 0.18 m/s results in decrease of CO and H_2 composition in the product gas and increase of N_2 composition
- With the increase in biomass moisture content from 4% to 15% (wet basis) CO composition decreases and N_2 composition increases
- Reduction zone length variation from 5 cm to 10 cm resulted in increase in CO and H_2 composition, however at 15 cm the composition of CO and H_2 decreased

Funding

I would like to thank Department of Science and Technology (DST) Govt. of India for funding me to attend and present my paper at AIChE Annual Meeting 2017 Minneapolis, MN, USA.

THANK YOU

BITS Pilani, Deemed to be University under Section 3 of UGC Act, 1956

498686 Site-Directed Mutagenesis of Family 64 CBM Provides Insights into the Anomalous Binding Interactions with Pretreated Cellulose during Biomass Saccharification

Abstract:

Lignocellulosic biomass, composed of cellulose, hemicellulose and lignin, is broken down into simple fermentable sugars in nature by cellulolytic enzyme cocktails secreted by various microbes. Enzymatically-catalyzed breakdown of cellulose into sugars like glucose has critical significance towards cost-effective production of cellulosic biofuels. Lignocellulosic biomass pretreatment methods can increase the enzymatic accessibility of cellulose microfibrils and/or change the crystalline ultrastructure of cellulose that can also impact enzyme activity (1). Cellulose III, an unnatural allomorph of cellulose, is formed during anhydrous liquid ammonia based pretreatment of the native crystalline cellulose I allomorphic state (*1-2*). Cellulose III has already been shown to have anomalous binding interactions with carbohydrate-binding modules and cellulolytic enzymes (e.g., cellulases) in our previous studies *(3)*. Here, we report results from our site-directed mutagenesis study of a novel carbohydrate-binding module (CBM) belonging to family 64, followed by detailed binding characterization of the mutant CBM library with both cellulose I and cellulose III. These studies shed light into the structure-function relationships guiding the binding of native and engineered CBMs to unnatural cellulose III, as opposed to native cellulose I. This work has direct implications on the cost-effective enzymatic hydrolysis of biomass into fermentable sugars in lignocellulosic biorefineries employing pretreatments that can modify the cellulose ultrastructure.

1. Chundawat, S. P. S. et al. Restructuring the crystalline cellulose hydrogen bond network enhances its depolymerization rate. J. Am. Chem. Soc. 133, 11163â€"11174 (2011).

2. da Costa Sousa, L. et al. Next-generation ammonia pretreatment enhances cellulosic biofuel production. Energy Environ. Sci. 9, 1215â€"1223 (2016).

3. Gao, D. et al. Increased enzyme binding to substrate is not necessary for more efficient cellulose hydrolysis. Proc. Natl. Acad. Sci. 110, 10922â€"10927 (2013).

Direct Succinic Acid Production from Non-Hydrolyzed Biomass Using Sequential Solid-State and Slurry Fermentation with Mixed Fungal Cultures

Jerico Alcantara [a,b], Andro Mondala [a], Logan Hughey [a], Shaun Shields [a]
jerico.alcantra@wmich.edu

[a] Western Michigan University
Department of Chemical and Paper Engineering

[b] University of the Philippines Los Baños
Department of Chemical Engineering

Succinic Acid

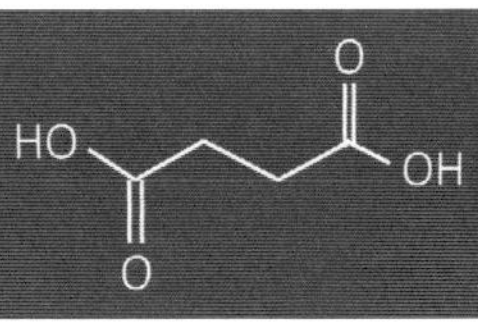

- a four-carbon 1,4-dicarboxylic acid
- one of US Department of Energy's top 12 value-added chemicals from biomass

Uses

Platform chemical	Specialty chemical
1,4-butanediol	Detergent and surfactants
Adipic acid	Ion chelators
2-pyrrolidone	Food ingredients
Tetrahydrofuran	Pharmaceuticals

Zeikus et al. (1999)

2013 Succinic acid market shares

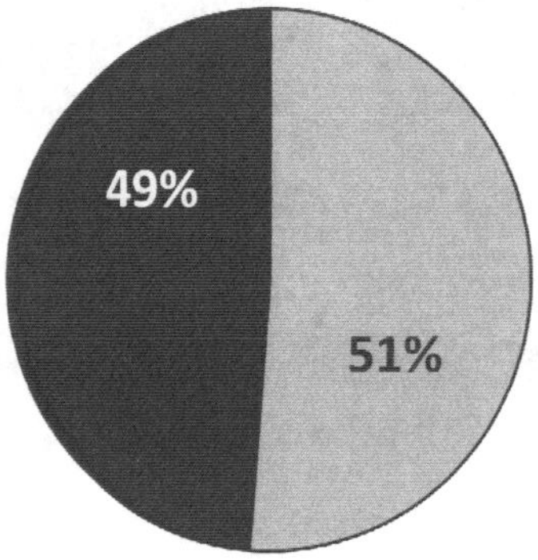

Bio-based
Fermentation
- yeast
- bacteria

38 ktons
$2,860/ton

Fossil-based
Maleic anhydride from butane
40 ktons
$2,500/ton

E4tech, RE-CORD and WUR (2015)

Metabolic Pathway for Bacterial Succinic Acid Fermentation

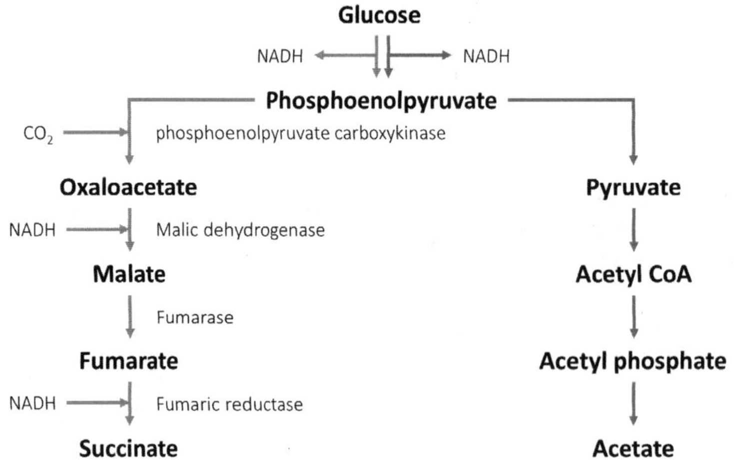

van Heerden and Nicol (2013)

Bacteria vs Fungi

	Bacteria	Fungi
Feedstock	Corn syrup Starch Sugar from energy rich crops	Non-pretreated plant biomass
Operational Requirement	Carbon dioxide/carbonate salt	not costly
Nutrient Requirement	Yeast extract Biotin	not costly

Jansen and van Gulik (2014)

Werpy and Petersen (2004)

Two-step fermentation process

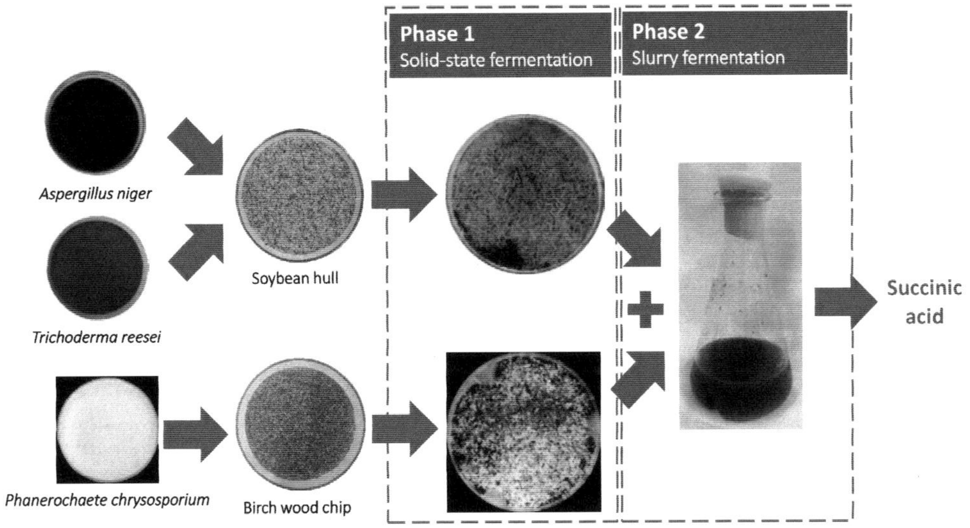

Solid-state Fermentation Pre-culturing

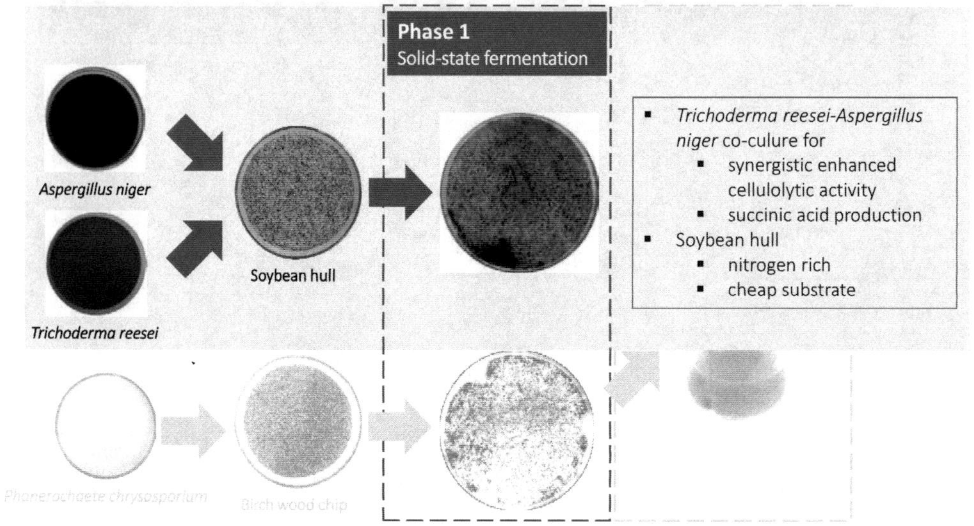

Solid-state Fermentation Pre-culturing

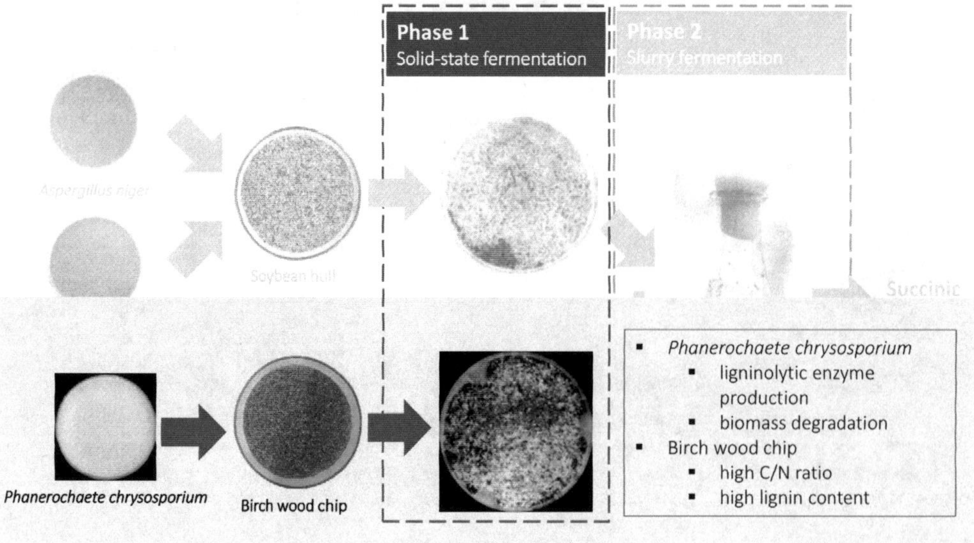

Slurry Fermentation

- Solid-state fermentation setups were combined and submerged in buffer with constant mixing
- Mixed fungal culture for simultaneous
  - lignocellulolytic enzyme production
  - biomass degradation and enzymatic saccharification
  - succinic acid production

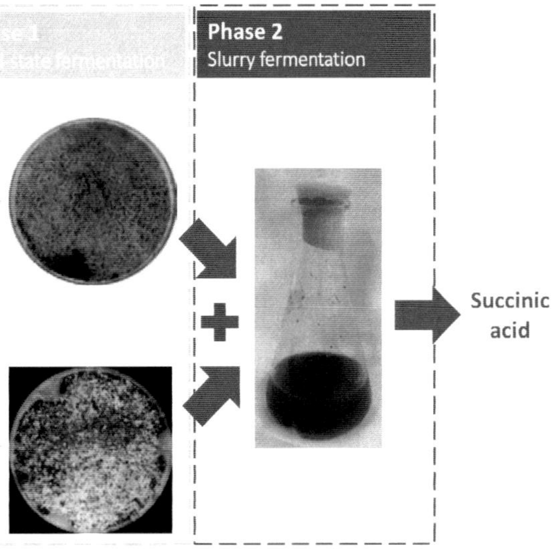

Methodology

Biomass preparation
(0.50 – 1.00 mm)

Inoculum preparation
(2 days)

Solid-state fermentation
(30 °C, 7 days, 95% RH, 70% moisture)

Slurry fermentation 1 (SF1)
(30 °C, 150 rpm, pH 4.8 0.05 M acetate buffer 20% solids)

3:1 BWC to SH ratio

Slurry fermentation 2 (SF2)
(30 °C, 150 rpm, pH 4.8 0.05 M acetate buffer 20% solids)

4:1 BWC to SH ratio

non-pretreated wood chip
(< 0.5 mm)

Slurry Fermentation 1 (SF1)

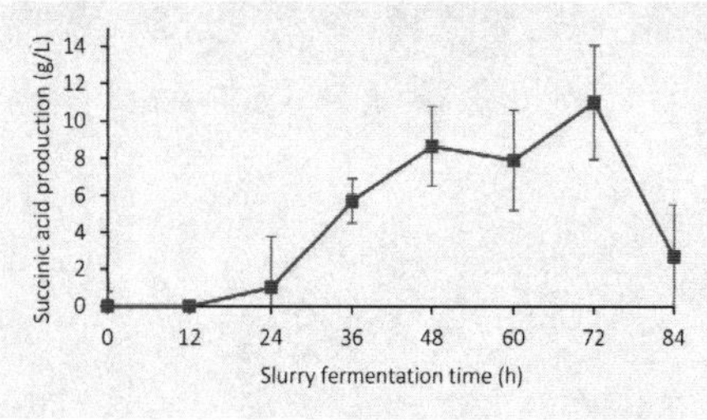

Succinic acid production

- Succinic acid was the main product of slurry fermentation
- Oxalic acid and citric acid were detected in small quantities (< 1 g/L)
- Little to none of succinic acid produced within 24 h
- Maximum succinic acid production of 11.01 g/L after 72 h
- Volumetric productivity of 0.15 g/L·h

Slurry Fermentation 1 (SF1)

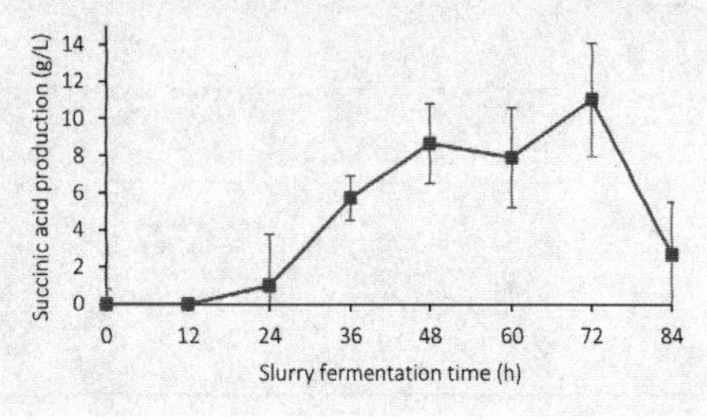

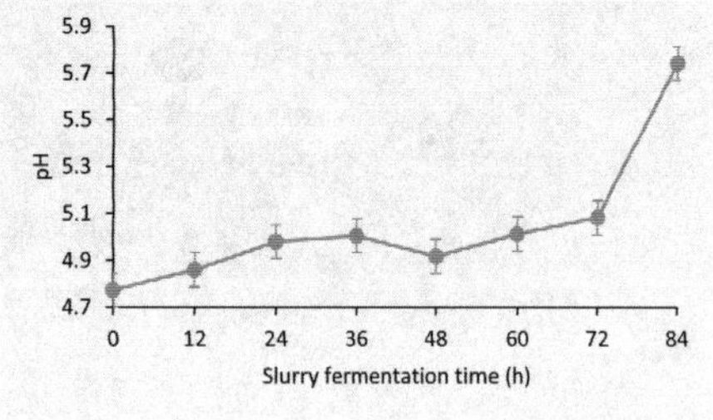

pH

- Initial pH was set at 4.8 for optimal cellulolysis
- Succinic acid production occurred within pH range of 4.77 to 5.08
- High succinic acid present in the slurry mixture resulted to low pH
- After 72 h, succinic acid production dropped resulted to drastic increase in pH

Slurry Fermentation 1 (SF1)

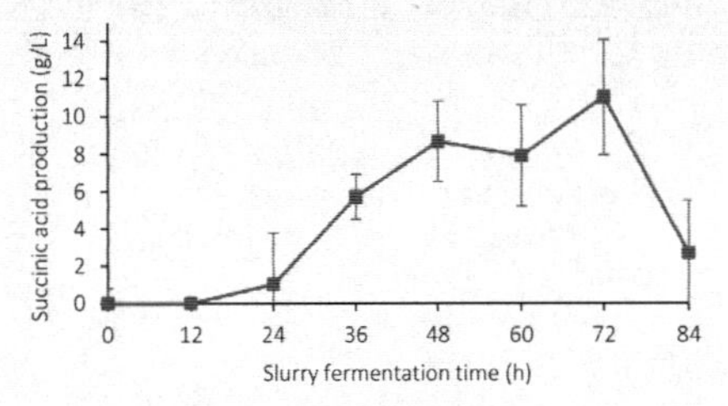

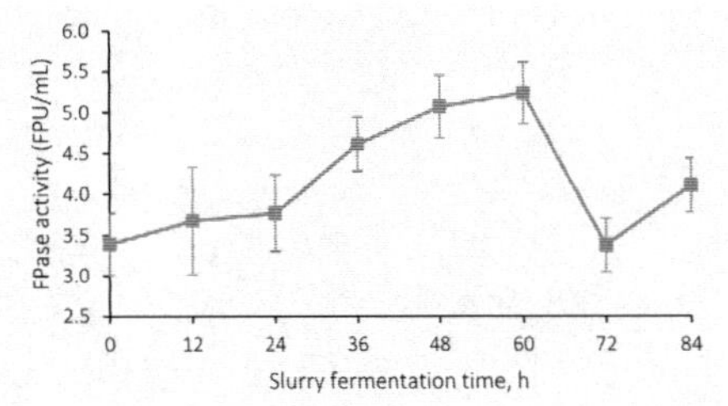

Cellulolytic enzyme activity

- Succinic acid concentration increases with increasing cellulose activity
- Sugars were released from the lignocellulosic matrix resulting to increase in succinic acid production
- After 60 h, cellulase activity decreased, high succinic acid concentration may inhibited cellulolytic enzyme activity
- After 72 h, cellulase activity started to increase again as succinic acid concentration decreased.

Slurry Fermentation 1 (SF1)

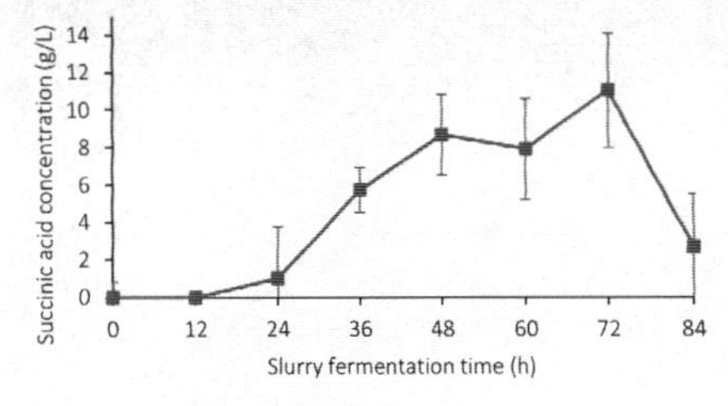

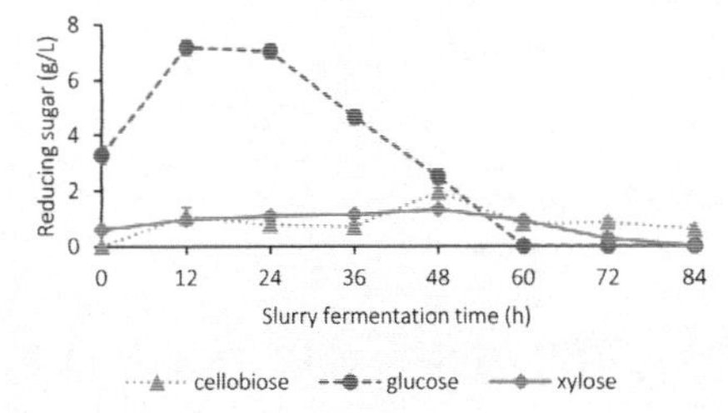

Reducing sugar

- Residual glucose was doubled within 12 h
- After 24 h, glucose started to decline
- Succinic acid increases as residual sugar decreases
- Residual sugars decreased with decrease in cellulase activity

Slurry Fermentation 2 (SF2)

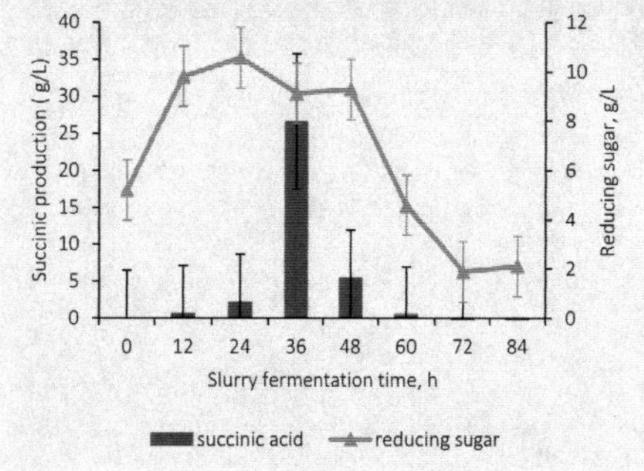

Addition of non-pretreated wood chip

- Non-pretreated birch wood chip was added at the start of slurry fermentation
- Additional carbon rich substrate resulted to higher reducing sugars liberated during cellulolysis
- Maximum succinic acid production of 26.62 g/L succinic acid after 36 h
- Time to reach maximum production shifted
- Volumetric productivity increased to 0.74 g/L·h
- Productivity increased 5 folds

Batch fermentation for succinic acid production

	Microorganism	Substrate	Time (h)	SA ($g L^{-1}$)	Productivity ($g L^{-1} h^{-1}$)	References
Bacteria	*A. succinogenes*	glucose	78	105.8	1.34	Guettler et al. (1996)
Bacteria	*A.succiniciproducens*	glucose	19	34.2	1.8	Lee et al. (1999)
Bacteria	*A.succiniciproducens*	glucose + corn steep liquor	13.5	17.8	1.32	Lee et al. (2000)
Bacteria	*A. succinogenes*	wheat bran	54	64	1.19	Du et al. (2008)
Bacteria	*A. Succinogenes*	sugarcane bagasse	121	52	0.43	Borges and Pereira (2011)
Bacteria	*A.Succiniciproducens*	wood hydrolysate + corn steep syrup	32	23.8	0.74	Lee et al. (2003)
Fungi	*A. Niger*	molassses			0.14	Ikram-ul et al. (2004)
Fungi	Fungal mixed culture	birch wood chip + soybean hull (3:1)	72	11.01	0.15	this study
	Fungal mixed culture	birch wood chip + soybean hull (4:1)	36	26.62	0.74	this study

Summary

Highlights

- Maximum production of 26.62 g/L after 36 h of slurry fermentation
- Maximum volumetric productivity of 0.74 g/L·h
- Productivity increased 5 folds when fresh carbon rich substrate was added

Advantage

- Much cheaper substrate
- No chemical pretreatment
- Minimal operational and nutrient requirement

Disadvantage

- Biological pretreatment (i.e., solid-state fermentation) is relatively slow

Engineering implication

- Results of this study can be used for future scale up of this consolidated bioprocessing technology

References

Borges, Pereira Jr., E. R. N., 2011. Succinic acid production from sugarcane bagasse hemicellulose hydrolysate *by Actinobacillus succinogenes*. J. Ind. Microbiol. Biotechnol. 38, 1001–1011.

Du, C., Lin, S. K. C., Koutinas, A., Wang, R., Dorado, P., Webb, C., 2008. A wheat biorefining strategy based on solid-state fermentation for fermentative production of succinic acid. Bioresour. Technol. 99, 8310-8315.

E4tech, RE-CORD and WUR, 2015. From the Sugar Platform to biofuels and biochemicals. Final report for the European Commission, contract No. ENER/C2/423-2012/SI2.673791

Guettler, M.V., Jain, M.K., Rumler, D., 1996. Method for making succinic acid, bacterial variants for use in the process, and methods for obtaining variants. US patent 5,573,931

Ikram-ul, H., Ali, S., Qadeer, M. A., Iqbal, J., 2004. Citric acid production by selected mutants of *Aspergillus niger* from cane molasses. Bioresour. Technol. 93 (2), 125-130.

Jansen, M. L. A., van Gulik, W. M., 2014. Towards large scale fermentative production of succinic acid. Curr. Opin. Biotechnol. 30, 190-197.

Lee, P. C., Lee, W. G., Lee, S. Y., Chang, H. N., 1999. Effects of medium components on the growth of Anaerobiospirillum succiniciproducens and succinic acid production. Process Biochem. 35, 49-55.

Lee, P. C., Lee, W. G., Lee, S. Y., Chang, H. N., Chang, Y. K., 2000. Fermentative production of succinic acid from glucose and corn steep liquor by Anaerobiospirillum succiniciproducens. Biotechnol. Bioprocess Eng. 5, 379-381.

Lee, P. C., Lee, S. Y., Hong, S. H., Chang, H. N., Park, S. C., 2003. Biological conversion of wood hydrolysate to succinic acid by *Anaerobiospirillum succiniciproducens*. Biotechnol. Lett. 25, 111–114.

van Heerden, C. D., Nicol, W., 2013. Continuous and batch cultures of Escherichia coli KJ134 for succinic acid fermentation metabolic flux distributions and production characteristics. Microbial Cell Factories, 12(80).

Werpy, P., Petersen, G., 2004. Top value added chemicals from biomass: Volume I - Results of screening potential candidates from sugars and synthesis gas. US Department of Energy.

Zeikus, J. G., Jain, M. K., Elankovan, P., 1999. Biotechnology of succinic acid production and markets for derived industrial products. Appl. Microbiol. Biotechnol. 51, 545-552.

494849 Enhanced Alcohols, Ketones and Organic Acids Production Via Co-Fermentation of Sugars and Gases

Abstract:

Butanol is a better fuel than ethanol due to its compatibility with gasoline and existing fuel infrastructure. Butanol can also be converted into drop-in diesel and jet fuel by a simple hydrogenation step. Butanol has been produced biologically by the traditional acetone-butanol-ethanol (ABE) fermentation using molasses, starches and, recently, lignocellulosic biomass. The ABE process suffers from low conversion yields that hinder its commercialization. More than 50% of the carbon in sugars is wasted in producing H_2 and CO_2. Increased conversion efficiency of renewable raw materials to butanol could make this butanol production viable. The present study reports on a novel process for biological production of alcohols, organic acids and ketone. This process consists of a co-fermentation of sugars and gaseous substrates for ABE production in a single fermentation system while converting the off gases containing H_2 and CO_2 to produce ethanol and acetic acid. Depending on the microorganism and reactor configuration used, the novel process can enhance alcohol yield by more than 25%. Media containing pure sugar and redcedar hydrolyzate were used in this study. The use of two-stage reactors resulted in a 19% improvement in ABE yield from sugar by *Clostridium acetobutylicum* ATCC 824 in the first stage with 60 g/L glucose and the conversion of generated H_2 and CO_2 by *Clostridium ragsdalei* to ethanol and acetic acid in the second reactor. The efficiency of H_2 and CO_2 conversion by *C. ragsdalei* was 72% and 16%, respectively. The total organic acid production in the co-fermentation process in the two-stage reactors was also increased by 141%. In addition to reducing CO_2 emissions, the co-fermentation of sugars and gases has the potential to enhance the feasibility of ABE fermentation by producing more alcohols from an otherwise waste gas stream. The results demonstrate that co-fermentation of sugars and gases to various products is feasible, and can be exploited to enhance a bioprocessing economy.

502541 Attainable Region Analysis for Production of Butyl Butyrate Via Biochemical Route

Abstract:

EcoLodge aims at providing a proof-of-concept for a new, integrated biotechnological production process for C8 ester butyl butyrate, a promising supplement/substitute for diesel and jet engine fuels, from lignocellulose derived C5 and C6 sugars. Previous studies suggest that butyl butyrate has excellent properties both as gasoline and as a diesel component [1]. Butanol and butyric acid are the process intermediates produced via anaerobic fermentation, which undergo enzymatic esterification yielding the product, butyl butyrate.

The attainable region concept employs a computational/geometrical procedure for determining the boundaries of the region that include all possible reactor products for a known feed condition [2]. In this paper, we applied the attainable region concept to study different process configurations for biochemical production of butyl butyrate. These are i) butyl butyrate reactor with the feed from the butanol and butyric acid reactors, ii) combined butanol and butyl butyrate production with the feed from butyric acid reactor, and iii) combined butyric acid and butyl butyrate production with the feed from butanol reactor. We used mathematical models and model coefficients from the literature for cell growth, butyric acid and butanol fermentations, and their esterification for butyl butyrate production. Models include cell growth inhibition due to butanol and butyric acid, and enzyme inhibition due to butanol, butyric acid and glucose [3-5].

In this paper, we demonstrated the capabilities of the attainable region analysis as a tool to assess different process configurations for butyl butyrate production. Process configuration ii) showed the highest capability for butyl butyrate production, since the butyric acid enhances the butanol production, and butanol esterification decreases butanol inhibition on the cell growth. The results of this study will be employed in the experimental phase of the project.

1. Lange, J. P., Price, R., Ayoub, P. M., Louis, J., Petrus, L., Clarke, L., & Gosselink, H. (2010). Valeric biofuels: a platform of cellulosic transportation fuels. Angewandte Chemie International Edition, 49(26), 4479-4483.
2. Milne, D., Glasser, D., Hildebrandt, D., & Hausberger, B. (2004). Application of the attainable region concept to the oxidative dehydrogenation of 1-Butene in inert porous membrane reactors. Industrial & engineering chemistry research, 43(8), 1827-1831.
3. Yang, X., & Tsao, G. T. (1994). Mathematical modeling of inhibition kinetics in acetone-butanol fermentation by Clostridium acetobutylicum. Biotechnology progress, 10(5), 532-538.
4. Varma, M. N., & Madras, G. (2008). Kinetics of synthesis of butyl butyrate by esterification and transesterification in supercritical carbon dioxide. Journal of chemical technology and biotechnology, 83(8), 1135-1144.
5. Song, H., Eom, M. H., Lee, S., Lee, J., Cho, J. H., & Seung, D. (2010). Modeling of batch experimental kinetics and application to fed-batch fermentation of Clostridium tyrobutyricum for enhanced butyric acid production. Biochemical Engineering Journal, 53(1), 71-76.

500314 Biomass Conversion to Functionalized Carbon

Abstract:

Biomass Conversion to Functionalized Carbon

Masoudeh Ahmadi, Jacek Jasinski, Jagannadh Stayavolu, Mahendra Sunkara*

Conn Center for Renewable Energy Research, University of Louisville, Louisville KY 40292, United States

*Corresponding Author: mksunk01@exchange.louisville.edu

In this research emphasis is placed on the applied methodology to produce functionalized carbon from hemp agriculture by products. We have studied an effective, economic, and novel technology to transform lignocellulosic biomass into the functionalized carbon. The effects of various parameters on the preparation such as carbonization methodology and impregnation ratio were studied and their effects on the textural properties such as surface area, pore volume and crystalline structure were discussed. There has been a large amount of research and interest in the conversion of biomass to carbons for various applications. However, the challenges remain both in terms of processing time scale and also with the ability to tune the resulting carbonÕs properties. In addition, the functionalization schemes for using the resulting carbons toward catalyst schemes require additional development. The challenges in utilizing carbon supports for catalysts include uniform dispersion of active catalyst metals and metal oxide clusters and improved catalyst and support interactions.

In this study, Hemp fibers were mixed with KOH solution with different impregnation ratio. Then carbonization was done in a horizontal tube furnace under nitrogen flow at 700ûC for one hour. Textural and chemical properties for the samples were investigated by means of XRD, SEM, BET, FTIR and XPS. The textural properties of carbon were found to be strongly dependent on the carbonization method and impregnation ratio. Reduction technique was used to load Pt particles on to the resulting carbon structures using tetraammine platinum nitrate as Pt precursor. The carbon supports were first impregnated with tetraammine platinum nitrate to produce 5%wt Pt/C. The mixture was further dried overnight at 100 ¡C followed by reduction under flowing hydrogen at 450¡C for 2 h using tube furnace to remove the ligands of the precursor and to reduce the metal to its active elemental state. In this presentation, various mechanistic models will be presented on the carbonization of biomass using different means.

501921 Synthesis of Water-Soluble Oligosaccharides As Potential Prebiotics Via Non-Enzymatic Sugar Glycosylation

Abstract:

Bioactive oligosaccharides have found promising applications as functional food, prebiotic compounds, immunostimulating agents, and pharmaceuticals. Oligosaccharides synthesis prevalently relies on biological methods such as enzymatic transglycosylation and controlled hydrolysis of polysaccharides. In this study, we demonstrated a novel non-enzymatic glycosylation method to produce water-soluble oligosaccharides in acidic lithium bromide trihydrate from monomeric sugars, in which new glycosidic bonds were formed between the C1 oxocarbenium on a protonated sugar and the hydroxyl groups (C2, C3 or C6) of another sugar. The oligosaccharide products were analyzed using HPIC, MALDI-TOF/MS, and NMR. The results indicated that up to 70% of the glucose could be converted to gluco-oligosaccharides. The resultant gluco-oligosaccharides were linked predominantly through Î±-1,6-glycosidic bonds with small qantity ofÂ Î²-1,6-, Î±-1,2-, and Î±-1,3-glycosidic bonds. Preliminary evaluation indicated that the oligosaccharides could be metabolized by gut microbiota; therefore, the oligosaccharides have the potential as prebiotics.

499334 Resonant Soft X-Ray Scattering of Cellulose Microstructure in Plant Cell Walls

Abstract:

Lignocellulose is the most abundant raw material for biofuels and the most important bio-renewable resource for building materials, paper, and textiles. A detailed understanding of the structure at the nano- and meso-scale could accelerate the development of an efficient biofuel conversion method. The plant cell wall is a heterogeneous mixture of various complex biopolymers including cellulose, hemicellulose, and pectin. The structure and assembly of these constituents in the plant cell wall is still not well understood. It is difficult to resolve the different elements through conventional hard X-ray scattering as they have similar electron density and hence, do not generate sufficient scattering contrast. The chemical specificity of resonant soft X-ray scattering (RSOXS) allows contrast to be generated based on differences in the chemistry of the different polysaccharides. By varying incident X-ray energies, we have achieved increased scattering contrast between cellulose and other polysaccharides from onions, such that features of the network structure of the cell wall are resolved. Thus, RSOXS reveals the packing distance of cellulose microfibrils embedded in the polysaccharide network.

499032 Lignocellulosic Biomass Pretreatment Using Radical Species

Abstract:

Currently, much effort is directed at moving away from fossil fuel based economy towards renewable feedstocks and the development of biorefineries. The major challenge for refining the complex chemical structure of biomass to biofuel is the separation of lignin from cellulose and hemicellulose. Wood-eating termites provide an excellent model for degradation of biomass and separation of lignin from cellulose and hemicellulose by the presence of radical species and hydroxyl radical intermediates. In order to validate the suspected effect of radical on degrading lignin and carbohydrate, activated persulfate systems were applied to produce different reactive oxygen species. Persulfate system is a combination of powerful oxidative agents, reductive species, and nucleophilic ones, which all in all make this system outstanding to attack lignin, and makes noteworthy changes which are ideal for biomass pretreatment.
The base-activated and heat-activated persulfate were used to treat 60 mesh wheat straw particles for different reaction time (30min-7 h), and the formation of radicals was confirmed by chemical probes including nitrobenzene and anisole. The results from FTIR spectra, SEM images, compositional analysis, and enzymatic hydrolysis showed that the lignin structure was removed and in somehow modified, leading to higher sugar yield from 20% in raw material to 40% and 70% corresponding to heat-activated persulfate and base-activated persulfate, respectively. The better sugar yield obtained in base-activated persulfate was correlated to the capability of this system to generate an array of different radicals such as sulfate radical anion, hydroxyl radical, superoxide radical anion and hydroperoxide anion. These diverse groups of radicals make persulfate system a widely-reactive oxidant that can potentially penetrate in biomass pores to remove lignin and make cellulose accessible.

496239 Stabilization of Carbohydrates with Formaldehyde during Integrated Biomass Depolymerization

Abstract:

Stabilization of Carbohydrates with Formaldehyde during Integrated Biomass Depolymerization

Biomass-derived carbohydrates are important platform molecules for the production of renewable fuels and chemicals. The production of carbohydrates from lignocellulosic biomass requires the extraction of lignin and the cleavage of ether bonds in hemicellulose (mostly xylan) and cellulose chains while minimizing further degradation of the resulting carbohydrates.(*1*) Current methods lead to incomplete biomass depolymerization (producing only polysaccharides) and high process costs due to mineral acid recovery and enzyme production.(*2*) Lowering acid use to improve process economics requires the use of higher temperatures and generally leads to significant sugar degradation and low yields, which is why these strategies have generally been difficult to implement.

Figure 1. (A) Xylose conversion to diformylxylose and furfural during biomass pretreatment (B) Glucose conversion to diformylglucose isomers and 5-hydroxymethylfurfural (5-HMF) during cellulose acid depolymerization.

Recently, we have discovered that formaldehyde (FA) could be used to stabilize lignin and facilitate the conversion of extracted lignin to monomers at high yields (up to 97% of theoretical yield).(*3*) This method could also prevent xylan degradation by producing a stable xylose-derived molecule that we refer to as diformylxylose. In the current work, we study the stabilization of carbohydrates by the addition of FA during integrated biomass depolymerization. The low water content and the acidic environment in the biomass pretreatment allow FA to react with xylose, forming diformylxylose (Figure 1-A) at yields above 90% and minimizing xylose degradation into furfural. In comparison, reactions without formaldehyde lead to almost full xylose degradation into furfural with only 16% xylose recovery. Diformylxylose could be used as is or converted back to xylose at high yields in aqueous environments. A similar process is observed with glucose during cellulose acid depolymerization, forming diformylglucose (DG). The presence of FA led to the formation of two DG isomers from glucose by forming 1,3-dioxolane and 1,3-dioxane structures (Figure 1-B). As with DX, these structures stabilize glucose after depolymerization at conditions that were previously unfavorable due to carbohydrate degradation. Future efforts include the characterization of protected sugars and the removal of their protective groups.

References:

1. J. S. Luterbacher *et al.*, Nonenzymatic Sugar Production from Biomass Using Biomass-Derived γ-Valerolactone. *Science*. **343**, 277–280 (2014).

2. L. Shuai, Y. M. Questell-Santiago, J. S. Luterbacher, A mild biomass pretreatment using γ-valerolactone for concentrated sugar production. *Green Chem.* **18**, 937–943 (2016).

3. L. Shuai *et al.*, Formaldehyde stabilization facilitates lignin monomer production during biomass depolymerization. *Science*. **354**, 329–333 (2016).

500555 Milling, Crystallinity and Pyrolysis Rates for Cellulose and Whole Biomass

Abstract:

The effect of milling and cellulose crystallinity on pyrolysis rates of microcrystalline cellulose (MCC) and two forms of whole biomass (switchgrass and tall fescue) were studied using a combined experimental and computational approach. The crystallinity and particle size of milled samples were carefully determined along with the temperature at which the peak rate of decomposition occurs ($T^{DTG}{}_{max}$). A statistical analysis of the data shows that over a broad range of particle sizes from about 15 to 750 μm, only the crystallinity has a statistically significant effect on the $T^{DTG}{}_{max}$. Mirrored real-time high-temperature X-ray (RTHT-XRD) diffraction and thermalgravimetric analysis (TGA) show that while the onset of crystallinity loss is abrupt, it tends to lag the onset of weight loss as a function of temperature. The RTHT-XRD data shows that crystallinity of gently milled samples decrease by as much as 10% between room temperature and about 100 ^{o}C. Thereafter, crystallinity remains about constant until at between 200 and 250 ^{o}C a quantifiable recrystallization occurs followed by the sharp onset of crystallinity loss, lagging weight loss by as much as 25 ^{o}C. The notable recrystallization suggests high molecular mobility and the possibility of a liquid phase. Careful mirrored experiments using MCC suggest that crystallinity is retained until between 30 and 40% of the mass has been volatilized. Molecular dynamics simulation for cellulose deploymerization to form levoglucosan (LVG) assuming random dissociation of the Î²-glycosidic bond and bond rearrangement are consistent with the experimental observations. Upon removal of between 20 to 30% of the thus transformed glucose units, the simulated cellulose structure abruptly transitions from crystalline to amorphous. Notably, the retention of at least a portion of the formed LVG (as a condensed phase) appears to help extend crystallinity until somewhat higher levels of conversion.

501557 Effect of Pretreatment on Biomass Pyrolysis for Better Quality of Bio-Oil

Abstract:

Effect of Pretreatment on Biomass Pyrolysis for Better Quality of Bio-oil

Yuan Xue, Joel Braden, Xianglan Bai

Biomass is recognized as an important sustainable resource for energy and chemicals. Typical biomass, such as agricultural wastes and forest residues, can contain up to 12% ash as received. As the major components of ash, alkali and alkaline earth metals (AAEM) naturally presenting in biomass are known to have strong detrimental effect on biomass pyrolysis by promoting char and gas formations. Washing biomass by diluted acid or directly infusing acid into biomass prior to pyrolysis can increase bio-oil yield and greatly improve sugar production. It has been considered that acid pretreatments improve biomass pyrolysis either by removing AAEM as acid soluble or passivating catalytic effect of AAEM. On the other hand, it has also been reported that pyrolysis of acid pretreated biomass reduces the recovery of lignin-derived phenolic monomers and oligomers. The acid pretreated biomass (either by acid-washing or acid-infusion) also easily agglomerates due to the formation of unknown melts during pyrolysis. Although various theories have been suggested, the effects of acid treatment on lignin and carbohydrates in AAEM containing biomass are still not well understood. In the present study, biomass is subjected to various types of pretreatments, including acid treatment, oxidative treatment and alkaline delignification to decouple the effects of AAEM on lignin and cellulose in biomass. The biomass pretreated under various conditions were characterized and further fast pyrolyzed. The overall goal of this study is to increase the quality of bio-oil from biomass pyrolysis through effective biomass pretreatment.

IOWA STATE UNIVERSITY
Department of Mechanical Engineering

Experimental Multiphase Flow Laboratory

An Examination of Ash from Fast Pyrolysis

Dane R. Erickson, Tannon J. Daugaard, Mark Mba-Wright, and Theodore (Ted) J. Heindel
Department of Mechanical Engineering
Iowa State University
Ames, Iowa

AIChE Annual Conference
31 October 2017

IOWA STATE UNIVERSITY
Department of Mechanical Engineering

Presentation Outline

Project Objectives

1. Determine optimal conditions for heat carrier regeneration for a fast pyrolysis process
2. Design a collection cyclone to capture elutriated ash during the process of regeneration
3. Analyze the content of ash that is elutriated

Conclusions

1. Ash can efficiently elutriate from a fluidized bed regenerator
2. A collection cyclone can capture a significant amount of elutriated ash-sized particles
3. Elutriation efficiency and the amount of elutriated ash in general increases as the sweep gas flow rate through the fluidized bed regenerator increases
4. Optimal heat carrier regeneration conditions can be determined once data is collected at additional temperatures and flow rates

IOWA STATE UNIVERSITY
Department of Mechanical Engineering

Introduction

Fast Pyrolysis

- A process in which biomass is rapidly heated to a temperature of approximately 500°C in the absence of air
- An earlier stage of this project involved fast pyrolysis within an auger reactor
- Auger reactor fast pyrolysis is limited by the economic need for continuous operation

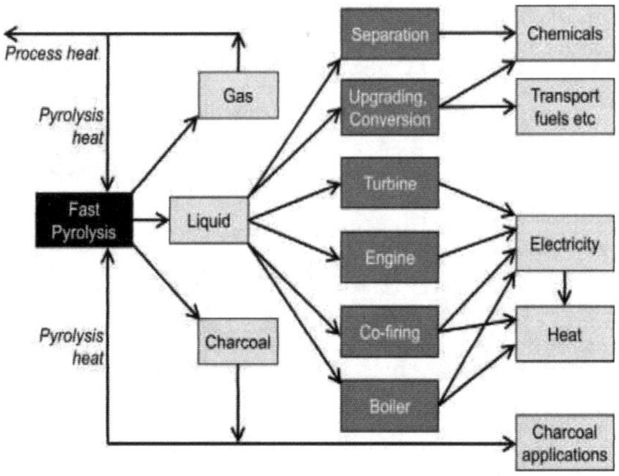

Illustration of the fast pyrolysis process [1].

[1] Bridgwater, A.V. "Review of Fast Pyrolysis of Biomass and Product Upgrading." *Biomass and Bioenergy*, ScienceDirect, 3 Mar. 2011, www.sciencedirect.com/science/article/pii/S0961953411000638.

IOWA STATE UNIVERSITY
Department of Mechanical Engineering

Motivation

Clean Energy

- Climate change is an impending threat
- Clean energy solutions should be developed in a timely manner

Improving Economic Feasibility

- In order for fast pyrolysis to be affordable on a large scale, the process must be run continuously
- Heat carrier must be regenerated
- Ash from regeneration must be removed in order for continuous operation to be possible
- Room for improvement by recycling energy from the downstream end of the process

Red Oak (300-710μm) used in pyrolysis experiments.

IOWA STATE UNIVERSITY
Department of Mechanical Engineering

Regenerator Setup

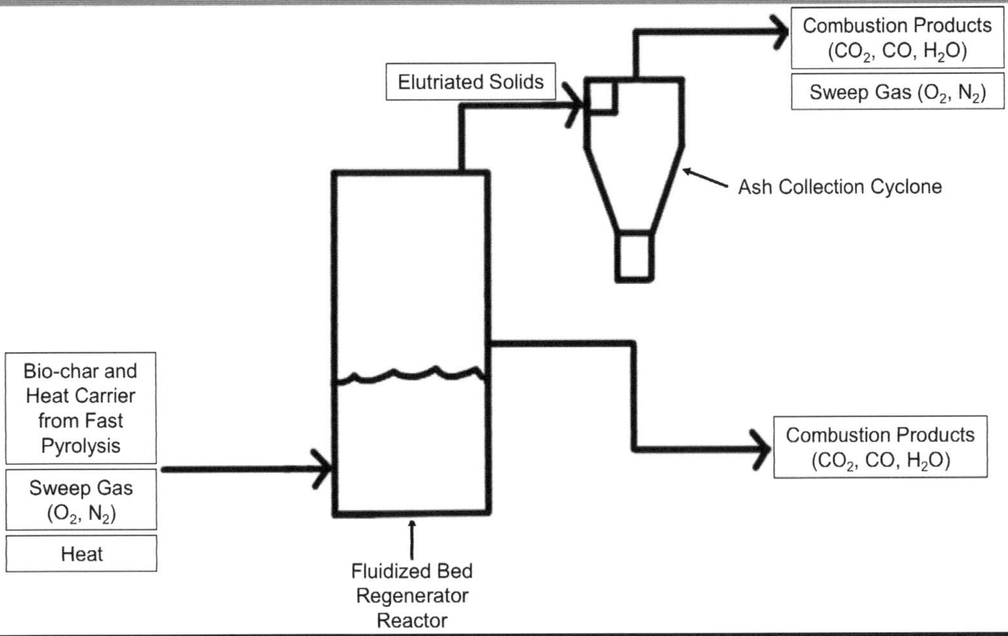

IOWA STATE UNIVERSITY
Department of Mechanical Engineering

Cyclone Modeling

Barth Model

Assumptions [2]:

- Ash particles are approximately spherical
- The sweep gas has a radial velocity equal to zero
- Stokes Law can adequately represent the radial force on the ash particle
- Adjacent particles to not affect the motion of other particles
- Force balances are sufficient in estimating cyclone dimensions

Parameters:

- **D**: Body diameter
- **D_x**: Diameter of the gas outlet
- **D_d**: Diameter of the dust outlet
- **H**: Total height
- **H_c**: Height of the conical portion
- **a** and **b**: Dimensions of the inlet port
- **S**: Length of the vortex finder

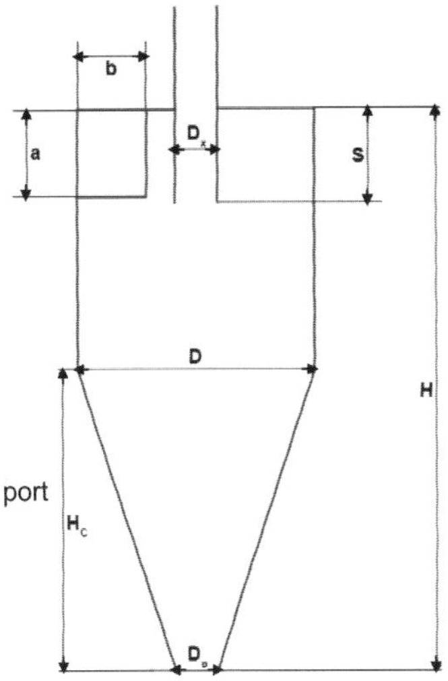

Generic cyclone drawing with dimensions.

Barth model design literature-based correlations [3].

	a	b	D_x	D_d	H	S	H_c
Barth	0.53D	0.13D	0.33D	0.33D	2.58D	0.73D	1.89D

[2] Altmeyer, S. *Comparison of Differential Models of Cyclone Prediction Performance for Various Operating Conditions Using a General Software.* 2003.
[3] Crowe, C., E. Michaelides, and J. Schwarzkopf, *Multiphase Flow Handbook.* 2016, Taylor & Francis/CRC: Boca Raton, FL. p. 846.

IOWA STATE UNIVERSITY
Department of Mechanical Engineering

Cyclone Modeling

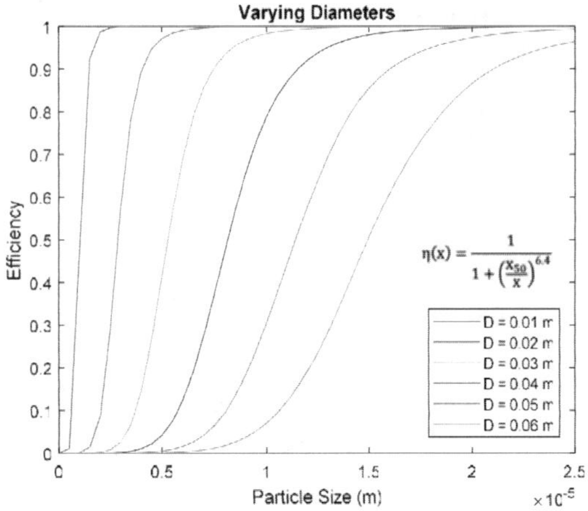

Examination of cyclone separation efficiency as a function of cyclone diameter.

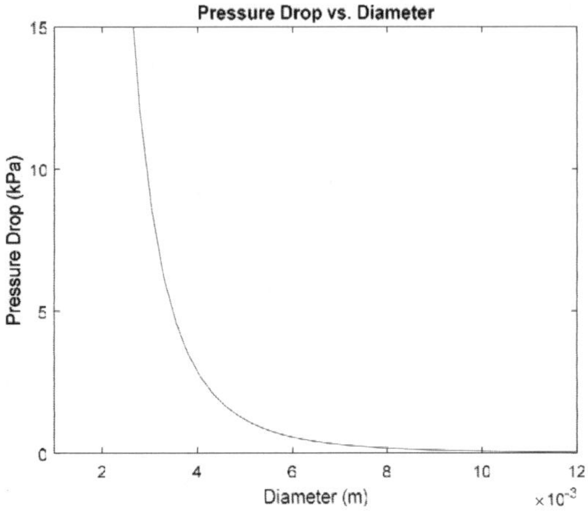

Examination of cyclone pressure drop as a function of cyclone diameter.

IOWA STATE UNIVERSITY
Department of Mechanical Engineering

Cyclone Design

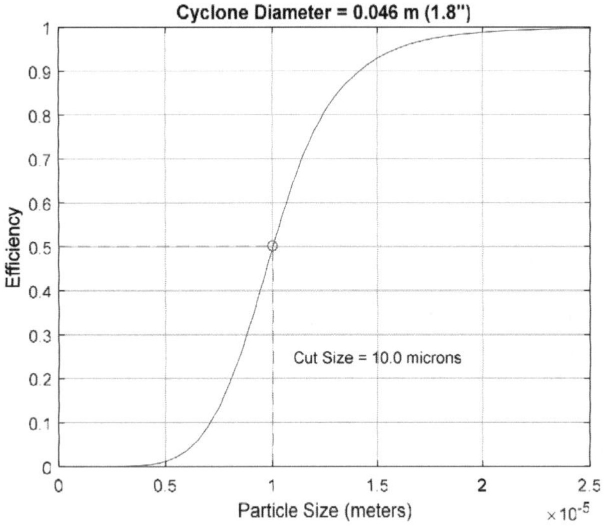

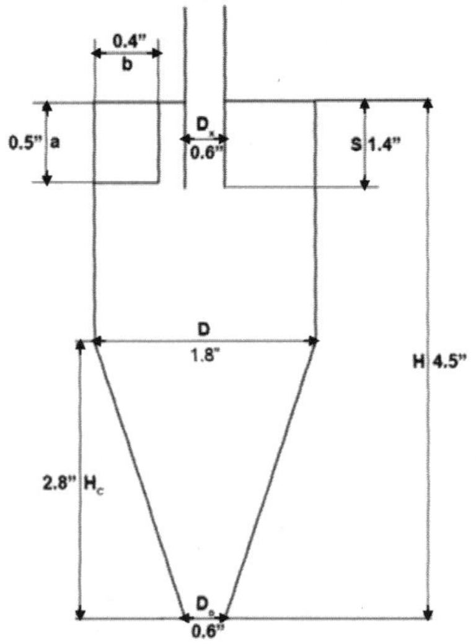

Cyclone design dimensions for a 50% cut-size of 10.0 microns. Design assumes air as the sweep gas.

IOWA STATE UNIVERSITY
Department of Mechanical Engineering

Selected Cyclone

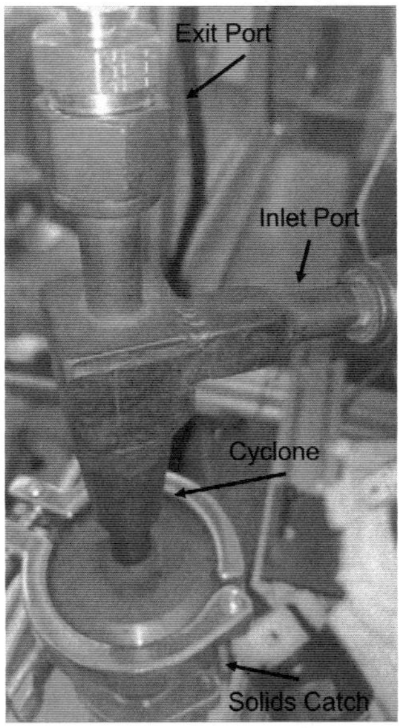

In-house cyclone utilized for experiments.

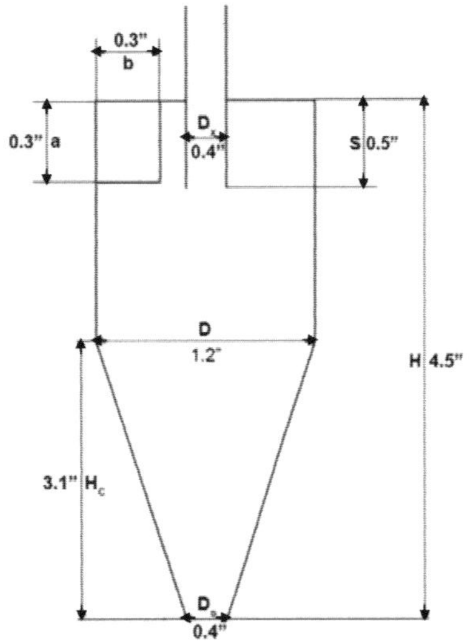

In-house cyclone dimensions.

Knowledge. Innovation. Leadership. www.me.iastate.edu 9

IOWA STATE UNIVERSITY
Department of Mechanical Engineering

Experimental Operation

- Regenerator operated at temperatures ranging from 450°C to 750°C
- At each temperature, air sweep gas flow rates are applied ranging from the minimum fluidization velocity to 2.5x the minimum fluidization velocity
- Samples are collected from the cyclone catch as well as from the fluidized bed following each reactor run
- Samples consistently contain a combination of ash, char, and heat carrier (silica)

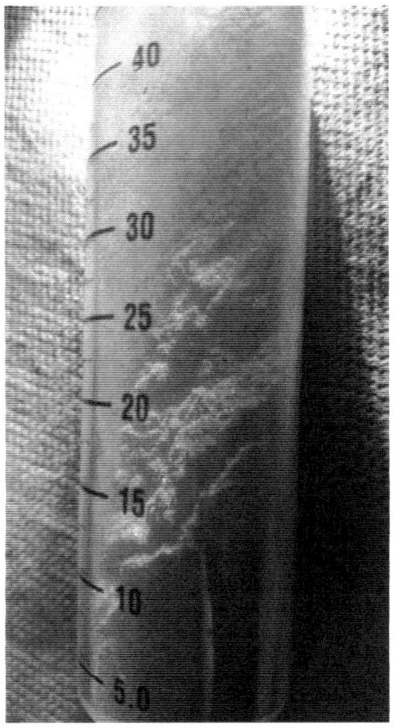

Cyclone catch sample.

Knowledge. Innovation. Leadership. www.me.iastate.edu 10

IOWA STATE UNIVERSITY
Department of Mechanical Engineering

Microscopic Imaging

Magnification of a cyclone catch sample collected after a 550°C, 16 SLPM (1x u_{mf}) reactor run.
(Image Credit: Meng Han, PhD Candidate at Iowa State University)

Knowledge. Innovation. Leadership. www.me.iastate.edu 11

IOWA STATE UNIVERSITY
Department of Mechanical Engineering

Solids Catch Separations

- In order to analyze the ash content, first need to distinguish it from the rest of the solids catch
- Accomplished using combination of Thermogravimetric Analysis (TGA) and acidic digestion via Inductively Coupled Plasma Mass Spectrometry (ICP)

$$m_{Cyclone} = m_{Sand} + m_{Char} + m_{Ash}$$

$$m_{Sand} = (m_{Cyclone})(\ wt\%_{Sand})$$

$$m_{Char} = (m_{Cyclone})(wt\%_{Volatiles} + wt\%_{Fixed\ Carbon} + wt\%_{Moisture})$$

$$m_{Ash,\ Actual} = m_{Total\ Ash,\ Cyclone} - (m_{Char})\ (wt\%_{Ash,\ Char\ Feedstock})$$

Knowledge. Innovation. Leadership. www.me.iastate.edu 12

IOWA STATE UNIVERSITY — Department of Mechanical Engineering

TGA Example

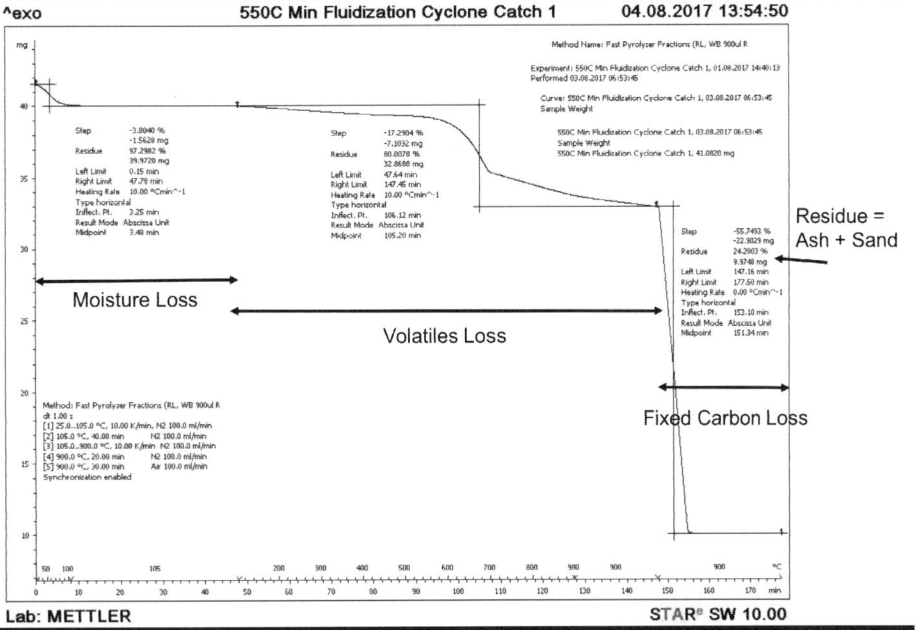

IOWA STATE UNIVERSITY — Department of Mechanical Engineering

Ash Elutriation Results

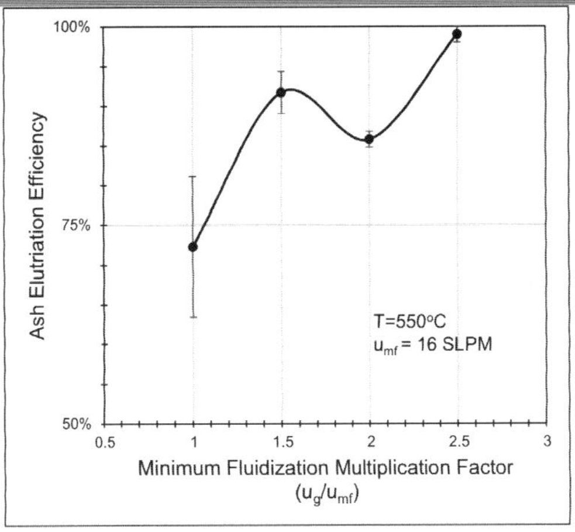

Change in ash elutriation efficiency as a function of sweep gas velocity.

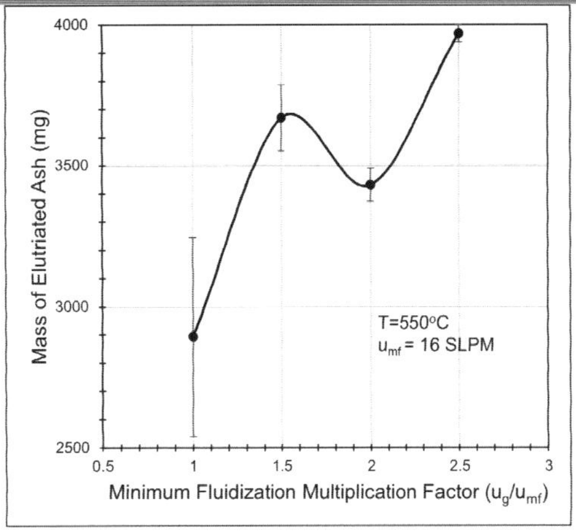

Change in mass of elutriated ash as a function of sweep gas velocity.

$$\eta_{Ash\ Elutriation} = \frac{m_{Actual\ Ash\ Elutriated}}{m_{Theoretical\ Ash\ Elutriated}}$$

In general, increasing the sweep gas velocity increases the amount of elutriated ash.

IOWA STATE UNIVERSITY
Department of Mechanical Engineering

Elemental Ash Analysis

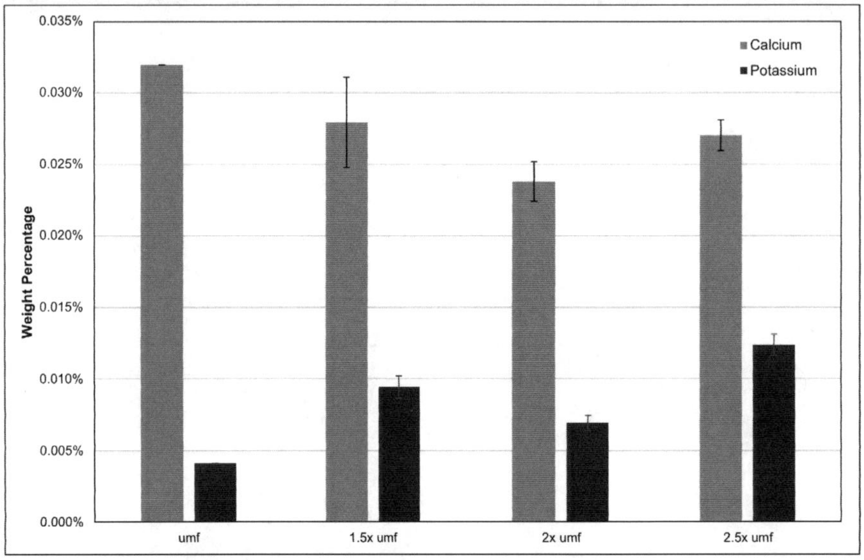

NOTE
Other elements, including **iron**, **magnesium**, **manganese**, **phosphorus**, and **aluminum** have also been identified in trace amounts.

Calcium and potassium weight percentages in ICP cyclone catch samples. Samples taken from reactor runs at 550ºC and varying flow rate. (u_{mf} = 16 SLPM)

Knowledge. Innovation. Leadership. www.me.iastate.edu 15

IOWA STATE UNIVERSITY
Department of Mechanical Engineering

Final Remarks

Project Objectives

1. Determine optimal conditions for heat carrier regeneration for a fast pyrolysis process
2. Design a collection cyclone to capture elutriated ash during the process of regeneration
3. Analyze the content of ash that is elutriated

Conclusions

1. Ash can efficiently elutriate from a fluidized bed regenerator
2. A collection cyclone can capture a significant amount of elutriated ash-sized particles
3. Elutriation efficiency and the amount of elutriated ash in general increases as the sweep gas flow rate through the fluidized bed regenerator increases
4. Optimal heat carrier regeneration conditions can be determined once data is collected at additional temperatures and flow rates

Knowledge. Innovation. Leadership. www.me.iastate.edu 16

Acknowledgements

Finally, a huge thanks to the following for their help in making this project possible:

- Project Funding:
  - **Phillips 66**
- Laboratory Assistance:
  - **Dr. Patrick Johnston**
- Undergraduate Students:
  - **Angela Wagner**
  - **Zach Paskach**
  - **Omer Mahmoud**

An Examination of Ash from Fast Pyrolysis

Dane R. Erickson, Tannon J. Daugaard, Mark Mba-Wright and Theodore (Ted) J. Heindel
Department of Mechanical Engineering
Iowa State University
Ames, Iowa

AIChE Annual Conference
31 October 2017

494900 Introducing Catalytic Diversity into Single-Site Zeolites of Fixed Composition Via Synthetic Control of Active Site Proximity

Abstract:

Introducing Catalytic Diversity into Single-Site Zeolites of Fixed Composition via Synthetic Control of Active Site Proximity

John R. Di Iorio, Claire T. Nimlos, Rajamani Gounder*

Charles D. Davidson School of Chemical Engineering, Purdue University, 480 Stadium Mall Drive, West Lafayette, IN 47907, USA

Distinct Brønsted acid site (H^+) ensembles within a given zeolite arise from differences in the arrangement (Al-O(-Si -O)$_x$-Al) of framework Al atoms between isolated (x = 3) or paired configurations (x = 1, 2), the latter defined functionally by their ability to exchange extraframework divalent cations. The effects of proton proximity on hydrocarbon catalysis have been previously studied in MFI zeolites,[1-2] but the concomitant changes in Al distribution among different crystallographically distinct tetrahedral sites (T-sites) and void environments (i.e., straight and sinusoidal channels, and their intersections) have precluded unambiguous kinetic assessments of proton proximity in MFI. Here, we use chabazite (CHA) zeolites, a high-symmetry framework with only 1 unique lattice T-site, to develop a predictive relationship that describes how synthetic methods can be developed to prepare zeolites, of fixed topology and elemental composition, with systematically different framework Al arrangements. Such arrangements, in turn, generate different proton active site ensembles that markedly influence turnover rates for the Brønsted acid-catalyzed dehydration of methanol to dimethyl ether, a known probe reaction of solid acids.[3] Specifically, we extend previously reported synthetic procedures[4] to crystallize CHA zeolites at similar composition (Si/Al = 15), but with fractions of paired Al sites that systematically varied from the limit of site isolation (0% paired Al) to nearly half of the sites (44% paired Al) in paired arrangements. This series of CHA zeolites is used to demonstrate that first and zero-order rate constants (415 K, per total H^+) of methanol dehydration are one order of magnitude larger on paired than on isolated protons. *In situ* IR spectra measured during steady-state dehydration catalysis are used to observe the formation of surface methoxy species at paired Al sites in CHA, which are not observed on isolated protons in CHA or MFI, providing evidence that alternate dehydration pathways are accessible on paired protons in CHA zeolites. These results represent a step toward the synthetic control of active site atomic arrangement and establish a predictive synthesis-structure-function relation for Brønsted acidic CHA zeolites, demonstrating that catalytic diversity can arise solely from differences in the atomic arrangement of active sites, even among single T-site zeolites of fixed composition.

References:

1. Janda, A.; Bell, A. T. *J. Am. Chem. Soc.* **2013**, *135*, 19193-19207

2. Bernauer, M.; Tabor, E.; Pashkova, V.; Kaucký, D.; Sobalík, Z.; Wichterlová, B.; Dedecek, J. *J. Catal.* **2016**, *344*, 157-172.

3. Carr, R. T.; Neurock, M.; Iglesia, E. *J. Catal.* **2011**, *278*, 78-93.

4. Di Iorio, J. R.; Gounder, R. *Chem. Mater.* **2016**, *28*, 2236-2247

493983 Probing the Site-Pairing Preference of Framework Aluminum in ZSM-5 with $Ga(CH_3)_3$

Abstract:

Probing the Site-Pairing Preference of Framework Aluminum in ZSM-5 with $Ga(CH_3)_3$

Kyle Groden[a], Zachary Jones[b], Renqin Zhang[a], Susannah Scott[b], Jean-Sabin McEwen[a]

[a] Washington State University, WA, 99164;

[b] University of California, Santa Barbara, CA, 93106;

Figure 1. Comparison of the computed and the simulated XANES for methylene bridged Ga and how the Al distribution influences the Ga-Ga distance within a zeolite framework. The computed Ga-Ga distance is 2.96 Å, in strong agreement with experimental findings.

ZSM-5 is a zeolite catalyst marked with great versatility in catalytic application as a result of its flexibility in cation incorporation. Due to this property, the structure of ZSM-5 has been under question for its ability to accept divalent cations, implying that the substituted aluminum sites exist in paired configurations at specific locations within the zeolite framework. Trimethylgallium was used to probe these sites, as the dimethyl gallium fragments that exist in these locations following impregnation were thought to bind to nearby fragments via terminal methyl groups. Experimental extended x-ray absorption fine structure (EXAFS) spectroscopy data supports this "bridging" configuration due to the presence of potential gallium-gallium scattering signals. Zeolite cluster models were created containing potential sites for aluminum pairings with dimethyl gallium fragments, with optimized geometrical conformations obtained from first principles calculations based on density functional theory (DFT) implemented in the VASP program.[1,2] As methyl-bridged structures were unfavorable, additional cluster and periodic models were constructed to explore the possibility of carbon-hydrogen bond cleavage due to the reactivity of the methyl groups bound to the gallium centers. This reaction results in a methylene (CH_2) group joining the gallium centers in contrast with the methyl bridges previously hypothesized. X-ray absorption near edge spectra (XANES) were computed using CASTEP[3], another DFT program, for many of these configurations examined, which allows for a direct comparison with the synchrotron-based experimental results. As can be seen from Figure 1, the agreement between the gallium-gallium bond distances and spectral results from our calculations and those measured experimentally imply that paired aluminum sites are indeed present within the ZSM-5 architecture.

References:

[1] Kresse, G.; Furthmüller, J.*Phys. Rev. B* **1996**, *54* (16), 11169.

[2] Kresse, G.; Hafner, J. *Phys. Rev. B* **1993**, *47* (1), 558.

[3] Clark, S. J.; Segall, M. D.; Pickard, C. J.; Hasnip, P. J.; Probert, M. I. J.; Refson, K.; Payne, M. C. *Zeitschrift fur Krist.* **2005**, *220* (5-6), 567–570.

496765 Quantitative Analysis of Catalysis in Hierarchical Zeolites and Surface Topology Effect Based on MFI and MWW Frameworks

Abstract:

Recent advances in hierarchical zeolites have demonstrated the possibility of synthesizing zeolites with microporous domains down to single-unit-cell. The improved catalytic performances of such zeolites were mostly associated with the reduced diffusion limitation in microporous regions, and the better access to active sites on external surface. Here, liquid-phase alkylation, self-etherification and de-etherification reactions are performed at mild conditions using two 10MR acid zeolites, H-self-pillared-pentasil zeolite (H-SPP) and H-MCM-22. A kinetic study with quantitative analysis of above reactions shows the dominant contribution to catalytic performance from external active sites in comparison to their 3D counterparts, and therefore the increasing importance of surface topology effect in such hierarchical zeolites.

496772 Multi-Faceted Approach to Optimize ZSM-11 Catalysts for Methanol-to-Hydrocarbon Reactions

Abstract:

ZSM-11 (MEL) is an emerging catalyst for a wide variety of applications. The MEL framework is very similar to ZSM-5 (MFI), which is one of the most widely used zeolite catalysts in industry. Recent studies have shown that ZSM-11 has better catalytic performance in several different reactions compared to ZSM-5 [1-3], possibly due to improved intra-crystalline diffusion within the less-tortuous straight channels of the MEL topology; however, there are relatively few systematic studies of ZSM-11 synthesis, specifically with respect to tailored physicochemical properties such as size, shape, and Si/Al ratio (SAR). Here, we will present a systematic study of ZSM-11 catalyst preparation and testing for C_1 upgrading reactions. Through the judicious selection of synthesis parameters, we successfully developed a facile protocol to generate a library of ZSM-11 catalysts with tunable physicochemical properties (e.g., size, shape, and SAR). These methods for tailoring zeolite crystallization will be discussed along with catalytic studies of methanol-to-hydrocarbon (MTH) reactions. Collectively, our findings are part of a broader initiative to establish quantitative structure-performance relationships as a platform for optimizing zeolite catalysts.

[1] Kustova, M. Y.; Rasmussen, S. B.; Kustov, A. L.; Christensen, C. H., *Applied Catalysis B-Environmental* **2006**, *67* (1-2), 60-67.

[2] Zhang, L.; Liu, H.; Li, X.; Xie, S.; Wang, Y.; Xin, W.; Liu, S.; Xu, L., *Fuel Processing Technology* **2010**, *91* (5), 449-455.

[3] Bleken, F.; Skistad, W.; Barbera, K.; Kustova, M.; Bordiga, S.; Beato, P.; Lillerud, K. P.; Svelle, S.; Olsbye, U., *Physical Chemistry Chemical Physics* **2011**, *13* (7), 2539-2549.

500785 Atomic-Level Insights into Molecular Adsorption and Reaction Properties at Distinct Heteroatom Sites in Aluminosilicate Zeolite Catalysts

Abstract:

Heteroatom-containing zeolites are of technological interest because they exhibit high surface areas, highly uniform sub-nanometer pore dimensions, and high catalytic activities for diverse applications, including hydrocarbon conversion (e.g., fluidized catalytic cracking, FCC, and methanol-to-olefin, MTO, processes). Heteroatoms, such as B or Al, in zeolite frameworks are associated with molecular adsorption and reaction sites, and therefore the types and distributions of heteroatom environments strongly influence the macroscopic reaction properties of zeolite catalysts. However, measurements of local heteroatom environments, their distributions in zeolite frameworks, and their interactions with adsorbed molecular guest species, such as structure-directing organic molecules, reactants, or catalytic reaction products, have been extremely challenging. This has been due, in part, to non-stoichiometric zeolite framework compositions and distributions of both framework and extra-framework species that lack long range order, which are often exacerbated by post-synthesis treatments, such as calcination, steaming, or exposure to reaction conditions. Nevertheless, detailed atomic-level insights on heteroatom environments and their interactions with adsorbed molecular guest species are provided by solid-state nuclear magnetic resonance (NMR) spectroscopy, correlated with complementary transmission electron microscopy, X-ray diffraction, and catalytic reaction analyses. Notably, two-dimensional dynamic-nuclear-polarization (DNP)-enhanced NMR techniques provide substantially increased NMR signal sensitivity (ca. 100x) and resolution, which enables detection and correlation of specific heteroatom environments with the adsorbed molecules with which they interact. Such measurements have, until now, been infeasible by scattering or other spectroscopic techniques. Results will be presented for methanol, toluene, and carbonaceous deactivation products adsorbed on aluminosilicate zeolites Y and SSZ-13, which are associated with technologically important cracking or methanol-to-olefin reactions. These analyses identify heteroatom sites in the framework and at exterior surfaces that have distinct adsorption and reaction behaviors, providing useful new insights for the design of highly effective and stable catalysts.

500318 Synthesis and SCR Testing of SSZ-39

Abstract:

Synthesis and SCR Testing of SSZ-39

Ross Ransom[1], Jonathan Coote[1], Roger Moulton[2], Feng Gao[2], and Daniel F. Shantz[1]*

[1]Tulane University, New Orleans, Louisiana 70118 (USA)

[2]SACHEM, Inc., Austin, Texas 78704 (USA)

SSZ-39, the aluminosilicate anologue of the AEI ($AlPO_4$-18) topology, can be prepared in a straightforward manner using faujasite as an aluminum source[1]. This material shows great promise as a catalyst for deNOx catalysis,[2,3] and thus a practical, simple, and scalable synthesis is something that would be highly desirable. A variety of organic SDAs have been shown to afford SSZ-39 at various levels of efficacy, including N,N-dimethyl-3,5-dimethylpiperidinium, the organic SDA employed here. Prior work has shown that not only does this SDA lead to SSZ-39 formation, but has suggested that the isomer identity has a significant effect on growth kinetics. [4]

This talk will describe how the synthesis conditions used impact the growth kinetics and composition of the final material obtained from synthesis. There are four main findings from our synthesis study.[5]Â The first is that using a faujasite with a Si/Al = 2.7 it is possible to form phase pure SSZ-39 across a wide gel Si/Al ratio (15-90), something not previously reported. Second, the yield from these syntheses changes consistently and clearly shows that the aluminum is the limiting reagent, and in all syntheses the yield based on aluminum is over 90%. Third, the trans isomer effect reported previously in the literature is confirmed and quantified across a wide range of cis/trans ratios in this work. Fourth, at an Si/Al = 45 one observes a clear increase in the Si/Al of the material obtained (from 8.6 â€“ 10.5) as one increases the trans isomer content of SDA.

The talk will then describe our initial results in evaluating this material for SCR, showing how the SSZ-39 properties such as Si/Al, which can be controlled based on our synthesis studies, impact the catalytic properties of the SSZ-39 samples. These results will be put into the context of the voluminous literature around SSZ-13.

References

1. Maruo, T., Yamanaka, N., Tsunoki, N., Sadakane, M., Sano, T. . *ChemLett* **2014**, *43*, 302.
2. Albarracin-Caballero, J. D., Khurana, I., Di Iorio, J. R., Shih, A. J., Schmidt, J. E., Dusselier, M., Davis, M. E., Yezerets, A., Miller, J. T., Ribeiro, F. H., Gounder, R. *Reaction Chemistry and Engineering* **2017**, *2*, 168.
3. Moliner, M., Franch, C., Palomares, E., Grill, M., Corma, A. *ChemComm* **2012**, 8264.
4. Dusselier, M., Schmidt, J.E., Moulton, R., Haymore, B., Hellums, M., Davis, M.E. *Chem. of Materials* 27, 7 (2015).
5. Ransom, R., Coote, J., Moulton, R., Gao, F., Shantz, D. F. *Ind. Eng. Chem. Res.* ASAP Article.

500518 Determining Siting Preference of Exchanged Fe Ions in Fe-SSZ-13 Zeolite through Density Functional Theory andÂ Ab Initio Molecular Dynamics

Abstract:

Fe-exchanged zeolites are known to be catalytically active for various reactions, for example, N2O decomposition, paritial methane oxidation, selective catalytic reduction of NOx, etc. In spite of great efforts spent on understanding catalytic mechanisms of these reactions over Fe-zeolites, identification of available Fe species and their siting preference, even under ex situ conditions, is still under debate. Several factors complicate the site identification, including: (1) potential coexistence of Fe with different nuclearities, ranging from monomeric cations to oligomeric Fe complexes and Fe oxides, (2) coexistence of Fe with oxidation states (Fe2+ and Fe3+) and (3) uncertainties in interpretations of experimental characterization results. The available literature [1] suggests that in low-loaded SSZ-13 samples, each Fe cation charge-compensate single framework Al, and that these cations are present primarily as isolated and dimeric species.

Here we present a theoretical approach to determine Fe siting preference in Fe-SSZ-13 zeolite using density functional theory (DFT) and ab Initio molecular dynamics (AIMD). Relative stabilities of exchanged Fe monomers in isolation and dimers ligated by arbitrary combinations of O and OH ligands at isolated Al and different paired Al sites are determined by construction of reactions that interconvert these species. AIMD and time-dependent DFT are combined to compute ensemble-averaged UV-visible spectra of different Fe species. We show that both Fe2+ and Fe3+ have strong tendency to form hydroxo-bridged dimers at paired Al sites, and OH-ligated Fe3+ monomer and dimers have qualitatively similar computed UV-visible spectra, suggesting previous UV-vis based assignments of isolated Fe3+ cation preference may need to be revisited. Through first-priciples thermodynamic analysis, we identify potential active sites by determining abundant Fe species at different Al sites (either isolated or paired) under conditions of catalyst activation for partial methane oxidation and other reactions. These findings help in better understanding of relevant active sites as well as redox mechanism in Fe-SSZ-13 catalysis.

Reference

[1] Gao, F., Zheng, Y., Kukkadapu, R.K., Wang, Y., Walter, E. D., Schwenzer, B., Szanyi, J. and Peden, C. H. G. ACS. Catal. 6, 5 (2016)

485484 Fabrication of Bifunctional Acid-Base Catalyst By Organic Pillared MFI Nanosheets

Abstract:

Organic pillared MFI zeolite has been synthesized by detemplation of organic templates and intercalation of arylic silsesquioxane (1,4-bis(triethoxysilyl)benzene, BTEB) molecules between multilamellar MFI nanosheets. The acid extraction and UV light radiation were sequentially employed for removal of organic templates located not only between MFI layers but also in the zeolite micropores. The removal of organic templates by non-thermal calcination method prevents the condensation of external silanol groups of zeolitic nanosheets stacked next to each other, which allows intercalation of BTEB molecules between the zeolitic nanosheets and their successive reaction with the silanol groups to form the organic pillared zeolite structure. A post-treatment amination of the resultant zeolite sample introduced amino groups in the organic BTEB pillars. The acid sites from the zeolite framework aluminium and basic sites from the amino groups in BTEB pillars endow organic pillared MFI bifunctionality for cascade catalytic reactions. The synthesis of organic pillared MFI is a new addition to the structural modifications of lamellar zeolite with more open structures for processing bulky molecules. The resultant inorganic-organic hybrid zeolite structure creates new opportunities for potential applications of lamellar zeolites with tailored properties.

490696 Production of Furandicarboxylic Acid (FDCA) Using Biomass Derived Î³-Valerolactone

Abstract:

The demand for bio-based polymers is increasing globally due to their lower carbon footprint and independence from fossil fuels. 2,5-Furandicarboxylic acid (FDCA) is a potential bio-based alternative to terephthalic acid for the production of bio-based polymer polyethylene furanoate (PEF). However, the economical production of FDCA from carbohydrates must address technological challenges, such as instability of hydroxymethylfurfural (HMF), low solubility of FDCA, and requirement of stoichiometric amounts of base for the oxidation of HMF to FDCA.

We will report a process for production of FDCA from fructose which utilizes a solvent system comprising of Î³-valerolactone (GVL) and water to address the aforementioned challenges. In our process, a high concentration of fructose (15 wt%) is dehydrated to HMF at high yield (>70%), and HMF is then oxidized to FDCA over a heterogeneous catalyst (Pt/C) without intermediate separation. We demonstrate that the GVL/H_2O solvent system has high FDCA solubility which not only leads to high FDCA yield but also eliminates the need of a homogeneous base during HMF oxidation. We further show that FDCA can be separated from the GVL/H_2O solvent system by crystallization, and a highly pure (>99% purity) FDCA product is obtained. Furthermore, we demonstrate that FDCA is an effective catalyst for fructose dehydration in GVL/H_2O, which obviates the need of a mineral acid catalyst, thereby improving both the economic and the environmental impact of the process. We estimate through preliminary techno-economic modeling that the overall process is cost-competitive to state of art FDCA production processes.

492378 Spectroscopic and Modeling Study of Aluminum Active Species in Glucose Isomerization

Abstract:

The need to reduce greenhouse gas emissions and our dependence on fossil fuels has prompted considerable research in the production of fuels and chemicals from lignocellulosic biomass.[1] Among the top biomass-derived chemicals, 5-hydroxymethyl furfural (HMF) has gained great interest as a platform chemical.[2] HMF synthesis includes the hydrolysis of cellulosic biomass to glucose, glucose isomerization to fructose, and fructose dehydration to HMF.[1] Currently, HMF production is not commercial because industrial glucose isomerization uses enzymes, which are costly and incompatible with the Brønsted acidity necessary for fructose dehydration.[3] To overcome this challenge, homogeneous metal chlorides (e.g., $CrCl_3$, $AlCl_3$) have demonstrated effectiveness in direct glucose conversion to HMF in a single pot.[4] This rather unexpected result was rationalized by tandem catalysis where the metal species catalyzes glucose isomerization via Lewis acidity and hydrolyzes to produce Brønsted acid sites (H_3O^+) necessary for dehydration.[4] Direct, in situ experimental evidence in support of multiple catalytic species has though been lacking. This in turn hinders our ability to develop molecular descriptors and ultimately to predict better homogeneous and heterogeneous catalysts. In this work, we investigate the effect of combining Lewis ($AlCl_3$) and Brønsted (HCl) acids in glucose isomerization. For the first time, we demonstrate the interplay between the acids and from a practical position, optimize the concentration of the Lewis and Brønsted acids to maximize the formation of fructose. Simultaneously, the aluminum speciation in water is explored using in situ ^{27}Al nuclear magnetic resonance (NMR), dynamic light scattering (DLS), and inductively coupled plasma-mass spectrometry (ICP-MS), and the glucose-to-fructose isomerization is correlated with the various Al species. An $AlCl_3$ speciation model combined with kinetic results indicates that the hydrolyzed Al(III) complex $[Al(H_2O)_4(OH)_2]^{1+}$ is the active species in glucose isomerization and acts as a Lewis acid-Brønsted base bifunctional site.

References:

[1] van Putten, R. J., et al., *Chem Rev* **2013,** *113*(3), 1499-1597.

[2] Corma, A., et al. *Chem Rev* **2007,** *107*(6), 2411-2502.

[3] Nguyen, H., Vlachos, D., et al., *Acs Catal* **2016,** *6*(3), 1497-1504.

[4] Choudhary, V., Vlachos, D., et al., *J Am Chem Soc* **2013,** *135* (10), 3997-4006.

495639 Hydrodeoxygenation of Furfural over Multifunctional Catalysts

Abstract:

Hydrodeoxygenation (HDO) of biomass-derived platform molecules, such as furfural and hydroxymethylfurfural (HMF) is required for their utilization as fuels or chemical precursors. In our previous work [1,2], we have shown the utility of trifunctional Ru/RuO_x catalysts for the HDO of furfural and HMF. Based on theoretical and experimental studies, we have shown that these reactions require the presence of a metal and an oxide and proceed through a Lewis-acid-mediated transfer hydrogenation from isopropanol to furfural over the RuO_2, followed by the deoxygenation of furfuryl alcohol over O vacancies on the RuO_2 surface. These vacancies are generated in the reaction of surface oxygen with hydrogen produced in situ from the dehydrogenation of isopropanol over the metal sites [3].

However, the 2-methylfuran (2MF) product may further react over Ru, giving undesirable 2-methyltetrahydrofuran (MTHF) and 2-pentanol, and consuming valuable hydrogen [4]. Moreover, RuO_2 is reduced to metallic Ru under reaction conditions, resulting in the suppression of the HDO reaction. For these reasons, we have developed an alternative catalyst for the HDO of furfural to 2MF. To do this, we correlate the intrinsic HDO rates and the catalyst stability over a series of oxides with bulk and surface DFT descriptors and we choose an active and stable oxide. Extensive catalyst characterization, including microscopy, XPS, XRD, and XAS, is conducted. In a parallel effort, we correlate the selectivity of monometallic and bimetallic catalysts with the descriptors using correlations for thermochemistry and kinetics, such as group additivity and Polanyi relations. These studies demonstrate volcano like behavior and reveal best materials for HDO of biomass molecules.

References:

1. Panagiotopoulou, Vlachos. Appl. Catal A: General, 480 (2014) 17-24

2. Jae, et al. ChemCatChem, 6 (2014) 848-856

3. Mironenko and Vlachos. J. Am. Chem. Soc. (2016) 2016, 138, 8104â€"8113

4. Gilkey, et al. (2016) ChemSusChem, 9, 3113-3121

498198 Functionalization of 5-Hydroxymethylfurfural By Selective Etherification

Abstract:

There is a desire to use biomass to replace oil as the feedstock for modern chemicals and products. In this regard, polymers represent an attractive class of potential biobased materials. In this study, we seek to create tunable polymers using 5-hydroxymethylfurfural (HMF) as a renewable platform chemical. The properties of the final polymer can be tuned by modifying the side-chain of the monomer, which is added by the etherification of HMF with an alcohol. The R-group of the alcohol is then carried through the subsequent reactions to become the side-chain of the monomer. Here the etherification reaction was investigated first, by evaluating several catalysts with a variety of active sites and pore structures for their activity in this reaction and second by evaluating the interplay of R-group identity and active site structure on the etherification kinetics.

Several catalysts were evaluated including, H-FAU, H-MFI, H-BEA, H-MOR, amorphous SiO_2-Al_2O_3, Amberlyst-15, hydrotalcite, tungsten oxide, and Î³-Al_2O_3. It was found that catalysts with BrÃ¸nsted acid sites (*i.e.*, H-FAU, H-MFI, H-MOR, SiO_2-Al_2O_3, and H-BEA Zeolites, and Amberlyst-15) showed the highest selectivities to the ether product at 70-80% HMF conversion. These same catalysts also showed high production rates, most notably Amberlyst-15 and H-BEA. Catalysts possessing either basic sites (e.g., hydrotalcite) or Lewis acid sites (e.g., WO_3 and Î³-Al_2O_3) were an order of magnitude less active and less selective than those possessing BrÃ¸nsted sites. The ether production rates using H-BEA and Amberlyst-15 are very similar; however, H-BEA is ~40% more selective than Amberlyst-15. The improved selectivity observed for H-BEA-Zeolite, compared with Amberlyst-15, is hypothesized to be due to the ability of the zeolite to stabilize reaction intermediates. To further understand the capability of H-BEA several different SiO_2:Al_2O_3 ratios were evaluated for the etherification, revealing that the turnover frequency for HMF etherification is approximately constant and suggesting the absence of strong transport limitations with this zeolite morphology.

In the second portion of this study ethanol, butanol, phenol, and cyclohexanol were used to evaluate the production of a variety of ether products. Increasing the size of the alcohol (*e.g.*, cyclohexanol or phenol vs. ethanol or butanol) was found to have little effect on the ether production rate, suggesting that this rate is governed by the interaction between HMF and the catalyst. Based on these results, we hypothesize that the structure of H-BEA zeolite leads to improved activity for etherification, possibly due to the size compatibility between HMF and the pores of H-BEA. We suggest that Brønsted acidity is needed for etherification, and catalyst morphology may be highly important for obtaining high selectivity.

498584 Inhibition of Xylene Isomerization in the Production of Renewable Aromatic Chemicals from Biomass-Derived Furans

Abstract:

High selectivities (>90%) have been achieved with an H-BEA catalyst in the Diels-Alder cycloaddition, and subsequent dehydration, of 2,5-dimethylfuran (DMF) and ethylene to produce *p*-xylene.[1-4] This success is unexpected due to the ability of xylene to isomerize in the presence of zeolite acid catalysts at similar conditions. In this work, three possible mechanisms describing the conversion of 2,5-dimethylfuran to *p*-xylene were evaluated, where the dominant mechanism was indicated as the homogeneous, uncatalyzed cycloaddition of DMF and ethylene, followed by an acid catalyzed dehydration (Figure 1a).[5] The inhibition of xylene isomerization was attributed to large active site coverage of 2,5-hexanedione, the ring opened product of DMF. This is supported by the similar adsorption enthalpies of 2,5-hexanedione and the cycloadduct versus the less favorable adsorption enthalpy of *p*-xylene (Figure 1b). Though all reaction components were independently examined, only the presence of 2,5-dimethylfuran and 2,5-hexanedione, even at 1/50th the concentration of xylene, inhibited transalkylation and methyl shift reactions of *p*-xylene on H-Y (Si/Al 2.6) and H-BEA (Si/Al 12.5) zeolites. A combination of experimental techniques (thermogravimetric analysis, TGA, diffuse reflectance infrared Fourier transform spectroscopy, DRIFTS, and ^{27}Al NMR) and ONIOM computational method allowed for the investigation of adsorption energetics of relevant species on zeolite acid active sites.

Figure 1. Proposed mechanisms of reaction of 2,5-dimethylfuran and ethylene to *p*-xylene and adsorption enthalpies of compounds in reaction mixture. A) Proposed catalytic cycles converting dimethylfuran to *p*-xylene. Brønsted acid sites within H-BEA and H-Y zeolite catalyze conversion of DMF and ethylene to *p*-xylene and water. B) Heat of adsorption versus proton affinity in H-Y zeolite. The heat of adsorption correlates linearly with proton affinity above 200 kcal mol^{-1}. Deviation from the linear proton affinity prediction exists below 200 kcal mol^{-1}.

[1] Williams, C. L.; Chang, C.-C.; Do, P.; Nikbin, N.; Caratzoulas, S.; Vlachos, D. G.; Lobo, R. F.; Fan, W.; Dauenhauer, P. J. *ACS Catal.*, 2012, 2, 935-939.

[2] Chang, C.-C.; Green, S. K.; Williams, C. L.; Dauenhauer, P. J.; Fan, W. *Green Chem.* 2014, 16, 585-588.

[3] Patet, R. E.; Nikbin, N.; Williams, C. L.; Green, S. K.; Chang, C.-C.; Fan, W.; Caratzoulas, S.; Dauenhauer, P. J.; Vlachos, D. G. *ACS Catal.*, 2015, 5, 2367-2375.

[4] Williams, C. L.; Vinter, K. P.; Chang, C.-C.; Xiong, R.; Green, S. K.; Sandler, S. I.; Vlachos, D. G.; Fan, W.; Dauenhauer, P. J. *Catal. Sci. Tech.*, *6*, 178-187.

[5] Williams, C. L.; Vinter, K. P.; Patet, R. E.; Chang, C.-C.; Nikbin, N.; Feng, S.; Wiatrowski, M. R.; Caratzoulas, S.; Fan, W.; Vlachos, D. G.; Dauenhauer, P. J. *ACS Catal.*, 2016, 6, 2076-2088.

501076 Renewable Butadiene Production from Tetrahydrofuran over HZSM-5

Abstract:

Butadiene is an indispensable monomer to the production of synthetic rubber.[1] It is also used in other important copolymers with unique characteristics, such as styrene-butadiene rubber (SBR) and the impact-resistant ABS (acrylonitrile-butadiene-styrene) plastic. In the classic petrochemical approach butadience is recovered after separation of the C4 cut from naphtha crackers, where ethylene is the main product. However, the recent shale gas revolution increases reliance on non-conventional, shale-derived feeds for ethylene production, which leads to a shortage in C3/C4 olefins including butadiene.[2] Therefore, new, on-purpose butadiene production routes that are not coupled to ethylene production need to be developed.

In this work, we present a bio-based route to butadiene from tetrahydrofuran (THF), which can be produced by decarbonylation and reduction of biomass-derived furfural. The Brønsted-acidic zeolite HZSM-5 has shown good selectivity for THF dehydration to butadiene at low conversions, but the selectivity decreases at high conversions, due to byproducts such as formaldehyde and propene.[3] In this paper, we present different decomposition pathways of THF on HZSM-5 using density functional theory (DFT) calculations, and elucidate reaction mechanisms for the formation of butadiene, formaldehyde and propene. . In addition, we develop a microkinetic model in order to identify key surface intermediates and the rate-limiting step and compare with experimentally obtained apparent activation energies and kinetics.

References:

1. Wagemann, K., *Chemie Ingenieur Technik* **2014,** *86* (12), 2115-2134.

2. Bruijnincx, P. C. A.; Weckhuysen, B. M., *Angewandte Chemie-International Edition* **2013,** *52* (46), 11980-11987.

3. Abdelrahman, O. A.; Park, D. S.; Vinter, K. P.; Spanjers, C. S.; Ren, L.; Cho, H. J.; Vlachos, D. G.; Fan, W.; Tsapatsis, M.; J., D. P., *ACS Sustainable Chemistry and Engineering* **2017**, DOI: 10.1021/acssuschemeng.7b00745.

502462 Reaction Mechanisms Responsible for the Selective Vapor Phase Rearrangement of Furanics

Abstract:

The conversion of the abundant but highly reactive furfural to more stable fuel precursors is an important challenge in the conversion of biomass to higher value products. Cyclopentanone is a stable building block that can serve both as a commodity chemical and as a building block for diesel and jet fuels. The ring rearrangement of furfural to cyclopentenone in the aqueous phase in the presence of an acid catalyst is a well-studied reaction, commonly referred to as Piancatelli rearrangement, that is believed to require the presence of a condensed aqueous phase to proceed. Here we report alternative mechanisms for the conversion of furfural to cyclopentenone and cyclopentanone in a single vapor phase upgrading step using multifunctional catalysts. Direct vapor phase upgrading enables catalytic strategies that can be coupled with pyrolysis and torrefaction steps that donâ€™t require a condensed phase to proceed, where undesirable humin formation may occur. We illustrate how selective sites present at the interface of Ru nanoparticles on a TiO_2 support can facilitate this rearrangement in the vapor phase at temperatures as high as 400°C in a flow reactor, both with model compounds and real biomass vapors. We further show how water molecules can interact with these interfacial sites to modify the chemistry, which we propose is due to a different mechanism than the established Piancatelli rearrangement. We show that acidic coadsorbates that promote this reaction in the condensed aqueous phase, such as acetic acid, inhibit the vapor phase rearrangement by competing for active sites. This behavior is expressed in the context of a kinetic model, with a mechanism proposed involving binding to the metal as well as interaction with the interfacial site. The requirement of an interfacial site for this reaction is contrasted over supports with varying acidity, such as TiO_2 vs. Al_2O_3. We compare this chemistry over Ru with that over Pd and Ni to generate an overall correlation with ring rearrangement rate with the metal-carbon bond strength.

Interfacial sites and oxide overlayers manipulate selectivity for selective C-O cleavage and structure sensitive decarbonylation reactions as well. By systematically controlling the degree of oxidation while maintaining nanoparticle size, we illustrate the importance of partially oxidized sites on the Ni surface, and how these sites can influence the rate of C-O cleavage and decarbonylation reactions.

498959 Implications of High Accuracy Thermochemical Kinetics for H + CH_3 (+M) â‡„ CH_4 (+M) on Combustion Models

Abstract:

CH_3, the simplest alkyl radical, is an extremely stable intermediate that persists in large concentrations even in high temperature chemically reacting systems such as flames. Radical-radical recombination with H-atoms (R1: H + CH_3 â‡„ CH_4) is the dominant process for CH_3 removal in high temperature combustion and flame processes. As such, the net result of this reaction is to terminate reactive H-atoms and CH_3 radicals in flames. Numerous studeies have been conducted to directly measure k_1, however, k_1 has only been measured at low temperatures (< 700 K) while all data at high temperatures (> 1300 K) relevant to the pre-flame regime have been measured in the reverse direction. Hence, reconciling k_1 and k_{-1} requires accurate equilibrium constants spanning an extended range of temperatures. In the present work, we include all relevant anharmonic corrections to calculate an accurate Active Thermochemical Tables (ATcT) partition function for CH_3. The resulting nonrigid-rotator-anharmonic-oscillator partition function is used to determine thermochemical parameters (heat capacity, entropy, and enthalpy increment values) for CH_3 and K_{eq} values for reaction R1 over a wide temperature range (200-6000 K). With the newly evaluated equilibrium constant, literature experiments and theory (based on two-dimensional master equation calculations with a first principles energy and angular momentum transfer kernel) are used to obtain an accurate representation of the kinetics for the title reaction. In particular, the present work provides fits to k_1(T,P) for N_2, Ar, and He as colliders for direct use in combustion modeling. The results of the new rate constant evaluation for R1 has a dramatic impact on combustion modeling. In particular, we assess the impact of incorporating anharmonic thermochemistry for CH_3 and the updated fits for k_1 (and k_{-1}) in current widely used literature models for simulations of CH_4/air and n-heptane/air laminar flame speeds and find that flame speed predictions can be affected by as much as 50%.

This work was supported by the Office of Basic Sciences, Division of Chemical Sciences, Geosciences, and Biosciences, U. S. Department of Energy, under Contract No. DE-AC02-06CH11357.

496829 Estimation of Thermodynamic Properties of Polycyclic Molecules By a Linear Regression Model

Abstract:

Though recent advances in *ab initio* methods have opened the door toward accurate calculations of molecular thermochemistry, large-scale theoretical studies are still often limited, at least initially, to the use of empirical methods to rapidly screen out unimportant species, so only the important species are the subject of CPU-time intensive quantum chemistry calculations. Among the empirical methods developed in the past decades, the Benson group additivity scheme is one of the quickest and most convenient methods to determine thermodynamic properties of molecules without needing the 3D molecular structures. It has achieved great success in accurate prediction of thermochemistry of small molecules and has been widely adopted in modeling software for on the fly prediction of thermodynamic parameters. However, because this additivity scheme simply uses the properties of individual chemical groups independently to calculate the entire property for a molecule, the contribution of the overall molecular structure to the property is usually not taken into account. This problem manifests itself for strained structures and can cause significant errors for polycyclic molecules with fused rings. Therefore, the application of the additivity method is often restricted to simple chemical systems without the presence of polycyclic species.

To address this issue, we have developed an improved model to predict the thermochemistry of polycyclic species.Â A regularized linear model was chosen to resemble the simplicity and interpretability of the additivity method. However, instead of following the definitions of the chemical groups of Bensonâ€™s scheme, we generated a comprehensive list of identifiers containing local (atoms, bonds, and angles) and/or nonlocal (rings) structural information and used these identifiers to compose feature vectors for molecules.Â The identifiers carrying nonessential information were eliminated by L1 regularization during model training so that those remained in the final model represent an optimum set of chemical units for the calculation of the thermodynamic property of interest with the contribution of the cyclic structures. These chemical units are human interpretable and conceptually equivalent to the chemical groups defined in Bensonâ€™s scheme but are selected by the model objectively without human intervention.

For a training set of 25,716 cyclic and polycyclic organic molecules made up of C, H, and O atoms, 408 identifiers were found to be needed for the calculation of formation enthalpies. The transferability of the trained model was validated on an independent test set of 2,858 molecules with a mean absolute error of 1.88 kcal/mol.

490830 Hydrogen Binding Versus Dissociation during Soot Formation: Insights from Reactive Molecular Dynamics

Abstract:

It has been proposed that two reactions are prominent in the earliest stages of soot nanoparticle formation; hydrogen abstraction via collision between a growing PAH (polyaromatic hydrocarbon) cluster and a hydrogen atom (yielding H_2 and a dehydrogenated cluster) and carbon addition via the binding of an alkane/alkene/alkyne chain to the cluster. We examine the first part of this process via reactive molecular dynamics using REAXFF C-H potentials. In simulations, cluster structures determined using the previously developed Atomistic Model for Particle Inception (AMPI) code of Violi and coworkers are equilibrated (at a predefined temperature) and a hydrogen atom with a specific velocity and impact parameter is driven towards the cluster. Depending upon the velocity, impact parameter, and collision orientation, we find that pure hydrogen abstraction is rare and instead at low velocities hydrogen binds to clusters. At increasing velocities, hydrogens rebound from the cluster surface and can drive hydrogen facilitated collision induced dissociation (CID) of the cluster. CID events yield smaller alkane/aklene/alkyne chains, which may serve to bind other clusters. Integration over all impact parameters and velocities for collisions (in accordance with the Maxwell Boltzmann distribution) yields the hydrogen binding rate coefficient and hydrogen induced dissociation rate coefficient for a given cluster structure. We discuss how these rates vary with temperature and cluster size.

499595 Effect of Composition Variation of Syngas Mixtures on NO_x Formation at High Pressure

Abstract:

Increasing interest in syngas combustion in gas turbines has led to highlighting the environmental feasibility, in particular the control of nitrogen oxide (NO_x) emissions. Trace amounts of NO_x species may be present during combustion and lean premixed (LPM) combustion method is one of the effective NO_x control technologies for gas turbines. Despite the proven feasibility of syngas, these fuels face variation in their composition due to different feedstock and production methods. Syngas can contain small amounts of hydrocarbons, such as methane, ethane and ethylene, and the presence of hydrocarbon species in the fuel can affect the NO_x emissions in the post-flame region. A detailed investigation of high-pressure syngas combustion is required to provide both practical information regarding NO_x emissions from gas turbine systems and fundamental understanding of reaction pathways to verify and develop chemical kinetic models.

In this study, a high-pressure burner facility capable of operating at pressures up to 15 bar, was designed and fabricated to simulate syngas combustion in a gas turbine. Syngas consisting mainly of H_2 and CO was blended with components such as nitric oxide (NO), nitrogen dioxide (NO_2) and trace amount of hydrocarbons prior to combustion at high pressure. Hot combustion products were supplied to the high pressure vessel by a flat flame burner operating with lean, premixed syngas-air mixtures with various H_2/CO ratios ranging between 0.25-1.0 and equivalence ratios between 0.5-1.0. Detailed NO_x speciation measurements in the post-flame region were conducted for different fuel compositions at various pressures up to 10 bar using Fourier transform infrared (FTIR) spectrometer. The experimental data generated from this project can be used to validate/develop kinetic models to predict NO_x emissions.

Numerical modelling of industrial burners for reduction of NO_x emissions using flamelet methods in combination with a newly developed postprocessor for fast and accurate emission prediction

DDI Werner Pollhammer, K1-MET GmbH, Linz, Austria
DI Dr. Christoph Spijker, Montanuniversitaet, Leoben, Austria
DI Michael Koller, Ebner Industrieofenbau GmbH, Leonding, Austria
Univ.-Prof. DI Dr. Harald Raupenstrauch, Montanuniversitaet, Leoben, Austria

Abstract

Current trends of research on combustion systems show increased interests on prediction of pollution. To meet the demands of modern combustion systems such as industrial burners detailed studies of their behaviour is needed. One of these pollutants which are commonly of interest is NO_x. The aim of the work is to develop a numerical toolbox that is able to predict NO_x emissions of industrial burners by numerical simulation using Computational Fluid Dynamics (CFD).

A set of numerical equations is solved for flow, temperature, pressure and species. Using CFD it is possible to model a combustion system using different approaches like flamelet methods or detailed chemical reaction systems. To predict the emission of NO_x from a flame a detailed chemical reaction system is necessary such as the GRI 3.0. mechanism [1]. The problem of using this mechanism is that it can be applied only for small geometries to avoid long calculation time or processing costs. Using a detailed chemical reaction system for every species a few chemical reactions have to be solved. Using the GRI 3.0. mechanism 325 reactions for 53 species have to be solved [2]. Also with modern computers it is not affordable to model industrial burners using this approach which would calculate NO_x emissions well.

The second main possibility for simulation of burners is to use flamelet methods [3][4]. Choosing this approach the combustion system is pre-calculated and reaction data will be stored in so called lookup tables. During simulation this reaction data will be accessed by reading them from the lookup tables. This approach allows fast calculation of temperature and flow field of gas burners as well as fast chemical reactions like formation of CO_2, H_2O or CO. This is the standard procedure for simulation of industrial burners for analysis of flame shape and temperature fields. The problem is that the described approach is not applicable for slow chemical reactions such as formation of NO_x. Therefor a detailed chemical reaction system is necessary. The problem is that on the one hand there is the detailed chemical reaction system which would predict NO_x emissions very well but takes too long for calculation. On the other hand there is the flamelet method which is fast but delivers only for flow fields, temperature fields as well as some species fields right results but models the formation of NO_x not very well. The idea now is to combine the benefits of both approaches and to postprocess the results out of the flamelet calculation using a detailed chemical reaction system. This means temperature and flow field are calculated by the fast flamelet calculation. After convergence of this calculation the solution for the flow field and temperature field as well as for some species fields are known. Subsequently for slow chemical reactions such as NO_x production, which cannot be predicted by the flamelet method, the postprocessor calculates the reactions using the already existing temperature and flow field from the flamelet calculation. This combination enables a much faster calculation of NO_x formation in the flame and combines the speed of flamelet method with the accuracy of using detailed chemical reaction systems with reasonable calculation costs [5].

Model description

For a more precise description of radiation effects it was necessary to implement the Discrete Ordinate Model (DOM) into the flamelet solver of OpenFOAM [6]. The standard radiation model is only valid for free stream flames with a constant surrounding wall temperature. The radiation model was used in combination with a Weight Sum of Gray Gases Model (WSGGM) which calculates the optical density of the gas by applying a mixture of temperature dependent absorption coefficients for calculation of the radiation beams [7].

For flamelet calculation a steady laminar flamelet for non-adiabatic combustion was used out of literature [3][4]. By defining a range for enthalpy defects in flamelet generation heat transfer in the system and out of the system can be taken into account. Flamelets were generated for different scalar dissipation rates which takes mixing time scales into account and for different enthalpy defects.

For modelling of turbulent effects the k-ε-model was used with a slight change according to literature data. For a more precise prediction of flame acceleration the constant $C_{2\varepsilon}$ was modified to 1.8. [8]. Out of data from Sandia Flame D boundary conditions for turbulence were generated for turbulent kinetic energy k and dissipation of it named ε [1].

The NO_x postprocessor uses a detailed chemical reaction approach the so called partially stirred reactor model implemented in OpenFOAM for calculation of chemical reaction [9]. It takes chemical reaction kinetics, for instance out of GRI 3.0., as well as turbulent mixing effects into account [2]. The solver uses a conservation equation for each species and chemical reaction is implemented as a source term in every species equation. This means for chemical reaction in every mesh cell the whole reaction mechanism consisting of hundreds of Arrhenius equations has to be solved.

To reduce the consumption of calculation power a special strategy of solving chemistry was implemented. As a matter of fact that the system seeks for a steady state solution that means after a certain time species composition will not chance any more in the cells. While post processing there is no change of temperature and turbulence because these fields are calculated by the previous flamelet calculation and they are constant during post processing. The reaction rates of chemistry then only depend on species concentration. If steady state is nearly reached in a cell and the species concentration varies just a bit there is no big difference between reaction rates of the previous and actual time step. An algorithm which analyses these changes was implemented. Using this algorithm the chemical reaction source terms were tabled from time step to time step. If the change in species is lower than 2 percent of the relative concentration of every species in a cell then the source terms of the previous time step is used. Using this algorithm about 95 percent of chemistry can be accessed by tabling and only 5 percent of the cells have to be calculated. This leads to huge saving of computational power and to a big speed up in simulation.

To avoid unphysical states there is also a check if one of the species concentrations goes under zero which is physically incorrect. If this occurs then chemistry will be calculated in the next time step. To avoid pseudo steady state solutions every 200 time steps the whole domain will be calculated.

The model validation was done by means of a comparison with measurements on the Sandia Flame D [1]. Therefor a two dimensional mesh consisting of 200 000 hexahedra was generated using blockMesh and analysed in terms of mesh independency. The benefit of validation on Sandia Flame D is that temperature, velocity, turbulence and species decomposition is also known at certain points in the flame. This means it is feasible to analyse the whole flame shape and its position dependent characteristics and not only the properties at the outlet. For reliable results it is important to be able to predict the temperature maximum very well. Comparison of species decomposition shows very good agreement in terms of oxidizer consumption and CO_2 formation [1]. Using the newly developed postprocessor a very accurate description of NO_x formation is possible which fits well to literature data [1].

Model validation on Sandia Flame D and results

First the flamelet model was used to describe the temperature, velocity and turbulence field of the flame. The following Figure 1 shows the solution of the Sandia Flame D case and its temperature field.

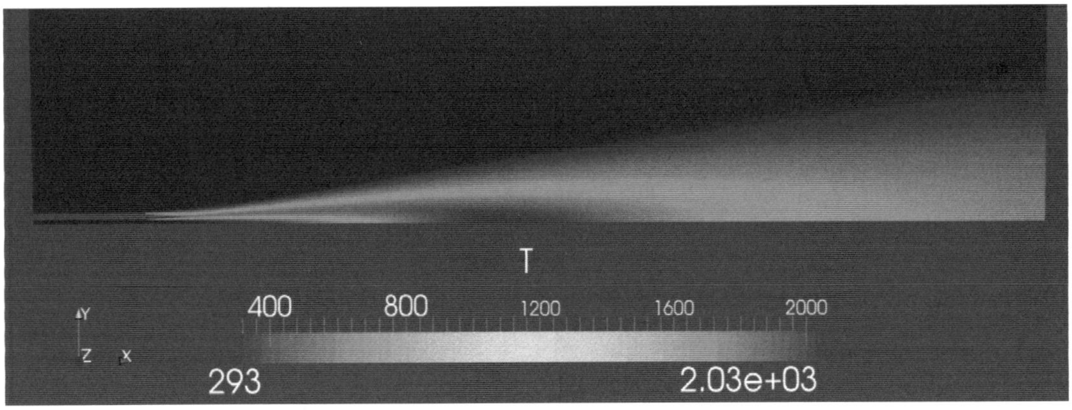

Figure 1: Temperature distribution in Sandia Flame D calculated using flamelet model

This simulation results were then compared to literature data [1] which is displayed in Figure 2. As it can be seen is that the temperature along the rotational axis fits very good to measurements. This is very important for an accurate prediction of nitrogen oxide formation.

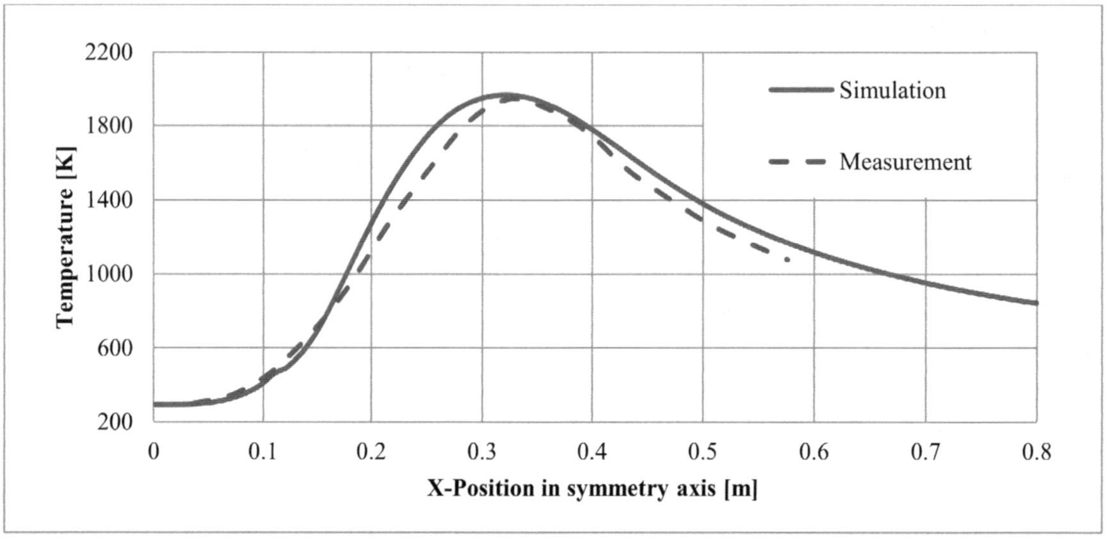

Figure 2: Plot of flame temperature comparing simulation and measurement data from Sandia Flame D [1]

Using this data for temperature, velocity, turbulence and species which can be predicted by flamelet the post processing was performed to calculate species like NO, N_2O, NO_2 which cannot be predicted accurate enough by flamelet. For this species a detailed chemical reaction mechanism was used. In this case the simulations were performed using GRI 3.0. [2]. The following Figure 3 shows the distribution of NO in the Sandia Flame D with three sections through the flame. At these positions a comparison with measurement data was made which is displayed in Figure 4.

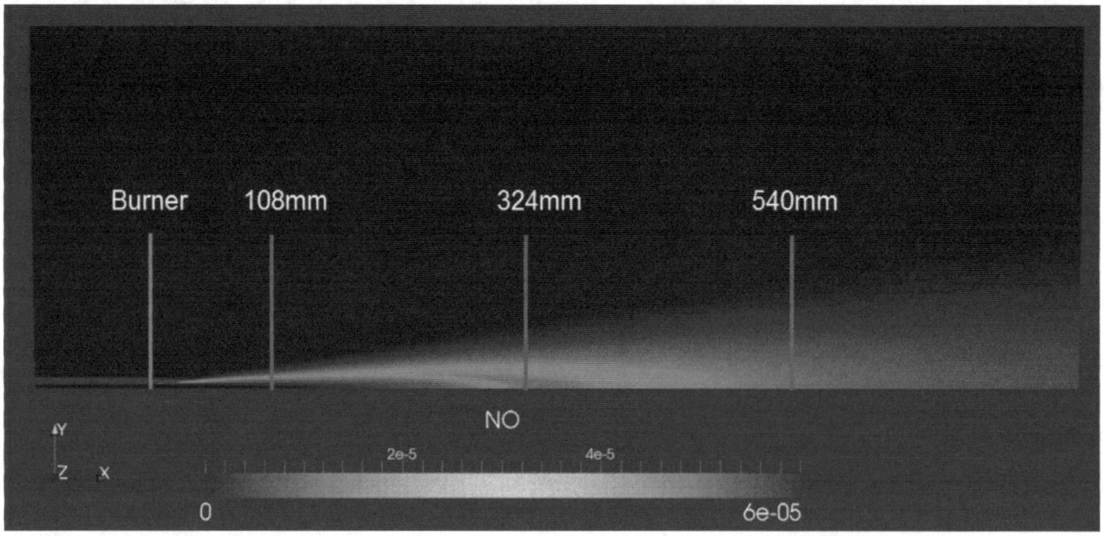

Figure 3: NO field of Sandia Flame D after convergence of postprocessor

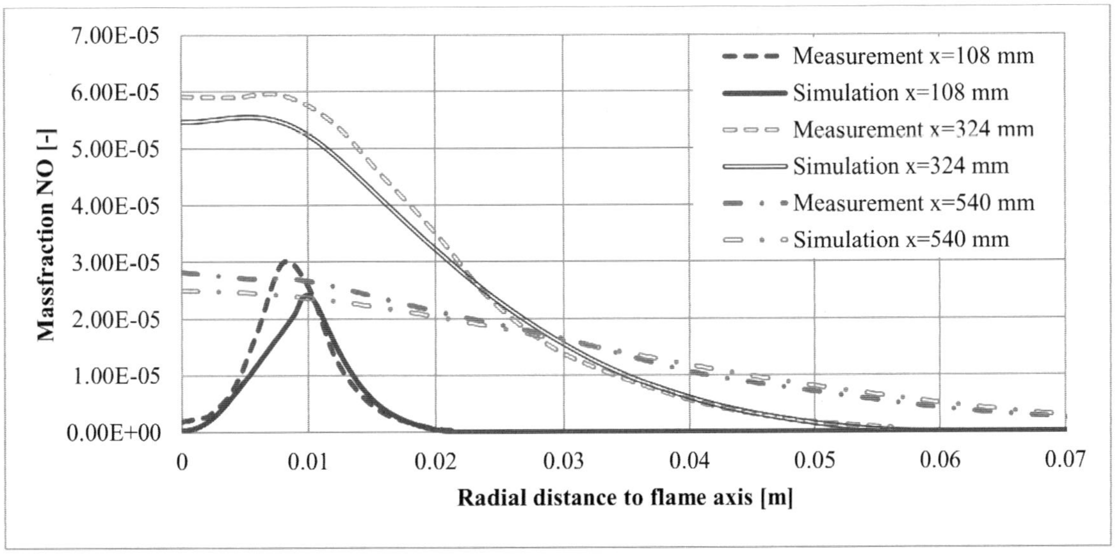

Figure 4: Comparison between simulation and measurement at three cuts through Sandia Flame D as shown in Figure 3 [1]

The result of the first validation is that the generated postprocessor is able to calculate nitrogen oxide formation in Sandia Flame D very well. It predicts slightly lower concentration but the accordance is good in the whole flame. So the developed model can be used for detailed investigations on nitrogen oxide formation in industrial burners.

Model validation on test chamber and results

A second model validation was done by means of an industrial test where a burner was operated in a test chamber. Measurements of wall temperatures by thermocouples and a detailed exhaust gas analysis were performed. Out of the measurements of 20 thermocouples at different positions on the furnace wall a temperature profile was generated to implement it as boundary condition for the flamelet calculation. This lab test was modelled using the developed simulation toolbox. A mesh of the whole test geometry was generated using cfMesh [10]. Out of mass flow and temperature measurements boundary conditions for simulation were calculated. The first validation for the model was a comparison between exhaust gas temperature in the lab test and simulation. These data were then used to start the NO_x post processing. Therefor the developed transient postprocessor was used to calculate the formation of NO_x and various other species which are not predicted well by the flamelet method. At that point it is only necessary to calculate the chemical reaction source terms without solving flow and temperature fields which are now constant in this calculation and already known from the flamelet calculation. This way the simulation is much faster and calculation of NO_x formation now is possible in affordable time using the detailed chemical reaction mechanism with the benefit of a much more accurate description of NO_x formation. For the whole geometry a mesh independency study was performed on the one hand for the flamelet model and on the other hand for the postprocessing. The mesh was generated with a cell size which delivers mesh independent results. For the test chamber with a height of 2.3, a width of 1.7 and a depth 2.5 meters the mesh consists of 5.26 million polyhedral cells.

Figure 5 shows the result of the post processing of the test chamber simulation.

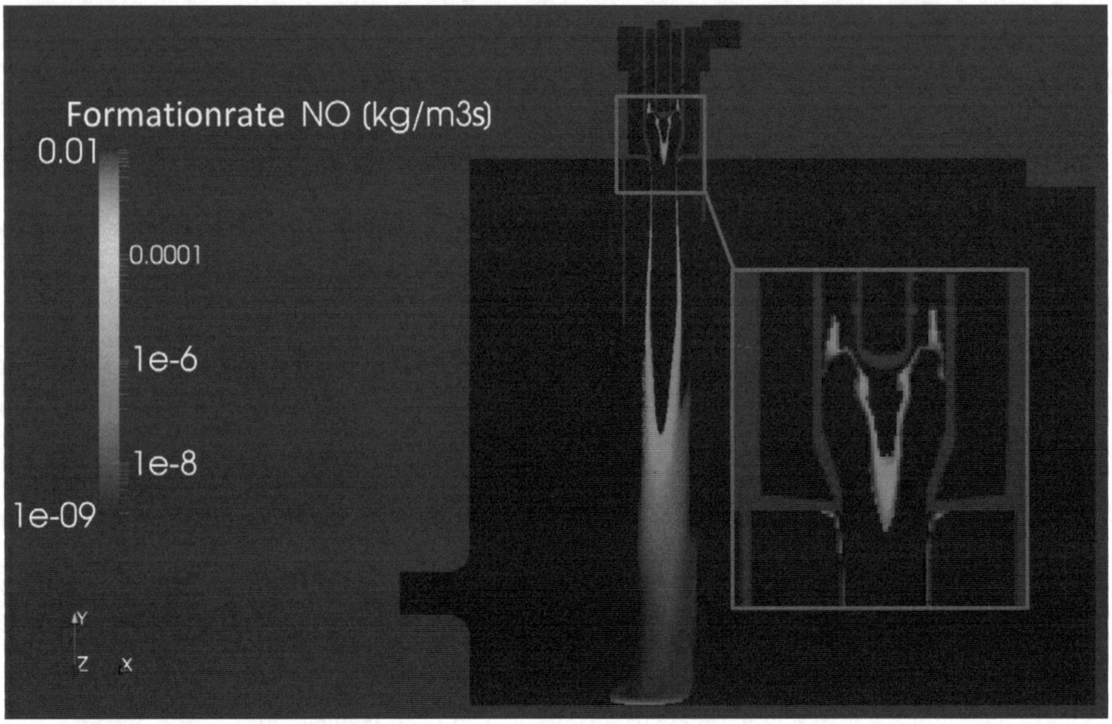

Figure 5: Areas of NO formation in test chamber, calculated using developed postprocessor

This test was performed twice. One test was made with air at ambient temperature and one with preheated air with 199°C.

Table 1: Comparison between simulation results and measurements in test chamber

Formation of NO_x	Simulation [ppm]	Measurement [ppm]
Cold burner air with 28°C	63	65
Preheated burner air with 199°C	78	92

As it can be seen in Table 1 there is also a good accordance for the postprocessor with measurements in the big test chamber. The case with air at 28°C shows the best accordance. The second lab test using preheated air shows the right trend by increasing NO_x formation in terms of the cold air case but shows lower NO_x emissions like the measurement at the test chamber. This effect of lower NO_x concentrations can also be seen in Figure 4 that the model predicts lower nitrogen oxide emissions than the measurements. Important is that the model predicts relative improvements and trends. So now the model was used to perform a geometry study and to test different improvements of the burner geometry by changing the geometry and using the postprocessor for analyzation.

Outlook

Now the model is used to analyse the formation zones of nitrogen oxides in the burner. The next steps are further improvements of the code efficiency and to find more strategies to reduce the calculation time. After some improvement steps of simulation a prototype of the new developed burner will be build a new chamber test will be performed to check these improvements also in a real test.

Acknowledgements

The authors gratefully acknowledge the funding support of K1-MET GmbH, metallurgical competence center. The research program of the competence center K1-MET is supported by COMET (Competence Center for Excellent Technologies), the Austrian program for competence centers. COMET is funded by the Federal Ministry for Transport, Innovation and Technology, the Federal Ministry for Science, Research and Economy, the province of Upper Austria, Tyrol, and Styria and the Styrian Business Promotion Agency.

References

[1] Barlow R, Frank J. *Piloted CH4/Air Flames C, D, E, and F – Release 2.1*. Available at: http://www.sandia.gov/TNF/DataArch/FlameD/SandiaPilotDoc21.pdf; 2007 [accessed 07.04.2017].

[2] Smith GP, Golden DM, Frenklach M, Moriarty N, Eiteneer B, Goldenberg M, Bowman T, Hanson RK, Song S, Gardiner WC, Lissianski, Jr., Vitali V., Qin Z. *GRI Mech 3.0*. Available at: http://www.me.berkeley.edu/gri_mech/; 1999 [accessed 11.04.2017].

[3] Cuoci A, Frassoldati A, Faravelli T, Ranzi E. Formation of soot and nitrogen oxides in unsteady counterflow diffusion flames. *Combustion and Flame* 2009;156:2010–22, doi:10.1016/j.combustflame.2009.06.023.

[4] Cuoci A, Frassoldati A, Faravelli T, Ranzi E. Kinetic Modeling of Soot Formation in Turbulent Nonpremixed Flames. *Environmental Engineering Science* 2008;25:1407–22, doi:10.1089/ees.2007.0193.

[5] Spijker C. *Unsteady Laminar Flamelet Modellierung zur Beschreibung von Mündungsmischbrennern*. Masterthesis, Montanuniversitaet. Leoben; 2010.

[6] Thynell ST. Discrete-ordinates method in radiative heat transfer. *International Journal of Engineering Science* 1998;36:1651–75, doi:10.1016/S0020-7225(98)00052-4.

[7] Barlow RS, Karpetis AN, Frank JH, Chen J-Y. Scalar profiles and NO formation in laminar opposed-flow partially premixed methane/air flames. *Combustion and Flame* 2001;127:2102–18, doi:10.1016/S0010-2180(01)00313-3.

[8] Pfeiler C, Spijker CJ, Raupenstrauch H. CFD als Werkzeug in der Industrieofentechnik. *Berg Huettenmaenn Monatsh* 2011;156:347–52, doi:10.1007/s00501-011-0018-z.

[9] OpenFOAM. *OpenFOAM 2.4.0*. Available at: https://openfoam.org/; 2017.

[10] cfMesh. *Professional Meshing Solutions - cfmesh.com*. Available at: http://cfmesh.com/; 2017 [accessed 11.04.2017].

491775 A CFD Model-Based Optimization of a Low NO_x Burner in Integrated Process System

Abstract:

Although process burners are frequently used in many industries such as refeinring, petrochemical and chemical processes, there is a serious concern about emission control of these equipment. Recently, the deveopment of new numerical approaches such as computational fluid dynamics (CFD) has led to significant improvements in emission control of industrial processes. The proper analysis and testing leads to more efficient combustion, less physical testing and lower cost for full scale equipment burners.

In this study, we used CFD simulations to design, develop and optimize a low NOx burner which produces heat for methane steam reforming. The CFD method helped to design a burner with angled air holes arranged in a swirl pattern. This optimized design increased the mixing near the burner plate that led to lower peak flame temperature and therefore lower NOx emissions. The designed burner was built and experimentally tested and showed a major NOx reduction in steam reforming hydrogen generation process. In addition, reformer performance improved significantly due to the better temperature distribution by the new CFD optimized burner.

496078 Mapping the Stable Region of Operation of Homogenously Charged Compression Ignition (HCCI) Engine for Methane and DMF Fuel Blend

Abstract:

Abstract:

The emission legislations are transmuting to be rigid in accordance to reduce the emissions from the IC engines. Research has been focused to ameliorate both Spark Ignition and Compression Ignition engine in order to make them more efficient. Over the previous decade, major research has been focused on Homogenous Charge Compression Ignition (HCCI) engines. This engine has the potential to consummate the eminent feature of Spark Ignition engine, with no emission of NOx and Soot leading to cleaner combustion, the best of CI engine with higher efficiency and low fuel combustion along with lower carbon dioxide emissions[1].

The focus of this work is to map the stable regions of operation of HCCI engine using Methane and Dimethyl Furan blend as fuel. DMF is comparable to gasoline since it has an energy density of 40% higher than ethanol. Also, Burning DMF requires 33% less air compared to the same amount of gasoline since its stoichiometric air to fuel ratio is 10.72. The Simulations of HCCI engine were carried out in CHEMKIN, where the model used was a closed transient CSTR[2]. The main objective of the study was to depict the stable regions of operation of HCCI engine with and without Exhaust gas recirculation. EGR refers to a part of the exhaust stream recycled back to the engine inlet along with the fuel mixture[3].

Parameters such as EGR Temperature, EGR percentage and equivalence ratio were varied to study the effect on the sustainable emissions at the exhaust for CO, CO_2 and NOx. Detailed Kinetics of DMF obtained from NCBI was used. Initial understanding of the working of HCCI engine was done using Methane as source fuel, keeping EGR temperature and HCCI engine temperatures as a constant at 600K and 447K and the equivalence ratio were varied from 0.5 to 0.7 and the EGR percentage were varied from 0 to 30%. The in-cylinder temperature, pressure and species profile were obtained as a function of Crank Displacement angle (CAD).

The Figure (1) shows the effect of equivalence ratio on the in-cylinder temperature, pressure and species profile for a engine temperature of 447K and EGR temperature of 600K for 30% EGR. Increase in equivalence ratio states the presence of higher amount of fuel where as smaller ratios represent deficit of fuel in the mixture. It could be observed that Peak temperature and pressure increases with increase in equivalence ratio. The mole fraction of CO increase with increase in equivalence ratio, this is because of the lack of oxygen available for complete combustion. In case of Thermal NO production rate, it increases with increase in equivalence ratio due to increase in temperature which leads to increased production of NO at higher temperatures. It was observed from the initial study using Methane as a fuel that, for Lesser exhaust emissions, the fuel must be lean with a higher percentage of EGR.

Fig 1: Variation of In-cylinder temperature, pressure, mole fraction of CO, CO_2 and Thermal NO production rate with Equivalence ratio

The study was further extended to DMF as an additive to Methane and the parameters such as EGR percentage, Equivalence ratio and EGR temperature were varied to find the stable region of operation of HCCI engine for Methane and DMF blend. The effect of Equivalence ratio on the in-cylinder temperature, pressure and species properties for 0% EGR for an engine temperature of 447K and EGR temperature of 600K were studied initially. The figure (2) shows the effect of Equivalence ratio on Peak CO_2 mole fraction in the cylinder for different percentage of Methane and DMF blend. It could be observed that the peak mole fraction of CO_2 increases with increase in Equivalence ratio and also increases with increase in blend percentage. The latter is due to the increase in the carbon availability in the cylinder with increase in fuel.

Fig 2: The Variation of Peak Mole Fraction of CO_2 with percentage of the DMF blend for different Equivalence ratios

As a next step, it is intended to study the effect of EGR percentage and EGR mixing temperature on Methane and DMF blend as shown in the Table 1 and analyse the Peak in-cylinder temperature, pressure and species properties for the different parameters combinations and map the stable region of operation of HCCI engine for Methane and DMF fuel blend.

Table 1: Parameters to be estimated for the Blend

Blend	**Parameters**
Methane & DMF blend	EGR Percentage – 0%, 10%, 20% & 30% EGR temp – 400K, 500K & 600K Equivalence ratio – 0.5, 0.6 & 0.7 Blend Ratio – 0%, 50% & 90%

Keywords:

Exhaust Gas Recirculation, DMF, Crank Rotation angle.

References:

1. Harrison R. M. ed. (1996) Pollution, Causes, Effects and Control. 3rd Ed. Cambridge, The Royal Society of Chemistry.

2. Xu H., Liu, M., Gharahbaghi, S., Richardson, R., Wyszynski, M., Megaritis, T., “Modelling of HCCI Engines: Comparison of Single-zone, Multi-zone and Test Data”, SAE 2005-01-2123

3. Heywood JB, Internal Combustion Engine Fundamentals, Mc Graw Hill; 1988

THE UNIVERSITY *of* EDINBURGH

SCCS

Towards multiscale screening of porous materials for carbon capture

<u>S. Krishnamurthy</u>, A. H. Farmahini, D. Friedrich, S. Brandani and L. Sarkisov

School of Engineering, University of Edinburgh, UK

s.krishnamurthy@ed.ac.uk

Presentation for the
2017 AIChE Annual Meeting, Minneapolis, USA
31/10/17

Introduction

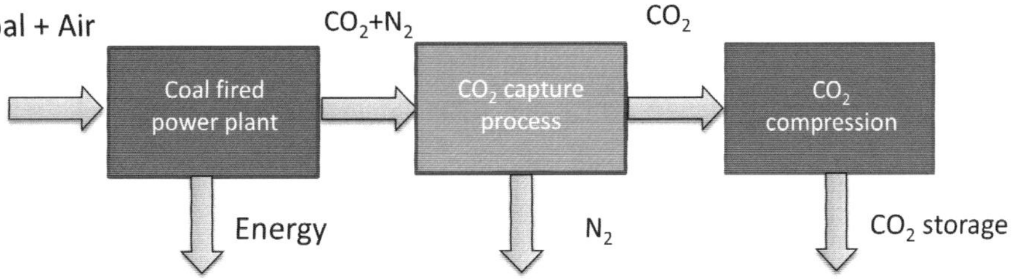

Typical flue gas composition 15% CO_2, 85% N_2 assuming dry flue gas

Challenge: Capturing 90% CO_2 at 95% purity with minimum energy penalty

SCCS

Adsorption processes for CO_2 capture

- Two modes
  - Pressure/Vacuum swing adsorption (PSA/VSA)
  - Temperature swing adsorption (TSA)
- TSA
  - Long cycle times affect throughput
- PSA
  - Pressurizing flue gas with large N_2 content not attractive
- Vacuum swing adsorption (VSA)
  - Adsorption at atmospheric pressure and desorption at vacuum
  - Current benchmark adsorbent: Zeolite 13X

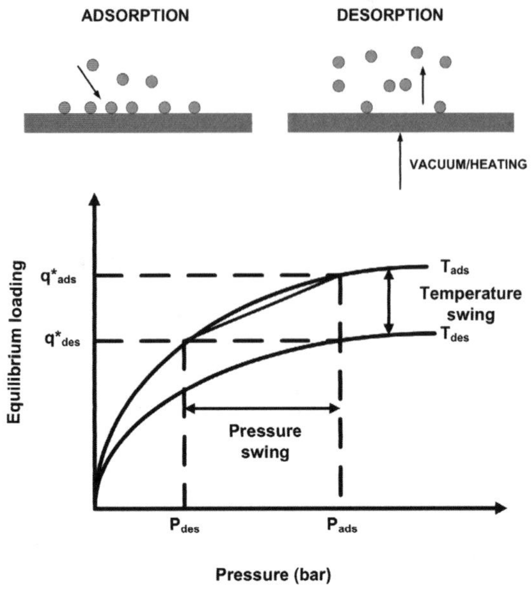

SCCS

Adsorbents for CO_2 capture

Adsorbents

- Zeolites [1,2]
- Carbons [3,4]
- Metal organic frameworks (MOFs)[5,6]
- Zeolite Imidazolate frameworks (ZIFs)[7]
- Amine impregnated Silica[8]
- Porous polymer networks (PPNs)[9]
- Coordinated organic frameworks (COFs)[10]

How do we select the right material for carbon capture processes ?

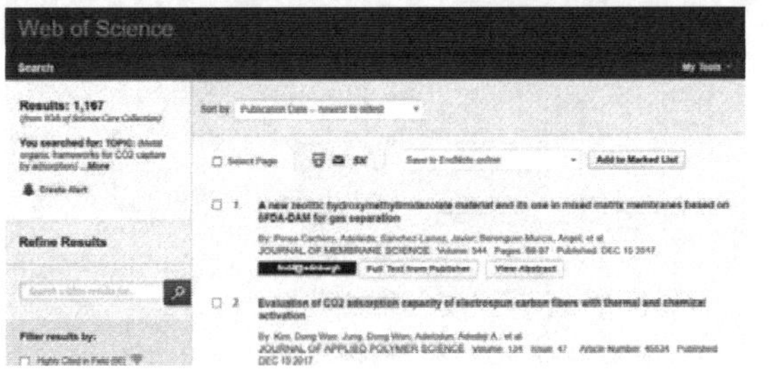

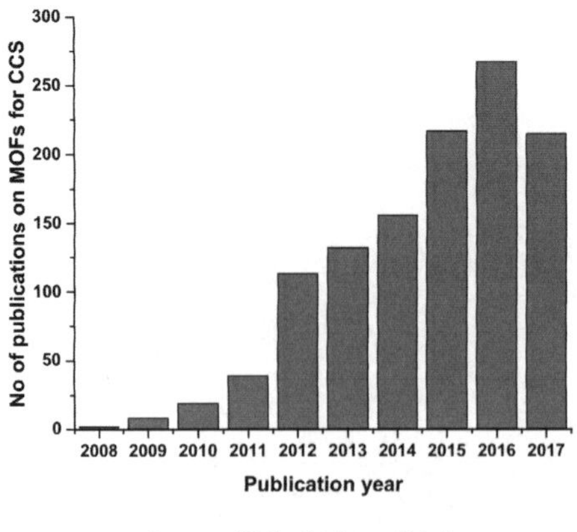

Source : Web of science database

1. Lee et al., J. Chem. Eng. Data 2002,47, 1237-1242
2. Pham et al., Langmuir, 2013, 29 (2), 832–839
3. Ribero et al., J. Chem. Eng. Data, 2008, 53 (10), 2311–2317
4. Shirani and Eic, Can. J. Chem. Eng. 94, 2016 2023–2034
5. B. Pato-Doldán, et al., ChemSusChem 2017, 10, 1710-1719.
6. Queen et al., Chem Sci 2014, **5**, 4569-4581
7. Zhang et al., J. Phys. Chem. C, 2014, 118 (17), 8788–8794
8. Belmabkhout et al. Chem Eng Sci 65 (2010) 3695–3698
9. Lu et al., J.Phys.Chem.C 2013, 117, 4057-4061
10. Zheng et al., Adv. Mater. **2016**, 28, 2855–2873

SCCS

Material ranking : Adsorption isotherms

Experiments

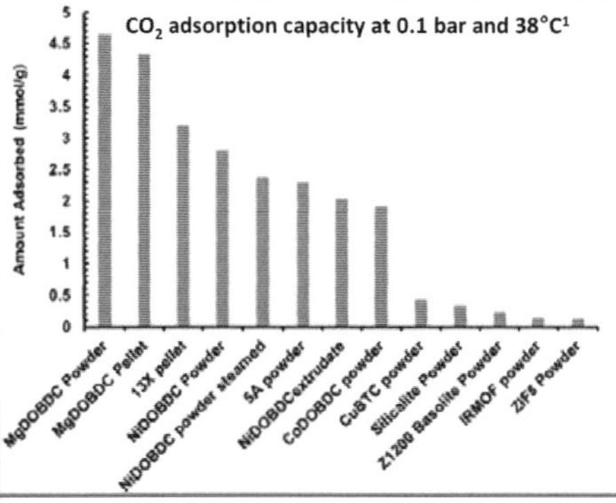

Molecular simulations

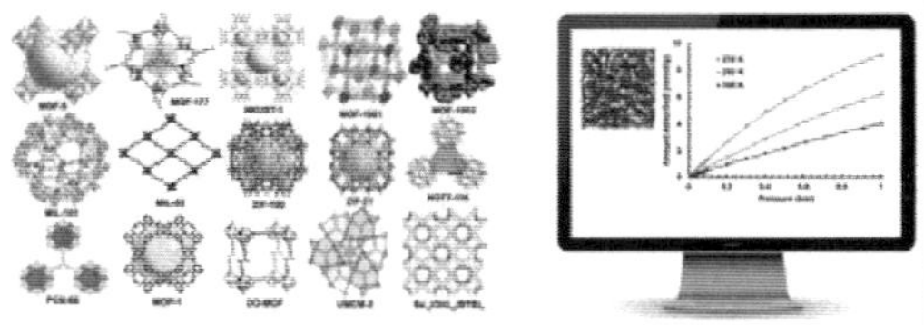

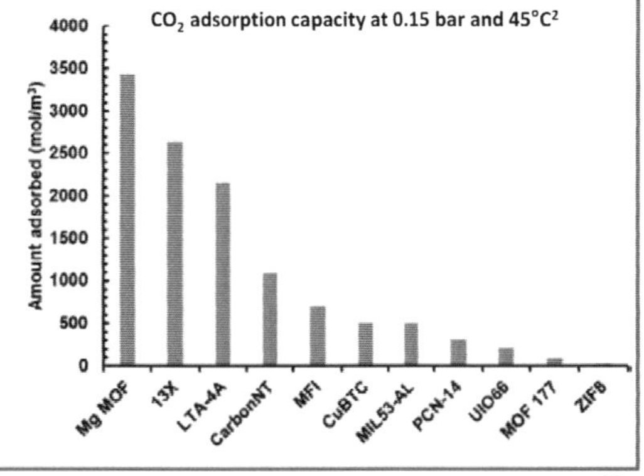

1. Hu et al., Ind. Eng. Chem. Res. 2015 54 6772–6780
2. Bahamon, D. Vega, L.F. Chemical Eng J 2016 284 438–447

SCCS

Process based material ranking

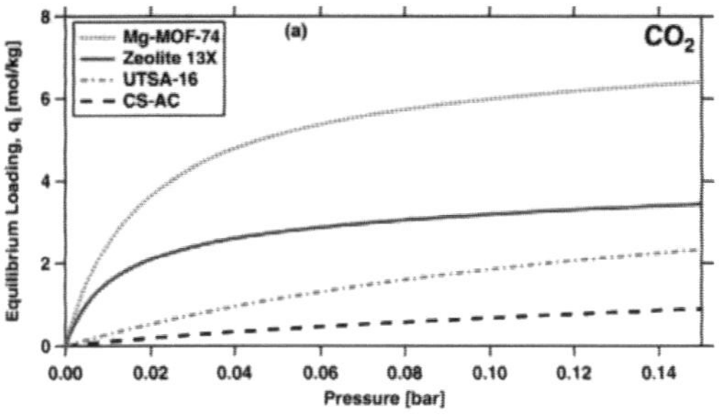

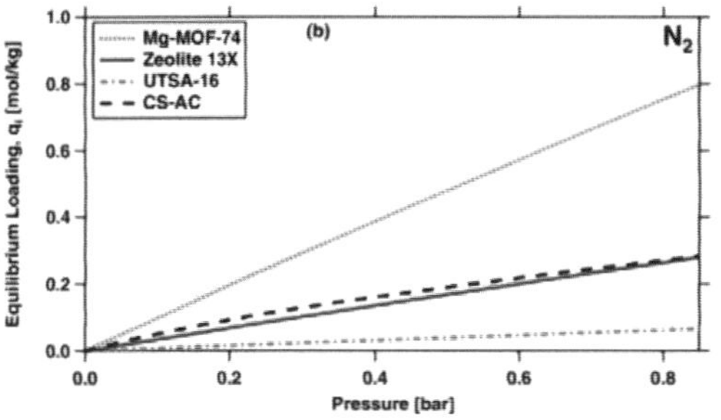

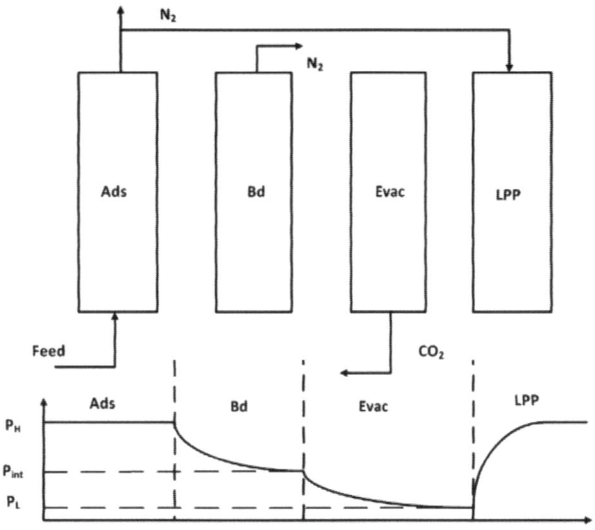

Rajagopalan, A.K, Avila, A.M, Rajendran, A.. *Int J Greenhouse Gas Control.* 2016;46:76-85.

SCCS

Process based material ranking

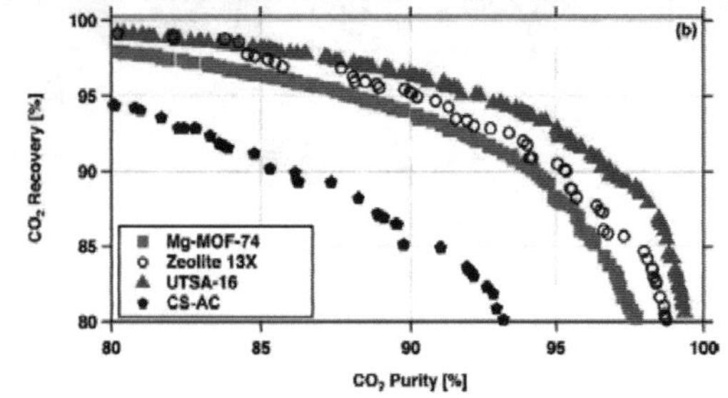

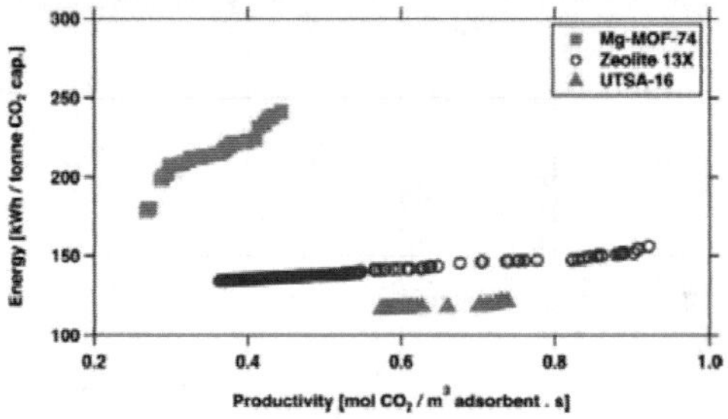

Maximise Purity & Recovery

Maximise Productivity
Minimize Energy

Main conclusions from this work

1. Detailed process optimization provides a more reliable picture than ranking materials based on adsorption capacity
2. Nitrogen adsorption plays an important role in determining the process performance

SCCS

Rajagopalan, A.K, Avila, A.M, Rajendran, A.. *Int J Greenhouse Gas Control.* 2016;46:76-85.

Multiscale screening of adsorbents

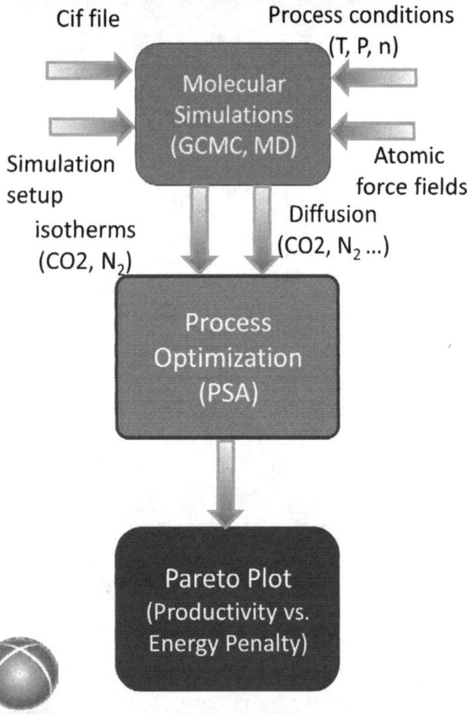

Scope of this work

- Predict isotherms on well known materials and compare with experimental data
- Process optimization with experimental and simulation isotherms
- Materials : Zeolite 13X and CPO-27-Ni

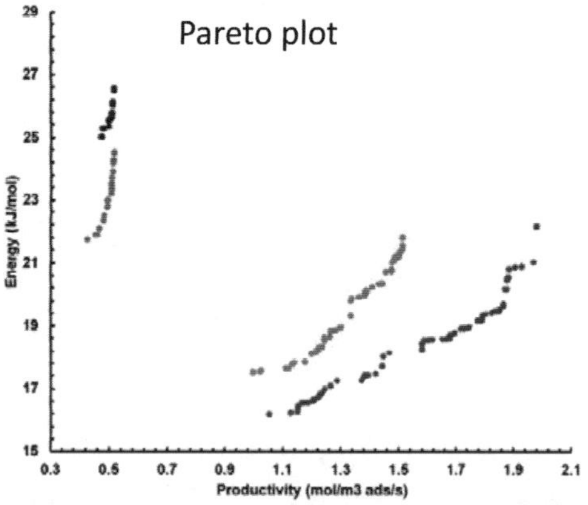

SCCS

Experimental isotherms: 13X

Zeolite 13X

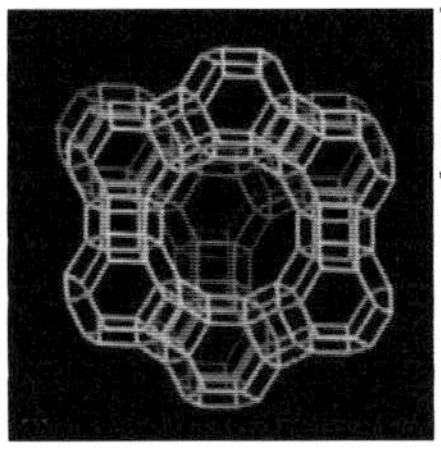

Source : IZA database

Ideal selectivity= K_{CO2}/K_{N2}: 1126

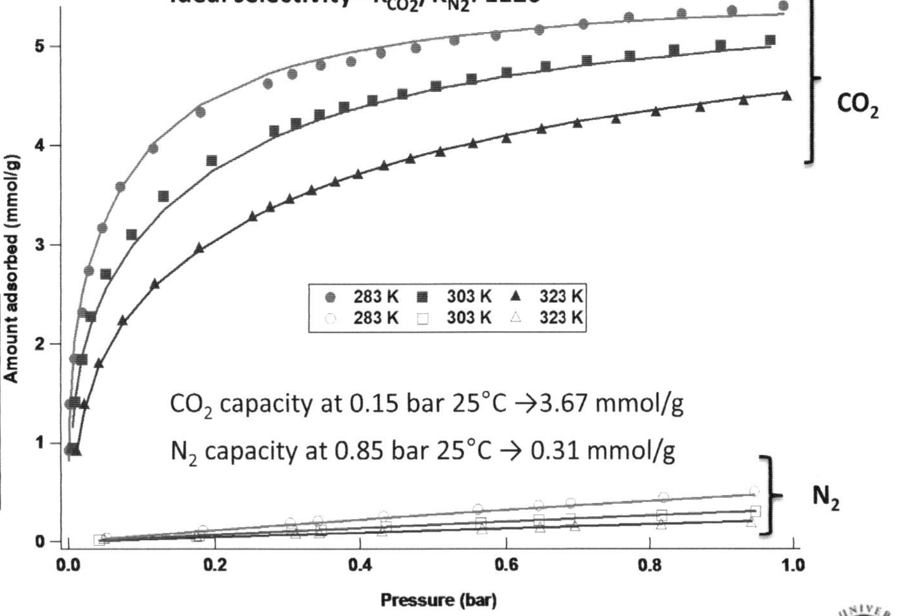

SCCS

Experimental isotherms: CPO-27-Ni

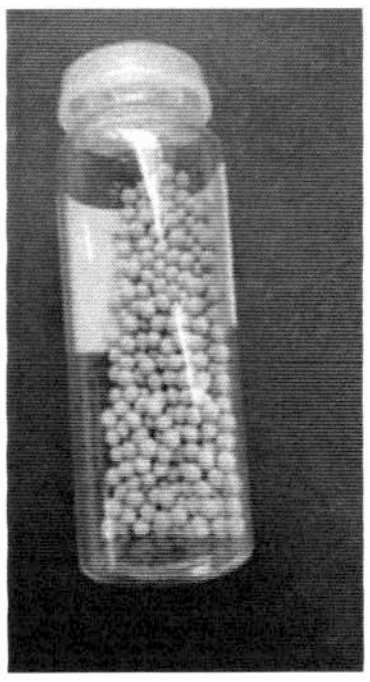

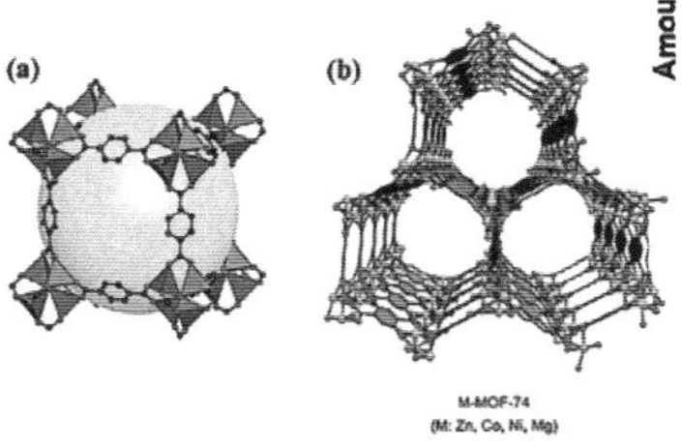

Ideal selectivity= K_{CO2}/K_{N2}: 90

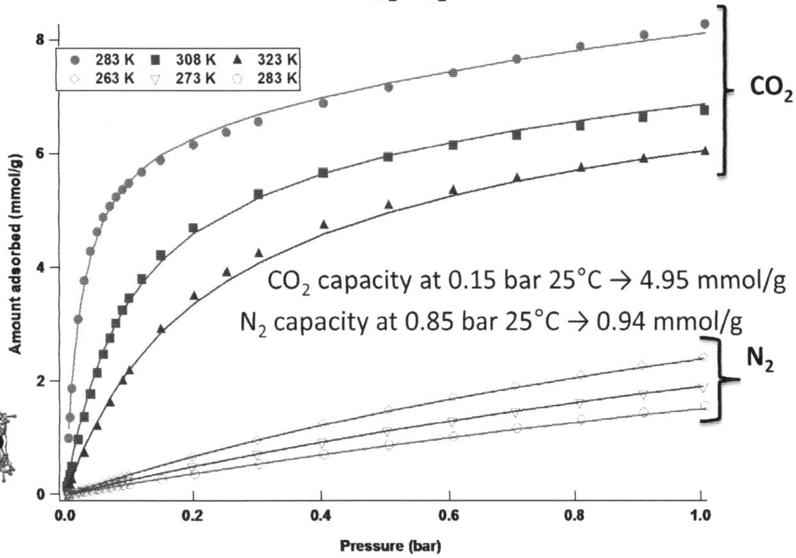

SCCS

Liu et al., **Chem. Soc. Rev.**, 2012,41, 2308-2322

Molecular simulations: Zeolite 13X

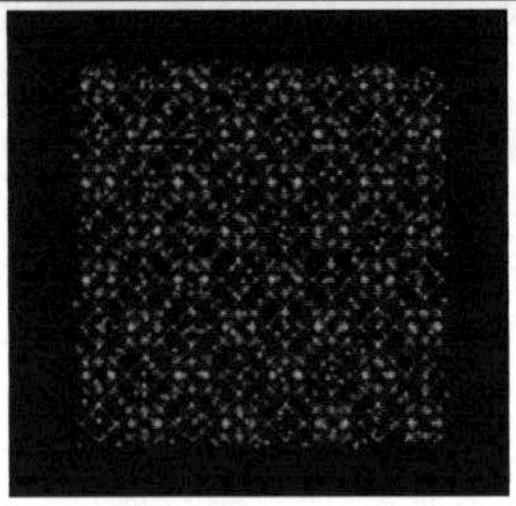

Zeolite 13X

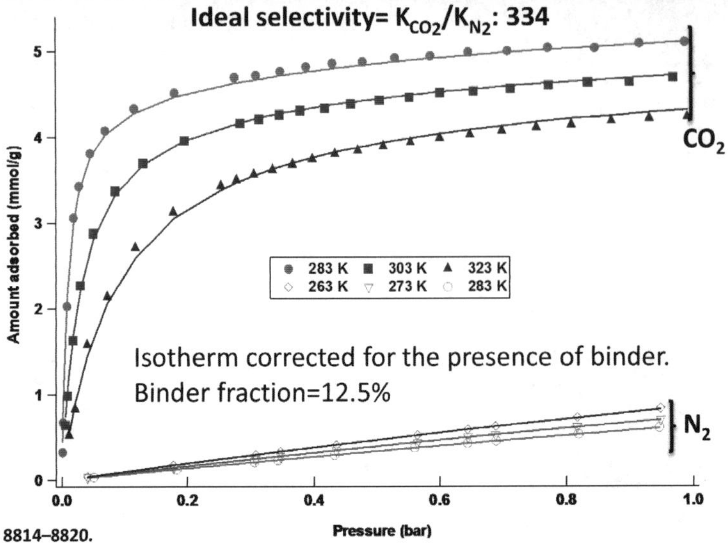

Crystal structure
D. H. Olson, Zeolites 15 1995 439-443.
Force fields
Garcia-Sanchez et al. J., Phys.Chem.C 2009 113 8814–8820.
Vujić, B., Lyubartsev, A. Modelling Simul. Mater. Sci. Eng. 2016
Jiang. J., Sandler, S. I., J. Am. Chem. Soc. 2005 127 11989-11997

SCCS

Experiment vs molecular simulation: 13X

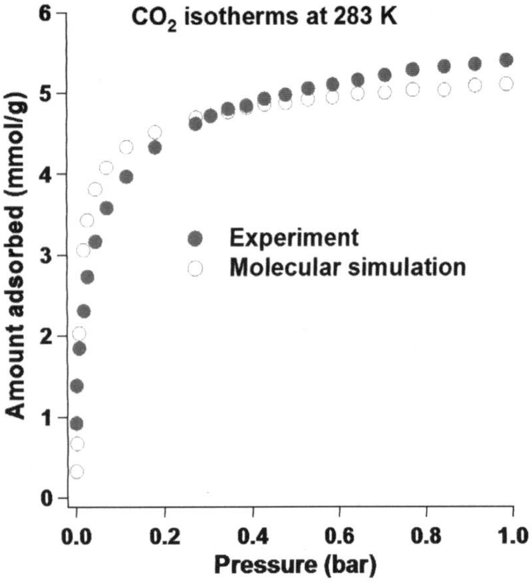

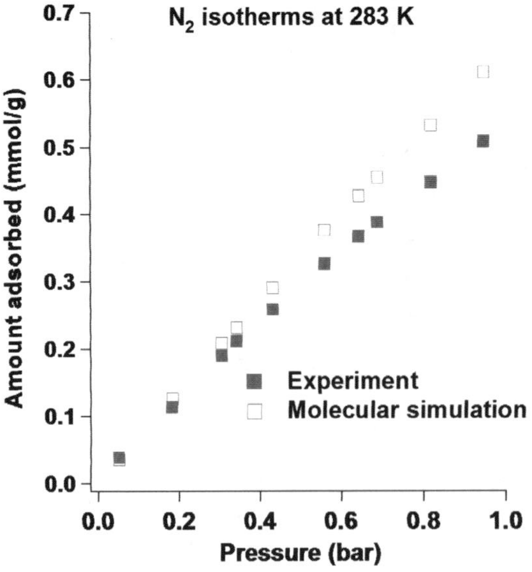

CO_2 capacity at 0.15 bar 25°C → 3.97 mmol/g

N_2 capacity at 0.85 bar 25°C→ 0.44 mmol/g

SCCS

Molecular simulation: CPO-27-Ni

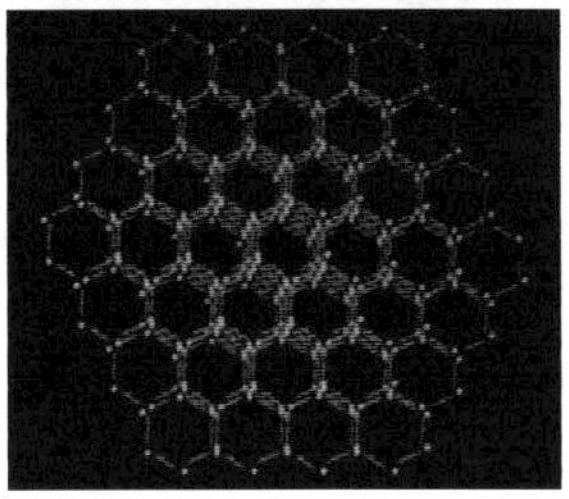

CPO-27-Ni

Ideal selectivity= K_{CO2}/K_{N2}: 145

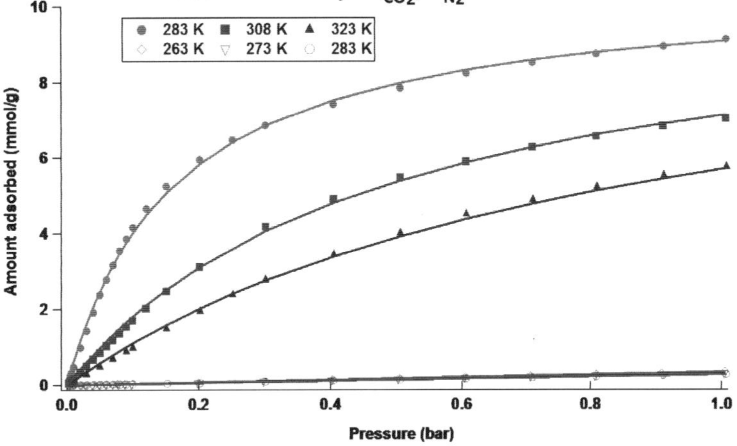

Crystal structure
Lee et al., *Chem. Mater.*, 2015 27 668–678.
Force fields
Mercado et al., *J. Phys. Chem C* 2016 120 12590–12604.
Bai et al., *J.Phys.Chem C* 2013 117 24375–24387
Rappe et al., J. Am. Chem. Soc. 1992 114 10024–10035

Isotherm corrected for the presence of binder.
Binder fraction=12.5%

SCCS

Molecular simulation: CPO-27-Ni

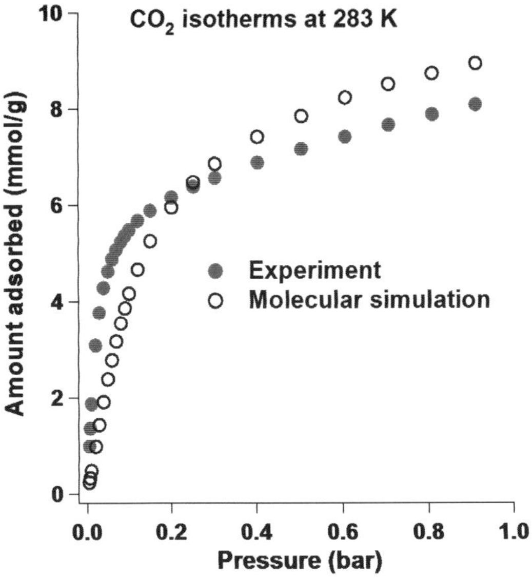

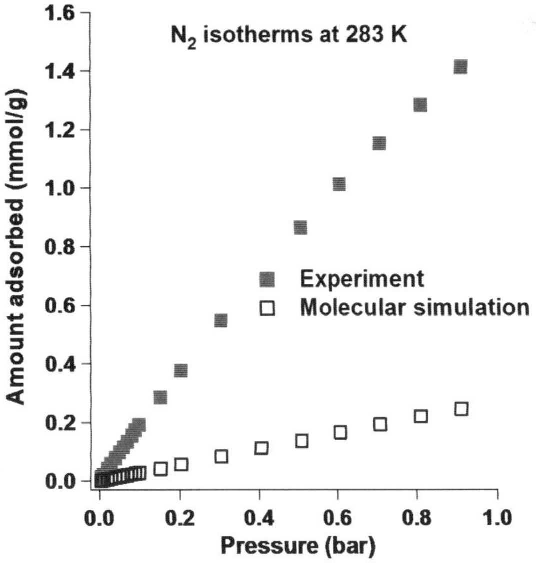

CO_2 capacity at 0.15 bar 25°C → 3.39 mmol/g

N_2 capacity at 0.85 bar 25°C → 0.19 mmol/g

SCCS

Process simulation

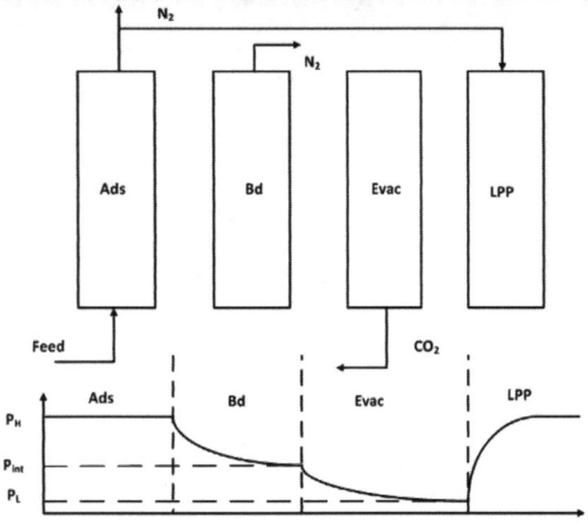

4-step VSA cycle with LPP [1,2]

- CySim adsorption simulator[3]
- Feed 15% CO_2 , 85% N_2 at 1atm and 25°C
- Adsorption isotherms
  - Pellet isotherms from experiment
  - Crystal isotherms from molecular simulations corrected for binder
- Same pellet size for the two materials
- Competitive dual-site Langmuir model for binary adsorption equilibrium

$$CO_2 \text{ purity} = \frac{\text{moles } CO_2 \text{ in evac}}{\text{total moles evac}}$$

$$CO_2 \text{ recovery} = \frac{\text{moles } CO_2 \text{ in evac}}{\text{moles } CO_2 \text{ in feed}}$$

$$\text{Specific energy} = \frac{\text{Compression+Evacuation energy}}{\text{moles } CO_2 \text{ in evacuation}}$$

$$\text{Productivity} = \frac{\text{moles } CO_2 \text{ in evacuation}}{\text{volume of adsorbent X cycle time}}$$

SCCS

1. Haghpanah et al AIChE J (2013)
2. Krishnamurthy et al AIChE J (2014)
3. Dang et al., Energy Procedia (2013)

Process optimization

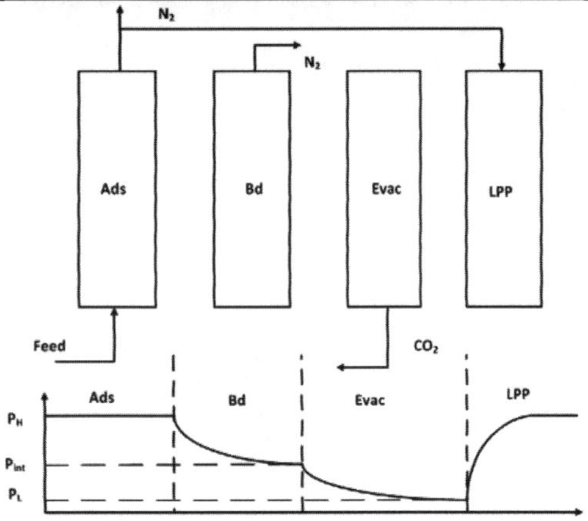

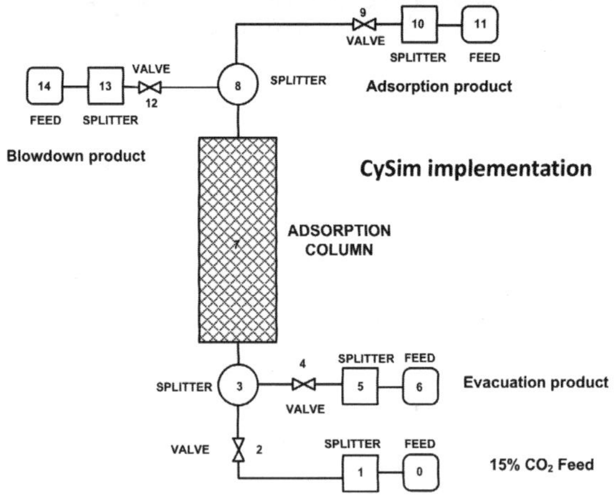

Decision variables

Adsorption, blowdown, evacuation step times

Blowdown & evacuation pressures

Feed flow rate

valve coefficients of blowdown and evacuation steps

GA based optimization (NSGA-III)

Minimize energy and maximize productivity

Constraints : Purity > 95% Recovery >90%

SCCS

Performance analysis : 13X

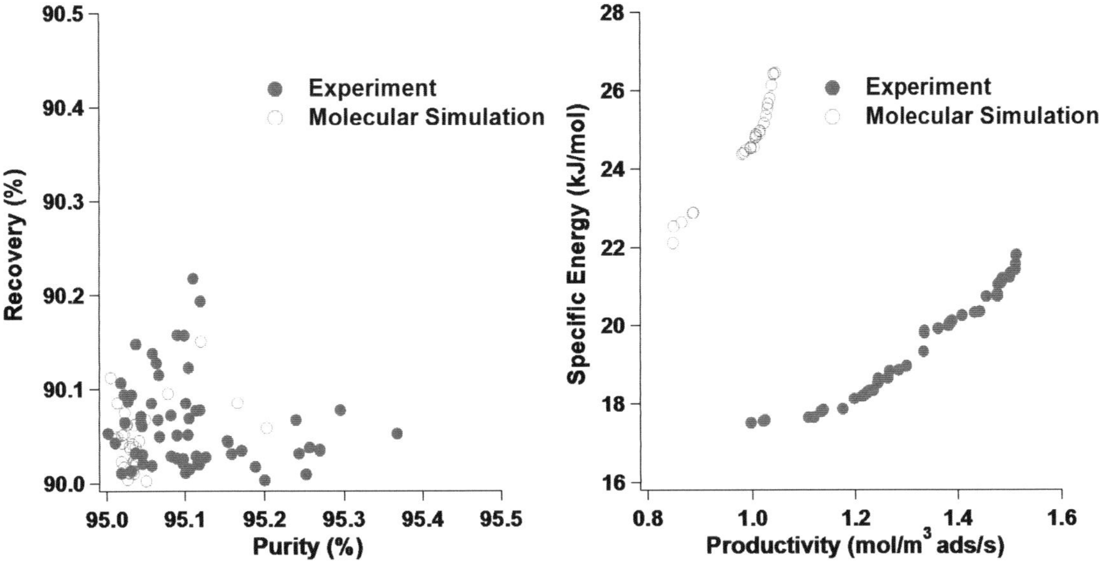

Minimum energy 17.51 kJ/mol (experiment isotherm) < 22.6 kJ/mol (simulated isotherm)

SCCS

Experiment vs Simulation: 13X

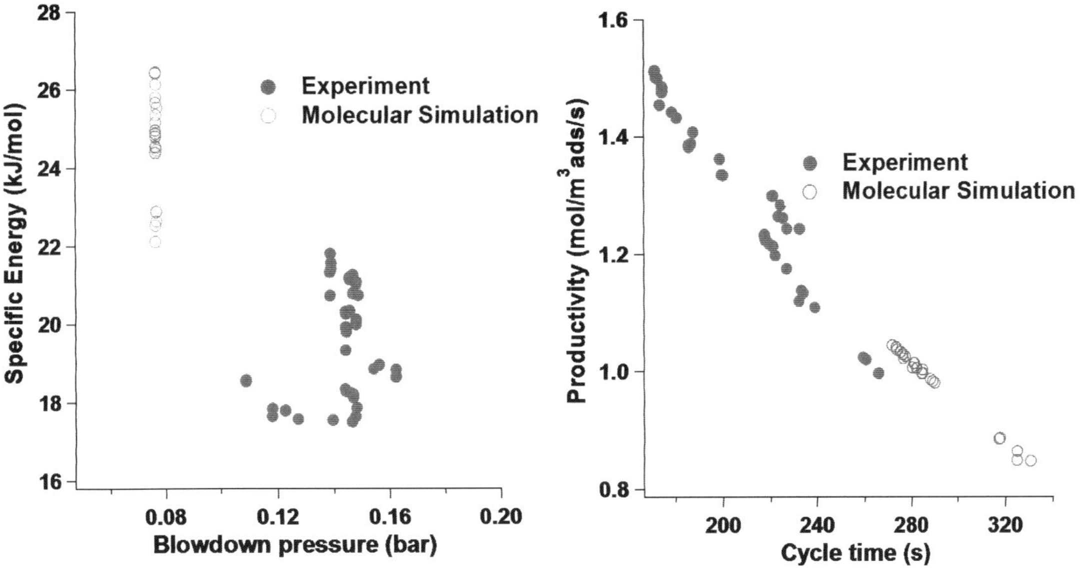

SCCS

Performance analysis: CPO-27-Ni

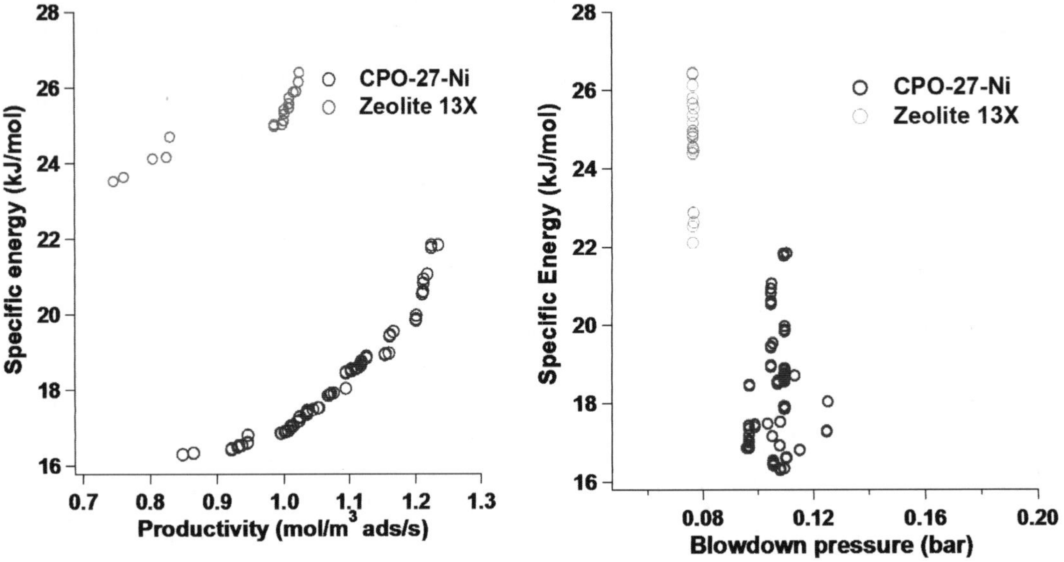

95% Purity and 90% Recovery constraints not satisfied with experimental isotherm

Minimum specific energy 16.37 kJ/mol < 13X (22.6 kJ/mol)

SCCS

Conclusions

- Multiscale screening of adsorbents from molecular scale to process optimization: first step in a toolbox for screening materials for CCS.
- Simulated isotherms compared with experimental isotherms:
  - Shape of CO_2 isotherm varied in both the samples
  - CO_2 capacity at 0.15 bar: over-predicted in 13X and underestimated in CPO-27-Ni
  - N_2 isotherms overestimated in 13X and underestimated in CPO-27-Ni
- Detailed process optimization carried out for two materials
- Prediction of N_2 isotherm has a profound impact on process performance
- Preliminary conclusion: accurate prediction of equilibrium properties is crucial for process optimization
- Yet, it still remains a substantial challenge for molecular simulations. Although predicted adsorption isotherms may look reasonable, they differ in essential features such as capacity in low pressure region
- Nitrogen adsorption largely ignored in simulations: needs to be addressed

SCCS

Acknowledgements

EPSRC
Engineering and Physical Sciences Research Council

Grant Ref : EP/N007859/1

SINTEF
Materials and Chemistry

17 AIChE
Annual Meeting, Minneapolis, MN

SCCS

501445 Evaluation of Diamine Appended Metal-Organic Frameworks for Post-Combustion CO_2 Capture Using a Vpsa

Abstract:

The recent invention of diamine appended metal organic frameworks [1] that show unusual step-shaped isotherms have opened up the possibility to tailor-make adsorbents for post-combustion CO2 capture. By varying the metal, it was possible to position the step at various pressure levelsl. The characteristic CO2 uptake of these materials indicate a possibility to obtain larger theoretical working capacities when compared to typical classical materials such as Zeolite 13X that show classical Langmuirian-type isotherms. The performance of five different materials with step-shaped isotherms was evaluated for Vacuum swing adsorption(VSA)-based post-combustion CO2 capture from a dry flue gas. In order to evaluate the true potential of these materials, a rigorous optimization based on detailed models that take into account mass+heat transfer, cycle configuration was performed. Using Genetic algorithms the operating space was throougly scanned to obtain the Pareto curves for Purity and Recovery for five diamine-appended MOFs. The optimization results revealed that some of these materials show promise as they are able to outperform the current benchmark material, i.e., Zeolite 13X. The optimization results also demonstrated that the flue gas temperature plays a critical role in deciding which of the MOFs should be used and provide intersting heuristics for choosing operating conditions depending on the type of the MOF and the flue gas temperature. Finally, detailed optimization to reduce parastic energy showed that some of the MOF s are capable of providing lower energy consumption compared to Zeolite 13X for CO2 purity> 95% and Recovery > 90%, making them attractive materials for further study.

1.MMcDonald, Thomas M., Jarad A. Mason, Xueqian Kong, Eric D. Bloch, David Gygi, Alessandro Dani, Valentina Crocella et al. "Cooperative insertion of CO2 in diamine-appended metal-organic frameworks." *Nature* 519, no. 7543 (2015): 303-308.

495090 Sub-Ambient Pressure-Swing Adsorption for Enhanced Post-Combustion Carbon Dioxide Capture Via Phase-Change Materials Application

Abstract:

Sub-ambient pressure-swing adsorption for enhanced post-combustion carbon dioxide capture via phase-change materials application

Héctor Octavio Rubiera Landa, Stephen J. A. DeWitt, Matthew J. Realff, Ryan P. Lively, and Yoshiaki Kawajiri.

School of Chemical & Biomolecular Engineering, Georgia Institute of Technology,

311 Ferst Drive N. W., Atlanta, GA, 30332-0100, USA.

As a transition step towards a carbon-free society, a number of process technologies are currently being investigated and developed. Amongst these, adsorption-based separations for carbon capture and storage (CCS) offer an attractive alternative for this global challenge **[1]**. Critical aspects to successful implementation of adsorption-based CCS technologies for post-combustion carbon capture include the application of high-performance adsorbents, suitable for high carbon dioxide recovery from flue gas streams.

As demonstrated experimentally in **[2]**, and analyzed via numerical simulations of breakthrough curves at sub-ambient conditions **[3]**, the application of a phase-change material (PCM) embedded in fiber adsorbents has the potential to enhance the performance of adsorption cycles by providing suitable heat management, therefore aiming at lower plant footprint and enabling a quasi-isothermal operation with optimal adsorptive capacity usage. In combination with tailored metal-organic frameworks operating at sub-ambient conditions, wherein significant adsorption of carbon dioxide occurs, 'PCM-assisted' cyclic adsorption processes promise to enable higher productivity adsorption systems.

The present work discusses firstly simple 'PCM-assisted' modeling and numerical simulation for sub-ambient pressure-swing adsorption cycles (PSA/VPSA), whereby a smooth-interface phase-change model is applied to represent melting/freezing transitions of the PCM, occurring by the intrinsic non-isothermal, periodic process operation. Next, as preliminary step for further process optimization work, we conduct a parametric study, useful in obtaining insight into the process dynamics governing 'PCM-assisted' PSA/VPSA cycles.

References

[1] J. Wilcox, R. Haghpanah, J. He, K. Lee, E. Rupp. Advancing Adsorption and Membrane-Based Separation Processes for the Gigaton Carbon Capture Challenge. *Annual Review of Chemical and Biomolecular Engineering* **2014**, *5(1)*, 479-505. http://dx.doi.org/10.1146/annurev-chembioeng-060713-

040100.

[2] R. P. Lively, N. Bessho, D. A. Bhandari, Y. Kawajiri, W. J. Koros. Thermally moderated hollow fiber sorbent modules in rapidly cycled pressure swing adsorption mode for hydrogen purification. *International Journal of Hydrogen Energy* **2012**, *37(20)*, 15227-15240. http://dx.doi.org/10.1016/j.ijhydene.2012.07.110.

[3] H. O. Rubiera Landa, M. J. Realff, R. P. Lively, Y. Kawajiri. *2016 AIChE Annual Meeting* **2016**, San Francisco, CA, USA.

497486 Pressure Swing Adsorption Cycle Synthesis Utilizing Artificial Neural Networks As Surrogate Models

Abstract:

Pressure Swing Adsorption Cycle Synthesis Utilizing Artificial Neural Networks as Surrogate Models

K.T. Leperi,[1] F. You,[2] R.Q. Snurr[1]

[1]Northwestern University, Evanston, IL, USA

[2] Cornell University, Ithaca, NY, 14850

With fossil fuels expected to be a significant portion of the world's energy mix for the near future, it is important to minimize CO_2 emissions from current power plants through carbon capture and sequestration (CCS).[1] In post-combustion CCS, CO_2 is separated from the power plant flue gas emissions, containing mainly N_2 and CO_2. Of the technologies currently available for CCS, pressure swing adsorption (PSA) is one of the more promising due to low energy requirements and short cycle times compared to other adsorption based technologies. However, one challenge that exists in using PSA for CCS is designing the cycle to match newly developed materials for this application. Although the steps in all PSA cycles can be classified into six different possibilities (pressurization, feed, depressurization, light reflux, heavy reflux and pressure equilibration), the arrangement of the steps and interactions between steps lead to hundreds of potential different combinations. The objective of this work is to develop a new approach that is capable of synthesizing PSA cycles to capture the CO_2 from flue gas at the required purities and recoveries while minimizing energy requirements and maximizing adsorbent productivities.

In this work, we present a new framework for synthesizing the PSA cycle with the lowest predicted CO_2 capture costs. In this framework, we train artificial neural networks (ANNs) using Bayesian regularization methods as surrogate models for the various PSA steps. The ANNs are trained on simulation data collected from our PSA model consisting of a system of partial differential algebraic equations incorporating mass and energy balances, pressure drop across the column, competitive multi-site Langmuir isotherms and the linear driving force model.[2] With the ANN surrogate models, we propose a mixed integer nonlinear programming (MINLP) model to determine the ideal ordering and duration of steps in order to minimize the energy requirements and maximize the adsorbent productivity. We evaluate this model with several adsorbents, including Ni-MOF-74, UTSA-16 and zeolite 13X in order to compare the adsorbents under optimized cycle conditions.

1. MIT. *The Future of Coal*; 2007. http://web.mit.edu/coal/.

2. Leperi KT, Snurr RQ, You F. Optimization of Two-Stage Pressure/Vacuum Swing Adsorption with Variable Dehydration Level for Postcombustion Carbon Capture. *Ind. Eng. Chem. Res.* 2016;55:3338-3350.

501398 Process Flow Sheet Optimization and Techno-Economic Assessment of Post-Combustion CO_2 Capture Using Heat Integrated, Sub-Ambient PSA

Abstract:

Solid sorbents are a promising low-energy approach to post-combustion CO2 capture. However, the relatively low swing capacities (i.e., ~ 1 mmol CO2/g) of most solid sorbent systems drives up both capital and operating costs. These challenges can be overcome by an unorthodox approach: flue gas cooling and compression matched with sorbent materials that have large isotherm steps. These two factors enable high swing capacity (>10 mmol CO2/g) sorbent that can drive down capital costs relative to other state-of-the-art sorbent systems by many folds. Further, the sorbents are embedded within hollow fiber modules that have been shown to be 3-5x more capital and energy efficient than traditional pellet-packed fixed beds. Critically, the energy associated with flue gas cooling and compression system is largely recouped by employing sub-ambient heat exchange and power recovery from the clean N2 exhaust. The goal of this study is to show the viability of this heat exchange/power recovery approach for our sub ambient pressure swing adsorption (PSA) system.

Flue gas compression and cooling is major cost elements and tradeoffs between the capital cost of the heat exchangers and compressor train and utility use are explored. In the optimization model, the major components are considered including the gas-gas heat exchanger to recover the cold utility from the CO2 stripped flue gas and the work recovery from the stream for gas compression. The optimization problem is solved by considering a superstructure that embeds the different options of multistage compression and cooling. Some optimal designs are considered as case studies to perform high-level technoeconomic analysis. After the technoeconomic analysis, the optimization models discussed earlier are refined and updated, thus making the entire effort iterative. To estimate the capital cost, only major components are considered such as fiber modules, compressors, turbines, and heat exchangers utilizing the existing models. The specifications given by NETL for a 550MW (net) power plant is used and our initial studies show the CO_2 capture cost to be below $25.0/ton.

494609 Fast Sorbent Mediated Water-Gas Shift (C-SHIFT) Process for Pre-Combustion CO_2 Capture

Abstract:

Fast Sorbent mediated water-gas shift (C-SHIFT) process for pre-combustion CO_2 capture

M.E. Boot-Handford, P.S. Fennell, J.A.Z. Pieterse, P.D. Cobden, J. Boon, M.G. Sceats, B.N.C. Sweeney, B. Arstad, R. Blom

The carbonated shift (C-SHIFT) process is part of the ASCENT project aiming at proof of concept of three different advanced CO_2 capture process. C-SHIFT, a sorption enhanced water-gas shift process, uses a highly innovative fluidized bed reactor system to match the heat requirements of reaction and regeneration of medium temperature carbon dioxide acceptor materials (300-600°C) with pressure swing technology. The aim is to produce a fuel stream consisting of high temperature H_2 and H_2O, at a pressure above the compressor temperature of a gas turbine, with a CO_2 capture efficiency greater that 90%. Alkali carbonate promoted Mg/Al-hydrotalcites (HTCs) and magnesium oxides were considered the most promising of the bifunctional CO_2 sorption and water-gas shift reactive materials reviewed. In particular, high MgO containing Mg/Al-hydrotalcites impregnated with K_2CO_3 and high magnesium contents were identified as being strong candidates for high pressure, mid-temperature CO_2 capture in the presence of large quantities of steam. However, in the context of C-SHIFT, the cyclic performance information provided thus far is limited. There is a significant discrepancy between the long carbonation periods under which these materials have been investigated and the short residence time that the CO_2 sorbent material is expected to reside within the proposed C-SHIFT carbonator (< 60 s). At present there are no published studies reporting either the uptake capacity or kinetics for the reaction between CO_2 and HTC-derived materials over time scales relevant to C-SHIFT. The work performed within ASCENT project aims to address this gap in the knowledge and provide information relating to the kinetics and capacities for CO_2 uptake by K_2CO_3-promoted HTC-derived materials at conditions relevant to a C-SHIFT process. This has been achieved using a pressurized spouted fluidized bed and fixed bed reactor and for pressure-swing operation at pressures up to 10-20 bar with large concentrations of steam. Fluidized beds are particularly suited for kinetic studies based on their superior heat and mass transfer characteristics when compared with other reactor configurations such as fixed beds and thermo-gravimetric analysis (TGA). The reactor has now been commissioned for use at pressures up to 10 bar with steam concentrations up to 40 % v/v. Modifications to the water trap design and drainage protocol that enables continuous removal of water from the system without disruption to the system pressure stability. Results from initial testing of the cyclic performance of the selected material over carbonation/ calcination cycles are also provided serving as a proof-of-concept for the pressurized fluidized bed reactor testing. The aim of this initial cyclic test was to provide empirical measurements of the carbonation and calcination kinetics and cyclic capacities of K_2CO_3-promoted HTC-derived mixed metal oxide materials at conditions and within the time period relevant to an industrial C-SHIFT process. The capabilities of the pressurized fluidized bed reactor to test the cyclic performance of candidate mid-temperature CO_2 sorbents with pressure-swing regeneration was demonstrated with carbonation at 10 bar 400 °C, 20 % v/v CO_2 and 40 % v/v steam and regeneration at 1 bar, 400 °C under mild condition (100% N_2). The presence of steam was also found to dramatically enhance both the carbonation rate and short-term carbonation capacity. Furthermore, the presence of H_2S or operation with syngas did not affect the cyclic performance of either of the sorbent materials tested. The materials reacted reversibly with H_2S presenting the possibly of a consolidating CO_2 and sulfur gas removal into a single step in a sour sorption enhanced water gas shift reaction.

Acknowledgements

The presented work is funded within the ASCENT project as part of the European Union's Seventh Framework Programme (FP7/2007-2013) under grant agreement nº 608512. Note: "The present publication reflects only the authors' views and the European Union is not liable for any use that may be made of the information contained therein".

502557 Numerical Evaluation of a Fluid Catalytic Cracking Unit with Internal Baffles

Abstract:

Fluid Catalytic Cracking (FCC) units play a very important role in an oil refinery, because they convert heavy fractions (vacuum distillates or some vacuum residues) to gasoline, C3-C4 cuts and petrochemicals. These units and the catalysts used are in continuous evolution. They must adapt to market changes: gasoline yield and/or quality maximization, petrochemicals production, conversion of residues, environmental requirements. New riser configurations require that the phases have an increasingly short contact time, being separated after 1 s or 2 s of flow. Regenerators operating at a higher combustion temperature allow a higher recovery of the catalyst. The catalyst entering the reactor warmer requires that the mixture between the phases be extremely efficient, since a poor thermal exchange between the phases results in high temperature points and, consequently, the favoring of thermal cracking reactions. Products formed by thermal cracking have lower value than those formed by catalytic route, and therefore regions with temperatures too high must be avoided. Once the gas-solid flow in risers can develop a "core-anulus" pattern, in which catalyst particles are more concentrated near the walls of the reactor, the inclusion of internals can enhance the phases mixture. However, it can also increase its pressure drop, which is a limiting factor to the process. Therefore, the aim of this study is to evaluate the performance of an industrial riser type reactor, with the inclusion of airfoil-shaped ring-type baffles in order to improve the gas-solid flow. Computational fluid dynamics (CFD) techniques are applied to simulate in a three-dimensional geometry, in which the turbulent three-phase flow was calculated in a transient formulation. Gases (air and steam) and catalyst particles were the two Eulerian phases simulated, and a third Lagrangian phase was composed of liquid oil droplets which were evaporated during the process. Results indicate that it is possible to enhance the gasoline production without causing an expressive increase in the pressure drop of the unit.

17AIChE
Annual Meeting, Minneapolis, MN

A Group Contribution Method for the Prediction of the Mid Infrared (MIR) Absorption Spectra of Species Involved in Fluid Catalytic Cracking (FCC)

Sandra López-Zamora, Alejandro Molina, Hugo de Lasa

UNIVERSIDAD NACIONAL DE COLOMBIA SEDE MEDELLÍN
Western UNIVERSITY · CANADA
FRUN Flujos Reactivos de la Universidad Nacional de Colombia

Outline

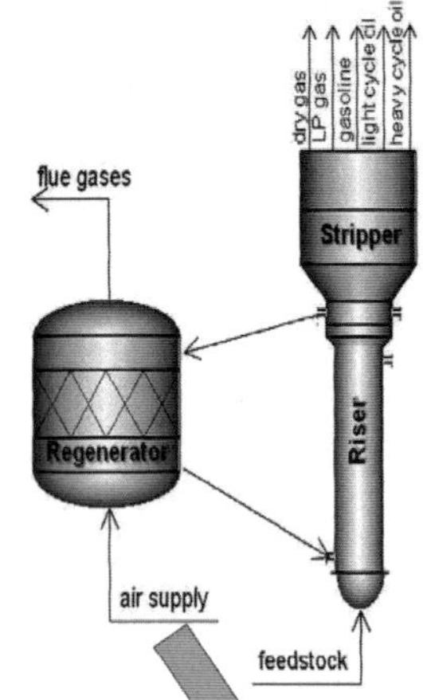

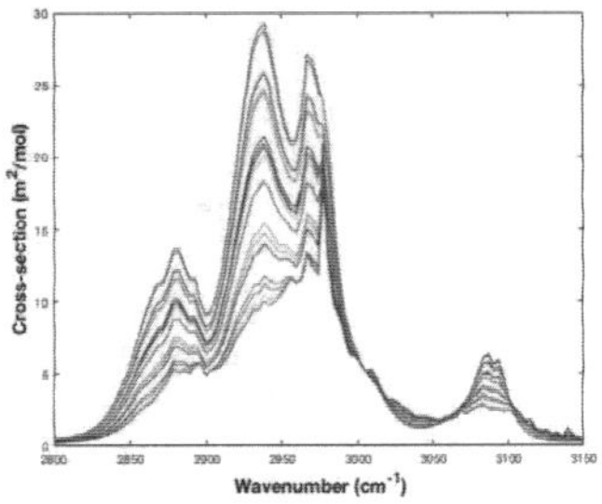

Validation of the Group contribution method

Introduction

Effect of C-H bonds number in absorbance

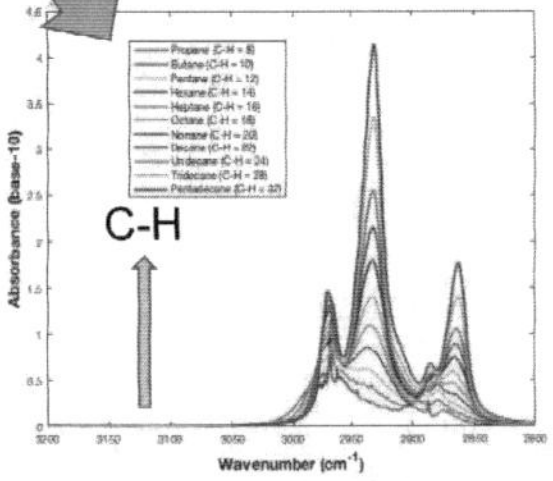

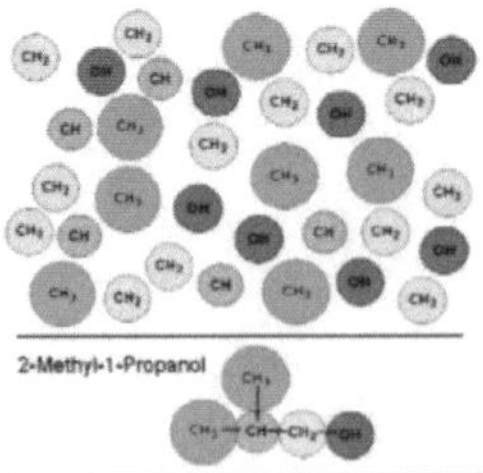

Group contribution method

UNIVERSIDAD NACIONAL DE COLOMBIA SEDE MEDELLÍN
Western UNIVERSITY · CANADA
FRUN Flujos Reactivos de la Universidad Nacional de Colombia

Fluid Catalytic Cracking (FCC)

- Fossil fuels are extensively employed in transportation, industry and electricity generation
- Over 95% of the energy in the transportation sector is derived from petroleum
- Fluid catalytic cracking is one of the most important and profitable process technologies in the refinery

Image from: elfinanciero.com.mx & 123rf.com

UNIVERSIDAD NACIONAL DE COLOMBIA SEDE MEDELLÍN — Western UNIVERSITY · CANADA — FRUN Flujos Reactivos de la Universidad Nacional de Colombia

Fluid Catalytic Cracking (FCC)

- The feed to the riser is basically composed of C30−C40 hydrocarbons that are cracked to more value molecules
- The reaction zone in FCC involves:
  - ✓ Catalyst stripper
  - ✓ Two coupled reactors: Riser (or downer) and regenerator
- Operating conditions
  - ✓ Riser: 3 atm, Pressure drop: 35-62 kPa, Temperature: 550 °C
  - ✓ Downer:2.5 atm, Pressure drop: 20 kPa, Temperature: 550 °C

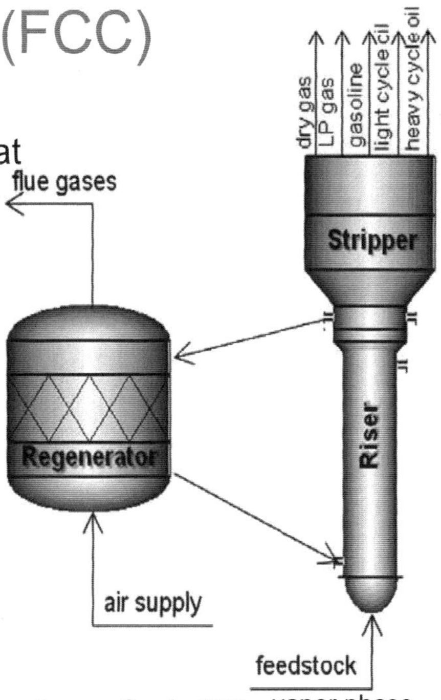

Image from: Jimenez-García, 2011

UNIVERSIDAD NACIONAL DE COLOMBIA SEDE MEDELLÍN — Western UNIVERSITY · CANADA — FRUN Flujos Reactivos de la Universidad Nacional de Colombia

FCC Reactions

- The feed to the riser is basically composed of C30−C40 hydrocarbons that are converted to smaller molecules.
- Cracking reactions are fast (<10 s).
- Reactants and products can be divided in: paraffins, olefins, naphthenes and aromatics
- Main reactions:
  - ✓ Cracking
  - ✓ Cyclization
  - ✓ Isomerization
  - ✓ H transfer
  - ✓ Dehydrogenation

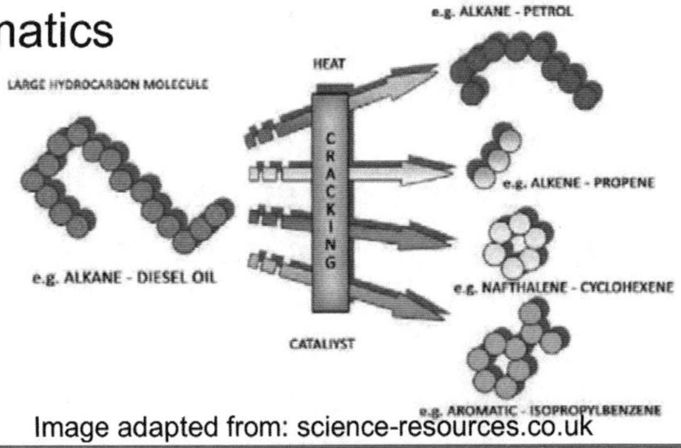

Image adapted from: science-resources.co.uk

UNIVERSIDAD NACIONAL DE COLOMBIA SEDE MEDELLÍN Western UNIVERSITY · CANADA FRUN Flujos Reactivos de la Universidad Nacional de Colombia

FCC reactions

- FCC experiments miss the first seconds of the reaction.
- Some experiments have resolved reaction data down to seconds by quenching the reaction
- Short reaction time: 0.5 – 10 s
- Reaction characterization is made ex-situ with long analysis time >20 min

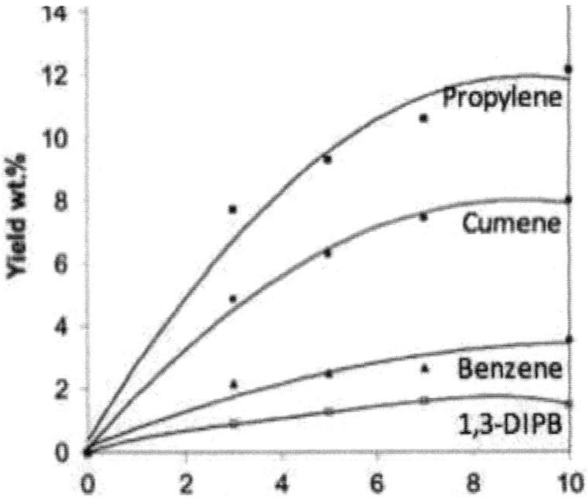

Catalytic cracking of 1,3,5-TIPB: change in the product yield with reaction time (Taken from Al-Khattaf et al. 2002)

UNIVERSIDAD NACIONAL DE COLOMBIA SEDE MEDELLÍN Western UNIVERSITY · CANADA FRUN Flujos Reactivos de la Universidad Nacional de Colombia

Reactions characterization

Conventional techniques

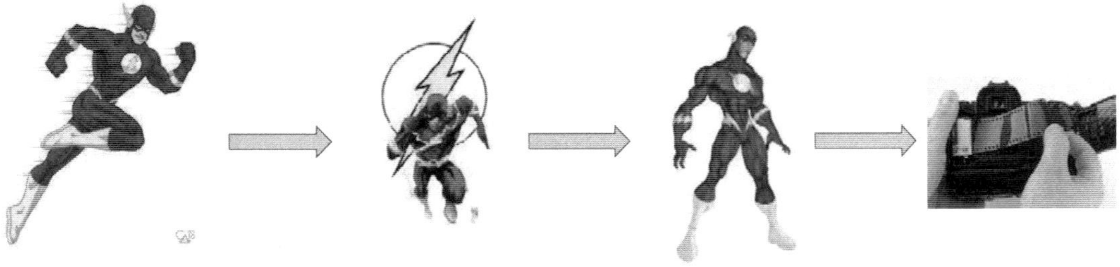

Laser techniques

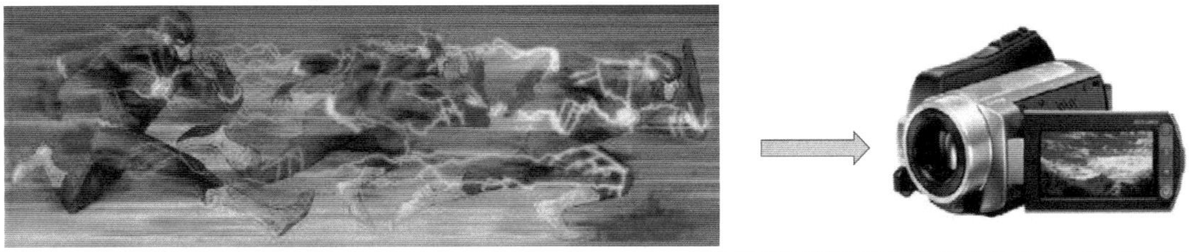

Reactions characterization

Laser techniques

Useful for:

- Higher number of data during the first seconds of reaction
- Process control
- Computational fluid dynamics (CFD) validation → Time-resolution

UNIVERSIDAD NACIONAL DE COLOMBIA SEDE MEDELLÍN
Western UNIVERSITY · CANADA
FRUN Flujos Reactivos de la Universidad Nacional de Colombia

Previous works

➢ Several studies present the evaluation of hydrocarbons by the use of laser techniques

PARAFFINS

- C-H stretch vibrational absorption can be exploited.
- n-Heptane mole fraction → IR HeNe laser absorption
- Iso-octane → 2948 cm^{-1} HeNe laser
- N-dodecane and n-decane → Tunable mid-infrared laser using difference frequency-generation (DFG) near 2941cm^{-1}

OLEFINS

- Ethylene and propylene → widely-tunable rapid-scanning Mid-IR Laser Spectrometer: 3283 - 2812 cm^{-1}
- Ethylene → CO_2 gas laser near 943 cm^{-1}
- Propene → External cavity quantum-cascade laser (ECQCL) system
- 1,3-butadiene and propylene → Tunable diode laser (TDL): 891 cm^{-1}

AROMATICS

- Toluene → Differential-absorption LIDAR, third harmonic of the Nd:YAG laser
- Aromatic compounds and particles in an atmospheric-pressure opposed-flow flame of ethylene → Laser Induced Emission
- polycyclic aromatic hydrocarbons (PAH) → Laser-induced fluorescence, Nd:YAG Laser in the fourth harmonic (266 nm) and a nitrogen laser at 337 nm

➢ In the referred literature no study applied laser diagnostics to a mixture such as the one for FCC process

UNIVERSIDAD NACIONAL DE COLOMBIA SEDE MEDELLÍN

FRUN Flujos Reactivos de la Universidad Nacional de Colombia

Molecules interaction with light

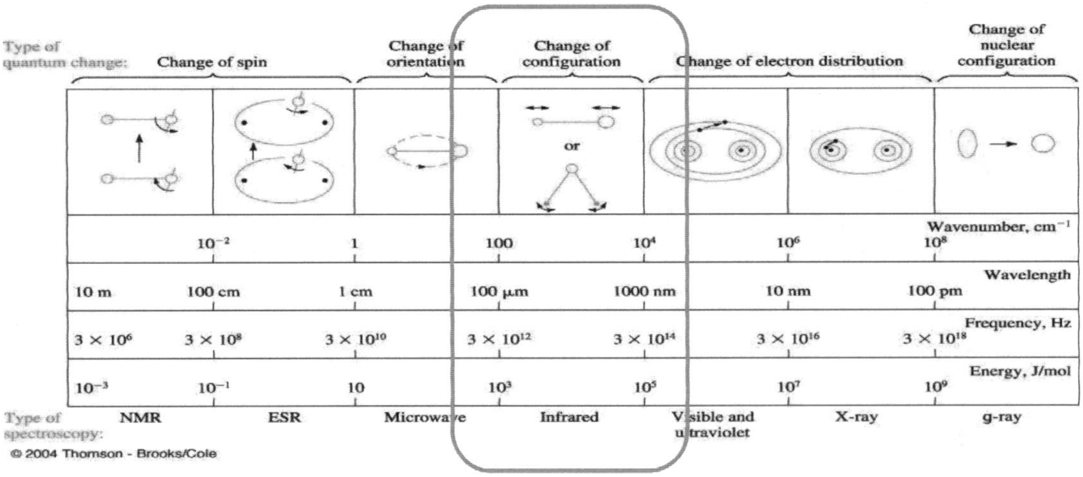

Image from: en.wikipedia.org/wiki/Infrared_spectroscopy

UNIVERSIDAD NACIONAL DE COLOMBIA SEDE MEDELLÍN

Western UNIVERSITY · CANADA

FRUN Flujos Reactivos de la Universidad Nacional de Colombia

IR spectra

- A change of configuration in the molecule bonds occurs
- Molecules absorb specific frequencies that are characteristic of their structure: resonant frequencies
- The absorption of IR radiation can be used to identify and study chemicals
- The behaviour of hydrocarbons in mid-infra red (2.5 - 40 μm) is of interest

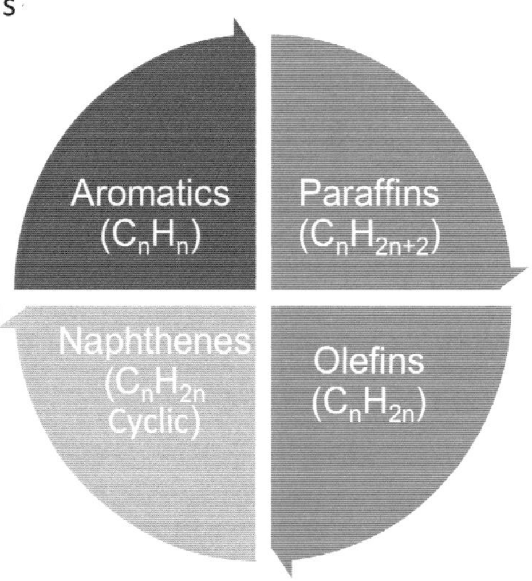

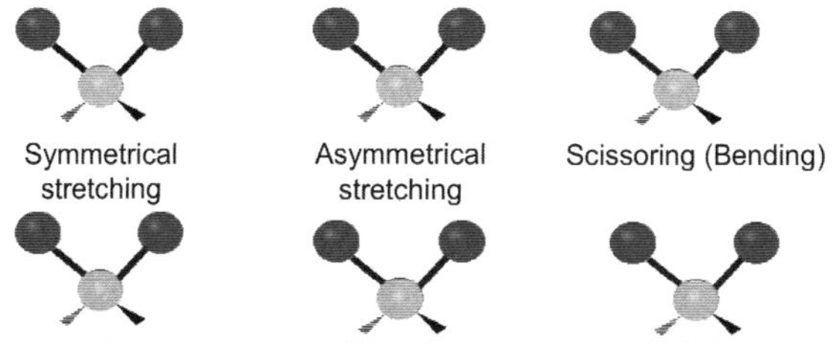

Image from: en.wikipedia.org/wiki/Infrared_spectroscopy

UNIVERSIDAD NACIONAL DE COLOMBIA SEDE MEDELLÍN
Western UNIVERSITY · CANADA
FRUN Flujos Reactivos de la Universidad Nacional de Colombia

Absorbance: Beer law

According with Beer law, cross section can be related to absorbance:

$$\alpha = -ln\left(\frac{I}{I_0}\right) = \sigma c L \qquad \sigma = \frac{\alpha}{cL}$$

where I_0 is light intensity, I is light intensity after interaction with molecules, σ is cross section, c is concentration and L the pathlength

In a given region the integrated band intensity can be obtained:

$$\psi = \int_{\lambda_0}^{\lambda} \sigma d\lambda$$

UNIVERSIDAD NACIONAL DE COLOMBIA SEDE MEDELLÍN
Western UNIVERSITY · CANADA
FRUN Flujos Reactivos de la Universidad Nacional de Colombia

MIR characterization of hydrocarbons

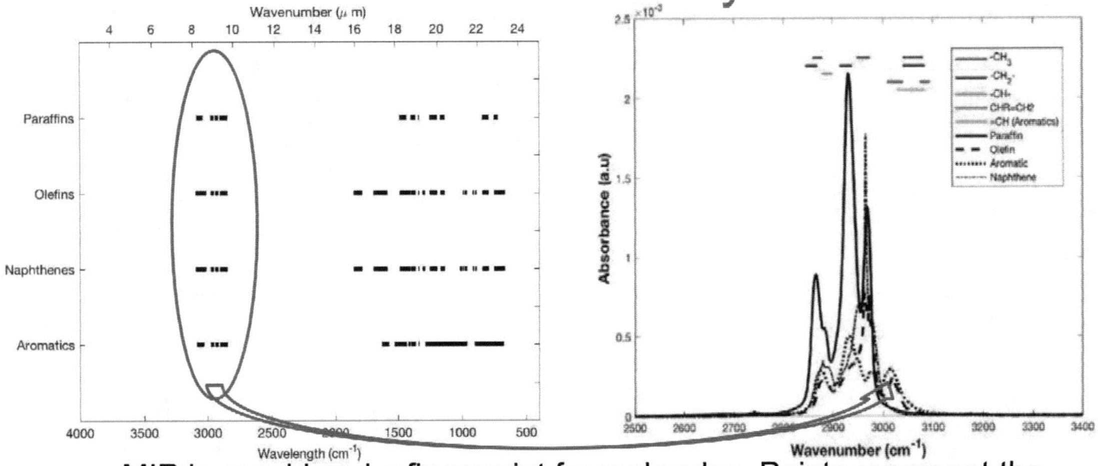

MIR is considered a fingerprint for molecules. Points represent the interaction with light of different functional groups. It is possible to use the region 3200-2800 cm^{-1} to characterize the functional groups present in hydrocarbons: paraffins, olefins, naphthenes and aromatics according to the C-H bonds

UNIVERSIDAD NACIONAL DE COLOMBIA SEDE MEDELLÍN

Western UNIVERSITY · CANADA

FRUN Flujos Reactivos de la Universidad Nacional de Colombia

MIR behaviour: 3200-2800 cm^{-1}

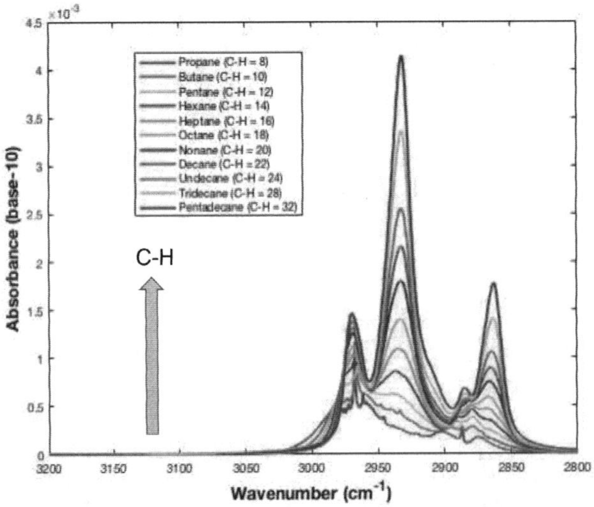

The region 3200 - 2800 cm^{-1} presents a relation between the number of C-H bonds and the absorption.

Spectra are taken from PNNL laboratory (Sharpe, 2004)

UNIVERSIDAD NACIONAL DE COLOMBIA SEDE MEDELLÍN

Western UNIVERSITY · CANADA

Sharpe (2004), Klingbeil (2007), Alfarae (2013), Campbell (2014)

FRUN Flujos Reactivos de la Universidad Nacional de Colombia

Group contribution method (GCM)

- Group contribution methods use the chemical structure of components to predict the properties of a given compound
- Since Lyndersen (1955) and Joback (1987), group contribution methods have been employed for the prediction of thermodynamic properties
- We proposed to use a group contribution method for the prediction of cross section in the region of interest.

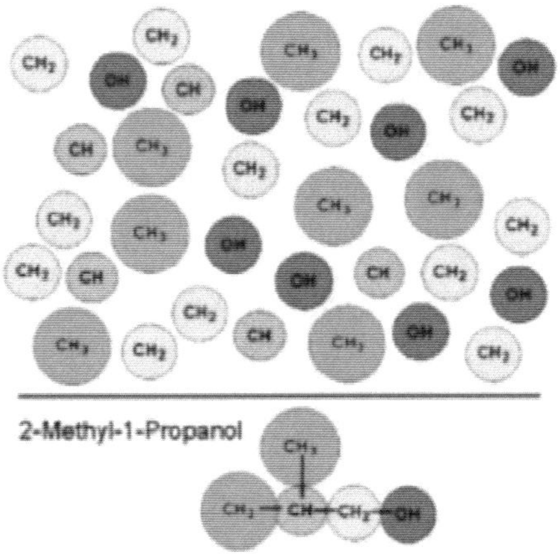

Graphical representation of a group contribution method

Image from:
https://en.wikipedia.org/wiki/Group_contribution_method

UNIVERSIDAD NACIONAL DE COLOMBIA SEDE MEDELLÍN
Western UNIVERSITY · CANADA
FRUN Flujos Reactivos de la Universidad Nacional de Colombia

Group contribution method (GCM)

- Mathematical model
- Linear behaviour of functionalities
- Statistical parameters should be consider
- Validation and applications
- Experimental setup

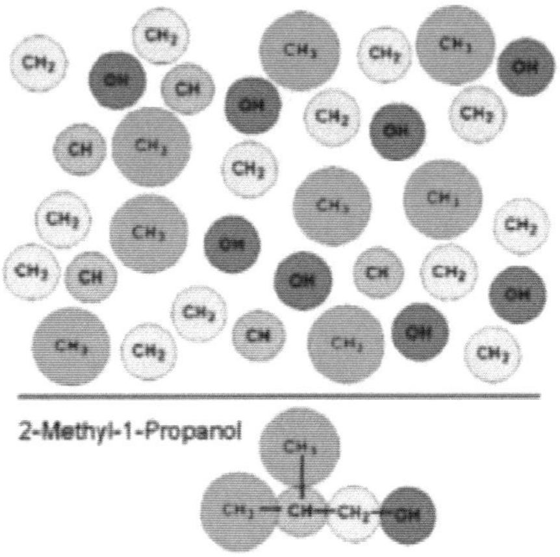

Graphical representation of a group contribution method

Image from:
https://en.wikipedia.org/wiki/Group_contribution_method

UNIVERSIDAD NACIONAL DE COLOMBIA SEDE MEDELLÍN
Western UNIVERSITY · CANADA
FRUN Flujos Reactivos de la Universidad Nacional de Colombia

Group contribution method (GCM)

- 15 parameters related to the kind of C-H bond are employed for the prediction of cross section
- Hydrocarbons present in PNNL database (Sharpe, 2004) were employed: 65 compound

9 parameters:

- ✓ $-CH_3$
- ✓ $-CH_2-$
- ✓ >CH-
- ✓ >C<

Simple bond ➡ 6 parameters related to α type carbons:

- ✓ $-CH_3$ (α)
- ✓ $-CH_2-$ (α)

- ✓ $=CH_2$
- ✓ =CH-
- ✓ =C<

Double bond ➡

- ✓ $-CH_3$ (α)
- ✓ $-CH_2-$ (α)

- ✓ a=CH
- ✓ a=C<

Aromatic ring ➡

- ✓ $-CH_3$ (α)
- ✓ $-CH_2-$ (α)

$$\sigma = \sum_{i=1}^{15} n_i \gamma_i$$

UNIVERSIDAD NACIONAL DE COLOMBIA SEDE MEDELLÍN — Western UNIVERSITY · CANADA — FRUN Flujos Reactivos de la Universidad Nacional de Colombia

Relationship between C-H number and cross section

A linear relationship between cross section (σ) and C-H number at HeNe wavelength (3.39μm) for paraffins, olefins, naphthenes and aromatics can be observed.

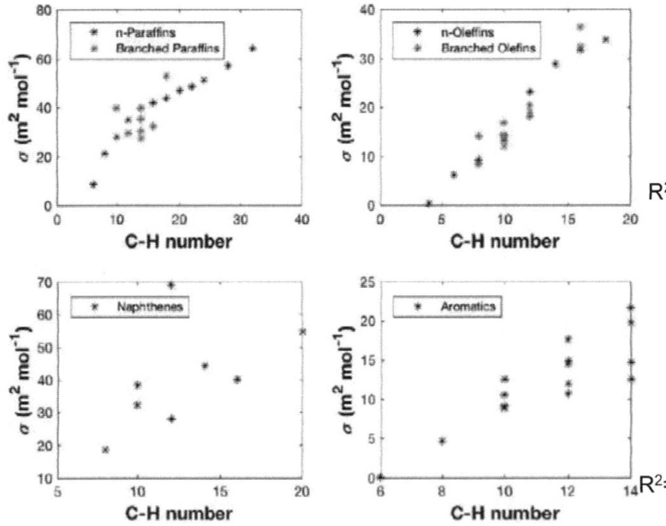

R^2=0.858

R^2=0.945

R^2=0.288

R^2=0.768

UNIVERSIDAD NACIONAL DE COLOMBIA SEDE MEDELLÍN — Western UNIVERSITY · CANADA — FRUN Flujos Reactivos de la Universidad Nacional de Colombia

Relationship between C-H number and integrated band intensity

Linear behaviour increases for integrated band intensity (Ψ) between 3200-2800 cm^{-1} and C-H number

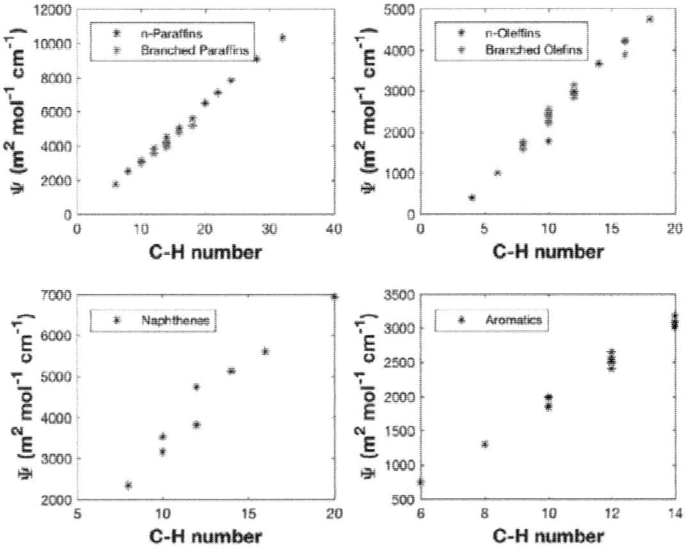

R^2=0.959 R^2=0.977

R^2=0.954 R^2=0.988

UNIVERSIDAD NACIONAL DE COLOMBIA SEDE MEDELLÍN — Western UNIVERSITY · CANADA — FRUN Flujos Reactivos de la Universidad Nacional de Colombia

GCM validation: pure components

The group contribution method (GCM) was employed for the prediction of paraffins, olefins, naphthenes and aromatics.

A good prediction for the spectra was obtained. Statistical moments of the different order (statistical moments of order one, two and three) had a R^2 of 0.97, 0.90, 0.78 respectively.

The prediction for integrated band intensities had a R^2=0.978

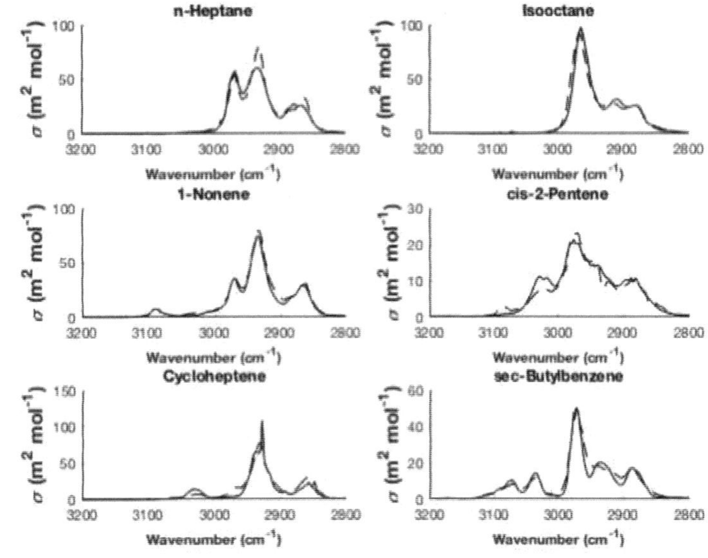

Cross section prediction.
-- Experimental (PNNL)
- Predicted

UNIVERSIDAD NACIONAL DE COLOMBIA SEDE MEDELLÍN — Western UNIVERSITY · CANADA — FRUN Flujos Reactivos de la Universidad Nacional de Colombia

GCM validation: Gasolines

Group Contribution Method was used for the prediction of absorption spectra of 21
The statistic summary for the prediction of absorption coefficients in the case of regular and premium gasolines present $R^2>0.97$ and RER>15

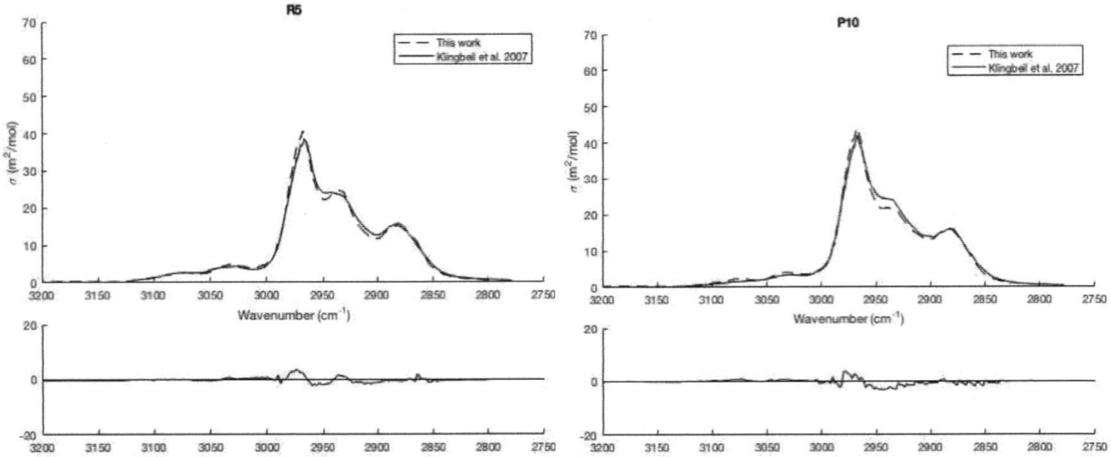

(a) Regular gasoline comprised of 20.6% n-paraffins, 32.8% isoparaffins, 18.6% olefins, 27.9% aromatics, 0% ethanol (b) Premium gasoline comprised of 8% n-paraffins, 49% isoparaffins, 4.7% olefins, 27.2% aromatics, 11.2% ethanol.

UNIVERSIDAD NACIONAL DE COLOMBIA SEDE MEDELLÍN

Western UNIVERSITY · CANADA

FRUN Flujos Reactivos de la Universidad Nacional de Colombia

GCM validation: PIOA and RON

PIOA prediction for premium gasolines was possible with $R^2=0.936$, RER=13,56.
RON prediction for premium gasolines was possible with $R^2=0.958$, RER=14.11.

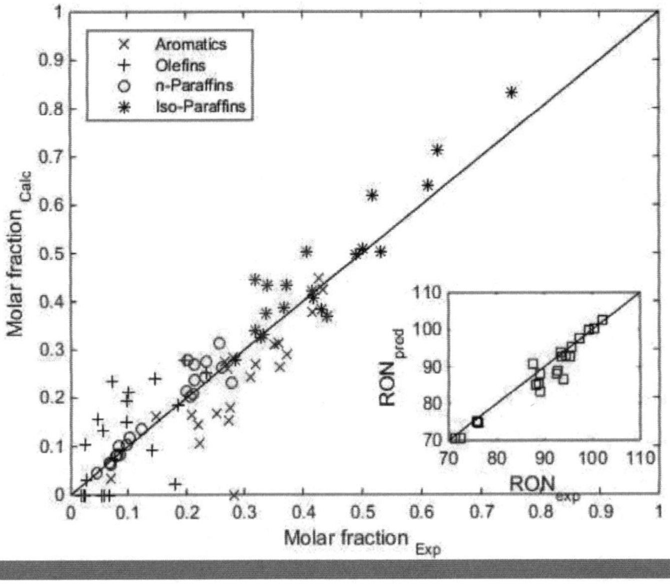

UNIVERSIDAD NACIONAL DE COLOMBIA SEDE MEDELLÍN

FRUN Flujos Reactivos de la Universidad Nacional de Colombia

GCM application: Reaction progress

It is possible to employ the linear relationship between the C-H number and the integrated band intensity for calculating the reaction progress of catalytic cracking reaction

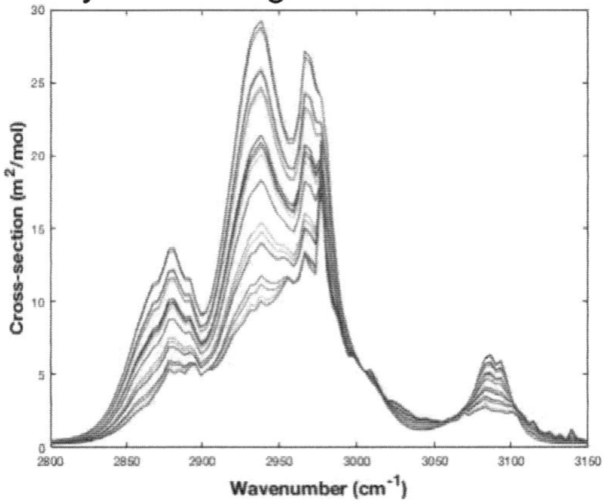

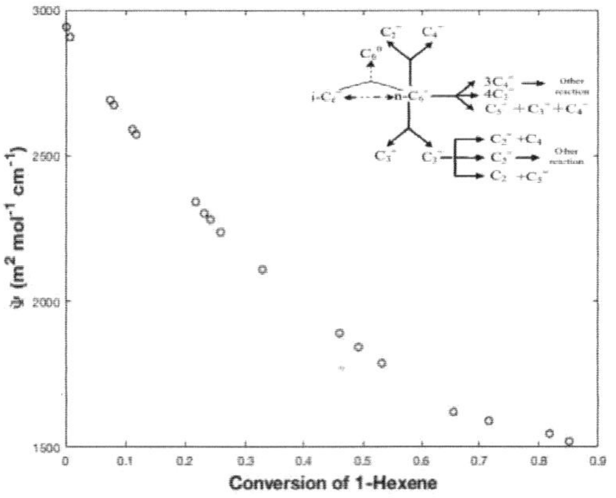

Model reaction: Catalytic cracking of 1-Hexeno. a) change in spectra with time b) Relationship between integrated band intensity (Ψ) and conversion

UNIVERSIDAD NACIONAL DE COLOMBIA SEDE MEDELLÍN

Western UNIVERSITY · CANADA

FRUN Flujos Reactivos de la Universidad Nacional de Colombia

GCM application: Reaction progress

It is possible to employ the linear relationship between the C-H number and the integrated band intensity for calculating the reaction progress of catalytic cracking reaction

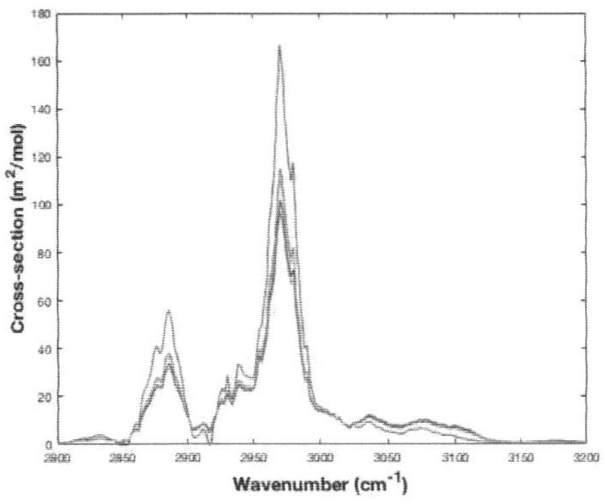

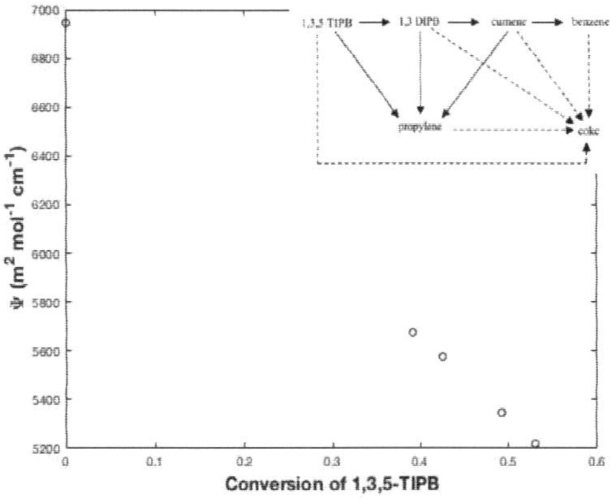

Model reaction: Catalytic cracking of 1,3,5-Triisopropylbenzene. a) change in spectra with time b) Relationship between integrated band intensity (Ψ) and conversion

UNIVERSIDAD NACIONAL DE COLOMBIA SEDE MEDELLÍN

Western UNIVERSITY · CANADA

FRUN Flujos Reactivos de la Universidad Nacional de Colombia

GCM application: Reaction progress

In laboratory, the use of 3.39 µm HeNe laser to estimate the reaction conversion is possible

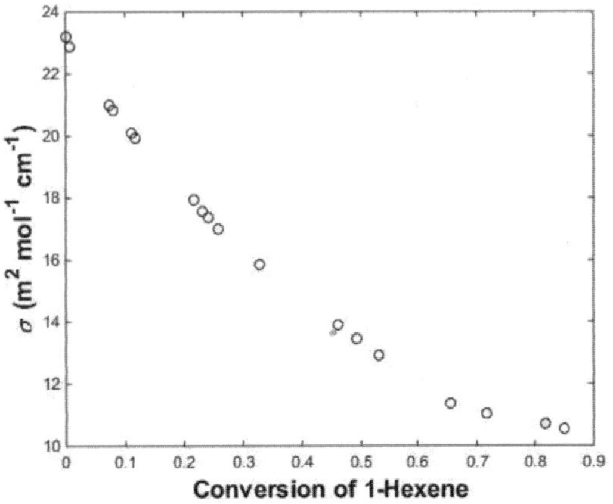

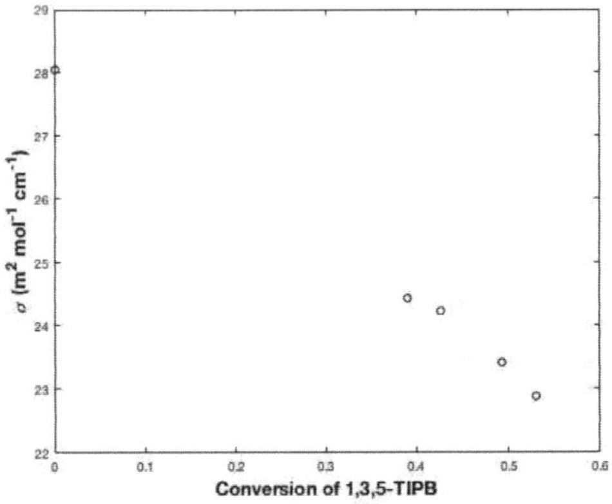

Model reactions: Catalytic cracking of a) 1-Hexeno and b) 1,3,5-TIPB in the HeNe wavenumber

UNIVERSIDAD NACIONAL DE COLOMBIA SEDE MEDELLÍN Western UNIVERSITY · CANADA FRUN Flujos Reactivos de la Universidad Nacional de Colombia

Experimental setup

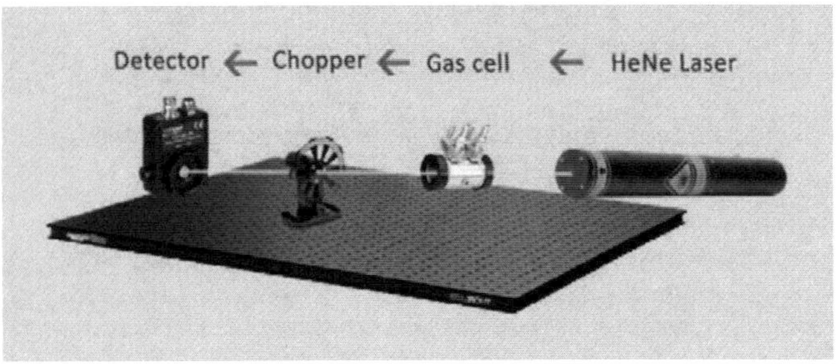

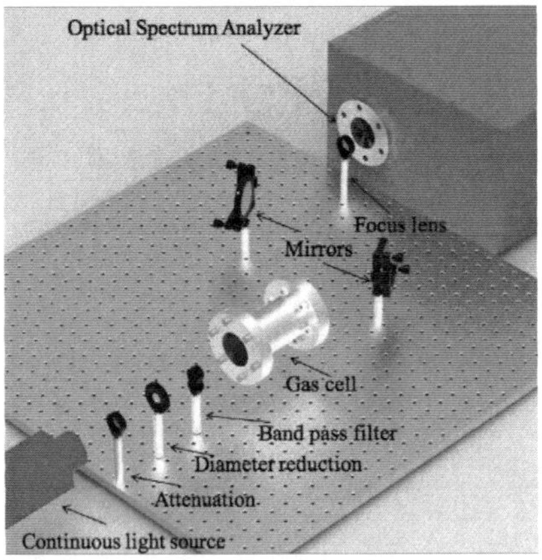

- ✓ A 3.39µm HeNe laser and a 1.5- 4.8 µm detector between were employed for the measurement
- ✓ The measurement in the whole region (3200-2800 cm^{-1}) could be possible with a high energy continuous light source and an optical spectrum analyzer

UNIVERSIDAD NACIONAL DE COLOMBIA SEDE MEDELLÍN Western UNIVERSITY · CANADA FRUN Flujos Reactivos de la Universidad Nacional de Colombia

Experimental results

- HeNe was employed for the measurement of absorption cross section of pure compounds
- It is possible to validate GCM predictions for 3.39 μm wavelength

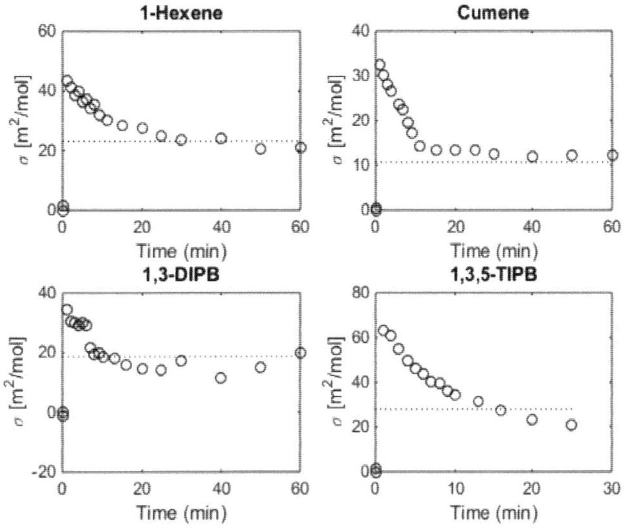

Experimental results for absorbanc at HeNe wavelengnt (Dotted line corresponds to theoretical value)

UNIVERSIDAD NACIONAL DE COLOMBIA SEDE MEDELLÍN Western UNIVERSITY · CANADA FRUN Flujos Reactivos de la Universidad Nacional de Colombia

Future work

- Measurement of synthetic mixtures representing the catalytic cracking reaction
- In situ measurement of the reaction progress

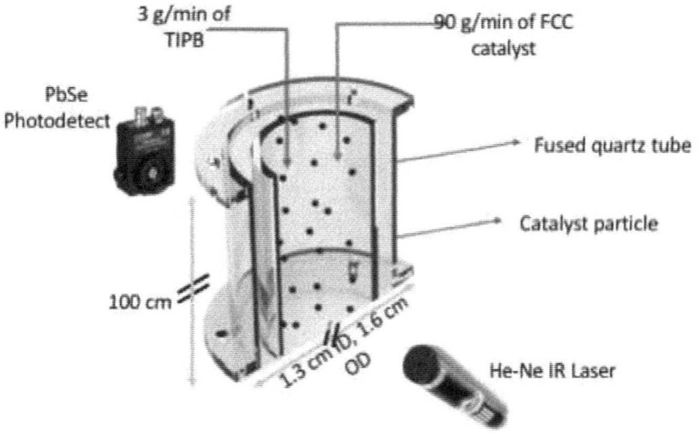

UNIVERSIDAD NACIONAL DE COLOMBIA SEDE MEDELLÍN Western UNIVERSITY · CANADA FRUN Flujos Reactivos de la Universidad Nacional de Colombia

Conclusions

- A group contribution method for the prediction of absorption spectra of paraffins, olefins, naphthenes and aromatics in the region 3200-2800 cm^{-1} was proposed.
- A good prediction for the paraffins, olefins, naphthenes and aromatics by using the GCM was obtained.
- Validation of GCM by gasoline spectra prediction was presented obtaining $R^2 > 0.97$ and RER>15
- Application of GCM for PIOA and RON prediction was possible
- The linear behaviour of absorption coefficient and C-H number was used for the conversion prediction of catalytic cracking reaction
- An experimental setup for the use of this methodology was presented

UNIVERSIDAD NACIONAL DE COLOMBIA SEDE MEDELLÍN — Western UNIVERSITY · CANADA — FRUN Flujos Reactivos de la Universidad Nacional de Colombia

Acknowledgements

The authors would like to acknowledge

- The financial support of the Colombian Science Foundation (COLCIENCIAS), the Colombian State Oil Company (ECOPETROL S.A.) under contract No. 0423 2013

- Emerging Leaders in the Americas Program (ELAP) Canadian Exchange Program.

UNIVERSIDAD NACIONAL DE COLOMBIA SEDE MEDELLÍN — Western UNIVERSITY · CANADA — FRUN Flujos Reactivos de la Universidad Nacional de Colombia

UNIVERSIDAD NACIONAL DE COLOMBIA SEDE MEDELLÍN — Western UNIVERSITY · CANADA — FRUN Flujos Reactivos de la Universidad Nacional de Colombia

References

S. Al-Khattaf, The role of diusion in alkyl-benzenes catalytic cracking, Applied Catalysis A: General 226 (2002) 139-153.

M.A. Alrefae, Mid-IR Absorption Cross-Section Measurements of Hydrocarbons, King Abdullah University of Science and Technology, 2013.

M.F. Campbell, K.G. Freeman, D.F. Davidson, R.K. Hanson, FTIR measurements of mid-IR absorption spectra of gaseous fatty acid methyl esters at T=25-500 C, J. Quant. Spectrosc. Radiat. Transf. 145 (2014) 57–73.

G. Jiménez-García, R. Aguilar-López, R. Maya-Yescas, The fluidized-bed catalytic cracking unit building its future environment, Fuel 90 (2011) 3531–3541.

Klingbeil, A. E. (2007). MID-IR Laser absorption diagnostics for hydrocarbon vapor sensing in harsh environments. Standford University.

Sharpe, S. W., Johnson, T. J., Sams, R. L., Chu, P. M., Rhoderick, G. C., & Johnson, P. A. (2004). Gas-phase databases for quantitative infrared spectroscopy. Applied Spectroscopy, 58(12), 1452–61. http://doi.org/10.1366/0003702042641281

UNIVERSIDAD NACIONAL DE COLOMBIA SEDE MEDELLÍN — Western UNIVERSITY · CANADA — FRUN Flujos Reactivos de la Universidad Nacional de Colombia

Previous works

➢ Qualitative or quantitative determination of hydrocarbons

- Methane (CH_4) (Sepulveda-Jauregui et al. 2012)
- Ethane (C_2H_6) (Serdioutchenko et al. 2004)
- Ethylene (C_2H_4) (Bamford et al, 2007)
- Acetylene (C_2H_2) (Serdioutchenko et al. 2004)
- Propylene (C_3H_6) (Harward et al. 2004)
- 1,3-butadiene (C_4H_6) (Harward et al. 2004)
- Toluene (Porter 2009)
- n-dodecane (Porter 2009)
- n-decane (Porter 2009)

➢ Only a small number of experiments reported the successful evaluation of hydrocarbon with optical techniques

➢ In the referred literature no study applied laser diagnostics to a mixture such as the one for FCC process

UNIVERSIDAD NACIONAL DE COLOMBIA SEDE MEDELLÍN — Western UNIVERSITY · CANADA — FRUN Flujos Reactivos de la Universidad Nacional de Colombia

References

A. Sepulveda-Jauregui, K. Martinez-Cruz, A. Strohm, K. M. Walter Anthony, F. Thalasso, A new method for field measurement of dissolved methane in water using infrared tunable diode laser absorption spectroscopy, Limnology and Oceanography: Methods 10 (2012) 560-567.

D. J. Bamford, D. J. Cook, S. J. Sharpe, A. D. Van Pelt, Widely tunable rapid scanning mid-infrared laser spectrometer for industrial gas process stream analysis, Applied Optics 46 (2007) 3958.

C. N. Harward, R. E. Baren, M. E. Parish, Determination of molecular parameters for 1,3-butadiene and propylene using infrared tunable diode laser absorption spectroscopy, Spectrochimica Ada Part A: Molecular and Biomolecular Spectroscopy 60 (2004) 3421-3429.

A. Serdioutchenko, I. Moller, H. Soltwisch, Infrared diode laser absorption Spectroscopy of C_2H_2 and C_2H_6 in capacitively coupled methane RF discharges, Spectrochimica Acta Part A: Molecular and Biomolecular Spectroscopy 60 (2004) 3311-3318.

J. M. Porter, Laser Based Diagnostics for Hydrocarbon Fuel in the Liquid and Vapor Phases, Ph.D. thesis, Stanford University, 2009.

UNIVERSIDAD NACIONAL DE COLOMBIA SEDE MEDELLÍN — Western UNIVERSITY · CANADA — FRUN Flujos Reactivos de la Universidad Nacional de Colombia

Statistical parameters

Average absolute error

$$\varepsilon = \frac{1}{N}\sum_i \left|\sigma_{Pred} - \psi_{Exp}\right|$$

Average signed error or BIAS

$$\varepsilon_{sgn} = \frac{1}{N}\sum_i \left(\psi_{Pred} - \psi_{Exp}\right)$$

Average relative absolute error

$$\xi = \frac{1}{N}\sum_i \left|\frac{\psi_{Pred} - \psi_{Exp}}{\psi_{Exp}}\right|$$

Average relative signed error

$$\xi_{sgn} = \frac{1}{N}\sum_i \left(\frac{\psi_{Pred} - \psi_{Exp}}{\psi_{Exp}}\right)$$

Root mean squared error of prediction

$$RMSEP = \sqrt{\frac{1}{N}\sum_i \left(\psi_{Pred} - \psi_{Exp}\right)^2}$$

Standard Error of Prediction

$$SEP = \sqrt{\frac{1}{(N-1)}\sum_i \left(\psi_{Pred} - \psi_{Exp} - BIAS\right)^2}$$

Range Error Ratio

$$RER = \frac{Range}{SEP}$$

$RER \geq 4$, the calibration is acceptable for sample screening; $RER \geq 10$, it is acceptable for quality control; and $RER \geq 15$, it is good for quantification (Hayes2015)

UNIVERSIDAD NACIONAL DE COLOMBIA SEDE MEDELLÍN

Western UNIVERSITY · CANADA

FRUN Flujos Reactivos de la Universidad Nacional de Colombia

GCM validation: Gasolines

The statistic summary for the prediction of absorption coefficients in the case of regular and premium gasolines present R^2>0.97 and RER>15

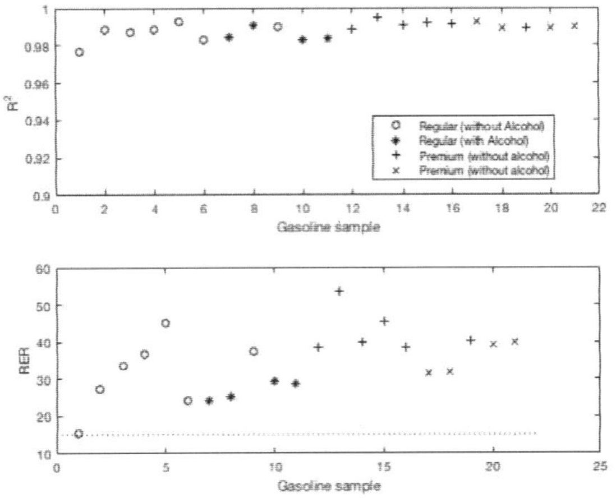

UNIVERSIDAD NACIONAL DE COLOMBIA SEDE MEDELLÍN

FRUN Flujos Reactivos de la Universidad Nacional de Colombia

504589 Fundamental research on phase equilibrium of extraction desulfurization

Abstract:

Fundamental research on phase equilibrium of extraction desulfurizaiton of FCC gasoline

Yuhao Zhang, Liang Zhao, Jinsen Gao, Chunming Xu

State Key Laboratory of Heavy Oil Processing, China University of Petroleum (Beijing), Beijing, China, 102249

Due to the increasing awareness of protecting environment, the great demand for ultralow sulfur contents in motor gasoline has become increasingly significant. Nowadays, many technologies have the ability to produce clean gasoline to meet the requirement of sulfur content (<10 ppm). Approaches of desulfurization could be divided into hydrodesulfurizaiton and non-hydrodesulfurization. Wherein, the hydrodesulfurization processes were well developed and widely applied in the whole world, such as ISAL, Octgain, Prim-G^+, SCAN Fining, and CDTech [1-5] processes. Part of non-hydrodesulfurization processes were also gathered much attention, such as adsorptive desulfurization, extraction desulfurization, oxidative desulfurization, and biological desulfurization[6-12]. Wherein, extraction desulfurization process was well developed because advantages of variety of solvents, high selectivity, low operating price, low energy consumption, eco-friendly, et al. Commonly, it was divided into solvent extraction[7,13] and extractive distillation[14-17] depending on the mechanism, such as Shen reported a gasoline desulfurization using solvent extraction[7] and GTC-BTX Plus technology[14] by extractive distillation. In spite of the theory of these two methods were different, the interaction of phase equilibrium between solvents and key components of FCC gasoline still plays a significant role to impact the effect of separation which could be express through liquid-liquid equilibrium or vapor-liquid equilibrium experiment.

In this research, the composition of different distillation fractions of FCC gasoline were analyzed and the feedstock of solvent extraction and extractive distillation were selected carefully. And then, particular investigated the properties of different kinds of single solvents using liquid-liquid equilibrium and vapor-liquid equilibrium experiment to found the fundamental of extraction desulfurization of FCC gasoline. Furthermore, an optimal compound solvent were selected using phase equilibrium experiment.

In conclusion, the effect of separation performance of different solvents were studied in several ways. Sulfolane shown a great interaction of aromatics and sulfides. Furthermore, a best compound solvent was determined which the sulfur separation rate was 75 % and has the ability to separate C6~C8 olefins into the raffinate containing a small quantity of aromatics. The results were shown in Figure 1 to Figure 3.

Figure 1 Selectivity of olefins and aromatics using different single solvents

k_A shows the ability to dissolve aromatics; θ shows the ability of selectivity between aromatics and olefins

Figure 2 Sulfur separation rate and raffinate yield using compound solvents

Figure 3 Olefin and aromatics distribution in raffinate using optimal compound solvent

Acknowledgement

This research is supported by National Natural Science Foundation of China (Grants 21336011, 21476260 and 21236009). And also supported by the Science Foundation of China University of Petroleum-Beijing (Grants 2462015YQ0311).

References

[1] Martinez, N.P. Meeting gasoline pool sulfur and octane targets with ISAL process. *NPRA Annual Meeting*, **2000**, San Antonio.

[2] Shih, S. S., Owens, P. J., Palit, S. Mobil's Octgain Process: FCC gasoline desulfurization reaches a new performance level. *NPRA Annual meeting*. **1999**, San Antonio.

[3] Baco, F., Debuisschert, Q. Prime-G^+ process, desulfurization of FCC gasoline with minimized octane loss. *Fifth International Conference on Refinery Processing, AIChE 2002 Spring National Meeting*. **2002**, New Orleans.

[4] Kaufmann, T., Kaldor, A., Stuntz G. Catalysis science and technology for cleaner transportation fuels. *Catalysis Today*. **2000**, 62 (1), 77-90.

[5] Babich, I.V., Moulijn, J.A.. Science and technology of novel processes for deep desulfurization of oil refinery streams: a review. *Fuel*. **2003**, 82 (6), 607-631.

[6] Khare et al. Transport desulfurization process utilizing a sulfur sorbent that is both fluidizable and circulatable and a method of making such sulfur sorbent. **2002**, US 5914292.

[7] Haitao Shen, Benxian Shen, Hao Ling. Desulfurization of fluid catalytic cracking gasoline by extractive distillation coupled with hydrodesulfurization of heavy fraction. *Energy & Fuels*. **2013**, 27, 5153-5160.

[8] Ke Ming, Zhou Aiguo, Zhao Zhensheng, Jiang Qingzhe, Song Zhaozheng. Alkylation of thiophenic sulfur technology in the production of clean fuel. *Chemical Industry and Engineering Progress*. **2006**, 25 (4), 357-361.

[9] Qiu Shi, Wu Liguang, Shi Guozhong, et al. Preparation and pervaporation property of chitosan membrane with functionalized multiwalled carbon nanotubes. *Ind Eng Chem Res*. **2010**, 49 (22), 11667-11675.

[10] Waserscheid P, Keim W. Ionic liquids new solutions for transition metal catalysis. *Angew Chem*. **2000**, 39 (21), 3772-3789.

[11] Xu Hang, Zhang Dandan, WU Fengmin, Cao Renqiang. Deep oxidative desulfurization of fuels based on [C(4)mimCl] CoCl2 ionic liquid oxone solutions at room temperature. *Fuel*. **2017**, 208, 508-513.

[12] Zhang Liang, Lin Xiaojuan, Wang Jinting, Jiang Feng, Wei Li, Chen Guanghao, Hao Xiaodi. Effects of lead and mercury on sulfate-reducing bacterial activity in a biological process for fuel gas desulfurization wastewater treatment. *Scientific Reports*. **2016**, 6, 1-10.

[C_8mim][BF_4] ionic liquid. *Separations*. **2007**, 53, 3108-3115.

[14] GTC Technology Corporation. Process of removing sulfur compounds from gasoline. US 6551502, **2003**.

[15] Shen H, Mei Z, Shen B, Ling H. The desulfurization of fluid catalytic cracking gasoline by extractive distillation. *Energy Sources Part A Recovery Utilization & Environmental Effects*. **2012**, 34 (2), 187-196.

[16] Tang Xiaodong, Wen Xiaohong, Wang Rui, Nie Tongyuan. Study on desulfurization of FCC light gasoline by extractive distillation. *Petroleum Processing and Petrochemicals*. **2013**, 44 (12), 77-80.

[17] Xiaochuan Jiang; Yi Nie; Chunxi Li; Zihao Wang Imidazolium-based alkylphosphate ionic liquids-A potential solvent for extractive desulfurization of fuel. *Fuel*. **2008**, 87, 79-84.

500365 Rational Design and Synthesis of Surfactant-Free Pt-Based Ternary Nanoalloys As Superior Oxygen Reduction Reaction Electrocatalysts

Abstract:

Efficient and low cost electrocatalysts are indispensable for electrochemical oxygen reduction reactions (ORR) during proton exchange membrane fuel cell (PEMFC) operations. Yet, the challenge for solution phase synthesis of nanocatalysts with low precious metal loading, high durability and catalytic activities remains in tailoring their architectures and compositions without using unwanted chemicals and surfactants/ligands that are known to impede their active catalytic sites. We present to synthesize ternary nanoalloys (NAs) of Pt with transition metals (Co, Cu, Ni, Ti, Ru, Mn) as ORR electrocatalysts using our recently developed laser ablation synthesis in solution-galvanic replacement reaction (LASiS-GRR) technique as a facile and surfactant-free nanomanufacturing route. The specific choice of the elemental compositions is driven by the respective target metal/metal salt redox potential gaps as well as the target metal and/or, metal oxide solubility in desired acids. The high-energy thermodynamics of the LASiS-GRR process enables control of the sizes, elemental compositions and distributions of the ternary NAs through systematic tuning of initial metal salt concentration, pH, laser fluence and ablation time. Specifically, the PtCuCo NAs synthesized with an elemental composition of 72:12:16 (Pt:Co:Cu) indicates the best ORR catalytic activity. The NAs largely possess shell-core structure with the shell comprising mostly of Pt and minor amount Cu, along with a uniformly alloyed PtCuCo core. Mass and specific activities for ORR performance of the NAs indicate a 4 and 6.5-fold improvements respectively over the corresponding activities of commercial Pt/C. We attribute the enhanced activity to 1) our surfactant/ligand-free synthesis technique that prevents catalytic site degradations; and 2) minor alloying of the second transition metal Cu that shifts back the Pt d-band center to an optimal position between that of Pt and the PtCo binary NAs, thereby tuning their binding affinities for both oxygen and oxygenated species. Finally, this work establishes the versatility of the LASiS-GRR technique through the synthesis of other ternary NAs (PtRuNi, PtCoMn, PtNiTi) that also exhibit reasonably good ORR activities.

496706 Improved Activity and Durability of Ionic Liquid Composite Nanoporous Nanoparticle Electrocatalysts for Oxygen Reduction Reaction

Abstract:

The development of efficient and stable catalysts for the cathodic oxygen reduction reaction (ORR) is one of the greatest challenges for the exploitation of polymer electrolyte membrane fuel cells (PEMFCs). Dealloying has shown increasing utility in the field of electrocatalysis as a tool for the synthesis and development of nanoporous materials possessing high surface-to-volume ratios with controlled morphology and compositional gradient (core-shell structure) [1]. After electrochemical dealloying, the open, bicontinuous, three-dimensional nanoporous nanoparticle electrocatalysts exhibit dramatically enhanced electrocatalytic properties [2,3]. We have taken advantage of the free volume within these nanoporous nanoparticles and created a composite catalyst architecture through the incorporation of a hydrophobic, protic ionic liquid (IL), specifically [MTBD][beti]. Modification of the interface through the formation of IL thin film on the surface of an electrocatalyst yields enhanced catalytic rates by chemically biasing reactants to the surface and changing the degree of solvation/stabilization of adsorbed intermediates.

In the development of efficient electrocatalysts for ORR, durability is too often ignored in the pursuit of higher activities. Here we studied the manner by which ILs impact the stability of 3-dimensional, nanoporous nanoparticle electrocatalysts. In addition to the standard mechanisms of catalyst degradation including Pt dissolution/Ostwald ripening and coalescence/aggregation, nanoporous materials can lose electrochemically active surface area (ECSA) through surface smoothening driven coarsening. We use a combination of in-situ and ex-situ experimental techniques, under accelerated degradation conditions, to develop insight into the structural and compositional evolution of nanoporous PtNi nanoparticles (np-NiPt) formed through the dealloying of $Pt_{20}Ni_{80}$precursor nanoparticles, and demonstrate that facilitated diffusion of stable surface atoms under relevant electrochemical conditions will drive evolution of the particle morphology from a porous to a nonporous structure. With a better understanding of the interplay between nanoporous structure coarsening and transition metal loss, we have developed strategies to mitigate coarsening through incorporating ILs at the metal/electrolyte interface to directly mitigate electrochemically enhanced surface diffusion, which leads to ECSA loss. The hydrophobicity of these ILs is known to change the interaction of water with the surface, positively shifting the onset potential of Pt oxidation [3]. The interfacial IL acts to limit the charge dependent formation of a surface metal/electrolyte anion complex which is responsible for the potential dependent enhancement in surface diffusion. The oxygen reduction reaction activity, ECSA and residual transition metal content of these complex composite nanoporous electrocatalysts was tracked as a function of cycle number during accelerated degradation testing providing insight into the mechanism of catalyst degradation and the effect of ILs on catalyst stability.

References

[1] Snyder, J.; McCue, I.; Livi, K. & Erlebacher, J., *J. Am. Chem. Soc.* **134,**8633-8645 (2012).

[2] Snyder, J.; Fujita, T.; Chen, M. W. & Erlebacher, J., *Nature Mater.* **9,**904-907 (2010).

[3] Snyder, J.; Livi, K. & Erlebacher, J., *Adv. Funct. Mater.* **23,** 5494-5501 (2013).

489132 Particle Atomic Layer Deposition for Stabilization of Oxygen Reduction Catalysts

Abstract:

As a highly efficient, portable, non-greenhouse gas emitting source of electrical power, the proton exchange membrane (PEM) fuel cell holds great potential as a replacement for the internal combustion engine. One of the most significant development challenges in the commercialization of the PEM fuel cell is the long-term durability of the oxygen reduction reaction (ORR) catalyst. Cathode potential cyclingâ€"resulting from the variable voltage loads imposed during vehicular operationâ€"is known to promote agglomeration and growth of the Pt nanoparticle catalyst, which in turn degrades the power output of the fuel cell. In this work, atomic layer deposition (ALD) was used to deposit protective TiO_2 films onto both commercial and ALD-synthesized Pt/C catalyst in order to increase its electrochemical durability. Deposition half-reactions during fluidized bed particle ALD were observed via in-situ mass spectrometry, and chemisorption analysis was used to quantify changes in Pt surface area following TiO_2 deposition. Rotating disk electrode analysis indicated that electrochemical surface area (ECSA) and activity were slightly depressed by the addition of TiO_2 ALD layers. Activity retention during electrochemical durability testing was modestly improved, with TiO_2-coated catalysts retaining upwards of 60% of their initial activity. Examination of catalysts before and after testing also showed that TiO_2 ALD barriers were effective at preventing Pt agglomeration under fuel cell operating conditions. Further testing of the catalysts in membrane electrode assemblies was used to quantify effects of TiO_2 ALD on the hydrophobicity of the catalysts, which has been previously demonstrated as an important factor in fuel cell device operation with ALD-fabricated catalysts.

500977 A Mechanistic Insight into the Electrocatalytic Behavior of Mixed Oxide Cathode Catalysts for Li-O_2 Batteries

Abstract:

The recent advancements in renewable energy generation will require investments in the development of energy storage technologies. Generally, the generated renewable energy can be stored in multiple forms, such as chemically (i.e., water splitting to form hydrogen fuel), mechanically (i.e., flywheel that converts electrical to rotational energy), or electrochemically (i.e., batteries that can store electrical energy). Among the many different electrochemical energy storage technologies, lithium-oxygen (Li-O_2) batteries exhibit the highest theoretical energy density (11,680 Wh/kg) [1]. Unlike the current state-of-the-art Li-ion batteries, where the technology has been explored for many years and reached a ceiling performance, Li-O_2 batteries are still far from reaching the theoretical energy density, thus are still under wide investigation [2]. In order to improve the performance of these systems, a fundamental understanding of the reaction mechanism involved in both the oxygen reduction reaction (ORR) and oxygen evolution reaction (OER) during discharge and charge respectively, at the cathode of these systems is necessary. It is in part due to the reported discharge product, Li_2O_2, being a high bandgap insulator, thereby making the electron transfer during OER extremely challenging [1]. We have shown previously that the performance of the Li-O_2 battery cathodes can be improved upon addition of non-precious metal oxide catalysts, known as Ruddlesden-Popper (R-P) oxides [3]. To optimize the performance of these catalytic materials, a fundamental understanding of the mechanism that leads to this improvement is required. In this contribution, we combine electrochemical and spectroscopic techniques to unravel the complex chemistry involved in Li-O_2 cathodes containing R-P oxide electrocatalysts. The effect of the nature of the discharged product species during ORR and its effect on lowering the overpotential losses during OER will be discussed. We finalize by devising ways of enhancing the performance of R-P oxides for oxygen reduction and evolution in Li-O_2 batteries.

References

1. Girishkumar, G., et al., *Lithium - Air Battery: Promise and Challenges.* Journal of Physical Chemistry Letters, 2010. **1**(14): p. 2193-2203.

2. Armand, M. and J.M. Tarascon, *Building better batteries.* Nature, 2008. **451**(7179): p. 652-657.

3. Nacy, A., X.F. Ma, and E. Nikolla, *Nanostructured Nickelate Oxides as Efficient and Stable Cathode Electrocatalysts for Li-O-2 Batteries.* Topics in Catalysis, 2015. **58**(7-9): p. 513-521.

494499 Characterization of Feni Alloy Nanoparticles for Oer As a Function of Iron-Nickel Composition

Abstract:

Iron-doped nickel hydroxide catalysts are now recognized as the leading catalyst composition for alkaline electrochemical water splitting, and in particular, the anodic reaction of oxygen evolution [1]. The iron-nickel bimetallic composition of these hydroxide catalysts has been demonstrated to be critical for optimal catalyst performance [2]. However, few studies thus far have investigated if the same trend in catalyst activity is observed when catalyst morphology is scaled from bulk films to the nanoscale [3], where the overall field of iron-nickel hydroxide/oxide catalysts is in its infancy in terms of developing, synthesizing, and fully characterizing these nanoparticulate bimetallic hydroxides. Further, preliminary stability studies performed on iron-nickel [4] and iron-cobalt [5] thin films suggests that an evaluation of the stability of the bimetallic catalysts as a function of composition will be critical to identifying the best-performing catalyst compositions to pursue. In particular, Burke et al. report that while iron atomic compositions of less than 54 % resulted in high catalytic activities, the activities of these iron-cobalt bimetallic catalysts decreased by up to 62 % over two hours [5]. In the iron-nickel catalyst field, little work has evaluated how catalyst stability is affected by bimetallic composition in either thin film or nanocatalyst morphologies. Thus far, most studies have primarily focused on how the iron-nickel bimetallic composition affects catalyst performance from the perspective of lowering the overpotential, thereby lowering the kinetic limitation to the oxygen evolution reaction [2, 6].

In this talk, our ongoing work to develop and characterize a suite of bimetallic iron-nickel alloy hydroxide nanoparticle catalysts for the oxygen evolution reaction will be discussed. In particular, results will be presented for three iron-nickel compositions (1:5, 1:1, and 5:1 mol:mol Fe:Ni) and trends in both overpotential and stability will be discussed. Nanoparticle characterization via high resolution transmission electron microscopy, elemental analysis, and synchrotron-based x-ray absorption spectroscopy will be presented and discussed as a function of bimetallic composition. The role of nanoparticle structure in controlling activity and stability will be discussed, and electrochemical performance data will be used to demonstrate that both activity (i.e., overpotential) and stability (i.e., degradation rate as mV/hr) are critical to evaluate as we develop these catalysts for water splitting.

References

[1] D. Friebel, M.W. Louie, M. Bajdich, K.E. Sanwald, Y. Cai, A.M. Wise, M.-J. Cheng, D. Sokaras, T.-C. Weng, R. Alonso, R.C. Davis, J.R. Bargar, J.K. Norskov, A. Nilsson, A.T. Bell, Identification of highly active Fe sites in (Ni,Fe)OOH for electrocatalytic water splitting, J. Am. Chem. Soc., 137 (2015) 1305â€"1313.

[2] L. Trotochaud, S.L. Young, J.K. Ranney, S.W. Boettcher, Nickel-iron oxyhydroxide oxygen-evolution electrocatalysts: The role of intentional and incidental iron incorporation, J. Am. Chem. Soc., 136 (2014) 6744-6753.

[3] J.A. Bau, E.J. Luber, J.M. Buriak, Oxygen Evolution Catalyzed by Nickelâ€"Iron Oxide Nanocrystals with a

Nonequilibrium Phase, ACS Applied Materials & Interfaces, 7 (2015) 19755-19763.

[4] M. Gong, Y. Li, H. Wang, Y. Liang, J.Z. Wu, J. Zhou, J. Wang, T. Regier, F. Wei, H. Dai, An Advanced Niâ€"Fe Layered Double Hydroxide Electrocatalyst for Water Oxidation, Journal of the American Chemical Society, 135 (2013) 8452-8455.

[5] M.S. Burke, M.G. Kast, L. Trotochaud, A.M. Smith, S.W. Boettcher, Cobalt-iron (oxy)hydroxide oxygen evolution electrocatalysts: The role of structure and composition on activity, stability, and mechanism, J. Am. Chem. Soc., 137 (2015) 3638-3648.

[6] M.W. Louie, A.T. Bell, An investigation of thin-film Ni-Fe oxide catalysts for the electrochemical evolution of oxygen, J. Am. Chem. Soc., 135 (2013) 12329-12337.

495038 Active Sites in Nitrogen-Doped Carbon Nanostructures for Oxygen Reduction and Oxygen Evolution Reactions

Abstract:

Regenerative polymer electrolyte membrane (PEM) fuel cell technology has a promising future for sustainable power generation. In regenerative PEM fuel cells, oxygen reduction (ORR) and oxygen evolution (OER) are two important reactions that take place in the fuel cell mode and electrolyzer mode, respectively. State-of-the-art catalysts used for these reactions – Pt for ORR and Ir (or Ru) for OER – are expensive and found in limited reserves. Also, both these catalysts are active only for one of the two reactions, thus necessitating the development of cheaper materials that can serve as efficient bifunctional catalysts for both ORR and OER in regenerative PEM fuel cells. We have reported nitrogen doped carbon nanostructures, denoted as 'CN_x' [1], to be active, selective and stable catalysts for ORR in acidic medium. Our current focus is to further evaluate the applicability of CN_x as bifunctional ORR and OER catalysts, and also understand the nature of active sites in these materials.

Electrocatalytic activity measurements using rotating ring disk electrode (RRDE) show significant ORR and OER activity of CN_x compared to state-of-the-art Pt and Ir-based catalysts. Analysis of bifunctional electrocatalytic activity demonstrates much better bifunctional characteristics of CN_x compared to Ir/C and Pt/C [2].

Use of poisoning probes is a useful methodology to systematically examine the nature of active sites in catalytic materials. However, due to the absence of a metal centered active site, CN_x materials are not susceptible to poisoning by CO, H_2S or CN^- [3, 4]. We have recently identified phosphate anions ($H_2PO_4^-$) as probe molecules for poisoning CN_x catalysts in an acidic ORR environment [5]. In-situ RRDE measurements show a linear decrease in ORR activity with increasing $H_2PO_4^-$ anion concentration. Transmission IR spectroscopy and Raman spectroscopy reveal the presence of $H_2PO_4^-$ species on H_3PO_4–soaked CN_x. A linear decrease in ORR kinetic current with a decrease in pyridinic nitrogen content obtained using X-ray photoelectron spectroscopy (XPS) is observed. This poisoning phenomenon is consistent with two possible active site models: (i) pyridinic N as the active site which is rendered inactive by protonation [6, 7], and (ii) carbon sites adjacent to pyridinic N as the active site (i.e. pyridinic N as a marker for the active site) wherein poisoning is caused by a site-blocking effect due to adsorption of $H_2PO_4^-$ species on carbon. OER activity has also been found to increase with increasing pyridinic nitrogen site density in CN_x samples.

References

[1] P.H. Matter, U.S. Ozkan, Non-metal catalysts for dioxygen reduction in an acidic electrolyte, Catal. Lett., 109 (2006) 115-123.

[2] K. Mamtani, D. Jain, D. Dogu, A. C. Co, U. S. Ozkan, Insights into Oxygen Reduction Reaction (ORR) and Oxygen Evolution Reaction (OER) Active Sites for Nitrogen-doped Carbon Nanostructures (CNx) in Acidic Media, Applied Catalysis B: Environmental, Under Review, (2017).

[3] D. von Deak, D. Singh, J.C. King, U.S. Ozkan, Use of carbon monoxide and cyanide to probe the active sites on nitrogen-doped carbon catalysts for oxygen reduction, Appl. Catal. B-Environ., 113-114 (2012) 126-133.

[4] D. von Deak, D. Singh, E.J. Biddinger, J.C. King, B. Bayram, J.T. Miller, U.S. Ozkan, Investigation of sulfur poisoning of CNx oxygen reduction catalysts for PEM fuel cells, J. Catal., 285 (2012) 145-151.

[5] K. Mamtani, D. Jain, D. Zemlyanov, G. Celik, J. Luthman, G. Renkes, A.C. Co, U.S. Ozkan, Probing the Oxygen Reduction Reaction Active Sites over Nitrogen-Doped Carbon Nanostructures (CN x) in Acidic Media Using Phosphate Anion, ACS Catalysis, 6 (2016) 7249-7259.

[6] G. Liu, X. Li, P. Ganesan, B.N. Popov, Studies of oxygen reduction reaction active sites and stability of nitrogen-modified carbon composite catalysts for PEM fuel cells, Electrochim. Acta, 55 (2010) 2853-2858.

[7] X. Li, G. Liu, B.N. Popov, Activity and stability of non-precious metal catalysts for oxygen reduction in acid and alkaline electrolytes, J. Power Sources, 195 (2010) 6373-6378.

497178 Atomic Iron-Dispersed Electrocatalysts Derived from Metal-Organic Framework for Oxygen Reduction

Abstract:

To significantly reduce the cost of proton exchange membrane (PEM) fuel cells, current Pt must be replaced by platinum-metal-group (PGM)-free catalysts for the oxygen reduction reaction (ORR) in acid. We report here a new class of high-performance atomic iron-rich catalyst via chemical doping of iron ions into zinc-zeolitic imidazolate framework (ZIF), a type of metal-organic framework (MOF). The novel synthetic chemistry enables accurate size control of Fe-doped ZIF catalyst particles with a wide range from 20 to 1000 nm without changing chemical property. This provides a great opportunity to increase the density of active sites. An optimal size at 50 nm exhibited the highest activity; further reducing size to 20 nm results in particle agglomeration with decreased active sites. In addition to the size effect, we carefully explored the thermal activation mechanism to provide insight into the origin of active site formation. A temperature of 800oC was identified as the critical point to generate active sties with new bond formation associated with nitrogen doping, carbon, and iron. 1100 oC was found for the best activity with maximum density of active sites along with optimal porosity, graphitic nitrogen doping, and carbon structures. We also found the control of optimal iron doping is crucial for performance enhancement without the formation of inactive iron cluster. Overall, the best performing catalyst is enriched in atomic iron sites embedded into carbon structures, achieving a new performance milestone for the ORR including a half-wave potential of 0.84 V vs RHE in 0.5 M H_2SO_4 and only 20 mV loss after 10,000 cycles in O_2 saturated electrolyte, which has been approaching to Pt in the challenging acidic media. Thus, the chemically Fe doped MOF catalysts are able to further narrow down the performance gap with Pt catalysts in more challenging acidic media, holding great promise to advance PGM-free catalyst as a replacement of Pt for PEM fuel cells.

494356 IGCC Process Intensification with the Reforming Technologies for Enhanced Power Generation with CCS Technology

Abstract:

IGCC (Integrated gasification and combined cycle) is a pre-combustion technique which represents higher thermal efficiency with large scale implementation of CO_2 capture. On the other hand, NGCC (Natural gas combined cycle) power plants offers higher process performance compared to IGCC process with CCS technologies but the fluctuating and comparatively higher costs of natural gas limits its extensive implementation over coal based power generation systems. Therefore, process intensification of coal based gasification processes with the natural gas reforming technology has been proposed in this study to enhance the overall process performance while limiting the natural gas consumption. In this study, two IGCC based process models have been developed and evaluated in terms of both the process performance and economics with CO_2 capture. Case 1 is taken as the conventional IGCC process, whereas, case 2 presents an idea of integrating methane reforming process with an IGCC technology. Coal is used as primary fuel in both the cases for the coal gasification process, however, small amount of natural gas is used as secondary fuel only in the case 2 design. The high enthalpy steam generated during coal slurry gasification process in case 2 is used to assist the methane reforming to generate additional H_2. The integration of IGCC with methane reforming process not only supplies the heat required for the endothermic reforming process but also increases the heating value of the resulting syngas. Moreover, the H_2O generated in the gasification process in case 2 can be consequently utilized in the natural gas reforming process without utilizing any extra steam. This concept also provides an opportunity for the process economics since shared water gas shift reactors and CO_2 capture units will suffice the process needs. The net electricity generation capacity and efficiency for case 1 and case 2 is calculated as (375.08 MW_e, 472.92 MW_e) and (35.93%, 40.70%), respectively. While comparing the results, it has been seen that case 2 design offers nearly 4.77% higher efficiency compared case 1 design with CO_2 capture. The process economics analysis showed that the case 2 design requires a higher CAPEX and OPEX throughout the project life compared to the case 1 design. However, due to the higher power generation capacity of case 2 design, it has a potential to reduce the LCOE (levelized cost of electricity) by 5.81% compared to the case 1 design. Moreover, case 2 design not only requires the less CO_2 capture and avoidance costs compared to the case 1 but also exhibits higher rate of return on the investment with the less payback time. With the higher efficiency and least case 2 has been considered as the most feasible option for electricity production at an economical price compared to the case 1 design.

500239 Vegetable Oil Process Intensification with Sterically Hindered Alcohols to Biofuels and Biochemicals

Abstract:

Vegetable oils, refined to different degrees, are the raw material to manufacture biodiesel. Commercial batch times can exceed 8 h due to poor mass transfer and miscibility between oil and methanol. Moreover, free fatty acids methyl esters (FAME) have a cloud point (CP) and pour point (PP) ranging from −3 °C to 12 °C and −15 °C to 10 °C, respectively [1], i.e. poor cold flow properties unsuitable for most climates unless blended with petro-diesel. Substituting methanol with branched alcohols to transesterify or esterify vegetable oil matrices improves the cold flow properties of the fatty alkyl esters. For example, the CP and PP of isopropyl esters range from −14 °C to 0 °C and −20 °C to −5 °C, respectively [1]. We propose a fluidized catalytic bed and ultrasound-based processes to either transesterify vegetable oils with alcohols other than methanol or esterify free fatty acids with a trimethylolppropane to produce biolubricants.

A fluidized catalytic bed concurrently transesterified and cracked canola oil with iso-propanol (iPrOH) to fatty acid isopropyl esters in the vapour phase. We alternated the injection of vegetable oil + iPrOH (reaction stage) and air (regeneration stage) to remove the coke from the CaO/Al_2O_3 catalyst. The reaction stages and regeneration stages lasted 2 min. We varied the iPrOH:oil ratio from 35 to 51 (mol:mol) and the air flow for during the regeneration step. The coke yield ranged from 3.5% to 34% and the isopropyl esters yield from 2% to 31%. The highest hydrocarbons yield was 42% (GC analysis). Fluidized beds are excellent candidates for biodiesel process intensification because as little as 10 t of catalyst may produce 80 kt/y of biodiesel, which corresponds to a reactor size at least 10X lower than the current commercial processes [2-4].

We propose a new ultrasonic-assisted mixing device to transesterify vegetable and waste cooking oil with ethanol and iso-propanol in a continuous process. We adapted a 13 mm diameter tip ultrasound probe to a media mill, where a rotating basket circulated the oil: alcohol mixture at 1200 rpm in the vicinity of the tip [5]. The alcohol:oil ratio was stoichiometric and we adopted KOH as basic catalyst. Ultrasound operated at 20 KHz and 500 W nominal power and with pulses 1 s on and 1 s off. After 1 min reaction, 40 %, 55 % and 90 % oil converted to isopropyl, ethyl and methyl esters, respectively. In absence of ultrasound, the same conversions were reached after 2 hours. Ultrasound creates a very fine oil: alcohol emulsion that increases the contact among the reagents and the catalyst.

Vegetable oils as biolubricants suffer from poor oxidative stability because of the hydrogen in b position to the ester groups, which have the tendency to give b-elimination [6]. Trimethylolpropane (TMP) esterify free fatty acids to biodegradable and stable tri-esters whose application is as biolubricants. We apply ultrasound to esterify oleic acid with TMP with heterogeneous catalysts including Al_2O_3-supported $H_3PW_{12}O_{40}$ and sulphated ZrO_2 systems. We vary the power amplitude (20 to 60 % of 500 W) and adopt different ultrasound pulses lengths. Sulphated zirconia systems convert over 80 % of substrate in 2 hours, while Al_2O_3-supported $H_3PW_{12}O_{40}$ is mostly inactive.

References

[1] Joshi R.M., Pegg M.J., Flow properties of biodiesel fuel blends at low temperatures, *Fuel*, 2007, 86:143–151

[2] Boffito D.C., Galli F., Pirola C., Patience G.S., CaO and isopropanol transesterify and crack triglycerides to isopropyl esters and green diesel, *Energ. Conv. Mangement.*, 2017, 139, 71-78

[3] Boffito D.C., Neagoe C., Edake M., Pastor-Ramirez B., Patience G.S., Biofuel synthesis in a capillary fluidized bed, *Cat. Today*, 2014, 237, 13–17

[4] Boffito D.C., Blanco-Manrique G., Patience G.S., One step cracking/transesterification of vegetable oil: Reaction–regeneration cycles in a capillary fluidized bed, *Energ. Conv. Manage.*, 2015, 103, 958–96

[5] Boffito D.C., Galli F., Martinez P.R., Pirola C., Bianchi C.L., Patience G.S., Transesterification of triglycerides in a new ultrasonic-assisted mixing device, *Chem. Eng. Trans.*, 2015, 43, 427-432.

[6] Zaccheria F., Mariani M., Psaro R., Bondioli P., Ravasio N., Environmentally friendly lubricants through a zero waste process, Appl. Catal. B., 2016, 181, 581–586

487349 Targeting Maximum Energy and Water Efficiencies for the Sustainable Total Textile Waste Refinery

Abstract:

The global textile industry is among the top ten manufacturing economic activities attaining a market value of €0.82 trillion. In 2008, 24 million tonnes of cellulosic and 39 million tonnes of synthetic fibres were sold globally (Oerlikon report, 2009). Despite industry's great economic value, the textile industry is among the most polluting ones. To only cultivate 1 kg of cotton; 16 g of pesticides, 460 g of fertilizer and 22 $m^3$of water is consumed (Saleem et al., 2010). In EU alone, 10 million tonnes of textile waste are produced each year (EASME, 2015). Specifically, in central Europe, Nordic countries, USA and Japan, the total post-consumer textile waste going to landfill is estimated 0.6, 1.9, 0.6, 0.1, 0.4, 0.9 million tonnes respectively. The present work coordinates established and original Process Integration technologies to the solution of a new problem; the design of the sustainable Total Textile Waste Refinery (TTWR); a new biorefinery type, that it is being currently developed in the course of the European Horizon2020 research project (RESYNTEX, 2015). It uses the complete spectrum of waste textiles as feedstock, aiming to provide a holistic response to the largely overlooked waste textile management problem. TTWR is now just before the phase of pilot construction and process integration technologies are currently implemented to accelerate the design solution.

TTWR consists of gate links between textile materials with commodity and specialty products and chemicals. Feedstock includes the entire range of textile fibres, consisting mostly of cotton, polyethylene terephthalate (PET), polyamides (PAs), wool but also other material in smaller quantities like acrylics and elastane. Glucose, amino-acid solutions, and PET and PA monomers and oligomers are the gate chemicals produced by the decomposition of the feedstock before the process eventually produces bio-ethanol, adhesives used in the wood-based panel production, PET bottles and PA-derived value added chemicals (caprolactam, amylamine, hexanoic acid etc.). The decomposition of natural and synthetic textile fibres is realised using patented chemistries with emphasis on chemical and enzymatic hydrolysis. The overall process consists of 20 sections dedicated to pretreatment for the reduction of the feedstock size, discolorations for the removal of dyes and impurities, depolymerizations for the decomposition of material and the production of gate chemicals, separations and purification of gate chemicals, reactions and separations for the production of end products and finally solid and liquid waste treatment processes. At particular stages (discolorations, separations and purifications), the TTWR process can be considered as the reverse process of textile production (pretreatment, dying, polishing etc.) and therefore, it has similar characteristics regarding water consumption. Without any water integration attempts the water consumption is as high as 52 tonnes per tonne of feedstock, and it is shared among reaction and separation stages (mostly washing and filtration steps) at a 2:1 ratio. Contrary to the textile fabrication processes which uses an average of 30 MJ/kg of fabric (Kalliala and Nousianinen, 1999), TTWR is more energy intensive, requiring more than 53 MJ/kg of feedstock. The majority of the energy requirements (~40%) is used in reactions, which are mainly neutral thermochemical hydrolyses at high temperatures for PAs and PET, and biochemical hydrolyses at mild temperatures. Considering that water is used at high ratios compared to the feedstock material in all reactions, it is also the main energy carrier in the reactors. This evidence indicates clear scope for a combined energy and water integration study, since water is used both as a cleaning agent for removing dyes and impurities as well as the main heating medium.

The TTWR process design procedure benefits from the rich technical know-how that has been accumulated so far for the design of conventional chemical processes. The present work takes advantage and improves on the significant contributions in process synthesis and integration (Mountraki et al., 2011a, 2011b; Tsakalova et al., 2012) made in the course of the European research project BIOCORE (Biocore, 2014) dedicated to the evolution of 2ndgeneration biorefineries.

Novel process integration tools have been developed for assessing alternative process integration scenarios and achieving maximum efficiencies in energy and water use in modern biorefineries (Koufolioulios et al., 2014; Nikolakopoulos and Kokossis, 2016). Also, total waste management and recycle design technology has been produced to enable the development of designs for processes with minimum environmental impact (Mountraki et al., 2016; Nikolakopoulos et al., 2016a).

Water and energy Integration is applied with a combined use of process modelling targeting, process synthesis and mathematical optimization. Targeting sets the scope to match and improve efficiencies for energy and water (Pinch, Water Pinch), explaining efficiency bottlenecks, modifications required to improve targets, means to re-use and regenerate water and solvents, as well as means and technologies appropriate to treat waste.

A new tool is proposed In this paper for simultaneously targeting maximum energy and water efficiencies in the TTWR. It is developed by integrating a mathematical programming heat cascade sub-model for targeting and optimally allocating heating and cooling utilities (Koufolioulios et al., 2014) with a two-step iterative procedure to coordinate the targeting of minimum fresh water requirements, the calculation of recycle and treatment flows and the selection of water treatment processes (Nikolakopoulos et al., 2016b).

For the first model, combinations of utilities are built by matching pseudo-cooling utilities with steam levels so as to permit alternation of the purpose of the utility from cooling to heating and thus exploit excess of heat. This actually defines a dual function for a subset of cold utilities. The method calculates for each combination of utilities the minimum utility requirements through an expanded transhipment model of (Papoulias and Grossmann, 1983). The model incorporates the utilities as process streams with the objective of minimizing total utility cost. Heat receivers (i.e. cool streams matched against steam at the same level) are being maximized by being assigned a negative sign. Hot and cold utilities at the same level are assigned different costs and prices respectively. The model calculates the utility combination that minimizes the annualized total cost and maximizes profit.

The second model, consists of two interlinked transhipment models, where successively increasing recycle flows are introduced until convergence to the optimal recycle flow value based on a total cost criterion. The first is a transhipment model used to identify the flowrates and concentrations of the wastewater streams at the effluent water mains. The second model, originally intended for targeting treatment flowrates (Nikolakopoulos et al., 2014) is extended in screening treatment technologies and suggesting processes to integrate. The models demonstrated cascade characteristics by allowing the transfer of contaminant between concentrations intervals.

The application of the proposed tool in the design of the TTWR is able to produce significant energy and water saving targets, that may reach up to 50% for energy and 70% for water. An additional advantage of the procedure is the possibility to detect critical parameters of the system, where retrofitting changes improve on total cost. These are the number and order of washing and filtering stages and their concurrent or countercurrent mode of operation. The results of total water network suggest the use of centralized treatment using ultimately ultra and nano-filtration preceded by simple pretreatment filtering stages.

Acknowledgement

This project has received funding from the *European Union's Horizon 2020 research and innovation programunder* grant agreement No 641942.

References

Biocore, 2014. EU research project FP7-241566, http://www.biocore-europe.org

Kalliala, E.M., Nousianinen, P., 1999. Life Cycle Assessment Environmental Profile of Cotton and Polyestercotton Fabrics. AUTREX Res. J. 1, 8–20.

Koufolioulios, D., Nikolakopoulos, A., Pyrgakis, K., Kokossis, A., 2014. A mathematical decomposition for the synthesis and the application of total site analysis on multi-product biorefineries, Computer Aided Chemical Engineering.

Mountraki, A., Tsakalova, M., Panteli, A., Papoutsi, A.I., Kokossis, A.C., 2016. Integrated Waste Management in Multiproduct Biorefineries: Systems Optimization and Analysis of a Real-Life Industrial Plant. Ind. Eng. Chem. Res. 55, 3478–3492.

Mountraki, A.D., Nikolakopoulos, A., Mlayah, B.B., Kokossis, A.C., 2011a. BIOCORE- A systems integration paradigm in the real-life development of a lignocellulosic biorefinery. Comput. Aided Chem. Eng. 29, 1381–1385.

Mountraki, A.D., Nikolakopoulos, A., Pyrgakis, K.A., Mlayah, B.B., Kokossis, A.C., 2011b. BIOCORE - On a paradigm for the design and synthesis of real-life biorefineries, in: 11AIChE - 2011 AIChE Annual Meeting, Conference Proceedings.

Nikolakopoulos, A., Faskiotis, D., Kokossis, A., 2016a. Integrated Transhipment Models for Synchronous Screening of Treatment Technologies and Targeting of Fresh Water and Recycle Flows, Computer Aided Chemical Engineering.

Nikolakopoulos, A., Faskiotis, D., Kokossis, A., 2016b. Cascade Models for Targeting and Synthesis of Total Water Networks. Chem. Eng. Trans. 52, 943–948.

Nikolakopoulos, A., Kokossis, A., 2016. A Problem Decomposition Approach for Developing Total Water Networks in Lignocellulosic Biorefineries. Process Saf. Environ. Prot. in press.

Nikolakopoulos, A., Thomaidis, L., Kokossis, A., 2014. Mathematical Programming Shortcut Screening Models for the Design of Integrated Waste Treatment Systems. Comput. Aided Chem. Eng., Computer Aided Chemical Engineering 34, 381–386.

Papoulias, S.A., Grossmann, I.E., 1983. A structural optimization approach in process synthesis—II. Comput. Chem. Eng. 7, 707–721.

Tsakalova, M., Nikolakopoulos, A., Kokossis, A., 2012. On the systematic synthesis screening and integration of real-life biorefineries, in: AIChE 2012 - 2012 AIChE Annual Meeting, Conference Proceedings.

495001 Advances in Process Intensification: Using Reactive Distillation for the Conversion of Crude Glycerol

Abstract:

Advances in Process Intensification: Using Reactive Distillation for the Conversion of Crude Glycerol

Obakore Agbroko and Tracy J Benson (Lamar University)

With the rise in biodiesel production over the last couple of decades has come the abundance of glycerol, a byproduct from biodiesel manufacturing. A typical biodiesel process includes the transesterification of triglycerides to produce fatty acid methyl esters, followed by fuel clean up steps: product separation, water washing, and acidulation. In short, the glycerol stream also includes water and methanol and is typically considered â€œcrude glycerol.â€ In this study, crude glycerol was converted to propanediols using isopropanol as a hydrogen donor molecule in a reactive distillation column.

Combining reaction engineering and distillation into a single unit operation presents some interesting challenges, yet, reduces the operational footprint as well as the operating and capital costs of manufacturing. Operational and design challenges include mismatched temperatures between the separating components and the reacting components, unfavorable volatilities, and slow reaction rates generally due to poor reactant/catalyst/product mass transfer rates. Reactive distillation is a likely pathway towards process intensification to minimize overall energy consumption, particularly since distillation accounts for 20 % of total energy used within the U.S. manufacturing sector. Increased reflux ratios typically yield higher product purities but at the cost of reboiler heat duties. In comparison with conventional approaches of reaction followed by distillation, the proposed scheme holds great potential due to its ability for lowering both operating and capital costs.

In this study, Aspen Plus was used to model the conversion of glycerol to 1,3-propranediol and 1,2-propanediol while being separated from methanol, water, and other inorganic components that are commonly found in waste glycerol from biodiesel production processes. Using the Non-Random Two Liquid activity coefficient model, the number of trays, conversion, and product selectivities were estimated. To verify simulation results, laboratory experiments were conducted using a 1-inch ID, packed distillation column. Column temperature profiles as well as reboiler and condenser heat duties were determined. The complete reactive distillation analysis will be presented as well as the extension of this technology into other areas of the chemical process industry.

495999 Process Design and Techno-Economic Assessment of a Seaweed-Based Biorefinery: Integration of Thermochemical and Biochemical Conversion Processes

Abstract:

Biofuel production using carbohydrate rich brown seaweed is being intensely researched in the last few years as a viable replacement for 1st and 2nd generation biomass. In this study, for the first time, integrated pyrolysis and anaerobic digestion is simulated on an industrial scale using Aspen Plus software and experimental data, and technoeconomical assessment is made. Main points of integration include the pyrolytic conversion of the digestate, instead of raw seaweed and further digestion of the carbohydrate rich aqueous phase, which is a byproduct of both pyrolysis and catalytic upgrading. The synergistic effect of this integrated process should yield a more economically and environmentally viable biofuel production, compared to standalone pathways.

491067 Process Intensification and Modularization for Sustainability

Abstract:

We will describe various concepts for intensifying catalytic reactions and reactors, including bifunctional materials and cascade reactions, fairly mature by now microtechnology, hybrids of reactors and separations, and multifunctional reactors including static micromixers and microheat exchangers (spatial domain), periodic operation (temporal domain), and use of alternate energy forms such as microwaves. These concepts will be elucidated with examples taken from biomass, shale gas, and ammonia chemistry as means of sustainable chemical processing.

503175 Keynote - Jet Fuel from H_2O, CO_2, and Solar Energy

Abstract:

The entire production chain for renewable kerosene obtained directly from sunlight, H_2O, and CO_2 has been experimentally demonstrated. The key component of the production process is a high-temperature solar reactor containing a reticulated porous ceramic (RPC) structure made of ceria, which enables the splitting of H_2O and CO_2 via a 2-step thermochemical redox cycle. In the 1st endothermic reduction step, ceria is reduced using concentrated solar radiation as the energy source of process heat. In the 2nd exothermic oxidation step, nonstoichiometric ceria reacts with H_2O and CO_2 to form H_2 and CO – syngas – which is finally converted into kerosene by the Fischer-Tropsch process. The RPC features dual-scale porosity for enhanced heat/mass transport and rapid redox kinetics, while 500 consecutive redox cycles further validate material stability and structure robustness. We further report on the thermochemical splitting of CO_2 into separate streams of CO and O_2 with 100% selectivity, 83% molar conversion, and 5.25% solar-to-fuel energy efficiency.

References:

Marxer D., Furler P., Scheffe J., Geerlings H., Falter C., Batteiger V., Sizmann A., Steinfeld A., "Demonstration of the entire production chain to renewable kerosene via solar thermochemical splitting of H_2O and CO_2", *Energy & Fuels* 29, pp. 3241-3250, 2015.

Marxer D., Furler P., Takacs M., Steinfeld A., "Solar thermochemical splitting of CO_2 into separate streams of CO and O_2 with high selectivity, stability, conversion, and efficiency", *Energy & Environmental Science*, 2017.

503174 The Role of Charge Compensating Dopants in Decreasing the O-vacancy Formation Energy in ceria

Abstract:

Solar thermochemical water and carbon dioxide splitting provide a pathway to high efficiency solar fuels. To-date, ceria remains the benchmark material due to its stability and rapid kinetics. Of the many doping strategies investigated, only Hf- and Zr-doping has shown increased performance over undoped ceria. Here we use paired charge compensating dopants (PCCD) to improve the performance of ceria. Specifically, we use density functional theory calculations explore the reduction behavior of ceria co-doped with tri- and pentavalent dopants. We show that this strategy increases the range of available reduction energies available in ceria and provide a fundamental understanding for the found behavior. The identity of the trivalent determines the reduction energy of the system while pentavalent dopants only have a minor effect, due to their individual mechanisms of action. The pentavalent dopants cluster and cause large local restructuring which highly suppresses the O-removal energy in its vicinity to level below that needed to drive gas splitting. The trivalent dopants however weaken the bonds with their coordinating O anions through decreased ionic interaction; this interaction only slightly decrease the O-vacancy formation energy, making these sites more redox active. Overall we predict relative reduction energies of PCCD ceria of Hf doped ceria (Hf-CeO_2) < ScNb-CeO_2 < YNb-CeO_2 < LaNb-CeO_2 < CeO_2. Experimental studies using thermos-gravimetric analysis confirm the predicted reduction energies. Although these materials are investigated in light of solar thermochemistry, their behavior is of potential interest to the metal oxide catalytic community as a whole.

496839 Redox-Active $Y_{0.5}Ba_{0.5}CoO_{3-\delta}$ As Thermochemical Oxygen Pump for O_2-Inert Gas separations

Abstract:

To develop a thermochemical oxygen pump driven by solar-thermal process heat, we screen perovskites for their oxygen reduction- and oxidation-activities using electronic structure computations. Materials with high predicted redox-activity, $La_{0.5}Ba_{0.5}CoO_{3-\delta}$, $Pr_{0.5}Ba_{0.5}CoO_{3-\delta}$ and $Y_{0.5}Ba_{0.5}CoO_{3-\delta}$, are synthesized and evaluated for reversible oxygen exchange in up to 50 redox cycles. Each cycle consist of endothermic perovskite reduction, that absorbs heat from 1 bar air at 500°C to form oxygen vacancies and O_2, and exothermic perovskite oxidation, that refills oxygen vacancies with O_2 stripped from O_2-Ar mixtures at 0.01 bar O_2, 1 bar, and 290°C. We demonstrate as well producing pure O_2 via pressure-temperature swing from 1 bar O_2 at 600°C to 1 bar air at 300°C. Evaluating mixed $Y_{0.5}Ba_{0.5}CoO_{3-\delta}$ powder/SiO_2 sand and porous $Y_{0.5}Ba_{0.5}CoO_{3-\delta}$ granules at these conditions in a packed bed reactor gives oxygen exchange capacities of 64 mmol lattice oxygen per mol perovskite for decreasing pO_2 from 0.01 to 0.001 bar and 114 $mmol_O$ $mol_{perovskite}^{-1}$ for increasing pO_2 from 0.2 to 1 bar, exchanged at about 12-19 μmol_{O2} $g_{perovskite}^{-1}$ min^{-1}. $Y_{0.5}Ba_{0.5}CoO_{3-\delta}$ thereby outperforms the CuO/Cu_2O benchmark, active in reduction with air only above 1030°C, and $SrCoO_{3-\delta}$, reaching 88 $mmol_O$ $mol_{perovskite}^{-1}$ in air at 627°C. With the measured heat capacities and reduction enthalpies we solve the energy balance for the Ar purifcation and O_2 production modes, outlining the conditions at which thermochemical oxygen pumping might be a viable alternative to cryogenic distillation and vacuum pumping.

497841 Reduction Kinetics of Hercynite ($FeAl_2O_4$) and Hydrogen Production for Solar Thermochemical Water Splitting

Abstract:

The solar thermochemical water splitting process (STWS) is a promising technology for producing renewable H_2. A metal oxide is first reduced with high temperature concentrated sunlight to produce O_2. Then, the reduced material is exposed to steam to reoxidize it and produce H_2. Un-doped hercynite ($FeAl_2O_4$) is shown to be a promising and low-cost iron and aluminum based solar thermochemical water splitting material [1]. The reduction kinetics of $FeAl_2O_4$ was studied using a thermogravimetric analysis (TGA) technique. Isoconversional methods and XRD analysis show that hercynite is first formed by reacting aluminum oxide with iron oxide as shown in R1. Then, hercynite is further reduced, forming O-vacancies, while maintaining the spinel structure as shown in R2. Apparent activation energies for both reactions have been determined. This is the first kinetics study to be reported for the total reduction process to form herycnite and then to form the spinel, and to identify H_2 production from oxidizing the reduced hercynite with H_2O as shown in R3.

$Fe_2O_3 + 2Al_2O_3 \rightarrow 2FeAl_2O_4 + 0.5\ O_2$ R1

$FeAl_2O_4 \rightarrow FeAl_2O_{4-\delta} + \delta/2\ O_2$ R2

$FeAl_2O_{4-\delta} + \delta H_2O \rightarrow FeAl_2O_4 + \delta H_2$ R3

1. Muhich, C.L., et al., *Predicting the solar thermochemical water splitting ability and reaction mechanism of metal oxides: a case study of the hercynite family of water splitting cycles.* Energy & Environmental Science, 2015. **8**(12): p. 3687-3699.

492700 Near-Isothermal on-Sun Demonstration to Split Water

Abstract:

A sustainable hydrogen economy based on the low-cost production of H2 from renewable resources has the potential to drastically transform the energy sector. While solar energy is the most abundant renewable energy resource, the capture, storage, and distribution of it remains a challenge. Solar thermochemical water splitting (STWS) provides a promising route for efficient utilization of this dispersed resource because it allows for use of the entire solar spectrum to convert water to a renewable energy carrier and fuel, H_2. Simple two-step reduction/oxidation (redox) cycles are preferred because the production of oxygen and hydrogen can be spatially or temporally separated, eliminating the need for high-temperature gas separation, while also avoiding the design complications and inefficiencies of multistep processes. In this process, oxygen is generated by the reduction of a metal oxide during a high temperature step in an inert atmosphere. In the second step, the reduced metal oxide is exposed to high temperature steam to reoxidize the material and produce hydrogen fuel. Although the STWS reaction schemes appear straightforward, their implementation in an economically viable process is extremely demanding. In order to commercially realize this technology, several challenges must be overcome including an energy efficient reactor design, stable reactor and heat exchanger materials, and robust redox active particles. In this work we present results showing near-isothermal on-sun hydrogen production utilizing dual lab-scale fluidized beds. The use of fluidized beds that are cycled between oxidation and reduction negates the need for energy intensive moving solids. Active hercynite particles were redox cycled at the High Flux Solar Furnace at the National Renewable Energy Laboratory and production rates of H_2 exceeding 500 umol/g were measured. The effect of environmental variability on the incident solar radiation and hydrogen production is evaluated during on-sun testing. This work demonstrates the viability of solar thermal water splitting leveraging continuous operation of both reduction and oxidation at near-isothermal conditions.

500324 Solar Thermochemical Fuel Production in a Microreactor

Abstract:

Harvesting and utilizing solar energy offers a sustainable way to reduce the use of fossil fuels and meet the increasing demand of energy. High temperature thermochemical reactions using concentrated solar power are of great interest due to the stored energy density associated with chemical fuels. The aim of this project is to develop technology suitable for the direct utilization of solar thermal energy through chemical processes. Both high temperature (ca. 1400°C) thermochemical water/carbon dioxide splitting cycle and moderate temperature (600-700°C) dry reforming of methane are considered. The critical component of this project is mesoscale-based reactor for the purpose of solar fuel generation. The mesoscale-based reactor, which offers advantages such as fast heat and mass transfer as well as flexibility in capacity and in design, was designed, built and tested. The reactive materials (perovskites) for high-temperature thermochemical splitting and catalysts for dry reforming of methane were synthesized, characterized and tested in the reactor.

Saroj Bhatta and Juan Pablo Trelles

Motivation: Solar Chemical Synthesis

Solar Chemical Synthesis:
- Solar for targeted chemical reactions
- Value-added products (e.g. solar fuels)

Activation Energy

Thermo-chemical:

Reactants + Thermo-catalyst —*Solar Intensity*→ Products

- High temperature → high reactivity
- ~ Independent of spectra

+

Photo-chemical:

Reactants + Photo-catalyst —*Solar Spectra*→ Products

- Low temperature → low throughput
- Depend on solar spectrum (e.g. UV)

=

Photo-Thermochemical:

Reactants + High-temperature photo-catalyst —*Solar Intensity + Spectra*→ Products

- High temperature, but potentially lower than thermochemical
- Higher yield than photochemical

This research

R·Eng | Lab 1 UMASS LOWELL

Reactor Design Rationale

Reactor – Simulator Operation

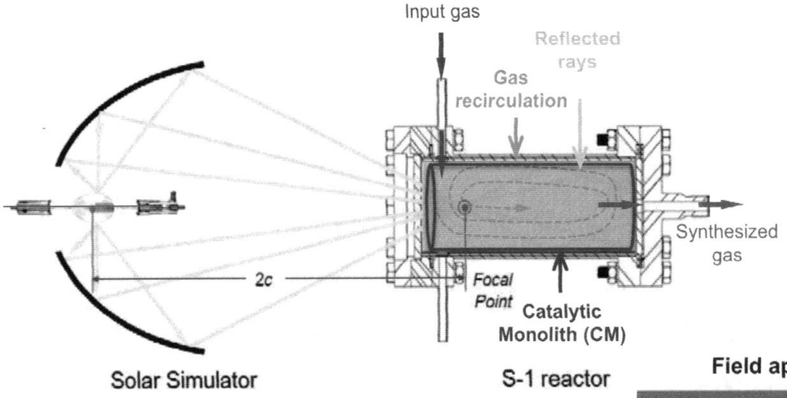

Rationale

- Optical cavity:
  - Light trapping
- Residence time:
  - Flow recirculation
- Catalytic monolith:
  - "Optically-tuned" high temp photo-catalyst

Field application

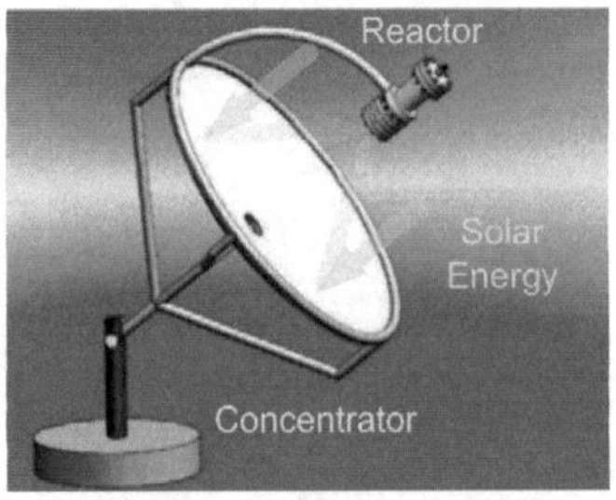

➢ High temp. photo-catalyst
+ High gas recirculation
+ High photons absorption
= Promote photo-thermochemical processes

S. Bhatta, D. Nagassou, J. P. Trelles, "*Solar Photo-Thermochemical Reactor Design For Carbon Dioxide Processing*", Solar Energy (2017), Vol. 142, pp. 253–266.

R·Eng | Lab 2 UMASS LOWELL

Photo-Thermochemical Reactor

S-1 Reactor

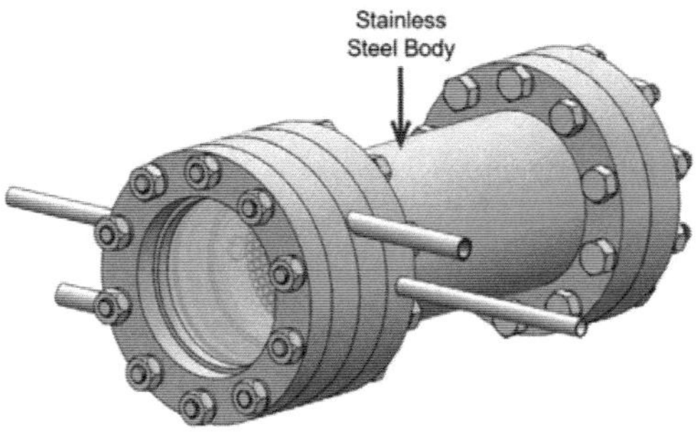

Characteristics:

- Stainless steel body:
  (Design Operating Temperature ~ 1000 [°C])

S. Bhatta, D. Nagassou, J. P. Trelles, "*Solar Photo-Thermochemical Reactor Design For Carbon Dioxide Processing*", Solar Energy (2017), Vol. 142, pp. 253–266.

Photo-Thermochemical Reactor

S-1 Reactor

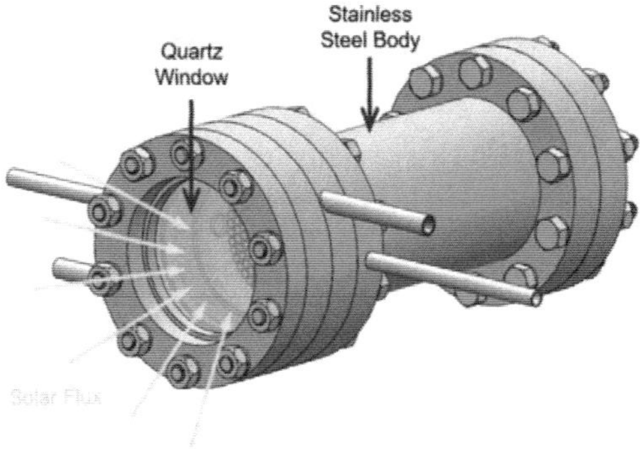

Characteristics:

- Stainless steel body:
  (Design Operating Temperature ~ 1000 [°C])
- Quartz window:
  (Solar Flux Inlet)

S. Bhatta, D. Nagassou, J. P. Trelles, "*Solar Photo-Thermochemical Reactor Design For Carbon Dioxide Processing*", Solar Energy (2017), Vol. 142, pp. 253–266.

Photo-Thermochemical Reactor

S-1 Reactor

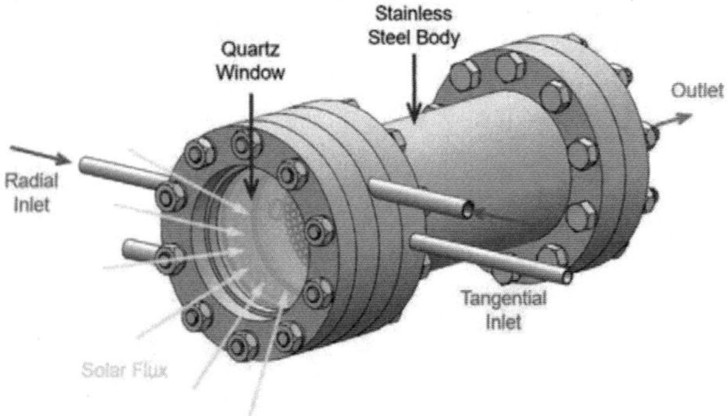

Characteristics:

- Stainless steel body:
  (Design Operating Temperature ~ 1000 [°C])
- Quartz window:
  (Solar Flux Inlet)
- Multiple flow inlets:
  (Radial & Tangential Inlets)

ORꞏEng | Lab

S. Bhatta, D. Nagassou, J. P. Trelles, "*Solar Photo-Thermochemical Reactor Design For Carbon Dioxide Processing*", Solar Energy (2017), Vol. 142, pp. 253–266.

5

UMASS LOWELL

Photo-Thermochemical Reactor

S-1 Reactor

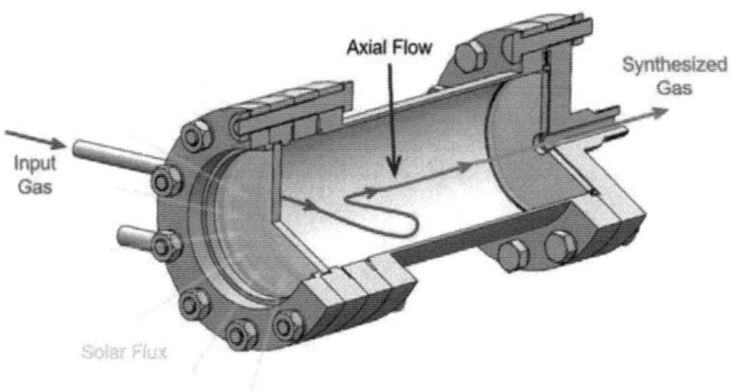

Characteristics:

- Stainless steel body:
  (Design Operating Temperature ~ 1000 [°C])
- Quartz window:
  (Solar Flux Inlet)
- Multiple flow inlets:
  (Radial Inlets → Axial flow)

ORꞏEng | Lab

6

UMASS LOWELL

Photo-Thermochemical Reactor

***S*-1 Reactor**

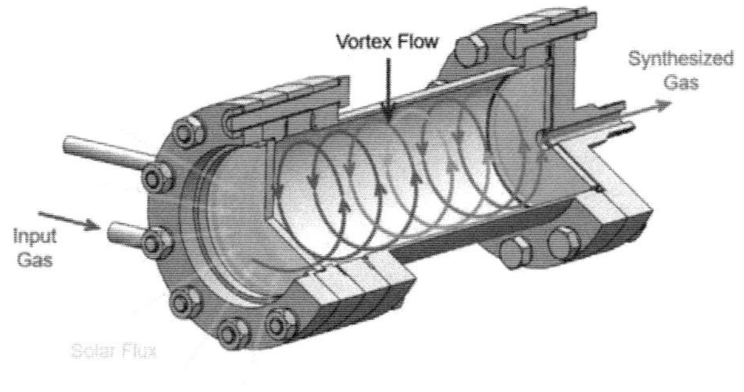

Characteristics:

- Stainless steel body:
  (Design Operating Temperature ~ 1000 [°C])
- Quartz window:
  (Solar Flux Inlet)
- Multiple flow inlets:
  (Tangential Inlets → Vortex flow)

R·Eng | Lab

UMASS LOWELL

Photo-Thermochemical Reactor

***S*-1 Reactor**

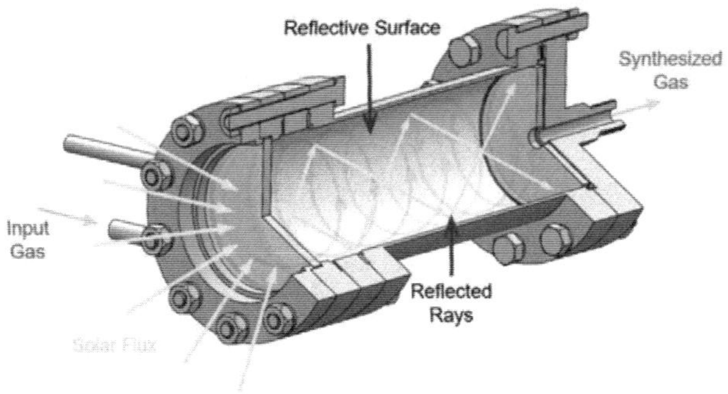

Characteristics:

- Stainless steel body:
  (Design Operating Temperature ~ 1000 [°C])
- Quartz window:
  (Solar Flux Inlet)
- Multiple flow inlets:
  (Radial & Tangential Inlets)
- Reflective interior
  (Light Trapping Cavity)

R·Eng | Lab

S. Bhatta, D. Nagassou, J. P. Trelles, "*Solar Photo-Thermochemical Reactor Design For Carbon Dioxide Processing*", Solar Energy (2017), Vol. 142, pp. 253–266.

UMASS LOWELL

Photo-Thermochemical Reactor

S-1 Reactor

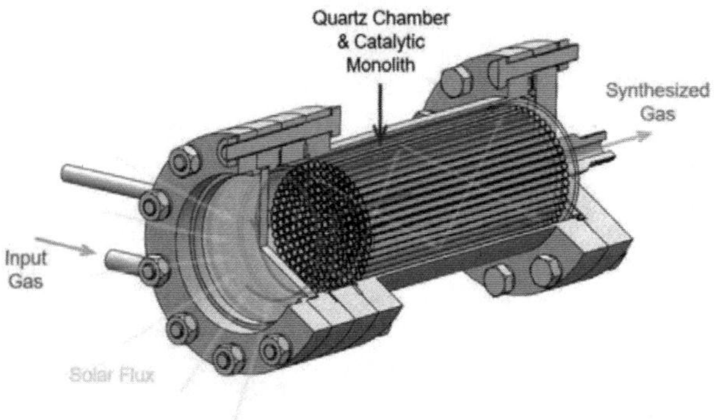

➢ Increase interaction among:
(1) photons, (2) feedstock, and (3) high temp. photo-catalyst
→ Enhance photo-thermochemical processes

Characteristics:

- Stainless steel body:
  (Design Operating Temperature ~ 1000 [°C])
- Quartz window:
  (Solar Flux Inlet)
- Multiple flow inlets:
  (Radial & Tangential Inlets)
- Reflective interior
  (Light Trapping Cavity)
- Quartz reaction chamber
  (Reactor Integrity & Catalytic Monolith)

S. Bhatta, D. Nagassou, J. P. Trelles, "*Solar Photo-Thermochemical Reactor Design For Carbon Dioxide Processing*", Solar Energy (2017), Vol. 142, pp. 253–266.

Catalytic Monolith

Zirconia Foam Monolith (ZrFM)

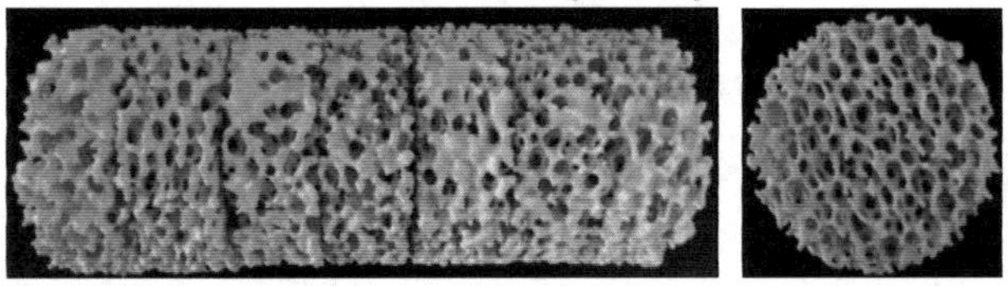

- Absorb incident radiation
- Better for thermo-chemical processes

Light Transmission

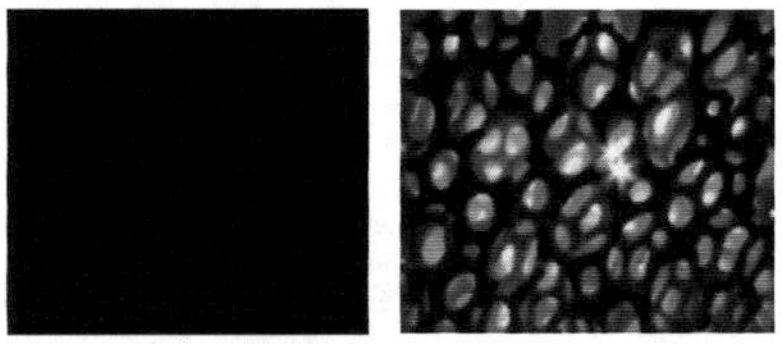

Full monolith — Two zirconia foams

Optically thick

Tubular Quartz Monolith (TQM)

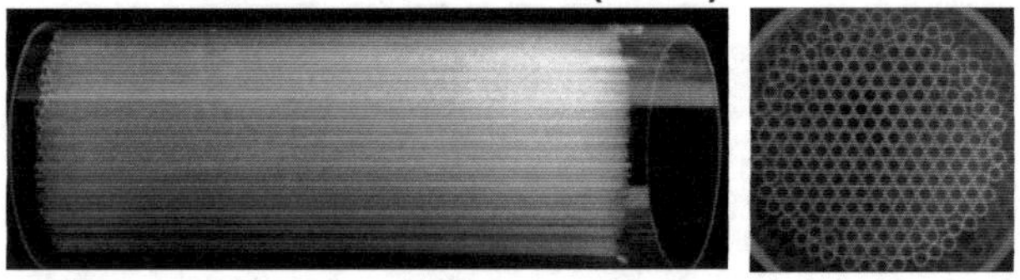

- Absorb + transmit incident radiation
- Better for photo-thermo-chemical processes

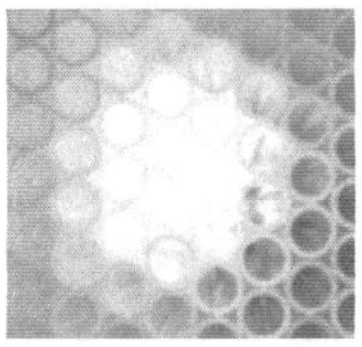

Full monolith

Optically thin

➢ if yield: ZrFM > TQM → higher thermo-catalysis
ZrFM < TQM → photo-thermo-catalysis

Catalytic Coating

- **Sol-gel process**
  - ZrFM & TQM as substrate
  - Dip coated (5 layer) → 1% wt. and 2% wt. Cu/TiO_2
  - Coating wt. ZrFM (~ 3.3 gm) → 3X more than QTM

Before heating **After heat @ 550 °C**

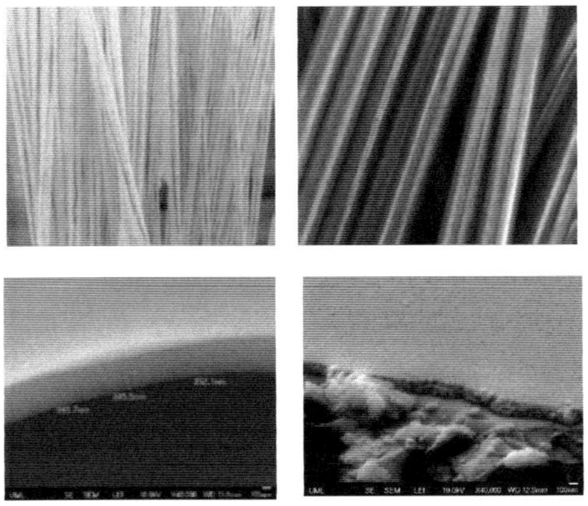

Sample SEM Images

Catalytic monolith

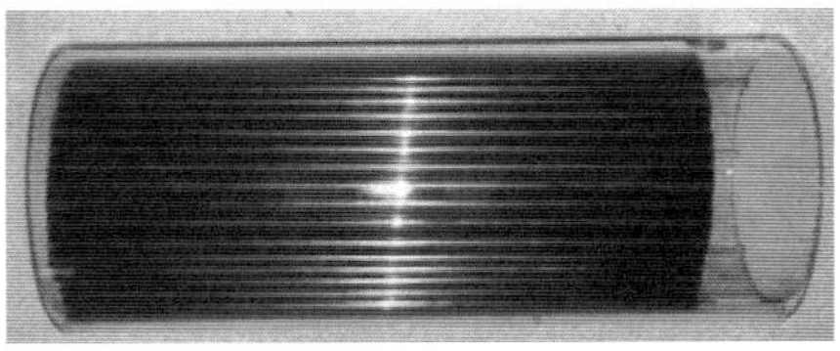

EDS Analysis

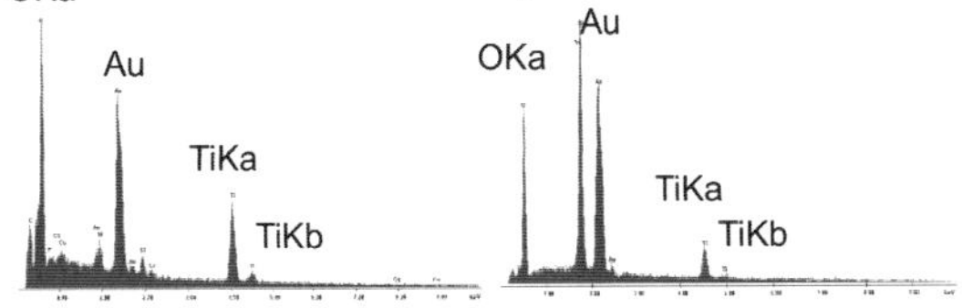

Before heating **After heating**

➢ ZrFM → higher coating mass → larger surface area (estimated 3X)

Experiment Results: $CO_2 + H_2O$

- Catalytic monolith → 1% wt. Cu/TiO_2 & 2% wt. Cu/TiO_2 coated
  - ZrFM and TQM
- Flow rates → 4, 3, 2, 1 & 0.5 slpm CO_2 + bubbler
- Run time → at least 20 min for each flow after steady state

Sample results: 2% Cu/TiO_2 coated ZrFM

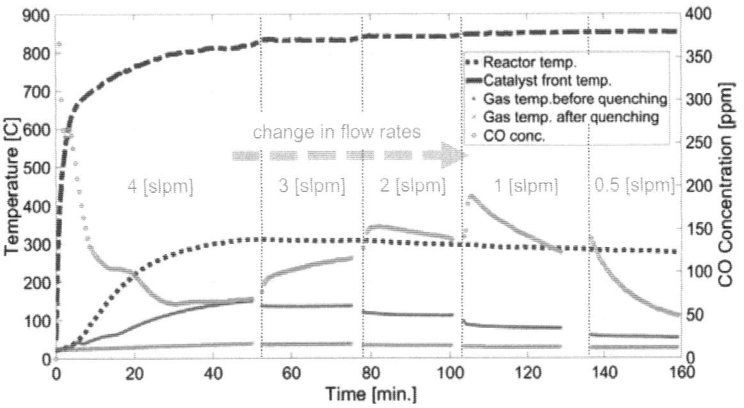

➢ Different catalytic monoliths @ different flowrates tested

Experiment Results: $CO_2 + H_2O$

- Gas temp. @ outlet for different flowrates
- Gas temp. changes with flow
  - Low flow → Low temp.

- 1 % wt. Cu/TiO_2 coated TQM → **20% higher temp.**

Gas temp. @ reactor outlet

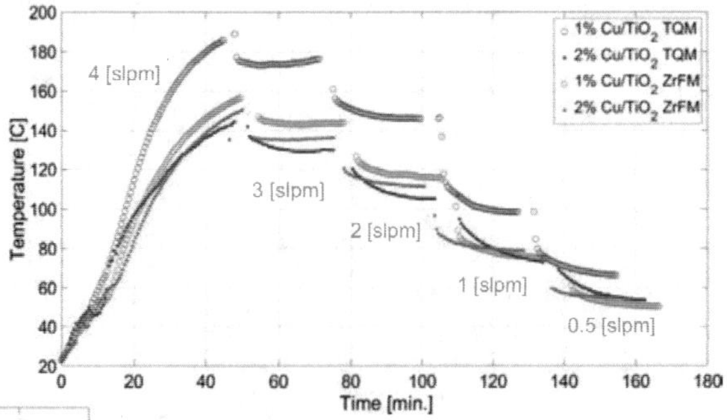

Catalytic monolith front temp.

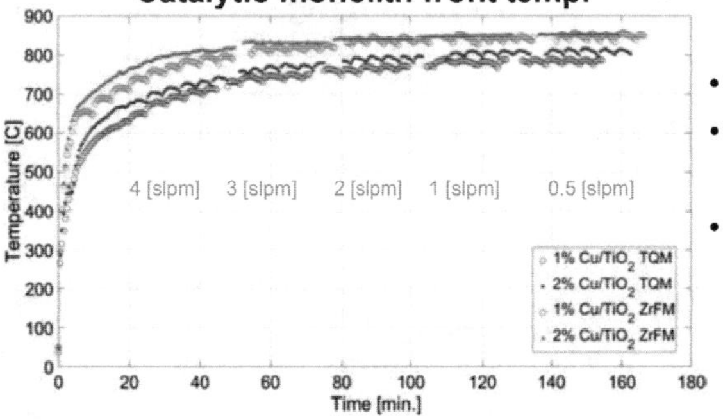

- Catalytic monolith temp. near quartz window
- No change with change in flow rates

- ZrFM higher temp. than TQM
  ~ 10 % higher than TQM

➢ ZrFM: Higher surface temp. but lower gas temp.

R·Eng | Lab

UMASS LOWELL

Experiment Results: $CO_2 + H_2O$

CO conc. @ reactor outlet for different flowrates

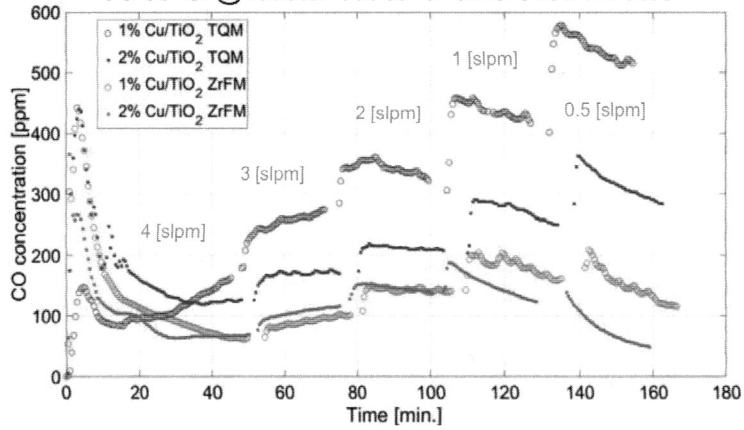

- Low flow rate → high CO conc.
- CO conc. TQM > ZrFM (~50 - 400 %) → 1 % wt. Cu/TiO_2 coating TQM

➢ ZrFM: higher surf. temp + ~ 3X catalyst surface area
→ Yet lower conversion

➢ Optically thin catalytic monolith + optimized flow rate
→ better performance

R·Eng | Lab

UMASS LOWELL

Experimental Results: $CO_2 + CH_4$

- CO_2 (Tangential inlet) : CH_4 (Radial inlet) = 2:1

Dissociation of CO_2 & CH_4 into CO and carbon black

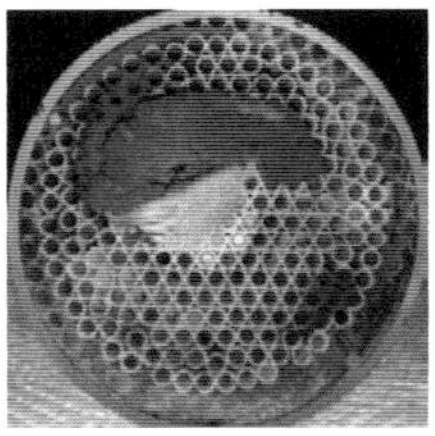
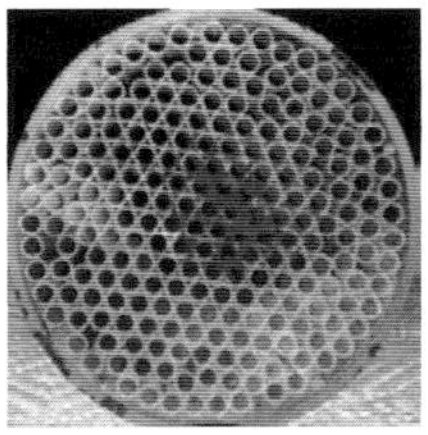
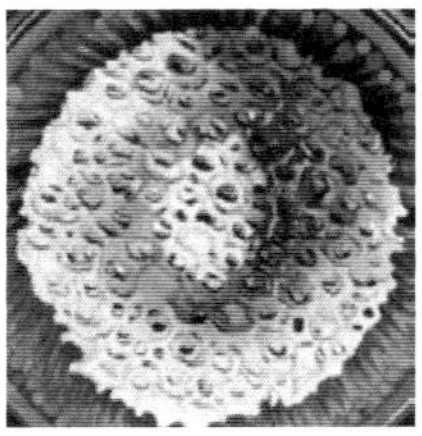
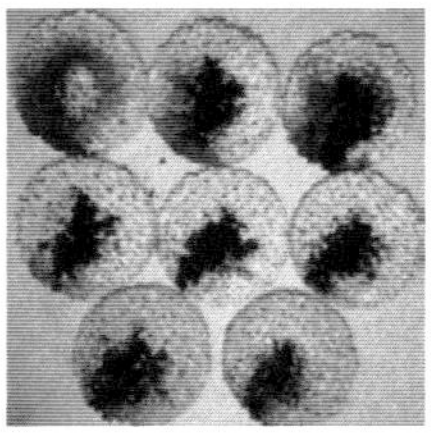

▪ 2 % wt. Cu/TiO_2 coated TQM	**Catalytic monolith**	▪ Uncoated Zirconia foam monolith
▪ 1.63 area % of CO	**GC result**	▪ 6.39 area % of CO

- Issue: Carbon deposition on catalytic monolith → deactivate catalyst

➢ Thermal decomposition

R·Eng | Lab

UMASS LOWELL

Deposition Analysis

EDS analysis of coat before experiment

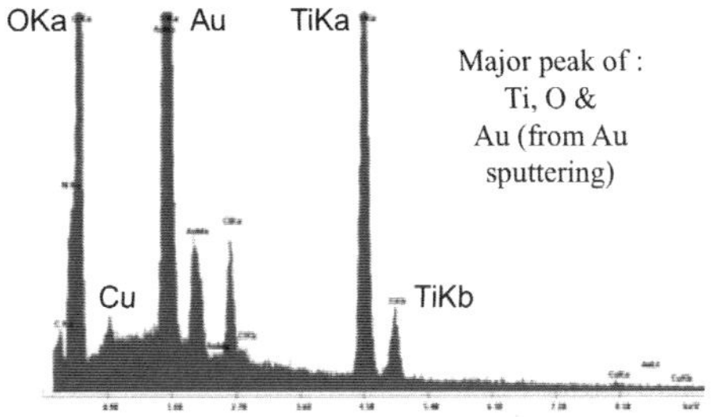

EDS analysis of coat after experiment

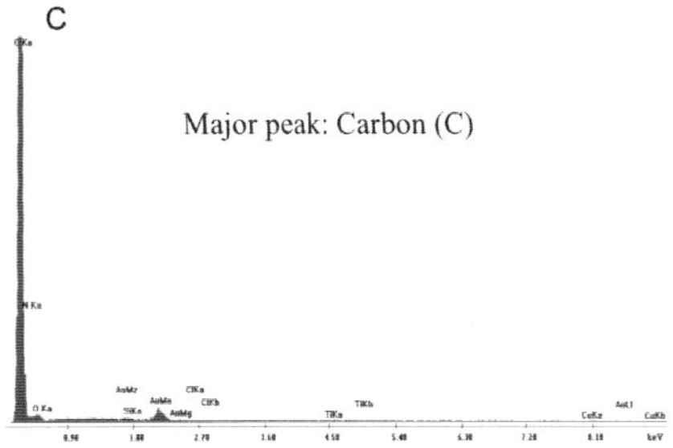

SEM images of the deposition on the quartz monolith

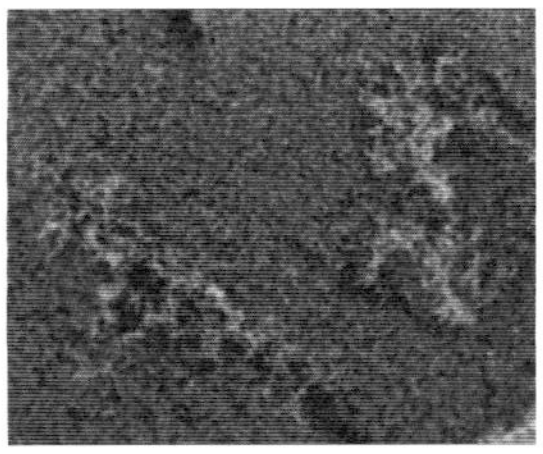

➢ Carbon black → catalyst deactivation due to deposition

R·Eng | Lab

UMASS LOWELL

Computational Model

- 3D CFD model in ANSYS Fluent
- Multiple species, incompressible ideal gas
- Discrete ordinate radiation transport model
- Species transport w. heterogeneous surface chemistry

3D Reactor Geometry

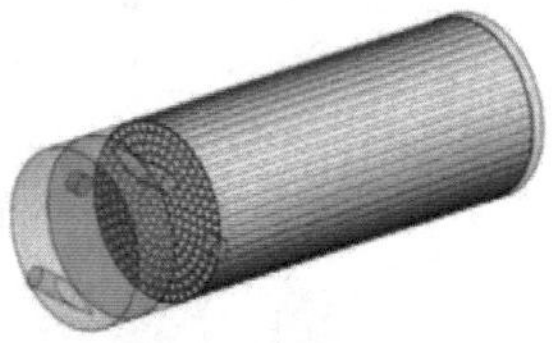

Governing equations

Equation	Variable	Expression
Continuity	$\mathbf{v}(\mathbf{x})$	$\nabla \cdot \mathbf{v} = 0$
Momentum	$p(\mathbf{x})$	$\nabla \cdot (\rho \mathbf{v}\mathbf{v}) = -\nabla p + \nabla \cdot \tau$
Energy	$T(\mathbf{x})$	$\nabla \cdot (\mathbf{v}(\rho E + p)) = \nabla \cdot (k_{eff} \nabla T - \sum_s h_s \mathbf{J}_s) + S_h$
Radiation	$I(\mathbf{x},\mathbf{s})$	$\mathbf{s} \cdot \nabla I + aI = \pi^{-1} a \sigma T^4$
Species	$Y_s(\mathbf{x})$	$\nabla \cdot (\rho \mathbf{v} Y_s) = -\nabla \cdot \mathbf{J}_s + R_s + S_s$

Mass Fraction

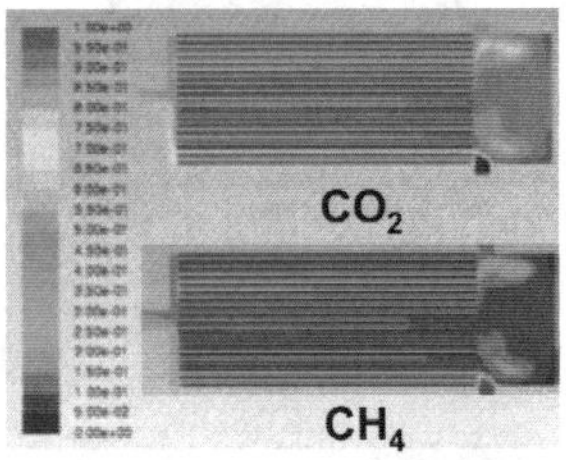

Temperature Contour

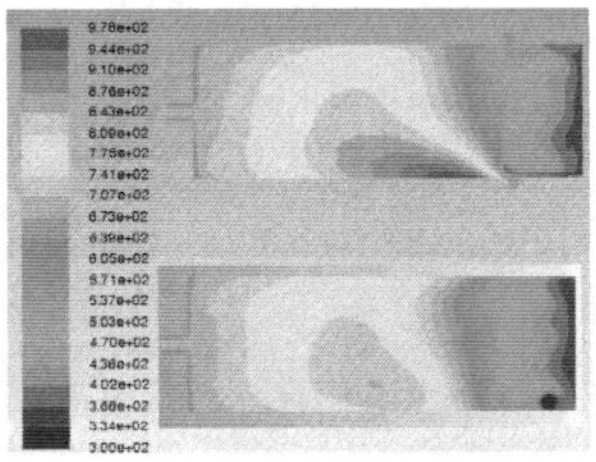

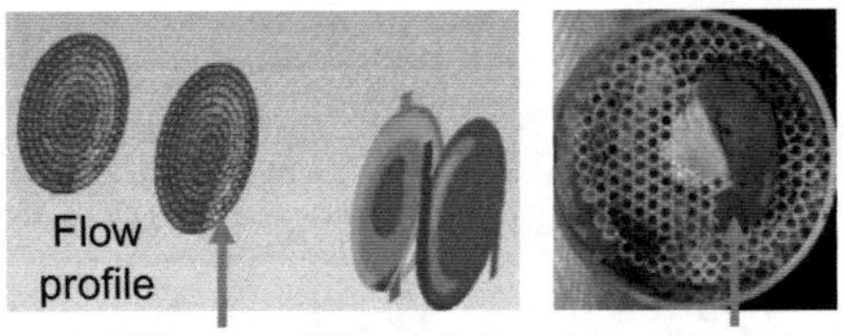

Flow profile matches to carbon deposition profile

R·Eng | Lab

UMASS LOWELL

Computational Model

Heterogeneous surface reaction model

Reaction	Rate
$TiO_2 + h\upsilon \rightarrow TiO_2 + e_{cb}^- + h_{vb}^+$	r_g
$e^- + h^+ \xrightarrow{k_2,(TiO_2)} heat$	$k_2[e^-][h^+]$
$2H_2O + 4h^+ \xrightarrow{k_3,(TiO_2)} O_2 + 4H^+$	$k_3[H_2O]^2[h^+]^4$
$CO_2 + 2H^+ + 2e^- \xrightarrow{k_4,(TiO_2)} CO + H_2O$	$k_4[CO_2][H^+]^2[e^-]^2$

J. Marugán et al. / Catalysis Today 240 (2015) 9–15

O. Ola, M.M. Maroto-Valer / Journal of Photochemistry and Photobiology C: Photochemistry Reviews 24 (2015) 16–42

Proposed photo-thermochemical reaction rate (r_g)

$$r_g = \alpha G e^{(-\frac{E}{RT})}$$

Thermal response (exponential term)

Photolytic response (G)

α → proportional to quantum yield

G → Incident radiation flux

E/R → Constant

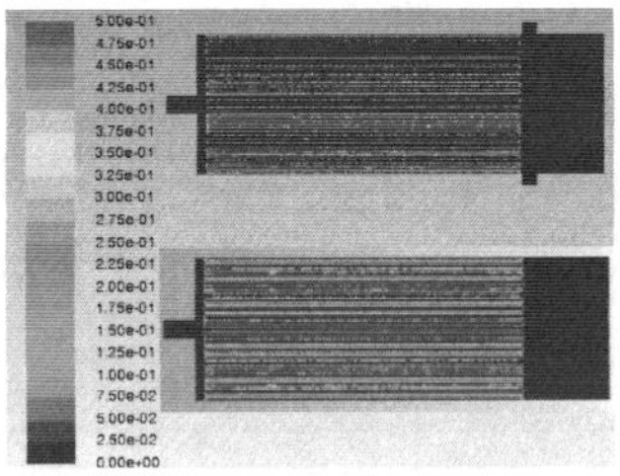

Surface coverage of [e⁻] on the catalytic monolith surface

> Parameters α and E to be found from experimental result → model calibration

R·Eng | Lab

UMASS LOWELL

Summary and Conclusion

- **Rationale:**

  Solar Photo-Thermochemical Process → combine advantages of high-temperature reactivity + photo-catalytic activation

- **Photo- Thermo-chemical Reactor Design:**

  Optical cavity + flow recirculation + "Optically-tuned" catalytic monolith → increase interaction among photons, gas & catalyst surface

- **Experiments:**

$CO_2 + H_2O$

  - ZrFM: higher surface temp. + high surface area but optically thick → lower conversion than TQM
  - Optically thin catalytic monolith + optimized flow → higher CO_2 reduction

$CO_2 + CH_4$

  - Carbon black deposition throughout the monolith
  - Thermochemical decomposition of CO_2 assisted by CH_4

R·Eng | Lab

Thank You

19

UMASS LOWELL

497085 Mechanistic Insight for Oxidative Coupling of Methane on Mg6MnO8-Based Redox Catalysts in a Chemical Looping System

Abstract:

Methane activation has been the topic of interest for many decades for its upgradation to value added chemicals. With the high availability of shale gas due to advancements in fracking technology, understanding this C-H bond activation becomes important as it is a key reaction step in the direct synthesis of chemicals from methane. Traditionally, methane is first converted to syngas, which can then be processed and used for producing chemicals and fuels. This indirect method is both process and energy intensive and thus incentivizes direct conversion of methane to chemicals. One such method for direct conversion is oxidative coupling of methane (OCM), which provides a single step path towards synthesizing higher hydrocarbons from methane, especially ethylene. A popular way for methane oxidation in OCM is through co-feeding molecular oxygen with methane over a catalyst. Conversely, the chemical looping method utilizes the lattice oxygen in a catalytic oxygen carrier (COC) for methane oxidation. A Mg_6MnO_8-based redox system was studied under the chemical looping method for understanding the mechanism of OCM. The atomistic thermodynamics methods and density functional theory (DFT) calculations were carried out to elucidate the lattice oxygen diffusion and vacancy formation mechanism during the oxidation and reduction process.

Additionally, experiments in a fixed bed reactor were performed to validate the computational findings, by running cyclic redox reactions on a fixed amount of COC. Fixed bed experiments were performed with methane as the reducing gas and air as the oxidizing gas. Several cofeed experiments were also performed in addition to the redox experiments to get a holistic view of the process. These results were coupled with thermogravimetric analysis (TGA) to investigate the effect of temperature on the OCM reaction. A dopant screening was also investigated computationally, to potentially increase the selectivity towards the desired hydrocarbon products. Several analysis techniques such as X-ray diffraction (XRD), X-ray photoelectron spectroscopy (XPS) were used to characterize the COC in its oxidized and reduced form. It was observed that the advent of carbon monoxide in conjunction to the desired hydrocarbons as the OCM reaction progresses gives an insight into the importance of oxygen vacancy formation on the COC surface.

496130 Gold-Palladium Particle Size Effect on Methane Oxidation Activity

Abstract:

Effect of Gold-Palladium Particle Size on Methane Oxidation Activity

Christopher Williams*[a], James Carter[a], Nicholas F. Dummer[a], Robert Armstrong[a], Sara Yacob[b],David J. Willock[a], Randall J. Meyer[b], Stuart H. Taylor[a], Graham J. Hutchings[a]

[a]Cardiff Catalysis Institute, School of Chemistry, Cardiff University, CF10 3AT

[b] ExxonMobil Research and Engineering Company, Corporate Strategic Research, Annandale, NJ 08801, USA

*Email:WilliamsCP1@cf.ac.uk

Natural gas, of which methane is a major component (approx. 70-90%), is a naturally abundant gas with reserves of ca. 190 trillion cubic meters.[1] Both affordable and readily available, natural gas may provide an alternative feedstock from the traditional petroleum reserves in addition to providing a cleaner source of fossil energy. Directly upgrading methane to oxygenates presents a global challenge complicated by the chemistry of methane. Current utilization involves indirect routes via the formation of synthesis gas (CO and H_2), requiring costly and energy intensive processes. A direct methane-to-methanol process at low temperatures provides an attractive and efficient method of producing an energy-dense fuel and important chemical feedstock.

In addition to requiring the use of preformed reactive oxygen species, current systems such as Periana[2] or Palkovits[3] catalysts have required highly acidic reaction media for the oxidation of methane. Alternatively, Hutchings et al. have demonstrated the use of H_2O_2 as an efficient oxidant for the low-temperature oxidation of methane, with noticeable productivities with the application of zeolite catalysts and supported gold-palladium nano-particulate catalysts.[4,5] Gold-palladium catalysts have also been shown to activate C-H bonds present in toluene[6] and in benzyl alcohol[7], in addition to the direct synthesis of H_2O_2 from H_2 and O_2.[8] The in situ generation of H_2O_2 provides a catalytic system for the oxidation of methane.

Figure SEQ Figure * ARABIC 1 : Particle size distribution of sol immobilised AuPd catalysts: Dried (blue) vs. Calcined at 800 °C (red).

It has previously been reported that gold-palladium catalysts, prepared by incipient wetness, produce materials with a bimodal distribution of particles ranging from 5-20 nm and 100-200 nm.[6,8] These 1- wt% AuPd/TiO_2 catalysts gave turnover frequencies of 6.5 h^{-1} and methanol and oxygenate selectivities of 12.1% and 85.4%, respectively.

Furthermore, methanol selectivity was improved with increasing metal weight loading to 5 wt%, whilst retaining high oxygenate selectivity. The bimodal particle size observed with incipient wetness is thought to inhibit reactivity due to (i) small particles that are highly activation.

In this work, a particle size effect is observed for methane oxidation using added H_2O_2 at 50 °C. 1 wt% AuPd/TiO_2 catalysts prepared by sol immobilization initially show relatively low rates of methane activation, forming CO_2 only and decomposing all H_2O_2. The AuPd catalysts are composed of a uniform size distribution ranging 4-8 nm, with mean particle size of 4.9 nm. Heat treatment of the dried catalyst produced a catalyst 3.6 times more active for the oxidation of methane, with 80% oxygenate selectivity and 50% of added H_2O_2 remaining after 0.5 h. Analysis by TEM showed that heat treatment at 800 °C yielded materials with increased AuPd particle sizes, increasing mean particle size from 4.9 nm to 20.3 nm (Fig.1). This particle size increase correlated with an increase in TOF (1.3 h^{-1} to 4.8 h^{-1}) and oxygenate selectivity (0% to 80%). These catalysts demonstrated high selectivity towards methyl hydroperoxide (64%), a key intermediate towards the formation of methanol.

The effect of particle size is an important consideration for the efficient utilization of H_2O_2. Increased particle size results in reduced H_2O_2 decomposition rates which facilitate the selective oxidation of methane to valuable oxygenates. Further work to optimize the synthesis of gold-palladium particles of larger size will be investigated to further improve the oxygenate selectivity.

References/ Bibliography

(1) BP Statistical Review of World Energy June 2016, bp.com/statisticalreview

(2) Periana, R. A.; Taube, D. J.; Gamble, S.; Taube, H.; Satoh, T.; Fujii, H. *Science (80-.).* 1998, *280* (5363), 560–564.

(3) Palkovits, R.; von Malotki, C.; Baumgarten, M.; Müllen, K.; Baltes, C.; Antonietti, M.; Kuhn, P.; Weber, J.; Thomas, A.; Schüth, F. *ChemSusChem* 2010, *3* (2), 277–282.

(4) Hammond, C.; Forde, M. M.; Ab Rahim, M. H.; Thetford, A.; He, Q.; Jenkins, R. L.; Dimitratos, N.; Lopez-Sanchez, J. A.; Dummer, N. F.; Murphy, D. M.; Carley, A. F.; Taylor, S. H.; Willock, D. J.; Stangland, E. E.; Kang, J.; Hagen, H.; Kiely, C. J.; Hutchings, G. J. *Angew. Chemie - Int. Ed.* 2012.

(5) Ab Rahim, M. H.; Forde, M. M.; Jenkins, R. L.; Hammond, C.; He, Q.; Dimitratos, N.; Lopez-Sanchez, J. A.; Carley, A. F.; Taylor, S. H.; Willock, D. J.; Murphy, D. M.; Kiely, C. J.; Hutchings, G. J. *Angew. Chemie - Int. Ed.* 2013.

(6) Kesavan, L.; Tiruvalam, R.; Rahim, M. H. A.; bin Saiman, M. I.; Enache, D. I.; Jenkins, R. L.; Dimitratos, N.; Lopez-Sanchez, J. A.; Taylor, S. H.; Knight, D. W.; Kiely, C. J.; Hutchings, G. J. *Science (80-.).* 2011, *331* (6014), 195-199.

(7) Pritchard, J.; Kesavan, L.; Piccinini, M.; He, Q.; Tiruvalam, R.; Dimitratos, N.; Lopez-Sanchez, J. A.; Carley, A. F.; Edwards, J. K.; Kiely, C. J.; Hutchings, G. J. *Langmuir* 2010, *26* (21), 16568–16577.

(8) Edwards, J. K.; Solsona, B. E.; Landon, P.; Carley, A. F.; Herzing, A.; Kiely, C. J.; Hutchings, G. J. *J. Catal.* 2005, *236* (1), 69–79.

490965 Printable Bioreactors for Bioconversion of Methane to Value-Added Chemicals

Abstract:

An industrial process for the selective oxidation of methane under mild conditions would be highly valuable for controlling emissions to the environment and for utilizing vast new sources of natural gas. Methanotrophic bacteria utilize methane under nearly ambient conditions, suggesting a biochemical process for methane conversion could be more efficient, less capital intensive, and more scalable compared to current industrial practice. However, the industrial use of bioconversion for large-scale applications, especially those involving gas-to-liquid reactions, has been limited by poor mass transfer within stirred-tank reactors. We have embedded the enzyme particulate methane monooxygenase (pMMO) within a 3D printed hydrogel to create a mechanically robust and gas permeable material that overcomes mass transfer limitations to convert methane to methanol. Building up those findings, we also explored encapsulation of whole cells within our material to utilize the entire cell proteome to further oxidize methanol into additional chemicals and expand our production capabilities.

Methods: Biocatalysts including enzymes and/or dehydrated cells were immobilized in poly(ethylene glycol) (PEG) hydrogel during crosslinking by UV-initiated photopolymerization. The utilization of methane by the encapsulated biocatalysts was measured by an infrared-based methane detector over time and high pressure liquid chromatography was used to detect production of lactate and methanol. The stability of the encapsulated biocatalysts was investigated for at least 72 hours.

Results and Discussion: We demonstrated the ability to encapsulate biocatalysts within a gas-permeable hydrogel material while retaining catalytic activity. Encapsulation within our material allowed an increased surface area to volume ratio, thereby overcoming mass transfer limitations and improving methane utilization. We investigated the stability of the encapsulated biocatalysts over time as well as the effect of loading, hydrogel weight fraction, and surface area to volume ratio on biocatalyst activity. We look forward to improving the activity of the encapsulated biocatalyst, and eventually curing the hydrogel material within a poly(dimethyl siloxane) scaffold as part of a flow-through reactor, shown in Figure 1, with which we have demonstrated continuous methane conversion using encapsulated pMMO.

Figure 1: Schematic of enzyme immobilization within hydrogel in a structured polymer scaffold to create a gas-permeable biocatalytic material in a flow-through bioreactor.

This work was performed under the auspices of the U.S. Department of Energy by Lawrence Livermore National Laboratory under Contract DE-AC52-07NA27344.

493072 Influence of Platinum on the Activity of Ga/HZSM-5 Catalyst in Ethane Dehydroaromatization Reaction

Abstract:

Gallium-platinum promoted HZSM-5 is found to be a promising catalyst for ethane dehydroaromatization reaction. In the present work, the influence of Pt as a promoter on the activity of Ga/HZSM-5 catalyst for ethane dehydroaromatization has been investigated. Also comparative study is performed between bimetallic Ga-Pt based (GaPt/HZSM-5) and Mo based HZSM-5, where the GaPt/HZSM-5 showed better aromatic and hydrogen selectivity. Pt promoted Ga/HZSM-5 catalyst exhibited higher activity compared to pure Ga/HZSM-5 catalyst. The presence of platinum in the gallium zeolite considerably accelerates dehydrogenation step in ethane aromatization. In addition, GaPt/HSZM-5 deactivates significantly slower than well-known dehydroaromatization catalyst such as Mo/HZSM-5 and Ga/HZSM-5. TPO study of spent catalysts [Figure 1] revealed that carbonaceous deposit on GaPt/HZSM-5 catalyst were burnt off at lower temperature compared to pure Ga/HZSM-5 catalyst. This happened because the presence of Pt metal on the catalyst facilitates hydrogen spillover resulting in the hydrogenolysis of coke and hence it can moderate the coke deposition. This property of Pt improves the stability of the GaPt/HZSM-5 catalyst. The reaction mechanism associated with aromatic formation is postulated based on the correlation between catalytic performance and surface characterization.

Figure 1: TPO profiles of spent catalysts after running the reaction for 5 h.

496112 Effect of Catalyst Composition and Reaction Mechanism Study on Non-Oxidative Methane Conversion into Higher Hydrocarbons

Abstract:

Direct non-oxidative methane conversion to higher hydrocarbons is a potential approach to convert natural gas into valuable hydrocarbons.[1] Zeolite supported molybdenum (Mo) catalyst is the most effective catalyst reported for direct non-oxidative methane conversion into higher hydrocarbons in which HZSM-5 & HMCM-22 are the two well-known zeolites as support.[2] Mo/HMCM-22 catalyst was reported for effective catalytic performance due to unique pore structure and framework of HMCM-22 zeolite support.[3] Different factors such as reaction parameters (T, P and GHSV), catalyst formulation, such as Mo loading over HMCM-22 zeolite support, and SiO_2/Al_2O_3 ratio (SAR) of HMCM-22 zeolite support significantly affects the reaction. In this regard we have studied here the effect of metal (Mo) loading over HMCM-22 (SAR-55) support and space velocity (GHSV) for non-oxidative methane conversion reaction. Firstly, effect of different Mo loading (2, 5 and 10 wt%) was analyzed and observed that 5 wt% Mo loading is suitable for benzene formation, major product of the reaction as shown in Figure 1(a). Effect of GHSV for the reaction has been shown in Figure 1 (b) for 5wt% Mo loaded HMCM-22 (SAR-55) catalyst, and it was observed that lower space velocity (720 ml/gm-hr) is effective for aromatics (benzene) formation compared to higher space velocity (1200 ml/gm-hr). Higher space velocity gives more ethylene compared to benzene due to lower residence time. Theoretical calculations were performed using density functional theory (DFT) for the Mo/Zeolite catalyst. Molybdenum carbide formed during the reaction has been proposed to be the active phase for C-H activation of methane, C_2 coupling and aromatic formation reactions.[4] Mo_xC_y[5] type nanocluster were supposed to be suitable form of molybdenum carbide species for methane activation and C_2 coupling. In this reference, Mo_4C_2 nanocluster grafted over Al and Si site of zeolite channel was reported for C-H bond activation and a 112 kJ/mol energy barrier was observed when grafted over Al site.[6] In the study we have investigated the C-H bond activation and ethylene (primary intermediate of benzene) formation via C_2 coupling over Mo_4C_2 nanocluster. Schematic reaction paths for methane activation, ethane and ethylene formation over the Mo_4C_2 cluster and also the potential energy diagram for all these steps have been shown in Figure 2(a and b). Methane molecule gets activated over the Mo_4C_2 nanocluster with an activation barrier 112.8 kJ/mol to form CH_3 and H adsorbed on the cluster. In the next step the second CH_4 molecule dissociated over the same site with 115.5 kJ/mol activation barrier to form two CH_3 groups attached to the same Mo atom, which then undergo C-C coupling reaction with an activation barrier of 150 kJ/mole which is comparatively high to form ethane. The energy barrier for C-H bond activation of ethane (89 and 94 kJ/mol) is relatively low and forms ethylene (C_2H_4), which is the major primary unit of benzene formation, over the catalyst as also observed experimentally, in Figure 1(a) and (b).

Figure 1. (a) Product distribution for different Mo loading over HMCM-22 (SAR-55) catalyst (b) Product distribution for 5wt%Mo/HMCM-22 (SAR-55) catalyst at two different GHSV (720 & 1200 ml/gm-hr)

Figure 2. (a) Schematic paths for methane activation and C-C coupling over Mo_4C_2 nanocluster (b) Potential energy diagram for C-H activation and C_2 coupling over Mo_4C_2 nanocluster

References

1. S. Majhi, P. Mohanty, H. Wang, K. K. Pant, *Journal of Energy Chemistry*, 2013, 22, 543-554.
2. S. Ma, X. Guo, L. Zhao, S. Scott, X. Bao, *Journal of Energy Chemistry*, 2013, 22, 1-20.
3. Y. Shu, D. Ma, L. Xu, Y. Xu, X. Bao, *Catalysis Letters*, 2000, 70, 67-73.
4. D. Zhou, S. Zuo, S. Xing, *J. Phys. Chem.* C 2012, 116, 4060-4070.
5. J. Gao, Y. Zheng, G. B. Fitzgerald, J. D. Joannis, Y. Tang, I. E. Wachs, S. G. Podkolzin, *J. Phys. Chem.* C 2014, 118, 4670-4679.
6. J. Gao, Y. Zheng, J. M. Jehng, Y. Tang, I. E. Wachs, S. G. Podkolzin, *Science*, 2015, 348, 62-35.

Acknowledgement

We would like to acknowledge Gas Authority of India Limited (GAIL) for financial support

490990 CO_2 Activation By Methane in a Dual-Bed Configuration Via Methane Cracking and Iron Oxide Lattice Oxygen Transport – Concept and Materials Development

Abstract:

In this work, we explore a novel process configuration for CO_2 activation into CO by methane in a dual-bed reactor configuration by circulating supported iron oxide/iron particles with deposited carbon between two reactors. Iron was chosen as the active phase due to its ability to effectively crack methane at relevant temperatures, its favorable properties for lattice oxygen transport between the two reactors, and its low cost. In reactor 1 Fe_3O_4 is reduced to metallic iron by H_2 that is produced by catalytic cracking of methane fed to the bottom of the reactor. The reduced iron particles and the carbon deposited on them from methane cracking are then introduced to reactor 2, to which CO_2 is fed. There, the carbon is gasified by the reverse Boudouard reaction and the iron is reoxidized back to Fe_3O_4, thereby activating the CO_2 into CO. The merit of cracking the methane in reactor 1 instead of fully combusting it lies in the concentration of the entire carbon fed to the system in the product gas stream from reactor 2. This further allows oxidation of some hydrogen in reactor 1 with air to provide heat for this otherwise highly endothermic process without generating any CO_2 that is diluted with nitrogen from air.

To demonstrate the feasibility of this concept, we pursued a two-fold approach. First, the theoretical potential of the process was evaluated by thermodynamic equilibrium calculations. For both reactors, moving bed configurations with counter-current gas-solid flow and well-mixed fluidized bed configurations were investigated. Moving beds were approximated with a multi-stage equilibrium model, whereas fluidized beds were modeled as CSTRs, with two separate stages for reactor 1 (a cracking and an iron oxide reducing stage, respectively). Results indicated that moving beds offer significant advantages over well-mixed fluidized beds for both reactors. With both reactors realized as counter-current moving bed reactors, about 2.7 mol CO_2 per mol of CH_4 could be potentially activated while providing a product stream of about 82 mol-% CO, with the remainder being CO_2, at a moderate reactor temperature of 850°C.

Second, suitable materials for this process were developed. To prevent sintering of the iron/iron oxide and to facilitate fast methane cracking, reduction and oxidation kinetics, it needs to be combined with a suitable support material. Several potential support materials were investigated based on their electronic, ionic and protonic conduction properties, and cermet materials with 30 wt-% Fe_2O_3 and 70 wt-% support were produced. The reduction kinetics with H_2, the kinetics for methane cracking and the reoxidation kinetics with CO_2 were then evaluated in a TGA in a cyclic manner. Materials were further characterized by X-ray diffraction and SEM. Iron oxide reduction kinetics were fastest with $BaZr_{0.9}Y_{0.1}O_{3-\delta}$ and $SrFe_{0.5}Ti_{0.5}O_{3-\delta}$ as support material. $BaZr_{0.9}Y_{0.1}O_{3-\delta}$ support also exhibited favorable properties for methane cracking, while $SrFe_{0.5}Ti_{0.5}O_{3-\delta}$ proved to be an ineffective support for methane cracking.

In conclusion, our results indicate the high potential of this process for CO_2 activation as evidenced by the high achievable ratio of CO_2 activated per mol of CH_4 in a dual moving-bed configuration. Iron/Iron oxide supported on Zr-based ceramics, in particular $BaZr_{0.9}Y_{0.1}O_{3-\delta}$, exhibited highly promising properties for this application.

504631 Development and Demonstration of Advanced Supply Chain Equipment for Harvest, Delivery, and Processing of Herbaceous Biomass

Abstract:

Results to date and future plans from several multi-year cooperatively-funded industry/government development and demonstration projects will be presented. These projects have been co-funded by the U.S. Department of Energyâ€™s Bioenergy Technologies Office to develop and demonstrate innovative new equipment and methods for round- and square-baled herbaceous biomass, and to develop and demonstrate new methods for sustainably building and harvesting future herbaceous biomass supplies. A primary project objective is to demonstrate an advanced biomass supply chain for delivering high impact, high quality feedstocks from the field to the throat of a biorefinery. Photos, video, and measured performance results from the projectâ€™s demonstrations will be provided. Among the developments discussed will be: a new header that will enable efficient variable-rate sustainable harvest of corn stover; a new header enabling a 2x to 3x reduction of field passes for baling equipment; a self-propelled baler operating with crop-adjustable headers and a 6-pack bale accumulator; a bale gathering vehicle capable of collecting 36 bales per trip to the field, self-loading and unloading bale trailers that can reduce load/unload times to 5 minutes per truckload; new automated receiving, handling, pre-processing equipment for round and square bales for biorefinery or biomass depot operations; a near-infrared bale probe that enables instantaneous determination of moisture, ash, glucan, and xylan content in the field or a process line; and harvest and processing operations to help reduce dirt content in biomass delivered to the process. Dirt content in delivered corn stover has been a significant challenge for initial commercial-scale biorefineries. Harvest operations and results from corn stover harvests from one to over two tons per acre, and from switchgrass harvests of up to eight tons per acre will be presented and discussed. Preliminary results of pre-processing testing at Idaho National Labâ€™s Process Demonstration Unit will also be presented and discussed.

504343 Feedstock-Driven Operational Challenges Facing Pioneer Biorefineries in the Emerging Biofuel Industry

Abstract:

Recent evidence indicates that the operational performance of DOE's 1st-plant Integrated Biorefineries (IBRs) are significantly impacted by feedstock variability. This is a high impact problem because several have already failed due in part to these unexpected operational issues. The remaining IBRs are working to solve the issues, however, solving one invariably creates other issues because of interactions across the system.

Conventional feedstock supply systems (CFSS) provide biomass to existing markets that have less stringent quality requirements than do biorefineries. They are vertically integrated, designed around limited markets, single biomass types, single biorefineries and a limited supply radius. In contrast, an Advanced Feedstock Supply System (AFSS) dynamically would provide a "conversion-ready" engineered feedstock from a myriad of diverse raw biomass sources, decoupling the supply of this engineered feedstock from the biorefinery infeed. "Conversion-ready" implies that the engineered feedstock meets all requirements for the conversion process, including physical requirements such as flowability and particle size distribution, as well as more conventional specifications such as convertible composition, convertibility and ash content.

Typically CFSS and AFSS are compared on an nth-plant delivered feedstock cost, which ignores the operational impacts of constant variation in moisture, ash and composition of the biomass entering preprocessing for introduction to the conversion process. A discrete event simulation model was constructed to examine the impacts of reduced throughput and equipment downtime on feed handling and preprocessing systems, replicating the operations and flow of material to the reactor throat of the pretreatment reactor of an integrated corn stover-to-ethanol biorefinery. Results indicate that neither CFSS nor AFSS were capable of achieving design capacity, due primarily to moisture and fines generated during processing. A significant result was the identification of the plug screw feeder to the pretreatment reactor as the primary cause of down time in the modeled pioneer biorefinery scenario, specifically as a result of fines (60-70% of total down time). Moisture-caused failure of grinders far exceeded any other failure, but accounted for only 20-30% of total down time. Additionally, decoupling the feedstock supply from the reactor throat with a five day feedstock buffer only marginally improved minimum scale throughputs for both CFSS and AFSS.

Perennial solutions to annual problems?
Emily Heaton[1], Elke Brandes[2], Gabe McNunn[1,3], Mauricio Tejera[1], Andy VanLoocke[1], and Lisa Schulte[4]
[1]Dept. of Agronomy, Iowa State University, Ames, IA, USA
[2]Thüenen Institute, Braunschwieg, Germany
[3]AgSolver, Inc., Ames, IA, USA
[4]Dept. of Natural Resources, Ecology and Management, Iowa State University, Ames, IA, USA
Photo: Jim Meade

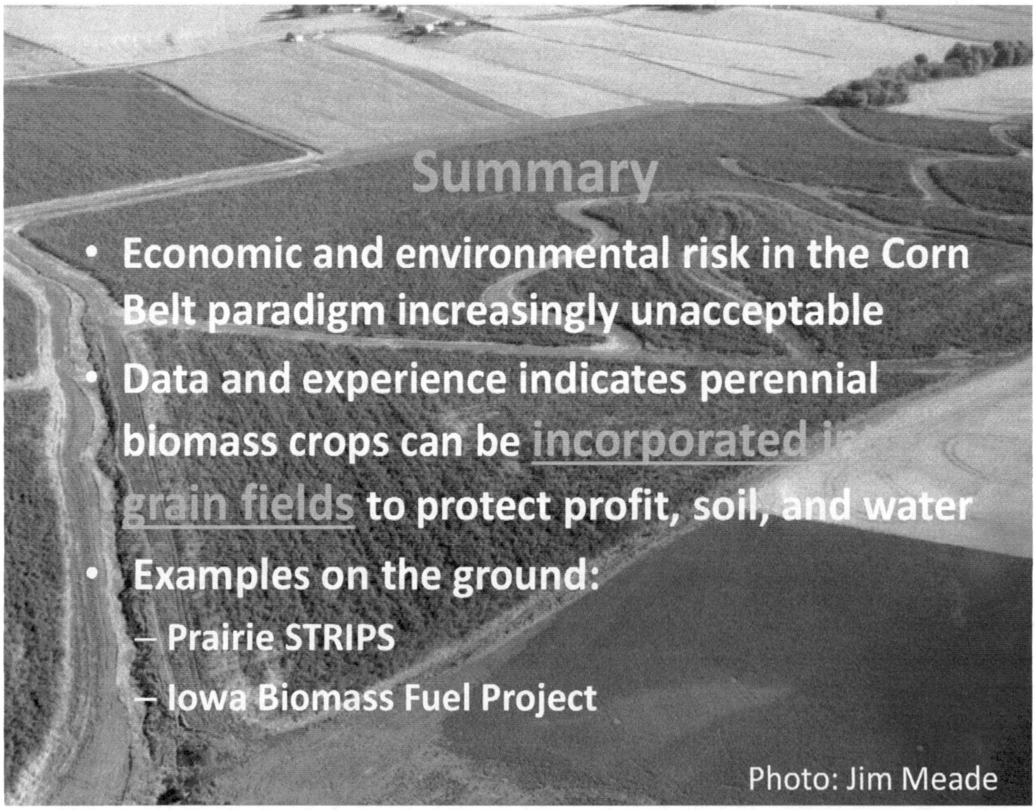
Summary
• Economic and environmental risk in the Corn Belt paradigm increasingly unacceptable
• Data and experience indicates perennial biomass crops can be incorporated into grain fields to protect profit, soil, and water
• Examples on the ground:
– Prairie STRIPS
– Iowa Biomass Fuel Project
Photo: Jim Meade

HARVEST PUBLIC MEDIA

Covering Our Food System From The Heartland

KCUR 89.3 · KUNC · NET · KBIA · IOWA PUBLIC RADIO

Donate

Home · Food · Energy · In The Field · In Depth · Video · About · Search

Farm Economy Downturn Prompts Fears Of A 'Crisis'

By PEGGY LOWE • JUL 23, 2017

Twitter
Facebook
Google+
Email

Image: Sac County, Tim Youngquist

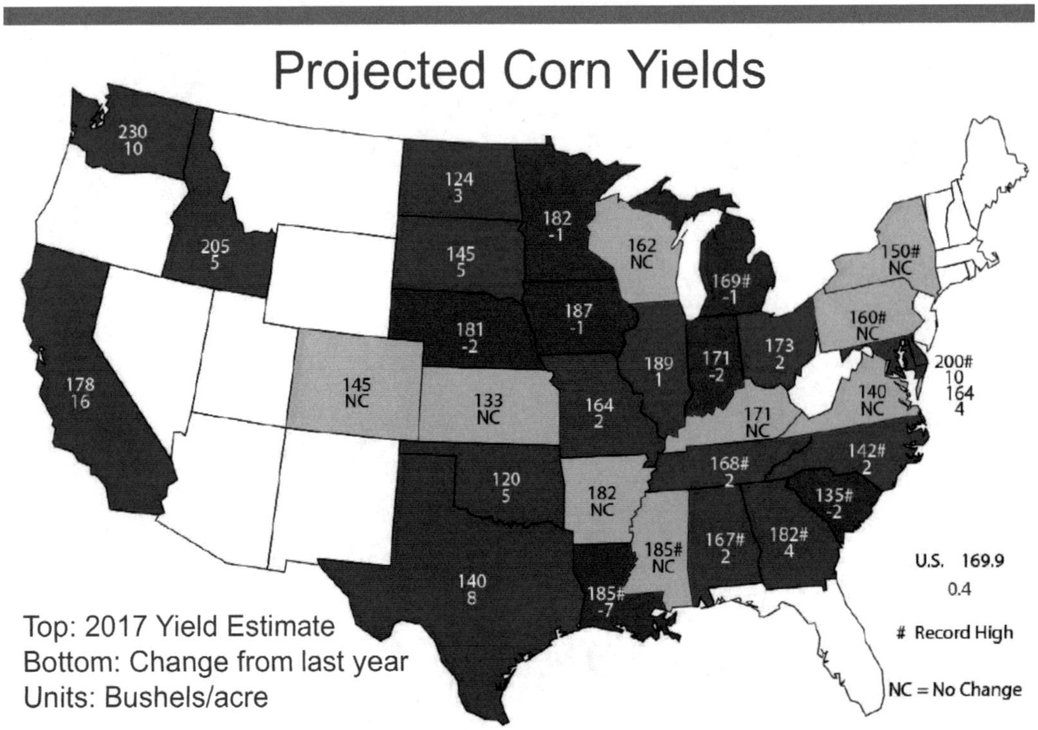

Source: USDA-NASS; slide courtesy of Chad Hart

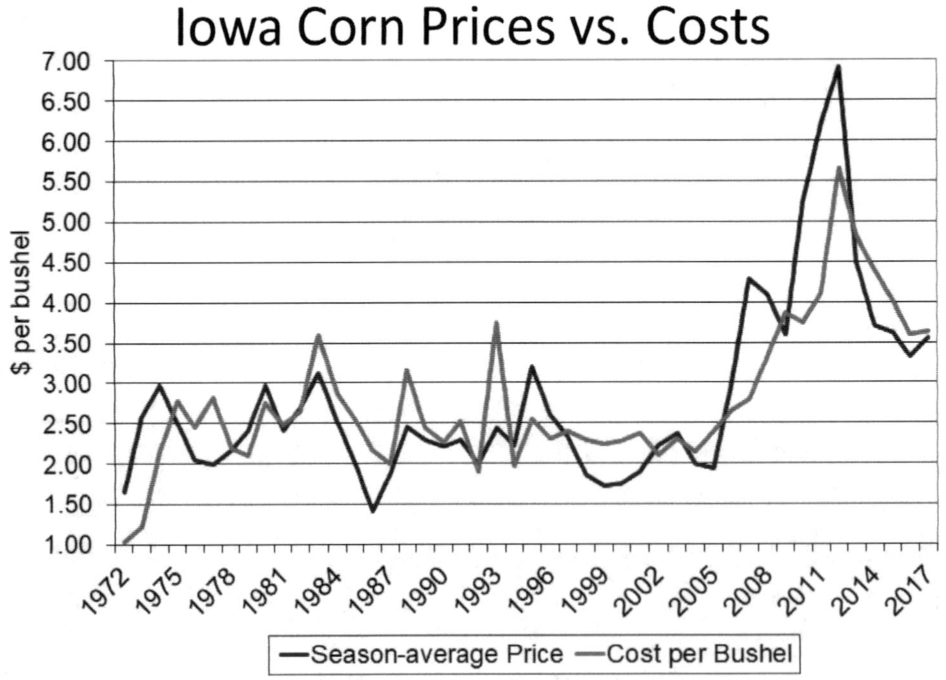

Source: ISU Ag Decision Maker; slide courtesy of Chad Hart

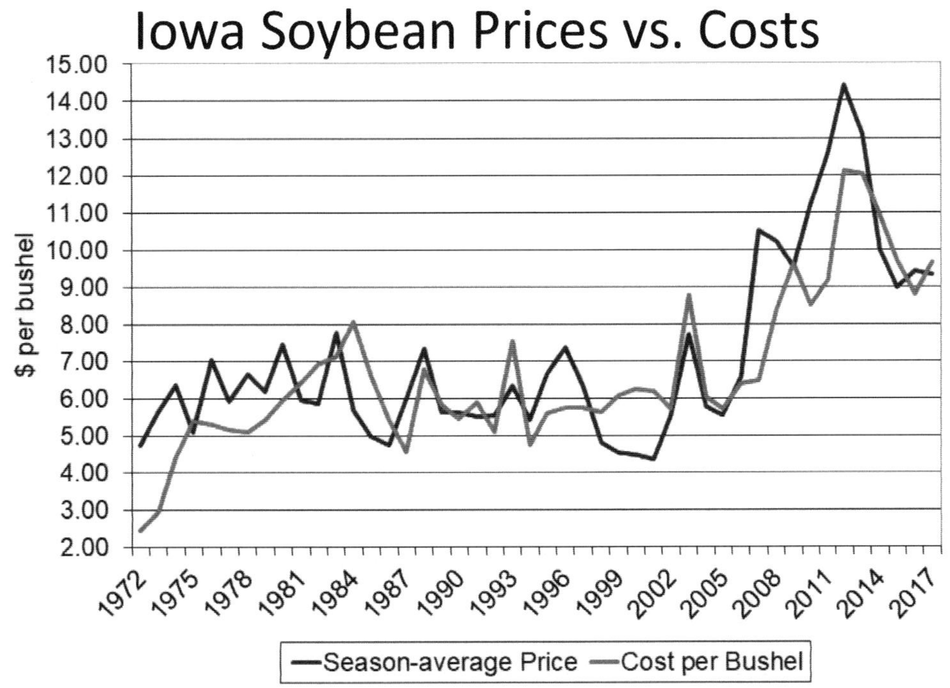

Source: ISU Ag Decision Maker; slide courtesy of Chad Hart

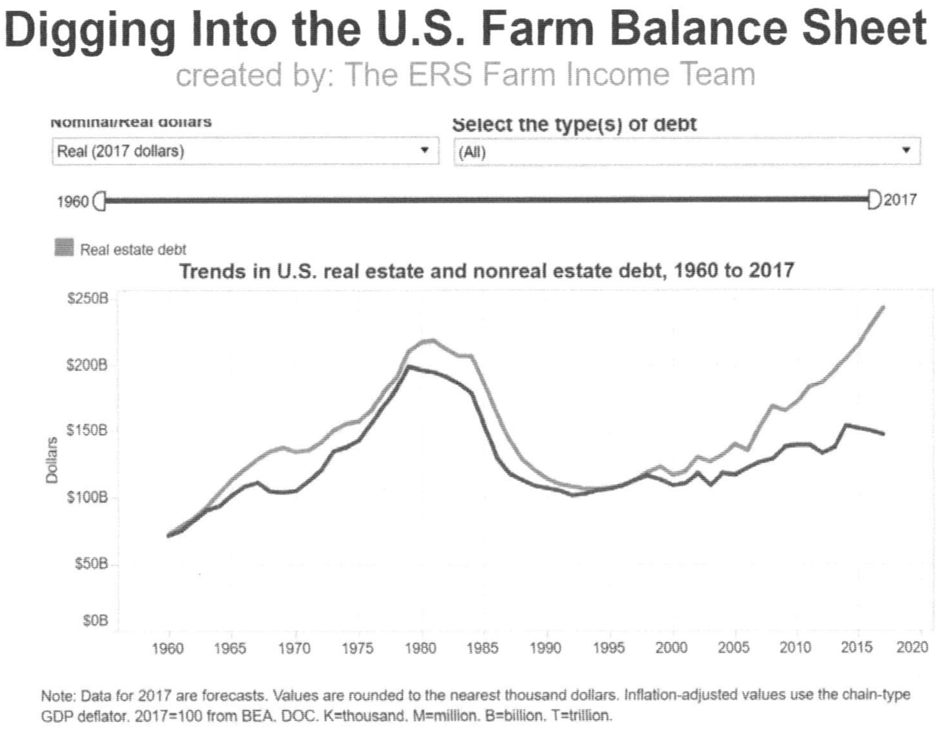

https://www.ers.usda.gov/data-products/farm-income-and-wealth-statistics

Image: Boone Co.; Todd Ontl

Image: Story Co., Drake Larsen

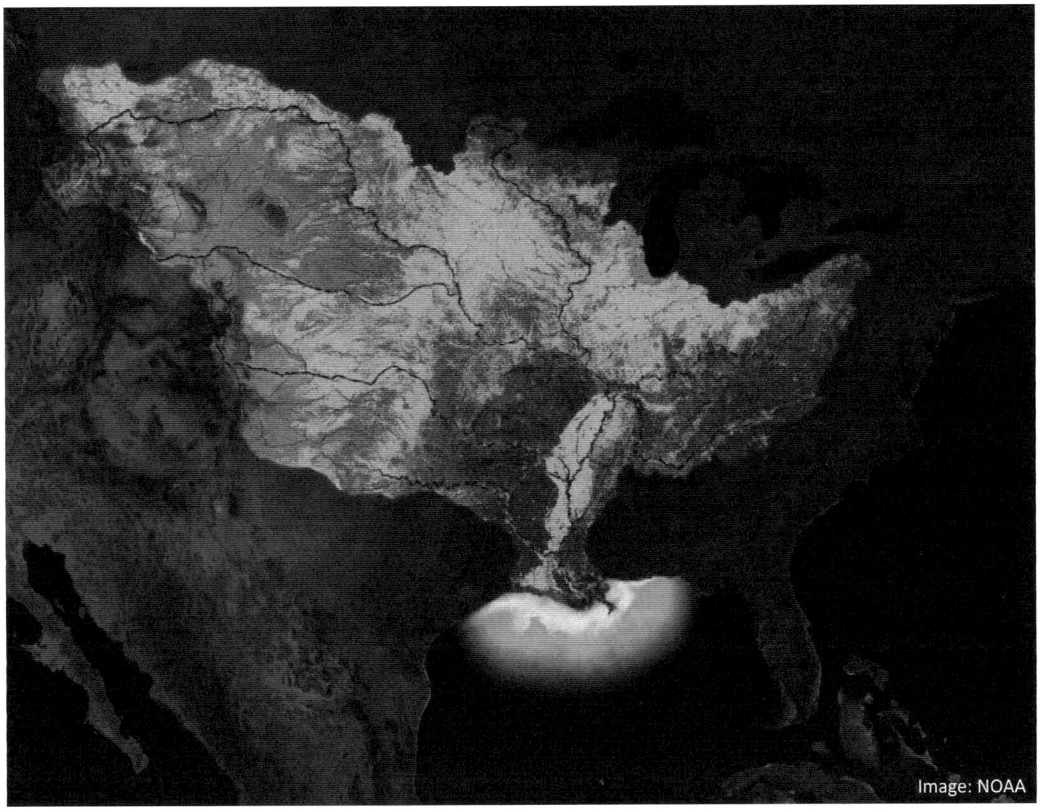
Image: NOAA

Image: Mark Licht

Precision Business Planning Workflow – Alternative Systems as a Business Decision

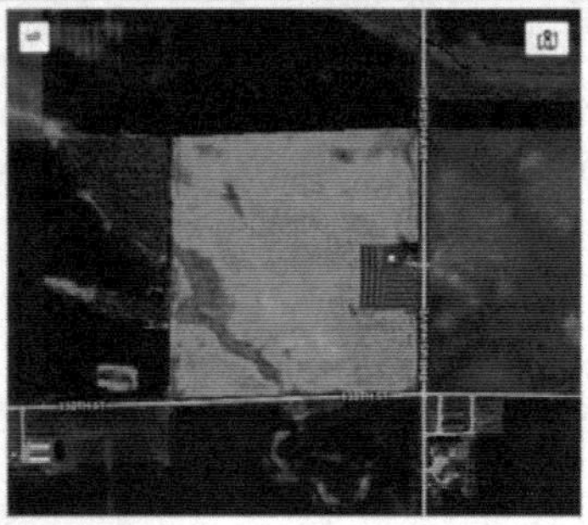

Scenario: Actual Production

Parameter	Value
Field Acreage	143.3 ac
Average Yield	170.2 bu/ac
Profit	$49.63/acre
ROI	6.2 %
Production Efficiency	212.4 bu/$1000
Acreage Opportunity Ratio	23 %
Working Capital Opportunity	$25,973.83
Total Field Expenses	$114,800.50
Total Field Revenue	$121,912.06
Total Field Profit	$7,111.56

Scenario: Conservation-Final

Parameter	Value
Field Acreage	143.3 ac
Average Yield	179.2 bu/ac
Profit	$93.85/acre
ROI	12.6 %
Production Efficiency	239.7 bu/$1000
Acreage Opportunity Ratio	22 %
Working Capital Opportunity	$19,494.23
Total Field Expenses	$107,085.95
Total Field Revenue	$120,534.99
Total Field Profit	$13,449.04

turning data into decisions for agriculture™

Large Scale Project: What is the economic story?

Parameter	Total
number of fields	3,796
total acres	207,937
total acre-years analyzed	641,998
average years of data per field	3.1
average acres per field	54.8
total profit	$ 5,703,472
per field total profit (all years)	$ 1,502
average per acre profit (all years)	$ 27.43
average max annual per acre profit	$ 128.23
average min annual per acre profit	$ (147.78)
total revenue	$ 405,192,182
total expenses	$ 399,488,710
total annualized ROI	0.46%

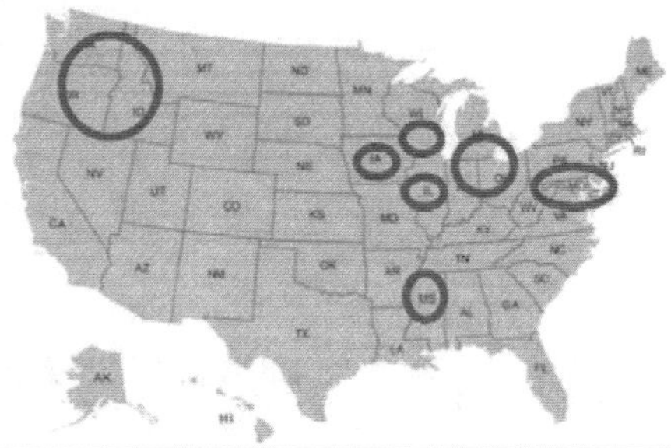

turning data into decisions for agriculture™

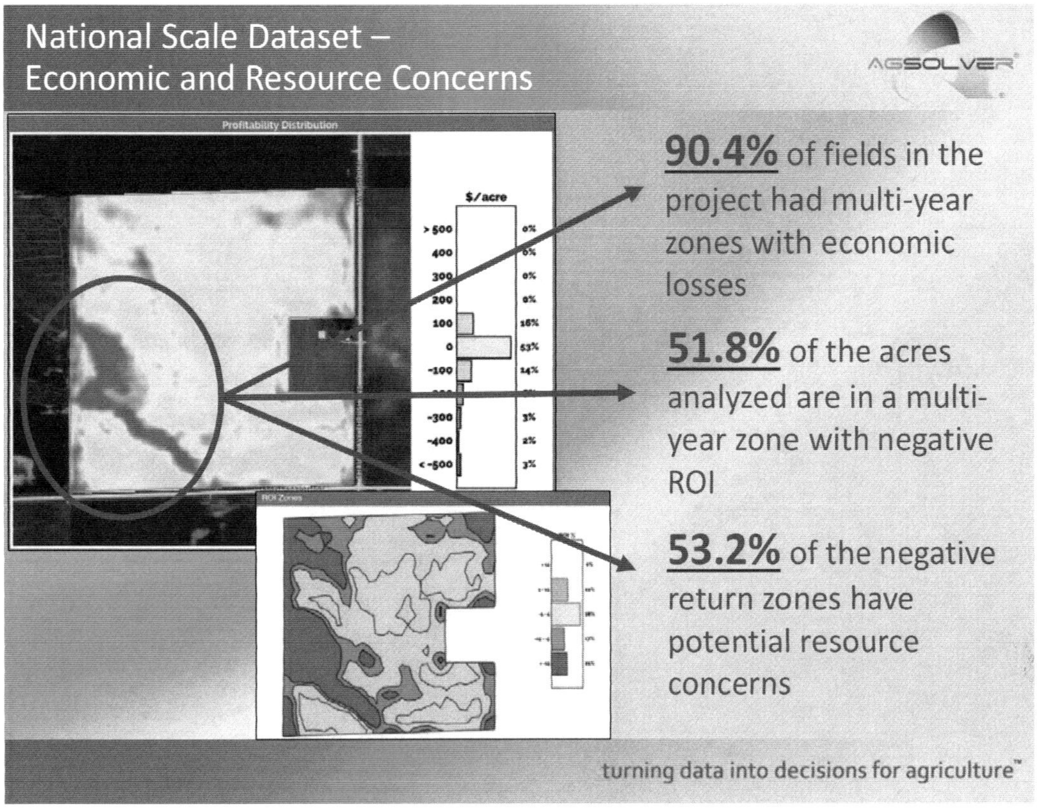

Is there an economic rational to grow

something other

than

corn or soybean in Iowa?

Modeling Framework

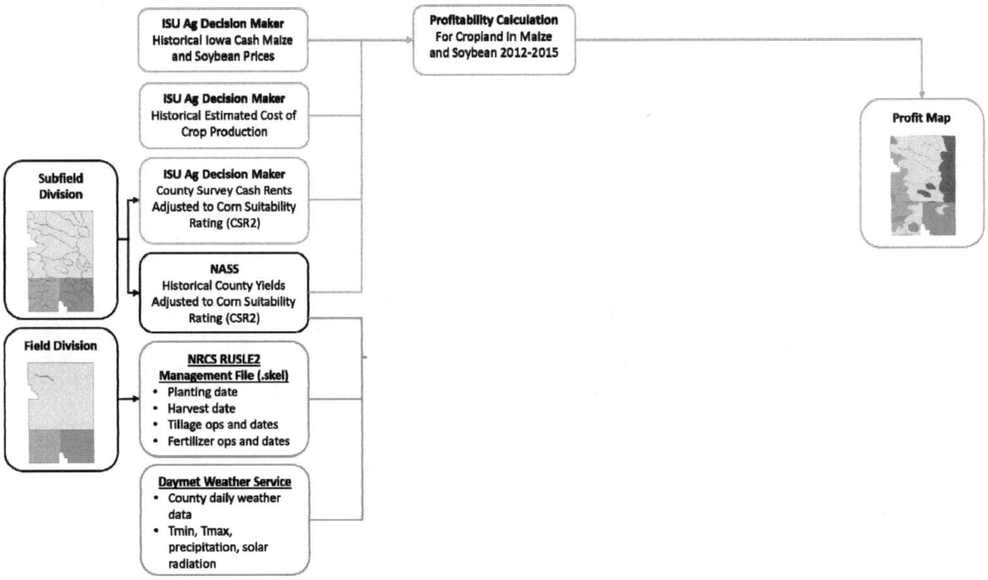

Source: Brandes et al. 2017 Global Change Biology Bioenergy

Corn and soy profitability is down in Iowa.

Some parts of fields are less profitable than others.

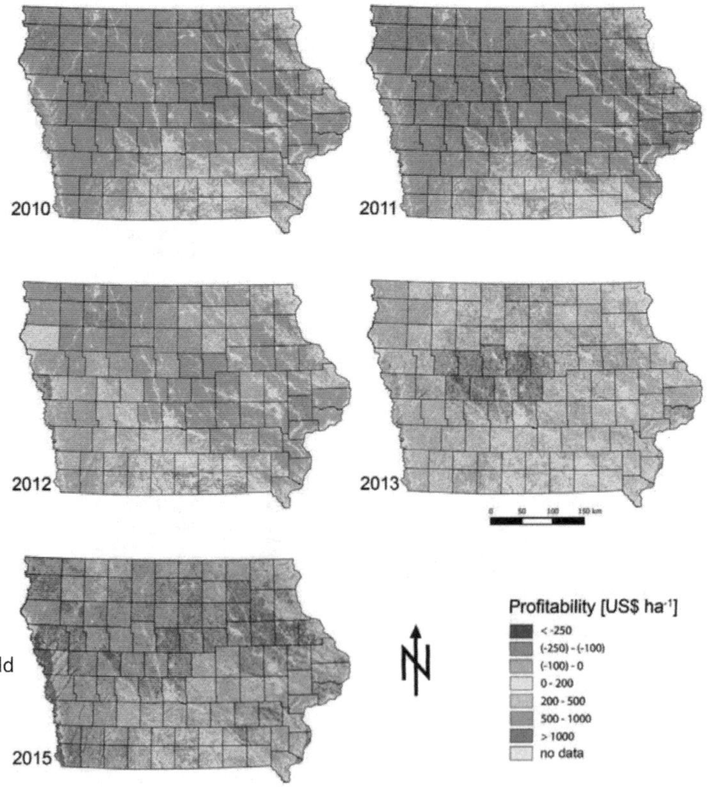

Brandes E, McNunn GS, Schulte LA, Bonner IJ, Muth DJ, Babcock BA, Sharma B, Heaton EA (2016) Subfield profitability analysis reveals an economic case for cropland diversification. *Environmental Research Letters*, **11**, 014009.

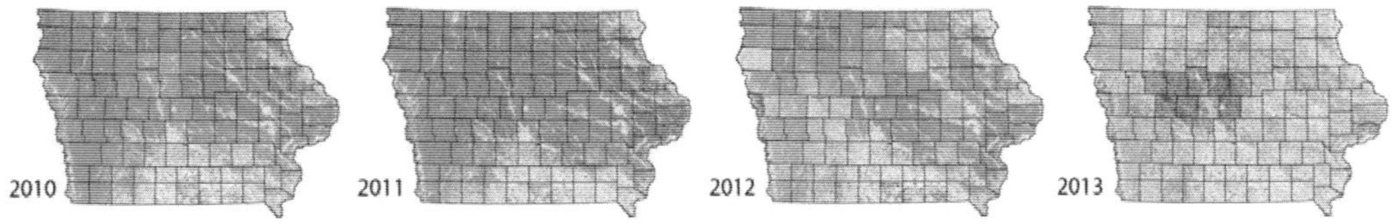

Zoom in to learn more at:
http://mesonet.agron.iastate.edu/GIS/apps/profit/

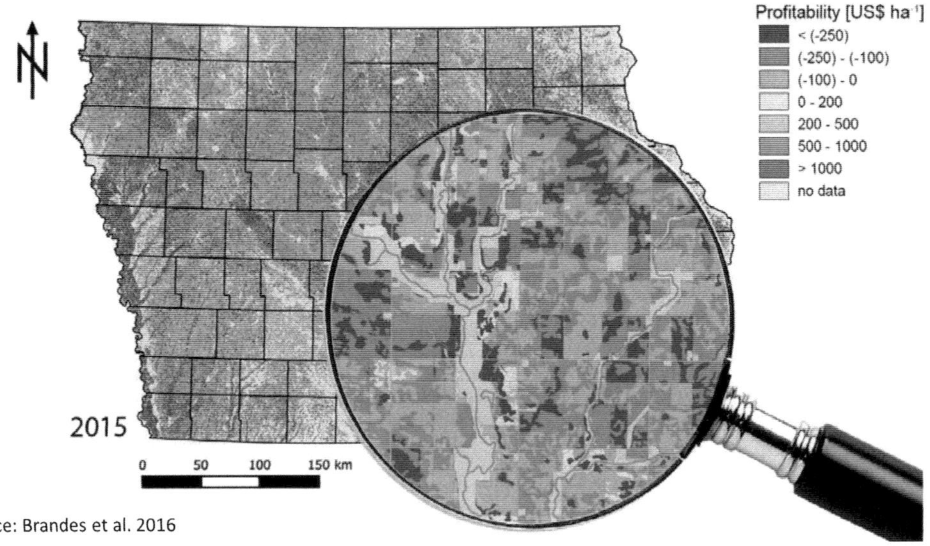

Source: Brandes et al. 2016

Corn/Soy Profitability (avg. 2012-2015)

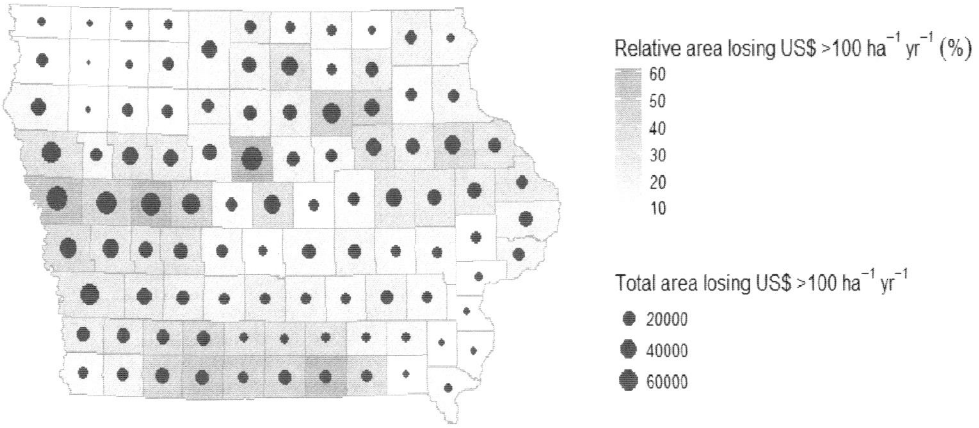

Source: Brandes et al. 2017, in press

What is the impact of replacing
annual crops...

with a high yielding
perennial crop?

Why switchgrass?

1. Year-long activity & potentially cover
2. Deeper, abundant roots
3. 'Liberty' variety trialed nationally by USDA ARS and NIFA CenUSA CAP

Modeling Framework

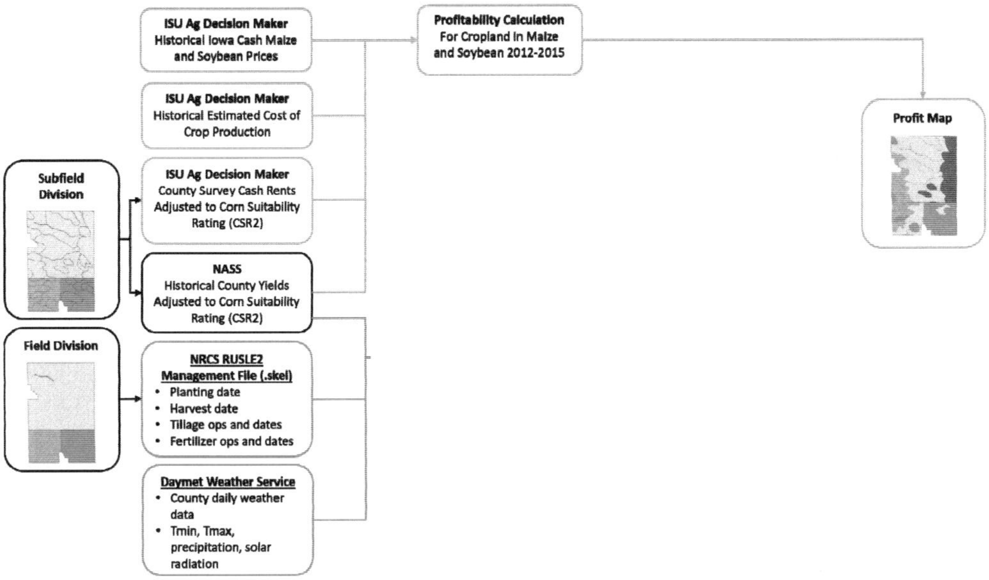

Source: Brandes et al. 2016

3 scenarios

Baseline:

Cropland in corn and/or soybeans between 2012 and 2015

Conservative:

Converting the most unprofitable cropland leaching the most nitrate to switchgrass

Nutrient reduction:

Converting unprofitable cropland with moderate nitrate leaching potential to switchgrass

Baseline nitrogen and profit loss in corn/soy fields

Subfield-scale profitability on land continuously in corn/soy during 2012-2015.

Brandes et al., (2016) *Env. Res. Let.*

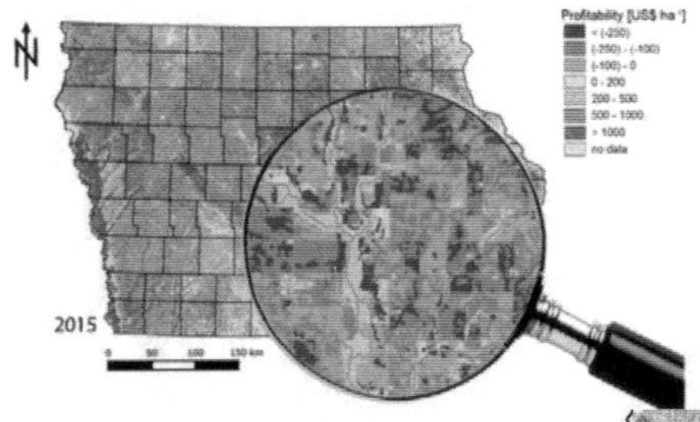

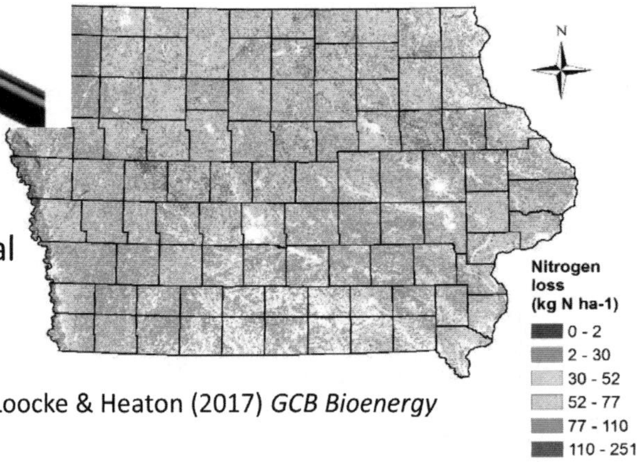

Subfield-scale NO_3-N leaching on corn/soy cropland. Values are annual rates averaged over the years 2012-2015.

Brandes, McNunn, Schulte, Muth, VanLoocke & Heaton (2017) *GCB Bioenergy*

3 scenarios

Baseline:

Cropland in corn and/or soybeans between 2012 and 2015

Conservative:

Converting the most unprofitable cropland leaching the most nitrate to switchgrass

Nutrient reduction:

Converting unprofitable cropland with moderate nitrate leaching potential to switchgrass

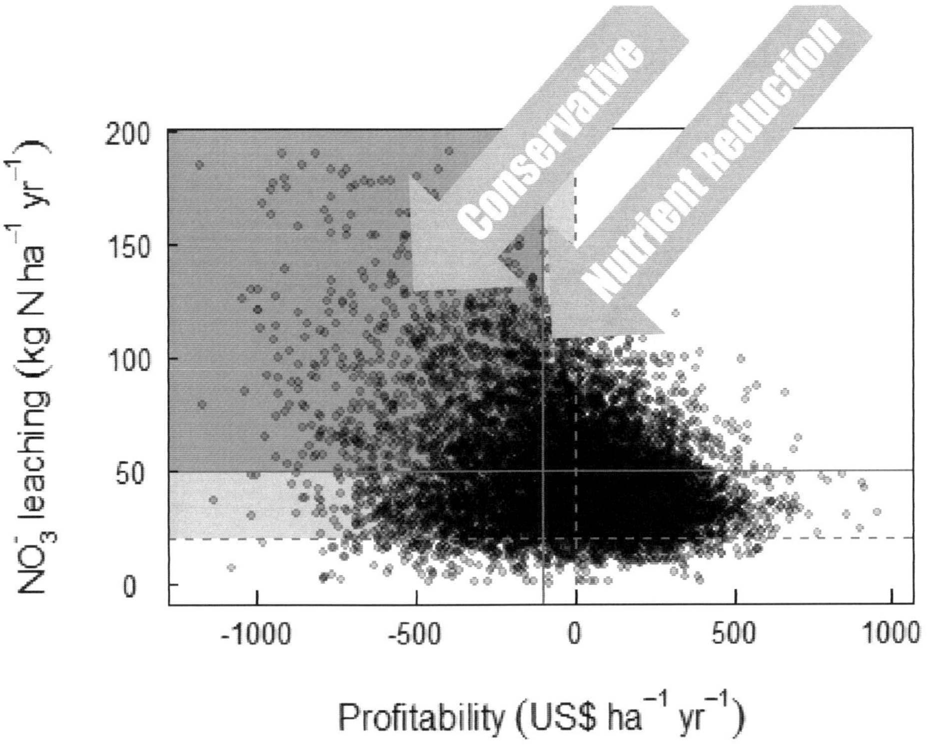

Brandes, McNunn, Schulte, Muth, VanLoocke & Heaton (*in press, GCB Bioenergy*)

Integrating perennials on unprofitable parts of fields can meaningfully increase profitability *and* retain N

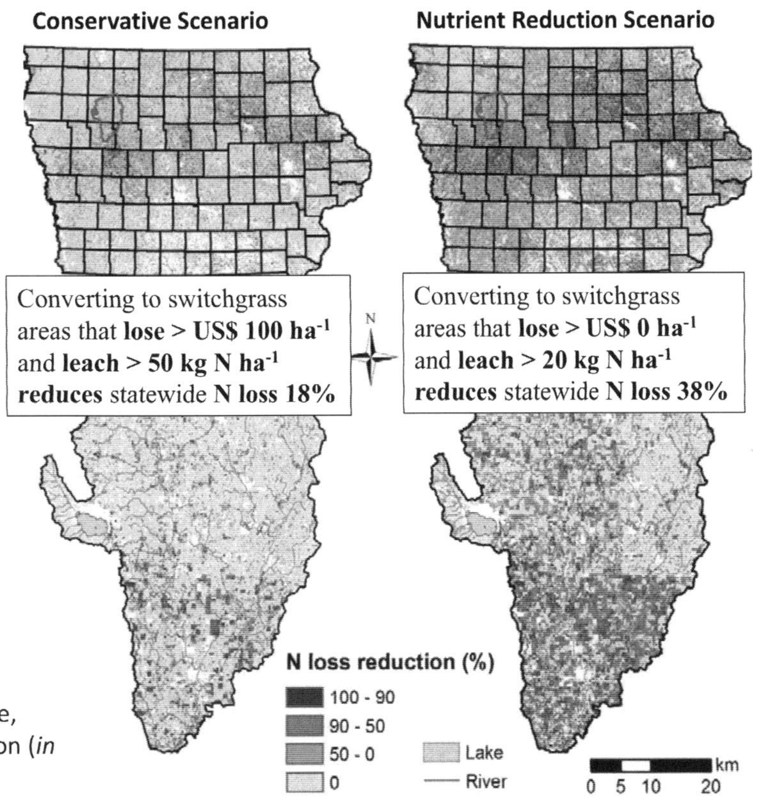

Brandes, McNunn, Schulte, Muth, VanLoocke & Heaton (*in press, GCB Bioenergy*)

Tradeoffs and Biomass

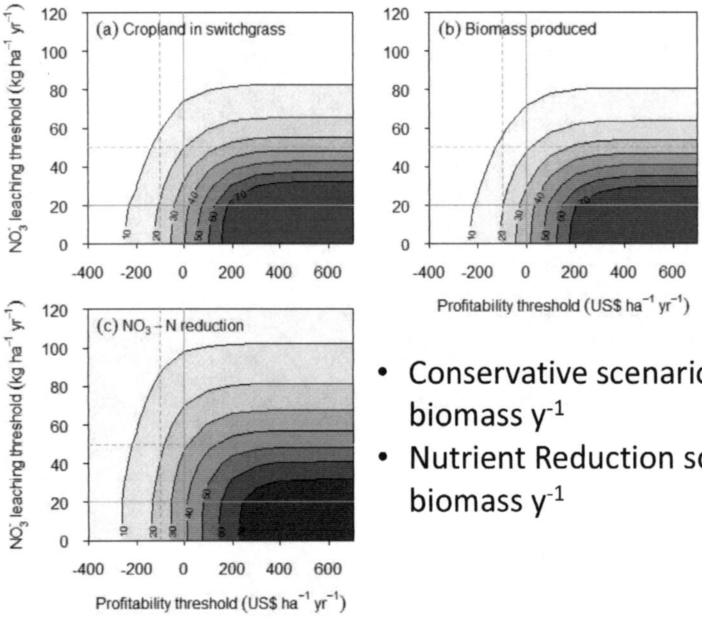

- Conservative scenario: ~10 M Mg biomass y^{-1}
- Nutrient Reduction scenario: ~30 M Mg biomass y^{-1}

Brandes, McNunn, Schulte, Muth, VanLoocke & Heaton (*in press, GCB Bioenergy*)

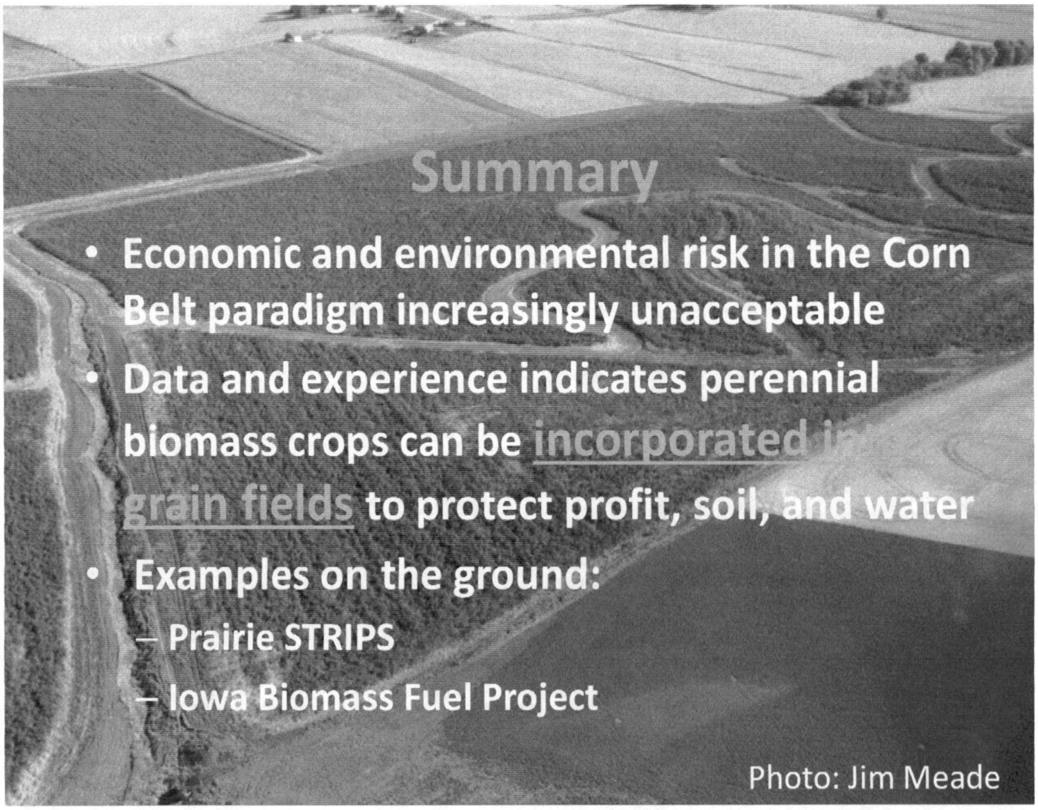

STRIPS
Prairie Strips
Collaborator
www.prairiestrips.org
Slide courtesy Lisa Schulte Moore
Photo: Tama Co., Tim Youngquist

Slide courtesy Dr. Lisa Schulte Moore
Photo: Wright Co., Lynn Betts

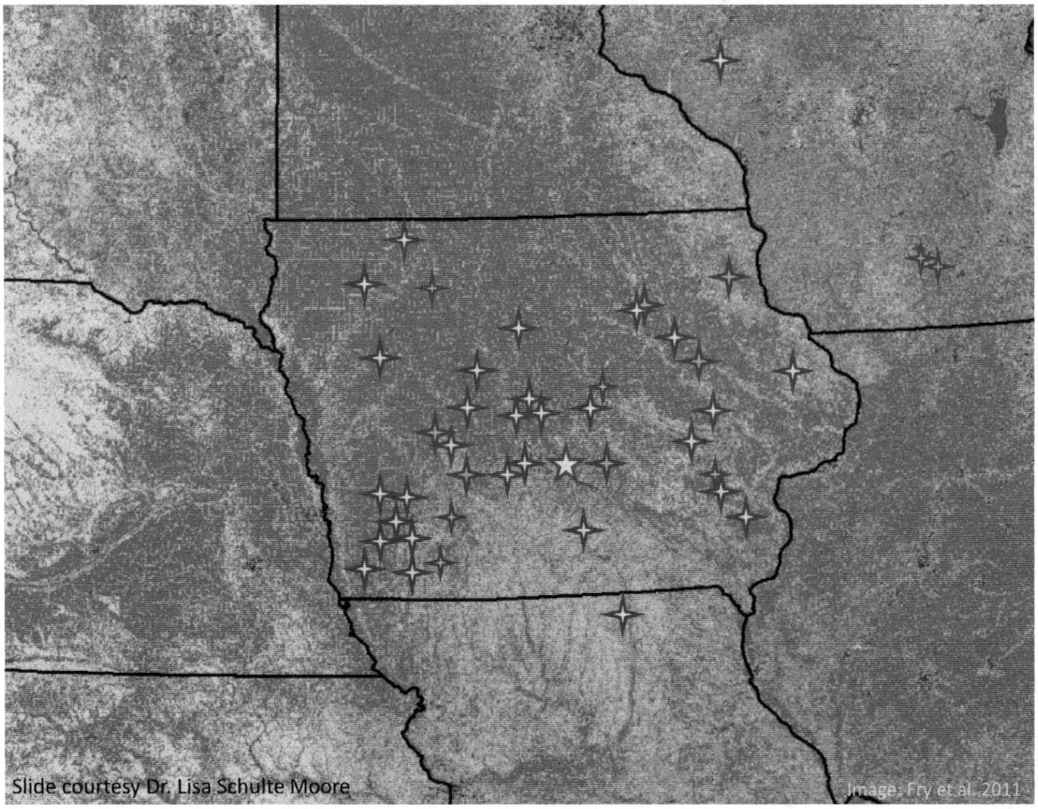

100% crops	90% crops: 10% prairie	100% prairie

Sources: Zhou et al. 2012, Helmers et al. 2012, Hernandez-Santana et al. 2013, Iqbal et al. 2014, Mitchell et al. 2014, Zhou et al. 2014
Slide courtesy Dr. Lisa Schulte Moore

Strategically adding ~10% prairie to annual row crop fields...

1. 44% reduction in water runoff
2. 95% reduction in soil loss through runoff
3. 90% reduction in phosphorus runoff
4. 84% reduction in nitrogen runoff and 70% reduction in subsurface nitrate loss (not tiled)
5. 2-3 times more beneficial insects and birds
6. No reduction in per acre yields
7. Costs less than terraces; comparable to cover crops

Source: Schulte et al. (2017) PNAS. Data collected by STRIPS team, 2007-2014 at Neal Smith National Wildlife Refuge

Slide courtesy Dr. Lisa Schulte Moore

Pigs & Prairie Goal

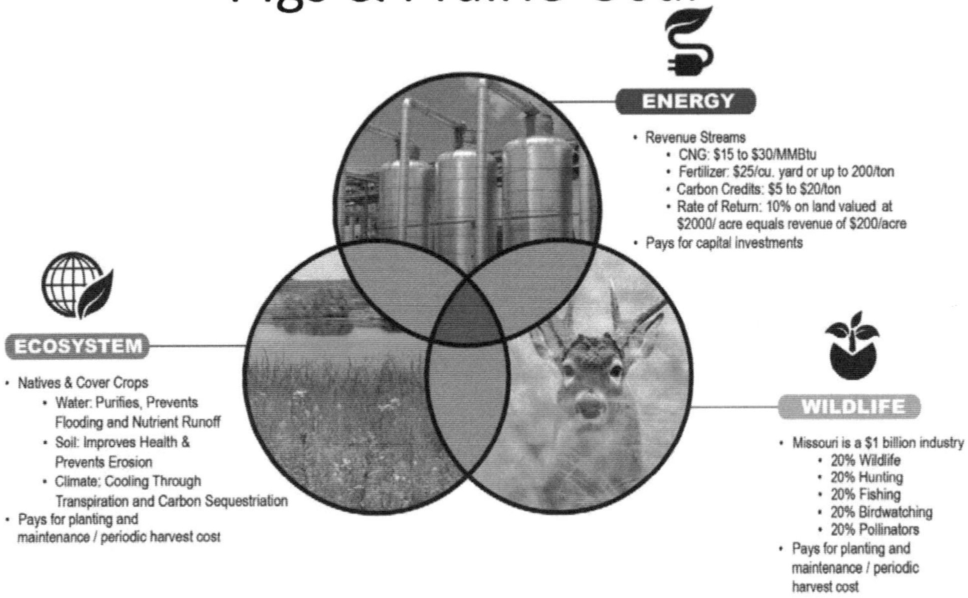

...producing pipeline-quality natural gas ultimately with the energy equivalent of 17 million gallons of diesel fuel each year.

http://sustainability.uiowa.edu/biomass

USDA
United States Department of Agriculture
National Institute of Food and Agriculture

Facilities Management
The University of Iowa

Conversion
- (processing)
- combustion
- heat & power

AGgrow TECH

Agricultural Services
- planting
- management
- harvest

Iowa State University
Department of Agronomy

Agronomy
- research
- extension
  - nitrogen, water, temperature

iowa energy center

LEOPOLD CENTER FOR SUSTAINABLE AGRICULTURE

IOWA EPSCoR

NEW HOLLAND AGRICULTURE

Current Biomass Fuels at U Iowa

- Total fuel budget = $14.6M

- Biomass demand by 2025: 110,000 t y[1]

BOILER	FUEL	CAPACITY (LB/hour)
7	Gas	140,000
8	Gas	145,000
10	Coal, Biomass, & Gas	170,000
11	Coal & Biomass	170,000

Images courtesy Ingrid Gronstal Anderson and Erin Hazen

Giant miscanthus (*Miscanthus × giganteus*)

Looking ahead:
- Densification - blend into Energy Pellets
- Optimize Storage & Logistics
- Focus on minimum time from field-to-boiler

- Crop selected & managed in close partnership with Iowa State University and AgGrow Tech (formerly Repreve Renewables)
- Perennial, grows well in marginal soil
- Displaces 4 tons coal per planted acre
- Co-fired in stoker boiler at 5% by heat (37.5% Volume)
- Over 800 acres planted since 2013
- Target 2500 acres (22,500t) = 10% of fuel mix

University of Iowa Miscanthus Business Plan Review

		Manageable Risk Profile
TECH	Miscanthus Combustibility	low
	Compatible Spec	mid
	Technology to Produce Spec	high
	Technology Development	high
OPERATIONAL	**Land**	
	Participation in Project Area	low
	Land Suitability	low
	Crop Establishment	
	Planting Stock / Material Source	low
	Planting Schedule	mid
	Planting Method	low
	Successful Crop Establishment	mid
	Annual Yield Expectations	mid
	Winterkill / Crop Loss	mid
	Invasive potential	low
	Supply Chain	
	Harvesting	mid
	Storage	low
	Collection / Transportation	low

Slide courtesy Ferman Milster

Long Term Assessment of Miscanthus Productivity and Sustainability (LAMPS)

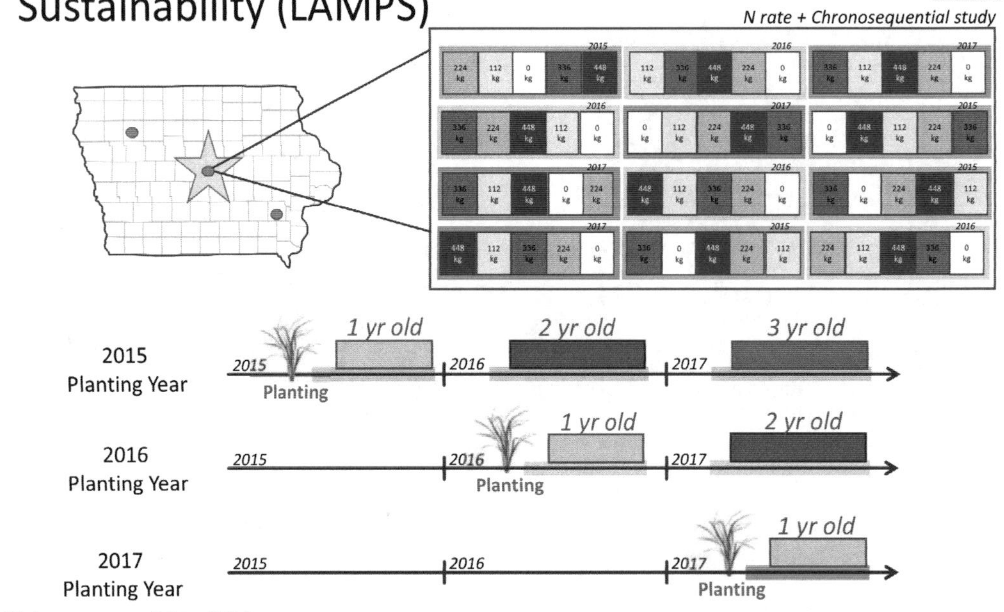

Slide courtesy Mauri Tejera

IMPACT

By burning Miscanthus instead of coal the UI will annually:

- keep **\$10 million y^{-1} in Iowa** by buying biomass instead of out-of-state coal
- Reduce **nitrate leaching and soil loss by ~90%** while building soil and providing wildlife habitat
- Provide farmers **~\$200/acre**, similar to Conservation Reserve Program (CRP payments)
- directly displace **>50,000 tons of fossil CO_2**, (>20 tons of CO_2 per acre from >4 tons of coal per acre)

References and more information at:

- http://faculty.agron.iastate.edu/heaton/
- Facebook and Twitter: @ISUBiomass
- www.prairiestrips.org
- http://sustainability.uiowa.edu/biomass
- www.agsolver.com
- www.newbio.psu.edu

Questions?
WORKMASTER 55
LAND

504321 Methanol to Propylene Process: An Integration of Catalyst Development and Reactor Engineering

Abstract:

Methanol, which can be produced from coal or natural gas, is one of the most important C1 chemicals. The utilization of methanol to make light olefins including ethylene and propylene has been recently commercialized. However, in many applications, it is desired to maximize the yield of propylene, and thus methanol to propylene (MTP) process becomes increasingly interesting. It is normally hard to achieve high selectivity to propylene in methanol conversion. In this work, we have developed a multi-functional catalyst, over which three reactions, i.e. methanol conversion, ethylene alkylation, and C4+ cracking, can be catalyzed. By integrating these three reactions, we attempted to develop a MTP process with the selectivity to propylene higher than 75%. Evaluation of this catalyst has been carried out in laboratory-scale micro-fluidized bed reactor. It shows that the selectivity to propylene of 50% has been achieved in the methanol conversion reaction. We also studied the ethylene alkylation reaction and C4+ cracking reaction, and identified the optimal operation conditions. Based on the results obtained in the micro-fluidized bed reactor, we proposed an integrated process to maximize the selectivity of propylene. In achieving this goal, a turbulent fluidized bed reactor, in which the optimal operation conditions suitable for both methanol conversion and ethylene alkylation have been considered, has been designed and constructed in our pilot plant. The reactor can be operated with 300t/a methanol feed rate. Pilot plant experiments have been conducted for validating the process concept and scaling up the reactor. It shows the selectivity to propylene of 75%, in addition to a selectivity of ethylene of 10%, can be achieved in the pilot plant. This MTP process can be seen as a successful example of the integration of catalyst development and reactor engineering.

504320 Direct Production of Value-Added Chemicals Via Fischer-Tropsch Synthesis

Abstract:

Fischer-Tropsch synthesis (FTS) is a heterogeneous catalytic process for sustainable production of fuel and value-added chemicals via syngas derived from coal, biomass and natural gas. It is considered as an attractive alternative non-petroleum-based production route. FTS is a very complicated catalytic reaction process with co-production of paraffin, olefin, alcohol and other products. The direct synthesis of high value-added olefins and higher alcohols from syngas has drawn particular attention due to due to its process simplicity, low energy consumption and clean utilization of carbon resource. For the purpose of high selectivity to specific olefins or higher alcohols, it is very desirable to regulate the reaction network for certain reactions to take place preferentially. Obviously, it is necessary to control CO activation and then both carbon chain growth and termination for high selectivity to target product by development new catalyst systems and optimization of reaction conditions.

Reaction network during syngas conversion was firstly investigated by probe molecules techniques [1]. Alcohol, aldehyde, olefin as probe molecules were in-situ introduced into the reaction system. Several parallel reactions and second reactions were clarified. The existence of surface acyl species indicated that the insertion of CO rather than hydroxyl was the underlying mechanism for higher alcohols synthesis over modified FT catalysts along with the support from surface simulation. In addition, the dual active sites for higher alcohols formation was further investigated [2]. Several kinds of dual active sites (Cu/FeC_x, CuCo, and Co/Co_2C) were controllably prepared for higher alcohol synthesis. Both Cu/FeC_x and Co/Co_2C could afford high selectivity for long chain alcohols, while only short chain alcohols were produced for CuCo. The difference of the detailed surface structure and the corresponding catalytic performance was discussed in detail.

Fischer-Tropsch to olefins (FTO) by using new active phase was also investigated. Co_2C nanoprisms with exposed facet of {101} and {020} were prepared and tested for syngas conversion [3]. It is widely accepted the formation of Co_2C is considered to be one of the main reasons for deactivation of the Co-based FT reaction because it possesses very low activity for CO hydrogenation, with methane and CO_2 as the main products. However, due to the strong facet effect of Co_2C, Co_2C nanoprisms afforded very high selectivity for the production of lower olefins (60.8 C%), while little methane (about 5.0 C%) was generated with the ratio of olefin/paraffin amongst the C_{2-4} products being as high 30. The product distribution deviates markedly from the classical ASF distribution with the highest selectivity to propylene. The effect of several promoters and the model for modified ASF distribution were discussed [4]. In addition, the nanoeffects of Co_2C were studied in details to reveal the structure-performance relationship.

References

[1] *Catalysis Science & Technology*ï¼Œ2015, 5, 4224

[2] *Catalysis Science & Technology*ï¼Œ2013, 3, 1591

[3] *Nature*, 2016, 538, 84..

[4] *ACS Catalysis*, 2017, 7, 3622.

504319 An Investigation into the Effect of Li and Mn Promotions on the Activity and Selecivities to Olefins and Alcohols of Co@Co_2C/Activated Carbon (AC) for Fischer-Tropsch Reaction

Abstract:

Abstract:

Utilization of non-precious transition metals for high alcohols and olefins synthesis is of a great importance in heterogeneous catalysis. We synthesized successfully Li and Mn promoted AC supported cobalt metal-carbide (Co@Co_2C/AC) catalysts, which present remarkable activity of more than 200g/kg-cat.h and selectivities of ca. 65 C% both for alpha-alcohols and alpha-olefins via the Fischer-Tropsch reaction at the mild conditions. The formation of the stable cobalt carbide and the Co@Co_2C interface are found essential for the observed reactivity. The results of characterization from In-situ EXAFS, in-situ XRD, HRTEM, XSP, and chemo-adsorption of H_2 and CO experiments showed that Mn promotes the formation of Co_2C species during Fischer-Tropsch reaction and Li increased the CO uptake, while both promoters have little impact on the amount of H_2 uptakes. Density functional theory calculations show that Co_2C is highly efficient for CO non-dissociative adsorption, behaving as noble-metal-like, whereas the Co metal is highly active for CO dissociative adsorption and the subsequent carbon-chain growth. The interface between the cobalt metal and its carbide phase as well as the dual sites available at the interface for facile CO insertion to C_nH_m-groups formed on the Co sites. The remarkably high alpha-alcohol and alpha-olefins selectivities from the Li and Mn promotion effects, namely the formation of more interfacial active sites between Co and Co_2C and enhancement of the ratio for chemo-adsorption CO to H_2 uptake on the surface of Co@Co_2C/AC catalysts, paves a new way for the design of the new catalysts for synthesizing various high value-products in syngas applications.

504154 Direct Production of Gasoline-Range Hydrocarbons from Carbon Dioxide over Iron-Based Multifunctional Catalysts

Abstract:

Converting CO_2 from a detrimental greenhouse gas into value-added liquid fuels not only contributes to mitigating CO_2 emissions, but also reduces dependence on petrochemicals. Unfortunately, the activation of CO_2 and its hydrogenation to hydrocarbons or other alcohols are challenging tasks because CO_2 is a fully oxidized, thermodynamically stable and chemically inert molecule. Most research to date, not surprisingly, is focusing on selective hydrogenation of CO_2 to short-chain products, while few studies on long-chain hydrocarbons, such as gasoline-range (C_5–C_{11}) hydrocarbons. The key to this process is to search for a high efficient catalyst.

Herein, we have succeeded in preparing high efficient, stable, and multifunctional Na–Fe_3O_4/Zeolite catalysts for the direct production of gasoline-range (C_5–C_{11}) hydrocarbons from CO_2 hydrogenation. This catalyst displayed record selectivity towards C_5–C_{11} hydrocarbons (78%) as well as low CH_4 and CO selectivity under industrial relevant conditions, and gasoline fraction are mainly isoparaffins and aromatics thus favouring the octane number. Moreover, the composition of C_5–C_{11} can be tuned by the choice of zeolite type and the integration manner of multifunctional catalyst. When Na–Fe_3O_4 nanocatalyst was matched with HZSM-5 zeolite and integrated by granule-mixing manner, up to 63% of aromatics in gasoline fraction were produced. While Na–Fe_3O_4 was matched with HMCM-22 zeolite, up to 60% of isoparaffins in gasoline fraction were obtained under the dual-bed configuration. Not only that, the multifunctional catalyst exhibited a remarkable stability for 1,000 h on stream, which definitely has the potential to be a promising industrial catalyst for CO_2 utilization to liquid fuels.

In-depth characterizations indicate that, during CO_2 hydrogenation reaction, this catalyst enables RWGS over Fe_3O_4 sites, olefin synthesis over Fe_5C_2 sites, and oligomerization/aromatization/isomerization over zeolite acid sites. The concerted action of the active sites calls for precise control of their structures and proximity. This study paves a new path for the synthesis of liquid fuels by utilizing CO_2 and H_2. Furthermore, it provides an important approach for dealing with the intermittency of renewable sources (sun, wind and so on) by storing energy in liquid fuels.

References

1. G. A. Olah, A. Goeppert and G. S. Prakash, *J. Org. Chem.*, 2008, **74**, 487-498.

2. R. W. Dorner, D. R. Hardy, et al., *Energy Environ. Sci.*, 2010, **3**, 884-890.

3. W. Wang, S. Wang, X. Ma and J. Gong, *Chem. Soc. Rev.*, 2011, **40**, 3703-3727.

4. F. Jiao, J. Li, X. Pan, et al., *Science*, 2016, **351**, 1065-1068.

5. P. Zhai, C. Xu, R. Gao, et al., *Angew. Chem. Int. Ed.*, 2016, **55**, 9902-9907.

6. J. Wei, Q. Ge, R.Yao, et al., *Nat. Commun.*, 2017, **8**, 15174.

Innovative Syngas Production Catalyst for Utilization of CO_2 (CT-CO_2AR™)

Takenori Kanda, Osamu Hirohata, Tomoyuki Mikuriya and Fuyuki Yagi
Chiyoda Corporation, Kanagawa, Japan

1. Introduction

Production of synthesis gas from natural gas and conversion of synthesis gas to fuels and chemicals have become increasingly attractive from the viewpoint of energy resource diversification. The technology of synthesis gas production by the reforming of natural gas with CO_2 (hereafter referred to as "CO_2-reforming") is an effective solution for the issues of efficient energy resource utilization and environmental preservation.

Chiyoda has developed and commercialized a new noble metal based reforming catalyst named CT-CO_2AR™ which has a higher resistance to carbon formation and produces synthesis gas more effectively compared with commercial reforming catalyst.

In this paper, we will elaborate on the characteristics and the competitiveness of the CT-CO_2AR™ catalyst and the results of the operation of the commercial plant.

2. Characteristics of CT-CO_2AR™ catalyst

(1) The characteristics of CO_2/steam reforming

Synthesis gas is used for hydrogen production and as the feed stock of chemicals, and the H_2/CO ratio of synthesis gas depends on the purpose.

For example, hydrogen production requires an H_2/CO ratio of more than 3.0. For ethylene glycol production, the required H_2/CO ratio is 2.0. For chemical production through oxo synthesis, the required H_2/CO ratio is 1.0. For the feedstock of acetic acid or MMA requires only CO or as low as possible H_2/CO ratio.

When hydrocarbons such as natural gas or naphtha and steam are used as reformer feedstocks, the steam reforming reaction shown in Eq. 1 is normal and the H_2/CO ratio of produced synthesis gas becomes more than 3.0.

$$CH_4 + H_2O \Leftrightarrow 3\ H_2 + CO \qquad (1)$$

However, when CO_2 and steam are used as reforming agents, the CO_2 reforming reaction shown in Eq. 2 occurs.

$$CH_4 + CO_2 \Leftrightarrow 2\ H_2 + 2\ CO \qquad (2)$$

Therefore, by changing the ratio of feed CO_2 and steam, various H_2/CO ratios of synthesis gas can be produced.

However, during the reforming reactions, the carbon formation reaction shown in Eq. 3 occurs on the catalyst surface. Deposited carbon causes the deactivation of the catalyst and the occlusion of the catalyst tube[1] leading to an unstable operation.

$$CH_4 \Leftrightarrow C(s) + 2\,H_2 \qquad (3)$$

To prevent carbon formation for a stable operation, commercial Ni based reforming catalysts require steam and CO_2 greater than the stoichiometric ratio.

Fig. 1 shows the relation between the operation condition and risk of carbon formation[2]. The bold line in the center represents carbon limit line at 600 ℃. In the black area to the left of this line, the process gas has enough high thermodynamic potential for carbon formation at 600 ℃. Operation around this line cannot be carried out with commercial Ni based reforming catalysts. They are usually operated within the excess steam carbon ratio (S/C) zone (blue area).

On the other hand, our new commercialized noble metal based reforming catalyst (CT-CO_2AR™) enables a stable operation under a lower S/C condition (red area) because it has a higher resistance to carbon formation.

In next section, the advantages of a lower S/C operation will be shown.

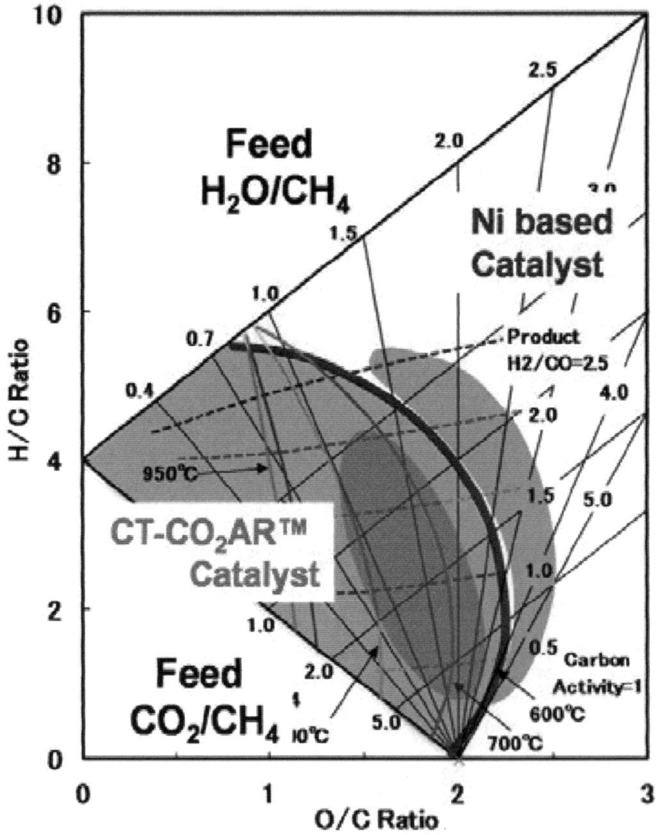

Fig. 1. Carbon Limit Diagram

(2) The advantages of a lower S/C ratio operation

In the case of the synthesis gas production process shown in Fig. 2, the advantages of a lower S/C operation are as follows:

1. Since the amount of steam and CO_2 is reduced, the heat duty to heat steam and CO_2 can be reduced. Therefore, the fuel consumption of the reformer can be reduced.
2. Fig. 3 shows the comparison of synthesis gas concentration at the reformer outlet between a high S/C (=2.2) operation and a low S/C (=1.0) operation in the case of H_2/CO=1.0
   In the case of a low S/C operation, the amount of CO_2 can be reduced and as a result, the system volume of CO_2 removal is downsized.
   In addition, because of lower CO_2/C, the capacity of the CO_2 compressor is downsized.
   Therefore, both the CAPEX and the OPEX of CO_2 removal and CO_2 compressor are reduced.
3. Fig. 3 shows that a lower S/C operation can reduce the amount of H_2O at the reformer outlet and it is possible to reduce the amount of process condensate.
   In the case of a low S/C operation, the total amount of produced steam is less than that of a high S/C operation. However, the amount of steam uses as the reforming agent in a low S/C operation is significantly less than that in a high S/C operation.
   As a result, a low S/C operation can increase the amount of export steam.

From these reasons, the CT-CO_2AR™ catalyst is preferable to a commercial Ni based catalyst.

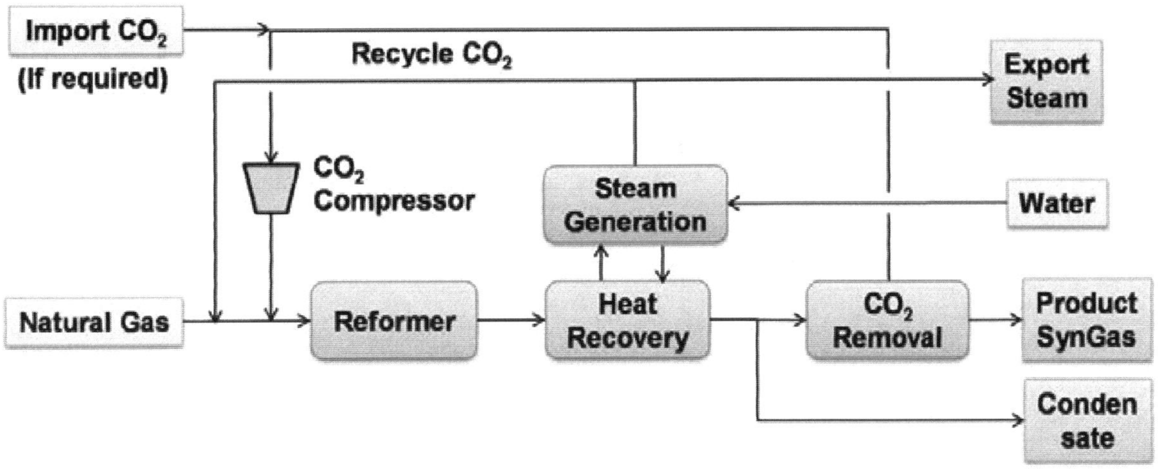

Fig. 2. Block Flow Diagram of Syngas production

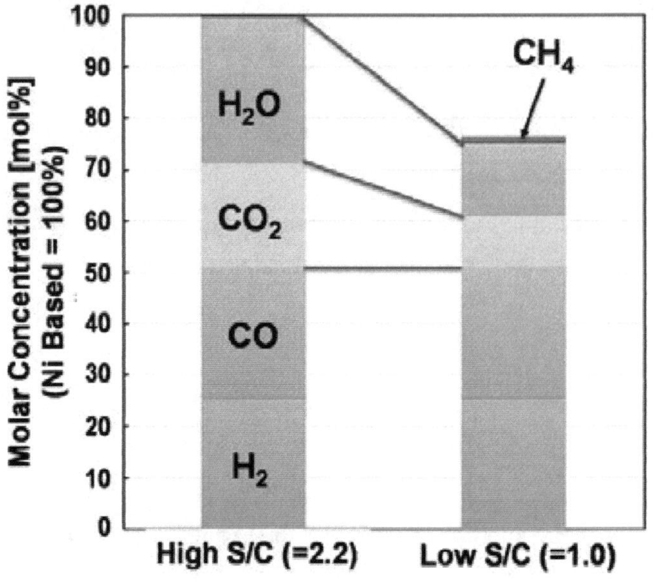

Fig. 3. Typical Example of synthesis gas composition in $H_2/CO=1.0$

(3) The characteristics of the CT-CO_2AR™ catalyst

Fig. 4 shows the result of EPMA (Electron Probe Micro Analyzer) of the CT-CO_2AR™ catalyst. Generally, the surface reaction on the reforming catalyst is the rate-determining step. Therefore, to reduce the amount of active metal, the active metal is supported on only the outer surface of the carrier like egg-shell.

Fig. 5 shows the accelerated experiment result. The CT-CO_2AR™ catalyst keeps its activity due to its high resistance to carbon formation.

We built the reaction kinetics model shown in Fig. 6 to estimate the behavior of the CT-CO_2AR™ catalyst.

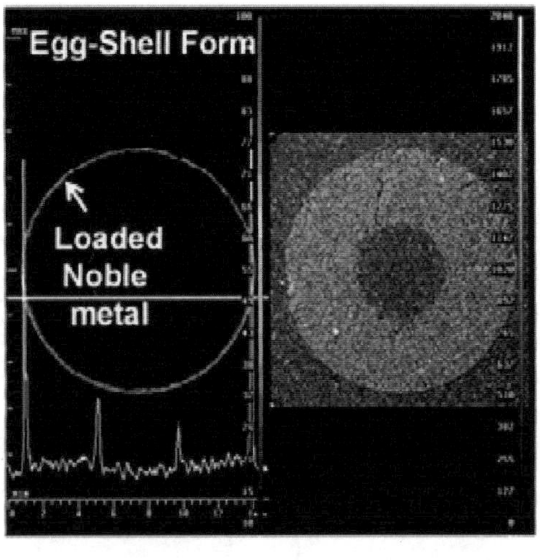

Fig. 4. EPMA result of CT-CO_2AR™ Catalyst

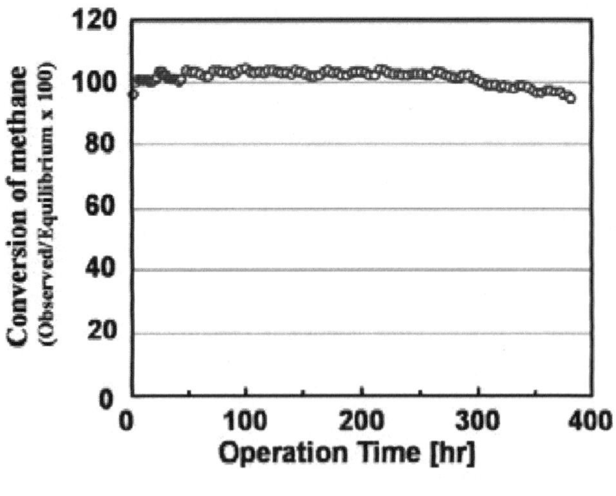

Fig. 5. Result of accelerated experiment

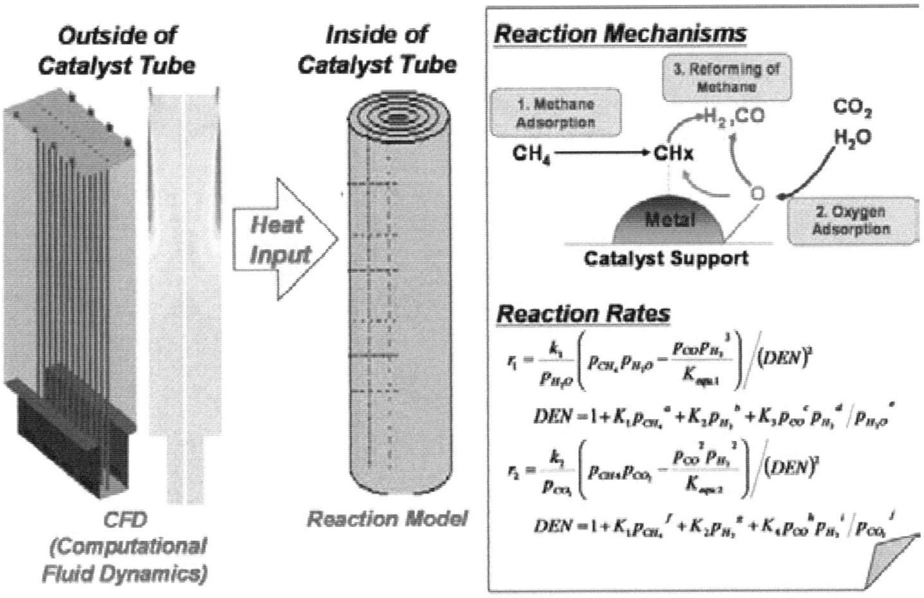

Fig. 6. Reaction model of CT-CO_2AR™ catalyst

3. Advantages of CT-CO_2AR™ catalyst

(1) Study Basis

Regarding the competitiveness of the CT-CO_2AR™ catalyst compared to a commercial Ni based reforming catalyst, we have compared the CT-CO_2AR™ catalyst and Ni based catalysts from the following three points of view based on the block flow shown in Fig. 2.

1. Total energy input (feed + fuel hydrocarbon + utility consumption)
2. Manufacturing costs of synthesis gas
3. CO_2 Emission per final products

[CO_2 Emission] = [CO_2 in Flue and Utility] - [Import CO_2] + [CO_2 in downstream process to produce the final products]

In this study we considered cases with H_2/CO ratios of 1.0 and 0.5. The conditions and estimated final product of each case are listed in Table 1.

(2) Study Result

From Fig.7-9 show the results of the process study. In the case of H_2/CO=1.0, compared to Ni based catalysts, the CT-CO_2AR™ catalyst can cut total energy input by 11% (Fig. 7) manufacturing cost by 12% (Fig. 8).Regarding CO_2 emission, the CT-CO_2AR™ catalyst can achieve a reduction of 22% compared with a Ni based catalyst (Fig. 9). In the case of H_2/CO=0.5, CT-CO_2AR™ catalyst is more competitive. Approximately 30% of both total energy input and manufacturing cost can be reduced and 76% of the amount of CO_2 emission can be reduced. From this study, the CT-CO_2AR™ catalyst is competitive compared with Ni based catalysts both in terms of CAPEX and OPEX.

Table 1. Process study conditions

H_2/CO Ratio	Catalyst	Feed S/C	Amount of (H_2+CO)	Final Products	Capacity
[-]	[-]	[-]	[Nm3/h]	[-]	[ton/hr]
CO (0.5)	Ni Based	3.3	37,500 (25,000 as CO)	Acetic Acid	500,000
	CT-CO_2AR™	0.4			
1.0	Ni Based	2.2	25,000	n-butanol (oxo synthesis)	270,000
	CT-CO_2AR™	1.0			

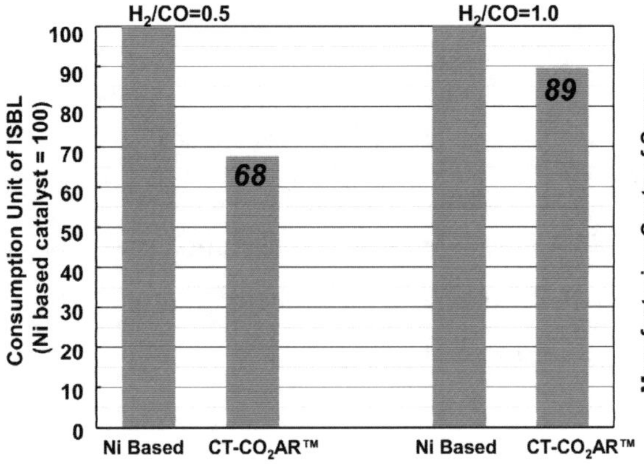

Fig. 7. Total Energy Input

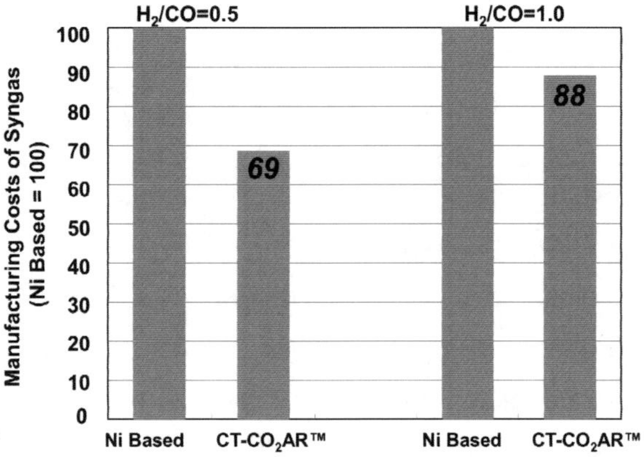

Fig. 8. Manufacturing Costs of Syngas

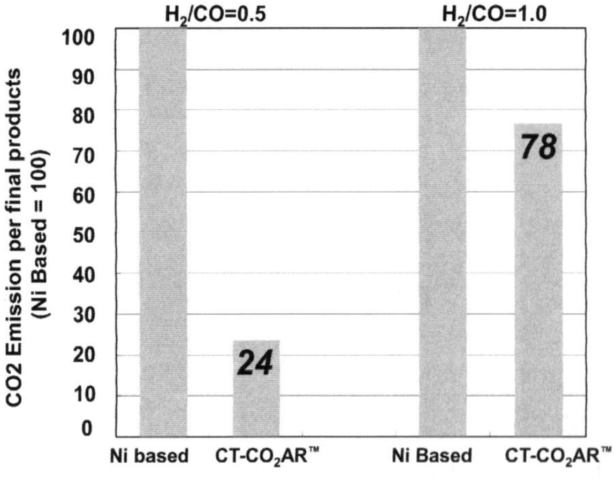

Fig. 9. CO_2 Emission per final products

4. Application to Commercial Plant

The CT-CO_2AR™ catalyst has been operated in a commercial plant in Japan since 2014. The information of the commercial plant is listed in Table 2. In this project, the reforming catalyst has been changed to the CT-CO_2AR™ catalyst and the S/C ratio at the inlet of the prereformer has reduced from 2.2 to 1.4 without changing the amount and specification of the produced synthesis gas.
Regarding the comparison between analysis result using the reaction model and the operation result, a pressure drop through the reformer tube is shown in Fig. 10 and the methane concentration of the synthesis gas is shown in Fig.11.
Since a significant increase in the pressure drop and methane concentration was not observed, a stable operation was achieved during the whole operation time.
In addition, our reaction model can accurately predict the performance of the CT-CO_2AR™ catalyst.

Table 2. Information about the commercial plant

Client's Name :	KH Neochem Co., Ltd
Plant Location :	Yokkaichi, Japan
Product :	Synthesis gas H_2/CO ratio=1.0
Product capacity :	5,000 Nm^3/hr (as H_2+CO)
Project Scope :	Catalyst supply to the existing plant
Feed Hydrocarbon :	Natural gas or Naphtha

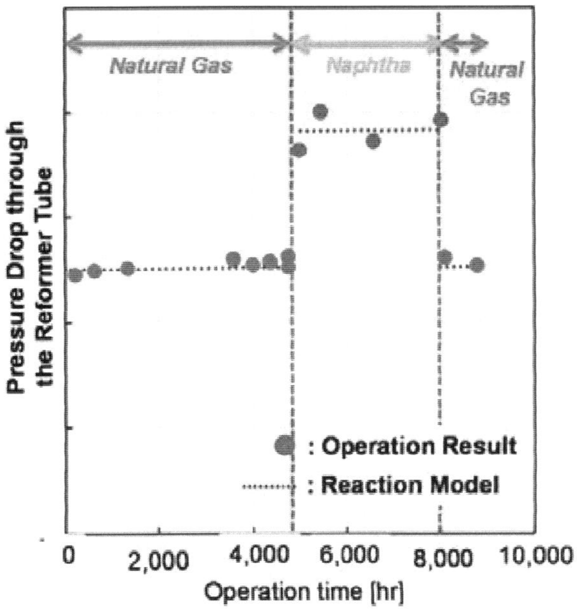

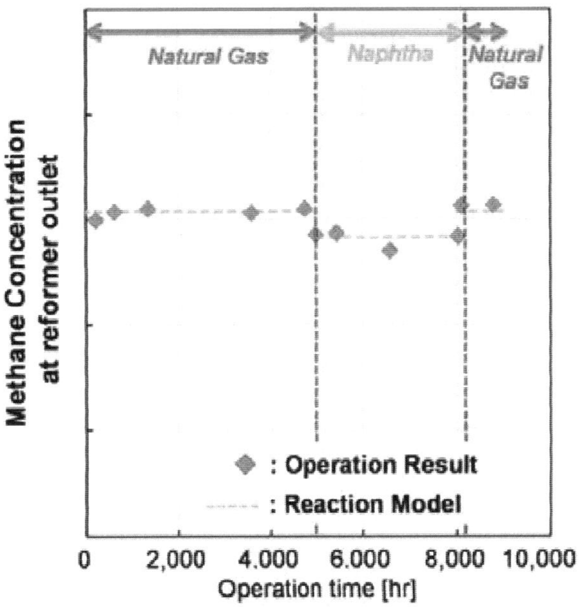

Fig. 10. Pressure difference of a reformer tube   Fig. 11. Methane concentration at reformer outlet

Fig. 12 shows the comparison of the reformer duty and the amount of input natural gas between before and after replacing the CT-CO_2AR™ catalyst. By replacing the CT-CO_2AR™ catalyst, 10% of the natural gas input is reduced and this result matches the result of process study shown in section 3 (Advantages of the CT-CO_2AR™ catalyst).

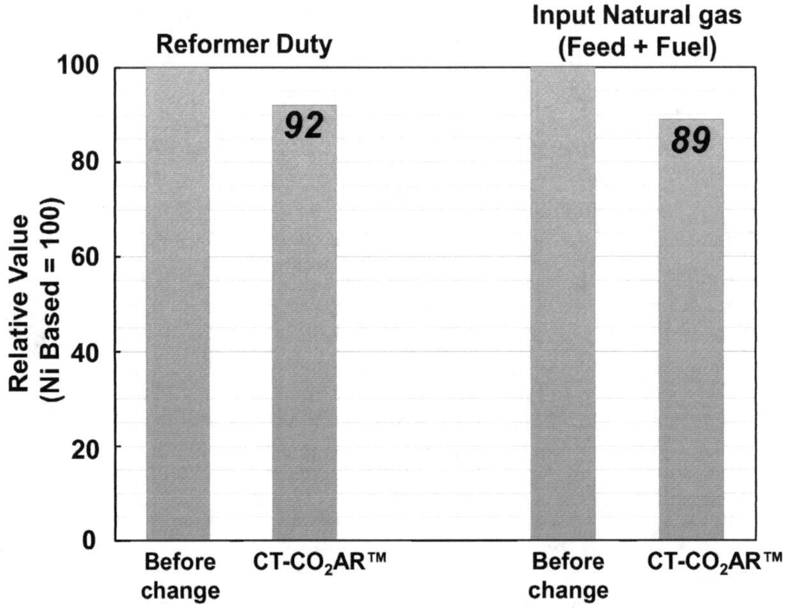

Fig. 12. Comparison result of commercial plant operation

5. Conclusion

The CT-CO_2AR™ catalyst enables a lower S/C condition operation because of its higher resistance to carbon formation.

Compared with a commercial catalyst, the CT-CO_2AR™ catalyst demonstrates high competitiveness in both manufacturing cost and CO_2 emission in the case of H_2/CO ratio = 0.5 and 1.0, especially, 0.5. The CT-CO_2AR™ catalyst was installed in a commercial plant in 2014 and has been operating stably. 10% reduction of natural gas input was achieved in the commercial operation.

6. References

1) J.R. Rostrup-Nielsen, J. H. B. Hansen, *J. Catal.*, **144**, 38 (1993)

2) N.R.Udengaard, J.-H. Bak Hansen and D.C.Hanson, Oil & Gas J., Mar.9(1992)

17AIChE
Annual Meeting, Minneapolis, MN

Solar Thermochemical Splitting of H_2O and CO_2 using Nonvolatile Metal Oxides

(Principal Author)
Rahul Bhosale
Department of Chemical Engineering
College of Engineering
Qatar University, Doha, Qatar
rahul.bhosale@qu.edu.qa

(Other Authors)
Gorakshnath Takalkar
Department of Chemical Engineering
College of Engineering
Qatar University, Doha, Qatar

Anand Kumar
Department of Chemical Engineering
College of Engineering
Qatar University, Doha, Qatar

Fares A. AlMomani
Department of Chemical Engineering
College of Engineering
Qatar University, Doha, Qatar

Prepared for Presentation at
American Institute of Chemical Engineers
2017 Fall Meeting
Minneapolis, MN, USA
October 29 – November 3, 2017

Solar Thermochemical Splitting of H_2O and CO_2 using Nonvolatile Metal Oxides

Rahul R. Bhosale, Anand Kumar, Gorakshnath Takalkar, Fares A. AlMomani

Department of Chemical Engineering
College of Engineering
Qatar University, Doha, Qatar

Keywords: Thermochemical Reactions, Metal Oxides, Solar Fuels, H_2O/CO_2 Splitting

Abstract

In this study, a Co ferrite/ZrO_2 ceramic nanomixture was synthesized by the propylene-oxide-assisted sol–gel method. Cobalt, iron, and zirconium nitrate salts were used as the metal precursors, and ethanol and propylene oxide were used as the solvent and gelation agent, respectively. The prepared gels were aged, dried, and annealed at 1000°C for different dwell times. The annealed Co ferrite/ZrO_2 powder was analyzed by powder X-ray diffraction (PXRD), Brunauer–Emmett–Teller surface area analysis (BET), and energy-dispersive spectroscopy (EDS),. The characterization results are reported in detail.

1. Introduction

Metal-oxide-based solar thermochemical water and carbon dioxide splitting is considered a promising technology to satisfy future energy demands [1]. Zinc oxide [2-5], iron oxide [5-6], tin oxide [7-8], doped iron oxides (ferrites) [9-14], ceria-based redox materials [15-19], and perovskites [20-24] are used for splitting. Among these oxides, ferrites have been used extensively for solar thermochemical water splitting; however, their use is extremely limited for CO_2 splitting.

However, the major issue limiting the use of ferrites in solar thermochemical cycles is the increased size of the ferrite grains obtained as a result of high-temperature processing. In this study, a Co ferrite/ZrO_2 ceramic mixture was synthesized by the propylene-oxide-assisted sol–gel method [25-27], exhibiting mixing of Co ferrite and ZrO_2 with a uniform distribution. The effect of annealing time on the phase and chemical composition, crystallite size, specific surface area (SSA), pore volume, and morphology of the Co ferrite/ZrO_2 ceramic nanomixture was investigated, and the results were reported in detail.

2. Experimental

2.1 Sol–Gel Synthesis

Metal precursors of cobalt, iron, and zirconium were first mixed in a specific weight ratio and dissolved in ethanol using a sonication bath. Once a visibly clear solution was obtained, a predetermined amount of propylene oxide was slowly added, affording a gel. The gel was aged, dried, and annealed at 1000°C (at various annealing times) in air. Fig. 1 shows the flow chart for the synthesis of the Co ferrite/ZrO_2 ceramic nanomixture.

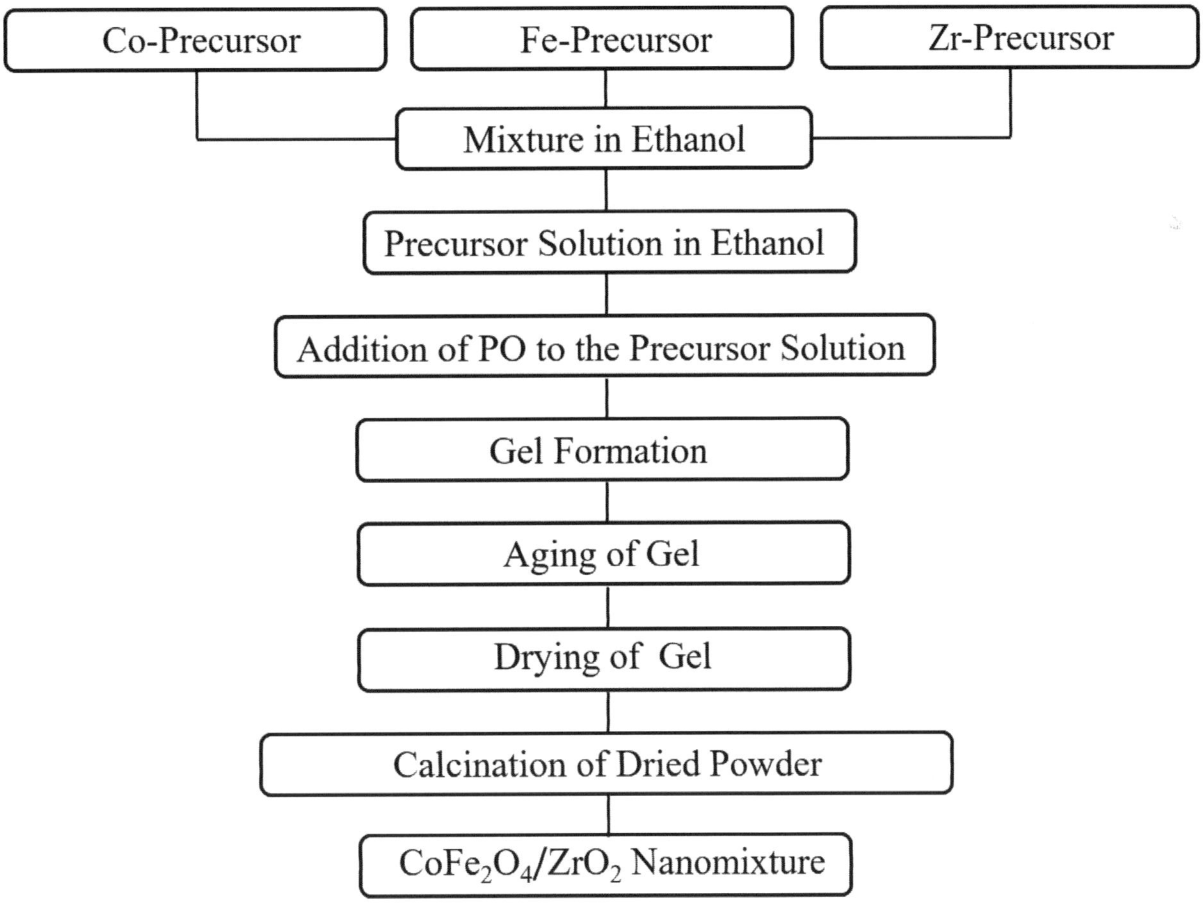

Fig. 1. Flow chart for the synthesis of the Co ferrite/ZrO_2 ceramic nanomixture by the sol–gel method.

2.2 Characterization

To investigate the effects of annealing on the physiochemical properties of the Co ferrite/ZrO_2 ceramic nanomixture, the Co ferrite/ZrO_2 powder obtained after annealing was characterized by powder X-ray diffraction (PXRD), Brunauer–Emmett–Teller (BET) surface area analysis, scanning electron microscopy (SEM) with a field-emission gun, X-ray energy-dispersive spectroscopy (EDS), and high-resolution transmission electron microscopy (HR-TEM).

3. Results & Discussion

The effect of annealing time on the physicochemical properties of the sol–gel-derived Co ferrite/ZrO_2 ceramic nanomixture was investigated. For this purpose, the Co ferrite/ZrO_2 gels (prepared using 20 mL of propylene oxide, aged for 24 h, and dried at 100°C) were annealed at 1000°C for 1–5 h. The calcined powders thus obtained were characterized by PXRD and EDS to determine the phase and chemical composition. The results shown Fig. 2 clearly indicated that the phase and chemical composition of the sol–gel-derived Co ferrite/ZrO_2 ceramic nanomixture is unaffected by the increase in the annealing time.

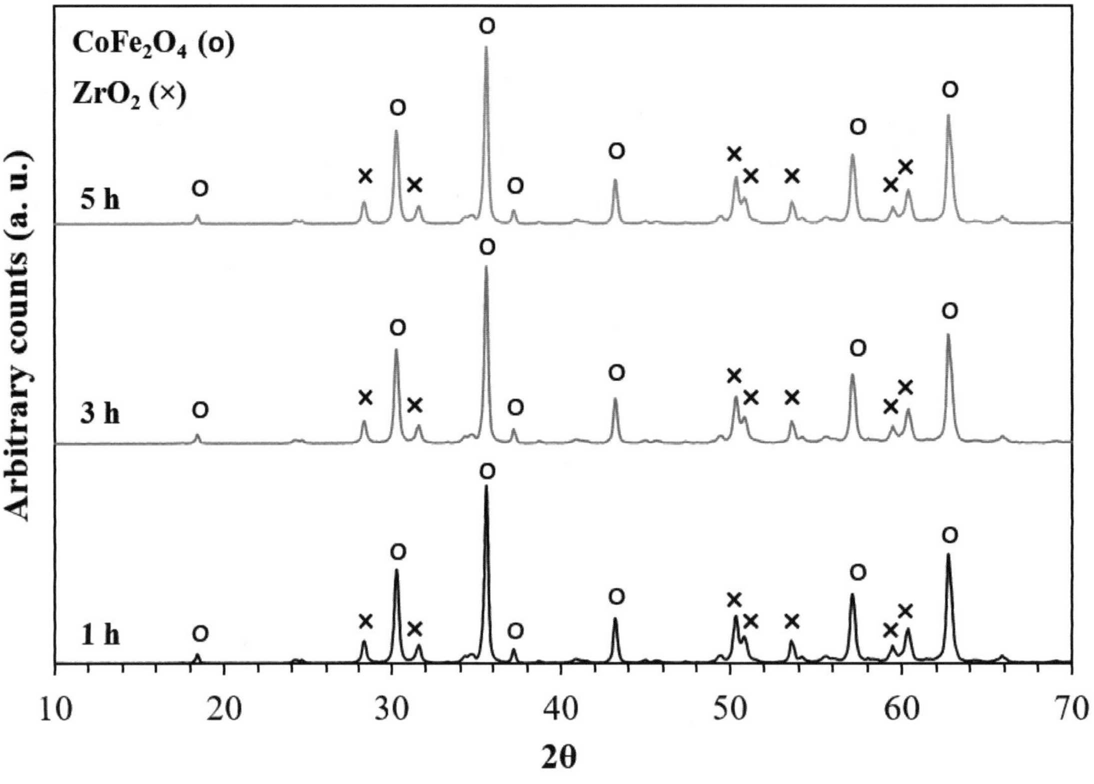

Fig. 2 – Effect of the annealing time on the phase composition of sol–gel-derived Co ferrite/ZrO_2 ceramic nanomixture.

The effect of the annealing time on the crystallite size of the sol–gel-derived Co ferrite/ZrO_2 ceramic nanomixture was also determined. Fig. 3 shows the results obtained. With the increase in the annealing time from 1 to 3 h, the crystallite size increases by 14.86 nm. With the further increase in the annealing time from 3 to 5 h, the crystallite size further increases by 22.83 nm. In summary, with the increase in the annealing time from 1 to 5 h, the size of the Co ferrite/ZrO_2 crystallites increases from 53.66 to 91.35 nm.

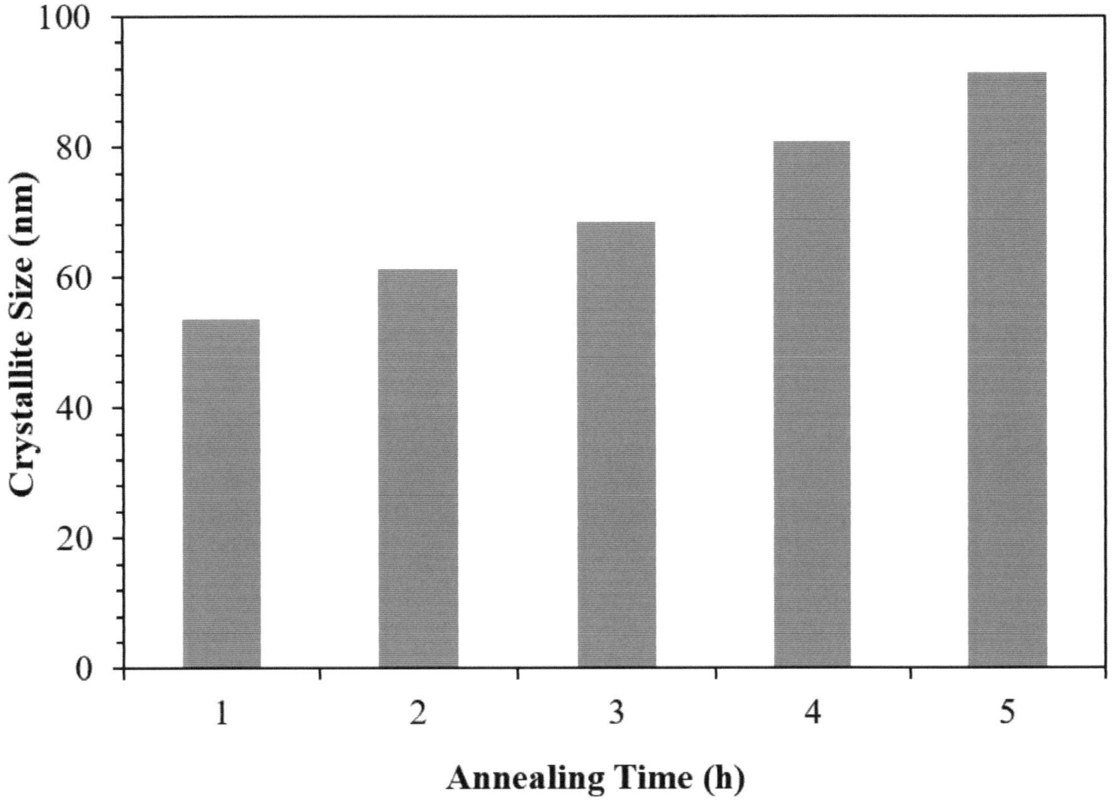

Fig. 3 – Effect of the annealing time on the crystallite size of the sol–gel-derived Co ferrite/ZrO_2 ceramic nanomixture.

In addition to the composition and crystallite size, the SSA and pore volume are also crucial parameters to be investigated. The results obtained by BET analysis indicated that with the increase in the annealing time from 1 to 3 h, the SSA decreases by 14.33%. Similarly, with the increase in the annealing time from 3 to 5 h, the SSA decreases by 13.57%. With the increase in the annealing time from 1 to 5 h, the pore volume decreases from 1.52×10^{-3} cm^3/g to 1.03×10^{-3} cm^3/g.

4. Conclusion

In this paper, a Co ferrite/ZrO_2 ceramic nanomixture, with a pure phase composition, high SSA, pore volume, nanoparticulate morphology, and a uniform distribution of Co ferrite and ZrO_2 in the mixture, was synthesized by the propylene-oxide-assisted sol–gel method. The phase and chemical composition of the Co ferrite/ZrO_2 nanomixture was not affected under all of the annealing conditions. With the increase in the annealing time from 1 to 5 h at an annealing temperature of 1000°C, the crystallite size increased by 37.69 nm. with the increase in the annealing time from 1 to 5 h, the SSA and pore volume decreased by 2.59 m^2/g and 0.484 × 10^{-3} cm^3/g, respectively.

5. Acknowledgments

This publication was made possible by the NPRP grant (NPRP8-370-2–154) from the Qatar National Research Fund (a member of Qatar Foundation). The statements made herein are solely the responsibility of author(s).

References

[1] J.R. Scheffe, A. Steinfeld, Oxygen exchange materials for solar thermochemical splitting of H_2O and CO_2: a review, Materials Today 17 (2014) 341-348.

[2] P.G. Loutzenhiser, A. Steinfeld, Solar syngas production from CO_2 and H_2O in a two-step thermochemical cycle via Zn/ZnO redox reactions: thermodynamic cycle analysis, International Journal of Hydrogen Energy 36 (2011) 12141-12147.

[3] A. Stamatiou, P. Loutzenhiser, A. Steinfeld, Solar syngas production from H_2O and CO_2 via two-step thermochemical cycles based on Zn/ZnO and FeO/Fe_3O_4 redox reactions: Kinetic analysis, Energy & Fuels 24 (2010) 2716-2722.

[4] R. R. Bhosale, A. Kumar, F. AlMomani, R. B. Gupta, Solar thermochemical ZnO/$ZnSO_4$ water splitting cycle for hydrogen production, International Journal of Hydrogen Energy (2017) in-press.

[5] P. Loutzenhiser, M. Gálvez, I. Hischier, A. Stamatiou, A. Frei, A. Steinfeld, CO_2 splitting via two-step solar thermochemical cycles with Zn/ZnO and FeO/Fe_3O_4 redox reactions II: Kinetic analysis, Energy & Fuels 23 (2009) 2832-2839.

[6] R. R. Bhosale, A. Kumar, van den Broeke, Leo JP, S. Gharbia, D. Dardor, M. Jilani, J. Folady, M.S. Al-Fakih, M.A. Tarsad, Solar hydrogen production via thermochemical iron oxide–iron sulfate water splitting cycle, International Journal of Hydrogen Energy 40 (2015) 1639-1650.

[7] P. Charvin, S. Abanades, F. Lemont, G. Flamant, Experimental study of SnO_2/SnO/Sn thermochemical systems for solar production of hydrogen, AIChE Journal 54 (2008) 2759-2767.

[8] R. R. Bhosale, A. Kumar, P. Sutar, Thermodynamic analysis of solar driven SnO_2/SnO based thermochemical water splitting cycle, Energy Conversion & Management 135 (2017) 226-235.

[9] R. R. Bhosale, A. Kumar, F. AlMomani, U. Ghosh, P. Sutar, G. Takalkar, A. Ashok, I. Alxneit, Effectiveness of Ni incorporation in iron oxide crystal structure towards thermochemical CO_2 splitting reaction, Ceramics International 43 (2017) 5150-5155.

[10] C.C. Agrafiotis, C. Pagkoura, A. Zygogianni, G. Karagiannakis, M. Kostoglou, A.G. Konstandopoulos, Hydrogen production via solar-aided water splitting thermochemical cycles: Combustion synthesis and preliminary evaluation of spinel redox-pair materials, International Journal Hydrogen Energy 37 (2012) 8964-8980.

[11] R. R. Bhosale. A. Kumar, F. AlMomani, I. Alxneit, Propylene oxide assisted sol–gel synthesis of zinc ferrite nanoparticles for solar fuel production, Ceramics International 42 (2016) 2431-2438.

[12] M. Roeb, N. Gathmann, M. Neises, C. Sattler, R. Pitz-Paal, Thermodynamic analysis of two-step solar water splitting with mixed iron oxides, International Journal of Energy Research 33 (2009) 893-902.

[13] R.R. Bhosale, R.V. Shende, J.A. Puszynski, Thermochemical water-splitting for H_2 generation using sol-gel derived Mn-ferrite in a packed bed reactor, International Journal of Hydrogen Energy 37 (2012) 2924-2934.

[14] J.R. Scheffe, J. Li, A.W. Weimer, A spinel ferrite/hercynite water-splitting redox cycle, International Journal of Hydrogen Energy 35 (2010) 3333-3340.

[15] R.R. Bhosale, A. Kumar, F. AlMomani, I. Alxneit, Sol–gel derived CeO_2–Fe_2O_3 nanoparticles: Synthesis, characterization and solar thermochemical application, Ceramics International 42 (2016) 6728-6737.

[16] R.R. Bhosale, A. Kumar, F. AlMomani, U. Ghosh, S. Al-Muhtaseb, R. Gupta, I. Alxneit, Assessment of $Ce_xZr_yHf_zO_2$ based oxides as potential solar thermochemical CO_2 splitting materials, Ceramics International 42 (2016) 9354-9362.

[17] J.R. Scheffe, A. Steinfeld, Thermodynamic analysis of cerium-based oxides for solar thermochemical fuel production, Energy & Fuels 26 (2012) 1928-1936.

[18] W.C. Chueh, C. Falter, M. Abbott, D. Scipio, P. Furler, S.M. Haile, A. Steinfeld, High-flux solar-driven thermochemical dissociation of CO_2 and H_2O using nonstoichiometric ceria, Science 330 (2010) 1797-1801.

[19] P. Furler, J.R. Scheffe, A. Steinfeld, Syngas production by simultaneous splitting of H_2O and CO_2 via ceria redox reactions in a high-temperature solar reactor, Energy & Environmental Science 5 (2012) 6098-6103.

[20] J.R. Scheffe, R. Jacot, G.R. Patzke, A. Steinfeld, Synthesis, Characterization, and Thermochemical Redox Performance of Hf^{+4} , Zr^{+4} , and Sc^{+3} Doped Ceria for Splitting CO_2, The Journal of Physical Chemistry C 117 (2013) 24104-24114.

[21] J.R. Scheffe, D. Weibel, A. Steinfeld, Lanthanum–strontium–manganese perovskites as redox materials for solar thermochemical splitting of H_2O and CO_2, Energy & Fuels 27 (2013) 4250-4257.

[22] A.H. McDaniel, E.C. Miller, D. Arifin, A. Ambrosini, E.N. Coker, R. O'Hayre, W.C. Chueh, J. Tong, Sr-and Mn-doped $LaAlO_{3-\delta}$ for solar thermochemical H_2 and CO production, Energy & Environmental Science 6 (2013) 2424-2428.

[23] M. Gálvez, R. Jacot, J. Scheffe, T. Cooper, G. Patzke, A. Steinfeld, Physico-chemical changes in Ca, Sr and Al-doped La–Mn–O perovskites upon thermochemical splitting of CO_2 via redox cycling, Physical Chemistry Chemical Physics 17 (2015) 6629-6634.

[24] R. R. Bhosale, A. Kumar, A. Ashok, P. Sutar, G. Takalkar, M. Khraisheh, F. AlMomani, U. Ghosh, La-Based Perovskites as Oxygen-Exchange Redox Materials for Solar Syngas Production, MRS Advances (2017) in-press.

[25] R. R. Bhosale, R. Khadka, J. Puszynski, R. V. Shende, H_2 generation from two-step thermochemical water-splitting reaction using sol-gel derived $Sn_xFe_yO_z$, Journal of Renewable & Sustainable Energy 3 (2011) 063104-1 – 063104-12.

[26] R. Bhosale, R. Shende, J. Puszynski, H_2 generation from thermochemical water splitting using sol-gel derived Ni-ferrite, Journal of Energy and Power Engineering 4 (2010) 27-38.

[27] R.R. Bhosale, R.V. Shende, J.A. Puszynski, H_2 Generation From Thermochemical Water-Splitting Using Sol-Gel Synthesized Zn/Sn/Mn-doped Ni-Ferrite, International Review of Chemical Engineering 2 (2010) 852-862.

500493 Critical Issues in the Development of Commercial Natural Gas to Industrial Chemicals Bioprocesses

Abstract:

The low cost and abundant supply of natural gas make it America's greatest untapped energy resource, leading to the development of technologies for natural gas conversion to compounds of greater value. Intrexon has developed the first natural gas-to-liquids bioconversion platform powered by methantrophic fermentation, which applies synthetic biology to program bacteria and produce higher value materials. Our innovative genetic toolbox allows us to engineer the biology of the methanotroph to more efficiently convert natural gas into a wide range of valuable chemicals and fuels. These applications include isobutanol, farnesene, 1,4-butanediol, 2,3-butanediol, isoprene and isobutyraldehyde – the building blocks for industrial and consumer products like fuels, synthetic rubber, acrylics, resins and spandex.

Using a methanotrophic bioprocess to produce fuels and chemicals has several advantages. Natural gas bioconversion offers scaling opportunities beyond traditional gas-to-liquid (GTL) facilities, as well as biomass feedstock facilities. The theoretical yields of methanotroph/methane exceed that of yeast/sugar approaches by a wide margin. Biocatalytic processes that achieve high yields under milder processing conditions allow for reduced capital and operation expenditures, offering greater economic viability. To successfully bring this technology to commercial scale, consideration must be given not only to optimizing the bacterial metabolic pathway, but to key issues in the fermentor design, downstream processing, scale-up and overall plant economics.

At any scale, aerobic natural gas fermentation must overcome the challenge of gas-to-liquid transfer of low solubility gases to increase bioavailability of gas substrates to the bacteria. A low gas-to-liquid mass transfer rate reduces the bacterial productivity and process yield, which in turn increases fermentor size, capital cost, operating costs and may add additional equipment necessary to recycle gases. During commercial scale-up of gas fermentation processes, these factors can become cost prohibitive. Standard stirred tank reactors become unfeasible for use at larger scales, as the agitator power input and size become excessive. There are many fermentor designs currently available to increase the overall mass transfer coefficient of these gases. However, capital and operating expenditures of the reactor are critical metrics for evaluating commercialization feasibility. Unique fermentor configurations can greatly increase capital cost by raising the cost of fabrication and decreasing the ability to receive competitive bids. These unique fermentors also may not create a homogenous, well-mixed broth – which increases the cost of controls and instrumentation. The economic feasibility of gas transfer designs must be balanced with the biologic effects of reactor forces on the bacteria. Methods of increasing mass transfer often include increasing turbulence in the reactor, which can create detrimental effects on the bacteria through high shear forces, high pressure and vacuum areas, and dissolved gas substrate inhibition. Careful equipment selection and pilot trials are necessary to determine the various effects of gas transfer methods on mass transfer rates and cell viability.

Downstream process selection for commercialization is a balance of biologic limitations, technical feasibility, capital cost, operating cost and the risk of cutting-edge over well-established techniques. The developing bioprocess must be iterated upon the limitations of strain engineering and practicality of downstream processing techniques. Fermentation broth concentrations may be low to product inhibition on the cell, but this increases the size and energy requirement of downstream equipment. Cutting-edge technologies can increase the efficiency of low titer separation and reduce energy inputs, which can be capital intensive and unproven at commercial scales. Genetic engineering tools can increase the tolerance of the bacteria to products. However, the directed or natural evolution process requires time.

A minimum broth titer must be determined through techno-economic modeling of processes at each stage of process development and refinement. Bioprocesses also create metabolic byproducts which can add unit operations, and therefore cost, to the downstream purification process. Some byproducts may originate from the genetic pathway or from commercial feedstocks, as they are often lower in purity and have higher variability than those at the research scale. Additional downstream steps must be evaluated to determine if they are technically and economically feasible to meet final product specifications for sale, or if the byproduct genetic pathways can be manipulated to terminate production and not add cost to the overall process.

Furthermore, certain tools that aid in genetic engineering and strain development at the research and development scale must be eliminated or retooled to create commercially viable strains. During development, production pathways are often introduced to the bacteria through plasmids which carry antibiotic resistance. To avoid the material cost and downstream separation of the antibiotic at commercial scale, the genetic material must be integrated into the chromosome. This genomic integration may affect the strain's capability to produce product at previously observed yields, titers and rates. Another common development tool is that production pathways may be turned on or off using an externally applied chemical inducer. At scale, these chemical inducers can add material cost as well as affect downstream processes and product purity. Induction systems may be completely removed for commercial scale or based off an easily controllable environmental switch.

Favorable plant economics are necessary to move forward with commercialization of a technology. Scale is carefully selected based on a comparison of the production rate, fermentation productivity and yield targets to revenue, capital and operating costs. The plant location is an important selection factor, as costs are affected by utility rates in the area and/or co-location of the plant with shared infrastructure.

Traditional engineering workflows, specifically front-end loading strategy, were used to shape the commercial scale process design of Intrexon's natural gas to chemical plants. The biologic, technical and economic commercial feasibilities of the end-to-end natural gas to isobutanol process were critical to Intrexon's plant design. Laboratory scale learnings were applied to select technology that was commercially feasible at demonstration scale. This demonstration scale design was scaled down to pilot scale, where the operation will yield learnings into refinement of at-scale design, equipment selection, empirical data on heating and cooling, and controls strategies and feedback to further iterations on genetics.

495124 Direct, Single-Pass Thermocatalytic Upgrading of Biogas and Landfill Gas into Renewable Natural Gas over the Ultra-Low Loading Ru/ Î³-Al_2O_3 Catalyst

Abstract:

Direct, single-pass thermocatalytic upgrading of biogas and landfill gas into renewable natural gas over the ultra-low loading Ru/g-Al_2O_3 catalyst

Yichen Zhuang and David Simakov

Department of Chemical Engineering, University of Waterloo, Waterloo, ON N2L 3G1, Canada

Converting CO_2 into synthetic renewable fuels is an attractive pathway to decrease CO_2 emissions and to reduce our dependence on fossil fuels. Thermocatalytic hydrogenation provides advantages of fast reaction rates and high conversion efficiencies (as compared to electro-catalytic and photo-catalytic routes), thus allowing for compact, high-throughput operation. One option is to convert CO_2 into synthetic methane, so-called renewable natural gas (RNG) via the Sabatier reaction ($CO_2 + 4H_2 = CH_4 + 2H_2O$). Water electrolysis using renewable or surplus electricity can be used to provide a non-fossil source of H_2 required for the reaction. There are multiple sources of concentrated CO_2-containing streams including biogas and landfill gas (up to 50 mol% CO_2). It is highly desirable to develop a technology that can convert such CO_2-rich streams directly, without costly upstream CO_2 separation.

One of the most fundamental challenges is the development of a catalyst, which is highly active, selective to CH_4 formation, stable against coking, and low-cost at the same time. The selectivity becomes a major challenge when a mixed feed (biogas or landfill gas contain up to 50 mol% CH_4 in addition to CO_2) is used instead of purified CO_2 feed. In this study, a highly selective Ru-based catalyst with ultra-low Ru loadings (less than 0.5 wt%) is reported. A series of catalysts with Ru loadings ranging from 0.05-0.5 wt% were synthesized and evaluated for their catalytic activity in selective conversion of CO_2 in the mixed CH_4/CO_2 feed. CO_2 conversions up to 80% were achieved with nearly complete selectivity to CH_4 generation (left plot in the figure below). Importantly, the catalysts showed good stability, without any notable decline in performance over 60 hours on stream at very high space velocity of GHSV = 90,000 1/h (right plot in the figure below). A variety of analytical techniques were employed in order to investigate the mechanism of the superior selectivity and activity, including electron microscopy, thermal gravimetric analysis, temperature programmed reaction, chemisorption, and X-ray photon spectroscopy. Our results point out at the strong structure-activity relationship and the importance of the active phase interaction with the support.

492217 Energy Quality Factor and Exergy Destruction Processes Analysis for a Proposed Polygeneration System

Abstract:

A new polygeneration system was established through combining coal pyrolysis with semicoke combustion, semicoke gasification and a steam Rankine cycle, and then the calculation methodology was developed to calculate the system. The energy and exergy efficiencies of the system are 65.9 % and 75.1 %, respectively; and the standard energy quality factor Î±Î¸ values of the involved 17 components and the energy quality factor Î± of each system stream were obtained. By comparing the differences among the standard energy quality factor Î±Î¸ values of the related 17 components, the features of the energy grade of the components at the standard conditions were elucidated. Then the effect of streamÂ¢s composition and temperature on the actual energy quality factor was discussed. Combining the concept of energy quality factor, the processes of exergy destruction in system blocks were visually expressed via a kind of diagram; and the reasons of exergy destruction were tried to be explained from the perspective of energy quality factor.

491941 Mechanistic Implications of Low-Pressure Feeds for Methanol-to-Olefins Conversion on MFI

Abstract:

The production of methanol and its subsequent conversion on zeolites to produce important petrochemical precursors, such as ethylene and propylene, provides a route for the commercial-scale valorization of unconventional carbon sources such as coal, biomass, and natural gas. The complex network of reactions governing this chemistry can be summarized by a dual catalytic cycle description where olefins-based chemistries of methylation and β-scission are interconnected with the aromatics-based chemistries of methylation and dealkylation via hydrogen transfer and cyclization. Transient isotopic labeling studies implicate ethylene production predominantly from the aromatics-based cycle, thereby, suggesting that process or material parameters that result in the selective propagation of the olefins-based cycle would in turn result in high propylene-to-ethylene ratios. We implicate inlet methanol partial pressure as the crucial process parameter for preferential propagation of the olefins-based cycle over its aromatics counterpart, regardless of the reaction temperature, on H-ZSM-5. We report

(i) ~10-fold increase in the propylene-to-ethylene ratio (~2.7 – 27) as the inlet methanol pressure was decreased from 52 to 0.6 kPa; carbon selectivities towards propylene increased from ~29 to 41%C when compared at iso-conversion (~30 %C) levels. This effect persisted regardless of the reaction temperature (623 – 773 K). These results clearly suggest that low inlet pressures of methanol can effectively decouple the olefins- and aromatics-based cycles;

(ii) The previously unrecognized role of formaldehyde, obtained from persistent methanol dehydrogenation events, in initiating the aromatics-based cycle. A small co-feed of formaldehyde (~24 Pa) with 0.6 kPa of methanol resulted in ~15-fold increase in ethylene selectivity (from ~1.5 to 22 %C) while the propylene-to-ethylene ratio dropped from ~27 (in the absence of formaldehyde) to ~1 (in the presence of formaldehyde), thereby, advocating suppressed formaldehyde production at low inlet methanol pressures as the critical factor in selective propagation of the olefinic cycle.

These mechanistic studies provide guidance for process conditions to minimize the formation of undesirable products during methanol-to-hydrocarbons conversion on zeolites.

498789 Catalysis on Microporous Solid Acids: Mechanism and Catalyst Descriptors for the Coupling of Alkenes and Alkanones

Abstract:

The conversions of light alkenes and oxygenates on solid Brønsted acids are mediated by ion-pair transition states (TS) [1,2]. The stability (with respect to their relevant kinetic precursors) of these ion-pairs, consisting of organic cations and inorganic anions formed upon proton transfer, determine turnover rates for these reactions and also those for subsequent conversions of the primary products formed, such as isomerization, β-scission, hydride transfer, cyclization and the formation of unsaturated species that can render protons unreactive. The inorganic catalyst host affects TS stability by the ability of its anion to stabilize negative charge (acid strength) and of its voids to establish van der Waals contacts with the organic guest; such contacts depend sensitively on the matching of the size and shape of host and guest. The stability of the cationic moiety at the transition state also depends on its ability to accept the proton (proton affinity) and then to interact electrostatically with the localized negative charge at the framework and via dispersion forces through van der Waals contacts with the surrounding framework. These confinement and electrostatic effects are probed here for types of reactions mediated by TS structures that differ in size, shape, or charge from each other and from their relevant precursors on solids with diverse acid strength and confining environments. Density functional theory (DFT) methods that include pertinent dispersive interactions are used to provide insights that are inaccessible from experiment alone, such as the nature and magnitude of relevant energetic descriptors in the form of van der Waals confinement energies, proton affinities of organic molecules, and deprotonation energies.

In this work, we report turnover rates (per proton) for C_2-C_4 alkene coupling (503 K, 10- 500 kPa alkene) and for aldol condensation (473 K, 0.1-10 kPa acetone) reactions on acids with different acid strength and confining environments (H-Al-TON, H-X-MFI (X=Al, Ga, Fe, and B isomorphously substituted samples), H-Al-BEA, H-Al-FAU, H-Al-MCM-41, and supported Keggin polyoxometalate clusters ($H_3PW_{12}O_{40}/SiO_2$)). For both reactions, turnover rates are proportional to the pressure of the respective reactants on all solid acids and the first-order rate constants reflect an energy barrier between bimolecular C-C bond-forming TS and protons that are saturated with less-charged precursors, consistent with infrared spectra and DFT-simulations; these precursors include covalently-bound alkoxides during alkene coupling reactions and H-bonded acetone during condensation reactions.

The measured rate constants are sensitive to changes in acid strength, in the case of unconfined acid sites in polyoxometalates and silica-alumina and of similarly-confined sites within MFI frameworks with Al, Ga, Fe, or B heteroatoms; here, acid strength is described in terms of deprotonation energy, a theoretically accessible value. In both cases, rate constants decreased exponentially with increasing deprotonation energy, consistent with kinetically-relevant energy barriers between charged, ion-pair TS and less charged, precursors.

Measured condensation and alkene coupling rate constants on microporous and mesoporous aluminosilicates, which contain acid sites of similar strength, reach maximum values as the framework voids

and the TS structures become of similar size. The consequences of such "good fits" for reactivity are quantitatively described first based on spherical geometry by comparing the diameter of a sphere with volume equivalent to the van der Waals volume of the TS to a characteristic void size defined by the largest included diameter within the pore network. We find, however, that to describe these interactions both the size *and the shape* of the zeolite void and the TS must be characterized. The incomplete nature of such geometrical descriptors is resolved by using van der Waals interaction energies (E_{vdw}) obtained by ensemble-averaging Lennard-Jones interactions between each framework O-atom and each atom in the organic moiety (TS or precursor) at each crystallographically distinct T-site for each of the frameworks examined in this study.

The alkene coupling TS structures for alkenes of different chain length vary in size, shape, and proton affinity, thus allowing us to dissect effects of confinement and intrinsic TS stability. Their different size also show that oxide frameworks distort locally upon containment of TS structures and alkoxides as their size approaches that of the confining voids. These distortions are quantified on a geometric basis as the mean O-atom displacement and, energetically as E_{dist}; both metrics indicate that the distortions enhance van der Waals contacts for the host-guest structures. These distortions, however, bring forth energetic penalties that attenuate and ultimately negate the benefits of more effective van der Waals contacts, consistent with the repulsive component of E_{vdw} that dominates for the large isobutene coupling TS in the 10-membered ring channel of TON.

Here, we provide a framework via thermochemical cycles to understand the effects of these catalyst and mechanistic descriptors (DPE, E_{vdw}, E_{dist}, proton affinity) on the relevant energy barriers. The similar conclusions drawn for both alkene coupling and alkanone condensation reflect the resemblance of chemistry, i.e. C-C bond formation via bimolecular transition states, but, moreover, a commonality of the analysis using energetic descriptors for catalysis on microporous acids. This universality begets a tool to predict reactivity of catalysts and reactants not studied here.

[1] Sarazen, M. L., Doskocil, E. Iglesia, E., J. Catal. 344, 553 (2016)

[2] Herrmann, S., Iglesia, E., J. Catal. 346, 134 (2017)

Financial support from BP through the ICC Program and from the National Science Foundation Graduate Research Fellowship Programs. Computational resources from National Science Foundation's Extreme Science and Engineering Discovery Environment (XSEDE; ACI- 1053575) are gratefully acknowledged.

486644 Structure-Property Relationships for Unidimensional, Large and Extra-Large Pore Zeolites Using Alkane Hydrocracking and Hydroisomerization As Probe Reactions

Abstract:

Structure-Property Relationships for Unidimensional, Large and Extra-Large Pore Zeolites Using Alkane Hydrocracking and Hydroisomerization as Probe Reactions

Viktor J. Cybulskis[1], Stacey I. Zones[2], Tracy M. Davis[2], Cong-Yan Chen[2], Michael W. Deem[3] and Mark E. Davis[1]

[1]Chemical Engineering, California Institute of Technology, Pasadena, CA 91125, USA

[2]Chevron Energy Technology Company, Richmond, CA 94802, USA

[3]Bioengineering, Rice University, Houston, TX 77005 USA

High-silica zeolites (Si/Al > 30) are versatile catalytic materials in petroleum refining processes due to their thermal stability, shape selectivity, and chemical functionality. However, of the 232 known [1] and more than 330,000 predicted [2] zeolite structures, the global refining market is dominated by five main frameworks: FAU, MOR, MFI, *BEA and FER [3]. While these zeolites often deliver satisfactory commercial performance, the development of materials with improved function or entirely new structures for future applications requires a deeper understanding of how specific properties, such as composition, framework topology, and acid site function, can influence the product distribution for a given reaction.

For the present study, large and extra-large pore zeolites were prepared from the corresponding structure-directing agents (SDAs) shown in Table 1 to investigate the effects of the organo-cation guest molecules and synthesis conditions on zeolite crystallinity, microporosity, and heteroatom incorporation. Unidimensional (1D) structures were chosen so that the zeolite pore opening could be used to adequately describe the confining size. Hydrocracking and hydroisomerization of *n*-hexane (nC_6) and *n*-decane (nC_{10}) were used as probe reactions on these Pt-exchanged, 12- and 14-membered ring (MR) zeolites to examine the effect of pore structure on shape selectivity by comparing the distributions and yields of branched C_6 and C_{10} isomers. Elucidation of these structure-property relations support *de novo* computational design efforts to predict chemically-synthesizable SDAs and produce a novel, 1D zeolite framework with tailored catalytic properties.

Table 1. 1D, 12- and 14-MR zeolites synthesized from the corresponding SDAs

References:

[1] C. Baerlocher and L.B. McCusker, Database of Zeolite Structures, http://www.iza-structure.org/databases/. Accessed March 27, 2017 (2017).

[2] R. Pophale, P.A. Cheeseman and M.W. Deem, *Phys. Chem. Chem. Phys.* 13, 12407 (2011).

[3] W. Vermeiren and J.P. Gilson, *Top. Catal.* 52, 1131 (2009).

501594 Modeling Complex Reactions in Zeolites: Effects of Acid Site Location, Framework, and Reagent Structure on Methanol-to-Hydrocarbon Reactions

Abstract:

Acid-catalyzed reactions within proton-form zeolites are mediated by cationic transition states which (along with their precursors) are stabilized by electrostatic interactions with the conjugate base and are 'solvated' by the zeolite framework through van der Waals interactions. The diversity of zeolite frameworks has sustained decades of research into determining and synthesizing optimal zeolite materials for many reactions, including the conversion of methanol into hydrocarbons (MTH). The size and topology of zeolite pores can shift selectivity during MTH from light olefins (in TON zeolites) to gasoline-range hydrocarbons (in MFI zeolites), but zeolites relevant to industrial practice for MTH processes rapidly deactivate. Acid site locations within zeolites are often poorly controlled/understood, leading to difficulties in modeling zeolite-catalyzed reactions with theoretical tools, because of the configurational complexity of such reactions. Here, we investigate reactions within methanol-to-hydrocarbons reactions and show how they are affected by the local zeolite environment (by altering acid site location within a framework or by changing the framework entirely). Ultimately, the methods outlined here will enable rigorous screening of zeolite materials to identify synthesis-targets for controlling heteroatom location and new framework synthesis.

Acid site location directly impacts molecular adsorption energies, reaction energies, and activation barriers. Here, we examine the effects of acid site location using novel methods that rigorously examine molecular adsorption and reaction pathways at all acid sites within zeolite frameworks. Methanol adsorption free energies, calculated using dispersion-corrected energies within a periodic model of MFI, for example, vary from -126 kJ mol^{-1} to -16 kJ mol^{-1} at the 48 Al–O site pairs which define acid sites within MFI. Intrinsic free energy barriers for methanol dehydration to form surface methoxy species (CH_3-Z) similarly vary from 80 kJ mol^{-1} to 150 kJ mol^{-1}. Varying acid site location within MFI has the same impact on intrinsic rates as changing the entire zeolite framework, indicating the importance of NMR and synthesis techniques which attempt to locate and control the location of heteroatoms within zeolite frameworks. In this work, we will present methanol and ethanol dehydration pathways within all acid sites of TON, CHA, and MFI and use these data to demonstrate that rates, selectivities, and even mechanisms can be affected by the local zeolite environment by changing the framework or acid site location.

Methanol-to-hydrocarbon reactions are also impacted by the local zeolite environment, by certain environments permitting the formation of aromatic species which constitute a "hydrocarbon pool" of species which co-catalyze further reaction cycles—but can also grow into coke precursors, leading to catalyst deactivation. This "hydrocarbon pool" consists of methylated benzenes (from benzene to hexamethylbenzene, 13 species) and each of these species can 1) by methylated by methanol, 2) undergo intramolecular C–C bond rearrangements to form alkyl substituents, and 3) dealkylate to form light alkene products. The total number of elementary steps involved is enlarged by the large set of possible co-catalysts. Here, we use structural refinement methods to examine reaction energetics all co-catalysts for each elementary step. The free energy barrier for methylating an aromatic ring, for example, increases from 80 kJ mol^{-1} to form toluene to 140 kJ mol^{-1} to form hexamethylbenzene, indicating that methylation becomes more difficult as reagents become more sterically-crowded by methyl-substituents. These effects of co-catalyst structure, along with kinetic Monte Carlo methods, allow us to report not only the mechanism of light alkene formation during MTH, but also the hydrocarbon pool species predominantly responsible for light alkene production.

These two related topics focus on computational methods and strategies for examining reactions solvated by diverse zeolite environments (within one and across multiple frameworks) and for examining the effects of reagent structure through systematically varying –H and alkyl substituents in hydrocarbon pool species which facilitate methanol-to-hydrocarbon reactions. These approaches lay the groundwork for examining the complete MTH pathways in all known and predicted zeolite structures to ultimately use computation to direct zeolite synthesis efforts by identifying promising novel materials in which the hydrocarbon pool can be formed without concomitant coke buildup and catalyst deactivation.

497790 Simple Characterization of Solid Acid Catalysts By Reactive Gas Chromatography

Abstract:

One of the more common methods for characterizing solid acid catalysts is the temperature programmed desorption (TPD) of alkylamines, which selectively decompose to an alkene and ammonia over Brønsted acid sites. TPD is traditionally carried out either in a flow reactor or microbalance, coupled with a mass spectrometer for identification and/or quantitative purposes. Reactive gas chromatography is a new technique to measure Brønsted acid site densities of solid acid catalysts, which condenses the typical experimental setup required for a TPD into a single, fully automated gas chromatograph (GC). Reactions are carried out in a microcatalytic reactor, housed within a temperature controlled GC inlet liner, products are then separated in a GC column and quantified by a flame ionization detector (Figure 1a). This eliminates the need for more complex instrumentation such as on-line mass spectrometers or a flow reactor apparatus.

An investigation of the RGC method was carried out by characterizing ZSM-5, BEA and SPP zeolites of varying Si/Al ratios, utilizing alkylamine decomposition to selectively count Brønsted acid sites. A comparison between RGC measurements and conventional TPD methods show excellent agreement (Figure 1b), confirmed by comparison with pyridine in-situ titrations under Brønsted catalyzed reaction conditions. Reactive gas chromatography measurements performed using n-propylamine, isopropylamine and tert-butylamine were found to yield identical Brønsted acid site densities. Given the relatively low detection limits of a flame ionization detector, the method of reactive gas chromatography is found to be highly sensitive. Siliceous materials with Brønsted acid site densities of ~1 μmol g_{cat}^{-1} could be reliably measured.

Figure 1. Method of Reactive Gas Chromatography for Solid Acid Titration. A. Solid acid samples loaded into the inlet liner (red rectangle) of a gas chromatograph are titrated with amines, introduced via an automatic liquid sampler (ALS) syringe, adsorbed and reacted on solid acid sites to produce propylene and ammonia. Resultant products are quantified by a flame ionization detection (FID). **B.** Comparison of zeolite measured Brønsted acid site density between reactive gas chromatography and literature values. Catalysts include ZSM-5 (Si/Al 11.5, 40, 140: •) and BEA (Si/Al 12.5, 19: ¦).

497279 Understanding Effective Diffusion Length Theory in Nanoscale Zeolites

Abstract:

Understanding effective diffusion length theory in nanoscale zeolites

Xiaoduo Qi, Vivek Vattipalli, Paul Dauenhauer and Wei Fan

Department of Chemical Engineering, University of Massachusetts Amherst, 686 N Pleasant Street, Amherst, 01002

Zeolites represent a class of crystalline aluminosilicates with well-defined micropores in the framework. Due to their unique molecular sieving and catalytic properties, zeolites have been extensively utilized as catalyst to facilitate chemical reactions. However, as the micropore size is comparable to the molecular dimension, the catalytic performance of the zeolites is often limited by the slow intra-crystalline molecular transport within the micropore channel when bulky molecules try to access the active site. To overcome such a transport limitation, considerable efforts have been made with the aim of reducing the contribution of molecular transport to the overall rate of reaction.[1] Among all other methods, synthesizing zeolite nanoparticle has been considered as one of the most promising strategy due to simple and well controlled preparation procedure. As the radius of the zeolite crystal is minimized to nanometers, it is expected that the characteristic diffusion length of such material is extremely short, leading to a rapid micropore diffusion process.

However, diffusion studies done by ZLC technique have shown that for zeolite nanoparticles, the effective micropore diffusion length can be much longer than the radius of the crystal due to a secondary diffusion step on the zeolite surface which leads to the occurrence of pore re-entry[2]. Despite the fact that the effective diffusion length phenomena can greatly hinder the mass transport within zeolite nanocrystals, the origin of such negative effect has not been well studied. In this work, four adsorbates (Cyclohexane, Methylcyclohexane, Ethylcycllohexane, and cis-dimethylcyclohexane) with different adsorption energies to MFI zeolite were used as probing molecules to investigate the cause of the effective diffusion length effect. It shows that the significance of such effect strongly depends on the adsorption strength of diffusion molecule to the zeolite surface. More specifically, as the adsorption becomes stronger, it is harder for diffusing molecule to completely desorb from the zeolite surface after it exits the micropores. As a result, this molecule will continue to diffuse along the surface and repeatedly diffuse within the pore, leading to a longer diffusion length. This work provides solid experimental evidence suggesting that the effective diffusion length effect is strongly related to guest/host adsorption. This fundamental understanding of micropore diffusion process within zeolite nanoparticles also enables us to optimize the mass transport by enhancing the adsorptive property of the zeolite.

[1] Chang, C.-C.; Teixeira, A. R.; Li, C.; Dauenhauer, P. J.; Fan, W., Enhanced Molecular Transport in Hierarchical Silicalite-1. Langmuir 2013, 29, (45), 13943-13950.

<>[2] Vattipalli, V.; Qi, X.; Dauenhauer, P. J.; Fan, W., Long Walks in Hierarchical Porous Materials due to Combined Surface and Configurational Diffusion. Chemistry of Materials 2016, 28, (21), 7852-7863.

498345 3D-Printed Zeolitic Scaffolds forÂ Selective Transformation of Light Alcohols to Light Olefins

Abstract:

This study aims at developing novel structured zeolite monoliths with various compositions by using three dimensional (3D) printing techniques to promote the production of light olefins and catalyst stability in selective transformation of light alcohols. 3D-printed ZSM-5 and silicalite-1 monoliths were coated with a thin layer of SAPO-34 zeolite. Characterization results indicated that the 3D-printed monoliths were highly porous and their micropore surface area and pore volume further increased by consequent coating with SAPO-34 zeolite. The catalytic results showed that the thin layer of SAPO-34 on 3D-printed MFI structures impeded the gaseous mas transport but favored the production of light olefins. The total acidity of SAPO-34/3D-printed ZSM-5 structure was higher than both SAPO-34/3D-printed silica and uncoated 3D-printed ZSM-5 monolith which favored the light olefins production. Coating 3D-printed HZSM-5 monolith with SAPO-34 enhanced the ability to produce olefins while circumvented the quick coke formation. These results demonstrated that selectivity and reaction rate in dehydration of light alcohols to light olefins can be modulated by changes in catalyst topology and morphology.

494815 The Role of Surface Barriers As Dominant Transport Mechanism in Hierarchically Structured Zeolites-- Application to the Alkylation of Benzene with Ethylene

Abstract:

The meso-macropore network of a hierarchically structured zeolite catalyst is numerically optimized to maximize the volume-integrated reaction yield in the ethylation of benzene to produce ethylbenzene over zeolite H-ZSM-5. A hierarchical approach is used at multiple length scales to determine the optimal pore network properties. The maximum volume-integrated reaction yield of the hierarchically structured zeolite catalyst containing meso- and macropores is nearly twice the yield of a zeolite pellet containing only macropores, at the same macroporosity. To bridge the gap between modeling and experiments, a series of physical mixtures of ZSM-5 crystals and mesoporous silica, containing different weight fractions of zeolite is synthesized and used in fixed bed reactor experiments to determine the optimal pellet structure to maximize the conversion of ethylene. Comparison with reactor simulations of the zeolite composites shows that the performance of the zeolite composites might be limited by surface barriers at the external surface of the zeolite crystals, rather than by diffusion limitations within the meso-macropore network of the pellets.

References:

Rao, S. M.; Saraci, E.; Glaeser, R.; Coppens, M.-O. Chem. Eng. J. 2017 (http://doi.org/10.1016/j.cej.2017.04.015)

488270 Stabilizing Pd Particles on Nitrogen-Doped Carbon Supports Under Hydrothermal Conditions

Abstract:

Carbon coating has been applied to improve the hydrothermal stability of metal oxide supports for use in aqueous-phase biomass reaction systems. Previously, incorporation of oxygen heteroatoms in the carbon coating were found to provide some enhancement the hydrothermal stability of Pd particles on these carbon-coated materials. To further improve the hydrothermal stability of these Pd particles, nitrogen-doped carbon supports were synthesized. Compared with merely carbon supported Pd catalyst, better dispersion and stability of the particles were achieved with the nitrogen-doped carbon. Unlike with undoped carbon, the Pd particles had limited sintering after extended exposure to hydrothermal reactions conditions and Pd leaching was greatly reduced. Through detailed solid-state ^{13}C and ^{15}N NMR analysis, the carbon surface chemistry was investigated and systematically compared for varying degrees of N incorporation. Previous studies have suggested that heteroatom incorporation could improve stability, but the NMR characterization techniques used in this study allowed for more detailed and quantitative characterization of N species present in the carbon. Better stability was ascribed to nitrogen functional groups including pyridine, pyrrole and secondary amine groups. This work will help to develop a more systematic strategy for improving hydrothermal stability of metal particles by heteroatom doping in carbon supports.

502131 Reaction Mechanism for the Conversion of Î³-Valerolactone (GVL) over a Ru Catalyst

Abstract:

Conversion of biomass-derived gamma-valerolactone (GVL) to valuable chemicals has been studied extensively, and understanding the reaction mechanism is very valuable for improving turnover rates and selectivities. Recently, it was shown that, in an aqueous phase over a Ru/C catalyst, GVL could be converted to 1,4-pentanediol, 2-butanol and 2-pentanol in parallel paths[1]. Here, we report first-principles density functional calculations, through which we show in detail the reaction pathways of GVL conversion on a Ru(0001) surface[2]. We find that GVL undergoes a ring-opening reaction rather easily, and that the rate-limiting step towards the formation of 1,4-pentanediol and 2-pentanol is the hydrogenation step. The high energy barrier for this step is caused by a strong interaction between Ru and the unsaturated acyl intermediate that is formed after opening the ring. Among all the primary products, formation of the 2-butanol has the smallest activation barrier, while the slowest step is the C-C bond cleavage in the decarbonylation step. We further show that the same acyl intermediate after ring opening of GVL can also be formed by dehydrogenation of 1,4-pentanediol with a moderate activation barrier, which suggests both 2-pentanol and 2-butanol can also be produced in secondary steps.

Acknowledgements

The calculations were performed at NERSC and OU Supercomputing Center for Education and Research (OSCER). We appreciate support from the U.S. DOE through a DOE/EPSCOR grant (DESC0004600).

1. Rozenblit, A.; Avoian, A. J.; Tan, Q.; Sooknoi, T.; Resasco, D. E., Reaction mechanism of aqueous-phase conversion of Î³-valerolactone (GVL) over a Ru/C catalyst. *Journal of Energy Chemistry* **2016,** *25* (6), 1008-1014.

2. Bababrik, R. M.; Wang, B.; Resasco, D., Reaction mechanism for the conversion of Î³-valerolactone (GVL) over a Ru catalyst: a first-principles study. *Industrial & Engineering Chemistry Research* **2017,** *56* (12), 3217-3222.

495140 Hydrodeoxygenation of Biomass Derivatives on Metal-Modified Molybdenum Carbides

Abstract:

Hydrodeoxygenation of Biomass Derivatives on Metal-Modified Molybdenum Carbides

Weiming Wan, Jingguang G. Chen

Department of Chemical Engineering, Columbia University, New York, NY, 10027

Biomass has attracted significant attentions as an alternative resource for fuel and chemicals. However, biomass derivatives have a high O/C ratio and low energy density that make them unsuitable for use in the fuel and chemical sectors.[1] Thus, the hydrodeoxygen (HDO) reaction is necessary to upgrade these biomass derivatives to value-added products. For example, furfural and glycerol are two typical biomass derivatives. Through HDO reactions, furfural can be converted to 2-methylfuran, a promising fuel additive, and glycerol can be converted to propylene or propanediol, an important monomer.

Transition metal carbides are novella class of interesting materials that often demonstrate unique catalytic properties. Due to the high oxygen binding energy, which helps to break the C-O bond, many metal carbides have shown activities towards HDO reactions. Mo_2C is a typical transition metal carbide and it shows high HDO activity for oxygenates such as furfural and propanol.[2],[3] However, the high oxygen binding energy makes it hard to remove the surface oxygen to complete the catalytic cycle. Metal modifiers can be used to adjust the oxygen binding energy. The strong oxophilic metal modifiers, such as Co and Fe, have slightly lower oxygen binding energies than Mo_2C, and the modified Mo_2C surface could have a similar HDO selectivity with better stability. In contrast, the weak oxophilic metal modifiers, such as Cu and Pt, have much lower oxygen binding energies than Mo_2C, and the modified Mo_2C surface should have a different reaction pathway and selectivity.

The current work combines density functional theory (DFT) calculations and surface science experiments using temperature programmed desorption (TPD) and high-resolution electron energy loss spectroscopy (HREELS) under ultrahigh vacuum (UHV) conditions. The main objective is to study the effect of metal modifiers on the adsorption geometry and selective HDO of furfural and glycerol on the Mo_2C surface. The Mo_2C surface was prepared by cycles of ethylene treatment on the Mo(110) single crystal, and the metal layer was prepared by physical vapor deposition.

In the furfural HDO reaction, compared with the Mo_2C surface, furfural has a similar adsorption geometry on Co-modified Mo_2C, suggesting the same reaction pathway on both surfaces. The lower oxygen binding energy of the Co/Mo_2C surface improves the catalytic stability in the TPD experiments. On the other hand, Pt/Mo_2C has lower oxygen binding energy than Co/Mo_2C and the furan ring is flat on the Pt/Mo_2C surface. The flat furan ring and the low oxygen binding energy help produce more furan through decarbonylation.

In the glycerol HDO reaction, the Mo_2C surface can break all the C-O bonds to make propylene. Similarly, the Fe/Mo_2C surface can produce propylene from glycerol with a better stability. On the other hand, the Cu/Mo_2C surface can selectively break the C-O bonds. With a high Cu coverage, the Cu/Mo_2C surface can break one C-O bond to form acetol; with a low Cu coverage, the Cu/Mo_2C surface can break two C-O bonds to form ally alcohol.

This work provides guidance for designing metal-modified carbide HDO catalysts and a detailed discussion of the DFT calculations and the surface science experiments will be presented.

References:

[1] K. Xiong, W. Yu, D. G. Vlachos, J. G. Chen. *ChemCatChem* **2015**, *7* (9), 1402.

[2] K. Xiong, W. Lee, A. Bhan, J. G. Chen. *ChemSusChem* **2014**, *7* (8), 2146.

[3] H. Ren, W. Yu, M. Salciccioli, Y. Chen, Y. Huang, K. Xiong, D. G. Vlachos, J. G. Chen. *ChemSusChem* **2013**, *6* (5), 798.

496760 Structure-Activity Relations on Î³-Al_2O_3: From Alcohol Dehydration to Alkane Dehydrogenation

Abstract:

Structure-Activity Relations on Î³-Al_2O_3: From Alcohol Dehydration to Alkane Dehydrogenation

Pavlo Kostetskyy and Giannis Mpourmpakis

Department of Chemical Engineering, University of Pittsburgh, Pittsburgh, PA 15261, USA

Abstract:

Nonoxidative dehydrogenation of alkanes is an important chemical reaction, as it can be used for production of alkenes â€“ building blocks for a range of plastics and chemicals[1-2]. Several metal oxides, including gamma-alumina, have exhibited dehydrogenation activity, showing promise as candidates for the upgrade of alkanes. In this work we use computational chemistry methods to elucidate the surface mechanism of ethane, propane, n-butane and i-butane dehydrogenation pathways on gamma-alumina. Based on our calculations, it was shown that a concerted mechanism is energetically preferred for alkene formation, with a carbenium-ion-like transition state forming on surface acid-base pairs. A dehydrogenation model was developed as a screening tool for oxide catalysts, based on physicochemical properties of the oxide surface and stability of the reacting hydrocarbons. The dehydrogenation model was developed based on methodology previously applied in structure-activity relations (SAR) for alcohol dehydration on metal oxides[3-4]. The Carbenium Ion Stability (CIS), shown to be a descriptor in alcohol dehydration[3-5], was used as a quantitative descriptor in alkane dehydrogenation and was found to correlate with the calculated activation energy barriers for the hydrocarbons in question. Increased hydrocarbon substitution (branching) was found to decrease the calculated reaction barriers, based on the CIS at the transition state. Finally, SARs developed for alcohol dehydration on various metal oxides were found to accurately capture the trends in alkane dehydrogenation, accounting for catalyst acid-base surface properties and CIS of intermediates at the transition state.

References:

1. Coperet, C., *Chemical Reviews* **2010,** *110*, 656-680.

2. Joubert, J.; Delbecq, F.; Sautet, P., *Journal of Catalysis* **2007,** *251*, 507-513.

3. Kostestkyy, P.; Yu, J.; Gorte, R. J.; Mpourmpakis, G., *Catal. Sci. Technol.* **2014,** *4*, 3861-3869.

4. Kostetskyy, P.; Mpourmpakis, G., *Catal. Sci. Technol.* **2015,** *5*, 4547-4555.

5. Roy, S.; Mpourmpakis, G.; Hong, D.-Y.; Vlachos, D. G.; Bhan, A.; Gorte, R. J., *Acs Catalysis* **2012,** *2*, 1846-1853.

501410 Catalytic Non-Aqueous Processing and Hydrogenation of Chitin and Amino-Based CarbohydratesÂ Using Ru Carbene Organometallics Immobilized on Active Hydrotalcites

Abstract:

Although cellulose has garnered significant attention over the past decades in the search for renewable fuel sources that can replace petroleum fuels, chitin can be catalytically converted into similarly valuable products as well. Chitin is the second most abundant carbohydrate and only differs from cellulose by an acetyl amine on the C5 position of cellulose. In comparison to traditional mineral acids, tunable hydrotalcites (HT) promote glycosidic bond cleavage by its basic sites while dehydration of the monomer is catalyzed by its Lewis acid sites. Additionally, immobilization of homogenous catalysts onto HT create highly selective single site heterogeneous catalysts (SSHC-HT). The main challenge in the development of such a process was to pair a stable solvent that would dissolve chitin with an active heterogeneous catalyst. To overcome this challenge, a total of five systems were studied: gamma-valerolactone (GVL), GVL/levulinic acid and formic acid (as both solvent and acid), tetrahydrofuran, ionic liquids, and water. A 3:1 ratio of Mg:Al HT was used as a basis for all catalytic reactions while substitutions of Al with Fe, Cr, Ni, and Co were tested for improved catalytic activity. Hydrogenation of products were carried out using Ru and Pd HT. A wide array of products, as cited in literature under similar reaction conditions, can be generated, including 5-(hydroxymethyl)furfural, levulinic acid, amino furans, and ketones. Products were identified using LCMS, HPLC, NMR, GCMS, and FTIR.

495616 Selective C-C Bond Scission of Biomass-Derived Oxygenates Using Cerium Oxide-Supported Ruthenium Catalyst

Abstract:

Selective C-C Bond Scission of Biomass-Derived Oxygenates Using Cerium Oxide-Supported Ruthenium Catalyst

Tomoo Mizugaki,[1] Kohei Uesugi,[1] Kodai Nitta,1 Zen Maeno,[1] Takato Mitsudome,[1] Koichiro Jitsukawa,[1] and Kiyotomi Kaneda*[1,2]

1. Graduate School of Engineering Science, Department of Materials Engineering Science, Osaka University

2. Research Center for Solar Energy Chemistry, Osaka University

kaneda@cheng.es.osaka-u.ac.jp

Current requirements to reduce carbon dioxide emissions have led to the utilization of renewable resources, such as plant biomass, as carbon-neutral feedstocks for commodity chemicals. To date, much effort has been devoted to the direct C–O bond cleavage of high-oxygen containing biogenic polyols, such as glycerols, to produce valuable chemicals by hydrogenolysis and deoxydehydration. On the other hand, C–C bond cleavage has not yet been widely researched despite its great potential to extend the utility of biomass-derived chemicals. The existing C–C bond cleavage methods often suffer from low selectivity toward the desired chemicals and limited substrate scope. Therefore, the development of highly selective and versatile C–C bond scission catalysts able to work under mild conditions is highly desired to open new routes for industrially important chemicals from a wide range of biomass derivatives.

In this work, we chose levulinic acid as the model substrate, which is an important platform biomass-derivative leading to a variety of useful C5 chemicals such as g-valerolactone (GVL),[1] 1,4-pentanediol (1,4-PeD),[1] and 2-methyltetrahydrofuran (MTHF).[2] However, there are fewer examples of the transformation of LA into C4 chemicals through one-carbon scission reactions. We found that cerium oxide-supported ruthenium nanoparticles (Ru/CeO_2) efficiently promoted the selective one-carbon scission of levulinic acid to give 2-butanol in over 80% yield under 3 MPa of H_2 at 150 °C in water. The high generality of this method is demonstrated by its broad substrate scope, where the cleavage of C–C bonds chemospecifically occurs at positions adjacent to the carboxyl, ester, and hydroxymethyl groups of oxygenated compounds (Scheme 1).

References

[1] T. Mizugaki, Y. Nagatsu, K. Togo, Z. Maeno, T. Mitsudome, K. Jitsukawa, K. Kaneda, *Green Chem.* **2015**, *17*, 5136–5139.

[2] T. Mizugaki, K. Togo, Z. Maeno, T. Mitsudome, K. Jitsukawa, K. Kaneda, *ACS Sustainable Chem. Eng.* **2016**, *4*, 682–685.

500440 Large Eddy Simulation of Soot Formation in Oxy-Coal Combustion

Abstract:

Oxy-Coal combustion is a promising clean coal technology. Typical combustion burns coal in air, resulting in a product mixture of water, nitrogen, and carbon dioxide. Separating carbon dioxide from nitrogen is costly. In oxy-fired combustion, nitrogen is first separated from air and a mixture of coal, oxygen, and (optionally) recirculated flue gases is burned. The resultant products are water and carbon dioxide, which are easily separated for subsequent sequestration. Oxy-fired combustion often occurs at high temperatures and under flow and thermal conditions that differ somewhat from combustion in air. Radiative heat transfer is a key process in such combustion as it is a primary source of heat transfer in steam generation applications. Radiation is present from gaseous species, ash, coal char, and soot, with soot being a significant contributor. Unfortunately, soot formation is a complex process. While much research has been done on soot formation in general, relatively little research has been published on soot formation in coal combustion. In oxy-fire systems with high water and carbon dioxide concentrations, and at high temperature, soot consumption via gasification may be significant. (Normally so consumption is dominated by oxidation reactions.)

We present a series of large eddy simulations in an oxy-coal combustor in which sensitivity to the soot formation mechanism is presented. Differences is flow, temperature, and radiative heat transfer are shown. We compare results using a previously published coal soot model, as well as extensions that include optimized gasification and oxidation rates. An advanced physics-based soot formation from coal tar mechanism is presented.

Ash aerosol and ash deposit formation during high temperature oxy-combustion of petroleum coke

Yueming Wang [a], Xiaolong Li [a], Jost Wendt [a]

[a] Department of Chemical Engineering, University of Utah, Utah, USA

Background

What is petroleum coke

- Peroleum coke (petcoke) is a byproduct of petroleum refining process

Advantages

- Petcoke is available in large amounts with low price
- Petcoke has low ash content and high heating value

Challenges

- The volatile content is very low in petcoke making it hard to be burned efficiently
- The sulfur content is very high and can cause corrosion and pollution

Research Objectives

- ➢Assess the combustion efficiency of petroleum coke under pulverized coal furnace
- ➢Uncover the effect of flame temperature on combustion performance of petroleum coke
- ➢Study the formation mechanisms of ash aerosol and ash deposition (fouling)
- ➢Develop the relationship between ash deposition rates and concentration of the ash aerosol

THE UNIVERSITY OF UTAH

3

Oxy-Fuel Combustor (OFC)

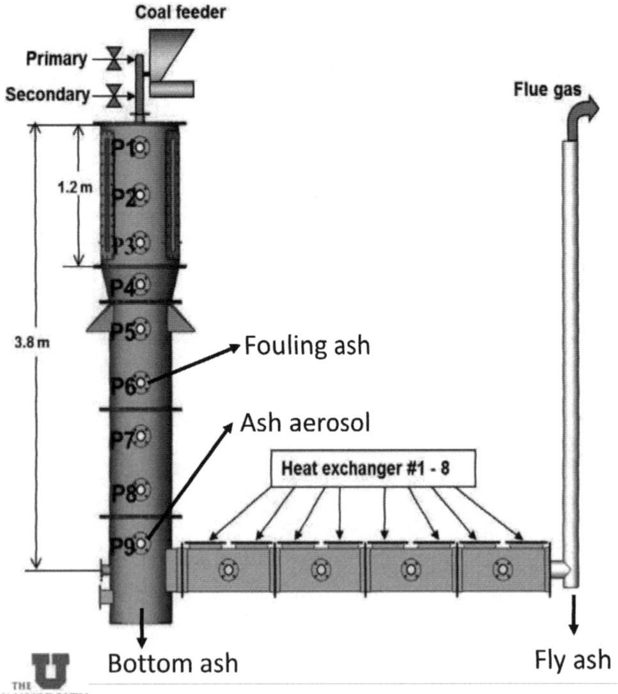

- ➢ 100 kW pilot scale
- ➢ Self-sustained down-fired
- ➢ Systematically controlled
- ➢ Temperature profile and particle concentration comparable with full scale units

THE UNIVERSITY OF UTAH

4

Fouling ash sampling system

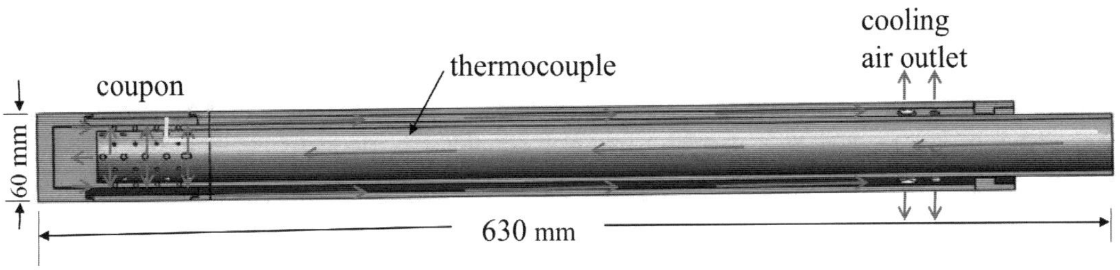

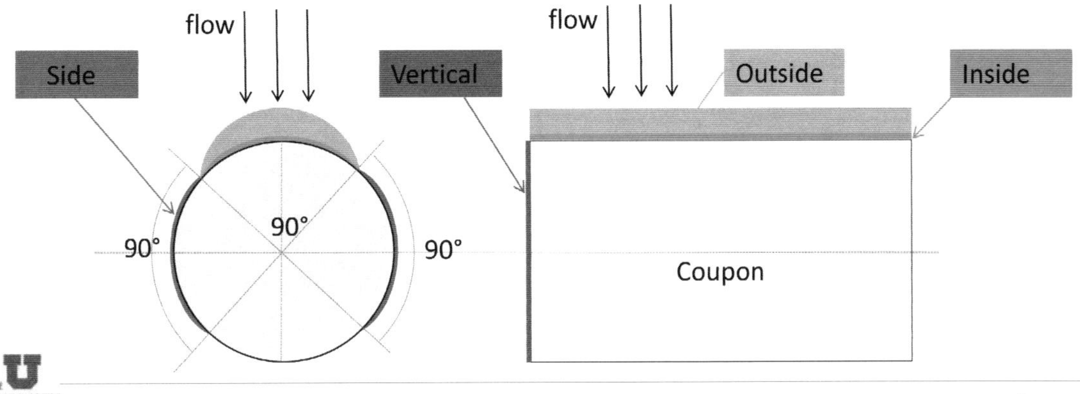

THE UNIVERSITY OF UTAH

Ash aerosol sampling system

BLPI:
0.0324 - 15.7 μm

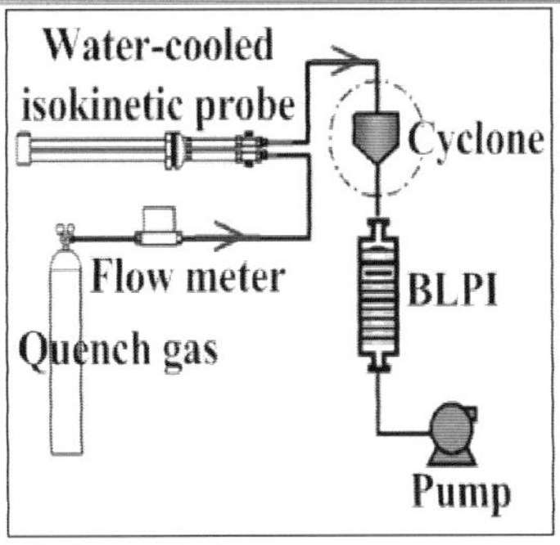

+

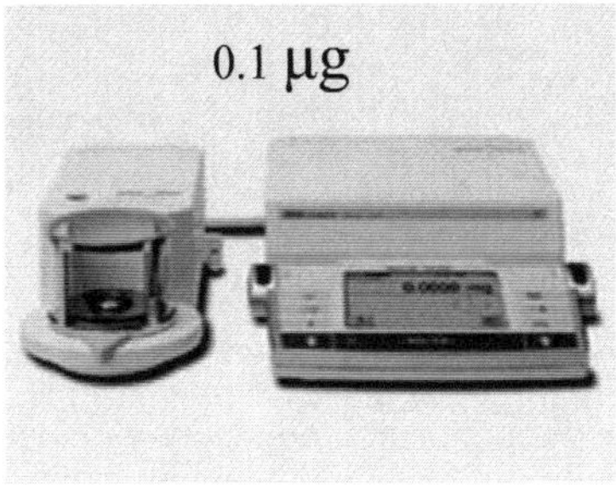

SMPS:
0.0143 - 0.6732 um
APS:
0.532 - 20 um

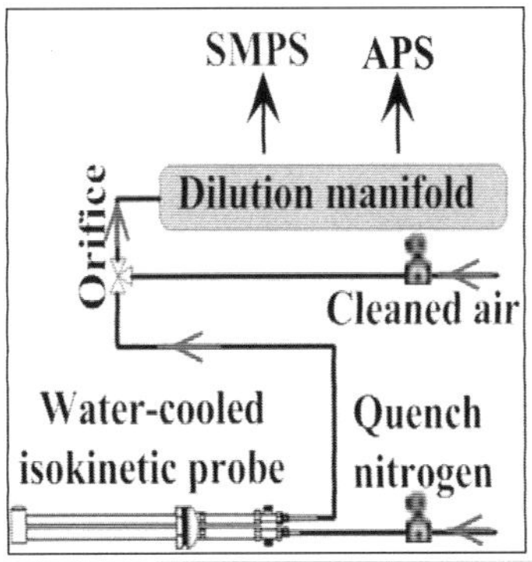

THE UNIVERSITY OF UTAH

Combustion parameters

Composition Analysis

ASH (%)	C (%)	H (%)	N (%)	S (%)	O (diff) (%)	H_2O (%)	Volatile (%)	Fixed Carbon (%)	HHV (kJ/kg)
2.99	82.51	6.02	1.71	5.65	0.49	0.57	10.18	86.26	34447.6

Al_2O_3 (%)	CaO (%)	Fe_2O_3 (%)	MgO (%)	K_2O (%)	SiO_2 (%)	Na_2O (%)	SO_3 (%)	TiO_2 (%)	NiO (%)	V_2O_5 (%)
19.4	4.22	7.02	0.66	1.17	46.7	0.72	3.77	0.63	1.26	8.24

Combustion conditions

Tests conditions	OXY70	Air
Feeding rate of petroleum coke (kg/h)	1.36	1.36
Feeding rate of natural gas (kg/h)	1.36	1.36
Excess O_2 (vol%_dry)	3	3
Adiabatic flame temperature (K)	3095	2312
Flow rate of flue gas (m^3/h_standard condition)	11.85	37.73

THE UNIVERSITY OF UTAH

7

Temperature profile of flue gas

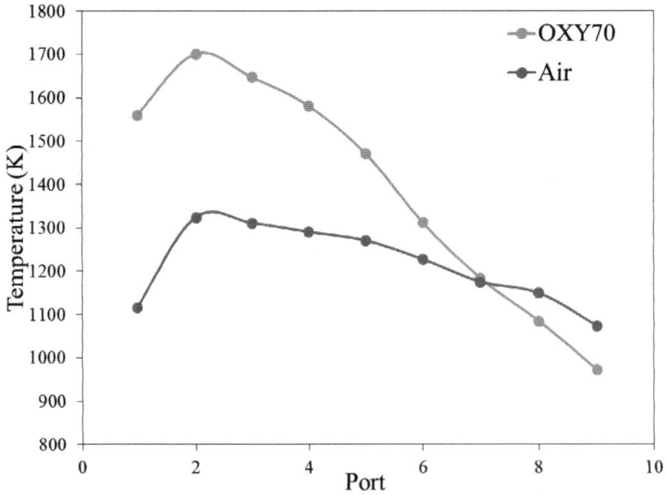

- Peak temperature of OXY70 (1701 K) is much higher than Air (1324 K)
- Peak temperatures for both cases are much lower than adiabatic flame temperatures.

THE UNIVERSITY OF UTAH

8

Aerosol particle size distribution (PSD)

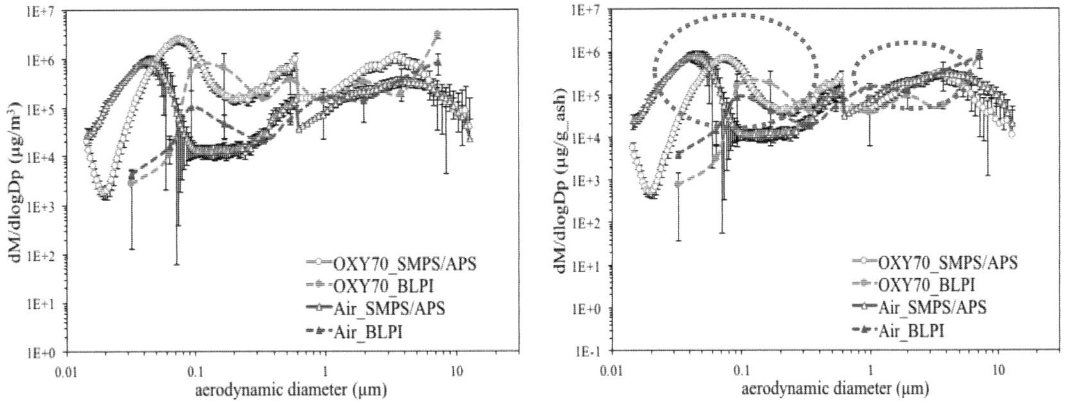

- BLPI data match with SMPS/APS data
- High temperature oxy-firing generates more submicron particles (99.1 mg/g_ash) than air-firing (72.4 mg/g_ash), but supermicron particles are not affected by temperature

THE UNIVERSITY OF UTAH

Aerosol size segregated composition

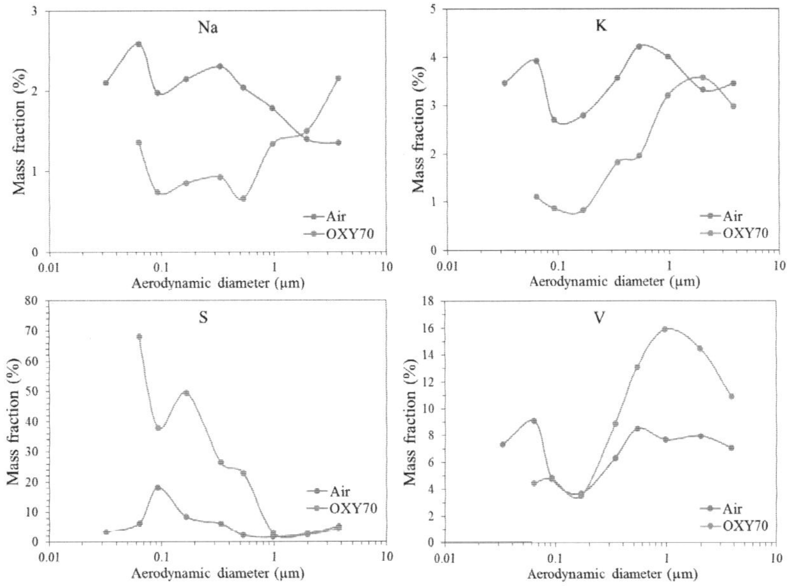

- More alkali metals are scavenged from submicron particles by alumina-silica under high temperature condition

THE UNIVERSITY OF UTAH

Composition of fouling ash

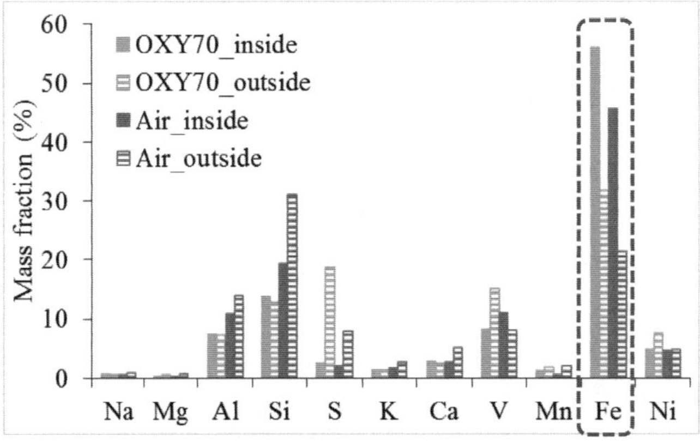

Al_2O_3 (%)	CaO (%)	Fe_2O_3 (%)	MgO (%)	K_2O (%)	SiO_2 (%)	Na_2O (%)	SO_3 (%)	TiO_2 (%)	NiO (%)	V_2O_5 (%)
19.4	4.22	**7.02**	0.66	1.17	46.7	0.72	3.77	0.63	1.26	8.24

➢ Composition of deposits are dominated by Fe

THE UNIVERSITY OF UTAH

SEM and EDS analysis for deposits

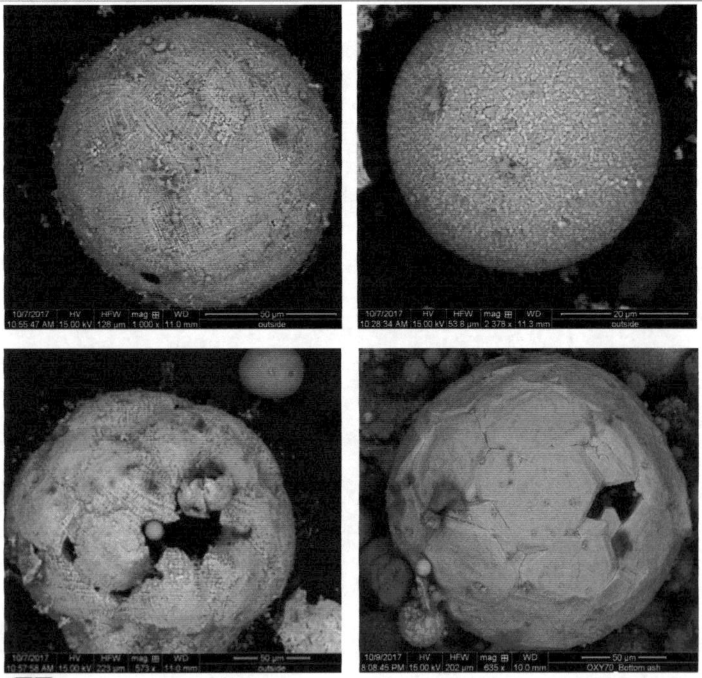

Na	0
Mg	0.17
Al	1.63
Si	2.30
S	1.05
K	0.33
Ca	0.82
V	1.33
Mn	0.61
Fe	**90.01**
Ni	1.75

THE UNIVERSITY OF UTAH

Composition of fouling ash

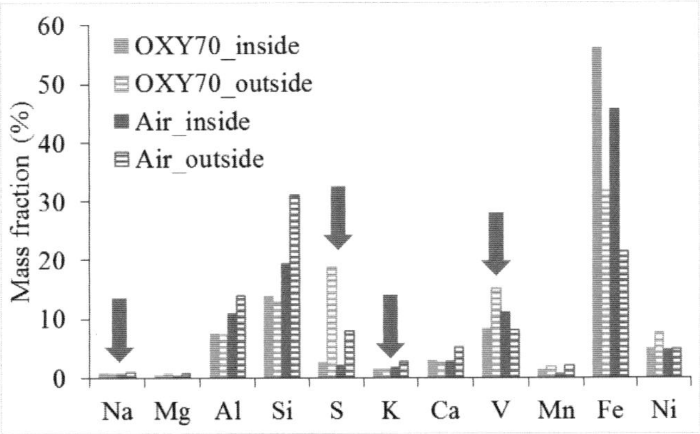

- Composition of deposits are dominated by Fe
- OXY70 deposits have more S and V, but less Na and K than Air, which are consistent with aerosol composition analysis

Sulfur and carbon analysis

Condition	OXY70		Air	
Content	S (%)	C (%)	S (%)	C (%)
outside deposit	3.21	79.0	1.13	28.0
fly ash	3.43	11.9	1.55	13.3
bottom ash	3.12	68.8	1.89	36.8

- Air firing seems surprisingly more efficient to burn petcoke than OXY70
- Sulfur content in OXY70 is higher than Air case

Correlation of deposition rates with PM1

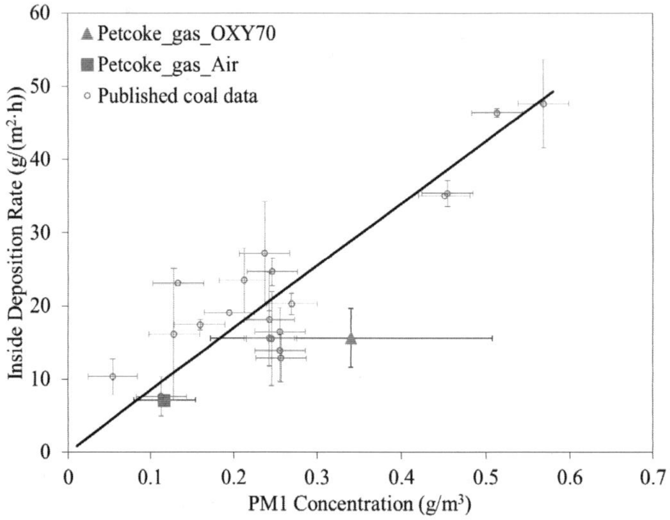

Published data from "Zhan, Z.; Wendt, J. O. L. Energy & Fuels **2017**, 31 (3), 2198–2202."

- Inside deposition rates of petcoke show positive relationship with concentration of PM1, which is consistent with previous findings in coal.

THE UNIVERSITY OF UTAH

Conclusions

- More alkali metals are scavenged from ultrafine fly ash by alumina-silica under high temperature condition
- The composition of fouling ash can be predicted by ash aerosol
- Fe is enriched in fouling ash, especially for high temperature condition
- Mixing between oxidant gas and petcoke is more important for carbon burnout than flame temperature
- Inside deposition rates can be predicted by the concentration of PM1

THE UNIVERSITY OF UTAH

Acknowledgements

- National Science Foundation under Award 1603249
- Praxiar for contributing the O_2 and CO_2 supply
- Brittany Walker, Dave Wagner, Paulo Perez and Brian van Devener

THE UNIVERSITY OF UTAH

17

Thank you!

THE UNIVERSITY OF UTAH

18

499074 Mercury Oxidation over Cu-SSZ-13 Catalysts under Flue Gas Conditions

Abstract:

Increasing concerns over mercury (Hg) emissions from coal-fired power plants have led the creation of new emission standards such as the recent Mercury and Air Toxic Standards, which requires a 90% reduction of Hg emissions. These new standards have led to the implementation of various novel technologies to control Hg emissions. Selective catalytic reduction (SCR) catalysts have shown promise as bifunctional catalysts for limiting both nitrogen oxide (NO_x) and Hg emissions through simultaneous NO_x reduction and Hg oxidation. The catalytic oxidation of elemental Hg, which is difficult to capture without dedicated Hg capture systems, to Hg^{+2} via HCl enables the capture of Hg^{2+} in the existing wet flue gas desulfurization unit as it is water soluble. Previous studies have investigated the possibility of utilizing current commercial SCR catalysts to fill this role, however these catalysts often have unsatisfactory performance when it comes to reducing Hg pollution to the levels required by current emission standards. To develop the next generation of SCR catalysts, a copper (Cu) exchanged small pore zeolite (Cu-SSZ-13) was investigated for its possible application in coal-fired power plants as a new bifunctional SCR and Hg oxidation catalyst. Cu-SSZ-13 has shown promise as a SCR catalyst for diesel exhaust, but has not been studied in coal-fired power plants.

In this study, the activity of Cu-SSZ-13 was investigated for its ability to oxidize Hg under a variety of flue gas conditions. The Cu-SSZ-13 catalyst was compared to SSZ-13 to understand the effect of Cu in the Hg oxidation mechanism. The Hg oxidation performances of both zeolites were compared to that of a commercial SCR catalyst (V-Ti-W), which served as a benchmark. All catalysts required hydrogen chloride (HCl) to exhibit any Hg oxidation with the zeolites outperforming the commercial catalyst. However, the activity of all catalysts significantly decreased in the presence of either sulfur oxides (SO_x) or SCR reactants (nitrogen oxide (NO)+ammonia (NH_3)) in the flue gas, with SO_x having a greater effect on Cu-SSZ-13 and SCR reactants having a greater one on the commercial catalyst. Additionally, it was determined that Cu was not the active site for Hg oxidation but was necessary for the SCR reaction. In fact, in the presence of SO_2, Cu actually contributed to the deactivation of the catalyst for Hg oxidation, likely due to the formation of sulfates as confirmed by XPS analysis.

502807 Mediated Oxycombustion with Integrated Uncoupled Oxygen Supply Â (MOBIUS)

Abstract:

Combustion systems that circumvent or reduce the energy penalty for carbon dioxide separation are being vigorously investigated. The approach we have pioneered aims at synergistically combining the basic objectives of three such technologies, namely oxy-combustion, chemical looping and liquid anode DCFC, and transforming them into a single lean, thoroughly optimized single unit process - Mediated Oxycombustion with Integrated Uncoupled Oxygen Supply (MOBIUS) - through innovations that overcome the bottlenecks of each of these technologies such as a) controlled combustion in a molten oxide to overcome the issues related to high temperatures (oxycombustion); b) obviating the need of solids transport between reactors (CLC); and c) avoiding operational issues such as electrical resistance, ash separation, impurity tolerance etc (DCFC). The system (simplistic concept in Figure 1) synergistically combines the effective aspects of the above mentioned technologies along with innovations to overcome the hurdles faced by these technologies. In MOBIUS, the solid fuel is combusted in the absence of any nitrogen as in oxy-combustion. However, the carbon is not combusted directly by oxygen but by a metal oxide as in a CLC. Unlike conventional CLC, the MeO_x and Me in the proposed Chemical Looping with Oxygen Uncoupling (CLOU) systems are in the liquid phase to improve the combustion rates and the mobility of the metal and the oxide. The metal is re-oxidized at the walls of the combustor with oxygen that is uncoupled from the reactor wall material. The individual subsystem studies will be presented and the integration of the various subsystems into a single reactor system will be discussed. Results from the subsystem will include the combustion rates, metal oxidation rates, oxygen uptake, transport and release rates under various conditions and composition of the membrane (during the membrane evaluation). The long terms operation of the integrated system will be demonstrated.

497932 Mobility Size and Effective Density of Carbonaceous Aerosols

Abstract:

Mobility Size and Effective Density of Carbonaceous Aerosols

Georgios A. Kelesidis, Eirini Goudeli, Sotiris E. Pratsinis

Particle Technology Laboratory, Institute of Process Engineering,

Department of Mechanical & Process Engineering, ETH Zurich, Switzerland

The mobility size estimation of evolving fractal-like soot aggregates is not trivial, as it depends on their structure as well as the number and size of their constituent primary particles. A scaling law describing the mobility size of agglomerates consisting of monodisperse primary particles in point contact (Sorensen, 2011) overestimates the mobility diameter of mature soot aggregates from diffusion flames for a given number of primary particles (Rissler et al., 2013). Rissler et al. (2013) attributed this difference to the primary particle polydispersity and chemical bonding induced by surface growth (aggregation), neglected in the above scaling law (Sorensen, 2011). Typically population balance models for soot formation estimate the mobility diameter assuming it is equal to the gyration diameter (Yapp et al., 2015).

Here, the Discrete Element Model of nascent soot agglomeration and surface growth by acetylene pyrolysis (Kelesidis et al., 2017) is extended to investigate the dynamics of mature soot formation at longer residence times and different flow regimes than nascent soot in the absence of soot oxidation. The evolution of soot mean gyration, mobility and primary particle diameters as function of the mean number of primary particles per aggregate are elucidated. The common assumption that the mobility is equal to the gyration diameter leads to 40 % underestimation of the nascent soot mobility size and 30 % overestimation of the mature soot aggregate mobility diameter.

Relationships for the soot mobility size and effective density are derived by following the evolution from nascent to mature soot morphology and compared to the correlation of Sorensen (2011) and mass-mobility measurements in quenched diffusion flames (Rissler et al., 2013; Yon et al., 2015). The scaling law for agglomeration in the absence of surface growth (Sorensen, 2011) overestimates the measured mature soot aggregate mobility size by 37 % and underestimates its effective density by 44 %. Accounting, however, for the primary particle polydispersity and chemical bonding or aggregation induced by surface growth reduces the deviation from the mass-mobility measurements down to 7 and 11 % for the mobility size and the effective density, respectively.

References:

Kelesidis, G.A., Goudeli, E., Pratsinis, S.E. (2017) *Proc. Combust. Inst.* **36**, 29-50.

Rissler, J., Messing, M.E., Malik, A.I., Nilsson, P.T., Nordin, E.Z., Bohgard, M., Sanati, M., Pagels, J.H. (2013)

Aerosol Sci. Technol. **47**, 792-805.

Sorensen, C.M. (2011) *Aerosol Sci. Technol.* **45**, 765-779.

Yapp, E.K.Y., Chen, D., Akroyd, J., Mosbach, S., Kraft, M., Camacho, J., Wang, H. (2015) *Combust. Flame* **162**, 2569-2581.

Yon, J., Bescond, A., Ouf, F.X. (2015) *J. Aerosol Sci.* **87** (2015) 28-37.

490602 Hybrid Geothermal Energy Conversion - a Potential Solution for Low-Temperature Geothermal Resources

Abstract:

Low- to moderate- temperature geothermal resources are typically used for direct heat energy applications, whereas deep, high-temperature, geothermal resources are often used for electricity generation. But deep geothermal resources usually require enhancement of the permeability of the hot dry rock at these depths (4-6 km), which can induce seismicity. Here we present an alternative geothermal energy utilization approach that uses shallower, lower-temperature, naturally permeable formations. Our approach can generate electricity while substantially reducing drilling costs and the risk of induced seismicity. This approach uses geothermal heat to supplement a secondary energy source, thereby potentially enabling the use of geothermal energy for electricity generation in many regions worldwide that are not viable with current approaches. We determine the net power output, energy conversion efficiencies, and economics of these hybrid power plants, where the geothermal power plant uses supercritical carbon dioxide ($scCO_2$) as the subsurface working fluid. When $scCO_2$ is used underground, the overall system stores 100% of the injected CO_2, while generating geothermal power, constituting a Carbon Capture Utilization and Storage (CCUS) system. Parameters varied include geothermal reservoir depth (2.5-4.5 km) and turbine inlet temperature (100-220°C). We find that hybrid power plants outperform two individual, i.e., stand-alone geothermal and waste-heat power plants. Furthermore, such hybrid power plants are more economical, and result in lower capital costs per electricity produced ($/kW) than separate power plants.

497836 On the Origin of Preferred Bicarbonate Production from Carbon Dioxide (CO_2) Capture into Aqueous 2-Amino-2-Methyl-1-Propanol (AMP)

Abstract:

Chemical absorption by aqueous amine-based solvents appears the most promising near-term solution for CO_2 capture from flue gas because of their ability to capture CO_2 at low pressure with adequate absorption/desorption kinetics.[1,2] Monoethanolamine (MEA) is one of the most widely used and studied alkanolamines, but its expansion for commercial use tends to be limited due to degradation problems and high parasitic energy consumption during solvent regeneration.[3-4] 2-amino-2-methyl-1-propanol (AMP) and other sterically hindered primary amines have been suggested as an alternative because they exhibit both high CO_2 loading capacities and relatively fast absorption rates, in addition to low degradation and corrosion rates.[5–8] However, the mechanism underlying the reaction between CO_2 and aqueous AMP still remains controversial.

It is now well adopted that CO_2 capture by most primary amines including MEA in an aqueous solution takes place via a two-step mechanism involving a zwitterion intermediate.[9-10] The zwitterion may subsequently undergo deprotonation by another amine to form a carbamate ion and a protonated amine, leading to a loading capacity of 0.5 mol CO_2/mol amine. Conversely, the primary products of the reaction of aqueous AMP with CO_2 have been found to be bicarbonate (HCO_3^-) and protonated AMP ($AMPH^+$), with carbamate ($AMPCOO^-$) but whose concentration is about an order of magnitude lower than HCO_3^-.[11] It is also known that bicarbonate formation predominantly takes place in aqueous tertiary amines such as methyldiethanolamine (MDEA), yielding the theoretical maximum capacity of 1 mol CO_2/mol amine.[12] In the tertiary amine system, bicarbonate can be formed by the base-catalyzed hydration of CO_2, e.g., MDEA + $H_2O + CO_2 \rightarrow MDEAH^+ + OH^- + CO_2$ ® $MDEAH^+ + HCO_3^-$.[13]

However, according to previous experimental observations, the absorption rate of CO_2 in aqueous AMP tends to be about two orders of magnitude faster as compared to the MDEA case.[14] On the basis of the much higher absorption rate of AMP relative to tertiary amines, it has been speculated that first AMP and CO_2 react to form carbamate ($AMPCOO^-$), which undergoes hydrolysis to form bicarbonate due to its possible instability (i.e., $AMPCOO^- + H_2O \rightarrow HCO_3^- + AMPH^+$).[5,15] However, previous theoretical studies have predicted free energy barriers for the hydrolysis of both $AMPCOO^-$ and $MEACOO^-$ to HCO_3^- to be high and also comparable, implying that not only $MEACOO^-$ but also $AMPCOO^-$ are relatively stable and may not easily undergo hydrolysis to form HCO^{3-}.[16–19] Rather, it would be more likely that the carbamates would revert back to the amine and CO_2.[20] This gives a hint that preferred bicarbonate formation in the AMP-H_2O-CO_2 system might not be due to carbamate being less stable than the MEA case.

In this work, we investigate the factors underlying the preferred production of bicarbonate over carbamate from CO_2 absorption in aqueous AMP, with comparison to the MEA case where carbamate is predominantly formed, by explicitly taking into account both thermodynamic and kinetic contributions. First, we evaluated the thermodynamic favorability for bicarbonate and carbamate formation by calculating the changes in free energy (and enthalpy) using static QM with an implicit solvent model (and AIMD with explicit solvent). Our calculations show that both CO_2 capture mechanisms are thermodynamically favorable with similar exothermicities in aqueous AMP and MEA solutions. In addition, AIMD simulations demonstrate that AMP carbamate ($AMPCOO^-$) can be as stable as MEA carbamate ($MEACOO^-$) in an aqueous solution, which is consistent with previous theoretical studies predicting large and similar energy barriers for carbamate hydrolysis to form bicarbonate in both cases[19,17], i.e., $AMPCOO^-$ ($MEACOO^-$) + $H_2O \rightarrow HCO_3^-$ + $AMPH^+$ ($MEAH^+$); the comparable stability between $AMPCOO^-$ and $MEACOO^-$ is also well demonstrated by analysis of the N-C (in CO_2) bonding interaction. This may suggest that the relative concentrations of carbamate and bicarbonate could not be predicted only by their thermodynamic favorability, especially since it may take too long to reach equilibrium at typical operating conditions.

From AIMD simulations, we observed that the AMP + $H_2O \rightarrow AMPH^+ + OH^-$ reaction frequently occurs, rather than carbamate formation (i.e., 2AMP + $CO_2 \rightarrow AMPCOO^- + AMPH^+$) which tends to occur readily in the case of MEA. If OH^- does not abstract a proton to form H_2O, bicarbonate can be formed via the OH^- + $CO_2 \rightarrow HCO_3^-$ reaction.[21,22] This can also occur in one step, via the amine-catalyzed hydration of CO_2 proposed by Donaldson and Nguyen[13], (i.e., AMP + H_2O + $CO_2 \rightarrow AMPH^+ + HCO_3^-$). These results suggest that bicarbonate formation may be more kinetically favorable relative to carbamate formation in aqueous AMP. The enhanced AMP protonation reaction is apparently related to its high basicity. Our electronic structure analysis shows that the steric hindrance of the CH_3 groups of AMP causes a more planar-like configuration around the central C atom (C_N) which is attached to N. The tendency of sp^2-like hybridization results in charge redistribution in which the C_N atom is more positively charged and the N atom is more negatively charged. This in turn strengthens the interaction between the N in AMP and surrounding H_2O molecules while suppressing the accessibility of CO_2 to the N site to form carbamate, relative to the MEA case, as confirmed by analysis of radial/spatial distribution functions from MD simulations.

Based on our calculation results, we attribute the preferential formation of bicarbonate in CO_2 absorption into an aqueous AMP solution largely to kinetic factors. As discussed above, the strong interaction between N (in AMP) and H (in H_2O) suppresses the reaction with CO_2 and promotes the protonation reaction, while the bicarbonate and carbamate reaction routes exhibit similar thermodynamic favorability. This study highlights that not only thermodynamic but also kinetic factors should be considered in describing CO_2 capture by aqueous amines.

References

1. Rochelle GT. Amine scrubbing for CO_2 capture. *Science*. 2009;325(5948):1652-4.

2. Figueroa JD, Fout T, Plasynski S, McIlvried H, Srivastava RD. Advances in CO_2 capture technology—The U.S. Department of Energy's Carbon Sequestration Program. *Int J Greenh Gas Control*. 2008;2(1):9-20.

3. Zhu D, Fang M, Lv Z, Wang Z, Luo Z. Selection of Blended Solvents for CO_2 Absorption from Coal-Fired

Flue Gas. Part 1: Monoethanolamine (MEA)-Based Solvents. *Energy & Fuels*. 2012;26:147-153.

4. Goff GS, Rochelle GT. Monoethanolamine Degradation: O_2 Mass Transfer Effects under CO_2 Capture Conditions. *Ind Eng Chem Res*. 2004;43(20):6400-6408.

5. Sartori G, Savage DW. Sterically Hindered Amines for CO_2 Removal from Gases. *Ind Eng Chem Fundam*. 1983;22(2):239-249.

6. Chakraborty AK, Astarita G, Bischoff KB. CO_2 Absorption in Aqueous Solutions of Hindered Amines. *Chem Eng Sci*. 1986;41(4):997-1003.

7. Lepaumier H, Picq D, Carrette P-L. New Amines for CO_2 Capture. I. Mechanisms of Amine Degradation in the Presence of CO2. *Ind Eng Chem Res*. 2009;48(20):9061-9067.

8. Lepaumier H, Picq D, Carrette P-L. New Amines for CO_2 Capture. II. Oxidative Degradation Mechanisms. *Ind Eng Chem Res*. 2009;48(20):9068-9075.

9. Caplow M. Kinetics of Carbamate Formation and Breakdown. *J Am Chem Soc*. 1968;90(24):6795-6803.

10. Danckwerts PV. The Reaction of CO_2 with Ethanolamines. *Chem Eng Sci*. 1979;34(4):443-446.

11. Ciftja AF, Hartono A, Svendsen HF. Experimental study on carbamate formation in the AMP–CO_2–H_2O system at different temperatures. *Chem Eng Sci*. 2014;107:317-327.

12. Jakobsen JP, Krane J, Svendsen HF. Liquid-Phase Composition Determination in CO_2 - H_2O - Alkanolamine Systems: An NMR Study. *Ind Eng Chem Res*. 2005;44(26):9894-9903.

13. Donaldson TL, Nguyen YN. Carbon Dioxide Reaction Kinetics and Transport in Aqueous Amine Membranes. *Ind Eng Chem Fundam*. 1980;19(3):260-266.

14. Vaidya PD, Kenig EY. CO_2-Alkanolamine Reaction Kinetics: A Review of Recent Studies. *Chem Eng Technol*. 2007;30(11):1467-1474.

15. Alper E. Reaction Mechanism and Kinetics of Aqueous Solutions of 2-Amino-2-methyl-1 -propanol and Carbon Dioxide. *Ind Eng Chem Res*. 1990;29(8):1725-1728.

16. Xie H-B, He N, Song Z, Chen J, Li X. Theoretical Investigation on the Different Reaction Mechanisms of Aqueous 2-Amino-2-methyl-1-propanol and Monoethanolamine with CO_2. *Ind Eng Chem Res*. 2014;53(8):3363-3372.

17. Davran-Candan T. DFT Modeling of CO_2 Interaction with Various Aqueous Amine Structures. *J Phys Chem A*. 2014;118(25):4582-4590.

18. Mehdizadeh H, Gupta M, Kim I, Da Silva EF, Haug-Warberg T, Svendsen HF. AMP–CO_2–water thermodynamics, a combination of UNIQUAC model, computational chemistry and experimental data. *Int J Greenh Gas Control*. 2013;18:173-182.

19. Yamada H, Matsuzaki Y, Higashii T, Kazama S. Density Functional Theory Study on Carbon Dioxide Absorption into Aqueous Solutions of 2-Amino-2-methyl-1-propanol Using a Continuum Solvation Model. *J Phys Chem A*. 2011;115(14):3079-86.

20. Da Silva EF, Svendsen HF. Computational chemistry study of reactions, equilibrium and kinetics of chemical CO_2 absorption. *Int J Greenh Gas Control*. 2007;1(2):151-157.

21. Stowe HM, Vilčiauskas L, Paek E, Hwang GS. On the origin of preferred bicarbonate production from

carbon dioxide (CO_2) capture in aqueous 2-amino-2-methyl-1-propanol (AMP). *Phys Chem Chem Phys.* 2015;17(43):29184-29192.

22. Nakai H, Nishimura Y, Kaiho T, Kubota T, Sato H. Contrasting mechanisms for CO2 absorption and regeneration processes in aqueous amine solutions: Insights from density-functional tight-binding molecular dynamics simulations. *Chem Phys Lett.* 2016;647:127-131.

493040 Evaluating the Performances of Reduction Process in the Solar Thermochemical Two-Step CO_2 Splitting Based on Ceria Redox Reactions

Abstract:

The emission of carbon dioxide is a severe problem in the current society. There are tremendous technologies that provide approaches to address the surplus CO_2. Solar thermochemical reactor provides an attractive approach that utilizes the most common solar radiation as the thermal driving force to motivate the reaction between CO_2 and metal oxides, which is so called metal oxide redox pair based thermochemical cycles. By two-step redox process, the incoming CO_2 can be transformed to CO for further applications. Among the diversified metal oxide redox pairs, CeO_2/CeO_{2-}î´ shows an anticipated performance, due to its fast redox kinetics, high crystallographic stability of a wide range of reacting oxygen non-stoichiometry, and relatively high oxygen solid-state conductivity.

In this work, a numerical analysis is applied to study on the performance of CeO_2 reduction process in a segment of cavity-receiver reactor. The CeO_2 is presented as catalyst particles in the domain to mimic the structure of porous medium. A three-dimension transient model is applied to simulate the momentum, heat, and mass transfer in the domain. CeO_2 particle size, void fraction of domain, purge gas velocity and radiation flux rate are four key parameters that are analyzed in this work. By investigating on the influences of each factor to the reduction reaction, an optimized reaction rate can be achieved, which is beneficial to improve the solar to fuel efficiency of the reactor.

494372 Mass Transfer Studies of Carbondioxide Absorption in Sodium Hydroxide in Millichannels

Abstract:

During the past several decades, CO_2 levels in the atmosphere have been dramatically increasing due to the man-made emissions such as fossil fuel-fired power plants. This results in global warming thus leading an increase in the need to study different ways to capture the emitted CO_2 directly from the exhausts of power plants or atmosphere. There are several ways to remove CO_2, such as absorption into a liquid solvent, adsorption into a solid, cryogenic separation, permeation through membranes and photochemical conversion. In most industries, the absorption of CO_2 in chemical solvents (in absorption towers) is used for CO_2 capture. In these towers, the mass transfer along with chemical reactions take place between the gas and liquid phase. It is important to understand these processes in detail. These flow patterns are difficult to maintain in large scale industrial absorbers. So to get accurate information controlled gas-liquid absorption experiments are carried out in milli-channels in this work under controlled atmosphere. The absorption experiments of CO_2 in varying concentrations of sodium hydroxide solution are carried out in T-junction glass milli-channels with a circular cross section (inner diameter of 2mm). The gas and liquid flow rates are controlled by a mass flow controller (MFC) and a Harvard syringe pump respectively. The slug flow in the channel is recorded using a camera and the videos are analysed. The gas slug of pure CO_2 is found to decrease in size along the length of the channel due to absorption of gas in the liquid. This is also captured with the model developed and the mass transfer characteristics are studied. The pressure drop across the channel is determined by sum of the pressure drops from the gas slugs and the liquid plugs. A dimensionless correlation for the mass transfer coefficient is developed in terms of Sherwood number and compared with the existing correlations in the literature. They are found to be in close agreement with each other. In this case, due to the presence of chemical reaction, the enhancement of mass transfer is obtained. This is quantified with the help of an enhancement factor.

Keywordsâ€" Absorption, Enhancement factor, Mass transfer coefficient, Sherwood number

495024 Design and Evaluation of Thermodynamic Conditions for an Off-Shore Topside CO_2 Injection System

Abstract:

Carbon capture and storage (CCS) is one of the dominant technologies to tackle the global warming issue. The transport of CO_2 for geological storage is economically feasible by ship when the storage site location is off-shore and installment of an off-shore pipeline requires a huge capital cost. Ship transportation requires the captured CO_2 to be in liquid phase under pressurized thermodynamic conditions. The injection of liquid CO_2 into the geological reservoir involves pressurization and heating in order to maintain the safe well head operating conditions. Most of the recent studies focusing on ship transportation of CO_2 set their research boundary upstream of an injection system. On the other hand, studies investigating the storage mechanism in the EOR field or other geological locations assumed the fixed CO_2 conditions from the literature. The purpose of this study is to evaluate the design of CO_2 injection system for an off-shore storage site where CO_2 is available in the liquid phase at the vessel. Many previous studies considered the liquid CO_2 at the thermodynamic conditions of 6-7 bara and -52 ^oC assuming large CO_2 transport volumes. However, currently there are no commercial ships which transport liquid CO_2 at such cryogenic conditions and only small size ships up to 1,500 m^3 are being used for CO_2 transport at -20 °C and 20 bar. This study evaluates and compares the two thermodynamic conditions (7 bara and 20 bara) considering all the design and operational constraints of a liquid CO_2 injection system. This study also aims to fill the study gaps from the previous researches. For example, all the previous studies assumed the CO_2 stream as a pure liquid CO_2, however, practically this may not be possible since some amount of water is left in the CO_2 stream posing a threat of hydrate formation. In this study, we incorporate the CO_2 vapor return line to the pressure vessel which is an important aspect from safety point of view. This paper also investigates the application of two-stage rankine cycle to utilize the cold energy available from liquid CO_2 which can help to improve the overall efficiency of the injection system compared to that of the base case design. The working fluid employed in both the stages of rankine cycle is ammonia because of its highest power generation performance, easy availability and environment friendly characteristics compared to other working fluids. A well head injection pressure and temperature of 65 bar and 10 ^oC is set as a base case value for a storage depth of 1000 m to avoid any hydrate formation in the injection well or at the well outlet. The results show that the net specific power consumption for the two conditions (7 bara and 20 bara) is 4.52 and 1.90 kWh per ton of CO_2 injected respectively. Economic comparison has been performed to investigate the impact of operating conditions on the CO_2 injection process. The results show that the CAPEX for both the conditions is almost same since equipment involved in both the processes is the same. However, the OPEX for 7 bara design condition is approximately 2.4 times high compared to that of the 20 bara conditions. The high OPEX is attributed from the high operational energy requirement of the process. For a project life of 10 years, the specific cost per unit ton of CO_2 injected for the two design conditions (7 bara and 20 bara) came out to be 1.20 \$ and 1.03 \$ respectively. Finally, a sensitivity analysis has been done in order to investigate the effect of some important variables in the study.

CHINA UNIVERSITY of PETROLEUM, BEIJING

2017 AICHE Annual Meeting

The Effect of New Nanocomposite on Fluidity of Waxy Crude Oil in Low Temperature

China University of Petroleum – Beijing (CUPB)

***Huirong Huang**, Wei Wang*, Yanfen Ding, Zeheng Peng, Jing Gong**

w.wang@cup.edu.cn

ydgj@cup.edu.cn

CHINA UNIVERSITY of PETROLEUM, BEIJING

Background

- ☐ Special problems:

  Waxy crude oil, liquidity deterioration, pipeline blockage

- ☐ Multi-phase mixed transport:

  Oil, Gas, Water three-phase flow

- ☐ Urgent need:

  Improve the safety of marine pipelines

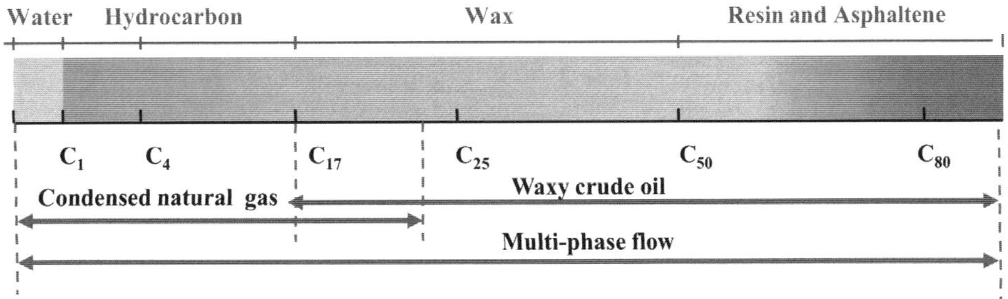

CHINA UNIVERSITY of PETROLEUM, BEIJING

How to solve this problem with efficient ways ?

Critical Technology

- ☐ physical solution

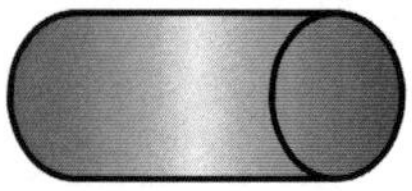

Heating

Dilution

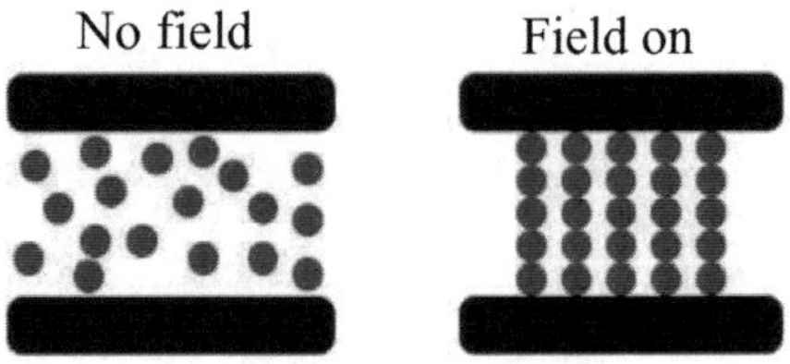

Electro-magnetic

- ☐ chemical additives

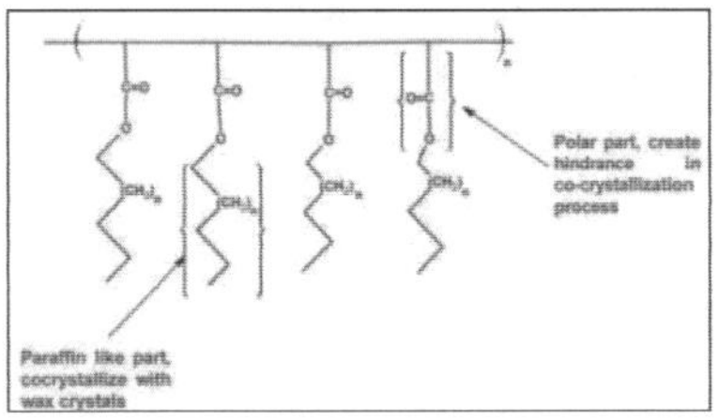

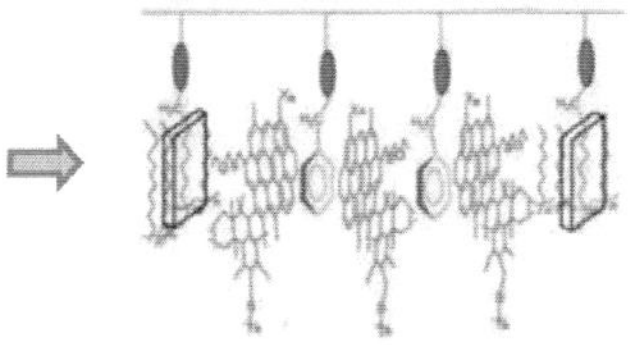

comb-like polymer

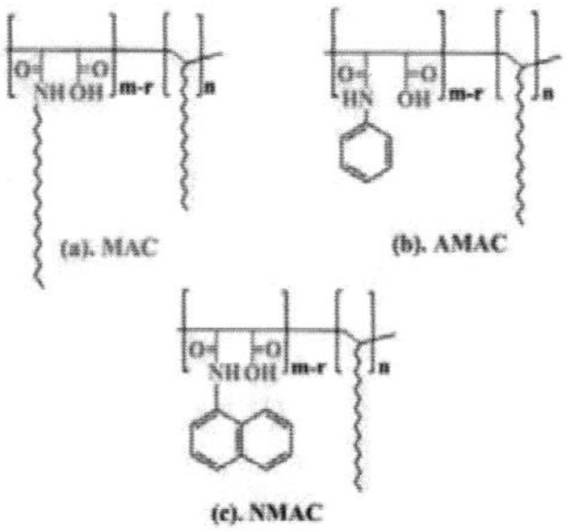

CHINA UNIVERSITY of PETROLEUM, BEIJING

How to solve this problem with efficient ways ?

Critical Technology

- ☐ chemical additives

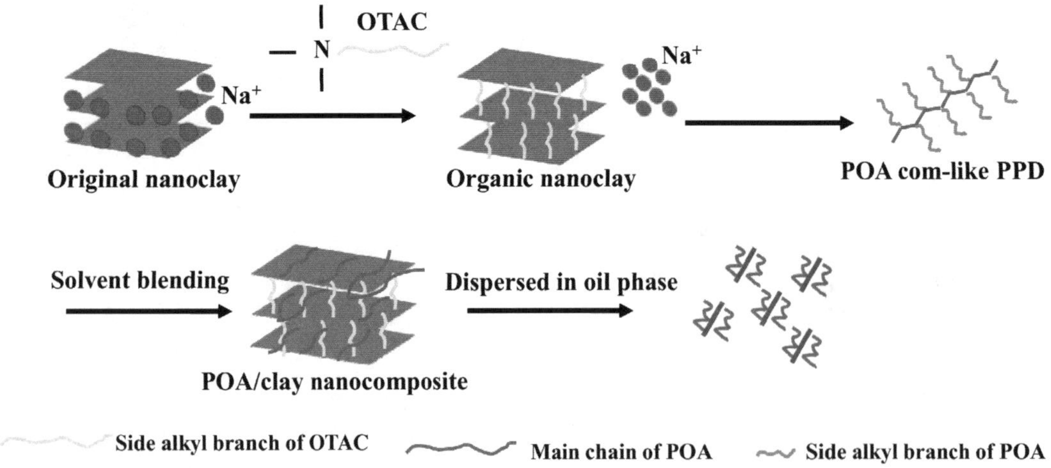

Critical Technology

□ Interaction Mechanism

Crystal nucleation, Co-crystallization, Adsorption, Solubilization

Microscopic | Mesoscopic | Macroscopic

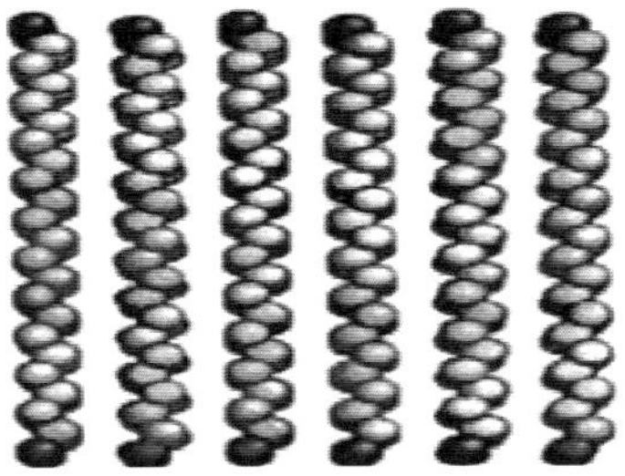

Wax crystals

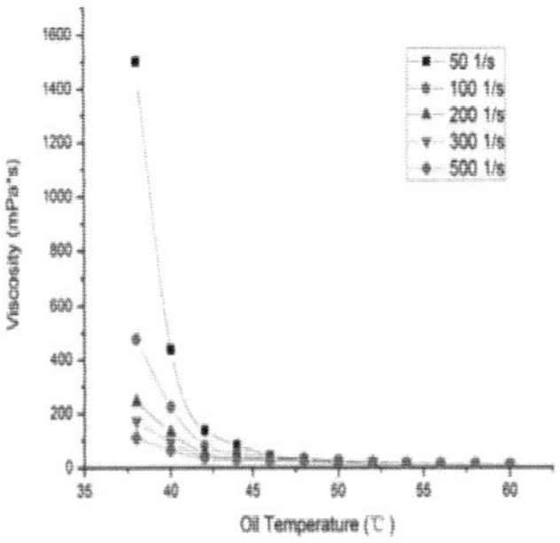

CHINA UNIVERSITY of PETROLEUM, BEIJING
厚积薄发 开物成务

Materials and Experimental Methods

□ Materials

a. **Mixed wax, D80, ethylene/vinyl acetate copolymers (EVA), Montmorillonite**

b. **The density of D80 is 0.795 kg/m^3 (15℃)**

c. **The viscosity is 2.09mm^2/s (25℃)**

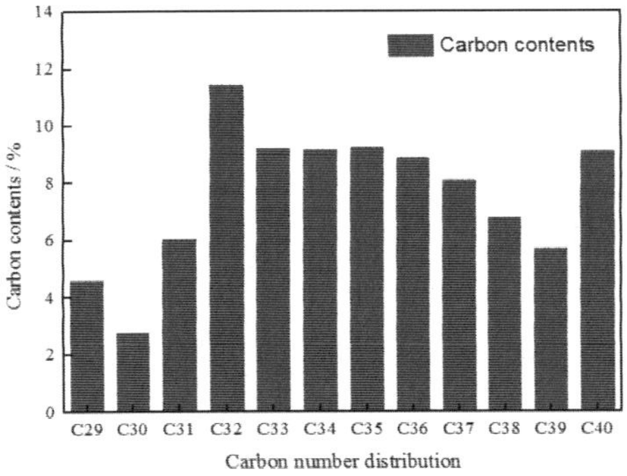

Carbon contents and distribution of mix wax

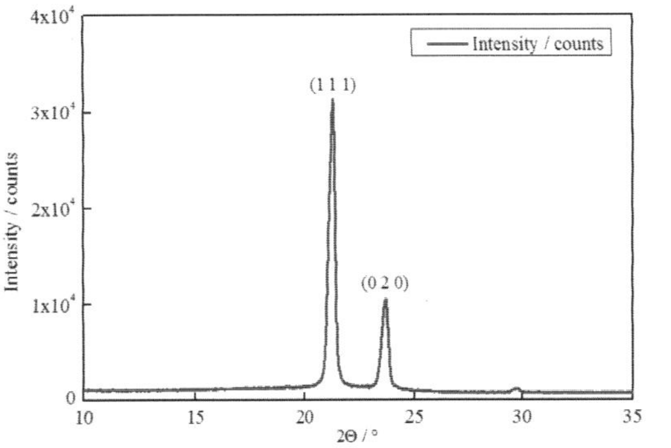

The XRD curve of mixed wax at room temperature

CHINA UNIVERSITY of PETROLEUM, BEIJING

厚积薄发 开物成务

Materials and Experimental Methods

☐ Materials

a. **Characterization of EVA and NPPD-2(composed of Montmorillonite and EVA)**

b. **The mass flow rate (MFR) of EVA and NPPD-2 is 20.5g/10min and 2.0g /10min**

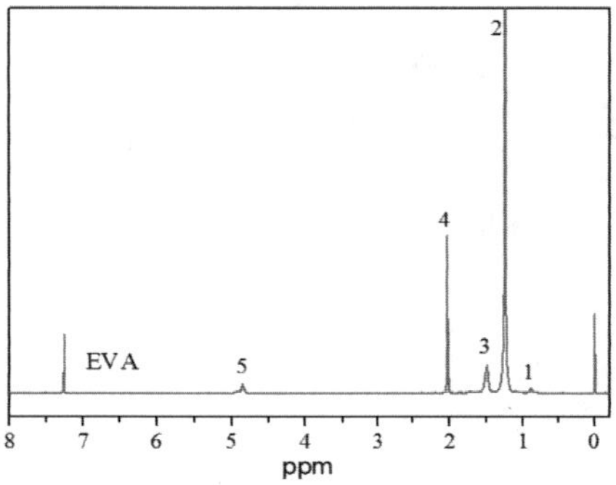

^{1}H NMR spectra of and EVA

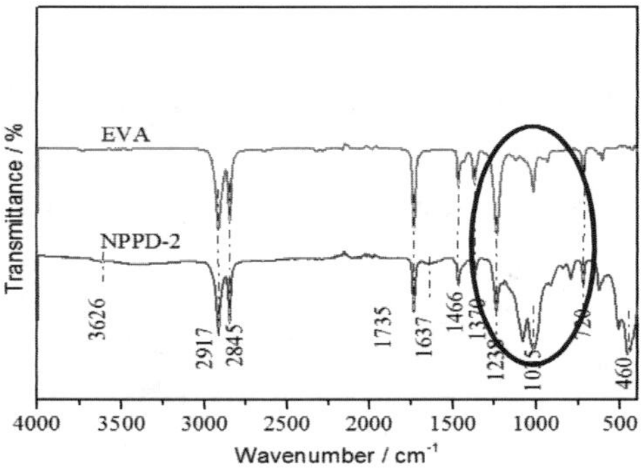

FTIR spectra of EVA and NPPD-2

CHINA UNIVERSITY of PETROLEUM, BEIJING

厚积薄发 开物成务

Materials and Experimental Methods

☐ Methods

a. **The rheometer (Gel point, Yield stress, Viscosity)** (Figure 1)

b. **The X-ray diffraction (Lattice parameters)** (Figure 2)

c. **The polarizing microscope BX51(crystal morphology)** (Figure 3)

Figure 1

Figure 2

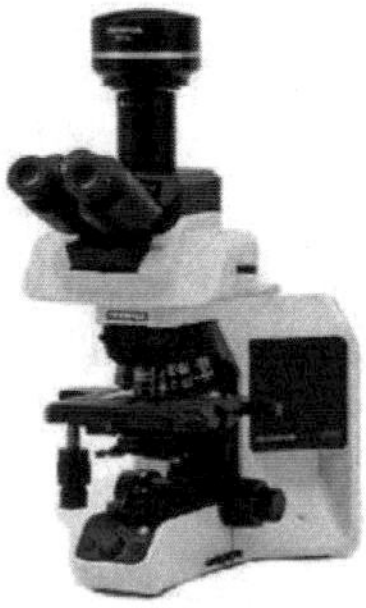

Figure 3

CHINA UNIVERSITY of PETROLEUM, BEIJING

The Effect of PPDs on Viscosity

PPD	shear rate/s^{-1}	Viscosity of model oil/ Pa.s	Doped oil 50ppm / Pa.s	Doped oil 100ppm / Pa.s	Doped oil 150ppm / Pa.s	Doped oil 200ppm / Pa.s
EVA	10.200	37.750	0.961	0.952	0.423	0.346
	50.000	1.888	0.082	0.073	0.060	0.056
	100.000	0.955	0.047	0.038	0.035	0.032
	150.000	0.735	0.038	0.029	0.029	0.026
	200.000	0.609	0.033	0.024	0.025	0.022
NPPD-2	10.200	37.750	0.349	0.418	0.299	0.290
	50.000	1.888	0.063	0.055	0.051	0.042
	100.000	0.955	0.044	0.039	0.034	0.025
	150.000	0.735	0.038	0.035	0.029	0.020
	200.000	0.609	0.034	0.032	0.026	0.017

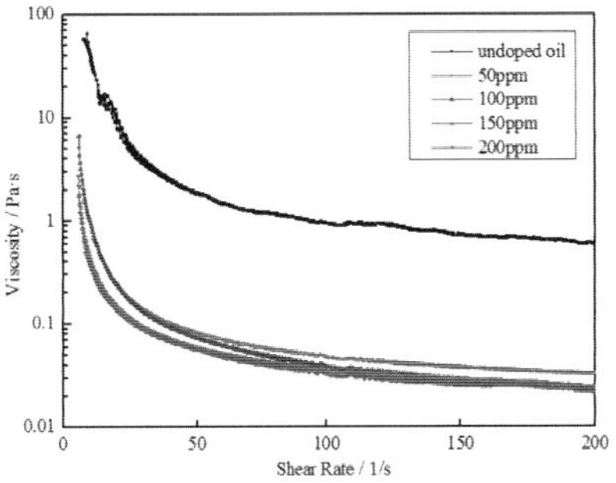

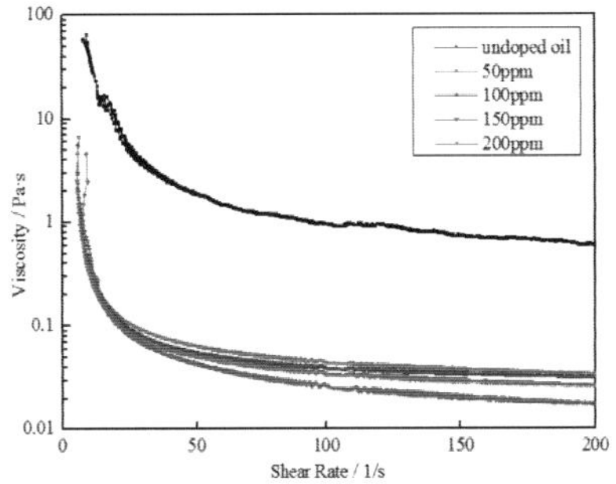

CHINA UNIVERSITY of PETROLEUM, BEIJING

The Effect of PPDs on Gelation

60℃ for 30 min

60℃ to 10 ℃(0.5℃/min), 0.01Pa

The gelation point represents the strength of the waxy network structure

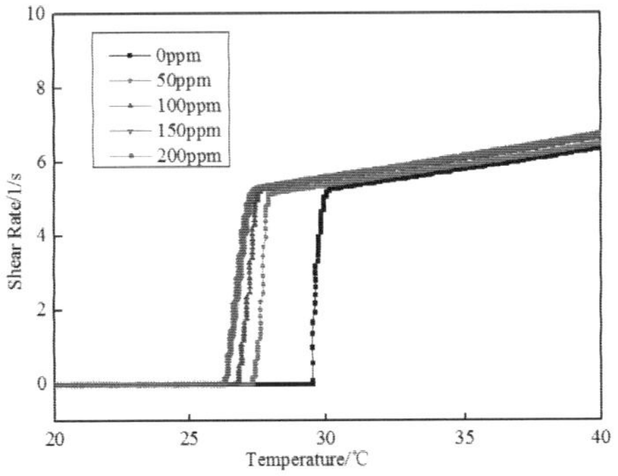

The gelation curves of undoped/doped oil with EVA

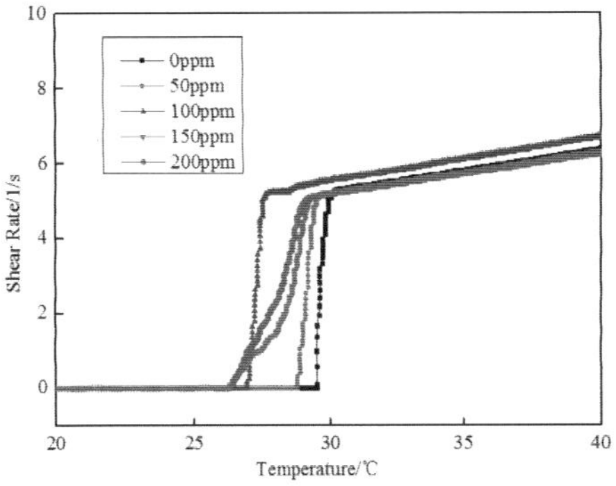

The gelation curves of undoped/doped oil with NPPD-2

The Effect of PPDs on Yield Stress

60℃ for 30 min → 60℃ to 10 ℃(0.5℃/min) → 0 to 600Pa at the rate of 5Pa / min

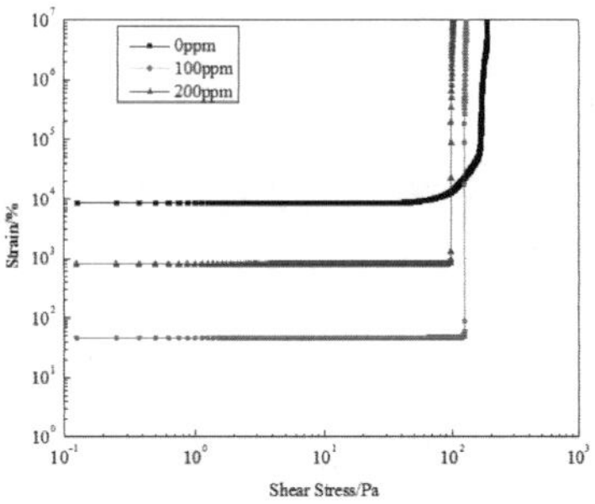

Yield curves（EVA）

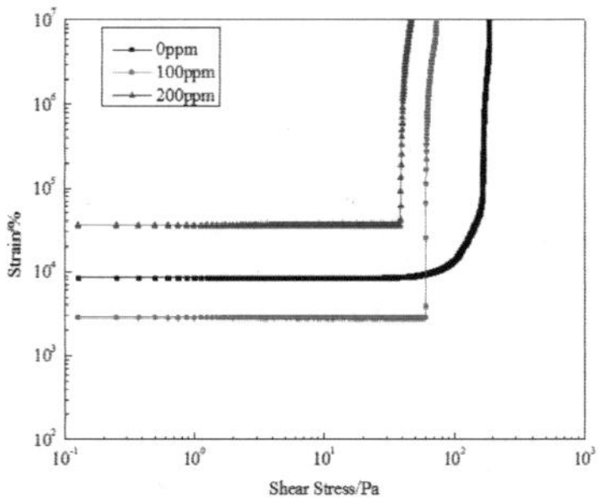

Yield curves（NPPD-2）

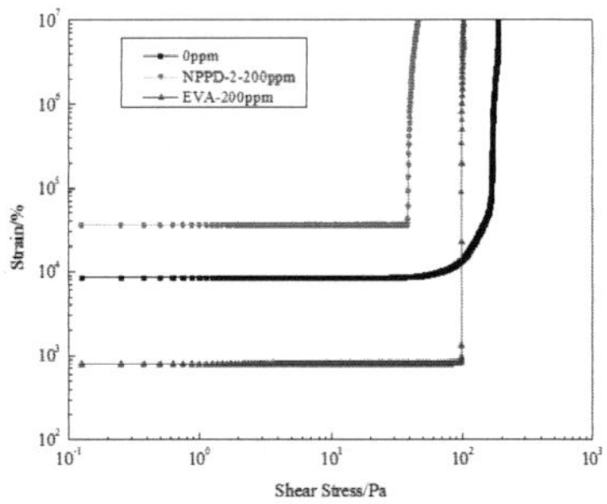

Yield curves（200ppm）

CHINA UNIVERSITY of PETROLEUM, BEIJING
厚积薄发 开物成务

The Effect of PPDs on WAT

$$r_c = \frac{2\gamma T^*}{\Delta H_f \Delta T}$$

γ is the interfacial tension, T^* is the solid-liquid equilibrium temperature expressed in kelvins, $\Delta T = T^* - T$ is the super-cooling, ΔH_f is the latent heat of fusion, and r_c is the radius of a critical nucleus.

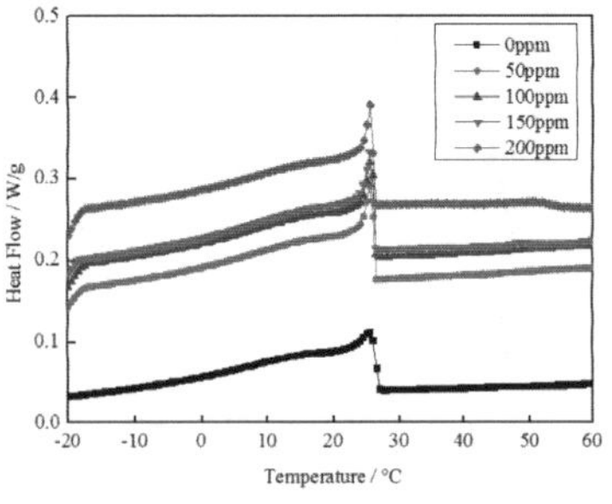

The WAT curves with EVA

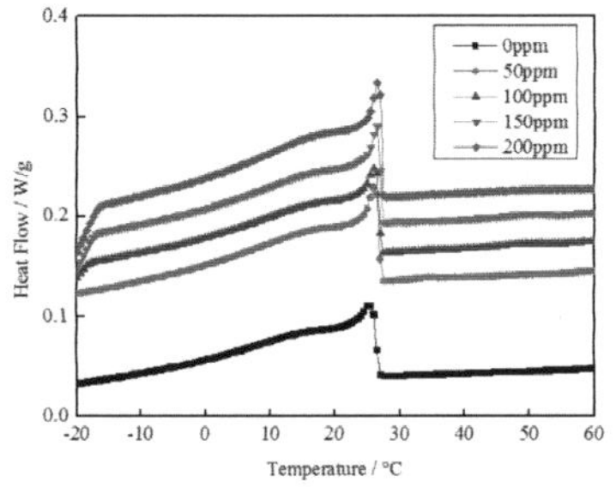

The WAT curves with NPPD-2

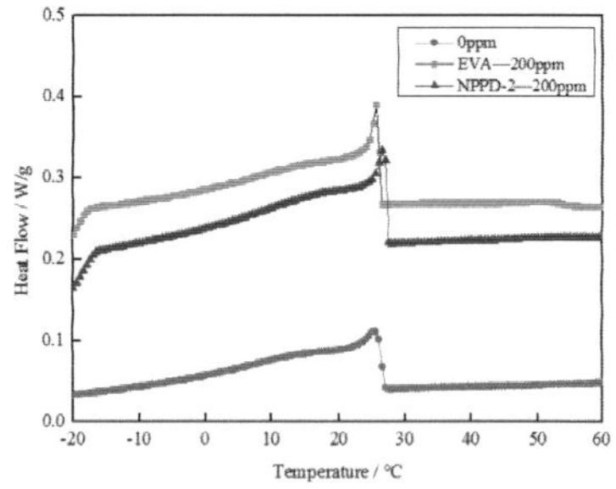

The WAT curves with at 200ppm

The X-ray Diffraction

$$2d_{hkl}\sin\theta_{hkl} = \lambda \qquad c = l*d_{00l} \qquad a = h*d_{h00} \qquad \frac{1}{d_{hkl}^2} = \frac{h^2}{a^2} + \frac{k^2}{b^2} + \frac{l^2}{c^2}$$

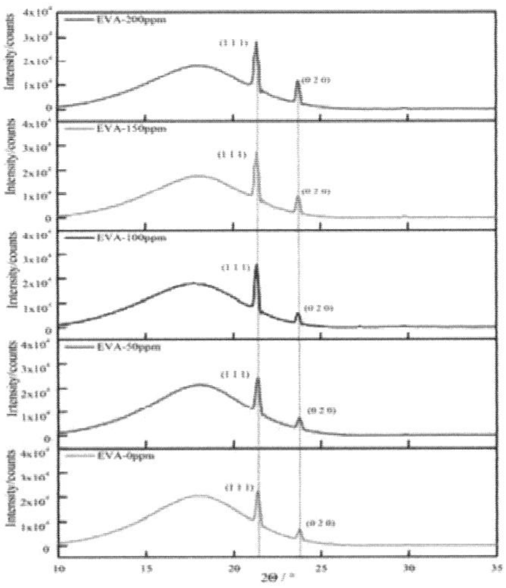

Doped with EVA

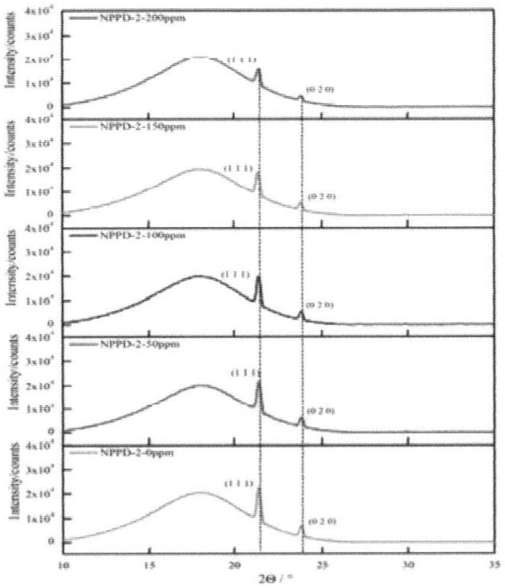

Doped with NPPD-2

CHINA UNIVERSITY of PETROLEUM, BEIJING
厚积薄发 开物成务

The X-ray Diffraction

	PPD concentration (ppm)	Peak (0 0 6)		Peak (110)		Peak (200)	
		2θ(°)	Intensity (counts)	2θ(°)	Intensity (counts)	2θ(°)	Intensity (counts)
undoped model oil	0	6.166	485	21.371	16919	23.748	5633
EVA	50	6.408	774	21.370	18325	23.748	5916
	100	6.446	857	21.370	22045	23.705	5287
	150	6.477	901	21.332	22004	23.708	8154
	200	6.486	879	21.330	23089	23.708	10322
NPPD	50	6.489	1312	21.373	16114	23.784	5011
	100	6.489	1367	21.373	14916	23.785	4698
	150	6.487	749	21.376	12125	23.784	4083
	200	6.485	814	21.373	5885	23.784	3216

(2)

The X-ray Diffraction

	PPD concentration(ppm)	d (006) (A)	d (110) (A)	d (200) (A)	a (A)	b (A)	c (A)
undoped model oil	0	14.3218	4.1543	3.7436	7.4872	4.9935	85.9308
EVA	50	13.7817	4.1545	3.7435	7.4870	4.9939	82.6902
	100	13.7016	4.1546	3.7503	7.5006	4.9900	82.2096
	150	13.6352	4.1618	3.7498	7.4996	5.0028	81.8112
	200	13.6153	4.1621	3.7498	7.4996	5.0033	81.6918
NPPD	50	13.6093	4.1539	3.7379	7.4758	4.9962	81.6558
	100	13.6100	4.1539	3.7379	7.4758	4.9962	81.6600
	150	13.6132	4.1534	3.7380	7.4760	4.9952	81.6792
	200	13.6193	4.1538	3.7381	7.4762	4.9959	81.7158

CHINA UNIVERSITY of PETROLEUM, BEIJING
厚积薄发 开物成务

The Observation of Polarized Microscope

- **Cooling the sample from 60℃ to 25℃ with natural and recording the morphology of wax crystals at room temperature.**

The picture of doped oil with NPPD-2

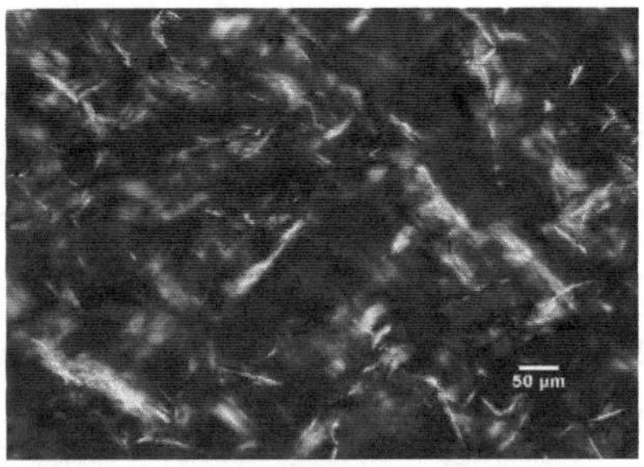

The picture of undoped oil

The picture of doped oil with EVA

CHINA UNIVERSITY of PETROLEUM, BEIJING

厚积薄发 开物成务

Interaction Mechanism

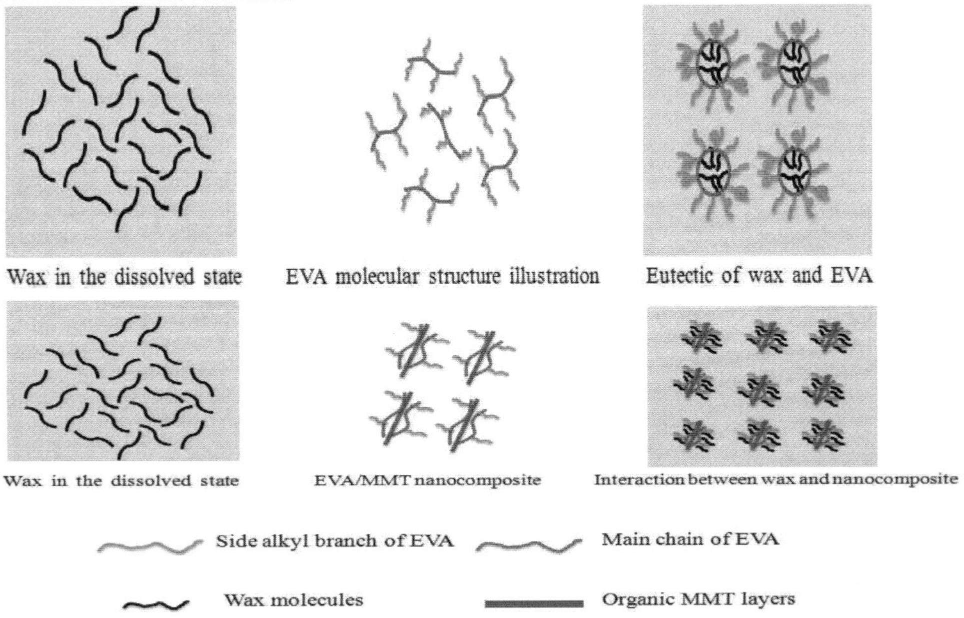

CHINA UNIVERSITY of PETROLEUM, BEIJING

厚积薄发 开物成务

Conclusion

- **The rheological parameters also demonstrate that NPPD-2 can reduce the gel point, and depress yield stress and viscosity of waxy crude oil more significantly than EVA.**
- **The WAT slightly decreases with the increase of EVA concentration, while the trend for NPPD is opposite.**
- **The X-ary diffraction displays that the intensity of the diffraction peak decreasing with the additive of NPPD-2 increasing, which indicates fewer wax crystals exit in doped oil.**
- **The interplanar spacing between the wax crystals decreases, which remarkably improves the low temperature fluidity of waxy crude oil.**
- **The morphology of the wax crystals observed by polarizing microscope shows fewer wax crystals and dispersed finely doped with NPPD-2.**

CHINA UNIVERSITY
of
PETROLEUM, BEIJING

Thanks for your attention!

488981 Co-Processing of Liquid Phase Pyrolysis Oil and Refinery Intermediates in a Continuous Hydrodeoxygenation Reactor

Abstract:

Introduction

Biomass pyrolysis is a suitable pathway for the production of 2nd generation biofuels.[1] During pyrolysis, one of the major products is pyrolysis oil. Due to its high water content, high corrosivity and other negative properties, pyrolysis oil needs intensive upgrading prior usage as fuel for combustion engines.[2] To achieve those standards, an upgrading step is necessary. One way to do so is hydrodeoxygenation (HDO).

BiomassPyrolysisRefinery[3]

An overall concept for the production of 2nd generation biofuels based on liquid phase pyrolysis (LPP) is the BiomassPyrolysisRefinery concept. In the first step, lignocellulosic biomass is pyrolysed in vacuum gas oil to produce an unpolar oil, pyrolysis oil and pyrolysis char through the bioCRACK process[4]. In the second step, the intermediate products are upgraded[5,6]. The bioCRACK process was successfully operated in pilot scale (100 kg/h lignocellulosic feed) over two years. HDO of LPP oil has already been performed continuously in the temperature range of 350°C to 400°C and at liquid hourly space velocities of 0.5 h^{-1} to 3 h^{-1}. To fulfill the requirements of a refinery integration, co-processing of LPP oil and refinery intermediates has to be investigated.

HDO of LPP oil and refinery intermediates

To perform co-processing in lab scale, a paraffin mix and LPP oil were used. HDO was carried out in a plug flow reactor of Parr Instrument Company on an inline-sulfided metal-oxide catalyst. Operating parameters were 121.1 bar hydrogen pressure and 400°C. Overall LHSV was 1 h^{-1}, with 75 wt.% paraffine mix and 25 wt.% LPP oil in the feed.

In Figure 1 the boiling ranges of the HDO products are compared to the boiling range of the paraffin feed. As the paraffin feed is already highly hydrogenated, only cracking reactions occurred. It can be seen, that 25 wt.% LPP oil didn't have a heavy impact on the product quality.

Figure 1: Boiling ranges of HDO products compared to the paraffin feed

1 A. Demirbas, *Appl. Energy*, 2011, **88**, 17–28.

2 Q. Zhang, J. Chang, T. Wang and Y. Xu, *Energy Convers. Manag.*, 2007, **48**, 87–92.

3 N. Schwaiger, R. Feiner, H. Pucher and L. Ellmaier, 2015, 1–8.

4 J. Ritzberger, P. Pucher and N. Schwaiger, 2014, **39**, 1189–1194.

5 H. Pucher, N. Schwaiger, R. Feiner, P. Pucher, L. Ellmaier and M. Siebenhofer, *Energy Res.*, 2014, **31**, 3205.

6 R. Feiner, N. Schwaiger, H. Pucher and L. Ellmaier, 2013, **43**, 1–6.

494343 Static and Dynamic Catalytic Adsorptive Desulfurization (CADS) of Real Diesel Using Low-Cost TiO_2/Fumed Silica

Abstract:

The catalytic adsorptive desulfurization (CADS) has been developed as a promising approach for ultra-deep desulfurization of diesel with low energy comsumption (Ren, X.; Miao, G.; Xiao, Z.; Ye, F.; Li, Z.; Wang, H.H.; Xiao, J. Catalytic Adsorptive Desulfurization Using TiO_2/SBA-15 under Mild Conditions. Fuel, *2016*, 174, 118-125.). In this work, low-cost fumed silica were screened as the supporting substrate for TiO_2 for effective CADS of real diesel (~35 ppm-S). Effect of TiO_2 loading and CADS temperature were optimized in batch reactor. The dynamic breakthrough performance was evaluated in fixed-bed reactor. The sorbent was regenerated by solvent wash followed by air oxidative treatment, and its regenerability was assessed in multiple adsorption-regeneration cycles. Results suggested that TiO_2/fumed silica (300-400 mesh) with higher surface area and larger amount of Si-OH groups showed higher desulfurization capacity. The TiO_2 loading was optimized to 13% with the maximized utilization of Ti catalytic sites and Si-OH adsorption sites on TiO_2/fumed silica for CADS. While the CADS temperature was optimized to 35°C with the balanced catalytic activity for the oxidation of dibenzothiophenes (and their derivatives, 4,6-DMDBT mainly) and minimized desorption of yielded dibenzothiophene sulfones. With a suitable O/S ratio of 5, and overall aspect ratio of 10, the TiO_2/fumed silica showed the maximal desulfurization capacity of 1.44 mg-S/g-A of real diesel at the breakthrough concentration of 5 ppm-S. Considering the effect of the dimension of fixed bed, the dynamic desulfurization performance with varied aspect ratio followed the order of H/R=10 > H/R=22 ~ H/R=5. The S-breakthrough concentration decreased with increased O/S ratio. The superior regenerability of TiO_2/fumed silica was suggested by no loss on desulfurization capacity was observed after six adsorption-regeneration recycles. The CADS using the low-cost TiO_2/fumed silica adsorbent with superior desulfurization capacity and remarkable stability provide an effective and economic approach for ultra-deep desulfurization of real diesel with low energy consumption.

Keyword: catalytic adsorptive desulfurization, TiO_2/fumed silica, diesel, dynamic adsorption.

494352 [O]-Induced Reactive Adsorptive Desulfurization of Liquid Fuel over Ag_Xo@SBA-15 Under Ambient Conditions

Abstract:

Development of clean energy production processes with low energy consumption is a central strategy for a sustainable and environmentally friendly planet. Here we report a novel [O]-induced reactive adsorptive desulfurization (RADS) approach for ultra-clean fuel production using multifunctional Ag_XO@SBA-15 adsorbent with and without the addition of air under ambient conditions. The adsorption capacity of Ag_XO@SBA-15 by RADS reached 4.4 times of that by conventional ADS at low feed sulfur concentration of 10 ppm-S; its adsorption selectivity was enhanced 2.4 orders of magnitude. Over 90% of sulfur was removed within 10 min (corresponds to the space volocity of 6 h^{-1}). The RADS mechanism was illustrated as a simultaneous oxidation (to sulfones) and selective adsorption of sulfones, where the nano-sized silver species detected at different valence states (AgO, Ag_2O and Ag) and the feed air played the versatile enabling roles as the oxidation catalyst and oxidant, respectively, and the surface silanol groups were suggested as the preferential adsoption sites.

Keywords: desulfurization, adsorption, fuel, silver, air

Pacific Northwest
NATIONAL LABORATORY
Proudly Operated by Battelle Since 1965

Modularized Production of Value-Added Products and Fuels from Distributed, Waste Carbon-rich feedstocks

Robert Weber[1], Johnathan Holladay[1], Cynthia Jenks[2]

1) Pacific Northwest National Laboratory; 2) Argonne National Laboratory

Topical Conference Session: Distributed Chemical and Energy Processes for Sustainability, AIChE National Meeting, October, 2017

Overview

Valorizing waste requires distributed facilities that, in turn, require new catalysts and new reactors to achieve economic, robust operation.

- We seek to locally convert waste carbon into chemical and fuel precursors, which aggregate to about 25% of US demand.
- Throughput will come from numbering up, not scaling up, which should accelerate deployment by attenuating technical and investment risk.
- The remoteness, dispersion and heterogeneity of the resources requires development of new catalysts, reactors and intensified processes.
- For example, electrocatalytic upgrading of liquefaction bio-oil could integrate upgrading with clean-up, further "preventing a waste and turning it to profit".
- This project probes both fundamental science and the practicality of electro-upgrading of the components of bio-oil *ex* hydrothermal liquefaction.

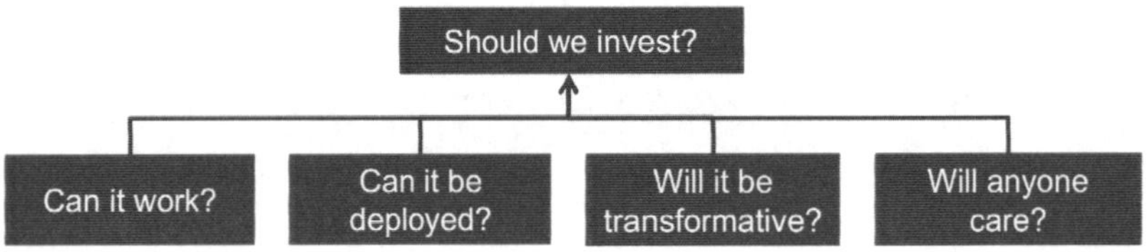

Will anyone care: *size of the opportunity*

We seek to locally convert waste carbon resources into chemical and fuel precursors, which sum to about 30% of US demand.

Type	Source	Availability (Mboe/y)	Site capacity (boe/d)	Possible Process	Tech. issues
Dry	MSW, ag & forest residue	1250	2-20	Pyrolyze, Hydrotreat	Hetero-geneity
Wet	Food, sludge manure	750	2-150	HTL, Hydrotreat	Product slate
Gas	Stranded CH_4 biogas, CO_2	50	70-1500 (CH_4) 1-10 (CO_2)	Reforming FT, Magic	Product slate
Total		**2000**			
US annual petroleum import		6900	130,000		
US annual net import of oil		1700			
Fuel demand of US cars		2800			
Energy demand of US homes		2000			

2/17

Will it be transformative: *benefits of modularity*

To attenuate technical and investment risk and to benefit from manufacturing efficiencies and linear scaling, we will number up rather than scale up.

- Replicating many (≳1000) small facilities promises more attractive investment and risk profiles than are typically found for scaling up for small scale.

Nominal risk profiles: most the risk is resolved in the first deployment for modular systems.

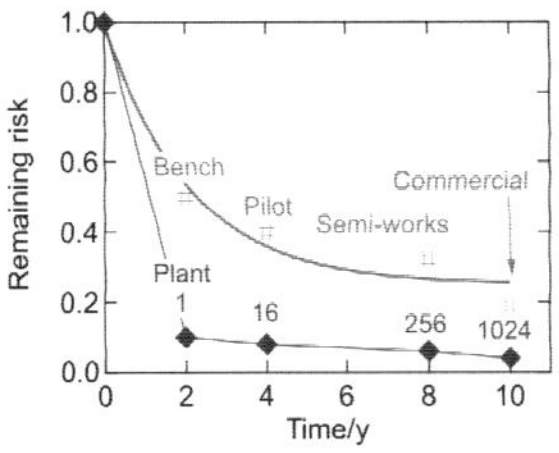

Nominal cost profiles for modular and stick-built plants each summing to 2000 bbl/day

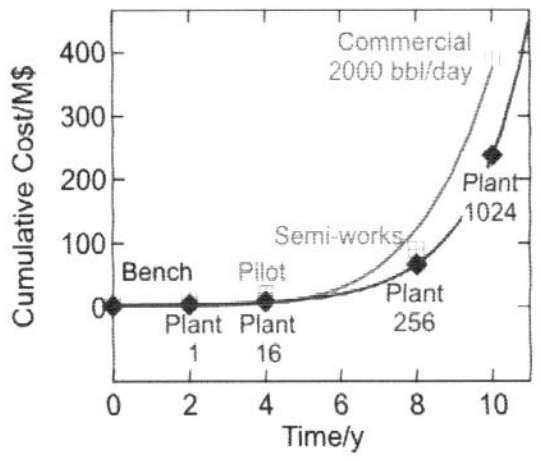

Having a large number of modular plants would benefit from economies of manufacture

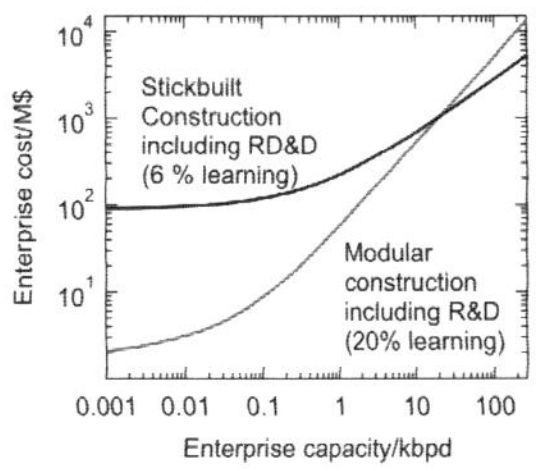

- The cross-over between scaling up & numbering up occurs at an aggregate throughput of >20 kbbl/day.
- Going from 2000 bbl/day to an aggregated 4000 bbl/day would cost ~$460 million to add 1000 more 2 bbl/day facilities; a second commercial plant (at a 6% learning rate) would cost "only" $280 million.

3/17

Can it work: *science and technology needs*

The remoteness, dispersion and heterogeneity of the resources requires development of new catalysts, reactors and intensified processes.

- The size of the facility will be dictated by the supply rate of feedstock, which will likely be neither homogeneous nor constant.
- Remoteness means we will not have a steam or H_2 grid or ways to valorize waste heat.
- The goal is to convert stranded carbon to fuel so we want to be as frugal as possible and not generate additional, even more dilute waste streams.

Challenge or technical need	Example of basic research
Separation processes and controls to facilitate conversion processes	Remote, online quantification of types and flow rates of solids and suspensions
Catalytic conversions with high efficiency, rates and selectivity under near ambient conditions	**Research on low temperature catalysis and chain reactions in liquid media, particularly in water**
Separation processes facilitating product valorization	Thermodynamic of and transport in complex mixtures
Water recovery and cleanup, again with high energy efficiency	Membrane science for separation of water from both polar and non-polar organics

4/17

Can it work: *implementing modularity*

To achieve the envisaged modularity requires changing from volume-based rates to area-based processes which downsize economically.

Modular facilities will incorporate Intensified unit operations, amenable to mass manufacture and which are cheaper than volume-based technologies at small scale

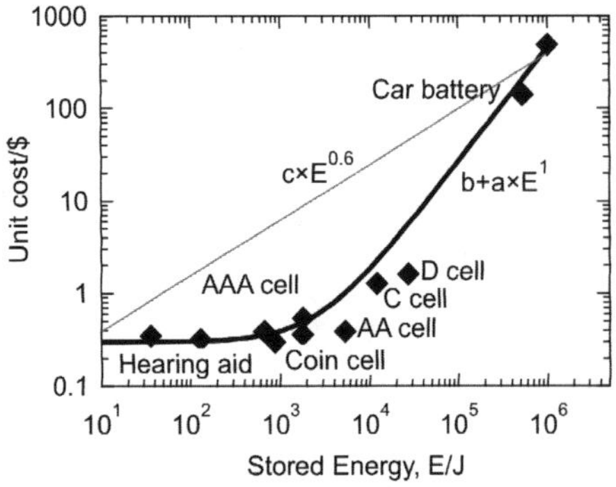

A similar scaling rule is observed for process heat exchangers (including consumer refrigerators) and other electrochemical reactors (fuel cells, flow batteries).

5/17

Can it work: *System architecture*

A complete system would include feed preprocessing, processing, upgrading and separations.

- Sites will be geographically distributed, likely without access to industrial utilities except electricity.
- The scale will be small (say, the equivalent to ~10 bbl oil/day).
- Operators may not be highly trained or closely supervised.
- The customers will be frugal.

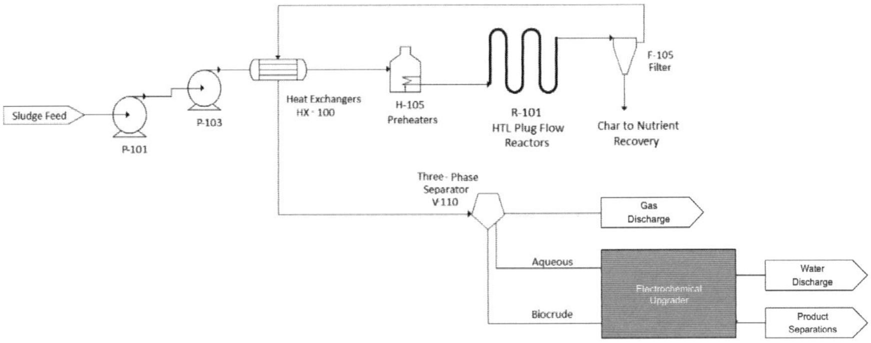

6/17

Can it work: *Scope*

We are now researching the downstream, upgrading and separations units to complement prior work at PNNL on the upstream conversion operations.

PNNL's Modular Hydrothermal Liquefaction

3 skids

- Feed prep and handling
- Hydrothermal liquefaction
- Biocrude recovery

Capability

- ~16 liters/h ≈ 2.4 bbl/day
- Solid-liquid separations at operating conditions
- Liquid-liquid separations for biocrudes that are heavier, or lighter, than water
- Product-feed heat exchange

7/17

Can it work: *electrolysis reactors*

We have constructed 3 scales of flow reactors that can provide both fundamental information and grow to practicable levels of capacity.

- ► Ag/AgCl reference electrode that can be positioned axially
- ► Carbon felt- or Carbon cloth-supported catalysts
- ► Automated data acquisition and control over E, q and T.

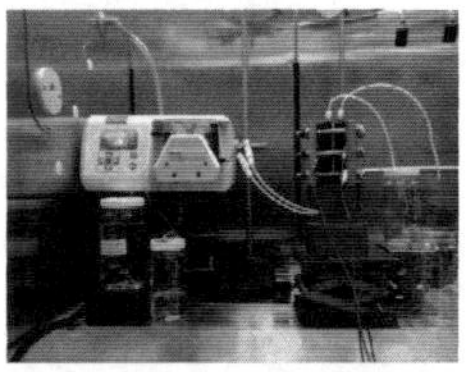
Bench: 2.4 cm × 4.2 cm

Pilot:, 5.7 cm × 10 cm

Demo: 10.7 cm × 18.7 cm

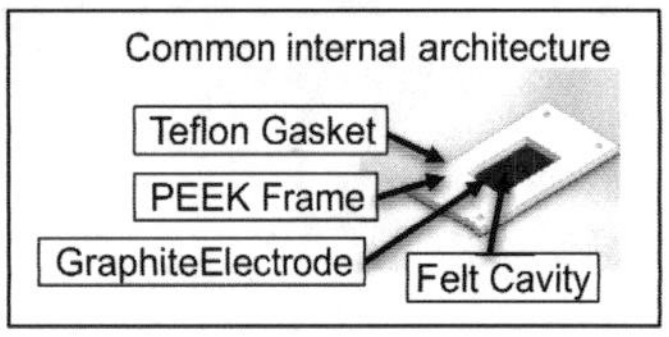

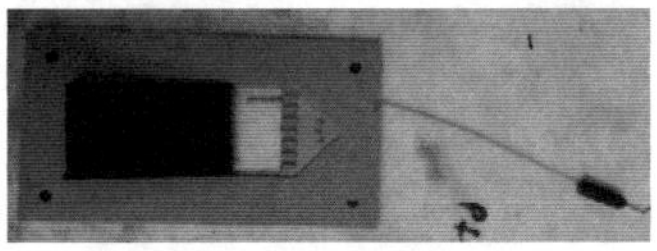
Reference electrode

Can it work: *electrolysis reactors*

We are in the process of characterizing, modeling & testing catalysts and reactors.

- ► The catalysts consist of redox sites and electrically conductive supports, which stabilize the redox sites and construct the solution environment.
  - ■ E.g., metals and alloys deposited on nanoporous and fuzzy carbons; akin to PEMFC catalysts but with *high* overpotentials for making H_2
- ► Characterization will entail site-counting as well as site description, in both cases at least *in situ* and better *in operando*, e.g., Raman, SFG, XAS, coulometry and (with care) electron microscopy
- ► Modeling needs to estimate ΔG, not just ΔE, of reactions in fields $\gtrsim 10^8$ V/m
- ► Testing will require reactors that afford intrinsic kinetics and that are scalable.

Visualization of flow in a lab-scale electrolysis reactor using Rhodamine 6G

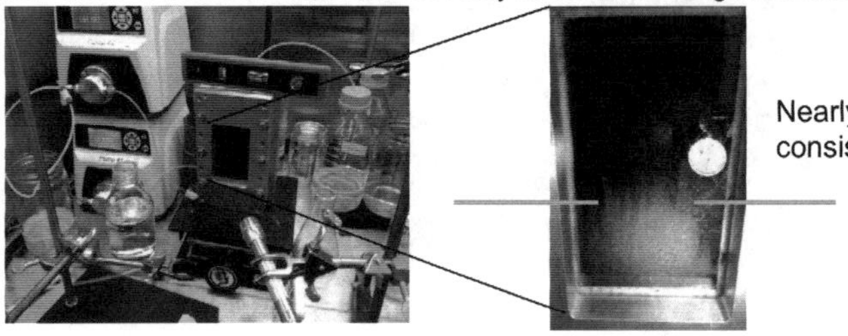

Nearly flat front is consistent with plug flow

Can it work: *Kinetics*

The rates of the model reactions correlate well with Butler-Volmer kinetics.

The reactors appear to be nearly plug flow; when the reaction rate laws accord with pseudo-first order kinetics, conversion, C_R/C_i = exp(-kτ), where the rate constant, k, derives from Butler-Volmer kinetics

$$C_R = \exp\left[-k_0\left(\left(\exp\left(\frac{-\alpha F\eta}{RT}\right)+\exp\left(\frac{(1-\alpha)F\eta}{RT}\right)\right)+C_i\exp\left(\frac{-\alpha F\eta}{RT}\right)\right)\tau\right]$$

Kinetics fit using the bench cell (V =5 mL)
$0.05 < C_0/M < 0.253$; $0.1 < E/V < 1.1$;
$150 < \tau/s < 450$; $0.01 <$ loading < 0.04

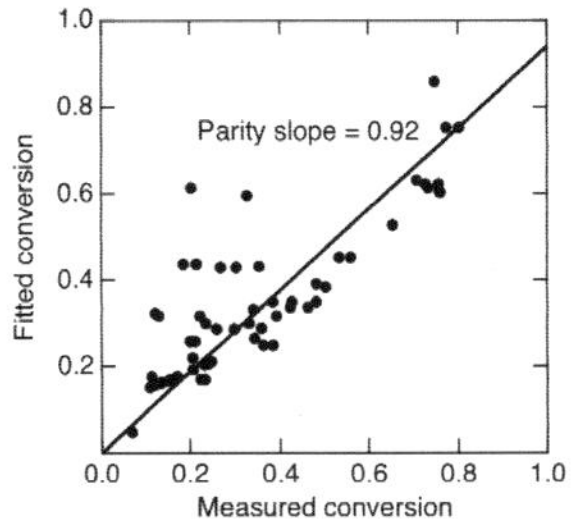

Comparison of the bench cell (5 mL) with the pilot cell (27 mL), each containing 4 wt% Cu/felt

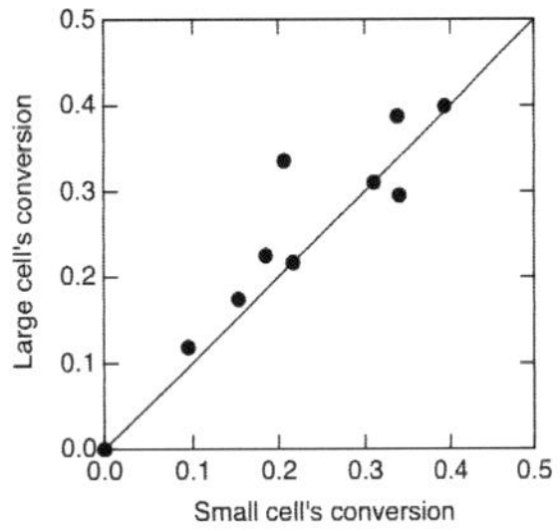

Can it work: *Cathode processes*

We have recently shown that we can electrochemically upgrade some of the functional groups present in neat bio-oil.

- Catalyst was 4 wt% Pd/C, 150 mL of whole bio-oil (ex pyrolysis of pine) recirculated for 20 h through a 6 mL cell at 1.5 V.
- The residence time τ = 20 h × 6 mL/156 mL = 0.8 h (*cf* 5 h for thermal hydrogenation).
- ^{13}C NMR and ^{1}H NMR indicate a significant decrease in carbonyl and aromatic species post electrolysis.

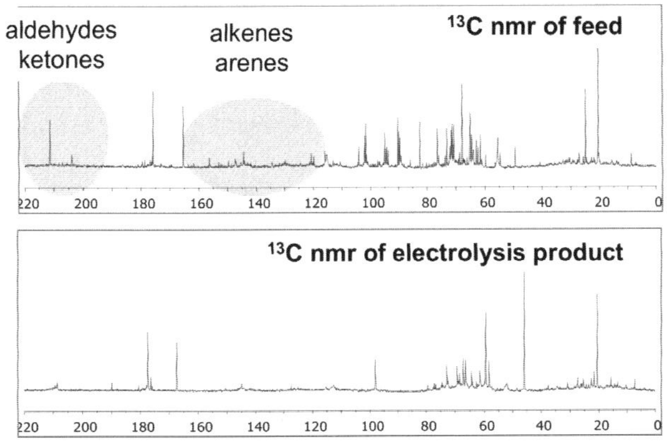

Can it work: *Anode processes*

The aqueous phase from liquefaction of wastes contains many compounds that can, in principle, be oxidized on the anode …

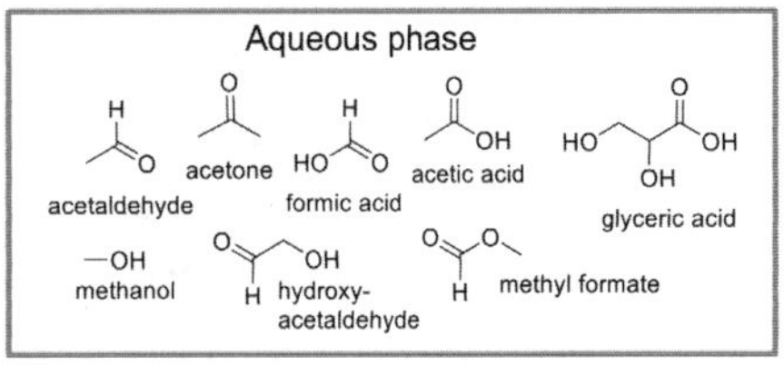

Stoichiometry of upgrading HTL bio-oil *ex* WWTP sludge (0.4 kg oil/kg sludge)

Stream	H^+ Equivalents/g_{sludge}
Oil	-0.03
Aqueous	+24

Estimated from Snowden-Swan et al., PNNL-25464 **2016**.

Can it be deployed: *process integration*

Electrocatalytic upgrading of the bio-oil could simultaneously remediate the wastewater further "preventing a waste and turning it to profit".[1]

- Electrochemistry provides two types of process intensification: high energy flux at ambient conditions (1 eV = 100 kJ/mol) and integrated operations.

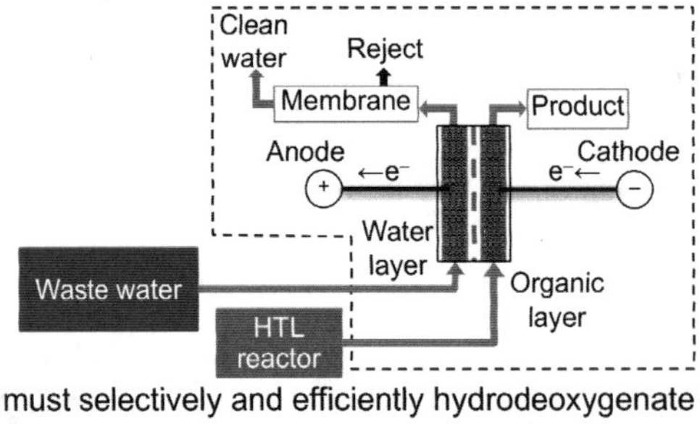

- The cathode catalyst must selectively and efficiently hydrodeoxygenate and hydrogenate the feed.
- The anode catalyst must preferentially and efficiently oxidize the dissolved polar compounds.
- The electrolysis reactor must be compact, robust and ensure good contacting between the liquid phases and their respective catalysts.

1) Arthur D. Little, *J Ind Eng Chem*, **1909**, *1*, 61-62."Every waste that is prevented or turned to profit, … and every more effective process which is developed, makes for better living in the material sense and for more wholesome living in the higher sense."

Can it be deployed: *Techno-economics*

Experience with redox flow batteries indicate that an electrochemical upgrader would cost about 1/3rd of the target CapEx = $50,000/bpd.

- Values for the electrochemical process were based on PNNL's analysis of the hardware costs for a redox flow battery.
  - Raising the heating value of 1 T/day of pyrolysis oil (~15 MJ/kg) to that of a fuel (~45 MJ/kg) requires adding 30 GJ in 86400 s = 350 kW
  - Estimated, near term CapEx for a redox flow battery, less electrolyte, = ~300 $/kW (exclusive of any PGM catalysts)
- An electrochemical upgrader based on similar technology would cost ~105 k$/T/day = $15,000/bpd.

CapEx estimate for a Redox Flow battery

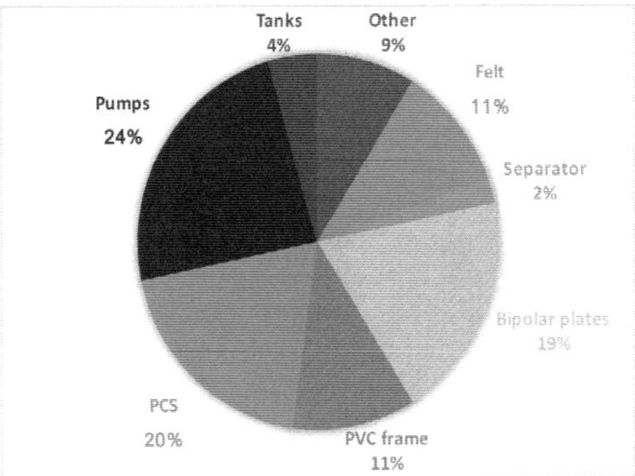

After V. Viswanathan et al., *J. Power Sources* **2014**, *247*, 1040

14/17

Summary

Downsized technology based on hydrothermal liquefaction and electrochemical upgrading appears to be a plausible way to valorize waste.

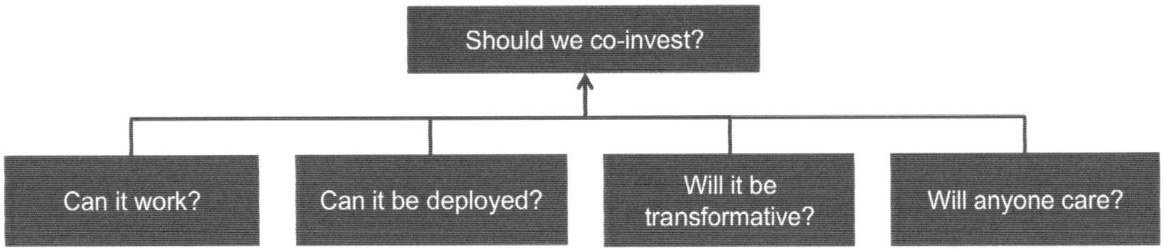

- We do need to devise new catalysts
- Reactors exist that could deploy them
- Such a new enterprise would be transformative: specifically it could convert a waste that now costs $60/ton to dispose and turn it into a profit stream.
- The new enterprise is being designed to appeal to waste-producing companies (e.g., agriculture), communities and product off-takers.

15/17

Acknowledgements

The research was supported by the Chemical Transformations Initiative at PNNL and was conducted under the Laboratory Directed Research and Development Program

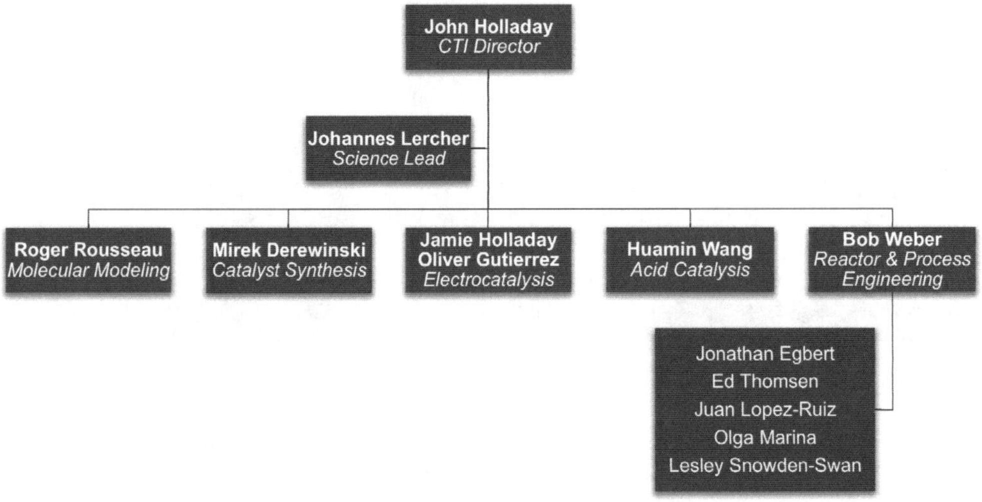

16/17

PNNL Overview

Pacific Northwest National Laboratory has been managed by Battelle for 50 years.

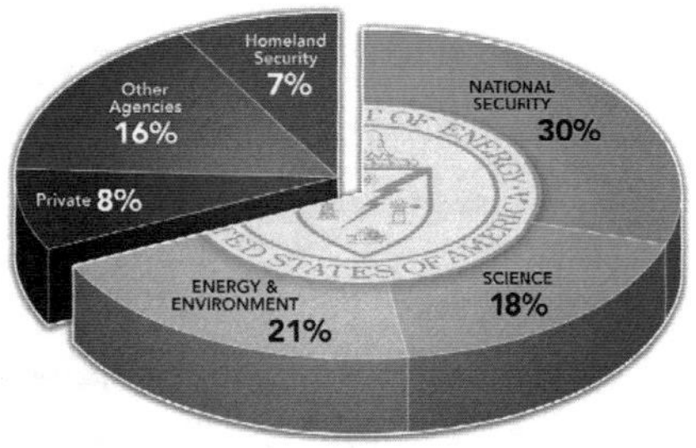

- Since 1965 operated by Battelle for US DOE
- Mission-driven collaborations with government, industry and academia
- Unique Science & Technology strengths and capabilities

- Operating budget of ~$1 B supporting a staff of 4,000
- ~900 peer-reviewed publications per year
- ~100 U.S. and foreign patents per year (2,400 since 1965)
- ~100 awards (R&D 100 and FLC)

17/17

TCP

Inclined Indirect Flaming Pyrolysis Rotary Gasifier

SERDP Project #WP-2211
Stephen Cosper, USACE-ERDC-CERL
Dr. Paul Amodeo, SUNY Cobleskill

SERDP DOD • EPA • DOE | ESTCP

Objective

- Create a design for a deployable waste-to-energy-system, with the following criteria:
  - Based on gasification
  - One to three tons/day capacity, with minimal sorting
  - Self-powered, with net-positive exportable electricity
  - Size to fit two, 20-ft containers

SERDP DOD • EPA • DOE | ESTCP

Technical Approach

Downdraft

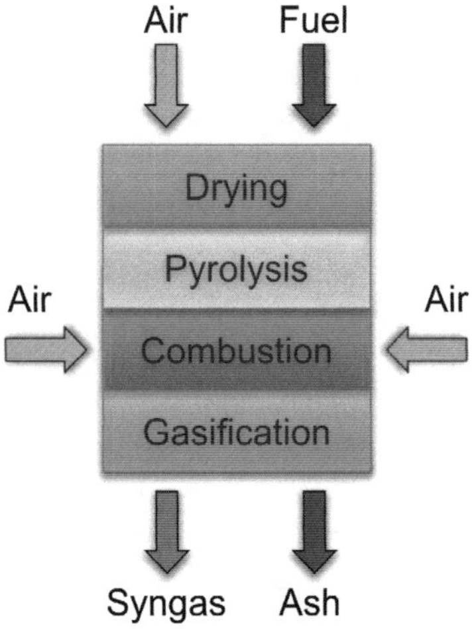

Updraft (Countercurrent)

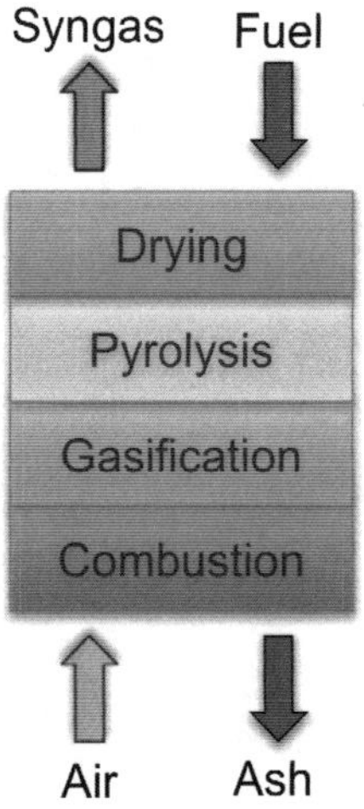

Countercurrent gasification has a number of inherent characteristics that would be beneficial for WTE systems:

- High thermal efficiency
- Less sensitive to fuel pretreatment
- Reduced particulate entrainment
- Low-reactivity ash produced from final-stage combustion

(Courtesy: UND-EERC)

3

SERDP DOD • EPA • DOE | ESTCP

Technical Approach – Rotary System

- System works like a modified updraft gasifier
- Waste fed from bottom, placed at hot, burning char layer
- Syngas collected through concentric tube, not directly exposed to char bed
- Rotates at ~1rpm

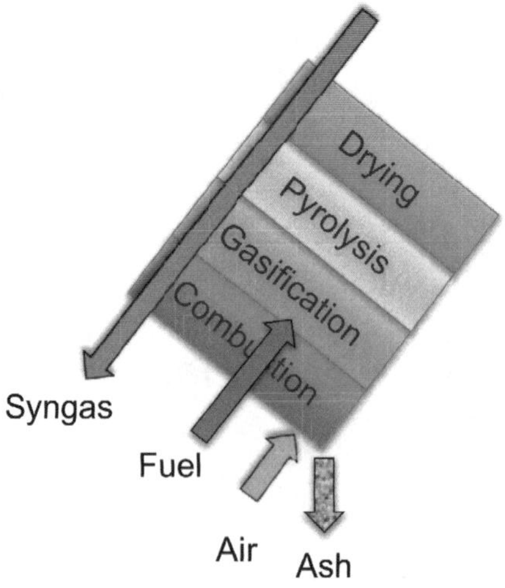

4

SERDP DOD • EPA • DOE | ESTCP

Initial design

- Initial hardware design and layout
- Experimental layout on two trailers to mimic portability requirements
- 1st trailer has hydraulic waste compression, gasifier, gas treatment
- 2nd has electronic control system, 60kW genset, hydraulic pump and controls, experimental electric load center

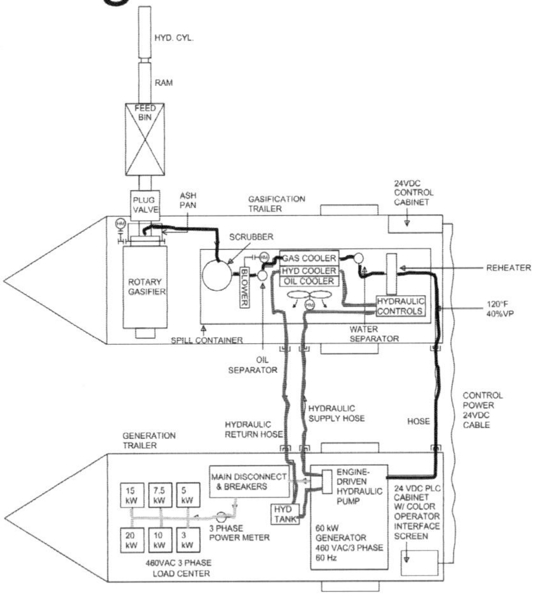

5

SERDP DOD • EPA • DOE | ESTCP

Waste Feed

6

Main Reactor

7

SERDP DOD • EPA • DOE | ESTCP

"Standard" waste composition

- Based on field surveys, 2013
- Variations included
  - 50% food
  - 50% plastics
  - 50% used oil

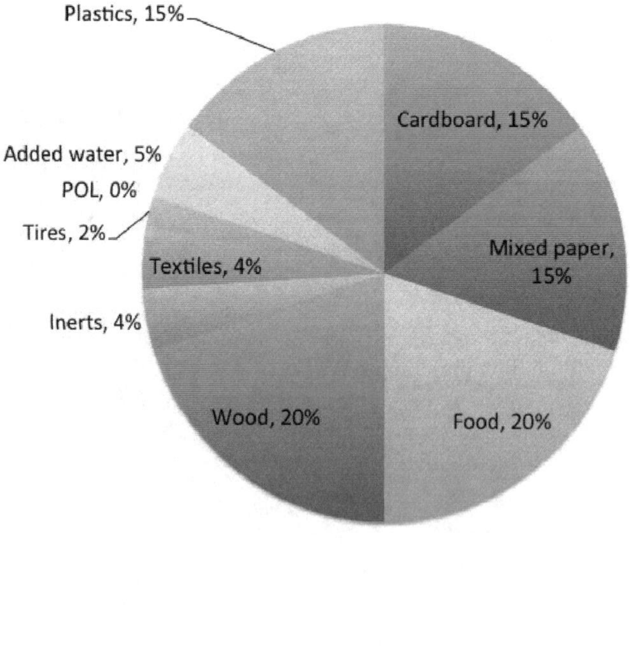

8

Temperature monitoring reactor

- Thermocouples down the length of reactor
- Steady state gas discharge temperature about 400°F
- Temperature peaks indicate time to add waste.
- High temps at end of run indicate, allowing waste layer to burn down.

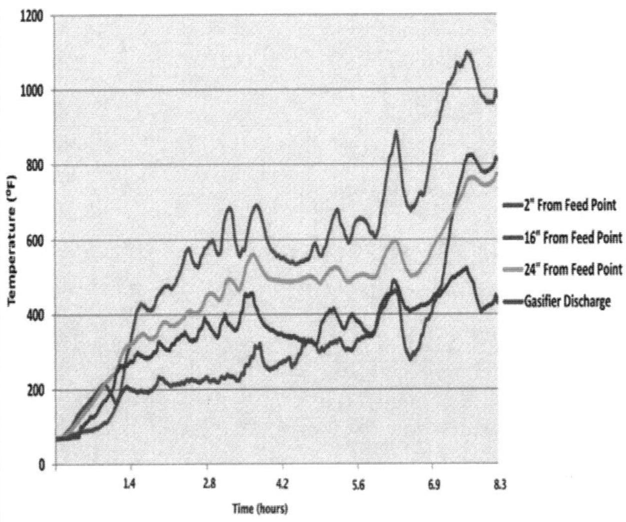

9

Quencher Tank / Pyro Oil Recycling

- Pyrolysis oils produced in reactor
- Accumulate in quencher tank
- Excess fed back into reactor
  - Eliminates a waste stream
  - Enriches syngas
  - ***Could be distilled to commercial fuels***

10

Quencher oil temperature

- Incoming syngas ~400°F
- Pyro-oil remains constant due to water emulsion
- Temp remains below boiling point of water

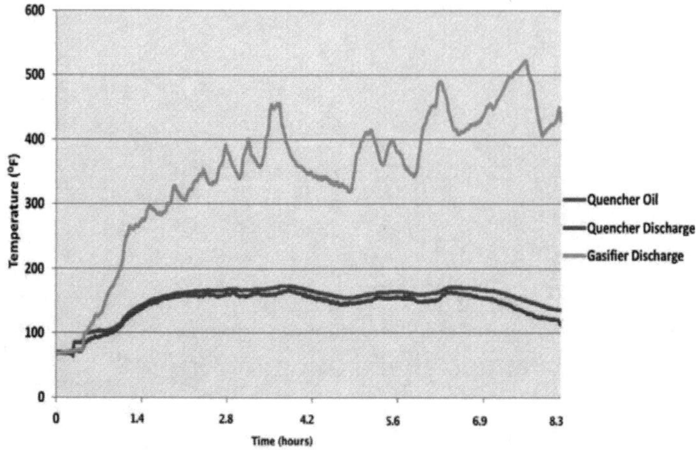

11

Comparison to Commercial Fuels

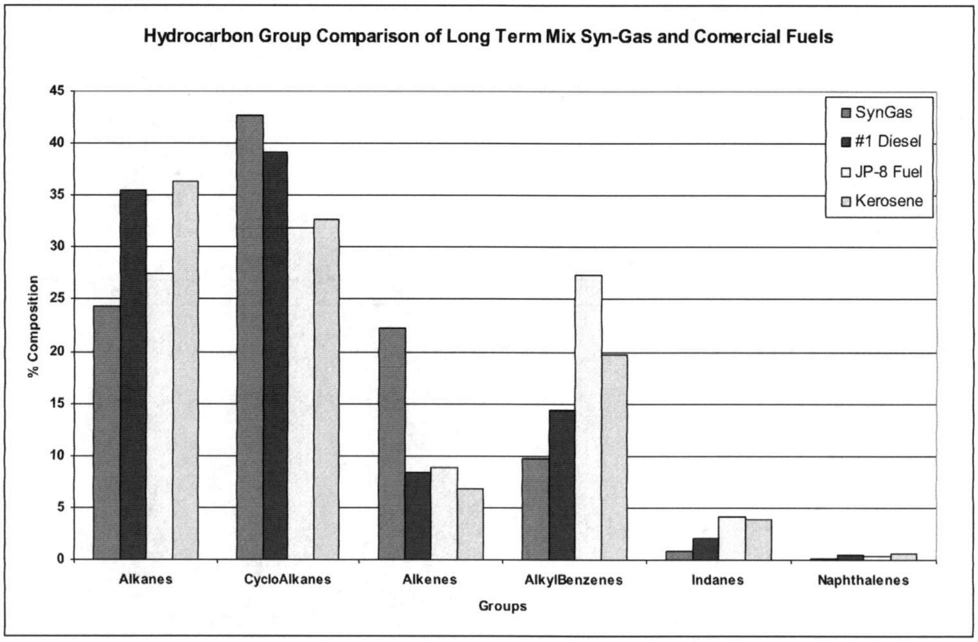

12

Exhaust gas comparison

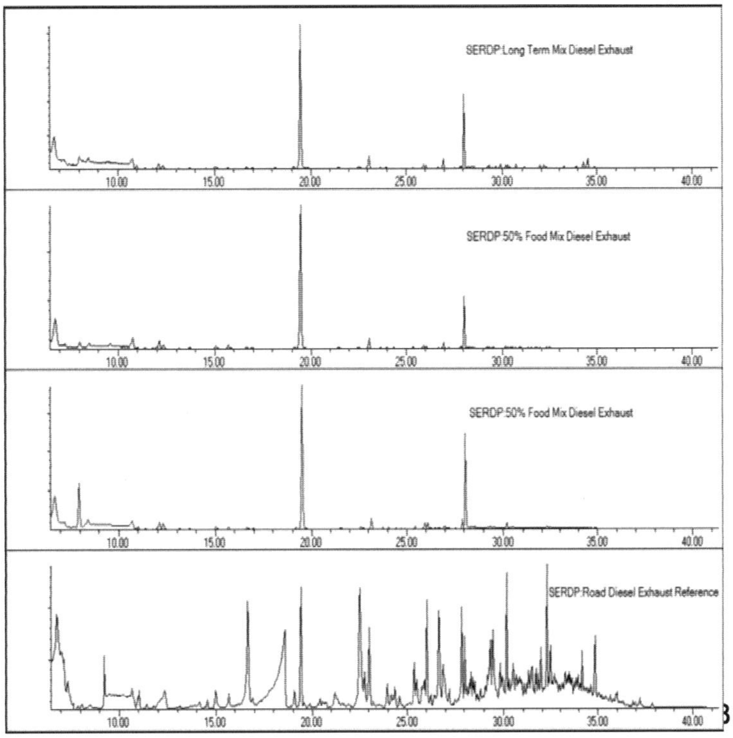

NO_X Emissions with Dual-fuel

- About 50% reduction in NO_X
- Due to:
  - less N_2 into engine
  - Cooler combustion temperatures

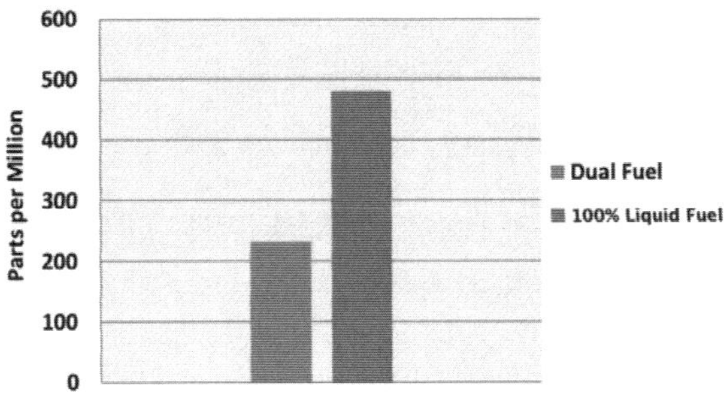

14

Status?

- Project complete
- Prototype designed, built, and tested on a variety of materials
- Final report published very soon
- Team is actively seeking funds for demonstration and/or to construct a "Mark II" automated, containerized version

15

Contact

Stephen Cosper
Army Corps of Engineers
Construction Engineering Research Lab
Champaign, IL
Stephen.D.Cosper@erdc.dren.mil
217-398-5569 M-Th
217-300-3721 F

Dr. Paul Amodeo
SUNY Cobleskill
518-255-5384
amodeopa@cobleskill.edu

https://www.serdp-estcp.org/Program-Areas/Weapons-Systems-and-Platforms/Waste-Reduction-and-Treatment-in-DoD-Operations/WP-2211/WP-2211

16

17AIChE
Annual Meeting, Minneapolis, MN

Monolithic Catalysts Coated with Hierarchical ZSM-5 for Fischer-Tropsch Synthesis

Chunxiang Zhu, David Gamliel, Julia A. Valla and George M. Bollas

Department of Chemical & Biomolecular Engineering
University of Connecticut

http://pdsol.engr.uconn.edu

πδσωλ

Motivation

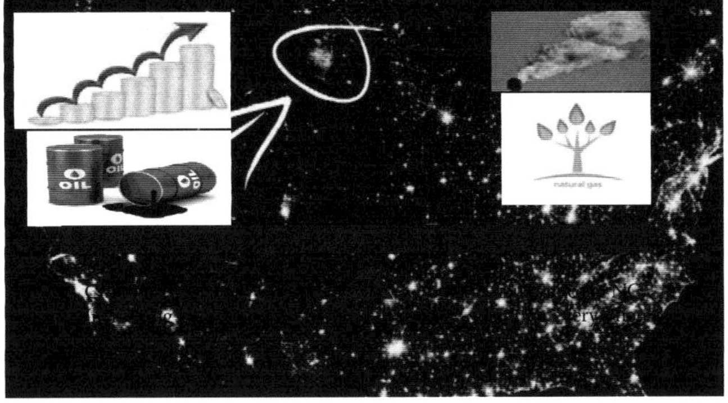

http://www.npr.org/sections/krulwich/2013/01/16/169511949/a-mysterious-patch-of-light-shows-up-in-the-north-dakota-dark
https://energyinfrapost.com/india-comfortable-crude-oil-prices-65barrel-cea/: oil barrel
https://www.123rf.com/photo_12683470_stacks-of-gold-coins-bullion-with-red-arrow-depicting-price-value-going-up-illustration.html : price goes up
https://www.google.com/search: nature gas

$\pi\delta\sigma\omega\lambda$

FTS Reaction Mechanism

Reaction	Stoichoimetry	Desirable?
Alkane Formation	$nCO + (2n+1)H_2 \rightarrow C_nH_{2n+2} + nH_2O$	YES
Alkene Formation	$nCO + 2nH_2 \rightarrow C_nH_{2n} + nH_2O$	YES
Water Gas Shift	$CO + H_2O \rightarrow CO_2 + H_2$	NO
Boudouard Reaction	$2CO \rightarrow CO_2 + C$	NO

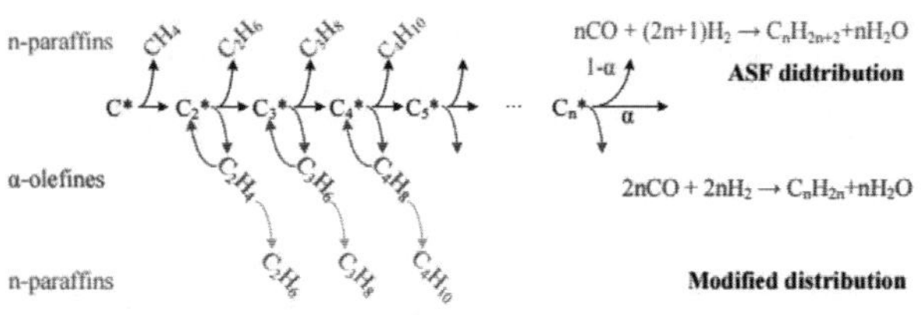

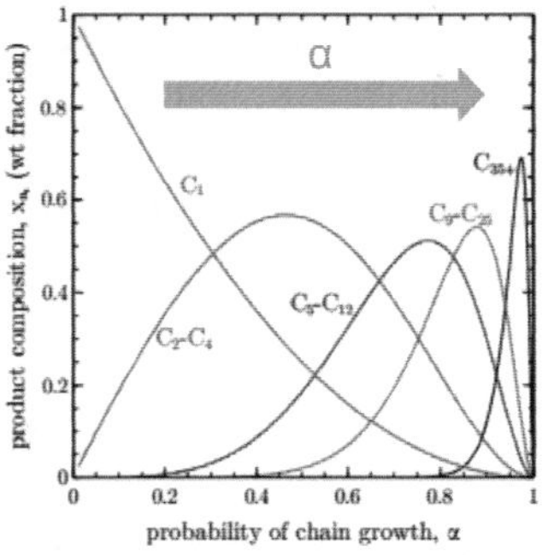

- **Low Temperature FTS (200-240 ºC)**
  - Co
  - Wax, heavy hydrocarbons
  - Low water gas shift activity

Source: F. Kapteijn, R.M. De Deugd, J.A. Moulijn, Fischer-Tropsch synthesis using monolithic catalysts, Catalysis Today. 105 (2005) 350–356.

$\pi\delta\sigma\omega\lambda$

GTL Plants

Existing GTL plants[1]

Location	Operator	Start-up	Capacity (b/d)	Note
Sasolburg, South Africa	Sasol	1955	5,600	Switched to nature gas in 2004
Mossel Bay, South Africa	PetroSA	1992	36,000	
Bintulu, Malaysia	Shell	1993	14,700	
Ras Laffan, Qatar	Oryx GTL	2006	34,000	Owned by Qatar Petroleum and Sasol
Ras Laffan, Qatar	Shell	2011	140,000	Owned by Qatar Petroleum and Shell
Escravos, Nigeria	Chevron	2013	34,000	Owned by NNPC and Chevron

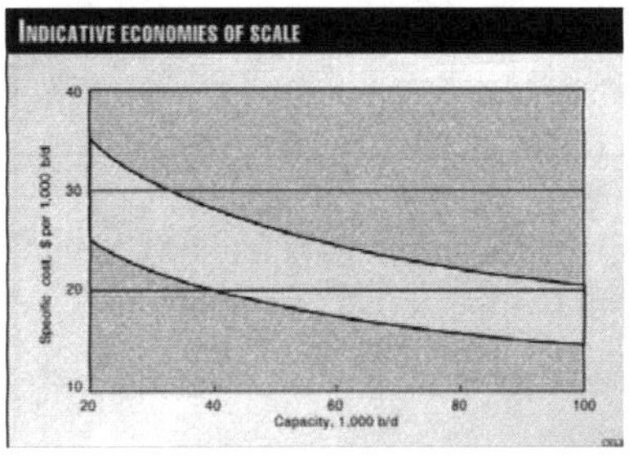

Product cost vs plant capacity.[2]

Less than 10% of the world's gas fields are large enough to sustain a 10,000+ bpd Fischer-Tropsch GTL plant[3]

1. https://www.eia.gov/todayinenergy/detail.php?id=28752
2. http://www.ogj.com/articles/print/volume-96/issue-38/in-this-issue/petrochemicals/gtl-technologies-focus-on-lowering-costs.html
3. Abstract, Process Economics Program Report 247 B, Small scale gas-to-liquids technology

Modular GTL Plants

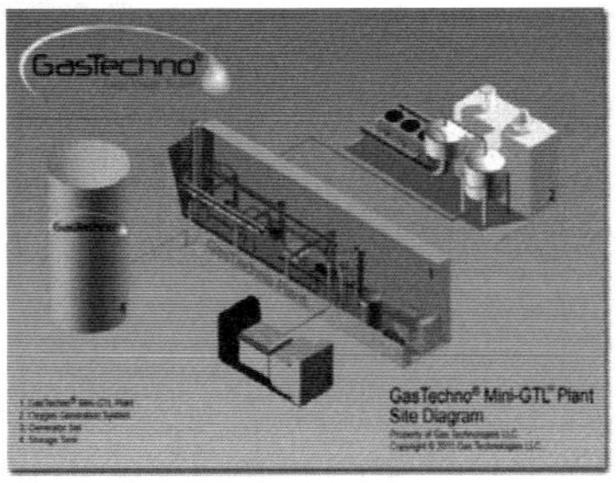

Modular GTL plant[1]

Numerous Companies Developing Mini-GTL Technology[2]

- **Reactor volume decrease**
- **High selectivity, Process equipment reduction**
- **Low capital investment, more profitable**

1. http://www.gastechno.com/gastechno-plant-scales.html
2. http://www.compactgtl.com/technology/petrobas

Objectives

- **Reactor volume decrease**
  - One pot catalysts with high catalyst activity
- **Process equipment reduction**
  - High gasoline selectivity
- **This catalyst would require:**
  - High surface area
  - Excellent heat transfer
  - Improved metal dispersion
  - Presence of zeolite film to enhance secondary cracking and shape-selectivity
  - Mass transfer limitation release

Component	Function
Monolith	Rigid structure, high thermal conductivity
Cobalt-Al_2O_3	Syngas to hydrocarbons
Zeolite ZSM-5	Secondary acid cracking, shape selectivity

πδσωλ

Properties of Mesoporous ZSM-5

Meso-ZSM-5 created using the **desilication** technique

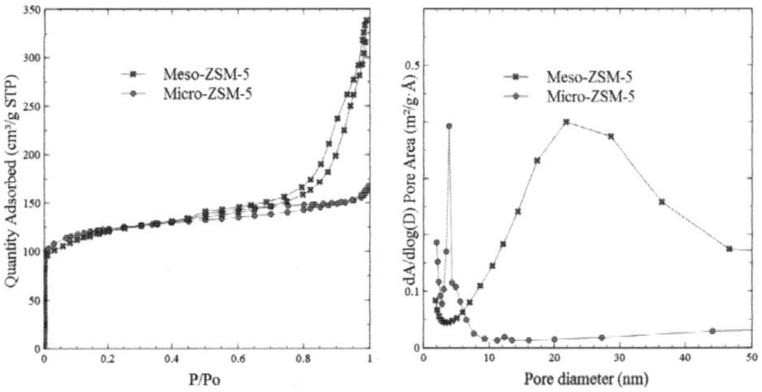

Sample	S_{BET} (m^2/g)[a]	V_{micro}[b]	V_{meso}[c]	$D_{average}$[d]	ΔV_{micro}	ΔV_{meso}
Micro-ZSM-5	418.00	0.14	0.08	6.00	--	--
Meso-ZSM-5	421.00	0.10	0.36	15.00	-27.71%	353.09%

[a] Surface area obtained from Brunauer-Emmett-Teller (BET) measurements (S_{BET}). [b] t-plot micro-volume (V_{micro}). [c] BJH adsorption pore volume 1.7-500 nm (V_{meso}). [d] BJH adsorption average pore diameter($D_{average}$).

πδσωλ

Proposed FTS Catalysis

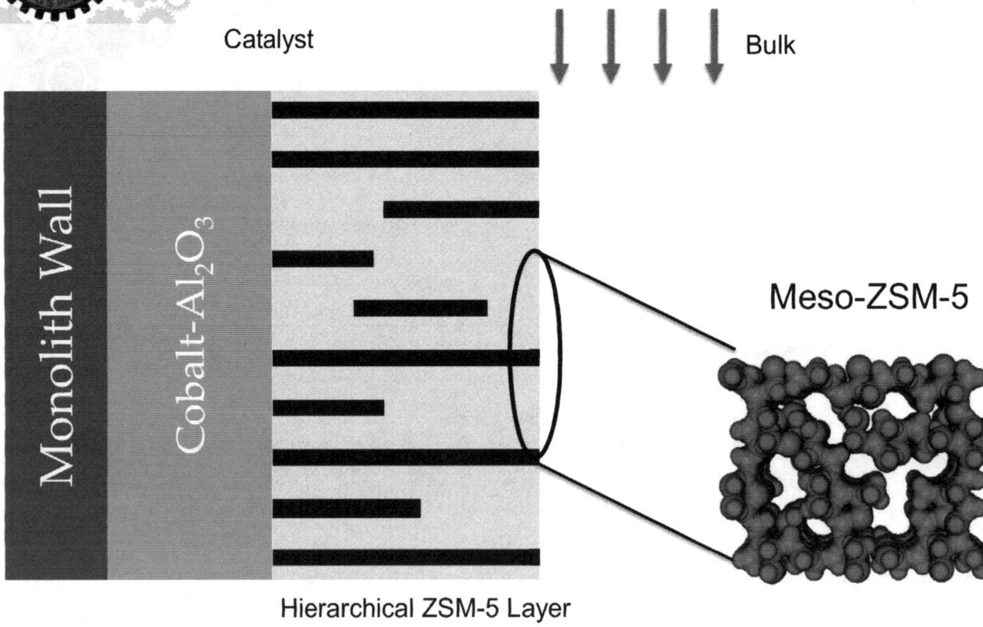

Proposed FTS Catalysis

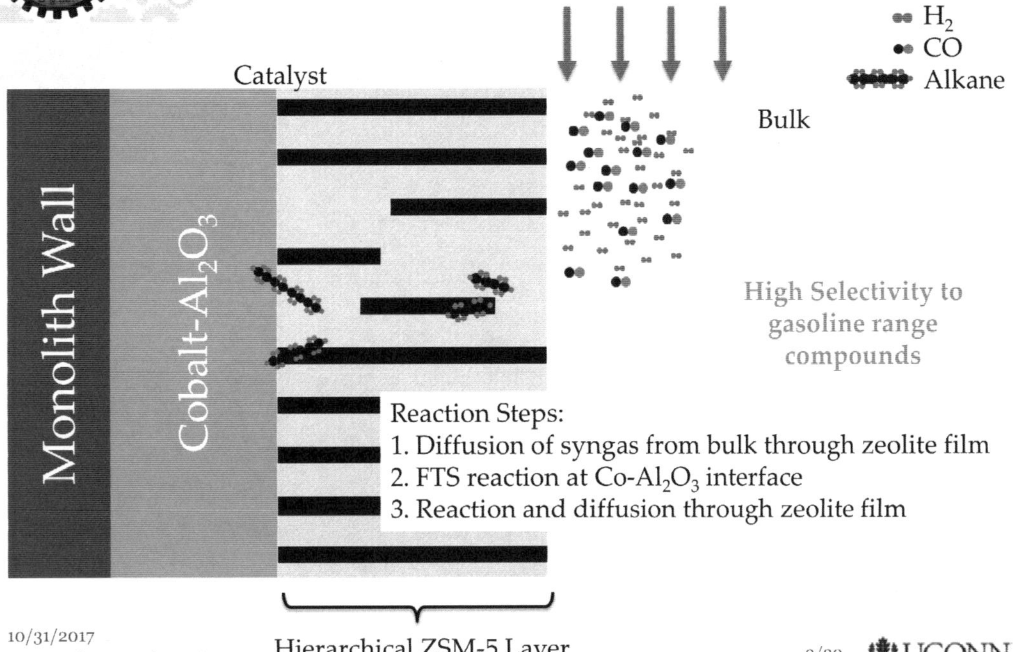

10/31/2017
2017 AIChE Annual Meeting

9/20 UCONN

Catalysts Preparation Method

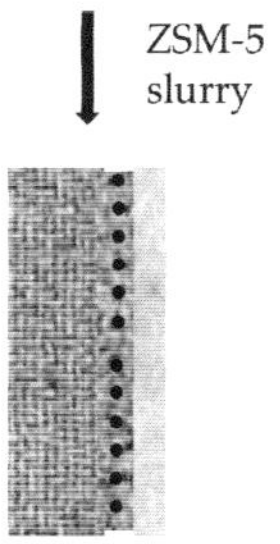

Catalysts were dried at 120 °C for 4 h and calcined at 400 °C for 12 h after each coating.

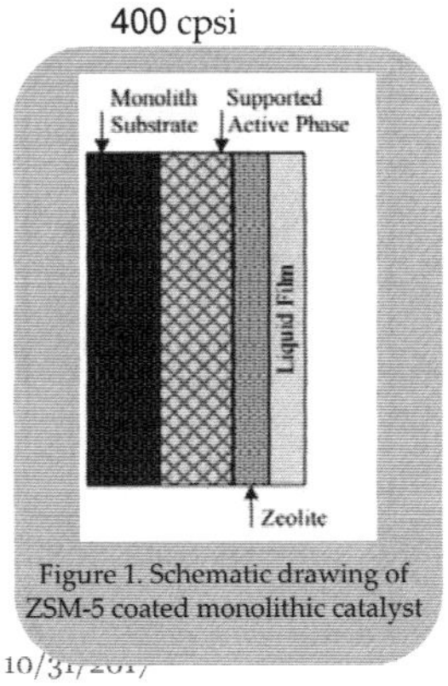

Figure 1. Schematic drawing of ZSM-5 coated monolithic catalyst

10/31/2017
2017 AIChE Annual Meeting

10/20 UCONN

πδσωλ

Equipment Set-up

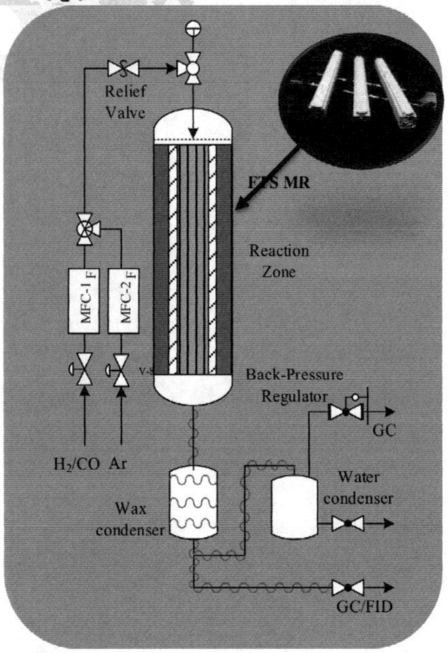

Catalysts Details

Catalyst	Catalyst weight(g)	Monolith (g)	Al_2O_3 (g)	Co_3O_4 (g)	ZSM-5 (g)
1.1 g-Micro	2.93 ± 0.03	1.42 ± 0.03	0.40 ± 0.08	0.13 ± 0.02	1.06 ± 0.04
1.1 g-Meso	3.11 ± 0.04	1.46 ± 0.03	0.50 ± 0.04	0.15 ± 0.02	1.07 ± 0.02
1.6 g-Meso	3.78 ± 0.01	1.45 ± 0.02	0.50 ± 0.01	0.16 ± 0.02	1.62 ± 0.03
1.9 g-Meso	4.01	1.43	0.53	0.16	1.89

- Premixed syngas (H_2:CO = 2:1).
- Ar was used as internal standard.
- Two-stage liquid trap. Wax trap was kept at 120 °C. Light oil and H_2O trap was at room temperature.
- Micro-GC 4900 with two columns and TCD for gases analysis.
- Agilent 6890 GC-FID with HP-5 column for wax and light oil analysis.

πδσωλ

FTS with Meso-ZSM-5 Coated Catalyst

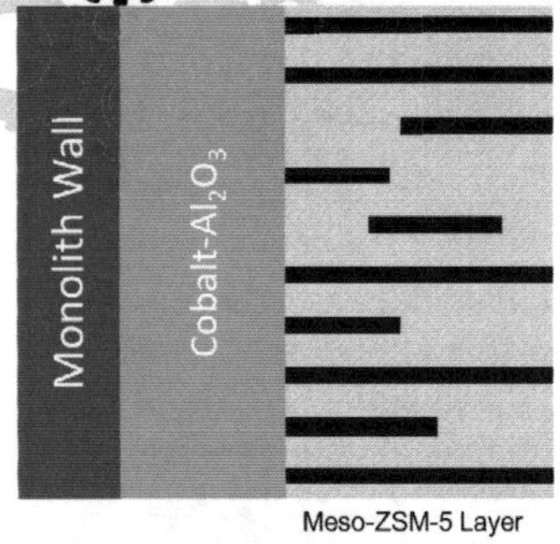

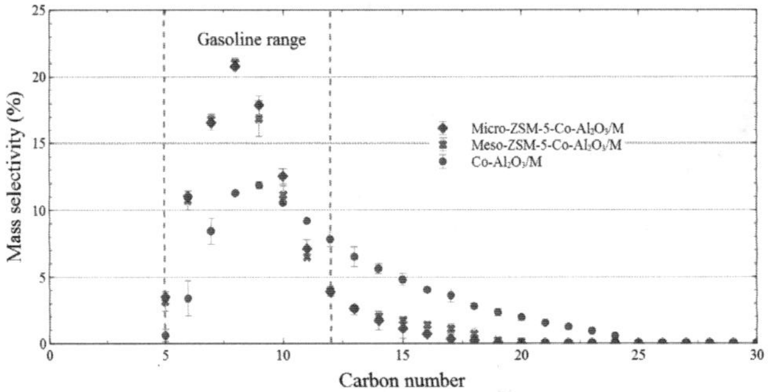

Component	CO Conv wt.%	Gas Selectivity wt.%	Liquid Selectivity wt.%	Mass Balance wt.%
$Co\text{-}Al_2O_3/M$	82	31.8	68.2	96
$ZSM\text{-}5/Co\text{-}Al_2O_3/M$	79	20.2	79.8	96
$Meso\text{-}ZSM\text{-}5/Co\text{-}Al_2O_3/M$	88	27.6	72.4	95

FTS with Meso-ZSM-5 Coated Catalyst

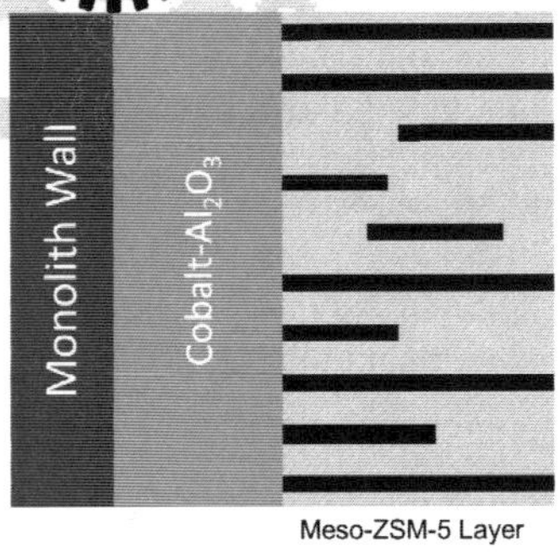

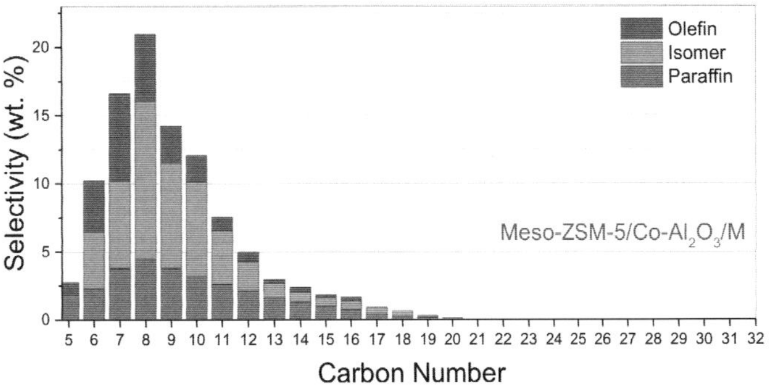

Component	Paraffin Selectivity wt.%	Isomer Selectivity wt.%	Olefin Selectivity wt.%
$Co\text{-}Al_2O_3/M$	69	26	5
$ZSM\text{-}5/Co\text{-}Al_2O_3/M$	29	50	21
$Meso\text{-}ZSM\text{-}5/Co\text{-}Al_2O_3/M$	29	47	23

Stability Comparison

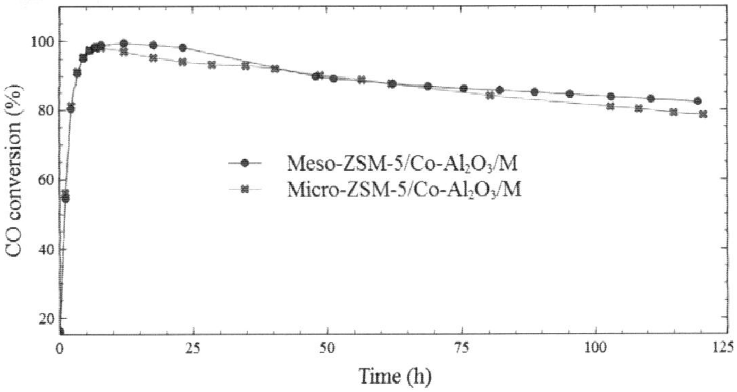

1.1 g-meso and 1.1 g-micro coated catalysts stability comparison within 120 h at 230 °C and 12 bar.

Catalysts coated with Meso-ZSM5 is more stable.

πδσωλ

Performance with Different Coating Thickness

Co-catalysts performance of 1.6 g meso-ZSM-5 at different pressures. Reaction condition: T: 230 °C; Syngas flowrate: 35 ml/min; Time on stream: 48 h

Catalyst	CO conv (%)	Carbon species selectivity (C%)		Oil phase selectivity (wt.%)			
		C_{5+}	C_5-C_{12}	C_{13+}	Paraffins	Isomers	Olefins
1.1 g-Meso	88.4 ± 0.6	77.3 ± 2.0	87.6 ± 1.3	12.4 ± 1.3	29.2 ± 1.8	44.9 ± 1.7	25.9 ± 0.5
1.6 g-Meso	87.8	80.2	93.2	6.8	27.5	50.6	21.9
1.9 g-Meso	85.1	81	92.9	7.1	26.1	50.2	23.7

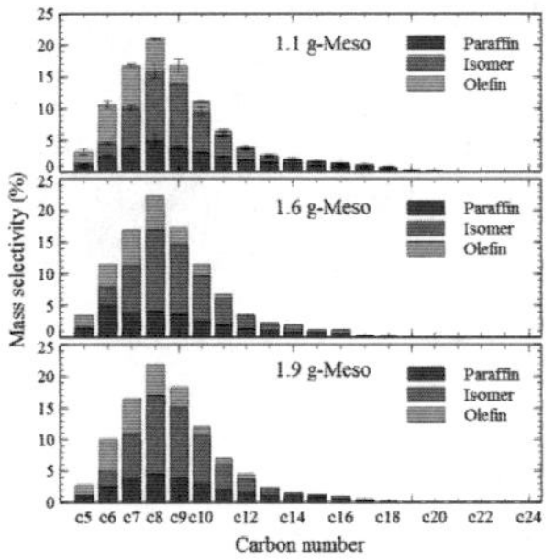

Liquid products distribution for different amount of meso-ZSM-5 coated co-catalysts.

- **CO conversion didn't change too much.**
- **Gasoline selectivity was above 90 wt%.**
- **Isomers selectivity was elevated with ZSM-5 coating**

πδσωλ

Performance at Different Pressures

Co-catalysts performance of 1.6 g meso-ZSM-5 at different pressures. Reaction condition: T: 230 °C; Syngas flowrate: 35 ml/min; Time on stream: 48 h

Catalyst	CO conv (%)	Carbon species selectivity (C%)		Oil phase selectivity (wt.%)			
		C_{5+}	C_5-C_{12}	C_{13+}	Paraffins	Isomers	Olefins
1.6 g -6 bar	55.3	84.6	88.9	11.1	33.1	44.3	22.6
1.6 g-12 bar	87.8	80.2	93.2	6.8	27.5	50.6	21.9
1.6 g-20 bar	92.8	78.4	92.4	7.6	28.1	50.1	21.8

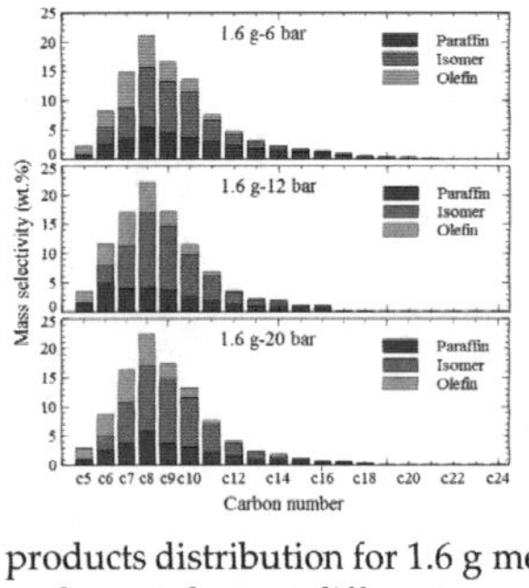

Liquid products distribution for 1.6 g meso-ZSM-5 coated co-catalysts at different pressures.

- **CO conversion increased with increase of pressure.**
- **Gasoline selectivity reached highest at 12 bar.**
- **Isomers selectivity was elevated with ZSM-5 coating.**

πδσωλ

Conclusions

- **Supported Zeolite/Co-Al_2O_3/monolith catalysts were created, characterized and evaluated for FTS**
- **Catalyst coated with Meso-zeolite ZSM-5 exhibits:**
  - High CO conversion
  - Excellent selectivity to gasoline range (C_5-C_{12}) compounds
  - Improved conversion to higher RON olefins and isomers
- **Recommended conditions for catalysts:**
  - Moderate loading of Meso-ZSM5: 1.6 g
  - Moderate pressure: 12 bar

πδσωλ

Acknowledgements

ACS-PRF Award No. 53648-DNI5

UConn Center for Clean Energy Engineering

Thank you!

Department of Chemical & Biomolecular Engineering
University of Connecticut

http://pdsol.engr.uconn.edu

499690 Valorization of Natural Gas Liquids from Shale Gas

Abstract:

Recent shale gas boom has transformed the US energy landscape. Most of the major shale basins are located in remote locations and historically non-gas producing region. Therefore, many major shale basins regions are lacking the infrastructure to distribute the extracted gas into the rest of the US and particularly the Gulf Coast region. Here, we propose a new shale gas valorization process that converts natural gas liquids (NGLs) in shale gas into higher molecular weight hydrocarbon, which is more valuable to transport from these remote basins than NGL or its constituents such as ethane and propane.

In this process, natural gas liquid is transformed to higher molecular weight hydrocarbon. Compared with the existing technologies, our new approach has the following advantages: (1) This process minimizes the production of by-products. (2) This process is amenable to size reduction.

We demonstrate that, through a case study for a 96 MMSCFD shale gas feed from the Bakken region, the total capital cost for the proposed process is much cheaper than existing Gas-to-Liquid (GTL) processes such as Fischer-Tropsch and Methanol-to-Gasoline process. Also, we estimate the break-even price for our process to be lower than GTL-Fischer-Tropsch process. Therefore, this new process is an economically attractive step to upgrade shale gas in remote shale basins.

496044 Energy Sustainability Analysis of H_2 Production

Abstract:

The perspective for H_2 as an energy carrier to decarbonize energy services brings about a more and more relevant role for technologies of hydrogen production. As matter of fact, the yearly majority (>95%) of H_2 is produced from fossil fuels by steam reforming or partial oxidation of methane with only a small quantity by other routes, such as biomass gasification or electrolysis of water, hence the needs to explore new routes to produce H_2 appears evident. Stringent is the necessity to explore different technologies even at the infancy state, to produce hydrogen using different feedstocks. In this respect the energy analysis of H_2-producing technologies aimed to evaluate the net energy produced or Energy Return Of Investment (EROI) and Energy Payback Time (EPT) in order to score them towards energy sustainability is utmost of importance. To this end, the present paper is aimed to test three of the most advanced processes: *i)* steam reformer of methane, *ii)* electrolysis of water by using photovoltaic (PV) electrical energy and *iii)* dark fermentation (DF) of organic refuses (OR) by mixed microorganisms Hydrogen Producing Bacteria (HPB) mainly *Clostridium spp.* The three technologies were tested at laboratory scale for about one year, including the steam reformer owing the necessity to test the hypothesis to produce hydrogen at "the point of use" rather than trying to distribute it. In the panorama of bioH_2 production, which include heterotrophic and autotrophic microorganisms, the choice of DF is based on many reasons: *i)* reactions do not require light, so hydrogen is produced throughout the day and night, *ii)* the liquid end products of DF can be used as feedstock for Anaerobic Digestion (AD) to produce CH_4 with additional energy produced, and *iii)* because the OR are so abundant (> 70 % of wastes) and almost uniformly distributed.The data acquired by experimental campaigns were used to evaluate and scoring the energy sustainability of the three process by the application of a previous candidate procedure (Di Addario et al.,2016). The procedure is recognised into three different steps: a first screening is performed by using the Energy Sustainability Index (ESI), which takes into account the total Produced Energy under form of hydrogen referred to the direct spent energy (heat and electricity). The second step is the Analogical Model, i.e. the quantification of each chemicals, materials and energy to estimate the Useful Energy with a Life Cycle Approach (LCA), taking into account all the Indirect Energy terms. At such a step the energy embedded in the feeds, i.e. methane and water were evaluated for the steam reformer and the electrolysis process, respectively. For the DF process instead, the energy embedded in OR was assumed to be equalto zero, following the allocation procedures in LCA analysis. The third step is the evaluation of two parameters: EROI and EPT, considering all the involved energy flows. The procedure is anuseful tool to score the three processes towards sustainability: higher score correspond to high EROI & low EPT, and this permits to select the most sustainable choice among the three analyzed H_2-producing processes from an energy point of view.

Di Addario, M., Malavè, A.C.L., Sanfilippo, S., Fino, D., Ruggeri, B., 2016. Evaluation of sustainable useful index (SUI) by fuzzy approach for energy producing processes. *Chemical Engineering Research and Design*, 107, pp.153–166.

501652 Chemical Looping Based Technology for High Efficiency Production of H_2 from Ammonia (NH_3)

Abstract:

Hydrogen (H2) is widely considered to be an integral part in solving the issues surrounding the consumption of fossil fuels. As a carbon neutral fuel, H2 can be used as an energy source with minimal emissions for a wide range of uses including chemical production and transportation. One of the most advantageous properties of using H2 as a fuel source lies in its high specific energy of 39.8 kWh/kg (HHV). Despite the advantages offered by the widespread utilization of H2, market penetration is prohibited by the high transportation costs associated with its low volumetric energy density of 1.55kWh/L and thus high compression losses in the storage of H2. In an effort to overcome these limitations, the use of ammonia (NH3) as a hydrogen carrier to minimize transportation and storage costs has been proposed. This vision for improving the market penetration of H2, is contingent upon development of high efficiency H2 from NH3 processes. OSU has developed a chemical looping based NH3 to H2 (ATH) process that can operate at high thermal and H2 production efficiencies. The OSU chemical looping approach consists of two reactors to convert NH3 to H2. The first reactor, namely the reducer reactor utilizes an intrinsic O2 gradient driven by the reduction potential of NH3 and a metal oxide like Fe3O4, ZrO2, WO3 to efficiently crack NH3 to a mixture of N2, H2 and H2O. The reduced metal-oxide from the reducer reactor is re-oxidized in an oxidizer using H2O as the oxygen source. This ensures ~1.5 moles of H2 production per mole of NH3 (nearly 100% H2 production efficiency). This presentation will initially focus on the thermodynamic rationale for developing a chemical looping based NH3 to H2 process. This will be followed by the heat balance approach to achieving the maximum H2 production efficiency while maintaining a high energy efficiency values. The relevant experimental data (using Thermo-gravimetric Analyzer and a fixed bed reactor) that identifies key design conditions associated with the kinetics of the reducer and the oxidizer reactor will be presented. A series of directed sensitivity studies that identify the trends associated with the chemical looping system and the design parameters that need to be de-risked in further scale up of this technology will be discussed.

496846 Distributed Ammonia Manufacturing

Abstract:

To eliminate the dependency on fossil fuel and the carbon dioxide emissions, carbon-free ammonia can be synthesized in small wind-powered plants, and from totally renewable resources: hydrogen from electrolysis of water and nitrogen from pressure swing adsorption of air. We are making carbon-neutral ammonia in the UMN-WCROC plant, located in the agricultural region of Morris MN.[1] While this process is working, the ammonia costs twice that made conventionally, which is due to downscaling (40,000 scaled down process.) To overcome the difficulties associated with the energy demand for ammonia synthesis, and to improve the economics of the small plant, we are seeking a more efficient process to synthesize carbon-neutral ammonia.

One strategy to improve this process is to design a new class of materials for the catalysis, that allows synthesis at lower temperature and pressure. However, a century of research confirms that this is not trivial and immediately viable.[2] Essentially with the conventional catalyst, high temperature is a must in order to obtain fast kinetics, while high pressure ensures the conversion. If we can efficiently remove low concentrations of ammonia formed in the reactor, then we can keep the reaction far from equilibrium, and intrinsically we obtain rates closer to forward reaction rates. Thus the constraints of equilibrium is surpassed. We have proposed the application of solid absorber for high temperature ammonia separations.[3,4] In this process, ammonia can be synthesized at much lower pressure, with rates comparable to high-pressure conventional processes. The capital and operating costs of low pressure reaction-absorption process is improved since the work of compressing the fresh feed is significantly reduced, and the need for heat exchange also drops.

The low pressure reaction-absorption process suggests a strategy for making ammonia fertilizer locally, using wind-generated electricity. Alternatively, this ammonia made in rural area, can be used as a fuel, or a hydrogen carrier for fuel cell use in more urban areas.

References:

(1) Reese, M.; Marquart, C.; Malmali, M.; Wagner, K.; Buchanan, E.; McCormick, A.; Cussler, E. L. Performance of a Small-Scale Haber Process. *Ind. Eng. Chem. Res.* **2016**, *55* (13), 3742.

(2) Schlögl, R. Catalytic Synthesis of Ammonia—A "Never-Ending Story"? *Angew. Chemie Int. Ed.* **2003**, *42* (18), 2004.

(3) Malmali, M.; Wei, Y.; McCormick, A.; Cussler, E. L. Ammonia Synthesis at Reduced Pressure via Reactive Separation. *Ind. Eng. Chem. Res.* **2016**, *55* (33), 8922.

(4) Wagner, K.; Malmali, M.; Smith, C.; McCormick, A.; Cussler, E. L.; Zhu, M.; Seaton, N. C. A. Column Absorption for Reproducible Cyclic Separation in Small Scale Ammonia Synthesis. *AIChE J.* **2017**.

499699 Deciphering the Chemical and Physical Mechanisms Active in Elevated Temperature Photocatalytic Synthesis of Hydrocarbons and Ammonia

Abstract:

The photocatalytic synthesis of hydrocarbons, oxygenates, and NH_3 from CO_2, N_2, NO/NO_2, and H_2O is a truly sustainable route to fuels and building block chemicals production. Unfortunately, the community still lacks the ability to rationally improve overall rates of reaction and control product distributions due to a lack of reaction mechanism understanding. Endemic issues with organic contamination in experimental studies has further contributed to this lack of progress. Our combined computational/experimental studies of photocatalytic CO_2 reduction by H_2O in organic-free reaction conditions at an appreciable temperature range (gas phase reaction at ~100-400°C+) have shed significant light upon the surface chemical and electronic properties that affect these reaction mechanisms. Moreover, studies have also isolated new H-transfer mechanisms. Both of these developments have the power to significantly aid in the design of these materials. The new H-transfer mechanisms and an understanding of the role of surface chemistry have further enabled control over the selectivity of surface-bound atomic H between hydrogenation and H_2 evolution -- a critical efficiency aspect of the system. Additionally, key carbon-containing reaction intermediates that allow for carbon-carbon coupling have been isolated as well. Catalysts shown to be active at elevated temperatures indicate semiconductors with a high Debye temperature, robust bulk bonding, and elevated surface chemical reactivity may lend additional tunable features for photocatalytic synthesis mechanisms. With the ultimate aim of producing higher hydrocarbons, oxygenates, and ammonia photocatalytically from simple thermodynamic-minimum reactants, these studies have mapped the course forward for materials selection and mechanistic understanding.

502400 A High-Throughput Computational Screening Approach for Solar Fuels Photoelectrocatalysis

Abstract:

In recent years, theoretical overpotential has become increasingly popular as a metric for characterizing the suitability of oxygen evolution electrocatalysts using density functional theory (DFT). In addition, DFT has become an invaluable tool for characterizing photoabsorbing materials in a number of high-throughput screening studies aimed at designing materials for solar fuels production from stability and bandstructure criteria. In this work, we present an approach which integrates calculations of theoretical overpotential with those of band-gap and band-edge positions on mixed metal oxides. Given that the prospect of discovering lone materials fulfilling the criteria of stability, catalytic activity, and photoabsorption efficiency, we also report on how materials with suitable surface chemistry may be more effectively paired with suitable photoabsorbing substrates using interfacial matching. Lastly, we benchmark our approach comparing potential combined metrics with experimentally measured photoelectrocatalytic activity.

499265 Computational Insights into Photo(electro)Chemical Nitrogen Fixation over Titania Catalysts

Abstract:

Photocatalytic nitrogen fixation is a form of artificial photosynthesis capable of enabling distributed production of fixed nitrogen at near-ambient conditions from the nitrogen, water, and photons that are readily available from the environment. The process has been demonstrated over titania catalysts through both reductive and oxidative pathways, although the currently observed nitrogen fixation rates are too low to be practical for fertilizer synthesis and there is little fundamental understanding of the reaction mechanism or chemical phenomena that enable the process. Interestingly, the process occurs under environmental conditions and has been speculated to play a significant role in the global nitrogen cycle owing to the prevalence of $TiO2$ in naturally occurring sands and man-made pigments. Despite the wide-ranging potential impacts of this reaction, first demonstrated nearly 75 years ago, the process has received relatively little attention and has not been studied with modern techniques. This talk will briefly review the historical context and current progress toward photocatalytic nitrogen fixation, followed by a theoretical perspective into the thermodynamic driving forces and likely reaction mechanisms, with an emphasis on the potential for developing photoelectrochemical processes. The computational results will be complemented by recent experiments demonstrating the feasibility of the reaction.

501308 Engineering the Interface to Improve the Efficiencies of Insulator-Protected Semiconductors for Photoelectrochemical Cells

Abstract:

Recent improvements in solar fuels production efficiencies have been enabled by the development of protection schemes for low bandgap semiconductors that would otherwise be unstable in the harshly oxidative environment of an electrochemical cell. One common protection strategy incorporates an insulator layer placed between a metal electrocatalyst and semiconductor. Recent research has shown that the characteristics of the insulator layer have a large effect on the overall efficiency of a photoelectrochemical cell. In this contribution, we will provide experimental evidence that demonstrates how by tuning the properties of the insulator layer we can better control charge transfer across the interface in order to improve efficiencies. We will also present a model to describe the insulator layerâ€™s effect on charge transfer mechanisms. Provided with these modeling insights, we will discuss which experimental systems can benefit from the most from insulator tuning in order toÂ expand the flexibility of material selection for a photoelectrochemical device.

492457 The Role of Surface-Bound Dihydropyridine Analogs in Pyridine-Catalyzed CO_2 Reduction over Semiconductor Photoelectrodes

Abstract:

The (photo)electro-catalytic reduction of CO_2 to useful fuels via energy from sunlight has received significant attention as a promising route for generating carbon-neutral fuels and value-added chemicals. Experimental evidence suggests that pyridine (Py) is an effective co-catalyst during CO_2 reduction over GaP, CdTe, and $CuInS_2$ semiconductor electrodes. Identifying the role of Py during catalytic reduction is essential for optimizing the design of such photocatalytic processes. We propose a general reaction mechanism for the Py-catalyzed reduction of CO_2 over GaP(111), CdTe(111), and $CuInS_2$(112) photoelectrode surfaces. This mechanism proceeds via formation of a surface-bound dihydropyridine (DHP) analog, which is a newly postulated intermediate in the Py-catalyzed mechanism. Using density functional theory, we calculate the standard reduction potential related to the formation of the DHP analog, which demonstrates that it is thermodynamically feasible to form this intermediate on all three investigated electrode surfaces under photo-electrochemical conditions. Hydride transfer barriers from the intermediate to CO_2 demonstrate that the surface-bound DHP analog is as or even more effective at reducing CO_2 to HCOO- as the $DHP_{(aq)}$ molecule in solution. This intermediate is predicted to be both stable and active on many varying electrodes, therefore pointing to a mechanism that can be generalized across a variety of semiconductor surfaces and explains the observed electrode dependence of the photocatalysis.

490917 Surface Plasmon-Assisted Photoelectrochemical CO_2 Reduction on Well-Defined Nanostructured Silver Electrodes

Abstract:

Plasmonic hot-carriers and the strong local electric fields produced by plasmon excitation may open new pathways for CO_2 reduction, resulting in increased product selectivity and lower overpotential. Plasmon decay generates â€œhotâ€ electrons which can be transferred selectively to a surface molecule for direct reduction. Simultaneously, the enhanced electric fields can alter the electronic coupling with surface adsorbed molecules, thereby changing the binding energy of these species and the catalytic properties of the plasmonic metals.[1-3] Plasmon-assisted photoelectrochemical CO_2 reduction was demonstrated when optically excited surface plasmons at the roughened surface of a Ag electrode in CO_2-containing solutions produced strong enhancement of the cathodic current.[4-5]

Here we reveal the mechanism of plasmonic hot-carriers interacting with CO_2 and associated adsorbates on plasmonically-active Ag nanopyramid catalysts. Photocurrent measurements and photoelectrochemical spectroscopy were performed on the plasmonic Ag nanopyramid photocatalysts in electrolyte saturated with CO_2 and Ar. We observed that the photocatalytic reaction is much stronger in the presence of CO_2 and is resonantly enhanced at a specific potential. In Ar-saturated electrolyte, no characteristic feature of photocurrent was observed. These results indicate that the hot-carriers created by the localized surface plasmon resonance effect are reacting specifically with CO_2 molecules and CO_2 reduction intermediates. Further, the photocurrent onset potential was significantly reduced with respect to the dark current. Our findings suggest that the plasmonic hot-carriers behave as an effective co-catalyst with Ag and may play a critical role in discovering a CO_2reduction process that forms readily usable hydrocarbon fuels.

Scanning electron microscopy (SEM) was used to confirm the shape, size, and distribution of the silver nanopyramids. Photocurrent measurements were performed in a custom glass photoelectrochemical cell with a quartz window that allows for front- or back-illumination of the electrode. The two electrode chambers are separated by a membrane and can achieve a gas-tight seal that allows for precise pressure measurements. The cell was coupled to a gas chromatograph (GC) that allows for detection and quantification of gaseous products. *Ex-situ* proton nuclear magnetic resonance spectroscopy (^1H NMR) and high performance liquid chromatography (HPLC) were used to quantify liquid products.

References

[1] A. O. Govorov, et. al., *Nano Today* 9, 85 (2014)
[2] R. Sundararaman, et. al., *Nature Comm.* 5, 5788 (2014)
[3] S. Mukherjee, et. al., *J. Am. Chem. Soc.* 136, 64 (2014)
[4] R. Kostecki, et. al., *J. Chem. Phys. Lett.* 194, 386 (1992)

[5] R. Kostecki, et. al., *J. Appl. Phys.* 77, 4701 (1995)

This material is based upon work performed by the Joint Center for Artificial Photosynthesis, a DOE Energy Innovation Hub, supported through the Office of Science of the U.S. Department of Energy under Award Number DE-SC0004993.
The work is supported by the National Science Foundation Graduate Research Fellowship under Grant No. 1106400. Work at the Molecular Foundry was supported by the Office of Science, Office of Basic Energy Sciences, of the U.S. Department of Energy under Contract No. DE-AC02-05CH11231.

490812 Artificial Leaf for Carbon Dioxide Photo-Reduction to Fuels

Abstract:

An artificial leaf that mimics photosynthesis process is a promising approach to directly convert solar energy into energy-rich chemicals[1–4]. Thus far, this concept has been successfully developed for the hydrogen generation process[2–4]. Here, we have developed an artificial leaf for the conversion of carbon dioxide (CO_2) into synthesis gas in an acidic electrolyte using the energy from a sun simulator. A configuration of the artificial leaf was achieved by using cobalt, as an oxygen-evolving catalyst and molybdenum disulfide (MoS_2) catalyst for CO_2 reduction reaction deposited on the illuminating and back-substrate surfaces of a photovoltaic cell, respectively. We have performed detail experimental studies such as linear sweep voltammetry (LSV) and chronoamperometry (CA), in-situ differential electrochemical mass spectroscopy (DEMS), and gas chromatography (GC) followed by some characterization techniques e.g. scanning electron microscopy (SEM), X-ray photoelectron spectroscopy (XPS) and energy dispersive spectroscopy (EDS) to elucidate the performance and stability of our developed artificial leaf. The experimental analyses indicate that our custom made artificial leaf cell can directly convert the solar energy to synthesis gas with a catalytic conversion efficiency of 75%. Our results reveal that the operation of the artificial leaf cell is not restricted by the CO_2 reduction reaction according to the remarkable activity of MoS_2 NFs catalysts[5]. However, it is rather limited by the oxygen evolution reaction (water splitting) at the anode side. This finding could potentially open a new route in the solar-driven CO_2 conversion field where the CO_2 reduction reaction is known to be the bottle neck of this process.

References:

1. Asadi, M. *et al.* Nanostructured transition metal dichalcogenide electrocatalysts for CO_2 reduction in ionic liquid. *Science.* **353,**467–470 (2016).

2. Kanan, M. W. & Nocera, D. G. In Situ Formation of an Water Containing Phosphate and Co^{2+}. *Science.* **321,**1072 (2008).

3. Reece, S. Y. *et al.* Wireless Solar Water Splitting Using Silicon-Based Semiconductors and Earth-Abundant Catalysts. *Science.* **334,**645–648 (2011).

4. Qiu, F. *et al.* Photocatalytic Hydrogen Generation by CdSe/CdS Nanoparticles. *Nano Lett.* **16,**5347–5352 (2016).

5. Abbasi, P. *et al.* Tailoring the Edge Structure of Molybdenum Disulfide toward Electrocatalytic Reduction of Carbon Dioxide. *ACS Nano.* **11**, 453–460 (2017).

500536 Understanding Photoelectrochemistry on Epitaxial Oxides through Surface Electronic Structure

Abstract:

The intermittent nature of renewable energy sources requires a clean, scalable means of converting and storing energy. One earth abundant storage option is water electrolysis, storing energy in the bonds of O_2 and H_2. Photoelectrochemical (PEC) cells based on semiconductor/liquid interfaces can convert sunlight to chemical fuels without external circuitry, such as "splitting" water into O_2 and H_2 upon illumination of a photoabsorber and catalyst. Tuning the location of the band edges of a semiconductor tailors the wavelength of photoabsorption and the charge transfer to adsorbed species. The efficacy of conversion depends in part on the rectifying properties of semiconductor–electrolyte junctions, as the band bending present in the semiconductor space-charge region from Fermi level equilibration drives the separation of electron–hole pairs. However, the relationship between band structure, band bending/flattening, photocatalytic activity, and surface functionality (such as hydroxylation properties and hydrophilicity) is not well understood. Improving our understanding of the semiconductor/electrolyte interface is crucial in the design of efficient photoelectrode materials for harvesting solar energy in a PEC system.

In the particular case of the semiconductor Ge, favorable conduction band alignment with the hydrogen evolution reaction (HER) suggests the potential for efficient solar-to-hydrogen conversion with visible light. However, material instability in aqueous environments and the potential for photocorrosion limits its utility in PEC applications. In order to protect the photoabsorber surface while maintaining favorable alignment of the conduction band, we epitaxially grow $SrTiO_3$ on top of p-Ge. We will present studies of such model oxide photoelectrodes grown by molecular beam epitaxy (MBE) on single crystal substrates that display a known crystallographic orientation, surface area, path for charge transport, and strain. Photoelectrochemical measurements on these heterostructures can establish the intrinsic activity of oxide catalysts in a way that cannot be realized with polydisperse nanoparticle systems. Insight into the band bending between the substrate and oxide overlayer, as well as at the semiconductor surface, can be obtained from X-ray photoelectron spectroscopy (XPS).[1] Measurement of XPS at ambient pressures (AP-XPS) can further elucidate the relationship between adsorbates and surface band bending.[2] This fundamental insight will build understanding necessary for the design of active, earth-abundant photocatalysts that can be integrated into PEC devices for efficient conversion of solar energy into chemical fuels.

References

1. S.A. Chambers, Y. Du, R.B. Comes, S.R. Spurgeon, P.V. Sushko, *Applied Physics Letters,* **110**, 082104 (2017).

2. K.A. Stoerzinger, R. Comes, S.R. Spurgeon, S. Thevuthasan, K. Ihm, E.J. Crumlin, S.A. Chambers. *J. Phys. Chem. Lett.* **8**, 1038 (2017).

486659 Separation of Anhydrosugars and Phenolic Species in a Fast Pyrolysis Aqueous Product Stream Using Resin Adsorbents and Simulated Moving Bed Technology

Abstract:

Anhydrosugars are a valuable product present in fast pyrolysis bio-oil with levoglucosan being the most abundant sugar. Appropriate acid pretreatment methods of lignocellulosic biomass before pyrolysis can yield 15 and 23 wt% levoglucosan on a biomass basis for switchgrass and red oak, respectively. A novel fractionating collection system developed at Iowa State University condenses and collects pyrolysis products according to their respective vapor pressures. The anhydrosugars condense in the first fraction alongside phenolic monomers, dimers and oligomers. Water extraction of the first fraction removes approximately 80 wt% of the anhydrosugars and a myriad of phenolic monomers up to their solubility limits in water, for example, 0.8 and 2.7 wt% for 4-vinylphenol and *o*-cresol, respectively, at 25 °C. The total average phenolics present in the water extraction is approximately 9 wt% and must be reduced to levels near 0.01 wt% to be below toxicity limits for fermentation. This work examines the use of hydrophobic polymeric resin adsorbents for the selective removal of phenolic species from the aqueous phase. Adsorption isotherms and kinetic adsorption parameters were determined for the anhydrosugars and phenolic species adsorption on commercially available hydrophobic polymeric resin adsorbents, SP207 and XAD-4. Isotherm data demonstrates an order of magnitude higher affinity of the resins for phenolic species over anhydrosugars. Although the anhydrosugars adsorb to a similar capacity on the resins as the phenolic species when adsorbing in the absence of phenolics, their presence competitively reduces anhydrosugar adsorption an order of magnitude. Chromatographic column experiments were performed and showed good agreement with breakthrough curves calculated using the adsorption parameters. A simulated moving bed chromatographic process was designed and modeled for the continuous production of a very high purity anhydrosugar stream and separate phenolics stream for a 50 ton per day pyrolysis bio-refinery.

497969 Use of Novel Reactor-Separator Combination (Membrane BioReactor) for the Enzymatic Hydrolysis of Waste Fines and Fiber Rejectsfrom Recycled Linerboard Paper Mills

Abstract:

Papermaking fibers, particularly waste fibers from recycling operations can be a valuable resource for biofuels (ethanol), and sustainable products such as bioplastics or biochemicals. The amount of lignin in such fibers is much lower than native wood or other plant based lignocellulosic resources eliminating the need for complex and expensive pretreatment processes. Moreover, the removal of hornified cellulosic fines from recycled pulps can accelerate water removal from the pulps on the papermachines and in the dryer sections, leading to significant productivity increases. Inter-fiber bonding in the paper sheet is improved due to a lower â€˜waste finesâ€™ content, reducing the use of strength agents like starch and improving sheet quality.

The present work develops a physico-chemical model for the kinetics of enzymatic hydrolysis of cellulosic pulp fines from recycled linerboard mills. Batch kinetics representing hydrolysis in a fixed agitated volume of the fines suspension is considered. Cellulosic fines are assumed to be spherical and porous, with the cellulase enzymes diffusing and adsorbing onto the interior pore surfaces. Enzymatic reactions result in the production of glucose which diffuses outward. The mechanistic role of inhibitory mineral (ash) fillers and the application of suitable surfactants to cover their surfaces and improve cellulose hydrolysis is considered.

Optimal usage of the enzyme catalyst is a critical part of this process. In this paper, microfiltration membranes are used to separate the enzymes and recycled so that the effective catalyst concentration is higher and overall enzyme usage is minimized. Experimental results showing the impact of flow rates, concentrations of substrate and concentrations as well as other variables such as pH and temperature are shown. Mathematical models for determining the optimal designs of both external and internal membrane bioreactors were set up for analysis.

501158 Evolving Topochemistry of Acer Saccharum Chips during Pretreatment Determined By Confocal Raman Microscopy

Abstract:

The utilization of dedicated biomass crops as a renewable resource of platform chemicals requires a thorough understanding of the physical and chemical processes inherent to the industry. Due to the complex network of physically and chemically recalcitrant polymers which comprise woody biomass, a pretreatment to deconstruct the material is a necessary preliminary step required for its use. Among the large variety of treatments that have been investigated, liquid hot water (LHW) pretreatment has been demonstrated to have some of the greatest potential. The technology balances low capital and operating costs with an acceptable increase in hydrolysis yield. It is well established that the primary effect of LHW pretreatment is the removal of hemicelluloses via a combination of autohydrolysis and dissolution (creating a sugar-rich side stream).As hemicelluloses are a large component of the inter-dependent network of polymers that is the cell wall, this creates more pore space and increases cellulose susceptibility to hydrolysis (enzymatic or otherwise). Also, while it is known that bulk lignin concentrations are largely unchanged, it is hypothesized that they undergo a process of plasticization and re-deposition. These facts have been corroborated by many gravimetric experiments, as well as visualization by electron and atomic force microscopy. However, in light of the as-yet unrealized potential of dedicated biomass crops, it is clear that a more fundamental knowledge describing these phenomena is required.

To this end the authors have made use of an analytical instrument finding increasing utility in the field, the confocal raman microscope (CRM). Acer sacchrum wood chips were subject to hot water pretreatment at 160° C for varying times. Samples from this treatment were transversely microtomed into 10Î¼m thick sections for analysis by CRM. This was coupled with gravimetric analyses as well as other spectroscopic techniques. Quantitative and qualitative analysis at the sub-cellular level for whole wood chips at real industrial reaction conditions is shown for the first time. A preferential de-lignification from the secondary wall to the compound middle lamella was observed. As pretreatment severity increased, it was seen that samples lost the clear delineation between morphological layers of the cell wall, becoming internally heterogeneous as compared to untreated samples. It is deduced this is due to limited mass transport capabilities within the wood chip, as previous works have shown that in non-mass transport limited scenarios the cell wall maintains its distinct, homogeneous layers. The quantitative data gained herein presents the possibility to increase our understanding of and model the physio-chemical processes occurring within the cell wall as a result of LHW and similar hydrolytic pretreatments.

CENTER FOR ADVANCES IN
WATER & AIR QUALITY
LAMAR UNIVERSITY
TEXAS
COLLEGE OF ENGINEERING
LAMAR UNIVERSITY®
MEMBER THE TEXAS STATE UNIVERSITY SYSTEM™
MICROWAVE ASSISTED SYNTHESIS OF SILVER NANOPARTICLES USING EXTRACELLULAR POLYMERIC SUBSTANCES (EPS) FROM *Cosmarium* sp.
Authors: Adarsh Bafana, Shishir V Kumar, Prasad P Pawar, Ashiqur Rahman, Si A. Dahoumane and Clayton S Jeffryes
Presenting Author: Adarsh Bafana
PhD. Candidate

COLLEGE OF ENGINEERING
Silver Nanoparticles (AgNPs): Applications
CURAD
silver
SoleFresh
Silver Nanoparticles
Reference: https://nanocomposix.com/pages/knowledge-base
http://waterislife.com/media/videos
LAMAR UNIVERSITY

Extracellular Polymeric Substances (EPS)

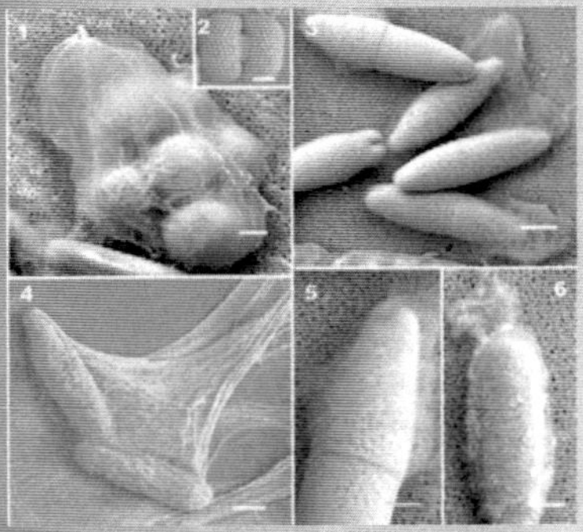

SEM image of *Cosmarium* cells covered in EPS (Kiemle et al. 2007)

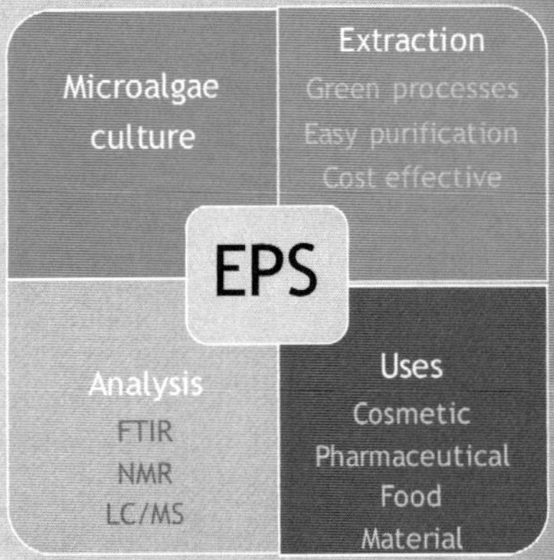

Cosmarium sp.

- *Cosmarium* is a non-motile, freshwater member of division Chlorophyta (desmids)
- Desmids are known to produce significant amounts of EPS which is utilized for different purposes.

Culturing :

- Flask scale cultures
- Culturing was done in Bold Basal medium (BBM)
- pH: 7.6 Light intensity: 30-40 μmoles m^{-2} s^{-1}

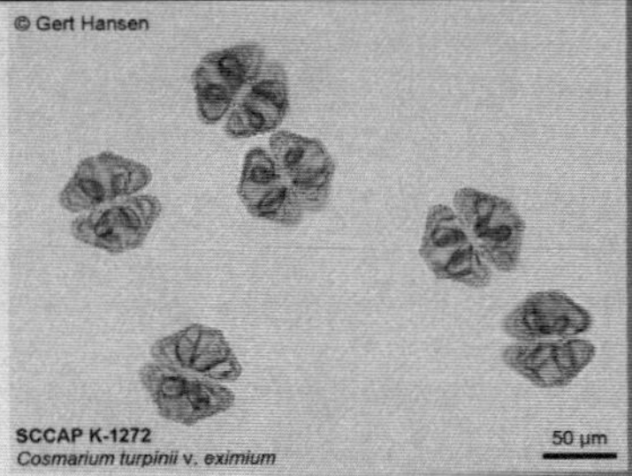

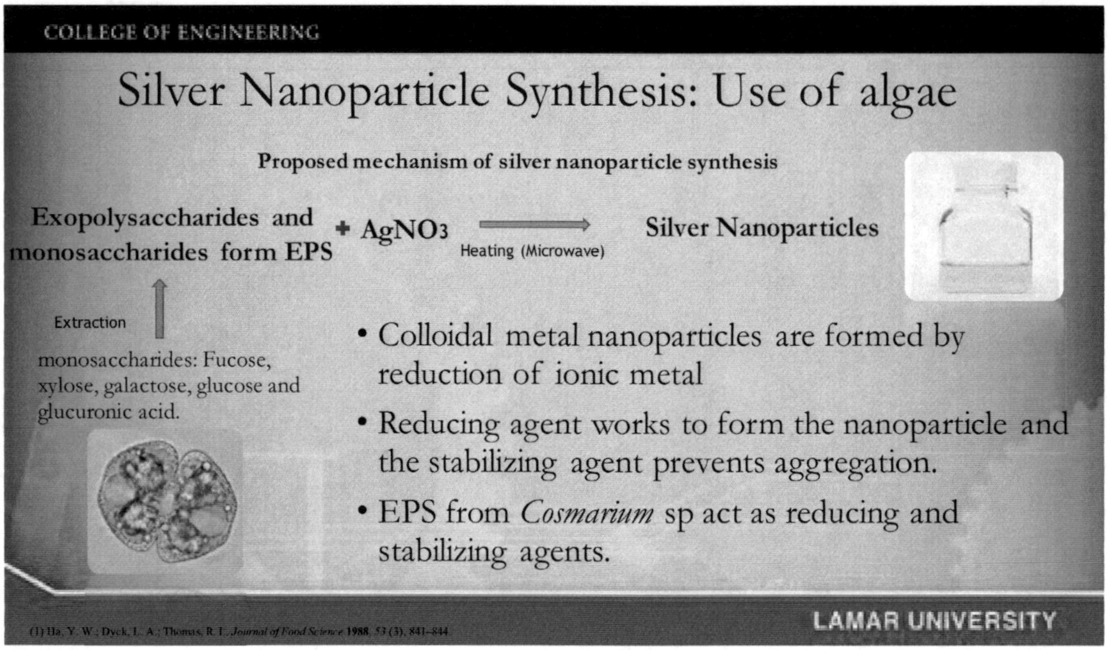
COLLEGE OF ENGINEERING
Silver Nanoparticle Synthesis: Use of algae
Proposed mechanism of silver nanoparticle synthesis
Exopolysaccharides and monosaccharides form EPS
+ AgNO3
Heating (Microwave)
Silver Nanoparticles
Extraction
monosaccharides: Fucose, xylose, galactose, glucose and glucuronic acid.
• Colloidal metal nanoparticles are formed by reduction of ionic metal
• Reducing agent works to form the nanoparticle and the stabilizing agent prevents aggregation.
• EPS from Cosmarium sp act as reducing and stabilizing agents.
(1) Ha, Y. W.; Dyck, L. A.; Thomas, R. L. Journal of Food Science 1988, 53 (3), 841–844.
LAMAR UNIVERSITY

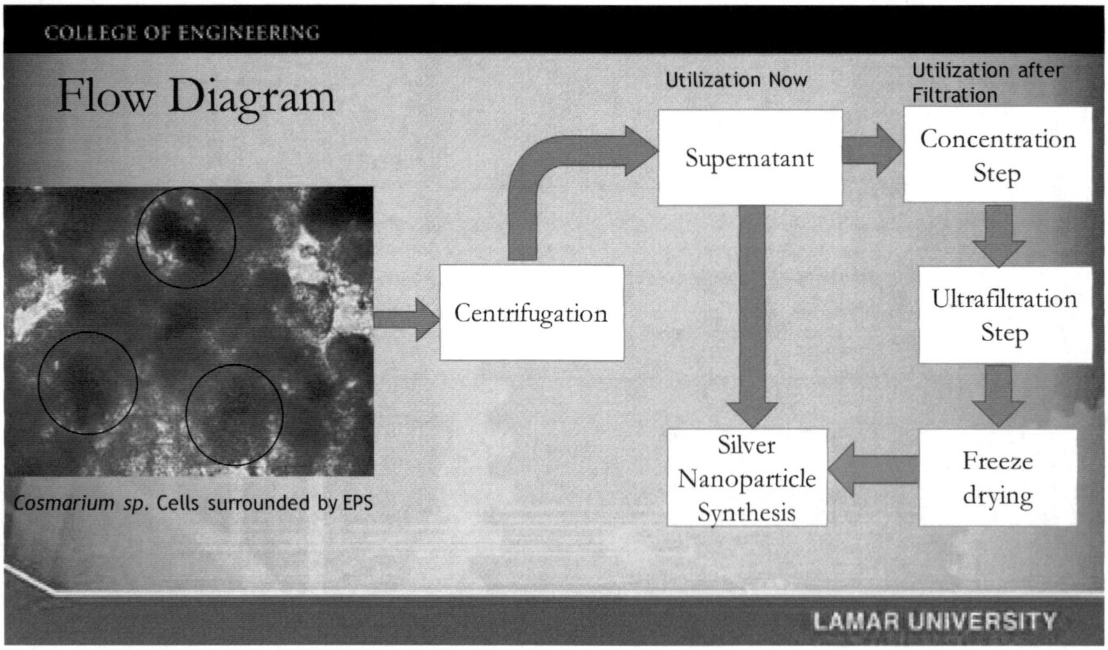
COLLEGE OF ENGINEERING
Flow Diagram
Utilization Now
Utilization after Filtration
Supernatant
Concentration Step
Centrifugation
Ultrafiltration Step
Silver Nanoparticle Synthesis
Freeze drying
Cosmarium sp. Cells surrounded by EPS
LAMAR UNIVERSITY

COLLEGE OF ENGINEERING

Harvesting and Separation of EPS

Fresh *Cosmarium* sp. culture and *Cosmarium* and EPS

- After about 1.5 month the culture turned pale yellow and were ready to harvest.
- Centrifugation was performed at 3300 rpm for 10 min.
- The supernatant was utilized for synthesis of silver nanoparticles.

LAMAR UNIVERSITY

COLLEGE OF ENGINEERING

Separation of EPS

- Rotary Evaporator was used to concentrate the sample.
- Ultrafiltration step was performed to remove the salts from the EPS which are in the media(more details)
- Freeze drying was done to utilize and analyze the EPS.

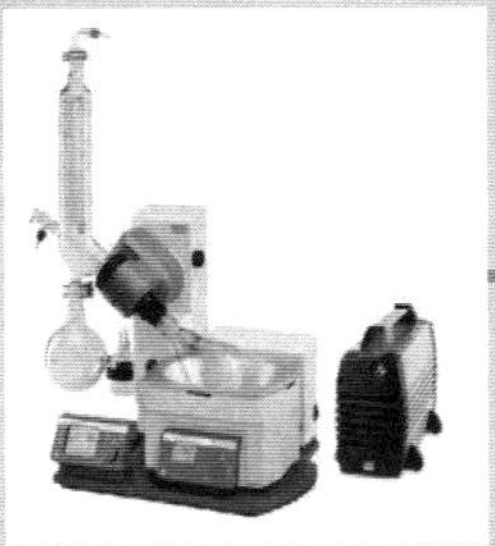

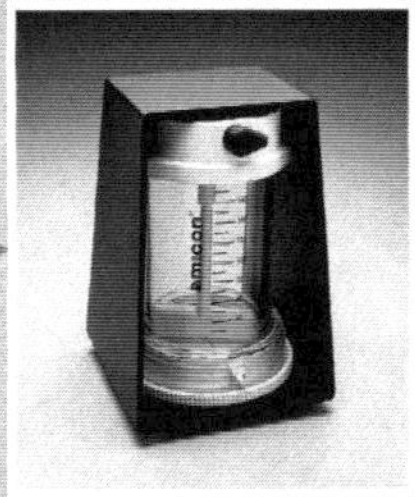

LAMAR UNIVERSITY

Organic content in the Supernatant

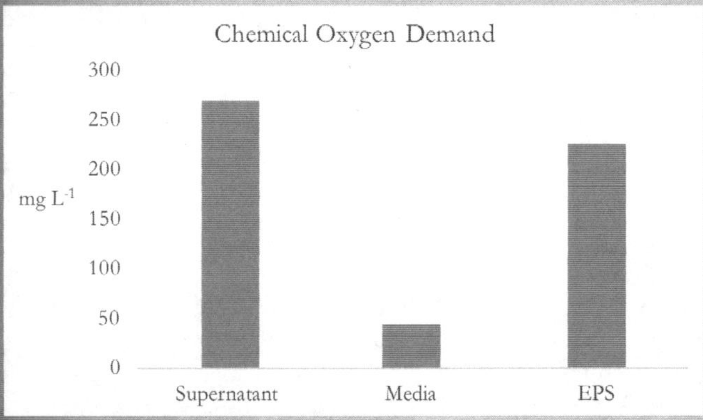

The COD results show that the amount of organics in the supernatant is 225 mg/l.

Freeze dried -EPS

FTIR to characterize the EPS

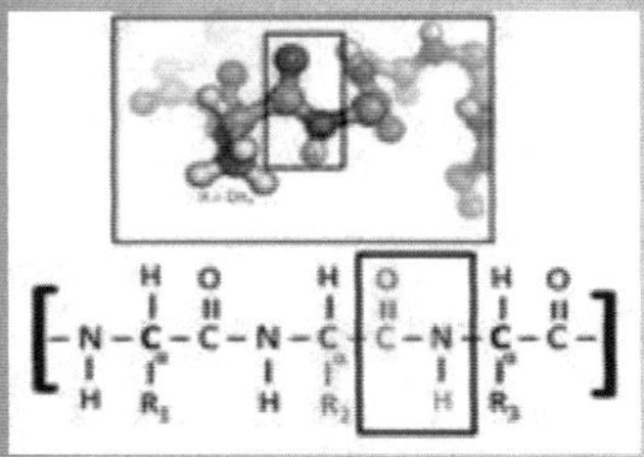

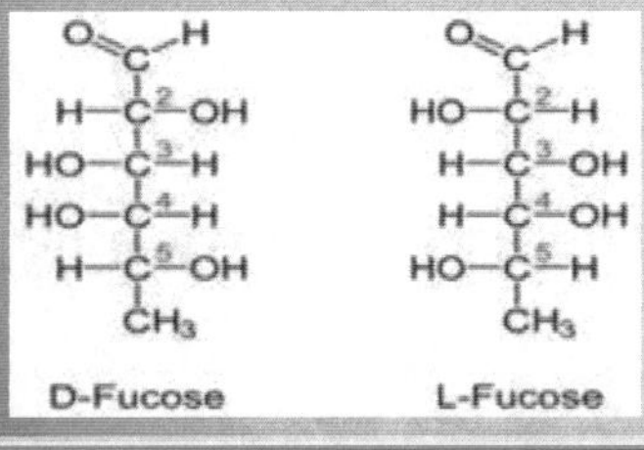

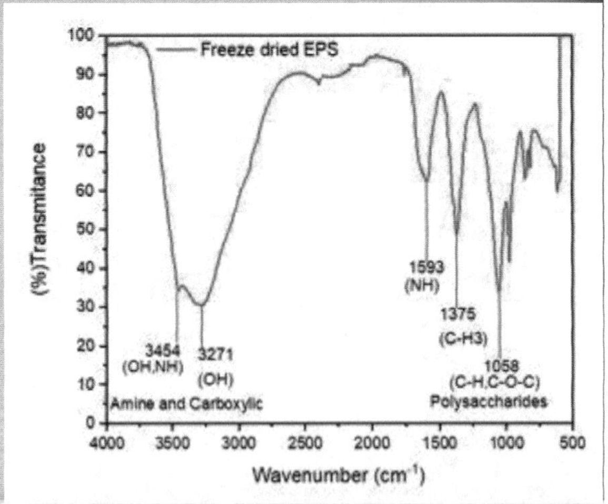

(1) Ha, Y. W.; Dyck, L. A.; Thomas, R. L. *Journal of Food Science* 1988, 53 (3), 841–844.

COLLEGE OF ENGINEERING

Synthesis method

- Materials: $AgNO_3$, *Cosmarium* sp. (supernatant EPS)
- Instrument: CEM MARS 5 microwave system
- Method: Power : 1200 W, Time: 120 s

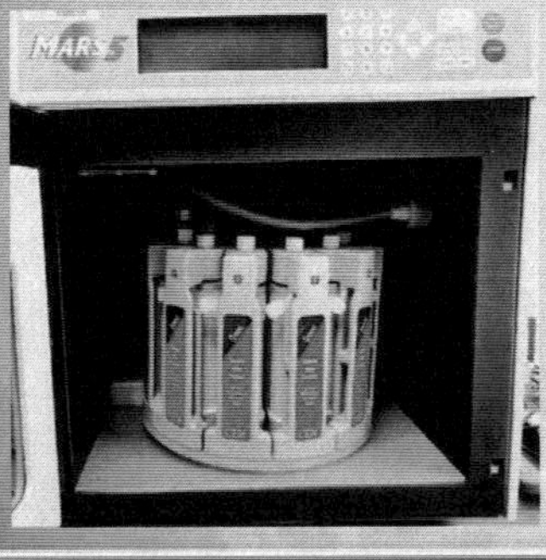

Sr. No	*Cosmarium sp.* supernatant (%) EPS	$AgNO_3$ (mL) 0.01 M
1	5	10
2	10	10
3	15	10
4	20	10
5	25	10

Total reaction volume in each vessel was 20 ml diluted with deionized water

LAMAR UNIVERSITY

COLLEGE OF ENGINEERING

Silver Nanoparticle Synthesis: Microwave assisted heating

Dielectric heating mechanism:

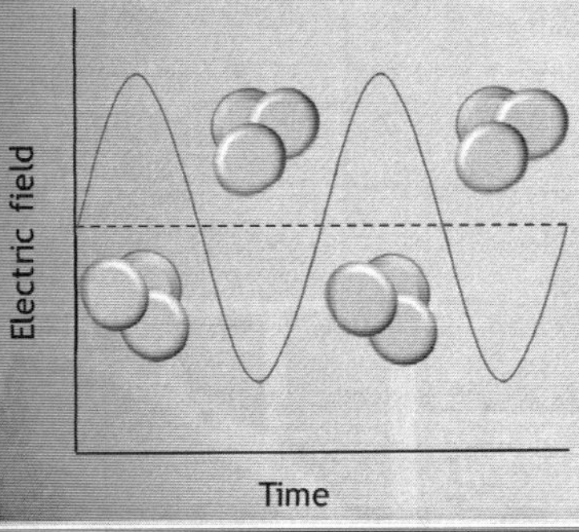

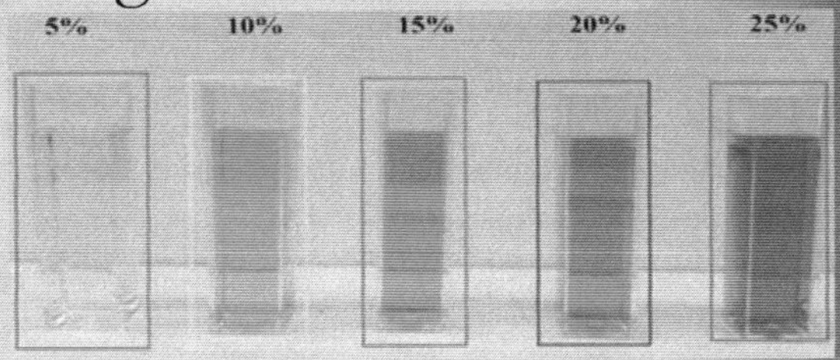

- Energy Efficient
- Biologically derived compounds
- Use of Aqueous solution
- Provision for tuning of parameters

LAMAR UNIVERSITY

COLLEGE OF ENGINEERING

RESULTS

LAMAR UNIVERSITY

COLLEGE OF ENGINEERING

Surface Plasmon Resonance :

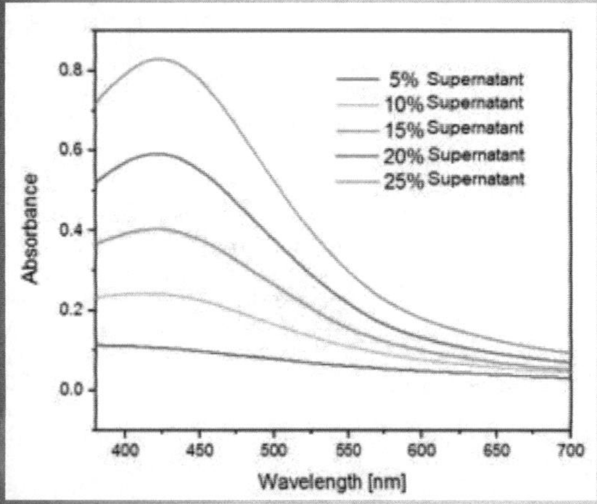

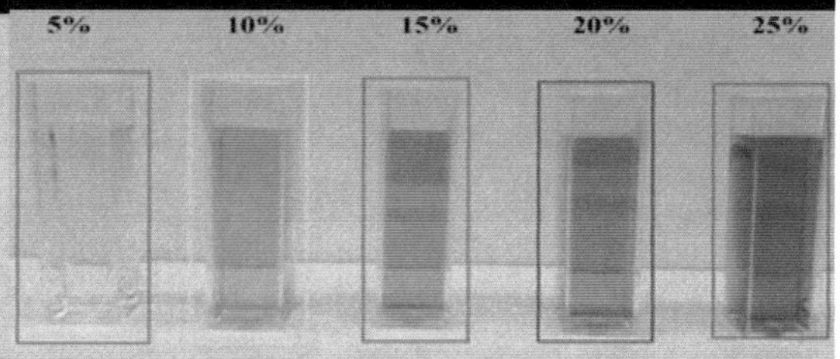

- Characteristic SPR curve for the AgNP is recorded
- The characteristic peak at 420-430 nm was observed

Pictures show formation of AgNPs

LAMAR UNIVERSITY

TEM Images of AgNPs

- The AgNPs formed are sphere-like in shape
- The AgNPs are coated with the biopolymers in EPS which has stabilized the particles

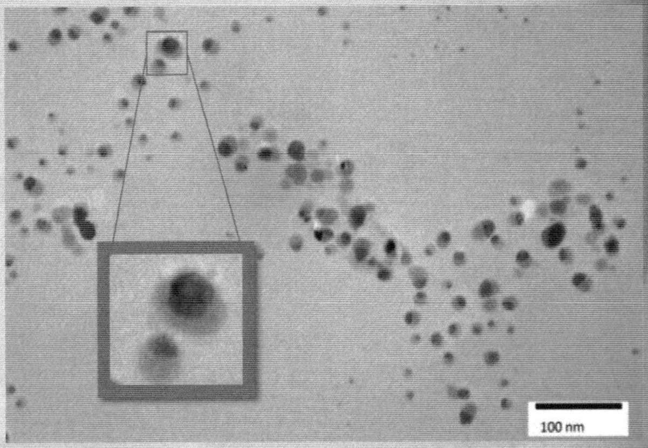

Electron Diffraction Pattern

The characteristic double rings for silver nanoparticles can be seen in the SAED.

COLLEGE OF ENGINEERING

Particle Size Distribution

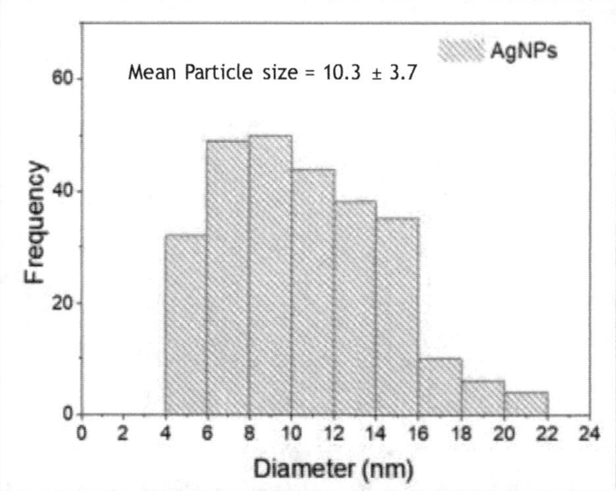

LAMAR UNIVERSITY

COLLEGE OF ENGINEERING

Conclusion

- Synthesized bio-organic capped AgNPs
- Confirmed the presence of AgNPs using SPR excitation, crystallographic techniques and morphological analysis
- Characterized and quantified the EPS
- Developed a novel technique using algal EPS of *Cosmarium* for AgNP synthesis

LAMAR UNIVERSITY

COLLEGE OF ENGINEERING

Future work

- Advanced separation techniques
- EPS characterization using HPLC.
- Synthesis of other metallic nanoparticles
- Study the effect of organic layer on the AgNPs.
- Application of synthesized silver nanoparticles for bio nanotechnologies

LAMAR UNIVERSITY

COLLEGE OF ENGINEERING

Thank you

LAMAR UNIVERSITY

References

- Kiemle, S.N., Domozych, D.S., Gretz, M.R., 2007. The extracellular polymeric substances of desmids (Conjugatophyceae, Streptophyta): chemistry, structural analyses and implications in wetland biofilms. Phycologia 46, 617–627. doi:10.2216/06-97
- Ha, Y. w., Dyck, L. a., Thomas, R. l., 1988. Hydrocolloids from the Freshwater Microalgae, Palmella texensis and Cosmarium turpinii. Journal of Food Science 53, 841–844. doi:10.1111/j.1365-2621.1988.tb08967
- Xiao, R., Zheng, Y., 2016. Overview of microalgal extracellular polymeric substances (EPS) and their applications. Biotechnology Advances 34, 1225–1244. doi:10.1016/j.biotechadv.2016.08.004
- Zhao, X., Xia, Y., Li, Q., Ma, X., Quan, F., Geng, C., Han, Z., 2014. Microwave-assisted synthesis of silver nanoparticles using sodium alginate and their antibacterial activity. Colloids and Surfaces A: Physicochemical and Engineering Aspects 444, 180–188. doi:10.1016/j.colsurfa.2013.12.008

488880 First Proof of Supported Hydrophobic Deep Eutectic Solvent Liquids Membranes for the Removal of Furfural and 5-(hydroxymethyl)Furfural from an Aqueous Environment

Abstract:

First proof of supported hydrophobic deep eutectic solvent liquids membranes for the removal of furfural and 5-(hydroxymethyl)furfural from an aqueous environment

Carin H.J.T. Dietz[a], Maaike C. Kroon[b], Martin van Sint Annaland[a], Fausto Gallucci[a]

[a] Chemical Process Intensification, Dept. of Chemical Engineering and Chemistry, Eindhoven University of Technology, PO Box 513, Eindhoven, Netherlands

[b] The Petroleum Institute, Chemical Engineering Department, P.O. Box 2533, Abu Dhabi, United Arab Emirates

Deep Eutectic Solvents (DESs) are designer solvents, which were first presented in literature by Abbot et al. in 2003.[1] In this publication a variation of amides and choline chloride, all in a 2:1 ratio, were mixed, where the combination of urea and choline chloride gave the lowest melting point (12 °C). A DES is in most cases a combination of at least one hydrogen bond donor (HBD) and one hydrogen bond acceptor (HBA). It is often claimed that DESs provide the same advantages as ionic liquids (ILs), where their easy preparation (mixing at moderate temperatures) is an advantage over ILs.[2] Until recently, most DESs have been hydrophilic due to the nature of the chosen constituents of the DESs. In 2015, the first hydrophobic DESs have been presented, which were a combination of decanoic acid and quaternary ammonium salts.[3]

In this work, the thermophysical properties (density, viscosity, glass transition temperature, decomposition temperature) of various newly designed hydrophobic DESs were measured first. The furfural (FF) and 5-(hydroxymethyl)furfural (HMF) extraction coefficients were determined in different hydrophobic DESs from an aqueous environment. Furthermore, liquid impregnated membranes were prepared for the first time from hydrophobic DESs and fully characterized. Finally, the diffusion of FF and HMF from an aqueous solution with different hydrophobic DESs liquid membranes was tested. Detailed results will be presented and discussed at the conference.

[1] A. P. Abbot, G. Capper, D. L. Davies, R. K Rasheed, V. Tambyrajah â€œNovel solvent properties of choline chloride/urea mixturesâ€ Chemical Communication, 2003, 70-71.

[2] M. Francisco, A. van den Bruinhorst, M.C. Kroon, Low-Transition-Temperature Mixtures (LTTMs): A New Generation of Designer Solvents, Angewandte Chemie International Edition, 2012, 52, 3074-3085.

[3] D. J. G. P. van Osch, L. F. Zubeir, A. van den Bruinhorst, M. A. A. Rocha, M. C. Kroon, Hydrophobic deep eutectic solvents as water-immiscible extractants, Green Chemistry, 2015, 17, 4518-4521.

Acknowledgements

This work was performed under the framework of InSciTe Horizontal project and with contributions from the European Regional Development Fund (ERDF) within the framework of OP-Zuid and with contributions from the province of Brabant and Limburg and the Dutch ministry of economy

500687 Use of Competitive Adsorption with 2,5-Dimethylfuran for the Inhibition of Isomerization and Dimerization ProductsÂ

Abstract:

The reaction of 2,5-dimethylfuran (DMF) with ethylene to form *p*-xylene has achieved high selectivities (>90%) using an H-BEA (Si/Al=12.5) zeolite acid catalyst.[1,2] As minimal xylene isomerization and transalkylation occurs under these conditions, this phenomenon was investigated and attributed to the surface site coverage of the ring opening product of DMF, 2,5-hexanedione.[3] Using alcohol dehydration as a probe chemistry, the ability of 2,5-hexanedione (always present with the addition of DMF to solution), to inhibit the reactivity of unsaturated bonds was investigated. In the absence of DMF cyclohexene was quickly produced followed by subsequent consumption of the cyclohexene product (Figure 1a). The addition of DMF increases the cyclohexene yield by almost 50% and inhibits its subsequent consumption through secondary pathways. Apparent activation energies of cyclohexanol dehydration appear to be identical in the presence and absence of DMF, indicating that the presence of DMF does not affect the energetics of the dehydration reaction (Figure 1b).

Figure 1. Cyclohexanol dehydration on H-BEA in the presence and absence of 2,5-dimethylfuran. A) Normalized concentration of cyclohexene to initial cyclohexanol concentration over time in the presence of and in the absence of 2,5-dimethylfuran. Reactions Conditions: 0.5 M cyclohexanol, 1.74 mM H-BEA (Si/Al 12.5), 160 °C, 200 psi N_2 added at 22 °C, 82 mL heptane, 2 mL tridecane (internal standard), and 0.74 M 2,5-dimethylfuran, when added. B) Arrhenius plot of apparent rate of cyclohexanol conversion versus inverse temperature for reactions including and excluding 2,5-dimethylfuran. Apparent activation energies are displayed and within error. Reaction Conditions: 135 – 200 °C, 0.5 M cyclohexanol, 1.74 mM H-BEA (Si/Al 12.5), 160 °C, 200 psi N_2 added at 22 °C, 82 mL heptane, 2 mL tridecane (internal standard), and 0.74 M 2,5-dimethylfuran, when added.

[1] Williams, C. L.; Chang, C.-C.; Do, P.; Nikbin, N.; Caratzoulas, S.; Vlachos, D. G.; Lobo, R. F.; Fan, W.; Dauenhauer, P. J. *ACS Catal.*, 2012, 2, 935-939.

[2] Chang, C.-C.; Green, S. K.; Williams, C. L.; Dauenhauer, P. J.; Fan, W. *Green Chem.* 2014, 16, 585-588.

[3] Williams, C. L.; Vinter, K. P.; Patet, R. E.; Chang, C.-C.; Nikbin, N.; Feng, S.; Wiatrowski, M. R.; Caratzoulas, S.; Fan, W.; Vlachos, D. G.; Dauenhauer, P. J. *ACS Catal.*, 2016, 6, 2076-2088.

503177 Keynote - Two-step Solarthermal Water Splitting – The Path Forward

Abstract:

Using concentrated sunlight to drive water splitting for renewable hydrogen production is the ultimate goal for a truly green energy world. The seminal paper, "Hydrogen and Oxygen from Water," *Science*, 197, 1050 (1977), by E.A. Fletcher and R.L. Moen ignited a flurry of activity in the field of solar thermalwater splitting. Focused research since then has resulted in substantial progress identifying particular chemical routes, active materials and various solarthermal reactor designs. This presentation will review key findings and will focus on the path forward emphasizing active materials and processes and identifying key areas of interest for future research.

491083 Syngas Production Via Solar Chemical-Looping Reformation of Methane in a Fixed-Bed Reactor

Abstract:

The production of syngas using concentrated sunlight via solar chemical-looping reforming (SCLR) is an approach to produce sustainable transportation fuels. SCLR is a two-step thermochemical cycle that produces separate streams of syngas and high-purity H_2. In the endothermic partial oxidation step, methane reacts with oxygen released from a metal oxide to produce syngas. In the exothermic water-splitting step, steam reacts with the reduced metal oxide to produce H_2. Solar energy is stored in the products, upgrading the calorific content of the feedstock by up to 28%. Combustion of the products emits 41% less carbon per kilojoule than combustion of the productions from non-solar CLR. The cycle is thermodynamically favorable at 1273 K and 1000 suns, conditions easily achievable by commercial solar concentrated optics used currently for electricity production. Based on a thermodynamic analysis of the process, the solar-to-chemical efficiency for SCLR, defined as the net increase in the higher heating value of the gases divided by the solar thermal input to provide process heat and parasitic work for gas pumping and separation, can reach 54% with complete conversion of CH_4 to syngas and H_2O to H_2.

While the net products and chemical energy requirements match that of solar steam reforming, decoupling the partial oxidation and H_2O-splitting steps provides numerous advantages. The products of SCLR are separate streams of 2:1 H_2:CO and high-purity H_2 compared to the single stream of 3:1 H_2:CO obtained from steam reforming. The 2:1 H_2:CO syngas matches the desired composition for the synthesis of liquid hydrocarbon fuels and H_2 is valuable as a fuel for electricity production via fuel cells or as a chemical feedstock in the synthesis of various chemical commodities such as ammonia. Furthermore, conventional steam reformers must operate with excess oxidizer (e.g. $H_2O/CH_4 = 3$) to prevent catalyst deactivation by carbon deposition. With solar CLR, carbon deposition can be prevented by limiting the reduction extent of the metal oxide.

In this paper, we present measured performance for SCLR in a prototype reactor that utilizes cerium dioxide (ceria) as the oxygen carrier material. The reactor is operated in the University of Minnesota high flux solar simulator. The reactor comprises two sections: a solar receiver/reactor and a heat recuperator. Concentrated sunlight enters the cylindrical receiver/reactor cavity section through an open circular aperture and heats a sealed assembly of concentric alumina tubes, referred to as the reactive element. The annulus of the reactive element is filled with 336 g of ceria particles with a bed void fraction of 45%. The particles are 5 mm (length and diameter), 78% porous and have a specific surface area of 0.114 $m^2\ g^{-1}$. The reactive element extends beyond the solar cavity to the heat recuperator. In the heat recuperator, the annulus and inner tube of the reactive element are filled with alumina reticulated porous ceramic that is 85-90% porous and has pore densities of 5 and 10 pores per inch in the inner tube and annulus, respectively.

The cycle is accomplished by alternating the gas feed to the reactive element between CH_4 and H_2O. During the endothermic step, CH_4 flows through the inner tube, reverses direction at the closed end, and flows over the ceria particles in the annulus. Sensible heat from the product gases is transferred to the inlet flow in the heat recuperator. At the end of the partial oxidation step, the gas flow is switched automatically to H_2O. The operating conditions (temperature, reactant feed rates, reaction times) are selected based on a thermodynamic process model of the reactor. The model predicts CH_4 and H_2O conversions and H_2 and CO selectivities data from prior benchtop experiments.

Measured results include radiative input, transient temperatures of the solar cavity, reactive element, and inlet and outlet gases, upstream and differential pressures across the reactive element, and product gas composition. The axial temperature distribution along the top surface of the reactive element in the receiver/reactor and surface temperature of the reactor cavity are measured by alumina-sheathed type-B (Pt-30% Rh/Pt-6% Rh) thermocouple probes. The axial temperature distribution along the reactive element in the heat recuperator section and the gas temperatures at the inlet and outlet of the active reactive element are measured with type-K (Ni-Cr/Ni-Al) thermocouples. The high-flux solar simulator is controlled to obtain a cycle-averaged reactive element temperature of 1273 K. The radiative solar input is measured using a water-cooled black body calorimeter. The product gas composition is measured at 1 s intervals using Ramen laser gas spectroscopy. Data are evaluated to determine the solar-to-chemical efficiency of the reactor. The reactor demonstration provides an important proof of concept for SCLR and a comparison of measured efficiency to prior theoretical predictions of efficiency.

486955 Solar Gasification in a Molten Salt Reactor for Continuous Production of Syngas

Abstract:

The design and operation of a concentrated solar reactor to gasify biomass in a ternary eutectic blend of alkali metal carbonate salts are presented. Solar gasification has numerous advantages over conventional gasification. The use of concentrated solar energy as the source of process heat eliminates the need for partial combustion of the feedstock. The product synthesis gas is not diluted by excess CO_2 and N_2 nor contaminated by other combustion by-products. There is no need for an economically and energetically expensive oxygen plant that otherwise is necessary for oxygen-based partial combustion. The energetic value of the product is greater than that of the feedstockâ€"the difference is equal to the storage of solar energy in the fuel. The benefits of the molten salt solar gasifier in comparison to other solar gasifiers are the thermal mass of the salt allows continuous operation during solar transients, provides rapid heat transfer, catalyzes the gasification reaction, and reduces the amount of tars and other contaminants in the product gas stream for an operating temperature of 1200 K. In prior work, utilizing cellulose as feedstock and carbon dioxide as oxidizer, the reactor achieved a solar efficiency of 30% and converted 47% of the carbon in a continuous process. The efficiency at 1218 K was comparable to the highest achieved in other solar gasifiers operated at higher temperature, but the pneumatic feed system was prone to blockage and required a stoichiometric oxidizer ratio of eleven. The use of excess carbon dioxide increased the sensible heating requirement and exacerbated the entrainment of fine char particles in the product gas stream prior to complete conversion.

In the present study, we present a model of an improved volumetric pumping feed system and provide operational data for gasification of cellulose in the improved reactor. The feed system consists of a single fluted screw within a straight barrel, similar to that used for polymer extrusion systems. The feed enters the reactor as a densified cylinder rather than as loose powder, encouraging the formation of larger char particles which have a reduced tendency to be entrained by the product gases leaving the reactor. The design of the feed delivery system is based in part on a numerical model of the transient heat transfer and kinetics of pyrolysis and gasification of the feedstock material during transport.

The reactor was operated for CO_2 gasification of cellulose in the University of Minnesotaâ€™s high flux solar simulator. The simulator consists of seven radiation units, each composed of a 6.5 kW xenon short arc lamp close-coupled to a custom reflector in the shape of a truncated ellipsoid of revolution. The total power delivered to the aperture of the reactor was measured using a water-cooled black body calorimeter of the same aperture diameter. The feedstock is ash-free microcrystalline cellulose ($C_6H_{10}O_5$) sieved to a particle size of 0.5 mm. Temperatures of the solar cavity and within the salt melt are measured using Type-K thermocouples. The product gas composition is measured using a Raman Laser Gas Analyzer. The downstream gas lines and HEPA filter are analyzed for residual secondary products. Tars are separated from solids using a solvent wash and the particle size distribution of the solid material is measured using laser diffraction.

The reactor performance is quantified by the solar to fuel efficiency defined in rate form as the ratio of the lower heating value of the useful products and the sum of the solar input and the LHV of the carbonaceous feedstock. Carbon conversion is given by the ratio of the net gaseous carbon released from the reactor to the feedstock carbon delivered to the reactor. The measured species flow rates are compared to the equilibrium flow rates predicted using Gibbs free energy minimization. The data demonstrate the improvements in performance for the reactor with the new feed system in terms of longer operation without blockage, and higher solar efficiency.

494979 An Evaluation of Feedback and Model Predictive Control for the Rejection of Weather Transients in Renewable High-Temperature Solar Thermal Chemical Processing on a Novel Hybrid Solar-Electric Reactor

Abstract:

Concentrated solar power, which has been commercialized for electrical generation, is now of interest as a means of directly driving chemical processes. Investigations have included the use of solar heat to produce metals from oxide ores, generate renewable hydrogen for use as fuel, and to gasify biomass. These processes often operate at high temperatures where reactors and receivers are built partly or wholly from ceramics, ceramics that are vulnerable to thermal shock and fatigue. Weather transients, which can deprive a facility of solar heat within seconds, must be controlled to insure the longevity of solar-driven chemical process equipment. Supplementary electrical and/or gas heaters can be used to provide auxiliary heat when transient cloud cover would otherwise interrupt solar thermal chemical production. We evaluate different control schemes on a new solar-electric reactor prototype at the University of Colorado Boulder. These include both feedback and model predictive control paradigms, whose performance was measured by integral squared absolute error over the course of programmed weather transients on a 45kWe high-flux solar simulator. Implications for the implementation of commercial solar thermal processes will be discussed.

494571 Solar Hybrid Photo-Thermochemical Sulfur-Ammonia Water-Splitting Cycle: Photocatalyst, Thermodynamics and Plant Analysis

Abstract:

The solar driven hybrid sulfur-ammonia water splitting cycle (HySA) is a promising technology for the environmentally sustainable production of hydrogen fuel. It is a state-of-the-art process that integrates a solar-photocatalytic hydrogen production step (driven by the photonic portion of solar irradiance) with a high-temperature solar thermochemical oxygen evolution step (driven by the thermal portion) and efficient thermal energy storage as an integral part of the cycle. This work summarizes the recent findings on the two subcycles and the overall process design and optimization, following detailed experimental and numerical calculations. First, multiple photocatalysts, for the hydrogen production step, were assessed and characterized, finally, a binary band gap photocatalyst based on ZnS-CdS core shell particles doped with Pt was selected. Second, the materials and their properties for the oxygen evolution step were verified using thermogravimetric experiments, that supported a detailed thermodynamic analysis for the selection and design of the overall process. Finally, all information was assembled in one process that takes advantage of the hybrid solar photo-thermo character of the cycle and the incorporation of a portion of the process-fluids to design an integrated thermal-storage cycle. The solar to H_2 efficiency of the process was estimated up to 25% depending on reactor configurations.

495376 On the Potential for Enhancement of Solar Thermochemical Synthesis Processes By Nonequilibrium Plasma

Abstract:

The sustainable synthesis of fuels and chemicals from abundant low-value feedstock, such as carbon dioxide, methane, and water, is a key technological and global challenge. Solar thermochemical synthesis processes directly utilize the most abundant form of renewable energy, but face severe challenges due to the low reactivity of most feedstock gases to solar radiation and the intermittency of the solar resource. These challenges often lead to limited process efficiency and resiliency, which translate into systems with higher capital and operating costs. The combination of a solar thermochemical process with a non-solar co-process can be an effective option to achieve continuous operation, improve resiliency, and reduce costs. Moreover, if the concurrent operation of the solar and non-solar processes leads to greater overall efficiency, then the combined approach could have greater viability than either process alone. This talk discusses the potential of the use of nonequilibrium plasma as a co-process to solar thermochemical synthesis. The combined solar-plasma approach, denoted as Plasma-Enhanced Solar Energy (PESE), has the potential to enhance the efficiency and resiliency, and consequently improve the viability, of solar thermochemical synthesis processes.

Nonequilibrium plasmas are partially ionized gases in which the temperature of the free electrons is significantly higher than that of the heavy-species (molecules, atoms, ions). The electron temperature in atmospheric-pressure nonequilibrium plasmas are typically on the order of 1 to 2 eV, commensurate with the threshold energies in molecular bonds, whereas the temperature of the heavy-species range between room temperature to a few thousand degrees. Therefore, electrons and heavy-species are not in state of Local Thermodynamic Equilibrium (LTE), i.e., they are in a non-LTE (NLTE) state. Plasma-based chemical synthesis processes use electricity to directly modify gas chemistry and produce activated species (radicals, excited states, ions). These processes can achieve high energy efficiencies by selectively transferring energy to different channels. For example, microwave plasmas have shown over 90% energy efficiency (i.e. electrical to chemical) for the decomposition of carbon dioxide thanks to effective step-wise vibrational excitation, the most effective channel for dissociation [1]. Moreover, the reliance of plasma processes on electrical energy makes them ideal for continuous operation, but also limits their economic viability and their net sustainability benefit, even if the required electricity is obtained from renewable energy sources, such as wind and solar.

PESE chemical synthesis processes seek to combine the advantages of solar thermochemical and plasma processes. The potential advantage of PESE to improve the efficiency of solar thermochemical synthesis is due to the fact that the molecular excitation produced by free electrons in non-equilibrium plasmas can significantly increase solar photon absorption leading to the enhancement of chemical reaction kinetics. Moreover, PESE chemical synthesis processes could rely on electrical energy to compensate for fluctuations in the solar energy input (during daytime) or even substitute it (at nighttime), and therefore ensure continuous operation.

Preliminary computational analyses of solar radiation absorption through a carbon-dioxide (CO_2) mixture in a state of NLTE reveal over two orders of magnitude increase in absorption compared to the same mixture in a state of LTE, achieving 50 to 80% radiative energy absorption efficiency. The calculations were performed with the line-by-line nonequilibrium radiation code SPARTAN [2, 3] using 32 types of atomic and molecular transitions. The results show that CO_2 in LTE absorbs negligible solar irradiance (mostly UV and far IR), whereas CO_2 in NLTE displays significantly more absorption, especially in the visible range, but also emission, which diminishes the overall amount of absorbed radiation. Both findings are consistent with established understanding of radiative transport; however, the novelty is finding that the net effect is a significant increase in solar energy absorption. The results also show that CO_2 tends to behave as a perfect absorber of solar energy for very high degrees of NLTE (i.e. larger differences between electron and heavy-species temperatures). This fact, although not intuitive given that plasmas are known to be good emitters (e.g., radiation from arc lamps), can be understood by considering that a medium is prone to absorb the same type of radiation it emits. Furthermore, it is to be noted that the concept of a blackbody, a perfect absorber and emitter or radiation, corresponds to a medium in LTE. In fact, the approximation of the solar spectrum as the emission from a blackbody at $\sim$ 5780 K is a consequence that the sun's photosphere is largely regarded as being in LTE. Therefore, a NLTE medium provides the possibility to control its radiative properties by tuning its degree of thermodynamic nonequilibrium.

To evaluate the potential of PESE, a solar receiver-reactor system equipped with two types of nonequilibrium plasmas, namely a gliding arc discharge and a microwave plasma, is being built at UMass Lowell. The PESE chemical synthesis reactor system is designed to evaluate comparable solar and electrical energy inputs at the 1 kW level, spanning the whole range of combined solar-plasma operation, i.e. from 100% solar to 100% electrical energy inputs. The microwave plasma has the potential for the greatest energy efficiency, whereas the gliding arc discharge has simpler operation, more stability, and greater potential for scalability. Design challenges being addressed include the establishment of a stable plasma volume within the focal point of the concentrated solar radiation from a high-flux solar simulator [4], as well as the maximization of the plasma volume and the gas-plasma interaction region, the incorporation of a downstream catalytic monolith, among others.

Acknowledgements

The authors gratefully acknowledge support from the U.S. National Science Foundation through award CBET-1552037.

References

[1] A. Fridman, *Plasma chemistry*, Cambridge University Press, New York (2008)

[2] M. Lino da Silva, *SPARTAN - Simulation of PlasmA Radiation in ThermdynAmic Nonequilibrium*, http://esther.ist.utl.pt/spartan/

[3] M. Lino da Silva, *An adaptive line-by-line—statistical model for fast and accurate spectral simulations in low-pressure plasmas*, J. Quant. Spectrosc. and Rad. Transfer 108 (2007) 106–125

[4] S. Bhatta, D. Nagassou, J. P. Trelles, *Solar photo-thermochemical reactor design for carbon dioxide processing*, Solar Energy 142 (2017) 253–266

496114 Direct Two Step Solar Metallothermic and Electrolytic Production of Rare Earth Elements from Oxides

Abstract:

Rare Earth Elements (REEs) exhibit a range of unique electronic, optical and magnetic properties which make them attractive for many niche applications, including metal alloys, catalysts, petroleum refining, glass additives, polishing compounds, and ceramics. The demand for rare earth elements is expected to grow in the coming years, especially in uses related to critical clean energy technologies such as PV devices, batteries, high-performance magnets, and phosphors for lighting and electronics. Cerium (Ce), lanthanum (La), neodymium (Nd), and praseodymium (Pd) account for ~86% of the total global REE consumption. The current electrolytic production route through molten chlorides is quite inefficient due to the stability of Re^{+2} ion in chloride melt, whereas the oxide-fluoride route suffers from low solubility of oxides in fluorides. Additionally, these extractive processes are currently driven by fossil fuel based energy sources with very high greenhouse gas impacts and resource depletion potential.

Energy consumption and environmental concerns, coupled with the fact that the current electrolytic production method must occur at high temperatures, provides significant motivation for investigating solar driven metallothermic/ electrolytic processing routes as alternatives to conventional REE production for providing â€œcleanâ€ high temperature process heat and electricity. In this work, a new two-step REE production route consisting of direct metallothermic reduction of RE-oxides using Ca and subsequent electrolytic recovery of metallic Ca in a mixed chloride bath has been explored from the theoretical energy consumption perspective. The new process is critically evaluated using thermodynamic analysis and compared with the state-of-the-art. The two-step process shows promise in terms of significant improvement in the energy efficiency compared to the current electrolysis routes. The potential technological challenges in scale-up and impact of solar variability will be discussed as well.

502531 Fossil Energy Policy in the Trump Administration

Abstract:

Fossil energy policy in the Trump administration

Those interested in a federal fossil energy policy are very alert to attempt to read the proper signals early in the Trump Administration. Several circumstances provide a sense of direction. Others will unfold over the June to January timeframe.

What we know now is:

- It appears that the Clean Power Plan is dead. This may be helpful because industry players perceive a somewhat arbitrary "number" representing emission reduction levels on a state-by-state basis. The energy industry is reducing emissions already and can likely meet climate goals more cost-effectively by not having to spend money on compliance costs and to use this capital to reduce emissions broadly instead on arbitrary emissions reduction targets.

- It appears that the Paris Accord is dead. Staying in or pulling out of the Paris Accord may be meaningless from an emission reduction standpoint. The U.S. energy industry is on track to realize the emission reductions represented in the accord. It can achieve emission reductions more cost-effectively without arbitrary reduction levels and regulatory compliance costs. Contributions that the United States would have made to the Green Climate Fund can be used for other purposes consistent with the President's "America First" campaign promise.

- And the U.S. Department of Energy Fossil Energy budget, this is trickier because while the "budget path" cuts fossil research it also cuts research in the energy efficiency and renewable energy budgets. The administration has strived to stress that the federal role should be basic research, not pre-commercialization research; which they suggest is more appropriately a private sector function.

Much more will be known over the next few months – impacting the fossil fuel research agenda perhaps for years.

502530 Emerging Technologies in Fossil Energy and Chemical Conversion

Abstract:

Advanced energy technologies are usually developed to meet emerging goals in stationary power, transportation, or aerospace applications. Research efforts typically focus on meeting energy goals such as efficiency, emissions, and cost. Less attention is given to possible "derivative" technologies where advances in energy conversion could also produce benefits for chemical conversion. This presentation will describe new developments in energy conversion that may also provide advantages for chemical production and process intensification. The talk will discuss emerging energy technologies—continuous detonation engines, cold plasmas, rotating fluid beds—and describe how they may benefit both energy and chemical conversion needs.

504579 University Coalition for Fossil Energy Research - a New Doe-NETL National Alliance for Clean Energy Solution

Abstract:

University Coalition for Fossil Energy Research - A New DOE-NETL National Alliance for Clean Energy Solution

Chunshan Song*

DOE-NETL University Coalition for Fossil Energy Research, the Pennsylvania State University, EMS Energy Institute, C211 Coal Utilization Lab, University Park, PA 16802

*E-mail: csong@psu.edu

The University Coalition for Fossil Energy Research (UCFER) was established in 2016 by the U.S. Department of Energy (DOE), National Energy Technology Laboratory (NETL) at Penn State after a nation-wide open competition in 2015. The Coalition brings together a multidisciplinary team of researchers from participating universities to address the fundamental challenges in developing clean carbon-based energy technologies. Currently there are 16 member universities in the DOE-NETL UCFER as of May 2017 including Carnegie Mellon Univ, Louisiana State Univ, MIT, Ohio State Univ, Penn State Univ, Princeton Univ, Texas A&M Univ, Univ of Kentucky, Univ of North Dakota, Univ of Pittsburgh, Univ of Southern California, Univ of Tulsa, Univ of Utah, Univ of Wyoming, Virginia Tech, and West Virginia Univ. Research performed by the coalition will directly support the mission of DOE NETL and NETLâ€™s Research and Innovation Center, by focusing on more efficient and more environmentally-friendly fossil energy utilization including diverse research programs such as advanced energy systems, CO_2 capture, storage and utilization, energy conversion, oil and gas technology, unconventional natural gas resources and infrastructure, crosscutting research and system analysis. The Coalition will facilitate basic and applied energy research and promote multidisciplinary collaboration among the member universities and NETL. The Coalition also seeks industry participation in research projects and promote technology transfer to the private sector. The research, completed throughout the six-year initiative, is expected to accelerate the development and deployment of fossil fuel-based technologies, enabling the continued use of our nationâ€™s abundant natural resources in a cost effective and environmentally responsible manner.

502533 Reliance Energy R&D

Abstract:

Reliance Energy R&D

Reliance industries is operating large scale refineries at its Jamnagar Manufacturing Complex. Reliance Industries has been continuously carrying out R&D work in Energy sector. We are at advance stage of implementation of Petcoke Gasification project at Jamnagar as value addition to Refinery and as a part of energy integration. Petcoke Gasification is carried out with entrained flow slagging Gasification technology.

Petcoke gasification is not a mature technology and a very few plants in the world are operating on 100% Petcoke. Therefore to understand the technology, Reliance has initiated an advance R&D projects in Petcoke Gasification mainly in the area of Slag Management, Petcoke Gasification Kinetics, Refractory Material Improvement in terms of life and alternate of Chrome Refractory.

Another area of research is recovery of V/Ni from Gasification Slag. Process has to be developed as there is no off-the- shelf Technology available for V/Ni recovery from Slag.

Reliance is also exploring next generation coal liquefaction technology and Bio conversion of syngas to higher hydrocarbons & petrochemicals.

Reliance scientists and technologist are engaged in indigenously developing Low Temperature & High Temperature PEM fuel cell.

Reliance is engaged with world class universities PSU, MIT, Frieberg, Kansas University to name a few working with us in field of Gasification, and environmentally beneficial technologies.

Reliance is carrying out research in the field of Bio-CBM i.e. bio-conversion of underground coal to methane, another potential breakthrough technology.

Another significant research is being carried out in the area of generating Oil from Algae (A2O).

501439 Short Contact Time Catalytic Partial Oxidation of Methane over Rhodium Catalysts Supported on 3-D Printed Ceramic Supports

Abstract:

Short contact time catalytic partial oxidation (SCT-CPO) is a process by which methane and other fuels can be converted to hydrogen and carbon monoxide at millisecond contact times without the addition of heat. To achieve such low contact times, structured supports with high void fractions. Since the reaction is limited by mass transfer, small pore diameters are desired at the cost of potential increases in pressure drop. 3-D printed ceramic supports (created via robocasting) can be produced with well defined and consistent pore structures and sizes with the desired high void fraction necessary for short contact time operation.

In this work, we have investigated the SCT-CPO of methane over 3-D printed ceramic supports experimentally demonstrating their ability to achieve high yields at low residence times and virtually no pressure drop. Supports of different compositions including alumina, silica, and zirconia have been investigated as well as differing pore size supports at constant void fraction. Experimental data at varying space velocities and feed ratios will be presented. Under the appropriate conditions, 3-D printed supports achieve methane conversions over 85% with hydrogen selectivities over 90% at space velocities over 400,000 hr-1.

499273 Characterization of Ni-Substituted Pyrochlore Catalysts for the Dry Reforming of CH_4

Abstract:

Pyrochlores are mixed metal oxides that have proven to be effective catalysts for the reforming of hydrocarbon fuels. These materials are thermally stable at the demanding conditions associated with fuel reforming, and have a modifiable structure which can incorporate various dopants to improve their catalytic activity. One example is a lanthanum zirconate pyrochlore (La-Zr). The Zr framework has been partially substituted by Rh to disperse the catalytic metal into small and stable clusters that are resistant to both carbon and sulfur poisoning [1]. Furthermore, alkaline metals (e.g. Ca) can be doped into the A-site for La to introduce oxygen vacancies which can increase lattice oxygen mobility, while also providing surface basicity that promotes the gasification of carbon.

Ni is an attractive catalytic reforming metal, because of its well-known specific activity, and low cost. However, Ni is problematic due to its low sintering temperature, and propensity to form carbon. This study will evaluate the substitution of different levels of Ni (1 vs 3 wt%) into a La-Zr pyrochlore to determine whether the pyrochlore structure is able to stabilize the Ni under conditions required for dry reforming of CH_4. The materials will be synthesized by a variation of the Pechini method [2], which is a well-known procedure to obtain mixed oxide solid solutions. The materials will be characterized by various methods to determine the location of the Ni, and how this will relate activity stability, and particularly carbon deposition. Traditional characterization methods like XRD, temperature programmed reduction, BET, and Raman are used. In addition, an analysis using by Cameca using a LEAP 5000 atom probe provided 3-D images with corresponding compositional analysis on each Ni containing material. The atom probe results (see Figure 1 for results on the 1 wt% substituted catalyst), and XRD results both show Ni is not soluble in the pyrochlore structure (for either weight loading). Instead, crystallization elicits the exsolution of Ni to the surface and grain-boundary regions of the pyrochlore, which forms NiO. At the elevated temperatures, and with high enough Ni loading (3 wt%), some of the Ni forms pocket of mixed-phases with the excess La (e.g. $LaNiO_3$). The Ni distribution at the surface and grain boundary regions suggests these enriched regions with metallic Ni would be the active component of the catalyst for dry reforming.

496600 Dry Reforming of Methane over Zr,Y-Modified Ni/Mg/Al Hydrotalcite Catalysts for Hydrogen and Carbon Monoxide Production

Abstract:

Dry reforming of methane over Zr,Y-modified Ni/Mg/Al hydrotalcite catalysts for hydrogen and carbon monoxide production

Katarzyna Swirk[a,b], Maria Elena Gálvez[b], Monika Motak[a], Teresa Grzybek[a], Patrick Da Costa[b]

[a] AGH University of Science and Technology, Faculty of Energy and Fuels, Al. A. Mickiewicza 30, 30-059, Krakow, Poland

[b] Sorbonne Universités, UPMC Paris 6, CNRS, UMR 7190, Institut Jean Le Rond d'Alembert, 2 Place de la Gare de Ceinture, 78210, Saint-Cyr-L'Ecole, France

1. Introduction

Dry reforming of methane (DRM) is a method of producing hydrogen and carbon monoxide while utilizing CO_2. The process has not been launched yet on an industrial scale due to lack of stable, active and selective catalyst [1]. Ni-based catalysts, used e.g. in steam reforming, seem to be a great alternative for expensive noble metal materials. However, the main drawbacks of nickel catalysts is their fast deactivation caused by carbon deposition and/or sintering [2].

The hydrotalcites (HTs, or layered double hydrotalcites LDHs, which exhibit brucite-like structure) appear to be promising support for catalysts due to their strong interrelated nanostructure which hinders Ni from sintering at high temperatures needed for DRM [3]. On the other hand, an addition of metallic promoters may positively influence catalytic activity and stability in DRM process [4].

Zr and Y species seems to be good components for the DRM catalyst, because only a slight addition of yttrium may stabilize the zirconia phase resulting in high thermal stability of catalysts [5]. Moreover, the yttrium as a promoter, enhances activity and stability of the used catalyst in DRM process, due to ability of oxygen vacancies creation [6].

Thus, the goal of this work was to examine the effect of the promotion of Ni/Mg/Al hydrotalcites with Zr and Y on their structure and performance in dry methane reforming.

2. Experimental

The HT sample was prepared by the co-precipitation method at pH of 10 ± 2, from the aqueous solution of nickel, magnesium and aluminum nitrates (Ni/Mg=0.33, M^{2+}/M^{3+}=3.0). Subsequently, the fresh HT material was modified with zirconium (5 wt.%) and yttrium (0.2 wt.%, 0.4 wt.%, or 0.6 wt.%), resulting in HT/Zr5 Y0.2, HT/Zr5 Y0.4, HT/Zr5 Y0.6 catalysts. The introduction of these two metals occurred through wetness impregnation method (co-impregnation). So synthesized materials were dried at 80°C overnight, and calcined in air at 550°C for 6 hours. The obtained catalysts were characterized by XRD, TPR-H_2, TPD-CO_2, and FTIR.

The dry reforming tests were carried out in a fixed-bed quartz reactor at temperature range of 850-600°C. Before each catalytic test, the samples were reduced in situ at 900°C for 1 hour with hydrogen (5% H_2 in Ar, flow 10°C/min). The gas composition was adjusted to 1/1/8=CH_4/CO_2/Ar, with total flow of feed gases of 100 cm^3/min (GHSV=20,000h^{-1}). The final products were analyzed by Varian GC-490 micro-GC.

3. Results and discussion

The X-ray analysis for the fresh samples confirmed successful synthesis of hydrotalcite-like materials (reflections at 2θ equal to ca. 11, 22 and 35°). Also, the periclase-like structure was registered after thermal treatment in air. On top of this, separate phase of Zr compound (ca. 31° and 51°) was observed. However, no separate phase was registered for yttrium in the XRD patterns of calcined and reduced samples.

The TPR-H_2 profiles show wide peaks with maximum value at ca. 830-860°C, which arise from reduction of Ni particles derived from mixed oxides in hydrotalcite structure. The reduction temperature is dependent on the presence of promoters. There was observed a shift of the wide peak towards higher temperatures in Zr, and Zr,Y-modified samples (847°C), with the highest temperature reached at 860°C for HT/Zr5 Y0.4 sample. This shift indicates decrease in reducibility, and stronger interaction of Ni with the support. The observed peaks at lower temperature, ca. 400°C, arise from NiO weakly bonded with the support.

The Zr,Y-modification influenced basicity of tested samples by expanding the number of medium sites while increasing Y content. Although, the total basicity decreased with promoters addition.

The FTIR measurements showed typical spectra for hydrotalcite-derived material.

The DRM tests (CH_4/CO_2/Ar=1/1/8) showed increase in CH_4 and CO_2 conversions while increasing temperature (Fig. 1 and Fig.2). The best results in the low temperature region (600-700°C) were observed for Zr,Y-modified samples. At 700°C the conversions obtained following sequence: HT/Zr5 Y0.6 > HT/Zr5 Y0.4 > HT/Zr5 > HT > HT/Zr5 Y0.2. Above this temperature the best performance was found for the catalyst impregnated only with Zr. The presence of side reactions, such as RWGS was deduced through CO_2 conversion greater than CH_4 conversion, and H_2/CO lower than 1.

Fig. 1 CH_4 conversion as a function of reaction temperature (DRM process CH_4/CO_2/Ar=1/1/8, GHSV=20,000h^{-1}) of hydrotalcite catalysts modified with Zr and Y

Fig. 2 CO_2 conversion as a function of reaction temperature (DRM process CH_4/CO_2/Ar=1/1/8, GHSV=20,000h^{-1}) of hydrotalcite catalysts modified with Zr and Y

4. Conclusions

All prepared catalysts were catalytically active in the DRM process. The presence of Zr and Y in Ni-containing hydrotalcite-derived catalysts resulted in a higher number of medium basic sites, which contributes to enhanced catalytic activities in DRM at low temperatures. The DRM tests show that the HT/Zr5 Y0.6 gives the highest CH_4 and CO_2 conversion, and the highest H_2/CO ratio in the temperature range of 600°C to 700°C. At temperatures higher than 700°C the presence of yttrium did not increase the catalytic activity of the Zr-modified sample. The effect of the reverse water gas shift reaction, on the hydrogen production and the molar ratio H_2/CO has been observed.

Acknowledgments

K. Swirk would like to acknowledge for financial support the French Embassy in Poland and InnoEnergy PhD School.

References

1. M. Rezaei et al., J Nat Gas Chem, 15 (2006) 327-334

2. A. S. A. A. Fatish et al., Appl Catal A, 364 (2009) 150-155

3. K. Mette et al., ChemCatChem, 6 (2014) 100-104

4. K. Asami et al., Catal Today 84, 27 (2003) 27-31

5. F. Legorreta-García et al., Bol Soc Esp Cerám V, 54 (2015) 119-123

6. Q. Wu et al., Fuel Process Technol, 89 (2008) 993-999

492067 A Multicore-Shell Catalyst Derived from Ni Phyllosilicate@SiO_2 for Low Temperature Dry Reforming of Methane: Confinement Effect Against Carbon Formation

Abstract:

Dry reforming of methane has been a popular research topic since it converts two â€œgreenâ€ gases into syngas (CO+H2). The main issue for Ni-based catalysts is the severe carbon formation which covers active sites, leading to deactivation. A lot of effort has been put to get stable Ni catalysts with less carbon formation, in which confinement effect has been proven to be a good strategy.

In this report, we manage to synthesize a novel multicore-shell catalyst. Firstly, Ni phyllosilicate is supported on silica nanospheres with ammonia evaporation method. Then Ni phylloslicate nanospheres are coated with a layer of mesoporous SiO2 to obtain a core-shell structure of Ni phyllosilicate@SiO2 via hydrolysis of TEOS. The thickness of shell can be tuned via varying the amount of TOES. After H2 reduction at high temperature, multiple small Ni nanoparticles (~3nm) are formed supported on the inner SiO2 core but also encapsulated within the outer mesoporous SiO2 shell.

This multicore-shell catalyst shows a high and stable conversion (~60%, GHSV=60,000ml/h g_{cat}) along 24h time of stream (TOS) at 600°C while the pristine Ni phyllosilicate suffers from heavy carbon formation and the reactor is blocked after 3h TOS. TGA and TEM of spent catalysts have shown that there is almost no carbon formation for this novel multicore-shell catalyst.

Compared with conventional Ni@SiO2 core-shell catalyst, our multicore-shell catalyst is much easier to synthesize and does not need any toxic organic solvent. And low reaction temperature (600°C) also helps to reduce the energy input. We believe this strategy of making multicore-shell catalyst can be extended and may be helpful to researchers in the same field.

489031 Mixed (dry & partial oxidation) Reforming of Methane with a New Catalyst Derived from a Negative Value Mining Residue Spinellized with Nickel

Abstract:

Mixed (dry & partial oxidation) reforming of methane with a new catalyst derived from a negative value mining residue spinellized with nickel

Frank DEGA(+), Mostafa CHAMOUMI, Nadi BRAIDY and Nicolas ABATZOGLOU(*)

Department of Chemical Engineering and Biotechnological Engineering, Université de Sherbrooke, Sherbrooke, QC, J1K 2R1, Canada

Key words: spinel, reforming, catalyst, nickel, carbon deposition, stability, selectivity.

(+) Presenting author: Frank.Blondel.Dega.Dongmo@USherbrooke.ca

(*) Corresponding author: Nicolas.Abatzoglou@USherbrooke.ca

This work belongs to a larger endeavor aimed at studying the performance and the stability of a hydrocarbons reforming spinellized nickel catalyst prepared from an ilmenite metallurgical residue. Said residue is a negative value upgraded slag oxide (UGSO) coming from a TiO_2 slag production unit, operated by Rio Tinto Iron & Titanium, Quebec, Canada. Recently published results from experiments carried out with this catalyst exhibited high hydrogen production selectivity during methane steam and dry (CO_2) reforming and stability over time-on-stream (TOS).

More specifically, this work report on the behavior of this catalytic formulation during methane partial oxidation (POX) regimes employing both CO_2 and O_2 as reactants. This has been done at 800-850°C, atmospheric pressure, molar ratios CH_4/CO_2 = 3, CH_4/O_2 = 1, 2, 3, and a space velocity between 4000 and 4600 $ml_{STP}/h/g_{cat}$. The conversion of CH_4 and CO_2 and the selectivity of $H_2$and CO have been assessed by means of mass balances whose closure is 100 +/-5%. The crystalline phases of the catalyst components as well as their changes over TOS in presence of different oxygen concentration have been studied. A first attempt to correlate the catalytic performances and the catalytic stability (mainly related to carbon deposition tendency) with the mechanisms and structural modifications of the catalyst over TOS is reported.

The promising results have shown that, within the experimental range tested, an optimum performance has been obtained at 850°C and at molar CH_4/CO_2 = 3, CH_4/O_2 = 2. At CH_4/O_2 lower and higher than 2 the catalytic

performance decreases. The catalyst exhibited a good stability for 2 days of reaction, with no detectable carbon deposition, high conversions (99% of CH_4, 62% of CO_2) and H_2 selectivity of 60%. The in situ X-ray diffraction (XRD) analysis showed that the crystalline structure of the catalyst is subject to modifications. The most significant observation is that the catalyst structure changes over, typically, the first 30min of TOS then stabilizes. This catalyst structure behavior has been correlated to its good stability, the absence of carbon deposition and the high methane conversion and hydrogen selectivity. This study sheds more light on the role of the reaction environment in the activity and the stability of this UGSO spinellized nickel catalyst.

500693 Design of Robust and Efficient Diesel Autothermal Reforming System Using Bi-Metallic Catalyst for the Mobile Applications

Abstract:

In contrast to the studies dealing with the hydrogen production from the light fossil fuels, the degradation problem of the catalyst for the diesel reforming has not been scrutinized thoroughly. Among the several reasons for the degradation problem, coke formation on the surface of the catalyst is known as the main cause of the catalytic failure and it encourages the production of the un-cracked hydrocarbons such as ethylene (C_2H_4), which may lead the severe poisoning problem on the anode of the fuel cell. In this study, bi-metallic catalyst consisted of platinum and ruthenium was introduced to prevent coke formation and ethylene production during the long-term autothermal reforming (ATR) of commercial diesel fuel. Every catalysts synthesized using a combustion method were demonstrated its long-term stability and catalytic activity over 200 hours. The rate of degradation for each time-on-stream experiment was evaluated by calculating its reforming efficiency containing the amount of hydrogen, carbon monoxide and methane produced from the diesel fuel. Through this study, we have found that the additive ruthenium performed as a great oxidation assistant on the catalytic surface and it could successfully oxidize the carbon species formed on the surface of the catalyst. Consequently, the prepared catalysts containing ruthenium showed great stability and the lowest rate of degradation among the samples. Several fresh and post-mortem characterizations were performed in terms of physical and redox properties using X-ray diffraction (XRD), N_2-isotherms, temperature programmed reduction (TPR) and temperature programmed oxidation (TPO) analyses. Based on the findings from the catalytic experiments and characterization, we could minimize the oxygen to carbon ratio (OCR) and steam to carbon ratio (SCR) for the diesel autothermal reforming reaction to enhance the overall efficiency without sacrificing the long-term stability. The bi-metallic catalyst containing ruthenium showed great performance and negligible degradation under the low OCR and SCR condition than conventional values with minimal production of ethylene species. The overall autothermal reforming efficiency was also dramatically enhanced without significant degradation for over 200 hours operation. Moreover, we suggested the total system design containing diesel reformer consisted of the bi-metallic catalyst and conventional solid oxide fuel cell using the commercial software and thermodynamic calculations. Finally, to demonstrate the possibility to use this new catalyst for the mobile application, we coated the bi-metallic catalyst on the cordierite monolith structure using a dip-coating method. The final structured catalyst containing only 0.4 g of active bi-metallic catalyst layer showed excellent reforming activity and efficiency near the thermodynamic limitation and presented no degradation signal over 1,000 hours under the commercial diesel autothermal reforming condition.

495591 Core-Shell Ni-Phyllosilicate@CeO2 Catalyst with Excellent Coke-Resistance in Steam Reforming of Biomass Tar

Abstract:

In this work, a highly active and coke-resistant core-shell structured Ni-SiO_2@CeO_2 catalyst has been developed for steam reforming of biomass tar. A major challenge in the production of hydrogen from biomass gasification is the removal of tar that is formed as a byproduct. Nickel-based catalysts have been shown to be highly active in catalytic steam reforming of tar, but suffer from a high tendency for forming coke that results in blockage of reactors and deactivation of catalyst. Here, we have explored novel core-shell structured catalysts with a Ni-phyllosilicate derived active core and a Cerium oxide shell, that shows high activity (> 60% conversion at 700°C) and stable performance in steam reforming of toluene as a model tar component. In a 20 hr catalytic test run, the Ni-SiO_2@CeO_2 catalyst showed negligible deposition of coke (ca 0.005 gC/ g-cat hr), whereas a reference Ni-SiO_2 catalyst without the CeO_2 shell showed extensive formation of filamentous carbon (ca 0.08 gC/g-cat hr) under the same reaction conditions, that resulted in blockage of the reactor. The excellent thermal stability and resistance to carbon formation of the Ni-SiO2@CeO2 catalyst is attributed to the high oxygen storage capacity and redox nature of CeO_2, and the confinement effect provided by the CeO_2 shell on the metallic nickel nanoparticles. The ceria shell hinders the spatial movement and sintering of Ni nanoparticle at the high reaction temperature, and thus reduces the tendency of carbon formation, which has been shown to be highly dependent on the Ni particle size. At the same time, the CeO_2 shell can also catalyze the oxidation of the coke precursors deposited on the nickel surface by supplying necessary oxygen species. Different synthesis routes have been explored and synthesis conditions have been optimized to tune the morphology of the core-shell catalyst and maximize the activity by increasing Ni dispersion and minimizing diffusion resistance through the shell. Characterization of catalyst has been done by TEM, XRD, XPS, TPR, and TGA analysis. The influence of reaction temperature on the catalyst performance has also been investigated. Due to its high activity and carbon resistance, the application of this catalyst can be extended to other significant H_2 producing reactions such as dry reforming of methane, steam reforming of ethanol etc.

500502 Improved Activity and Stability of Ni-based Anode for Direct Methanolâ€“Fueled Solid Oxide Fuel Cells

Abstract:

Ethanol steam reforming (ESR) is a promising route for sustainable hydrogen production. Recently, $LaNiO_3$ has received a lot of attention as the precursor of the catalyst in ESR reaction due to its interesting properties, such as good redox ability, accessibility and coke resistance. The major drawback of traditional $LaNiO_3$ is its low surface area, which limits its catalytic application. Therefore, it is highly desirable to develop an effective strategy for generating porous $LaNiO_3$. In recent years, three-dimensionally ordered macroporous (3DOM) perovskites have been reported to show better activity for soot combustion, methane combustion and CO oxidation. In this work, high-surface-area 3DOM $La_{1-x}Ce_xNiO_{3-\delta}$ (x=0, 0.1, 0.2, 0.3) (LCNO) catalysts are successfully synthesized and firstly used in ESR reaction. The 3DOM perovskite structure was examined in terms of the catalyst structure and reactivity. After the calcination at 850 ^oC, all the LCNO samples exhibit a 3DOM structure, whereas the $LaNiO_3$ sample shows an irregular morphology with a particle size of 200–600 nm. The incorporation of Ce has almost no effect on the morphology of the 3DOM structure. The 3DOM structure catalysts show significantly higher H_2 yields and lower C_2H_4 yield than that of LaNiO3 at 600 ^oC. In addition, the addition of Ce enhances the stability of the catalyst and effectively suppresses coking in the reaction.

495447 The Modeling and Validation of CO Preferential Oxidation (CO PROX) Multi-Reactor System Using CuO/Rare Earth-Doped Ceria Catalysts for Commercial Diesel Reforming Applications

Abstract:

The multi-reactor system using copper-ceria mixed metal oxide catalysts was proposed for the CO preferential oxidation (CO PROX) in H_2-rich streams. Although the CO PROX reactor system using precious metal-based catalysts has been widely adopted in commercial applications, it has crucial drawbacks such as severe exothermic hydrogen oxidation, unwanted methanation and narrow operating temperature window due to its intrinsic reaction mechanism. To solve the problems, researchers have introduced the redundant inter-coolers and liquid coolant pumps between the CO PROX reactors to keep the operating temperature for each reactor, however the additional heat exchangers lead the over-sized system. In this study, we demonstrated the possibility of copper-ceria catalysts for CO PROX system as an alternative of current precious metal-based catalysts and we suggested optimized multi-reactor design to reduce physical size of the CO PROX system without redundant heat exchangers and coolers. Rare-earth elements doped cerium oxides synthesized using a glycine nitrate combustion method were prepared as a support material for the catalytic system. Various rare-earth elements such as La, Pr, Nd, Sm, Gd, and Tb were introduced in the ceria lattice and the final catalysts containing copper oxides were demonstrated to find its operating temperature for CO PROX reactions. Several characterizations of the prepared catalysts in terms of structural and redox properties using inductively coupled plasma atomic emission spectrometer (ICP-AES), N_2-isotherms, X-ray diffraction (XRD), transmission electron microscope (TEM) and temperature programmed reduction (H_2-TPR) analyses. The industrial CO PROX reactants simultaneously containing H_2, CO, CO_2 and H_2O were blended to demonstrate the CO PROX activity and selectivity of the prepared catalysts. The introduction of rare-earth dopants in the ceria support showed significant effect on the CO PROX activity, selectivity and operating temperature. Consequently, we could utilize these various copper-rare earth doped ceria catalysts to construct the multi-reactor CO PROX system in proper temperature regions for each reactor. The computational fluid dynamics (CFD) modeling using commercial software was performed to build the complete multi-reactor CO PROX system with different copper-ceria catalysts for each reactor canister and laboratory scale validation was also conducted. Moreover, techno-economic evaluation was followed to compare the current precious metal-based CO PROX system with the newly proposed copper-ceria based system in the case of commercial diesel reforming application.

487887 Electrochemical Swing Process for Carbon Capture

Abstract:

Electrochemical Swing Process for Carbon Capture

The capture of carbon dioxide (CO_2) at low concentrations (<1%) is of interest for power generation and other industrial processes, as well as for ventilation of enclosed spaces. We report a solid-state Faradaic electrochemical cell that exploits the reductive addition of CO_2 to quinones for CO_2 capture. The behavior of the cell is modulated by the introduction of CO_2 which changes the reduction of quinone from two one-electron reductions to one two-electron reduction. This results in a change in the potential window of the cell and alters its charge-discharge behavior. An electrochemical cell with a cathode, comprised of a polyanthraquinone-carbon nanotube composite, captures CO_2 upon charging via the carboxylation of reduced quinones, and releases CO_2 upon discharging. The novel cell architecture sandwiches a polyvinylferrocene anode, which serves as both an electron source and sink for the reduction and oxidation of quinones respectively, with two polyanthraquinone cathodes, maximizing the cathode surface area exposed to gas, allowing for ease of stacking of the cells in a device. The bench-scale prototype made demonstrates capture of CO_2 from inlet streams of CO_2 concentrations as low as 0.5% (5000 ppm) at a Faradaic efficiency of >90% and a work of 40-90 kJ per mole of CO_2 captured. This makes it an efficiency improving addition to HVAC systems and cabin environmental control systems aboard aircraft and vehicles.

500966 Opportunities for Industrial CO_2 Capture and Utilization in the US

Abstract:

Carbon dioxide capture from flue gases has largely failed to gain traction due to their prohibitively high cost. One strategy to lower capture costs is to target capture from emissions with higher CO_2 concentration. This presentation focuses specifically on a sub-section of industrial CO_2 emissions, of which there are no carbon-free routes for the formation of products such as cement, glass, iron and other metals. Although these emissions do not represent the majority, unlike the electricity and transportation sectors, they are not replaceable with renewables or bio-energy routes, and in fact, their scale matches well with CO_2 utilization opportunities. A low-cost pathway, including separation, compression and transportation to currently available utilization opportunities has been identified. In particular, our cost model corrects for differences in exhaust composition, flow rate, and geo-specific utilization demand. A regional case-study for the US state of Pennsylvania reveals steel and cement manufacturing as the least cost options. Further, we find that transportation via trucks is generally the low-cost alternative compared to pipeline transport for small volumes on the order of 100 kt CO_2/a. These results are presented in the context of other complexities such as the relative feasibility of extracting CO_2 process emissions from combustion emissions, theoretical maximum commodity outputs by region, and competitive-based commodity trading (regional and international) inspired by localized CCUS efforts.

491206 Improvement of Oxy-Combustion Using Thermodynamic and Exergetic Analysis

Abstract:

Reducing carbon dioxide (CO_2) emissions to the atmosphere is has been the focus of many policy actions directed at mitigating the effects of climate change. Although renewable energies have been widely deployed, there is still a dependence on fossil fuels to satisfy the demands of the grid. Carbon capture and storage (CCS) is generally recognised as providing an essential role to reduce anthropogenic CO_2 emissions from burning fossil fuels for electricity generation. Several technologies have been developed to address this problem, such as oxy-combustion, which also possesses applications in the glass and steel industries. Oxy-combustion is a promising technology that enables the reduction of carbon dioxide (CO_2) emissions from combustion of fossil fuels, such as coal, by more than 90 %. However, the requirement of including an air separation unit (ASU) and a gas processing unit (GPU) introduces an energy penalty to the process, increasing the cost of electricity. This work aims at identifying the potential for improving this technology using a rational approach grounded on the laws of thermodynamics and techno-economic analysis.

Improving the efficiency of oxy-combustion by reducing the parasitic energy consumption of both ASU and GPU is a requirement for this technology to be more competitive in the market. This can be achieved by increasing the separation efficiency of the ASU and GPU. Another way of reducing losses in oxy-combustion is to improve the thermal efficiency of the plant, by increasing the operating conditions to the boiler. However, this would require the use of newer materials, increasing capital expenditure (CAPEX) and make it a riskier project.

Heat integration using low grade heat of compression from the ASU and GPU to pre-heat feedwater is another way of increasing efficiency. This strategy possesses the advantage of using the available resources more efficiently while reducing the number of unit operations, thus reducing CAPEX, and allowing more steam to be used for electricity generation. Acceptance of a technology is highly dependent on the levelised cost of electricity (LCOE) which is influenced by CAPEX and fuel price. This demonstrates that there must be an increase of process efficiency while keeping CAPEX reduced for technological innovation to occur.

An oxy-combustion comprised of a double column cryogenic ASU, an ultra-supercritical boiler with flue gas treatment, and a cryogenic distillation GPU was simulated using Aspen HYSYS. The model developed showed a good agreement with data from IECM program and the values reported by Callide oxyfuel project, with the exception for O_2 and fuel consumption. This difference was found due to the assumption of complete combustion, with no carbon monoxide formation and no heat losses to the environment.

The minimum thermodynamic separation work was calculated for both ASU and GPU and was combined with the theoretical Carnot efficiency to determine the maximum efficiency that an oxy-combustion could achieve. An exergy destruction analysis was performed on each unit operation to quantify the inefficiencies within the process, which would allow for a rational improvement of the process.

In conclusion, the process was improved based on inefficiency minimisation, and then a techno-economic analysis was performed to determine its economic viability. This allowed for an efficiency improvement of 1.4 % (gross) and 3 % (net) as well as a 15 % reduction of CAPEX while decreasing the parasitic power consumption of the ASU from 204 to 197 kWh/t_{O2} and the GPU from 140 to 137 kWh/t_{CO2}. Both reduction of CAPEX and power consumption leads to a decrease of the levelised cost of electricity (LCOE).

The Surface Layer Speciation of CO_2-loaded Aqueous Solutions of Alkanolamines

Naser S. Matin[1], Janice A. Steckel[2], Jesse Thompson[1], Moushumi Sarma[1], and Kunlei Liu[1]

1. Center for Applied Energy Research, University of Kentucky, 2540 Research Park Drive, Lexington, KY 40511, United States
2. National Energy Technology Laboratory, U.S. Department of Energy, Pittsburgh, PA, 15261

UKy-CAER

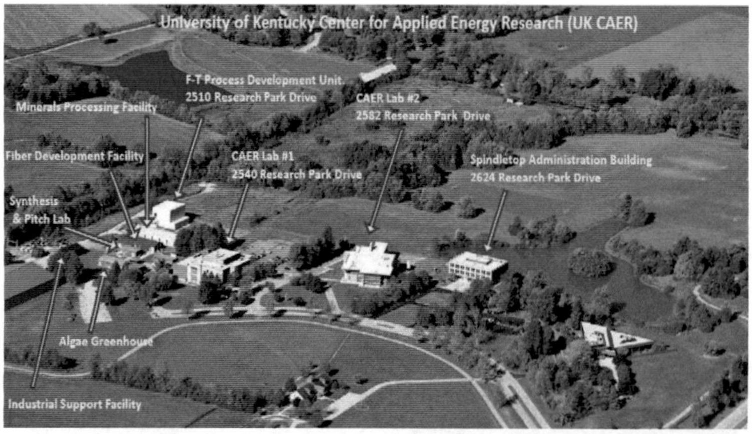

Outline

- ❑ Why Surface Speciation?
- ❑ Surface Behavior of Alkanolamine Solutions
- ❑ Thermodynamic Framework
- ❑ Bulk and Surface Speciation
- ❑ Conclusion

Why Surface Speciation?

Surface layer speciation;

- Has strong impact on the interfacial behavior of a solution.
- Provides deep knowledge and understanding about solutions surface properties.
- Helps in designing proper solvents.
- Is useful in designing proper gas-liquid contactors.
- Provides information about the link between liquid and surface properties.
- Has drawn less attention in amine based CO_2 capture processes.

CO_2 Absorption Rate

Liquid side mass transfer coefficient; $kg' = \{k_2 [Am]\ D_{CO2}\}^{0.5} / H_{CO2}$

To enhance the CO_2 Absorption rate:

- **The solvent with fast chemical reaction kinetics (k_2)**
  - High absorption kinetics
  - Proper mixed solvents
  - Application of a catalyst
- **Henry's law constant (H_{ij}) decreasing**
  - Different composition of a base solvent
- **Improvement of the CO_2 diffusion in liquid phase**
  - Solution viscosity reduction
  - Physical/mechanical/chemical approaches
- **Amine concentration increasing (optimization)**

The number of amine molecules available at the gas-liquid interface to react with CO_2
- Interfacial/surface properties enhancement

Surface Behavior of Alkanolamine Solutions

Unloaded alkanolamine solutions

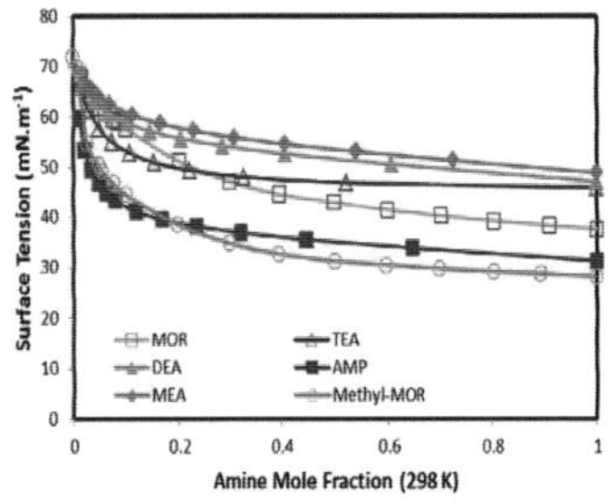

Loaded alkanolamine (MEA 30%)

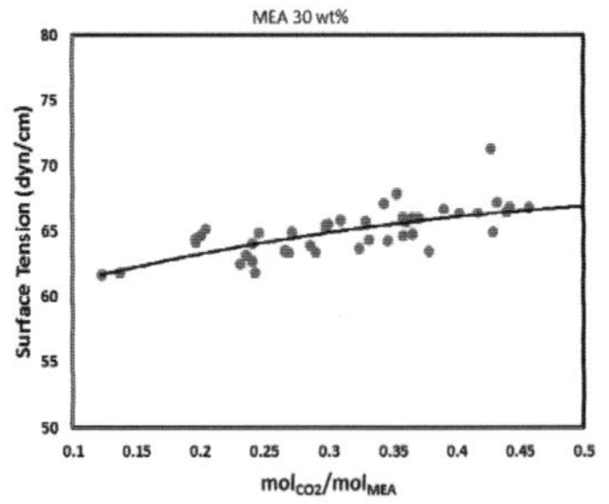

G. Vazquez, et al., J. Chem. Eng. Data 42 (1997) 57.
G. Vazquez, et al., J. Chem. Eng. Data 41 (1996) 806.
P. W. Derks, et al., J. Chem. Eng. Data 50 (2005) 1947.
Y. Maham, et al., J. Chem. Eng. Data 49 (2004) 411.
E. Alvarez, et al., J. Chem. Eng. Data 48 (2003) 32.

CO_2 + **Amine** (aq) →

- HCO_3^- (bicarbonate)
- $CO_3^{=}$ (carbonate)
- $AmCOO^-$ (carbamate)
- AmH^+ (protonated amine)
- Amine (free amine)

How the bulk concentrations of different species can affect the surface properties?

Surface and Bulk Concentrations

m^s : surface concentration

m^b : bulk concentration

Surface concentration (m^s) = *Bulk concentration* (m^b) ?

How can we determine the surface concentration (m^s) ?

Surface Concentration (Speciation) Determination ?

Measurable surface property ⇒ Surface Tension (σ) $\overset{?}{\Rightarrow}$ Surface Concentration

$\sigma = f(I, m_i^s, m_i^b, \Gamma_i)$: solution speciation

Γ_i : Excess of species "*i*" in a unit cross section

I: Ionic strength

m_i^s : concentration of species "*i*" in surface

m_i^b : concentration of species "*i*" in bulk solution

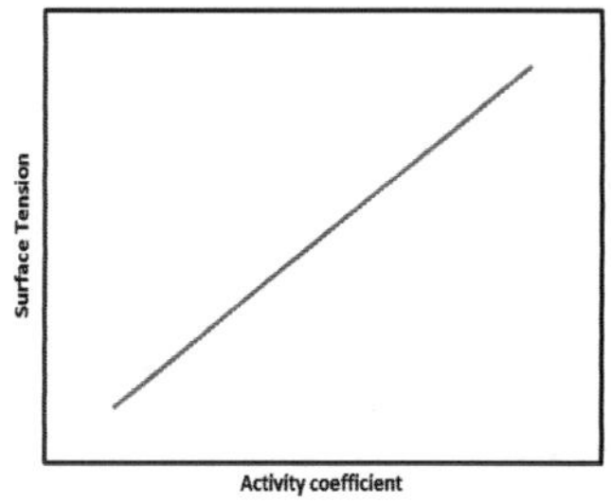

Z. Li, B. C.-Y. Lu, Chemical Engineering Science 56 (2001) 2879.
P. Wang, A. Anderko, and R. D. Young, Ind. Eng. Chem. Res. 2011, 50, 4086.
R. I. Slavchov, J. K. Novev, J. of Colloid and Interface Science 387 (2012) 234.

Theoretical Framework

$\sigma_{sln} = \sigma_w + f(\text{bulk \& surface compositions}, I, \gamma_i)$ **(1)**

$$\sigma_{sln} = \sigma_w + \frac{RT}{55.51\,A_w}(\phi^b \Sigma v_i m_i^b - \phi^s \Sigma g_i v_i m_i^b) \quad (2)$$

I : ionic strength

γ : activity coefficient

ϕ : osmotic coefficient

σ_w : water surface tension

A_w : partial molar area of water

g : partition coefficient ($g = m^s/m^b$)

Through equation (2), solution surface tension (σ_{sln}) is related to bulk and surface compositions.

Matin, et al., Ind. Eng. Chem. Res. 2017, 56, 5747.
Li, et al., Ind. Eng. Chem. Res. 1999, 38, 1133.

Surface Speciation Estimation

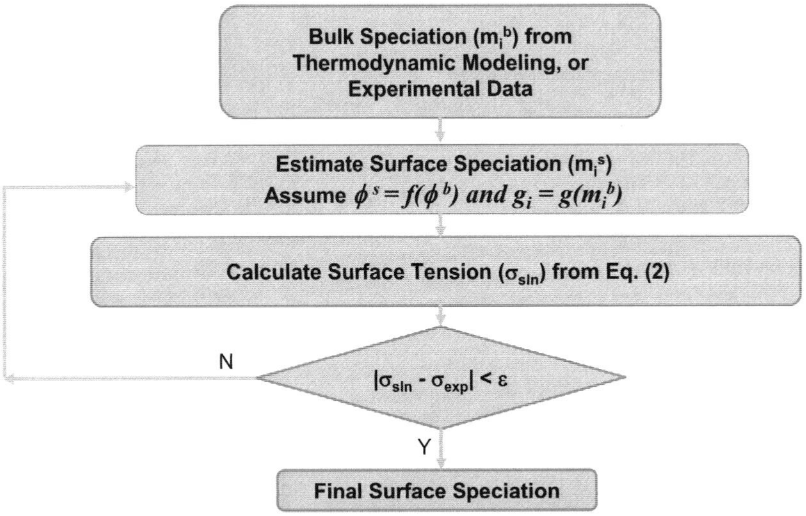

Predicted Surface Tension (MEA 30%, 25°C)

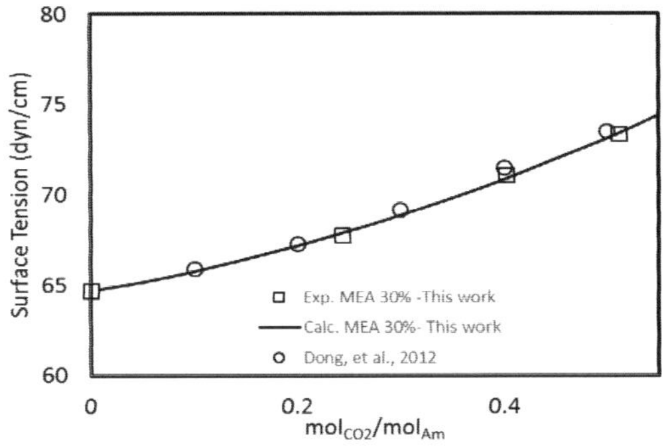

Dong, et al., Science China Chemistry, 55 (2012), 1467– 1473

Surface/Bulk Ratio Speciation
(MEA 30%, 25°C)

Surface accumulation

Surface depletion

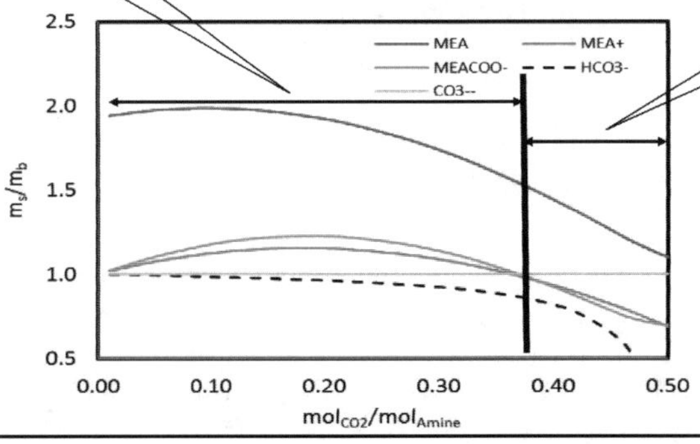

$m_s/m_b = 1$, evenly distribution between surface and bulk

$m_s/m_b < 1$, surface depletion

$m_s/m_b > 1$, surface accumulation

J. A. Steckel (NETL) and literature

Polarizability: Carbamate (10.2 A°³) > Carbonate (8.1 A°³) > Protonated Amine (5.4 A°³) > Bicarbonate (5.1 A°³)
Size: Carbamate > Protonated Amine > carbonate ≈ bicarbonate

Higher size and polarizability ⇒ higher surface tendency

Matin, et al., Ind. Eng. Chem. Res. 2017, 56, 5747.
Krei, et al., "Molecular Thermodynamics of Aqueous Solutions Containing Volatile Weak Electrolytes", Report, Lawrence Berkeley Laboratory University of California, September 1988.
dos Santos, et al., J. Phys. Chem. C 2008, 112, 19431.

Bulk and Surface Speciation
(MEA 30%, 25°C)

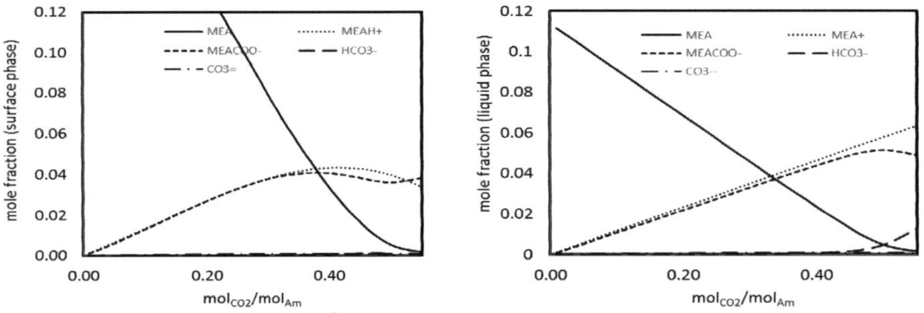

Ionic Strength and Surface Tension

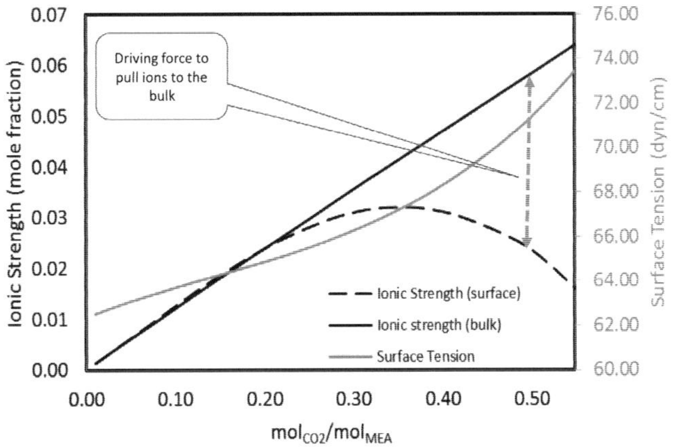

AMP and MEA solutions

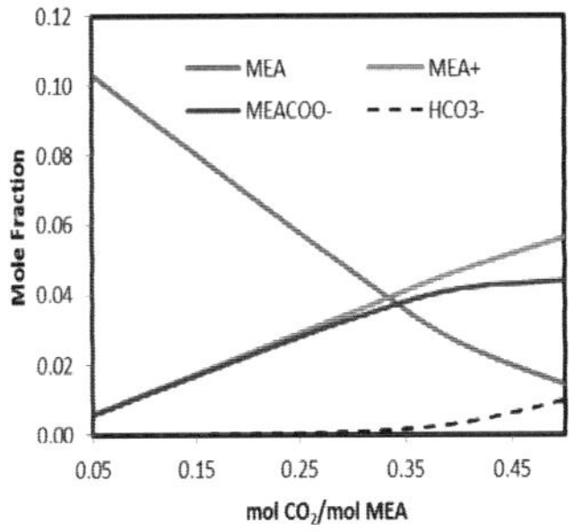

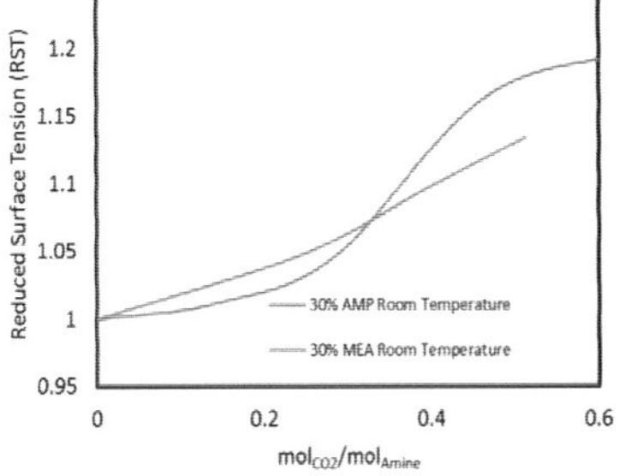

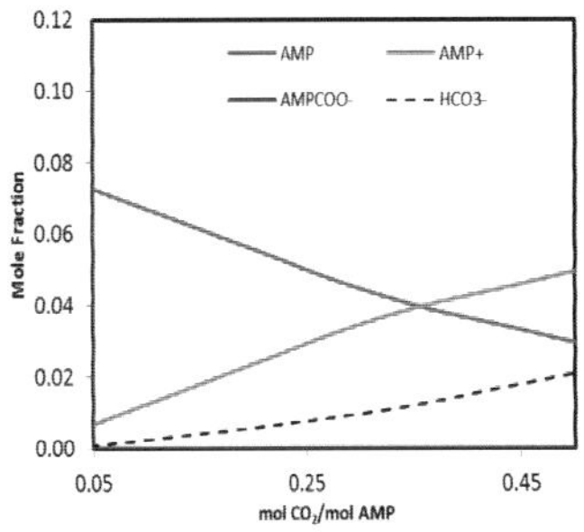

RST = (surface tension at loading α) / (surface tension of unloaded solution)

Solution Behavior Schematic (Theoretical Expectation)

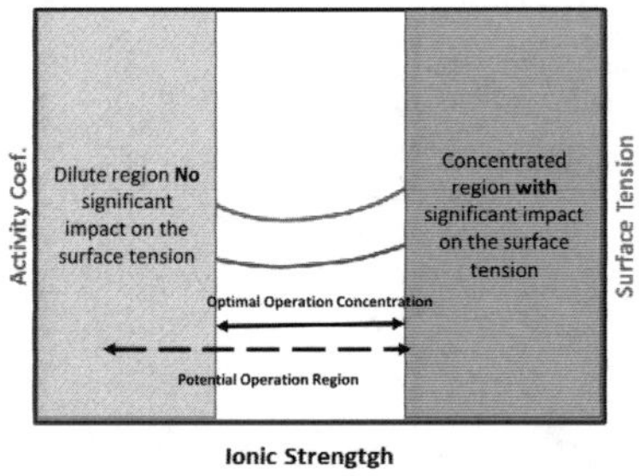

Conclusion

- Thermodynamic platform to estimate the surface phase speciation
- Bicarbonate ion has more impact on the surface tension increasing.
- At high CO_2 loadings all major species show depletion in the surface phase.
- Solvents with more bicarbonate route (hindered, tertiary, carbonated, ..) may need surfactant to decrease their surface tension.
- Higher size and polarizability mean higher surface tendency
- Carbonate polarizability > bicarbonate polarizability $\Rightarrow$ higher surface tendency for carbonate
- Size of MEA carbamate & protonated MEA > size of carbonate & bicarbonate ions $\Rightarrow$ higher surface tendency for carbamate and protonated MEA

Acknowledgement

Carbon Management Research Group Members:

DEDI

DUKE ENERGY

LG&E KU
PPL companies

AEP KENTUCKY POWER
A unit of American Electric Power

EPRI | ELECTRIC POWER RESEARCH INSTITUTE

Thank You

Questions?

Comments?

495771 Post-Combustion CO_2 Capture Using Hexamethylenediamine Activated Aqueous Sodium Glycinate Solvent

Abstract:

One of the most serious problems facing the world today is global warming and consequential climate change. Prevention of the catastrophic global worming requires an effective implementation of CO_2 capture strategy to the major point sources of the CO_2 emission. Among the major point sources, fossil fuel based power plants contributes almost 40% of the anthropogenic discharge. So the main target for implementing the CO_2 emission-reducing strategy is pointed towards this sector. Aqueous amine based absorption-regeneration process for removing CO_2 is a mature technology. Monoethanolamine (MEA), diethhanolamine (DEA) are the conventional solvents for rapid capture of CO_2. However, implementation of the conventional amine solvent for capturing CO_2 from power plant flue gas stream leads to high energy requirement, large solvent circulation rate and high solvent degradation rate, all of which in turn increases the operating and capital cost of carbon capture. These drawbacks are the main motivation for the investigation of the alternative solvents.

CO_2 absorption property of hexamethylenediamine (HMDA) activated aqueous sodium glycinate (SG) solvent is investigated in this work to explore its potential as an efficient alternative solvent for CO_2 capture. The ionic nature of aqueous SG makes it more resistant to oxidative and thermal degradation which is a desired property for the CO_2 capture from the oxygen rich power plant flue gas. It also provides certain other desirable physical properties such as low volatility, higher surface tension, etc. On the other, HMDA, a straight chain diamine (Mondal et al., 2017, 2015) will enhance the solvent characteristics due to its high kinetic and absorption capacity. All these characteristics of the component solvents motivated us to investigate aqueous (HMDA+SG) hybrid solvent.

In this work, the physico-chemical properties (density, viscosity, physical solubility), vapour-liquid equilibrium (VLE) and kinetics of CO_2 absorption in (5 mass% HMDA+ 25 mass% SG+ 70 mass% H_2O) solvent as well as (30 mass% SG+70 mass% H_2O) solvent are measured in the temperature range of 313-333K. These properties are required for the design of absorber-regenerator system for CO_2 capture from power plant flue gas. VLE of CO_2 is measured in a custom made stirred equilibrium cell set up using the method of Park and Sandall (2001). VLE data obtained in this work is presented in terms of CO_2 loading (mole CO_2 absorbed per mole total amine). The experimental VLE data shows a general trend of increasing CO_2 loading with the increase in CO_2 partial pressure. For both the solvent systems, loading capacity decreases with the increase in the temperature which indicate the exothermic nature of the chemical absorption. At 10kPa CO_2partial pressure and 313K temperature, comparison of the hybrid solvent (CO_2

loading = 0.616) with the 30% SG (CO_2 loading = 0.471) shows 20% enhancement of the loading capacity (due to addition of HMDA).

For the kinetic study, a reaction calorimeter (Model: RC1e, Mettler Toledo) set-up is used. Kinetics of CO_2 absorption is explored assuming pseudo-first-order reaction condition. Seconds order rate constant for CO_2 absorption in aqueous 30 mass% SG (592 $m^3mol^{-1}s^{-1}$) is estimated using zwitterion mechanism with fast deprotonation step and compared with the existing literature data (Lee et al., 2007) . For both the solvent systems, overall reaction rate constant is found to increase with the temperature. Overall reaction rate constant for the aqueous 30 mass% SG (2080s^{-1}) and (5mass% HMDA+25mass% SG) (10410 s^{-1}) are compared which shows 5 times enhancement of the absorption rate of the hybrid solvent compared to 30% SG solvent.

So, from the investigation of absorption properties, (HMDA+SG+H_2O) hybrid solvent appears to be a potential solvent with high CO_2 absorption rate and capacity which can be used for the CO_2 capture from the power plant flue gas.

References:

Lee, S., Song, H., Maken, S., Park, J., 2007. Kinetics of CO2 Absorption in Aqueous Sodium Glycinate Solutions. Ind. Eng. Chem. Res. 46, 1578–1583. doi:10.1021/ie061270e

Mondal, B.K., Bandyopadhyay, S.S., Samanta, A.N., 2017. Kinetics of CO2 absorption in aqueous hexamethylenediamine. Int. J. Greenh. Gas Control 56, 116–125. doi:10.1016/j.ijggc.2016.11.023

Mondal, B.K., Bandyopadhyay, S.S., Samanta, A.N., 2015. Vapor–liquid equilibrium measurement and ENRTL modeling of CO2 absorption in aqueous hexamethylenediamine. Fluid Phase Equilib. 402, 102–112. doi:10.1016/j.fluid.2015.05.033

Park, M.K., Sandall, O.C., 2001. Solubility of Carbon Dioxide and Nitrous Oxide in 50 mass % Methyldiethanolamine. J. Chem. Eng. Data 46, 166–168. doi:10.1021/je000190t

499551 Using Intellectual Property to Protect Carbon Capture Innovations

Abstract:

Carbon capture and storage (CCS) technology is increasingly an important tool to meet global emission reduction targets.[1] By the end of 2017, the number of operational large-scale CCS projects is set to rise to over 20, and the CO_2 capture capacity of operational projects will more than double since 2010.[2]

Fueling this growth, companies and universities alike are developing new and more cost-effective CCS technology.[3] Securing strong intellectual property (IP) rights is vital to protecting these innovations. In this session, IP attorneys Charles Collins-Chase and Lauren Dowty will explain what kinds of IP rights are available to protect carbon capture technologies, why those IP rights are valuable in the carbon capture field, how to obtain them, and common pitfalls to avoid. They will also analyze exemplary patents covering CCS technology and discuss case studies concerning CCS patents.

---

1 Global CCS Institute, *The Global Status of CCS 2016*, at 5, 11-12 (2016).

2 *Id.* at 11.

3 *See* Global CCS Institute, *The Global Status of CCS 2014*, at 113-14 (2014) ("Academia is the primary source of publications in general, but corporations are the driving force for patent applications in the area of carbon capture."); *see also ExxonMobil and FuelCell Energy progressing one-of-a-kind carbon capture fuel cell solution*, Energy Factor, https://energyfactor.exxonmobil.com/news/fuel-cell-site/ (Oct. 28, 2016) ("FuelCell Energy and ExxonMobil have announced the site of a pilot plant to test a novel research concept that uses fuel cell technology to capture carbon dioxide. This new technology has the potential to substantially reduce costs and lead to a more economical pathway toward large-scale application globally."); Suzanne McCarron, *Focus on Technology: Finding a way to capture carbon*, Energy Factor (Mar. 7, 2016), https://energyfactor.exxonmobil.com/perspectives/focus-technology-finding-way-capture-carbon/ ("The corporation has a working interest in more than one-third of the world's existing carbon-capture-and-storage capacity. Last year alone, we captured more than 6 million metric tons of CO_2 for storage. . . . In addition to our in-house programs, we have also supported carbon-capture-and-storage research efforts with leading universities such as Georgia Tech, MIT, Stanford, and the University of Texas. That's in addition to technical partnerships we have participated in with the U.S. Department of Energy and the European Commission to evaluate a range of carbon injection and storage technologies."). *But see* Shell, *Shell's Quest Carbon Capture and Storage Project reaches significant milestone*, Global CCS Institute (Oct. 4, 2016) https://www.globalccsinstitute.com/news/institute-updates/shell%E2%80%99s-quest-carbon-capture-and-storage-project-reaches-significant-milestone ("From the outset, any intellectual property or data generated by Quest has been publicly available, in collaboration with the governments of Alberta and Canada to help bring down future costs of CCS and encourage wider use of the technology around the world. This means that others can take the detailed engineering plans, valued at C$100 million, to help build future CCS facilities.")

UNIVERSITY OF BATH

Whole-Systems CO_2 Value Chain Modelling

A closer look at the pathways through syngas

Chris Quarton, c.j.quarton@bath.ac.uk
PhD Student
Department of Chemical Engineering
University of Bath

Supervisor: Dr Sheila Samsatli, s.m.c.samsatli@bath.ac.uk

AIChE Annual Meeting, Minneapolis
31 October 2017

UNIVERSITY OF BATH

A world-class University in a World Heritage City

50 UNIVERSITY OF BATH Established 1966

TOP 5 UK UNIVERSITY The Guardian 2018

TEF Gold

Department of Chemical Engineering

- **Top 3** for course satisfaction *(Guardian University Guide 2018)*
- **Top 5** in the UK for Chemical Engineering overall *(2018 Complete University Guide and Guardian University Guide 2018)*
- **Top 5** for teaching satisfaction *(Guardian University Guide 2018)*
- **89%** of our research is graded as either world-leading or internationally excellent *(Research Excellence Framework 2014)*

UNIVERSITY OF BATH

Outline

1. Context
2. Value Web Model
3. Modelling of UK CO_2 value chains
   - ➢ Model Setup
   - ➢ Results
4. Conclusion

UNIVERSITY OF BATH

Context

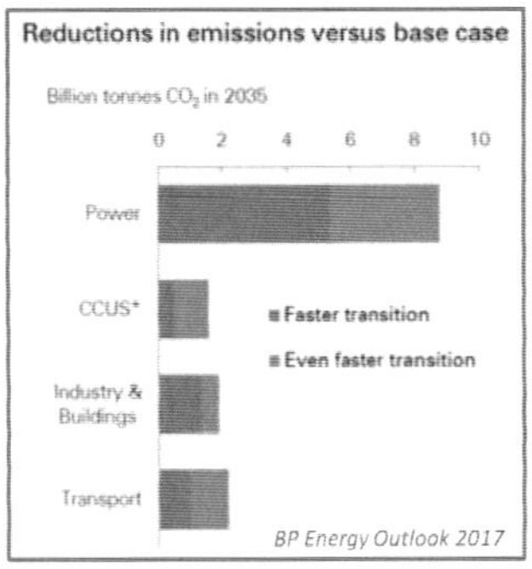

- Need to reduce carbon emissions
  - 2050 target of 20% of 1990 levels is enshrined in UK law
- This will require carbon capture
- CCS vs CCU debate is divisive
- Arguments focus around:
  - Scale
  - Sustainability
  - Economics and Efficiency
  - Infrastructure
  - Grid balancing
  - Diversification
- Aim: model both CCS and CCU on same terms to investigate conflicts, synergies, clear winners

UNIVERSITY OF BATH

VALUE WEB MODEL

S. Samsatli, N.J. Samsatli (2017). A multi-objective MILP model for the design and operation of future integrated multi-vector energy networks capturing detailed spatio-temporal dependencies, Applied Energy (in press).

UNIVERSITY OF BATH

Energy systems as value webs

Traditional supply chain model

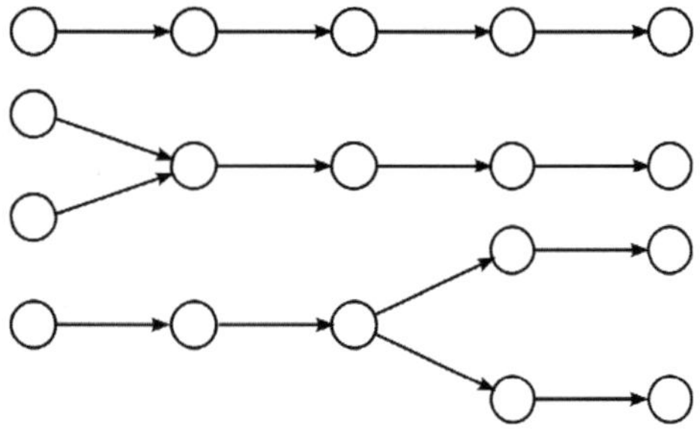

- Linear and unidirectional
- Framework used by many energy system models
- Very limiting for energy systems

Value web model

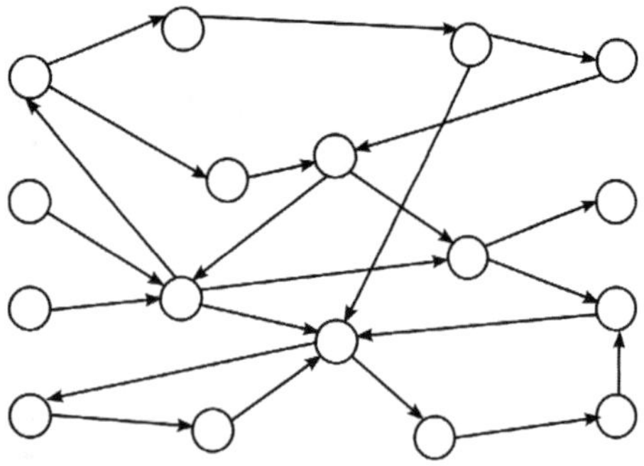

- Ideal framework for energy systems
- Interconvert energy between many different forms
- What is the best form of energy to transport and store?
- Multi-vector energy networks

UNIVERSITY OF BATH

Value Web Model

- Optimal design and operation of systems that utilise primary resources to deliver products and services
- Aids strategic planning and tactical operations (simultaneously)
- Detailed storage and transport representations
- Detailed spatial and temporal representations
- Technology-agnostic
- Data-driven framework (easy to add more networks by simply providing data)
- Can be applied to energy systems, water networks, supply chains for chemicals, pharmaceuticals, food etc., circular economy

UNIVERSITY OF BATH

Problem statement

Given:

- Spatio-temporal demands for resources and energy services
- Spatio-temporal availability of primary energy sources and raw materials
- Characteristics of each technology (e.g. CAPEX, O&M, efficiency, lifetime)

Determine:

- Network design
  - Location, number and capacity of generation/conversion and storage technologies
  - Structure of transport infrastructure network (transmission and distribution)
  - When and where to purchase/install
  - What interactions
- Network operation
  - Which resources to convert, store and transport (how much, where and when)
  - Which technologies to use at different times
  - Transport flows between different regions

UNIVERSITY OF BATH

Problem statement (cont...)

Subject to:

- Demand satisfaction
- Conservation and other physical laws
- Constraints on resources (e.g. land, water), costs and GHG emissions
- Technological constraints (e.g. tech. availability, build rates)
- Social and political constraints (e.g. siting of specific techs)

Objective:

- Minimise cost
- Minimise environmental impact (e.g. GHG emissions)
- Maximise value
- Any combination of the above

UNIVERSITY OF BATH

Temporal representation

- Long-term strategic decisions
- Short-term operational issues (intermittency, dynamics of energy storage)

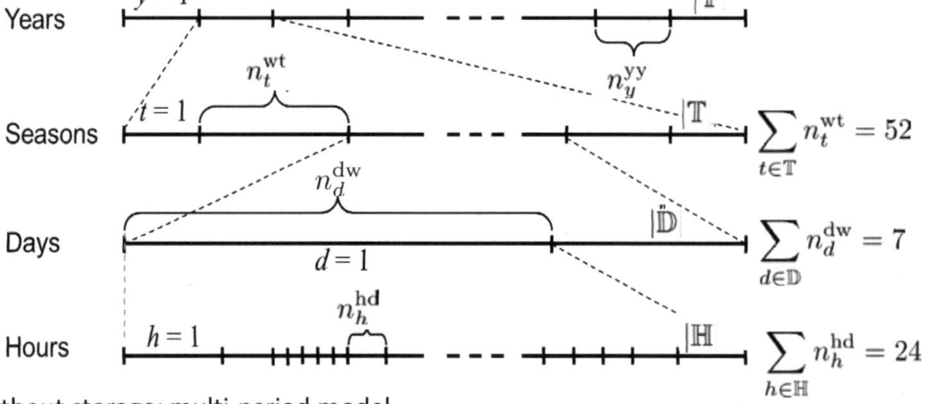

Without storage: multi-period model

With storage: dynamic model; extra variables for initial inventories; extra constraints to link inventories within and between time levels

S. Samsatli, N.J. Samsatli (2015). Computers & Chemical Engineering 80, 155-176, 0098-1354.

Spatial representation

- The region under study divided into a number of zones
- Each zone may:
  - Be of any shape and size
  - Have dynamic demands for various resources
  - Contain some resources that are available in varying quantities
  - Host technologies for conversion and storage of resources
  - Be connected with other cells via transport infrastructures
  - Import or export resources

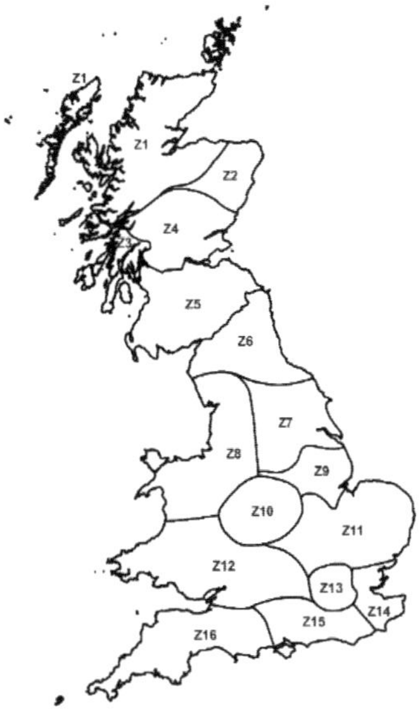

National Grid's SYS study zones

Integrated energy conversion pathways

- Represented using Resource-Technology Networks (RTN)
- Allows representation of "web" pathways
  - convert any resource to any other, including recycles (circular chains)
  - store and transport resources at any stage in the chain
- Consider a wide range of different feedstocks and technologies
  - different operating modes to generate different energy vectors
    - heat, electricity, transport fuel etc.

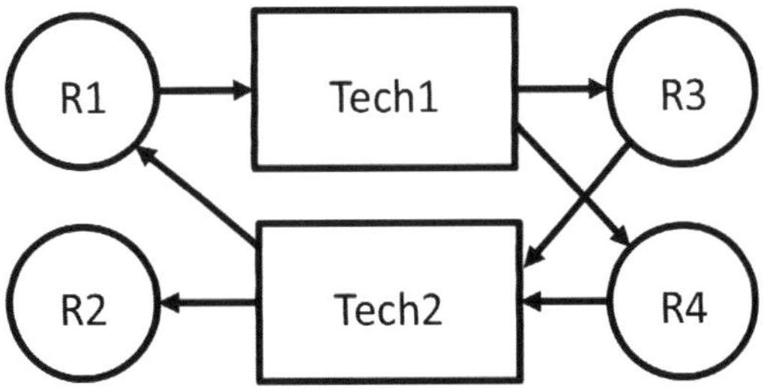

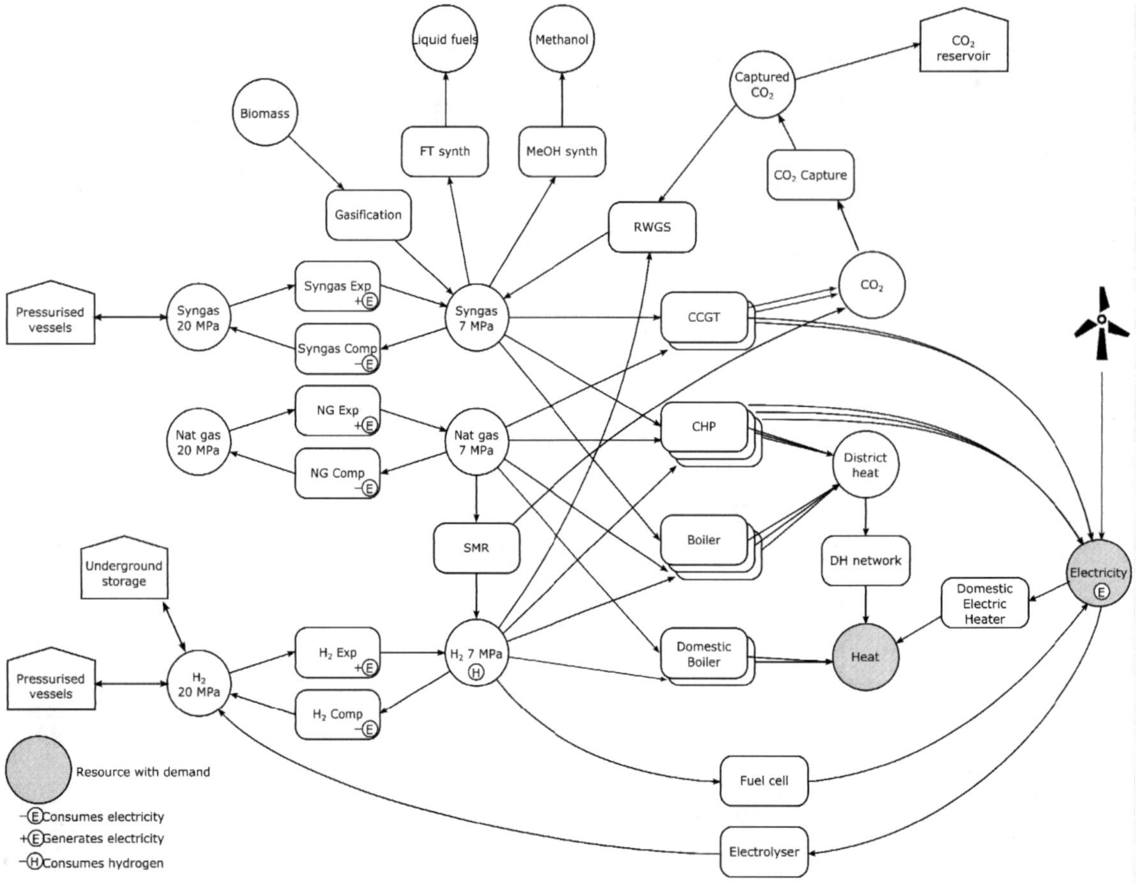

UNIVERSITY OF BATH

Modelling storage

Set of **storage operations** to store resource r_1.

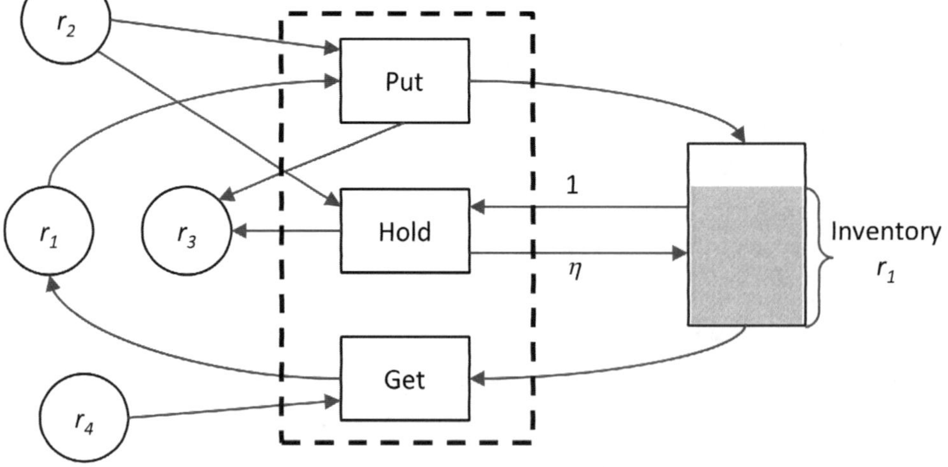

The "put" operation transfers r_1 from the zone to the store, requiring some r_2 and producing some wastes r_3 (e.g. CO_2). The "hold" operation maintains r_1 in storage which could be at less than 100% efficiency, the losses being converted to r_3 ; this operation may also require some r_2. Finally, the "get" operation retrieves r_1 from storage and delivers it to the zone, requiring some r_4.

S. Samsatli, N.J. Samsatli (2015). Computers & Chemical Engineering 80, 155-176, 0098-1354.

UNIVERSITY OF BATH

Modelling transport

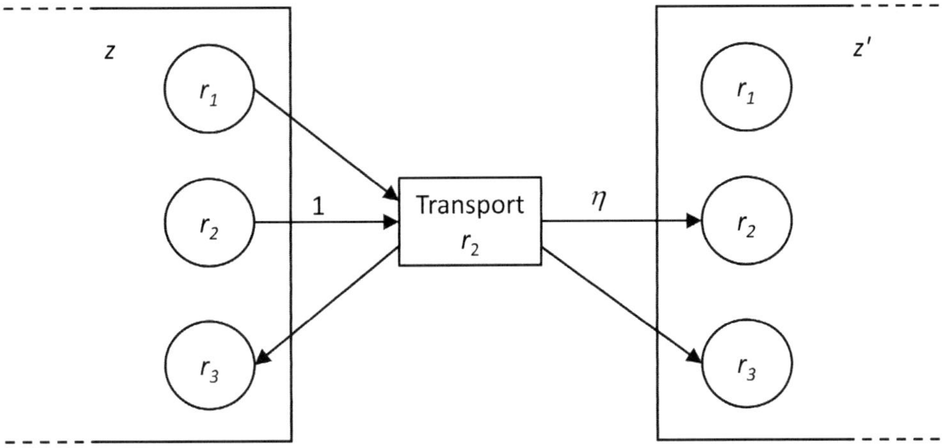

Resource r_2 is transported from zone z to zone z', which requires r_1 from zone z and results in waste r_3 being generated in both zones

S. Samsatli, N.J. Samsatli (2015). Computers & Chemical Engineering 80, 155-176, 0098-1354.

Transmission technologies connect the networks between different zones

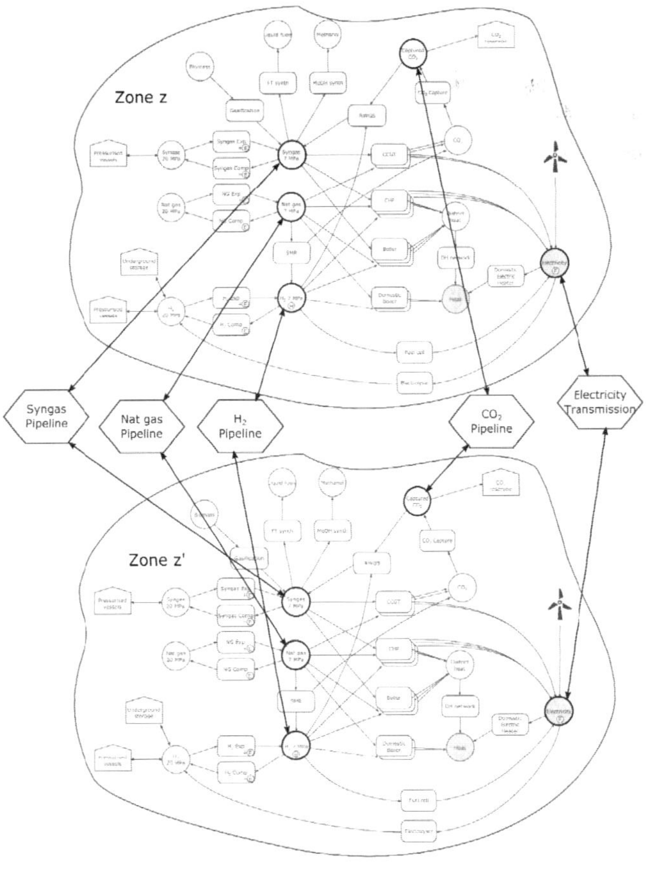

Resource balance and availability

Resource balance:

$$U_{rzhdty} + M_{rzhdty} + P^{\mathrm{h}}_{rzhdty} + P^{\mathrm{d}}_{rzdty} + S_{rzhdty} + Q_{rzhdty} \geq D_{rzhdty} + X_{rzhdty} \qquad \forall r \in \mathbb{R},\, z \in \mathbb{Z},\, h \in \mathbb{H},\, d \in \mathbb{D},\, t \in \mathbb{T},\, y \in \mathbb{Y}$$

Resource utilisation: $U_{rzhdty} \leq u^{\max}_{rzhdty} \qquad \forall r \in \mathbb{R},\, z \in \mathbb{Z},\, h \in \mathbb{H},\, d \in \mathbb{D},\, t \in \mathbb{T},\, y \in \mathbb{Y}$

Wind power potential: $u^{\max}_{\mathrm{Elec},zhdt} = 0.5 \times 10^{-6} \eta \rho^{\mathrm{air}} \pi N^{\mathrm{WT}}_{zy} \left(R^{\mathrm{WT}}\right)^2 v^3_{zhdty}$

Biomass potential: $\sum_{hd} U_{\mathrm{Bio},zhdty} n^{\mathrm{hd}}_h n^{\mathrm{dw}}_d n^{\mathrm{wt}}_t \leq A^{\mathrm{Bio}}_{zy} Y^{\mathrm{Bio}}_{zty} \qquad \forall z \in \mathbb{Z},\, t \in \mathbb{T},\, y \in \mathbb{Y}$

Land footprint constraints:

Wind turbines: $\pi \left(5R^{\mathrm{WT}}\right)^2 N^{\mathrm{WT}}_{zy} \leq A^{\mathrm{WT,max}}_{zy} \qquad \forall z \in \mathbb{Z},\, y \in \mathbb{Y}$

Biomass (each zone): $A^{\mathrm{Bio}}_{zy} \leq f_{zy} A^{\mathrm{Bio,max}}_{zy} \qquad \forall z \in \mathbb{Z},\, y \in \mathbb{Y}$

Biomass (GB): $\sum_z A^{\mathrm{Bio}}_{zy} \leq f^{\mathrm{GB}}_y \sum_z A^{\mathrm{Bio,max}}_{zy} \qquad \forall y \in \mathbb{Y}$

Constraints for conversion technologies

Net production of resource r:

$$P^{\mathrm{h}}_{rzhdty} = \sum_{p \in \mathbb{P}^{\mathrm{h}}} \mathscr{P}^{\mathrm{h}}_{pzhdty} \alpha_{rpy}$$

$$P^{\mathrm{d}}_{rzdty} = \sum_{p \in \mathbb{P}^{\mathrm{d}}} \mathscr{P}^{\mathrm{d}}_{pzdty} \alpha_{rpy}$$

Min/max rate of operation:

$$N^{\mathrm{P}}_{pzy} p^{\min}_p \leq \mathscr{P}^{\mathrm{h}}_{pzhdty} \leq N^{\mathrm{P}}_{pzy} p^{\max}_p$$

$$N^{\mathrm{P}}_{pzy} p^{\min}_p \leq \mathscr{P}^{\mathrm{d}}_{pzdty} \leq N^{\mathrm{P}}_{pzy} p^{\max}_p$$

Technology balance: $N^P_{pzy} = N^P_{pz,y-1} + NI^P_{pzy} - NR^P_{pzy}$

Retirements: $NR^{\mathrm{P}}_{pzy} = \sum_{y'} RF^P_{py'y} NI^P_{pzy'}$

Max build rate: $\sum_z NI^{\mathrm{P}}_{pzy} \leq BR_{py}$

- Number of domestic technologies is continuous instead of integers for computational efficiency (i.e. fraction of penetration)

UNIVERSITY OF BATH

Constraints for transport technologies

Net transport of resource r:

$$Q_{rzhdty} = \sum_{z'|\nu_{z'z}=1} \sum_{l\in\mathbb{L}} [(\bar{\tau}_{lr,\text{dst},y} + \hat{\tau}_{lr,\text{dst},y} d_{z'z}) \mathscr{Q}_{lz'zhdty}] + \sum_{z'|\nu_{zz'}=1} \sum_{l\in\mathbb{L}} [(\bar{\tau}_{lr,\text{src},y} + \hat{\tau}_{lr,\text{src},y} d_{zz'}) \mathscr{Q}_{lzz'hdty}]$$

Max rate of operation: $\mathscr{Q}_{lzz'hdty} \leq \sum_{b\in\mathbb{B}} q_l^{\max} N^{\text{B}}_{bzz'y}|_{LB_{lb}=1\wedge\nu_{zz'}=1}$

Max infrastructure capacity: $\sum_{l\in\mathbb{L}} \mathscr{Q}_{lzz'hdty} LB_{lb} \leq b_b^{\max} N^{\text{B}}_{bzz'y}$

Infrastructure balance: $N^{\text{B}}_{bzz'y} = N^{B}_{bzz',y-1} + NI^{\text{B}}_{bzz'y}$

Bi-directional links:

$$N^{\text{B}}_{bzz'y} = N^{\text{B}}_{bz'zy} \qquad \forall b \in \mathbb{B}|\beta_b = 1,\, z \neq z' \in \mathbb{Z}$$

UNIVERSITY OF BATH

Constraints for storage technologies

Max charging and discharging rates:

$$\mathscr{S}^{\text{put}}_{szhdty} \leq N^{\text{S}}_{szy} s_s^{\text{put,max}} a_{sz}$$

$$\mathscr{S}^{\text{get}}_{szhdty} \leq N^{\text{S}}_{szy} s_s^{\text{get,max}} a_{sz}$$

Net inflow of resource r due to storage operations:

$$S_{rzhdty} = \sum_s \left(\mathscr{S}^{\text{put}}_{szhdty} \sigma^{\text{put}}_{sr,\text{src},y} + \mathscr{S}^{\text{hold}}_{szhdty} \sigma^{\text{hold}}_{sr,\text{dst},y} + \mathscr{S}^{\text{get}}_{szhdty} \sigma^{\text{get}}_{sr,\text{dst},y} \right)$$

Inventory levels:

$$I_{szhdty} = n_h^{\text{hd}} \sum_r \left(\mathscr{S}^{\text{put}}_{szhdty} \sigma^{\text{put}}_{sr,\text{dst},y} + \mathscr{S}^{\text{hold}}_{szhdty} \sigma^{\text{hold}}_{sr,\text{src},y} + \mathscr{S}^{\text{get}}_{szhdty} \sigma^{\text{get}}_{sr,\text{src},y} \right)$$

Changes in inventory:

Daily: $\delta^{\text{d}}_{szdty} = I_{sz,|\mathbb{H}|,dty} - I^{0,\text{sim}}_{szdty}$

Weekly: $\delta^{\text{t}}_{szty} = \sum_d \delta^{\text{d}}_{szdty} n_d^{\text{dw}}$

Yearly: $\delta^{\text{y}}_{szy} = \sum_t \delta^{\text{t}}_{szty} n_t^{\text{wt}}$

Cyclic constraint: $\delta^{\text{y}}_{szy} = 0$

UNIVERSITY OF BATH

Constraints for storage technologies (cont...)

Link inventories between different time levels:

$$I^{0,\mathrm{act}}_{szdty} = I^{0,\mathrm{act}}_{sz,d-1,ty} + n^{\mathrm{dw}}_{d-1}\delta^{\mathrm{d}}_{sz,d-1,ty}$$

$$I^{0,\mathrm{act}}_{sz,1,ty} = I^{0,\mathrm{act}}_{sz,1,t-1,y} + n^{\mathrm{wt}}_{t-1}\delta^{\mathrm{t}}_{sz,t-1,y}$$

$$I^{0,\mathrm{act}}_{sz,1,1,y} = I^{0,\mathrm{act}}_{sz,1,1,y-1} + n^{\mathrm{yy}}_{y-1}\delta^{\mathrm{y}}_{sz,y-1}$$

Shift inventory levels to match average:

$$I^{0,\mathrm{sim}}_{szdty} = I^{0,\mathrm{act}}_{szdty} + \left[\left(n^{\mathrm{dw}}_{d} - 1\right)\delta^{\mathrm{d}}_{szdty} + \left(n^{\mathrm{wt}}_{t} - 1\right)\delta^{\mathrm{t}}_{szty} + \left(n^{\mathrm{yy}}_{y} - 1\right)\delta^{\mathrm{y}}_{szy}\right]/2$$

$$\mathscr{S}^{\mathrm{hold}}_{sz,1,dty} = I^{0,\mathrm{sim}}_{szdty}/n^{\mathrm{hd}}_{1} \qquad \mathscr{S}^{\mathrm{hold}}_{szhdty} = I_{sz,h-1,dty}/n^{\mathrm{hd}}_{h}$$

Storage capacity constraints:

$$s^{\mathrm{hold,min}}_{s} N^{S}_{szy} a_{sz} \leq I_{szhdty} \pm \left[\left(n^{\mathrm{dw}}_{d} - 1\right)\delta^{\mathrm{d}}_{szdty} \pm \left(n^{\mathrm{wt}}_{t} - 1\right)\delta^{\mathrm{t}}_{szty} \pm \left(n^{\mathrm{yy}}_{y} - 1\right)\delta^{\mathrm{y}}_{szy}\right]/2 \leq s^{\mathrm{hold,max}}_{s} N^{S}_{szy} a_{sz}$$

UNIVERSITY OF BATH

Constraints for CCS and CCU

Rate of utilisation or storage of CO_2:

$$CCUS_{zhdty} = -\sum_{p\in\mathbb{P}^{\mathrm{CCUS}}} \alpha_{p,\mathrm{CO}_2^{\mathrm{cap}}} \mathscr{P}^{\mathrm{d}}_{pzdty}$$

Total impact from carbon credits:

$$\mathscr{I}^{\mathrm{CCUS}}_{iy} = \varsigma \sum_{zhdt} D^{\mathrm{OM}}_{y} V^{\mathrm{CCUS}}_{iy} CCUS_{zhdty} n^{\mathrm{hd}}_{h} n^{\mathrm{dw}}_{d} n^{\mathrm{wt}}_{t}$$

$$V^{\mathrm{CCUS}}_{iy} = \begin{cases} \text{the } CO_2 \text{ price if } i \text{ is cost} \\ 1 \text{ if } i \text{ is } CO_2 \text{ emissions} \end{cases}$$

Objective function

$$\text{Minimise } Z = \sum_{iy} \omega_i \Big(\mathscr{I}_{iy}^{\mathrm{W}} + \mathscr{I}_{iy}^{\mathrm{P}} + \mathscr{I}_{iy}^{\mathrm{Q}} + \mathscr{I}_{iy}^{\mathrm{S}} + \mathscr{I}_{iy}^{\mathrm{w}} + \mathscr{I}_{iy}^{\mathrm{p}}$$
$$+ \mathscr{I}_{iy}^{\mathrm{q}} + \mathscr{I}_{iy}^{\mathrm{s}} + \mathscr{I}_{iy}^{\mathrm{r}} + \mathscr{I}_{iy}^{\mathrm{m}} + \mathscr{I}_{iy}^{\mathrm{x}} - \mathscr{I}_{iy}^{\mathrm{Rev}} - \mathscr{I}_{iy}^{\mathrm{CCUS}} \Big)$$

Impacts $\forall i \in \mathbb{I},\ y \in \mathbb{Y}$

	Capital	O&M
Wind turbines	$\mathscr{I}_{iy}^{\mathrm{W}} = \varsigma \sum_{z} D_y^{\mathrm{C}} C_{iy}^{\mathrm{WT}} NI_{zy}^{\mathrm{WT}}$	$\mathscr{I}_{iy}^{\mathrm{w}} = \varsigma \sum_{z} D_y^{\mathrm{OM}} F_{iy}^{\mathrm{WT}} N_{zy}^{\mathrm{WT}}$
Conversion techs	$\mathscr{I}_{iy}^{\mathrm{P}} = \varsigma \sum_{pz} D_{piy}^{\mathrm{C}} C_{piy}^{\mathrm{P}} NI_{pzy}^{P}$	$\mathscr{I}_{iy}^{\mathrm{p}} = \varsigma \sum_{pz} D_y^{\mathrm{OM}} F_{piy}^{\mathrm{P}} N_{pzy}^{P}$
Transport techs	$\mathscr{I}_{iy}^{\mathrm{Q}} = 0.5\varsigma \sum_{bzz'} D_{biy}^{\mathrm{C}} C_{biy}^{\mathrm{B}} NI_{bzz'y}^{B} d_{zz'})$	$\mathscr{I}_{iy}^{\mathrm{q}} = 0.5\varsigma \sum_{bzz'} D_y^{\mathrm{OM}} F_{biy}^{\mathrm{B}} N_{bzz'y}^{B} d_{zz'}$
Storage techs	$\mathscr{I}_{iy}^{\mathrm{S}} = \varsigma \sum_{sz} D_{siy}^{\mathrm{C}} C_{siy}^{\mathrm{S}} NI_{szy}^{S}$	$\mathscr{I}_{iy}^{\mathrm{s}} = \varsigma \sum_{sz} D_y^{\mathrm{OM}} F_{siy}^{\mathrm{S}} N_{szy}^{S}$

Primary resource production
$$\mathscr{I}_{iy}^{\mathrm{r}} = \varsigma \sum_{zhdt} D_y^{\mathrm{OM}} \left(RP_{izty}^{\mathrm{Bio}} A_{zy}^{\mathrm{Bio}} Y_{zty}^{\mathrm{Bio}} + \sum_{r \neq \mathrm{Bio}} RP_{rizhdty} U_{rzhdty} n_h^{\mathrm{hd}} n_d^{\mathrm{dw}} n_t^{\mathrm{wt}} n_y^{\mathrm{yy}} \right)$$

Export/import
$$\mathscr{I}_{iy}^{\mathrm{x}} = \varsigma \sum_{rzhdt} D_y^{\mathrm{OM}} V_{riy}^{\mathrm{X}} X_{rzhdty} n_h^{\mathrm{hd}} n_d^{\mathrm{dw}} n_t^{\mathrm{wt}} \qquad \mathscr{I}_{iy}^{\mathrm{m}} = \varsigma \sum_{rzhdt} D_y^{\mathrm{OM}} V_{riy}^{\mathrm{M}} M_{rzhdty} n_h^{\mathrm{hd}} n_d^{\mathrm{dw}} n_t^{\mathrm{wt}}$$

Revenue
$$\mathscr{I}_{iy}^{\mathrm{Rev}} = \varsigma \sum_{rzhdt} D_y^{\mathrm{OM}} V_{riy}^{\mathrm{Rev}} D_{rzhdty} n_h^{\mathrm{hd}} n_d^{\mathrm{dw}} n_t^{\mathrm{wt}}$$

UNIVERSITY OF BATH

MODELLING OF UK CO_2 VALUE CHAINS

Modelling of UK CO_2 Value Chains

Given:

- Spatio-temporal demands for domestic heat and electricity
- Spatio-temporal supplies of wind, natural gas, and biomass (Miscanthus)
- Carbon capture, storage and utilisation technologies
- What might the optimal 2050 network be?
- 12 timesteps per day, 1 day per season, 4 seasons per year, 1 year type to 2050

Carbon Pathways

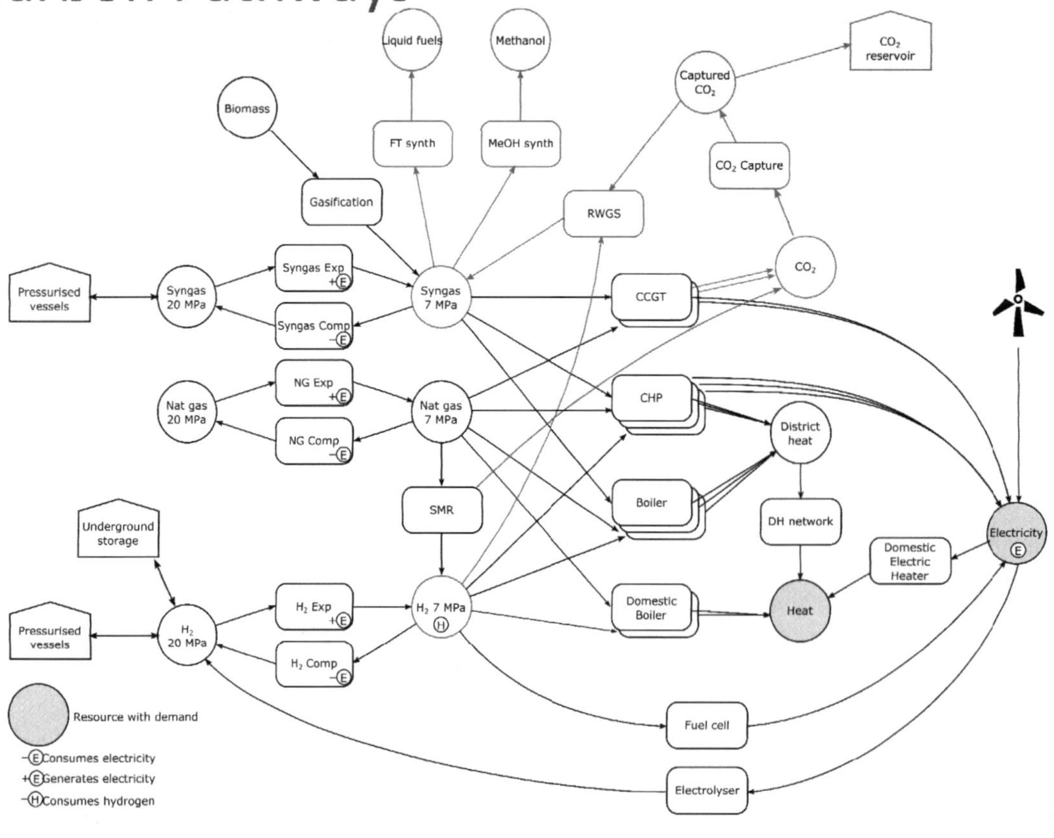

Carbon Storage

- 4 potential sequestration sites included
- Sites associated with nearest zone
- Model chooses whether to build CO_2 pipelines between zones

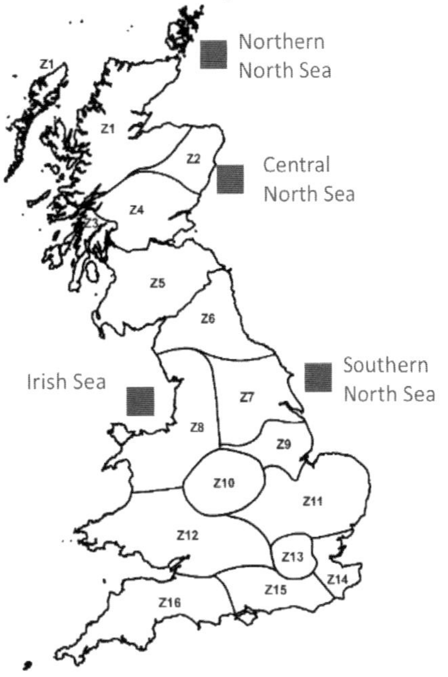

Baseline Case: Maximise NPV

- Wind turbines favoured over CCGTs due to long time horizon and production cost of natural gas
- Neither CCU nor CCS is profitable

NPV = £143bn (profit)
CO_2 = 1565 $MtCO_2$

Wind Turbine Installed Capacity (MW @10m/s)

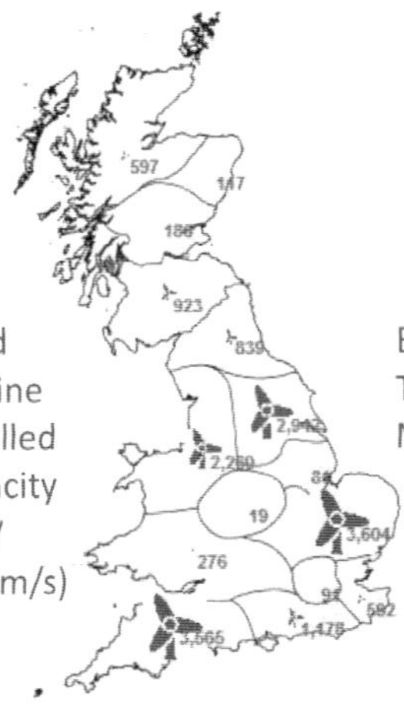

Electricity Transmission Network

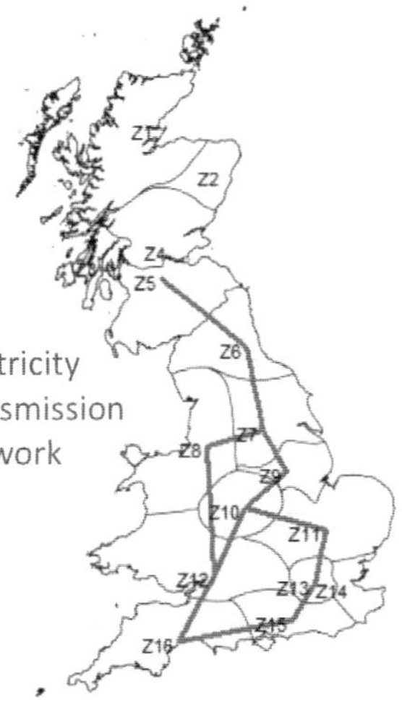

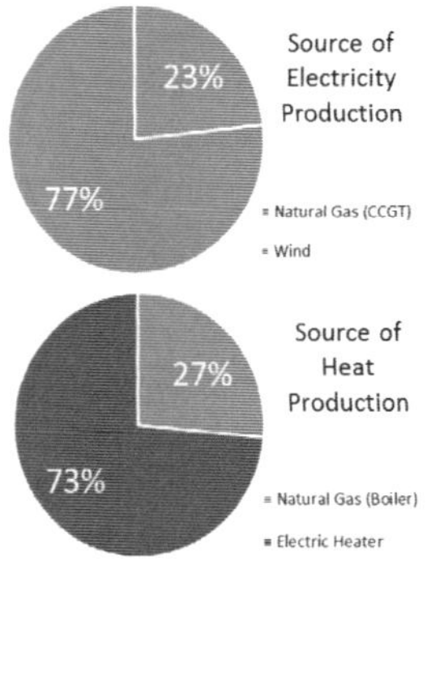

Constraint on Carbon Emissions

- CO_2 constrained to equivalent of 2050 CO_2 target (compared to baseline run)
- Model chooses CCU (MeOH synthesis) – no CCS
- All H_2 produced by electrolysis – some H_2 storage also built
- CO_2 pipeline network built
- CCU focussed in a few zones

NPV = £111bn (profit)
CO_2 = 508 $MtCO_2$

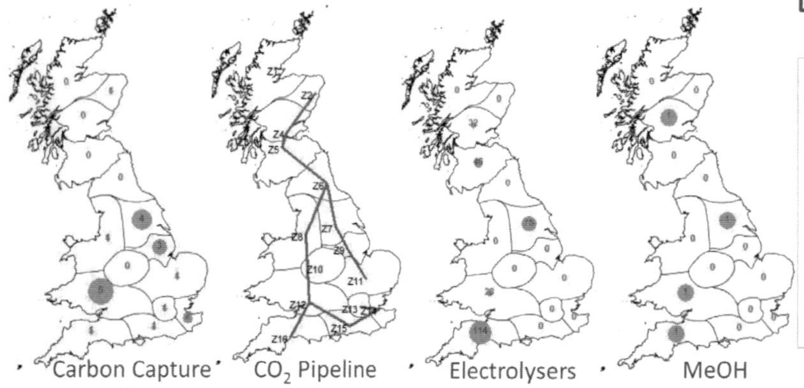

Carbon Capture Facilities · CO_2 Pipeline Network · Electrolysers (70MW) · MeOH Synthesis Plants

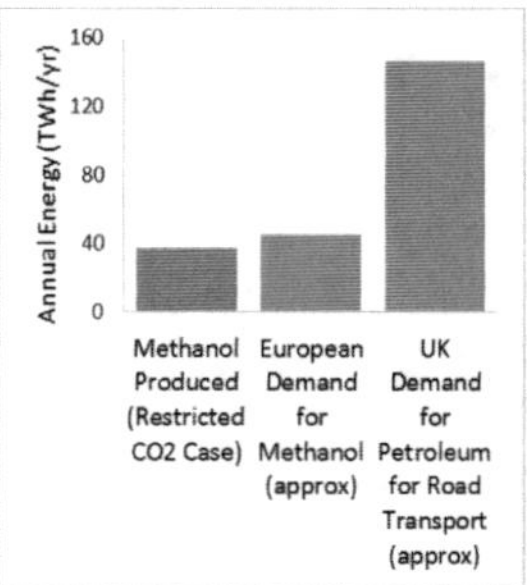

Carbon Credit

- No constraint on emissions
- £100 credit for each ton of CO_2 captured and utilised or stored
- CCS & CCU (MeOH synthesis) both used
- H_2 produced predominantly by SMR - CO_2 and H_2 pipeline networks both built
- Small overall CO_2 benefit given the large credit – due to SMR

NPV = £231bn (profit)
CO_2 = 1291 $MtCO_2$

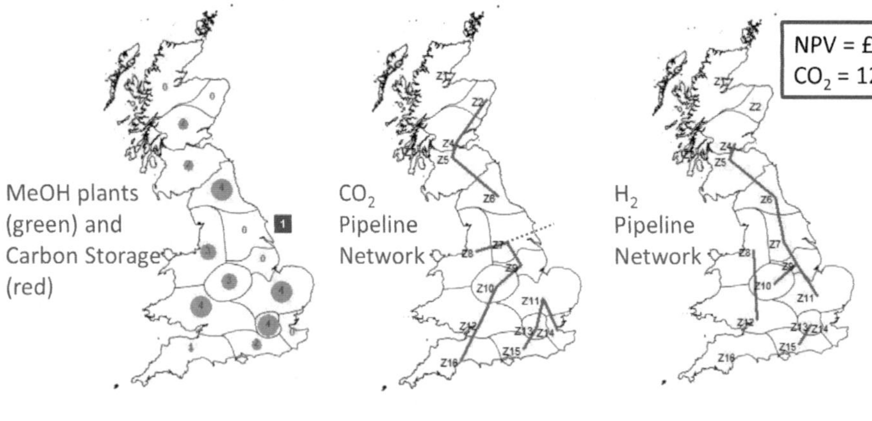

UNIVERSITY OF BATH

Inflated CCU Product Prices

- No carbon credit, no constraint on emissions
- MeOH & Synthetic Hydrocarbons prices doubled (compared to market price)
- CCU (MeOH synthesis) focussed in a few zones – no CCS
- SMR and electrolysis both used for H_2 production
- CO_2 levels are higher than the baseline – due to SMR

NPV = £167bn (profit)
CO_2 = 1617 $MtCO_2$

MeOH plants built

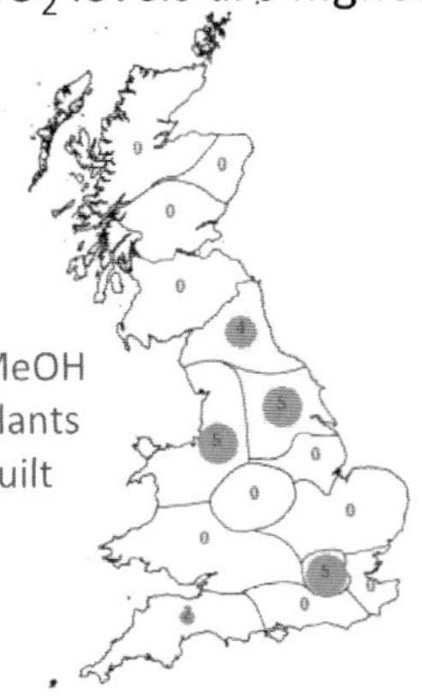

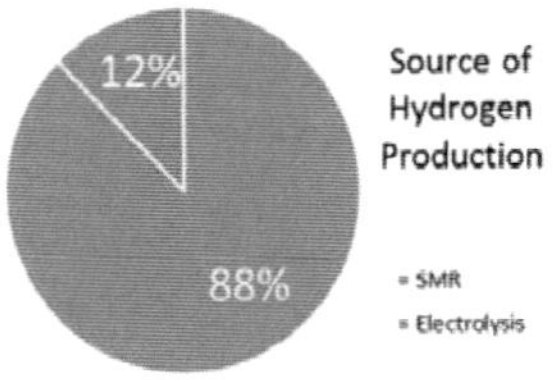

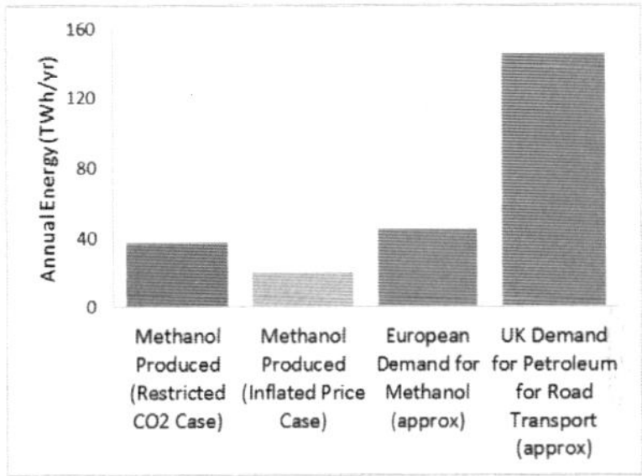

UNIVERSITY OF BATH

Conclusion

- Value Web Model a powerful framework for integrated multi-vector energy networks
  - Spatio-temporal nature matches up primary resource availability and energy service demands which are distributed in space and time
- CCU and CCS both have potential for carbon mitigation
  - CCU has economic advantage over CCS
  - CCU may become economically valid independent of environmental benefits
  - Current CCU products market may be too small but has expansion potential
  - Important to choose policy tools correctly to encourage genuine reductions in total carbon emissions

UNIVERSITY OF BATH

Acknowledgments

We would like to acknowledge the Department for Business, Energy and Industrial Strategy and the Engineering and Physical Sciences Research Council for their support and funding of this work.

Please note that the work presented here represents the views of the authors only and not necessarily those of the sponsors.

EPSRC
Engineering and Physical Sciences Research Council

Department for Business, Energy & Industrial Strategy

497306 Economic and Energetic Analysis of Biofuel Supply Chains

Abstract:

Economic and Energetic Analysis of Biofuel Supply Chains

Rex T. L. Ng[1,2] and Christos T. Maravelias[1,2]

[1]*Department of Chemical and Biological Engineering and* [2]*DOE Great Lakes Bioenergy Research Center, University of Wisconsin-Madison, Madison, WI*

The efficient and effective transportation of lignocellulosic biomass is one of the critical challenges in biofuel production, due to its low bulk density and durability properties. Generally, in a biofuel supply chain (SC), baled biomass is either transported to a biorefinery directly, or sent to a depot for densification prior to transporting densified biomass to a biorefinery [1]. Many studies show the installation of depots in biofuel SCs offer better economic performance in large scale biorefineries [2–4]. However, from an energy point of view, depots may be infeasible if the total primary energy input (e.g., biomass harvesting and collection, depot operation, and baled and densified biomass transportation) is higher than the energy content of biofuel.

To better understand what supply chain configurations are preferable under different scenarios, we develop systematic methods to compare the cost and energy efficiency of biofuel SC under different configurations. We define a configuration as a combination of depot selection (with and without depots) and transportation modes (truck and rail). The analysis aims to provide some insights into the following questions: *When should we ship biomass directly to the biorefinery? When depot is economically and energetically feasible? Which transportation mode yields low cost and high energy efficiency? What is the trade-off between the improvement of densification technology and cost or energy efficiency?*

In this analysis, we consider corn stover which is collected from 12 Midwest states in the US. We assume that corn stover productivity is on an annual basis. A high moisture pelleting process [5] is employed at the depot, while dilute acid pretreatment, separate hydrolysis and fermentation are employed at the biochemical biorefinery [6]. Note that the analysis can be repeated using the proposed method for any feedstock, technology, and type of productivity (annual, seasonal or monthly basis). We estimate the minimum ethanol selling price (MESP) and total energy input based assuming the feedstock collection area to be circular with the biorefinery or the depot located at its center. When a combination of two or more configurations leads to better economic or energy benefits (referred to in this work as 'hybrid configuration'), the average distance from depots in the annular area to the biorefinery at the center is estimated.

The results show what reductions in cost and energy input can be achieved through a hybrid configuration (Fig. 1A). From an economic point of view, direct shipment of baled corn stover (orange region) is preferred if the distance between biorefinery and harvesting site is lower than 155 km. Depots installation and truck transportation of densified corn stover (blue region) are favorable if the distance ranges between 155 to 260 km; while rail transportation of densified corn stover (red region) will only lead to better economic benefits at larger distance (=260 km). From an energy point of view, rail transportation of densified biomass is always better than truck transportation. The depot installations with rail transportation have lower total energy input values than those with truck shipment of baled biomass if the distance is greater than 170 km. We further discuss key cost and energy efficiency drivers: corn stover productivity, biorefinery size, and densification efficiency. The MESP decreases with the increase of productivity and biorefinery size (Fig. 1B). We will show how the improvements in densification technologies can reduce transportation cost and energy consumption.

Finally, we will briefly discuss how the finding of this study can be used to formulate tractable mixed-integer programming (MIP) models for biofuel SCs operational planning, which account for non-uniform biomass availability and other practical considerations. We will show results from two case studies in southern Wisconsin.

C2

Figure 1: (A) Hybrid configuration, and (B) MESP as a function of biomass productivity and biorefinery capacity.

References:

[1] R.T.L. Ng, C.T. Maravelias, Design of biofuel supply chains with variable regional depot and biorefinery locations, Renew. Energy. 100 (2017) 90–102.

[2] D.J. Muth, M.H. Langholtz, E.C.D. Tan, J.J. Jacobson, A. Schwab, M.M. Wu, et al., Investigation of thermochemical biorefinery sizing and environmental sustainability impacts for conventional supply system and distributed pre-processing supply system designs, Biofuels, Bioprod. Biorefining. 8 (2014) 545–567.

[3] S. Kim, B.E. Dale, Comparing alternative cellulosic biomass biorefining systems: Centralized versus distributed processing systems, Biomass and Bioenergy. 74 (2015) 135–147.

[4] P. Lamers, E.C.D. Tan, E.M. Searcy, C.J. Scarlata, K.G. Cafferty, J.J. Jacobson, Strategic supply system design – a holistic evaluation of operational and production cost for a biorefinery supply chain, Biofuels, Bioprod. Biorefining. 9 (2015) 648–660.

[5] J.S. Tumuluru, K.G. Cafferty, K.L. Kenney, Techno-economic analysis of conventional, high moisture pelletization and briquetting process, in: Am. Soc. Agric. Biol. Eng. Annu. Int. Meet. 2014, ASABE 2014, 2014: pp. 4177–4189.

[6] D. Humbird, R. Davis, L. Tao, C. Kinchin, D. Hsu, A. Aden, et al., Process Design and Economics for Biochemical Conversion of Lignocellulosic Biomass to Ethanol: Dilute-Acid Pretreatment and Enzymatic Hydrolysis of Corn Stover, Natl. Renew. Energy Lab. (NREL), Golden, CO. (2011).

489815 Synthesis of Sustainable Processing Networks: Location-Dependent Biorefinery Models

Abstract:

The current and projected increase in demand of commodities, water and energy caused by the growing world population motivates innovation in process synthesis-design towards more sustainable processes. This is achieved by using alternative (renewable) raw materials, such as biomass, incorporating new process technologies and satisfying new design objectives and constraints, including sustainability.

This work describes the development, modeling and optimization of a comprehensive biorefinery superstructure with multiple product and feedstocks. Over 100 processing alternatives for converting the mentioned feedstocks into the various products are included, leading to over a hundred trillion theoretical alternative solutions.

A systematic process synthesis framework for biorefineries has been developed (Bertran et al, 2017), alongside the necessary methods and tools (such as a comprehensive database of biomass, chemicals, reaction paths, technologies, prices and many more), and implemented in a software interface named Super-O. This framework has been applied in this contribution to solve various problems derived from the biorefinery superstructure, under different scenarios. These include: (i) selection of a product or a set of products given the feedstock mix; (ii) selection of a feedstock or set of feedstocks given the product mix; (iii) selection of the processing route for a given feedstock-product combination; (iv) selection of the location given the feedstock, product and route; (v) all possible combinations of the aforementioned options. Moreover, multiple location options are investigated, by incorporating transportation options between each location, to determine the optimal distributed location-configuration for the entire value chain. Using integer cuts, a number of top-ranked solutions are generated and analyzed.

In this contribution, a strong emphasis is made on the management of the problem data through a data structure implemented as a database. Upon a specific problem definition, the relevant data is retrieved from the database and the problem is then solved via optimization.

The production of bioethanol alongside co-products (such as succinic acid, lactic acid, and lysine) from various lignocellulosic feedstocks is used as a case study to determine the optimal feedstock mix, processing route, set of locations and configuration.

References:

- M.-O. Bertran, R. Frauzem, A.-S. Sanchez-Arcilla, L. Zhang, R. Gani, Comput. Chem. Eng. (2017) DOI: 10.1016/j.compchemeng.2017.01.030.

500097 Electrochemistry As a Sustainable Alternative for Distributed Processing of Biomass

Abstract:

Electrochemical processes use electrons, a green reagent, as main driving force for the conversion of chemical energy into electrical energy of vice versa. Conventional electrochemical technologies (i.e. fuel cells and batteries) are well developed and are ideal for distributed energy generation and storage. Recent developments have extended scope to include stabilization and energy densification of biomass feedstock to produce transportation fuels. Among the main advantage of electrochemical technologies for distributed processes are that it can features rapid start up and turn off, which enable them to operate either in batch or continues operations and be directly fed by intermittent renewable energy sources, such as solar and wind. In the specific case of distributed biomass processing, electrochemical processes can be implemented the conversion of generated CO_2, upgrading of lignocellulose biomass, and biomass stabilization.

In this work the applicability of electrochemical technologies on distributed biomass processing schemes will be contextualized. Current state of the art and potential impacts will be assessed based on technology readiness, roadblocks, and viability. Case studies will be presented for the electrochemical reduction of CO_2, and the electrochemical upgrading/stabilization of biomass derived streams.

AUTHOR INDEX

AUTHOR INDEX

AUTHOR INDEX

AUTHOR INDEX

AUTHOR INDEX

AUTHOR INDEX

AUTHOR INDEX

AUTHOR INDEX

AUTHOR INDEX

AUTHOR INDEX

AUTHOR INDEX

AUTHOR INDEX

AUTHOR INDEX

AUTHOR INDEX

AUTHOR INDEX

AUTHOR INDEX

AUTHOR INDEX

AUTHOR INDEX

International Congress on Energy 2017

Topical Conference at the 2017 AIChE Annual Meeting

Minneapolis, Minnesota, USA
29 October – 3 November 2017

Volume 2 of 2

ISBN: 978-1-5108-5776-6

Printed from e-media with permission by:

Curran Associates, Inc.
57 Morehouse Lane
Red Hook, NY 12571

Some format issues inherent in the e-media version may also appear in this print version.

Printed by Curran Associates, Inc. (2018)

For permission requests, please contact AIChE
at the address below.

AIChE
120 Wall Street, FL 23
New York, NY 10005-4020

Phone: (800) 242-4363
Fax: (203) 775-5177

www.aiche.org

Additional copies of this publication are available from:

Curran Associates, Inc.
57 Morehouse Lane
Red Hook, NY 12571 USA
Phone: 845-758-0400
Fax: 845-758-2633
Email: curran@proceedings.com
Web: www.proceedings.com

TABLE OF CONTENTS

VOLUME 1

VOLUME 2

Some abstracts in this book refer to images that were not made available by the publisher

499969 Assessing Fuel and Feedstock Energy Use in the U.S. Chemical Sector: A Supply Chain Analysis

Abstract:

Assessing Fuel and Feedstock Energy Use in the U.S. Chemical Sector: A Supply Chain Analysis Scott Nicholson Alberta Carpenter Rebecca Hanes

In chemical manufacturing, fossil fuels are consumed both as an energy source and as feedstocks that are converted to higher-value chemicals. This work expands on the U.S. Department of Energy's chemical manufacturing bandwidth study, which quantified the potential for reducing direct energy consumption in the chemicals sector through the adoption of the most energy efficient manufacturing technologies available [1]. The scope of the bandwidth study was limited to fuels consumed within the plant boundary and fuels consumed for electricity generation. Fuel consumed upstream in the supply chain and fuel consumed as feedstock (defined as fuels that are converted to chemical intermediates rather than being used as an energy source) were not quantified; by one estimate, 60% of the fossil fuels consumed in the U.S. chemical manufacturing sector are consumed as feedstocks [2]. The present study quantifies, in addition to process energy, energy consumed upstream in the supply chain and energy embedded in fossil feedstocks and identifies chemical supply chains in which an understanding of the supply chain can help to identify opportunities for reducing total fossil fuel consumption. Alternative biomass-based processes, which require no fossil feedstocks but do not necessarily consume less energy, are then analyzed to determine if the alternative processes can reduce the total fossil fuel consumption in the identified supply chains. Reducing the chemical industry's reliance on fossil fuels could lead to greater long-term profitability in the industry, as crude oil prices are projected to rise steadily over the coming decades [3].

The Materials Flows through Industry (MFI) supply chain modeling tool, developed at the National Renewable Energy Laboratory, is employed to quantify supply chain energy and feedstock requirements for the chemicals analyzed previously in the bandwidth study. The MFI tool is a flexible framework for modeling material and energy usage in the supply chains of commodities such as chemicals, primary metals, and paper products [4]. MFI contains U.S.-specific data on energy and material inputs to 1,469 manufacturing processes for 707 commodities. U.S. commodities produced via multiple technologies are modeled using weighted averages of the current production technology mix based on market shares reported in the literature [5]. For this study, energy values are calculated at both the immediate process level and as a cumulative total for 10 steps up a chemical's supply chain. In MFI, a "step" consists of the material and energy inputs to the intermediate products manufactured in the previous step, which then become the products of the next step in the supply chain. Material and energy inputs beyond the 10th step have been shown to be negligible compared to the first 10 steps, so the supply chain calculations are truncated at that step.

MFI is similar to standard life cycle assessment (LCA) software packages, but the tool goes beyond these platforms in several ways. First, MFI is adaptable; processes and process weightings can be adjusted individually to model existing or future manufacturing scenarios. Modeling alternative production technologies in a particular supply chain is therefore straightforward. MFI also has the capability to disaggregate supply chain energy requirements into three types: process fuel, energy in fuels used as feedstocks, and fuel for electricity generation. For each chemical in the current analysis, the MFI model is used to calculate process and feedstock energy on a per-kilogram basis. These per-kilogram values are multiplied by annual U.S. production quantities from 2014 or later [5] to determine the energy requirements needed to produce the given chemical at the national scale. These calculations are repeated for alternative and biomass-based pathways for chemicals with feedstock-intensive supply chains, in order to compare the total fossil fuel consumption in the conventional and alternative supply chains. When comparing alternative and conventional petroleum production routes, the former may use significantly less feedstock energy, particularly for biomass-based production routes. However, alternative production routes may or may not use less process energy; in the case of advanced, non-commercial technologies, process energy is likely to increase and this increase may outweigh any savings in feedstock consumption. Both feedstock consumption and energy use must be quantified to determine which, if any, alternative technologies offer a supply-chain-level reduction in total fossil fuel consumption over the conventional technologies.

One of the chemicals in the current analysis is 1,3-butadiene, a chemical intermediate used predominantly in the production of synthetic rubber [6]. Essentially all commercial-scale production of butadiene in the U.S. derives from the byproduct streams of ethylene cracking [7]. Since ethylene is derived from fossil feedstocks, there are significant feedstock-embedded energy inputs in the conventional butadiene supply chain. Preliminary results show that supply chain feedstock consumption for 1,3-butadiene is 220 GJ/tonne, which exceeds both supply chain scale process energy consumption (21 GJ/tonne) and the onsite energy consumption reported by the bandwidth study (18 GJ/tonne) by a full order of magnitude. The alternative manufacturing processes for 1,3-butadiene are all biomass-based and begin with the fermentation of sugars [8]. These processes, because they do not involve ethylene cracking, are expected to reduce feedstock consumption in the 1,3-butadiene supply chain significantly. However, these processes have yet to be optimized for commercial scale production, and may show an increase in both onsite and supply chain energy consumption. The other chemicals analyzed in this study are caprolactam, para-xylene, and propylene glycol, all of which have biomass-based alternative processes under development [8].

This analysis quantifies the technological potential of alternative chemical manufacturing processes to reduce fossil fuel consumption. The economic implications of transitioning chemical manufacturing to next-generation energy-efficient and/or biomass-based production technologies are outside the scope of this study, as are the logistical complexities involved in shifting from a fossil-based to a biomass-based chemical sector. Still, the results of this study could be used to inform decisions around the commercialization of advanced manufacturing processes. They can be used to determine whether it is more worthwhile, from the perspective of reducing fossil fuel consumption, to upgrade an existing manufacturing process or to move to another technology entirely. Future extensions of this work may explicitly analyze the process economics of these alternative production technologies by incorporating costs of material and energy inputs. If these results show that the bio-based technologies can both reduce feedstock consumption *and* be economically competitive, then a stronger argument can be made for commercialization in the near future. In contrast, an alternative process that is significantly more expensive would still require further development. Lastly, the MFI tool accounts for energy use from resource extraction through production (cradle-to-gate); it does not currently look at use-phase or end-of-life energy requirements (i.e. the full cradle-to-grave). These latter stages are beyond the scope of this chemical sector analysis, since the downstream products manufactured from these commodity chemicals are used across many sectors —not just the chemical sector. With additional data, MFI has the ability to capture use-phase and end-of-life, and this is another potential area of future research.

References

[1] Brueske, S., Kramer, C., & Fisher, A. (2015). *Bandwidth Study on Energy Use and Potential Energy Saving Opportunities in US Chemical Manufacturing* (No. DOE/EE-1229). Energetics. Retrieved from https://energy.gov/sites/prod/files/2015/08/f26/chemical_bandwidth_report.pdf

[2] U.S. Department of Energy, Energy Information Administration. Office of Energy Consumption and Efficiency Statistics. *2010 Manufacturing Energy Consumption Survey*. https://www.eia.gov/consumption/manufacturing/data/2010/

[3] U.S. Department of Energy, Energy Information Administration. *Annual Energy Outlook 2017 with projections to 2050* (No. DOE/EIA- 0383(2017)). Retrieved from http://www.eia.gov/forecasts/ aeo/pdf/0383(2017).pdf.

[4] Hanes, R.J. & Carpenter, A. (2017) Evaluating opportunities to improve material and energy impacts in commodity supply chains. *Environ Syst Decis 37: 6*. doi: 10.1007/s10669-016-9622-5

[5] IHS (2017). *Chemical Economics Handbook (CEH)*. https://www.ihs.com/products/chemical-economics-handbooks.html.

[6] White, W. C. (2007). Butadiene production process overview. *Chemico-biological interactions, 166(1)*, 10-14. doi: 10.1016/j.cbi.2007.01.009

[7] Sriram, P., Hyde, B., & Smith, K. (2016). Butadiene. *Chemical Economics Handbook*. IHS.

[8] Biddy, M. J., Scarlata, C., & Kinchin, C. (2016) *Chemicals from Biomass: A Market Assessment of Bioproducts with Near-Term Potential*. National Renewable Energy Laboratory (NREL). Retrieved from http://www.nrel.gov/docs/fy16osti/65509.pdf

501692 Integration of Fast Pyrolysis and Electrolyzer for Deoxygenation of Biomass

Abstract:

Domestically, the US is challenged with improving the environment, energy security, and energy productivity, thus many efforts are concerted in utilizing biomass as a viable source of renewable liquid transportation fuels and contributing to a robust carbon-neutral energy production. However, achieving a high carbon, hydrogen and overall energy efficiency is an ongoing challenge for current biomass upgrading technologies. The main approach of upgrading biomass to hydrocarbon fuels is by hydroprocessing, which relies heavily on large hydrogen consumption and on large, more centralized production scales. In this work, we present a different approach of using distributed pyrolysis unit scale with low hydrogen requirement by utilizing electricity to drive the deoxygenation of molecules present in the biomass. Currently, we have demonstrated the feasibility of electrochemical deoxygenation using solid oxide electrolysis cell in small cells, multi-cell stacks, and in an integrated test with a bench-scale fast pyrolysis unit. The electrolysis process removes oxygen from the oxygenated organic molecule as well as from steam to produce hydrogen in-situ allowing for a distributed, small scale integrated upgrading unit. Specifically, we will report recent work with a fast pyrolysis unit with 1 kg/h ponderosa pine feed, 480 deg C fast pyrolysis temperature and 550-600 deg C process at the electrolyzer. These process conditions allowed us to achieve an improved quality of the bio-oil. In this presentation, we will also discuss the present challenges of electrolyzer integration with a fast pyrolysis unit as well as the characterization of products using ^{13}C NMR, ultimate analysis (CHNOS), and GC-MS.

498622 Durable, Self-Hydrating, Tungsten Carbide-Based Composite Polymer Electrolyte Membrane Fuel Cells

Abstract:

Proton conductivity of the polymer electrolyte membranes in fuel cell (PEMFC) dictates the performance and requires sufficient water management. Here, we introduce a simple scalable method to produce well-dispersed transition metal carbide (TMC) nanoparticles with narrow size distribution (3-5 nm). This protocol can be extended to the synthesis of other early transition metal carbides. Extensive characterizations including HRTEM, STEM tomography, SEM, XPS, XRD and TG are conducted.

WC nanoparticles incorporated into a recast Nafion membrane demonstrate significant improvement of the PEMFC power density compared to the baseline Nafion membrane. Based on both theoretical and experimental results, the WC composite membrane catalyzes the reaction of crossover H_2/O_2 and provides sufficient water management within the membrane, improving the proton conductivity. Tomographic studies using FIB-SEM indicate large pinholes forming around Pt nanoparticles and provide support that the membrane/Pt catalyst facilitates the production of radicals that leach from Nafion. Superior to Pt, incorporated WC nanoparticles in the membrane result in considerable durability of the membrane. DFT calculations provide insights into the mechanisms by which WC improves performance.

492100 Teaching an Old Electrocatalyst to Do New Tricks

Abstract:

Electrocatalysis underlies several key clean energy technologies such as fuel cells and electrolyzers that are expected to be key components of a sustainable energy future. However, continued improvements in electrocatalyst activity, durability, and selectivity must be achieved in order to make this future a reality. In this study, we explore the use of electrocatalyst architectures based on metallic electrocatalysts that have been encapsulated by ultra-thin silicon oxide overlayers. These oxide layers are synthesized with a room temperature process and deposited as continuous layers onto metallic thin films and nanoparticles. The catalytic metal used in these studies is Platinum (Pt), an "old catalyst" material that is one of the most commonly used electrocatalysts for electrolyzers and fuel cells. Inspired by past studies on glass microelectrodes[1] and a recent study from our lab on oxide-coated nanoparticles,[2] we show that well-defined oxide overlayers deposited on smooth metal substrates can be highly permeable to certain electroactive species and thereby enable efficient and selective electrocatalysis at the metal oxide / metal interface. By systematically varying the thickness of the oxide overlayer, we explore its transport properties and influence on the performance of thin film hydrogen evolution electrocatalysts. Our studies show that the properties of an "old catalyst" like Pt can be greatly altered, and perhaps tuned, through the use of the ultra-thin metal oxide overlayers, which can be thought of as nanomembranes that can selectively control transport of reactants and products. These results highlight the potential of this membrane-coated electrocatalyst architecture to serve as a tunable, efficient, and stable platform for a variety of electrochemical reactions.

References

1. J. Velmurugan, D. Zhan, and M. V. Mirkin, *Nature Chemistry*, 2, 498-502, 2010.

2. N. Y. Labrador, X. Li, Y. Liu, J. T. Koberstein, R. Wang, H. Tan, T. P. Moffat, and D. V. Esposito, *Nano Letters*, 16, 6452-6459, 2016.

493787 Using Surface Chemistry to Understand Aqueous-Phase Thermal Catalytic and Electrocatalytic Hydrogenation of Bio-Oil Model Substrates

Abstract:

One of the major challenges of the future is matching the increasing demand for energy using renewable resources. The energy density of hydrocarbons makes them ideal for transportation fuels, especially for aviation, where electric vehicles are not available. As the amount of energy transitions to renewable energy such as solar and wind, which are intermittent and potentially more decentralized, developing ways to store this energy also becomes important. Catalysis can play an integral role in the chemical processes required to produce renewable transportation fuels and store renewable energy. In this talk, I will discuss developments in catalysts used in a method to convert waste biomass to renewable fuels. The feedstock of this process is bio-oil, a product of pyrolysis or hydrothermal liquefaction of biomass. Bio-oil (modeled here using phenol and benzaldehyde) can be hydrogenated and then deoxygenated to yield fuel grade hydrocarbons.[1] Although hydrogen generally comes from steam methane reforming, which emits carbon dioxide, renewable electricity can be used to drive carbon-neutral (i) electrochemical hydrogen production (from water) which is then used for thermal catalytic hydrogenation (TCH), or (ii) direct electrochemical hydrogenation (ECH) of the bio-oil. Electrochemical routes are more economically competitive in decentralized applications and can make use of intermittent renewable electricity. The challenge of both electrochemical hydrogen production coupled with TCH and ECH is developing hydrogenation catalysts with high rates for commercial applications. The target is to achieve comparable rates (at approximately 80 °C) to existing commercial electrochemical processes, on the order of hundreds of mA per square centimeter (geometric current density) or turnover frequencies near 10 s^{-1}. In this work we discuss some advancements in the understanding of the aqueous-phase hydrogenation of phenol and benzaldehyde on different platinum group metals, and how this leads to improved rates for ECH and TCH.

For phenol hydrogenation, ECH and TCH on rhodium or platinum have similar activation barriers due to a shared rate determining step of adsorbed H adatoms reacting with adsorbed phenol. Using a flow cell designed for X-Ray Absorption Spectroscopy (XAFS), we probe Pt nanoparticles under reaction conditions for both near-edge (XANES) and extended X-ray absorption fine structure (EXAFS). This allows us to look at the catalyst state (from XANES) and surface coverages (from Pt-C scattering that we show is due to the adsorbed reactant species) under operating conditions. This spectroscopy, coupled with kinetic measurements, affords us a better understanding of the surface chemistry and its dependence on hydrogen pressure and applied potential. Using this understanding, we demonstrate methods to prevent catalyst deactivation at temperatures at 80 °C.[2] We also show how modifying the adsorption energy of the adsorbed species controls the reaction rate, and discuss how other reaction conditions (ex. pH) affect the surface coverages and rates. These results lead us to propose material properties that may result in more active electrocatalysts for phenol TCH or ECH, and how to reach reaction rates that may be necessary for a commercial ECH electrolyzer system.

For benzaldehyde, on metals such as platinum and rhodium, TCH and ECH have different benzaldehyde rate order dependence, unlike what is observed for phenol. This may be due to a difference in the reaction mechanism for the ECH and TCH of benzaldehyde on Pt and Rh, or surface coverages under operating conditions. Using XAFS (as well as traditional cyclic voltammetry), we try to understand how the surface coverage of the species may be responsible for the observed reaction rates and orders. We connect this with kinetic measurements and adsorption energies in an attempt to understand how the surface chemistry controls the reaction rates that are observed.

References

(1) Zhao, C.; He, J.; Lemonidou, A. A.; Li, X.; Lercher, J. A. *J. Catal.* **2011**, *280*(1), 8.

(2) Singh, N.; Song, Y.; Gutiérrez, O. Y.; Camaioni, D. M.; Campbell, C. T.; Lercher, J. A. *ACS Catal.* **2016**, *6*, 7466.

498568 Design of Heterogeneous Catalysts for High Temperature Electrochemical Reduction of CO_2

Abstract:

The high levels of carbon dioxide (CO_2) emissions, consequential from the extensive use of fossil fuels, are major contemporary challenges. An approach to alleviate this process is to activate the reverse chemical pathways in which CO_2 is reduced into high-energy molecules. The electrochemical reduction of CO_2 to CO has attracted increasing attention, since CO represents a valuable intermediate to the production of synthetic fuels using established processes, such as Fischer-Tropsch. The use of solid oxide electrolysis cells (SOECs) to electrochemically reduce CO_2 to CO is an attractive technology, given that this solid-state electrochemical systems can, in principle, facilitate the CO_2 reduction with potentially very high rates due to the favorable kinetics at high operating temperatures. While electrochemical reduction of CO_2 using SOECs offers a great deal of promise, these systems are still limited by the high activation overpotential losses induced by the sluggish kinetics of CO_2 reduction at the cathode. In the present work, we combine experimental and theoretical techniques to understand the chemical/electrochemical process that governs CO_2 electrolysis, and develop a structure/performance relation in order to identify the optimum electrocatalyst for this process. Controlled electrochemical experiments show that upon modification of the electrode metal electrocatalyst, significant decrease in the overpotential is achieved.

REFERENCES

1. Ebbesen, S. D., Jensen, S. H., Hauch, A., & Mogensen, M. B. (2014). High temperature electrolysis in alkaline cells, solid proton conducting cells, and solid oxide cells. *Chemical reviews*, 114(21), 10697-10734.

2. Gu, X. K., Carneiro, J., Nikolla, E. (2017). Electrolysis on Carbon Dioxide. In Catalysis (pp. 94-116). The Royal Society of Chemistry.

3. Gu, X. K., & Nikolla, E. (2017). First-Principles Study of High Temperature CO_2 Electrolysis on Transition Metal Electrocatalysts. *Industrial & Engineering Chemistry Research*, under review 2017.

494155 Enhancing Methanol Electrooxidation Activity Using Double Oxide Catalyst Support

Abstract:

Electrooxidation of methanol was investigated on platinum catalyst supported on a double oxide nanocomposite of tin oxide clusters and carbon-doped titanium dioxide nanocoatings on carbon nanotubes. The electrocatalyst demonstrated a higher forward peak current density in methanol eectrooxidation than the electrocatalyst with a single tin oxide or just carbon as support. Formation kinetics of the oxygenated species (-OH) groups was dramatically enhanced in the catalyst with double oxide support. This leads to a much lower onset potential for adsorbed CO oxidation with an enhanced bifunctional effect in tin oxide. Doping with carbon in the titanium oxide enables backward electron donation, making supported platinum more electron negative and stable. Binding energy shifts in oxidation states was observed for the double oxide support catalyst and is attributed to strong metal support interactions. This study demonstrated a synergistic effect of combining bifunctional and electronic mechanisms with double oxide support in the enhancement of methanol electrooxidation.

501794 Study of Impact of Electrode Catalyst Loadings on Thin Flexible Fuel Cell (TFFC) Performance

Abstract:

Thin Flexible Fuel Cell (TFFC) is a flat and flexible device without any rigid support capable of being integrated into portable electronics. TFFC uses hydrogen and oxygen as fuel to provide high volumetric and gravimetric energy densities capable of competing with battery technology as a power source. The group is currently investigating a TFFC with an air-breathing design that maximizes the cathode interface with ambient air. The current challenges of making TFFC commercialized are the cost and performance. Enhanced cell performance is critical for a competitive cost of the device. The group investigated the impact of catalyst loading on the cell performance to increase the overall performance of the device. The catalyst loading of the cathode electrode was varied while keeping the anode catalyst loading constant and vice versa. The effects of the various catalyst loadings were characterized using various analytical techniques. Polarization curves and Electrochemical Impedance Spectroscopy (EIS) were used to separate and quantify the sources of losses for the TFFC. Power curves provided the maximum power density of the device. Cyclic voltammetry (CV) was implemented to characterize the electrochemically active surface area (ECSA) and utilization of the various catalyst layers. Rotating Disc Electrodes (RDE) tests were performed for a direct assessment of the catalyst. We will report the results of these assessments and summarize the resulting guidance for optimizing catalyst loading on similar fuel cell structures.

501059 Transition Metal Modified Graphitic Mesoporous Carbon for PEMFC Application

Abstract:

Proton exchange membrane fuel cell (PEMFC) has drawn considerable attention due to its high power density, zero emission and fast startup features. In PEMFC, platinum metal is the catalyst of choice in anode and cathode electrodes. The cost of Pt is a significant issue in commercialization of PEMFC. Further due to its sensitivity to CO, PEMFC requires high purity hydrogen as fuel. Conventional catalyst support (activated carbon) lacks in adequate porous structure, which results in underutilization of Pt metal particles. To better utilize the Pt-catalyst, we synthesized high surface area metal modified mesoporous carbon using hard template-based (SBA-15) method. Highly porous structure of this carbon allows well-dispersed Pt metal loading ensuring its proper utilization. Transition metals (Ni, Fe and Co) and Ni-Fe alloy were used for catalytic graphitization of the carbon. Resulting carbon showed graphitic structure in X-ray diffraction (XRD) and Raman spectroscopy. Electrocatalysts synthesized using these metal modified carbons were tested on the fuel cell. The structural characteristics of these graphitic mesoporous carbon and their performance in PEMFC will be presented in this paper.

488240 Mechanistic Modeling of Fast Pyrolysis of Hemicellulose

Abstract:

Hemicellulose is one of the major components of lignocellulosic biomass and has potential for the production of renewable drop-in transportation fuels and multiple commodity chemicals.[1] Previously, however, the successful modeling of fast pyrolysis of hemicellulose was hampered by lumped kinetic models, which failed to describe the heterogeneous structure of hemicellulose and cannot predict the bio-oil composition. In this work, a structural model for hemicellulose extracted from corn stover that captured the experimentally measured properties was proposed, and a mechanistic model for hemicellulose pyrolysis was constructed based on the reaction family approach that we used for cellulose pyrolysis[2]. The model described the decomposition of hemicellulose chains, reactions of intermediates, and formation of a range of low molecular weight compounds at the mechanistic level. Rate constants were specified for each elementary reaction with Arrhenius format. Overall, more than 500 reactions of 144 species were included in the mechanistic model for fast pyrolysis of hemicellulose. The mechanistic model was able to closely match experimental yields of all the major products (with yield ≥ 1 wt%) from hemicellulose pyrolysis at 500 °C reported by Shanks and coworkers[3]at Iowa State University. The model predicted about 4 s residence time for complete thermo-conversion of hemicellulose. Model results showed that both degree of polymerization (DP) and polydispersity index (PDI) of hemicellulose had an insignificant effect on the pyrolysis products distribution. The mechanistic model of extracted hemicellulose is extendable to fast pyrolysis of native hemicellulose, which will be also discussed in this work.

[1] X. Zhou, W. Li, R. Mabon, L. J. Broadbelt, *Energy Technol. 2016, 5, 52 –79.*

[2] a) R. Vinu, L. J. Broadbelt, *Energy Environ. Sci.* **2012**, *5*, 9808-9826; b) X. Zhou, M. W. Nolte, H. B. Mayes, B. H. Shanks, L. J. Broadbelt, *Ind. Eng. Chem. Res.* **2014**, *53*, 13274–13289; c) X. Zhou, M. W. Nolte, H. B. Mayes, B. H. Shanks, L. J. Broadbelt, *AIChE J.* **2016**, *62*, 766–777.

[3] J. Zhang, Y. S. Choi, C. G. Yoo, T. H. Kim, R. C. Brown, B. H. Shanks, *ACS Sustainable Chem. Eng.* **2015**, *3*, 293-301.

496698 Fast Pyrolysis Kinetic Study for Ten Ecuadorian Agricultural Residual Biomass Samples

Abstract:

Fast pyrolysis of biomass is a promising technology to produce fuels, commodity chemicals as well as fine chemical from renewable sources. Due to its CO2 neutrality replacing fossil resources by biomass helps to mitigate global warming. The thermochemical route of biomass conversion has the advantage that any type of organic matter can be used. Ecuador's economy is mainly based on the export of primary agricultural products, which creates large amounts of currently unused waste streams. In order to assess the feasibility of these waste materials for thermal conversion to valuable products, laboratory fast pyrolysis data are needed. Although numerous TGA studies are available in the literature, detailed product composition analyses obtained for pyrolysis at *high* heating rates are scarce. A new micro-pyrolysis setup was built to address this gap. The setup consists of a single-stage Frontier micro-pyrolyzer reactor which is connected to a comprehensive GCxGC coupled to a flame ionization detector (FID) and a time-off-flight mass spectrometer (TOF-MS). This configuration allows for the simultaneously identification and quantification of large molecules. In addition to the GCxGC, a customized GC is used to detect permanent gases, light oxygenated hydrocarbons, water and C2 hydrocarbon molecules.

The product spectra of fast pyrolysis of 10 biomass residues selected from the main agricultural processes in Ecuador are reported. These residues are: forestry residues from wood processing, palm oil production wastes (rachis, fiber and kernel shells), cocoa recollection (fresh and rotten cocoa kernel), coffee husks, banana stem, rice husk and sugar cane bagasse. The chosen set of biomass sample covers a wide range of cellulose, hemicellulose and lignin contents. Three pyrolysis temperatures are studied (500 ^{o}C, 550 ^{o}C and 600 ^{o}C) and quantitative products yields are obtained using either calibration factors of pure compounds or the effective carbon number approximation. The obtained experimental product distributions are compared to predictions using a literature biomass pyrolysis mechanism.

Acknowledgements

The SBO proposal "Bioleum" supported by the Institute for promotion of Innovation through Science and Technology in Flanders (IWT) is acknowledged. SG acknowledges the financial support from the Research Board of Ghent University (BOF). This research has been supported by the Belgian Development Cooperation through VLIR-UOS. VLIR-UOS supports partnerships between universities and university colleges in Flanders (Belgium) and the South looking for innovative responses to global and local challenges. Visit www.vliruos.be for more information. The research leading to these results has received funding from the European Research Council under the European Union's Seventh Framework Programme (FP7/2007-2013) / ERC grant agreement n° 290793.

494767 Biomass Pyrolysis: Can a Single Severity Factor Describe the Effect of Pyrolysis Conditions on the Final Biochar Product?

Abstract:

The amendment of soils with biochar has been heralded as a sustainable method for increasing crop yield and preventing pollution problems caused by fertilizer runoff. However, the available experimental results are highly variable. While some studies report significant beneficial effects with up to 60% higher yield after biochar addition, other studies report that biochar amendments either has no net effect or can even lower agricultural productivity by up to 30% [1]. A major reason for this variability is biochar heterogeneity. Numerous biomass feedstocks were used in these studies to produce biochars in multiple types of reactors under varying temperature and oxygen conditions, leading to thousands of biochars with widely varying properties.

Understanding the heterogeneity of biochars is a major challenge for this research. Although the maximum temperature is often used as a metric of the charring intensity or severity of slow pyrolysis conditions, the literature is full of studies reporting that multiple factors control the properties of biochars produced from the same feedstock. These factors include the peak pyrolysis temperature, the treatment time at that temperature, the heating rate, the particle size, the presence or absence of oxygen and others. Since not all these conditions are carefully controlled or even reported, it becomes almost impossible to compare biochars produced under different environmental conditions or in different laboratories. And, the problem becomes even more difficult when different feedstocks are used.

This study presents an approach that combines experimental results and reaction modeling to relate some key properties of biochars to charring intensity metrics that combine (a) the full pyrolysis temperature history and (b) the kinetics of pyrolysis reactions. To achieve very accurate control of the temperature history of our biochars, we used a thermogravimetric analyzer (TGA) to produce 71 biochar samples from slash pine wood under nitrogen and a variety of slow pyrolysis conditions. Two different particle sizes (210-250 and 850-1000 microns) were used and the initial amount of feedstock varied between 130 and 230 mg. Proximate analysis was used to determine the fixed carbon content, volatiles and ash of the produced biochars. Additional TGA experiments were carried out to identify the network of pyrolysis reactions and estimate the kinetic constants. After running a series of step isothermal and/or constant heating rate TGA runs, the weight vs. time data sets were post-processed with an optimization technique to estimate (a) the reaction network (i.e. number of parallel and sequential pyrolysis reactions), and (b) the kinetic constants (i.e. activation energy, pre-exponential factor and order) of each reaction.

As expected, the peak pyrolysis temperature was not a good metric for characterizing total biochar yield, volatiles or fixed carbon content. Fixed C showed a good correlation with peak temperature only for the longest processing times (3 hours) used in our experiments. The charring intensity metric (CITT) defined as the integral of the temperature history $T(t)$ of pyrolysis [2] also did a rather inadequate job in describing total biochar yield, volatiles or fixed carbon content. However, much better correlations were obtained when biochar properties were plotted vs. the charring severity function (CSF) defined as the integral over temperature of the kinetic rate constant of the reaction leading to the formation of char, the highly aromatic and crystalline biochar component that is the final product of the pyrolysis reactions.

Numerical simulations with the pine reaction network revealed that the same value of the CITT metric (the temperature-based charring intensity) can give us multiple concentrations of char and, thus, multiple total biochar yields, volatile contents etc. Clearly, we cannot infer the temperature history of a biochar from the properties of the final biochar product. We reached the same conclusion for the CSF metric. In the case of the CSF metric, however, our simulations revealed much smaller ranges of biochar properties for the same CSF value.

Our experiments also revealed the smaller particle size fraction (210-250 microns) contained a larger proportion of biomass that undergoes pyrolysis at lower temperatures. This was the result of the grinding. The different chemical composition of the two size fractions resulted in significant differences in the properties of the final biochar products, differences that could not have been predicted by the small variations in the effective heating rates resulting from the different particle sizes. The initial mass of the biochar sample also had a rather significant effect on the properties of the final biochar product due to the subtle temperature variations within the pyrolyzing biomass sample we have identified in earlier work.

References:

[1] Crane-Droesch, A; S. Abiven; S. Jeffery; M.S. Torn. "Heterogeneous global crop yield response to biochar: a meta-regression analysis." *Environmental Research Letters* **8** (2013) 044049.

[2] Pyle, L.A; W.C. Hockaday; T. Boutton; K. Zygourakis; C.A. Masiello, "Chemical and Isotopic Thresholds in Charring: Implications for the Interpretation of Charcoal Mass and Isotopic Data," *Environmental Science & Technology*, **49** (24), 14057-14064 (2015).

493304 Real-Time Mass Spectroscopy in Fast Pyrolysis Process

Abstract:

Biomass pyrolysis is inherently dynamic. Each individual particle entering an otherwise steady-state reactor must be heated to reaction conditions. Upon heating, regardless of the rate, the biomass particle will begin to pyrolyze under conditions of changing temperature. Thus, there are no true stead-state pyrolysis experiments from which kinetics can be extracted. In an effort to provide true real-time kinetic information from which reliable kinetic information can be extracted, a novel micro-reactor micro-particle technology was develop and coupled to a fast mass spectrometer. The newly developed micro-sphere micro-reactor (MSMR) was coupled with a fast mass spectrometer capable of sampling pyrolysis vapors at a rate of 100 Hz over a total reaction time of 100 to 10000 ms. A Q Exactive (QE) mass spectrometer at the Pacific Northwester National Laboratory (PNNL) Environmental Molecular Science Laboratory (EMSL) was used. In this study two types of biomass (tall fescue and switchgrass) and microcrystalline cellulose were formed into microspheres of different sizes and pyrolyzed at different temperatures to study the effect of particle size and process conditions on kinetics and chemistry of reaction. The results show clear dynamics even for fast pyrolysis conditions in which particles are complete pyrolyzed in less than 100 ms.

501350 Measuring the Torrefaction Products of Xylan and D-Xylose

Abstract:

Xylan torrefaction plays a crucial role in the pre-treatment of biomass, so this work examines its reactions by measuring the differences between torrefaction products of xylan vs. its monomer, D-xylose at 250° and 300°C. Xylan is a major component of hemicelluloses and one of the least thermally stable polymers in lignocellulosic biomass. While xylan pyrolysis at higher temperatures has received attention [1], its torrefaction kinetics and its differences with the torrefaction kinetics of D-xylose are largely unstudied.

A sample of 0.1 – 0.3 mg of xylan (Sigma Aldrich) or d-xylose (Acros Organics) is dried at 105°C for 15 mins before being flash-torrefied (Pyroprobe Model 5000, CDS) at the selected temperature and held for 30 s. Product analysis is by two-dimensional gas chromatography followed by time-of-flight mass spectrometry (Pegasus 4D, LECO). Products are considered to be identified if they have greater than 80% similarity with the library spectra and their S/N ratio is greater than 100. The products are divided into three categories: common products, products identified only in D-xylose torrefaction, and products identified only in xylan torrefaction. The % peak areas of common products are used to analyze the effects of temperature on the product distribution.

In both raw materials and at both torrefaction temperatures, the % total peak area of water is the largest among all the products. It is above 30% in each case, while the peak areas for the other products are less than 10 %. Interestingly, water % peak area significantly decreases at higher temperature (from 64% to 32 % in d-xylose torrefaction), possibly due to formation of additional products formation at higher temperatures. Water production due to water elimination reactions may play an important role in D-xylose and xylan torrefaction.

D-Xylose torrefaction yields higher % peak areas at 300°C than at 250°C for the most common products (CO_2, glycoladehyde, furfural, furanmethanol, acetol, acetic acid and formic acid), while for xylan, % peak area of CO_2 and formic acid decreased at 300°C. Furfural had the second largest peak area % among the other products of xylan and the highest % peak area for D-xylose (6.56% at 300°C). Another significant result is that methanol, ethanol, propanoic acid, and butyrolactone are identified as torrefaction products only from xylan. On the other hand, furancarboxaldehyde and 4-cylcopentene-1,3-dione are identified in only in D-xylose torrefaction.

[1] X. Zhou, W. Li, R. Mabon, L.J. Broadbelt, L. J., "A critical review on hemicellulose pyrolysis," *Energy Tech.*, 2016, 4, 1 –29.

494946 Stimulation and Inhibition of Organosolv Lignins on Enzymatic Hydrolysis Mediated By Zeta Potential and Hydrophobicity

Abstract:

Lignin typically inhibits enzymatic hydrolysis of cellulosic biomass, but certain lignin also enhances enzymatic hydrolysis. In this study, we examined the physicochemical properties and structures of ethanol organosolv lignins (EOL) from hardwood and softwood and ascertain the association between lignin properties and their inhibitory and stimulatory effects on enzymatic hydrolysis. The zeta potential and hydrophobicity of EOL lignin samples, isolated from organosolv pretreatment of cottonwood, willow, aspen, eucalyptus and loblolly pine, were determined and correlated with their effects on enzymatic hydrolysis of Avicel. EOLs from cottonwood (CW), black willow (BW) and aspen (AS) improved the 72 h hydrolysis yield by 8-12%, while EOLs from eucalyptus (EH) and loblolly pine (LP) decreased the 72 h hydrolysis yield by 6% and 16%, respectively. The results showed a vigorous correlation (y= 51.97-6.148 × hydrophobicity - 1.583 × zeta potential, r2=0.995) between the 72 h hydrolysis yield with hydrophobicity and zeta potential. The strong correlation indicated the hydrophobicity of EOL was a negative effect and the negative zeta potential of EOL was a positive effect. HSQC NMR spectra showed that Î²-O-4 linkages in lignin to react with ethanol to form an Î±-ethoxylated Î²-O-4Ê¹ substructure (AÊ¹) during organosolv pretreatment. Considerable amount of C2,6-H2,6 correlation in p-hydroxybenzoate (PB) units were observed for EOL-CW, EOL-BW and EOL-AS, but not for EOL-EH and EOL-LP. PB units could potentially contribute to the high zeta potential of EOL-CW, EOL-BW and EOL-AS, and give rise to the positive effect of these three lignins.

498402 Novel Inverse Isotope Effect for Cellulase Hydrolysis of Deuterated Switchgrass

Abstract:

Neutron scattering enables the study of structural and dynamic properties of lignocellulosic biomass at multiple length scales in a non-destructive manner. High levels of deuterium substitution are desired to enable contrast variation techniques which reduce background. However, changes in physical and chemical properties during growth and adaptation to D_2O solutions need to be identified in model plants used for neutron scattering. In this study, enzymatic hydrolysis of bacterial cellulose, and cellulose and holocellulose isolated from switchgrass showed the expected kinetic isotope effect, with deuteration lowering glucose yields by 17, 18 and 4% of theoretical yield, respectively. However, the opposite trend was found for deuterated switchgrass, in which glucose yield was 5% higher than that obtained with protiated switchgrass at lower enzyme loading. These data indicated that alterations to lignin might be responsible for this novel inverse isotope effect. Lignin content of deuterated switchgrass was about 2% higher than that of protiated switchgrass, while NMR spectroscopy indicated no significant structural differences in lignin, and Simonsâ€™ staining found comparable cellulose surface area. However, extensive CFM and TEM imaging showed that deuterated switchgrass had abnormal lignin distribution in some of its cell walls and many of them were collapsed possibly due to reduced rigidity, which would render them easier to deconstruct by cellulases. The changes in morphology resembled those of drought stressed plants, consistent with abiotic stress due to growth in deuterated media. These protio/deutero investigations clearly illustrate the key role of component distribution in the multilamellar cell wall architecture on biomass recalcitrance.

492208 Synthesis of High Biomass Content UV-Curable Epoxy Acrylate Oligomer with Cardanol

Abstract:

UV-curable polymerization is widely used in industrial applications for different purposes, such as decoration, protection, and some other specific functions.[1] Conventionally, petroleum-based materials are included in most UV-curing polymer.

This will be increasing environmental concerns and limit for fossil feedstock.[2] Much effort has been devoted to explore renewable materials to substitute non-renewable organic chemicals.[3] Among these, a cashew nut shell liquid (CNSL), is a kind of sustainable, low-cost and largely available natural resources, which is obtained as a byproduct of the cashew processing industry.[4] Its extracted product, cardanol, is a phenol derivative, which has a meta-substituent of a C_{15} unsaturated hydrocarbon chain with one to three double bonds.[5] This unique structure exhibits promising properties for making tailor-made polymers of high value.[6]

Cardanol can be reacted with epichlorohydrin (ECH) in order to introduce the epoxy group. Then the cardanol glycidyl ether is synthesized and used in the next step. The unsaturated double bonds in the molecular structure of the cardanol glycidyl ether can be oxidized by 3-Chloroperbenzoic acid to form epoxide groups. The epoxide groups of the product were reacted with acrylic acid in the presence of triphenyl phosphorus. High biomass content UV-curable epoxy acrylate oligomers were synthesized. Different contents of the modified Nano-silica were added into the epoxy acrylate oligomer to study the materials property variations.

Fig 1 Synthesis sequence adopted for the preparation of cardanol-based epoxy acrylate oligomers

The structure of the prepared products was analyzed by a nuclear magnetic resonance (NMR) and (FT-IR) spectroscopy. The biorenewable content of UV-curable epoxy acrylate oligomers is about 58.8% ~ 60.9%. The UV-cured coatings have high gel content and good flexibility. The hardness value of the samples is form 3H to 5H. With the increasing of the modified Nano-silica content, the contact angle values on the cured film surfaces and surface energy showed a great changes. The presences of 0.75% (relative to the total mass) of the modified Nano-silica could lead to a large decrease in the surface free energy. The scanning electron microscopy (SEM) images were evident that the solubility of the UV-cured coatings was improved with a small amount of modified Nano-silica addition. We expect that our results will provide insight into the utilization of a new epoxy acrylate oligomer for adhesive and coating applications.

[1] H. Wang, W. Liu, Z. Yan, J. Tan and G. Xia-Hou, RSC Adv., 2015, 5, 81838-81846.

[2] R. Liu, X. Zhang, J. Zhu, X. Liu, Z. Wang and J. Yan, ACS Sustainable Chemistry & Engineering, 2015, 3, 1313-1320.

[3] R. Liu, G. Zhu, Z. Li, X. Liu, Z. Chen and S. Ariyasivam, Green Chem., 2015, 17, 3319-3325.

[4] K. I. Suresh and V. S. Kishanprasad, *Ind. Eng. Chem. Res.*, 2005, 44, 4504-4512.

[5] J. Chen, X. Nie, Z. Liu, Z. Mi and Y. Zhou, ACS Sustainable Chemistry & Engineering, 2015, 3, 1164-1171.

[6] C. Voirin, S. Caillol, N. V. Sadavarte, B. V. Tawade, B. Boutevin and P. P. Wadgaonkar, *Polym. Chem., 2014, 5.* 3142-3162.

500782 Dissolution of Cellulosic Particles: Population Ensemble Modeling Informs Efficient Woody Biomass Processing

Abstract:

A major barrier to the efficient utilization of biomass is the recalcitrance to dissolution of semicrystalline cellulose. The present study addresses the kinetics of swelling and dissolution of cellulose particles in conditions emulating large-scale processing where the particles exhibit a distribution of size and degree of crystallinity. To this end, we have developed a model in which the behavior of a population of particles is obtained from an ensemble of individual cellulose particle dissolution models. The dissolution of individual solid cellulose particles is based on the relevant transport phenomena and kinetics and reveals decrystallization and disentanglement as two important and potentially rate-determinant steps in the process. [Ghasemi, M.; Alexandridis, P.; Tsianou, M., Cellulose dissolution: Insights on the contributions of solvent-induced decrystallization and chain disentanglement. Cellulose 2017, 24 (2), 571-590. DOI: 10.1007/s10570-016-1145-1] The average value or the number distribution of any intra-particle property captured by the individual particle model can be determined by simulation of a sufficient number of individual particles such that ensemble averages are independent of the particle numbers and the computed particle distributions are acceptably smooth. Using this population ensemble model, various cellulose particle size distributions and crystallinity distributions are analyzed for different dissolution parameters. The findings from this study would be useful for the rational design and optimization of pretreatment processes to reduce the particle size and degree of crystallinity, leading to enhanced woody biomass utilization. [Ghasemi, M.; Tsianou, M.; Alexandridis, P., Assessment of solvents for cellulose dissolution. Bioresource Technol. 2017, 228, 330-338. DOI: 10.1016/j.biortech.2016.12.049]

Growing Scientific Leaders With Concentrated Sunlight: A Tribute to E. A. Fletcher

Bob Palumbo
Prof. of Mechanical Eng.

Luke Venstrom
Asst. Prof. of Mechanical Eng.

VALPARAISO UNIVERSITY

Growing Scientific Leaders

By Developing Deep Science and Engineering Knowledge

By Developing Communication Skills

By Developing the Passion to Serve Others

VALPARAISO UNIVERSITY

Model for Engaging Undergraduate Students

Curricular

Senior Design:
- Two-semester course
- Design during fall semester
- Fabricate and test during spring semester
- Multi-disciplinary teams with faculty advisor

Credit:
- Research during academic year
- Students earn up to 3 credits towards mechanical engineering degree

Extra-curricular

Summer Research:
- 10 to 12 weeks over the summer
- Includes stipend and free housing in dorm
- Intentional cross-disciplinary interaction

VALPARAISO UNIVERSITY

Undergraduate Student Activity

VALPARAISO UNIVERSITY

Valpo Solar Furnace

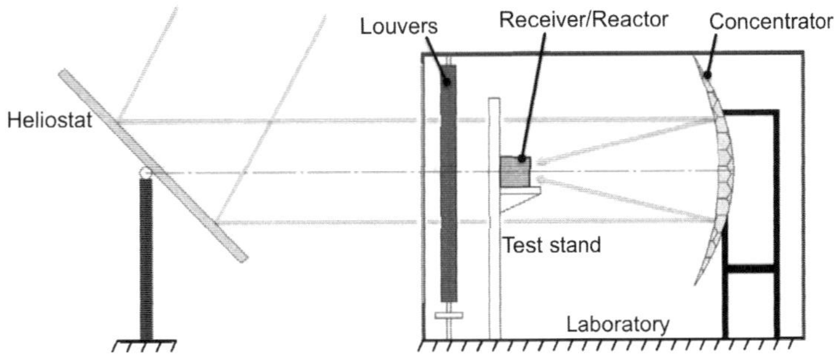

VALPARAISO UNIVERSITY

Valpo Solar Furnace

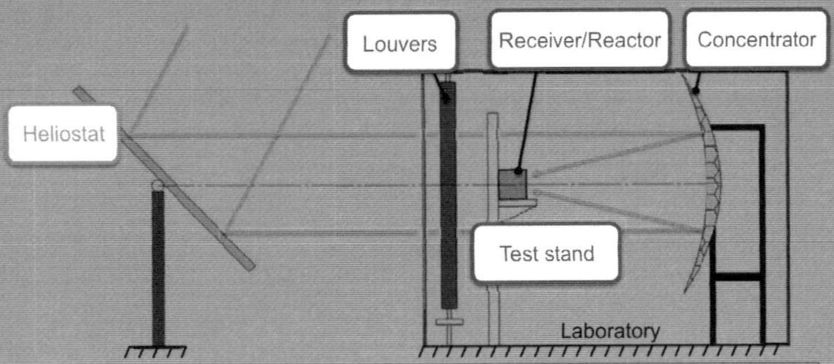

Designed and manufactured by undergraduate students

Undergraduate students contributed to design via analysis and construction of a prototype

VALPARAISO UNIVERSITY

Test Stand

Design Constraints:

- 3-axis motion accurate within .01"
- Support 250 lbf
- Integrate chilled water flow up to 10 gpm, pressurized air supply, and exhaust gas line
- Safety:
  - brake in event of power loss
  - emergency stop switches
  - laser safety scanners
- Able to hold numerous receivers:
  - Calorimeter
  - Lambertian target
  - Solar thermochemical reactors

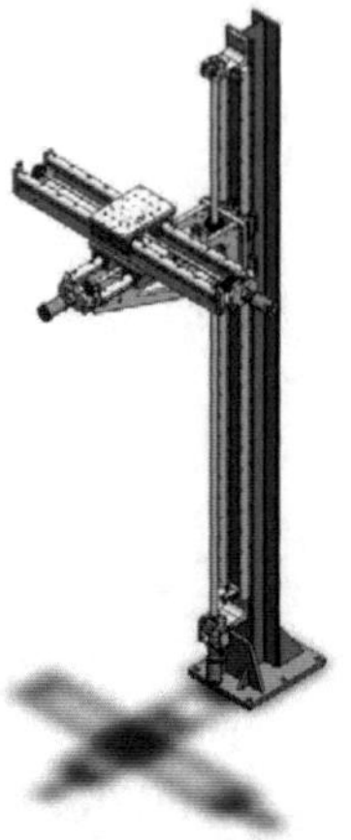

VALPARAISO UNIVERSITY

Test Stand

Analyzed by undergraduate students with FEA: Factor of safety of 11

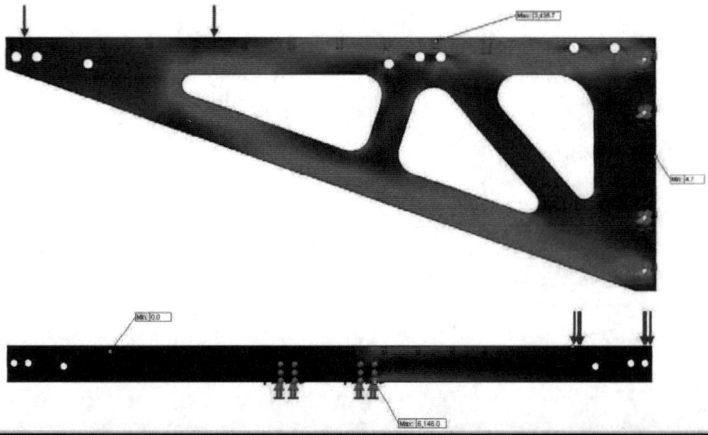

VALPARAISO UNIVERSITY

Test Stand

VALPARAISO
UNIVERSITY

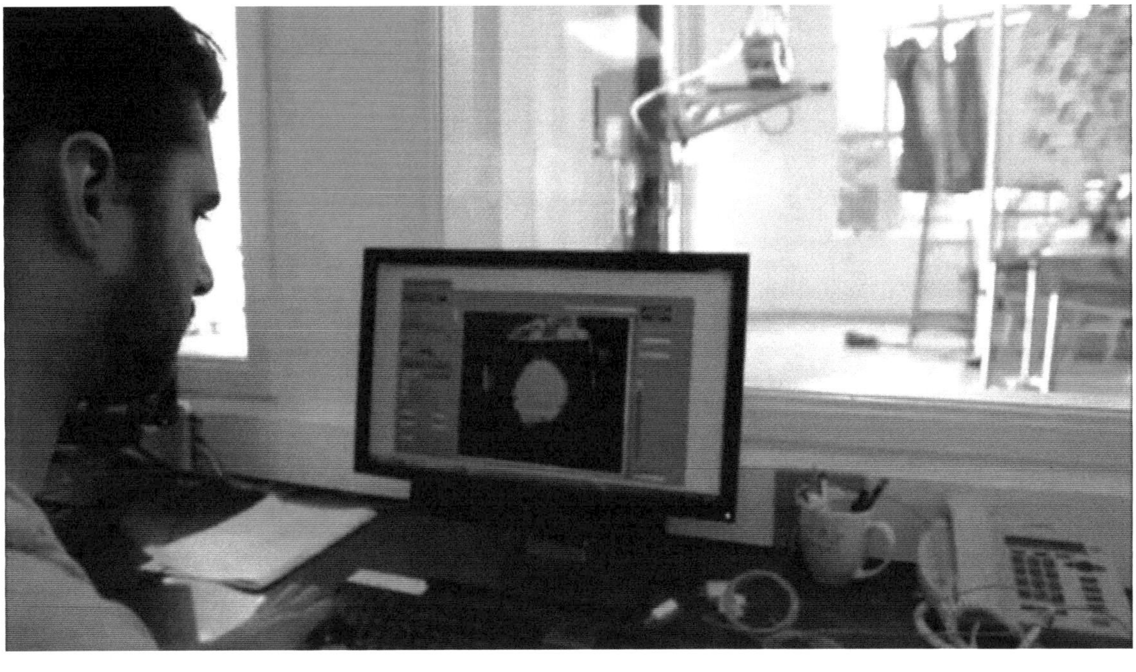

Solar Thermal Decoupled Electrolysis of Water

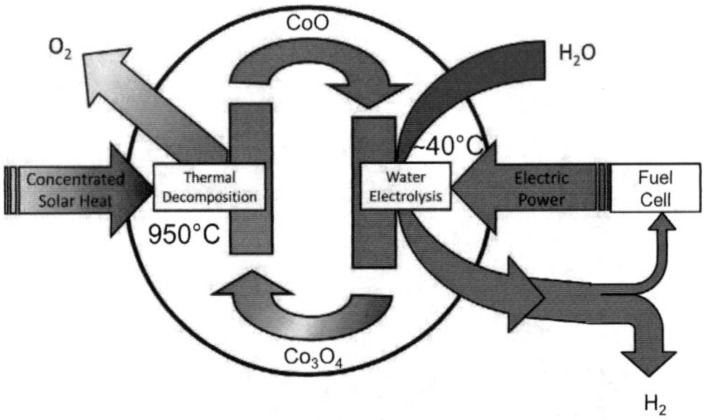

$$\eta_{cycle} = \eta_{abs} \frac{\Delta G_{H_2+0.5O_2=H_2O} - \Delta G_{Electrolysis}}{\Delta H_{Co_3O_4 \rightarrow 3CoO+0.5O_2}} = 38\%$$

Palumbo *et al.*, 2012, *Chem. Eng. Sci.*, **84**, 372–380.

VALPARAISO UNIVERSITY

Rotary Drum Reactor Concept

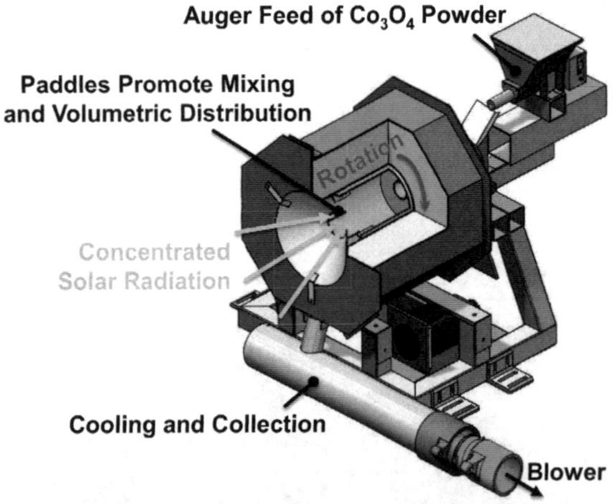

VALPARAISO UNIVERSITY

Rotary Drum Reactor Model

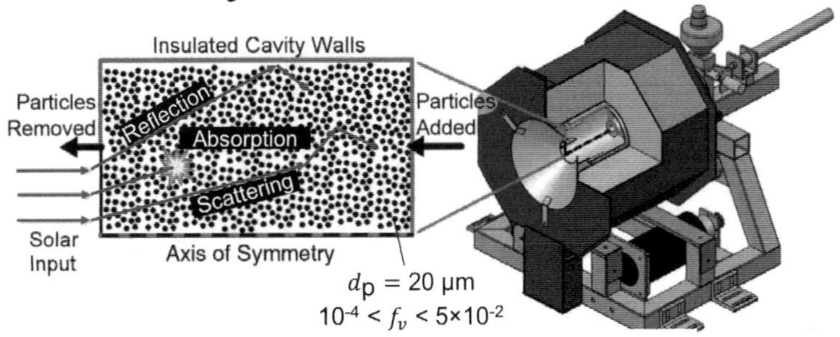

Conservation of Species (Reaction extent)

$$0 = -\nabla\cdot\left(\vec{G}''_{Co_3O_4}\right) - r'''$$

Particle motion

Rate of conversion of Co_3O_4 to CoO

Conservation of Energy (Temperature)

$$0 = -\nabla\cdot\left(-k_e\nabla T + \vec{G}''\hat{c}T\right) - q'''_{\mathrm{rad}} - r'''\Delta H_r$$

Conduction

Particle motion

Volumetric Radiation (participating media)

Rate of energy consumption from reaction of Co_3O_4

Equations solved using the Finite Volume Method

VALPARAISO UNIVERSITY

Model Results: Impact of Feed Rate

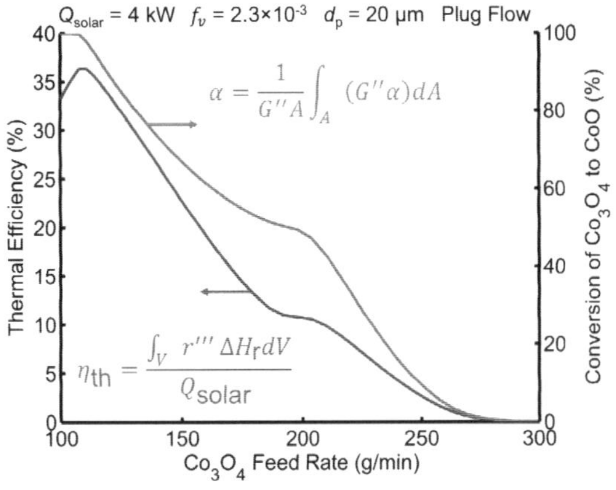

Maximum thermal efficiency achieved when the concentrated solar radiation is matched with the rate of the chemical reaction (α = 100%).

VALPARAISO UNIVERSITY

Reactor Testing

VALPARAISO UNIVERSITY

Reactor Operation

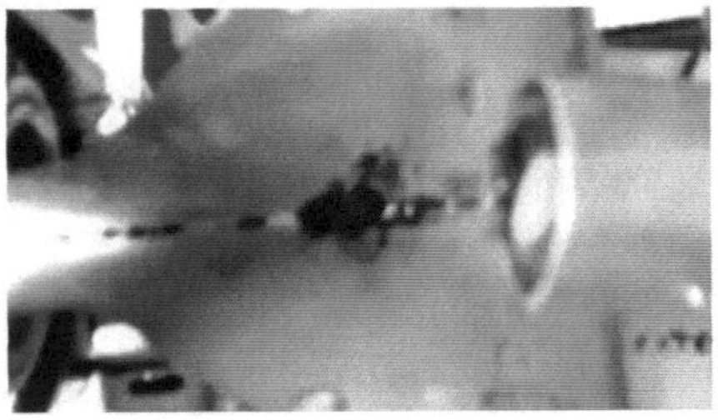

Reactor Inlet

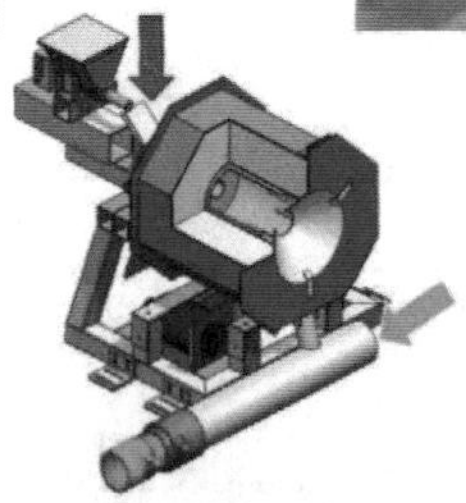

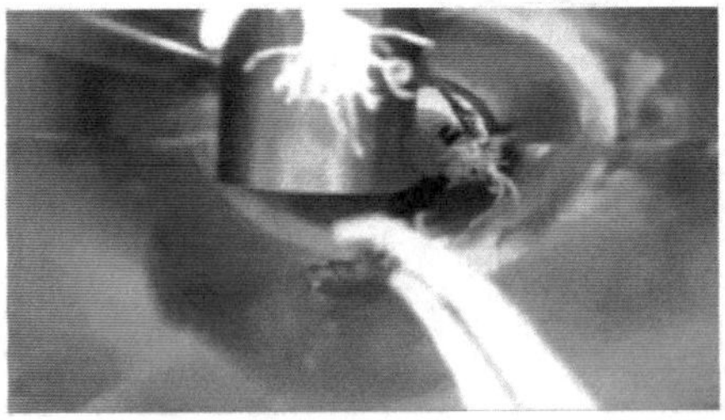

Reactor Outlet

VALPARAISO UNIVERSITY

Reactor Test Results

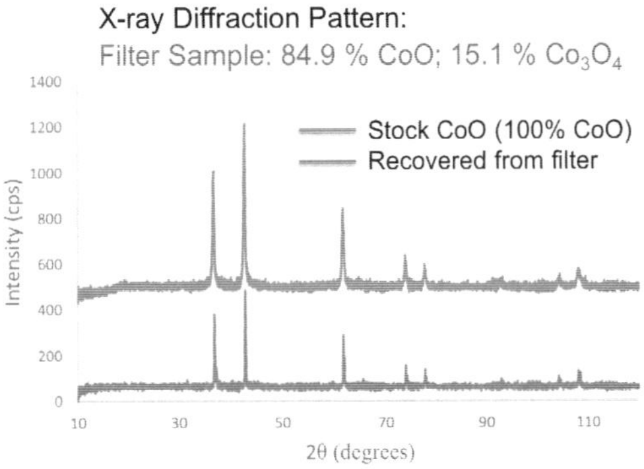

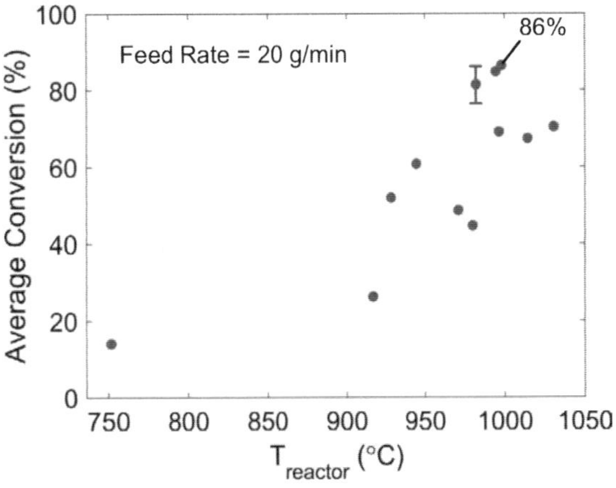

VALPARAISO UNIVERSITY

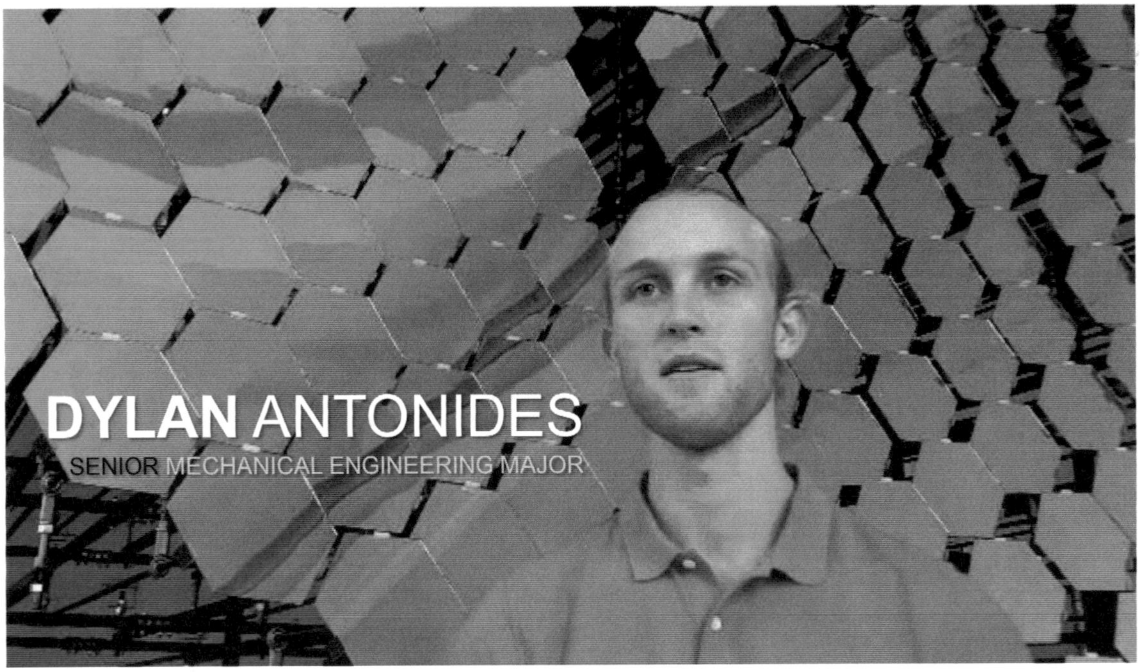

Project Roadmap ($2.3 Million Grant)

VALPARAISO UNIVERSITY

Project Roadmap ($2.3 Million Grant)

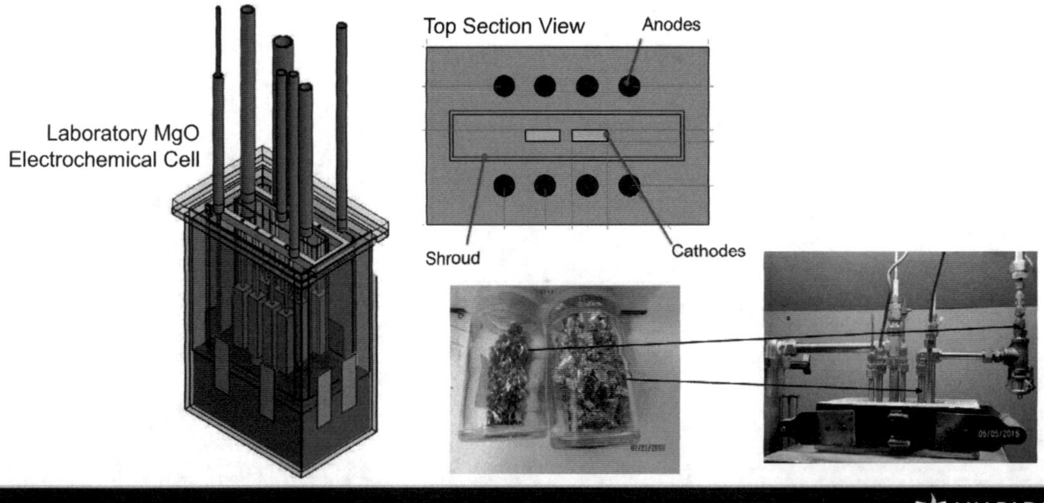

VALPARAISO UNIVERSITY

Claire Merrill

Fundamental Studies

VALPARAISO UNIVERSITY

Cathode Physics

Global Reaction: $Mg^{2+} + 2e^{-1} \leftrightarrow Mg\ (\ell)$

Reaction Kinetics: $$j = j_o \left[1 \cdot \exp\left(\frac{(1-\alpha)F\eta}{RT} \right) - \frac{C_{Mg^{2+}}}{C^{\infty}_{Mg^{2+}}} \exp\left(-\frac{\alpha F \eta}{RT} \right) \right]$$

Exchange Current Density (j_o); Transfer Coefficient (α)

Mass Transfer: $$0 = \frac{\partial}{\partial t}\left(C_{Mg^{2+}} \right) - D_{Mg^{2+}} \nabla^2 C_{Mg^{2+}}$$

Diffusion Coefficient ($D_{Mg^{2+}}$)

$$\frac{-j}{nF} = D_{Mg^{2+}} \nabla C_{Mg^{2+}}$$ At the cathode-electrolyte interface (Zero flux at all other boundaries)

VALPARAISO UNIVERSITY

Results from 3D Finite Element Model

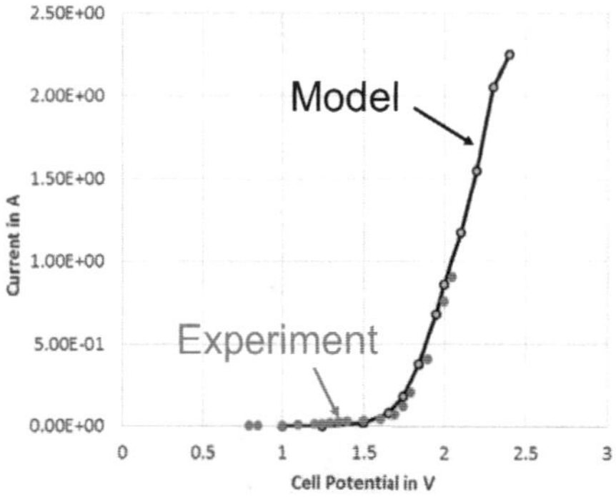

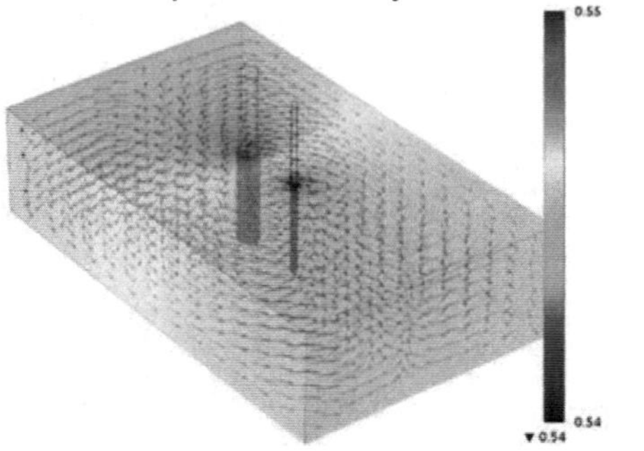

The cell is 13 cm x 7.5 cm x 2.54 cm

Anode = 4.0 cm^2 Cathode = 1.4 cm^2 Anode-Cathode Gap = 2 cm

VALPARAISO UNIVERSITY

Results from 3D Finite Element Model

Linear sweep voltammetry of Laboratory MgO Cell (20 A)

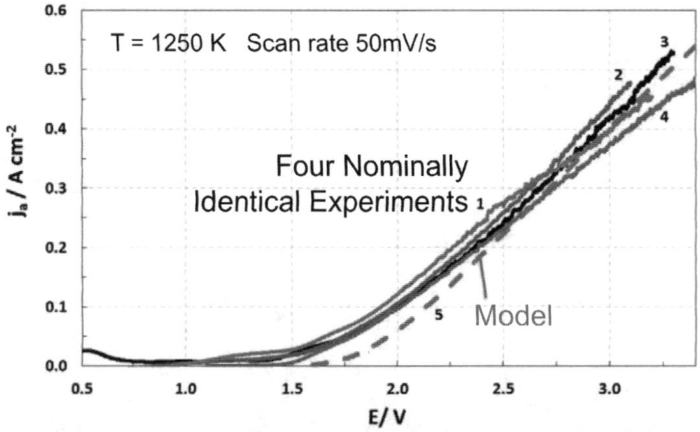

VALPARAISO UNIVERSITY

Towards Commercialization

VALPARAISO UNIVERSITY

Commercial Cell Concept

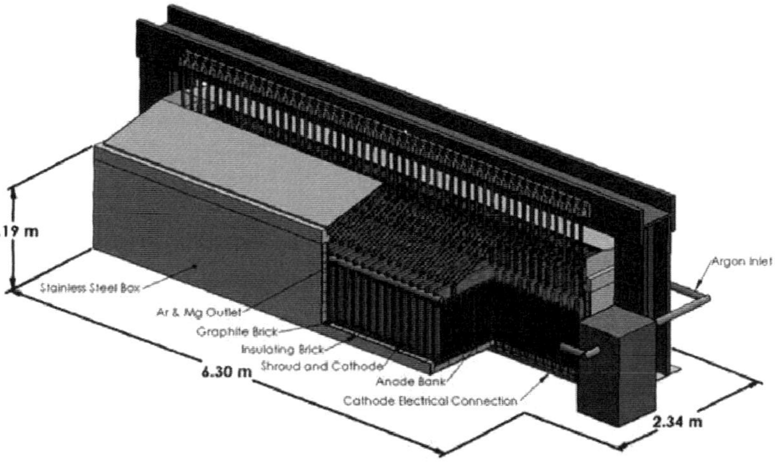

VALPARAISO UNIVERSITY

Finite Element Model of Commercial Cell Concept

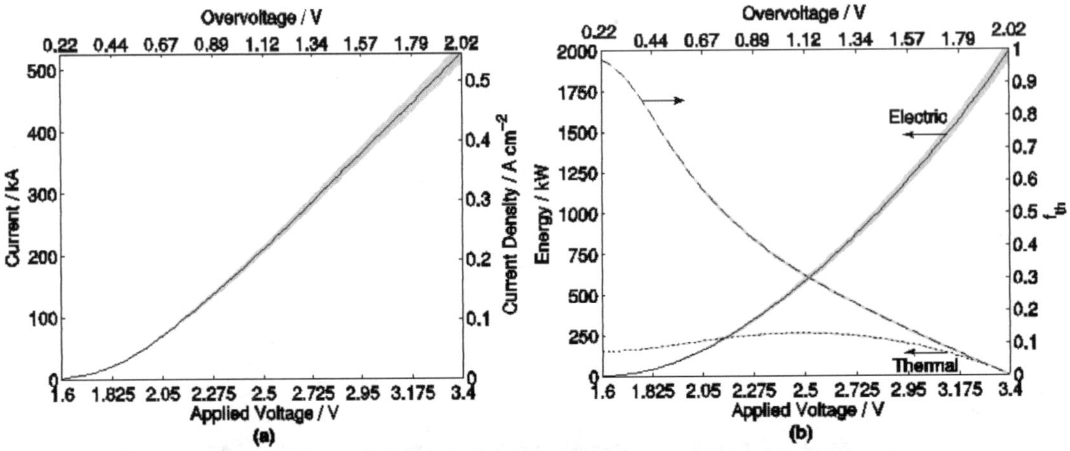

VALPARAISO UNIVERSITY

Visualizing the Solar-to-Commercial Cell Interface

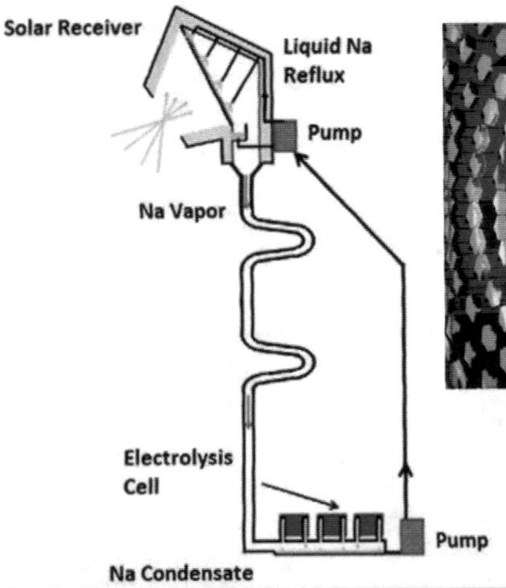

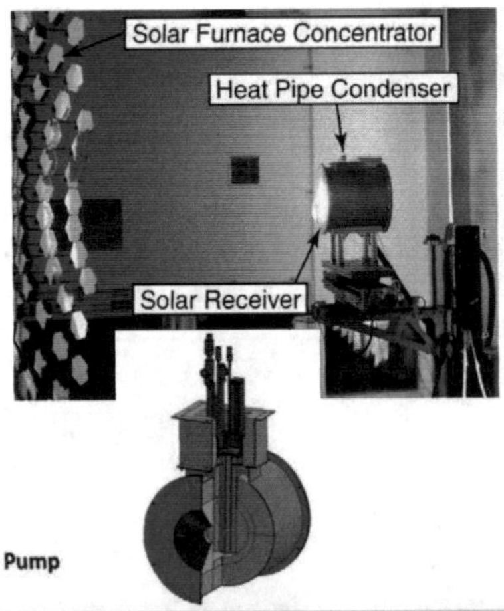

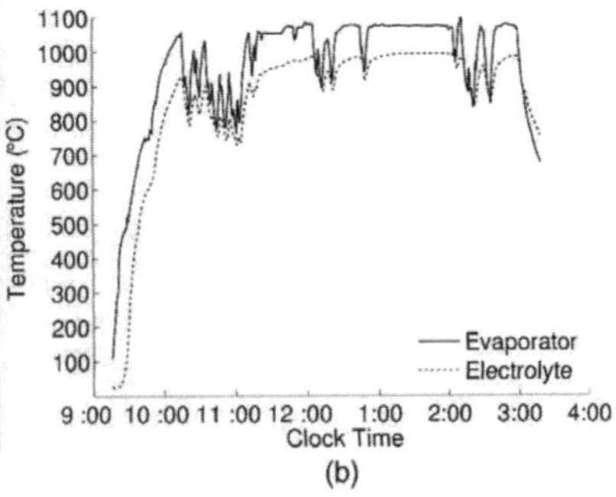

VALPARAISO UNIVERSITY

DR. NATHAN LEONARD
ELECTROCHEMICAL SCIENTIST AT GENERAL MOTORS

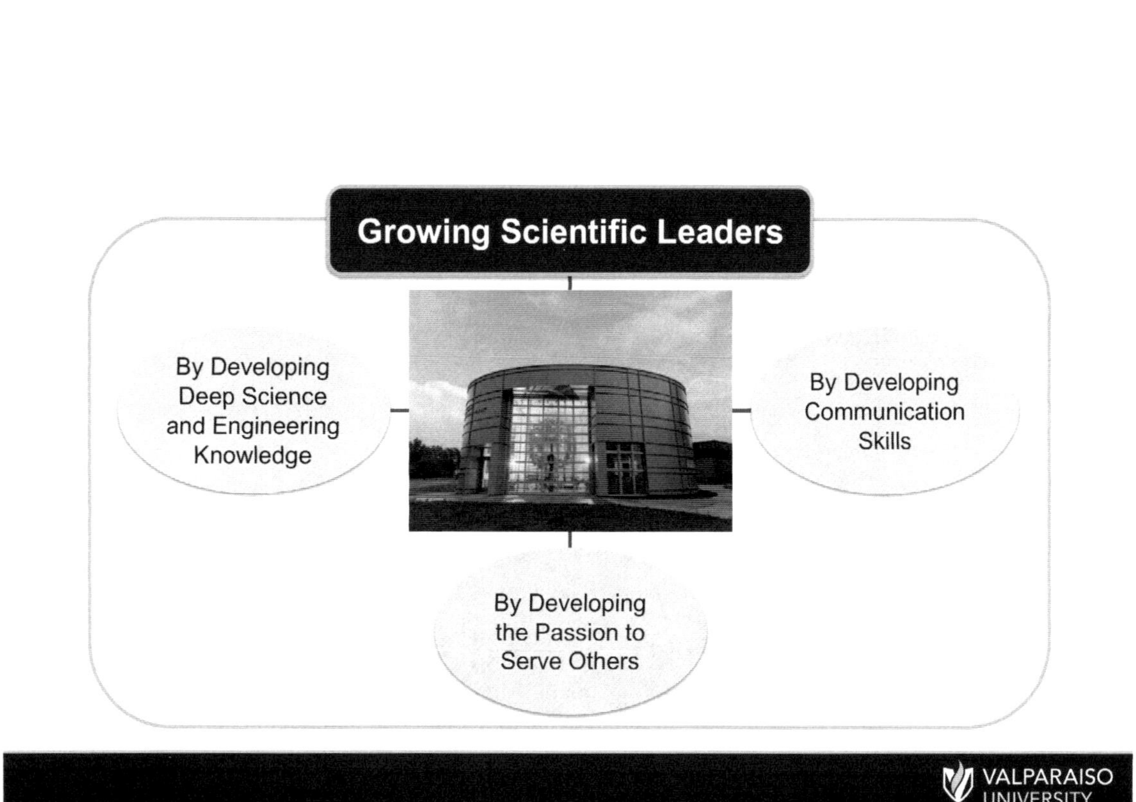
Growing Scientific Leaders
By Developing Deep Science and Engineering Knowledge
By Developing Communication Skills
By Developing the Passion to Serve Others
VALPARAISO UNIVERSITY

Undergraduate Student Co-authors

1) Leonard, N., Korenko, M., Larson, C., Blood, K., Venstrom, L.J., Nudehi, S., Duncan, S., Diver, R., Simko, F., Priscak, J. and Schoer, J., 2016, "The thermal electrolytic production of Mg from MgO: A discussion of the electrochemical reaction kinetics and requisite mass transport processes," *Chem. Eng. Sci.*, **148**, pp.155-169.
2) Palumbo, R., Korenko, M., Larson, C., Venstrom, L. J., Duncan, S., Nudehi, S., Schoer, J., Toberman, J., Prusinski, W., Johnson, D., Robinson, B., Barkely, S., Warren, K., Diver, R., Šimko, F., Boča, M., 2015, "Thermal Electrolytic Production of Mg from MgO: Reflections on Commercial Viability", *144th Annual Meeting & Exhibition of the Minerals, Metals & Materials Society TMS2015*, March 15-19, 2015, Orlando, FL.
3) Sheline, W., Matthews, L., Lindeke, N., Duncan, S., Palumbo, R., 2013, "An Exploratory Study of the Solar Thermal Electrolytic Production of Mg from MgO," *Energy*, **51**, pp. 163 – 170.
4) Palumbo, R., Diver, R.B., Larson, C., Coker, E.N., Miller, J.E., Guertin, J., Schoer, J., Meyer, M., Siegel, N.P., 2012, "Solar Thermal Decoupled Water Electrolysis Process I: Proof of concept," *Chem. Eng. Sci.*, **84**, pp. 372-380.
5) Schroeder, R., Matthews, L., Leatzow, D.,Kondratko, J., Will, J., Duncan, S., Sheline, W., Lindeke, N., Neves, P., and Palumbo, R., 2011, "Solar Thermal Electrolytic Process for the Production of Zn from ZnO: The Electrolysis of ZnO from 1275-1500 K," *J. Solar Energy Eng.*, **133**(4)
6) Krenzke, P., Krueger, K., Leonard, N., Duncan, G., and Palumbo, R. D., 2010, "A Solar Thermal Electrolytic Reactor for Studying the Production of Metals from Their Oxides," *J. Solar Energy Eng.*, **132**(3), pp. 34501-34505.
7) Venstrom, L., Krueger, K., Leonard, N., Tomlinson, B., Duncan, G., and Palumbo, R.D., 2009, "Solar Thermal Electrolytic Process for the Production of Zn from ZnO: An Ionic Conductivity Study," *J. Solar Energy Eng.*, **131**, pp. 1-9.
8) Freid, A.P., Johnson, P.K., Musella, M., M ller, R., Steinbrenner, J.E., and Palumbo, R.D., 2005, "Solar blind pyrometer temperature measurements in high temperature solar thermal reactors: a method for correcting the system-sensor cavity reflection error," *J. Sol. Energy Eng.*, **127**(1), pp.86-93.

VALPARAISO UNIVERSITY

Undergraduate Student Presentations

VALPARAISO UNIVERSITY

Student Outcomes: Graduate School

Graduate Schools Attended:

- Georgia Tech
- Penn State
- University of Florida
- University of Michigan
- University of Minnesota
- Harvard University
- University of Illinois at Urbana-Champaign
- Michigan State
- Stanford
- Vanderbilt
- Ohio State University
- University of Wisconsin
- Colorado State

64 percent enrolled in a graduate program (last 3 years)

4 NSF Graduate Research Fellowship winners and 4 honorable mentions (last 5 years)

VALPARAISO UNIVERSITY

A sustainability challenge: communicating science with the general public at an emotional level without compromising the science

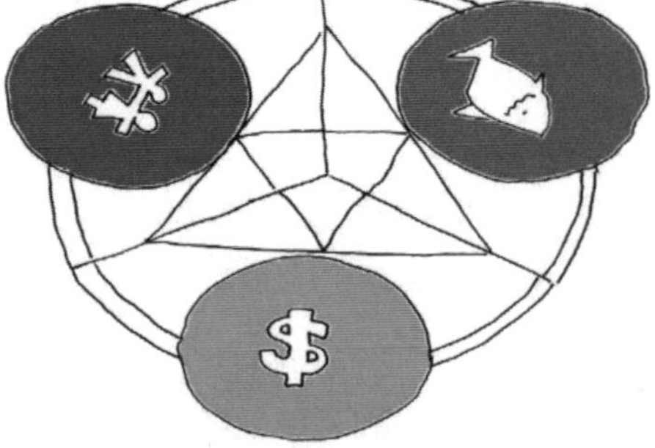

VALPARAISO UNIVERSITY

Thank you, Professor Edward Fletcher,
for helping us to Grow Future Scientific
Leaders with Concentrated Sunlight.
With Love, The VU Community
VALPARAISO
UNIVERSITY

495673 Metal Oxide Redox Materials for Solar Thermochemical Energy Storage

Abstract:

Thermochemical energy storage (TCES) systems are of great interest in concentrated solar power (CSP) applications. Storing sunlight as chemical energy during the day can enable power generation at night or during cloudy periods, effectively alleviating the inherent intermittency of solar sourced electricity. Metal oxides are among the most attractive TCES materials because they possess high-energy density and high reduction/oxidation temperatures (>800°C) suitable for driving high-efficiency thermodynamic power cycles. However, limitations involving high-temperature heat transfer, particle handling, sintering, and chemical performance of the metal oxide materials are still problematic in practical application.

Past works generally focus on evaluating pure metal oxide systems and investigate a select few mixed metal oxide systems in detail. In this work, we pre-selected over 100 high-potential mixed-metal oxide systems based on a comprehensive literature survey and then screened those materials for suitability in redox based high-temperature TCES using thermodynamic database software. The potential candidates were down-selected to the most promising contenders based on several selection criteria: high-temperature stability, large redox reaction enthalpy, and reaction temperatures of approximately 700-1100 Â°C for CSP applications. Ongoing and future work we will focus on experimental testing of the identified compositions to identify actual redox behaviours and to evaluate other thermochemical criteriaâ€”appropriate cycling physical stability, maintained chemical activity over many redox cycles, and acceptably fast reaction kinetics.

502849 SolPeD â€“ solar per demand for power generation and fuel production

Abstract:

We are challenged to develop an energy conversion method, which could gradually replace fossil fuel combustion on a worldwide scale, eventually leading to a renewabledominant global energy supply. Whereas other renewable energy resources can have important supporting roles, the only energy sources available on earth in large enough quantities to replace fossil fuels are nuclear and solar.
The talk will describe a solar energy conversion process based on several complementing novel technologies, and aimed to enable large-scale utilization of solar energy for yearround, 24/7 power generation and clean fuel production at high efficiency and affordable costs.
The major technological challenges and proposed solutions will be discussed, and ongoing effort of commercializing the technology will be reported.

498174 On the Role of Thermochemical Energy Storage in Concentrating Solar Power

Abstract:

Concentrating solar power (CSP) with thermal energy storage has the potential for grid-scale dispatchable power generation. Thermochemical energy storage (TCES), that is, the reversible conversion of solar-thermal energy to chemical energy, has high energy density and low heat loss over long periods. Previously, several reaction types have been studied as candidates, including dissociation of ammonia, hydrides, carbonates, oxides, hydroxides and methane reforming [1]. Although considerable work has been done, past research has mostly focused on experimental studies and reactor design. The limited system-level studies that are available in the literature are for specific reactions [2] and, furthermore, consider TCES in isolation, neglecting its interactions with the solar field and power generation system [3].

Accordingly, in this work, we establish a framework to systematically analyze and compare candidate reactions for TCES systems [4]. First, we develop a general process for CSP plants employing TCES systems. A general TCES system includes five tasks: reaction, storage, separation, material handling and heat recovery. We then formulate an optimization-based model for the process, which optimizes the plant design towards the minimum levelized cost of electricity (LCOE). We use this model to study ammonia and methane systems.

We identify that, for fluid phase TCES systems, the main cost driver is gas storage using pressure vessels, accounting for 77% of the TCES cost. This is because, even if reaction products are compressed and stored at 550 bar, both ammonia and methane systems require 65,000 m^3 of gas storage at a vessel price of 8900 $/$m^3$. The main energy driver is the electricity consumption for storage compression, which consumes 18% of the generated power. Adopting underground gas storage, where products are stored at lower pressure in much cheaper underground caverns, could significantly reduce the compression power consumption (from 18% to 3%) and improve the overall energy efficiency and thus cost. The methane system combined with underground storage is identified as the most cost-efficient alternative, achieving a 13% LCOE reduction over the current two-tank molten salt CSP plants.

Finally, we discuss the impact of the reaction temperature and operating pressure. The endothermic and exothermic reaction temperature determines not only the reaction conversion, but also the receiver and turbine efficiency. We find that both reaction temperature should be in the vicinity of the reaction turning temperature (at which the equilibrium constant K=1) and separation can increase the reaction conversion, allowing lower endothermic reaction temperature. The operating pressure, including process pressure, highest and lowest storage pressure, influences both the storage volume and compression work. For pressure vessel cases, the storage tank pressure varies in a wide range (150-550 bar), reducing the storage volume at the expense of more compression power consumption. When cheap underground storage is employed, a narrow storage pressure range (135-165 bar) leads to smaller electricity consumption and larger storage volume.

The proposed framework can be used to study a wide range of TCES systems, including solid-gas phase reactions, which will be the topic of future research.

References:

[1] P. Pardo, A. Deydier, Z. Anxionnaz-Minvielle, S. Rougé, M. Cabassud and P. Cognet, *Renew. Sustain. Energy Rev.*, 2014, **32**, 591–610.

[2] A. Luzzi, K. Lovegrove, E. Filippi, H. Fricker, M. Schmitz-goeb, M. Chandapillai and S. Kaneff, *Sol. Energy*, 1999, **66**, 91–101.

[3] C. Corgnale, B. Hardy, T. Motyka, R. Zidan, J. Teprovich and B. Peters, *Renew. Sustain. Energy Rev.*, 2014, 38, 821–833.

[4] X. Peng, T. W. Root and C.T. Maravelias, *Green. Chem*, 2017, DOI: 10.1039/c7gc00023e.

500484 Combined Thermochemical and Latent Heat Energy Storage for Low Temperature Residential Applications

Abstract:

Solar thermal energy has the potential to supply clean energy, but for space heating and cooking applications, peak usage times often are misaligned with peak solar hours. Solar thermochemical energy storage systems are one promising solution. Salt-hydrate reactions are well-suited to this purpose as they have low-decomposition temperatures and high energy densities. To enhance heat transfer, improve durability, and prevent conglomeration, a solid matrix material is used, typically made of porous substances such as activated carbon, vermiculite, or expanded graphite. Here, a matrix of expanded graphite reinforced with carbonized flour was used. Certain salt-hydrates have the potential to store both thermochemical and latent heat, and one such salt was investigated through TGA/DSC analysis. A functional eutectic was found at around 130°C and freezing/melting cycles were visible as well-defined peaks. The salt-matrix complex was also investigated through both TGA/DSC and realistic application testing. The success of this new chemistry sets the stage for a new dual-release approach to energy storage.

496055 Optical Losses from Bubble Evolution in Photo-Electrochemical Reactors

Abstract:

Solar fuels, such as the production of hydrogen by photo-electrolysis of water, have the potential to revolutionise the long term future of energy production as they can overcome the issue of intermittency. The potential of this technology can only be realised by scaling up current research and lab scale demonstrations and henceforth, design of photo-electrochemical (PEC) reactors will become increasingly important.

Successful PEC reactor design will minimise resistive losses in solution, ensure the separation of the photo-electrolysis products and permit maximal light absorption by the photo-electrode(s). Methodologies for simultaneously achieving these requirements have not been exhaustively studied and in particular a comprehensive quantification of optical losses in PEC is missing from the literature.

In this work we consider the optics of PEC reactor design and its ramifications, focusing on the optical losses of bubbles on the surface of the photo-electrode. Bubble evolution is a complex process of bubble nucleation, detachment and rise and not only detrimentally reduces the effective surface exposed to electrolyte but absorbs, reflects and scatters incoming photons.

We experimentally demonstrate this effect and quantify the optical losses due to bubble evolution using a transparent conductive oxide such as fluorine doped tin oxide (FTO) glass as an anode. The optical losses are correlated against current density (and hence gas evolution rate) and the ramifications of this discussed. From this relationship, a modified maximum Solar-To-Hydrogen efficiency for different semiconductors (Shockley-Queisser limit) is derived.

Opportunities for the elimination of optical losses from bubble evolution through bubble suppression are discussed where pressurisation and forced convection approaches are investigated. A multi-physics model is presented that demonstrates the operational requirements in order to a achieve bubble suppression in a novel membrane-less flow reactor design, which has been outlined in previous work. Experimental verification of bubble suppression through hydrodynamic control is given and compared to the theoretical results and the potential scalability of this technology explored.

500252 Acrylonitrile Production from Biomass-Derived Intermediates

Abstract:

Carbon fibers are key materials in critical energy applications including light-weighting of vehicles for improved fuel efficiency and enabling larger wind turbine designs for increased power generation. Carbon fibers are produced through heat treatment of polyacrylonitrile fibers. Today, the acrylonitrile monomer is made primarily via ammoxidation of propylene with ammonia. However, price fluctuations in the feedstocks and the energy-intensive processes use to make carbon fibers have resulted in low-volume and niche adoption of carbon fibers into automotive applications, motivating the drive towards carbon fibers produced cheaply and reliably from renewable biomass sources. This talk will describe efforts in the Renewable Carbon Fiber Consortium to pursue acrylonitrile production from biomass-derived sugars via hybrid biological and chemo-catalytic processes. In particular, bench-scale efforts in two separate process schemes will be described that proceed via biological production of 3-hydroxypropionic acid or lactic acid, followed by subsequent separation and catalytic conversion to acrylonitrile.

501349 Multistream Integrated Biorefinery (MIBR) for Carbon-, Road-, and Bio-Materials

Abstract:

The success of a modern biorefinery heavily depends on the availability of diverse product streams and utilization of all fractions of input material. Traditional lignocellulosic biorefining focuses on ethanol, which has limited by-product streams that add value to the biorefinery. Essentially all current bioconversion platforms lead to a lignin-containing waste stream that needs further processing into valuable products. Although a certain amount of lignin (~30â€“40%) is needed for the thermal requirements of biofuel production, a modern cellulosic processing plant will have ~60% excess lignin. The utilization of the lignin-containing biorefinery waste as feedstock for renewable products offers a unique opportunity to achieve a multi-stream integrated biorefinery (MIBR), where the lignin-containing biorefinery waste will be utilized for various carbon-, road- and bio- materials to maximize economic return and sustainability. Specifically, we have advanced the fractionation, conversion, and processing technologies to enable different bioproduct streams. An efficient enzyme-mediator system was developed to fractionate lignin into low- and high-molecular weight fractions with more homogenous chemical characteristics. We have demonstrated that these lignin fractions are more amenable to different applications in bioconversion, asphalt binder modifier, and carbon fiber. The low molecular weight fraction is more amenable to bioconversion into PHA for bioplastics. In addition, the low molecular fraction can be used as unique asphalt binder modifiers to improve both high and low temperature performance of road pavement. The high molecular weight fraction can be used to fabricate carbon fiber with significantly improved mechanical performance. Together, MIBR will reduce ethanol production cost through the creation of high value bioproducts, produce asphalt binder modifier with unique features, develop an innovative, new pathway to quality carbon fiber; and create a means to utilize all carbon in the feedstocks.

492807 Improving Economics of Cellulosic Biofuels: An Integrated Strategy for Co-Producing 1,5-Pentanediol and Bioethanol

Abstract:

Lignocellulosic biomass is an abundant and sustainable feedstock that can be used to produce biofuels and biobased commodity chemicals. However, the low oil price and the challenges in applying biorefinery technologies at scale hamper the commercial production of biofuels. In this presentation we will discuss a strategy to integrate the production of a high-value coproduct, 1,5-pentanediol (1,5-PDO), along with a typical bioethanol production process.

1,5-PDO and two other conventional α,ω-diols (e.g. 1,6-hexandiol and 1,4-butanediol), are building blocks for polyurethane and polyester plastics. These α,ω-diols are currently derived from petroleum and, as a group, represent an annual market over $6 billion. An economically viable approach of producing 1,5-PDO from biomass-derived furfural has been developed recently (Brentzel et al., 2017; Huang et al., 2017), which offers the potential to coproduce 1,5-PDO as a high value co-product of bioethanol manufacturing process to improve cellulosic biorefinery economics.

In this work, we propose a novel strategy that integrates lignocellulosic biomass fractionation with simultaneous conversion of cellulose and hemicellulose into ethanol and 1,5-PDO, respectively. First, we fractionate biomass into its three primary constituents— hemicellulose, cellulose, and lignin—using γ-valerolactone (GVL) as a solvent (Alonso et al., 2017), which also enables excellent yields of furfural by dehydration of xylose. The purified furfural is then used to produce 1,5-PDO (Huang et al., 2017), and the cellulose portion is converted to ethanol via enzymatic hydrolysis and fermentation (Humbird et al., 2011), while the lignin is combusted to generate steam and electricity. We have developed an experimentally based process simulation model (mass and energy balances) to determine the economic potential of the integrated strategy. If we consider a plant processing to 1,000 dry metric tonne per day white birch feedstock and producing 38,520 ton per year 1,5-PDO and 19.3 MMgal per year ethanol, the integrated strategy leads to a minimum ethanol selling price of $1.52 per gallon when 1,5-PDO is sold at $3,100 per ton (Figure 1). It is shown that this strategy could lower the cost of ethanol produced from biomass by more than two dollars per gallon. Finally, we perform sensitivity analysis to identify the major economic drivers of the process and suggest future research directions.

Figure 1. Minimum ethanol selling price as a function of 1,5-PDO selling price. The "green" region (1,5-PDO selling price > $2,775 per ton) indicates that the integrated strategy has favorable economics for ethanol production compared with the biochemical conversion process used in NREL 2011 report (Humbird et al., 2011).

References

1. Alonso, D. M.; Hakim, S.; Zhou, S.; Won, W.; Hosseinaei, O.; Tao J.; Garcia-Negron, V.; Motagamwala, A. H.; Mellmer, M. A., Huang, K.; Houtman C. J.; Labbé, N.; Harper, D. P.; Maravelias, C. T.; Runge T.; Dumesic, J. A. Increasing the revenue from lignocellulosic biomass: Maximizing feedstock utilization. *Science Advances* 2017.

2. Brentzel, Z. J.; Barnett, K. J.; Huang, K.; Maravelias, C. T.; George, W.; Dumesic, J. A.; Huber, G. W. Chemicals from Biomass: Combining Ring-opening Tautomerization and Hydrogenation Reactions to Produce 1,5-Pentanediol from Furfural. *ChemSusChem* 2017, 133, 12689.

3. Huang, K.; Brentzel, Z. J.; Barnett, K. J.; George, W.; Dumesic, J. A.; Huber, G. W.; Maravelias, C. T. Conversion of Furfural to 1,5-Pentanediol: Process Synthesis and Analysis. *ACS Sustainable Chemistry & Engineering* 2017.

4. Humbird, D.; Davis R.; Tao L.; Kinchin C.; Hsu D.; Aden A. Process Design and Economics for Biochemical Conversion of Lignocellulosic Biomass to Ethanol: Dilute-Acid Pretreatment and Enzymatic Hydrolysis of Corn Stover. *NREL/TP-5100-47764*. Golden, CO: National Renewable Energy Laboratory, 2011.

498859 High Value Products from a Photobioreactor-Based Biorefinery

Abstract:

Though Algenol continues research and development activities aimed at biofuels (ethanol and biocrude), increasing emphasis is being placed on high value products and co-products that can both provide sustainable income and serve as a market entry/risk reduction strategy for eventual development of biofuels. Algenol has successfully built and operated a large pilot scale (2 acre) integrated biorefinery (IBR) based on its proprietary photobioreactors (PBRs) and genetically modified cyanobacteria to produce fuel grade ethanol via an EPA approved pathway (70% carbon footprint reduction compared to gasoline). Ethanol productivities in the range of 4000-5000 gal/acre-yr (annualized) have been achieved. However, the system is not economic under current market conditions. The IBR and R&D support infrastructures are now being applied to higher value products, including a biocrude co-product. Though several products are under active consideration, algae based soil amendment is one application that is well advanced in our product development pipline. The soil amendment contains nitrogen-fixing cyanobacteria as the active component (based on *Anabaena* and/or *Nostoc* strains). The formulated product provides enhanced yields as a biofertilizer, can be used as a soil amendment for reclamation of degraded soil, and can result in high carbon storage levels in the soil.[1,2] This talk will use the soil amendment application as an example and provide a description of the selection process for product development (e.g., techno-economics, life cycle impacts, compatibility with PBR growth systems) and the development process from the lab, through outdoor testing, and then to pilot scale deployment. In particular, the alignment of indoor and outdoor performance will be demonstrated for several strains relevant to the above application, along with a modeling system that has been developed for productivity prediction for any rational position on the earth's surface, and for various configurations for Algenol's vertically-oriented PBR arrays. Harvesting and downstream processing will be described and experimentally demonstrated for several cyanobacterial strains, including strains designed for biocrude production. A comparison of the relative productivities and economics of PBR systems versus open pond systems will be discussed.

[1] Accelergy Corporation, *"Montana Integrated Carbon to Liquids Study,"* US DOE Award Number: DE-FE0003595, December 2013.

[2] US Patent2014/0345341, November 27, 2014; *"Integrated Process for Producing Fuels and BioFertilizers from Biomass and Products Produced."*

499440 A Highly Selective Dehydration of D-Xylose and Wheat Straw C_5-Sugars into Furfural By Using Supercritical CO_2 as Catalyst: A Green and Efficient Approach

Abstract:

Crude oil is the principal feedstock for production of liquid fuels and also affords a platform for synthesis of a wide range of added-value chemicals and materials. Due to price fluctuation of crude oil-derived products coupled to increasing concerns about the impact of fossil emissions on environment, society is becoming more environmentally conscious. Thus, traditional chemical technologies for production of fuels and chemicals from biomass are being replaced with innovations. This includes the use of novel and green solvents. In this work, the selective dehydration of both D-xylose and lignocellulose-derived C_5-sugars into furfural, using supercritical CO_2 as catalyst and tetrahydrofuran (THF) as extracting solvent, is for the first time reported. The scrutinised approach involves the benefits from the use of auto-generated carbonic acid from dissolution of CO_2 in aqueous phase, assisted by THF, guiding to the improvement of furfural yield and reaction selectivity. This novel approach allowed to achieve a xylose conversion into furfural above 83 %mol, furfural yield of 70 %mol and selectivity up to 84 %mol at 180 ºC for 60 min with 50 bar of initial CO_2 pressure.[1] At aforementioned operational condition, furfural yields and selectivities up to 40 %mol and 44 %mol with respect to wheat straw hemicellulose hydrolysate were reached.[2] First, the influence of various operating conditions on pentose conversion and furfural yields were investigated. Then, various volumetric ratios of water and THF (v/v) were investigated to help define reaction conditions that favour the highest yield of furfural. In an effort to understand how addition of CO_2 improves furfural yield and its impact by the presence of THF, different pressures of CO_2were also investigated.

Acknowledgments

IF/00471/2015 grants and by BBRI - Biomass and Bioenergy Research Infrastructure (ROTEIRO/0189/2013) and the Associated Laboratory for Sustainable Chemistry–Clean Processes and Technologies–LAQV which is financed by national funds from FCT/MEC (UID/QUI/50006/2013) and co-financed by the ERDF under the PT2020 Partnership Agreement (POCI-01-0145-FEDER-007265).

References

1. A. R. C. Morais and R. Bogel-Lukasik, *Green Chem.*, 2016, **18**, 2331-2334.

2. A. R. C. Morais, M. D. J. Matuchaki, J. Andreaus and R. Bogel-Lukasik, *Green Chem.*, 2016, **18**, 2985-2994.

493990 Process Design for the Synthesis of Biodiesel, Glycerol, 3-Hydroxypropionic Acid and 1,3-Propanediol

Abstract:

The current tendency towards the use of renewable biomass feedstocks instead of the use of non-renewable resources (for example, petroleum), has created the need for development of new, innovative production processes within the chemical and biochemical industries. Therefore, alternative solutions able to reduce the consumption of non-renewable resources and consequently minimize effects on climate change, waste, while, improving product yields, should be promoted. This replacement of petroleum with biomass as the primary material for fuel and chemical production is basically the idea behind the biorefinery concept.

This project is focusing on the ways that glycerol (an intermediate or by-product in a biorefinery) could be converted into value added platform chemicals so that its surplus production is not ending up as waste. The effective utilization of glycerol will ultimately contribute to the viability of biodiesel production since it will significantly lower its already high production cost. Thus, the design and development of a sustainable process to utilize this organic raw material becomes imperative.

The case of the simultaneous production of two platform chemicals with glycerol as a starting material has been investigated, namely of 3-hydroxypropionic acid (3HP) and 1,3-propanediol (1,3PDO). This process is part of a bigger one, where, by using vegetable oil as feedstock, the aim is to produce biodiesel. However, as glycerol is obtained as a by-product, it is then subjected to biotransformation with recombinant resting cells of *Lactobacillus reuteri* in an aqueous solution. The resting cells initially grow in the fermenter at 37°C and then the biotransformation of glycerol is performed in a stirred-tank reactor. This biotransformation has 3 main products, 3-hydroxypropionic acid (3HP) and 1,3-propanediol (1,3PDO), that are of interest as well as significant amounts of 3-hydroxypropionaldehyde (3HPA) as by-product (that is not of interest and needs to be reduced).

L.reuteri is selected for biotransformation for several reasons, such as, it is a safe bacteria, has a great potential as a biocatalyst, and it cannot be used as a carbon source by glycerol, unlike other bacteria. Last but not the least, apart from 3HPA, no other by-products are formed during the bioconversion of glycerol by *L.reuteri*, thereby making the downstream process much simplified.

The goal of the project is to investigate the optimal production process towards 3HP and 1,3PDO simultaneously, in comparison to the existing methods that are oriented to the production of either only 3HP [Ramakrishnan Gopi et al.,2015] or only 1,3PDO [Nakamura and Whited, 2014]. The research is orientated towards a setup of the production process that produces equal amounts of the desired products, is economically and environmentally feasible and sustainable. In addition, the existence of significant amounts of the aldehyde by-product (3HPA) needs to be avoided. Two scenarios are investigated; one where 3HPA is isolated, purified and sold in the market as a by-product, and the other, where it is recycled back to the biotransformation reactor such that the production of 3HP and 1,3PDO are

favoured.

According to the results regarding the production of 3HP and 1,3PDO, it is possible to determine the quantity of glycerol that needs to be allocated for the production 3HP and 1,3PDO and the remaining that will be allocated to the market. That is, the operation will depend on the function of the demand for glycerol, its current price, as well as the demands and prices of 3HP and 1,3PDO.

The presentation will provide the modelling and design details of the integrated process with a special focus on the increase of the atom efficiency of the processes involved, on the sustainability of the production process, the environmental impact versus the production and capital costs for different scenarios of product demand and price.

484549 Evaluation of Aromatics Formation in Thermal Cracking of Triglycerides

Abstract:

Liquid fuels produced by biomass conversion offer an alternative to our reliance on fossil sources. Among the process available in the literature, thermal cracking results on a liquid product (bio-oil) with similar characteristics to petroleum crude oil, especially when performed with triglyceride sources. However, undesired compounds, such as carboxylic acids and benzene might be formed as well. Although aromatics may have positive aspects (e.g. increasing the octane number on light fuels), they are toxic and, particularly benzene has its content restricted by local regulations. Moreover, benzene removal involves complex processes, being wiser to avoid its formation.

The aim of this work was to investigate one of the hypothesis for benzene formation, regarding a relationship between the degree of unsaturation of the biomass with the benzene content in bio-oil. Thermal cracking experiments were performed using triglyceride sources with three different degrees of unsaturation, in the same operational conditions. Commercial soybean oil and blends with 10% and 20% of hydrogenated were constituted as samples to be analysed. Density, acid index and iodine index were measured to evaluate the blending effect. The thermal cracking experiments were performed in triplicate at 525 ^{o}C of temperature and feed mass flow of 300 g/h in a continuous screw reactor operating over steady state conditions. The mass balance shows the liquid yield increased and coke formation decreased with for higher blends.

Benzene content decreased by 14% in the bio-oil produced by the biomass with the lowest degree of unsaturation. Also, the reactions for carboxyl group removal were favorable to the saturated compounds, reducing the acidity index in the biomass with the lowest unsaturation content. Notwithstanding, long-chain paraffinic compounds remained in the products, which suggests that saturated compounds does not crack easily.

500032 Tuning the Structure and Electronic Properties of Manganese Nitrides for Ammonia Production from First-Principles Modeling

Abstract:

The conventional Haber-Bosch ammonia synthesis process plays a significant role in the global food chain by feeding approximately 40% of worldâ€™s population, but is also extremely energy intensive, consuming 1-2% of the worldâ€™s energy supply. Ideally, simple technology enabling intermittent operation using renewable energy will be made available through step catalysis by exploiting manganese (Mn) based alloy materials to facilitate both nitrogen activation and ammonia production.

In this talk, a first-principles investigation based on periodic density functional theory (DFT) calculations has been carried out by focusing on the elucidation of molecular-level mechanisms for ammonia formation using a model based on the Mn_4N nitride phase, which can be readily obtained through the nitridation of metallic manganese. It is hypothesized that the reduction will follow a Mars-van Krevelan-type mechanism. Under a reducing environment at 1 atmospheric H_2 pressure, this computational study has shown that it is thermodynamically favorable to form NH utilizing the surface N species. Afterwards, the reaction becomes endothermic for ammonia evolution from NH. The diffusion of N in the nitride crystal lattice can influence the overall kinetics, and has been considered in this study as well. It turns out that both the overall thermodynamics and the N diffusion kinetics are the key for ammonia production and should be considered as the parameters to tune the materials performance. Also in this talk, the simple strategy to dope and alloy manganese with other transition metals to validate the research hypothesis has been explored computationally, and will be presented and discussed.

496330 Rapid and Extended Analysis of Syngas Using Micro GC Fusion

Abstract:

In 2012, the World Bank estimated that 140 billion cubic meters (bcm) of the 3370 bcm of natural gas produced worldwide was flared. Flaring burns surplus gas and is detrimental to the environment, contributing over 350 million tons of CO_2 in annual emissions. An alternative to flaring is to convert the associated natural gas into synthesis gas (syngas), which is an intermediate gas used to produce synthetic liquid fuel. Converting syngas to liquid chemicals through the Fischer-Tropsch (FT) process presents an economical and greener alternative to flaring. Syngas is composed predominantly of hydrogen and carbon monoxide, though it may also contain carbon dioxide, nitrogen, methane, ethane, ethylene, or H_2S in the high ppm to percent level. After catalytic conversion, determining the syngas composition is a critical step before using the gas in any downstream process. To ensure efficiency, the analysis of syngas must be analyzed quickly and accurately. Building on proven microelectromechanical systems (MEMS) technology, Micro GC Fusion is capable of providing rapid analysis of typical syngas components in a compact instrument. A method equipped with temperature programming provides an area %RSD of less than 1.0% and retention time %RSD of less than 0.05% repeatability. This accurate and repeatable analysis of syngas is performed in less than 50 seconds. Additionally, Micro GC Fusion is capable of performing analyses on more complex samples including C4-C5 olefins, and C1-C12 alkanes. This extended analysis of syngas is performed in less than 90 seconds.

490981 Dual-Alcohol Blending Effects on Gasoline Properties

Abstract:

A pure alcohol cannot provide the distillation temperature range required by spark ignition engines for smooth driveability. The front-end volatility in a distillation curve is of importance for cold start, engine warm-up, evaporative emissions, and vapor lock. Mid-range volatility is vital for warming up, acceleration, and cold weather performance. Finally, tail-end volatility is essential for running when the engine is hot, and it also indicates the propensity for combustion deposits and oil dilution. In addition, the high enthalpy of vaporization of alcohols is a prohibitive factor for good cold-start behavior. Therefore, both light and heavy hydrocarbon fractions of gasoline are needed to supply an ignitable mixture under these challenging conditions. In general, blends of gasoline with a single alcohol also have shortfalls. If lower alcohols (methanol and ethanol) are used, azeotropic behaviour due to the high polarity of these molecules, as well as their low calorific value, high enthalpy of vaporization and low stability would be problematic. If higher alcohols are used, achieving very low volatilities is challenging and can significantly limit the maximum volume that can be blended with gasoline. Thus, in order to increase the portion of bio-alcohols in the gasoline blend, another approach is needed. Using multi-alcohol-gasoline blends is an approach that might achieve a fuel with high portion of biofuel while meeting all standard specifications with no need for major changes in existing infrastructures and engines. Mixing a lower and a higher alcohol (e.g., ethanol and 1-butanol) in gasoline can provide a mixture that has the same vapor pressure as that of gasoline while reducing the evaporative emission on the fuel delivery system. However, other physicochemical and combustion related properties of dual alcoholâ€“ gasoline blends have not determined to fully characterize these fuel blends for a satisfactory performance in existing spark ignition engines. This presentation will describe the development of a database of all normal and cyclic C1-C10 alcohols and a procedure to downselect for a pair of alcohols that can be blended with gasoline in different target volumes from midrange (20 vol. %) to high range (40 vol. %). After selecting the desired pair of alcohols, a range of properties were measured to evaluate the fuel blends: distillation curve, calorific value, vapor pressure, vapor lock index, ignition quality, water tolerance and phase separation, density, and viscosity.

492614 Mechanism for Selective Fructose Etherification on Hierarchical Sn-SPP Zeolite

Abstract:

Isomerization of glucose to fructose is a key reaction for enabling the conversion of sugars into a wide range of biorenewable chemicals through furanic intermediates. The highest glucose to fructose yields on Sn-Beta are 33%[1], but a new catalyst, Sn-SPP can achieve a record 65% yield[2], even in excess of the glucose/fructose equilibrium (50%). Sn-SPP catalyzes a new reaction: fructose etherification (formally a ketalization) to ethyl fructoside, which reverts to fructose after water is added post-reaction (see Scheme 1). After these initial experimental results, many fundamental questions were unanswered, including how the structure of the active site enables this chemistry, why Sn-SPP performs ketalization while Sn-Beta does not, and why fructose ketalization is catalyzed but not glucose acetalization.

Scheme 1 - Reaction scheme for improving fructose yields from glucose using tandem reactions with fructose etherification.

We answered these questions using periodic DFT calculations and Monte Carlo simulations. A screening algorithm identified the most thermodynamically stable active site geometries; Sn-SPP has a "closed" Sn site with three nearby silanol groups, while Sn-Beta has a hydrolyzed "open" Sn site with a single, more distant silanol group. Identification of the most favorable ketalization mechanism revealed that the silanols in Sn-SPP facilitate ketalization through key H-bonding interactions at the transition state, particularly via a Sn-O-Si-OH moiety. However, the silanol in the Sn-Beta active site is not positioned to stabilize the TS, giving a larger barrier and indicating why Sn-Beta is not active for this chemistry. Analyzing glucose acetalization revealed differences in stability of the key oxonium intermediate at the respective transition states, indicating the reason for the remarkable selectivity of this process for fructose ketalization over glucose acetalization.

1. Moliner, M.; Roman-Leshkov, Y.; Davis, M. E. E.; Román-Leshkov, Y.; Davis, M. E. E. *Proc Natl Acad Sci U S A* **2010**, *107* (14), 6164–6168.

2. Ren, L.; Guo, Q.; Kumar, P.; Orazov, M.; Xu, D.; Alhassan, S. M.; Mkhoyan, K. A.; Davis, M. E.; Tsapatsis, M. *Angew. Chemie - Int. Ed.* **2015**, *54* (37), 10848–10851.

493506 Group IV and V Periodic Trends in Olefin Epoxidation: Effects of Local Environment and Electronic Structure

Abstract:

Hydrogen peroxide (H_2O_2) is a green oxidant that can epoxidize a variety of olefins over site-isolated group IV and V metal oxides (e.g., grafted onto mesoporous silica[1] or incorporated into zeolites[2,3]). These group IV and V metals incorporated into the *BEA framework (M-BEA) have been shown to activate H_2O_2 to form epoxidation-active hydroperoxide (M-OOH) or superoxide (M-$(O)_2^-$) (collectively referred to as M-(O_2)*) species, respectively. The reactivity and selectivity towards epoxidation is proportional to the functional Lewis acid strength of the catalyst (i.e., adsorption enthalpies of Lewis-acid bound deuterated acetonitrile) or the ligand-to-metal charge transfer (LMCT) energy of the reactive intermediate.[3] In this work, we explore the electronic properties of these activated metal complexes and extend our previous findings to include an analogous series of catalysts grafted onto mesoporous silica (M-SiO_2) via calixarene complexes. The differences in the reactivity towards olefin epoxidation over M-BEA and M-SiO_2 catalysts are attributed to two main effects. First, the microporous structure of the *BEA framework solvates, and thus stabilizes, the interaction between the olefin and M-(O_2)* intermediate in the transition state, which decreases the apparent activation enthalpies relative to SiO_2. Second, the coordination environment of the M-BEA or M-SiO_2 affects the electronic structure of the active intermediate (as detected via UV-vis spectroscopy), which results in a systematic decrease in the LMCT band energies for M-SiO_2 catalysts.

Turnover rates for styrene (C_8H_8O) oxide formation through primary reaction pathways as a function of styrene (C_8H_8) and H_2O_2 concentrations over M-BEA and M-SiO_2 (M = Ti, Nb, or Ta) reveal two distinct regimes, which correspond to a change in the most abundant surface intermediate (MASI). The observed dependencies are consistent with a rate expression that describes the irreversible activation of H_2O_2 to form a pool of M-(O_2)* intermediates, which then react with C_8H_8 or H_2O_2 via an Eley-Rideal mechanism to form C_8H_8O or H_2O_2 decomposition products, respectively. Time-resolved UV-vis measurements acquired *in situ* during reaction of cyclohexene with H_2O_2-activated M-BEA and M-SiO_2, in conjunction with molecular probe reactions with *cis*-stilbene, support the trend that the reactive intermediate on Ti-based catalysts is Ti-OOH (i.e., hydroperoxide), while Nb- and Ta-materials react through a M-$(O_2)^-$ (i.e., superoxide) intermediate.

Understanding of the fundamental properties of the catalysts and reactive intermediates that are responsible for the differences in apparent activation barriers for epoxidation require comparisons of these catalysts made under similar conditions (i.e., MASI). Activation enthalpies for C_8H_8O formation ($\Delta H^{\ddagger}_E$) are measured on similar M-(O_2)* saturated surfaces on all catalysts (shown by rates that are proportional to $[C_8H_8]$ and independent of $[H_2O_2]$), and correlate linearly with both the measured adsorption enthalpies of pyridine (ΔH_{Py}, determined via van't Hoff analysis of FTIR spectra of pyridine coordinated to the Lewis acidic Ti, Nb, and Ta atoms within the BEA framework or grafted onto SiO_2) and also to the LMCT band energy of the reactive intermediates on these materials. These comparisons reveal two critical differences among these materials. First, $\Delta H^{\ddagger}_E$ and ΔH_{Py} values for M-SiO_2 materials are systematically greater by 19 ± 2 and 11 ± 2 kJ mol^{-1} than M-BEA materials, respectively, which is consistent with the difference in bulk-averaged adsorption enthalpies for C_8H_8 and pyridine in the *BEA framework and in mesoporous SiO_2 (measured in the absence of metal atoms). This difference is attributed to the microporous environment of *BEA solvating the transition state (relative to the initial state) to a larger extent than SiO_2, which results in the systematic decrease of apparent activation and adsorption enthalpies between *BEA and SiO_2. Second, the linear relationship between $\Delta H^{\ddagger}_E$ and the LMCT band energy of the active intermediate possesses the same systematic offset in $\Delta H^{\ddagger}_E$ between *BEA and SiO_2 that matches the differences in adsorption enthalpies of C_8H_8 (*vide supra*). Interestingly, the M-SiO_2 catalysts possess systematically lower energy LMCT bands, which may be attributed to one of two phenomena, or both. First, the hydroperoxidolysis of a M-OSi bond in the zeolite framework leaves a vicinal SiOH group in a tetrahedral position that may contribute a small amount of electron density towards the metal atom, which decreases the Lewis acid strength of the metal atom. Second, the neighboring SiOH group in *BEA likely distorts the geometry of the M-(O_2)* intermediates, which will lower the extent of orbital overlap, lengthening the M-O bonds, and thus increase the LMCT band energy. The overarching trend relating $\Delta H^{\ddagger}_E$ and the LMCT band energy suggests that metal centers that are highly electron withdrawing (i.e., possess more highly polarized M-(O_2)* bonds) result in more electrophilic and reactive intermediates. To develop an intuition for the electrophilicity of the reactive intermediates, turnover rates for the epoxidation of *para*-substituted styrene (X-C_8H_8, where X = OMe, Me, H, Br, and NO_2) were measured under identical reaction conditions (3 mM X-C_8H_8, 0.01 M H_2O_2, 313 K). The corresponding Hammett plot reveals a reaction constant (ρ) of -0.9 ± 0.05 for all M-β and M-SiO_2, which definitively shows that the reactive intermediates implicated in olefin epoxidation have a strong preference to react with electron-rich olefins. Collectively, these data show that the rational design of next-generation epoxidation catalysts should seek to increase the Lewis acid strength of the metal center to increase the electrophilicity and reactivity of the active intermediates and leverage the decrease in apparent activation barriers that are attributed to the solvation of the transition states in microporous environments.

References:

1) Thornburg, N.E.; Thompson, A.B.; Notestein, J.M.; *ACS Catal.*, **2015**, *5*, 5077-5088.

2) Bregante, D.T.; Priyadarshini, P.; Flaherty, D.W.; *J. Catal.*, **2017**, *348*, 75-89.

3) Bregante, D.T.; Flaherty, D.W.; **2017**, *In Revision*

498047 Sulfur Deactivation Pathways in Cu-SSZ-13 Determined through First-Principle Modeling and X-Ray Spectroscopy

Abstract:

SSZ-13 zeolites with exchanged copper ions are used commercially in diesel exhaust treatment for the selective catalytic reduction (SCR) of nitrogen oxides into nitrogen and water because of their high activity and stability [1]. Exposure to sulfur oxides, which are produced from the combustion of sulfur in diesel fuel, causes a decrease in catalytic activity. The sulfur-containing species will accumulate in the catalyst in the long term and cause worse deactivation over time. Understanding of the poisoning intermediates and reaction pathway at the molecular level is insufficient due to the lack of theoretical studies. In this work we use first-principle density functional theory (DFT) calculations and ab initio molecular dynamics (AIMD) simulations to identify poisoning intermediates and examine molecular configuration at finite temperatures. These simulations provide a molecular interpretation of ex situ and in situ X-ray adsorption spectroscopy (XAS) experiments performed on poisoned Cu-SSZ-13 samples.

The interaction between Cu active sites and sulfur-containing species with variable O/H/N stoichiometry ($SO_xH_yN_z$) was explored by thermodynamic phase diagrams. We used AIMD simulation to explore molecular configurations of the adsorbate and picked lowest energy snapshots for structure optimizations. The potential of mean force method was used to calculate adsorption entropies of $SO_xH_yN_z$ on Cu sites. The phase diagrams show that ammonium sulfate and ammonium bisulfate are thermodynamically favorable below 600 K for a range of experimental conditions. Two Cu-SSZ-13 samples, containing either Z_2Cu or ZCuOH sites, were synthesized in lab [2] and exposed to dry SO_2 and O_2 at variable temperatures to simulate accumulated sulfur poisoning in actual catalytic converters. We performed ex situ and in situ X-ray spectroscopy on the sulfated samples and analyzed the near edge structure (XANES) region of the spectra. Oxidation states elucidated from both Cu K-edge XANES and S K-edge XANES were consistent with the conclusion from first-principle studies. The difference in poisoning species on the two types of Cu sites suggested the possibility of designing a more sulfur-resistant catalyst.

[1] Kumar, A., Smith, M., Kamasamudram, K., Currier, N.W., An, H., Yezerets, A.. *Catal. Today*. 231, 75 (2014).

[2] Paolucci, C., Parekh, A.A., Khurana, I., Di Iorio, J.R., Li, H., Albarracin Caballero, J.D., Shih, A.J., Anggara, T., Delgass, W.N., Miller, J.T., Ribeiro, F.H., *J. Am. Chem. Soc.* 138, 18 (2016).

486158 Catalytic Consequences of Framework Polarity for Ethanol Dehydration on Sn-Beta Zeolites

Abstract:

Catalytic Consequences of Framework Polarity for Ethanol Dehydration on Sn-Beta Zeolites

Jason S. Bates, Brandon C. Bukowski, Jeffrey P. Greeley, Rajamani Gounder*

Charles D. Davidson School of Chemical Engineering, Purdue University, West Lafayette, IN 47907

*rgounder@purdue.edu

Lewis acidic Sn-Beta zeolites containing different silanol defect density and framework metal site coordination (open: (HO)-Sn-$(OSi)_3$, closed: Sn-$(OSi)_4$) have been recognized to behave catalytically different in reactions of oxygenated compounds. In aqueous-phase glucose isomerization, titration and quantification of sites in Sn-Beta zeolites using pyridine and CD_3CN [1], together with theoretical estimates of 1,2-intramolecular hydride shift activation barriers [2], indicate that open sites are the dominant active sites. Furthermore, glucose isomerization *initial* rate constants (373 K, per open Sn) are >10× higher for open sites confined within Sn-Beta-F (fluoride-mediated, low-defect) than Sn-Beta-OH (hydroxide-mediated, high-defect) [1], which has been ascribed to the presence of extended hydrogen-bonded water structures that increase activation free energies in high-defect pores. Here, we employ gas-phase ethanol dehydration as a probe reaction to clarify the kinetic consequences of defect density in the absence of bulk solvent effects, and assess the influence of reaction conditions on active Sn site coordination.

Bimolecular ethanol dehydration rates to form diethyl ether (DEE) were measured over low-defect Sn-Beta-F and high-defect Sn-Beta-OH zeolites at 404 K, ethanol pressures between 0.5–35 kPa, and water pressures between 0.1–50 kPa. Rate data were interpreted using a mechanism-derived rate expression that describes measured reaction orders, supported by density functional theory calculations and microkinetic modeling to discriminate among candidate reaction pathways and surface intermediates. Infrared spectroscopic quantification of open and closed Sn sites using CD_3CN titrants before and after reaction indicated that Sn sites change structure dynamically during catalysis, leading to turnover rates (per total Lewis acidic Sn) that are invariant (within ~2–3× at 404 K) among six Sn-Beta-OH samples with varying Sn content (Si/Sn=30–95). Polar silanol groups located within the pores of Sn-Beta-OH preferentially stabilize, relative to Sn-Beta-F, dimeric intermediates that involve adsorption of a second polar water or ethanol molecule with an ethanol monomer, reflected in more negative apparent water orders at high water pressure, in lower apparent ethanol orders at high ethanol pressure, and in ~4× higher values of the regressed thermodynamic equilibrium constants (at 404 K) associated with ethanol-water dimer formation. These kinetic observations are consistent with single-component adsorption isotherms and two-component adsorption measurements that indicate higher uptakes of water relative to ethanol in high-defect micropores. Hence, differences in DEE formation rates (404 K) between the two materials arise solely from differences in prevalent surface coverages of inhibiting and reactive dimers rather than the intrinsic rate constants for DEE formation, which are insensitive to framework polarity.

[1] Harris et al., J. Catal. 335 (2016) 141–154.

[2] Bermejo-Deval et al., Proc. Natl. Acad. Sci. 109 (2012) 9727–9732.

497596 Size Selective Oxidation of Alkanes By Microporous Oxides

Abstract:

Introduction

MoVTeNb mixed oxides, with a particular crystal phase called the M1 phase, are among catalysts that provide the highest selectivity to desired C_2H_4 products in C_2H_6 oxidative dehydrogenation (ODH) with the least amounts of combustion products (CO, CO_2).[1] The origin of their selective catalysis is typically attributed to electronic properties of their lattice oxygens.[2] The M1 phase contains network arrangement of the corner and edge shared octahedra that form pentagonal, hexagonal and heptagonal channels. The heptagonal channels have dimensions comparable to kinetic diameters of linear alkanes (~ 0.4 nm).[3] Recently, Ueda et al. reported analogous crystalline materials containing only Mo and V elements and, by varying crystallite sizes, showed that C_2H_6 molecules react within heptagonal channels, while C_3 oxygenates react on external surfaces.[4] Here, we probe the origin of selective catalysis of these materials by examining oxidations ethane (C_2H_6) molecules that can enter the heptagonal channels, and cyclohexane (C_6H_{12}) molecules that are too big to enter these channels. We measure the C_2H_6/C_6H_{12} rate ratios and selectivity trends on MoVTeNb oxides and compare them with silica supported vanadium oxides (VO_x/SiO_2) without micropores. Such comparisons lead us to conclude that confinement within micropores plays a significant role in selective C_2H_6ODH catalysis.

Materials and methods

MoVTeNb mixed oxide samples were prepared by a hydrothermal method,[5] followed by thermal treatment in flowing Helium. The crystal structures were determined using X-ray diffraction (XRD) and microstructural details were examined using electron microscopy. VO_x/SiO_2 (41%wt.) samples were prepared by wet impregnation of ammonium metavanadate and followed by thermal treatments, flowing dry air. Surface areas and elemental compositions of oxides were measured using nitrogen physisorption methods and inductively coupled plasma atomic emission spectra, respectively. Samples were diluted with $SiO_2$to eliminate heat and mass transfer limitations during reactions. Catalytic alkane oxidation experiments were carried out in quartz plug flow reactors and the products were measured via online gas chromatography.

Results and discussion

The XRD pattern of MoVTeNbO is consistent with crystalline oxides with majority M1 phase. Electron micrographs reveal rod-like structures of crystalline oxides typical of the M1 phase (~1 μm length and 0.2-0.4 μm diameter) with a small fraction of much smaller particles. The catalytic sites of VO_x/SiO_2, a catalyst without micropores, are equally accessible to C_2H_6 and C_6H_{12} (Figure 1a) In contrast, the micropores of MoVTeNbO allow access to ethane, leading to rates from pores and external surfaces ($r1_{ext} + r1_{pore}$; Figure 1a), while restricting C_6H_{12} rates to external surfaces ($r2_{ext}$; Figure 1a). Therefore, the C_2H_6/C_6H_{12} ODH rate ratio on MoVTeNbO relative to that on VOx/SiO_2 reflects the contribution of micropores to C_2H_6 conversion. The measured rate ratio is 52 times higher for MoVTeNbO than that of VO_x/SiO_2, suggesting that most of the C_2H_6 ODH reactions on MoVTeNb occur within the pores (Figure 1b). The C_2H_6 oxidation product selectivity to C_2H_4, at similar conversions, is much higher on MoVTeNb than on VOx/SiO (Figure 2a), while both catalysts exhibit similar C_6H_{12} oxidation product selectivity to cycloalkenes and COx (Figure 2b). These trends confirm that only the pores, and not the electronic properties of external surfaces, are responsible to high alkene selectivity. The dependence of rates on reactant pressures are consistent with Mars-van-Krevelen redox cycles. Activation enthalpies for measured rate constants are lower for C_2H_6 than for C_6H_{12}, in spite of the weaker C-H bonds in the latter reactant, which is consistent with the role of van der Waals (vdW) interactions within pores in stabilizing the C_2H_6 activation transition states. We conclude that the high C_2H_4 selectivity arises from steric restriction to secondary reactions in the heptagonal channels. Our preliminary DFT calculations confirm that such vdW stablizations and steric restrictions are prevalent in the micropores.[6]

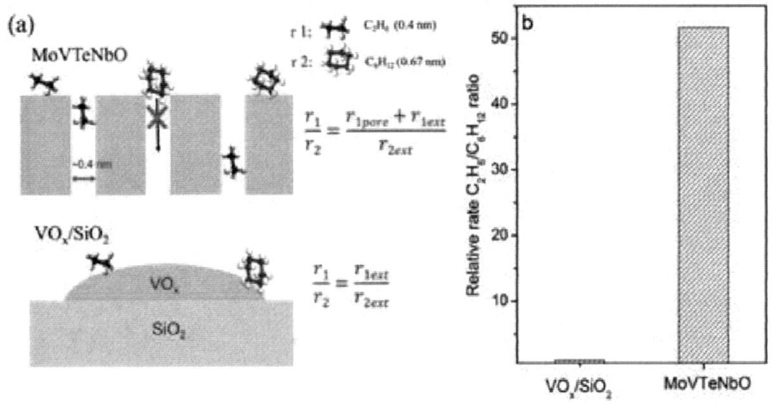

Figure 1. (a) Scheme representing size selectivity of alkane oxidation on MoVTeNb oxides and VO_x/SiO_2 using C_2H_6/C_6H_{12} rate ratios. (b) Measured rate ratios. [Reaction conditions: 30 cc/min; 3 kPa C_2H_6 or C_6H_{12}; 3 kPa O_2; 648 K; Conversions: VO_x/SiO_2 - 5% (C_6H_{12}) and 0.4 % (C_2H_6); MoVTeNbO - 8% (C_6H_{12}) and 1.7% (C_2H_6)]

Figure 2. (a) C_2H_6 and (b) C_6H_{12} product selectivity on the two oxides. [Reaction conditions: 30 cc/min; 3 kPa C_2H_6 or C_6H_{12}; 3 kPa O_2; 648 K; Conversions: VO_x/SiO_2 - 5% (C_6H_{12}) and 0.4 % (C_2H_6); MoVTeNbO - 8% (C_6H_{12}) and 1.7% (C_2H_6)]

Financial support from Tufts University and its Faculty Research Fund are gratefully acknowledged.

References

1. Botella, P.; López Nieto, J. M.; Solsona, B.; Mifsud, A.; Márquez, F. *J. Catal.* **2002,** *209*, 445-55.

2. Getsoian, A. B.; Zhai, Z.; Bell, A. T. *J. Am. Chem. Soc.* **2014,** *136*, 13684-97.

3. Ishikawa, S.; Ueda, W. *Catal. Sci. Technol.* **2016,** *6*, 617-29.

4. Ishikawa, S.; Yi, X.; Murayama, T.; Ueda, W. *Appl. Catal., A* **2014,** *474*, 10-17.

5. Gaffney, A. M.; Song, R, *US 7074956 B2*, B2, **2004**.

6. Cheng, M.-J.; Goddard, W. A. *Top. Catal.* **2016,** *59*, 1506-17.

500609 Impact of Hydrocarbon Trapping on Temperature Programmed Oxidation Â Â over a Pt/Pd/BEA Monolith Catalyst

Abstract:

Impact of hydrocarbon trapping on temperature programmed oxidation over a Pt/Pd/BEA monolith catalyst

Po-Yu Peng, Michael P. Harold, Dan Luss***

Department of Chemical and Biomolecular Engineering, University of Houston, Houston, TX 77204

*mharold@uh.edu, **dluss@uh.edu

The function of the diesel oxidation catalyst (DOC) comprising a washcoat of Platinum (Pt) and Palladium (Pd) (PGM = precious group metals) is to catalyze the oxidation of diesel effluent mixture of CO and hydrocarbons (HCs). A DOC operates at very high CO and HC conversion (>90%) after a sufficiently high operating temperature has been reached. However, during the "cold start" a fraction of the exhaust hydrocarbons may escape unreacted [1, 2]. A large-pore zeolite such as Beta (BEA) is added to the DOC to trap diesel exhaust hydrocarbons during the cold start. High molecular weight hydrocarbons are especially prone to trapping in BEA-modified PGM DOCs. Although extensive research has been conducted on the BEA storage capacity, there is a need to gain insight on how BEA affects the performance of HC light-off (LO) and of the HC oxidation transient performance [3-7]. We conducted experiments to gain a deeper understanding of functioning of the HC trap and to enable improved trap design and operating strategy. The insight gained from this study will help optimize the multi-functional and component catalyst formulation, reactor design and operating strategy for achieving high conversion of multiple hydrocarbon species.

In a typical experiment dodecane (C_{12}) is trapped on a 0.8 in. diameter by 3-in. long mixed washcoated Pt-Pd/BEA/Al_2O_3 400 cpsi cordierite monolith. The catalyst contains 60 g/ft^3 PGM in a Pt:Pd ratio of 2:1 by weight (~1:1 atomic ratio) on a 0.7 g/in^3 load of γ-Al_2O_3 support and incorporated with the BEA content (0 and 0.5 g/in^3) into the washcoat. The C_{12} and CO_2 gas concentrations were measured by spatially resolved mass spectrometry (SpaciMS, QMS-Omnistar GSD 300) and a FTIR. Spatial resolved temperature profiles were measured by a coherent optical frequency domain reflectometer (c-OFDR with a 1.26 m long gold-coated optical fiber placed in the center of monolith channel. The trapping experiments were conducted by flowing 160 ppm C_{12}, 10% O_2 and the balance Ar at 60 ^oC until the catalyst was saturated. This was followed by ~10 ^oC/min inlet temperature ramp. The total gas flowrate was 8 L/min (GHSV= 12,438 h^{-1}). During the dynamic hysteresis analysis, the saturated catalyst was heated up by 1 ^oC/min from 110 ^oC to 170 ^oC inlet temperature and then cooled by -1 ^oC/min from 170 ^oC to 110 ^oC. The total gas flowrate varies from 6 L/min (GHSV= 10,262 h^{-1}) and 8 L/min (GHSV= 12,438 h^{-1}). Adiabatic behavior was confirmed by measuring temperature profiles under non-reaction conditions.

Unusual CO_2 formation spikes were observed during the oxidation of HC that was pre-stored on the catalyst before exposure to the temperature ramp. Fig. 1 shows the effluent C_{12} concentrations along with the effluent generated CO_2 concentration during a linear feed temperature ramp. The C_{12} is completely trapped by C_{12} breakthrough with a slow approach to the feed concentration value at uniform temperature of ~60 ^oC. However, during the subsequent temperature ramp, in addition to the release of C_{12} at the 48 minute mark, two impressive spikes of CO_2 appeared between 50 and 57 minutes. Corresponding to the first two spikes are measureable spikes in the outlet temperature.

The c-OFDR measurements enabled construction of a 3-D contour plot of temperature versus time and position. (Figure 2) The temperature spikes correlate with the measured CO_2 yield spikes. The feed temperature increased linearly during the ramp. The local temperature sequentially increased and decreased three times at a nearly fixed position in the monolith. During the first and second cooling period, the temperature went back close to the inlet temperature. This behavior suggests that the oxidation stops abrubtly as the generated heat is quickly removed by the flowing gas stream. The drop in the CO_2 concentration between the first and second peaks in Fig. 1 also supports the stoppage of the HC oxidation. A mass balance analysis indicates that the fed C_{12} during this time period is released and re-trapped by the catalyst. The C_{12} continues to accumulate until that the catalyst saturates. This intermittent HC oxidation behavior continues until the HC trapping capacity becomes smaller at high temperature. The rapid propagation of the maximum temperature from the midstream to the downstream during the first peak also supports high heat convection by the gas flow. This rapid temperature peak propagation greatly accelerates the HC oxidation downstream.

A slow heating/cooling rate leads to a significant counter-clockwise hysteresis behavior, underscoring the competition between HC trapping and oxidation. Figure 3 shows the C_{12} and CO_2 concentrations during the hysteresis under the slow heating/cooling rate of ±1 ^oC/min. In the ramp-up period, the C_{12} (CO_2) gradually decreases (increases), indicating a delicate balance between trapping and oxidation. After ignition at ~159 ^oC, more impressive peaks occur. The occurrence of multiple peaks suggests that the C_{12} trapping occurs before the C_{12} oxidation. After a certain amount of C_{12} is trapped, the C_{12} oxidation reaction then triggers the release of stored-C_{12}. The HC trapping capability becomes smaller at high temperatures. During the ramp-down period, a high C_{12} conversion remains until the feed temperature is below ~136 ^oC. The reactor extinguishes at ~23 ^oC below the ignition, showing a strong ability to reduce HC emission when the engine stalls.

The non-isothermal nature of this adiabatic system suggests that HC trapping and oxidation dominates in the different regions in the catalyst. Since the downstream is much hotter than the upstream, the HC is re-trapped by BEA at the upstream and the HC is drastically oxidizes at the downstream. Both events greatly consume the C_{12}, helping to reduce C_{12} emission, and lead to a much lower extinction temperature. Even at such low heating/cooling rate of ±1 ^oC/min, a significant counter-clockwise hysteresis behavior still exists. This study is the first to provide insight into the spatiotemporal features of transient behavior of HC trapping and oxidation of BEA/DOC catalysts, and also the first discovery of multiple HC LO phenomena.

Figure 1. Dodecane and CO_2 concentration measured during adsorption followed by TPO with BEA-zeolite added $Pt/Pd/Al_2O_3$ monolith catalyst. Condition during saturation: 160 ppm dodecane and 10% O_2 balanced with Ar. Condition during desorption and TPO: 10% O_2 balanced with Ar.

Figure 2. The 3-D plot of the spatially-resolved temperature versus time in feed of 10% O_2 balanced with Ar. (Catalyst: D-05)

Figure 3. Experimental C_{12} and CO_2 concentration versus inlet fluid temperature in a dynamic hysteresis analysis. (C_{12} = 160 ppm, 10% O_2; space velocity of 12,438 h^{-1}; temperature ramp = ±1 °C/min).

References

1. Mukai, K., et al., *Adsorption and desorption characteristics of the adsorber to control the HC emission from a gasoline engine*. 2004, SAE Technical Paper.

2. AL-Harbi, M., et al., *Competitive no, co and hydrocarbon oxidation reactions over a diesel oxidation catalyst.* The Canadian Journal of Chemical Engineering, 2012. **90**(6): p. 1527-1538.

3. Li, H.-X., et al., *Application of zeolites as hydrocarbon traps in automotive emission controls.* Studies in Surface Science and Catalysis, 2005. **158**: p. 1375-1382.

4. Chang, H., et al., *Gasoline Cold Start Concept (gCSC™) Technology for Low Temperature Emission Control.* SAE Int. J. Fuels Lubr, 2014. **7**(2): p. 480-488.

5. Murakami, K., et al., *Development of a high performance catalyzed hydrocarbon trap using Ag-zeolite*. 2004, SAE Technical Paper.

6. Elangovan, S., et al., *A comparative study of zeolites SSZ-33 and MCM-68 for hydrocarbon trap applications.* Microporous and mesoporous materials, 2006. **96**(1): p. 210-215.

7. Elangovan, S., et al., *Silicoaluminophosphate molecular sieves as a hydrocarbon trap.* Applied Catalysis B: Environmental, 2005. **57**(1): p. 31-36.

495377 Consequences of Diffusion, Acid Strength, and Confinement on Bifunctional Reactions of Alkanes

Abstract:

Hydrocracking and hydroisomerization reactions convert large-molecular-weight alkane feedstocks to more valuable smaller and more highly branched molecules, respectively, through cascade reactions mediated by alkene intermediates [1]. Such reactions require presence of metal function—sufficient to establish alkane-alkene equilibrium for reactants and to scavenge intermediates and alkene products—that resides within diffusion distances from kinetically relevant acid function. These distances are seldom well-defined and are sensitive to synthetic and formulation protocols but, in part, define concentration gradients of reactant and product alkenes with significant consequences for rates and product selectivities, respectively, especially when acid sites reside within channels and voids of zeolite or zeotype materials (with characteristic sizes ~0.4-1.1 nm [2]). These apertures and voids restrict diffusion of reactants and products—and do so sensitively based on molecule size and shape—and also confer higher reactivity through stabilizing van der Waals and electrostatic interactions with carbocation moieties at ion-pair transition states. Resultant concentration gradients of product alkenes depend on the size of the acid domain and on the density of protons and lead to product selectivities reflecting multiple secondary interconversion events occurring before product alkenes exit acid domains and become hydrogenated at extracrystalline metal functions. Such diffusion-enhanced secondary reactions are rarely treated using rigorous kinetic-transport formalisms, and their effects on measured isomerization and β-scission rates and selectivities are typically attributed, unconvincingly and often inaccurately, to transition state selectivity or to effects of framework structure on acid strength.

Here, we dissect the effects of diffusion, acid strength, and transition state confinement to assess their respective catalytic consequences. Kinetic data, benchmarked against density functional theory, are reported for n-heptane (nH) and 2,4-dimethylpentane (24DMP) isomerization and β-scission on physical mixtures of Pt/SiO_2 and zeolites with diverse confining environments (MFI, MTW, BEA, SFH, FAU) but similar acid strengths and MFI zeotypes with different framework heteroatoms (Al, Fe, Ga, B) which contain H^+ of different acid strength within the same confining environment. To attain intrinsic product formation rates, we assess the effects of product-alkene concentration gradients within acid domains using rigorous reaction-transport formalisms that account for diffusion timescales independently measured from isochoric, isothermal transient uptake of 2,2-dimethylbutane. We additionally develop and report descriptors of confinement as van der Waals interaction energies, E_{vdw}, for transition states based on Lennard-Jones potential force fields [3] and of acid strength as deprotonation energy (DPE), defined as that required to heterolytically cleave the Brønsted O-H bond; these descriptors, together with thermochemical cycles providing the conceptual framework for interpreting activation barriers in terms of separable properties of the acid (E_{vdw} and DPE), the reactant (proton affinity), and the interacting moieties at the confined ion-pair transition state, permit systematic interpretation of the independent effects of acid strength and confinement on intrinsic isomerization and β-scission rates and selectivities.

n-Heptane and 2,4-dimethylpentane isomerization and β-scission turnover rates (per H^+; 548 K; 0.5-25 kPa alkane; 60-101 kPa H_2) on all physical mixtures are proportional to alkane/H_2 molar ratio (a surrogate for the prevalent pressure of heptenes) at low values of this ratio and approach constant values as molar ratio increases. First-order rate constants ($k_{isom,nH}K_{prot,nH}$ for n-heptane isomerization and $(k_{isom,24}+k_{\beta,24})K_{prot,24}$ for combined 2,4-dimethylpentane isomerization and β-scission) reflect the free-energy differences between their respective confined, ion-pair transition states and unconfined, neutral precursors (gas phase alkene & acid sites). Measured ratios of $k_{\beta,24}/k_{isom,24}$ reflect ubiquitous secondary reactions that occur within the acid domain before alkene products leave an acid domain and become alkanes. Similarly, on all acids and for both reactants, measured product selectivities (the ratio of the measured formation rate for a given product to the consumption rate of the reactant) were invariant with reactor residence time or alkene reactant pressure; these selectivities do not reflect the intrinsic values of primary formation events because primary alkene products undergo multiple secondary interconversions before they form their respective less-reactive alkane analogs. The extent of these diffusion-enhanced secondary reactions was assessed for each aluminosilicate by varying the intracrystalline residence time (t_{int}) through monotonic changes in the proton density ($[H^+]$) during desorption of weakly bound NH_3 titrants from the active protons. Figure 1 shows selectivities measured during NH_3 desorption as a function of t_{int} for (a) β-scission products and (b) isomer products during 2,4-dimethylpentane conversion on FAU-Pt/SiO_2 catalyst mixtures. Product selectivities at low values of t_{int} are invariant with t_{int}, indicating that such β-scission selectivities reflect primary chemical events. As t_{int} increased, β-scission, n-heptane, 2-methylhexane, and 3-methylhexane selectivities increased, while that for to 2,3-dimethylpentane decreases, indicating the prevalence of secondary alkene interconversions of primary 2,4-dimethylpentane products as $[H^+]$ increases.

Figure 1. Measured **(a)** β-scission and **(b)** 2-methylhexane (; left), 3-methylhexane (▲; left), n-heptane (¦; left), and 2,3-dimethylpentane (•; right) selectivities for reactant 24DMP as a function of the intracrystalline residence time (varied by desorption of NH_3) on physical mixtures of Pt/SiO_2 with FAU (548 K). Dashed lines are guides for the eye.

Rigorous reaction-transport treatments reveal that a Thiele parameter for the primary isoalkene products determines the extent of such diffusion-enhanced secondary reactions (, where $k_{i,iso}K_{prot,iso}$ is the first-order rate constant for secondary reaction i of the isoalkenes, D_{iso} is diffusivity of the isoalkenes, and R^2 is zeolite crystallite size). Figure 2 shows β-scission selectivities for nH reactant during NH_3 desorption on FAU, BEA, and MFI (as physical mixtures with Pt/SiO_2; BEA and MFI of different $[H+]R^2/D_{22DMB}$ are indicated, where D_{22DMB}/R^2 was measured using transient uptake of 2,2-dimethylbutane) as a function of the Thiele modulus for n-heptane ( ); the dashed curve is the regressed fit to the theoretical expression describing these secondary reactions (Fig. 2, inset equation, where and) with a single fitting parameter ( ) that has the same value for all aluminosilicates. β-Scission selectivities increased monotonically as increases and are a single-valued function of this parameter; selectivities for all aluminosilicates lie along the same curve, despite their very different void sizes, indicating that all frameworks stabilize the two transition states (for β-scission and isomerization) to the same extent.

Figure 2. Measured β-scission selectivity for reactant nH as a function of the n-heptane Thiele modulus (F_{nH}^2; varied by desorption of NH_3) on physical mixtures of Pt/SiO_2 with FAU, BEA, and MFI (548 K; 100 kPa H_2, 0.5 kPa nH). MFI and BEA of different [H^+] are indicated. Dashed line represents the regressed fit to the inset equation.

These confinement effects, however, lead to preferential stabilization of both β-scission and isomerization transition states with respect to their gas phase precursors; the stabilization of transition states by van der Waals interactions with the framework is reflected in E_{vdw}, where a better match between void size/shape and organic carbocation size/shape results in more negative values. First-order nH isomerization rate constants decrease exponentially with increasing E_{vdw}. The effects of E_{vdw} on activation barriers are attenuated by energy penalties brought forth by distortions of the framework (E_{str}) and the carbocation at the transition state in order to establish more effective van der Waals contacts, especially as the organic moiety approaches the size of the confining voids.

Such organic carbocations, together with the conjugate anion of the zeolite or zeotype, constitute ion pair transition states mediating isomerization and β-scission reactions. First-order isomerization and β-scission rate constants for 2,4-dimethylpentane, measured on MFI zeotypes with H^+ of different acid strength (by varying framework heteroatom, in order of decreasing acid strength: Al, Ga, Fe, B; DPE = 1226-1292 kJ mol^{-1}; in physical mixture with Pt/SiO_2), were corrected for diffusion-enhanced secondary reactions to obtain intrinsic values. These first-order rate constants decrease exponentially with increasing DPE, predominantly a reflection of the less stable conjugate anion at ion-pairs of weaker acids. The ratio of rate constants $k_{\beta,24}/k_{isom,24}$, however, increases with increasing DPE, indicating that β-scission occurs preferentially on weaker acids. Cations at transition states mediating β-scission resemble "harder" carbenium ions (according to classical acid-base formalisms [4]) compared to those for isomerization. These "harder" carbocations interact more effectively with the conjugate anion, recovering more of the energy required to deprotonate a weaker acid, and consequently are less sensitive to DPE. Thus, for transition state structures differing in their tendency to delocalize charge, acid strength can influence product selectivity through stronger effects of DPE for the "softer" transition state.

This work mechanistically assesses the effects of both confinement and acid strength on intrinsic isomerization and β-scission turnover rates and selectivities. Methods of untangling the dramatic and ubiquitous effects of diffusion-enhanced secondary reactions on measured product selectivities are also demonstrated. The insights gained from this work can be applied to design catalytic architectures to obtain desired selectivities and reaction rates.

References

[1] H.L. Coonradt, W.E. Garwood, Ind. Eng. Chem. Process Des. Dev. 3 (1964) 38–45.

[2] C. Baerlocher, L. McCusker, Database of Zeolite Structures, http://www.iza-structure.org/databases/.

[3] S. Herrmann, E. Iglesia, J. Catal. 346 (2017) 134–153.

[4] G. Pearson, J. Am. Chem. Soc. 85 (1963) 3533–3539.

The authors acknowledge technical discussions with Drs. Stacey I. Zones (Chevron) and William Knaeble (UC Berkeley), diffusion timescale measurements by Zhichen (Jane) Shi (UC Berkeley), financial support from Chevron Energy Technology Company, and computational resources from the National Science Foundation's Extreme Science and Engineering Discovery Environment (ACI-1053575).

500019 Hydrophobic Zeolites for Solketal Production from Crude Glycerol

Abstract:

Functionalizing zeolites (i.e., CBV600 with a SiO_2/Al_2O_3 ratio of 5.2) by organosilanes increase the degree of hydrophobicity. These hydrophobic zeolites can catalyze the solketal production from crude glycerol (i.e., glycerol containing moisture, methanol and salt impurities) and show the higher catalytic activity than the pristine zeolites. Glycerol is an under-utilized by-product from biodiesel industries. The global biodiesel consumption was ~34 billion liters in 2016 and is expected to reach 38 billion liters by 2020 from a rapid growth in fuel and power generation sectors. Hence, a large amount of crude glycerol will be available. Herein, crude glycerol and acetone were used for solketal production at low temperatures of 30-50 ËšC. Glycerol conversion and solketal yield were 84% and 80%, respectively. The physico-chemical characteristics of hydrophobic and pristine zeolites were characterized by N_2 physisorption, thermogravimetric analysis (TGA), X-ray diffraction (XRD), and infrared spectroscopy. The obtained characteristics were correlated with glycerol conversion and solketal yield. These results provide an understanding of how these hydrophobic zeolites function and how to tailor their catalytic activities.

497838 Transport Analysis of an Integrated Artificial Photosynthetic System for Direct Capture and Reduction of CO_2 from Air

Abstract:

The effect of anthropogenic emissions of CO_2 on the climate change is becoming progressively prevalent. The current concentration of CO_2 in the air is 409.11 ppm, which is increasing at a rate of ~1.9 ppm per year. There exist several technologies such as amine scrubbing, oxyfuel combustion, and calcium looping to capture CO_2 from flue gases, which are inherently energy intensive. However, there are only a few techniques such as moisture-swing adsorption that can capture CO_2 from the air but requires modulation of humidity and downstream separation. The captured CO_2 can be either sequestered at the storage site or chemically transformed into a usable product. Integrating CO_2 capture with reduction not only eliminates the need for additional transportation and storage but also increases the efficiency of CO_2 capture. Here we provide a conceptual design of an integrated system where an anion-exchange resin for CO_2 capture is directly connected to a photo/electrochemical cell for CO_2 reduction. The dry-side of the resin captures CO_2 from the air and produces a high concentration of HCO_3^- diffusing towards the wet-side of the resin, where HCO_3^- converts back to CO_2 at the attached cathode to produce fuels. The limiting current density of an integrated system as a function of resin properties, catalysts activity and selectivity, electrolyte pH, physical dimensions and relative humidity are simulated using COMSOL Multiphysics. The limiting current density increases with increasing the resin conductivity, and mass-transfer coefficient of CO_2 in air. The efficiency of such integrated system using air as well as flue gas will be discussed.

500306 Design of a 400 MW Carbon-Neutral, Coal-Fired Power Plant with Integration of Waste Heat and Solar Energy

Abstract:

Coal-fired power plants (CPP) consumes on an average of 221 tons h^{-1} of coal to generate 400 MW of electricity. The rate of emission of CO_2 from the state-of-the-art CPP operating at 60% efficiency is approximately 600 kg MW-h^{-1}, which is equivalent to 1.5 kmol s^{-1}. The amine-based scrubbers are known to effectively separate CO_2 from the flue gases which typically contains 12-15% CO_2. The captured CO_2 is then released using a regeneration process requiring an additional power of ~350 MW, which is typically supplemented by either burning natural gas or integrating with waste heat of ~250 MW from the CPP. Here we are proposing an electrochemical technique that directly coverts CO_2 absorbed in an amine-based solvent and water to either CO or HCOOH. This technique bypasses the need for regeneration process for amine-based solvent that saves ~350 MW of power, which can be used to conduct electrochemical reduction of CO_2. We found that continuous production of CO and HCOOH require lesser power as compared to other alcohols and hydrocarbons. For example, the minimum power required to produce either CO or HCOOH is ~585 MW (~380 MW is a thermodynamic limit) for 100% CO_2 Utilization Efficiency and Potential. The excess energy required for electrochemical conversion of CO_2 may be derived from the solar energy which requires a land area of 1.67 km^2 (38% of CPP land area), assuming a 20% effective efficiency of solar cells. Therefore, it seems plausible to support the energy requirement for electrochemical reduction of CO_2 using an integrated waste heat and solar energy management. The target performance of such system is ~1 A cm^{-2} of current density at ~2 V of cell voltage. Achieving such targets require development of a high surface area catalysts operating at total overpotential of <670 mV. This results in a carbon-neutral model that can be used as a basis for designing and adapting current and new conventional thermal power plants.

489902 The Energetics of Chemical Looping Processes for Post Combustion CO_2 Capture

Abstract:

The energetics of chemical looping processes for post combustion CO_2 capture

Hugo S. Caram*, Ramesh Gupta#, Fan Ni* and Hans Thomann#

(*) Department of Chemical and Biomolecular Engineering, Lehigh University, Bethlehem, PA

(#)ExxonMobil Research and Engineering Inc., Annandale, NJ

A number of processes have been discussed where the CO_2 is removed from flue gases by chemical reaction with a solid sorbent. While calcium looping is the most celebrated, many others including carbonate-bicarbonate, reactions with mixed oxides, treated hydrotalcites and organic impregnated solids, such as MOFS, have been proposed. The equivalent work requirement, that is the most important variable, power that will be drawn from the power plant are usually only examined when a detailed process design and pilot testing has been completed.

In this work, a simple thermodynamic tool is proposed that will allow determination of the

minimum equivalent work needed by the process. The tool only requires thermochemical properties at room temperature and evaluates major thermal and irreversible losses. It can be used to determine the effect of process parameters, equipment configuration (such as fluidized vs. fixed or moving bed operation), steam usage and other variables. Applications to some of the existing processes, where the method could apply, will be presented.

493365 A Technoeconomic Evaluation of Carbon Dioxide (CO_2) Electroreduction: Results and Implications

Abstract:

The electroreduction of carbon dioxide (CO_2) to value added C_1-C_2 chemicals (such as formic acid, carbon monoxide, methanol, ethylene, and ethanol) has been proposed as a possible technological solution that can utilize excess anthropogenic CO_2 emissions, thereby providing a sustainable pathway for closing the CO_2 emission cycle.[1, 2] However, even after a few decades of research, key questions regarding the technoeconomic feasibility of the process remain unanswered. These include: (i) what products of CO_2 electroreduction are the most economically viable to produce; and (ii) what performance targets such as maximum operating cell potential (V_{max}), minimum operating current density (j_{min}), Faradaic efficiency (FE), and catalyst durability (t_{catdur}), are required for an economically viable process.

In this talk, we will present results from a gross-margin based economic model that we recently developed to evaluate the technoeconomic feasibility of the CO_2 electroreduction process.[3] Using the model, we were able to quantitatively estimate the operating parameters (V_{max}, j_{min}, FE, t_{catdur}) required for an economically viable process of CO_2 electroreduction. A comparison of the operating parameters required for economic viability with the experimentally achievable performance of the state of the art electrochemical systems indicate that formic acid and carbon monoxide are the most economically viable CO_2 electroreduction products, and the process can be cost competitive with the existing industrial production methods for the same. However, to improve the economic viability of other products (such as ethanol, ethylene), alternate reaction chemistry/process designs with a lower electrical energy requirement need to be developed. This talk will cover some of our recent advances with respect to developing such alternate process design strategies as well.

References:

[1] A.M. Appel, J.E. Bercaw, A.B. Bocarsly, H. Dobbek, D.L. DuBois, M. Dupuis, J.G. Ferry, E. Fujita, R. Hille, P.J.A. Kenis, C.A. Kerfeld, R.H. Morris, C.H.F. Peden, A.R. Portis, S.W. Ragsdale, T.B. Rauchfuss, J.N.H. Reek, L.C. Seefeldt, R.K. Thauer, G.L. Waldrop, *Chem. Rev.* 2013, 113, 6621-6658.

[2] H.R.M. Jhong, S. Ma, P.J.A. Kenis, *Curr. Opin. Chem. Eng.* 2013, 2, 191-199.

[3] S. Verma, B. Kim, H.R.M. Jhong, S. Ma, P.J.A. Kenis, *ChemSusChem* 2016, 9, 1972-1979.

504355 Science with Purpose

Abstract:

As a company serving a multitude of industries in nearly every country on the planet, we recognize the opportunity we have: to grow our business while solving some of the worldâ€™s biggest problems. This requires collaboration with our customers, partners and stakeholders. Sustaining our business, and our planet, means protecting natural resources and empowering individuals and communities throughout the world to encourage progress. We use science to develop products that address challenges facing our customers and our global community, which include enhancing human health, promoting access to clean water, increasing energy efficiency and enabling safe work environments. We will share 3M examples and stories about who we are, how we work and what we create, and show how we use science with purpose every day.

504354 Some Thoughts on Sustainability: Food, Energy, and Water

Abstract:

The World is steadily moving into a major challenge. The challenge is how to provide for the human population within the limits of one planet, the Earth, over the long term. The effort is known as sustainability. To explore this challenge, I will focus on three closely linked issues of providing food, energy, and water. In addition, there is the issue of maintaining order and stability as necessary conditions for any kind of acceptable human existence. These issues, unfortunately, present critical but not well-bounded problems, and for that reason there are many more questions than answers. However, to explore the issues and their bounds, I will discuss what is known including some planetary limits and limits imposed by the known laws of Nature. I will lastly discuss where and how engineers can help in the effort to provide food, energy, and water for humanity. This will include some concrete examples where engineering design has helped to mitigate these challenges. The lecture will conclude with some thoughts on which approaches might stand the best chance of adoption and implementation under current and foreseeable future conditions.

Disclaimer: The views expressed in this presentation are those of the author and do not necessarily represent the views or policies of the U.S. Environmental Protection Agency.

504356 Industrial Sustainability Enhancement: A Control Perspective

Abstract:

Depletion of nonrenewable resources, global warming, air, water, and soil pollution, excessive population growth have challenged societies globally. This has demanded industries to seriously review engineering practice and find effective ways to ensure sustainable development. Over the past decades, a variety of sustainability assessment and enhancement methods and technologies have been developed for improving material and energy efficiency, environmental cleanness, human health, product life-cycle performance, social responsibility, eco-efficiency, etc. However, problem-solving solutions are mostly derived based on experience, which make them feeble in tackling complex problems involving multi-state, long-term strategic development under uncertainty. Fundamentally, industrial sustainability enhancement is a system science problem; it is featured by solutions for a series of industrial system state transitions toward sustainability. As such transitions occur in a multi-dimensional sustainability space, its dynamics and trajectory are difficult to predict, design, and control, especially when accessible data and available information are incomplete, imprecise, and uncertain. These challenges present a new opportunity for the study on system-transition-focused sustainability dynamics, assessment, and control. In this talk, we will discuss a number of fundamental aspects of sustainability control, including the controllability of system sustainability and transition. We will then introduce hierarchical control schemes for achieving sustainability goals under uncertainty. A few case studies will be presented and research directions will be explored. This talk will also discuss approaches for integration of sustainability elements and basic methods in undergraduate education, especially in the capstone process design and process control courses.

504357 Multi-Scale Life Cycle Optimization for Sustainable Engineering

Abstract:

Public interest in carbon and water footprints, ecosystem degradation and resource depletion, are just some of the environmental issues that drive the process industries to exercise greater responsibility for the environment in the design and operations of their manufacturing systems. Systems analysis and computational optimization methods from a life cycle perspective play an important role in the quest for sustainability. This talk will review some recent advances in the area of life cycle optimization for sustainability analytics. The focuses are on transformative modeling, theoretical and computational results that bridge the knowledge gaps between state-of-the-art life cycle assessment methods and emerging life cycle optimization tools. We will specifically discuss issues related to multi-scale system boundary selection, consequential perspective of the analysis, and potential presence of multiple non-cooperative decision makers in life cycle optimization for sustainable engineering problems. Applications of these general computational frameworks will be illustrated by case studies on bioenergy/biofuels, shale gas, and chemical manufacturing systems. We will conclude with a brief review of multi-scale applications of life cycle optimization methods from materials to the global scales.

499675 NO Uptake and Desorption on Fe-Modified Pd/ZSM-5:Â Impact of Pd and Fe Loading

Abstract:

NO Uptake and Desorption on Fe-Modified Pd/ZSM-5:

Impact of Pd and Fe Loading

Kyle Karinshak*, Michael P. Harold*, and Kiran Premchand+

*Dept. of Chemical & Biomolecular Engineering, University of Houston, Houston, Texas

+FCA US LLC Auburn Hills, MI 48236

* corresponding author: M.P. Harold (mpharold@central.uh.edu)

Reducing cold-start emissions, emissions under 200 °C, will play a significant role in meeting future EPA and EU emissions standards. By adsorbing NO and NO_2 at low temperatures and releasing these gases in temperatures regimes where traditional emission control catalysts can operate, Passive NOx Adsorbers (PNAs) augment existing emission control systems and are a promising new field of catalysis. The scant published literature has focused on Pd supported on various materials, including ceria or alumina. In contrast, the patent literature suggests the efficacy of ZSM-5 as a support as well as the addition of Fe to PNA compositions [1-6]. Our preliminary investigations into Palladium-exchanged MFI zeolites show significant NOx uptake capabilities and complex desorption behavior.

Catalyst samples were prepared through a wet impregnation protocol before being deposited onto monolith cores. Palladium(II) nitrate was used as the precursor and dissolved into water equal to the pore volume of the ZSM-5 zeolite (SAR 30). Alumina solution was added to the washcoating solution to improve binding. Four catalysts – 0 wt% Pd, 1wt% Pd, 1 wt% Pd/1 wt% Fe, and 2 wt% Pd, were washcoated onto a 0.25in diameter, 2in length, 600 cspi monolith cores at a fixed loading of 1.5 g/in^3. A bench-scale quartz reactor was utilized to simulate emissions and study uptake and desorption characteristics. DRIFTS (diffuse reflectance infrared Fourier transform spectroscopy) experiments of catalysts samples were performed to examine surface species formation and the mechanism through which NOx uptake occurs on Pd/ZSM catalysts.

Figure 1 shows uptake characteristics of NOx uptake at 50 ^{o}C over a five minute duration for the four catalyst. 400 ppm of NO was flowed through the catalyst accompanied by 2% O2 and balance Ar. Similar trends were observed at uptake temperatures of 80 ^{o}C and 150 ^{o}C with decreasing uptake quantities and breakthrough periods. NO adsorption was accompanied by noticeable NO_2 slip from the catalyst. The addition of Fe to the catalyst composition significantly improved the performance of NOx uptake.

Figure 2 shows the NOx desorption peaks from the same experiments. Over half of desorbed NOx occurred in the form of NO_2. As NO adsorbs as a mono nitrosyl palladium complex, this shows that NO is being converted into NO_2 through a zeolite-surface reaction and likely storing on the Pd-less acid sites of the zeolite and binder. Unlike Murata et al, who reported a single NOx desorption peak during TPD experiments [7], our results reveal multiple desorption peaks. Three distinct peaks are identified in Pd-containing samples whereas only two exist in the Pd-less sample. A preliminary assignment of these peaks include acid sites of the binder, the acid sites of the zeolite, and the Pd nitrosyl complexes.

Figure 3 shows DRIFTS measurements of three samples: 0 wt% Pd, 1 wt% Pd/1 wt% Fe, and 2 wt% Pd. Experiments were conducted in the presence of oxygen. Peak analysis matches that reported by Descorme and Gelin, confirmed the NO adsorption mechanism [8]. The Pd/Fe sample did not show the peaks associated with NO adsorption onto Fe complexes, confirming that the role of Fe in the catalyst composition is ensure a high dispersion of Pd [9]. Experiments conducted without oxygen (not shown) display peaks associated with NO adsorption onto Pd but do not show peaks associated with NO_2 formation and storage.

Additional TPD experiments (not shown) have shown that the presence of water significantly suppresses NOx uptake and desorption. The introduction of water inhibits the surface-reaction formation of NO_2, likely through the suppression of zeolitic acid sites. Ongoing experiments include powdered catalyst TPD experiments as well as additional DRIFTS experiments to better understand and quantify the effects the addition of binder.

Ongoing work will include modifications to the catalyst compositions as well as the addition of a modelling component for both uptake and desorption. Additionally, concurrent experiments are being carried out to understand the effects of adding hydrocarbons to the gas environment.

Figure 3: Evolution of DRIFTS data over time from three samples.

488343 Insights into Influencing Factors on Oxygen Storage Characteristics for Ceria-Zirconia Catalysts

Abstract:

Ceria systems have been investigated extensively in the field of automotive catalysis as oxygen storage materials in order to compensate for deviating stoichiometric operation during exhaust aftertreatment. The ability of ceria to cycle between Ce^{4+} and Ce^{+3} provides an attractive characteristic desirable for reactions such as water-gas shift, hydrocarbon oxidation, and hydrocarbon reforming [1]. Often, an additional dopant is added into the ceria lattice in order to increase oxygen storage capacity (OSC) and oxygen mobility, with ceria zirconia (CZO) being the most common system used [2]. The oxygen release rate (ORR) by these materials can be further enhanced by depositing platinum group metal (PGM) such as Pd or Rh, which provides beneficial metal-support interface in order for rapid oxygen transfer to occur [3]. While these catalytic materials have been in use, additional fundamental insight in the mechanism of OSC and the effect of PGM dispersion, CZO surface area, material pre-treatment, and varying testing conditions on catalyst performance are explored in this study.

A series of Pd/CZO catalysts was prepared by incipient wetness impregnation with different loadings and aged in redox conditions at high temperatures. Characterization by BET, H_2 chemisorption, and XRD will be discussed. OSC measurements were conducted via lambda probes before and after the catalyst as a function of temperature, while ORR were evaluated by the production of CO_2 as measured by FTIR as a function of time. Structure-activity relationships were developed to determine what factors were most important in achieving high OSC and fast ORR, particularly after high temperature aging. In addition, catalysts were evaluated for CO oxidation and water-gas shift to establish the importance of OSC in either reaction. Results from this study will help provide guidelines in the design of future oxygen storage materials in order to meet future emission standards.

[1] R. Gorte, AIChE Journal 56 (2010) 1126-1135

[2] E. Mamontov, et al., J. Phys. Chem. B 104 (2000) 11110-11116

[3] H.W. Jen, et al., Catal. Today 50 (1999) 309-328

Pd doped $CaCo_xZr_{1-x}O_{3-\delta}$ perovskites for automotive emissions control

Nov 01, 2017, AIChE Annual Meeting

<u>Qinghe (Angela) Zheng</u> and Marty Lail*, PhDs
Energy Technology Division, RTI International
Research Triangle Park, NC 27709
qizheng@ri.org

Conventional supported PGM-based TWC *vs.* Perovskite-TWC

Conventional PGM-based TWCs
($PdRh/\gamma Al_2O_3$ (CeO_2-ZrO_2))

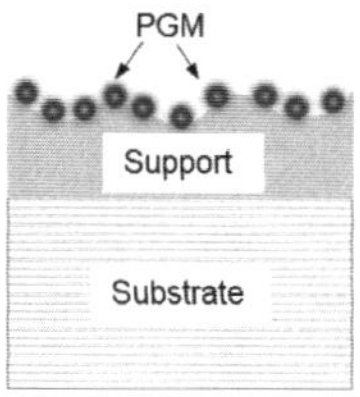

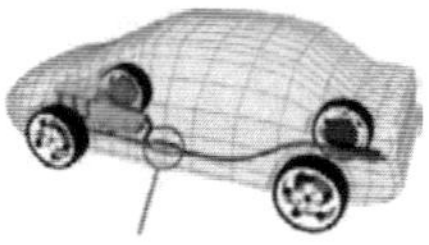

Three Way Catalytic (TWC) Converter

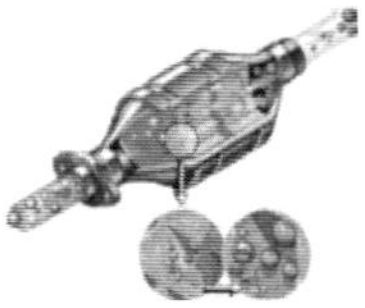

$ABO_{3-\delta}$ Perovskites

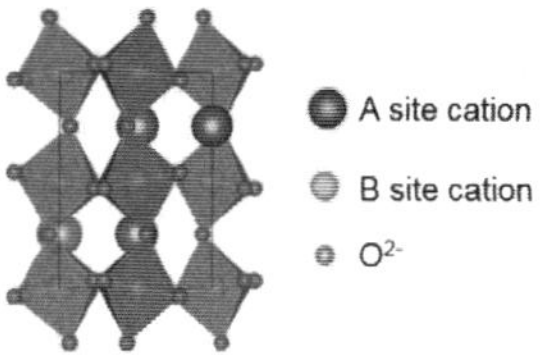

$$ABO_{3-\delta} + \frac{1}{2}\Delta\delta\, CO \rightarrow ABO_{3-\delta-\Delta\delta} + \frac{1}{2}\Delta\delta\, CO_2$$

$$ABO_{3-\delta} + \Delta\delta\, C_xH_y \rightarrow ABO_{3-\delta-(2x+\frac{1}{2}y)\Delta\delta} + \Delta\delta\,(x\, CO_2 + \frac{1}{2}y\, H_2O)$$

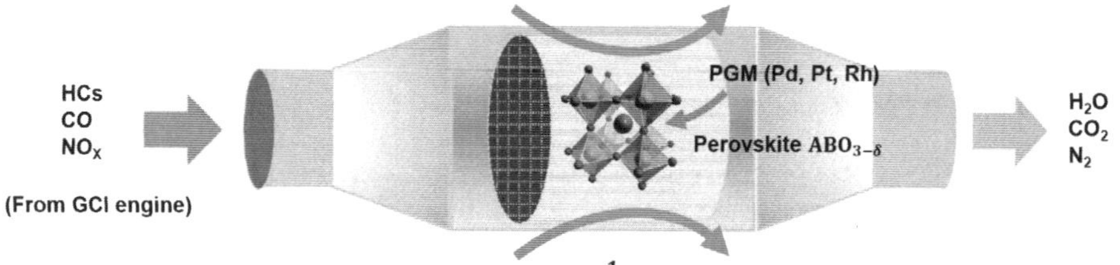

$$ABO_{3-\delta-\Delta\delta} + \frac{1}{2}\Delta\delta\, O_2 \rightarrow ABO_{3-\delta}$$

$$ABO_{3-\delta-\Delta\delta} + \Delta\delta\, NO \left(or \frac{1}{2}\Delta\delta\, NO_2\right) \rightarrow ABO_{3-\delta} + \frac{1}{2}\Delta\delta\, N_2\ (or \frac{1}{4}\Delta\delta\, N_2)$$

Conventional supported PGM-based TWC *vs.* Perovskite-TWC

RTI INTERNATIONAL

Conventional PGM-based TWCs
($PdRh/\gamma Al_2O_3$ (CeO_2-ZrO_2))

- ✓ High activity (@ stoichiometric)
- ✓ Good performance with CeO_2 added
- ✓ High surface area support
- o Low thermal stability (@ fuel cutoff)
- o High cost of PGM
- o Market control of Ce resources

Three Way Catalytic (TWC) Converter

$ABO_{3-\delta}$ Perovskites

- ✓ Improved thermal stability
- ✓ Low cost
- ✓ Tunable compositions
- ✓ Large availability
- o Low surface area
- o Less activity

$$ABO_{3-\delta} + \frac{1}{2}\Delta\delta\, CO \rightarrow ABO_{3-\delta-\Delta\delta} + \frac{1}{2}\Delta\delta\, CO_2$$

$$ABO_{3-\delta} + \Delta\delta\, C_xH_y \rightarrow ABO_{3-\delta-\left(2x+\frac{1}{2}y\right)\Delta\delta} + \Delta\delta\,(x\, CO_2 + \frac{1}{2}y\, H_2O)$$

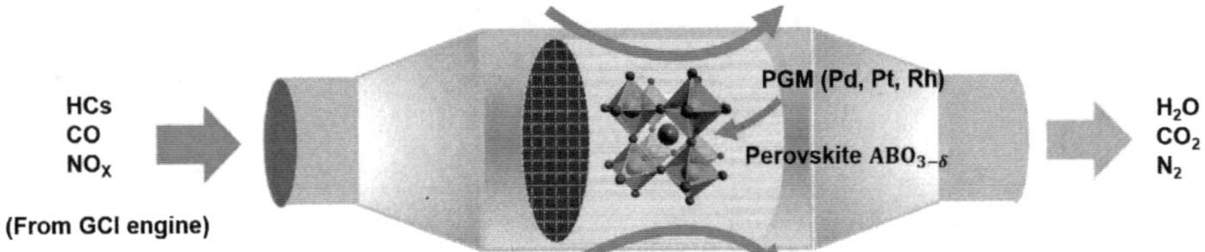

$$ABO_{3-\delta-\Delta\delta} + \frac{1}{2}\Delta\delta\, O_2 \rightarrow ABO_{3-\delta}$$

$$ABO_{3-\delta-\Delta\delta} + \Delta\delta\, NO \left(\text{or} \frac{1}{2}\Delta\delta\, NO_2\right) \rightarrow ABO_{3-\delta} + \frac{1}{2}\Delta\delta\, N_2\ (\text{or} \frac{1}{4}\Delta\delta\, N_2)$$

3

$ABO_{3-\delta}$ perovskite structure

RTI INTERNATIONAL

$ABO_{3-\delta}$ perovskite structure

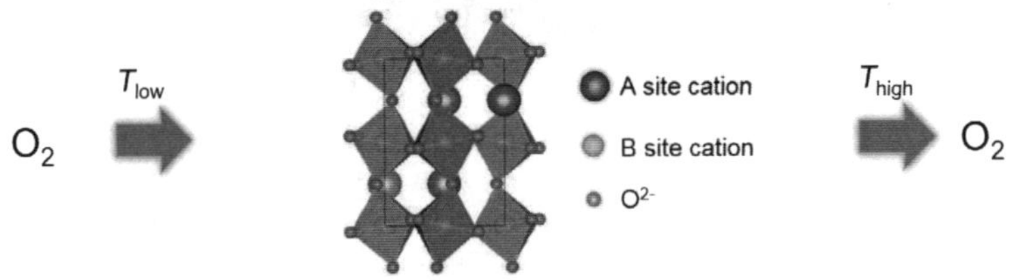

A site cation: alkaline earth, alkaline, and lanthanide ions in a 12-fold coordination (e.g. A= La, Sr, Ca, Ba)

- Important in perovskite structural formation
- Influence oxygen vacancy and structural defects density
- Influence charge-valence balance

B site cation: smaller transition metal ions in a 6-fold coordination with oxygen ions (e.g. B=Co, Mn, Fe, Cu)

- Perovskite oxygen storage capacity depends on the change of oxidation state of the B-site cation
- Perovskites with Co at B site was reported to show best activity in HC and CO oxidation
- Co-based perovskites have low thermal stability due to decomposition at high T
- Improved perovskite properties can be achieved by B-site doping

Oxygen non-stoichiometry δ: oxygen anion deficiency or excess depending on the total valence amount of A-site and B-site cations

4

Oxygen mobility within $ABO_{3-\delta}$ perovskite lattice

RTI INTERNATIONAL

- $ABO_{3-\delta}$ perovskite can exhibit significant swings in δ through the reaction with gas phase O_2.
- δ can be tailored by doping A- or B-sites with appropriately selected cations
- δ may vary with T and P_{O2}
- Adsorption of superficial [O] at T<600 °C
  Activation of interfacial lattice [O] at T>600 °C } Mars-van-Kreveken mechanism

RTI developed novel $CaCo_xZr_{1-x}O_{3-\delta}$ type perovskites wi or w/o PGM doping

(Patent application RTI 16P027-PROV)

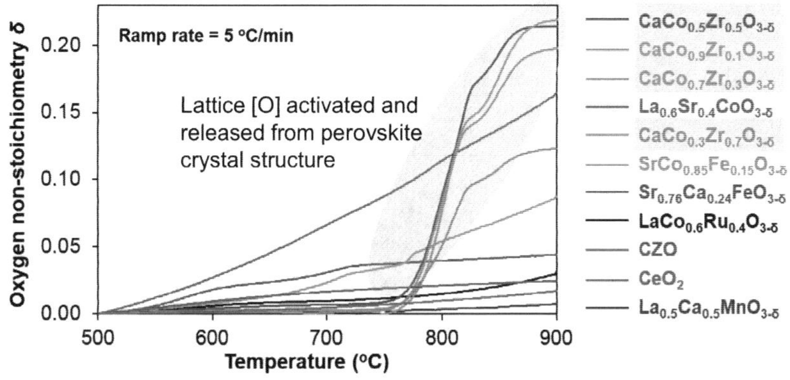

Fig. 1. O_2 non-stoichiometry as a function of sample temperature (500 ~ 900 °C) during TGA temperature programmed study of $ABO_{3-\delta}$ perovskite materials.

5

Perovskite with high OSC as potential TWC

RTI INTERNATIONAL

CeO_2-ZrO_2 as a support for TWC

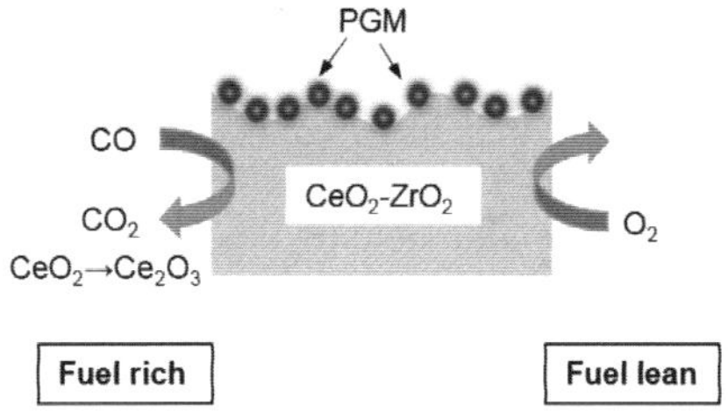

$ABO_{3-\delta}$ perovskite-type TWC

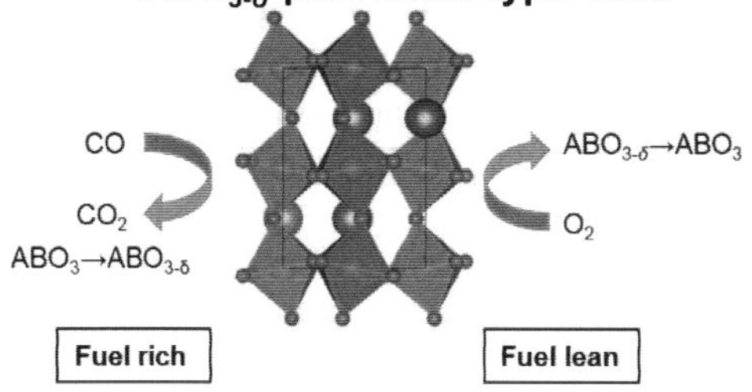

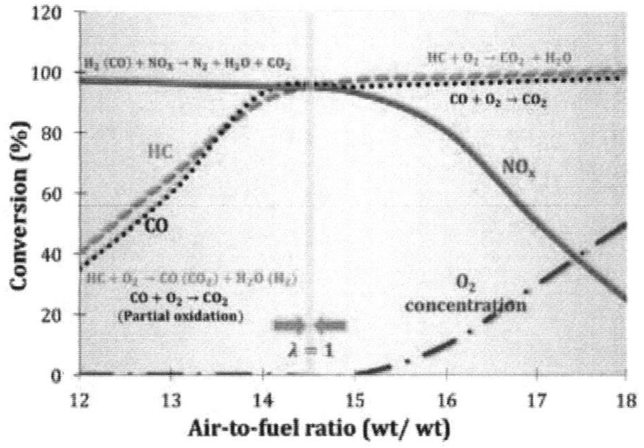

6

Figure 2. The TWC conversion profile as a function of air-to-fuel ratio (λ).

Experimental methodology

RTI INTERNATIONAL

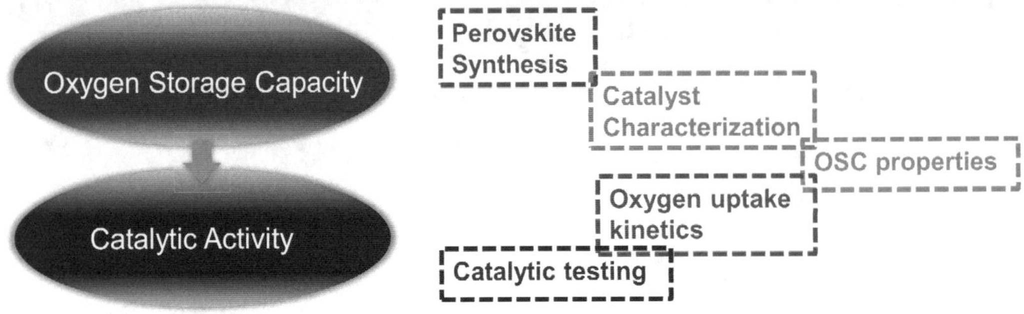

Synthesis Hood

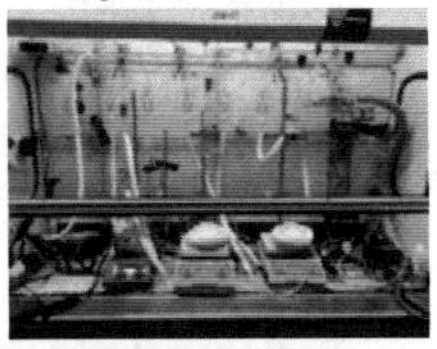

ASAP 2020 BET

TGA

Microreactor

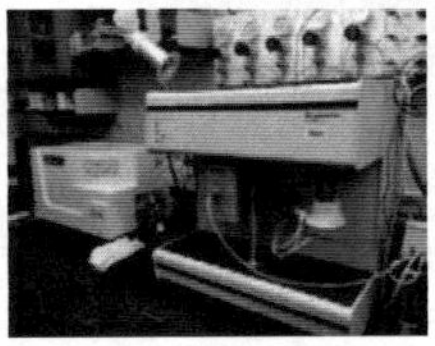

AutoChem reactor

XRF

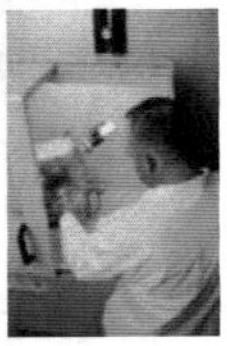

XRD

7

Synthesis and characterization of $CaCo_xZr_{1-x}(Pd)O_3$ perovskites

RTI INTERNATIONAL

Table 1. Analytical result of as-synthesized (Pd-doped) $CaCo_xZr_{1-x}O_3$ perovskites.

#	Theoretical composition	Actual composition (XRF)	Tolerance factor, τ
CCZPd2	$CaCo_{0.4}Zr_{0.4}Pd_{0.2}O_3$	$CaCo_{0.4}Zr_{0.3}Pd_{0.2}O_3$	0.81
CCZPd1	$CaCo_{0.5}Zr_{0.4}Pd_{0.1}O_3$	$CaCo_{0.5}Zr_{0.3}Pd_{0.1}O_3$	0.82
CCZPd05	$CaCo_{0.55}Zr_{0.4}Pd_{0.05}O_3$	$CaCo_{0.55}Zr_{0.3}Pd_{0.06}O_3$	0.83
CCZ91	$CaCo_{0.9}Zr_{0.1}O_3$	$CaCo_{1.0}Zr_{0.1}O_3$	0.86
CCZ73	$CaCo_{0.7}Zr_{0.3}O_3$	$CaCo_{0.8}Zr_{0.2}O_3$	0.85
CCZ55	$CaCo_{0.5}Zr_{0.5}O_3$	$CaCo_{0.5}Zr_{0.5}O_3$	0.84
CCZ37	$CaCo_{0.3}Zr_{0.7}O_3$	$CaCo_{0.3}Zr_{0.6}O_3$	0.82
CZ10	$CaZrO_3$	$CaZrO_3$	0.80

- **Perovskite synthesis:** Pechini method, calcination @ 1200 °C;
- **Perovskite target compositions:**
  $CaCo_xZr_{1-x}O_{3-\delta}$ (x=0, 0.3, 0.5, 0.7, and 0.9),
  Pd doped $CaCo_{0.6-y}Zr_{0.4}Pd_yO_{3-\delta}$ (y=0.05, 0.1 and 0.2);

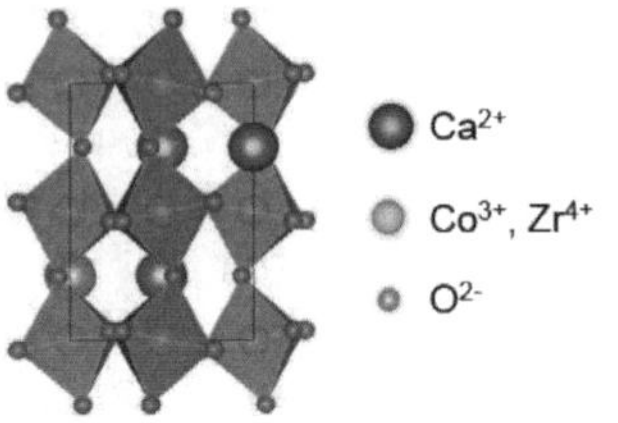

8

Synthesis and characterization of $CaCo_xZr_{1-x}(Pd)O_3$ perovskites

RTI INTERNATIONAL

Table 1. Analytical result of as-synthesized (Pd-doped) $CaCo_xZr_{1-x}O_3$ perovskites.

#	Theoretical composition	Actual composition (XRF)	Tolerance factor, τ
CCZPd2	$CaCo_{0.4}Zr_{0.4}Pd_{0.2}O_3$	$CaCo_{0.4}Zr_{0.3}Pd_{0.2}O_3$	0.81
CCZPd1	$CaCo_{0.5}Zr_{0.4}Pd_{0.1}O_3$	$CaCo_{0.5}Zr_{0.3}Pd_{0.1}O_3$	0.82
CCZPd05	$CaCo_{0.55}Zr_{0.4}Pd_{0.05}O_3$	$CaCo_{0.55}Zr_{0.3}Pd_{0.06}O_3$	0.83
CCZ91	$CaCo_{0.9}Zr_{0.1}O_3$	$CaCo_{1.0}Zr_{0.1}O_3$	0.86
CCZ73	$CaCo_{0.7}Zr_{0.3}O_3$	$CaCo_{0.8}Zr_{0.2}O_3$	0.85
CCZ55	$CaCo_{0.5}Zr_{0.5}O_3$	$CaCo_{0.5}Zr_{0.5}O_3$	0.84
CCZ37	$CaCo_{0.3}Zr_{0.7}O_3$	$CaCo_{0.3}Zr_{0.6}O_3$	0.82
CZ10	$CaZrO_3$	$CaZrO_3$	0.80

- **Perovskite synthesis:** Pechini method, calcination @ 1200 °C;
- **Perovskite target compositions:**
  $CaCo_xZr_{1-x}O_{3-\delta}$ (x=0, 0.3, 0.5, 0.7, and 0.9),
  Pd doped $CaCo_{0.6-y}Zr_{0.4}Pd_yO_{3-\delta}$ (y=0.05, 0.1 and 0.2);
- **Perovskite actual compositions** confirmed by XRF;

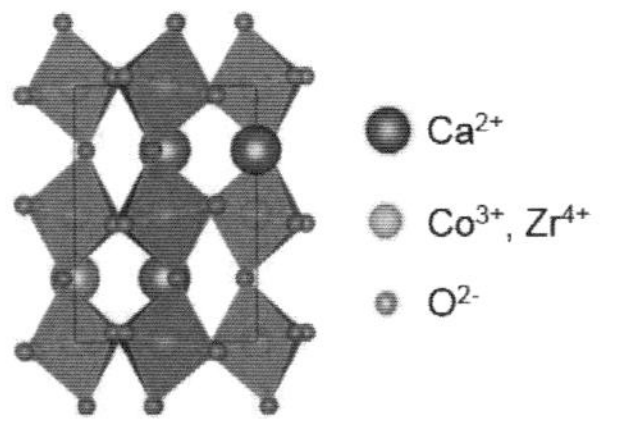

9

Synthesis and characterization of $CaCo_xZr_{1-x}(Pd)O_3$ perovskites

RTI INTERNATIONAL

Table 1. Analytical result of as-synthesized (Pd-doped) $CaCo_xZr_{1-x}O_3$ perovskites.

#	Theoretical composition	Actual composition (XRF)	Tolerance factor, τ
CCZPd2	$CaCo_{0.4}Zr_{0.4}Pd_{0.2}O_3$	$CaCo_{0.4}Zr_{0.3}Pd_{0.2}O_3$	0.81
CCZPd1	$CaCo_{0.5}Zr_{0.4}Pd_{0.1}O_3$	$CaCo_{0.5}Zr_{0.3}Pd_{0.1}O_3$	0.82
CCZPd05	$CaCo_{0.55}Zr_{0.4}Pd_{0.05}O_3$	$CaCo_{0.55}Zr_{0.3}Pd_{0.06}O_3$	0.83
CCZ91	$CaCo_{0.9}Zr_{0.1}O_3$	$CaCo_{1.0}Zr_{0.1}O_3$	0.86
CCZ73	$CaCo_{0.7}Zr_{0.3}O_3$	$CaCo_{0.8}Zr_{0.2}O_3$	0.85
CCZ55	$CaCo_{0.5}Zr_{0.5}O_3$	$CaCo_{0.5}Zr_{0.5}O_3$	0.84
CCZ37	$CaCo_{0.3}Zr_{0.7}O_3$	$CaCo_{0.3}Zr_{0.6}O_3$	0.82
CZ10	$CaZrO_3$	$CaZrO_3$	0.80

- **Perovskite synthesis:** Pechini method, calcination @ 1200 °C;
- **Perovskite target compositions:**
  $CaCo_xZr_{1-x}O_{3-\delta}$ (x=0, 0.3, 0.5, 0.7, and 0.9),
  Pd doped $CaCo_{0.6-y}Zr_{0.4}Pd_yO_{3-\delta}$ (y=0.05, 0.1 and 0.2);
- **Perovskite actual compositions** confirmed by XRF;
- **Goldschmidt tolerance factor:**
  $\tau = \frac{r_A + r_O}{\sqrt{2}(r_B + r_O)}$
  (r_A, r_B, and r_O are the ionic radii for the A^{n+}, B^{m+}, and O^{2-});
- All the τ values are within the range of 0.8 ~0.9:
  Suggest orthorhombic structures
- **BET surface areas: <10 m²/g**

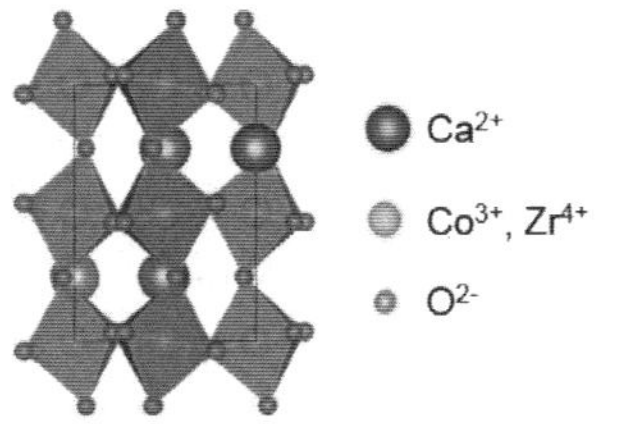

10

XRD: Partial B-site substitution by Zr enhances crystallinity Pd dopant type includes surface Pd^0 and bulk Pd^{2+}

RTI INTERNATIONAL

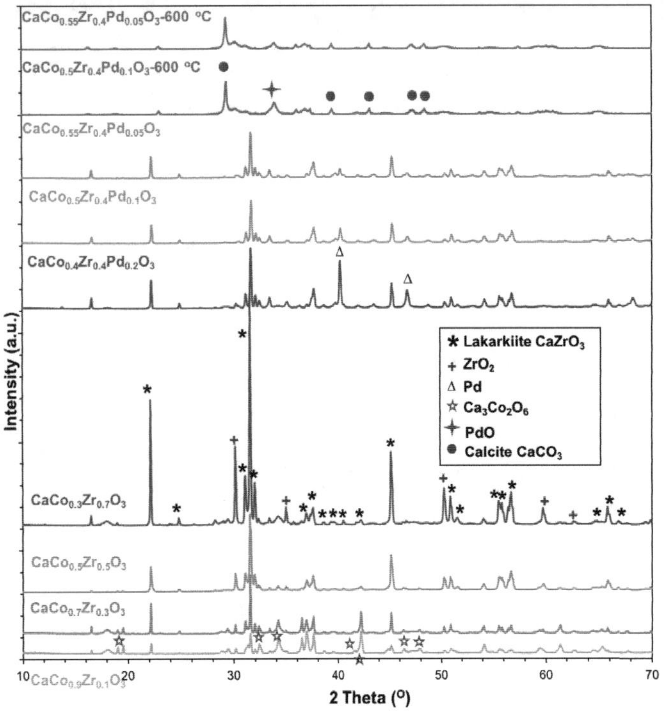

Fig. 3. XRD patterns of the $CaCo_xZr_{1\text{-}x}O_3$ (x = 0, 0.3, 0.5, 0.7 and 0.9) and Pd-doped $CaCo_{0.6\text{-}y}Zr_{0.4}Pd_yO_3$ (y = 0.05, 0.1 and 0.2) perovskite samples.

- **For $CaCo_{0.6\text{-}y}Zr_{0.4}Pd_yO_3$ (calcined at low temperature 600 °C):**
  - Perovskite structure not formed
  - Palladium presents as PdO
- **For $CaCo_{0.6\text{-}y}Zr_{0.4}Pd_yO_3$:**
  - Main phase remained $CaCo_{0.5}Zr_{0.5}O_3$ perovskite;
  - Doped Pd includes bulk Pd^{2+} and surface Pd^0.
- **For $CaCo_xZr_{1\text{-}x}O_3$:**
  - Retain Lakarkiite $CaZrO_3$ main structure;
  - Partial substitution of Co by Zr enhances the structural crystallinity of $CaCoO_3$;
  - When x=0.5, optimum crystallinity and minimum impurity.

11

H_2-TPR: Enhanced Co reduction by Zr substitution and Pd doping

RTI INTERNATIONAL

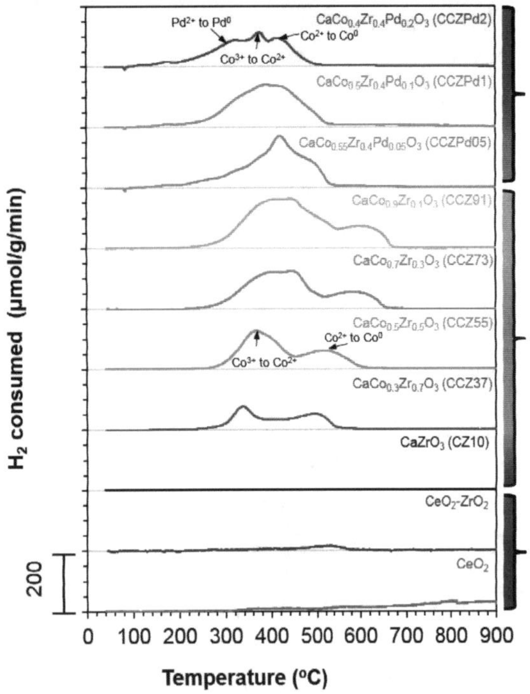

Fig. 4. H_2 consumption as a function of reaction time on stream and temperature of studied samples during H_2-TPR measurements (temperature ramp rate at 5 °C/min).

- **For $CaCo_{0.6\text{-}y}Zr_{0.4}Pd_yO_3$:**
  - Pd doping also facilitates the reduction of Co species;
  - $CaCo_{0.4}Zr_{0.4}Pd_{0.2}O_3$:
    Bulk Pd^{2+}→Pd^0 (T_{max}=320 °C)
    Co^{3+}→Co^{2+} (T_{max}=390 °C)
    Co^{2+}→Co^0 (T_{max}=420 °C);
  - Possible H_2 dissociation on surface Pd^0 particles followed by the successive spill-over of dissociated H atoms to the Co species.
- **For $CaCo_xZr_{1\text{-}x}O_3$:**
  - Co^{3+}→Co^{2+} (280-500 °C)
    Co^{2+}→Co^0 (450-680 °C);
  - B-site substitution by Zr enhances the reduction of Co species by lowering the reduction T;
  - No reduction peak observed with $CaZrO_3$.
- **Perovskite vs. CeO_2 and CeO_2-ZrO_2 (CZO):**
  $CaCo_xZr_{1\text{-}x}(Pd)O_3$ perovskites have significantly higher reducibility/OSC than CeO_2 and CZO.

12

Redox cyclic tests: $CaCo_{0.5}Zr_{0.5}O_3$ showed optimum OSC and stability

RTI INTERNATIONAL

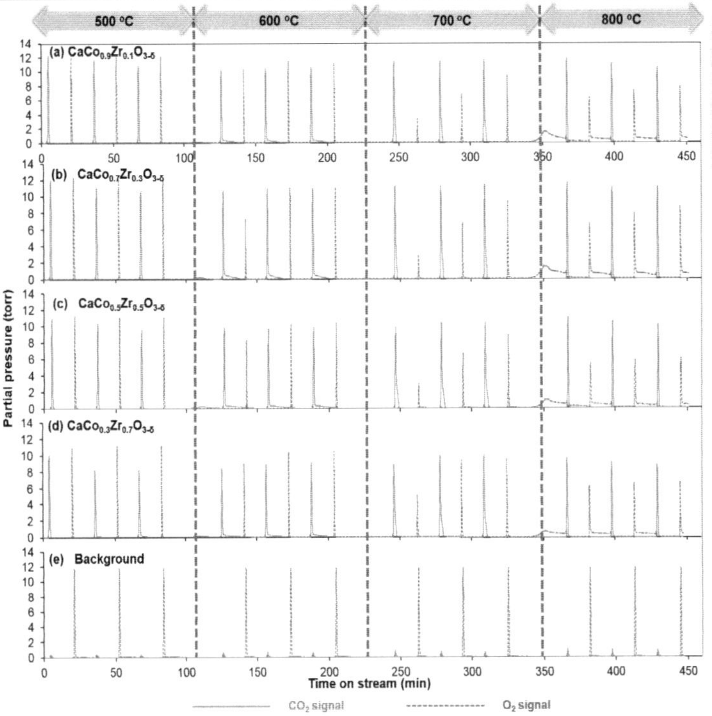

Reduction-Oxidation (3 cycles)

Reduction by 20% CO/He

Oxidation by air

Temperature: 500, 600, 700 and 800 °C

- Perovskite OSC increases with increasing T;
- Perovskite OSC increases with increasing *x* (Co content);
- Stable oxygen storage performance by Zr substitution;
- Optimum perovskite OSC and stability with $CaCo_{0.5}Zr_{0.5}O_3$.

Fig. 5. CO_2 production and O_2 intake vs. reaction time on stream (TOS) of $CaCo_xZr_{1-x}O_{3-\delta}$ (x=0.3, 0.5, 0.7, or 0.9) perovskite samples, as represented by MS partial pressure signals during isothermal CO reduction-air oxidation cyclic tests at 500, 600, 700, and 800 °C.

13

Oxygen intake kinetics measurements

RTI INTERNATIONAL

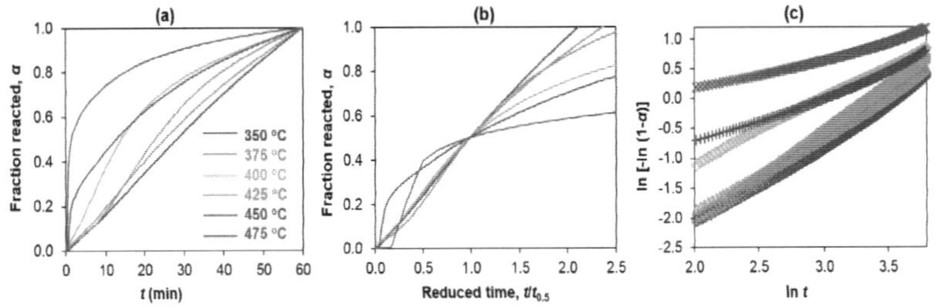

$$\alpha = \frac{m_0 - m_t}{m_0 - m_{final}}$$

$$\alpha \in [0, 1]$$

Fig. 6. Plots for oxygen-intake kinetic calculation of $CaCo_{0.5}Zr_{0.5}O_{3-\delta}$ perovskite showing (a) Fraction reacted α with time variation, (b) Fraction reacted α against reduced time $t/t_{0.5}$, and (c) $\ln[-\ln(1-\alpha)]$ as a function of $\ln t$ (with t in min) for the determination of reaction mechanisms at temperatures varied from 350 °C to 475 °C.

Table 2. List of solid-state rate equations with suggested *m* values studied for the oxygen-intake reaction for $CaCo_{0.5}Zr_{0.5}O_{3-\delta}$ and $CaCo_{0.9}Zr_{0.1}O_{3-\delta}$ perovskite samples. *m* values: slopes in Fig.6(c) at each temperature.

Eq. #	Function	Equation	m	Reaction mechanism
1	D_1 (α)	$\alpha^2 = kt$	0.62	Diffusion controlled
2	D_2 (α)	$(1-\alpha)\ln(1-\alpha) + \alpha = kt$	0.57	Diffusion controlled
3	D_3 (α)	$[1-(1-\alpha)^{1/3}]^2 = kt$	0.54	Diffusion controlled
4	D_4 (α)	$1 - 2\alpha/3 - (1-\alpha)^{2/3} = kt$	0.57	Diffusion controlled
5	F_1 (α)	$-\ln(1-\alpha) = kt$	1.00	First-order kinetics
6	R_2 (α)	$1-(1-\alpha)^{1/2} = kt$	1.11	Surface reaction controlled
7	R_3 (α)	$1-(1-\alpha)^{1/3} = kt$	1.07	Surface reaction controlled
8	Zero order	$\alpha = kt$	1.24	Zero-order kinetics

14

Oxygen intake kinetics measurements: First order kinetic model

RTI INTERNATIONAL

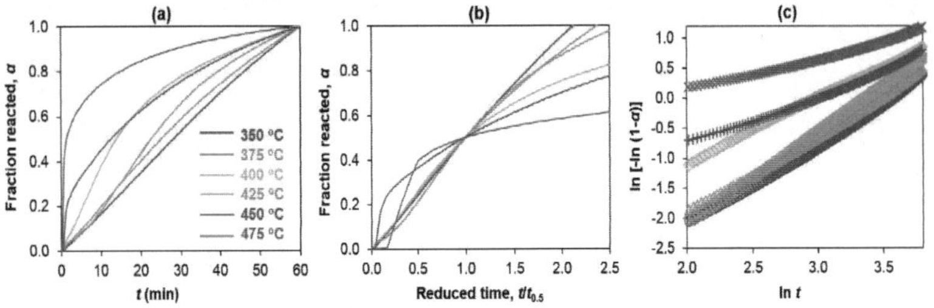

$$\alpha = \frac{m_0 - m_t}{m_0 - m_{final}}$$

$$\alpha \in [0, 1]$$

Fig. 6. Plots for oxygen-intake kinetic calculation of $CaCo_{0.5}Zr_{0.5}O_{3-\delta}$ perovskite showing (a) Fraction reacted α with time variation, (b) Fraction reacted α against reduced time $t/t_{0.5}$, and (c) $\ln[-\ln(1-\alpha)]$ as a function of $\ln t$ (with t in min) for the determination of reaction mechanisms at temperatures varied from 350 °C to 475 °C.

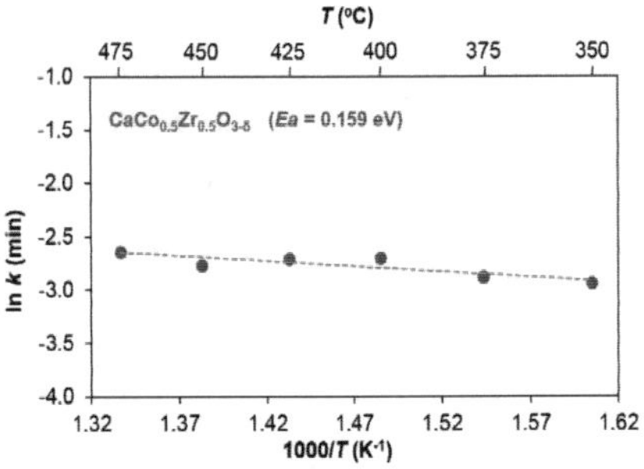

$-\ln(1-\alpha) = kt$ m = 0.6~1.3 First-order kinetics

Fig. 7. Temperature dependent-reaction rate constants k in natural log scale as a function of inverse temperature for the calculation of reaction activation energy for $CaCo_{0.5}Zr_{0.5}O_{3-\delta}$ perovskite. The k values were achieved by fitting reaction data with various kinetic models, and first-order kinetic model was selected.

15

Catalytic oxidation activity of Pd doped $CaCo_{0.6-y}Zr_{0.4}Pd_yO_3$

RTI INTERNATIONAL

Simulated dry exhaust feed: 0.060-0.066% C_3H_6, 0.750-0.825% CO, 0.675-0.750% O_2, 19.5% CO_2, N_2 bal.

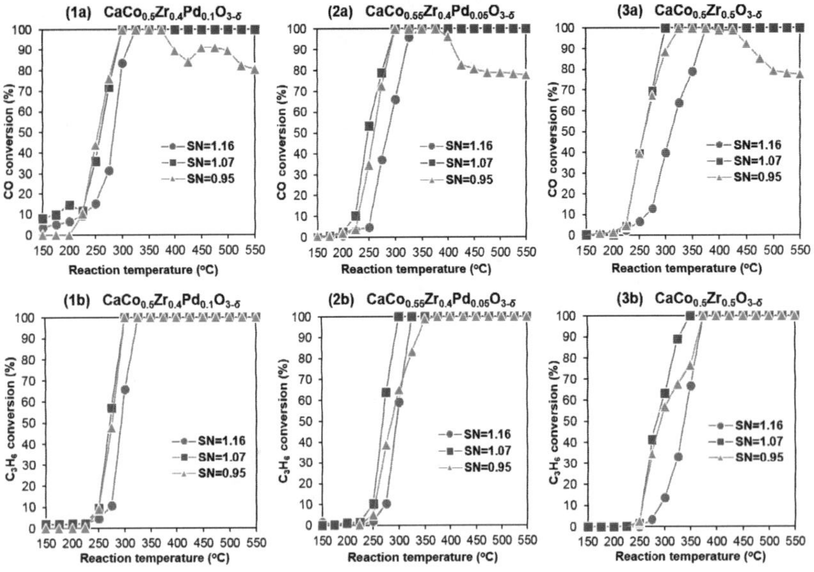

Fig. 8. (1) CO and (2) C_3H_6 conversions during catalytic oxidation tests over (a) $CaCo_{0.5}Zr_{0.4}Pd_{0.1}O_{3-\delta}$, (b) $CaCo_{0.55}Zr_{0.4}Pd_{0.05}O_{3-\delta}$ and (c) $CaCo_{0.5}Zr_{0.5}O_{3-\delta}$ perovskite with simulated exhaust feed **at stoichiometric numbers (SNs) of 1.16, 1.07 and 0.95. SN=2[O_2]/([CO]+9[C_3H_6])**

- Complete conversions of CO and C_3H_6 below 350 °C;
- Pd-doped samples showed higher oxidation activities than Pd-free one at all three conditions;
- Higher conversions at richer condition (SN=0.95) for Pd samples.
- "Pd regeneraetion" effect at rich condition (Tanaka *et al*)

16

Catalytic oxidation activity with addition of H_2O in the feed

RTI INTERNATIONAL

Simulated wet exhaust feed: 0.056% C_3H_6, 0.702% CO, 0.648% O_2, 17.55% CO_2, 10% steam, N_2 bal.

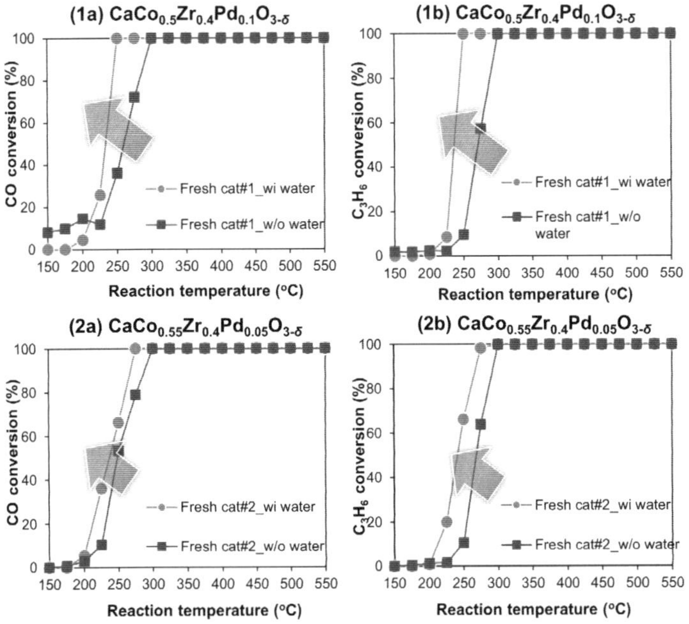

Fig. 9. (1) CO and (2) C_3H_6 conversions during catalytic oxidation tests over fresh and aged (a) $CaCo_{0.5}Zr_{0.4}Pd_{0.1}O_{3-\delta}$, (b) $CaCo_{0.55}Zr_{0.4}Pd_{0.05}O_{3-\delta}$ perovskites **at SN of 1.07, with presence of 10 vol-% H_2O in the feed.**

Enhancements in both CO and C_3H_6 conversions when water (steam) was added to the feed;

17

Catalytic oxidation activity of Pd doped $CaCo_{0.6\text{-}y}Zr_{0.4}Pd_yO_3$

RTI INTERNATIONAL

Table 3. Main reaction pathways during catalytic conversions of CO and C_3H_6 over Pd doped $CaCo_xZr_{1\text{-}x}O_3$ perovskites by feeding simulated engine exhaust at SN=1.07, when water is present in the feed.

Eq. #	Reaction pathway	Reaction	ΔH^0_{298} (kJ/mol)
(1)	Oxidation of C_3H_6	$C_3H_6 + 4.5\ O_2 \xrightarrow{\text{Perovskite}} 3CO_2 + 3H_2O$	-2058
(2)	Oxidation of CO	$CO + 0.5\ O_2 \xrightarrow{\text{Perovskite}} CO_2$	-283
(3)	Steam reforming of C_3H_6	$C_3H_6 + 6\ H_2O \xrightarrow{\text{Perovskite}} 3CO_2 + 9H_2$	+110
(4)	Water gas shift	$CO + H_2O \xrightarrow{\text{Perovskite}} CO_2 + H_2$	-41
(5)	Combustion of H_2	$H_2 + 0.5\ O_2 \xrightarrow{\text{Perovskite}} H_2O$	-286

When water was added to the simulated exhaust feed:

- Reaction pathways other than oxidation reactions were favored;
- Conversions of C_3H_6 and CO were enhanced respectively by catalytic SR and WGS on the same catalyst ;
- H_2 generated by the two reactions was readily consumed by stoichiometric amount of O_2 in the feed.

18

Catalytic oxidation activity of perovskites after accelerated aging

Simulated wet exhaust feed: 0.056% C_3H_6, 0.702% CO, 0.648% O_2, 17.55% CO_2, 10% steam, N_2 bal.

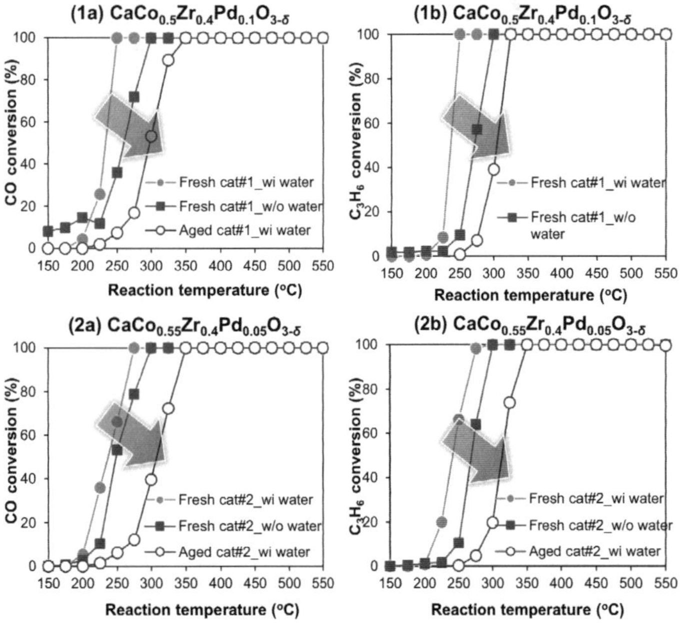

Certain degree of catalyst deactivation observed after accelerated aging in air at 1000 °C for 50 hrs, due to extensive catalyst sintering.

Fig. 10. (1) CO and (2) C_3H_6 conversions during catalytic oxidation tests over fresh and aged (a) $CaCo_{0.5}Zr_{0.4}Pd_{0.1}O_{3-\delta}$, (b) $CaCo_{0.55}Zr_{0.4}Pd_{0.05}O_{3-\delta}$ perovskites with at SN of 1.07, with presence of 10 vol-% H_2O in the feed. **Aging condition: 1000 °C in air for 50 hrs.**

19

Summary

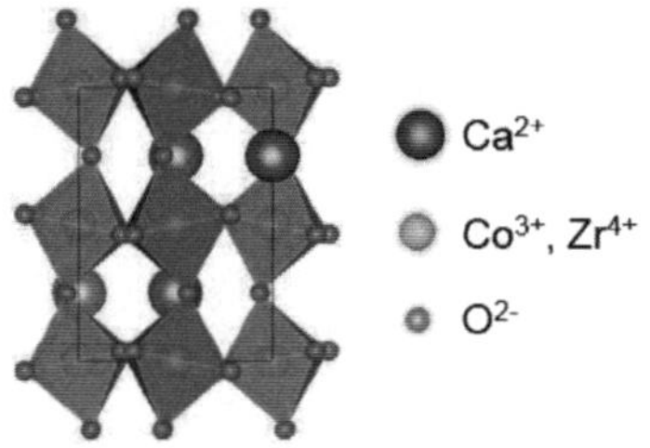

- Novel $CaCo_xZr_{1-x}O_{3-\delta}$ (x = 0, 0.3, 0.5, 0.7 and 0.9) and Pd-doped $CaCo_{0.6-y}Zr_{0.4}Pd_yO_3$ (y = 0.05, 0.1 and 0.2) perovskites were synthesized;
- The studied perovskites retain their main phase of orthorhombic Lakargiite $CaZrO_3$ structure, partial substitution of Co by Zr at B sites enhances the perovskite structural crystallinity;

20

Summary

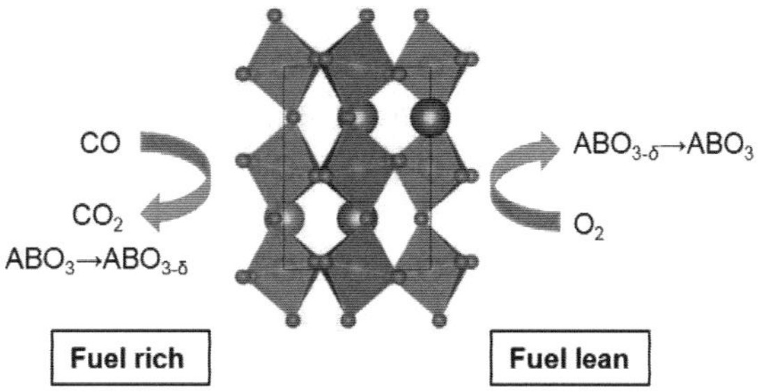

- Novel $CaCo_xZr_{1-x}O_{3-\delta}$ (x = 0, 0.3, 0.5, 0.7 and 0.9) and Pd-doped $CaCo_{0.6-y}Zr_{0.4}Pd_yO_3$ (y = 0.05, 0.1 and 0.2) perovskites were synthesized;
- The studied perovskites retain their main phase of orthorhombic Lakargiite $CaZrO_3$ structure, partial substitution of Co by Zr at B sites enhances the perovskite structural crystallinity;
- The studied perovskites showed improved redox property and OSC compared to the ceria-based oxygen storage materials (OSMs).

21

Summary

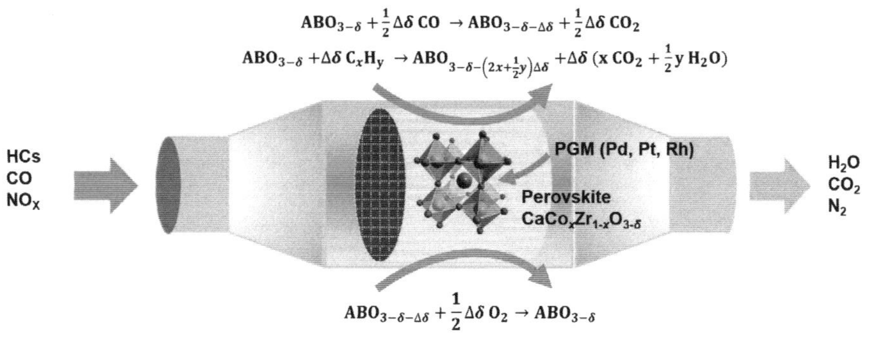

- Novel $CaCo_xZr_{1-x}O_{3-\delta}$ (x = 0, 0.3, 0.5, 0.7 and 0.9) and Pd-doped $CaCo_{0.6-y}Zr_{0.4}Pd_yO_3$ (y = 0.05, 0.1 and 0.2) perovskites were synthesized;
- The studied perovskites retain their main phase of orthorhombic Lakargiite $CaZrO_3$ structure, partial substitution of Co by Zr at B sites enhances the perovskite structural crystallinity;
- The studied perovskites showed improved redox property and OSC compared to the ceria-based oxygen storage materials (OSMs).
- Pd-doped samples showed promising catalytic activity towards C_3H_6 and CO oxidation under simulated exhaust conditions

22

Selected references

RTI INTERNATIONAL

[1] A. Simson, K. Roark, R. Farrauto, A feasibility study of niobium-containing materials for oxygen storage in three way catalytic converters, Appl. Catal. B Environ. 158–159 (2014) 106–111. doi:10.1016/j.apcatb.2014.04.006.

[2] A. Schön, C. Dujardin, J.-P. Dacquin, P. Granger, Enhancing catalytic activity of perovskite-based catalysts in three-way catalysis by surface composition optimisation, Catal. Today. 258, Part 2 (2015) 543–548. doi:10.1016/j.cattod.2014.11.002.

[3] Q. Zheng, R. Farrauto, M. Deeba, I. Valsamakis, Part I: A Comparative Thermal Aging Study on the Regenerability of Rh/Al2O3 and Rh/CexOy-ZrO2 as Model Catalysts for Automotive Three Way Catalysts, Catalysts. 5 (2015) 1770–1796. doi:10.3390/catal5041770.

[4] R. You, Y. Zhang, D. Liu, M. Meng, L. Zheng, J. Zhang, T. Hu, YCeZrO Ternary Oxide Solid Solution Supported Nonplatinic Lean-Burn NOx Trap Catalysts Using LaCoO3 Perovskite as Active Phase, J. Phys. Chem. C. 118 (2014) 25403–25420. doi:10.1021/jp505601x.

[5] Q. Zheng, R. Farrauto, M. Deeba, Part II: Oxidative Thermal Aging of Pd/Al2O3 and Pd/CexOy-ZrO2 in Automotive Three Way Catalysts: The Effects of Fuel Shutoff and Attempted Fuel Rich Regeneration, Catalysts. 5 (2015) 1797–1814. doi:10.3390/catal5041797.

[6] S. Royer, H. Alamdari, D. Duprez, S. Kaliaguine, Oxygen storage capacity of La1−xBO3 perovskites (with A′ = Sr, Ce; B = Co, Mn)—relation with catalytic activity in the CH4 oxidation reaction, Appl. Catal. B Environ. 58 (2005) 273–288. doi:10.1016/j.apcatb.2004.12.010.

[7] A. Klimkowicz, K. Świerczek, A. Takasaki, B. Dabrowski, Oxygen storage capability in Co- and Fe-containing perovskite-type oxides, Solid State Ion. 257 (2014) 23–28. doi:10.1016/j.ssi.2014.01.018.

[8] A. Mishra, N. Galinsky, F. He, E.E. Santiso, F. Li, Perovskite-structured AMnxB1−xO3 (A = Ca or Ba; B = Fe or Ni) redox catalysts for partial oxidation of methane, Catal. Sci. Technol. 6 (2016) 4535–4544. doi:10.1039/C5CY02186C.

[9] B. Levasseur, S. Kaliaguine, Effects of iron and cerium in La1−yCeyCo1−xFexO3 perovskites as catalysts for VOC oxidation, Appl. Catal. B Environ. 88 (2009) 305–314. doi:10.1016/j.apcatb.2008.11.007.

[10] G. Perin, J. Fabro, M. Guiotto, Q. Xin, M.M. Natile, P. Cool, P. Canu, A. Glisenti, Cu@LaNiO3 based nanocomposites in TWC applications, Appl. Catal. B Environ. 209 (2017) 214–227. doi:10.1016/j.apcatb.2017.02.064.

[11] S. Sartipi, A.A. Khodadadi, Y. Mortazavi, Pd-doped LaCoO3 regenerative catalyst for automotive emissions control, Appl. Catal. B Environ. 83 (2008) 214–220. doi:10.1016/j.apcatb.2008.02.014.

[12] Z. Say, M. Dogac, E.I. Vovk, Y.E. Kalay, C.H. Kim, W. Li, E. Ozensoy, Palladium doped perovskite-based NO oxidation catalysts: The role of Pd and B-sites for NOx adsorption behavior via in-situ spectroscopy, Appl. Catal. B Environ. 154–155 (2014) 51–61. doi:10.1016/j.apcatb.2014.01.038.

23

Acknowledgement

RTI INTERNATIONAL

Material Chemistry Manager: Dr. Marty Lail

Material Group: Dr. Mustapha Soukri
Dr. Jianping Shen
Dr. Ignacio Luz
Dr. John Thompson
Kelly Amato
Timothy Bellamy
Devaughn Body

U.S. DEPARTMENT OF ENERGY

Thank you and please feel free to ask any question!

One patent application (RTI 16P027-PROV) on perovskite material development and applications.

One manuscript just accepted by *Catalysis Today*.

25

500801 Probing the Oxidation and Reduction Halves of the Cu Redox Cycle during Standard NH_3-Selective Catalytic Reduction of NO_x on Cu-SSZ-13

Abstract:

The selective catalytic reduction of NO_x (x=1,2) using NH_3 as a reductant (NH_3-SCR) is a leading strategy for NO_x abatement in oxygen-rich automotive exhaust to meet increasingly stringent emissions constraints. It was shown that low temperature (473 K) NOx SCR on Cu-SSZ-13 occurs via a Cu(II)/Cu(I) redox cycle [1] on two types of isolated Cu sites, both divalent Cu^{2+} cations compensating paired framework Al atoms and monovalent $[CuOH]^+$ complexes at isolated framework Al atoms, which are solvated by NH_3 during reaction [2]. Two different kinetic regimes that are characterized by different apparent orders, activation energies, and steady-state Cu(I)/Cu(II) distributions in the Cu(I)/Cu(II) redox cycle have been identified under standard SCR conditions, in which O_2 is used as an oxidant. The rate-limiting steps during standard SCR depends on the Cu density (determined by the Si/Al and Cu/Al) of the Cu-SSZ-13 catalysts. One regime corresponds to SCR cycles limited by Cu(I)->Cu(II) oxidation with O_2 and the other regime corresponds to cycles limited by Cu(II)->Cu(I) reduction, which we probe in *operando* (varying Cu density, gas pressures and temperature) and in transient XAS experiments. O_2-oxidation during low temperature SCR requires two Cu(I) sites to be oxidized, in contrast to reduction occurring at single, isolated Cu(II) sites. Therefore, standard SCR rates depend on Cu cation density and distribution in Cu-SSZ-13 zeolites when Cu(I) oxidation steps are kinetically-relevant, but are independent of Cu density otherwise. We probe the nature of oxidation and reduction half-cycles independently by measuring their intrinsic kinetics. In-situ monitoring of Cu(I)/Cu(II) fractions under oxidation and reduction-limited conditions showed the existence of two different pools of Cu sites, those that are within diffusion distances of another Cu cation and can be oxidized with O_2 in standard SCR cycles, and others that are stranded and isolated from other Cu.

[1] C. Paolucci, A.A. Verma, S.A. Bates, V.F. Kispersky, J.T. Miller, R. Gounder, et al., Isolation of the Copper Redox Steps in the Standard Selective Catalytic Reduction on Cu-SSZ-13, (2014) 12022â€“12027. doi:10.1002/ange.201407030.

[2] C. Paolucci, A.A. Parekh, I. Khurana, J.R. Di Iorio, J. Albarracin, A. Shih, et al., Catalysis in a Cage: Condition-Dependent Speciation and Dynamics of Exchanged Cu Cations In SSZ-13 Zeolites, J. Am. Chem. Soc. (2016) 1â€“63. doi:10.1021/jacs.6b02651.

493807 Sulfur Oxidation Studies with BimetallicÂ Pd/Pt Catalysts

Abstract:

Bimetallic Pd/Pt catalysts are good automotive-engine oxidation aftertreatment catalyst candidates due to their improved activity and potential sintering resistance relative to the monometallic formulations. Any of these catalyst systems have to endure extended times on stream and tolerate exposure to high temperatures, water, and trace sulfur. Of note, the literature states that the Pd/Pt bimetallic benefits only exist in the absence of sulfur. Previous studies with bimetallic Pd/Pt catalysts show that sulfur sorption and oxidation characteristics are influenced by both Pd:Pt mole ratio and precious metal particle size. Here, studies were conducted with mono- and bimetallic Pd/Pt/Al_2O_3 catalysts to investigate more specifically how SO_2 oxidation activity is influenced by the Pd:Pt mole ratio. Diffuse reflectance infrared Fourier transform spectroscopy (DRIFTS), temperature-programmed oxidation, and steady-state oxidation studies were conducted on catalysts with different precious metal composition but similar precious metal crystallite particle sizes. During surface studies, it was found that catalysts with a higher Pd content tended to have greater activity for oxidizing sulfur species at low temperatures and as a result formed more SO_3 and SO_4 species. However, reactor studies showed that catalysts with a lower Pd content appeared to have greater activity for oxidizing sulfur species at a low temperature because more conversion of SO_2 to SO_3 species was detected at the reactor outlet. Due to this obvious disconnect, studies are now focused on investigating how the oxidation products formed on the catalyst surface influence the catalystâ€™s apparent activity for the SO_2 oxidation reaction. This work provides evidence that the SO_2 oxidation characteristics, as well as the potential for sulfur decay, for mono- and bimetallic Pd/Pt catalysts are heavily influenced by the Pd:Pt mole ratio.

495480 Nature of SO_2 Poisoned Cu-SSZ-13 Catalysts Under Ammonia Selective Catalytic Reduction (NH_3-SCR) Conditions

Abstract:

Nature of SO_2 poisoned Cu-SSZ-13 Catalysts under Ammonia Selective Catalytic Reduction (NH_3-SCR) Conditions

Arthur J. Shih[1], Hui Li[2], Ashok Kumar[3], Juan M. Gonzalez[1,4], Ishant Khurana[1], Christopher Paolucci[2], Jonatan D. Albarracin Caballero[1], Atish A. Parekh[1], Aída Luz Villa[4], W. Nicholas Delgass[1], Rajamani Gounder[1], Aleksey Yezerets[3], William F. Schneider[2], Jeffrey T. Miller[1], and Fabio H. Ribeiro[1]

[1]Davidson School of Chemical Engineering, Purdue University, West Lafayette, IN

[2]Department of Chemical and Biomolecular Engineering, University of Notre Dame, IN

[3]Cummins Incorporated, Columbus, IN

[4]Chemical Engineering Department, Universidad de Antioquia, Medellín, Colombia

Reduction of NO_x emissions from diesel engine exhaust is an environmental issue driven by increasingly stringent emission regulations. Cu-SSZ-13 catalysts have been shown to be promising for this application due to their hydrothermal stability and selectivity in reducing NO_x to N_2 using NH_3 compared to other zeolite catalysts. SO_x formed from the combustion of ppm levels of sulfur in diesel fuel deactivates the Cu-SSZ-13 catalyst. However, its effect on Cu^{2+} and CuOH active sites and the SCR mechanism is unclear and under debate.[1-4]

Two Cu-SSZ-13 catalysts, one with only Cu^{2+} (3.8 Cu wt%, 100% Cu^{2+}) and another with primarily $[CuOH]^{1+}$ (1.5 Cu wt%, 80% CuOH) active sites were synthesized, hereby denoted as Z_2Cu and ZCuOH, respectively.[5] Each catalyst was sulfated with dry SO_2 at 200°C and 400°C, resulting in sulfur contents of 0.7 wt% and 1.0 wt%, respectively. Sulfation at 400°C decreased the standard SCR rate by 90% for the ZCuOH catalyst but by only 60% for the Z_2Cu catalyst. In addition, the apparent activation energy for ZCuOH sites decreased from 70 kJ mol^{-1} to 11 kJ mol^{-1} with increasing extents of sulfur poisoning but remained constant at 70 kJ mol^{-1} for the Z_2Cu catalyst. These kinetic results suggest two different deactivation mechanisms for the Z_2Cu and ZCuOH sites. The decrease in apparent activation energy for the ZCuOH catalyst is attributed to changes in the binding energy of NH_3 to the Cu active sites, corroborated by evidence from UV-Visible where sulfur-Cu ligands were observed on ZCuOH but not on Z_2Cu. The unchanged apparent activation energy and reactant orders for the Z Cu catalyst after sulfation are attributed to pore-

2

filling by ammonium sulfate salts; this reduces the total number of accessible Cu, as supported by argon micropore measurements.

Z_2Cu sites are preferred over ZCuOH sites due to the lower impact of sulfur poisoning on the rate and apparent activation energy. These results predict that synthesizing catalysts with higher fractions of Z_2Cu sites will lead to improved emission control catalysts for commercial application.

References:

1. Jangjou, Y.; Wang D.;Kumar, A.; Li J.; Epling, W.S. *ACS Catal.* **2016**, *6*, 10.
2. Shen, M.; Wen, H.; Hao, T.; Yu, T.; Fan, D.; Wang, J.; Li, W.; Wang, J. *Catal. Sci. Technol.* **2015**, *5*, 1741.
3. Zhang, L.; Wang, D.; Liu, Y.; Kamasamudram, K.; Li, J.; Epling, W. *Appl. Catal. B: Environ.* **2014**, *156-157*, 371.
4. Luo, J.; Wang, D.; Kumar, A.; Li, J.; Kamasamudram, K.; Currier, N.; Yezerets, A. *Catal. Today* **2016**, *267*, 3.
5. Paolucci, C.; Parekh, A.A.; Khurana, I.; Di Iorio, J.R.; Li, H.; Albarracin Caballero, J.D.; Shih, A.J.; Anggara, T.; Delgass, W.N.; Miller, J.T.; Ribeiro, F.H.; Gounder, R.; Schneider, W.F. *J. Am. Chem. Soc.* **2016**, *138*, 18.

496535 Measurement of the Oxygen Storage Capacity and Its Role in the Catalytic Oxidation

Abstract:

CeO_2, thanks to its unique redox property of oxygen storage and release under lean and rich conditions, respectively, has been widely used as the catalyst support component. This property of oxygen storage and release can usually be quantitatively measured, called the oxygen storage capacity (OSC), via a reduction method using different techniques. The amount of OSC measured depends on the type of reducing agent and the technique used. Furthermore, the OSC is also influenced by the synthesis method of CeO_2, the modification by other elements like ZrO_2, the physical and geometric properties, and the metal supported on CeO_2-based materials. On the other hand, it is well accepted that the catalytic activity benefits from the unique characteristic of oxygen storage and release for reactions carried out over the CeO_2-based catalysts. Unfortunately, the significant enhancement of catalytic activity is still poorly correlated by the OSC, or the role of OSC in favoring the catalytic reaction is still poorly understood. We establish the measurement of OSC by comparing the use of different reducing agents (CO and H_2) and the use of different techniques (TCD and TGA). The effects of synthesis method, high temperature aging, precious metal loading, and the structure modification by ZrO_2 (i.e. CZO) and other stabilizer elements on CeO_2 have been also investigated. The OSC measured by different methods and the influencing factors are analyzed with emphasis on to the catalytic activity like CO oxidation for the objectives to establish a correlation between OSC and catalytic activity.

495203 Activation of Pd/SSZ-13 in Low Temperature NO Adsorption for Cold Start Application

Abstract:

As the regulation on NOx emission in exhaust gas has been globally issued and enforced, the purification of NOx from diesel engine is regarded as a great challenge. Although selective catalytic reduction (SCR) catalyst effectively reduce NOx in exhaust gas, this system cannot operate properly below 200 Â°C (i.e. during cold start period, initial ca. 100 â €“ 200 sec of engine operation). Therefore, during the warm-up of NOx reduction catalyst, the considerable amount of NOx is emitted into atmosphere without any treatment [1]. Recently, cold start NOx trap catalysts attract significant attention with a concept of storing NOx at low temperature (i.e. during cold start) and releasing them at elevated temperature, where SCR operates. In this work, we tried to apply various Pd/SSZ-13 catalysts with different Pd loading method and investigate the effects of hydrothermal aging (HTA) treatment on the NOx adsorption ability and the physicochemical change of Pd/SSZ-13.

504060 Continuous Distillation of Fast Pyrolysis Bio-Oils

Abstract:

Integration of pyrolysis-based biorefineries into petroleum refinery infrastructure will require novel strategies for bio-oil separations, including but not limited to distillation. Traditionally, distillation of fast pyrolysis bio-oils has been regarded as futile due to the thermal instability of the bio-oil. However, with developments in advanced pyrolysis techniques such as catalytic fast pyrolysis (CFP) and tail-gas reactive pyrolysis (TGRP) that reduce oxygen content and increase the thermal stability of bio-oils, distillation can proceed with significant yields that concentrate valuable chemicals into usable cuts. Still, little data on continuous separations has been offered. This talk will focus on preliminary data obtained from a continuous flash distillation system for pyrolysis bio-oil separations. The distillation system can be operated as either a one-stage or two-stage flash distillation. Based on our previous distillation studies, the first stage was operated at 120 ^{o}C, while the second stage can be operated between 250 and 300 ^{o}C. For one-stage distillation, the higher temperature set point was used. Yields and compositions of the fractions are analyzed and compared with results from batch experiments.

495467 Lignin Filers in Polylactic Acid Composite: The Effects of Impurities in Lignin and Lignin Type

Abstract:

Lignin filers in polylactic acid composite: The effects of impurities in lignin and lignin type

Yiwei Gao, Wangda Qu, Xianglan Bai

Lignin is second largest biopolymer on Earth. Lignin byproduct is abundantly available from paper and pulping industries, as well as emerging biorefineries. Due to its low cost, aromatic polymer structure and carbon rich nature, utilization of lignin for value-added products has been of great interest. For example, lignin filer is blended with other polymers to improve the properties of the composite materials, lower the product cost and enhance biodegradability. It is known that isolated lignin often contains considerable amount of inorganic impurities, and the impurities could affect the properties of lignin. Although the impurities are removable, purifying lignin can be a costly process. In the present study, the effect of impurities on the performance of lignin as the filer was studied by blending polylactic acid (PLA) with corn stover organosolv lignins with different amounts of impurities. The impurity content in the lignin was controlled by acid washing raw lignin with different numbers of washing and rinsing cycle. The PLA-lignin composites were characterized for their thermal, chemical and mechanical properties. Overall, blending corn stover lignin with PLA reduced the tensile strength of the composite but increased its modulus compared to pure PLA. When the ratio of lignin to PLA was fixed, using higher ash containing lignin resulted in the composite with better tensile strength. However, the composite with highest modulus was obtained with the lignin filer containing an intermediate amount of ash. For the lignin filer with fixed ash content, the composite containing 10% lignin filer had both highest tensile strength and modulus. It was also found that using raw lignin containing 7% of ash reduced thermal stability of the composite due to catalytic effect of the impurities for thermal decomposition. However, the lignins with lower ash contents did not significantly affect the thermal stability of the composites. A biorefinery produced hardwood lignin was also used as the filer in PLA to investigate the effect of lignin structure on the performance of lignin filer.

495477 Effect of Polyethylene Terephthalate in Producing Carbon Fiber from Pyrolytic Lignin

Abstract:

Pyrolytic lignin is a mixture of lignin-derived phenolic monomers and oligomers. Our previous study showed that repolymerized pyrolytic lignin could be used as a precursor to produce carbon fiber with enhanced mechanical properties. In the present work, pyrolytic lignin was mixed with different ratios of polyethylene terephthalate (PET) to extend the chain length of the precursor and control the molecular orientation. The FTIR spectra of the pyrolytic lignin and PET treated together at 200 ^{o}C were significantly different from the spectra of the physical mixture of individually treated pyrolytic lignin and PET, suggesting pyrolytic lignin and PET chemically reacted to form a new polymer structure. It is likely that pyrolytic lignin first caused chain scission of PET at the low temperature and then attached to the PET backbone through reacting with the carboxylic acid and vinyl groups on the PET fragments. The new precursors were further characterized using GPC, DSC, TGA, CHN and rheology analysis. The TGA analysis has shown that mixing pyrolytic lignin with 5% of PET increases carbon residue upon heating, thus potentially improving carbon fiber yield. The precursors were subjected to melt-spun at 200 ~ 210 ^{o}C and the spun fibers were oxidatively stabilized from room temperature to 250 ^{o}C using a heating rate of 0.3 ^{o}C/min. Subsequently, the stabilized fibers were carbonized at 1000 ^{o}C under inert environment. The produced carbon fibers were further characterized for mechanical properties and morphology.

501555 Sustainable Graphenic Fixed-Bed Hybrid Adsorbents: Green Synthesis and Application

Abstract:

Organic dyes are one of the most common types of pollutants found in urban waterbodies, with potentially serious health hazards. While various technologies including adsorption, precipitation, separation, amalgamation, and ion-exchange are employed to remove such contaminants from water, adsorption proved to be the most economical and efficient. Due to the high surface area and benign nature of activated carbon (AC), it is heavily used in the removal of various pollutants. However, increasing amounts of anthropogenic wastes necessitates the discovery of adsorbents with higher remediation efficacy. Graphene, the newest member of carbon-based nanomaterials, composed of sp2 hybridized carbons arranged in a perfect planar honeycomb lattice, have the highest surface area among the carbon family. Recent research indicated the large utility of graphenic materials in environmental remediation applications. However, the dependence on petroleum-based or high-value precursors for graphene synthesis is one of the challenges in its large-scale utilization for applications such as water purification. In addition, this dependence also limits the sustainability of such adsorbents. Hence, we report, *for the first time*, an immobilized graphenic adsorbent system prepared from a highly sustainable, naturally abundant, and underutilized bio-product, Lignin. Lignin, the most abundant natural polymer with aromatic monomers, was coated onto inexpensive sand particles. The lignin coated sand was thermochemically converted to graphene coated sand (GCS), leveraging the inherent high density of aromatic rings in lignin. The as-synthesized GCS with a large number of defects or void spaces, that act as the active adsorption sites, were investigated for their utility in removing a variety of dyes from simulated contaminated water. By controlling the process parameters, we engineered the microstructure of the adsorbent to derive highest possible adsorption capacity. To further embellish the adsorption efficacy, a metal-graphene hybrid adsorbent was prepared via a green reduction process. The hybrid system, due to enhanced surface area and activity, demonstrated improved adsorption capacity. It is expected that the completely green process employed for the synthesis and the highly sustainable nature of the product will attract widespread attention and help the push towards the application of graphenic materials for largescale applications such as water purification.

502694 Large-Scale Implementation of a New Technique for Sugar Extraction and Conversion to Furans from Biomass Hydrolysates

Abstract:

Lignocellulosic biomass is a sustainable feedstock for production of furans – 5-hydroxymethy furfural and furfural – versatile platform molecules that can be converted to drop-in fuels and precursors for synthetic materials. Lack of efficient pathways from lignocellulosic sugars constitutes a major barrier to the industrial production of furans. We have recently achieved high yields of furans from biomass hydrolysates using a novel hybrid enzyme- and chemo-catalytic technology. The presentation will provide proof-of-concept (bench-scale) experimental data on this novel technology as well as results from larger-scale implementation in process-modules, using biomass hydrolysates generated via AFEX (Ammonia Fiber Expansion) pretreatment.

496167 Microwave Assisted Depolymerization of Lignin to Phenolics in Polar Solvents

Abstract:

Lignin, a natural polymer is the third major constituent of vascular plants and provides structural support. It is separated from cellulose and hemicelluloses during the Kraft process in paper industry. An estimated 50 million tonnes of lignin is produced annually worldwide which is burned to generate process heat within the pulping industry [1]. The structure and aromatic units present in lignin makes it a valuable source of wide range of products ranging from macromolecules to phenolics. The conventional methods of lignin conversion require higher temperature and pressure conditions. Non-conventional methods such as microwave energy provide alternate ways of lignin conversion which are less energy intensive. Microwaves are electromagnetic radiation in the frequency range of 300MHz to 300 GHz. The interaction of microwaves with the reaction media is based on the mechanism of dipole rotation and ionic conduction. In this study, microwave assisted conversion of lignin was carried out in solvents, with high to medium polarity such as ethylene glycol (EG), dimethyl sulfoxide (DMSO) and dimethyl formamide (DMF). These solvents also have the ability to dissolve alkali lignin completely at 100°C. Experiments were carried out in a microwave reactor at different temperatures (100-140°C) and reaction time (0-80min). The effect of microwave treatment on the molecular weight distribution (MWD) and average molecular weights of lignin were analyzed using gel permeation chromatograph. The phenolics produced by lignin conversion were identified using MALDI-TOF mass spectrometry and high pressure liquid chromatography. In high polarity solvent such as EG, maximum reduction in weight average molecular weight (M_w) was observed at 120 °C, while in medium polarity solvents such as DMSO and DMF, M_w decreased with increase in temperature. The highest yield of lignin oligomers (127.74 mg g^{-1}_{lignin}) and phenolic monomers (133.7 mg g^{-1}_{lignin}) containing acetosyringone, guaiacol, syringaldehyde, anisole was obtained at 100Â°C in DMF and DMSO respectively. The possible reactions for the formation of phenolic monomers and oligomers occur by the cleavage of etheric Cî²-O and Cî±-Cî² bond induced by solvent interactions. The molecular weight distribution of lignin shifted to larger molecular weights at higher temperatures and reaction time indicating the involvement of recombination reactions.

References

1. Tolbert, A.; Akinosho, H.; Khunsupat, R.; Naskar, A.K.; Ragauskas, A.J. Characterization and analysis of the molecular weight of lignin for biorefining studies. *Biofuels Bioprod. Bioref.* **2014**, *8*, 836â€“856.

501832 Microwave-Responsive CNT/Polyaniline Core/Shell Structured Catalysts for High Energy Efficiency in Saccharide-HMF Conversion

Abstract:

Microwave absorption and catalytic activity have been synergistically integrated into a core/shell structured CNT/polyaniline(PANI) hybrid catalyst that enables localized heating on the catalyst surface and uplifts energy efficiency in the dehydration reaction of saccharides to 5-hydroxymethylfurfural (HMF). Results reveal that the integration of CNT and polyaniline significantly improves the overall saccharide conversion and HMF yield. The PANI shell thickness and morphology are controlled by both reactant feeding ratio and polymerization reaction rate. It is found that a thinner and rougher PANI shell is favourable to enhance the overall microwave absorption, improve heat transfer efficiency to catalyst surface and thus reaction efficiency. Moreover, the packing of polyaniline shell on CNT core facilitates its separation from reaction mixture and effectively prevents catalyst loss. These catalysts show excellent recyclability that the catalyst activity can be completely recovered by a simple re-doping process in 1.0 M sulfuric acid. This work opens up a new window in microwave responsive catalyst design that enhances energy efficiency in saccharide-HMF conversion reactions and promotes the economic feasibility of such reactions in practice.

493480 Base-Free Aerobic Oxidation of 5-Hydroxymethylfurfural to 2, 5-Furandicarboxylic Acid over Nanoscale Pt Catalysts Prepared By Atomic Layer Deposition

Abstract:

Pt/ZrO_2 catalysts synthesized by atomic layer deposition (ALD), incipient wetness impregnation (IW) and co-precipitation (CP) were employed for aerobic oxidation of 5-hydroxymethylfurfural (5-HMF) to produce 2, 5-furandicarboxylicacid (2, 5-FDCA) in aqueous phase without using alkali or acid. Under a mild reaction condition (T = 100 °C, P_{O2} = 0.4 MPa, time = 12 h), the Pt/ZrO_2 synthesized by ALD of 30 Pt deposition cycles gave a complete 5-HMF conversion and 2, 5-FDCA yield of 97% compared with the Pt/ZrO_2 synthesized by IW method (a 2, 5-FDCA yield of 9.4 %) and CP method (a 2, 5-FDCA yield of 3.9 %) with the same Pt loading of 3.23 wt%. The three catalysts were characterized by ICP, XRD, TEM and CO-TPD, and the results indicated that high Pt dispersion and uniform Pt particle size distribution on the ZrO_2 surface were obtained by ALD and the performance of C=O adsorption was strong. While the growth of Pt particles synthesized by incipient-wetness impregnation was not controllable resulting in the metal particles non-uniformed and the performance of C=O adsorption was weak. The Pt particles synthesized by co-precipitated were wrapped in the ZrO_2 and the catalytic activity of the Pt are completely eliminated due to the difficulty of accessing the Pt active surface for reactant and intermediate products. The ALD produce uniform Pt particles to ensure the unitive active sites for the aerobic oxidation of HMF to FDCA. As the deposition cycles of Pt for Pt/ZrO_2 (ALD) increased, the average size of the Pt nanoparticle increased but the dispersion did not change, and the Pt/ZrO_2 with ALD 30 deposition cycles exhibited the highest TOF. This work has high commercial value for developing efficient heterogeneous catalysts using the cheap and available metals oxide that with low surface area but with high catalytic active which have potential applications in many heterogeneous reactions and maximize the atom efficiency of noble metals.

Mo@Pt Overlayer as Effective Catalysts for Hydrodeoxygenation of Guaiacol and Anisole

Joseph H. Holles
Qinghua Lai

Department of Chemical Engineering
University of Wyoming
Laramie, WY 82071

UNIVERSITY OF WYOMING

1

Hydrodeoxygenation (HDO) of Bio-oils

Biomass can be converted into crude bio-oil via thermochemical treatment such as pyrolysis or liquefaction.

Drawbacks caused by high oxygen content of crude bio-oils:
low heating value, poor stability, poor volatility, high viscosity and corrosiveness.

Upgrading of crude bio-oils is necessary

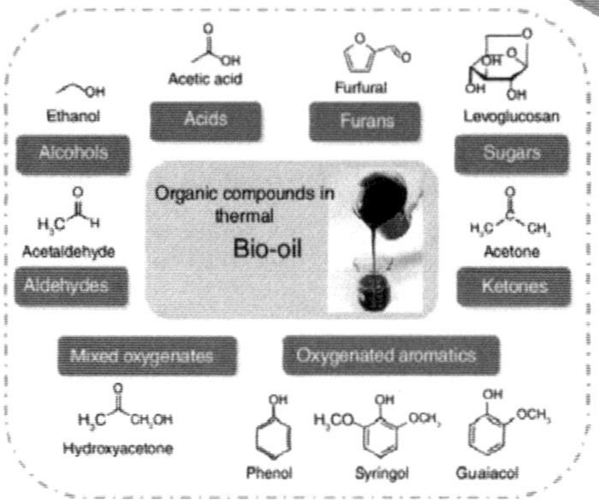

Fermoso J.; Wiley Interdisciplinary Reviews: *Energy Environ.* **2017,** 6:e245

Model compounds for lignin-derived bio-oils:

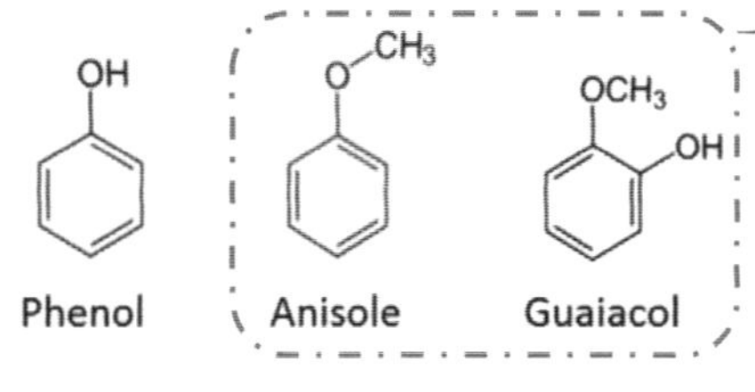

One or two types of C–O bond

Prototype compound to represent lignin-derived bio-oils

Strong surface-reactant binding has been shown to inhibit HDO reaction

Resasco D.; *Energy Fuels* **2011,** 25, 4155-4162

2

Bimetallic Overlayer Catalysts

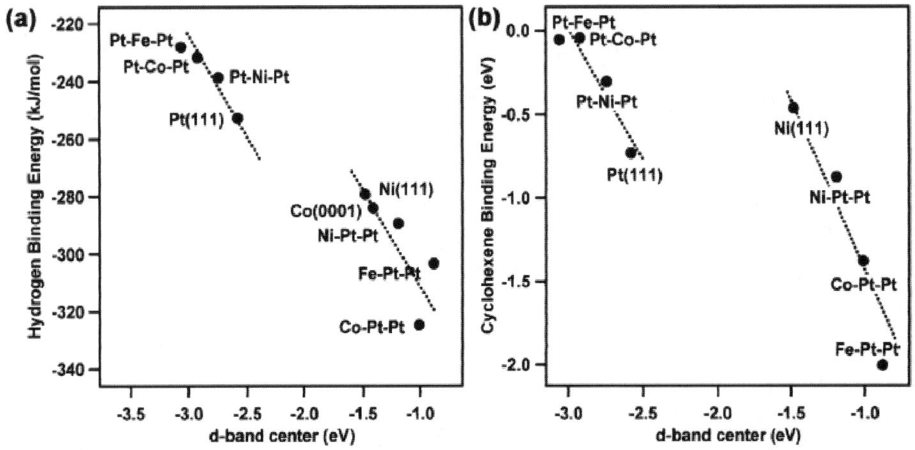

DFT calculations of binding energies of hydrogen and cyclohexene on Pt−3d−Pt(111) and 3d−Pt−Pt(111) surfaces.

- Overlayer structure could change the d-band center energy;
- Changing d-band center energy would alter bonding characteristic to adsorb stronger/weaker as desired;

Yu et al., *Chem. Rev.* **2012**, *112*, 5780−5817

3

Mo@Pt Overlayer Core@Shell Catalyst

- Mo@Pt/SiO_2-Al_2O_3 for Hydrodeoxygenation:
  - Decreased adsorbate binding strength
  - Mo favors the C-O bond cleavage compared to Pt only catalyst
  - Pt facilitates the dissociation of H_2 in comparison to Mo catalyst

- Objective:

Study how synthesis of Mo@Pt overlayer affects surface properties and activities in HDO of guaiacol and anisole

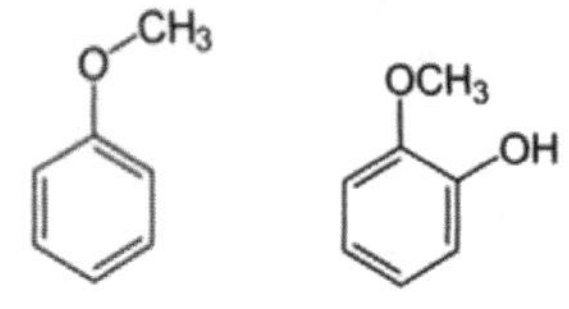

Anisole Guaiacol

4

Catalyst Synthesis

- Silica-Alumina (Alfa Aesar, grade 135) used as support for all synthesized catalysts.
- Mo/SiO_2-Al_2O_3, Mo-Pt/SiO_2-Al_2O_3, and Pt/SiO_2-Al_2O_3 prepared by using incipient wetness impregnation :
  - $Pt(acac)_2$ /toluene slurry.
  - Incipient wetness impregnation with nickel (II) nitrate hexahydrate precursors.
- The Mo@Pt overlayer catalysts used were prepared by using the directed deposition technique.

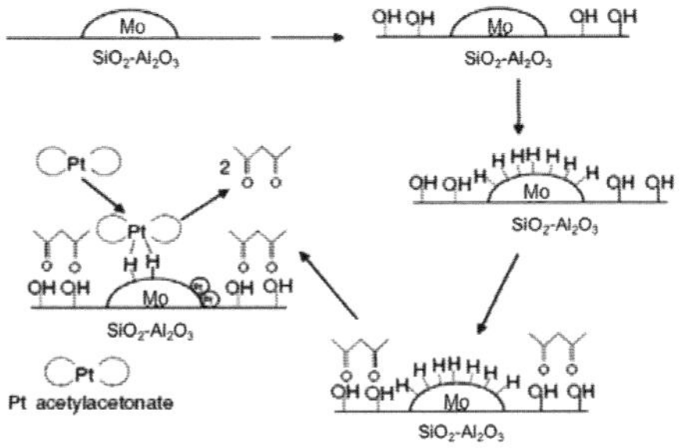

5

Physicochemical Properties

Sample	Metal loading [wt%]		Dispersion [%]	Overlayer coverage [%]	S_{BET} [m^2g^{-1}]	Pore volume [cm^3g^{-1}]
Mo/SiO_2-Al_2O_3	4.98	Mo	5.78		339	0.58
Mo@Pt SD	0.171	Pt	5.45	29	335	0.55
Mo@Pt DD	0.303	Pt	5.75	52	319	0.53
Mo-Pt Bimetallic	4.97	Mo	6.10		304	0.56
	0.285	Pt				
Pt	4.96	Pt	12.9		409	0.64

Typically, additional depositions lead to an increase in Pt loading

With similar S_{BET} and pore volume

6

XRD Characterization

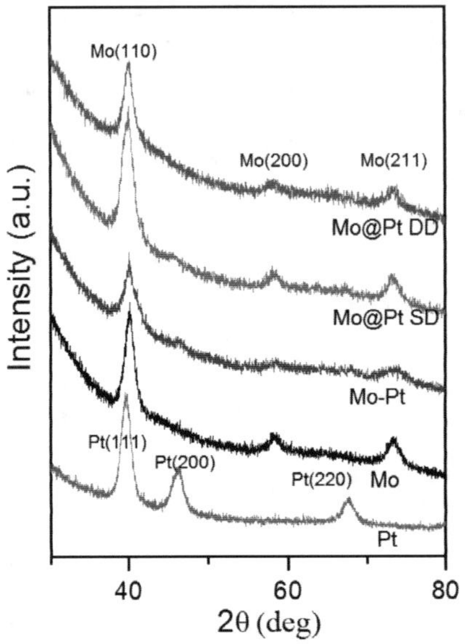

Metallic Pt peak was not observed in overlayer catalysts.

7

H_2 Temperature Programmed Reduction

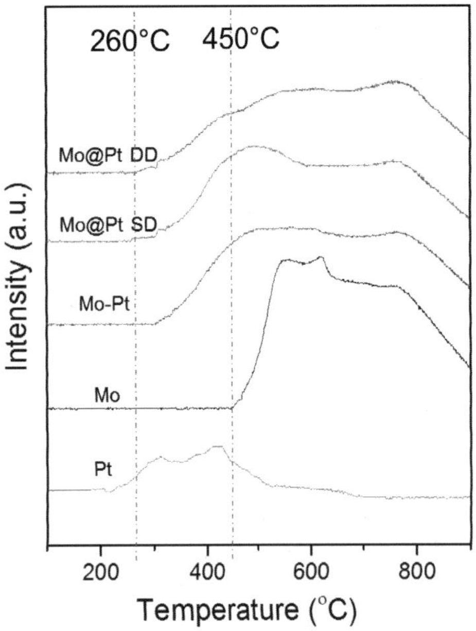

Deposition of Pt on Mo enhances the reducibility of molybdenum oxides

Indicates a close interaction between Pt and Mo species

8

H_2 and CO Heat of Adsorption

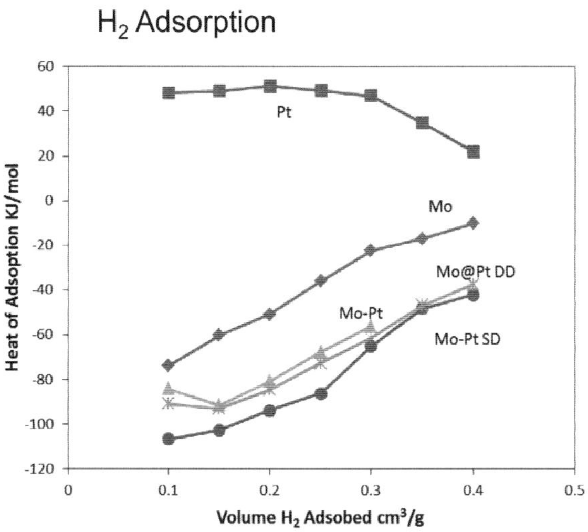

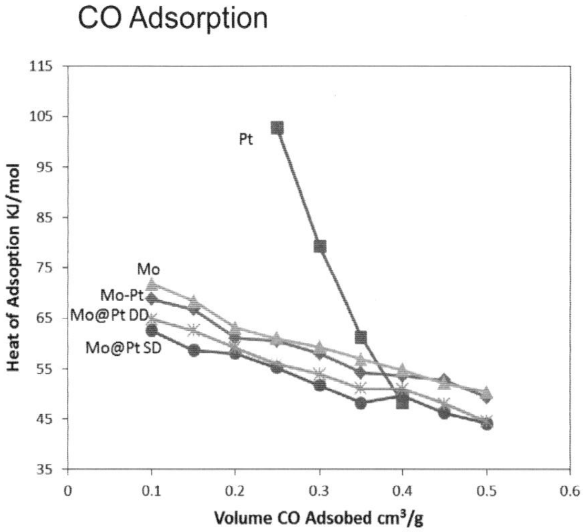

- Overlayer synthesis results in lower heat of adsorption.
  - Lower than Mo and Pt.
  - Similar to computational predictions.

9

Ethylene Hydrogenation Experiments

$$C_2H_4 + H_2 \rightarrow C_2H_6$$

- ~50 mg catalyst (<100 mesh) diluted in 450 mg silica gel (Alfa Aesar, 70-150 mesh).
- Reactivity experiments utilize ethylene hydrogenation in an 8 mm tubular glass reactor.
  - Ceramic furnace fitted around reactor to maintain temperature (30-200°C).
- Electronic mass flow controllers used to control inlet gas concentrations/flow rates:
  - Standard conditions: 3%C_2H_4/20%H_2/N_2 at 750 mL/min
    - Inlet concentrations consistent with literature
- Thermo-Finnigan Trace GC/ISQ GC/MS system.
  - 25 m Varian CP-Al_2O_3/KCl PLOT column.

10

Ethylene Hydrogenation Activity

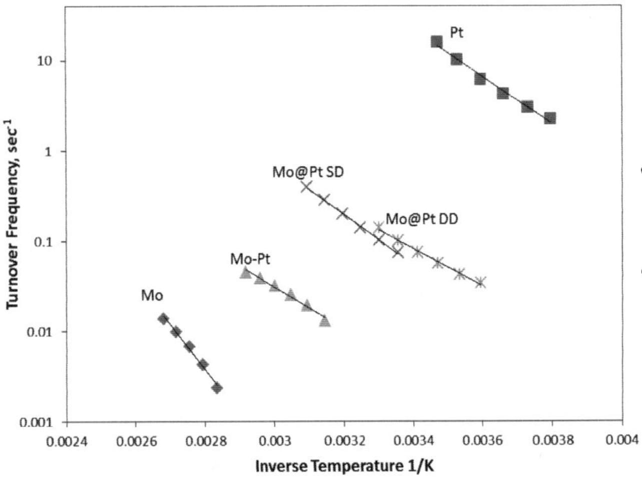

- Ethylene hydrogenation sensitive to hydrogen surface coverage
- Decreased binding strength decreases coverage and reaction rate

- All overlayers show reduced activity compared to pure Pt.

  Overlayer catalysts have lower hydrogen surface coverage and reduced binding strength

Guaiacol HDO Result

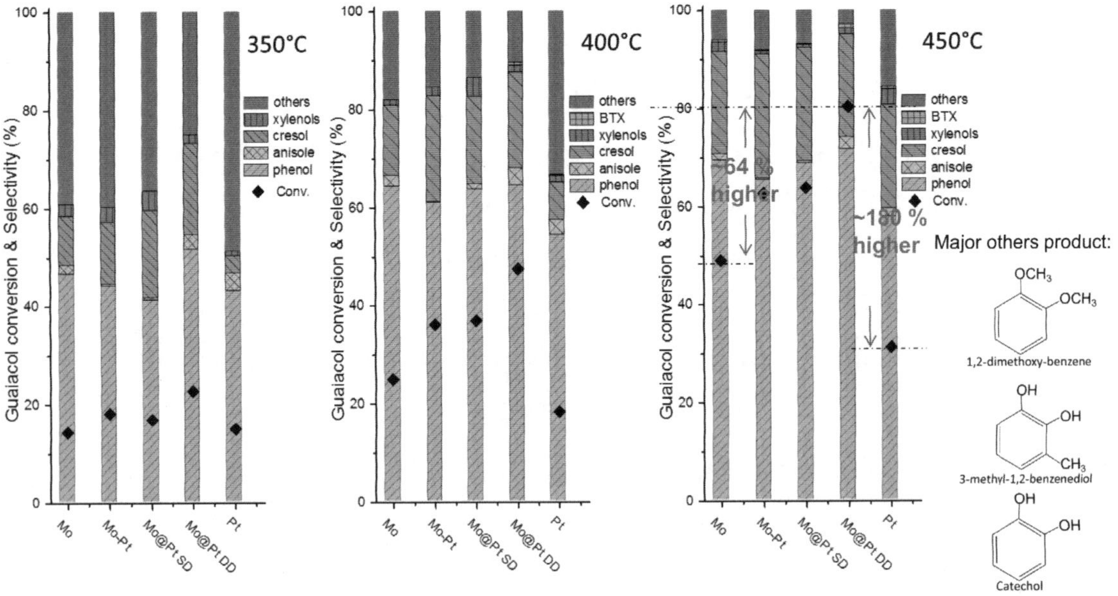

Mo@Pt DD catalyst showed much higher guaiacol conversion than other samples

Guaiacol HDO Result

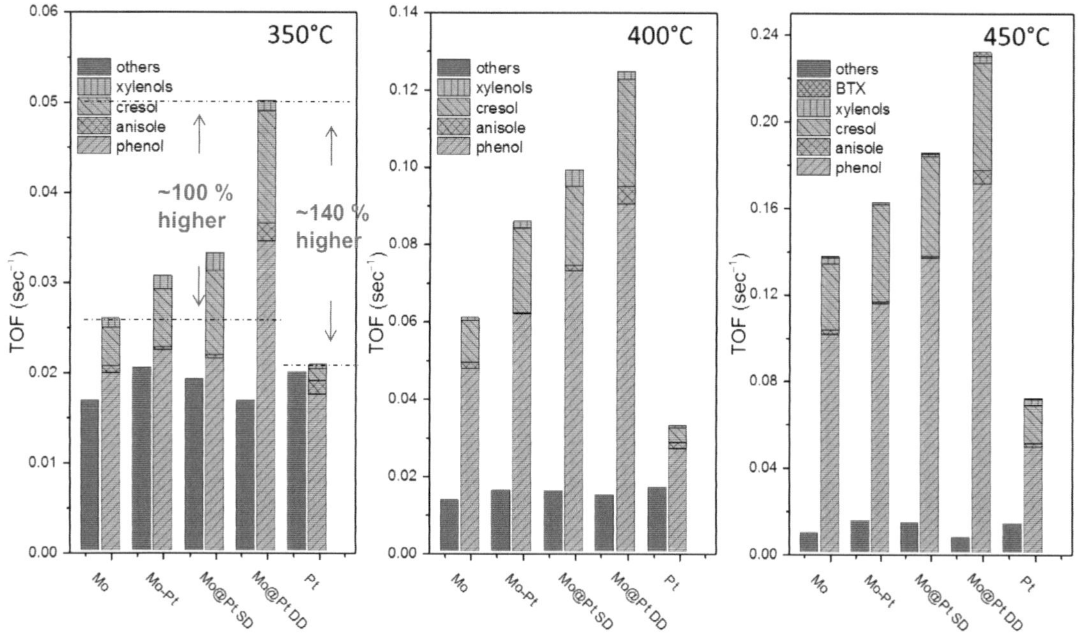

Mo@Pt overlayer catalysts showed much higher activity than other samples

13

Guaiacol HDO Result

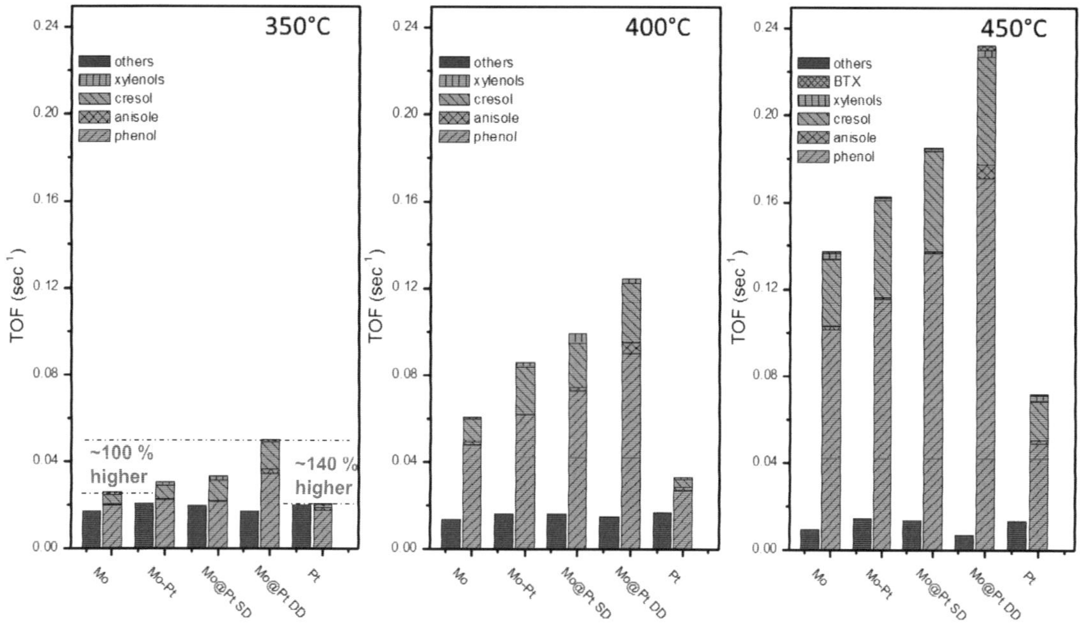

Mo@Pt overlayer catalysts showed much higher activity than other samples

14

Guaiacol HDO with Varying W/F

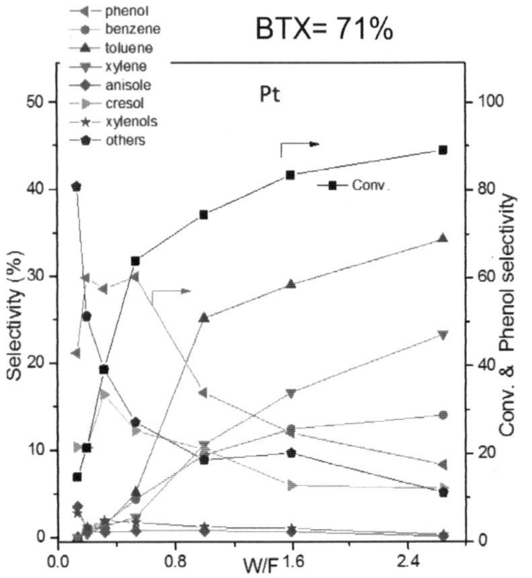

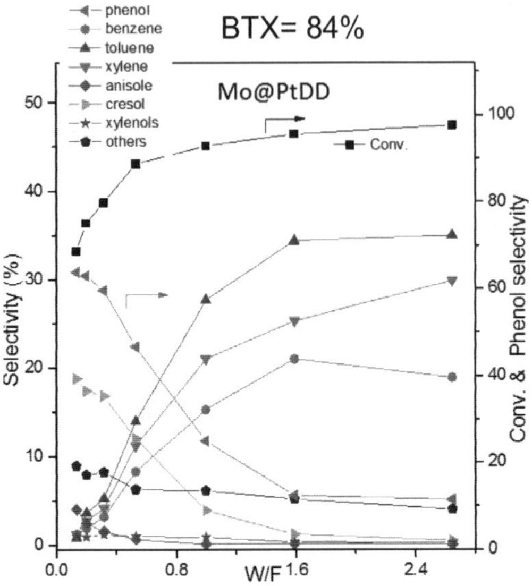

A BTX selectivity up to 84% over Mo@Pt DD at W/F of 2.65 h.

15

Anisole HDO Result

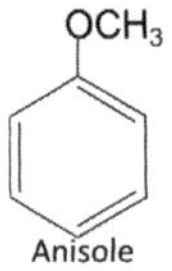

Non-deoxygenation products:

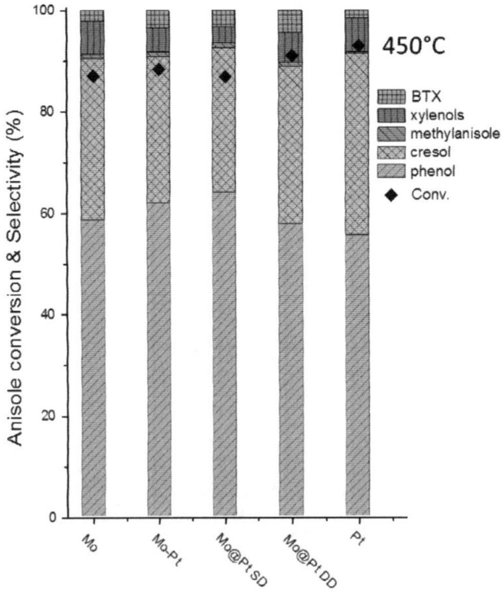

Deoxygenated products:

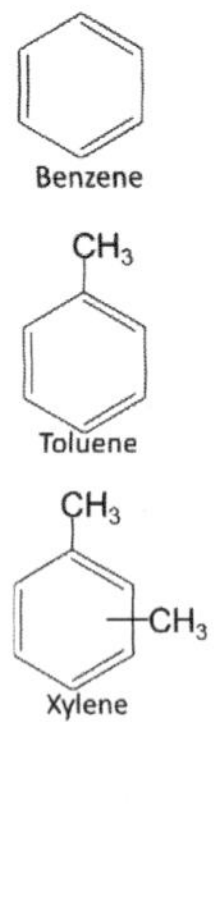

Mo@Pt overlayer catalysts showed a higher selectivity for BTX than Mo and Pt

16

Anisole HDO Result

TOF of Deoxygenation

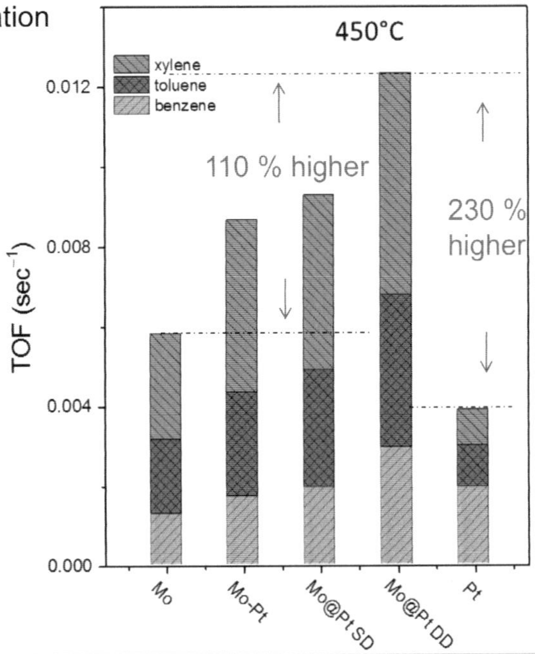

Mo@Pt overlayer catalysts showed higher deoxygenation activity than other samples

17

Anisole HDO with Different W/F

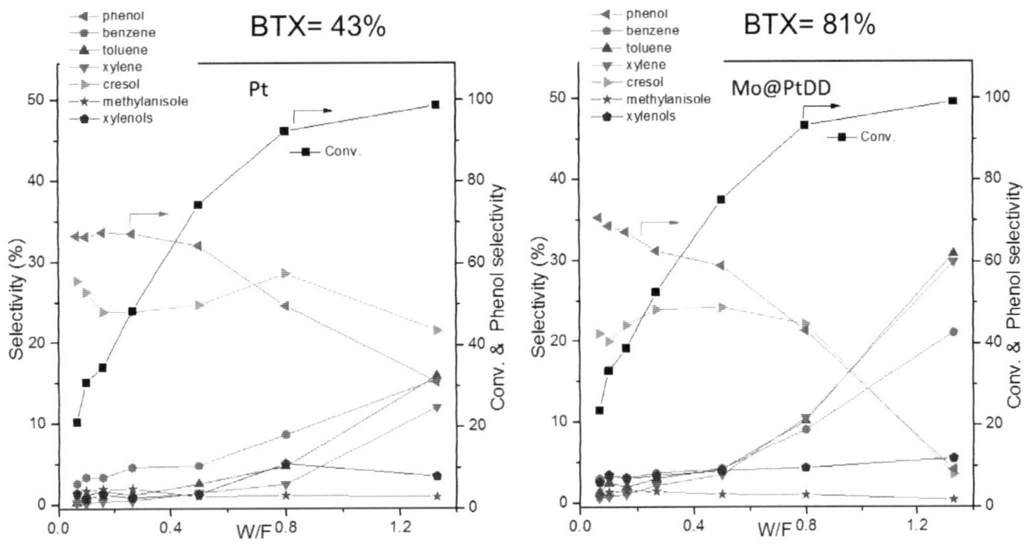

A BTX selectivity of 81% was achieved over Mo@Pt DD at W/F of 1.33 h

18

Conclusion

- Mo@Pt overlayer catalysts have demonstrated weaker H_2/CO adsorption than Pt, which is consistent with computational study.
- Mo@Pt overlayer show significantly increased turnover frequencies for deoxygenation product in both guaiacol and anisole HDO, confirming that the decrease in binding strength increases reactivity in HDO reaction.
- A selectivity up to 80% for BTX could be achieved over Mo@Pt DD for HDO of guaiacol and anisole.

19

Acknowledgements

Funding provided by National Science Foundation, Chemical, Bioengineering, Environmental and Transport Systems
[CBET-0933017]

National Science Foundation
WHERE DISCOVERIES BEGIN

Financial support from University of Wyoming School of Energy Resources through its Graduate Assistantship program.

School of Energy Resources
UNIVERSITY OF WYOMING

Research Group Members
Allen R. Morris
Chen Zhang

20

HDO Experimental

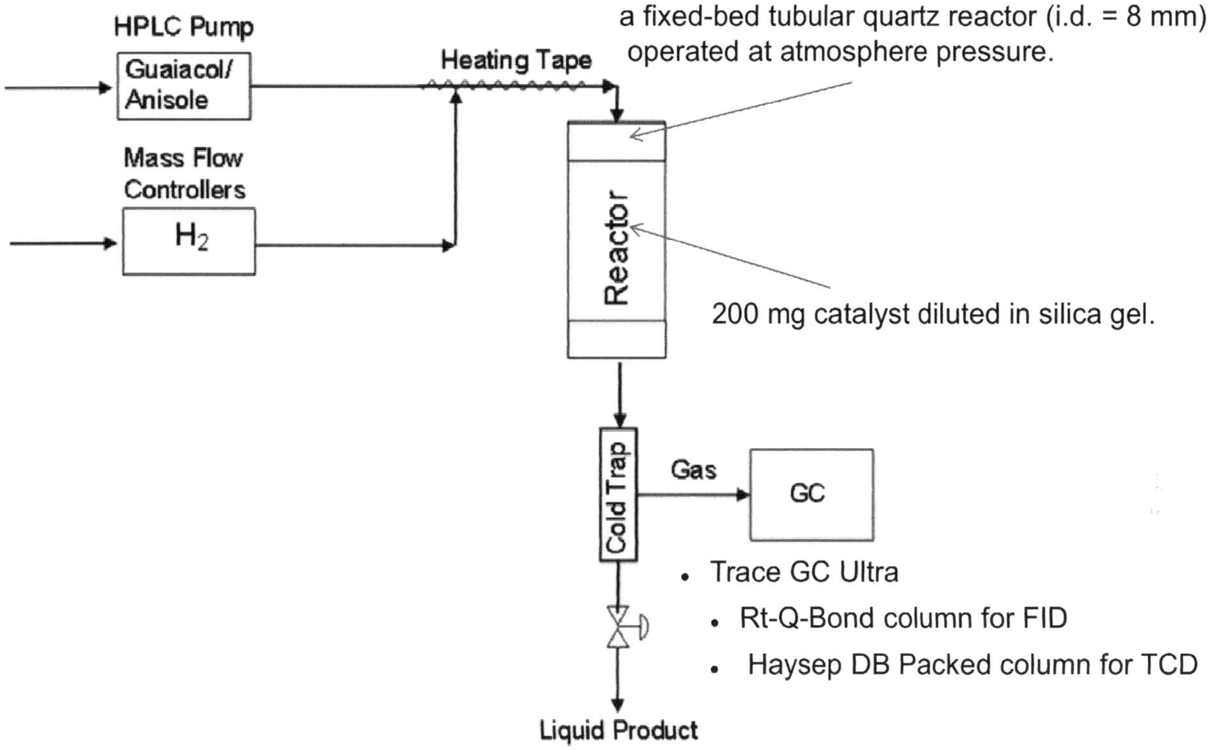

Schematic Diagram of Guaiacol/Anisole HDO Set-up

21

497210 Reaction Engineering for Bio-Renewable Furan-Derived Surfactants

Abstract:

Despite years of technology development, most large-volume surfactants are made from petrochemical sources, while efforts to make renewable surfactants are focused on making existing surfactant structures from renewable sources. In this work, we demonstrate a new surfactant based on the natural structure and chemistry of plant-based oils and sugars with superior function and suitability as a replacement to petrochemicals. Furans obtained from lignocellulosic biomass can be acylated with natural oils-derived fatty acids and anhydrides in the presence of a heterogenous zeolite catalyst. The results obtained for the reaction of lauric anhydride with furan show that different pore sizes, structures and acidity of zeolites result in varying acylation activity. Following acylation, the molecule can be upgraded via several independent and sequential chemistries such as etherification, hydrogenation and aldol condensation and finally subjected to sulfonation to yield surfactant molecules termed as oleo-furan sulfonates (OFS) in high yield. Evaluation of surfactant performance of OFS revealed hundredfold better detergency and stability in hard water conditions in comparison with petroleum-derived counterparts. The synthesis of OFS molecules is highly tunable and selective where the number of carbon atoms in the linear or branched chain can be easily varied without compromising on reaction yields to achieve desired surfactant properties.

References

[1] Joseph, K. E., Park, D. S., Koehle, M., Krumm, C., Ren, L., Damen, J. N., Shete, M. S., Lee, H. S., Xiao, Z., Lee, B., Fan, W., Vlachos, D. G., Lobo, R., Tsapatsis, M., Dauenhauer, P. J. ACS Cent. Sci. **2016**, 2, 820

501750 Upgrading of in-Situ Catalytic Fast Pyrolysis Bio-Oil to Hydrocarbon Fuels

Abstract:

Domestically, the US is challenged with improving the environment, energy security, and energy productivity, thus many efforts are concerted in utilizing biomass as a viable source of renewable liquid transportation fuels and contributing to a robust carbon-neutral energy production. However, achieving an economically competitive process is an ongoing challenge for current biomass upgrading technologies. Specifically, catalyst cost is one of the primary driver of thermochemical conversion cost associated with biomass upgrading to biofuels. The current state of technology for fast pyrolysis route includes a bio-oil stabilization step, which uses catalysts incorporating precious metals such as ruthenium to conduct hydrogenation of bio-oil at low temperatures. However, precious metal catalysts are orders of magnitude more expensive than conventional hydrotreating catalysts and very sensitive to sulfur contamination which limits there lifetime significantly. Catalytic fast pyrolysis (CFP) provides hope for a competitive production of an improved bio-oil. This bio-oil is also stable and higher in quality, and is also more readily hydroprocessed to liquid hydrocarbon fuels. Specifically, In-situ catalytic fast pyrolysis (in-situ CFP) uses a circulating fluidized-bed reactor which offers a straightforward, cost effective application of the technology. This presentation will report our recent research on in situ CFP and produced bio-oil hydrotreating. CFP bio-oils produced by using a low cost red mud catalyst (recovered waste from the Bayer process) and different reaction conditions from different biomass feedstocks were directly hydrotreated without a stabilization step. The impact of CFP parameters and feedstocks on hydrotreating will be elucidated. The overall biomass to fuel yield and cost information based on technoeconomic analysis will be reported. It showed that in situ CFP with an appropriate catalyst, such as red mud, has great potential for an efficient and economic viable route to convert biomass to fuels.

502374 Bio-Terephthalic Acid Synthesis from Cross Metathesis of Unsaturated Carboxylic Acids and Consecutive One-Pot Cycloaddition and Aromatization Reactions

Abstract:

Introduction

Terephthalic acid in its pure form (PTA) is essential for the preparation of a variety of polyester fibers and is largely produced via the catalytic oxidation of petroleum-derived p-xylene [1]. The ongoing exchange of naphtha crackers with shale gas crackers is causing a shortage in some aromatic species. As a consequence, alternative and sustainable routes to produce terephthalic acid from biomass derivatives, are highly desirable [2]. PTA precursors can be prepared via Diels-Alder cycloaddition with ethylene from bio-derived chemicals, like furnaic compounds or carboxylic dienes. These synthetic pathways are, however, limited by the low yields of either the starting chemicals or the final product [3-5]. In the last decade, Ru-catalyzed olefin metathesis, and in particular cross-metathesis, has become an important tool for the formation of carbon–carbon bonds and the formation of valuable organic molecules [6]. Di-ester muconates, in particular, can be synthesized from cross-metathesis of bio-sourced olefins. This route has never been considered before for the synthesis of bio-sourced terephthalates.

Results and discussion

We have investigated the ruthenium-catalyzed cross metathesis of sorbates with acrylates to form di-ester muconates. The effect of alkyl substituents and solvents was investigated over various organometallic ruthenium-based catalysts to optimize the conversion and selectivity towards the muconates. Conversions of up to 60% and 40% yields of the muconates were achieved within the first hour of reaction with low amounts (3 mol.%) of ruthenium-based catalyst containing a N-heterocyclic carbene (NHC) ligand. With the aim to develop a greener process, the acrylate reactant was used in excess with no other solvent added. In a later step, a one pot synthesis of diester terephthalates from the as-prepared di-ester muconates was achieved. First, the Diels-Alder cycloaddition between muconate esters and ethylene forms a cyclohexene dicarboxylic compound in >95% yield, and a consecutive aromatization on the same reaction batch in the presence of Pd/C catalyst affords the formation of terephthalic acid diesters (~ 40% yield). Upon further development, this route can provide a sustainable alternative for the industrial production of biomass derived polyethylene terephthalate (PET).

References

[2] Roger A. Sheldon. Green Chem. 16 (2014) 950.

[3] C. Luke Williams, Chun-Chih Chang, Phuong Do, Nima Nikbin, Stavros Caratzoulas, Dionisios G. Vlachos, Raul F. Lobo, Wei Fan, and Paul J. Dauenhauer. ACS Catal. 2 (2012) 935.

[4] S. Berard, C. Vallee, D. Delcroix, Ind. Eng. Chem. Res. 54 (2015) 7164-7168.

[5] J.W. Frost, Preparation of trans,trans muconic acid and trans,trans muconates, Draths Corporation, U.S. Patent US2010314243-A1, 2010

[6] Georgios C. Vougioukalakis, and Robert H. Grubbs. Chem. Rev. 110 (2010) 1746

499097 Conceptual Design and Techno-Economic Analysis of a Solar Hybrid Sulfur Process

Abstract:

The hybrid sulfur (HyS) cycle is the only practical two-step thermochemical water-splitting cycle with all fluid reactants. It is also one of the most advanced of the thermochemical water-splitting cycles, with both reaction steps having been experimentally validated and with detailed system designs having been proposed for hydrogen production from a continuous heat source like an advanced, high-temperature reactor. This presentation describes the conceptual design of a HyS process based on a variable heat source, specifically concentrated solar energy, which is only available for part of the day. The presentation will address how the diurnal variation was accommodated by the design of the plant and will lay out a techno-economic analysis for a hypothetical 50-MT/d solar hydrogen plant in the southwestern US. It will conclude with a development path toward hydrogen production at $2/kg H_2, the renewable hydrogen production cost target set by the DOE Office of Energy Efficiency and Renewable Energy.

NREL | 40

NATIONAL RENEWABLE ENERGY LABORATORY | 1977 - 2017

Potential Size of and Value Proposition for H2@Scale Concept

Mark F. Ruth, Paige Jadun,

Amgad Elgowainy,* Bryan Pivovar

2017 AIChE Annual Meeting

November 1-3, 2017

Minneapolis, Minnesota

NREL/PR-6A20-70457

* Amgad Elgowainy is an employee of Argonne National Laboratory

H2@Scale Energy System

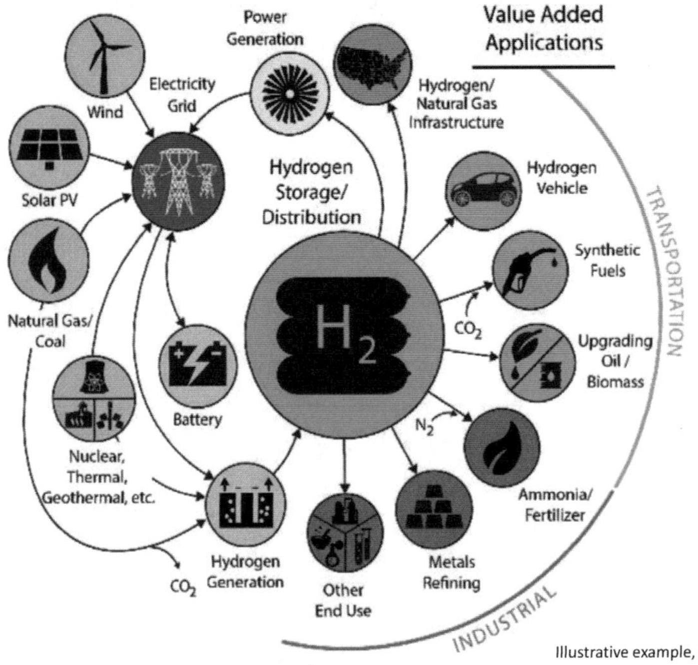

Illustrative example, not comprehensive

Technical Potential: Definition

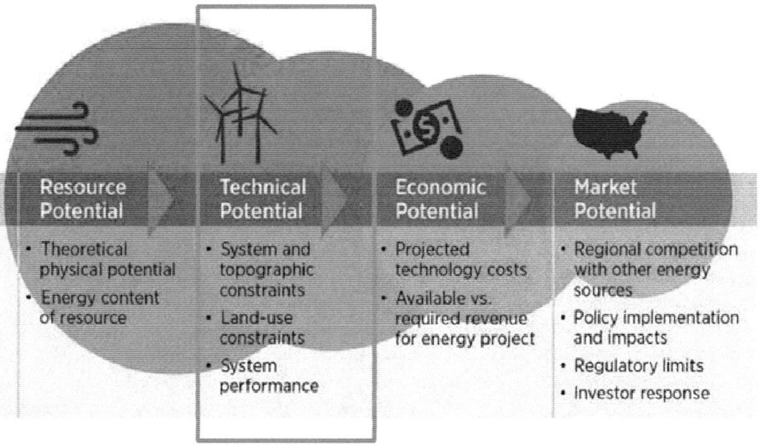

Technical potential is the subset of the available resource potential that is constrained by real-world geography and system performance but not economics

Brown, A., P. Beiter, D. Heimiller, C. Davidson, P. Denholm, J. Melius, A. Lopez, D. Hettinger, D. Mulcahy, and G. Porro. 2015. *Estimating Renewable Energy Economic Potential in the United States: Methodology and Initial Results.* Golden, CO: National Renewable Energy Laboratory. NREL/TP-6A20-64503

Technical Potential Hydrogen Demand

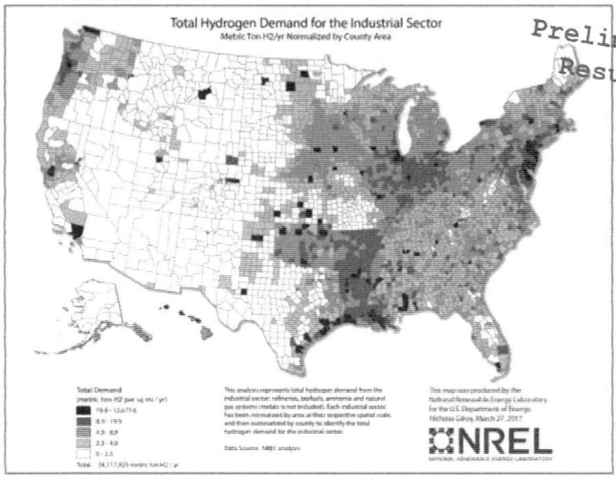

Use	Market potential (million metric tonne H_2 / year)
Industrial Use	
Refineries & CPI[§]	8*
Metals	5
Ammonia	5
Natural Gas	7
Biofuels	4
Light Duty Vehicles	28
Other Transport	3
Total	**60**

Total market potential: 60 MMT/yr

Current U.S. market: ≈ 10 MMT/yr

Near-term Outlook for Hydrogen Production Volume: 5% CAGR (2014-2019)[1]

§ CPI: Chemical Processing Industry not including metals, biofuels, or ammonia
* Current potential used due to lack of consistent future projections
Light duty vehicle calculation basis: 190,000,000 light-duty FCEVs from http://www.nap.edu/catalog/18264/transitions-to-alternative-vehicles-and-fuels
1. Global hydrogen Generation Market by Merchant & Captive Type, Distributed & Centralized Generation, Application & Technology- Trends & Forecasts (2011-2016)

Technical Potential: Impact on Renewable Resources

	EIA 2015 current consumption (quads/yr)	Required to meet demand of 60 MMT / yr (quads/yr)	Technical Potential (quads/yr)
Solid Biomass	4.7	15	20
Wind Electrolysis	0.7	9	170
Solar Electrolysis	0.1	9	1,364

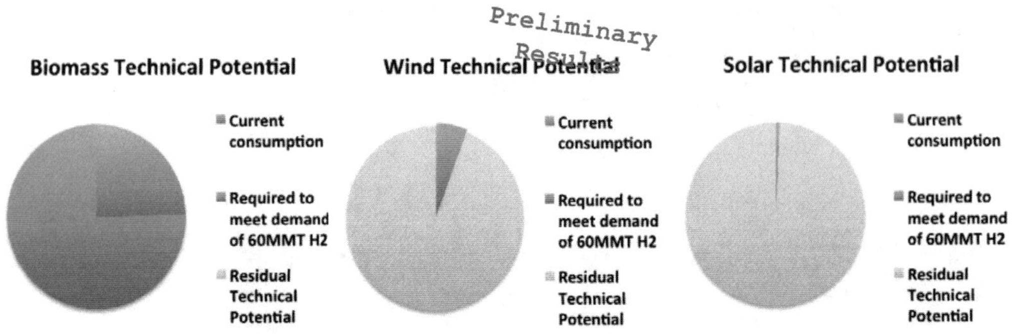

Total demand including hydrogen is satisfied by ≈6% of wind, <1% of solar, and ≈100% of biomass technical potential

Technical Potential: Impact on Fossil & Nuclear Resources

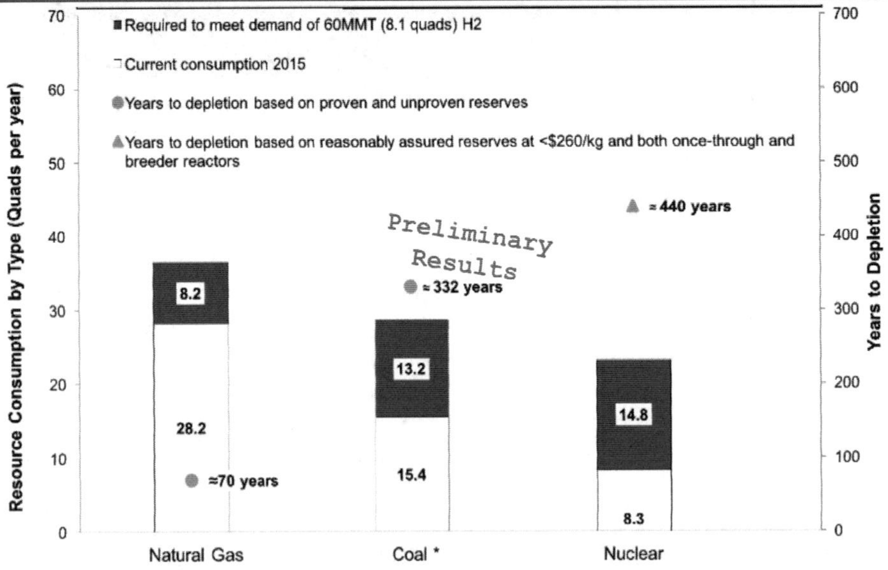

Hydrogen can be produced from diverse domestic resources to meet aggressive growth in demand

* Based on estimated recoverable reserves

Technical Potential: Impacts on Resources

Use	H_2 Consumed	*Resource Savings*		Emissions Reduction
	MMT / yr	Petroleum (bbl/yr)	Natural Gas (quad btu/yr)	CO2 (million metric ton/yr)
Refineries	8	900,000	1.332	87
Metals	5	0	0.365	78
Ammonia	5	500,000	0.833	54
Natural Gas System	7	700,000	0.923	63
Biofuels§	4	77,500,000	-0.026*	28
Light Duty Vehicles	28	1,017,600,000	0.629	469
Other Transport	3	113,400,000	0.051	50
Total	**60**	**1.2 Billion bbl**	**4.1 Quads**	**830 Million MT**
		~17% of U.S. petroleum consumption in 2016	~14% of U.S. natural gas consumption in 2016	~16% of U.S. energy-related emissions in 2016

Preliminary Results

Growth in electrolytic hydrogen using renewable electricity can reduce petroleum and natural gas utilization by ≥15%

*Negative values represent increase in use due to fertilizer production

§ 12% of the benefits of hydrogenated biofuels are credited to hydrogen

Economic Potential: Definition

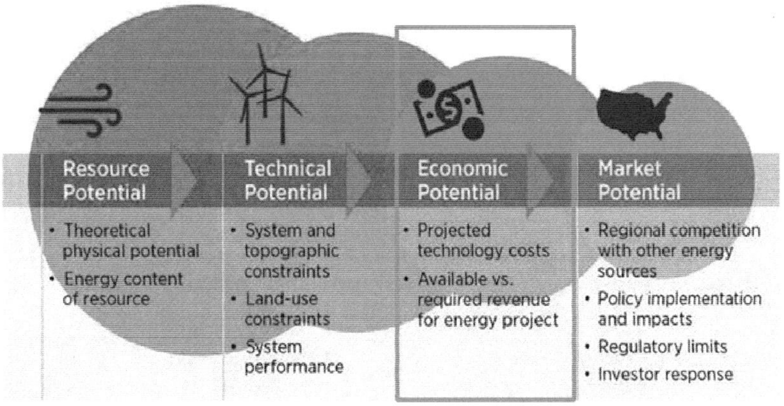

Economic potential is the **subset of the technical potential** where the **cost required** to produce hydrogen **is below** the **revenue available**

Brown, A., P. Beiter, D. Heimiller, C. Davidson, P. Denholm, J. Melius, A. Lopez, D. Hettinger, D. Mulcahy, and G. Porro. 2015. *Estimating Renewable Energy Economic Potential in the United States: Methodology and Initial Results.* Golden, CO: National Renewable Energy Laboratory. NREL/TP-6A20-64503

Improvements Enabling Use of Low-Cost Electricity

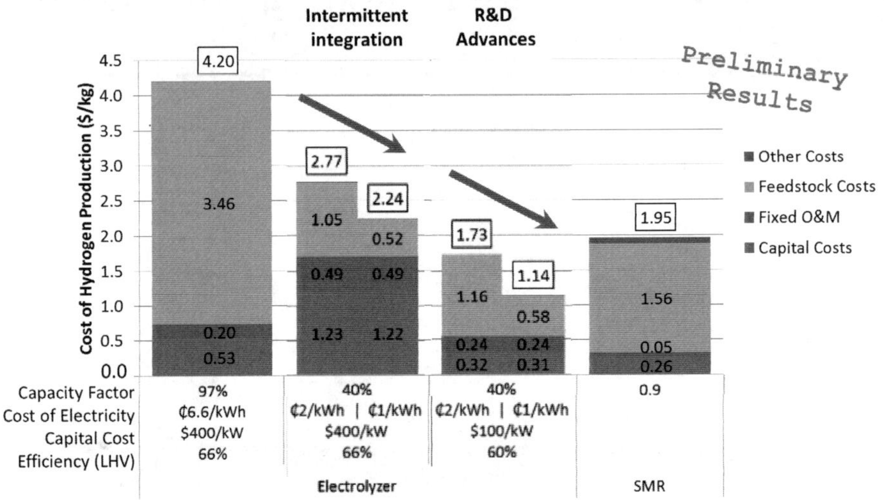

Leveraging of intermittent low-cost electricity can enable low-cost hydrogen production and also support grid stability.

Methodology for Estimating Otherwise Curtailed Electricity

Value of Otherwise Curtailed Electricity

Scenario Assumptions

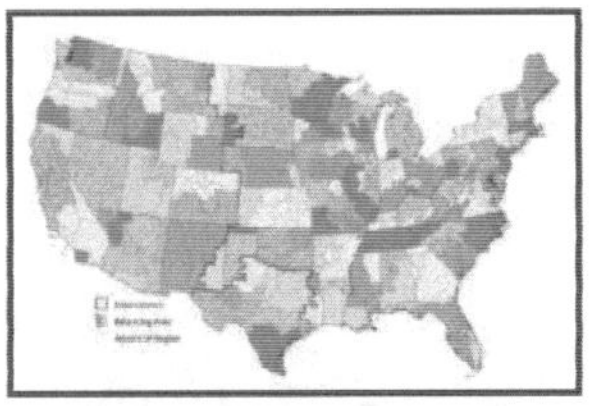

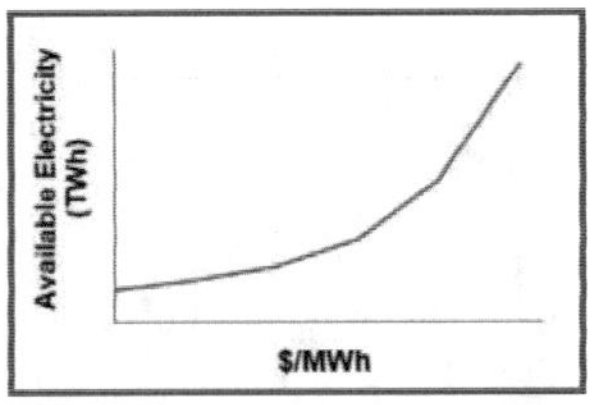

Key Assumptions

- ReEDS input parameters to achieve high renewable energy penetration:
  - Lower renewable energy technology costs
  - Higher natural gas prices

ReEDS

- Capacity expansion model
- Competes conventional, renewable, and storage technologies
- Projects capacity, generation, and transmission portfolio for 136 U.S. balancing areas out to 2050

Supply Curve

- Value of otherwise curtailed electricity vs. amount of curtailed electricity calculated by ReEDS
- Converted to a low-temperature electrolysis hydrogen supply curve using H2A

Economic Potential: Supply Curves

Developed supply curves using three resources

- Steam methane reforming (SMR)
- Nuclear generation
- Otherwise curtailed electricity with high penetrations of variable renewable generators on the grid and no transmission costs

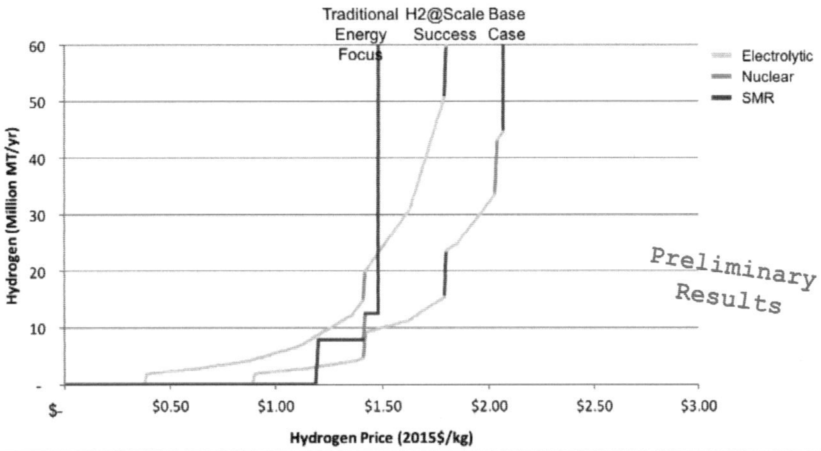

Economic Potential: Demand Curves

Developed demand curves under three paradigms

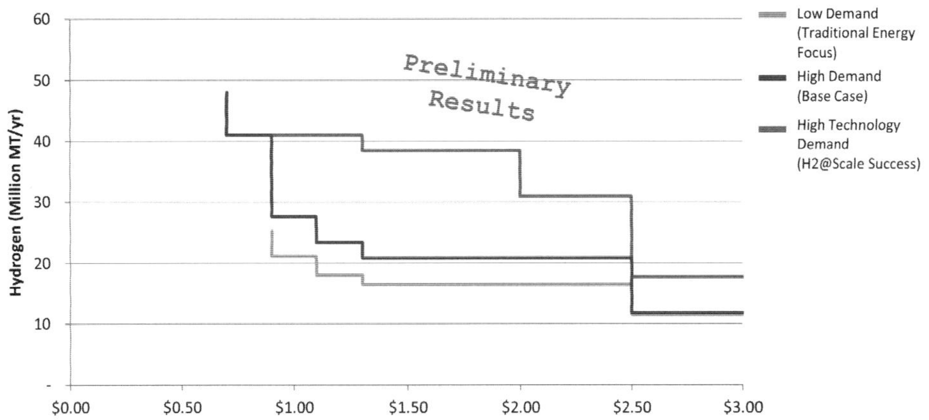

Demand	Base Case (High)	H2@Scale Success (High Tech.)	Traditional Energy Focus (Low)
Metals Reshoring	Economically competitive	Willingness to pay for H2 for metals	Economically competitive
LDV	Economically competitive	Full potential at $2.50/kg	Economically competitive
Synthetic Fuels	Economically competitive	Full potential at $2.00/kg	Economically competitive

Economic Potential: Scenario Summary

Preliminary Results

	Base Case	H2@Scale Success	Traditional Energy Focus
H_2 Use	21 MMT/yr	38 MMT/yr	17 MMT/yr
H_2 Price	$1.80/kg	$1.70/kg	$1.49/kg
Demand (MMT/yr)	• Refining (8), • Ammonia (3), • Synthetic fuel (1), • LDVs (9)	• Refining (8), • Ammonia (3), • Synthetic fuel (9), • Metals (6), • LDVs (13)	• Refining (8), • Ammonia (3), • Synthetic fuel (1), • LDVs (5)
Supply (MMT/yr)	• Low-temperature electrolysis (11), • Existing nuclear plants (5), • Existing NG reforming (5)	• Low-temperature electrolysis (33), • Existing nuclear plants (5)	• Existing nuclear plants (5) • NG reforming (8 MMT/yr from existing and 4 MMT/yr from new)
Electrolysis	10% curtailment, $19/MWh wholesale price	25% curtailment, $26/MWh wholesale price	No grid electrolysis

Economic Potential: Energy Use and Emissions Summary

H2@Scale can reduce emissions by up to 20% on top of baseline electricity sector emission reductions

Preliminary Results

Reduction Metric	Base Case	H2@Scale Success	Traditional Energy Focus
NO_X (Thousand MT)	130 (1%)	230 (2%)	61 (1%)
SO_X (Thousand MT)	33 (1%)	170 (5%)	13 (0%)
PM_{10} (Thousand MT)	10 (0%)	59 (2%)	4.0 (0%)
Crude Oil (Million Barrels)	470 (7%)	800 (12%)	280 (4%)
CO_2 (Million MT)	280 (9%)	590 (19%)	110 (4%)

Conclusions and Future Work

H2@Scale can transform our energy system by providing value for otherwise-curtailed electricity and a clean feedstock for numerous industries

- Technical potential:
  60MMT H_2/ yr can reduce emissions by 15%
- Economic potential:
  17-38 MMT H_2/ yr can be produced, given R&D advancements and access to low-cost intermittent power

Further analysis into regional issues and temporal issues is warranted to better quantify H2@Scale potential

www.nrel.gov

NREL
NATIONAL RENEWABLE ENERGY LABORATORY

NREL is a national laboratory of the U.S. Department of Energy, Office of Energy Efficiency and Renewable Energy, operated by the Alliance for Sustainable Energy, LLC.

Economic Potential Analysis Caveats

- Ideal markets
- Hydrogen storage and transportation requirements and costs are negligible
- Technology improvement and market assumptions in ReEDS result in a high VRE penetration and quantity of curtailed electricity
- Electrolyzer capacity factors are estimated using coarse temporal resolution of ReEDS (16 time slices representative of entire year)
- Low-temperature electrolyzer system costs achieve a target of $100/kW
- Nuclear-generated hydrogen costs do not include potential value to the grid in capacity and/or flexibility
- Economic rebound effects are negligible
- Additional hydrogen markets outside of our analysis may develop

504346 An Overview of H2@Scale: Energy System-Wide Benefits of Increased H2 Implementation

Abstract:

Supplying all of societyâ€™s growing energy demands is hard and has been limited to date to the fossil fuel based energy system in place since the start of the industrial age. Hydrogen, as a flexible, clean energy carrying intermediate, has the potential to be a centerpiece of a future energy system where aggressive market penetration of renewables (wind and solar) are coupled with renewable hydrogen production to meet societyâ€™s energy demands using clean, renewable resources and processes.

A number of key recent changes are bringing hydrogen back into the spotlight: 1) Renewable energy is getting cheaper, and penetration levels are increasing at an exponential pace; 2) the costs of GHG emissions (climate change) and criteria pollutants (health concerns) are being acknowledged and are reaching a point where society is demanding change (renewable portfolio and zero emission vehicle standards); and 3) commercial viability of fuel cell vehicle technology has been demonstrated (commercial vehicles being sold). Hydrogen, as an area of R&D interest (from either the federal or private sectors) has had its ups and downs over the past few decades. Hydrogen has suffered at times from being over-hyped or misunderstood. Claims of fuel cell commercialization, in particular, were in the past (early 2000â€™s) overly optimistic to be polite. The safety concerns of hydrogen have also often been poorly understood or overblown. These situations have arisen, in part, due to a lack of appreciation for the science and data of hydrogen and hydrogen systems.

This presentation will focus on the role of hydrogen at (grid-)scale and the efforts of a large, national lab team assembled to evaluate the potential of hydrogen to play a critical role in our energy future. Facts about hydrogen will be shared, as will the vision of how it will fit into our future energy system.

502524 Highly Efficient Visible Light Photocatalytic Production of H_2

Abstract:

Intensive effort has been made to develop photocatalytic processes to convert solar energy to clean fuels. So far, however, energy conversion efficiency for the visible-light photocatalytic splitting of water to hydrogen is less than 1%. In this talk, I will highlight recent findings for visible light photocatalysis in my group, including: (1) demonstrating that surface-diffuse-reflected-light can be 2 orders of magnitude more efficient than incident light for photocatalysis, (2) revealing that the inefficiency of absorbed visible light for the photocatalytic H2 production from water with a sacrificial agent is due to its kinetic limitation, and (3) inventing a novel temperature-induced visible light photocatalytic H2 production from water steam with a sacrificial agent to reach a high photohydrogen yield of 497 mmol/h/gcat with a large apparent quantum efficiency (QE) of 65.7% for entire visible light range.

498247 Dynamic Simulation of a Solar Powered Hybrid Sulfur Process for Hydrogen Production

Abstract:

Hydrogen can be produced using a thermochemical Hybrid Sulfur (HyS) Process. Using solar energy as the high temperature energy source allows for efficient environmentally-friendly production of hydrogen. The proposed route is an alternative to traditional photovoltaic plus electrolysis for hydrogen production. Making the process economically competitive is a major challenge. Operating the process with changes in the availability of solar energy also increases process complexity. The HyS flowsheet was modeled in Aspen Plus. The process includes electrolyzers, H2S04 decomposition, and O2 separation of the SO2/O2 product from the H2SO4 decomposition. Flowsheets for this process have demonstrated a thermal efficiency in excess of 50% higher heating value basis. The current work will discuss the process flowsheet, present results from a dynamic process simulation, and consider hydrogen production economics.

491322 Hydrogen Generation from Hydrogen Sulfide Decomposition

Abstract:

We propose a new method for the processing of hydrogen sulfide to create elemental sulfur and hydrogen gas. The process is based on a thermochemical cycle that leverages differences in temperature and reaction conditions to accomplish the thermodynamically unfavorable hydrogen sulfide decomposition over two reaction steps. First, hydrogen sulfide is reacted with iodine in aqueous conditions to produce dissolved hydroiodic acid and elemental sulfur. The sulfur is filtered out and the concentrated hydrogen iodide and water mixture is vaporized. The hydrogen iodide is then catalytically decomposed to hydrogen gas and iodine, which are flash separated. Overall, the two reaction and separation steps combine to accomplish the decomposition of hydrogen sulfide to hydrogen and elemental sulfur. This proposed process is similar to ideas put forth in the 1980s and 1990s by Kalina, Chakma, and Oosawa, but makes use of more recently discovered thermochemical hydrogen iodide decomposition methods and catalysts rather than electrochemical or photoelectrochemical methods.

This process is especially useful when the hydrogen sulfide was created by the reaction of hydrogen with sulfur-rich hydrocarbons during hydrodesulfurization. By recovering and recycling the hydrogen gas originally used to desulfurize the hydrocarbon, the amount of hydrogen gas needed to treat a sulfur-rich hydrocarbon is dramatically reduced. As larger quantities of sulfur-rich resources are extracted, methods to efficiently desulfurize hydrocarbons are increasingly important. Further, this process is inherently carbon neutral which is an advantage over steam reforming, the current main source of hydrogen gas.

A process model was created to explore the viability of the process. The model suggests that hydrogen could be profitably generated in processes with a maximum temperature of just 300 degrees Celsius. This implies that the heat required for the decomposition of hydrogen sulfide could be supplied by solar concentrators or heat integration with other process units. Process units in the model were defined by literature and experimental results.

Adsorptive on-Board Desulfurization of Liquid Fuels: High Efficiency in Desulfurization and Full Thermal Regeneration Via Hot Exhaust Gas

Raphael Neubauer, Institute of Thermal Engineering, Graz University of Technology, Austria
Norbert Kienzl, Bioenergy 2020+, Graz, Austria
Christoph Hochenauer, Institute of Thermal Engineering, Graz University of Technology, Austria

Introduction

Desulfurization of fuels is mandatory to fulfill environmental regulations and to protect the environment. Sulfur levels of liquid fuels are in the range of 10 to 5000 ppm and depend on the type of fuel and their correlated legal requirements (1). Hydrodesulfurization (HDS) is used for crude oil desulfurization to achieve the country-specific legal requirements. This HDS is the most prevalent and industrially relevant desulfurization process which operates at temperatures above 300 °C and more than 35 bar of H_2 operating pressure. The HDS efficiency decreases with increasing molecular size of the refractory sulfur compounds and thus is inefficient for removing benzothiophene (BT) and especially dibenzothiophene (DBT) and its derivatives.

Fuel cells are one of the most effective tools to convert chemical to electrical energy. This technology is not only suitable for stationary approaches but is also a promising source of on-board electricity supply for all kinds of vehicles, ships, and aircraft. The advantages of fuel cells are in particular higher efficiency, reduced emissions, and lower noise generation in comparison to conventional combustion engines. This technology becomes very attractive when on-board fuel is used to produce hydrogen-rich syngas via reforming, which is subsequently fed to the fuel cell. However, the sulfur threshold limit for the reformer and the fuel cell are < 50 and < 1 ppm of total sulfur, respectively (2, 3). Consequently, fuel cells as one of the most efficient tools for energy conversion are excluded from using the most common types of fuels.

Adsorptive desulfurization is a promising approach to provide the possibility of on-board desulfurization and thus operating fuel cells with commercial fuels. Several adsorbents have been investigated and published in recent years. However, adsorbents regeneration is still a major issue; especially after adsorption of dibenzothiophene (DBT). The aim of this work is to identify and prepare a highly thermal stable adsorbent for thermal on-board regeneration via hot exhaust gases.

Experimental Section

A silver based adsorbent was prepared by incipient wetness impregnation of γ–Al2O3. The impregnated support was further dried and calcined in order to get Ag-Al2O3. This adsorbent was characterized via different methods including N2-physisorption, XRD, SEM, EDX, and ICP-AES. The adsorbent was further used in different experiments including equilibrium saturation, kinetics, and breakthrough experiments. The adsorptive desulfurization performance of Ag-Al2O3 was tested under different conditions and fuels such as jet fuel and diesel fuel. These fuels contained different concentrations of poly aromatic sulfur heterocycles (PASHs) in the range of 300 to 900 ppmw of total sulfur. In situ regeneration experiments were carried out with preheated air and simulated exhaust gas from a solid oxide fuel cell driven auxiliary power unit

Experimental Section

The overall adsorption mechanisms of PASHs were studied by relating experimental data with theoretical principles. High breakthrough adsorption capacities were observed for different PASHs like BT, DBT, and 4,6-dimethyldibenzothiophen dissolved in commercial fuels. The breakthrough adsorption capacity was in the range of 0.9 to 2.3 mg/g at 20 °C and atmospheric pressure depending on the type of PAHS. Within these investigations, the role of the acid base interaction (S-H) on the overall adsorption mechanism was identified. The results showed, that the weak S-H interaction is a third mechanism beside the known π-complexation (π-Ag) and the direct metal sulfur interaction (S-Ag). Both π-Ag and S-Ag interactions have a selectivity order of DBT > BT which is caused by a higher adsorption energy of DBT in comparison to BT. The S-H interaction is thus responsible for the shift of the final selectivity order to BT > DBT at equilibrium saturation conditions. This result is confirmed by density functional theory (DFT) calculations reported in the literature as well as by the hard and soft acid base (HSAB) theory proposed by Pearson. Consequently, π-Ag and S-Ag interactions are the main adsorption mechanisms at breakthrough. In these two interactions, the silver cation (soft acid) interacts with the thiophenic ring or directly with the π lone-pair of electrons of the sulfur atom (soft base) (4).

Additional equilibrium saturation experiments were carried out at -10 and 60 °C. The adsorption temperature showed no influence on the equilibrium saturation capacity in the case of DBT. For BT, the equilibrium saturation capacity at -10 °C stayed at the same level as at 20 °C. At 60 °C, the equilibrium saturation of BT decreased to the level of DBT and thus confirmed that the weak S-H interaction is only relevant for BT at equilibrium stage. At increased temperatures (60 °C) the S-H interactions is to week to contribute in the overall adsorption mechanism of BT. The stronger adsorption of DBT in particular makes it difficult to achieve full thermal regeneration without thermal destruction of the adsorbent. These are important findings in order to develop highly efficient regeneration strategies.

The lack of solvents requires thermal regeneration of the adsorbent in the case of on-board desulfurization units. This is the reason, why we studied the desorption mechanisms of PASHs from Ag-Al_2O_3 in the presence of air; and in particular of N_2, O_2, CO_2, and H_2O. In these investigations, thermal regeneration was carried out in a preheated gas mixture with increasing temperature. A final temperature of 450 °C was used in the first regeneration experiments. This temperature was sufficient for full thermal regeneration in flowing air after adsorption of BT on Ag-Al_2O_3. The same experiment was carried out with DBT and showed only 60% recovery of the breakthrough capacity after thermal regeneration. This result confirms the stronger adsorption of DBT in comparison to BT. Additional regeneration experiments were carried out at a final temperature of 525 °C. The results showed full thermal regeneration in flowing air after adsorption of DBT. This is an excellent result as full thermal regeneration after DBT adsorption has not been reported so far. The regenerated Ag-Al_2O_3 was further analyzed and showed no thermal induced destruction.

One problem of thermal regeneration is the source of hot regeneration gas as electrical heating would be inefficient and cost-intensive. To overcome this drawback, hot exhaust gases were identified as a source of hot regeneration gases. This is the reason why regeneration experiments were performed by using hot off-gas from a solid oxide fuel cell based auxiliary power unit; this gas is referred to as APU. The flow sheet of this novel regeneration strategy is illustrated in Fig. 1 (right).

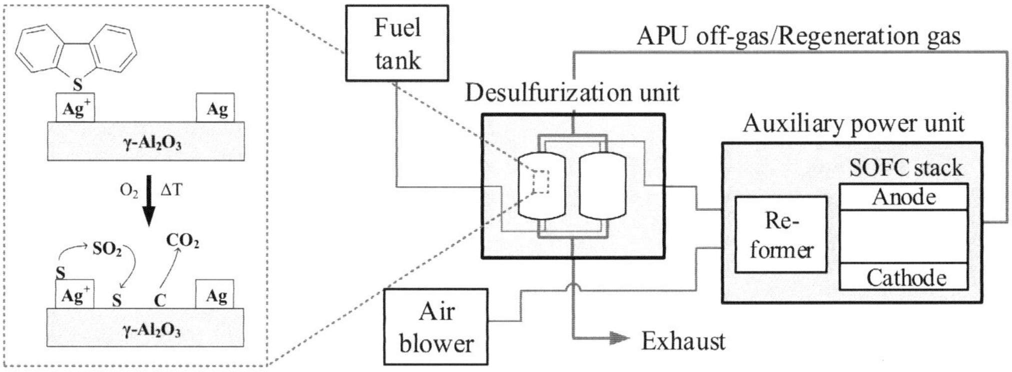

Fig. 1 Flow sheet diagram of novel regeneration strategy using hot APU off-gas for thermal regeneration of the adsorbent (right) and simplified scheme of the overall desorption mechanism of PASHs (left).

Excellent results were obtained with APU as regeneration gas. In these experiments, 100% regeneration was achieved over 5 cycles at a final temperature of 500 °C. The identified desorption mechanisms (Fig. 1 left) showed that CO_2 in the regeneration gas has no negative influence on the overall regeneration experiment as it desorbs already at temperature above 100 °C. In addition, H_2O was identified to have a positive influence on the thermal decomposition of adsorbed PASHs. This positive effect was proven by achieving full thermal regeneration at a final temperature of 450 °C with increased H_2O content in the regeneration gas (12 vol.%). These findings even highlight the possibility to use hot exhaust gas from a combustion engine as regeneration gas in a hybrid vehicle based on low temperature fuel cells.

Conclusion

Adsorptive desulfurization of liquid fuels is mandatory to provide the possibility to operate fuel cell with commercial fuels. $Ag-Al_2O_3$ was identified as excellent adsorbent for on-board desulfurization units. Investigations on the overall adsorption mechanism showed that the silver cations are the main active adsorption sites. Adsorption of DBT via π-Ag and S-Ag interactions is stronger in comparisons to BT. Consequently, regeneration of these silver sites after adsorption of DBT is essential to achieve full regeneration.

Hot exhaust gas was identified as ideal regeneration gas for thermal in situ and on-board regeneration. The identified desorption mechanism showed that CO_2 has no negative influence on the regeneration performance. Full thermal regeneration was reported for the first time after adsorption of DBT at a final temperature of 500 °C over 5 cycles. The final regeneration temperature could even be reduced to 450 °C by increasing the H_2O content in the regeneration gas. All these findings showed excellent adsorption and regeneration performances of $Ag-Al_2O_3$ under a wide range of conditions including different fuels. Full on-board regeneration via hot exhaust gas, as reported, highlights the possibility to operate fuel cell with commercial sulfur containing fuels and thus achieve high efficiency in energy conversion combined with lowest emissions.

References

1. A. Samokhvalov, B.J. Tatarchuk, Catal. Rev. 52 (2010) 381-410. http://dx.doi.org/10.1080/01614940.2010.498749.

2. X. Xu, S. Zhang, P. Li, Y. Shen, Fuel Process. Technol. 124 (2014) 140–146. http://dx.doi.org/10.1016/j.fuproc.2014.03.001.

3. C. Hulteberg, Int. J. Hydrogen Energy 37 (2012) 3978-3992. http://dx.doi.org/10.1016/j.ijhydene.2011.12.001.

4. R. Neubauer, M. Husmann, C. Weinlaender, N. Kienzl, E. Leitner, C. Hochenauer, Chem. Eng. J. 309 (2017) 840-849. http://dx.doi.org/10.1016/j.cej.2016.10.094.

504791 Neuronal Biosensors

Abstract:

Biomimetics has stimulated innovation in developing medical devices, no less in the field of biosensing. The most exquisite sensors have already been created with, for example, crustacean antennae useful in detecting femtomolar (10^{-15} M) concentrations or less through neuronal chemoreceptors. Over the past 30 years our team has sought to understand and mimic neuron-like sensing. We began studying living neuron action potentials as a function of species concentrations. We turned to synthetic analogues isolating ion channels in bilayers and modifying structures to make responses sensitive to specific proteins. To further reduce these systems to practice we pivoted to inexpensive ionophore preparations in micro- and nanoscale ion selective electrodes. Finally, to make these systems practical for detecting cells and proteins we are developing a dual ionophore biosensing concept which ultimately mimics the neuronal system where we began. This plenary session will take the attendee from the fundamental physiology, through Nicolsky-Eisenman and Goldman-Hodgkin-Katz modeling, to synthetic practicality in the neuronal biosensor area, and finally to simple learning tools for teaching neuronal biomimicry.

505102 Mix (and Unmix) It Up with Biomembranes

Abstract:

Since the invention of the optical microscope in the 17th century it has been observed that biological membranes compartmentalize the cellular machinery of life. Then, nearly 100 years ago, it was concluded, through surface science, that the cell membrane is a lipid bilayer. Since that time, many observations and theories have been put forward to try to explain how complex behavior emerges in living systems from the ubiquitous lipid bilayer structure and its integrated proteins and carbohydrates. We are now in a period of time when some of the most important paradigms posited in the last several decades for emergence of complex behavior in living cell membranes are being tested and questioned. These include the membrane (or lipid) raft hypothesis and mechanisms for generation of membrane curvature and compositional asymmetry. An important outcome of this work may be the design of new micro- and nano-scale self-assembled systems and composites. This talk will focus upon our work in lipid demixing, crowding-induced mixing, sorting by nanoscale curvature, and bionanocomposites that may contribute toward the design of new biological membrane-inspired medicine and technology.

504790 Bridging the Gap Between Viral and and Nonviral Gene Vectors

Abstract:

Our understanding of disease, both genetic and acquired, continues to progress rapidly and provides insight into how most ailments can be treated at the genetic level. In spite of our improved understanding of genetic treatments, the field of gene therapy has failed to keep pace and advance human health. The lack of progress can be attributed primarily to the obstacles associated with delivery of genetic material. We been working to develop novel gene therapy vectors that bridge the gap between viral and nonviral vectors. We are working to combine the advantages associated with each respective vector while reducing or eliminating common drawbacks to gene delivery. We have primarily done so by replacing viral proteins with synthetic materials, such as polymers, lipids, or peptides. By working to understand how these materials interface with viruses and how they change the delivery pathway, we are aiming to gain knowledge necessary for designing novel materials that will help us progress step-wise from viral vectors to nonviral vectors.

497735 Acid Catalyzed Production of 1,3-Butadiene from Biomass Derived Tetrahydrofuran

Abstract:

Butadiene is an industrially relevant precursor in the production of polymers, where it is typically polymerized with itself or other monomers to produce synthetic rubber. Polymerizing butadiene with styrene and acrylonitrile produces the plastic acrylonitrile butadiene styrene (ABS), while polymers of butadiene and styrene are also used in the production of automobile tires.

Here we present a renewable pathway to produce butadiene from biomass based tetrahydrofuran (THF), proceeding via a dehydrative ring opening of THF (Dehydra-Decyclization). A screening of various solid acid catalysts reveals that an all silica phosphorous containing zeolite (Phosphrous-self pillared pentasil, P-SPP) provides the highgest selectivity to butadiene of ~ 95%.[1] The most significant undesired reaction pathway is the conversion of THF to propene and formaldehyde, which proceeds via a retro-prins condensation reaction. Increased methyl substitution at the 2 and 5 carbon positions of THF leads to significant increase in the rate, without change in selectivity to dienes, while substitution at the 3 carbon position leads to no change in rate. This suggests that a carbenium ion formed at the 2 and 5 carbon positions, is a kinetically relevant surface intermediate. A combination of in-situ titrations of pyridine and a sterically hindered base, 2,6-ditertbutyl pyridine, suggest that the surface chemistry is pre-dominantly Brønsted acid catalyzed.

Figure 1. Dehydra-decyclization in a catalytic packed-bed flow reactor. Reaction of tetrahydrofuran to butadiene at 250-350 °C, WHSV = 1.0 h^{-1} and 5 Torr with catalysts: P-SPP (red), Sn-BEA (blue), amorphous silica-alumina (SiAl, yellow), and ZSM-5 (Si:Al = 140, green). Conversion for each experiment noted numerically within each bar.

[1] Abdelrahman, O. A.; Park, D. S.; Vinter, K. P.; Spanjers, C. S.; Ren, L.; Cho, H. J.; Vlachos, D. G.; Fan, W.; Tsapatsis, M.; Dauenhauer, P. J. *ACS Sustainable Chem. Eng.* **2017**, In Press.

502430 External Surface and Pore Mouth Catalysis in Hydrolysis of Inulin over Zeolites with Different Micropore Topologies and Mesoporosities

Abstract:

Hydrolysis of inulin over zeolite catalysts with various micropore topologies (FER, MFI, MOR, BEA, MWW and FAU) and mesoporosities (pillared MFI (PMFI) and pillared MWW (PMWW)) was studied. The reaction includes cleavage of four types of glycosidic bonds: the terminal glucosyl to fructosyl bond to produce glucose, the terminal sucrosyl to fructosyl bond to form sucrose, and the terminal fructosyl to fructosyl bond and internal fructosyl bonds within the polymer chain to generate fructose. Inulin conversion has shown an initially slow rate followed by pseudo first-order kinetics. Fructose production occurred at a much faster rate than that of sucrose followed by glucose. Rigorous kinetic data analysis showed that the reaction was inclined to proceed on the external surface acid sites of zeolites with cleavage of terminal sucrosyl to fructosyl and terminal fructosyl to fructosyl bonds. The increase in the micropore size in zeolites promoted pore mouth catalysis for the cleavage of the terminal fructosyl to fructosyl bond and terminal glucosyl to fructosyl bond. The mesoporosity in PMFI and PMWW zeolites enhanced external surface and pore mouth catalysis compared to those of their microporous analogues, but did not enable new types of catalytic events. The measured kinetic data were interpreted using a mathematical model based on a network involving parallel and series reactions. Inulin hydrolysis was probed in the transition from external surface to pore mouth catalysis depending on the zeolite topology and mesoporosity in bulky biomass processing. The present study provides guidelines for the utilization of zeolites with variable topologies and porosities for processing inulin and other biomass feedstocks for food and energy applications.

501211 On the Role of Na Cations and Solvent in Glucose Isomerization and Epimerization in Sn-BEA

Abstract:

The efficient conversion of glucose to platform chemicals such as 5-hydroxymethylfurfural (HMF) is crucial in establishing new green chemistry routes to useful products ranging from transportation fuels to chemical intermediates. Since the direct conversion of glucose to HMF is challenging, isomerization of glucose to fructose provides an alternative path that leverages the more favorable dehydration of fructose to HMF. Davis and coworkers[1, 2] have recently demonstrated that the hydrolyzed metal site of the Lewis-acidic zeolite Sn-BEA is particularly selective to glucose-fructose isomerization. However, the selectivity tilts in favor of the epimerization reaction to mannose in Na-exchanged Sn-BEA (Na-Sn-BEA). The reasons behind the selectivity switch have been the source of speculation and debate. Using DFT calculations, we have been able to uncover for the first time the reasons underlying the change in selectivity. Specifically, we shall show that selectivity in Na-Sn-BEA is effectively determined by H_2O molecules vicinal to the active site.

The glucose-fructose isomerization proceeds via a 1,2-hydride shift (HS) from the C2 to the C1 carbon atom, while there are two pathways for mannose formation: a direct pathway via an intramolecular 1,2-carbon shift (CS) from C2 to C1 and an indirect pathway via a second 1,2-HS from C1 to C2 through fructose. We have compared all three reaction pathways both in the absence and presence of water molecules in the vicinity of the active site. In Sn-BEA, the isomerization to fructose is the kinetically dominant pathway, while mannose is formed via the indirect epimerization pathway. In Sn-BEA, water does not influence the selectivity, but it does open the energy gap between the isomerization and epimerization pathways, making the latter even less favorable on account of the extensive solvent re-organization that it entails. In Na-Sn-BEA the lack of water molecules at the active leads to an incorrect prediction that favors isomerization. However, calculations with water at the active site clearly demonstrate that epimerization via the direct 1,2-CS pathway is favored over isomerization, in complete agreement with experiment. Upon coordination to Na^+, the polar H_2O molecules screen the strong electrostatic interaction between the C3 carbon of GLU and Na^+ and thus facilitate the 1,2-CS, which is now the preferred pathway. Our calculations resolve for the first time the role of water and Na cations in the chemistry while rationalizing the experimental data.

References

[1] Bermejo-Deval, R.; Orazov, M.; Gounder, R.; Hwang, S. J.; Davis, M. E., *ACS Catalysis* **2014,** *4* (7), 2288-2297.

[2] Hwang, S. J.; Gounder, R.; Bhawe, Y.; Orazov, M.; Bermejo-Deval, R.; Davis, M. E., *Topics in Catalysis* **2015,** *58* (7-9), 435-440.

Strategies for Improved Design of Nb-Silicates for Propylene Epoxidation

Swarup K. Maiti[1]
Anand Ramanathan[1]
Bala Subramaniam[1, 2]

[1] Center for Environmental Beneficial Catalysis, Univ. of Kansas (Lawrence, KS)
[2] Chemical & Petroleum Engineering, Univ. of Kansas (Lawrence, KS)

Center for Environmentally Beneficial Catalysis CEBC

THE UNIVERSITY OF KANSAS

AIChE Annual Meeting on "Catalysis with Microporous and Mesoporous Materials", Minneapolis, MN, October 29 - November 3, 2017

Why Propylene Oxide (PO)?

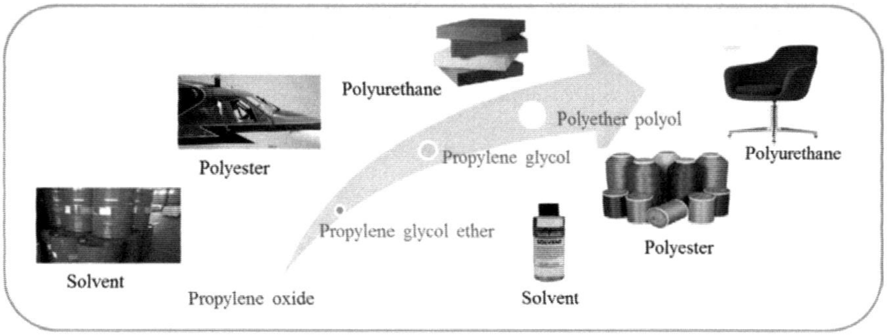

- PO is an important chemical intermediate
- Global PO demand is ~10 million metric tons/y (2013)
- Market is growing @ 3-4 %/y

www.ihs.com., published in March **2016**

Industrial Propylene Epoxidation Processes

Process	Reagents	Conditions	PO Yield (Y) /Selectivity (S)	Disadvantages
Chlorohydrin process	Cl_2, caustic, lime	45-90 °C; 1.1-1.9 bar	S = 88-95 %	• Adverse environmental impact
Hydroperoxide process	Ethylbenzene O_2, W, V, Mo	100-130 °C; 1.1-1.9 bar	Y = 95% S = 98%; 2h	• Dependent on stable byproduct value
DOW/BASF HPPO process	H_2O_2, TS-1	40-60 °C; ~20-25 bar	S = 95%	• PO selectivity ≤95% • TS-1 is expensive • H_2O_2 not fully utilized

Lee et al., *Chem. Eng. Sci.*, **2007**, 62, 7282
Nijhuis et al., *Ind. Eng. Chem. Res.*, **2006**, 45, 3447
Lu et al., *J. Catal.*, **2016**, 342, 173

Nb-based Silicates as Epoxidation Catalysts

Catalyst	Substrate	T	t	Conversion	$S_{Epoxide}$
		°C	h	%	%
Nb-MMM-E	Cyclohexene	50	0.5	20	40
#Nb/SiO_2	Cyclohexene	reflux	1	53	94
Nb-β	Cyclohexene	40	0.5	0.12	96
#Nb/SiO_2	Limonene	reflux	1	52	78
Nb-EISA	Cyclohexene	50	6	45.7	77.1
Nb-SBA-16	Cyclohexene	50	2	47.1	61.6
Nb-KIT-5			2	13.2	48.5
Nb-KIT-6			2	26.2	56.4

Solvent in all cases is acetonitrile, # Catalysts prepared by grafting/impregnation

1. Ivanchikova et al., J. Catal., **2015**, 332, 138–148.
2. Tiozzo et al., Catal. Today, **2014**, 235, 49-56.
3. Bregante et al., J. Catal., **2017**, 348, 75-89.
4. Tiozzo et al., Inorg. Chem. Acta, **2015**, 443, 190-196.
5. Ramanathan et al., Ind. Ing. Chem. Res., **2015**, 54, 4236-4242.
6. Ramanathan et al., *Microporous Mesoporous Mater.* **2017**, in press.

Goal

- Investigate Nb-based mesoporous silicates for liquid-phase propylene epoxidation under similar operating conditions to those of DOW/BASF HPPO process.
  - Develop Nb-silicate catalyst to maximize PO selectivity and H_2O_2 utilization efficiency.

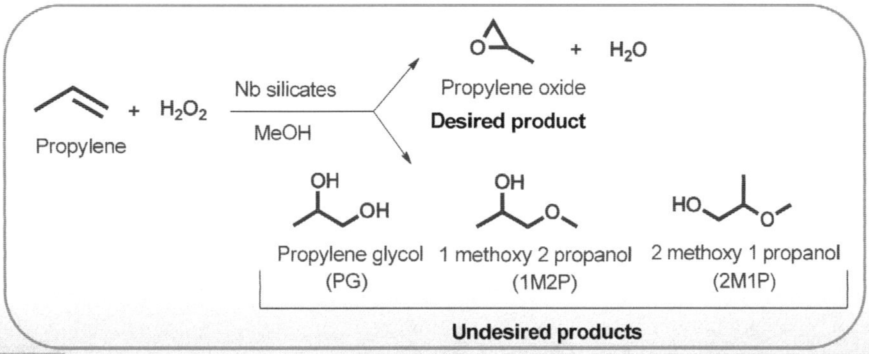

Catalyst Performance Evaluation

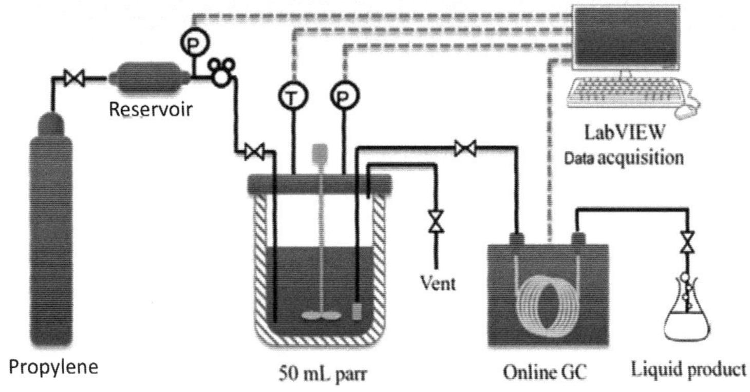

H_2O_2 utilization efficiency

$$U_{H_2O_2} = \frac{Epoxidation\ product\ (moles)}{H_2O_2\ consumed\ (moles)} \times 100\%$$

PO productivity

$$P_{PO} = \frac{m_{PO}}{(time)(m_{metal})}$$

Compressed Propylene "Expands" Reaction Mixture

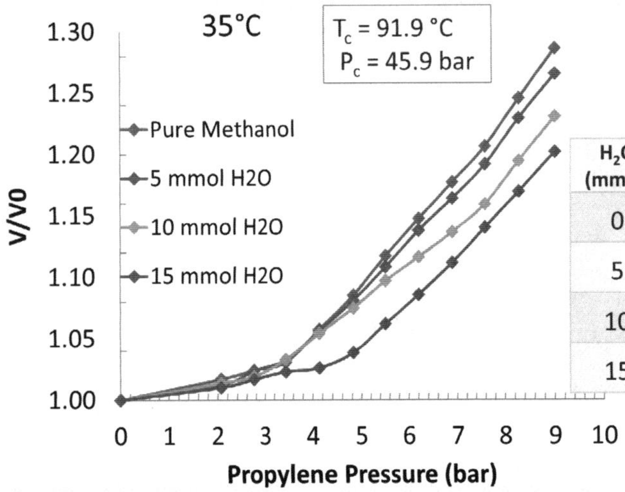

At 9 bar

H_2O (mmol)	Volume expansion
0	29%
5	27%
10	23%
15	20%

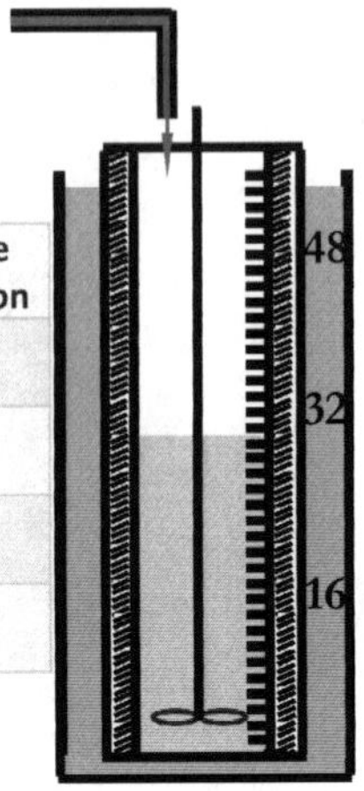

View Cell

Nb-TUD-1 Shows Promise

Entry	Nb wt. %	P_{PO} (mg/h/g)	S_{PO} (%)	U_{H2O2} (%)	X_{H2O2} (%)	Nb leaching %
1	0.8	11680	>99	80.1	17	24.8
2	3.9	4075	>99	89.4	24	20.9

Reaction conditions

- MeOH = 20 g (625 mmol), Acetonitrile = 0.1 g (2.4 mmol),
- H_2O_2 = 10 mmol, catalyst = 0.3 g,
- Propylene pressure = 9.5 bar (maintained constant), T = 35°C, t = 3h,
- Stirrer speed =1400 rpm, catalyst size ~100 μm

Maiti *et al., Ind. Eng. Chem. Res.,* **2017**, 56, 1999-2007
Yan *et al., J. Catal.,* **2016**, 336, 75-84

H_2O_2 Interaction with Simplest Niobyl Silicate

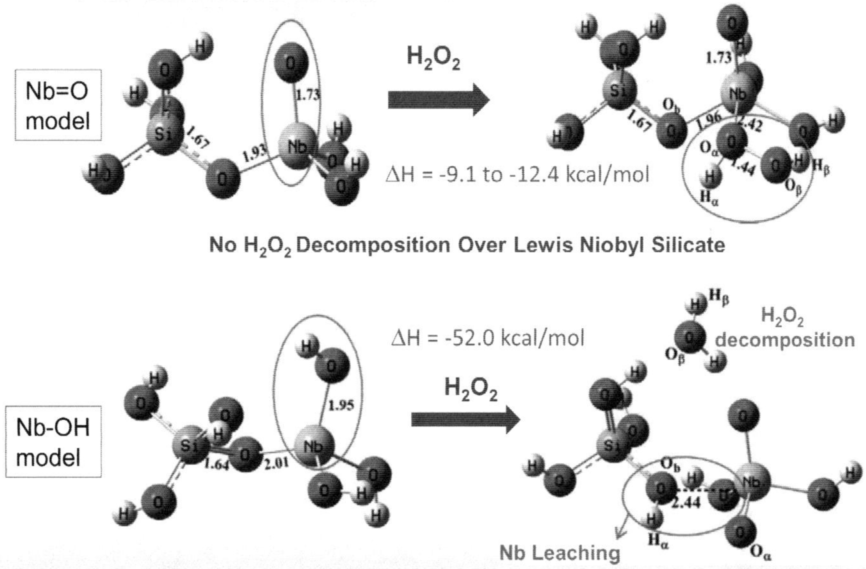

H_2O_2 Decomposition and Nb Leaching Are Tied to Brønsted acidity

Yan *et al., J. Catal.,* **2016**, 336, 75-84

Desired Catalyst Attributes

- Predominantly Lewis acid sites
- Lower Brønsted acid sites
- Better dispersion of metal species

Strategy

- Prepare and evaluate Nb-Silicates prepared by Evaporation Induced Self Assembly (EISA)

Yan *et al., J. Catal.,* **2016**, 336, 75-84.
Maiti *et al., Ind. Eng. Chem. Res.,* **2017**, 56, 1999-2007.
Wu *et al., J. Catal.,* **2017**, 350, 182-188.

W-EISA Catalysts Enhance Metathesis Activity

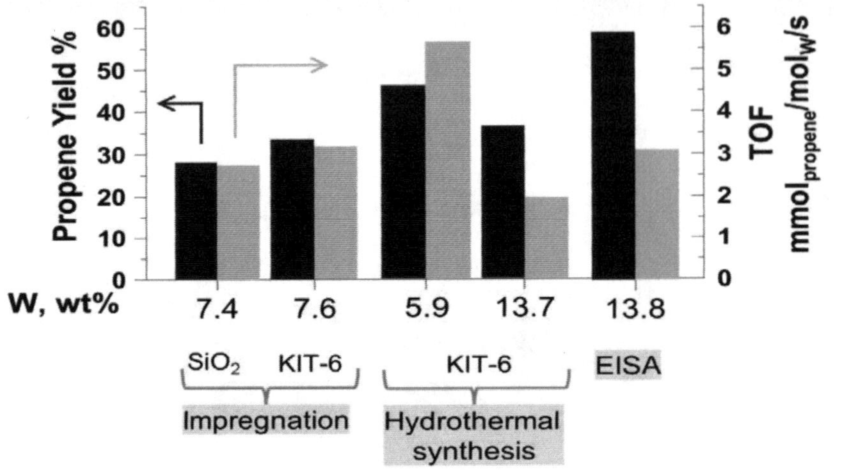

Wu et al., *J. Catal.*, **2017**, 350, 182-188
Yan *et al.*, *J. Catal.*, **2016**, 336, 75-84

The University of Kansas CEBC

11

Preparation of Nb-EISA: General Synthesis Procedure

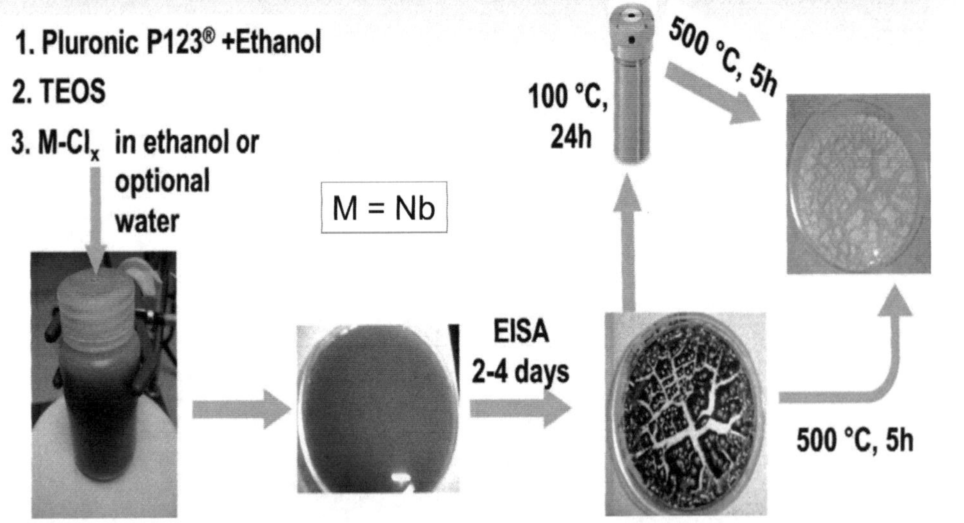

Wu et al., *J. Catal.*, **2017**, 350, 182-188.
Ramanathan et al., *Microporous Mesoporous Mater.*, **2017**, 245, 118-125.
Zhu et al., *Microporous Mesoporous Mater.*, **2016**, 223, 46–52.

The University of Kansas CEBC

12

Nb-EISA Materials are Mesoporous with No Detectable Bulk Oxides

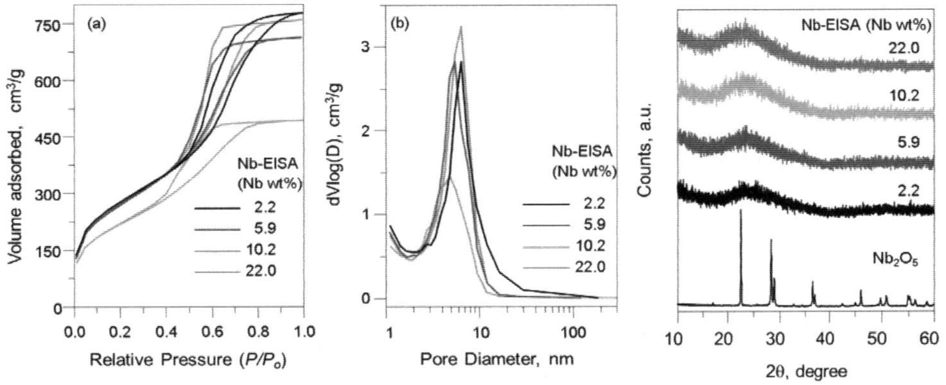

- Type IV isotherm with average pore dia 3.9 - 4.9 nm
- No bulk oxide even at very high loading of 22 wt%

The University of Kansas CEBC

KU THE UNIVERSITY OF KANSAS

DR-UV-Vis Reveals Isolated Tetrahedral (T) and Polymeric Octahedral (O) Sites

Diffuse Reflectance UV-Vis

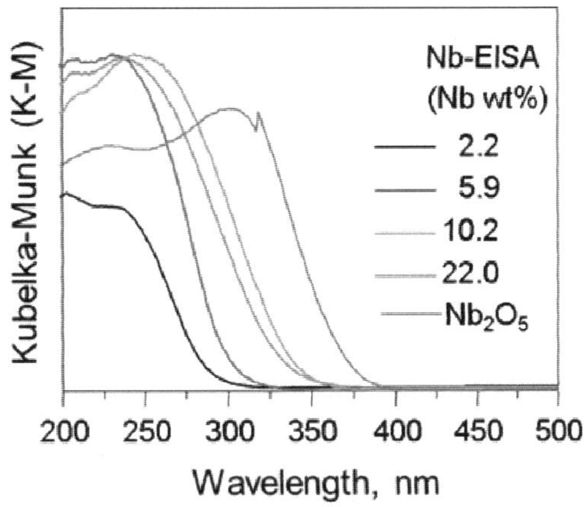

EISA	Isolated	Oligomeric
	Tetrahedral (T, nm)	Octahedral (O, nm)
Nb	220	270

The University of Kansas CEBC

KU THE UNIVERSITY OF KANSAS

STEM Images Reveal Uniformly Dispersed Nb-O_x Species

Scale bar 50 nm

Scale bar 20 nm

Scale bar 20mm

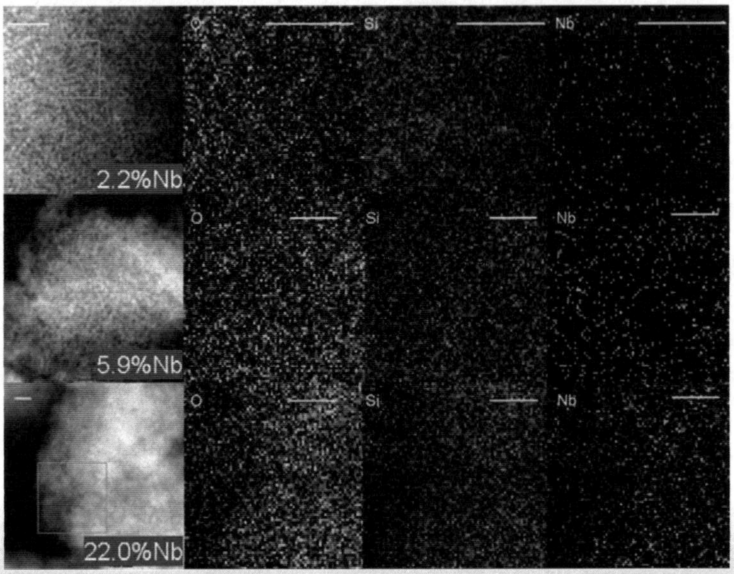

Relatively Low Brønsted Acidity of Nb-EISA Materials

Pyridine FT-IR spectra

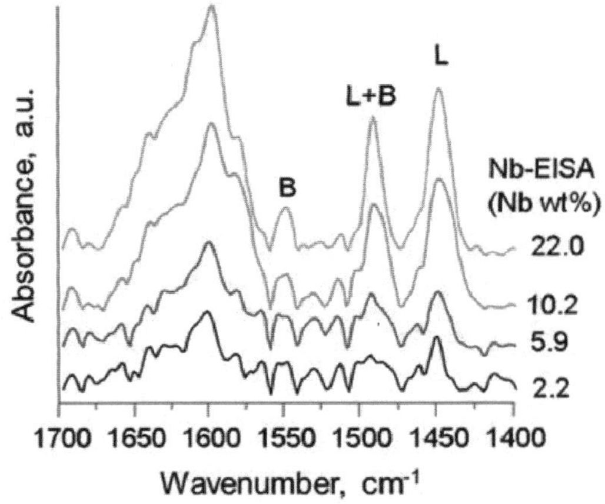

- An increase in Lewis acid sites (1450 cm^{-1}) at higher Nb loading.
- Relatively stable and low Brønsted acid sites (1550 cm^{-1}) at higher Nb loading.

Nb-EISA Performs Better than Nb-TUD-1

Entry	Catalyst	Nb wt.%	S_{PO} (%)	U_{H2O2} (%)	P_{PO} (mg/h/g)	% Nb leaching (mg)
1	Nb-TUD-1	3.9	>99	89	4075	20.9 (2.50)
2	Nb-EISA	1.9	>99	>99	3174	6.3 (0.36)
3	Nb-EISA	4.7	>99	95	2071	5.4 (0.50)
4	Nb-EISA	9.3	>99	>99	1235	3.1 (0.86)

Parametric Study

[Reaction conditions: T = 35°C, P = 9.5 bar, 10 mmol H_2O_2, 0.3 g Nb-EISA (1.9 wt. % Nb)]

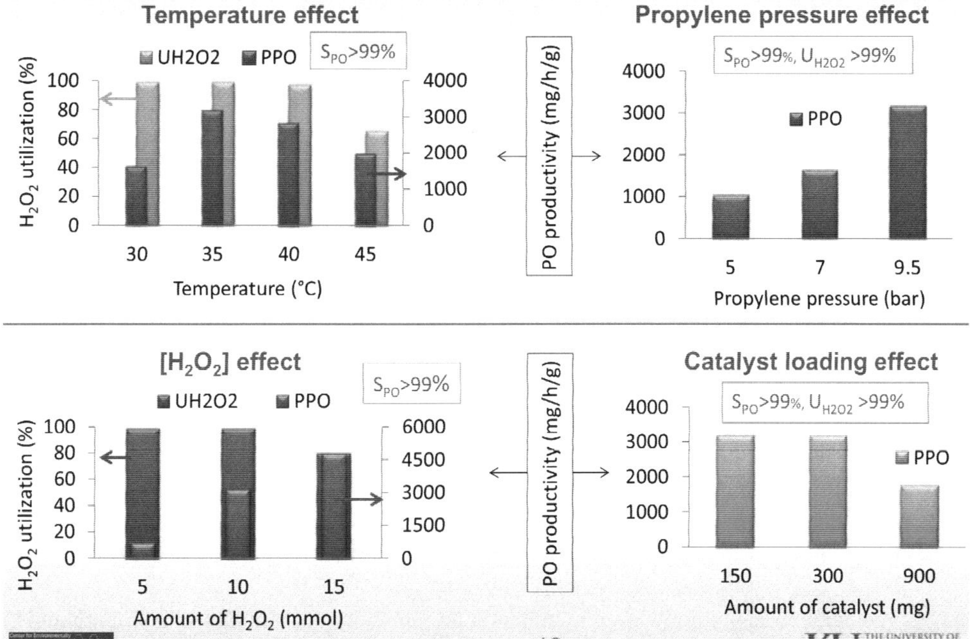

Nb-EISA (1.9 wt.% Nb) Shows Relatively Stable Performance on Recycling

Entry	Y_{PO} (mmol)	P_{PO} (mg/h/g)	S_{PO} (%)	U_{H2O2} (%)	Nb leaching %
Fresh	1.88	3274	>99	>99	6.3
Recycle-1	1.66	4682	>99	98	6.7
Recycle-2	1.55	4960	>99	>99	5.7
Recycle-3	1.46	4525	>99	>99	5.2
Recycle-4	1.38	4279	>99	>99	5.8

CEBC 19 KU THE UNIVERSITY OF KANSAS

Conclusion

- Nb-EISA shows promise as propylene epoxidation catalyst at commercial process conditions
  - >99% epoxide selectivity/H_2O_2 utilization, <5% Nb leaching
- Computational chemistry calculations implicate Brønsted acidity for H_2O_2 decomposition and metal leaching
  - Provide valuable guidance for catalyst design/selection

Path Forward

- Prepare and evaluate NbS-1 catalysts (similar to TS-1 support) to enhance catalyst activity.
- Perform intrinsic kinetics studies with Nb-EISA catalysts for rational reactor design and scale-up.

CEBC 20 KU THE UNIVERSITY OF KANSAS

Acknowledgments

Funding

NSF

EPA
United States
Environmental Protection
Agency

Current CEBC industry partners

ADM

Chevron Phillips Chemical Company LP

GRACE
Talent | Technology | Trust™

INVISTA™

Chemours™

DUPONT®

Reliance
Industries Limited

UOP
A Honeywell Company

Thank you!

21

The University of Kansas CEBC

KU THE UNIVERSITY OF KANSAS

497710 Catalytic Activity of Unique Base Sites Formed Via Nitridation and Subsequent Methylation of Mesoporous Silica SBA-15

Abstract:

Ordered micro and meso porous silica-based materials (e.g., mesoporous silica and zeolite) have attracted wide attention because of their unique structures and attended high surface area, besides their low toxicity, and high chemical and thermal stabilities. Among their applications, their use as heterogeneous base catalysts is one of a challenge. Nitridation of silica materials (i.e., substitution of N-atom with framework O-atom) can create Lewis base sites. Although they have been known to catalyze basic reaction such as Knoevenagel condensation, their basicity is relatively lower than conventional homogeneous catalysts and applicable reactants have been limited.

Here, we demonstrate the preparation of methylated nitrogen substituted mesoporous silica SBA-15 (Me-NSBA-15), and confirmed that the basicity and also nucleophilicity were enhanced to catalyze several reactions.[1),2)] Methylation of N-substituted mesoporous silica SBA-15 was achieved by reacting methyl iodide under reflux condition, and the functionalization was confirmed by elemental analysis and FTIR spectroscopy. Powder XRD showed the ordered porous structure was maintained. The catalytic test was performed with a batch reactor and formed products were analyzed with GC-FID. The Me-NSBA-15 catalyzed the Knoevenagel condensation of benzaldehyde and diethyl malonate (2 mmol C_7H_6O, 2 mmol $C_7H_{12}O_4$, 2.13 mL ethanol, 20 mg catalyst, 353 K for 24 h), which has never been catalyzed with simple N-substitution SBA-15. This suggests that the methylation enhanced the basicity of Lewis base sites, and allowed to deprotonate a higher pK_a reactant (i.e., pK_a of diethyl malonate is 16.4, whereas the pK_a of ethyl cyanoacetate is 13.1). Methylation also enhanced the nucleophilicity. In Morita-Baylis-Hillman reaction of formaldehyde and methyl acrylate (4mmol CH_2O, 2 mmol $C_4H_6O_2$, 2.0 mL 1,4-dioxane, 20 mg catalyst, 423 K for 48 h), the yield of the product (64% at 24 h) was comparable with that of conventional homogeneous catalyst, DABCO (69 % at 24 h), and higher than that of the catalyst without methylation, NSBA-15 (30% at 24 h).

1) Y. Furukawa, M. Ogura, J. Am. Chem. Soc. 136, 2014, 119.

2) K. Sugino, N. Oya, N. Yoshie, and M. Ogura, J. Am. Chem. Soc. 133, 2011, 20030.

486833 The Effect of Water on the Reusability of Aminated Mesoporous Silica Catalysts for Aldol Condensations

Abstract:

Aldol condensations are important reactions which are used in a wide range of processes to produce new carbon-carbon bonds, for example in the pharmaceutical industry for the production of chalcones and in the fine chemical industry for the production of 2-ethylhexanol for PVC plasticizers [1]. A potential novel application in the bio-based industry is the production of liquid fuels from lignocellulosic biomass by upgrading smaller furanic molecules to heavier hydrocarbons [2]. Currently, homogeneous bases such as KOH or NaOH are typically used as aldol condensation catalysts. Despite their excellent activity, these catalysts are dangerous to handle, pose an environmental risk, and are difficult to separate from the final product stream [1]. Hence, innovative heterogeneous catalysts are pursued to establish a more sustainable alternative.

Due to their large pore sizes, large surface areas and the easy incorporation of different types of active sites, mesoporous silicas are promising supports for many types of catalysts. By grafting amine groups on a fraction of the weakly acidic surface silanols, an acid-base aldol condensation catalyst can be obtained. Previously, it has been demonstrated that the residual silanols enhance the catalytic activity of the amine active site by promoting the attack of its nitrogen lone electron pair on the carbonyl group of the reactants [3]. Previous research on acid-base mesoporous silica catalysts has shown that the highest catalytic activity can be achieved when a methyl substituted secondary amine is attached with a propyl linker to the support, and is fully promoted by surface silanols [4]. When the amines are randomly grafted on mesoporous silica, full promotion is achieved when at least 1.7 silanols are available per amine [5].

Although catalyst lifetime is an important factor for the industrial application of a material, no systematic investigation of the deactivation behavior of these cooperative acid-base catalysts has been performed, yet. In this work, a detailed investigation is performed on the factors which induce catalyst deactivation. Using this information, complete catalyst reusability has been achieved.

First, a heterogeneous acid-base cooperative catalyst has been synthesized by grafting of N-methylaminopropyltrimethoxysilane (MAPTMS) on a mesoporous silica gel. The amine density is kept low in order to ensure full promotion by an excess of weakly acidic surface silanol groups. An unpromoted amine catalyst is obtained by, subsequently, end-capping the promoting silanol groups using 1,1,1,3,3,3-hexamethyldisilazane (HMDS). Both materials are characterized by means of nitrogen adsorption-desorption measurements and elemental (CHN) analysis. These catalysts were then employed in the aldol condensation reaction of 4-nitrobenzaldehyde with acetone at 55 °C in a batch-type reactor, using DMSO as a solvent and methyl-4-nitrobenzoate as internal standard. The results of these experiments show that the reaction rate remains constant throughout an experiment, i.e., the slope of the conversion versus batch time graph remains constant. However, subsequent reuse of the spent catalyst yields a lower, albeit also constant, reaction rate. Hence, it seems that catalyst deactivation has occurred upon unloading the catalyst from the reactor. The ratio of the turnover frequency in the second run to that of the first run is used as a measure for the catalyst reusability. For the cooperative acid-base catalyst this ratio amounts to 38 ± 10%, while with the unpromoted amine catalyst a reusability of 77 ± 10% is achieved.

To investigate the nature of the deactivation, the spent catalysts have been characterized via nitrogen adsorption-desorption and CHN analysis. No evidence of pore-blocking was observed. Interestingly, after reaction, the amount of nitrogen in the samples increased with 60% for the cooperative catalyst and with 10% for the unpromoted amine catalyst. This indicates the presence of reaction intermediates on the surface. Raman spectroscopy confirms the presence of aromatic rings, conjugated with double bonds and nitro groups, covalently attached to the active site.

Figure 1: catalytic reaction cycle of the aldol condensation of acetone with 4-nitrobenzaldehyde, catalyzed by a secondary cooperative aminated silica catalyst [3-5]

Possible intermediate species that comply with the organic groups identified using Raman spectroscopy are displayed in Figure 1 as species (VII), (V) and (VI). Species (VII) is the iminium ion of a 4-nitrobenzaldehyde molecule. However, this undesired side-reaction is suppressed by using a large excess of acetone. The other potentially remaining species are the product iminium ion (V) and the product carbinolamine (VI). This iminium ion can undergo hydration to form the carbinolamine and subsequently desorb from the catalytic site. However, when drying the catalyst after a catalytic run, iminium ions that are still present on the surface could dehydrate and form a stable conjugated compound that acts as a site-blocking species, as displayed in Figure 2.

Figure 2: the iminium intermediate species (a) from the catalytic cycle is in equilibrium with its enamine form (b) and can further dehydrate into a stable conjugated species (c) which acts as a site blocker to the amine active site

Adding extra water to the reactor could increase the water-assisted desorption step of the iminium intermediate, as displayed in Figure 1, and thereby reduce its surface coverage upon unloading the catalyst from the reactor. In this work, 0.5 wt% to 3 wt% water is co-fed to the batch reactor. The first and second run turnover frequencies of both the acid-base catalyst, as well as the unpromoted amine catalyst, as a function of the water amount in the reactor, are displayed in Figure 3. When adding at least 0.5 wt% water, the unpromoted amine catalysts are completely reusable, exhibiting a turnover frequency ratio of 92 ± 10%. The maximum reusability achieved with the cooperative acid-base catalysts amounted to 72 ± 10%. This indicates that the promotional ability of the silanol groups irreversibly decreases due to the presence of water, while the activity of the amine sites can be fully recovered.

Figure 3: turnover frequency (s^{-1}) for the aldol condensation of 4-nitrobenzaldehyde with acetone as a function of water in the reactor. Aminated silicas promoted by silanols (red) and unpromoted (blue), empty bars are a first run, full bars are a second run. Error bars indicate a 95% confidence interval. Karl-Fischer titrations have been performed to correct for water in the organic solvents.

Additionally, Figure 3 shows that a small amount of water has a beneficial effect on the reaction rate. In contrast, a high amount of water has a negative effect on the reaction rate. This is directly related to the effect of the water concentration in the catalytic cycle, as shown in Figure 1. Firstly, the enamine (species IV) can react with water leading to the desorption of acetone, which decreases the overall reaction rate. Secondly, by adding water, the rate of the water-assisted desorption step of the iminium ion (V) increases, thereby freeing up catalytic sites and increasing the overall reaction rate.

In this work it has, thus, been shown that the catalyst reusability can be drastically increased with the addition of water, causing a reduction in surface coverage of iminium species that could otherwise dehydrate to a stable site blocking species when the catalyst is dried. For the unpromoted amine catalyst, 0.5 wt% additional water already causes the catalyst to be fully reusable. However the cooperative acid-base catalysts cannot be made fully reusable due to a loss of the promotional ability of the silanol groups. The applicability of aminated mesoporous silica catalysts for continuous aldol condensation processes has, hence, been confirmed in this work by proving that a stable catalyst regime can be obtained by co-feeding a small amount of water. Further testing of these catalysts in a continuous-flow reactor will yield more information about the deactivation behavior of the catalyst with time-on-stream.

[1] G. Kelly, F. King, and M. Kett, "Waste elimination in condensation reactions of industrial importance," *Green Chemistry,* vol. 4, no. 4, pp. 392-399, 2002.

[2] Q. Deng *et al.*, "Efficient synthesis of high-density aviation biofuel via solvent-free aldol condensation of cyclic ketones and furanic aldehydes," *Fuel Processing Technology,* vol. 148, pp. 361-366, 2016.

[3] N. A. Brunelli, K. Venkatasubbaiah, and C. W. Jones, "Cooperative catalysis with acid–base bifunctional mesoporous silica: impact of grafting and co-condensation synthesis methods on material structure and catalytic properties," *Chemistry of Materials,* vol. 24, no. 13, pp. 2433-2442, 2012.

[4] J. Lauwaert, E. De Canck, D. Esquivel, P. Van Der Voort, J. W. Thybaut, and G. B. Marin, "Effects of amine structure and base strength on acid–base cooperative aldol condensation," *Catalysis Today,* vol. 246, pp. 35-45, 2015.

[5] J. Lauwaert, E. De Canck, D. Esquivel, J. W. Thybaut, P. Van Der Voort, and G. B. Marin, "Silanol-Assisted Aldol Condensation on Aminated Silica: Understanding the Arrangement of Functional Groups," *ChemCatChem,* vol. 6, no. 1, pp. 255-264, 2014.

500731 Novel Method for Synthesizing Non-Leaching Functionalized Mesoporous-Silica for Acid Catalyzed-Alkylation Reaction

Abstract:

Novel method for synthesizing non-leaching functionalized mesoporous-silica for acid-catalyzed alkylation reaction

*Santiago J. Umbarila[§], Tuong V. Bui[§], Tawan Sooknoi, Daniel E. Resasco**

[§]*Both authors contributed equally to this work*

School of Chemical, Biological and Materials Engineering, University of Oklahoma, 100 East Boyd Street, Norman, Oklahoma, 73019, United States

**Corresponding Author: resasco@ou.edu*

ABSTRACT

The ordered mesoporous silica materials such as MCM-41 and SBA-15 have attracted special attention because of its chemical and thermal stability[1]. Several studies have been reported modification method of the silica surface through conventional grafting (CG) and co-condensation (CC) functionalization[1–6]. The CG modifies the surface via sylilation of the free silanols with organosilanes. In turn, the CC method uses the TEOS as the a building block for generating the silica structure in the one-pot mixture with organosilanes, which act as functional moieties. The resulting hybrid materials has the characteristic mesoporous hexagonal structure together with organic functional groups anchoring on the surface.

It has been reported that the Si-O bonds obtained from conventional functionalization process are thermally stable, but only under rather mild conditions, which in turn do not cause any loss of active moieties via the leaching or decomposition. There are only a few studies that report the leaching under more severe chemical conditions in either non-polar [5–7] or polar solvents [8]. In this study, the leaching process will be investigated at the molecular level by utilizing a combination of techniques including Fourier-Transform Infrared Spectroscopy (IR), Transmission Electron Microscopy (TEM), Solid-State Nuclear Magnetic Resonance (NMR) of ^{29}Si, ^{1}H and cross polarization ^{1}H-^{29}Si, X-ray Diffraction (XRD).

In addition, a new synthesis method that we identify as "dry-grafting" will be proposed. This method provides a novel and simple way to synthesize stable non-leaching materials useful as catalysts in acid-catalyzed reactions, even under severe conditions (e.g. in polar solvents, at high temperature, etc). The catalytic stability is examined for the alkylation reaction between cyclopentanol (CPOL), acting as the alkylation agent and m-cresol as a substrate. This model compound reaction has practical relevant to biomass conversion and upgrading of pyrolysis vapors into chemicals and drop-in transportation fuels.

Key words: Dry grafting, Leaching, Mesoporous silica, Alkylation and Biomass conversion

(1) Hoffmann, F., et al., *Silica-Based Mesoporous Organic–Inorganic Hybrid Materials.* Angewandte Chemie International Edition, **2006**, 45(20): p. 3216-3251

(2) Huang, Y. Functionalization of Mesoporous Silica Nanoparticles and Their Applications in Organo-, Metallic and Organometallic Catalysis, Iowa University, **2009**.

(3) van der Graaff, W. N. P.; Olvera, K. G.; Pidko, E. A.; Hensen, E. J. M. Stability and Catalytic Properties of Porous Acidic (Organo)silica Materials for Conversion of Carbohydrates. *J. Mol. Catal. A Chem.* **2014**, *388*, 81–89.

(4) Hoffmann, F.; Cornelius, M.; Morell, J.; Fröba, M. Silica-Based Mesoporous Organic-Inorganic Hybrid Materials. *Angew. Chem. Int. Ed. Engl.* **2006**, *45* (20), 3216–3251.

(5) García, N.; Benito, E.; Guzmán, J.; Tiemblo, P.; Morales, V.; García, R. A. Functionalization of SBA-15 by an Acid-Catalyzed Approach: A Surface Characterization Study. *Microporous Mesoporous Mater.* **2007**, *106* (1), 129–139.

(6) Gholami, M.; Talaie, M. R.; Aghamiri, S. F. CO2 Adsorption on Amine Functionalized MCM-41: Effect of Bi-Modal Porous Structure. *J. Taiwan Inst. Chem. Eng.* **2015**.

(7) Testa, M. L.; La Parola, V.; Venezia, A. M. Transesterification of Short Chain Esters Using Sulfonic Acid-Functionalized Hybrid Silicas: Effect of Silica Morphology. *Catal. Today* **2014**, *223*, 115–121.

(8) Canilho, N.; Jacoby, J.; Pasc, A.; Carteret, C.; Dupire, F.; Stébé, M. J.; Blin, J. L. Isocyanate-Mediated Covalent Immobilization of Mucor Miehei Lipase onto SBA-15 for Transesterification Reaction. *Colloids Surfaces B Biointerfaces* **2013**, *112*, 139–145.

500634 Catalytic Conversion of Biomass Derived Ethanol to Advanced Hydrocarbon Fuels and Valuable Chemicals

Abstract:

Global ethanol production is increasing rapidly and anticipated to reach more than 33 billion gallons by 2017. The ethanol market is at a saturation point in transportation sector since its use is limited by the 10-15% blend-wall due to technological and infrastructure constraints. Therefore, it is expected that excess ethanol will be available as a platform molecule for the production of advanced hydrocarbon fuels and value-added chemicals. We will present our findings in exploring conversion of ethanol into hydrocarbon fuels and valuable co-products (e.g., benzene, toluene and xylenes (BTX)). Selective conversion of ethanol over Ga-ZSM-5 will be presented. At 450 °C and 0.4 h^{-1}, Ga-ZSM-5 produces 55% BTX which is two-fold increase over H-ZSM-5 (26% BTX). The BTX yield increases with increase in temperature (300-500 °C) and Ga loading (0.5% to 6.2%). Enhanced hydrogen production and reduced light paraffins (propane and butane) formation are also observed with the increase of Ga loading. The role of Ga in promoting BTX formation and molecular hydrogen production will be discussed. The active gallium species will be further presented from the studies of physical mixture experiments and extensive catalyst characterizations. We will also discuss our findings in alternative zeolite structures for ethanol conversion, and the effect on product selectivity and catalyst stability.

491403 Continuous Hydrogenation of Furfural to Cyclopentanone Using Activated Carbon Monolith Catalysts

Abstract:

Cyclopentanone, defined as a cyclic ketone, plays a key role in synthesis of pharmaceuticals, rubber chemicals, flavors, fragrances, and aviation fuel. The current industrial process in production of cyclopentanone involves a petroleum-based feedstock. During this process, adipic acid undergoes catalytic cyclization at high temperatures to form cyclopentanone. Adipic acid is a petroleum-based chemical that is obtained from oxidization of a cyclohexanol and cyclohexanone mixture in presence of excessive amounts of nitric acid. Generating large amounts of N_2O as by-product is one of the major drawbacks of this adipic acid production process.

Furfural is a green chemical that can be hydrogenated to form cyclopentanone. Acid hydrolysis of hemicellulose-rich biomasses such as corn stover, wheat bran, or corncobs results in xylose. In a second step, furfural can be obtained through catalytic dehydration of xylose. This work focuses on continuous hydrogenation of furfural to cyclopentanone using activated carbon monolith catalysts. Activated carbon monolith catalysts (ACMC), derived from woody biomass, are impregnated with precious metals and used to form cyclopentanone from furfural in the presence of hydrogen. Multiple advantages of using ACMC, in addition to low pressure drop, high mass transfer rates and surface area to volume ratios, are first, the monolith is made from renewable carbon sources, second, high stability in aqueous/acid/base reaction medium and finally, the precious metal can be recovered after combustion of the catalyst. The kinetics of cyclopentanone formation from furfural using ACM impregnated with precious metals will be studied over temperatures of 100 to 180 °C, hydrogen pressures of atmospheric to 300 psig, liquid flow rates of 0.5 to 4 mL/min and hydrogen flow rates of 40 to 100 mL/min. The immediate goal of this work is to develop a kinetic model and rate law, perform comprehensive catalyst characterization correlating structure with function, and determine the longevity of the catalyst.

496742 Soluble Sugars from Cellulose Using Polyvinyl Chloride Derived Hydrochloric Acid As Catalyst

Abstract:

Municipal Solid Waste (MSW) contains a large portion of organic material that has the potential to be converted into valuable fuels through thermochemical conversion. In particular, cellulosic waste-derived sugars are important platform chemicals of biofuels and chemicals. In addition to organic wastes, MSW also contains significant quantities of halogenated plastics like polyvinyl chloride (PVC) that are difficult to recycle using traditional methods. When thermally converted (e.g., pyrolysis, gasification, and incineration), PVC produces corrosive hydrochloric acid (HCl) that can react with organic waste-derived products to form organohalogens and dioxins that create various environmental and health hazards. In the current work, we co-converted cellulose and PVC in tetrahydrofuran (THF) at elevated temperatures and pressures. The reactions were carried out in miniature Swagelok reactors with reaction temperature, time and PVC concentration as the experimental variables. The liquid product was analyzed and quantified with GC/MS/FID, and the solid residue was characterized by pyrolysis using a Frontier micropyrolyzer coupled to a GC/MS/FID. The results showed that HCl released during the thermal decomposition of PVC acts as an effective catalyst in the polar aprotic solvent to promote cellulose decomposition to soluble sugars. Levoglucosan was found to be the most abundant product. The highest monomeric sugar yield was 42% of cellulose at 335 ºC and 8 minute reaction time with 2.5% cellulose weight equivalent PVC. The liquid products also contained up to 46% of soluble oligomeric sugars and small amounts of hydroxymethylfurfural and furfural. Pyrolysis of the solid residue also produced a high yield of levoglucosan (~22%, residue weight basis), indicating the residue is unconverted or partly decomposed cellulose. Chlorinated organics were nearly absent in both the liquid product and the pyrolyzates of the solid residue, suggesting the chloride content in PVC is converted to the solvent-dissolved HCl. The present study aims to explore an environmentally friendly approach to convert mixed solid wastes into value added products.

495204 Effect of Temperature and Vapor Residence Time on Product Distribution of High Density Polyethylene Fast Pyrolysis

Abstract:

In search of more sustainable options for plastic waste disposal, thermochemical conversion to produce alternative fuels or chemicals appears as an outstanding alternative due to its environmental benefits such as having a lower carbon footprint than fossil fuel extraction.[1] One method of thermochemical conversion is fast pyrolysis, the thermal degradation of plastic waste in the absence of oxygen. Fast pyrolysis at high temperatures, with reaction times shorter than 10 s, has been used to produce hydrocarbon products from plastic waste that can be converted into fuel or chemical intermediates. This study analyzes the degradation products of high density polyethylene (HDPE), the most abundant plastic waste, with respect to temperature and vapor residence time. Through the use of a novel, high-temperature reactor extension to an existing micro-pyrolysis unit (CDS Analytical 5200 HP Pyroprobe), the residence time at each temperature was varied by adjusting the helium carrier gas flowrate. Residence times were varied from ~0.0 to 5.6 s (25mL/min He to 100mL/min He). The temperature of the micropyrolysis unit and the reactor extension was established at three different temperatures (650°C, 675°C, and 700°C) for each set of residence times. The pyrolysis products were analyzed in a ThermoFisher K8880181 Trace Gas Chromatograph coupled to a DSQ II mass spectrometer. The observed chemical species were alkanes, alkenes, alkadienes and aromatics and the carbon content of the pyrolysis products ranged from 2 to 32 atoms per molecule. At constant reactor temperature of 700°C, the slight variation in residence time from ~0.0 to 1.3 s greatly shifted the product distribution towards lower molecular weight compounds (LMWC, C15 and below) as opposed to higher molecular weight compounds (HMWC, C16 and above). Without extended reaction, 25% of the products were LMWC, while with extended reaction, 96% were LMWC. When the pyrolysis temperature was lowered to 675 °C the amount of LMWC decreased 6% to 90%. These results illustrated the trend that as temperature and residence time increased, there was a greater abundance of lower molecular weight compounds in the product distribution, mostly alkenes with some alkanes but almost lacking dienes. Changes in chemical species distribution were also observed. An increase in the composition of aromatic compounds was exhibited at longer residence times and higher temperatures whilst the opposite trend occurred for high molecular weight alkenes and alkadienes. These results show a promising pathway for HDPE waste as a potential source for hydrocarbon fuel fractions and chemical intermediates.

References:

1. Gracida-Alvarez, U.R.; Keenan, L.M.; Sacramento-Rivero, J.C. & Shonnard, D.R. Resource and Greenhouse Gas Assessments of the Thermochemical Conversion of Municipal Solid Waste in Mexico. *ACS Sustainable Chemistry and Engineering* **4**, 5972-5978 (2016).

497522 Effects of Acid-Pretreatment on Co-Pyrolysis of Biomass and Plastic

Abstract:

Over 521 million pounds of agricultural plastics are annually used in US farm field to protect plants and improve crop yield. Due to the non-biodegradability of the plastics, disposal of the used plastics is problematic. The agricultural waste plastics are mostly burned in farm land, buried or sent to landfill, which cause many environmental problems and health hazards. Since waste plastics and crop residues are often present together in the farm field, co-pyrolyzing the two streams of materials could be an attractive solution to address the environmental issue while promoting energy recovery from waste plastics. In the present study, the methods for improving the co-pyrolysis efficiency of biomass and plastic are explored. It is known that the inorganic impurities presenting in cellulosic biomass favor undesired ring opening and charring reactions to reduce bio-oil yield. Acid washing and acid infusion of biomass prior to pyrolysis are known as effective methods to overcome the detrimental effect of inorganic impurities to increase bio-oil yield. The effect of acid pretreatment on co-pyrolysis of biomass and plastic is studied. Corn stover biomass pretreated with sulfuric acid (both acid washed and acid infused) were co-pyrolyzed with polyethylene in a tandem micropyrolyzer with and without zeolite catalyst. During non-catalytic pyrolysis, levoglucosan yield increased from 0.5wt% with non-treated corn stover to 9.5wt% in acid treated biomass, which is expected. It was found that co-pyrolysis of acid-treated corn stover with PE can further increase the yield of levoglucosan, as well as the yields of phenolic monomers due to synergistic effect. The stronger synergistic effects were also observed when the acid treated corn stover and PE were co-pyrolyzed using HZSM-5 zeolite catalyst in comparison to co-pyrolyzing non-treated corn stover and PE using same catalyst. Due to the synergistic effects, the amount of total aromatic hydrocarbons and the selectivity of mono-ring aromatic hydrocarbons both increased significantly.

497745 Effect of Dielectric Oxides on Microwave Assisted Deconstruction of Lignin Â

Abstract:

Lignin, a natural polymer is a potential source of fuels and platform chemicals due to its aromatic structure. The conventional industrial processes which utilize plant polysaccharides, burn lignin to generate process heat. With the depletion of fossil fuels, cellulosic biorefineries would generate surplus lignin than required to fuel the biomass conversion process [1]. Conventional lignin conversion methods require temperatures greater than 200°C and longer reaction time. A non-conventional source of energy such as microwave reduces the reaction time and gives higher heating rates. The heating of reaction medium by microwaves is through the mechanism of dipolar polarization and ionic conduction. In this study, alkali lignin was solubilized in 1:1 Dimethyl Sulfoxide (DMSO)/water mixture. The DMSO/water mixture has a high polarity (0.7) and solubilizes lignin completely. Metal oxides such as lanthanum ferrate ($LaFeO_3$), iron oxide (Fe_2O_3) and zirconium oxide (ZrO_2) were synthesized by the combustion method. Experiments were carried out in a microwave reactor at different temperatures (80-100°C) and samples were withdrawn at regular intervals between 0-120min. The catalysts were suspended in the reaction media and the solution vessel was continuously stirred for the entire reaction time. The microwave treated samples were analyzed using Gel Permeation Chromatography to study the effect of catalysts on reduction in the average molecular weight of lignin. Liquid chromatography Mass Spectrometry was used for the identification of phenolic monomers produced. The formation of products such as vanillin, acetoguaiacone, homovanillic acid and coniferaldehyde were observed at temperatures as low as 80°C while using $LaFeO_3$ and Fe_2O_3 as catalysts. The effect of different catalyst properties such as structure, dielectric constant and surface area, on the yield of monomers was studied and possible reaction mechanisms were proposed.

References:

1. A.J. Ragauskas *et.al.*, Lignin valorization:improving lignin processing in the biorefinery. *Science*, 344 (2014).

495589 Thermodynamic Simulations and Techno-Economic Analysis on the Utilization of CO_2Â and a Novel Modularization Strategy for Chemical Looping Based Gtl Processes

Abstract:

The utilization of carbon dioxide (CO_2) as a feedstock in a chemical looping process that converts methane to syngas and a variety of downstream products has the ability to be impactful in the field of carbon capture, utilization, and sequestration (CCUS). The Ohio State University Chemical Looping methane to syngas (MTS) processâ€™ novelty comes from the coupling of carbon capture and utilization strategy, which captures and utilizes CO_2 to create a synergy atypical to many processes that are designed to deal with the challenges of carbon constraints. CO_2 utilization allows for increased flexibility of H_2:CO molar ratios that may be required for variations in the downstream processing while producing a comparable quantity of syngas with reduced natural gas input. In this study, thermodynamic models for chemical looping reactor systems are examined using ASPEN Plus. The chemical looping reducer reactor considered employs steam, natural gas and CO_2 as feedstock along with metal oxide oxygen carriers. This presentation will initially present system optimization based on thermodynamic Gibbs free energy minimization for the desired syngas composition required for downstream gas to liquid (GTL) processes. With CO_2 co-injection in the MTS process, the thermodynamic optimization indicates a natural gas savings of more than 20% over that of the baseline partial oxidation processes for GTL applications. Experimental verification using Thermo-gravimetric analyzer and bench-scale moving bed reactor that compared well with the thermodynamic model simulations will also be presented. A comprehensive techno-economic analysis on an energy, balanced (water balance, steam balance, heat and electricity balance) thermodynamic performance model of the MTS process with and without CO_2 utilization & modularization strategy was developed. The use of techno-economic analysis model for studying sensitivities associated with parameters including capital charge factor, first year versus leveled costs of fuel, crude oil equivalent prices and cost of chemical looping reactors will be presented. The comprehensive techno-economic analysis findings exemplify the importance of a CO_2 utilization and modularization strategy in chemical looping GTL processes and encourage further optimization in natural gas to commodity chemical conversion processes including gas to methanol, gas to ammonia, and gas to ethanol, amongst others.

534C:

Gasification of Residue from the Solvent Extraction of Powder River Basin Coal

William C. Schaffers
Ying Wang
David A. Bell (speaker)

American Institute of Chemical Engineers National Meeting

November 1, 2017

UW UNIVERSITY OF WYOMING

Shenhua Direct Coal Liquefaction (DCL)

- Commercial plant in China that makes liquid fuels by hydrogenating coal.
- Residue from plant contains undigested coal + 1st stage catalyst
- This project:
  - Investigate feasibility of gasifying residue
    - Provide hydrogen for plant
    - Eliminate residue waste stream
  - Sponsored by U.S./China Clean Energy Research Center for Coal

UW

Project Assumptions

- Unable to contact Chinese team
- Assumed:
  - Shenhua DCL plant or similar built in U.S.
  - Powder River Basin coal feed.
  - Need to recommend gasification technology for residue.

UW

Fluidized bed gasifiers

- Lower cost than entrained flow gasifiers
- Requires reactive feedstock
  - Biomass
  - Lignite
  - Sub-bituminous coal, such as PRB
- Bituminous coals require more severe conditions in entrained flow gasifier.

UW

Experimental strategy

- Approximate coal liquefaction process by extraction with tetralin, a hydrogen donor solvent

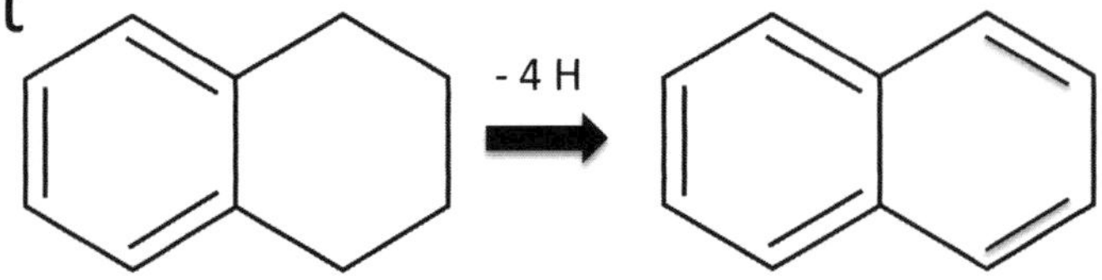

- Compare gasification reactivity of extraction residue to parent coal.

UW

UW

Extraction at 360 C

Run #	As Received Weight (g)	Dry Weight (g)	Moisture Loss (wt %)	Extraction Residue Weight (g)	Material Extracted (wt %)
3	54.38	46.54	14.42	25.31	45.62
5	57.88	50.19	13.29	30.01	40.21
6	60.27	51.38	14.75	28.39	44.75

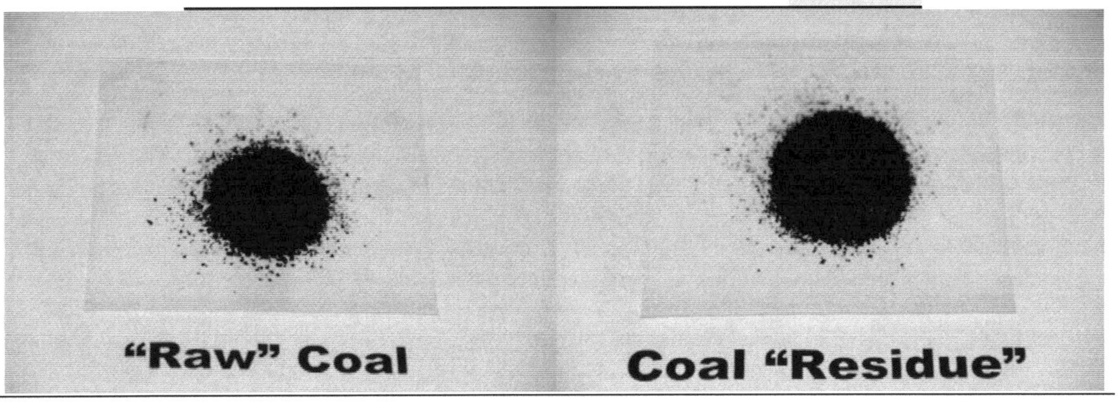

Proximate & Ultimate Analyses

	As Received Basis (wt %)		Moisture Free Basis (wt %)		Moisture & Ash Free Basis (wt %)	
	Decker Coal	***Extraction Residue***	***Decker Coal***	***Extraction Residue***	***Decker Coal***	***Extraction Residue***
Proximate Analysis (Method: ASTM D-5142; D-7582)						
Moisture	24.29	*3.27*	****	****	****	****
Ash	3.62	*9.78*	4.78	*10.11*	****	****
Volatile Matter	29.42	*31.48*	38.86	*32.54*	40.81	*36.20*
Fixed Carbon	42.67	*55.47*	56.36	*57.35*	59.19	*63.80*
Total	**100.00**	*100.00*	**100.00**	*100.00*	**100.00**	*100.00*
Ultimate Analysis (Method: ASTM D-5142/5373; D-7582/4239)						
Moisture	24.29	*3.27*	****	****	****	****
Hydrogen	2.88	*3.80*	3.80	*3.93*	3.99	*4.38*
Carbon	56.48	*68.28*	74.59	*70.59*	78.33	*78.53*
Nitrogen	0.87	*1.05*	1.15	*1.09*	1.21	*1.21*
Sulfur	0.34	*0.28*	0.45	*0.29*	0.47	*0.32*
Oxygen	11.52	*13.54*	15.23	*14.00*	16.00	*15.57*
Ash	3.62	*9.78*	4.78	*10.11*	****	****
Total	**100.00**	*100.00*	**100.00**	*100.00*	**100.00**	*100.00*

UW

Gasification tests

- Modified drop tube reactor
- Preheated, gas flow established
- ~1 g of coal or residue dropped into reactor
- Lands on glass frit in center
- Effluent gases measured using mass spectrometer

UW

Typical mass spectrometer results

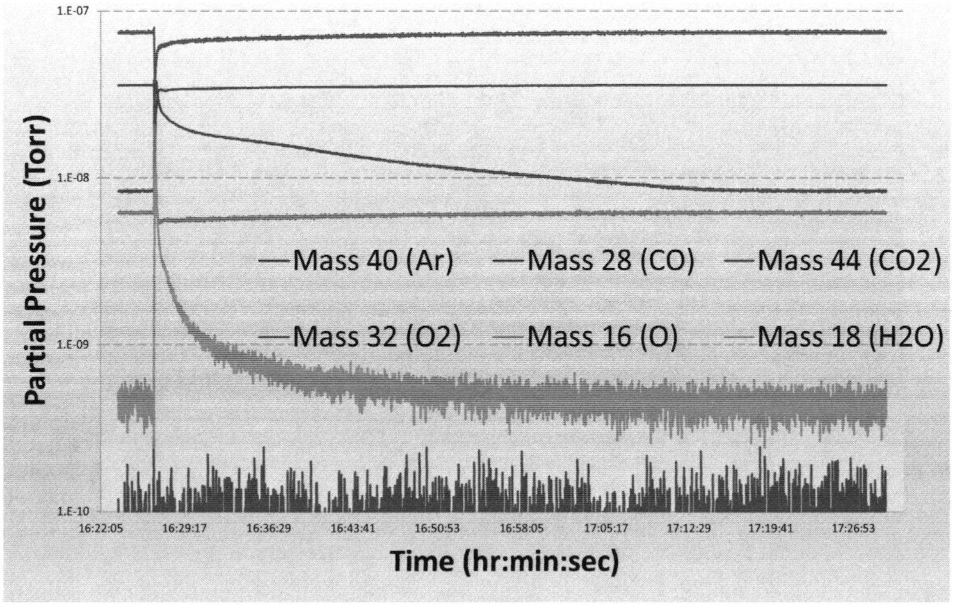

UW

Data fit

- Reaction complete when mass spectrometer signal returns to baseline
- Integration of signal versus time used to measure extent of reaction, *X.*
- Data fit to random pore model (assumes reaction on pore walls):

$$dX/_{dt} = \boldsymbol{k_p}(1 - X)\sqrt{1 - \psi ln(1 - X)}$$

Where $\boldsymbol{k_p}$ is the rate parameter, ψ is dimensionless structural parameter.

UW

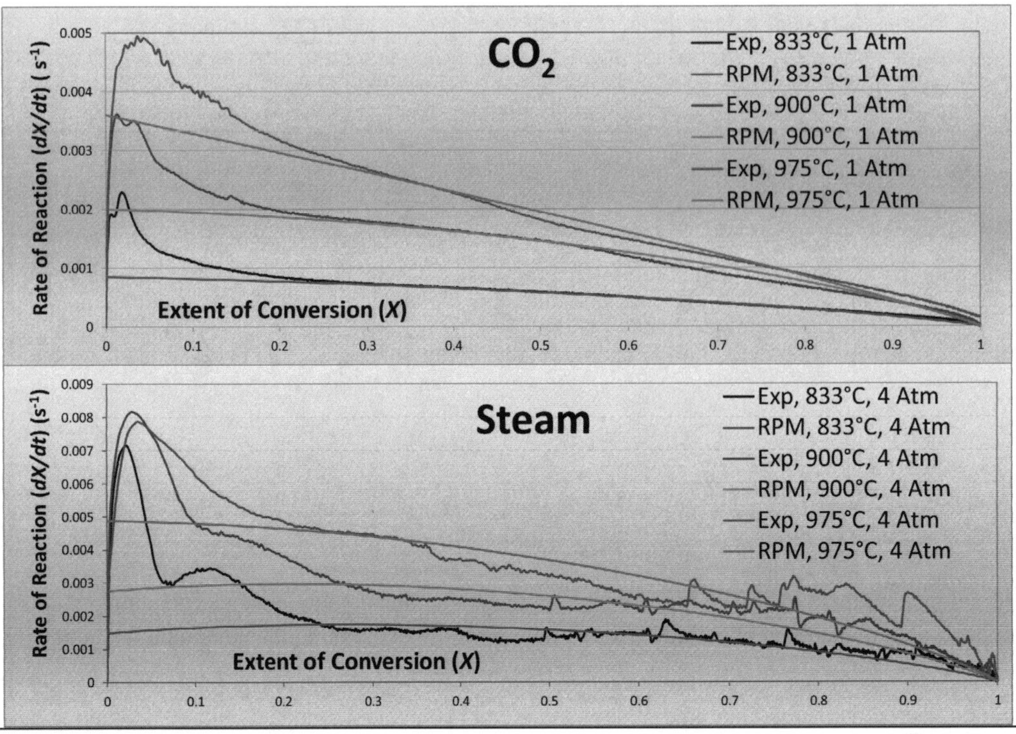

UW

Coal & Residue Gasification in CO_2

Pressure	Temp.	Coal Time (hr:min:s)	Residue Time (hr:min:s)	*Coal* $k_p (s^{-1})$	*Residue* $k_p (s^{-1})$	*Coal* ψ	*Residue* ψ
1 atm	833 °C	1:06:12	0:47:45	4.10E-04	8.75E-04	3.37	1.50
	900 °C	0:25:00	0:22:52	1.20E-03	1.81E-03	3.12	1.51
	975 °C	0:12:06	0:15:28	2.75E-03	3.44E-03	2.45	0.63
4 atm	833 °C	1:01:36	1:00:02	5.00E-04	8.17E-04	2.65	0.92
	900 °C	0:23:00	0:28:36	1.28E-03	1.95E-03	2.63	0.28
	975 °C	0:11:00	0:17:11	2.50E-03	3.88E-03	3.25	0.10

Time is time to complete gasification.

Residue gasification is usually faster than gasifying parent coal.

UW

Coal & Residue Gasification in Steam

Pressure	Temp.	Coal Time (min:s)	Residue Time (min:s)	*Coal* $k_p (s^{-1})$	*Residue* $k_p (s^{-1})$	*Coal* ψ	*Residue* ψ
1 atm	833 °C	17:41	13:19	1.28E-04	2.01E-03	5.25	4.69
	900 °C	9:26	9:34	2.75E-03	3.95E-03	5.25	2.18
	975 °C	4:31	7:07	6.30E-03	4.67E-03	3.50	3.39
4 atm	833 °C	19:53	19:19	1.28E-04	1.43E-03	3.63	4.27
	900 °C	13:15	10:02	2.60E-03	3.00E-03	2.95	3.65
	975 °C	6:47	07:31	5.85E-03	4.89E-03	2.23	2.88

Steam gasification faster than CO_2 gasification.

Residue gasification usually faster than coal gasification.

UW

Surface Areas of Chars from Coal & Residue

Method	Surface Area (m^2/g)					
	Coal Char (833°C)	Residue Char (833°C)	Coal Char (900°C)	Residue Char (900°C)	Coal Char (975°C)	Residue Char (975°C)
BET (N_2) (Mesopores)	0.36	10.81	1.49	24.39	34.54	40.25
D-R (CO_2) (Micropores)	408.51	382.13	341.44	382.61	344.68	404.13

UW

Calcium and Sodium on Char Particles

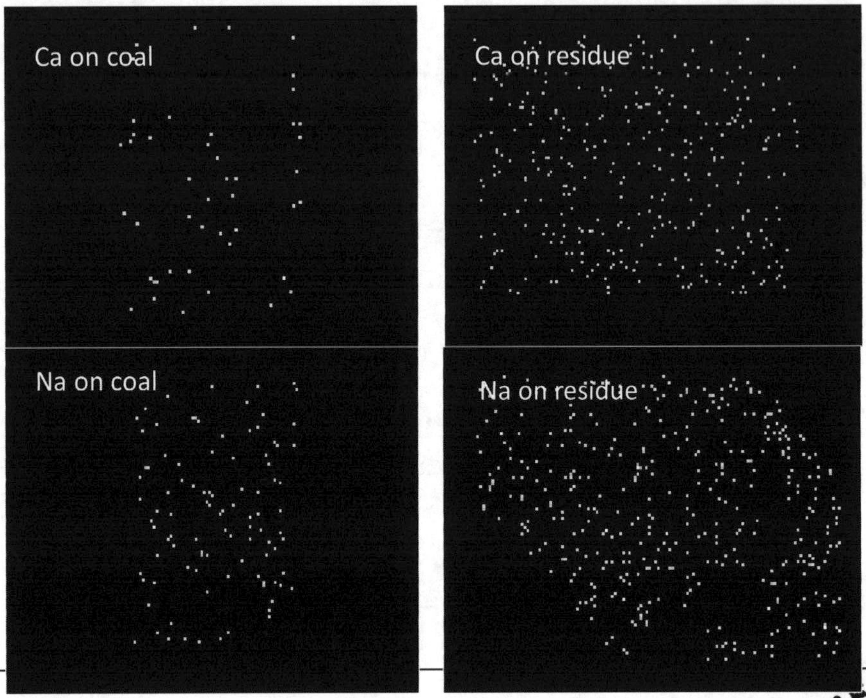

Conclusions

- Residue is more reactive than parent coal
- Effect may be due to
  - Increase in mesoporous surface area
  - Concentration of naturally-occurring catalysts, sodium and calcium
- Residue can be gasified in fluidized bed gasifier.

Center for Applied Energy Research

Development of a Modular Coal/Biomass to Liquids Research Facility

DOE Project Number: FE0010482

Andy Placido

University of Kentucky, Center for Applied Energy Research

http://www.caer.uky.edu/powergen/home.shtml

caer.uky.edu

2017 AIChE Annual Meeting

November 1st, 2017

Center for Applied Energy Research

Facility Overview

- Integrated coal/biomass-to-liquids (CBTL) facility at a capacity of 1 bbl./day
- Purposely designed for maximum flexibility: modular, skid-mounted, anticipating frequent change-outs; "plug and play;" and future re-purposing

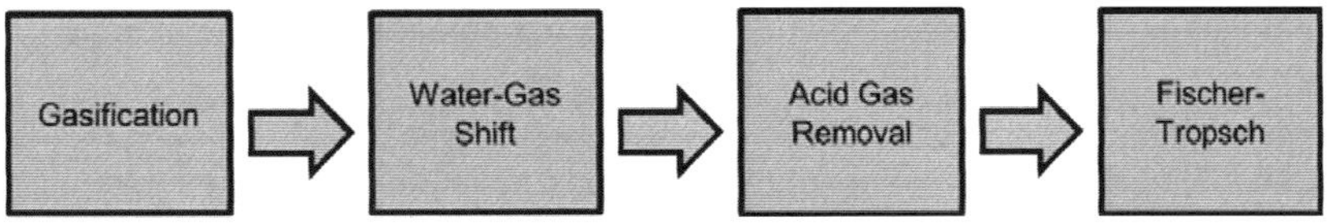

caer.uky.edu

2017 AIChE Annual Meeting

November 1st, 2017

Center for Applied Energy Research

caer.uky.edu

OMB Gasification System

- Entrained flow gasifier
  - Coal/water slurry
  - Oxygen blown
  - Molten slag
  - 1 ton/day coal consumption
- Includes biomass/coal feed preparation tower
  - Hydro-vibrating Ball mill

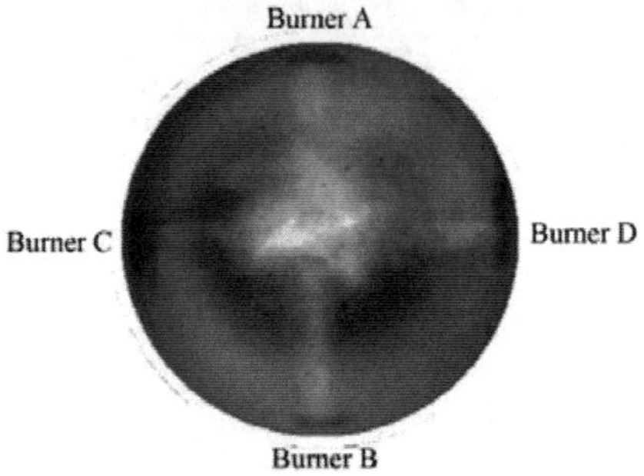

2017 AIChE Annual Meeting

November 1st, 2017

Center for Applied Energy Research

caer.uky.edu

Feed Preparation

- Ash fusion temperature generally limiting factor of feed
- Capable of utilizing wide range of coals and blends
- Designed to handle torrefied biomass
- Important Final Slurry Properties
  - Particle size
  - Viscosity

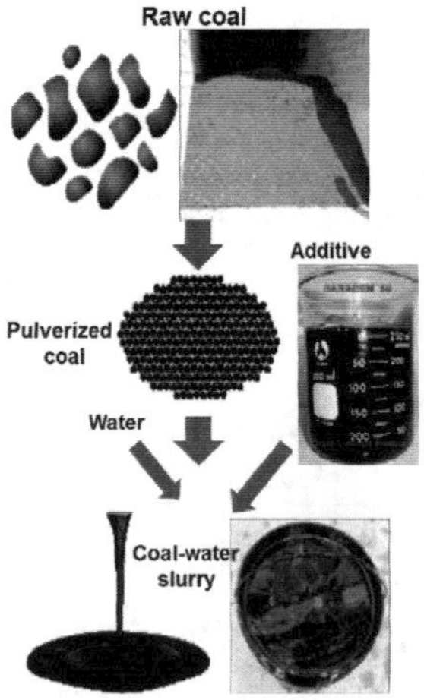

Typical Properties of CWS

Average particle size (μm)	Mass concentration	Viscosity (mPa·s)
<50	<60%	<250

2017 AIChE Annual Meeting

November 1st, 2017

Center for Applied Energy Research

caer.uky.edu

WGS

$CO + H_2O \Leftrightarrow CO_2 + H_2$ Tunable H_2/CO ratio

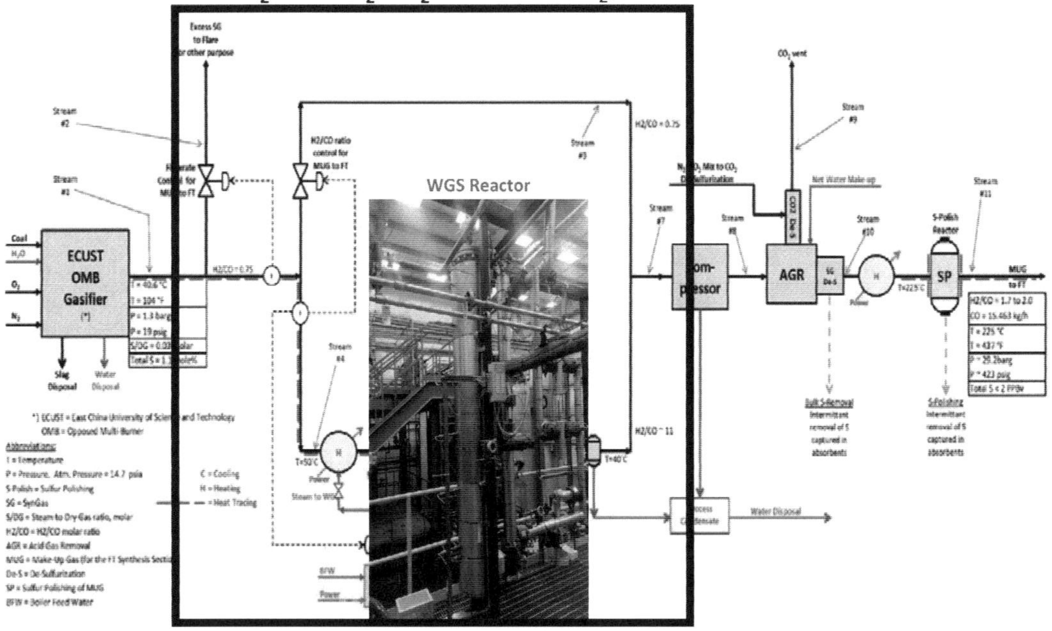

Center for Applied Energy Research

caer.uky.edu

Acid Gas Removal

Aqueous Solvent System

- CAER has 10+ years Experience with Similar Technology
- Thermal and Pressure Swing Absorption
- Howden Burton Corblin metal diaphragm compressor
  - Pressures up to 450 psi
- Wide range of Amine Solvents
- Sulfur Treatment on Both Rejected Acid Gas and Treated Gas
  - Activated carbon
- 95+% removal of CO_2
- 99.99% removal of sulfur (less than 1ppm)

Center for Applied Energy Research

caer.uky.edu

Downstream - FT

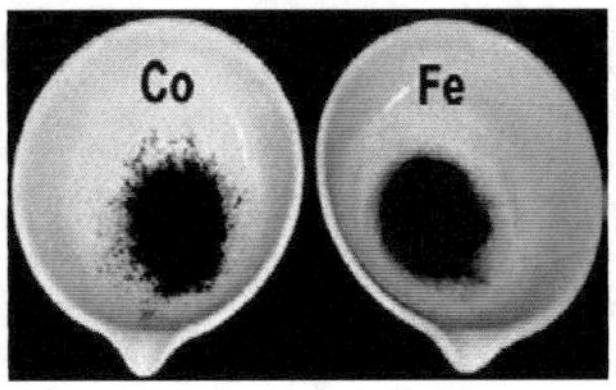

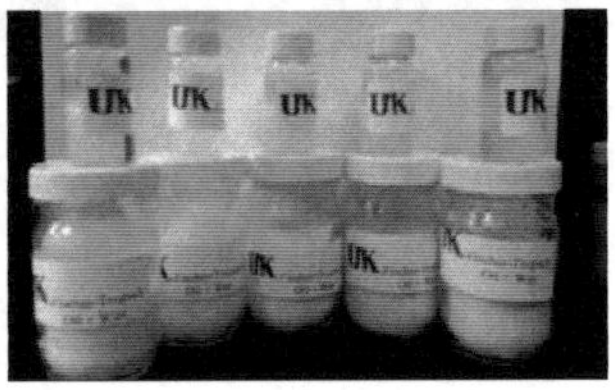

- Chart Energy microchannel reactor with production capacity of 1bbl./day
- FT Reactor is aluminum heat exchanger inside a steel kettle
- Capable of utilizing either Fe or Co based catalysts
  - Difference in H_2/CO molar ratio
- Wide ranging fuel production
- Water Gas Shift and FT modules were designed and built by Zeton Inc. with assistance from Stovlbaek Consulting

Center for Applied Energy Research

caer.uky.edu

Balance of Plant

UK Center for Applied Energy Research

caer.uky.edu

Experimental Data from Coal Only

- So far Gasifier test operation only
- Typical Operation data for 24 hour period
- Operationally stable for entire testing run
- Feed details
  - 30-40% coal/60-70% water
  - Co-fire small amount of natural gas (~15% heat input) to maintain heat

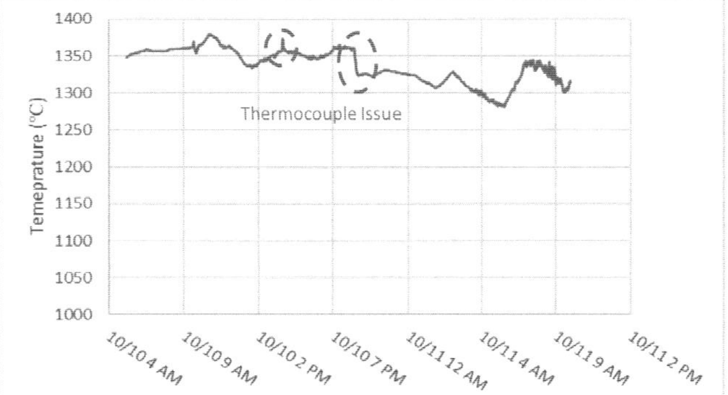

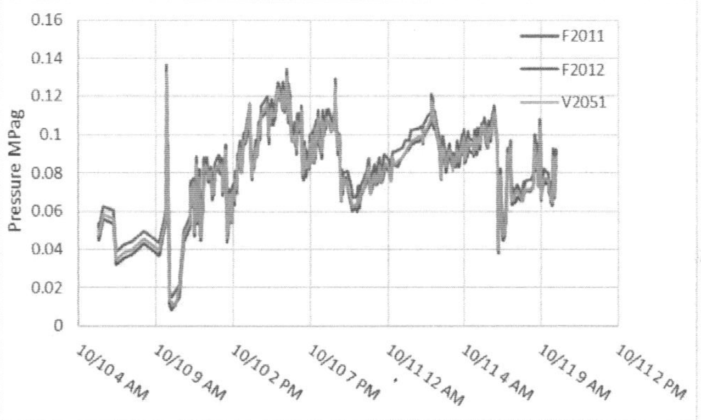

2017 AIChE Annual Meeting

November 1st, 2017

UK Center for Applied Energy Research

caer.uky.edu

Experimental Data from Gasifier Operation

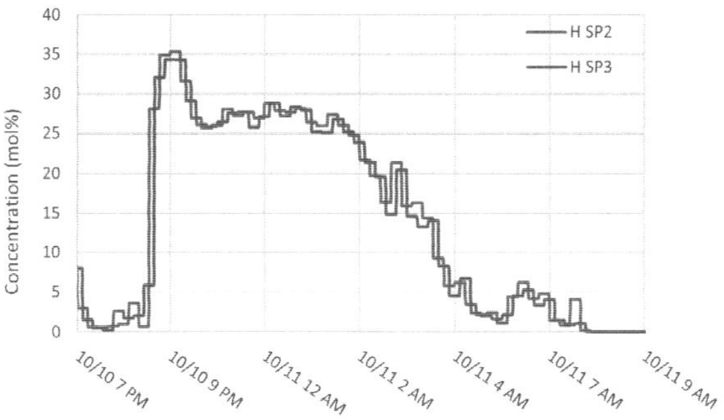

- Overall availability ~50%
  - First operation for syngas quality control testing
  - GC learning curve (filters clogging, learning software and hardware)
- H_2/CO important syngas quality measure for downstream
  - Ratio stable around 1.1
  - Higher than expected due to co-firing with natural gas

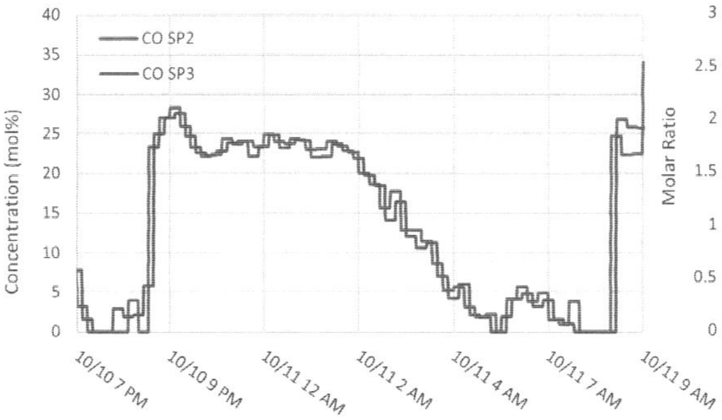

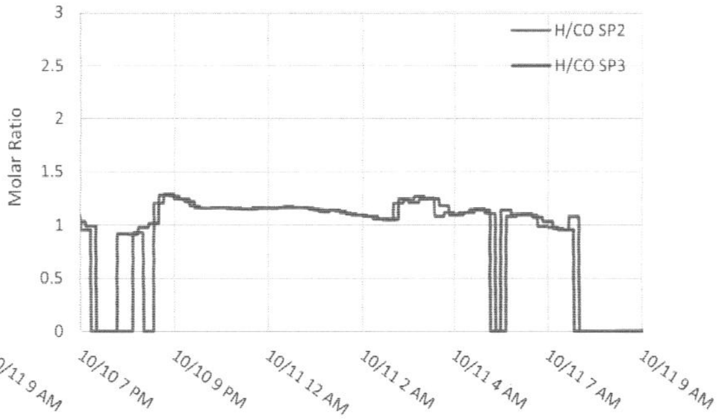

2017 AIChE Annual Meeting

November 1st, 2017

Center for Applied Energy Research

caer.uky.edu

Experimental Data from Gasifier Operation

- Coal Conversion
  - CO and CO_2 produced
  - Total carbon fed (coal and natural gas)
- Low Conversion
  - First trial run
- Burner Atomization
- Slurry Concentration

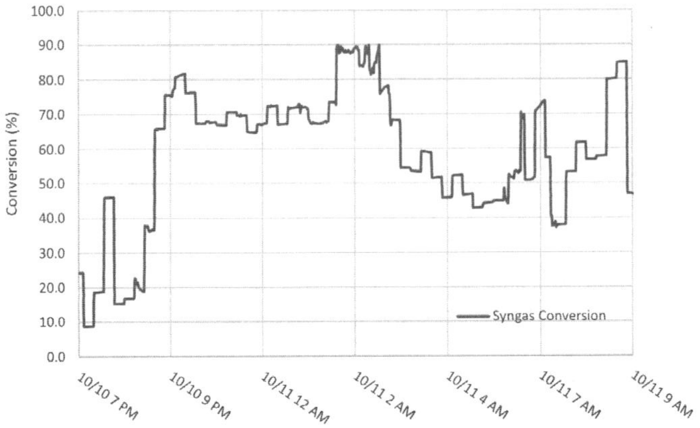

2017 AIChE Annual Meeting

November 1st, 2017

Center for Applied Energy Research

caer.uky.edu

Lessons Learned

- Finding pilot scale process equipment within budget and scope often require exhaustive and creative search
- The Project Benefited Greatly from Strong Collaborations in China in the area of Coal Gasification while also posing some unique challenges
- Slurry concentration has a dramatic affect on gasifier operation and gas composition
  - Challenge to maintain constant concentration (density)
  - Blend based on density and determining acceptable density range limits
  - Additive for settling and pumping
- Refractory on this small scale has caused some integrity issues with thermocouples
  - Thermal expansion/contraction
  - Wrong initial design
  - New thermocouples to meet our gasifier needs
  - Development/utilization of temperature probes without physical insertion (IR, Pyrometer, etc)

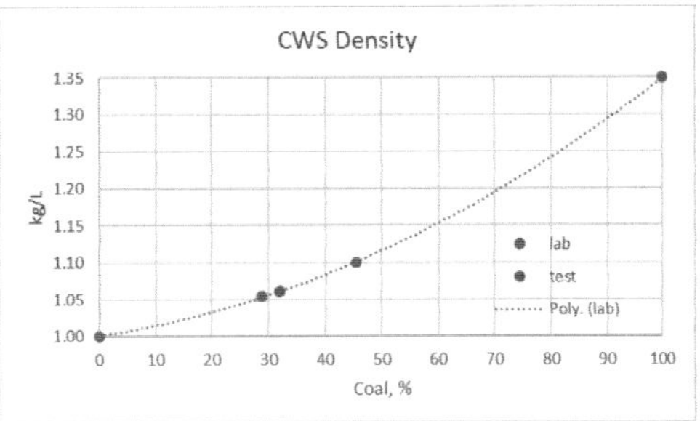

Gasifier Thermocouple Well

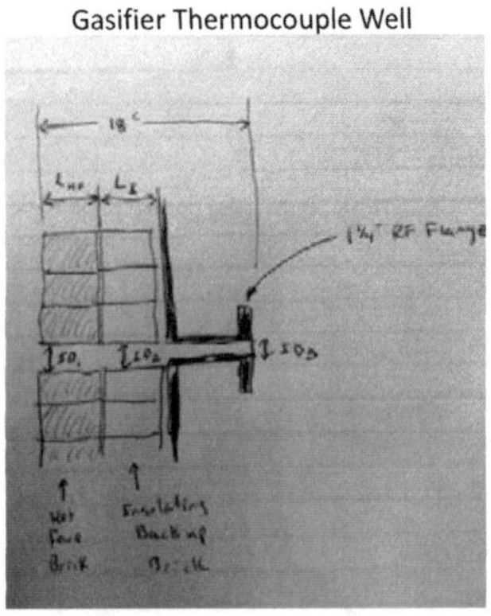

2017 AIChE Annual Meeting

November 1st, 2017

Summary

- Modular nature of the facility provides wide range of capabilities and flexibility for future use
- Whole facility ready for operation
- Facility is available for testing

Acknowledgements

- DOE-NETL: Arun Bose, Jenny Tennant and Jason Hissam
- UK: Rodney Andrews, Don Challman, Kunlei Liu, Burt Davis, Dennis Sparks, Steve Summers, Tate VanHoose, Len Goodpaster, Otto Hoffman, Marshall Marcum, Brad Irvin, Jimin Zeng and Jonathan Pelgen
- Zeton: Ravin Dave, Danielle De Sousa, Erik Gaspar
- Stovlbaek Consulting: Peter Pedersen
- ECUST: Qinghua Guo, Yan Gong, Jianliang Xu
- CEEDI: Wang Qiuming and Tao Li

507270 Production of Fuels from Coal-Biomass Feedstock using Fischer-Tropsch

Abstract:

The Fischer-Tropsch synthesis is actually composed of four sub-processes: gasification, gas clean-up, Fischer-Tropsch synthesis and product upgrading. The impact of feedstock (coal, natural gas, biomass) and mixtures of these feedstocks upon each sub-process will be discussed. The major impact of the feedstock is on the first two sub-processes: gasification and gas clean-up.

499845 Coal Power Plants with Enhanced Profitability and No Carbon Dioxide Emissions

Abstract:

Coal Power Plants with Enhanced Profitability and No Carbon Dioxide Emissions

A zero carbon dioxide emission process is developed, in which electrical power and formic acid are coproduced from coal. The overall reaction for the proposed process uses coal, oxygen, and water as reactants to generate formic acid as the only product. A thermodynamic analysis is carried out, so that feasible operating limits for the proposed process are identified. Since the process cannot be realized within one single step, a reaction cluster is proposed which consists of feasible chemical reactions. These are: gasification of coal to syngas; CO_2 hydrogenation to formic acid; hydrogen combustion; and CO combustion. The principal technologies employed in this process network include combustion, gasification of coal to syngas, gas compression, supercritical fluid pumping, hydrogen separation using pressure swing adsorption, and formic acid production from CO_2 and H_2. The software UniSim [TM, Honeywell Inc] is employed to simulate the flowsheet. Two Gibbs reactors act as the gasification reactor and a combustor, while a conversion reactor carries out the formic acid reaction. Coal, water, and part of oxygen are fed into the gasification reactor; off-gas from the hydrogen PSA and oxygen are fed into the combustor; pure hydrogen from hydrogen PSA and CO_2 which is compressed to supercritical phase are fed into the formic acid reactor. By adjusting the oxygen delivery ratio, the coal gasification is designed to be energetically self-sufficient, so that the whole process produces the largest amount of power. Heat exchangers, phase separators, compressors, pumps, etc. are also employed to realize the process. Temperature and heat load information for each process stream is obtained from the software simulation and used in carrying out a heat and power integration study. Through the use of heat engine and heat pump subnetworks, a feasible operating envelope for the considered process is identified. A preliminary technicoeconomic analysis is carried out, which establishes the superiority of the proposed process to current coal power plant technology.

503477 Engineering Materials and Processes for Flexible Electronics

Abstract:

Printed electronics is a new technology envisioning the fabrication of electronic devices using solution processable materials and printing methodologies. In this presentation I will describe the materials synthesis, development, and process engineering enabling the fabrication of unconventional electronic and opto-electronic devices, such as transistors, circuits and solar modules, all on flexible foils using several new materials. Materials development include new synthetic, green routes to semiconducting polymers. Examples of unconventional electronic materials include organic small molecular and polymeric semiconductors, metal chelates and complexes, and hybrid organic-inorganic metal oxides. Thus, we will show the fabrication of polymer based complementary circuits with excellent performance (field-effect mobilities ~ 1-6 cm^2/Vs), flexible amorphous oxides TFTs (field-effect mobilities ~ 5-40 cm^2/Vs) as well as polymeric donor-fullerene solar cells with efficiencies approaching 11% in inverted architectures.

503526 Spectroelectrochemistry of Conjugated Radical Polymers

Abstract:

Conjugated radical polymers are an interesting marriage of a conjugated backbone with radical pendant groups. This brings new opportunities for organic electronics, energy storage, and thermoelectrics. The challenge is to understand how the singly occupied molecular orbital (SOMO) of the radical interacts with the properties of the conjugated backbone, as this can strongly affect the resultant conductivity, mobility and redox activity. Here, a series of polythiophenes with pendant TEMPO (2,2,6,6-Tetramethylpiperidin-1-yl)oxyl) groups of varying alkyl spacer length are compared in the context of electrochemical energy storage. The conjugated radical polymers are examined as cathodes for nonaqueous batteries. Energy storage could potentially occur at both the radical group and the conjugated backbone. However, the overall capacity of the conjugated radical polymers is lower than expected, as is the conductivity. The explanation for this reveals an important design criteria for conjugated radical polymers in that the redox potential (or energy level) of the radical group and the conjugated backbone should be finely tuned so as to control electron exchange between the two moieties. This is further evidenced by in situ spectroelectrochemistry, in which the oxidation state of the polythiophene backbone is observed to change with time as electrons are transferred from one moiety to the other.

503527 Conjugated Copolymers Based on Purines: Linking Monomer Design and Macromolecular Properties

Abstract:

Purines seem to be compelling building blocks for conjugated polymers: They are susceptible to a wide range of site-specific chemical transformations that affect energy levels and emission wavelengths, and they possess the ability to self-assembly through non-bonded interactions. Motivated by these ideas, we describe the preparation of â€œpoly(purine)â€ copolymers in which purine-derived monomers are introduced directly into a polymer backbone via Stille cross-coupling polymerization with benzodithophene (BDT), as well as purine-based donor-acceptor small molecules. Variations in the nature of the BDT comonomer and substitution pattern of purinyl comonomer lead to changes in thermal and charge-transfer characteristics. Specifically, UV-Vis and fluorescence spectroscopies show that poly(purine)s exhibit intramolecular charge-transfer characteristics and display distinct changes in emission wavelengths based on purinyl comonomer design. These efforts are extended toward fully conjugated poly(purine)s through the development of small molecule donor/acceptor chromophores in which purines behave either as electron-donating or electron-accepting unit, which tunes the charge transfer behavior, excited state lifetime, and quantum yield of the building block. These demonstrations of main chain poly(purine)s and pi-conjugated, purine-based chromophores highlight the potential of purines as useful motifs in conjugated polymers for optoelectronic applications.

503541 Pinch-Off Dynamics and Printability of Semi-Flexible and Rigid Rod Polymer Solutions

Abstract:

Understanding and controlling drop formation and liquid transfer is crucial for designing printable photovoltaic and electronic devices using semi-flexible and rigid rod polymer solutions. Applications involving jetting, coating, spraying, atomization and various printing techniques (ink-jet, gravure printing, roll-to-roll) all involve formation of liquid necks that undergo capillary-driven thinning and pinch-off. Stream-wise velocity-gradients associated with extensional flows arise in thinning liquid necks, and as the response of polymer solutions to extensional flows influences both their processability, as well as final orientation and morphology. Characterizing and controlling the extensional viscosity that quantifies the resistance to stream-wise gradients and extensional relaxation time of polymer solutions that determines the pinch-off time are of critical importance for selecting and optimizing semi-flexible and rigid rod polymer solutions used in functional inks. However, due to the low viscosity and low elasticity of most printable inks, the quantitative analysis of pinch-off dynamics and extensional rheology response is beyond the measurement rage of commercially extensional rheometer called Capillary Break-up Extensional Rheometer (CaBER). In this contribution, we employ the recently developed dripping-onto-substrate (DoS) rheometry protocols for visualizing and analyzing capillary-driven thinning and pinch-off dynamics of a columnar neck formed between a nozzle and a sessile drop for typical functional inks including aqueous PSS-PEDOT solutions, aqueous NaPSS solutions and polythiophene solutions in organic solvents. We elucidate the influence of flexibility and extensibility on pinch-off dynamics and processability, emphasizing how printability can be quite distinct even if different polymer solutions exhibit a dynamically similar response in shear rheology characterization.

501384 Surface-Directed Multi-Scale Assembly of Conjugated Polymers

Abstract:

Organic electronic and photoelectronic materials that are light-weight, flexible and can be manufactured using energy-efficient and high-throughput methods. The solution printability at near ambient conditions enables deposition on flexible polymer substrates to create wearable, stretchable, imperceptible electronic devices for use in applications unimagined before. On the other hand, key challenges remain: how does molecular assembly proceed during solution printing and how to control the resulting thin film morphology? The significance of this challenge lies in the fact that charge transport in printed thin films is highly sensitive to their morphological parameters from molecular, mesoscopic to device scale. Addressing this challenge can open up new avenues for attaining high electronic performances, facilitating the much needed structure-property relationship studies in polymer-based electronic devices.

In this talk, I will present a new strategy we recently developed for controlling multi-scale assembly of conjugated polymers that are directly compatible with solution printing. Central to our method is the design of surfaces for overcoming the barrier to polymer nucleation, thereby directing the nucleation-triggered multiscale assembly process during printing. Using this method, we achieved high degree of global and local alignment over large area. In certain cases, we even observed small-molecule-like morphology for high molecular weight conjugated polymers, which has been rarely observed before. With high degree of control over thin film alignment and molecular packing, we correlate these morphological characteristics with anisotropic charge transport properties towards establishing structure-property relationships. By systematically tuning the degree of alignment, we show that the charge transport anisotropy can be switched to favor either transport along the polymer backbone, or along the pi-pi stacking direction.

503523 Interpenetrating Networks for Flexible Bulk Heterojunction Opvs

Abstract:

Bulk heterojunction organic photoltaic (OPV) devices can be fabricated fully though solution processing on a variety of substrates, but the active layer typically exhibits very poor flexibility and robustness in response to strains and deformation. Here, we present a general approach to fabricating flexible bulk heterojunction OPVs through the incorporation of an interpenetrating polymer network. We show that this provides P3HT/PCBM bulk heterojunction OPVs that are fully flexible and able to sustain stresses in excess of 50 %. Small molecular OPV blends are more robust, and we flexible OPV devices with power conversion efficiencies exceeding 5 % power conversion efficiencies.

500100 Development of an Anisotropic Coarse-Grained Conjugated Polymer Model for Optoelectronic Applications

Abstract:

Conjugated polymers have attracted attention for their uses in a variety of optoelectronic applications; their electronic properties can be altered to meet a variety of needs by fine-tuning both molecular structure and processing conditions. Developing models for simulating conjugated polymers that are fast yet contain the fine structural details necessary for examining the influence of morphology on electronic structure are crucial for understanding how to optimize conjugated polymer processing conditions. Specifically, conventional coarse-graining techniques fail to address the anisotropic interactions of the conjugated polymer backbone fundamental to both polymer pi-stacking and optoelectronic transport. In order to address this problem, we have developed a coarse-grained conjugated polymer model wherein monomers are represented as disks that interact through an anisotropic Gay-Berne potential and the backbone is modeled as a twistable wormlike chain. These two features allow this coarse-grained model to capture the anisotropic pi-stacking and semi-flexibility properties that are crucial to understanding these polymers, while substantially simplifying computations. We have parameterized the coarse-grained model from an atomistic model of a thiophene-containing conjugated polymer through free energy surface calculations and persistence length fitting. The anisotropic structural detail, including dihedral degrees of freedom, maintained by our coarse-grained model allows for the straightforward computation of x-ray scattering profiles and other relevant correlation functions, without the need for atomistic backmapping. Moreover, the form of our coarse-grained model allows for a simple mapping to an electronic tight-binding hamiltonian wherein the electronic structure of the valence bands may be evaluated. We perform structural and optoelectronic benchmarking of our coarse-grained model and compare these results to atomistic simulations and experimental data.

500244 Automated Quantitative Analysis of Oriented Morphologies and Inter-Grain Connectivity in Conjugated Polymers

Abstract:

A wide range of conjugated polymers form thin films with oriented morphologies, typically characterized with an imaging technique such as AFM. To provide quantitative metrics to describe and compare these morphologies, we developed an automated image analysis software package called GTFiber. GTFiber was originally designed for the high-throughput extraction and measurement of fibers from images, and has enabled the determination of quantitative process-structure-property relationships for poly(3-hexylthiophene)-based field effect transistors. These included both the effect of the spin- and blade-coating flow fields on fiber alignment during thin film deposition as well as the correlation between fiber alignment and field effect mobility. Data visualizations produced using GTFiber revealed striking nanofiber bundling behavior during solution processing. More recently, we have demonstrated GTFiber's generalizability to fibrillar systems in other materials, as well as the oriented morphologies observed in thin films of PNDI or strain-stretched thin films of conjugated polymer blends.

The extraction and vectorization of fibers from images also enabled us to simulate tie chain behavior on a length scale unattainable by molecular dynamics. By performing a Monte Carlo simulation of P3HT chain packing overlaid on the real, extracted fiber backbones, we were able to define a quantitative measure of grain interconnectivity: the tie-chain length density. Tie chain length density quantifies the expected number of tie chain connections to a given grain per unit lenth, and is sensitive to fiber packing, orientational order, and the assumed molecular weight distribution. While experimental observation of tie chains remains difficult, our simulation provides an upper bound on grain interconnectivity, while laying the quantitative groundwork necessary to describe this important feature of conjugated polymer thin films.

494252 Spray Coating Chitin & Cellulose Nanomaterials for Enhancement of Barrier Properties

Abstract:

Chitin is an abundantly occurring biopolymer like cellulose. The annual natural chitin production is estimated to be around 10^{10} - 10^{12} tons per year (Barikani et al., 2014) and is about the same order of magnitude as cellulose, whose annual bio-production by photosynthesis is estimated to be of the order of 10^{11}-10^{12} tons per year (Klemm et al.,2004). Similar to cellulose, chitin can also occur as chitin nanofibers (ChNFs) and when it is subjected to high shear treatment, nanofibers of 10-20 nm diameter and several microns in length are obtained (Wu et al., 2014; Ifuku & Saimoto, 2012).Â Wu et al. (2014) were able to produce chitin nanofibers in suspension from purified crab Î±-chitin, and when these suspensions were cast into films the resultant films had O_2 & CO_2 gas permeabilities of 0.006 & 0.018 barrer respectively. These good barrier properties as well as chitinâ€™s natural bacterial resistance make chitin an excellent candidate for sustainable barrier packaging applications. But packaging materials are generally multilayered, and a good method for incorporating these chitin films in continuous multilayer packaging processing needs to be developed. Spray coating is one of the most versatile methods of applying liquid onto surfaces. Due to its frequent use in painting, spray coating processes are widely encountered in everyday life (McKeen, 2016). One of the main features of a spray coating process is the contactless delivery of the material to be coated onto the surface. This implies that the actual delivery process is not influenced by the surface topography and thus uneven, or delicate surfaces can be coated. The delivered load of coated material is independent of the surface and the lack of a metering gap leads to decreased sensitivity to defects and especially helps in reduction of web breaks in roll to roll process (Czerwonatis, 2008). Beneventi, et al. (2014) were able to coat slurries of micro fibrillated cellulose (MFC) onto wet highly porous papers allowing complete retention of MFC on the surface of the substrate and improved barrier properties of the final product. In the present study we demonstrate the successful spray coating of chitin nanofiber & cellulose nanocrystal (CNC) suspensions onto polylactic acid films. The spray coating of these materials was found to lead to a reduction in the oxygen barrier properties of the final composite film by as much as 75 % of its original value depending on the loading and the material used (CNC, Chitin or both). Unlike the study by Beneventi, et al. the substrate in this case is non porous and the suspensions do not wet the surface. The resulting de-wetting problems were successfully solved by heating the substrate surface. This process could open up a path for the use of more sustainable polymers with otherwise excellent properties but are lacking in barrier properties. Addition of a ChNF or CNC layer could also lead to a reduction in the amount of the more expensive polymer in the multilayer (with the sole function of providing oxygen barrier) all the while increasing the sustainability.

500752 Development of Low-Concentration Alkaline Treatment Method to Produce Thermally Stable Cellulose Nanofibrils (CNFs)

Abstract:

The thermal stability of cellulose nanofibrils (CNFs) can be improved by converting cellulose crystal structure from cellulose I to cellulose II using alkaline pretreatment method. The conventional pretreatment method requires 17.5% or higher concentrations of sodium hydroxide and large volume of water for neutralization to produce thermally stable, but interdigitated or aggregated CNFs. The main objective of this study was to develop a low concentration alkaline pretreatment method to produce thermally stable and highly dispersible CNFs. Cellulose nanofibrils with improved thermal stability were successfully prepared by treating cellulose powder with 2wt.% NaOH solution at low temperature (below 0), followed by mechanical disintegration through a French cell press with only ten passes. The thermally stable CNFs had an average diameter of 90 nm with the mean onset and maximum thermal decomposition temperature of 305 and 343 respectively, compared with those of 283 and 310 respectively, for the CNFs prepared without the NaOH treatment. The thermally stable CNFs can create new opportunities for the development of CNFs based packaging composites and high-end electronic applications.

491943 Tailored and Integrated Production of Carboxylated and Lignin Containing Cellulose Nanocrystals and Nanofibrils for Composite Applications

Abstract:

Here we demonstrate di-carboxylic acid hydrolysis for the integrated production of lignin containing cellulose nanocrystals (LCNC) and nanofibrils (LCNF) using two unbleached mixed hardwood chemical pulps of lignin contents of 3.9 and 17.2%. Acid hydrolysis experiments used maleic acid solution of 60 wt% concentration at 120°C for 120 min under ambient pressure. Yields of LCNC were low of less than 6% under this set of condition. The higher lignin content samples produced LCNC with greater height (diameter) of 25 nm but similar length of approximately 230 nm to that from the lower lignin content fibers (height of 20 nm). Interestingly, the higher lignin content sample resulted in LCNF with smaller height (diameter) of 7 nm but longer length of > 1 μm, or greater aspect ratio than the LCNF from the lower lignin fibers of height 10 nm and length < 1 μm. Lignin protected cellulose from esterification which resulted in LCNC and LCNF that was less carboxylated compared to those lignin-free CNC and CNF and therefore had lower charges. However, lignin is more hydrophobic and thermally stable than carbohydrates therefore LCNC and LCNF are favorable for composite applications.

504076 Nano Carbon Structures from Cellulosic Biomass for Use As Functional Materials Â

Abstract:

We report herein the production of hierarchical nanocarbon materials from cellulose fibers. The carbonized cellulosic fibers with branching out CNTs (CC-CNT) offer interesting surface and electrochemical properties that are of interests for chemical processing and biosensing applications. In our tests for chemical adsorption tests, CC-CNT offered highly selective adsorption of hydrocarbons from gas mixtures, especially chemicals with C=C bond and aromatic structures. More interestingly, the chemicals absorbed and accumulated can be quickly released from the surface of the materials upon application of modest electrical current. Such an electrically switchable adsorption behavior may promise applications in a broad areas including air quality control, chemical sensing and fuel storage.

491886 Reforming of Hydrocarbons for Hydrogen Generation - Hyres

Abstract:

Steam methane reforming (SMR) is a commercial technology that is used at large scales to generate syngas and hydrogen, which can be used as fuel directly, or used as a chemical feedstock in petroleum refining and chemicals production. TDA Research is developing a reforming process capable of converting nearly any hydrocarbon feedstock into H_2. Our process has the potential to use as little as 1/4 of the energy required per pound of H_2 produced (energy required to operate the process) compared to SMR. We call our process *HyRes (Hydrogen from Residuum),* and we can steam-reform anything from shale derived natural gas to bottom-of-the-barrel atmospheric and vacuum residuum at low steam to carbon ratios. *HyRes* uses circulating fluidized bed reactors, one for steam reforming and one for catalyst regeneration. This reactor configuration permits us to maintain catalyst activity by continuously removing carbon deposited on the catalyst during the reforming process.

In addition to using less energy, the use of circulating fluidized beds for the reformer and regenerator means that the process no longer requires the use of the extremely expensive high alloy tubes such as those used in conventional fixed bed steam reforming. Also, because the catalyst is regenerated before enough carbon can build up to damage the catalyst, there is no irreversible catalyst deactivation due to coking or sulfur deposition. We discuss results where we have demonstrated our fluidized bed reforming process at the bench scale with several catalysts and hydrocarbon feedstocks.

497979 Syngas Production on Thermally Conductive SiC Catalyst

Abstract:

A new catalyst support technology has been developed to increase the thermal conductivity of a steam reforming catalyst system. The technology (*HeatPath*^{TM}) is designed to improve the heat transfer of the catalysts during the production of hydrogen from hydrocarbons. By allowing heat to move easily into the reaction zone, the new catalyst system reduces the amount of heat that must be generated, significantly lowering the cost of hydrogen produced.

The core technical achievement of this study is the feasibility validation of the high thermal conductivity supports with protective coating, and a washcoated catalyst layer on an application-suitable geometry. Packed-bed silicon carbide (SiC), along with structured metal foams and metal monoliths were all evaluated in the early stage to confirm the suitability of the coating processes and performance as steam reforming catalysts. Both metal and SiC supports showed promise for steam reforming reactions, though SiC has significant advantages in cost, weight and mechanical strength. Nexceris has optimized the integration of washcoated catalysts and protective layers on high thermal conductivity SiC supports. Testing has been performed to assess long term performance stability and to confirm agreement between the experimental and modeling results.

Through computational simulation of steam reformer operation, Nexceris determined that the best path for and reducing cost of syngas from natural gas is to reduce heat loss in the reformer. The identified approach was to establish coating technology that will enable the use of high thermal conductivity catalyst supports. Combining a protective coating technology and catalyst formulations designed for the reforming of natural gas, the *HeatPath*^{TM} platform offers three times better heat transfer than conventional catalyst support materials, which will reduce energy consumption during syngas production by more than 15 percent.

Results of our modeling and experiment work suggest that these newly developed supported catalysts will provide a leap forward in the design of hydrogen production systems.

497796 Thermodynamic Simulations, Experiments and Techno-Economic Analysis of Iron-Based Chemical Looping Systems for H_2 Production

Abstract:

The iron-based moving bed based chemical looping technology developed at the Ohio State University (OSU) is a promising candidate to yield cost benefits in natural gas utilization for a carbon-constrained scenario. OSU has developed three commercially relevant, distinct approaches for high efficiency H2 production from natural gas[1,2,3]. The first approach focuses on a three-reactor scheme that produces CO2 and H2 in separate reactors, providing process intensification benefits[1]. The second approach utilizes a membrane based separation scheme to surpass the thermodynamic limits associated with the H2 production system in the first approach[2]. The third approach retrofits the methane to syngas based co-current reaction system in a steam methane reformer and considers co-feed of steam and CO2 for high efficiency H2 production[3]. This presentation will initially examine the thermodynamic limits using Gibbs Free Energy minimization the three approaches. The thermodynamic performance will be further quantified by a cold-gas efficiency and an energy efficiency comparison between the three approaches and a baseline Steam Methane Reforming case. Relevant experiments that include high-pressure kinetic data and sub-pilot scale experimental demonstration will be used as an input for a base techno-economic analysis model for the three approaches. Sensitivity studies for each approach that identify the critical parameters to be de-risked for being competitive or better than steam methane reforming plants will be presented.

References

1. Kathe, M. V., Empfield, A., Na, J., Blair, E., & Fan, L. S. (2016). Hydrogen production from natural gas using an iron-based chemical looping technology: thermodynamic simulations and process system analysis. *Applied Energy*, *165*, 183-201.
2. Nadgouda, S. G., Kathe, M. V., & Fan, L. S. (2017). Cold gas efficiency enhancement in a chemical looping combustion system using staged H 2 separation approach. *International Journal of Hydrogen Energy*,Â *42*(8), 4751-4763.
3. Kathe, M., Empfield, A., Sandvik, P., Fryer, C., Zhang, Y., Blair, E., & Fan, L. S. (2017). Utilization of CO 2 as a partial substitute for methane feedstock in chemical looping methaneâ€"steam redox processes for syngas production.Â *Energy & Environmental Science*.

496174 Development of Thermo Neutral Reforming Catalyst for on-Board Hydrogen Production

Abstract:

This study investigated the development of a reforming catalyst for on-board generation of hydrogen for transportation applications. The development of a new reforming catalyst was aimed for high throughput system for syngas production for on-board transportation applications such as in auxiliary power units. A new reforming catalyst using a monolith support was prepared for liquid hydrocarbons based fuel processor. For mobile applications, liquid fossil fuels like diesel are suitable due to their high energy density, their easy storage and their already existing distribution infrastructure. However, diesel reforming is known as a major challenge due to its high carbon content, aromatics, and sulfur, which play a major role in catalyst deactivation. In this study, a new fuel delivery technique (atomization) was investigated for diesel reforming. An ultrasonic injector (instead of using conventional pre-heating) was used to mix all feedstock (fuel, air and steam) before it reaches the catalyst bed. The new method has improved the thermo neutral reforming (TNR) catalyst performance and stability with diesel. Monolith-based TNR catalysts of various sizes were prepared and their performance was evaluate in a fixed bed reactor with heavy naphtha and diesel as feedstocks. These catalysts were prepared by successive impregnation of active metals on a monolith with a wash coat of alumina on monolith. Compared to any other catalyst structure such as pellets or extrudates, a significantly lower pressure drop is observed when monolithic catalysts are used. The newly developed TNR catalyst showed stable performance with heavy naphtha as well as diesel for hydrogen production.

500939 1 kW_e Diesel Autothermal Reformer Integrated with Catalytic Burner for Exhaust Gas Oxidation

Abstract:

Heavy-duty trucks and recreational vehicles have to idle their engines when extended duration of using energy is required. However, idling causes the environmental issues and performance degradation of vehicles. Auxiliary Power Unit (APU) is a device on a vehicle that provides deficient energy for functions other than propulsion using the fuel from the vehicle. Solid Oxide Fuel Cell (SOFC) based auxiliary power units are expected to be about 15-20% more efficient compared to commercially available diesel based APUs resulting in significantly lower CO_2 footprint and lower emissions of criteria pollutants with low noise. The fuel reforming process, which is integrated with a SOFC for APU application, also plays an important role in an APU system. The synthesis gas consisting mainly of hydrogen, carbon monoxide, water and carbon dioxide generated as a result of hydrocarbon (i.e., diesel) reforming is sent to an SOFC stack for electrochemical oxidation operating at temperatures around 800°C. Moreover, the recent studies have reported a possibility to use SOFC as a CO_2 conversion system. Diesel autothermal reforming (ATR) is used to produce hydrogen-rich gas for SOFC application. 1 kW_e diesel autothermal reforming system is developed and tested for SOFC APU application with consideration of heavy-duty truck integration.

A catalytic burner is integrated into 1 kW_e diesel autothermal reformer to oxidize exhaust gases from the reformer. During the reformer operation including startup process, produced hydrogen and other combustible gases such as carbon monoxide must be fully oxidized before exhaust. Also, heat produced from the burner during the oxidation is utilized into the system to heat up the cathode air supply. In this study, the air supply rate of the catalytic burner is controlled to ensure full oxidation of the exhaust gases. With the burner temperature limitation of 800°C, hydrogen and carbon monoxide are fully oxidized with the catalytic burner. Generated heat is also utilized to increase temperature of cathode input air. As a result, stable startup and operation profile with integration of the catalytic burner can be obtained.

497030 CFD Study of a Pilot-Scale Multi-Tube Palladium Membrane Module for Hydrogen Separation

Abstract:

CFD Analysis of Pilot-Scale Multitube Pd Membrane Module for Hydrogen Purification from Coal-Derived Syngas

Rui Ma[1], Bernardo Castro-Dominguez[2], Ivan P. Mardilovich[1], Nikolaos K. Kazantzis[1], Anthony G. Dixon[1], Yi Hua Ma[1]

[1.] Center for Inorganic Membrane Studies, Department of Chemical Engineering, Worcester Polytechnic Institute, 100 Institute Road, Worcester, MA 01609, USA.

[2.] University of Limerick, Sreelane, Castletroy, Co. Limerick, Ireland

Syngas production and hydrogen separation technologies are very mature, and also extremely important for the energy and chemical industries. Enhancing or intensifying these mature technologies is extremely challenging, but on the other hand would have tremendous impact on performance and economics of many processes. At this moment, most research on Pd membrane technology has been conducted at lab scale, and membrane module scale up and commercial unit design are still challenges that need to be addressed in order to commercialize this promising technology. Compared with the one shell, one membrane tube module, many questions need to be answered for a pilot scale, multitube membrane module, such as the study of tube to tube variation in hydrogen permeation due to the position in the module, the exploration of the optimal shell side operating conditions and the evaluation of mass transfer limitations such as concentration polarization. For this purpose, a three dimensional CFD simulation model was developed and used to study the aforementioned aspects of a pilot scale membrane module.

The geometry of a seven-tube module with a total permeable area of 1050 cm^2 is shown in Figure 1a; coal-derived syngas was introduced to the shell side (retentate side) and pure hydrogen was collected from the tube side (permeate side). The asymmetrical fluid flow (Fig 2a) resulting from the outlet location leads to an uneven hydrogen permeation rate in each membrane tube (Fig 3). By adding a manifold for the membrane tubes in between the membrane tubes and the outlet (Fig1b), fluid flow tends to be more symmetrical (Fig 2b.), hydrogen permeance flux of each outer tube appears to be more even (Fig 3), thus the membranes are used more efficiently. In addition, due to the competition effect of the center tube with the six tubes around it, the center tube has the lowest hydrogen permeance flux. This phenomenon was addressed more in detail by the study of a further scaled-up module with 19 membrane tubes and a total permeable area of 2850 cm^2 (Fig 4). Concentration polarization was also observed from the model and indicated by the second Damkohler Number. Furthermore, hydrogen recovery appears to be favored by high pressure and low temperature while membrane usage benefits from moderate pressure and high temperature, the trade-off between these

two most important parameters of membrane modules is investigated and a thorough process optimization study was carried out using Box-Behnken method.

Figure1. Geometry of the 7 membrane tube module: (a) without manifold (b) with manifold

Figure2. Velocity field of the 7 membrane tubes module: (a) without manifold (b) with manifold

Figure 3. Hydrogen permeance flux of the outer tubes with different geometry

Figure 4. Cross sectional view of the 19 membrane tubes module

495931 Novel SAPO-34 Hollow Fiber Membrane for H_2 Separation Under Various Gas Atmospheres at High Temperature

Abstract:

Abstract

In the current research, a novel SAPO-34 hollow fibre membrane for hydrogen separation was fabricated using seeded secondary growth method. As compared to the conventional tubular membranes, hollow fibre membranes have a great advantage of higher surface to volume ratio, higher gas permeation due to reduced wall thickness etc. To obtain high quality SAPO-34 hollow fibre membranes, zeolite seeding and growing conditions were optimized to achieve a thin and defect-free zeolite layer for high gas permeability and selectivity. In addition, it was found out that membranes calcined under flowing oxygen had significantly higher gas permeability than those calcined under stagnant air. In this research, the membranes were subjected to permeation test of binary gas mixtures such as H_2/CH_4, H_2/CO_2, and H_2/N_2 at both room temperature and elevated temperatures ranging from 25 to 500 â„ƒ. Remarkable H_2 separation performance was achieved in separating a H_2/CH_4 mixture through the SAPO-34 hollow fiber membrane with a maximum H_2 selectivity of 45 at 100 â„ƒ and a H_2 permeation flux of $10.1*10^{-8}$ mol. $m^{-2}.s^{-1}.Pa^{-1}$. In addition, H_2 permeation flux increased to $28.9*10^{-8}$ mol. $m^{-2}.s^{-1}.Pa^{-1}$ with at a temperature of 500â„ƒ while the selectivity dropped to around 25. Due to the good H_2 permeation performance at high temperature, SAPO-34 hollow fiber membrane can be integrated with catalysts to form catalytic membrane reactors for various H_2 production reaction such as water gas shift (WGS) and dehydrogenation of hydrocarbons like propane. This novel catalytic membrane reactor can simultaneously remove hydrogen product, and increase the productivity of the reaction beyond the equilibrium yield.

492929 Thermochemical and Catalytic-Upgrading System Design to Convert Biomass to Liquid-Fuel Using a Superstructure-Based Approach

Abstract:

Nowadays, liquid fuels have become an essential commodity for day-to-day living, and this reliance is expected to increase in the future with increasing population, and raising standards of living. With dwindling petroleum resources and raising environmental concerns, biomass is considered to be a sustainable feedstock for liquid fuels. While multiple approaches to convert biomass to liquid fuels exist, we focus on the design and optimization of a biorefinery based on a multistage thermal decomposition [1,2] strategy.

The general multistage thermal decomposition process comprises of either one, or two, or three thermal decomposition stages for biomass pretreatment, each at different temperatures. While the streams leaving each of these stages, in practice, contain a wide variety of species, here, for the purpose of tractability, we represent all species using a fixed set of model-compounds, namely, acetic acid, acetol, furan, furfural, levoglucosan, toluene, guaiacol, char, carbon-di-oxide and water. The composition of these model-compounds in each stream leaving the thermal decomposition stages is dependent on the number of thermal decomposition stages used during pretreatment, which accordingly influences the sequence of downstream catalytic-upgrading chemistries to be used. The catalytic-upgrading steps are needed to increase the carbon-chain-length of the molecules, and reduce/eliminate oxygen-content so as to finally obtain fuel-like molecules. In this work, we consider the following kinds of chemistries for catalytic-upgrading as they are applicable to a variety of organic-molecules: acylation, aldol-condensation, alkylation, hydroxy-alkylation, ketonization, hydrogenation, oxidation and hydrodeoxygenation. With so many options for both thermal decomposition and catalytic-upgrading, it is non-trivial to identify cost-effective pathways to convert biomass to liquid fuel. To make the process-layout decisions and design the biorefinery in a systematic manner, we use a superstructure-based approach where we develop a mixed-integer nonlinear programming (MINLP) model.

We first develop a novel superstructure for a torrefaction-based biorefinery. The superstructure consists of reactors that are linked by a network of connecting streams associated with heat exchangers for temperature-conditioning. Further, we devise connectivity rules between the reactors. Some of these rules eliminate cyclical-paths and symmetrical solutions. Thereafter, we present strategies to model the individual elements of the superstructure.

If a reactor is present in the final solution, it is assigned one of many possible tasks. Each task in the thermal decomposition section is associated with a specific thermal decomposition temperature, while a task in the catalytic upgrading section is associated with a specific catalyst carrying out a specific kind of chemistry (listed earlier). For example, if acylation is the task assigned to a catalytic upgrading reactor, then, all molecules in the inlet stream to this reactor amenable for acylation react. So, multiple reactions can take place inside a reactor. To model the reactivities of the various molecules for a specific task, we use conversion data from literature and experiments. We also develop new methods to handle parallel reactions that any molecule can participate inside a reactor, alongside identifying the limitting reactant species of a reaction, which is unknown because stream flows are unknown. Furthermore, for each

reactor, we determine the heating/cooling requirement based on the extents of reactions that occur in them. A new approximate method is proposed for the calculation of heating and cooling duties for phase- and temperature-conditioning the streams that connect reactors. While the overall heating/cooling duties along with the biomass and hydrogen usage contribute to the overall operating costs, capital costs are governed by the reactors used and the tasks that are assigned to them.

Using this general framework, we identify cost-optimal pathways to covert biomass to liquid fuels based on multistage thermal decomposition, and study the sensitivity of the solutions to different parameters like biomass costs, hydrogen costs, and fuel-range carbon-yield.

References:

[1] M.J. Prins, K.J. Ptasinski, F.J.J.G. Janssen, More efficient biomass gasification via torrefaction, Energy. 31 (2006) 3458â€"3470.

[2] J.A. Herron, T.Vann, N.Duong, D.E.Resasco, S. Crossley, L.L. Lobban, C.T. Maravelias, A Systems-Level Roadmap for Biomass Thermal Fractionation and Catalytic Upgrading Strategies, Energy Technology. 5 (2017) 130â €"150.

493434 Kepler Workflow in a Cloud Infrastructure for Temperature Balancing in a Steam Methane Reformer Furnace Using a Computational Fluid Dynamics and a Data-Driven Optimization Approach

Abstract:

Steam methane reforming (SMR) is widely used in industrial-scale hydrogen production process because of its relatively low cost compared to other methods such as electrochemical process. Even though the industrial SMR process is relatively well optimized, it still has room for improvement in the energy efficiency where industry can save hundreds of thousands of dollars in operation cost. SMR is a net endothermic process which requires heat to be supplied by the furnace. This, in turn, drives the reaction of methane and super-heated steam at approximately 1100 K to produce hydrogen. Hot spots can cause critical failure, and dramatically reduce the lifespan of the tubes. Prolonged increase of temperature even by 20 K can decrease the lifespan of the tubes by half [1], [2]. Therefore, balancing the temperature gradients inside the furnace and across all tube walls is important.

This was the motivation for our previous work to develop a computational fluid dynamics (CFD) model [3], [4] and a balancing algorithm [5] for a top-fired furnace with 336 tubes and 96 burners. This procedure requires an expert knowledge in CFD and mathematical models to implement, but we want this to be used by non-experts such as plant operators. This is also the vision of the Department of Energy Smart Manufacturing Project where studies such as this one, needs to be implemented in a reusable fashion across the industry.

In this study, we are using ANSYS Fluent to create an accurate CFD model of the reformer and a reduced order model (ROM) of temperature distribution and burner rate distribution is used to balance the furnace. To create the ROM, we are using IPOPT and CPLEX software packages. To build a converged ROM, we composed a mini workflow within Ansys Fluent itself using the user defined function (UDF) feature of Fluent and integrated the code for ROM as a shared library. In this way Ansys Fluent will be called in a loop as many times as required until the results from ROM calculation and CFD calculation are within acceptable tolerance. A typical CFD calculation takes 1 to 3 days on 80 cores of distributed computing power on a high-performance computing cluster and a ROM calculation can be carried out within minutes on a single core. So, in a real-time furnace temperature balancing process, a ROM calculation is quicker and practical.

Since we are designing this procedure for the automatic operation by plant operators. We also created a scientific workflow using Kepler software [6] to integrate and run this model in a cloud computing environment. We tested this procedure on a computing platform where the input data from a plant is being received through a Historian [7]. We used this input to run CFD and ROM calculations. Outputs of the calculations containing updated optimized valve positions to adjust the fuel flow rate through the burners and predicted temperatures are sent back to the plant operator. In a typical scenario, a plant operator needs to just visit the user interface and start the workflow. The workflow will then automatically perform a series of file transfer and computation processes. The entire process appears as a black box to the operator as all steps inside the workflow are already composed by an expert in the field.

[1] Latham D. Masterâ€™s Thesis: Mathematical Modeling of an Industrial Steam Methane Reformer. Queenâ€™s University, 2008.

[2] Pantoleontos G, Kikkinides ES, Georgiadis MC. A heterogeneous dynamic model for the simulation and optimisation of the steam methane reforming reactor. International Journal of Hydrogen Energy. 2012;37:16346-16358.

[3] Lao L, Aguirre A, Tran A, Wu Z, Durand H, Christofides PD. CFD modeling and control of a steam methane reforming reactor. Chemical Engineering Science. 2016;148:78-92.

[4] Aguirre A, Tran A, Lao L, Durand H, Crose M, Christofides PD. CFD modeling of a pilot-scale steam methane reforming furnace. In: Kopanos G, Liu P, Georgiadis M (Eds.). Advances in Energy Systems Engineering. Springer, Switzerland, 2017, pp. 75-117.

[5] Tran A, Aguirre A, Crose M, Durand H, Christofides PD. Temperature balancing in steam methane reforming furnace via an integrated CFD/data-based optimization approach. Computers & Chemical Engineering. in press.

[6] Korambath P, Wang J, Kumar A, Hochstein L, Schott B, Graybill R, Baldea M, Davis J. Deploying Kepler workflows as services on a cloud infrastructure for smart manufacturing. Procedia Computer Science. 2014;29:2254-2259.

[7] Korambath P, Wang J, Kumar A, Davis J, Graybill R, Schott B, Baldea M. A smart manufacturing use case: Furnace temperature balancing in steam methane reforming process via Kepler workflows. Procedia Computer Science. 2016;80:680-689.

496021 A Multiparametric C.F.D. Analysis of Multiphase Annular Flows for Oil and Gas Drilling Applications

Abstract:

The increasing global energy demand has pushed the oil industry towards developing more intelligent and advanced methods of enhancing oil recovery even under more unfavourable technical and environmental conditions. The severity of many operational problems affecting the drilling and production of oil and gas wells is worsened by inaccessibility; hence remedial efforts must be implemented from afar [1]. One of these problems is ensuring efficient removal of formation rock cuttings with a suitable fluid, whose rheology is often complicated. Furthermore, pressure losses along the annular geometry involved, and decreased Rates of Penetration (ROP) due to accumulated drill cuttings downhole, constitute significant portions of the total energy to be supplied [2]. Thus, the application of sophisticated modelling techniques with credible elucidation of the phase distributions (solids, liquids and gas) [3] and prevalent flow regimes becomes essential, if adequate and economic design of a drilling program is desired [4].

The advent of Computational Fluid Dynamics (CFD) and the growth in the available computational power to support it have provided an unprecedented opportunity to simulate and understand complex real flows especially when experimental methods may be too demanding [3]. The application of CFD for the analysis, design and optimisation of drilling programs [5] has mostly focused on vertical and concentric annular geometries, with few studies focusing on the intricacies that evolve due to hole inclination and eccentric geometric configurations [6]. Another complication characterising the modelling process is the occurrence of multiple phases in the flow domain, which requires robust multiphase closure equations and appropriate fluid-particle tracking models to be incorporated with the principal flow equations [7]; thus, enhancing solvability and accurate prediction of the drilling variables of interest.

The choice of the multiphase flow tracking scheme in turn significantly depends on the governing particle driving force during flow (drag, lift or collision). Nonlinearity of these multiphase interactions yield a variety of flow phenomena which are modelled using two major approaches. These include; the Lagrangian tracking of computational particles coupled with the Eulerian flow description of the continuous phase and the Eulerian-Eulerian description in which the solid particles are represented as a random field in the Eulerian reference frame [5]. Accuracy of mathematical representation, consistency of accompanying closure models and numerical stability are the attributes that make these techniques very applicable. However, there are still opportunities for improvement in these models as far as the particle-particle interactions and limits on solid phase concentrations are concerned. The limitation of the Lagrangian-Eulerian approach in handling flow systems involving high solids concentration [2] (>12%) makes the Eulerian-Eulerian approach more suitable for this computational analysis.

The present study employs the tool of Computational Fluid Dynamics to simulate a two-phase Solid-Liquid (SL) flow in an annulus, by elucidating multiple state variable distributions (namely cuttings concentration, pressure drop profiles, axial fluid and solid velocities) as a function of several drilling parameters: hole eccentricity, inclination, ROP, drill-pipe rotation and fluid rheology. Anon-Newtonian (power law) fluid model

with well described flow parameters [8] has been implemented, considering a uniform cuttings size distribution (3 mm). A commercial CFD software suite (ANSYS FLUENT 17.1) has been used: the descriptive and predictive potential of the CFD software has been confirmed on account of the reasonable agreement with previously published experimental data (a relative error of less than 7% is achieved), as illustrated by the corresponding sensitivity plots. This multi-parametric CFD analysis study of multiphase cutting transport during drilling applications has confirmed that fluid velocity, hole inclination and annular eccentricity are the most influential factors governing the cuttings transport efficiency.

LITERATURE REFERENCES

1. Chin, W. C. (2001). *Computational Rheology for Pipeline and Annular Flow: Non-Newtonian flow Modeling for Drilling and Production, and Flow Assurance Methods in Subsea Pipeline Design,* Gulf Professional Publishing.
2. Pereira, F. A. R., Ataide, C. H., & Barrozo, M. A. S. (2010). A CFD Approach using a discrete phase model for annular flow analysis. *Latin American applied research*, **40**(1), 53-60.
3. Abdulkadir, M. (2011). *Experimental and Computational Fluid Dynamics (CFD) studies of gas-liquid flow in bends,* PhD Thesis, University of Nottingham, UK.
4. Gerogiorgis, D. (2013). Isothermal CFD modeling of annular multiphase flows during Underbalanced Drilling (UBD) in oil reservoirs, paper presented at the 2013 AIChE Annual Meeting, San Francisco, CA, USA.
5. Subramaniam, S. (2013). Lagrangian–Eulerian methods for multiphase flows. *Progress in Energy and Combustion Science*, **39**(2), 215-245.
6. Li, Y., & Kuru, E. (2003). Numerical modelling of cuttings transport with foam in horizontal wells. *Journal of Canadian Petroleum Technology*, **42**(10) (PETSOC 03-10-06).
7. Ofei, T. N., Irawan, S., & Pao, W. (2014). CFD method for predicting annular pressure losses and cuttings concentration in eccentric horizontal wells. *Journal of Petroleum Engineering*, **486423**, 1-16.
8. Rooki, R., Ardejani, F. D., Moradzadeh, A., & Norouzi, M. (2014). Simulation of cuttings transport with foam in deviated wellbores using computational fluid dynamics. *Journal of Petroleum Exploration and Production Technology*, **4**(3), 263-273.

498294 Development of a 3D Computational Fluid Dynamics Model for Microbial Fuel Cells

Abstract:

10D00 Modeling and Computation in Energy and Environment

Development of a 3D Computational Fluid Dynamics Model for Microbial Fuel Cells

Xianhua Li, Robert Ferrari, Jaclyn Guglielmi, Zuyi (Jacky) Huang

Department of Chemical Engineering, Villanova University, PA, 19085

Abstract:

Wastewater is an important new source of substantial energy. The energy potential contained in wastewater and its biosolids exceeds by 10 times the energy used to treat it [1]. However, conventional wastewater treatment currently consumes about 3% of all electrical power produced [2] and costs more than $25 billion annually in the United States [3]. Therefore, cost-effective methods are in high demand for converting the chemical energy contained in wastewater into bioenergy to address the energy crisis nowadays. Microbial fuel cell (MFC) is a bio-electrochemical system that is capable of producing electricity from wastewater under ambient conditions through the biodegradation of organic compounds by exoelectrogenic microorganisms [4]. Specific species of bacteria such as Geobacter and Shewanella spp. can transfer electrons directly to the anode by membrane-bound cytochromes through anaerobic respiration [5]. The generated electron flow can be used by an external electrical circuit as power supply. In the meanwhile, protons migrate to the cathode and react with oxygen to form water. While bench-scale MFCs have been studied for decades, for the practical implementation of MFCs in wastewater treatment plants, it is still challenging to scale up MFCs [6]. Regarding the complex multi-layer structure, fuel composition and electrochemical reactions in MFCs, a robust computational model will be helpful to direct the scaling up of MFCs.

This study aims to investigate the performance and scalability of MFCs with a three-dimensional computational fluid dynamics (CFD) model. In the CFD model, dynamic electricity generation data that was used for model validation was obtained from experiments conducted in different scales of single-chamber air-cathode MFCs with working volumes of up to one liter. Based on existing one-dimensional mathematical models that have been published [7-9], an integrated and advanced three-dimensional model with complex geometry specifications has been developed for the first time in CFD. User defined functions have been developed and compiled to CFD to accurately analyze process properties such as mass transfer rates, diffusion rates, electrochemical reactions, biofilm formation, bacterial endogenous metabolism, internal and external ohmic resistance. As a result, the 3D CFD model can quantify and visualize the MFC performance by showing electric potential, electrode surface potential, current density and species concentration distribution in any designated cross section across a MFC. From the model, the influence of current density, substrate concentration and conductivity on the long-term dynamics of power generation and coulombic efficiency has been revealed, thus optimal operational conditions have been identified. The developed three-dimensional MFC CFD models will be of great benefit for predicting and evaluating the scaling up of MFCs in the future.

References:

[1] W. E. R. Foundation, "Energy management exploratory team report executive summary," March 23, 2011.

[2] B. E. L. a. K. Rabaey, "Conversion of wastes into bioelectricity and chemicals by using microbial electrochemical technologies," *Science,* vol. 337, pp. 686-690, 2012.

[3] R. R. Hong Liu, and Bruce E. Logan, "Production of electricity during wastewater treatment using a single chamber microbial fuel cell," *Environ. Sci. Technol.,* vol. 38, pp. 2281-2285, 2004.

[4] "Exoelectrogenic bacteria that power microbial fuel cells," *Nature Reviews Microbiology,* vol. 7, pp. 375-381, 2009.

[5] Liang Shi, David J. Richardson, S. N. K. Zheming Wang, J. M. Z. Kevin M. Rosso, and J. K. Fredrickson, "The roles of outer membrane cytochromes ofÂ shewanellaÂ andÂ geobacterÂ in extracellular electron transfer," *Environmental Microbiology Reports,* vol. 1, pp. 220â€"227, 2009.

[6] B. E. Logan, "Scaling up microbial fuel cells and other bioelectrochemical systems," *Appl Microbiol Biotechnol,* vol. 85, pp. 1665-71, Feb 2010.

[7] R. P. Pinto, B. Srinivasan, M. F. Manuel, and B. Tartakovsky, "A two-population bio-electrochemical model of a microbial fuel cell," *Bioresour Technol,* vol. 101, pp. 5256-65, Jul 2010.

[8] C. Picioreanu, I. M. Head, K. P. Katuri, M. C. van Loosdrecht, and K. Scott, "A computational model for biofilm-based microbial fuel cells," *Water Res,* vol. 41, pp. 2921-40, Jul 2007.

[9] C. Picioreanu, K. P. Katuri, M. C. M. van Loosdrecht, I. M. Head, and K. Scott, "Modelling microbial fuel cells with suspended cells and added electron transfer mediator," *Journal of Applied Electrochemistry,* vol. 40, pp. 151-162, 2009.

500050 Optimal Resource Expansion and Placement in Smart Grid Communications Networks

Abstract:

With concerns about climate change, the adoption of new renewable standards and accelerated cost reductions are driving sharp growth in renewable energy technologies. In particular, the number of distributed solar photovoltaic (PV) installations is growing rapidly. However, as distributed solar power becomes an increasing fraction of total energy generation, the electric power grid must ensure continued reliability and cost-effectiveness. The development of the Smart Grid, as a natural evolution of the electric power grid, seeks to incorporate new technologies in order to support these distributed generation sources.

New grid devices, such as synchrophasors and Smart Meters, will be built to provide the existing Supervisory Control and Data Acquisition (SCADA) systems with vital power flow information. These SCADA systems allow operators to monitor and control transmission and maintain operational efficiency. But since network deployments can involve thousands of devices, manual reconfiguration and tuning of these devices are impractical. Therefore, automatic placement and optimization of network topology is critical to providing a high-performing, reliable, secure, and scalable communication network in support of the Smart Grid evolution.

In this work, we develop a procedure to optimally place communication hardware within a distributed PV network, while taking advantage of existing communications technologies to greatly reduce costs and maintaining quality of service for consumers. We propose a mixed-integer linear programming approach for optimal placement of communication nodes to ensure communication between Photo-Voltaic (PV) inverters and transmission substations in a multi-hop network via relay nodes, aggregation points, and various ad-hoc communication systems. We also consider uncertainty in link parameters, such as packet loss, latency, and bandwidth. We show results that are robust to changes in economic considerations and future technological improvements.

501771 Quantification of Parameter Space Regions Consistent with System Models and Associated Experimental Data

Abstract:

CO_2 adsorption on Hydrotalcite is an increasing field of study, as CO_2 is emitted in a variety of energy producing reactions and an increasing need for reduction of greenhouse emissions. With an ever-increasing demand for energy, accurate predictive models are imperative in order for lab to plant scale up. The Linear Driving Force (LDF) model is used to predict the concentration of adsorbed CO2 at equilibrium with parameters inputs from an isotherm. Additionally, we present a novel Two Step Adsorption mechanism that employs several parameters.

It is with this in mind that we also present a method for determining parameters for a model from experimental data. These models are typically comprised of ordinary differential equations that cannot be analytically integrated. In such cases, local non-linear regression techniques are employed in the solution. There are however shortcomings in local non-linear regression. The first is that it fails to guarantee the global optimality of the solution. Additionally, it also fails to ascertain whether or not the proposed problem parameterization is appropriate for the available data. In this work, we have addressed these issues by proposing a novel method for parameter identification. This method will be applied to the already established LDF, and compared with the feasibility of the Two Step Model. This method is guaranteed to identify the global optimum of the non-linear regression problem. In addition, given a predetermined level of accuracy, it also has the ability to define a range of suitable values for each parameter.

500661 Design and Intensification of Integrated Carbon Capture and Conversion to Chemicals

Abstract:

To address the rising CO_2 levels in the atmosphere, research efforts are directed towards making widespread deployment of carbon capture, utilization and sequestration (CCUS) cost-effective. CO_2 utilization can be implemented to produce as much as 2 Gt/yr. of fuels and 200 Mt/yr. of chemicals.[1] Utilization of CO_2 to deliver high-value products such as fuels and chemicals is an attractive and promising option. Existing standalone capture and utilization technologies perform energy-intensive CO_2 capture from dilute streams such as flue gas and then convert the captured CO_2 to syngas which is again energy-intensive. To intensify the capture and utilization steps to reduce energy inputs and hence costs, we design and optimize modular and multi-functional processes for direct utilization of CO_2 from flue gas to value-added fuels and chemicals.

We propose one such intensified process exploiting an integrated adsorption-purge-reaction system[2], which operates cyclically involving CO_2 adsorption from flue gas in the first step, followed by purging due to methane rich feed in the second step. The desorption of CO_2 from the column is achieved by a concentration driving force difference brought about by introduction of a second feed with a composition very different from that of flue gas. The N_2 separated from the column is vented out, while the CH_4/CO_2 purge mixture becomes the feed for dry reforming of CO_2 subsequent reactor section. A variety of methane rich feeds such as natural gas, fuel gas, unused methane, biogas etc. can be used by the process. The novel aspect of the process is the elimination of pressure and temperature swings employed in every cycle in conventional standalone CO_2 capture and integrating it with reactor feed pre-mixing which results in reduction in the overall cost of CO_2 capture and utilization. The process is integrated with existing CO_2 capture and utilization technologies.

The entire process is simulated using high-fidelity non-linear algebraic partial differential equation (NAPDE) model to identify the key decision variables, process metrics and the interactions between the adsorption and reaction sections. While it is important to operate the process such that it meets constraints pertaining to greenhouse gas emissions and product quality, it is only possible to obtain the values of these constraints after performing computationally intensive simulations. This poses a challenge in identifying the feasible and optimal operating conditions of the process. A data-driven constrained grey-box optimization framework is employed to obtain feasible operating conditions and to proceed towards the optimal objective desired. Applying the optimization framework, we are able to achieve a maximum of 99.67% net overall CO_2 utilization while accounting for auxiliary CO_2 emissions at costs ranging from \$110-\$130 per ton of syngas. About 14.6% of the total CO_2 input to the process is captured with low cost â€œdirectlyâ € from flue gas while still maintaining 91% overall CO_2 utilization. The rest of the CO_2 input is sourced from existing capture plants integrated with the developed process.

Using the above design, intensification and optimization framework, we also develop a novel cyclic process for the direct utilization of CO_2 to methanol. This process again uses a two-step process and exploits synergies between CO_2 capture, CH_4/CO_2 mixing, syngas production and syngas conversion to methanol. In this presentation, we will further elucidate the techno-economic feasibility of the methanol process.

References:

(1) Quadrelli, E. A.; Centi, G.; Duplan, J.-L.; Perathoner, S. Carbon Dioxide Recycling: Emerging Large-Scale Technologies with Industrial Potential. *ChemSusChem* **2011**, *4* (9), 1194â€"1215.

(2) Iyer, S. S.; Bajaj, I.; Balasubramanian, P.; Hasan, M. M. F. Modular Process Intensification of Carbon Capture and Conversion to Syngas. *Submitted.*

489213 Response Surface Models for SET POINT Determination of Smokeless Steam and Air Assisted Flares

Abstract:

Electronic code of Federal Regulations Â§63.670 requires flare operators to specify smokeless design capacity for flares and operate flares with no visible emissions when regulated vent gasÂ is sent to flares with flow rate less than the smokeless design capacity. Alternatively, 96.5% combustion efficiency (CE) or 98% destruction efficiency (DE) must be achieved using site specific operating limits to demonstrate the compliance in lieu of threshold values of minimum combustion zone net heating value (NHV_{cz}) as 270 BTU/scf during any 15â€"minute average period, maximum flare tip velocity (V_{tip}) less than 400 ft/s and net heating value dilution parameter (NHV_{dil}) as 22 BTU/ft^2. Site specific operating limits include threshold values for NHV_{cz}, V_{tip} and NHV_{dil} as well as vent gas flow rate, vent gas composition and steam/ air assist flow rate. This study aims at developing robust response surface models to express CE and opacity as a function of operating variables and to determine the set points for steam/air and make up fuel under smokeless flaring conditions. Multivariate response surface models were developed for steam and air assisted flares burning propylene, propane, natural gas, methane, ethylene and typical refinery vent gas mixtures. In this study, soot emissions, visible emission scale, DE, and CE test data from 1983 to 2014 (including 1983/1984/1985 EPA, 2010 TCEQ, 2009/2010 Marathon Detroit/Texas City, and 2014 Carleton University) were analyzed. The input variables include measurable variables like velocity, lower heating value, flare tip diameter, vent gas composition (Ï€-bond, carbon to hydrogen molar ratio and molecular weight), and operating variables (NHV_{cz}, NHV_{dil}, steam/air assists, and fuel). Minitab and Polymath were used to develop models and solve them to determine set points. Response surface models developed for CE and opacity showed high coefficient of determination (R^2) values 0.90 and 0.94 respectively for steam assisted and 0.91 and 0.93 respectively for air assisted flare test data.

502033 Kinetics of the thermal decomposition of sulfur compounds

Abstract:

Abstract

In this contribution the kinetics of the thermal decomposition of sulfur compounds will be discussed both experimentally and theoretically under typical steam cracking and pyrolysis conditions. The experimental investigation has been conducted in a new 85 cm long annular quartz flow reactor, having an outer and inner tube diameters of 2.1 and 1.5 cm respectively. Both pure sulfur compounds pyrolysis in nitrogen (0.04 mol% S in N2) as well as in the presence of hydrocarbons (0.04 mol% S and 13.7 mol% of HC in N2) were studied covering a wide temperature range of 600 to 850 °C, at a fixed pressure of 1.5 bara and residence time of ~0.5 s. The reaction effluent is sampled on-line at 300 °C. The unique on-line sampling system consists of valve-based sampling manifold and uniformly heated transfer lines connected to a comprehensive two-dimensional gas chromatograph (GC × GC) equipped with a flame ionization detector (FID) and a sulfur chemiluminescence detector (SCD). This dedicated set-up allows to detect and quantify both pure (with the FID) and sulfur-containing hydrocarbons (with the SCD) using heptane and 3-chlorothiophene as internal standards, respectively. The GC × GC-FID/-SCD settings are optimized for trace analysis of sulfur compounds in olefin-rich (ethylene- and propylene-) hydrocarbon matrices produced during steam cracking.

To understand the underlying decomposition mechanism of sulfur compounds in the presence of a complex hydrocarbon matrix formed by its decomposition, a group additive kinetic model was constructed via an in-house automated reaction mechanism generation software called Genesys [1]. It relies on ab-initio CBS-QB3 based kinetic coefficients for the reference/ parent reactions for each elementary reaction family, and for the others, contribution of groups to the single event Arrhenius parameters. No adjustment of the Arrhenius parameters to the experimental data is performed. The kinetic model is validated by comparison with the set of experimental data gathered in the annular quartz flow reactor by providing a satisfactory representation of the latter.

[1] Nick M. Vandewiele, Kevin M. Van Geem, Marie-Françoise Reyniers, Guy B. Marin, Genesys: Kinetic model construction using chemo-informatics, Chemical Engineering Journal, Volumes 207-208, 2012, Pages 526-538.

500300 A Cylindrical Formulation of the One Dimensional Turbulence (ODT) Model for Turbulent Jet Flames

Abstract:

The One Dimensional Turbulence (ODT) model has been successfully applied to a wide range of turbulent flows including homogeneous turbulence, jets, wakes, mixing layers, channel flows, Rayleigh-Taylor, buoyant wall flows, and others. ODT solves a flow by temporally-evolving one-dimensional transport equations for mass, momentum, and (optionally) reacting scalars. Advection is modeled through stochastic eddy events that are implemented as triplet maps. The size, location, and rate of these maps are drawn from an eddy rate distribution in a manner that this consistent with turbulent inertial range advection. The model is an extension to the Linear Eddy Model (LEM), and differs from LEM in that, in ODT, momentum components are solved in addition to scalar quantities, and these momentum components are used to define eddy rate distribution function. Hence, ODT is a dynamic turbulence model and can accurately capture velocity and mixing profiles in nonuniform turbulence. Because the model is one dimensional, it is limited to boundary-layer like flows, but such flows are important and widely studied, with a wealth of experimental data available. ODT is unique in that it is able to resolve a full spectrum of turbulent time and length scales (in one dimension). Only direct numerical simulation is able to resolve all scales of a turbulent flow, but this is done at a computational cost that is prohibitive for engineering flows at realistic Reynolds numbers. Large Eddy Simulation directly captures large eddies, but models small subgrid scale motions. Conversely, ODT resolves fine scale structures, allowing diffusion processes in the physical space coordinate, but models the large-scale advection. Often the fine scales are more difficult to model since nonlinear multi-component diffusion and reaction processes occur at dissipation scales. Large-scale advection is often not too difficult to model in boundary layer flows.

ODT has been recently validated quantitatively against several DNS in compatible configurations, namely, temporal, planar jets. This has been done for soot formation, and flame extinction and reignition using ethylene and syngas fuels. Premixed flames have also been studied. As the model has been extended to more and more complex flows, there has been a desire to use ODT as a surrogate DNS, due to its low cost compared to DNS. Such validation studies comparing ODT and DNS are promising in this area. However, DNS is limited to relatively few simulations and relatively low Reynolds number. Application of ODT to experimental flows is limited by the configuration. Experimental jets are usually cylindrical and evolve spatially, whereas existing ODT simulations are planar and (normally) evolve temporally. (There is a spatial formulation of the model in which a one-dimensional flow evolves spatially in a down-stream direction like a parabolic boundary layer formulation.)

We extend the ODT model to a cylindrical formulation. The formulation is presented, along with challenges and interpretations. We present comparisons of the cylindrical ODT formulation to round pipe flow, round nonreacting jets at high Reynolds number, and to reacting nonpremixed jet flames. The new formulation is expected to improve previous predictions of jet flows, including, for instance, particle flows. We discuss such expected improvements.

AUTOMATED TRANSITION STATE THEORY CALCULATION OF HYDROGEN ABSTRACTION FROM NOVEL BIOFUELS

Nathan D. Harms, Richard H. West
Northeastern University, Boston MA 02115

Abstract

Understanding and predicting the combustion properties of novel fuels allows for development of greener and more efficient fuels and engines. To predict the detailed combustion kinetics of a single novel fuel requires the construction of a reaction mechanism consisting of thousands of elementary reactions and their kinetics. The reaction rates can be determined by performing transition state theory (TST) calculations on each elementary reaction. However, with dozens of novel biofuels being considered, and each fuel having its own reaction mechanism, the TST calculations must be automated. We present a fully-automated method to determine reaction rates via automated Transition State Theory (AutoTST) calculations, coupled to the Reaction Mechanism Generator (RMG) [1, 2]. We focus here on the abstraction of hydrogen by six common radicals present in combustion ($HO\cdot$, $H\cdot$, $HOO\cdot$, $O\cdot$, $H_3C\cdot$, and O_2) from several biofuels being considered by the U.S. Department of Energy "Co-Optima" program. Hydrogen abstraction reactions are one of the primary reaction families that facilitate combustion and is arguably the most important class of reactions in the combustion process; therefore, accurately determining kinetic parameters for these reactions is crucial. Our approach uses RMG to determine the possible hydrogen abstraction reactions for each fuel, and the reaction kinetics are determined using our AutoTST method. The AutoTST method estimates the transition state geometry using a group contribution and distance geometry method, performs DFT geometry optimization, performs intrinsic reaction coordinate calculations to verify the transition state, and uses CanTherm [3], a canonical transition state theory calculator, to determine kinetic parameters. This method has previously been applied to hydrogen abstractions and results have been compared to published values and benchmark calculations [4] The application of AutoTST is broad and can be applied to all families of combustion reactions given proper resources.

Introduction

Long term energy, environmental, and economic needs have motivated the study of renewable energy sources to replace fossil fuels. Currently, the United States is responsible for 25 % of the world's CO_2 production despite the size of its population. [5] In order to limit the United State's production of CO_2 from motor vehicle transport, the Department of Energy (DOE) developed the "Co-Optima" program to simultaneously optimize the selection of novel biofuels and the design of efficient engines, hoping co-optimization of both will maximize performance, efficiency, and adoption. Designing an engine for a fuel that has not yet been made in significant quantities, however, requires a *predictive* kinetic model, as does evaluating a fuel's performance in an engine that does not yet exist.

From this investigation, 23 biofuels were identified as potential replacements for fossil fuels or blend-stocks (Figure 1). Few of these fuels have accurate kinetic models.

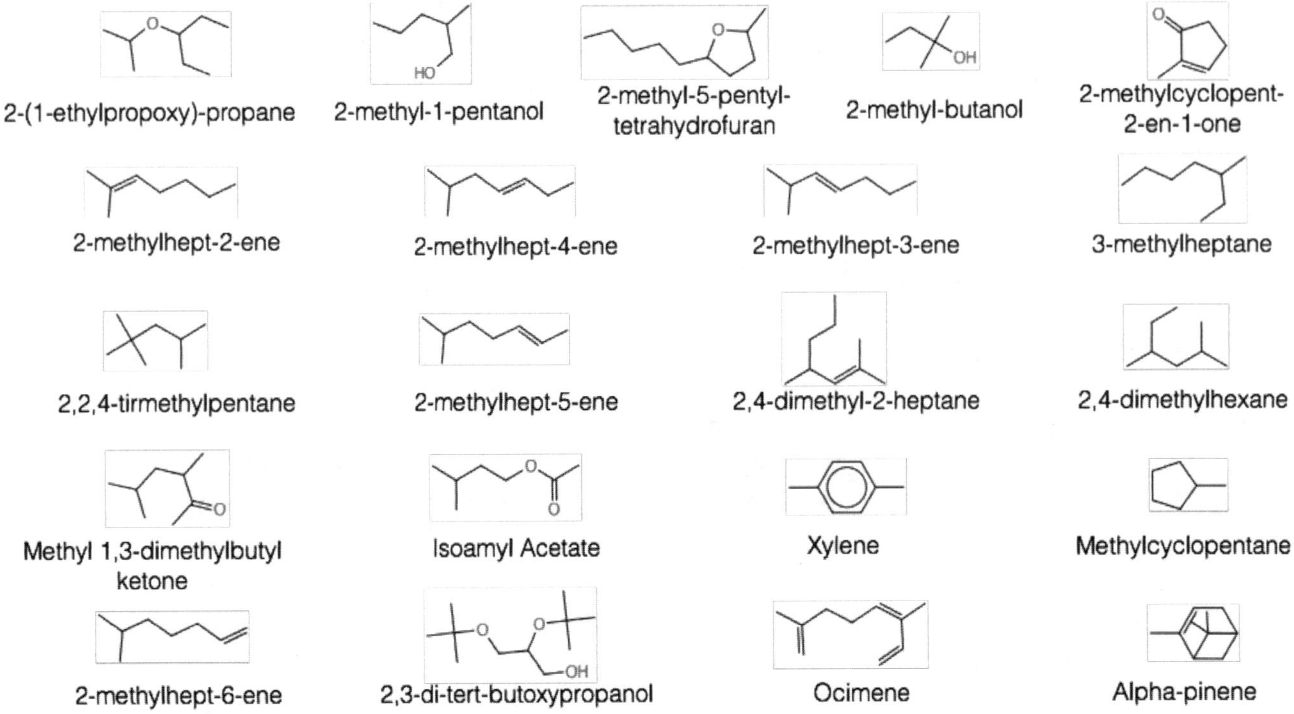

Figure 1: *The 23 novel biofuels being investigated in this study.*

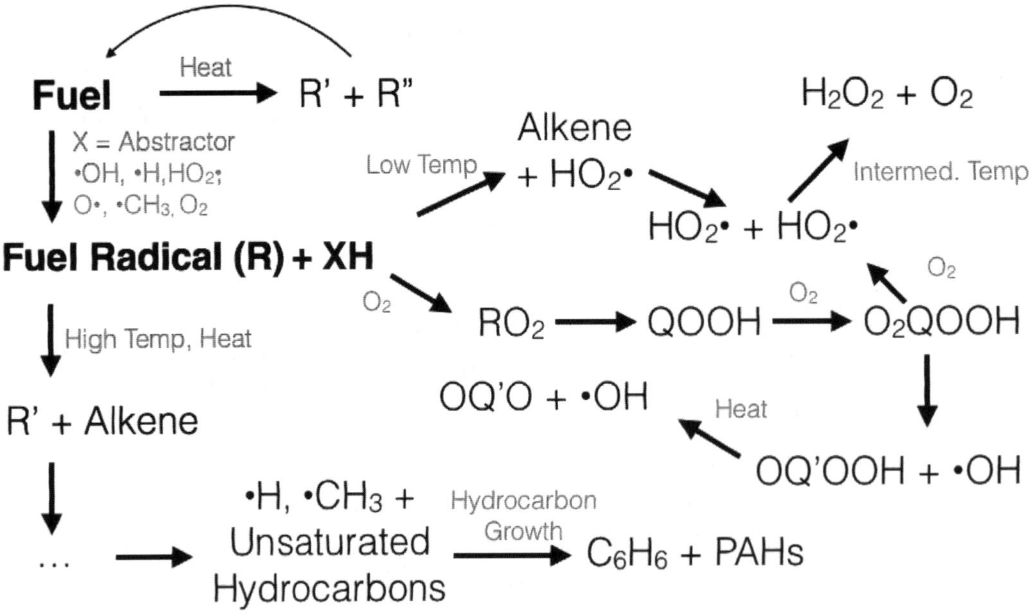

Figure 2: *Overview of a generic combustion process*

Reaction mechanisms often follow a typical road map outlined in Figure 2. For this work, we focus on the abstraction of hydrogen by six common radicals present in combustion: $HO\cdot$, $H\cdot$, $HOO\cdot$, $O\cdot$, $H_3C\cdot$, and O_2. Hydrogen abstraction reactions are some of the most important reactions in the combustion of fuel, and abstractions from the fuel molecule are the first in a set of thousands of reactions that allow for the rest of combustion to occur.

When constructing a reaction mechanism, rates are usually generated in three ways: estimates and intuition, rate rules, or transition state theory on the bases of quantum chemistry calculations [6]. Decent estimation require decades of prior knowledge in the combustion field and, even then, these estimates will be just that – estimates. Rate rules are an efficient method to determine kinetics for reactions, essentially codifying the estimates an experienced scientist might make. These methods are orders of magnitude faster than quantum calculations but their accuracy suffers when applied outside of the range of "chemical space" where the rules were developed and trained: These rate rules can be misapplied (i.e. applied to a reaction that is too dissimilar to the reactions from which it was derived), or the chemistry of a reaction might not follow the rules. Because of this, quantum calculations are often used to develop more accurate reaction rates.

Although experiments are essential to test and validate an overall kinetic model, it is often challenging to isolate an individual elementary reaction in an experimental study, and with the latest computational techniques some reactions can be calculated *ab initio* with an accuracy exceeding experiments, and at a fraction of the cost. Because of this, theoretical calculations are often utilized to determine kinetics in combustion mechanisms.

Utilization of quantum packages like Gaussian [7] allows one to optimize geometries of reactants and products by walking down the potential energy surface (PES) until a minimum energy has been found. This has been automated for on-the-fly calculation of thermochemical properties of intermediate species in RMG [8]. For transition state geometries, an initial guess of the transition state will undergo a similar search, but the program will search for a saddle point on the PES instead: a stationary point with negative curvature in one direction (the normal mode corresponding to the reaction coordinate) and positive curvature in all other directions. The search algorithm is usually only successful if the initial guess is close enough to the correct saddle point geometry that the direction of negative curvature is correct and pointing towards the saddle point. In other words, it needs a very good initial guess, which often requires significant input from the user. In order to automate the search for saddle points, Bhoorasingh et al. developed a procedure to provide a guess for the transition state geometry based on the reactants, products, and the type of reaction [9]. The algorithm used group-additivity to determine inter-atomic distances for the atoms participating in the reaction, and distance geometry to convert these into a 3D structure. Successful optimization of the TS geometry is validated using Intrinsic Reaction Coordinate (IRC) calculations. These calculations use the TS as a starting point and walk down the PES until a local minimum energy is attained. If the reactants and products of these local minima match the actual reactants and products of the reaction, the TS is validated.

Successful optimization of these geometries allows one to determine the the canonical partition functions from the results of the DFT calculations and subsequently kinetics using transition state theory (equation 1).

$$k(T) = \frac{k_B T}{h} \frac{Q^{TS}}{\Pi_i^n Q_i} \exp\left(\frac{-\Delta E_0}{k_B T}\right) \tag{1}$$

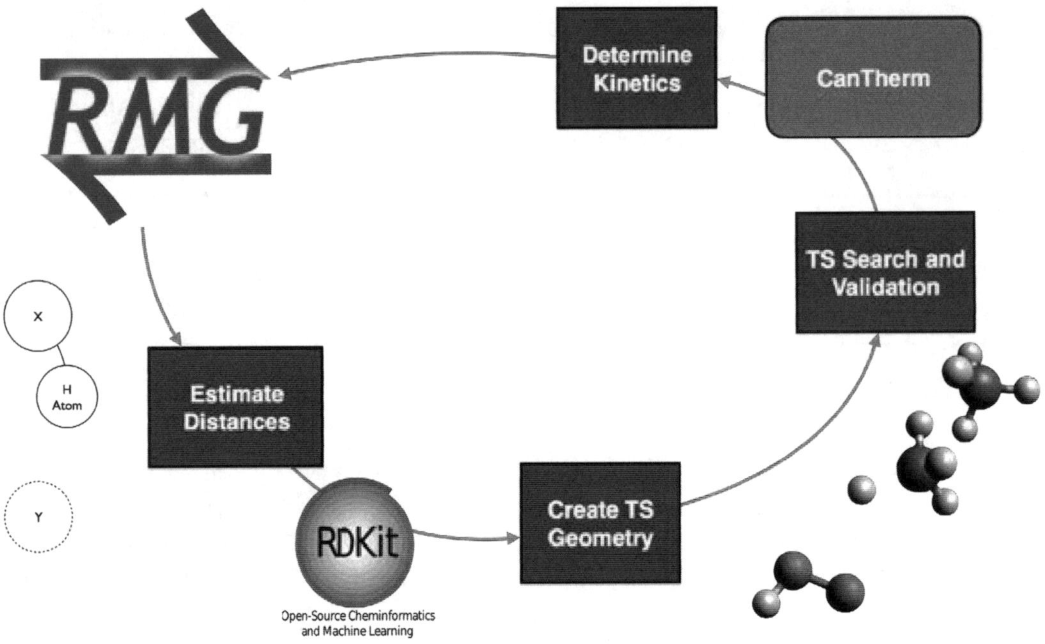

Figure 3: *Overview of the AutoTST process developed by [10]*

$k(T)$ represents the reaction rate, k_B is the Boltzmann constant, T is temperature, h is Plank's constant, Q^{TS} is the canonical partition function for the TS, $\Pi_i^n Q_i$ represents the product of the canonical partition functions for the reactants and ΔE_0 is the activation energy. This highlights that the temperature-dependent reaction rate can be determined by the partition functions and activation energy, which can all be determined from the electronic structure or DFT calculations of the reactants and transition state.

This geometry-producing algorithm was coupled to a canonical TST calculator to provide a fully automated AutoTST workflow (procedure outline in Figure 3) [4, 10].

Methods

Identification of possible hydrogen abstraction reactions

Identification of the hydrogen abstraction reactions of each of the proposed biofuels was accomplished by using Reaction Mechanism Generator (RMG) [1]. RMG identified 692 hydrogen abstraction reactions that were subsequently run using AutoTST.

Perform automated transition state theory calculations

The overall workflow is described in the recent paper by Bhoorasingh et al. [4] but a brief overview is given here. Reactant and product geometries were created using algorithms in RMG developed by Magoon and Green [8]. Transition state geometries were estimated using a group contribution method developed by Bhoorasingh and West [9]. This group contribution method identifies functional groups in each of the reactants and products, which are used to estimate the distances between atoms at the reacting center using previously identified transition states as training data. The transition state geometry estimate was optimized to an energy minimum with the reaction

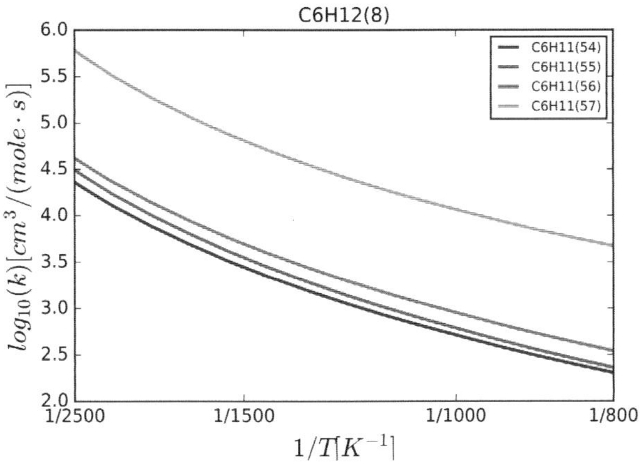

(a) *Arrhenius plot of HOO· abstracting a hydrogen from methylcyclopentane. Colors correspond to subfigure (b)*

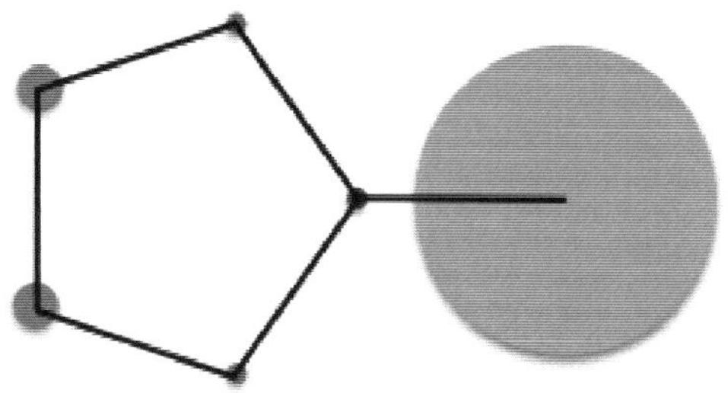

(b) *Depiction of methylcyclopentane. Colors correspond to subfigure (a). Area of each circle proportional to H abstraction rate at that site by HOO·*

Figure 4: *An Arrhenius plot and heat map of the hydrogen abstraction reactions from methylcyclopentane by HOO·.*

center distances frozen followed by a saddle point search with all distances frozen except for the reaction center. The resulting geometry is then fully optimized to a saddle point. If a saddle point is found, an intrinsic reaction coordinate (IRC) calculation is performed to verify the reactants and products. If the reactants and the products from the IRC calculations, determined from the Cartesian coordinates using a connect-the-dots algorithm, match the reactants and products from the initial reaction, the transition state validation is complete. All the geometries were optimized and validated with the M06-2X DFT functional with the MG3S basis set in the Gaussian 09 quantum chemistry package.

Following this validation, the CanTherm software package [3] was used to determine kinetic parameters using TST. Symmetry numbers for the rate calculations were determined using the SYMMETRY software [11]. SYMMETRY takes the optimized transition state geometry, allows for small deviations, and calculates the point group. Product properties were also calculated so that tunneling corrections could be applied via the Eckart model.

Results

Through our presented work flow, we have identified 692 hydrogen abstraction reactions on 23 novel biofuels, started calculations on 540 of these reactions, and successfully calculated kinetics for 106 of them. At time of preparation of this manuscript, the completion rate is 19.6%, which is substantially lower than the success rate previously reported for this method of $\sim$70% [4].

Although the current success rate of this work flow is low, we have successful identified the reaction rates of HOO· abstracting a hydrogen from methylcyclopentane at all unique sites, for example. In figure 4, we see an Arrhenius plot of the hydrogen abstraction rate by HOO· and a corresponding "heat map" of methylcyclopentane. This has told us that the methyl group of this proposed biofuel is most likely to lose a hydrogen.

Total Reactions Identified	692
Reaction Calculations Started	540
Overall Successes	106
Success Rate	19.6%

Table 1: *Success rates at time of manuscript preparation.*

Conclusions

We are still investigating the currently low success rate. Some failure modes include: not finding the lowest energy conformer for a reactant so that the barrier height is negative; failed TS optimizations in Gaussian; converging upon a saddle point that is not the required transition state (eg. a conformational change); and long computational time associated with IRC calculations leading to interrupted calculations. Many abstraction reactions will succeed in one direction but not the other, which is not reflected in these results.

In order to obtain the lowest energy conformers, we intend to implement Fafoom, a conformer searcher that utilizes a genetic algorithm technique [12]. We have tested this approach on reactant and product molecules, and plan to extend the approach to transition states.

A number of our calculations resulted in a non-existent barrier height. This could have occurred because tolerances on calculated energies of reactants, products, and TSs can overlap if the optimization is not exact enough. We propose running a stricter calculation on these geometries to minimize their tolerances and obtain barrier heights.

Some of our TS optimizations failed to converge as well. This could be because the guess provided by AutoTST was too far from the actual TS. One method to try and correct this is to retrain the data which the distance geometry estimates in AutoTST are based on. This will result in more accurate reaction coordinates for the TS guess and should give a higher success rate. Another avenue to explore is performing a looser optimization on the geometry in order to get convergence, then performing a tighter optimization on the loosely optimization geometry.

Finally, we would like to find a faster method of validation for AutoTST calculations. Currently, the IRC calculations are the most computationally expensive calculation. We have identified Quick Reaction Coordinate calculations [13] and nodal analysis as possible avenues to explore.

References

[1] Gao CW, Allen JW, Green WH, West RH. Reaction Mechanism Generator: Automatic construction of chemical kinetic mechanisms. *Comput Phys Commun*. 2016;203:212–225. https://doi.org/10.1016/j.cpc.2016.02.013.

[2] Green WH, Allen JW, Barlow J, Bhoorasingh PL, Buesser BA, Class CA, Gao CW, Han K, Khanshan FS, Lambert V, Merchant SS, Slakman BL, Troiano S, Vandeputte A, Vandewiele NM, West RH, NathanYee. Reaction Mechanism Generator;.

[3] Allen JW, Green WH. CanTherm: Open-source software for thermodynamics and kinetics. Included in: Reaction Mechanism Generator, v2.0.0. 2016.
URL http://reaction.mechanismgenerator.github.io

[4] Bhoorasingh PL, Slakman BL, Seyedzadeh Khanshan F, Cain JY, West RH. Automated transition state theory calculations for high-throughput kinetics. *J Phys Chem A*. 2017; https://doi.org/10.1021/acs.jpca.7b07361.

[5] U.S. Energy Information Administration - EIA - Independent Statistics and Analysis. URL https://www.eia.gov/tools/faqs/faq.php?id=32&t=6

[6] Van de Vijver R, Vandewiele NM, Bhoorasingh PL, Slakman BL, Seyedzadeh Khanshan F, Carstensen HH, Reyniers MF, Marin GB, West RH, Van Geem KM. Automatic Mechanism and Kinetic Model Generation for Gas- and Solution-Phase Processes: A Perspective on Best Practices, Recent Advances, and Future Challenges. *Int J Chem Kinet*. 2015;47(4):199–231. https://doi.org/10.1002/kin.20902.

[7] Frisch MJ, et al. Gaussian 09. Gaussian, Inc., Wallingford, CT. 2009.

[8] Magoon GR, Green WH. Design and implementation of a next-generation software interface for on-the-fly quantum and force field calculations in automated reaction mechanism generation. *Comput Chem Eng*. 2012;52:35–45. https://doi.org/10.1016/j.compchemeng.2012.11.009.

[9] Bhoorasingh PL, West RH. Transition state geometry prediction using molecular group contributions. *Phys Chem Chem Phys*. 2015;17(48):32173–32182. https://doi.org/10.1039/C5CP04706D.

[10] Bhoorasingh P, Slakman B, Khanshan FS, Cain J, West R. Kinetic data for manuscript describing the AutoTST algorithm for automated Transition State Theory calculations of chemical reaction rates. *figshare*. 2016;https://doi.org/10.6084/m9.figshare.4234160.v1.

[11] Patchkovskii S. SYMMETRY software, University of Calgary (2003). 2003. URL http://www.cobalt.chem.ucalgary.ca/ps/symmetry/

[12] Supady A, Blum V, Baldauf C. First-Principles Molecular Structure Search with a Genetic Algorithm. *J Chem Inf Model*. 2015;55(11):2338–2348. https://doi.org/10.1021/acs.jcim.5b00243.

[13] Goodman JM, Silva MA. QRC: a rapid method for connecting transition structures to reactants in the computational analysis of organic reactivity. *Tetrahedron Letters*. 2003; 44(45):8233 – 8236. https://doi.org/10.1016/j.tetlet.2003.09.074.

490269 First Principles Based Automated Kinetic Model Generation Using on-the-Fly *Ab Initio* Calculations

Abstract:

Introduction

Many kinetic models for combustion and pyrolysis processes have been published in the past decade, and a common challenge in most work is often being stressed, i.e. the shortage of accurate rate coefficients for a large number of elementary steps. Currently, many rate coefficients are calculated based on rough approximation methods, and their accuracy can be questioned when these approximations are used outside of the original application domain of the applied method. Furthermore, building these approximation methods requires a complete training and test set of reactions of which the rate coefficients are known from either experiments or theory. Manually obtaining all these rate coefficients is time consuming, and this work is aiming at automatically calculating a large number of rate coefficients using quantum chemistry, minimizing the manual work. The developed methods can furthermore be used after the model generation, i.e. when not all the kinetics are sufficiently accurate.

This work has been performed in the framework of Genesys[1], an automatic kinetic model generation tool developed at the Laboratory for Chemical Technology which employs both the rule-based and the rate-based termination criteria. The latter can in particular benefit from the automatic *ab initio* calculations in the search for more complete, more accurate and predictive kinetic models.

Methodology

Genesys offers a representation of chemical reactions, supported by the Chemistry Development Kit[2]. Molecules, radicals and transition states are treated as mathematical graphs, in which the atoms are nodes and the bonds are edges of the graph. This is a very powerful molecular representation to convert reactants into products, to find sub-molecular patterns, to search in databases for appropriate data, etc. However, it is insufficient to perform *ab initio* calculations, where the coordinates of the atoms in the 3D space are required. The conversion of connectivity information to 3D coordinates is already well established for molecules and radicals, for which rule-based heuristic methods and numerical methods exist. "Distance geometry" is a method belonging to the latter group in which the coordinates are estimated by minimizing an error function describing all the interatomic distances compared to expected distances, which are tabulated. Additional optimization steps, among others using the Merck Molecular Force Field (MMFF)[3], allow the generate a meaningful set of 3D coordinates, directly applicable for *ab initio* calculations. This has been implemented in Genesys.

To obtain accurate rate coefficients, quantum chemical calculations of transition states are also required. The 3D coordinates of atoms in transition states are more complex to deduce and cannot be summarized in the same simple rules used for molecules and radicals. Since Genesys has been programmed to offer the end-user flexibility in terms of which reaction families to use and how to constrain them, this philosophy is further employed for the generation of 3D coordinates for transition states. The end-user has to possibility to guess the interatomic distances of the atoms in the transition state that change in connectivity throughout the reaction, called the reactive atoms. This information, the so-called "template" is used in Genesys to obtain 3D coordinates of all the atoms via "distance geometry", i.e. similar to stable molecules and radicals. The template can be built from a small set of manually calculated transition states or from the vast literature data on quantum chemical calculations. Refining the "distance geometry" output to a more meaningful set of coordinates is also done. For this, the MMFF is used in which equilibrium distances for the reactive atoms originate from the template, and the force constants are taken from analogous stable molecules and radicals.

Once initial 3D coordinates are available for molecules, radicals and transition states, quantum chemical calculations are performed using the Gaussian09 suite of programs[4] in several steps. First, the geometry is energy-optimized using fast semi-empirical calculations such as PM3. Using the resulting geometry, the conformational space is scanned by rotating around single bonds in a systematic and exhaustive manner. Each of these conformers are optimized using DFT methods, and the geometries belonging to the lowest 5 kJ mol^{-1} are used for high-level optimization, i.e. CBS-QB3 calculations. Finally, hindered rotor potentials are obtained to correct for the harmonic oscillator approximation.

For transition states, it is necessary to verify if the imaginary frequency indeed corresponds to the expected reaction. This is done in two steps. First, Genesys controls if the final geometry is close to the user-provided template. Second, the normal mode corresponding to the imaginary frequency is analyzed and it is evaluated whether the bonds that are broken or formed have a large contribution to this normal mode. For this, the normal mode in Cartesian coordinates is transformed into a normal mode in internal coordinates.

Via statistical thermodynamics, standard enthalpies of formation, standard entropies and heat capacities of all the molecules, radicals and transition states are calculated. The frequencies that resemble internal rotations are automatically searched for via an analysis of the normal mode. Their contributions to the vibrational partition function are automatically replaced by the hindered rotor partition functions. Also, the total rotational symmetry number of the species is automatically calculated and used to obtain correct entropy values. In the case where stereocenters are present, several approaches are possible. In Genesys, it is possible to explicitly track each stereoisomer as a separate species and to calculate separate thermodynamic data. However, when this is not necessary, they can also be lumped together, for which the entropy is then corrected by the number of optical isomers.

The thermodynamic data of reactants and the transition state of a reactions are subsequently used to obtain high-pressure limit rate coefficients. Tunneling corrections are automatically calculated using Eckart potentials, and Arrhenius parameters are determined by regression of the rate coefficients.

Results

The new methods have been illustrated by calculating rate coefficients for 8 reactions, each belonging to a different reaction family, i.e. inter- and intramolecular hydrogen abstractions, inter- and intramolecular radical additions, a substitution, a dehydration, a dehydrogenation and a retro-ene reaction. For each reaction family, the template was created based on known geometries of transition states from the literature or from manual calculations. The resulting Arrhenius parameters and rate coefficients at 1000K are given in Table 1.

These rate coefficients have been compared to experimental and theoretical values in literature and an excellent agreement is found, which is the same agreement found when calculating rate coefficients from CBS-QB3 calculations "manually". Most of the rate coefficients are within a factor of 2 of experimental and theoretical literature rate coefficients.

Table 1: Arrhenius parameters and rate coefficients at 1000K of the eight reactions automatically calculated using

Genesys. The first line of each reaction corresponds to the forward reactions, the second line to the reverse reaction. The pre-exponential factor and rate coefficients are expressed in s^{-1} for monomolecular reactions and $m^3\ mol^{-1}\ s^{-1}$ for bimolecular reactions. The activation energies are expressed in kJ mol^{-1}.

Conclusions

New methods have been developed to automatically calculate rate coefficients from *ab initio* calculations. For this, Genesys is used, which provides a connectivity based representation of the species, transition states and reactions. This information is used to create 3D coordinates of the atoms in the species and transition states. These coordinates are inputted to quantum chemistry calculations, yielding high level thermodynamic and high pressure limit kinetic data for gas-phase reactions.

The new methods are of great value to calculate a large number of rate coefficients from *ab initio* calculations without the need for manual interventions or user knowledge. This is especially of great interest for (1) building approximation methods for rate coefficient calculations such as group additivity models and (2) increasing the accuracy of important rate coefficients in a kinetic model.

1. Vandewiele NM, Van Geem KM, Reyniers MF, Marin GB. Genesys: Kinetic model construction using chemo-informatics. *Chem Eng J.* 2012;207:526-538.

2. Steinbeck C, Han Y, Kuhn S, Horlacher O, Luttmann E, Willighagen E. The Chemistry Development Kit (CDK): An Open-Source Java Library for Chemo- and Bioinformatics. *Journal of Chemical Information and Computer Sciences.* 2003;43(2):493-500.

3. Halgren TA. Merck molecular force field .1. Basis, form, scope, parameterization, and performance of MMFF94. *Journal of Computational Chemistry.* 1996;17(5-6):490-519.

4. *Gaussian 09* [computer program]. Wallingford, CT, USA: Gaussian, Inc.; 2009.

492655 Quantifying the Catalytic Effect of Calcium on Initiation during Cellulose Pyrolysis

Abstract:

During high temperature biomass pyrolysis, biopolymers undergo rapid decomposition to form hundreds of organic compounds which condense to form bio-oil and can be subsequently upgraded to fuel and chemicals. Inorganic metals naturally present in lignocellulosic biomass such as calcium and magnesium have been shown to catalyze pyrolysis chemistry[1]. However, little information is available that quantifies the catalytic behavior of these metals on pyrolysis chemistry. This work focuses on identifying the fundamental chemical mechanisms and kinetics involved in this catalytic chemistry. Using Î±-cyclodextrin as a cellulose surrogate[2] and a novel reactor system PHASR (Pulsed Heated Analysis of Solid Reactions)[3-4], time-resolved conversion data is obtained between 370 and 430 ^{0}C. To study the effect of calcium catalysis, varying concentrations of $CaNO_3$ (0.4 to 2.0 wt%) were mixed with Î±-cyclodextrin to simulate calcium concentrations in lignocellulosic biomass[1]. Cellulose conversion exhibits first order like kinetic behavior among all temperatures and calcium concentrations. A catalyzed kinetic rate constant was extracted and exhibits similar energetics compared to a high-temperature non-catalyzed intra-chain scission regime determined in an earlier study[4].

References:

1. Zhu, C.; Maduskar, S.; Paulsen A. D.; Dauenhauer, P. J., *Alkaline-earth-metal-catalyzed thin film pyrolysis of cellulose*. ChemCatChem, **2016**, 8, 818-829.
2. Mettler, M. S.; Mushrif, S. H.; Paulsen, A. D.; Javadekar, A. D.; Vlachos, D. G.; Dauenhauer, P. J., Revealing pyrolysis chemistry for biofuels production: Conversion of cellulose to furans and small oxygenates. *Energy & Environmental Science* **2012,** 5, (1), 5414-5424.
3. Krumm, C.; Pfaendtner, J.; Dauenhauer, P.J., Millisecond pulsed films unify the mechanisms of cellulose fragmentation. *Chemistry of Materials* **2016**, 28, 3108-3114.
4. Zhu, C.; Krumm, C.; Facas, G. G.; Neurock, M.; Dauenhauer, P.J., Energetics of cellulose glycosidic bond cleavage. *Reaction Chemistry and Engineering* **2017**, 2, 201-214.

486729 Artificial Neural Networks for Flare Modeling and Set Point Determination

Abstract:

The refineries and chemical industries are mandated to meet the EPAâ€™s regulations for flare performance, i.e., smokeless flaring while maintaining a minimum combustion zone heating value of 270 BTU/scf and NHV_{dil} >22 Btu/ft^2 for all steam/air/non-assisted flares. While field tests of the many flare operation parameters are expensive, it is highly efficient to use mathematical models to analyse data and develop the right tools to make decisions in the real plant scenario.Â Modelling flare data using artificial neural networks to achieve the desired combustion efficiency (CE) and eliminate smoke can be done robustly. Selected flare data obtained from various flare tests having a wide range of exit velocity, heating value, and fuel composition have been analysed and the most influencing independent variables are chosen. Both steam and air assisted flares were modelled using the neural network toolbox in MATLAB with high correlation coefficients of greater than 0.95. Determining the set point (amount of steam/air/make-up fuel required) at the Incipient Smoke Point (ISP) and for smokeless flaring has been performed as a part of this study. Desirable operating inputs can be set for the ISP (NHV_{dil} >22 BTU/ft^2 & Opacity<Opacity $_{ISP}$) and for smokeless flaring with a high CE (>96.5).

486463 Preparation and Characterization of Co-Cu-Ga/ZrO_2-Al_2O_3 Catalyst for Higher Alcohol Synthesis from Syngas

Abstract:

A series of highly dispersed mesoporous ZrO_2-Al_2O_3 supported CoCuGa catalysts were prepared by incipient wetness impregnation. The catalytic performance was investigated for the synthesis of higher alcohols from syngas. The physicochemical properties of the catalysts have been characterized by TG-DTA, XRD, BET, H_2-TPR. The experimental parameters, i.e., reaction temperature, pressure and space velocity were also studied for this reaction. The experiment has been optimized process conditions: temperature 300 °C, pressure 4.0 MPa, H_2/CO=2, and space velocity 1200 h^{-1}. The mesoporous ZrO_2-Al_2O_3 served as reactors to ensure alcohol products easily diffuse into/out of its pores, as well as prevented hotpot formation, thus avoiding the further hydrogenation to hydrocarbons. As a result, the CoCuGa/ZrO_2-Al_2O_3 catalyst showed the highest conversion of CO and selectivity of total alcohols and C_2^+ alcohols. It was concluded that the catalytic performance was improved by adding Ga into the CoCu catalyst. Appropriate amount of Ga improved metal dispersion and CO uptake, thus leading to the enhancement of catalytic activity. This catalyst with significantly improved catalytic behavior can be potentially used in CO hydrogenation to produce higher alcohols.

504294 Conversion of Gluconate and Glycerol to Ethnaol Using the Recombinant Klebsiella Oxytoca Strains

Abstract:

Gluconate could potentially be directly produced from cellulosic biomass and is an attractive substrate for fuels and chemicals production. While gluconate was used as the substrate for ethanol production, a small fraction of the acetate has to be produced along with ethanol due to the fact that gluocnate was an more oxidized substrate than glucose. This study explore the co-fermentation of gluconate and glycerol, which is a more reduced substrate than glucose and a biodiesel byproduct for balanced ethanol production. Strain engineering strategies were applied to improve the overall ethanol yield by deleting the competing pathways.

487282 Ammonia Stripping for Enhanced Biomethanization of Chicken Manure

Abstract:

Paper submitted to 2017 AIChE Annual Meeting, which will be held at Minneapolis, USA within October 29-November 3, 2017

Ammonia stripping for enhanced biomethanization of chicken manure

Kun Li[1,2], Ronghou Liu[1,2]*, Shaofeng Cui, Ruijie Ma

1. Biomass Energy Engineering Research Centre, School of Agriculture and Biology, Shanghai Jiao Tong University, 800 Dongchuan Road, Shanghai 200240, P.R. China

2. Key Laboratory of Urban Agriculture (South), Ministry of Agriculture, 800 Dongchuan Road, Shanghai 200240, P. R. China

Abstract

Anaerobic digestion of animal manure is always with problems of ammonia inhibition, although it is characterized with high moisture, buffering capacity, organic matter and wide variety of nutrients. Ammoniacal nitrogen is generated by the biological degradation of nitrogenous matter. Optimal ammonia content guarantees sufficient buffer capacity of methanogenic medium in anaerobic digestion, thus improving the stability of the digestion process. However, high ammonia is the primary reason of digester failure because of its direct inhibition of microbial activity. In physical stripping, the desorbed component is more soluble in the gas than in the liquid state. The ammonia in gas is absorbed by sulphuric acid solution and ammonium sulphate is formed when ammonia is bubbled through sulphuric acid. In this research, the application of ammonia stripping to the digestate was tested with chicken manure. Four 2.5-L anaerobic digesters fed on chicken manure slurry were coupled to batch ammonia stripping columns and were run semi-continuously over 110 days at 37 . The results of performance and stability were compared with control without stripping. Stripping medium was air, and the stripping columns were operated at three temperature (35, 55, 70). All the pH adjusted to 10, and retention time was 2 days. Results showed that temperature of the stripping column as high as 70 at a pH value of 10 was needed to maintain a stable anaerobic digestion of chicken manure. Under these conditions, 24.5% of the total ammonia nitrogen(TAN) was removed in two days. And the highest specific methane production (0.199 m^3 CH_4/kg VS) and volumetric methane production rate (1.693 L/L•d) were obtained at 9 g VS/L•d. The reason for the good performance of digester R3 might be that optimal TAN concentration in R3 (coupled to column with a stripping temperature of 70) guarantees an adequate buffer capacity which maintains a healthy pH and allows the system to recover after a VFA shock caused by acetogenic-methanogenic imbalance. The research can provide guidance for the operation of anaerobic digesters running on chicken manure slurry.

Keywords: Ammonia removal; Batch ammonia stripping; Anaerobic digestion; Chicken manure

Kun Li , Ph. D student, majoring ecology,

Email:53772765xz@sina.com,

Phone: 0086 15021165113

Ronghou Liu, corresponding author

Professor, and Director of Biomass Energy Engineering Research Centre,

School of Agriculture and Biology,

Shanghai Jiao Tong University, 800 Dongchuan Road, Shanghai 200240, P.R. China

Email: liurhou@sjtu.edu.cn; liurhou@gmail.com; Phone: 0086 21 34205744

500860 Role of O in Ru Catalyst in Hydrodeoxygenation of Furfural

Abstract:

Role of O in Ru Catalyst in Hydrodeoxygenation of Furfural

Weiqing Zheng,[1] Konstantinos A. Goulas, [1] Ayman Karim,[2] Prashant Kumar,[3] Andre Mkhoyan,[3] Michael Tsapatsis,[3] Dionisios G. Vlachos[1]

1. Catalysis Center for Energy Innovation, University of Delaware, 221 Academy St. Newark, DE 19716

2. Department of Chemical Engineering, Virginia Polytechnic Institute and State University, Blacksburg, VA

3. Department of Chemical Engineering and Materials Science, University of Minnesota, Minneapolis, MN

Recent theoretical investigations in our group has shown the importance of surface oxygen vacancies in the furfuryl alcohol hydrodeoxygenation (HDO) reaction over ruthenium oxide structures.[1] We have also demonstrated the critical roles of metallic Ru and RuO_2 (physical mixture) in HDO of hydroxymethylfurfural (HMF) and furfural.[2,3] On a single metallic surface, the surface and subsurface oxygen species play a significant role in HDO. In this work, we present evidence of the coexistence of metal and oxide (nonstoichiometric Ru_xO_y) on Ru catalyst surface by atomic resolution high angle annular dark field scanning transmission electron microscopy (HAADF-STEM) at lower electron beam potential. This structure is further confirmed by X-ray absorption spectroscopic analyses of fresh, used and *in situ* reduced Ru/C catalyst. The results show that the fresh catalyst is predominately RuO_2 with small quantities of metallic Ru.

In order to directly correlate the nonstoichiometric Ru_xO_y structure to the reaction of furfural HDO we performed in *operando* X-ray absorption spectroscopy at beamline X18B, NSLS, BNL. All the experiments were performed with a custom-made XAS cell, which was charged with 1wt% furfural-isopropanol and pressurized with 400 psi He, heated up to defined temperatures (140 °C, 160 °C and 180 °C). Based on both near edge structure (XANES) and extended X-ray absorption fine structure (EXAFS) analyses of the data, we conclude that the confirmed nonstoichiometric Ru_xO_y structure can be directly correlated to the 2-methylfuran (MF) production from furfural via catalytic transfer hydrogenation. However, once the surface is dominated by metallic Ru, MF formation is suppressed.

Reference

1. A. Mironenko and D. Vlachos. *J. Am. Chem. Soc.* **2016**, *138*, 8104
2. J. Jae, W. Zheng, R. Lobo, D. Vlachos, *ChemSusChem*, **2013**, 7, 1158
3. J. Jae, W. Zheng, A. Karim, W. Guo, R. Lobo, D. Vlachos, *ChemCatChem*, **2014**, *6*, 848

504121 Optimization for Technological Conditions of Magnetic Ferric Oxide/SO_4^{2-} Biomass-based Solid Acid Catalyzed Hydrolysis Corn Straw to Prepare Levulinic Acid Using Response Surface Methodology

Abstract:

The optimization of the process conditions for preparing levulinic acid which used magnetic ferric oxide/SO_4^{2-} biomass-based solid acid as catalyst to was investigated using response surface methodology (RSM). Four parameters, namely those of Quality of Liquid-Solid Ratio (QLSR), Amount of Catalyst (AC), Hydrolysis Temperature (HT) and Hydrolyzation Duration (HD) were chosen as variables. TheÂ mathematical model of quadratic polynomial on the levulinc acid of P value was less than 0.0001, the loss of quasi P value is less than 0.01, the considerably high R^2 value (0.9969) indicated the statistical significance of the model and reached extremely significant level; $CV = 3\ \% < 10\ \%$, the experiment had most high credibility and accuracy. The optimal process conditions for preparing the levulinic acid yield (LAY) were determined by analyzing the response surfaceâ€™s three-dimensional surface polt and contour polt, and by solving the regression model equation with Design Expert software. The Box-Behnken design (BBD) was used to optimize the process conditions, which showed that the most favorable dosages of QLSR, AC, HT, and HD were 17.2:1 (V/W, mL:g), 3 g, 249 °C, 67.3 min, respectively. Under the optimized conditions, the maximum LAY was 15.82 %, and the predictive value was 15.81 %, which the relative deviation between them was 0.01 %, this model is good repeatability. This shows that using this model to optimize the technology of magnetic ferric oxide/SO_4^{2-} biomass-based solid acid catalyst corn stalk of hydrolysis to prepare of levulinic acid is feasible.

Keywords: Magnetic Ferric Oxide/SO_4^{2-} Biomass-based Solid Acid; Levulinic Acid; Optimization; Response Surface Methodology; Box-Behnken design

510667 Heavy Metal Removal with Lignin

Abstract:

The lead removal from the wastewater system with Kraft lignin was investigated in this study. The Kraft lignin was characterized through NMR analysis that ion-exchanging functional groups such as phenolic hydroxyl groups and carboxyl groups were observed. The aromatic structure in considered as another adsorption functional group due to its electron quenching ability. Due to the existence of multiple functional groups, a non-ideal adsorption isotherm model was proposed considering the different interaction between Pb(II) and different functional groups. Each two adjacent function groups were grouped up as one active site that is able to adsorb one lead cation. Experimental data performed at different pH and temperature were interpolated with the proposed model. Based on the parameters obtained from the modeling, it was concluded that the ionic bindings of Pb(II) on PhOH and â€“COOH were not impacted by the changes of temperature. However, the interaction between Pb(II) and aromatic rings was significantly weakened with the increase of temperature.

2017 AIChE Annual Meeting, Oct. 28-Nov. 2, Minneapolis, USA

Conversion of Biomass to Microbial Lipids for Biofuels (生物质制微生物油脂和生物燃料)

Zongbao K. Zhao
(赵宗保)

Dalian Institute of Chemical Physics, CAS
(中国科学院大连化学物理研究所)
E-mail: zhaozb@dicp.ac.cn

- **Background**
- **Progresses**
- **Prospects**

2

Biodiesel

The definition:

Biodiesel is a renewable, DROP-IN biofuel composed of <u>long-chain fatty acid esters</u>, usually made from triacylglycerols (TAG)

The chemistry:

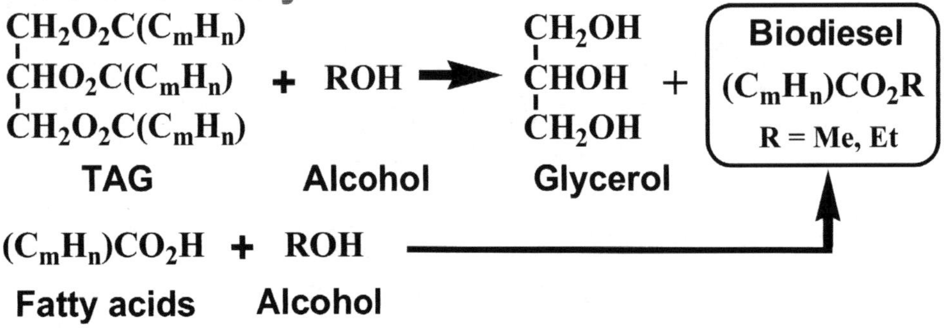

3

Lipids (Triacylglycerols)

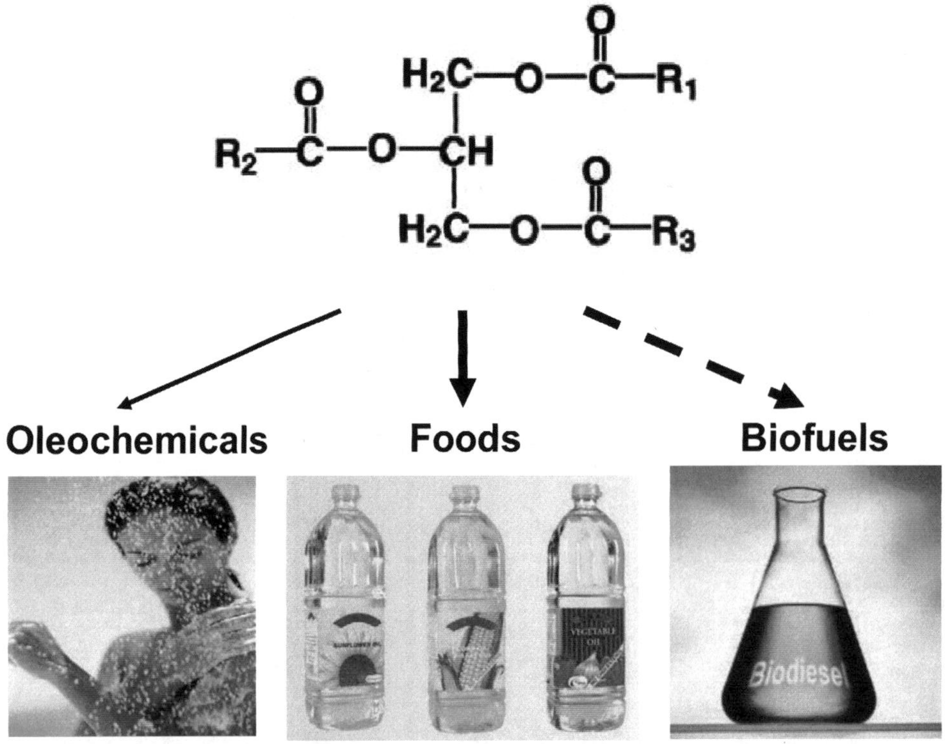

4

Resources for Lipids

Annual lipids production: ~ 200 MTons (2016')

Soybean

Rapeseed

Palm

Animals

5

Limitation of Current Lipid Technology

Yield

Productivity

Value

Structural diversity

New lipid technology required !

6

Lipid-Producing Microbes

Oleaginous microbes can take sugars and related organic materials, accumulate lipids up to 70% of their cellular dry weight

(FYI: oil seeds contain 20–50 wt% lipids)

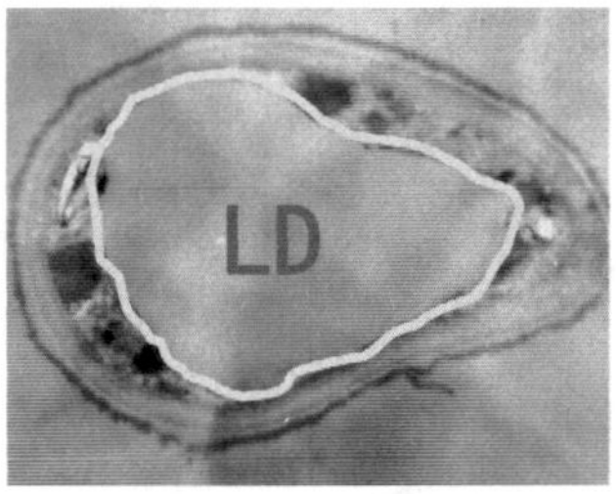

☜ TEM picture of red yeast contained 72% lipid

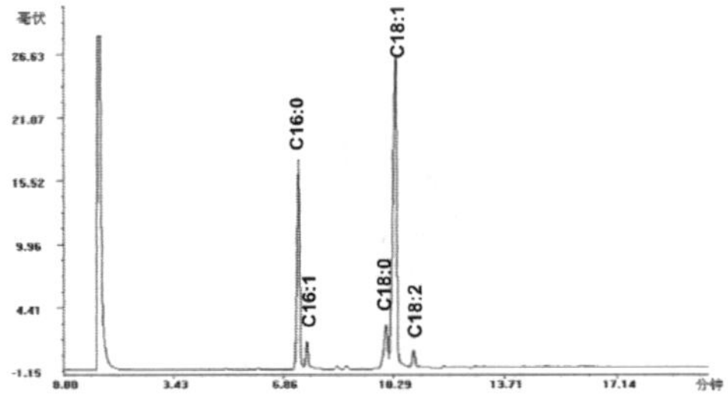

Fatty acid composition

To explore microbial oleaginicity!

7

Fatty Acid Compositional Profiles

Oil source	C16:0 Palmitic	C16:1 Palmitoleic	C18:0 Stearic	C18:1 Oleic	C18:2 Linoleic	C16 + C18
1# strain	27	3	7	60	2	99
2# strain	29	5	1	60	2	97
3# strain	21	2	4	44	27	98
4# strain	27	1	3	57	9	97
Rapeseed	2.5-6	< 0.6	1-2	50-66	18-30	>95
Olive	8-16	-	~3	70-85	5-15	>90
Soybean	10-12	-	~3	20-30	50-56	>95

Li YH, Liu B, Bai FW, Zhao ZB. *China Biotechnol.* 2005, 25(12), 39

8

High-Cell-Density Lipid Production

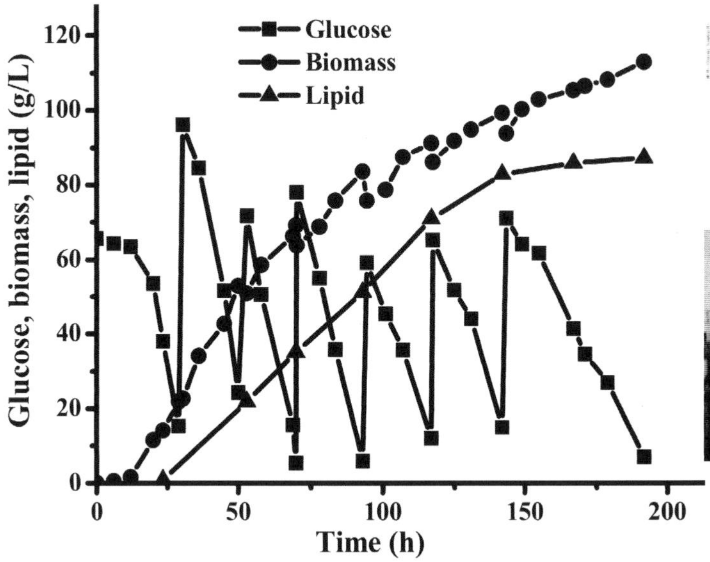

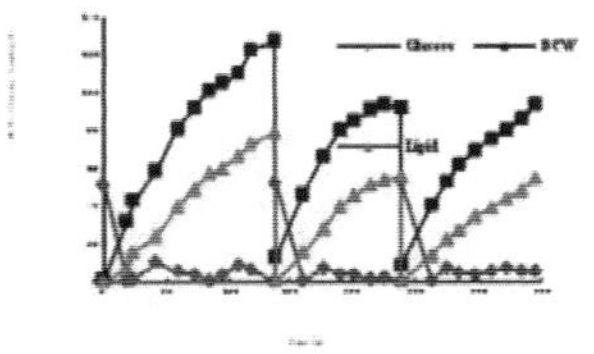

R. toruloides being cultivated at 30 °C in 30-L bioreactor.

Cell mass 110 g/L; lipid content: 77%; productivity: 0.45 g/L//h; titer: 83 g/L

Li YH, Zhao ZK, Bai FW. *Enzyme Microb Technol* 2007, 41, 312

Converting Yeast Lipids into Biodiesel

Direct methanolysis of lipid-rich yeasts

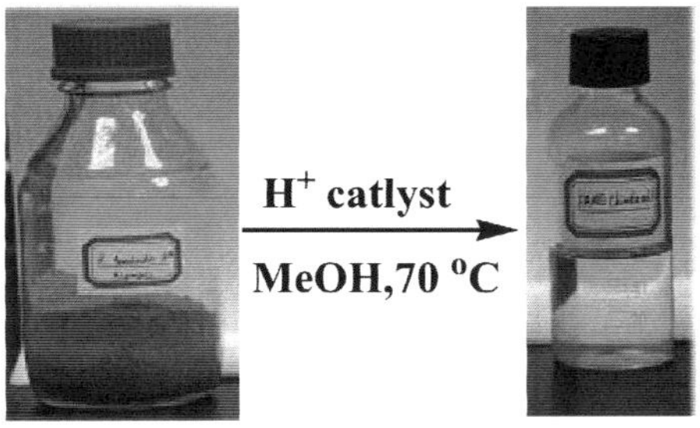

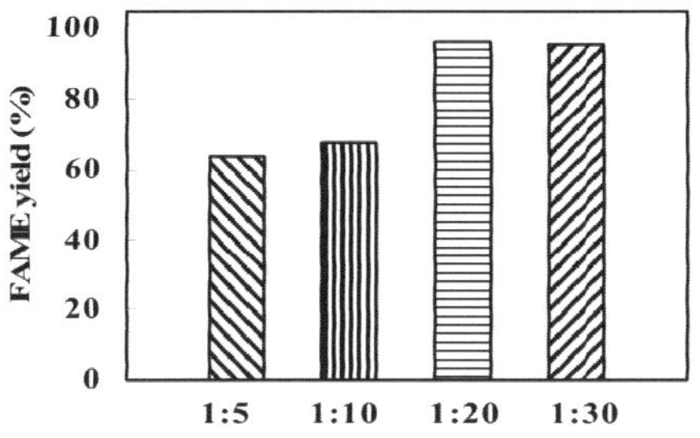

Enzymatic conversion of microbial oil

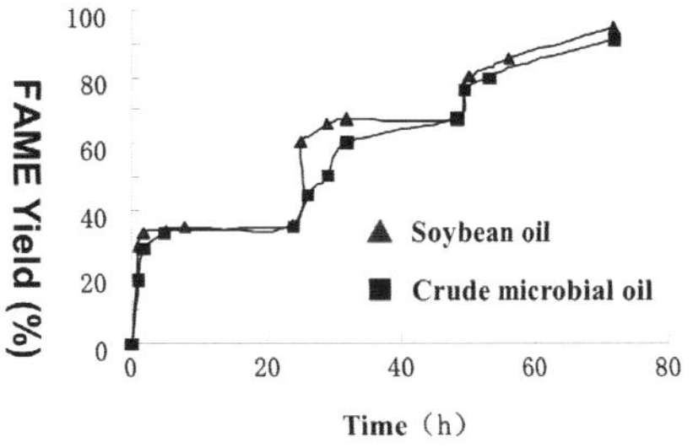

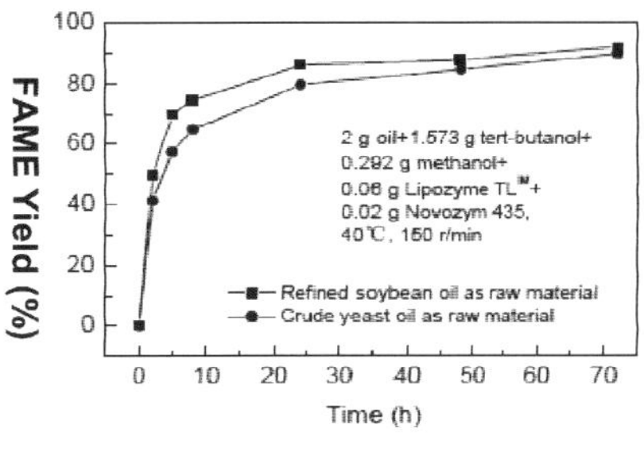

Liu B, Zhao ZK. *J Chem Technol Biotechnol* 2007, *82(8)*, 775.
Li W, et al. *Chin J Process Eng* 2007, *7(1)*, 137.

Biomass to Microbial Lipids for Biofuels

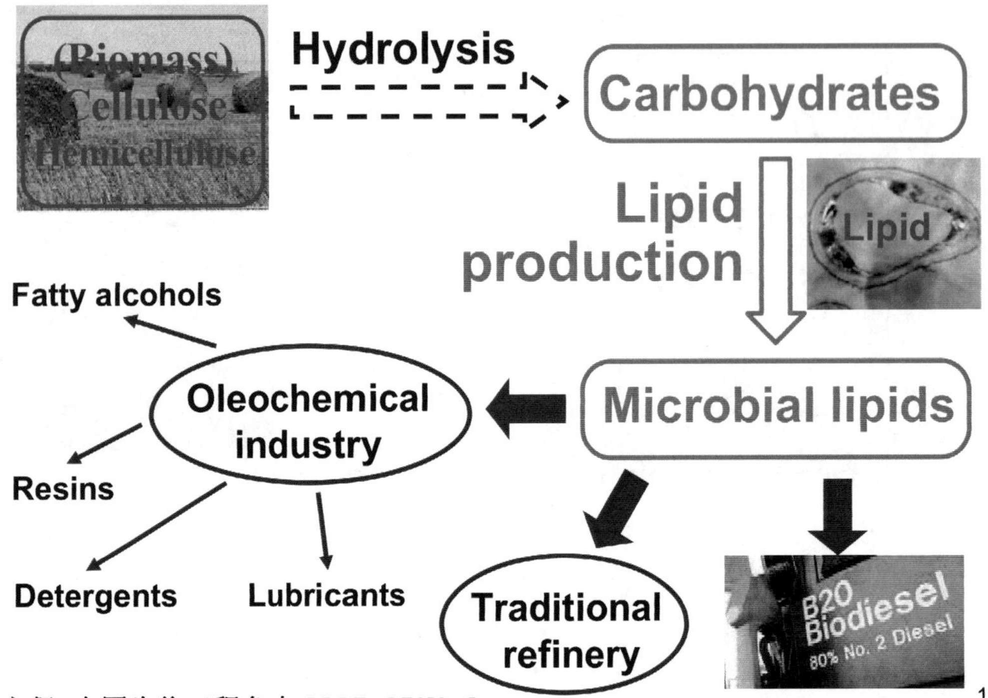

赵宗保 *中国生物工程杂志* **2005, 25(2), 8.**

11

- ➢ Background
- ➢ **Progresses**
- ➢ Prospects

12

Biomass De-Polymerization

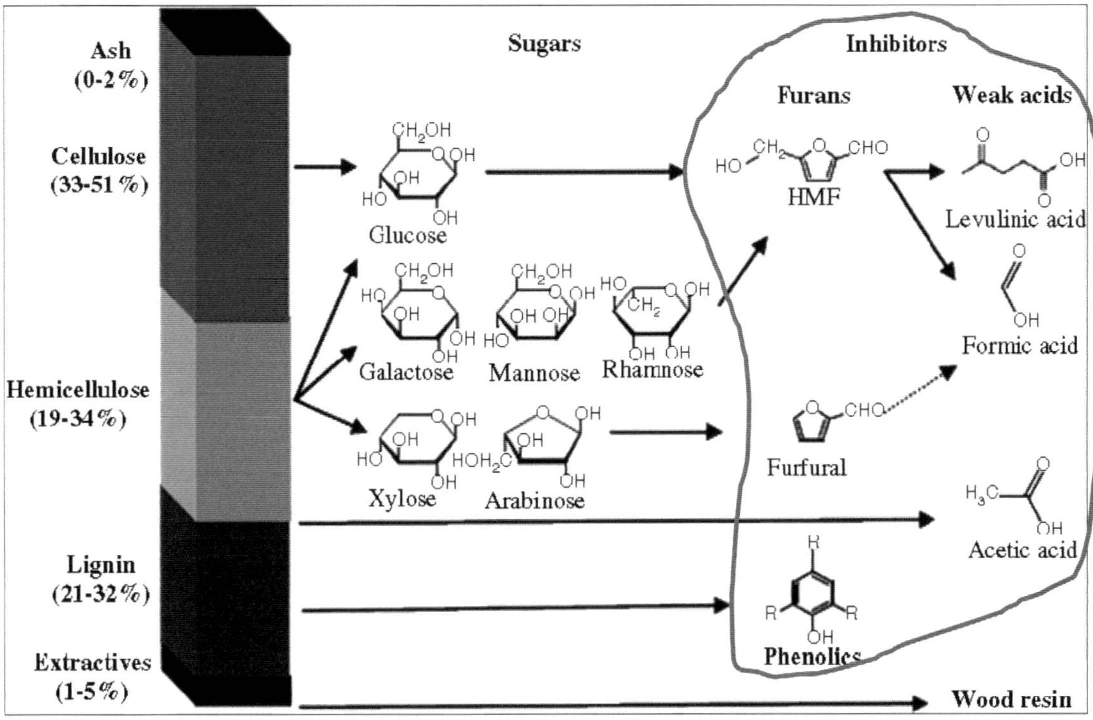

Lignocellulosic Biomass Hydrolysates Are Challenging Feedstock

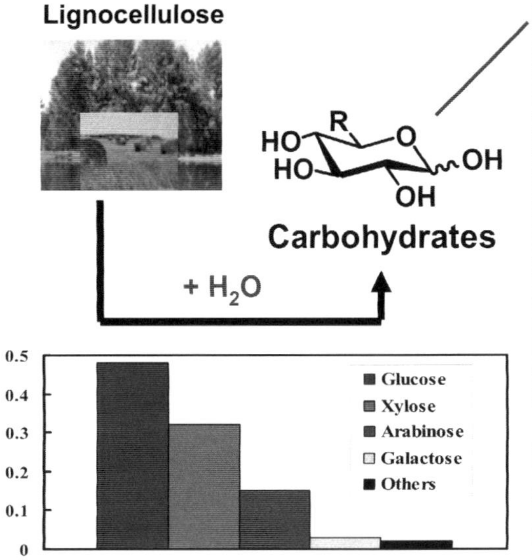

Sugar distribution profile of corn stalk hydrolysates prepared in ionic liquids

Notably different from starchy materials (e.g., corn), because of the following two major features:

- **C_6/C_5 co-presence**

  Major challenge for effective microbial use of C_6/C_5

- **Inhibitors**

Li CZ, et al. *Green Chem* 2008, 10, 177.

14

Robust Lipid Producer

Lipid production by *R. toruloides* in the presence of 6 inhibitors*

	Biomass (g/L)	Lipid content (%)	Lipid yield (g/g)
Control	15.8±0.3	65.2±0.1	0.214±0.001
Inhibited**	13.2±0.1	65.9±0.1	0.211±0.002

*Conditions: glucose 70 g/L, 30 °C

** Acetate 52.2 mM, furfural 10.2 mM, 5-hydroxymethylfurfural 15.0 mM, p-hydroxybenzaldehyde 0.44 μM, vanillin 1.0 μM, syringaldehyde 0.12 μM

▸ Effective lipid production in the presence of biomass-derived inhibitors

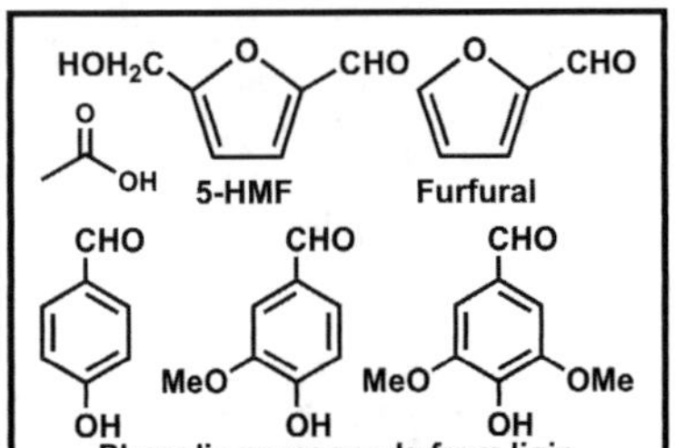

Phenolic compounds from ligin

Inhibitor found in biomass hydrolysates

Hu CM, et al. *Bioresour Technol*, 2009, *100*, 4843.

15

Glucose/Xylose Co-Utilization

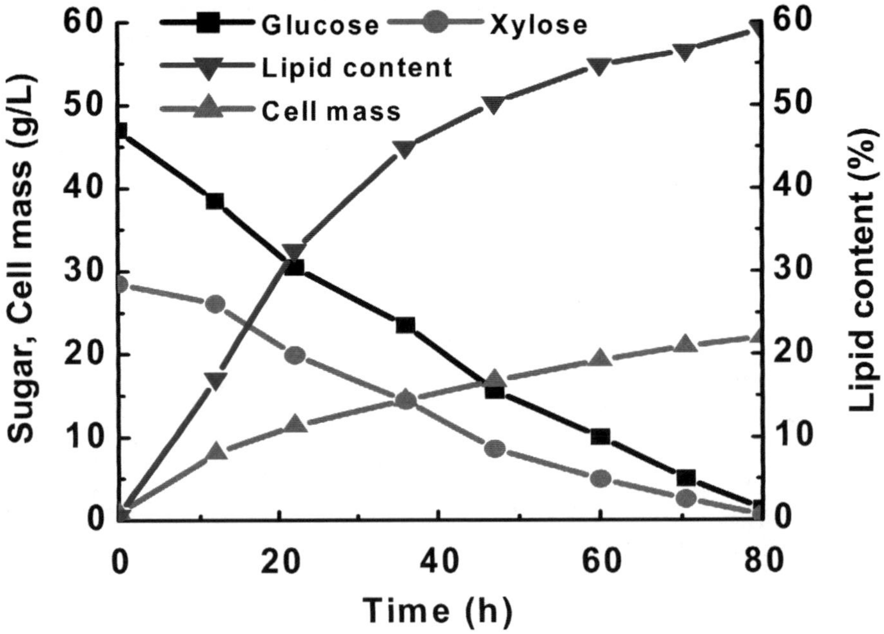

▸ Key to use cellulosic biomass hydrolysates

Hu CM, et al. *Biotechnol Biofuels*, 2011, *4*, 25.

16

Closed-Cycle Lipid Production

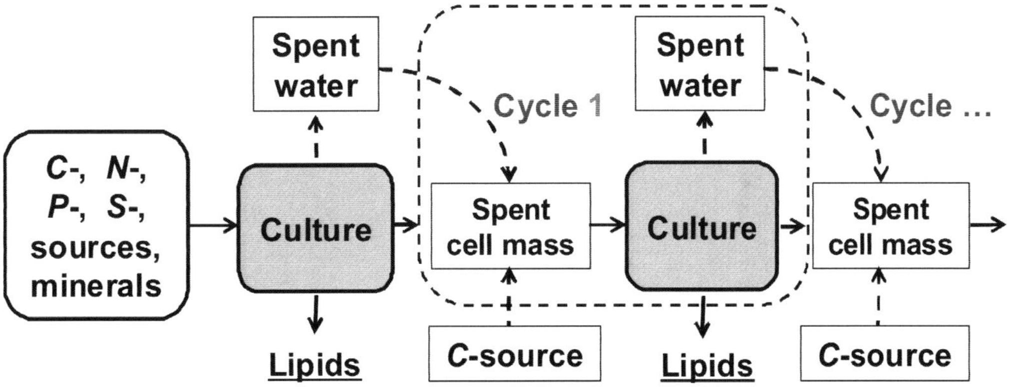

- *R. toruloides* was used as lipid producer
- Both cell mass and spent water recycled
- Lipid titre remained constant for over 3 cycles
- Significantly reduce waste discharge and nutrient costs

Yang XB, et al. *Bioresour Technol* 2015, *175*, 91.

17

Process Integration

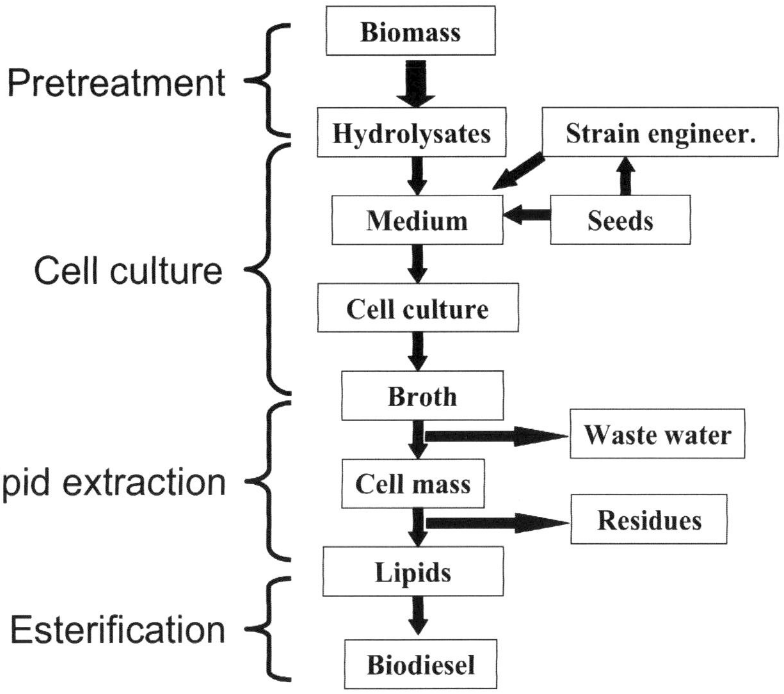

18

Lipid Production on Lignocelluloses

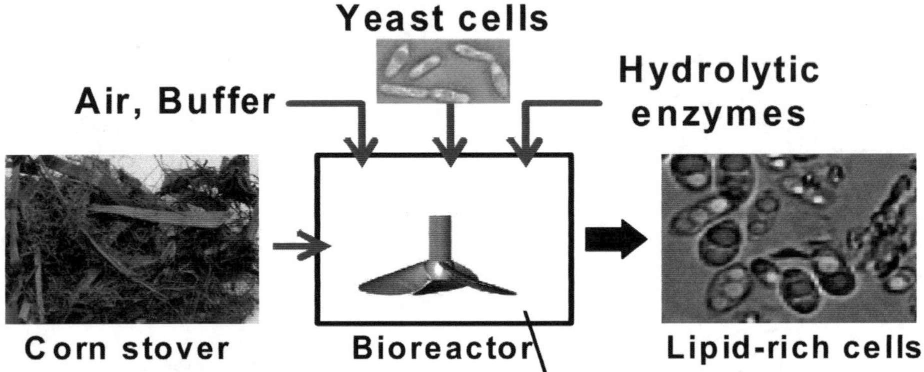

➢ "One-step" process

➢ Co-utilization of C6/C5 sugars

➢ Yield: 0.159 g/g corn stover

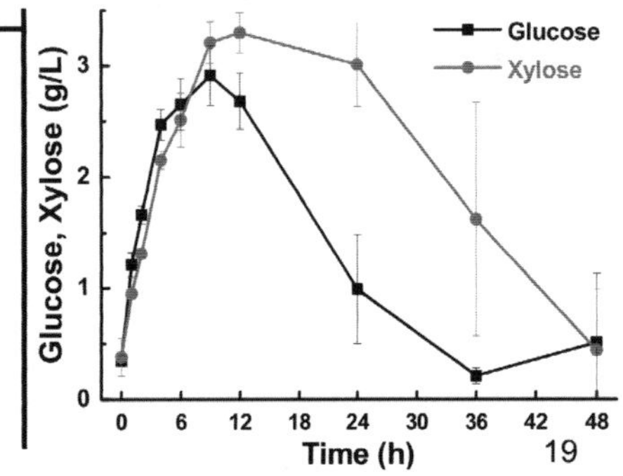

Xie HB, et al. *Green Chem* 2012, *14*, 1202.
Gong ZW, et al. *Biotechnol Biofuels* 2013, *6*, 36.
Gong ZW, et al. *Biotechnol Biofuels* 2014, *7*, 158.

19

Enzyme-Assisted Direct Lipid Extraction

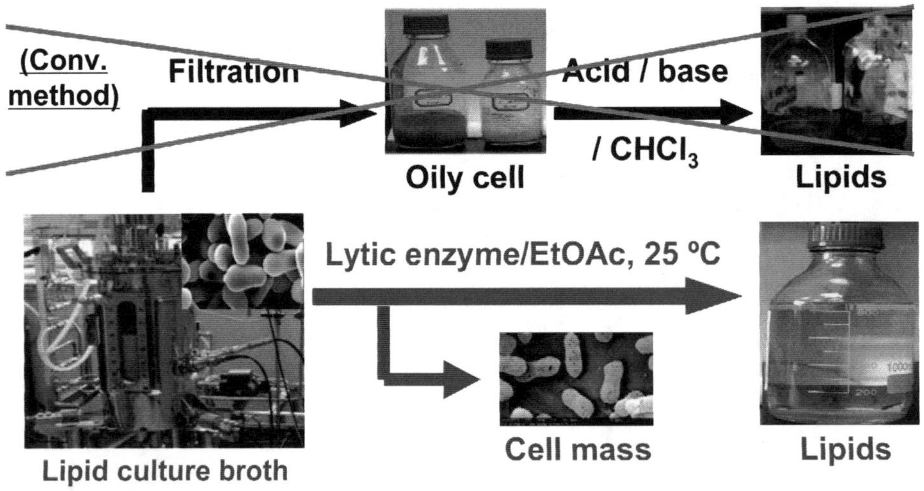

➢ Extraction from culture broth without cell harvest

➢ Working with ethyl acetate at room temperature

➢ Lipid recovery yield >95%

Jin GJ, et al. *Bioresour Technol* 2012, *111*, 378.

20

In-situ Trans-Esterification

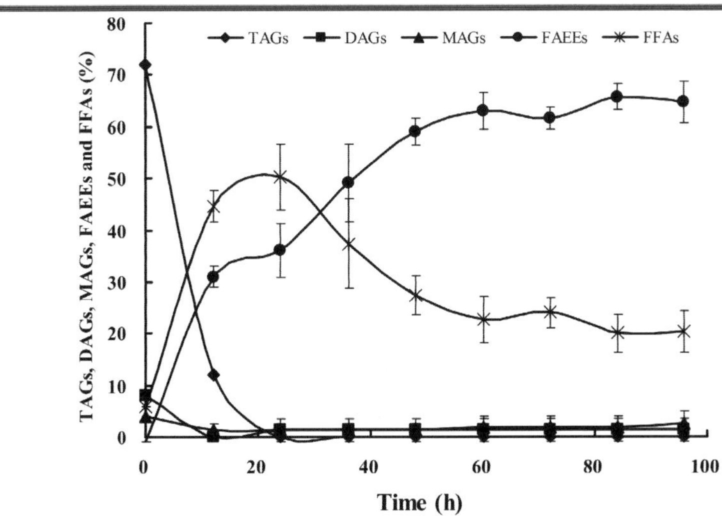

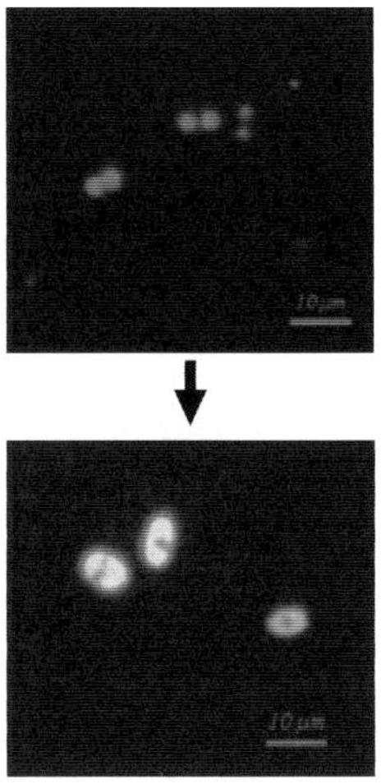

- Direct formation of fatty acid ethyl esters by lipid-rich yeasts in aqueous environment, yield up to 73%
- Cost saving, acid/base free and simpler process

Jin GJ, et al. *Bioresour Technol* 2013, *150*, 266.

21

Multi-Omic Analysis of Lipid Producer

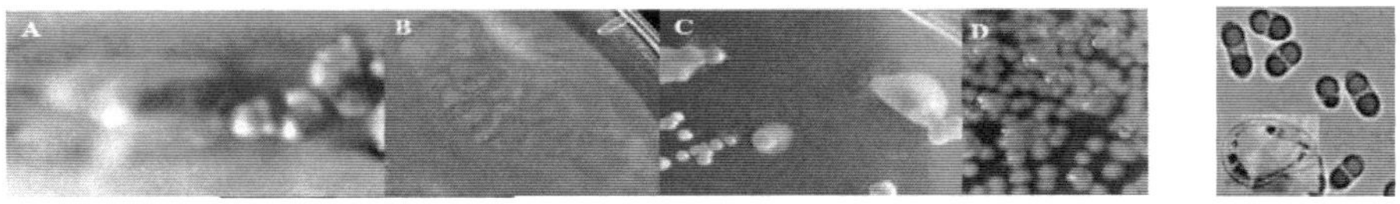

R. toruloides: Basidiomycota yeast

Accumulates lipids up to 70 wt%

Genomic, transcriptomic, and proteomic analysis

- 8171 protein-coding gene anonated

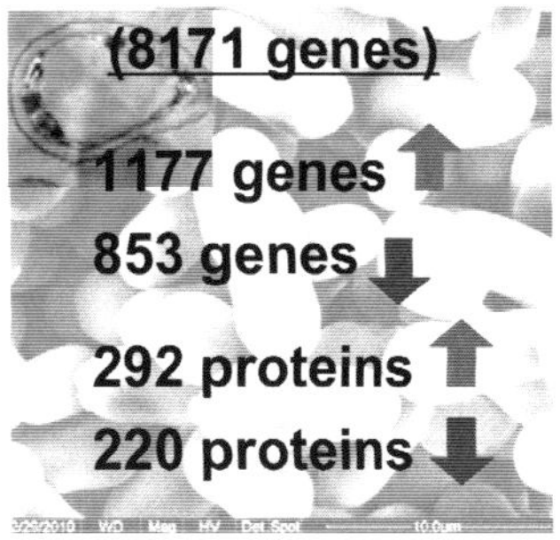

Rhodosporidium toruloides

- Key genes for lipid production identified
- The 1st multi-omic analysis of oleaginous microbe, provided rich and solid infomation for the community

Zhu ZW, et al. *Nat Commun* 2012, 3, 1112

22

Lipid Droplet Proteomics

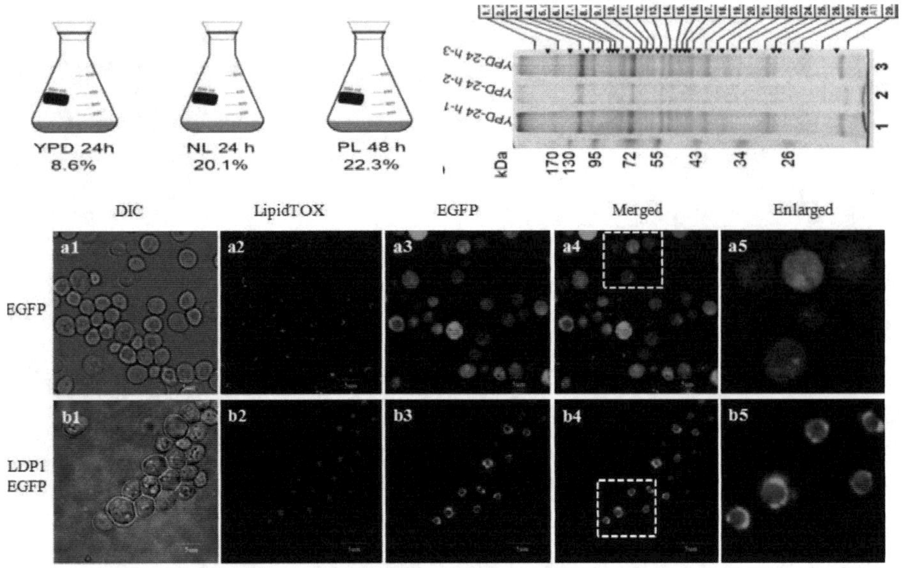

- Identification 226 LD-associated proteins
- Ldp1 faciliates lipid production by *S. cerevisiae*

Zhu ZW, et al. *Eukaryot Cell* 2015, *14*, 252.

23

Genetic Tools

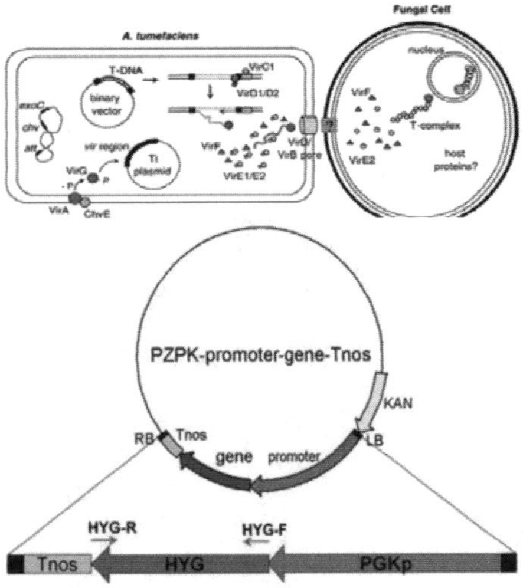

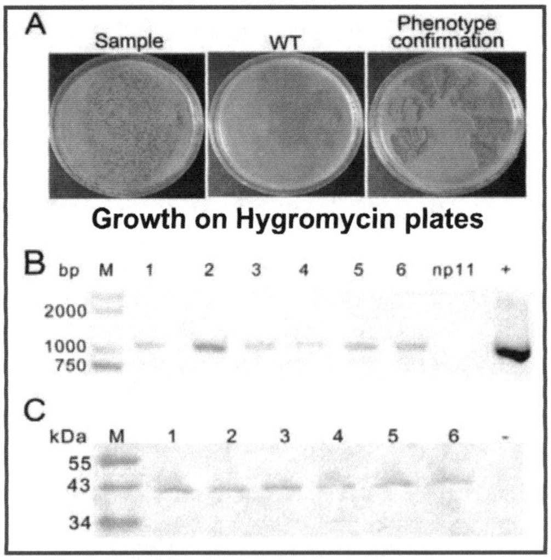

- Establish transform methods and new strains

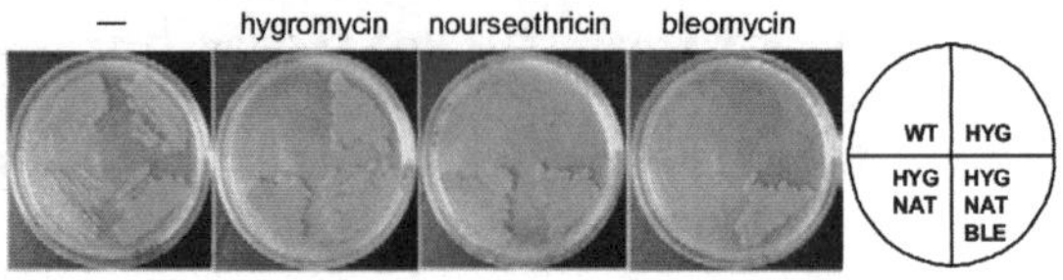

Lin XP, et al. *FEMS Yeast Res* 2014, *14*, 547; Wang YN, et al. *Yeast* 2016, *33*, 99.

24

Biomass-to-Biodiesel

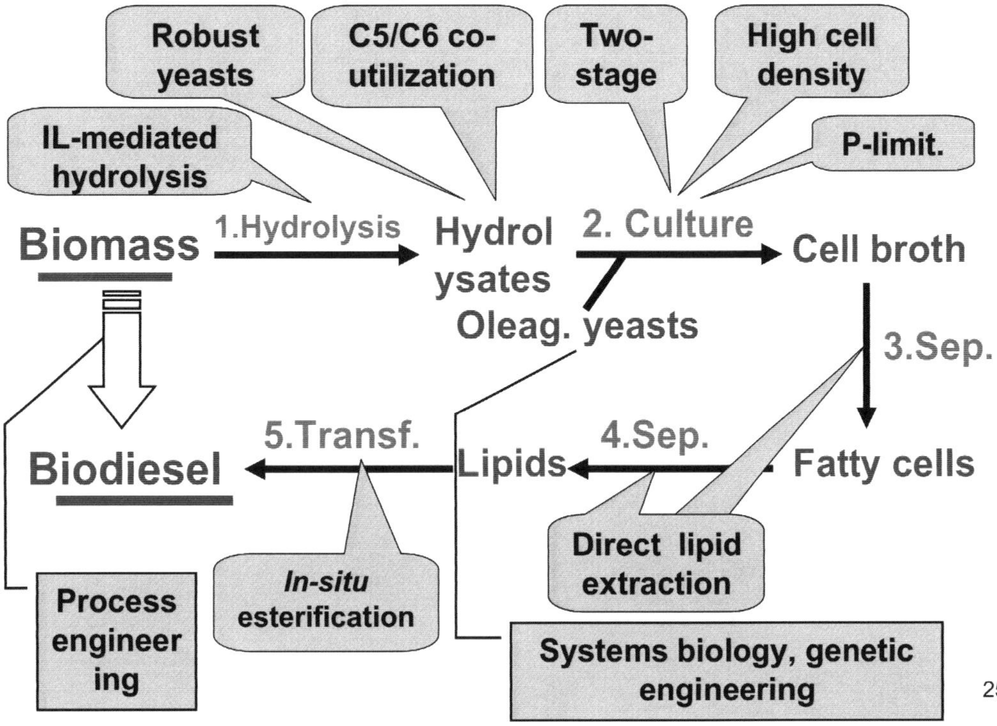

25

Over 10 Years Work on Microbial Lipids

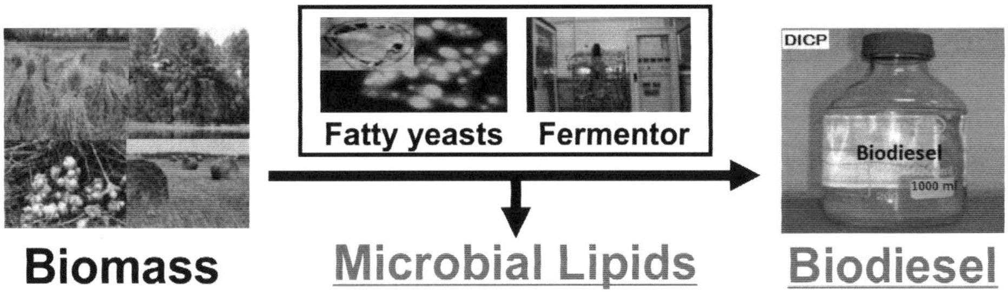

- **Key data: lipid content > 60 wt%; product titre > 100 g/L; lipid yield > 0.20 g/g sugar; productivity > 0.7 g/L/h**
- **>80 papers, 19 patents**

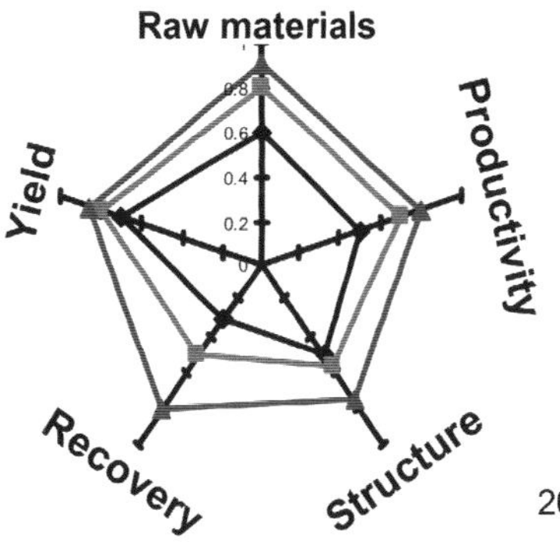

Enzyme Microb Technol. 2007, 41, 312 (259 citation)
J Chem Technol Biotechnol. 2007, 82, 775 (131 citation)
Bioresour Technol. 2009, 100, 4843 (106 citation)
Bioresour Technol. 2010, 101, 6124 (64 citation)
Bioresour Technol. 2011, 102, 1803 (59 citation)
Nat Commun. 2012, 3, 1112 (51 citation)
Biotechnol Biofuels. 2013, 6, 36 (25 citation)
Biotechnol Biofuels. 2014, 7, 158 (8 citation)
Biotechnol Biofuels. 2015, 8, 189

26

➢ Background

➢ Progresses

➢ **Prospects**

27

Fatty Acid Pathway Engineering

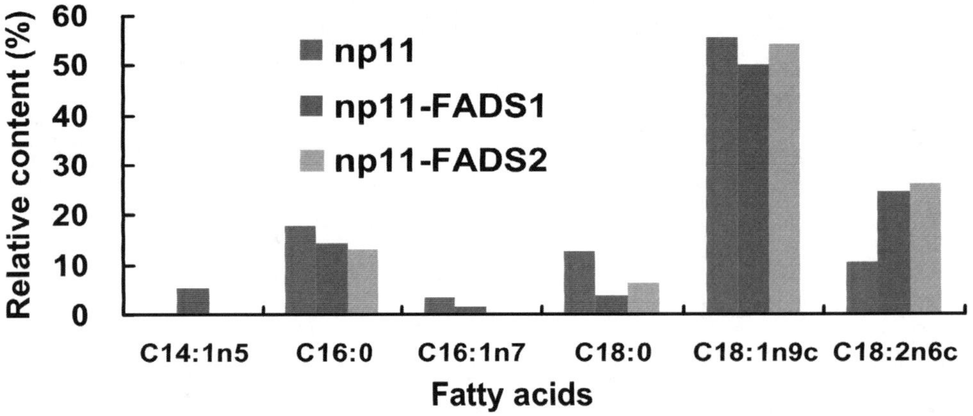

- Engineering *R. toruloides* for unsaturated fatty acids, ie. linoenic acid

Wang YN, et al. *Appl Biochem Biotechnol* 2016, *180*, 1497

28

Engineering Fatty Acid Synthase

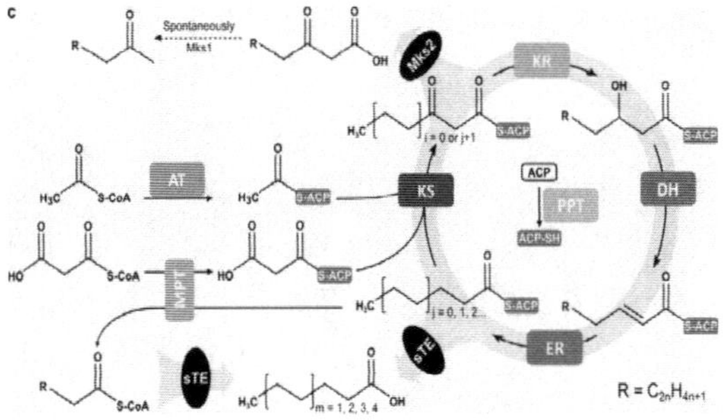

- ➢ Novel FAS with duplicated ACP identified
- ➢ Replace one ACP with sTE or Mks2, led to non-natural FAS, for the production of medium chain fatty acids or methylketones
- ➢ Enriched FAS product portfolio

Zhu ZW, et al. *Nat Chem Biol* 2017, *13*, 360

29

Future Works

- ➢ **Scale-up microbial lipid production technology**
- ➢ **Integrating chemical and biochemical approaches for biomass-to-biodiesel**
- ➢ **Engineering oleaginous yeasts to produce high-valued molecules**

30

Acknowledgement

NSFC

MOST

CAS

DICP

British Petroleum Co. Plc

Tianguan Group

LanzaTech Co.

Evonik Degussa

31

502578 Process Kinetics of Mixed Bacteria Photosynthetic Hydrogen Production

Abstract:

The process kinetics of mixed bacteria photosynthetic hydrogen production is studied by experimental research and model fitting. The results showed that growth cycle of photosynthetic hydrogen production mixed bacteria is the law of the nonlinear. The fittingÂ curve of Logistic model reveal the maximum mixed bacteria dry weight is 0.4323 mg/mL, specificÂ growthÂ rate is 0.1077mg/h and the correlation coefficient is 0.9207. The modified Logistic model is used to fit the substrate consumption. The substrate consumption utilization degree of hydrogen production system which substrate concentration is 5 and 10 mg/mL is higher, but the substrate largest consumption of hydrogen production system of substrate concentration 20mg/mL of is 17.0146 mg/mLã€,When substrate concentration is 5, 10 and 20 mg/mL, the correlation coefficient of the Gompertz model fitting hydrogen production are greater than 99%, and hydrogen production rate respectively reach the maximum in 95, 101, 107 h. The substrate consumption and hydrogen production mainly concentrated in the logarithmic phase and the early stage of stable phase which is 48-120 h.

502553 5-Ethoxymethylfurfural Production from Cellulose Catalyzed By Ultra Low Mass Concentration Sulfuric Acid in One-Pot Reaction

Abstract:

5-Ethoxymethylfurfural (EMF) is a new kind of liquid biofuel,which has a high energy density and excellent combustion properties, and has been considered as one of the promising next-generation biofuels. In this paper, ultra low concentration of sulfuric acid (0.001mol L^{-1}) was used as catalyst to catalyze cellulose and ethanol to produce EMF in one-pot reaction. The effects of temperature,reaction time and substrate concentration on EMF yield were firstly studied and then the response surface methodology experiments were designed to optimize the reaction conditions. The interactions of factors and the optimum reaction conditions were obtained. The results showed that the reaction time and reaction temperature were the factors that mostly impacted the one-pot production from cellulose and ethanol catalyzed by ultra low mass concentration sulfuric acid. The interactions of two factors on the EMF yield were all significant ($P<0.005$). The maximum mean EMF yield of 14.35% was obtained under the optimum reaction conditions: reaction temperature 200 ,reaction time 60min and substrate concentration 20g/L, with the prediction error of 5.35%. This study could provide research foundation for the EMF one-pot production from much cheaper and more abundant biomass resource catalyzed by ultra low mass concentration sulfuric acid.

492488 Rapid Estimation of Life Cycle Inventories

Abstract:

Growing chemical use is a challenge to both chemical manufacturers and regulators given the emerging desire to evolve into a sustainable society. Sustainable chemicals provide benefits to society through their designated functions while minimizing their associated environmental and economic risks related to items such as resource consumption, health and safety, and profitability. As part of alternative chemical assessments, the methodological framework known as life cycle assessment (LCA) accounts for resource use and releases from a network of processes within the life cycle of a product, including extraction, production steps, use, and disposal / recycling. However, a significant challenge to using LCA is the extensive material and energy flow data needed to support assessments. This poster describes methods that can be used to determine the life cycle inventory (LCI) of necessary data to enable LCA and alternative chemical assessments.

The approach to developing LCI data is comprised of four efforts. The first effort focuses on Top-Down Data Mining, which collects data for facilities from EPA sources: Chemical Data Reporting (CDR), National Emissions Inventory (NEI), Toxics Release Inventory (TRI), Discharge Monitoring Report (DMR), etc. Steps for application of data mining and subsequent manipulation are illustrated in Cashman et al. [1]. The second effort applies Bottom-Up Simulation, where design and simulation are used to develop energy and material flows of the system. Additional techniques for estimating uncontrolled emissions from chemical processing equipment are then applied to obtain a detailed emission profile for the process [2]. This simulation method is refined by applying pollution controls to update resource use and controlled emissions. The third effort determines the LCI Chemical Lineage, which quickly determines the chemicals involved in an LCA study, defining the lineage of parents and children in the supply chain of a chemical. Finally, the fourth effort is LCI Reconciliation, which, through the combination of methods, aims to produce inventories that are qualitatively and quantitatively better than the individual methods.

[1] Cashman, S.A., Meyer, D.E., Edelen, A., Ingwersen, W., Abraham, J., Barrett, W., Gonzalez, M., Randall, P., Ruiz-Mercado, G., Smith, R.L. (2016). "Mining Available Data from the United States Environmental Protection Agency to Support Rapid Life Cycle Inventory Modeling of Chemical Manufacturing," *Environ. Sci. Technol.*, 50, 9013-9025.

[2] Smith, R.L., Ruiz-Mercado, G.J., Meyer, D.E., Gonzalez, M.A., Abraham, J.P., Barrett, W.M., Randall, P.M. (2017). "Coupling Computer-Aided Process Simulation and Estimations of Emissions and Land Use for Rapid Life Cycle Inventory Modeling," *ACS Sustainable Chem. Eng.* (Accepted).

The views expressed in this presentation are those of the authors and do not necessarily represent the views or policies of the U.S. Environmental Protection Agency.

494887 Aspects of Sustainable Production of Palm Oil in the Municipality of Teapa in Tabasco, Mexico: Evaluating the Current and Future Use of Palm Oil

Abstract:

Palm has been considered an advantageous food and bioenergy crop, due to its high fruit yields and the fruit bunchesâ €™ high oil content, approximately 40% by mass. Currently in the municipality of Teapa in the state of Tabasco, Mexico several palm plantations exist and are being expanded to accommodate the growing demand for vegetable oil markets and potential biodiesel market. This study site evaluates several different aspects of sustainability when considering palm as a feedstock as a food crop along with the future use of producing palm biodiesel. Social, economic and environmental indicators were used to evaluate the multidimensional understanding of sustainability.Â 55 palm plantations were evaluated and the data aggregated to evaluate the overall sustainability of palm cultivation. Data such as soil C, wild pollinator count, fertilizers and insecticides use, palm fruit yield, etc., were collected from a questionnaire of local plantation owners along with measurements from field ecologists. A life cycle assessment (LCA) was conducted to evaluate overall greenhouse gas emissions, water consumption and releases and air quality of the current production of edible palm oil along with the future use of palm as a renewable energy feedstock. The LCA includes direct soil C land use change, land preparation and plantation, fruit harvest and transportation and oil extraction when evaluating the current use of palm oil. When looking to future use, the LCA includes processing of the palm oil into biodiesel and final combustion within vehicles. Socioeconomic conditions were evaluated within the state of Tabasco to look at social and economic impacts of expanded palm production for palm oil and/or palm biodiesel. Indicators such as working conditions, income, and employment were evaluated within the current use of the palm oil and were extrapolated to evaluate future production of palm biodiesel. This study takes a multidimensional approach to sustainability and recommends best practices for the continued expansion of palm oil and/or palm biodiesel production.

501534 Harvesting and Extraction Technologies Contributing to Algae Biofuel Environmental Viability-Life Cycle Analysis of Naabb Developed Novel Technologies

Abstract:

There are many challenges and technical barriers to implementation of a thorough life-cycle assessment of algae-based biofuels. Biofuels made from algae must typically go through a complicated cultivation, harvesting, dewatering, oil extraction, and conversion process, involving many unit operations. Feedbacks between unit operations often depend on the particular operations being performed, so selection of a particular technology will have upstream as well as downstream consequences in the value chain. The National Alliance for Advanced Biofuels and Bioproducts (NAABB) project was established to produce and evaluate new technologies that can be implemented by the algal biofuel industry and establish the overall process sustainability. The goal of our LCA is to determine greenhouse gas (GHG) emissions and fossil energy demand associated with harvesting and extraction of algae. Four harvesting technologies and two extraction technologies within the NAABB group were evaluated. These include: Chitosan Flocculation, Electrolytic Flocculation, Membrane Harvesting, Acoustic Harvesting, Acoustic Extraction, and Wet fractionation. The scope will be limited in the sense that our focus is only on inputs of harvest and extract technologies, not including the materials and chemicals used to cultivate algae. The functional unit for harvesting technologies study is 1 kg dry algae output. Harvesting technologies will be normalized to this unit, while extraction technologies will be normalized on the basis of 1 kg algae oil output. Input/output data on the industrial operations being investigated was used to construct LCA models of the unit operations. Using the input data from multiple resources, the harvesting and extraction processes were all placed on an equivalent basis for comparison and the GHG emissions impacts were analyzed using SimaPro LCA software. In order to assess the magnitude of each technology in a similar fashion as the other unit operations, the gate-to-gate results for these unit operations were placed in the context of our previous LCA baseline modeling and the results for these unit operations are presented in g CO_{2eq} / MJ renewable diesel using the assumptions detailed in the Baseline Harmonized Report and GREET model.

495711 Comparative Techno-Economic Analysis of Algal Biofuel Production Via Hydrothermal Liquefaction: One Stage Versus Two Stages

Abstract:

Algal biofuels have been widely accepted as promising renewable energy resources and receive extensive investigations in four main research areas: algal feedstocks, cultivation methods, harvesting/dewatering technologies, and conversion processes. Conversion of algal biomass is the essential downstream step to realize high-efficient and cost-effective production of biofuels and associated co-products. Hydrothermal liquefaction (HTL) stands out in practically handling wet algal slurry and avoids toxic solvent usage as well as produces biofuels both in reasonable quantity and quality. Pacific Northwest National Laboratory (PNNL) developed a direct HTL (DHTL) and upgrading process to convert whole algal material to biofuels. This method could enable conversion of non-oil components in raw algae feedstocks such as polysaccharides and proteins to bio-crudes, thus improves overall bio-oil yield, which furthermore relieves constraints of strain-specific operation conditions. However, biofuels derived from DHTL contain high level content of nitrogen (1.5-6.0 wt.%) than crude oil (0.1 wt.%) operated at high temperature (350 â„ƒ) and pressure (20.9 MPa). SubsequentÂ denitrification process and high manufacture cost due to the sever reaction conditions would make DHTL process economicallyÂ unfeasible.

To address the drawbacks of DHTL, a novel sequential hydrothermal liquefaction (SEQHTL) for simultaneous extraction of co-products and bio-oil from algal biomass has been proposed by our group. This process highlights recovery of proteins and polysaccharides at first stage and extraction of bio-oils at second stage at much lower temperature and pressure. The pre-removal of protein substances at first stage substantially decreases nitrogen content of bio-oil obtained at second stage by 30-55% and enhanced fatty acids exaction yields additionally.

A comprehensive techno-economic analysis (TEA) of DHTL of algae associated with oil upgrading was performed by PNNL, resulting in minimum fuel selling price (MFSPï¼‰ $4.77 gal-1. However, this cost derived from DHTL cannot achieve the overall goal of 3 $/GGE established by the Bioenergy Technologies Office (BETO) by 2022. Based on sensitivity analysis, the total investment could play an important role in determining algal MFSP besides feedstock cost, thus decreases on investment cost could offer a desirable MFSP. In this case, SEQHTL process is expected to provide less investment cost due to the low operating temperature and pressure. Moreover, isolated proteins and polysaccharides can be regarded as co-products credits to further compensate production cost.

We have two phases analysis procedure. Our first goal is to compare our proposed SEQHTL process using *Chlorella* with DHTL process using another algae species, *Nannochlorpsis*. In this phase all the DHTL process data were retrieved from PNNL reports and SEQHTL data were retrived from pervious experimental work in our group. Preliminary results showed that HTL equipment cost of SEQHTL is only 60% of that in DHTL. The total capital investment is 150 million cheaper than DHTL. The sugar products indeed make some credits in operating cost. However, the final oil selling price of SEQHTL was still higher than DHTL because of the different strains applied (Oil conversion rate was 50% in PNNL, while our strain only have 30%). If we increase the oil yield up to the same value as DHTL, the selling price decreases to $4.21 gge. The second phase will be conducted to compare cost between SEQHTL and DHTL with same algal strain. Preliminay results showed that slimilar oil conversion rates were realized for these two methods, thus SEQHTL process is expected to be more economic to produce algal biofuels than DHTL in terms of capital costs, operating costs, and minimum fuel selling price.

In addition, sensitivity analysis indicated that four major factors that affecting MFSP in SEQHTL process were algal feedstock cost, bio-oil conversion rate, sugar selling price, internal rate of return. The cost of algae is the top reason that blocks the development of algal biofuels. If the cost could be lower to 70% of original price, the fuel selling price will be decreased by $1/gge. Realizing high bio-oil conversion yield is rather important, the more produced, the lower price could be sold. Additionally, the selling price of co-products plays an important role. If we could extract high value co-products as well as biofuels. The MFSSP could drop substantially.

The purpose of this study is to provide preliminary TEA result of SEQHTL process and demonstrate the potential of replacing conventional DHTL process with promising SEQHTL as a novel platform in conversion technology, especially with co-products extraction from natural biomaterials. Future R&D should be focused on enhancing bio-oil production yield and effective separation of co-products in the field of conversion technology as well as reducing the cost of algae feedstocks. Although uncertainty exists due to the lack of experimental data and scale-up experience of HTL facilities, SEQHTL process is a promising and competitive conversion technology to handle wet feedstocks and fractionate biomass into high valued co-products.

497701 Economic and Environmental Assessment for the Production of Propylene Glycol from Biodiesel Glyceroleconomic and Environmental Assessment for the Production of Propylene Glycol from Biodiesel Glycerol

Abstract:

Different countries around the world are implementing policies and legislations to promote the sustainable development of their economy (i.e. bioeconomy). Biofuels have been one of the most promoted areas in the transition towards sustainable schemes and many regulations seek their industrialization and commercialization. As a result, over the last years biodiesel production has gained track in the market, resulting in large amounts of waste glycerol being generated as low-cost and readily available feedstock.

Focusing on the enhancement of the bioeconomy and the valorization of biodiesel glycerol, we here address the economic and environmental assessment under uncertainty for different routes in the production of propylene glycol from biodiesel glycerol. More precisely, we compare three different hydrogenolysis routes based on the use of biodiesel glycerol between themselves and against the business as usual process, which uses oil-based propylene oxide as raw material. Using commercial simulators, an initial case for each alternative was first implemented and then optimized through standard heuristics, sensitivity analysis and heat integration. The economic performance was quantified using the economic potential and total annualized cost, while the environmental impact was assessed following Life Cycle Assessment principles using the methodology CML 2001. Finally, the main uncertainties affecting the calculations were evaluated via stochastic modelling.

The results revealed that there are glycerol-based alternatives that outperform the current propylene glycol process simultaneously in terms of profit and environmental impact (i.e. 90% increment in profit and 74% reduction in environmental impact at optimum process conditions).

Overall, our results aim to promote the development of sustainable biorefinery schemes to convert waste glycerol into high-value commodity chemicals, like propylene glycol, within the context of holistic bioeconomy frameworks.

489453 More Than Ethanol: A Techno-Economic Analysis of Corn Stover-Ethanol Biorefinery Integrated with Hydrothermal Liquefaction Process to Convert Lignin into Biochemicals

Abstract:

Energy being an â€œingredientâ€ for every economic activity presents challenges of providing qualitative and quantitative energy to the increasing global population. Biorenewable approach of producing energy in form of biofuels and biochemicals is the way to go. Hydrothermal liquefaction (HTL) process is an alternative to fast pyrolysis (FP) that can be used to thermal chemically convert feedstock into hydro-oil, gases, and solids. HTL compared with FP, has advantages such as low bio-oil oxygen content of 10 to 20 wt% compared to 40%, higher oil heating value of 35 MJ/kg compared to 16 -19 range MJ/kg of FP. HTL processes wet biomass whereas FP requires dry biomass thus saving on energy that would have been required for drying feedstock. Research is underway to improve biorefinery revenues by producing biochemicals and biofuels from lignin which accounts for 40% biomass energy, and 10 to 35% by weight. Besides lignin from paper and pulp industry, also ethanol plants produce lignin as a waste, thus, the objective of this work is to evaluate the techno-economics of integrating ethanol biorefineries with hydrothermal liquefaction of lignin to produce biochemicals. A 2000 metric ton per day (MTPD) corn Stover biorefinery produced 61 MMgal/yr. of ethanol and different yields of chemicals. The minimum ethanol selling price (MESP) accompanied with respective chemical were $2.70/gal (catechol), $3.17 (phenol), $3.27/gal (m-cresol), $2.80/gal (p-cresol), $3.22/gal (o-cresol), $2.91/gal (acetic acid), $2.91/gal (formic acid), $3.32/gal (furfural), and $3.36/gal (acetaldehyde). Total capital investment is M$ 479.6. The most influential factor on the minimum ethanol selling price is ethanol yield. Other critical factors are fixed capital investment, feedstock flowrate, and ethyl acetate price. Lignin derived chemicals using HTL and separation of the produced biochemicals is in the early stages of development thus more investment in terms of fixed and operating cost might be required for future commercial biorefinery plants than those obtained in this analysis.

500146 A Mixed Integer Programming Model for Municipal Waste Management and Landfill Gas-to-Energy Systems

Abstract:

The interest on the optimal management of municipal solid waste (MSW) has increased considerably due to technological and industrial developments. Several technologies have been developed as an attempt to address the problem of waste disposal from a waste-to-energy perspective. Apart from conventional problems associated to solid waste management (land space usage, high costs of transportation, groundwater pollution, etc.), methane emissions from landfill sites (Landfill Gas; LFG) represent an unresolved major problem. However, despite that environmental issue, landfilling is currently the best alternative for waste management in the developing world.

Mathematical programming approaches have been proposed to address many issues concerning waste management from various perspectives. As an extension to the existing approaches, this paper presents a mixed integer multiperiod approach to optimize a MSW network, considering waste reduction processes and landfilling. An original formulation is proposed to improve the economics of the landfill operation. The model incorporates an optimal LFG-to-energy design, which involves the installation of several harnessing technologies depending upon the annual increasing or decaying of the LFG flow rate. A direct-hauling system and an optimally located off-municipality transfer station are used to optimize the transportation of the MSW to the potential destinations. Actual data from a municipality in Mexico (Celaya, Mexico) has been used to define our case-study. The resulting Mixed-Integer Linear Programming (MILP) model has been solved through the GAMS® modeling environment. Our formulation has been assessed through several scenarios, including different LFG collection efficiencies and a bioreactor landfill operation. The minimum total cost of the MSW management is achieved through the installation of an LFG-to-electricity system and a materials recycling facility.

Keywords: *Municipal solid waste management, mathematical programming, Landfill gas-to-energy.*

496211 Techno-Economic Analysis of Integrated Solid Oxide Fuel Cell - Gas Turbine - Organic Rankine Cycle Powered By Seaweed Biogas

Abstract:

Due to escalating environmental issues, alternative energy sources are required instead of fossil fuels. One of the alternatives for stationary power generation are solid oxide fuel cells (SOFC) which electrochemically convert fuel into electricity and heat. By utilizing biogas from brown seaweed (*Saccharina japonica*) as a fuel for SOFC, power generation is considered carbon neutral.

Electrical efficiency of biogas-powered SOFC is calculated to be around 45 %, while it generates 2.6 MW of heat per MW of electricity. This makes it plausible for integration into a combined heat and power (CHP) generation process. Integration of SOFC with gas turbine (GT) and organic rankine cycle (ORC) to utilize the heat, can increase the efficiency by 10-15%.

In this work, energy and economic viability of integrated SOFC, Gas Turbine (GT), and Organic Rankine Cycle (ORC) are investigated for industrial size of stationary power generation. This system is profitable with the value of corresponding minimum power selling price in the range of US average power selling price of 2016.

494516 Photocatalytic Reforming of Biomass for Hydrogen Production

Abstract:

Photocatalytic reforming of biomass for hydrogen production

Renée M. Ripken, Vincent J.H.W. de Boer, Johannes G.E. Gardeniers and Séverine Le Gac

MESA+ Institute for Nanotechnology

University of Twente, The Netherlands

Here, we describe a novel microfluidic device to determine the required bandgap for the photocatalytic reforming of biomass model substrates (ethylene glycol, glycerol, xylose and xylitol) in water. Furthermore, this device is applied to eventually elucidate the reaction mechanism of aqueous photocatalytic reforming, which is currently still a matter of debate.

Combustion of fossil fuels is one of the main causes of the climate change, due to the production of harmful CO_2. As a consequence, significant scientific efforts aim at developing cleaner and more sustainable fuel resources. Hydrogen is a promising alternative, as hydrogen has a high energy density and no harmful products are formed during its combustion. Hydrogen is currently produced from natural gas or coal by gasification, liquefaction or steam reforming [1]. However, these methods are highly energy demanding and they still require natural resources that are not renewable. In contrast, PhotoCatalytic Reforming (PhCR) only uses solar energy to convert aqueous phase biomass waste streams into H_2 and CO_2. [2]

Although PhCR appears to be a clean alternative for hydrogen production, currently used catalysts are relatively inefficient as they only use the UV-band from the entire electromagnetic spectrum. Only UV-light has sufficient energy to activate the catalyst by means of creating electron-hole pairs. To improve the efficiency of PhCR, new catalysts must be identified that are activated with a wider range of wavelengths. Therefore, we have realized a microfluidic platform to determine electrochemically the minimum required energy potential to perform the photocatalytic reforming of four biomass model substrates, ethylene glycol, glycerol, xylose and xylitol. These compounds were dissolved in water and a small amount of H_2SO_4 was added to ensure enough conductivity of the solution.

The microfluidic device includes electrodes to which a potential (up to 3 Volt) corresponding to the bandgap of a photo catalyst can be applied, mimicking an electrochemical potential. Cyclic voltammetry is used to evaluate the potential at which degradation of the substrate occurs. To study the reaction mechanism, the gases produced at the cathode and the anode sides are kept separate by an array of small channels (length: 50 Î¼m, depth and width: 2 Î¼m, spacing between the channels: 3 Î¼m) as a result of capillary pressure. The gaseous products are next extracted from the aqueous phase by means of a PDMS membrane placed on the outlets and subsequently analysed by inline GC. The

liquid phase products are collected from the device and characterized offline by HPLC.

Recorded cyclic voltammograms clearly show that a reaction is taking place for ethylene glycol and xylitol, which is also confirmed by GC and HPLC analysis. Based on the identification of the reaction products, we suspect that both proposed reaction mechanisms in the literature are taking place. [3-5] However, deactivation of the electrode surface was also observed as a result of carbon deposits, as confirmed by SEM and XPS.

In a next step, this device will be applied to help identify the optimal band gap for the photo catalyst, which would significantly improve the efficiency of PhCR.

References

1. G. W. Huber, J. A. Dumesic, *Catal. Today* **2006**, *111*, 119-132.
2. M. de Oliveira Melo, L. Almeida Silva, *Theoretical and Experimental Chemistry,* **2009**, Vol. 45 No. 6, 343-348
3. M. de Oliveira Melo, L. Almeida Silva, *Chem. Soc. Chem.* **2011**. Vol. 22, 1399-1406
4. N. Stratakia et al. *Appl. Chem. Cat. B: Environ.*, **2007,** Vol. 77, (12) 184-189
5. M. Kawai et al. *Chem. Phys. Lett.* **1984** Vol. 110 (1) 58-62

Acknowledgements

The authors wish to thank Prof. Dr. Guido Mul (University of Twente, The Netherlands) and Fang Liu (Utrecht University, The Netherlands) for all the support and help with the analysis.

Furthermore, the authors thank Remco Sanders, Stefan Schlautmann and Sip-Jan Boorsma (University of Twente, The Netherlands) for the help with the development of the experimental set-up and microreactor fabrication.

This work was supported by the Netherlands Center for Multiscale Catalytic Energy Conversion (MCEC), an NWO Gravitation programme funded by the Ministry of Education, Culture and Science of the government of the Netherlands.

499837 Experimental Analysis of Catalytic Gasification of Waste

Abstract:

(1) Title:

Experimental Analysis of Catalytic Gasification of Waste Polymers

(2) [Undergraduate] Student Researchers:

- Samuel O. Sanya
- UC Obiako
- Aliandra Barbutti
- Stephen Reeves
- Eric M. Lange

(3) Faculty Mentor:

- Dr. Jorge Gatica

(4) Abstract

Catalytic gasification, or "gasification", is one effective method which can be applied to promote low-temperature conversion of solid waste to energy. This research focuses on advancing the knowledge of a catalytic gasification process as a potential in-situ resource utilization and waste management alternative. This research has significance in a variety of engineering applications, but it is of particular relevance towards reducing landfill waste or as an in-situ resource generation system for space exploration beyond LEO.

This process evolves through a reaction mechanism consisting of two liquid-phase oxidation reactions of long-chain polymers, complemented by two gas-phase reactions. Using a laboratory grade reactor, the gasification of two model substrates, polyethylene and cellulose, was examined in the presence of ruthenium and platinum catalysts.

Gaseous products consisting of hydrogen, methane, carbon monoxide, carbon dioxide, were quantified by Gas Chromatography (GC). Calibration of the GC using chromatographic standards enabled the formulation of an equation to calculate the composition of the gaseous products for different reaction times. Fundamental reactor design equations along with stoichiometry calculations were then used for determinations of percent polymer gasified as well as selectivity of the gasification process.

Additional characterization of the solid and liquid residuals was performed using Differential Scanning Calorimetry (DSC), Thermo-Gravimetric Analysis (TGA), and Scanning Electron Microscopy (SEM). Quantification of the DSC spectra was used to correlate the thermal characterization of the residues with the unconverted (or non-gasified) substrate after the reaction was quenched. The SEM provided information on the microstructure of the residues, their atomic composition, and preliminary assessment of the possibility of catalyst recovery.

Preliminary results suggest that catalytic gasification provides potential control for the liquid-phase oxidations. Under appropriate conditions, oxidation was found to follow the incomplete oxidation pathway which resulted in significant yields of hydrogen and methane, which are potentially high value products as they can be used for energy recovery.

Comparisons between the gasification rates for the two substrates are drawn. Similarly, comparative studies of different catalysts are also presented. The results are used in conjunction with a kinetic model for the gas-phase to formulate a methodological model for the gasification of solid waste. The results will be presented along with a preliminary extension for a continuous gasification process.

500200 Investigation of Closed Loop Bioponic Irrigation Systems for Urban Agriculture

Abstract:

Bioponics is a novel take on traditional hydroponics farming in which petrochemicals are replaced by naturally derived nutrients in the form of "teas". The irrigation system is virtually the same as in hydroponics farming minus the inclusion of petrochemical fertilizers. What is used instead is a biologically active, nutrient "tea" that is brewed by aeration of fresh compost and water and is then pumped through the irrigation system to crops. The main factor in the effectiveness of these fertilizer teas is the large population of active bacteria. These bacteria are the same species of bacteria found in healthy topsoil and there is thought to be many links between these bacteria and markedly improved crop yields within a bioponics system (Vessey, 2003).

Traditionally, a hydroponics system are a single pass system. Meaning that large amounts of water are passed through the system into the plant containers and then passed out of the system to not be used again. This causes a large amount of waste water to be produced which is difficult to process especially in urban settings where such systems may be used. This also poses numerous environmental problems since this waste stream contains unused fertilizers that pose a threat to vulnerable ecosystems if not properly treated. The model-based study of lab and pilot-scale systems could prove beneficial in the search for more sustainable options for low-impact urban farming or even large scale, soilless agriculture.

495345 A Mechanistic Model for the Product Distribution of Fast Pyrolysis of High Density Polyethylene Waste

Abstract:

Fast pyrolysis emerges as an interesting alternative for the recycling of plastic waste on an industrial scale, despite incomplete information about the fundamental principles governing this process. Among the most important research areas is the development of an accurate kinetic model to predict the formation of different hydrocarbon classes when plastic is subjected to high temperatures and different residence times in an inert atmosphere. Several researchers have developed kinetic models that attempt to estimate the pyrolysis products distribution at different temperatures and residence times. One of the first models was proposed by Bockhorn et al.[1], it predicted the thermal degradation of polyethylene and polypropylene by using mass spectrometry measurements of pyrolysis products under isothermal conditions. Costa et al.[2] established a simplified model by grouping the products of polyethylene pyrolysis into four classes. This study idealized a series of chemical reactions that gave a kinetic model with only nine elemental first order reactions. Other models attempted to predict the product distribution of over 100 intermediate and hydrocarbon products using over 10,000 reactions.[3,4] Nevertheless, these models require complex methods for solving the differential equations that are not easily accessible. Mastral et al.[5] developed the sole kinetic model for fast pyrolysis of polyethylene under short residence times (0.52-2.07 s) and high temperatures (500-1000 °C). This model predicted the production of 8 groups of linear hydrocarbon and aromatic products. Random chain scission has been proposed as the mechanism responsible for the degradation of polyolefins into shorter hydrocarbon molecules through a set of free radical-based chain reactions. In this mechanism, polymers are broken down through a series of initiation, propagation, and termination steps. The complexity of these sequential reactions suggests that new insights may be gained by isolating the primary reactions from the secondary reactions. Therefore, this project attempts to develop a kinetic model to estimate the product distribution of primary and secondary products of the fast pyrolysis of waste HDPE. Primary reactions are carried out in a micro-pyrolysis CDS Analytical 5200 HP unit while secondary reactions occur in reactor extension attached to the micro-pyrolyzer. Products of waste HDPE degradation will be tested at high temperatures (650 â€" 700 Â°C) and short residence times (~0.0 to 5.3 s) where no char is produced and be analyzed using a gas chromatography-mass spectrometry equipment. The developed model will be based on the random chain scission mechanism described previously and will describe the product distribution of four classes of linear hydrocarbon compounds and aromatics. Possible reactions that describe the formation of upcoming hydrocarbon species will be discussed.

References

1. Bockhorn H, Hornung A, Hornung U, Schawaller D. 1999. Kinetic study on the thermal degradation of polypropylene and polyethylene. Journal of Analytical and Applied Pyrolysis 48(2):93-109.
2. Costa PA, Pinto FJ, Ramos AM, Gulyurtlu IK, Cabrita IA, Bernardo MS. 2007. Kinetic Evaluation of the Pyrolysis of Polyethylene Waste. Energy & Fuels 21(5):2489-2498.
3. Németh A, Blazsó M, Baranyai P, Vidóczy T. 2008. Thermal degradation of polyethylene modeled on tetracontane. Journal of Analytical and Applied Pyrolysis 81(2):237-242.
4. Levine SE, Broadbelt LJ. 2009. Detailed mechanistic modeling of high-density polyethylene pyrolysis: Low molecular weight product evolution. Polymer Degradation and Stability 94(5):810-822.
5. Mastral JF, Berrueco C, Ceamanos J. 2007. Theoretical prediction of product distribution of the pyrolysis of high density polyethylene. Journal of Analytical and Applied Pyrolysis 80(2):427-438.

495712 A Model-Based Approach for Sustainability Assessment of Biomethane from Anaerobic Digestion of Food Waste and Manure Mixtures

Abstract:

Abstract:

Anaerobic digestion (AD) is a process involving a series of biological reactions (Hydrolysis, acetogenesis, acidogenesis and methanogenesis), which converts the biodegradable organic materials into biogas comprised mostly of CH_4 and CO_2. As an earlier study on lifecycle assessment of Bio-CH_4 has shown (Ankathi et al. 2017) [1], disposal of the food waste in landfills and conventional animal manure management has significant methane emissions contributing 20% to the U.S. national total. When these wastes are managed by AD, avoiding the landfilling and conventional manure management emissions, it is estimated that life cycle greenhouse gas emissions are -3.5 kg CO_2 eq./kg Bio-CH_4. The major advantage of AD on converting the food waste and manure into Bio-CH_4 is the displacement of fossil natural gas. On full-scale application of U.S. food waste / manure, AD saves 0.74 % of current annual natural gas consumption through large-scale Bio-CH_4 production.

A model-based approach involving AD modelling and Aspen process simulation for sustainability assessment is proposed based on an existing model by Batstone et al 2002 (ADM1) [2]. Aspen simulation includes different unit operations for Bio-CH_4 production; mixing, heating, anaerobic digestion, membrane separations, and gas compression. This kind of approach helps to study the effect of heat integration in the process and to predict the variations in the output concentrations of AD gases based on input compositions of food waste and manure. Through this approach, the economic costs and profits and environmental impacts will be analyzed and optimized.

Â References

1. Ankathi Sharath K., James S. Potter, David R. Shonnard, â€œCarbon footprint and energy analysis of Bio-CH_4 from a mixture of food waste and dairy manure in Denver, Colorado.â€ The manuscript is currently under review in *Environmental Progress and Sustainable Energy*. The manuscript number is EP-17-129.
2. Batstone, D. J., et al. (2002). "The IWA anaerobic digestion model no 1 (ADM1)." Water Science and Technology **45**(10): 65-73.

500196 Efficient Saccharification of Softwoods By an Integrated Thermochemical and Biological Process

Abstract:

Softwoods have great potentials as renewable feedstocks for biorefineries because of their high cellulose content. However, softwoods are more recalcitrant compared to herbaceous plants and hardwoods. Hence, the use of softwoods in biorefineries has been viewed as limited. To overcome their recalcitrance, we applied a cellulose solvent-based fractionation process on southern yellow pine and douglas fir at a mild condition: 50 ºC and 1 atm for 90 min. Pretreated softwood samples were hydrolyzed fast by cellulase enzyme (15 FPUs) and > 85% glucose yield was achieved by 24 h. Cellulose accessibility to cellulase of pretreated samples was characterized by the fusion protein and was found to increase by 39 folds, suggesting that enhanced surface accessibility via cellulose solvent-based fractionation was responsible for the high glucose yield. After fractionation, the degree of crystallinity of pretreated samples decreased and fibrous structure appeared to be disrupted as observed by solid-state 13C Cross-Polarization Magic Angle Spinning (CP/MAS) nuclear magnetic resonance (NMR) and scanning electron microscopy. No changes in lignin functionalities after fractionation were observed by 2D Heteronuclear Single-Quantum Correlation (HSQC) NMR, giving rise to the potential use of lignin as a co-product in biorefineries.

495604 Single-Objective Versus Multi-Objective Optimization of Integrated Fermentation and *in situ* Product Recovery Based on Time-Dependent Fermentation Models

Abstract:

Process simulation provides a platform that can be used to systematically analyze the process at hand, evaluate the feasibility of different process alternatives, and provide the most informative scenarios that can be experimentally investigated to aid in process design and optimization. The batch fermentation is characterized by low final product concentration, yield, and reactor productivities as a result of product toxicity to the microorganisms. Approaches to reduce product toxicity include integrated batch fermentation and *in situ* product recovery techniques, such as gas stripping, adsorption, membrane separation, etc. A broad range of experimental conditions in terms of gas recycle flow rate relative to fermentation volume, initial substrate concentration, gas stripping start time relative to fermentation start times, etc. have been investigated experimentally for fermentation and *in situ* gas stripping processes. Process optimization will reveal the optimal conditions of the process to guide the design of high performing integrated fermentation and *in situ* gas stripping processes across the broad range of experimental conditions investigated.

Typically, there exist a trade-off among the performance parameters of integrated fermentation and *in situ* gas stripping processes. For example, the total product produced increases with increasing gas flow rate whereas the concentration of products in the stripped stream (condensate) decreases with increasing gas flow rate in an integrated fermentation and *in situ* gas stripping process. The trade-off in these performance parameters stem from the fact that the parameters are conflicting; conflicting parameters mean as one parameter is enhanced, another parameter is simultaneously made worse. This phenomenon typically leads to conflicting objectives in optimization of the process. Single-objective optimization translate a multiple objective system into one objective system by either using relative weights to combine different objective functions into one objective function or choosing one objective function as the main objective function while transforming other objective functions into additional constraints. In contrast, multi-objective optimization is able to optimize multiple objectives simultaneously even in the presence of conflicting objectives, revealing multiple optimal solutions relative to the single-objective optimization that finds only one optimal solution.

To this end, this study will use and compare (a) single-objective optimization and (b) multi-objective optimization approaches for fermentation and *in situ* gas stripping processes under different conditions. An unsteady state batch fermentation and *in situ* gas stripping (simulated with a batch reactor in Aspen Plus linked to a Fortran user kinetics subroutine) will be connected to the multi-objective genetic algorithm in MATLAB for the multi-objective optimization. This study will highlight the advantages of multi-objective optimization to reveal and understand the trade-off that exist among conflicting objectives over traditional single-objective optimizati

496186 Enzyme Cocktail Design of β-Agarase Enzymes for Complete Hydrolysis of Agarose in Ionic Liquid-Pretreated *Gelidium amansii*

Abstract:

Red macroalgae *Gelidium amansii* is a prospective renewable resource for producing D-galactose and 3,6-anhydro-L-galactose due to its ubiquity and high carbohydrate content. Its main component agarose, is a linear polysaccharide with alternate repeating units of D-galactose and 3,6-anhydro-L-galactose linked via β-1,4- and α-1,3-glycosidic bonds. Enzymatic saccharification using β-agarase enzyme system with subsequent acetic acid or thermal pre-hydrolysis is currently the most popular method to produce the monomeric sugars from agarose. However, accumulation of 5-hydroxymethylfurfural (HMF) as sugar dehydration product, and generation of agarotriose as a byproduct that is recalcitrant to complete enzymatic hydrolysis, have been typically encountered.

Ionic liquid (IL) pretreatment with subsequent enzyme cocktail hydrolysis using β-agarases is a facile strategy to effectively deconstruct and hydrolyze the agar into its monomeric sugars, and to eliminate the generation of unwanted products. Moreover, the use of an IL system as an effective and eco-friendly solvent could alleviate enzyme activity losses in unfamiliar solvent phases. Four ILs namely [EMIM]Cl, [BMIM]Cl, [EMIM]OAc and [BMIM]OAc were employed to dissolve *G. amansii* from which [BMIM]OAc was found most suitable. The enzyme cocktail of β-agarase I (endo-acting), β-agarase II (exo-acting), and neoagarobiose hydrolase (NABH) were subsequently used for D-galactose and 3,6-anhydro-L-galactose production from the IL-dissolved *G. amansii*. Performance optimization of IL-enzyme cocktail system is currently under investigation. This work was supported by the Korea Research Fellowship Program through the NRF funded by the Ministry of Science, ICT and Future Planning (2015H1D3A1062172 and 2016R1C1B1013252) and the Ministry of Education (2009-0093816).

492357 Development of Bottom-up Life Cycle Inventory Methods for Chemical Reaction Systems

Abstract:

Bottom-up life cycle inventory (LCI) methods involve the use of process models to obtain the material and energy flows at various stages within an overall product life cycle. Process models of chemical reactions utilize the reaction rate, physical and chemical characteristics of the reaction mixture, and desired reaction extent to determine the reactor vessel configuration. Material flows of interest to life cycle inventory for the chemical reaction system include the reactant and product flows, byproducts, catalysts, and other reaction aids required. Energy balances for the reactor provide an estimate of the energy requirements. Energy flows include addition or removal of heat from the reaction system to maintain proper reaction conditions. The overall life cycle inventory must include an enumeration of the energy requirements and ancillary material flows, as well as the reactor infrastructure. Releases from the reactor, including fugitive emissions must be considered as part of the life cycle inventory. This presentation will focus on the development of life cycle inventory data from models of chemical reaction systems. The approach will include shortcut methods for sizing equipment and fugitive emission estimation.

UNIVERSITY OF LOUISVILLE.
CONN CENTER FOR RENEWABLE ENERGY RESEARCH

National Institute of Technology Tiruchirappalli
राष्ट्रीय प्रौद्योगिकी संस्थान तिरुचिरापल्ली

SIMULTANEOUS PRODUCTION OF BIOGAS AND HYDROGEN USING MICROBIAL ELECTROLYSIS CELL INTEGRATED WITH ANAEROBIC DIGESTER

Samsudeen N[a], Josh Spurgeon[b] and Jagannadh Satyavolu[b, *]
[a]Department of Chemical Engineering, National Institute of Technology, Tiruchirappalli, India
[b]Conn Center for Renewable Energy Research, University of Louisville, Kentucky, USA
Corresponding e-mail: Jagannadh.Satyavolu@Louisville.edu

Background

Microbial Electrolysis Cell (MEC) utilizing anaerobic digestion (AD) technology is a green technology for sustainable hydrogen production from wastewater. Simultaneous production of biogas as well as biohydrogen by integrating MEC in to an AD unit can increase the COD degradation rate in waste water as well as improve the overall economics of fuel production. The main aim of the present objective is to enhance the hydrogen and biogas production by integrating MEC with AD. The integrated system performance in terms of hydrogen and biogas generation and waste water treatment efficiency is investigated in this study

Reaction Mechanism

Biogas reaction $C_2H_4O_2 \rightarrow 2CO_2 + CH_4$

Anode : $C_2H_4O_2 + 2H_2O \rightarrow 2CO_2 + 8e^- + 8H^+$

Cathode: $8H^+ + 8e^- \rightarrow H_2$

Materials and Methods

Mixed inoculum was collected from MSD plant, Louisville, KY. The integrated system was designed with an anode chamber capacity of 450 mL and a cathode chamber capacity of 25 ml – both separated by a Nafion 117 membrane. Carbon cloth and Pt mesh used as anode and cathode electrode. Synthetic wastewater used as substrate in MEC. Potentiostat is used to apply potential to the MEC to produce hydrogen. Biogas and Hydrogen produced in MEC were analyzed using Gas chromatography.

Results

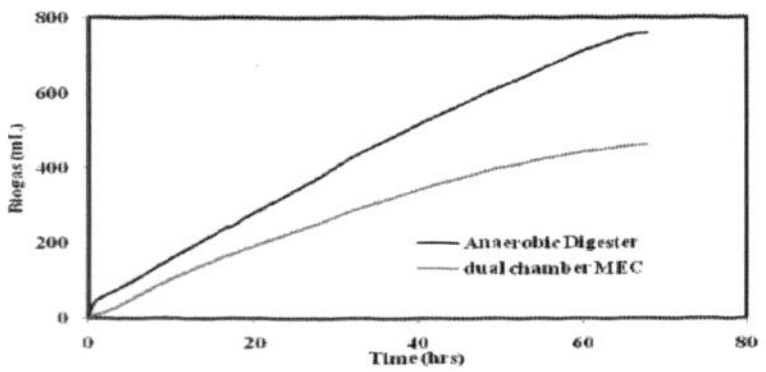

Fig. 1. Biogas production comparison in AD and MEC

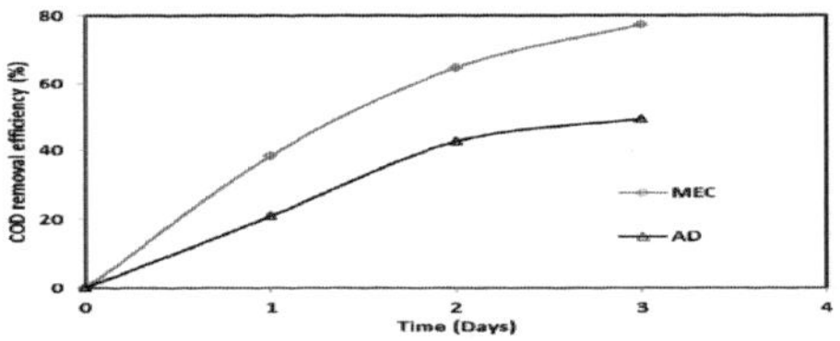

Fig.2. COD removal efficiency in AD and MEC

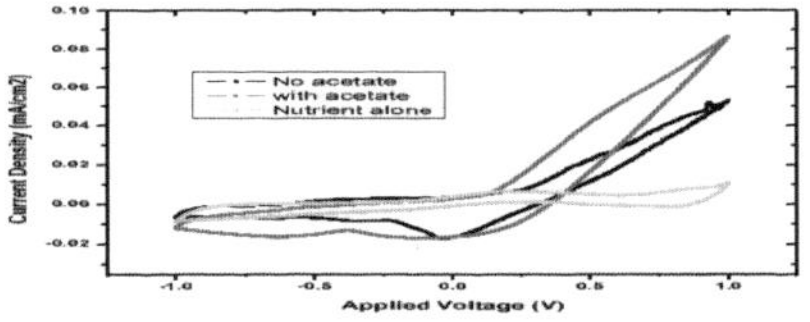

Fig. 3. CV analysis showed an effect of acetate in MEC

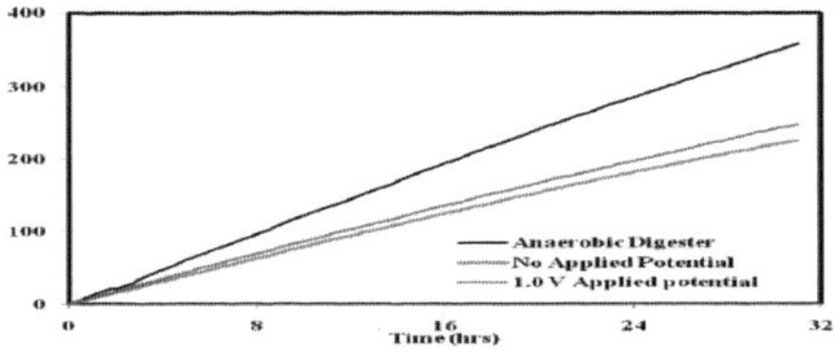

Fig. 4. Effect of voltage apply for biogas production

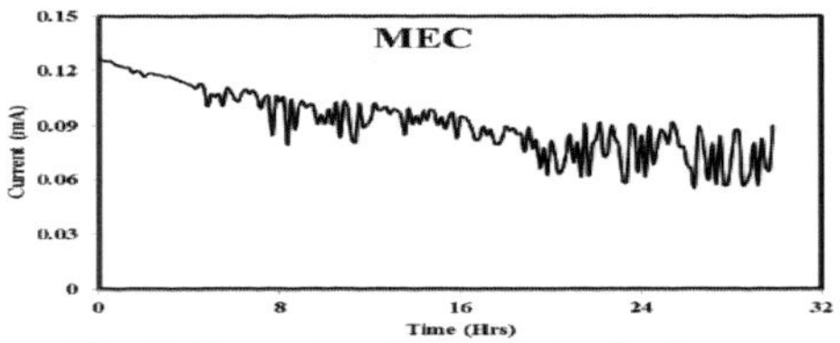

Fig. 5. Current variation during hydrogen production

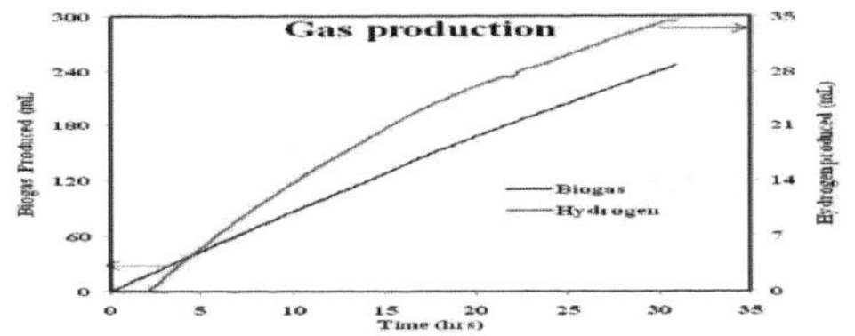

Fig. 6. Biogas and hydrogen production in MEC

Conclusions

- Biogas and hydrogen production by integration MEC into AD was studied.
- A maximum biogas and hydrogen of 232 mL and 32 mL were produced in 32 hours.
- Biogas and Hydrogen gas was confirmed by GC analysis.
- Cyclic Voltammetry (CV) showed a significant effect of presence and absence of acetate in MEC.

Acknowledgement

The authors thank Dr. Mahendra Sunkara of University of Louisville and Mr. Robert Bates of MSD treatment plant, Louisville, KY for their support during this work. Dr. Samsudeen also acknowledges the support from the Department of Biotechnology, Indo-US Science and Technology Forum (*IUSSTF*) for financial support for the B-ACER Award.

References

1. **Oh & Logan (2005).** Hydrogen and electricity production from a food processing wastewater using fermentation and microbial fuel cell technologies. *Water Research*, 39(19), 4673–4682.
2. **Kadier et al., (2016).** Recent advances and emerging challenges in microbial electrolysis cells (MECs) for microbial production of hydrogen and value-added chemicals. *Renewable and Sustainable Energy Reviews*, *61*, 501–525.

490378 Unconventional OIL and Natural Gas - Science & Technology Advancement

Abstract:

Despite the significant growth in the production of Unconventional Oil and Gas (UOG) resources during the shale revolution, very little is known in detail about recovery efficiency in these hydraulically-fractured reservoirs. The limited information that is available suggests that recovery factors (the ratio of produced resource to total in-place resource) are typically quite low: approximately 20 percent in gas-rich shale reservoirs and less than 10 percent in liquid-rich plays. By increasing the recovery factors of hydrocarbons, the environmental impact of these activities would also be reduced.

Therefore, the primary research challenges include, but are not limited to: 1) understanding the nano-scale structure of UOG reservoirs in order to more effectively create and control stimulated rock volume; 2) improving recovery efficiency to reduce the incidence of stranded/unproduced resources and the related lost economic and energy security benefits; 3) understanding the anticipated continued growth in recoverable resource volumes with technology advance; and 4) accurately assessing the potential nature, scale, and rate (intensity) of future development in various areas of the country to guide assessment of region-specific and environmental impacts.

Hydraulic and natural fracture diagnostic technologies have not advanced beyond microseismic monitoring techniques, and there is a need to improve the characterization of the far-field fracture geometry for recovery factor calculations and improvement. Advanced diagnostic and monitoring tools, as well as subsurface models, can be capable of predicting induced seismic events and may provide insight and benefit to our understanding and modeling of the stimulated reservoir volume during hydraulic fracturing operations.

Once the geometries of natural and induced fractures are better understood, new physical and chemical treatments can be formulated for each specific well to achieve maximum recovery factors. Advanced stimulation technologies, such as energized foams or other novel fluids could work in tandem with advanced fracture diagnostic tools to improve recovery efficiency and reduce potential environmental impacts.

Alternatives to Decline-Curve Models for Unconventional Reservoirs: A Case for Data-Driven Discovery of Natural Laws

17AIChE
Annual Meeting, Minneapolis, MN

Bharat Thakur & Michael Nikolaou
Petroleum Engineering
Chemical & Biomolecular Engineering
University of Houston, TX

Advances in Unconventional Oil and Gas Modeling

University of Houston

UNIVERSITY of HOUSTON

Decline-Curve Analysis (DCA): Conventional Reservoirs

- Decline curve analysis:
  - Estimate future production
  - From limited initial production data
- DCA for conventional reservoirs (Arps, 1944):

$$q = \frac{q_0}{(1 + bD_i t)^{1/b}}$$

- Arps curves:

$$P = P_0 e^{-t/a}$$

$$P = P_0 \left(1 + \frac{1}{a_0} t\right)^{-1}$$

$$P = P_0 \left(1 + \frac{b}{a_0} t\right)^{-1/b}$$

- Respective cumulative production:

$$Q = a(P_0 - P)$$

$$Q = a_0 P_0 \log\left(\frac{P_0}{P}\right)$$

$$Q = \frac{a_0 P_0^b}{1-b}\left(P_0^{1-b} - P^{1-b}\right)$$

17AIChE
Annual Meeting, Minneapolis, MN

Thakur
Nikolaou

Data-Driven Decline Curves

UNIVERSITY of HOUSTON

Introduction

Data

Data Modeling

Conclusions

3

17 AIChE
Annual Meeting, Minneapolis, MN

Decline-Curve Analysis (DCA): Unconventional Reservoirs

- DCA for unconventionals:
  - Arps not accurate
  - Alternatives cumbersome:
    - Different flow behaviors during:
      - ❑ Initial long transient,
      - ❑ Transition to boundary-dominated,
      - ❑ Boundary-dominated
    - Two models typically used
    - Unclear
      - ❑ when to switch models
      - ❑ which model to use in transition period

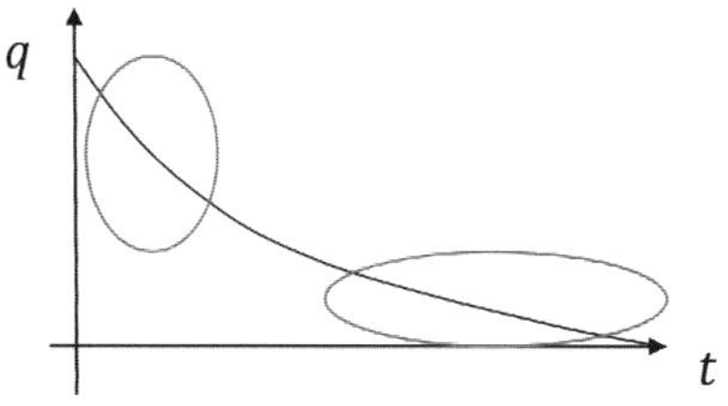

- Proliferation of Arps-like modified equations:
  - Fetcovich (1980)
  - Camacho-Velasquez (1987)
  - Robertson (1988)
  - Fulford & Blasingame (2013)
  - Duong (2011)
  - Valko (2009)

Thakur
Nikolaou

Data-Driven Decline Curves

UNIVERSITY of HOUSTON

Introduction

Data

Data Modeling

Conclusions

4

17 AIChE
Annual Meeting, Minneapolis, MN

Proposed Approach: Data-Driven DCA

- Data-driven discovery of governing equation:
  - Collect long-time production data
  - Develop Arps counterpart
    - Parsimonious

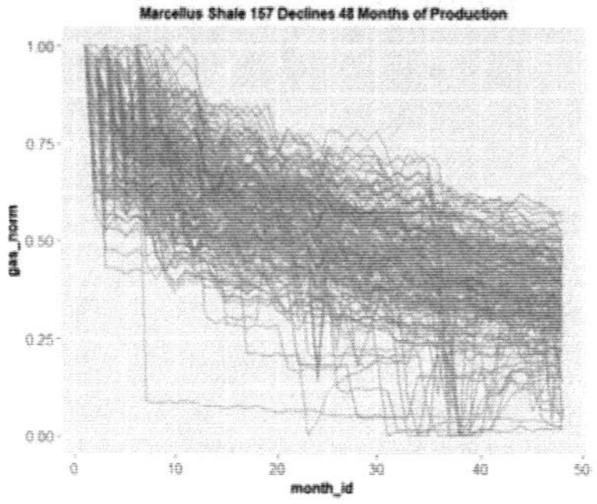

- Use Arps counterpart
  - Fit limited-time production with Arps counterpart
  - Extrapolate over time

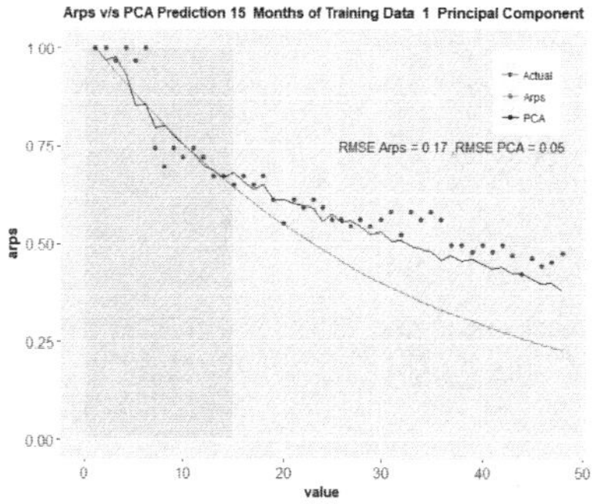

Thakur
Nikolaou

Data-Driven Decline Curves

UNIVERSITY of HOUSTON

Outline

Introduction

Data

Data Modeling

Conclusions

- Data of this study
- Modeling of data
- Conclusions

5

17 AIChE
Annual Meeting, Minneapolis, MN

Thakur
Nikolaou

Data-Driven Decline Curves

UNIVERSITY of HOUSTON

Data Analyzed

Introduction

Data

Data Modeling

Conclusions

- Lease Level Shale reservoir data analyzed:
  - Marcellus
  - Eagleford
  - Barnett

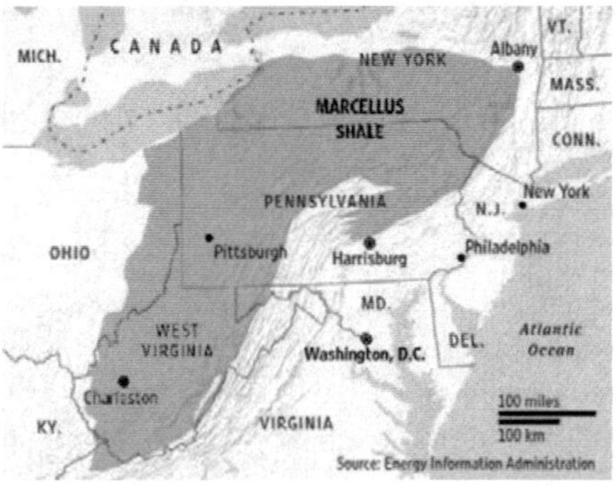

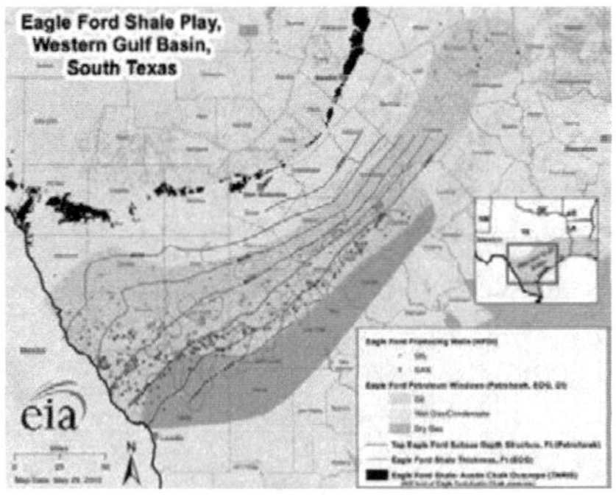

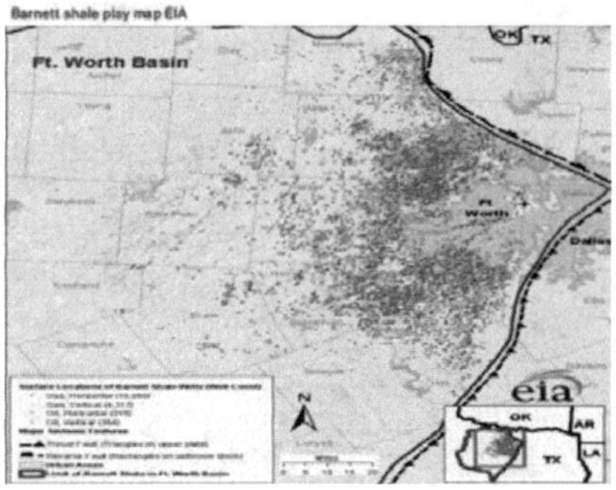

6

17 AIChE
Annual Meeting, Minneapolis, MN

Thakur
Nikolaou

Data-Driven Decline Curves

UNIVERSITY of HOUSTON

Introduction

Data

Data Modeling

Conclusions

7

17AIChE Annual Meeting, Minneapolis, MN

Data Analyzed

- Shale reservoir data analyzed:
  - Marcellus
  - Eagle Ford
  - Barnett

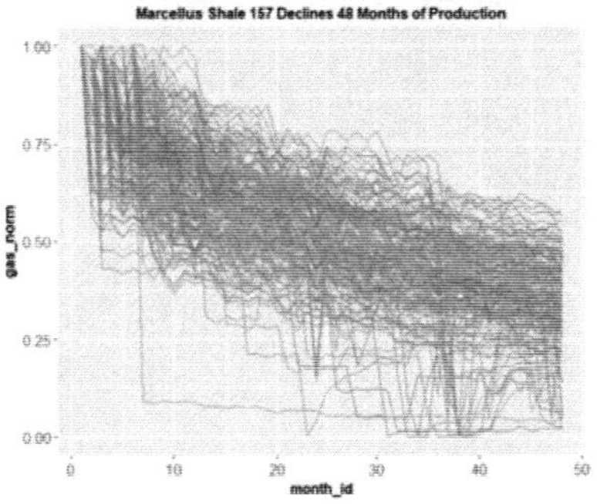

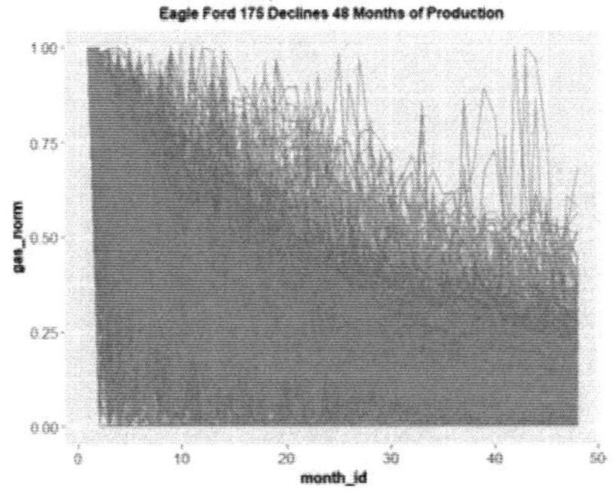

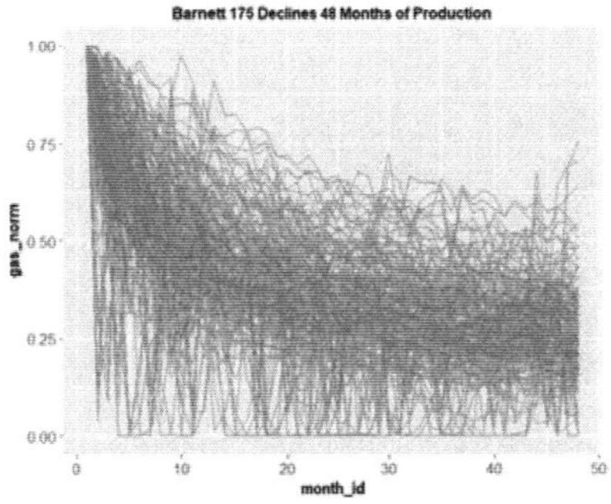

Thakur
Nikolaou

Data-Driven Decline Curves

UNIVERSITY of HOUSTON

Introduction

Data

Data Modeling

Conclusions

8

17AIChE Annual Meeting, Minneapolis, MN

Data Pre-processing

- Normalize the decline to 0-1 range with respect to Max Rate.
- Select decline curves with at least 48 months of production.
- Filter decline curves if average monthly CGR at least 1
- Remove if production increases for particular no of times and by particular percent
- From the Clean curves, subtract monthly average and perform PCA on deviation variable

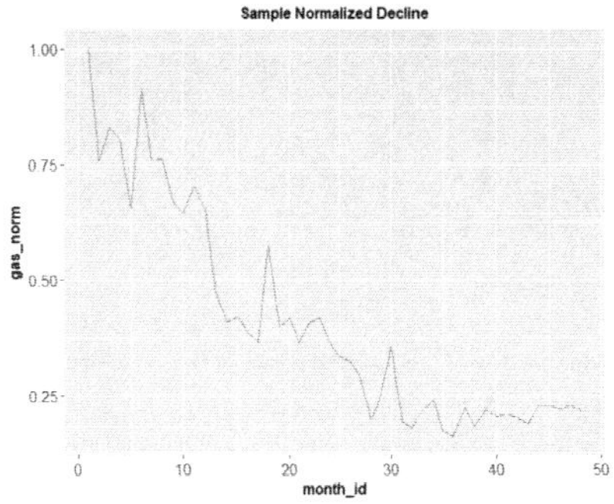

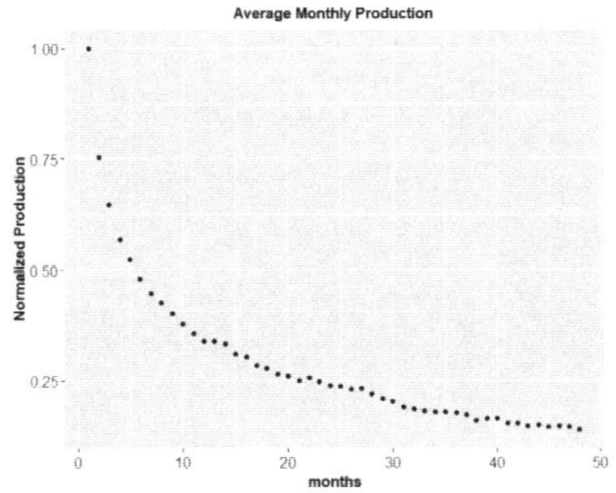

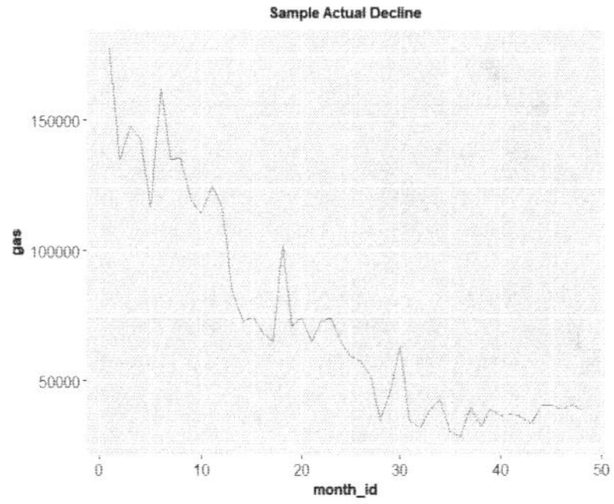

Thakur
Nikolaou

Data-Driven Decline Curves

UNIVERSITY of HOUSTON

Marcellus

Introduction

Data

Data Modeling

Conclusions

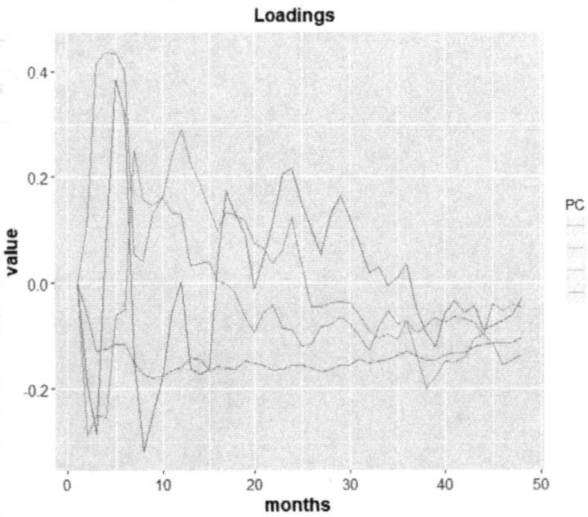

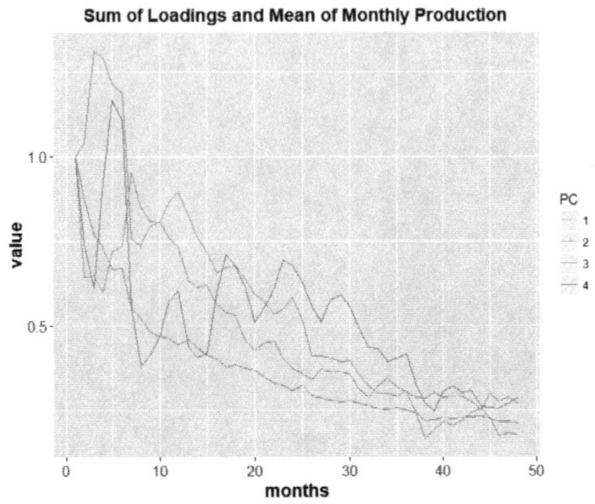

9

17 AIChE
Annual Meeting, Minneapolis, MN

Thakur
Nikolaou

Data-Driven Decline Curves

UNIVERSITY of HOUSTON

Eagle Ford

Introduction

Data

Data Modeling

Conclusions

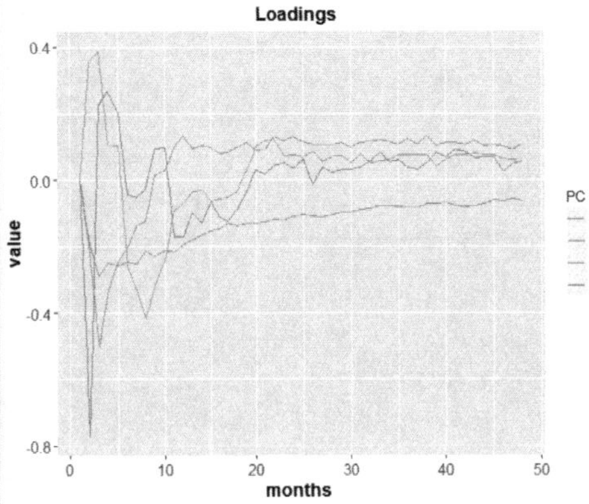

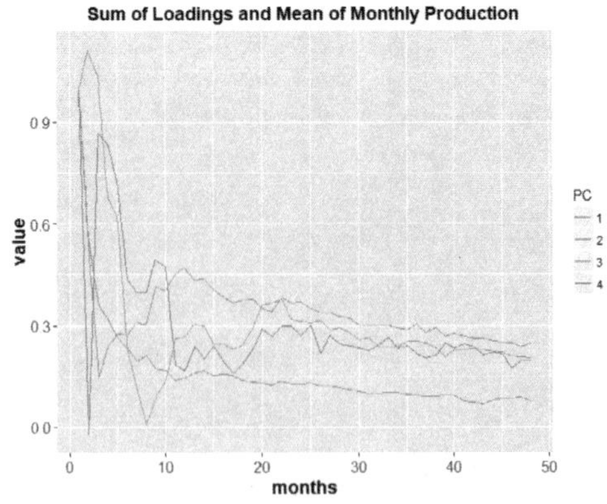

10

17 AIChE
Annual Meeting, Minneapolis, MN

Thakur
Nikolaou

Data-Driven Decline Curves

UNIVERSITY of HOUSTON

Barnett

- Introduction
- Data
- Data Modeling
- Conclusions

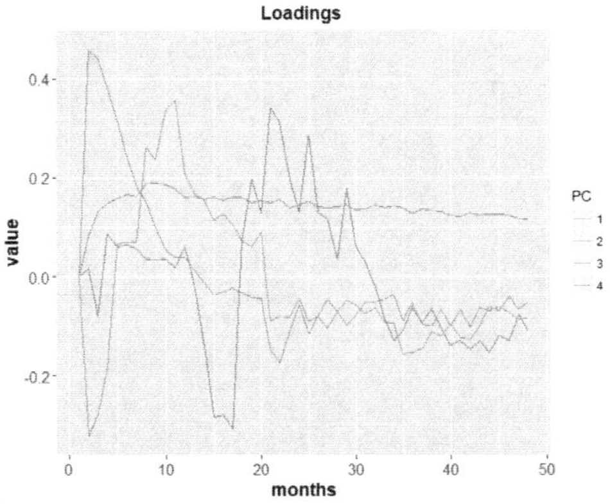

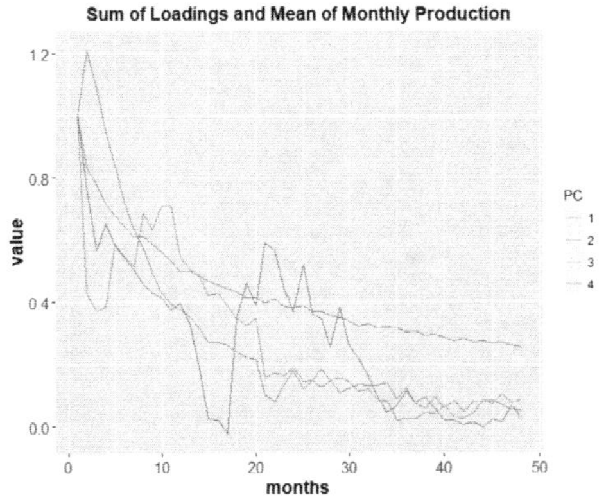

11

17AIChE
Annual Meeting, Minneapolis, MN

Thakur
Nikolaou

Data-Driven Decline Curves

UNIVERSITY of HOUSTON

Marcellus Variance Captured By PCs

- Introduction
- Data
- Data Modeling
- Conclusions

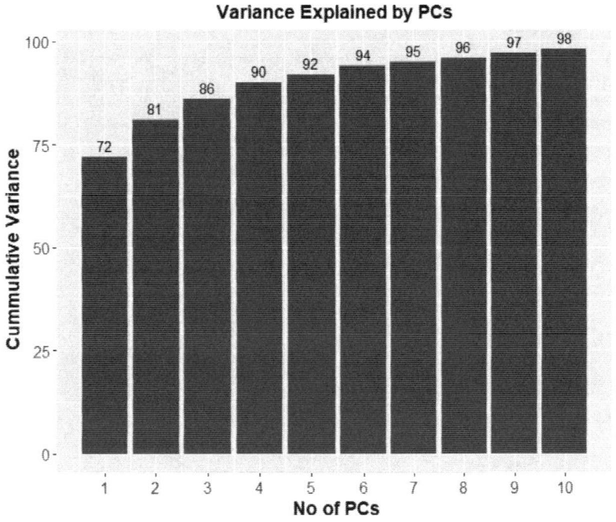

12

17AIChE
Annual Meeting, Minneapolis, MN

Thakur
Nikolaou

Data-Driven Decline Curves

UNIVERSITY of HOUSTON

Eagle Ford Variance Captured By PCs

- Introduction
- Data
- Data Modeling
- Conclusions

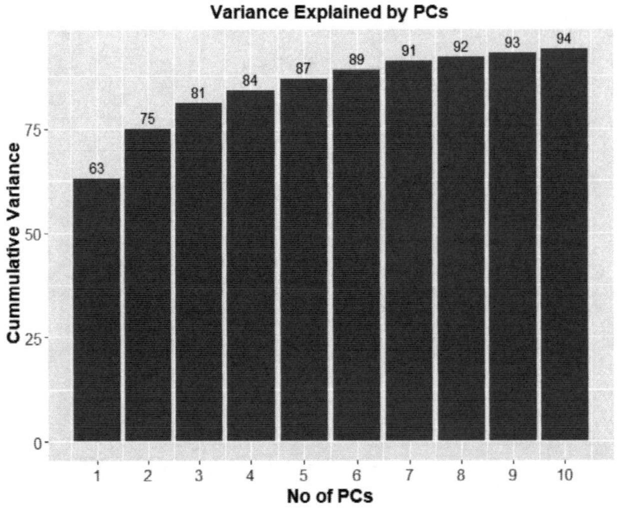

13

17AIChE Annual Meeting, Minneapolis, MN

Thakur
Nikolaou

Data-Driven Decline Curves

UNIVERSITY of HOUSTON

Barnett Variance Captured By PCs

- Introduction
- Data
- Data Modeling
- Conclusions

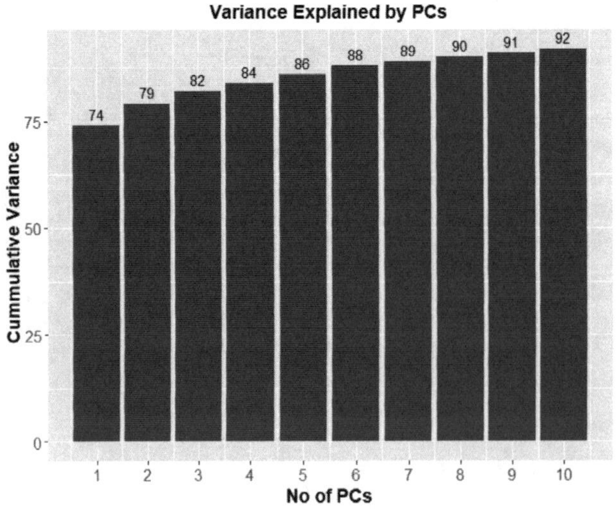

14

17AIChE Annual Meeting, Minneapolis, MN

Thakur
Nikolaou

Data-Driven Decline Curves

UNIVERSITY of HOUSTON

Deciding No of PC- Variance (Marcellus)

- Introduction
- Data
- Data Modeling
- Conclusions

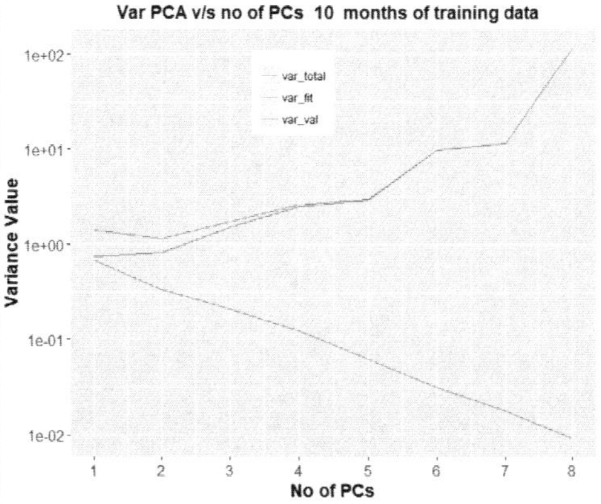

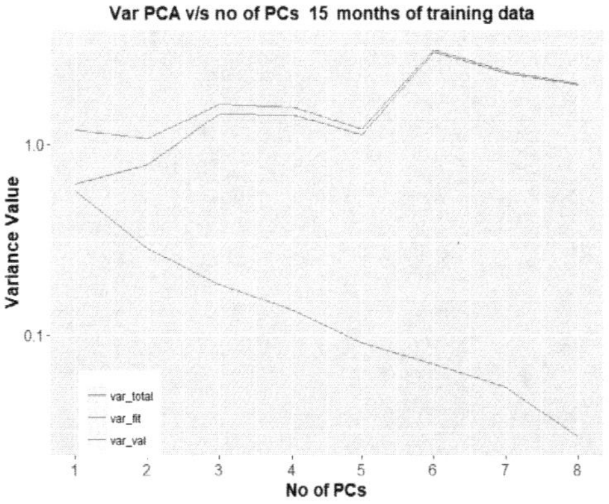

15

17AIChE
Annual Meeting, Minneapolis, MN

Thakur
Nikolaou

Data-Driven Decline Curves

UNIVERSITY of HOUSTON

Deciding No of PC-RMSE (Marcellus)

- Introduction
- Data
- Data Modeling
- Conclusions

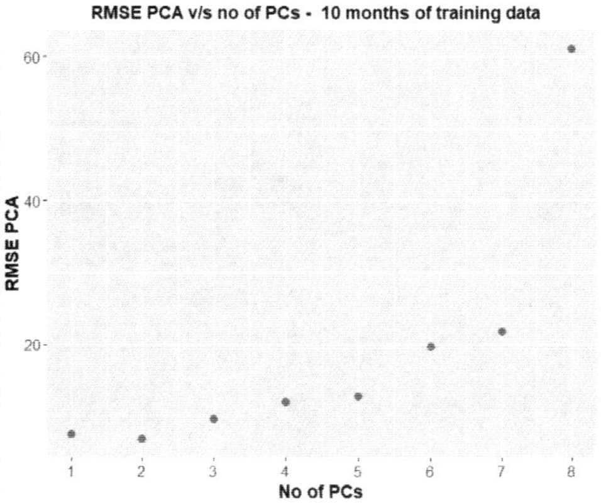

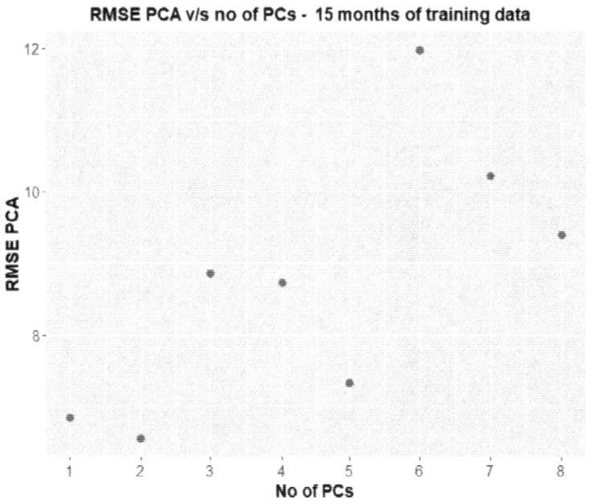

16

17AIChE
Annual Meeting, Minneapolis, MN

Thakur
Nikolaou

Data-Driven Decline Curves

UNIVERSITY of HOUSTON

Performance Overall - using 5 months of Training Data (Marcellus)

Introduction

Data

Data Modeling

Conclusions

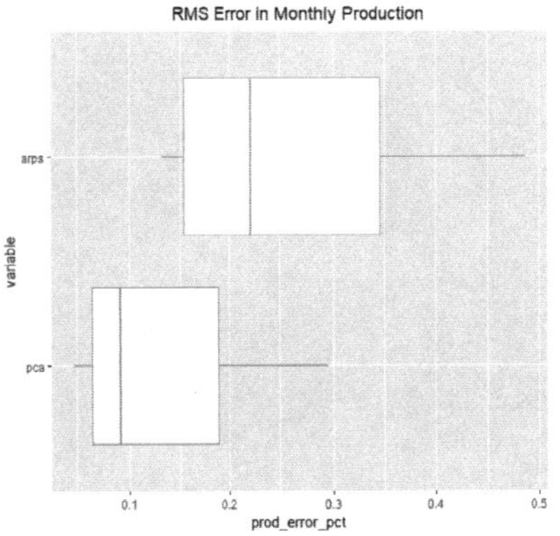

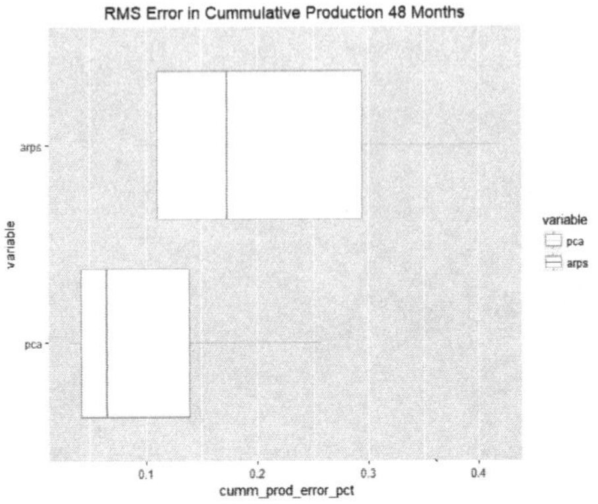

17

17AIChE
Annual Meeting, Minneapolis, MN

Thakur
Nikolaou

Data-Driven Decline Curves

UNIVERSITY of HOUSTON

Performance - using 10 months of Training Data (Marcellus)

Introduction

Data

Data Modeling

Conclusions

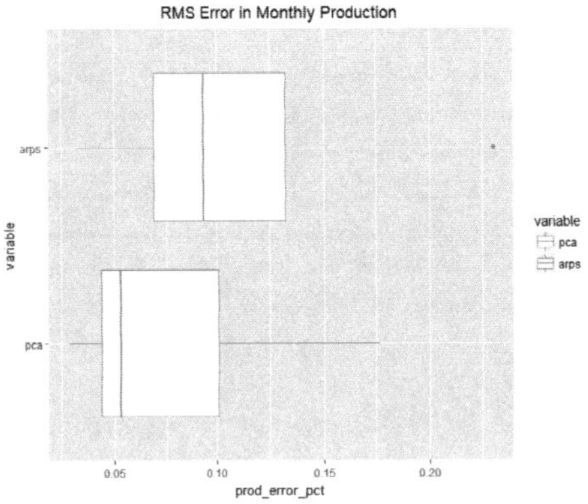

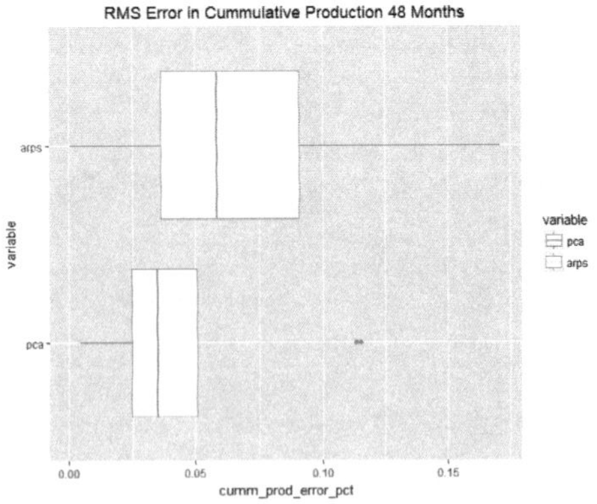

18

17AIChE
Annual Meeting, Minneapolis, MN

Thakur
Nikolaou

Data-Driven Decline Curves

UNIVERSITY of HOUSTON

Introduction

Data

Data Modeling

Conclusions

19

17AIChE
Annual Meeting, Minneapolis, MN

Performance - using 15 months of Training Data (Marcellus)

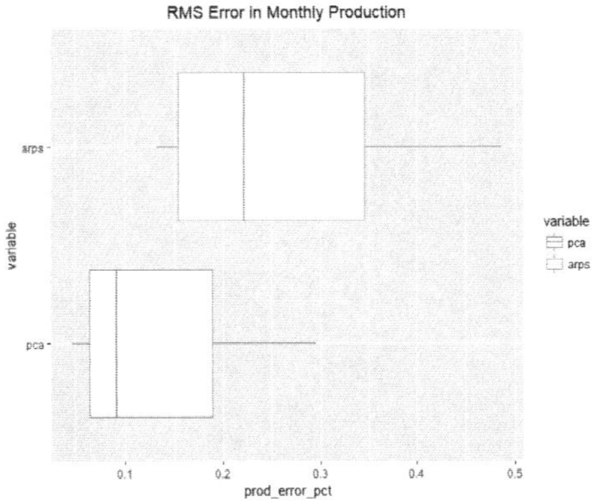

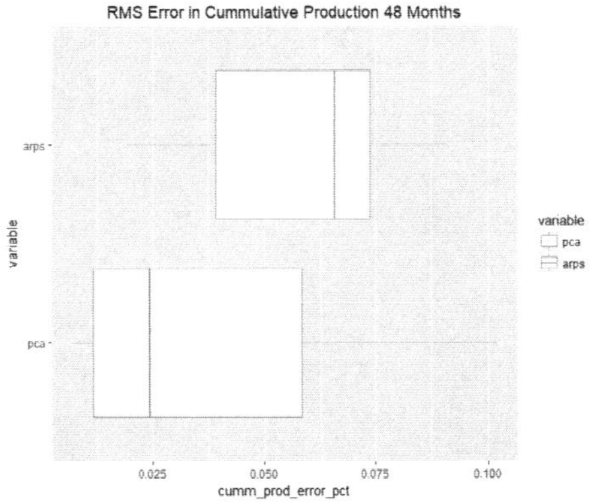

Thakur
Nikolaou

Data-Driven Decline Curves

UNIVERSITY of HOUSTON

Introduction

Data

Data Modeling

Conclusions

20

17AIChE
Annual Meeting, Minneapolis, MN

Effect of No of Training Months on Prediction (Marcellus)

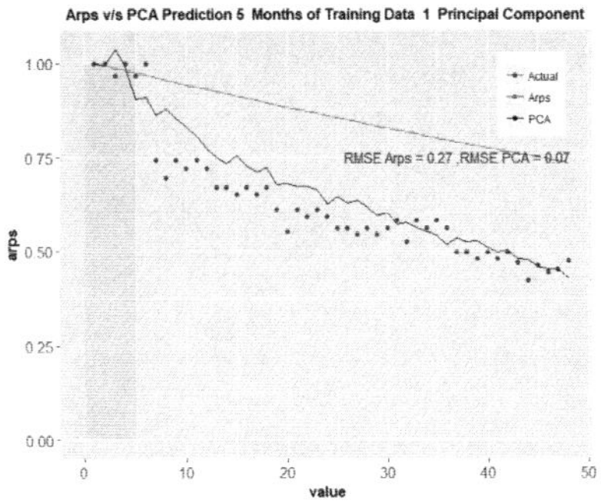

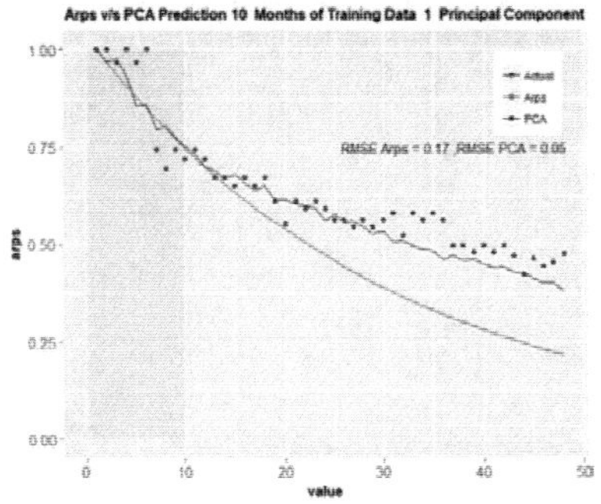

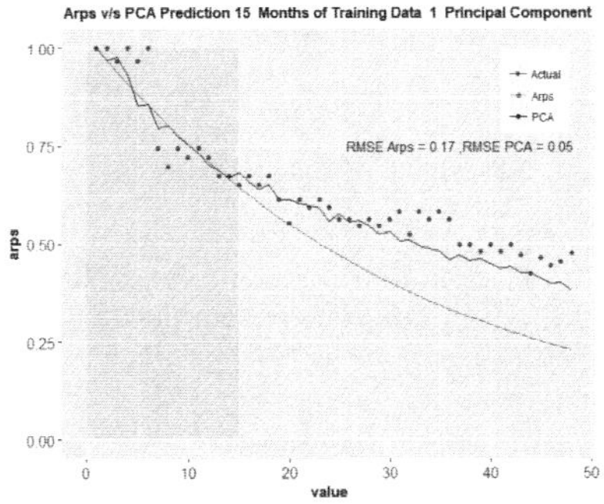

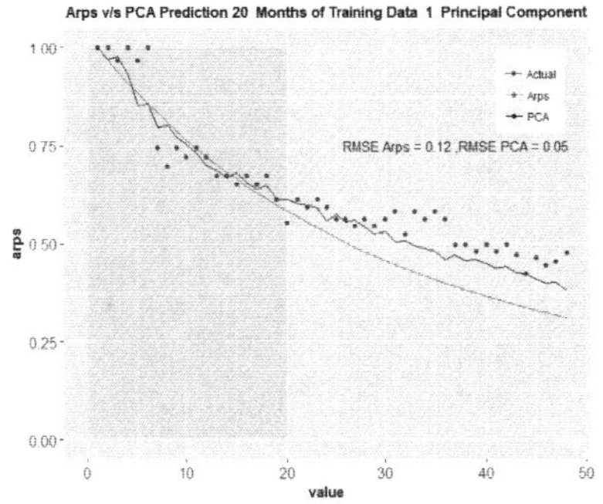

Thakur
Nikolaou

Data-Driven Decline Curves

UNIVERSITY of HOUSTON

Effect of No of Cs on Prediction (Marcellus)

Introduction

Data

Data Modeling

Conclusions

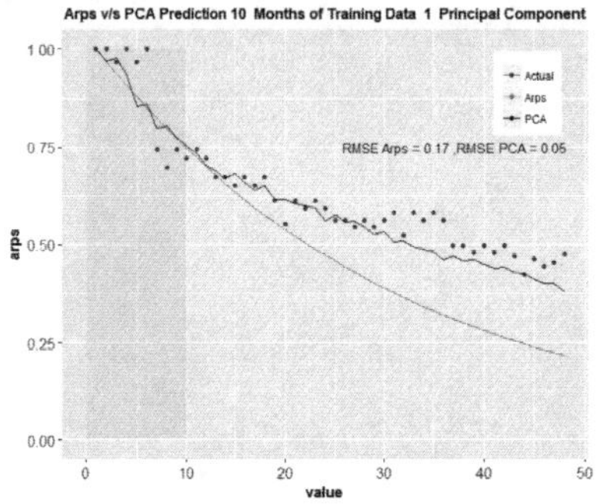

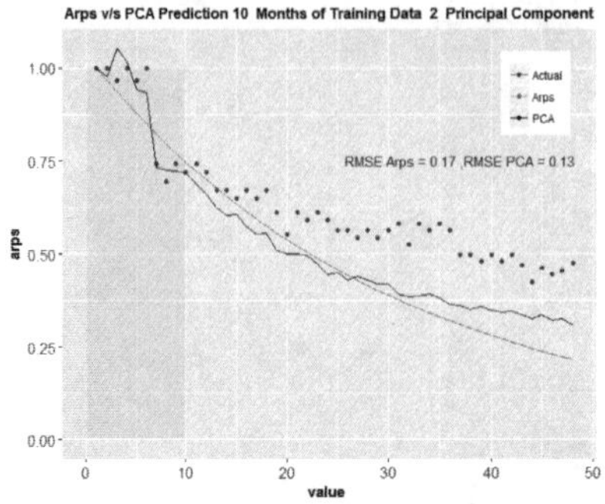

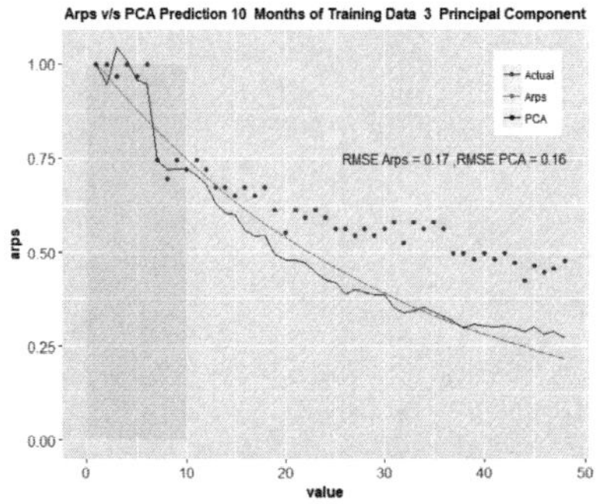

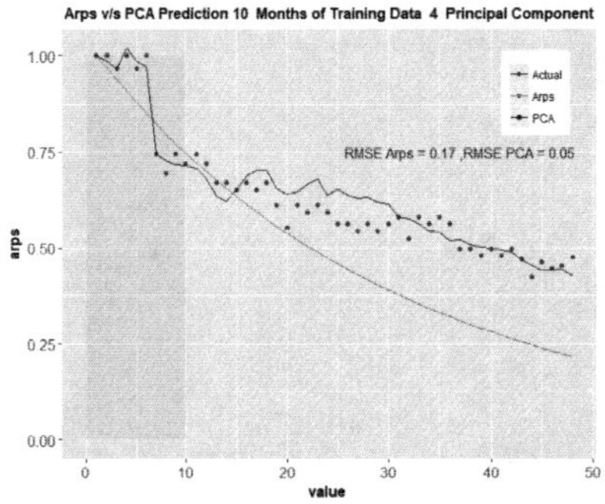

21

17 AIChE
Annual Meeting, Minneapolis, MN

Thakur
Nikolaou

Data-Driven Decline Curves

UNIVERSITY of HOUSTON

Confidence Intervals – PCA(using 1 PC) (Marcellus)

Introduction

Data

Data Modeling

Conclusions

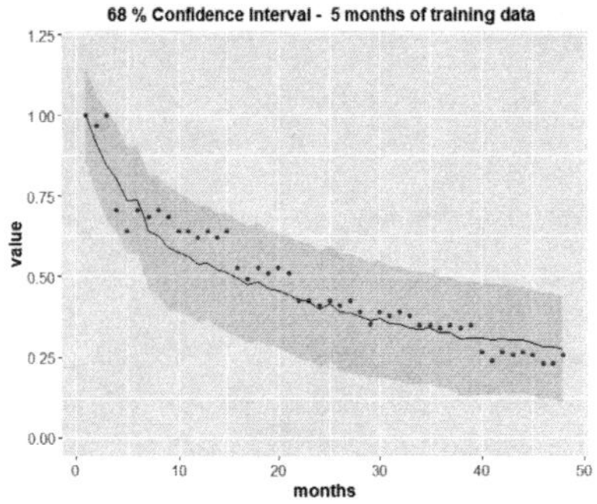

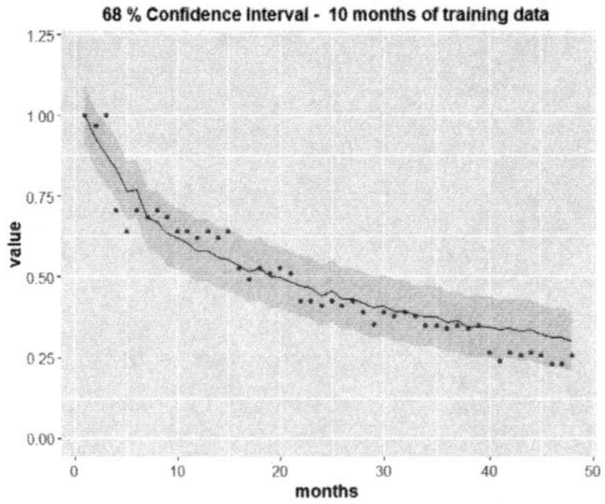

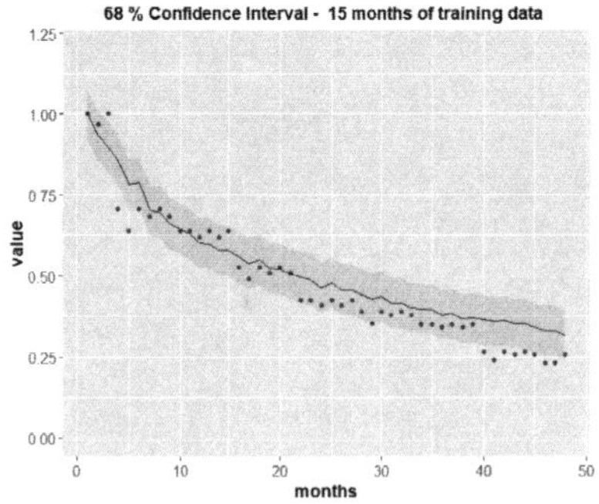

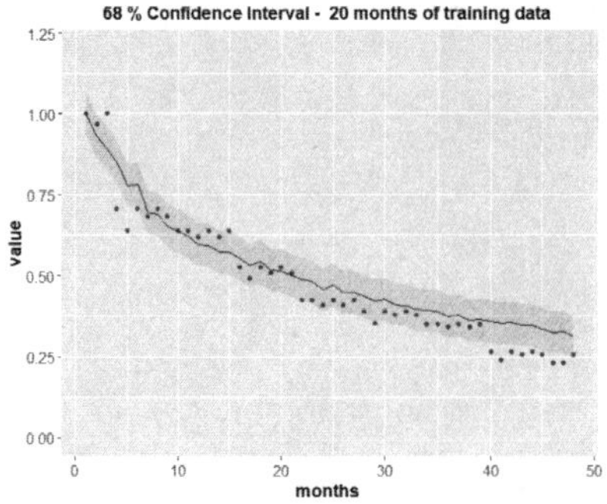

22

17 AIChE
Annual Meeting, Minneapolis, MN

Thakur
Nikolaou

Data-Driven Decline Curves

UNIVERSITY of HOUSTON

Introduction

Data

Data Modeling

Conclusions

Conclusions

- Data are equations
  - …if data is representative of what the equations try to predict
- Non-standard functions usable in simple "equations"

23

17AIChE
Annual Meeting, Minneapolis, MN

Thakur
Nikolaou

Data-Driven Decline Curves

UNIVERSITY of HOUSTON

Introduction

Data

Data Modeling

Conclusions

Acknowledgements

- Drilling Info
  - For access to data

24

17AIChE
Annual Meeting, Minneapolis, MN

Production Estimation and Well Classification for Hydraulically Fractured Horizontal Wells: A Data-Driven Model-Based Approach

17AIChE
Annual Meeting, Minneapolis, MN

Sunit Mathur and Michael Nikolaou
Petroleum Engineering
Chemical & Biomolecular Engineering
University of Houston, TX

Advances in Unconventional Oil and Gas Modeling

UNIVERSITY *of* HOUSTON

Mathur-Nikolaou

Predicting Production for Unconventionals

UNIVERSITY *of* HOUSTON

Production from Shale Gas Wells

- Introduction
- Data processed
- Modeling
- Results
- Conclusions

2

17AIChE
Annual Meeting, Minneapolis, MN

Mathur-Nikolaou

Predicting Production for Unconventionals

UNIVERSITY of HOUSTON

Introduction

Data processed

Modeling

Results

Conclusions

3

17AIChE
Annual Meeting, Minneapolis, MN

Predicting Shale Gas Well Production

- **Equations/simulator**:
  - Intense initial effort
  - Possibly inconclusive
- **Data-driven:**
  - Wealth of design and field data:
    - Exploration
    - ➢ Drilling job
    - ➢ Wellbore configuration
    - ➢ Fracture design
    - ➢ Historical production data

Mathur-Nikolaou

Predicting Production for Unconventionals

UNIVERSITY of HOUSTON

Introduction

Data processed

Modeling

Results

Conclusions

4

17AIChE
Annual Meeting, Minneapolis, MN

Outline

- Data processed
- Modeling approach
  - PLS and variants
- Results
  - Accuracy of predictions
  - Ranking of variables in importance
  - Comparisons with other methods:
    - Combined ANN/PCA and variants
    - Random forests for ranking variables
- Conclusions

Mathur-Nikolaou

Predicting Production for Unconventionals

UNIVERSITY of HOUSTON

Novelty of Work

Introduction

Data processed

Modeling

Results

Conclusions

- Previous works:

  Short-term production/completion data

  → predictions of future production

  → EUR

- This work:

$$\left\{\begin{array}{c}\text{Drilling job}\\ \text{Wellbore configuration}\\ \text{Fracture design \& construction}\\ +\\ \text{Historical production data}\end{array}\right\} \Rightarrow \left\{\begin{array}{c}\text{Production}\\ \text{in new wells}\end{array}\right\}$$

5

17 AIChE Annual Meeting, Minneapolis, MN

Mathur-Nikolaou

Predicting Production for Unconventionals

UNIVERSITY of HOUSTON

Data Processed

IPPTC

Introduction

Data processed

Modeling

Results

Conclusions

- Southwestern Energy acreage
- Fayetteville shale

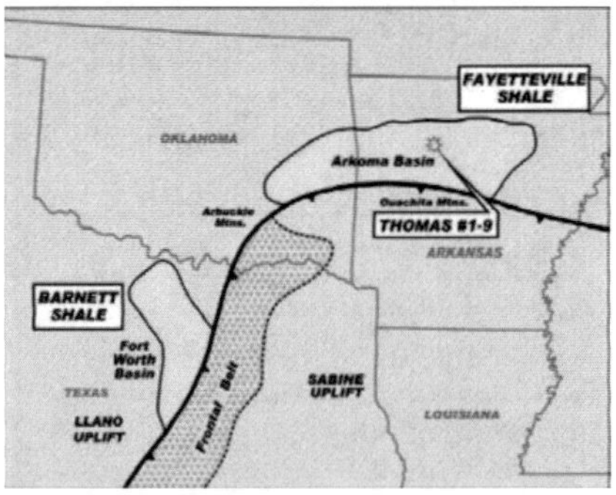

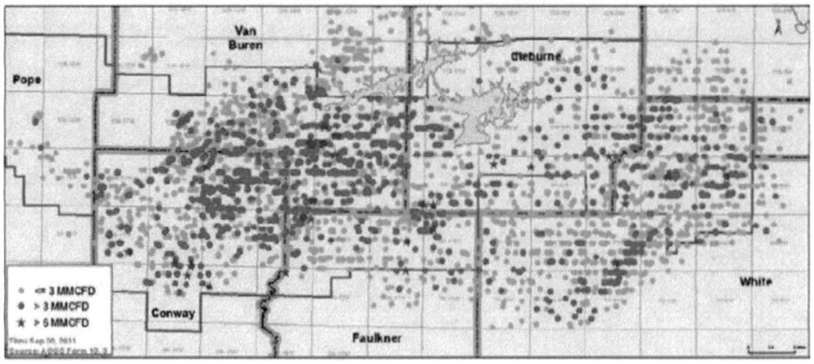

6

17 AIChE Annual Meeting, Minneapolis, MN

Mathur-Nikolaou

Predicting Production for Unconventionals

UNIVERSITY of HOUSTON

Introduction

Data processed

Modeling

Results

Conclusions

Data Processed

- Southwestern Energy
  - Drilling reports
  - Completion reports
  - Overall wellbook reports
- Automated workflow adopted for data capture:
  - **250** wells
  - **9** wellbore/fracture configuration data points per well
  - **26** production data points (monthly) per well
  - **14** data points per stage (hydraulic fracturing job)
  - Total **2649** stages
  - Overall **45,836** data points to be captured

7

17AIChE Annual Meeting, Minneapolis, MN

Mathur-Nikolaou

Predicting Production for Unconventionals

UNIVERSITY of HOUSTON

Introduction

Data processed

Modeling

Results

Conclusions

Snapshot of Hydraulic Fracturing Job Data

- Single stage fracture operation data
- Highlighted text important for this study

FRAC : Pressure test lines to 9,123 psi. Open well with 0,050 psi. Start down hole on stage 1 @ 5 bpm. Well broke @ 4,342 psi. Spearhead 1,500 gal of 15% HCl. Increase rate to 40 bpm, saw 1,386 psi drop from acid. Stepdown after acid hits as per procedure. Post-Acid ISIP was 2057 psi. Frac well as per procedure with SW fluid. Pumped a total of 9,596 bbls of clean fluid with 306,439 lbs of sand. Average rate/pressure was 66 bpm/6,426 psi. Max rate/pressure was 71 bpm/7,753 psi. ISIP: 2,415 psi. FG: 0.83 psi/ft

8

17AIChE Annual Meeting, Minneapolis, MN

Mathur-Nikolaou

Predicting Production for Unconventionals

UNIVERSITY of HOUSTON

- Introduction
- Data processed
- Modeling
- Results
- Conclusions

Modeling Overview

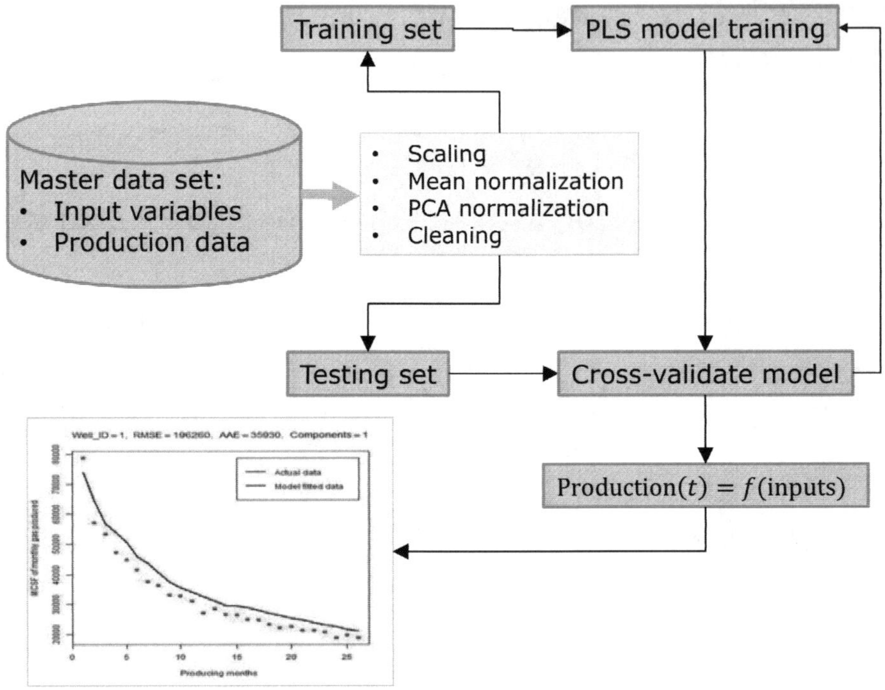

9

17AIChE
Annual Meeting, Minneapolis, MN

Mathur-Nikolaou

Predicting Production for Unconventionals

UNIVERSITY of HOUSTON

- Introduction
- Data processed
- Modeling
- Results
- Conclusions

Input Variables

- Pressure data (Hydraulic Fracturing)
  - Post Acid ISIP
  - Well Breaking Pressure
  - Opening Well Pressure
  - Pressure drop from acid
  - Pressure Test Line
  - Average Pressure during Injection
  - Maximum Pressure during Injection
  - ISIP
- Wellbore
  - Tubing Diameter
  - True vertical depth
  - Diameter of Horizontal section
  - Length of Horizontal section
- Fracture configuration
  - Number of cluster per stage
  - Number of perforations per cluster
  - Number of stages
  - Stage Length
  - Cluster spacing
  - Lbs of Sand added
  - Bbls of Clean fluid added
  - Bbls of HCl added
  - Increased Injection Rate
  - Average rate of injection
  - Maximum rate of injection

10

17AIChE
Annual Meeting, Minneapolis, MN

Mathur-Nikolaou

Predicting Production for Unconventionals

UNIVERSITY of HOUSTON

Introduction

Data processed

Modeling

Results

Conclusions

Output Variable

- Production over 26 months
- Derived output: EUR

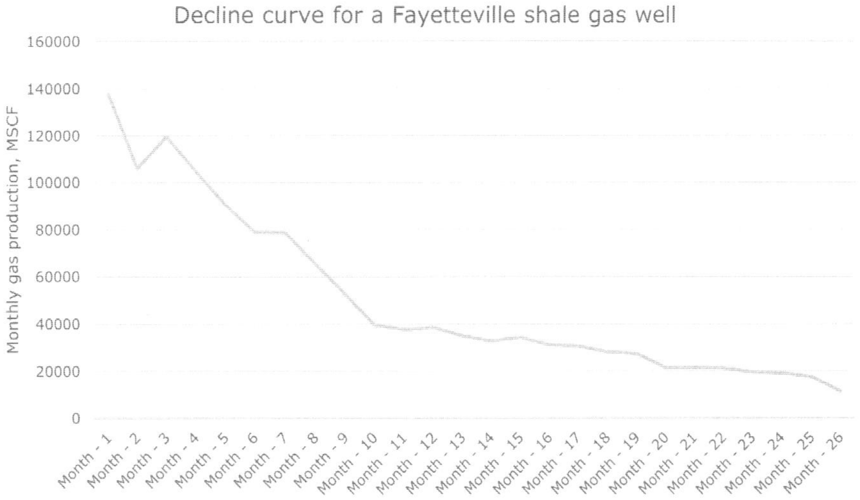

11

17AIChE

Annual Meeting, Minneapolis, MN

Mathur-Nikolaou

Predicting Production for Unconventionals

UNIVERSITY of HOUSTON

Introduction

Data processed

Modeling

Results

Conclusions

Testing Model Quality

- Cross-validation:
  - 80/20 split for training/validation

1	2	3	4	5
Train	Train	Validation	Train	Train

12

17AIChE

Annual Meeting, Minneapolis, MN

Mathur-Nikolaou

Predicting Production for Unconventionals

UNIVERSITY of HOUSTON

Introduction
Data processed
Modeling
Results
Conclusions

13

17AIChE Annual Meeting, Minneapolis, MN

Data Pre-processing

- Scaled:
  - 0 to 1
- Mean normalized
  - Deviation from average
- For ANN: PCA normalized
  - Principal components for 26 months of production
- Removal of outliers based on manual judgement:
  - Refractured wells
  - Wells temporarily closed

Mathur-Nikolaou

Predicting Production for Unconventionals

UNIVERSITY of HOUSTON

Introduction
Data processed
Modeling
Results
Conclusions

14

17AIChE Annual Meeting, Minneapolis, MN

Results

- PLS-based production curves for
  - Outlier points included or removed
  - Input/output scaled or unscaled
  - Input data with averages removed or not
- PLS-model comparisons with
  - ANN

17AIChE
Annual Meeting, Minneapolis, MN

Results

Decline curves PLS

UNIVERSITY *of* HOUSTON

Mathur-Nikolaou

Predicting Production for Unconventionals

UNIVERSITY *of* HOUSTON

PLS Uncleaned Unscaled 2 Latent Variables

- Introduction
- Data processed
- Modeling
- Results
- Conclusions

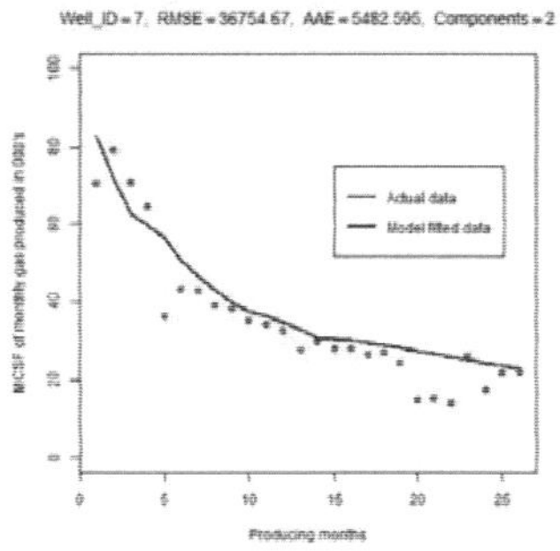

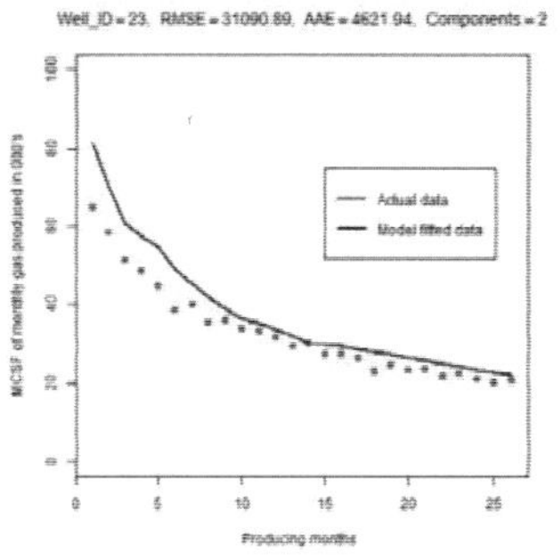

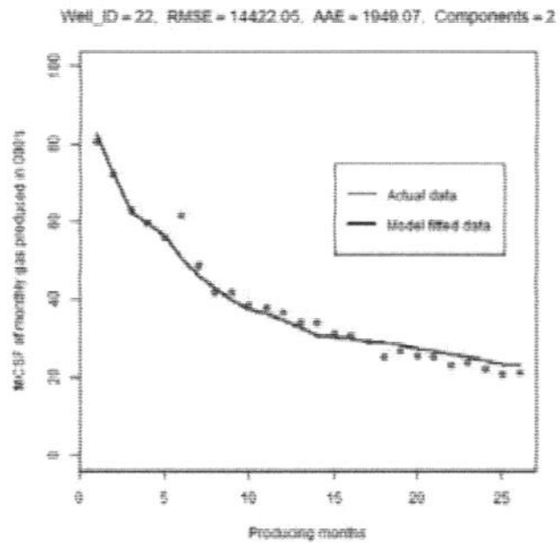

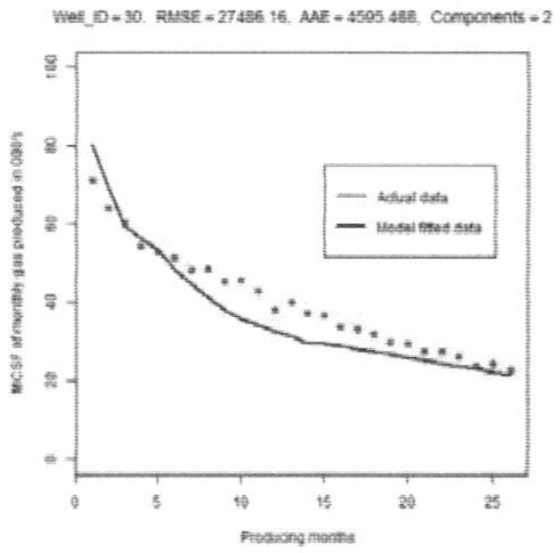

16

17AIChE
Annual Meeting, Minneapolis, MN

Mathur-Nikolaou

Predicting Production for Unconventionals

UNIVERSITY *of* HOUSTON

PLS Uncleaned Scaled 2 Latent Variables

Introduction

Data processed

Modeling

Results

Conclusions

17

17AIChE

Annual Meeting, Minneapolis, MN

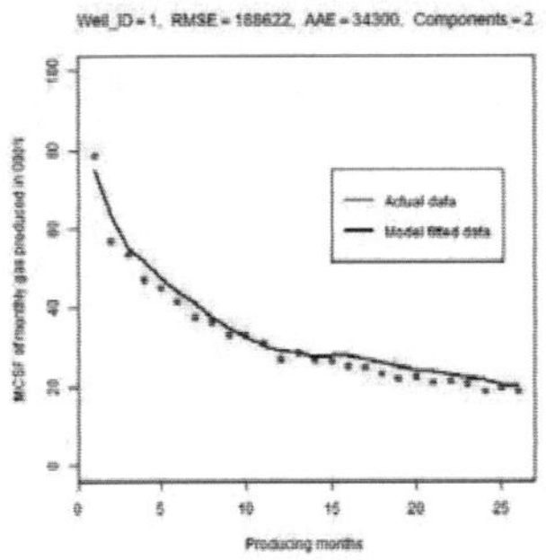

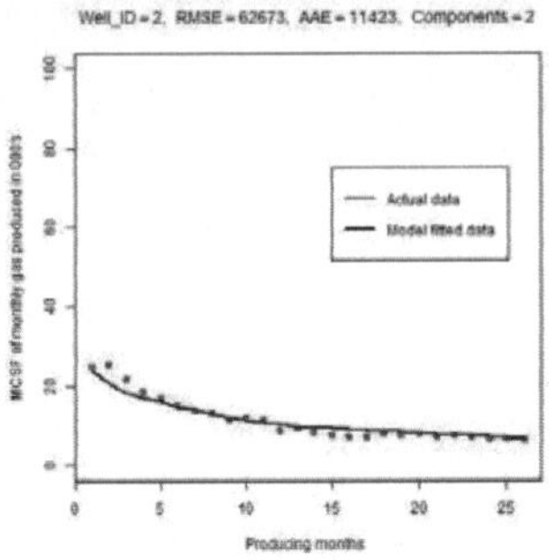

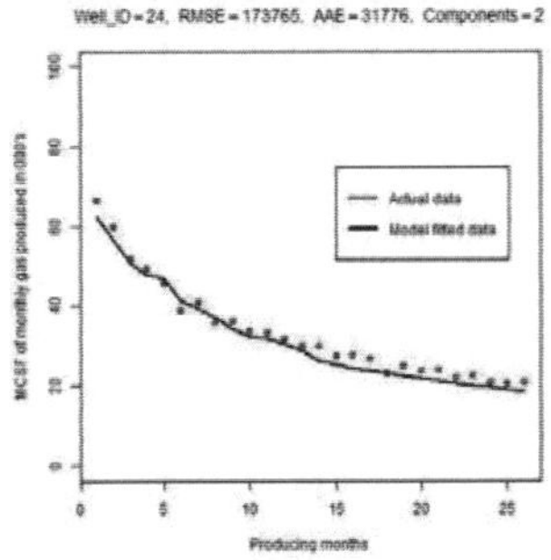

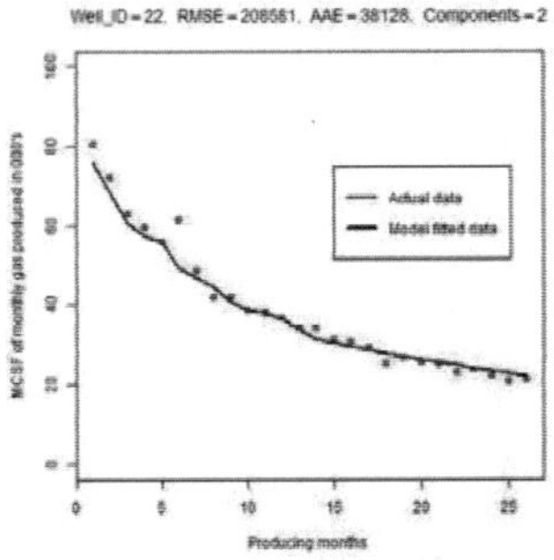

Mathur-Nikolaou

Predicting Production for Unconventionals

UNIVERSITY *of* HOUSTON

PLS Cleaned Average Corrected 2 Latent Variables

Introduction

Data processed

Modeling

Results

Conclusions

18

17AIChE

Annual Meeting, Minneapolis, MN

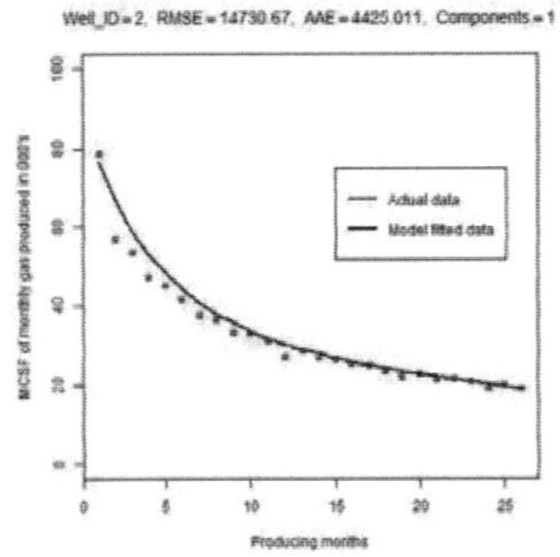

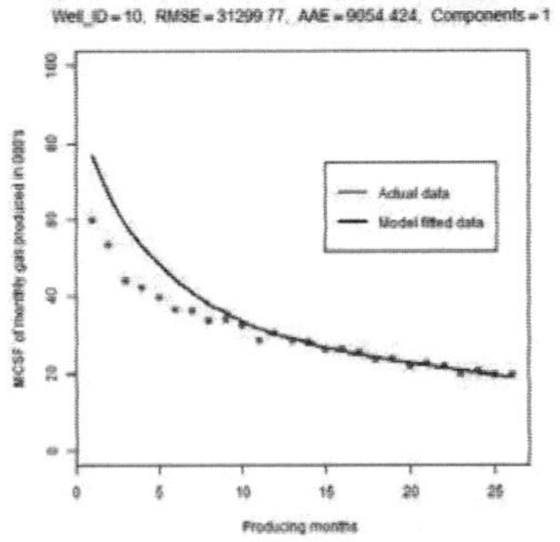

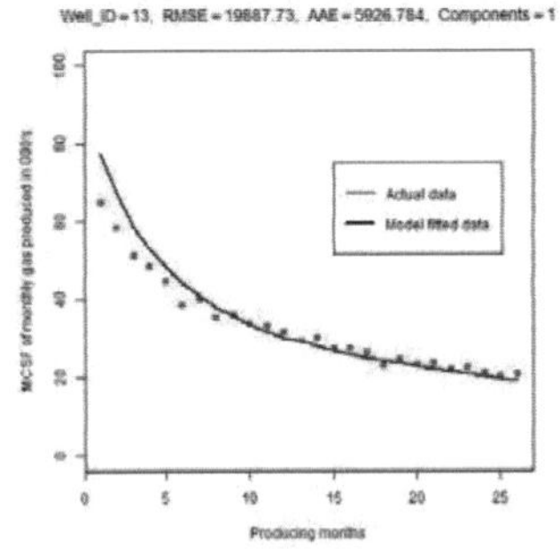

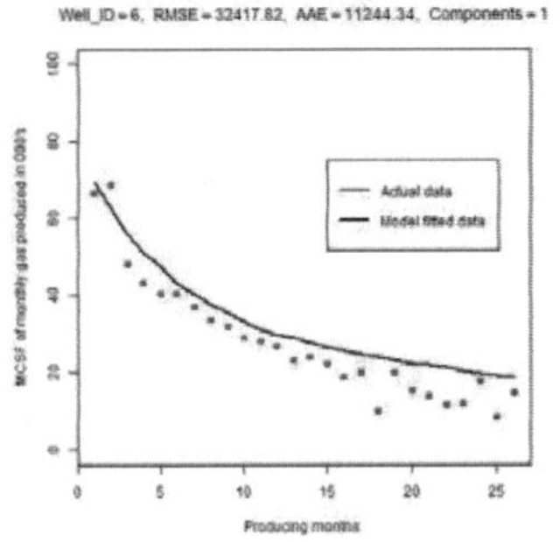

Mathur-Nikolaou

Predicting Production for Unconventionals

UNIVERSITY of HOUSTON

PLS Cleaned Unscaled 2 Latent Variables

- Introduction
- Data processed
- Modeling
- Results
- Conclusions

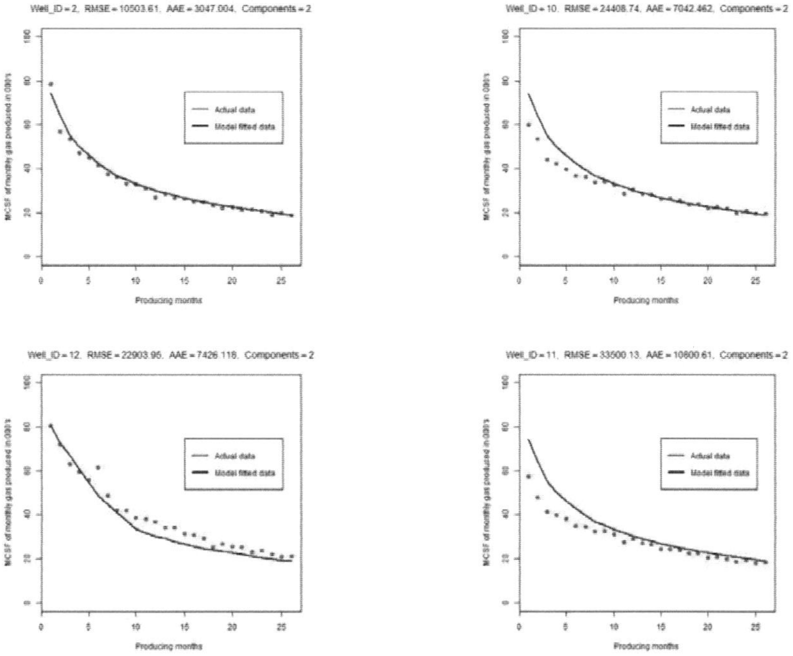

19

17AIChE Annual Meeting, Minneapolis, MN

Mathur-Nikolaou

Predicting Production for Unconventionals

UNIVERSITY of HOUSTON

PLS Cleaned Scaled 2 Latent Variables

- Introduction
- Data processed
- Modeling
- Results
- Conclusions

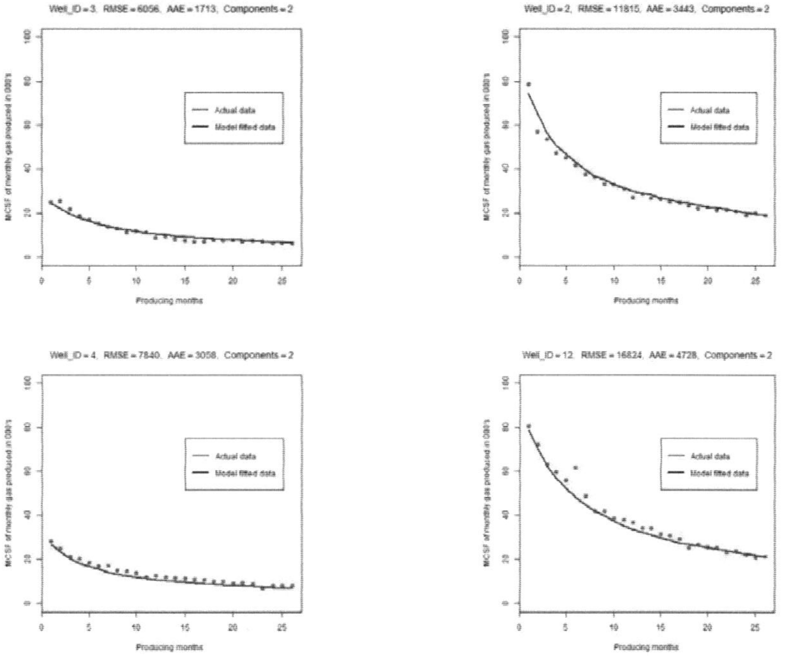

20

17AIChE Annual Meeting, Minneapolis, MN

Mathur-Nikolaou

Predicting Production for Unconventionals

UNIVERSITY *of* HOUSTON

Loading Weights for Uncleaned Scaled PLS

- Introduction
- Data processed
- Modeling
- Results
- Conclusions

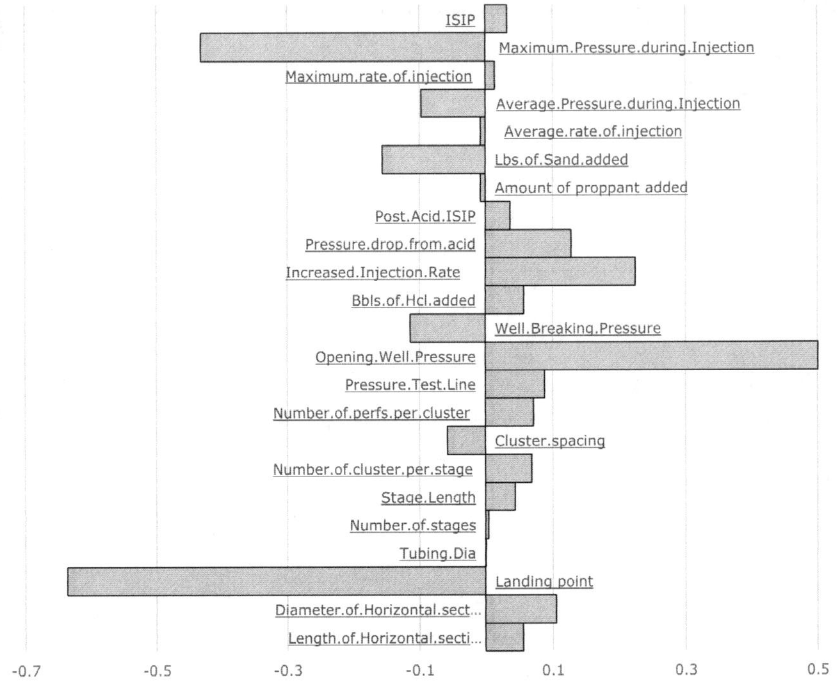

21

17AIChE Annual Meeting, Minneapolis, MN

17AIChE Annual Meeting, Minneapolis, MN

Importance of Inputs

Variable Importance in Projection (VIP) statistic

UNIVERSITY *of* HOUSTON

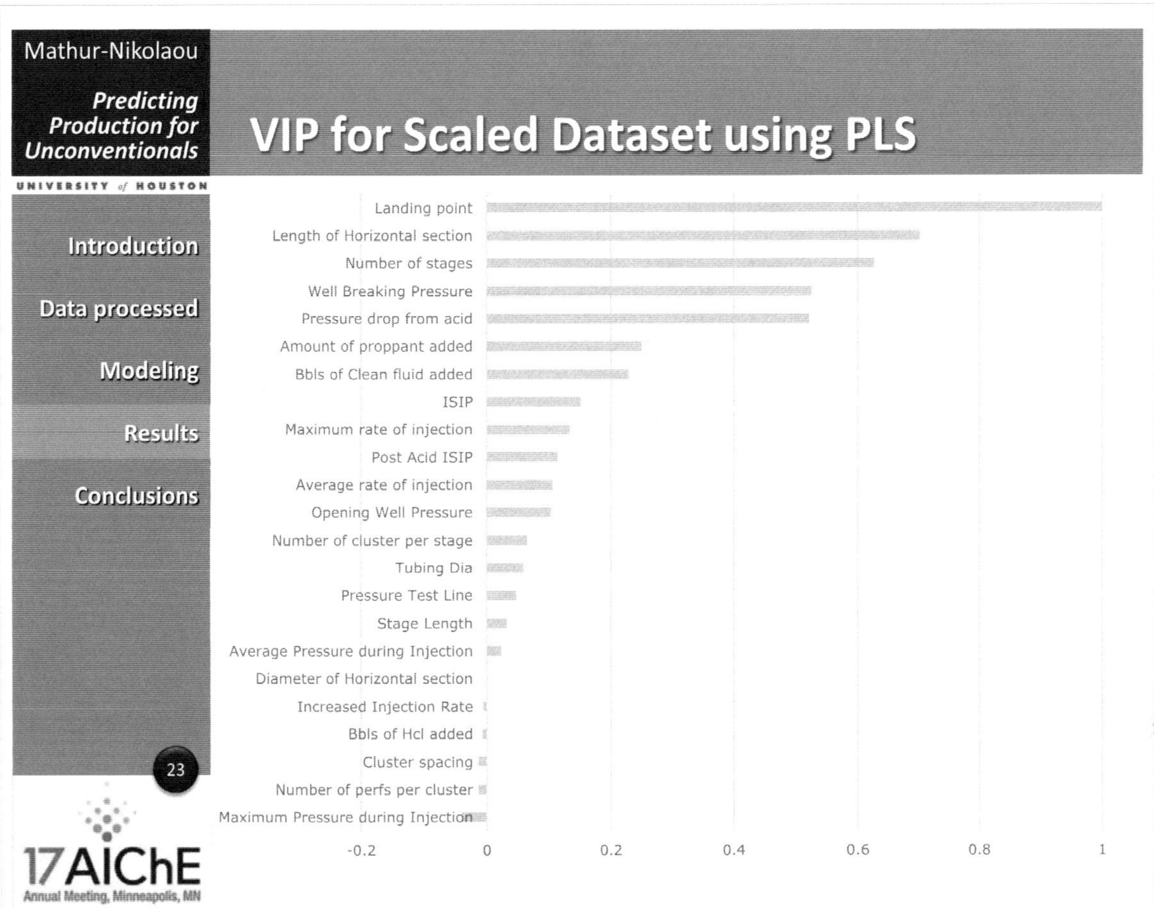
Mathur-Nikolaou
Predicting Production for Unconventionals
UNIVERSITY of HOUSTON
VIP for Scaled Dataset using PLS
Introduction
Data processed
Modeling
Results
Conclusions
23
17AIChE
Annual Meeting, Minneapolis, MN
Landing point
Length of Horizontal section
Number of stages
Well Breaking Pressure
Pressure drop from acid
Amount of proppant added
Bbls of Clean fluid added
ISIP
Maximum rate of injection
Post Acid ISIP
Average rate of injection
Opening Well Pressure
Number of cluster per stage
Tubing Dia
Pressure Test Line
Stage Length
Average Pressure during Injection
Diameter of Horizontal section
Increased Injection Rate
Bbls of Hcl added
Cluster spacing
Number of perfs per cluster
Maximum Pressure during Injection
-0.2
0
0.2
0.4
0.6
0.8
1

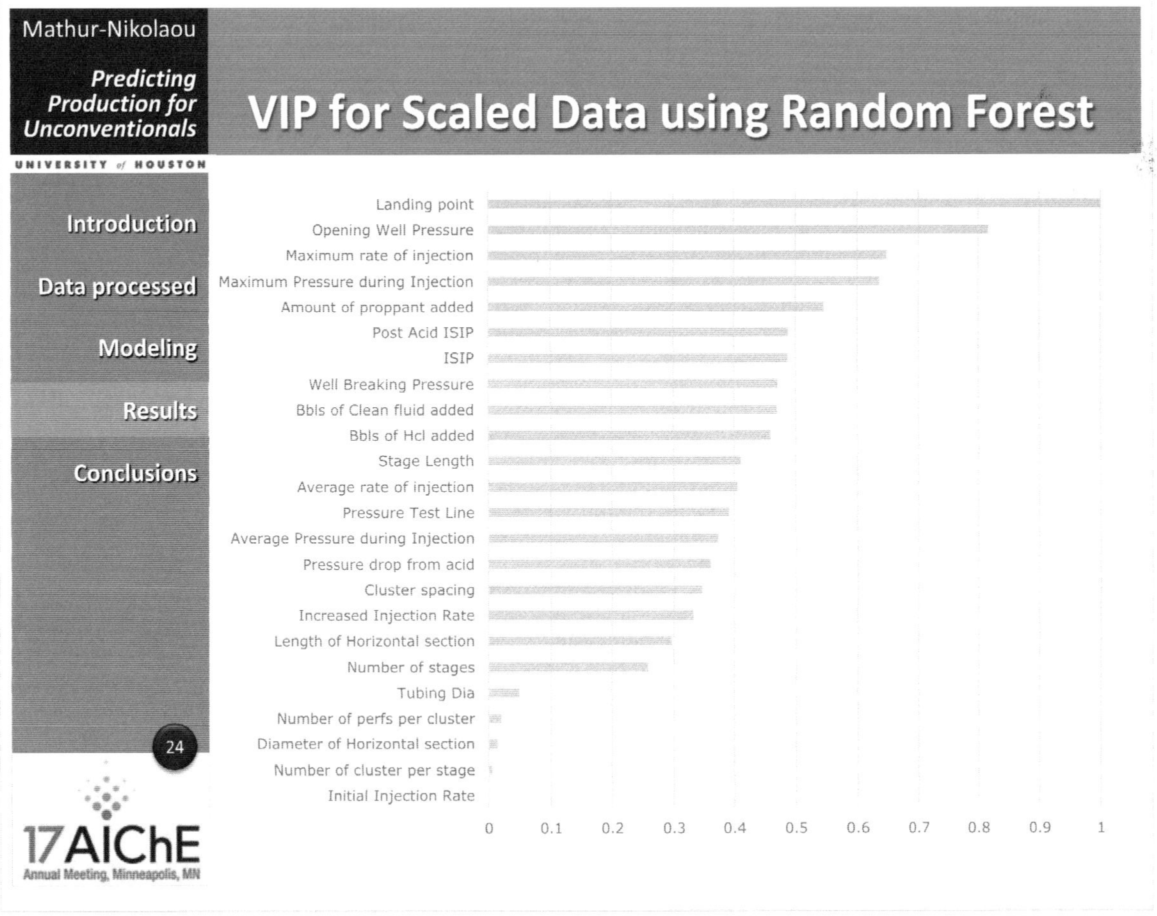
Mathur-Nikolaou
Predicting Production for Unconventionals
UNIVERSITY of HOUSTON
VIP for Scaled Data using Random Forest
Introduction
Data processed
Modeling
Results
Conclusions
24
17AIChE
Annual Meeting, Minneapolis, MN
Landing point
Opening Well Pressure
Maximum rate of injection
Maximum Pressure during Injection
Amount of proppant added
Post Acid ISIP
ISIP
Well Breaking Pressure
Bbls of Clean fluid added
Bbls of Hcl added
Stage Length
Average rate of injection
Pressure Test Line
Average Pressure during Injection
Pressure drop from acid
Cluster spacing
Increased Injection Rate
Length of Horizontal section
Number of stages
Tubing Dia
Number of perfs per cluster
Diameter of Horizontal section
Number of cluster per stage
Initial Injection Rate
0
0.1
0.2
0.3
0.4
0.5
0.6
0.7
0.8
0.9
1

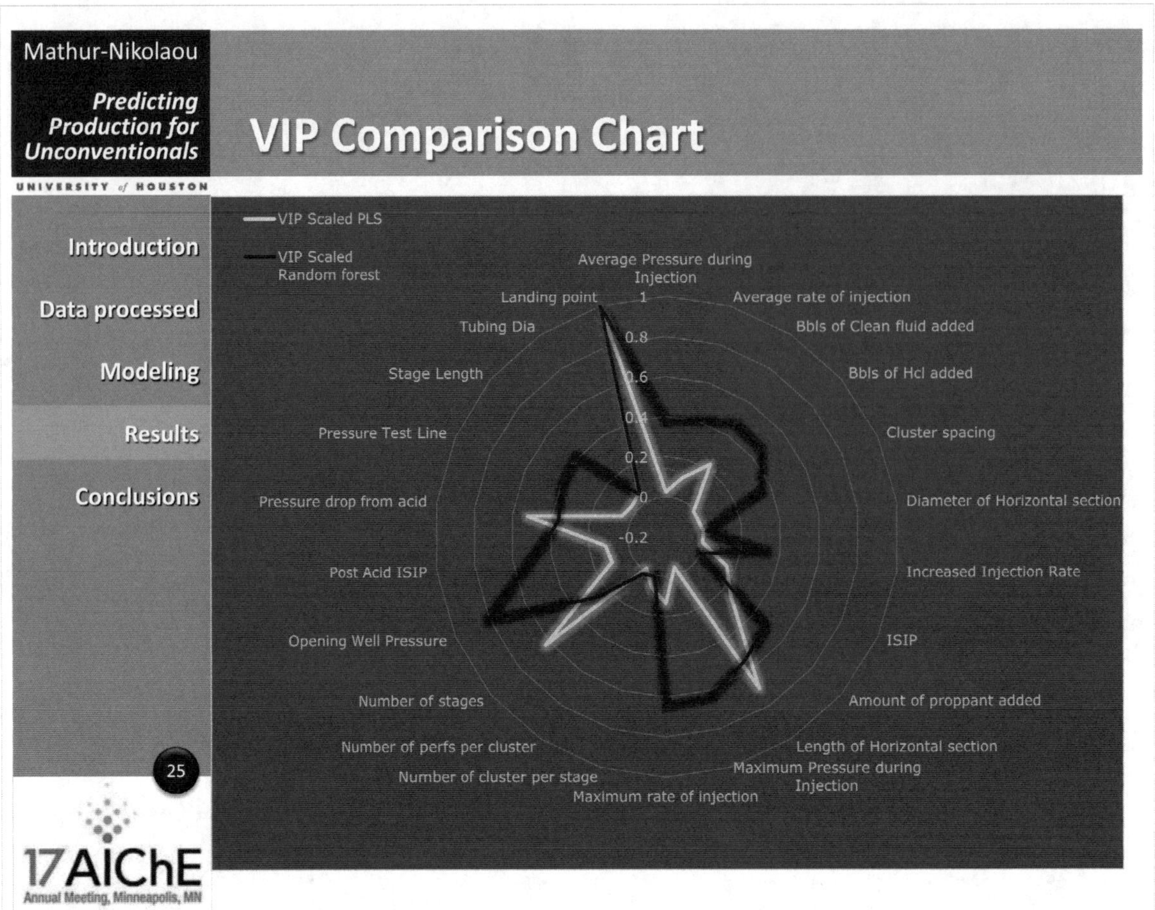

Mathur-Nikolaou
Predicting Production for Unconventionals
UNIVERSITY of HOUSTON
VIP Comparison Chart
Introduction
Data processed
Modeling
Results
Conclusions
25
17AIChE
Annual Meeting, Minneapolis, MN
VIP Scaled PLS
VIP Scaled Random forest
Average Pressure during Injection
Average rate of injection
Bbls of Clean fluid added
Bbls of Hcl added
Cluster spacing
Diameter of Horizontal section
Increased Injection Rate
ISIP
Amount of proppant added
Length of Horizontal section
Maximum Pressure during Injection
Maximum rate of injection
Number of cluster per stage
Number of perfs per cluster
Number of stages
Opening Well Pressure
Post Acid ISIP
Pressure drop from acid
Pressure Test Line
Stage Length
Tubing Dia
Landing point
1
0.8
0.6
0.4
0.2
0
-0.2

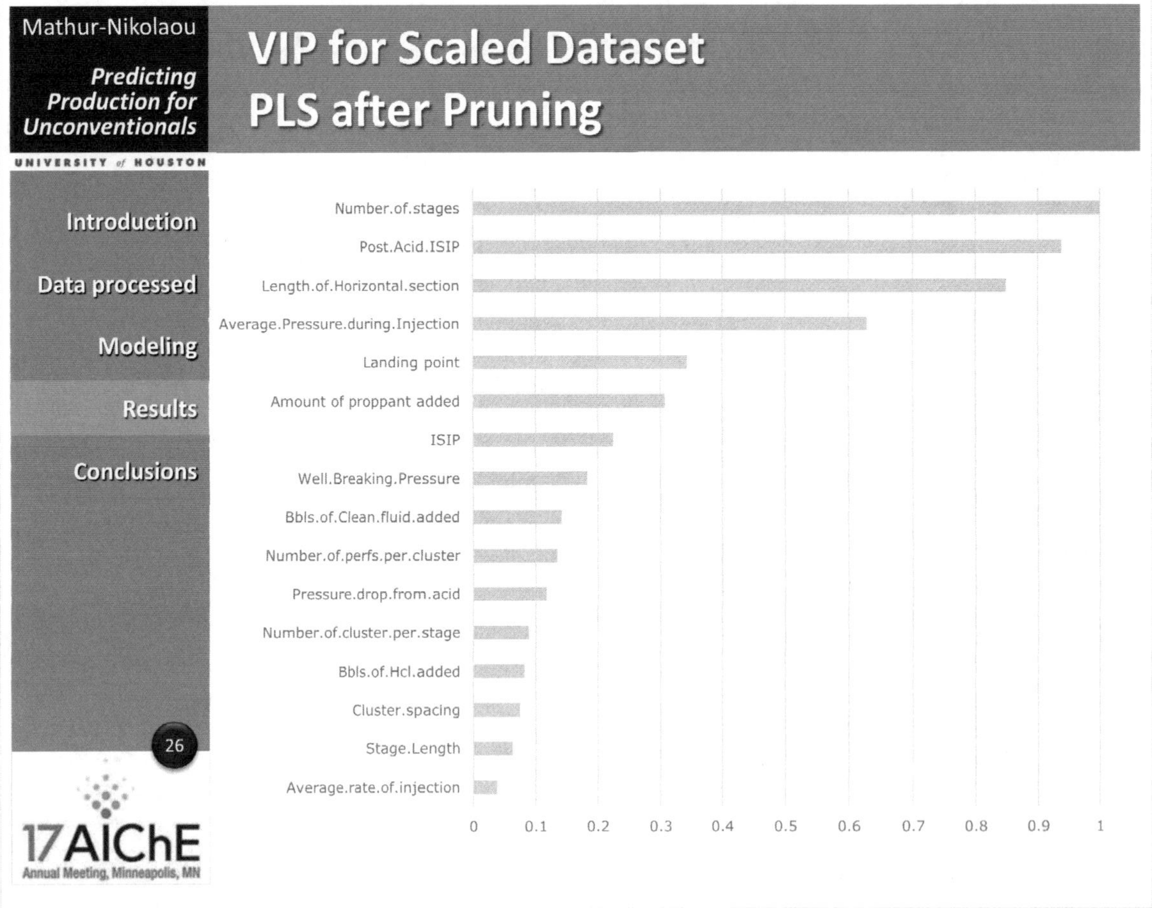

Mathur-Nikolaou
Predicting Production for Unconventionals
UNIVERSITY of HOUSTON
VIP for Scaled Dataset
PLS after Pruning
Introduction
Data processed
Modeling
Results
Conclusions
26
17AIChE
Annual Meeting, Minneapolis, MN
Number.of.stages
Post.Acid.ISIP
Length.of.Horizontal.section
Average.Pressure.during.Injection
Landing point
Amount of proppant added
ISIP
Well.Breaking.Pressure
Bbls.of.Clean.fluid.added
Number.of.perfs.per.cluster
Pressure.drop.from.acid
Number.of.cluster.per.stage
Bbls.of.Hcl.added
Cluster.spacing
Stage.Length
Average.rate.of.injection
0
0.1
0.2
0.3
0.4
0.5
0.6
0.7
0.8
0.9
1

Mathur-Nikolaou

Predicting Production for Unconventionals

UNIVERSITY of HOUSTON

VIP for Scaled Dataset Random Forest after Pruning

- Introduction
- Data processed
- Modeling
- Results
- Conclusions

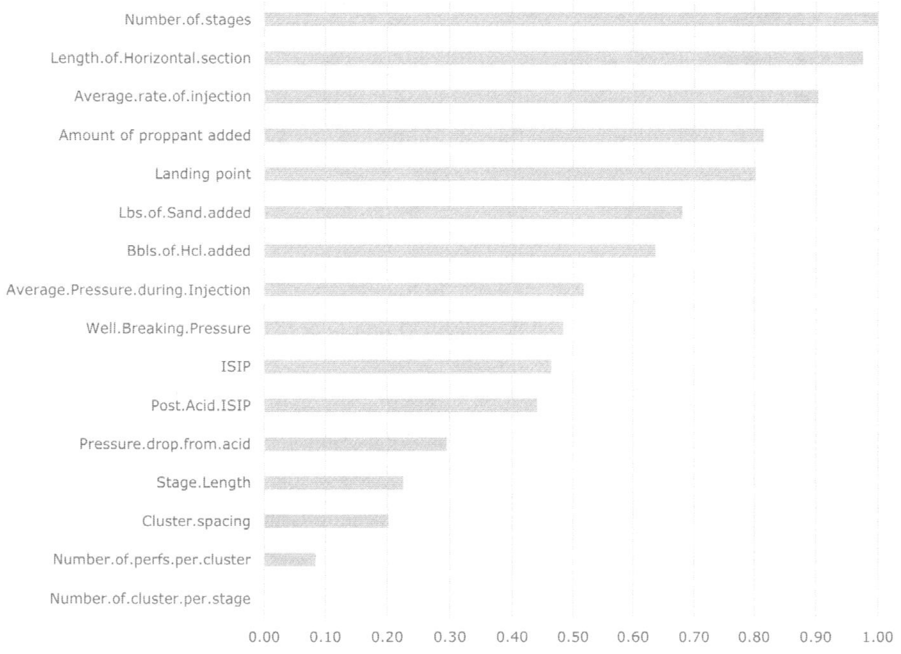

27

17AIChE Annual Meeting, Minneapolis, MN

Mathur-Nikolaou

Predicting Production for Unconventionals

UNIVERSITY of HOUSTON

VIP Comparison after Pruning

- Introduction
- Data processed
- Modeling
- Results
- Conclusions

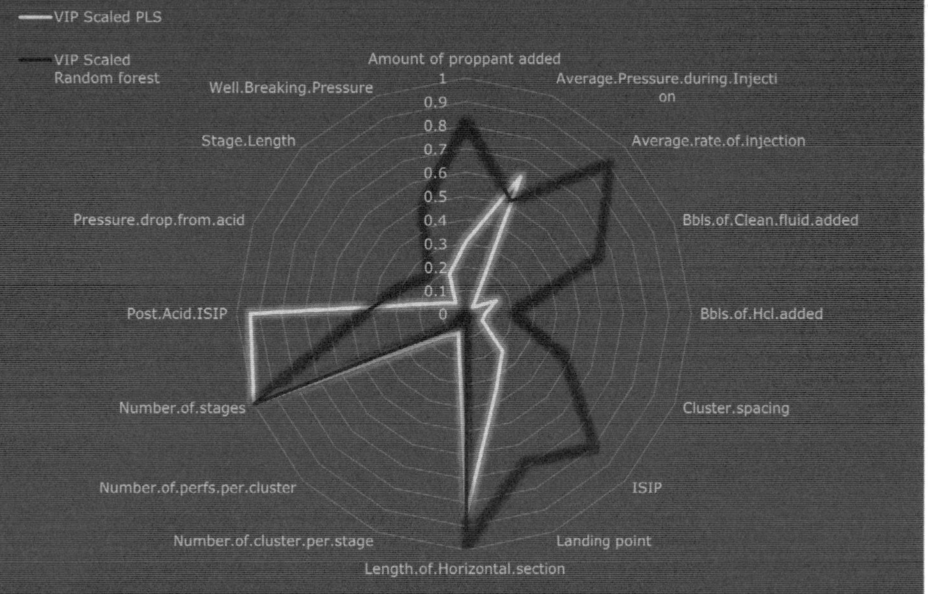

28

17AIChE Annual Meeting, Minneapolis, MN

Mathur-Nikolaou

Predicting Production for Unconventionals

UNIVERSITY of HOUSTON

Introduction

Data processed

Modeling

Results

Conclusions

29

17AIChE Annual Meeting, Minneapolis, MN

Comparison of Modeling Methods

- PLS
  - Scaled
  - Unscaled
  - Mean removed
- ANN
  - Scaled
  - Unscaled
  - PCA reduced

Mathur-Nikolaou

Predicting Production for Unconventionals

UNIVERSITY of HOUSTON

Introduction

Data processed

Modeling

Results

Conclusions

30

17AIChE Annual Meeting, Minneapolis, MN

RMSE comparison for PLS models

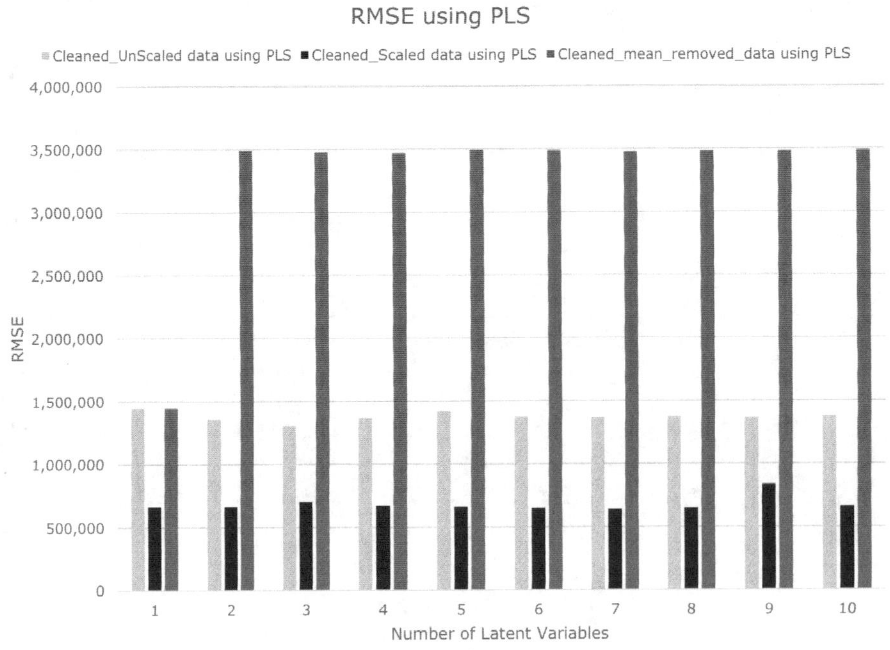

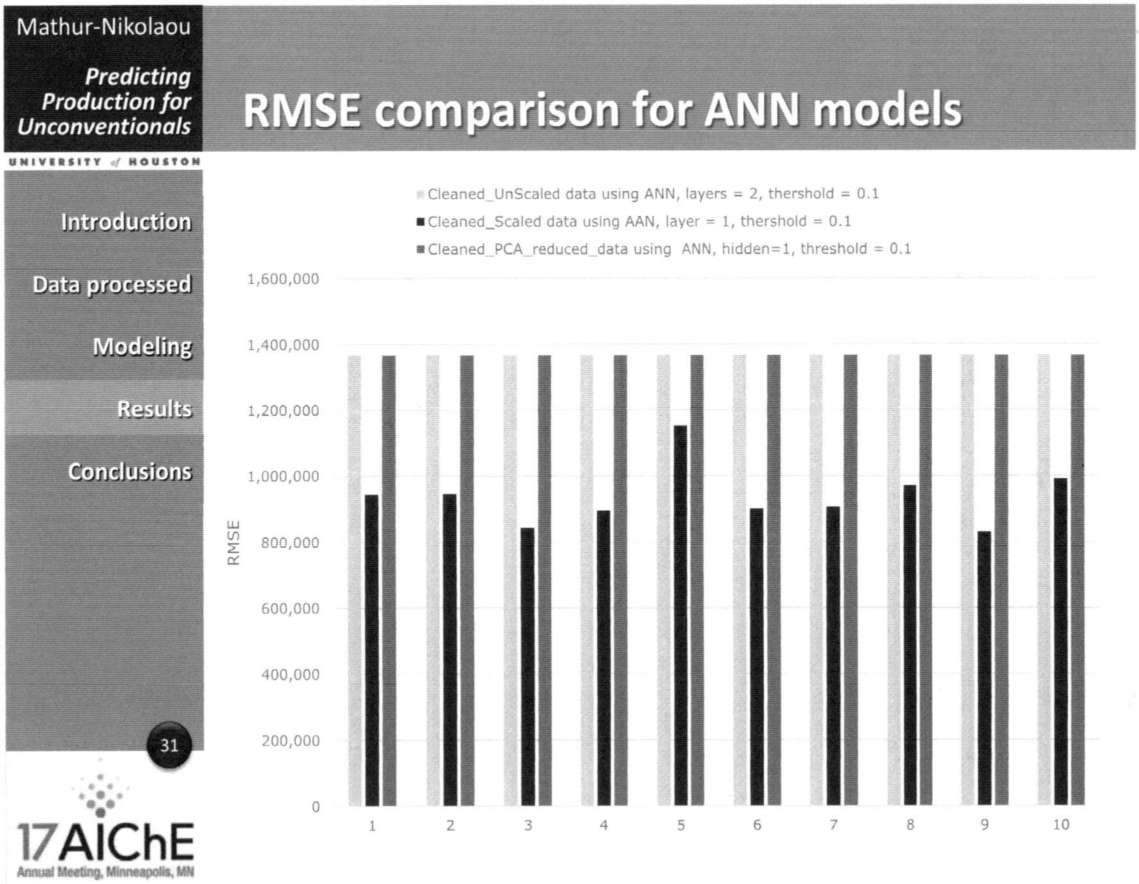

Mathur-Nikolaou
Predicting Production for Unconventionals
UNIVERSITY of HOUSTON
RMSE comparison for ANN models
Introduction
Data processed
Modeling
Results
Conclusions
31
17AIChE
Annual Meeting, Minneapolis, MN
Cleaned_UnScaled data using ANN, layers = 2, thershold = 0.1
Cleaned_Scaled data using AAN, layer = 1, thershold = 0.1
Cleaned_PCA_reduced_data using ANN, hidden=1, threshold = 0.1
RMSE
1,600,000
1,400,000
1,200,000
1,000,000
800,000
600,000
400,000
200,000
0
1
2
3
4
5
6
7
8
9
10

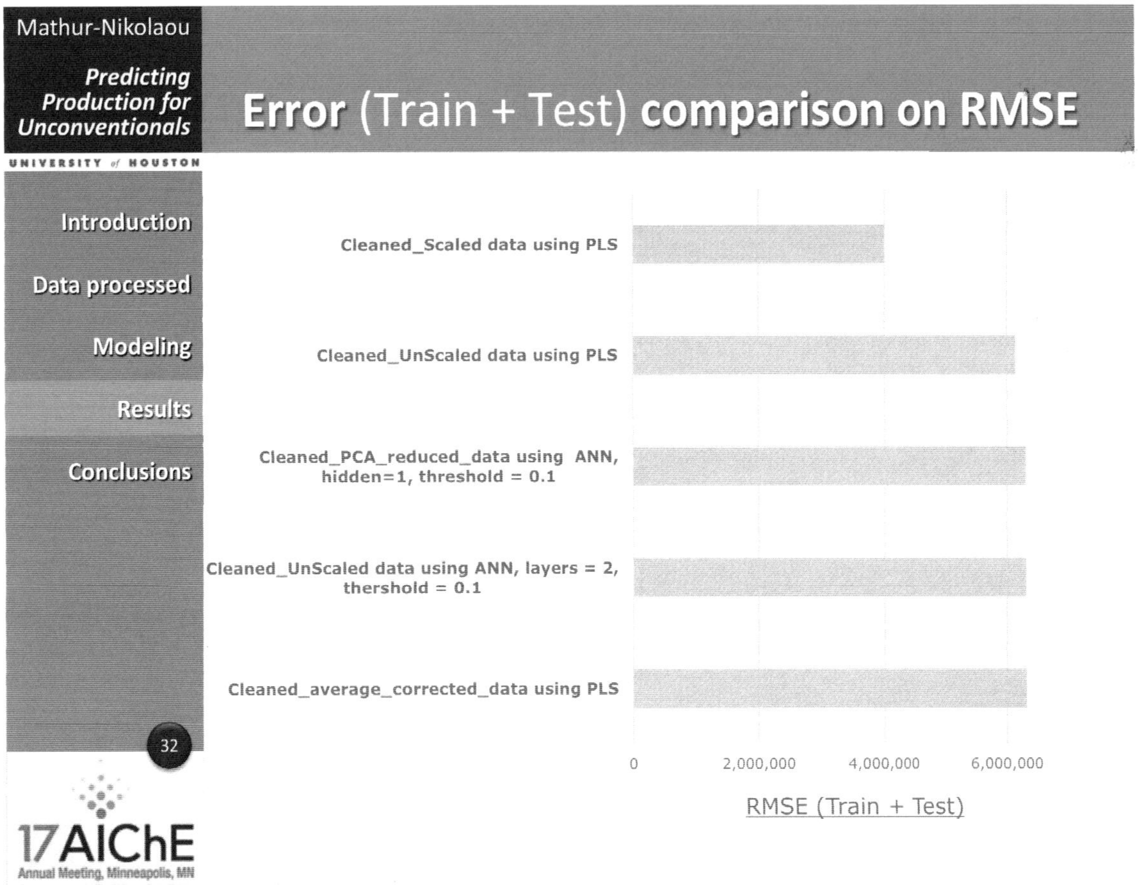

Mathur-Nikolaou
Predicting Production for Unconventionals
UNIVERSITY of HOUSTON
Error (Train + Test) comparison on RMSE
Introduction
Data processed
Modeling
Results
Conclusions
32
17AIChE
Annual Meeting, Minneapolis, MN
Cleaned_Scaled data using PLS
Cleaned_UnScaled data using PLS
Cleaned_PCA_reduced_data using ANN, hidden=1, threshold = 0.1
Cleaned_UnScaled data using ANN, layers = 2, thershold = 0.1
Cleaned_average_corrected_data using PLS
0
2,000,000
4,000,000
6,000,000
RMSE (Train + Test)

17AIChE
Annual Meeting, Minneapolis, MN

Results

Accuracy of Future Predictions

UNIVERSITY *of* HOUSTON

Mathur-Nikolaou

Predicting Production for Unconventionals

UNIVERSITY *of* HOUSTON

Introduction

Data processed

Modeling

Results

Conclusions

34

17AIChE
Annual Meeting, Minneapolis, MN

Error propagation with time on training data for uncleaned unscaled PLS

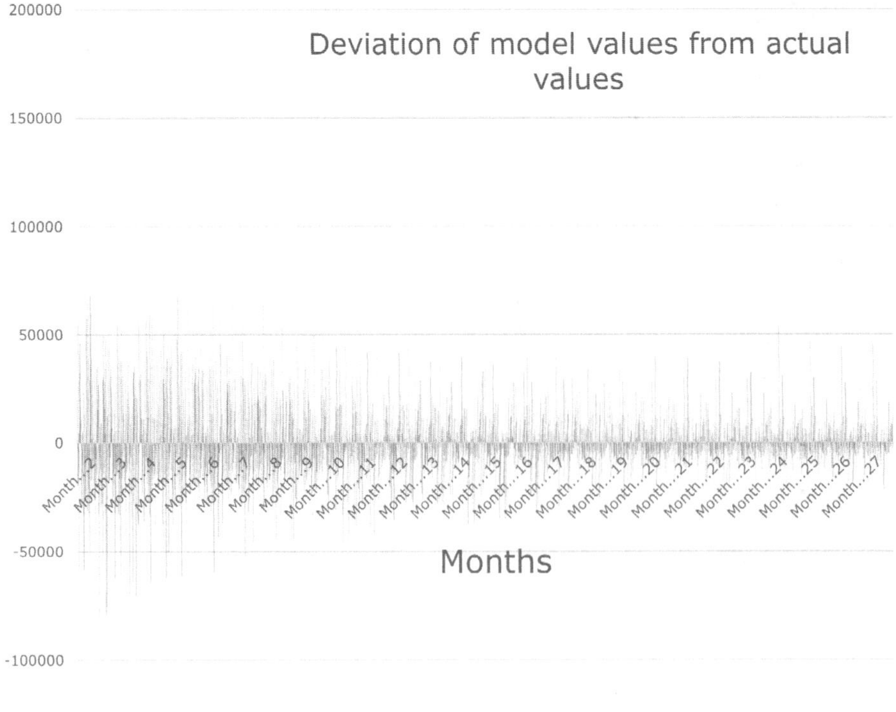

Mathur-Nikolaou

Predicting Production for Unconventionals

UNIVERSITY of HOUSTON

Error propagation with time on testing data for uncleaned unscaled PLS

- Introduction
- Data processed
- Modeling
- Results
- Conclusions

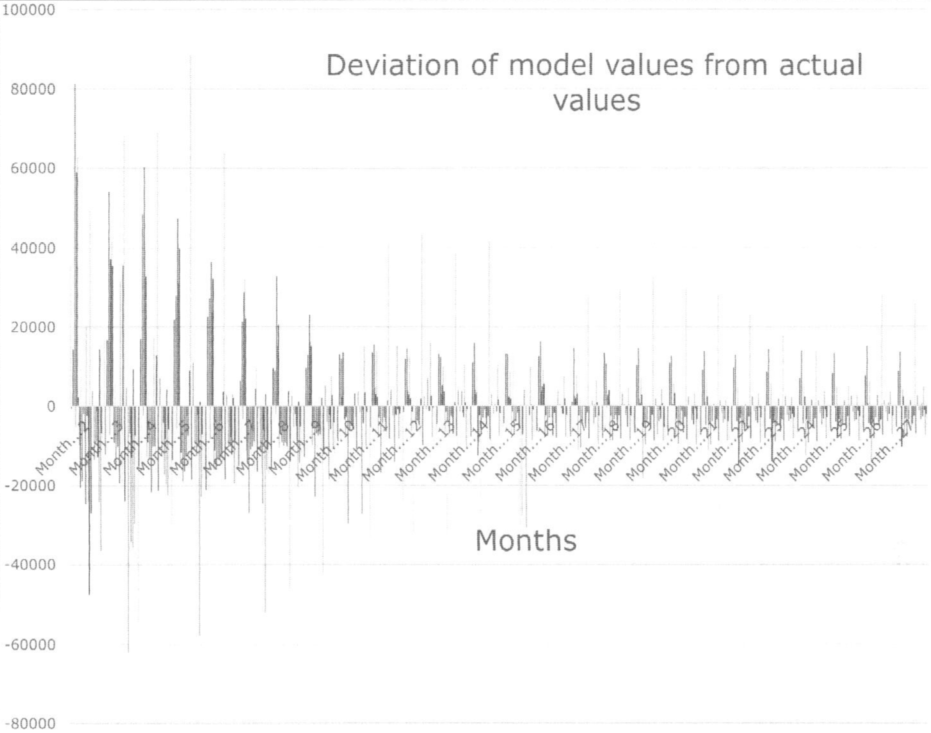

35

17AIChE
Annual Meeting, Minneapolis, MN

Mathur-Nikolaou

Predicting Production for Unconventionals

UNIVERSITY of HOUSTON

Conclusions

- Introduction
- Data processed
- Modeling
- Results
- Conclusions

- Data-driven modeling of shale gas production
  - Capitalizes on wealth of design and field data collected
    - Drilling job
    - Wellbore configuration
    - Fracture design
    - Historical production data
- Data-driven PLS model
  - Predicts future production in new wells remarkably well
  - Can rank variables in order of importance

36

17AIChE
Annual Meeting, Minneapolis, MN

Mathur-Nikolaou

Predicting Production for Unconventionals

UNIVERSITY of HOUSTON

Introduction

Data processed

Modeling

Results

Conclusions

37

17 AIChE Annual Meeting, Minneapolis, MN

Acknowledgements

- Southwestern Energy
- Dr. Matteo Marongiu-Porcu, Schlumberger

Mathur-Nikolaou

Predicting Production for Unconventionals

UNIVERSITY of HOUSTON

Introduction

Data processed

Modeling

Results

Conclusions

38

17 AIChE Annual Meeting, Minneapolis, MN

THANKS

493977 Measurement and Modeling of Competitive Sorption of Methane/Ethane Mixtures on Marcellus Shale â€“ Isotherms and Kinetics

Abstract:

As the primary mechanism of gas storage in shale, sorption phenomena of CH_4 and other hydrocarbons in the micropores and mesopores are critical to estimates of gas-in-place and of the long-term productivity from a given shale play. Since C_2H_6 is another important component of shale gas, besides CH_4, knowledge of CH_4-C_2H_6 binary mixture sorption on shale is of fundamental significance and plays a central role in understanding the physical mechanisms that control fluid storage, transport, and subsequent shale-gas production.

In this work, measurements of pure component sorption isotherms for CH_4 and C_2H_6 for pressures up to 114 bar and 35 bar, respectively, have been performed using a thermogravimetric method in the temperature range (40-60 ^oC). Sorption experiments of binary (CH_4-C_2H_6) gas mixtures containing up to 10% (mole fraction) of C_2H_6, typical of shale-gas compositions, for pressures up to 125 bar under the aforementioned temperature conditions have also been conducted.

To facilitate investigation of depletion dynamics (e.g. via application of the Extended Langmuir Model), excess sorption measurements must be converted to absolute sorption. This conversion requires accurate estimates of the adsorbate density. Here we investigate the performance of several of such methods proposed for evaluating the adsorbate density. We demonstrate that the equilibrium constant and hence the adsorption and desorption kinetic constants are sensitive to the choice of the adsorbate density model and conclude by providing recommendations for selection of adsorbate density modeling. Lastly, we apply the best-fit adsorbate density method to model the sorption dynamics of pure and binary gas mixtures on to the shale.

To the best of our knowledge, this is the first time that systematic measurements of CH_4, C_2H_6 pure and binary mixture sorption on the Marcellus shale have been conducted, thus providing a comprehensive set of CH_4-C_2H_6 competitive sorption data, which can help to improve the fundamental understanding of shale-gas storage mechanisms and its subsequent production.

SENSITIVITY STUDY OF FRACTURE PROPAGATION BY FOAMED FLUIDS AND SLICKWATER IN UNCONVENTIONAL RESERVOIRS

Fatick Nath, Chongwei Xiao, Texas A&M University-Kingsville, Kingsville, TX, USA

Abstract

Low water-content frac fluids are demanding for hydraulic fracturing application in order to improve oil and gas production from unconventional reservoirs. Non-Newtonian CO_2 foams stabilized by nanoparticles has been recently studied as a promising frac fluid. The proppant placement and fracture propagation are significantly affected by frac fluid's properties. High quality CO_2 foams can provide significant environmental-friendly benefits, such as low water usage and treatment, minimal effect to the formations, high-conductivity channels, deeper proppant placement. The viscosity of CO_2 foams, which is critical for the process, changes significantly along with gas quality and pumping rate. It is important to understand fracture propagation in unconventional reservoirs by using CO_2 foams under different foam quality and pumping rate.

This study investigated fracture propagation during hydraulic fracturing process by a 3D numerical model in an unconventional reservoir using a commercial fracturing simulator. The heterogeneous reservoir properties of a horizontal well perforated in Eagle Ford Shale formation of Zavala County, Texas were applied for the study. Effects of injection rate (20 ~ 40 BPM), CO_2 foam quality (50 ~ 80 %), incremental proppant distribution (0 ~ 5 PPA) on fracture propagation were investigated in multistage fracturing process. Fluid leakoff, which is important for the fracturing process, was investigated by pressure dependent leakoff method for different CO_2 foam qualities. The results showed that the viscosity of fracture fluid significantly influences fracture propagation. Along with the foam quality increased from 50% to 80%, the fracture width increased from 0.29 inches to 0.35 inches, whereas fracture half-length decreased from 780 ft. to 660 ft. The fluid leakoff rates decreased from 13 to 9 ml/cm^2 at 10 $\sqrt{\text{min}}$. Fracture propagation using CO_2 foams with varied injection rates compared to slick water fracturing. The results showed that CO_2 fracturing develop less fracture length compare to slick water fracturing but superior fracture width and height growth than slick water ranging from 15% to 20%. The fracture propagation by 2D and 3D numerical models were compared as well.

This study provides an insight and improved fracture treatments design by non-Newtonian frac fluids - CO_2 foams application with increased fracture conductivity and efficiency compare to slick water, which is vital for fossil fuel exploitation from unconventional reservoirs with reduced environmental impacts.

Keywords: Fracture propagation, CO_2 foam, Eagle Ford Shale, Fossil fuel, Slick water.

Introduction

Exploration and production from unconventional formations have increased in recent twenty years and the production from shale formation will last to rise due to two major technologies: horizontal drilling (Fisher et al., 2004) and hydraulic fracturing with pressurized frac fluids (Britt et al., 2006). Most of the fracturing operations used slick water or oil-based fluid as a base frac fluid to transport proppant (Barati and Liang, 2014; Schein, 2005). Carbon dioxide (CO_2) foams stabilized by nanoparticles have been recently studied as a low water content frac fluid to increase fracture conductivity (Fang et al., 2014; Xiao et al., 2017). The high-quality foams with water content less than 60% are eco-friendly as they reduce the potential damage for water sensitive formations and the volume of produced water. CO_2 foams have good mobility that result in better sweeping efficiency into formations and fast clean effect. The higher viscosity of CO_2 foams leads to better proppant transport than slickwater (Fang et al., 2014; Liu et al., 2014). Frac fluids play a key role in hydraulic fracturing treatment design (Barati and Liang, 2014) by extending fractures and transporting proppant within fractures. Some viability cases of supercritical CO_2 foams were studied from foam fracturing in high temperature and high pressure formations to control mobility (Harris, 1987; McAndrew et al., 2014; Alotaibi et al., 2016).

The mechanics of fracturing using CO_2 foams as a frac fluid needs more investigation, but precise prediction methods are still lacking. Typical 2D models are KGD (Geertsma and De Klerk 1969), PKN (Nordgren 1972), and radial model as shown in Figure 1. Both KGD and PKN models, deformation is assumed by increasing plane strain with fixed fracture height to the net pay and calculate width of the fracture by analytical solutions. The PKN models assumes the plain strain in vertical direction, which leads to an elliptical shape in cross section; whereas KGD model assumes the plain strain in horizontal direction, which leads to a rectangle shape in cross section. The radial model assumes propagation on a symmetrical plane, which leads to a circular shape in cross section for radial development. These models are found to be useful, but accuracy and simplification is the problem in most of the cases. KGD model is more appropriate for propagation of early time injection when height is larger than the length of the fracture ($x_f << h_f$), whereas PKN model for the late time of injection when length exceeds fracture height ($x_f >> h_f$). Now these classic 2D models are substituted by pseudo-3D models, but still aid as a validation tools for numerical models and providing insight of the fracturing process.

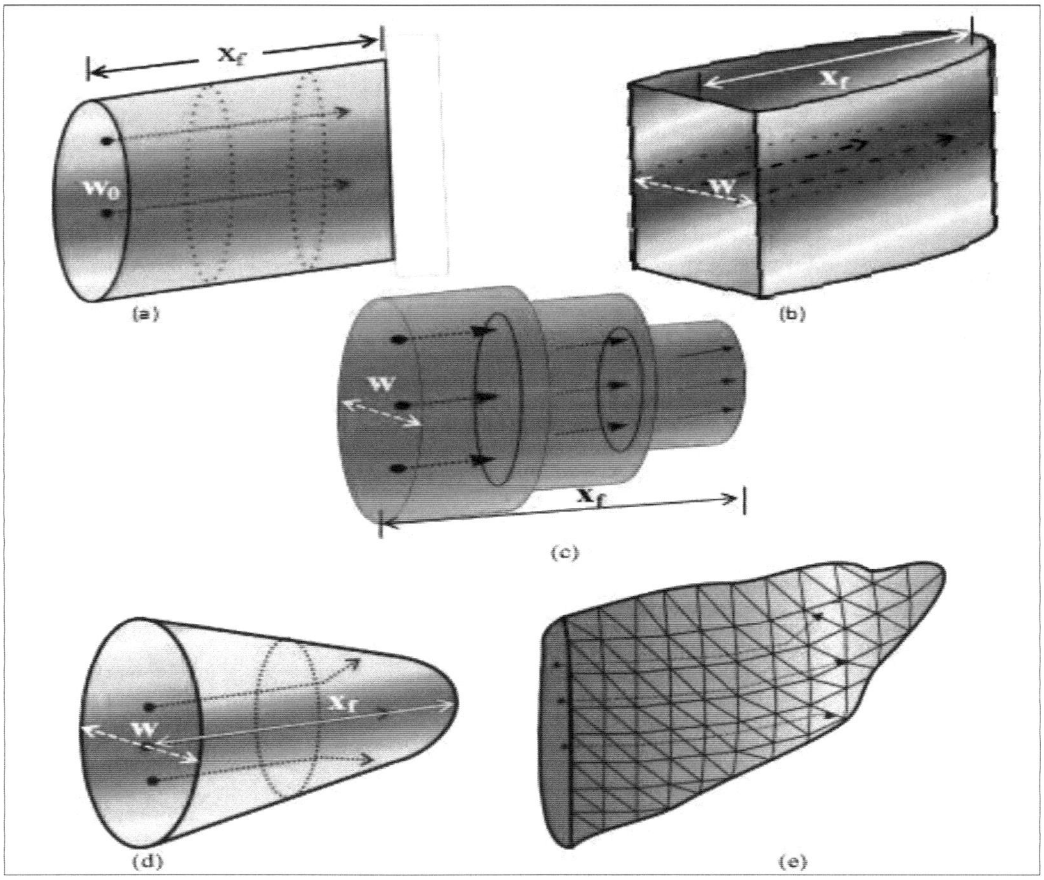

Figure 1. Fracture geometry

(a) PKN Model (b) KGD Model (c) P-3D Model (d) 3D-parametrised (e) Full 3D, gridded

The analytical models of hydraulic fracturing have greatly advanced from simple 2D to 3D models. Typical 2D models are focusing on the fracture width and length assuming constant height while the P-3D, full 3D and planar 3D models predict the height along with width and length variation. Based on the flexibility and advantages, different numerical models are used to simulate hydraulic fracturing process. The pseudo-three-dimensional (P-3D) and true 3D models were eventually established for height of fracture growth and to describe the insight of hydraulic fracturing process with enhanced accuracy. P-3D models are inexact but proficient to characterize the behavior of height growth of the fracture. A P-3D model demonstrates the height enlargements in a symmetric 3-layer formation investigating the material properties, in-situ stresses and pay zone pressure gradients and the fault of the formation (Simonson, Abou-Sayed et al. 1987). The P-3D model by Settari (Settari 1980) described a fluid flow in lateral direction and crack initiation for fracture and coupled with the development of fracture vertically. The typical methodology for true 3D models is based on 3D deformation of rock layers and 2D fluid flow within the fractures (Abou-Sayed, Sinha et al. 1984).

As CO_2 foams is emerging as a new frac fluid for unconventional resources development, it is important to study its effect on the hydraulic fracturing process. This research developed a 3D numerical model with mechanical heterogeneity in Eagle Ford shale formation for the fracturing process using CO_2 foams as the frac fluid. The combination of rock mechanics and two-phase fluid flow both in reservoir and inside the fractures will provide a guidance for future fracture treatment design. The sensitivity studies of fracture propagations affected by CO_2 foam quality, injection rate, and proppant concentration were analyzed accordingly. A comparison of fracture propagation from 3D model to 2D analytical models is performed to increase the understanding of the mechanism of hydraulic fracturing.

Fracture Propagation Model

The sensitivity studies is performed using finite element based numerical model and the model is based on a corner point gridding system on Cartesian plane and all parameters are defined at each corner point of grid blocks (GOHFER Manual). Each grid block is defined by all state variables - fluid composition, viscosity, pressure, rock mechanics, proppant concentration, etc.

The formation mechanical properties were derived from four different types of logs, including Gamma ray, Resistivity, Density Porosity, and Density Neutron, were acquired for the reservoir properties. Log-derived input data were analyzed by a built-in log analysis package and accurate lithologic description was generated, including lithology, in-situ stress and poroelastic properties of rock, etc. The in-situ stress profile was calculated from the pore pressure and poroelastic constant. The mechanical properties of the formation, including Young's modulus, Poisson's ratio, and Biot's constant, were determined by empirical correlations.
The complete stress equation is given by the closure pressure,

$$P_c = \sigma_h = \frac{\nu}{(1-\nu)}[\sigma_v - \alpha_v P_p] + \alpha_h P_p + \varepsilon_x E + \sigma_t \tag{1}$$

The Carreau rheology model (Elise, 1995) is used to describe the time-dependent fluid behavior in a different geometry including sphere motion, pipe flow, and plates.

$$\eta(\dot{\gamma}) = \eta_\infty + (\eta_o - \eta_\infty)\left[1 + (\dot{\gamma}\lambda)^2\right]^{\frac{n-1}{2}} \tag{2}$$

Where, $\eta(\dot{\gamma})$, Viscosity as a function of shear rate; η_o, zero shear viscosity, cp; η_∞: infinite shear viscosity, cp and λ : characteristic time in sec.

The fracture propagation was simulated using the fully 3D model based on finite difference scheme (Barree, 1983). Each grid node can be assigned an individual value of net stress, pore pressure, permeability, porosity, wall-building coefficient, rock strength, Poisson's ratio and Young's modulus and variables describing fracture-wall roughness and tortuosity. The fracture width equation used is the general formula given by Boussinesq (Sneddon, 1946) in terms of displacement for a semi-infinite pace acting by the distributed load.

In a dynamic flowing system, the fracture fluid pressure is associated with the apparent fluid viscosity, flow rate and fracture width at each point along the fracture –

$$Q = \frac{h}{12\mu_a} \frac{d}{dl} (P + \gamma h) w^3 \tag{3}$$

Where, h is the length of a grid block normal to the direction of flow, γ is the fracture gradient is the fracture width. For flow between two nodes harmonic average of the cube is used.

PKN model is applied when $x_f \gg h_f$. The maximum fracture width for non-Newtonian fluid is calculated by:

$$w_{max} = 12\left[\left(\frac{128}{3\pi}\right)(n'+1)\left(\frac{2n'+1}{n'}\right)^{n'}\left(\frac{0.9775}{144}\right)\left(\frac{5.61}{60}\right)^{n'}\right]^{1/(2n'+2)} \cdot \left(\frac{q_i^{n'} K' x_f h_f^{1-n'}}{E}\right)^{1/(2n'+2)} \tag{4}$$

$$\bar{w} = 0.625 w_{max} \tag{5}$$

Where, x_f : fracture half-length, ft
w_f: fracture width, in
h_f : fracture height, ft

KGD model is applicable where $h_f \gg x_f$. The average fracture width, for Newtonian fluid is:

$$\bar{w} = 0.34\left[\frac{q_i \mu x_f^2}{E' h_f}\right]^{1/4}\left(\frac{\pi}{4}\right) \tag{6}$$

For radial fracture geometry, the maximum fracture width is calculated by

$$w_{max} = 2\left[\frac{q_i \mu (1-\nu) R}{G}\right]^{1/4} \text{, Where, R is the fracture radius.} \tag{7}$$

Study Area: Eagle Ford Shale Formation

The hydrocarbon-rich Eagle Ford formation is located in South Texas. The productive zones vary from subsea 4,500 to 18,000 ft. (Shelley et al., 2012). with the thickness among 50 to 300 ft. The formation is thermogenically kerogen-rich and poroelastic properties are in attractive range including medium-ranged Young moduli, low Poisson's ratios, and least stress anisotropy. To represent formation heterogeneity, the rock mechanical properties were derived from well logs including Gamma Ray, Resistivity, Neutron Porosity, and Density Porosity logs, characterized by Young's modulus ($3\times10^6 \sim 6\times 10^6$ psi), Biot constant (0.6 ~ 0.8), and Poisson's ratio (0.2 ~ 0.4). The horizontal oil well (API: 42-507-32744) in Zavala County, TX was completed in Eagle Ford Shale formation in 15 March, 2010. The well was perforated in the interval of measured depth, 5567-10771 ft. The reservoir temperature is 150°F and formation pressure is 2300 psia with reservoir pressure gradient of 0.41 psi/ft.

Simulation Model Input

The simulation inputs of different properties were given in the Table 1 and fluid modeling procedure with determined parameters are given Table 2.

Table 1. Simulation Model Input

Rock Properties	
Young's modulus E, MMpsi	3.7-4.2
Poisson's ratio v,	0.27-0.32
Matrix Density, lb/ft3	165
Fracture Toughness, psi.(in) $^{0.5}$	1000
In-situ Stress data	
Vertical Stress, σ_v ,psi	~5200
Max. horizontal Stress, σ_H psi	~4200
Min. horizontal Stress, σ_H psi	~3200

Reservoir data	
Formation depth D, ft.	5300
Formation thickness h, ft.	300
Porosity ϕ,	0.08~0.1
Formation compressibility, C_t psi^{-1}	1E-6
Permeability k, md	~0.001
Water Saturation, Sw	0.32
Pore Pressure,psi	2286
Formation Tem., F	150
Fluid gradient, psi/ft.	0.41

Fracturing fluid Properties	
Fluid Name	CO_2 Foam
Foam Quality	0.5-0.8
Frac fluid density $\rho_L, lb/ft^3$	~65
Temperature, F	104
Power law exponent n',	0.5~0.7
Consistency index $K', lb_f sec^{n'}/ft^2$	0.0023-0.0286
Spurt loss coefficient	0
Proppant Data	
Proppant	100 mesh sand
Proppant porosity ϕ_p,	0.35
Proppant density $\rho_p, lb/ft^3$	165
EOJ slurry concentration c_t, ppg	0~5

Treatment Schedule

Each Stage	Pump Rate, BPM	Fluid Name	Stage Fluid Volume, Gal	Proppant Concentration, PPA per gal	Proppant Type + Mesh
Pad	20~40	CO_2 Foams	1000	0.00	-
Slurry	20~40	CO_2 Foams	1000	0.50	100 mesh sand
Slurry	20~40	CO_2 Foams	1000	1.00	100 mesh sand
Slurry	20~40	CO_2 Foams	1000	2.00	100 mesh sand
Slurry	20~40	CO_2 Foams	1000	3.00	100 mesh sand
Slurry	20~40	CO_2 Foams	1000	4.00	100 mesh sand
Flash	20~40	Slick H_2O	125	0.00	100 mesh sand

Table 2: Frac-fluid modeling using experimental data and parameters determination

	Viscosity, cp			
Shear Rate, 1/s	Foam quality 50%	Foam quality 60%	Foam quality 70%	Foam quality 80%
5950	1.45	2.02	2.78	3.49
8925	1.21	1.80	2.47	3.04
11900	1.06	1.65	2.28	2.76
14875	0.96	1.55	2.14	2.55
17850	0.88	1.46	2.03	2.40

Parameter	flowrate, ml/min	Foam Quality			
		50%	60%	70%	80%
n'	6	0.5506	0.7063	0.7156	0.6574
k', Pa.s^n	9	0.072	0.026	0.033	0.069
$\eta_{0, cp}$	12	1.80	2.50	3.35	4.25
$\eta_{\infty, cp}$	15	0.95	1.5	2.1	2.5
λ, s	18	0.0011	0.0012	0.001	0.001

Sensitivity Analysis of Fracture Propagation

During the process of CO_2 foams injection and flow in the unconventional reservoir, the effects of shear rates, elastic response of porous media, and leak-off on the fracture propagation were investigated. The effects of the fracture geometry, and non-Newtonian fluid flow were numerically analyzed.

Increasing injection flow rate augments the fracture length and width of the fracture. The viscosity of CO_2 foam is a dominant factor for fracture propagation. As the viscosity increases with varying foam quality (50%-80%), slurry fluid proppant-carrying capacity increases and average fracture width increases but the fracture extent reduces. Figure 1-4 showed the effects of CO_2 foam quality and injection rate on reservoir pressure and fracture propagation.

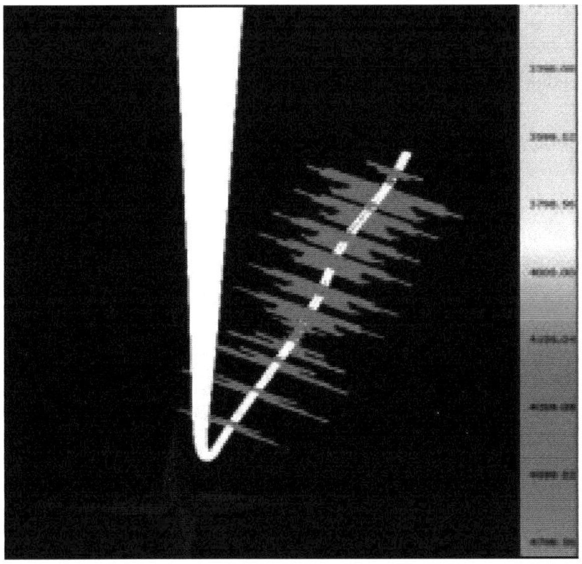

Figure 1. The profile of reservoir pressure during the process

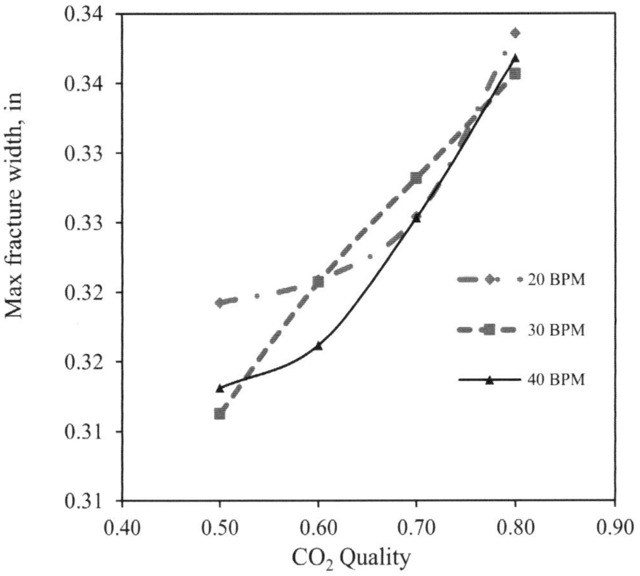

Figure 2. Effects of CO_2 foam quality and pumping rate on fracture width

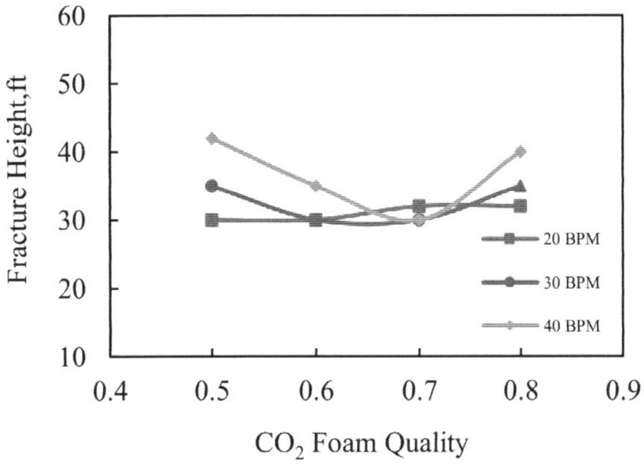

Figure 3. Effects of CO_2 foam quality and injection rate on fracture height

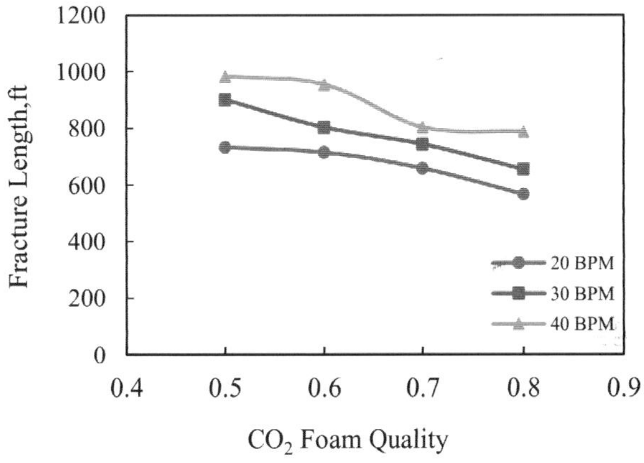

Figure 4. Effects of CO_2 foam quality and injection rate on fracture length

Fluid Leakoff

Fluid leakoff associated for different CO_2 foams quality (50% - 80%) was investigated. Pressure dependent leakoff method is used y to quantify the fluid loss during fracturing. The fluid leakoff rates along with resident time under different foam quality was shown Figure 5. The results showed that the fluid leakoff decreased with increasing CO_2 foams quality that leakoff rate reached the maximal 13 ml/cm^2 at 50% CO_2

foams quality and the minimal 9 ml/cm^2 at 80% of foam quality.

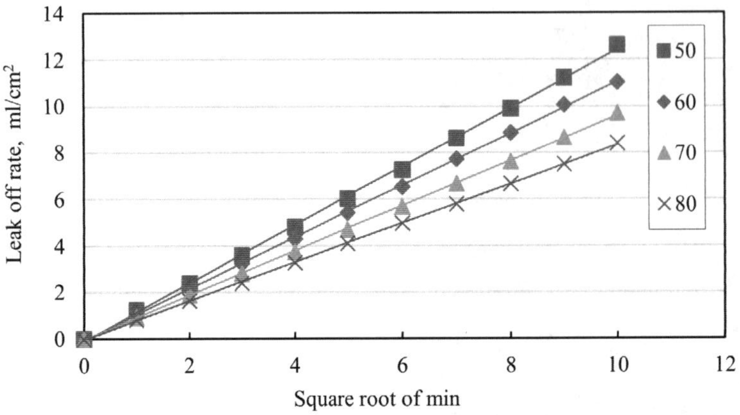

Figure 5. The fluid leakoff under different CO_2 foams quality

Comparison of fracture propagation with slick water

Fracture propagation using 60% quality CO_2 foams was compared (in 1-10 stages) to slick water fracturing and observed maximum 15% greater fracture width and height growth than slick water. The comparison of fracture width and height of CO_2 foams. Overall fracture propagation result is given in Table 4.

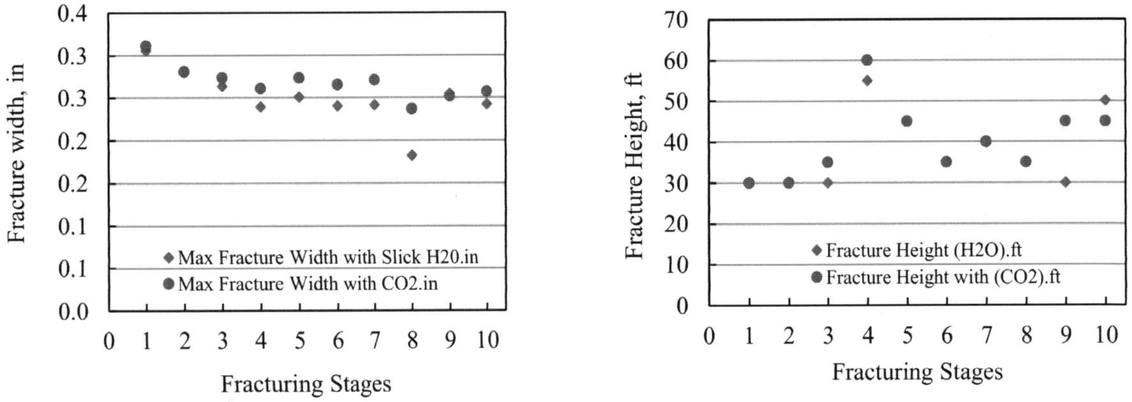

Figure 6. The comparison of fracture width and height fractured by CO_2 foams and slick water

Table 4. The comprehensive results of different treatment

Parameters	Slick H_2O	50% CO_2	60% CO_2	70% CO_2	80% CO_2
Half-length, ft.	852	743	612	560	524
Frac. height, ft.	35	30	35	35	40
Max. width, in	0.284	0.314	0.322	0.332	0.356
Avg.width, in	0.15	0.14	0.155	0.167	0.178

Comparison of fracture propagation among 3D and 2D numerical models

The fracture propagation was also studied with analytical 2D models. Fracture width from 3D model is compared for three different injection rates with 2D analytical model-PKN, KGD and radial model. The comparison shows a good agreement with fracture width variation. 3D model gives the highest width compared to the other models (Figure 7).

Fracture propagation by width is increased with increasing injection rate of CO_2 foams. For a given injection rate, 3D model considers the fundamental theories of linear elastic fracture mechanics with complex fluid flow behaviors inside the fractures whereas 2D models are not. In 3D model, fracture growth is simultaneously apply mixed-mode fracture propagation criterion which resulted the vertically and laterally fracture with highest extend. 2D models are good for quick checking the usual growth which gives a rough idea about the fracture propagation. 3D model is widely used in industry for commercial purposes as it incorporates complex behaviors of formation representing rock mechanical properties, fluid rheology, and complex stress distribution around the bore hole.

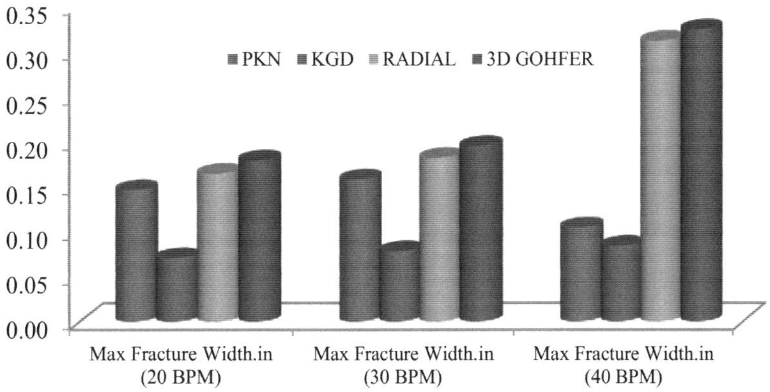

Figure 7. 3D model is compared with 2D analytical models

Conclusions

This study numerically simulated the process of hydraulic fracture propagation with the frac fluid - CO_2 foams in the Eagle Ford Shale formation with a 3D heterogeneous reservoir model.

The results showed that fracture propagation is obviously affected by the properties of frac fluid and injection rate. Along with increasing fluid viscosity and injection rate, both fracture length and width of the fracture augments increase; the proppant-carrying capacity of the slurry increases as well and proppant were placed deeper in fractures. However, the fracture extent reduces along with increasing fluid viscosity. Along with CO_2 foam quality increased from 50% to 80%, the fluid leakoff rates decreased from 13 to 9 ml/cm^2 and fluid efficiency increased from 85% to 96%. The fracture width and height growth using CO_2 foams is 15% to 20% greater than those of slick water fracturing. Finally, the comparison with 2D analytical model (PKN, KGD and radial) to 3D model showed a good agreement in fracture propagation.

Acknowledgement

The authors would like to acknowledge Houston Advanced Research Center (HARC) and the Department of Chemical and Natural Gas Engineering at Texas A&M University, Kingsville for their intellectual and financial support to conduct this research study. Authors also thankful to Barree & Associates, LLC to support this study by providing the simulator GOHFER.

References

1. Al-Dhamen, M., and Soriano, E., 2015. Increased Well Productivity from the Use of Carbon Dioxide to Foam Fracturing Fluids During a Refracturing Treatment in Saudi Arabia. Presented at the SPE Latin American and Caribbean Petroleum Engineering Conference, Quito, Ecuador, 18-20 November. SPE-177112-MS. http://dx.doi:10.2118/177112-MS.
2. Abou-Sayed, A. S., Sinha, K. P., and Clifton, R. J.,1984. Evaluation of the Influence of In-Situ Reservoir Conditions on the Geometry of Hydraulic Fractures Using a 3-D Simulator: Part 1-Technical Approach. Presented at the SPE Unconventional Gas Recovery Symposium, Pittsburgh, Pennsylvania, 13-15 May, . SPE-12877-MS. http://dx.doi.org/10.2118/12877-MS
3. Alotaibi, F., Zhou, X., Habah, A. et al 2016. EOR by CO_2 Emulsions Better Sweep Efficiency and Improved Recovery. Presented at the Abu Dhabi International Petroleum Exhibition and Conference, Abu Dhabi, UAE, 7-10 November.SPE-183014-MS. http://dx.doi:10.2118/183014-MS
4. Al-Qasim, A., AlOtaibi, F., Kokal, S. et al 2017. CO_2-Foam Rheology Behavior under Reservoir Conditions.Presented at the SPE KSA Annual Technical Symposium and Exhibition held in Dammam, Saudi Arabia, 24-27 April. SPE-188059-MS. http://dx.doi:10.2118/188059-MS
5. Barati, R. and Liang, J.T., 2014, A review of fracturing fluid systems used for hydraulic fracturing of oil and gas wells. J. Appl. Polym. Sci., 131, 40735, doi: 10.1002/app.40735
6. Barree, R. D., 1983. A Practical Numerical Simulator for Three-Dimensional Fracture Propagation in Heterogeneous Media. Presented at the Reservoir Simulation Symposium, San Francisco, CA, 15-18 November.SPE-12273-MS. http://dx.doi.org/10.2118/12273-MS.
7. Britt, L. K., Smith, M. B., Haddad, Z. A., et al., 2006. Waterfracs: We Do Need Proppant After All. Presented at the SPE Annual Technical Conference and Exhibition, San Antonio, TX, U.S.A. 24-27 Septermber. SPE-102227-MS. http://dx.doi.org/10.2118/102227-MS.
8. Economides, M., Oligney,R., and Valko, P., 2002. Unified fracture design: bridging the gap between theory and practice Orsa Press, TX,USA
9. Elise, A., 1995. An Improved Viscosity Equation to Characterize Shear-Thinning Fluids. Presented at the SPE International Symposium on Oilfield Chemistry, San Antonio, TX, U.S.A. 14-17 February. SPE-28973-MS. http://dx.doi.org/10.2118/28973-MS
10. Fang, C., Chen, W. and Amro, M., 2014. Simulation Study of Hydraulic Fracturing Using Super Critical CO_2 in Shale. Presented at the Abu Dhabi International Petroleum Exhibition and Conference, Abu Dhabi, UAE, 10-13 November. SPE-28973-MS. http://dx.doi.org/10.2118/172110-MS
11. Fisher, M. K., Heinze, J. R., Harris, C. D., et al., 2004. Optimizing Horizontal Completion Techniques in the Barnett Shale Using Microseismic Fracture Mapping. Presented at the SPE Annual Technical Conference and Exhibition, Houston, 26–29 September. SPE-90051-MS. http://dx.doi.org/10.2118/90051-MS
12. Fisher, M. K. and Warpinski, N. R., 2012. Hydraulic-Fracture-Height Growth: Real Data. SPE Production & Operations, February. SPE-145949-PA. http://dx.doi.org/10.2118/145949-PA
13. Geertsma, J. and De Klerk, F., 1969. A Rapid Method of Predicting Width and Extent of Hydraulically Induced Fractures. 10.2118/2458-PA
14. GOHFER-Manual, v8.3.0.184. GOHFER Manual.
15. Gupta, D. V. S., 2009. Unconventional Fracturing Fluids for Tight Gas Reservoirs. Presented at the SPE Hydraulic Fracturing Technology Conference, The Woodlands, Texas, USA, 19-21 January. SPE-119424-MS. http://dx.doi.org/10.2118/119424-MS
16. Gupta, A. P., Gupta, A., and Langlinais, J., 2005. Feasibility of Supercritical Carbon Dioxide as a Drilling Fluid for Deep Underbalanced Drilling Operation. Presented at the SPE Anuual Technical Conference and Exhibition,Dallas,Texas,USA, 9-12 October. SPE-96992-MS.http://dx.doi.org/10.2118/96992-MS

17. Gupta, S., Hlidek, B. T., Hill, E.S. W., and Dinsa, H.S., 2007. Fracturing Fluid for Low-Permeability Gas Reservoirs: Emulsion of Carbon Dioxide With Aqueous Methanol Base Fluid: Chemistry and Applications. Presented at the SPE Hydraulic Fracturing Technology Conference. College Station, Texas, U.S.A., 29-31 January.SPE-106304-MS. http://dx.doi.org/10.2118/106304-MS
18. Gupta, D. V. S., Leshchyshyn, T. T., and Hlidek, B. T., 2005. Surfactant Gel Foam/Emulsion: History and Field Application in Western Canadian Sedimentary Basin. Presented at the SPE Anuual Technical Conference and Exhibition,Dallas,Texas,USA, 9-12 October. SPE-97211-MS. http://dx.doi.org/10.2118/97211-MS
19. Harris, P. C., 1987. Dynamic Fluid-Loss Characteristics of CO2-Foam Fracturing Fluids. Production Engineering Journal, Issue. May, p. 89-94. SPE-13180-PA.. http://dx.doi.org/10.2118/13180-PA
20. McAndrew, J. J., Fan, R., Sharma, M., and Ribeiro, L., 2014. Extending the Application of Foam Hydraulic Fracturing Fluids. Presented at the Unconventional Resources Technology Conference,Denver,Colorado,USA,25-27 August. SPE-1926561-MS URTEC. http://dx.doi.org/10.15530/URTEC-2014-1926561
21. Geertsma, J., and De Klerk, F.,1969. A Rapid Method of Predicting Width and Extent of Hydraulically Induced Fractures. Journal of Petroleum Technology, Vol. 246 (01),Pg 1571-81. December. SPE-2458-PA. http://dx.doi.org/10.2118/2458-PA
22. Schein, G., 2005. The Application and Technology of Slickwater Fracturing. SPE-108807-DL SPE General. http://dx.doi.org/10.2118/108807-DL
23. Settari, A.,1980. Simulation of Hydraulic Fracturing Processes. Society of Petroleum Engineers Journal,Vol. 20,Issue 06,Dec 1980. SPE-7693-PA. https://doi.org/10.2118/7693-PA
24. Shelley, R. F., Saugier, L. D., Al-Tailji, W., Guliyev, N. and Shah, K., 2012. Understanding Hydraulic Fracture Stimulated Horizontal Eagle Ford Completions,Presented at the SPE/EAGE European Unconventional Resources Conference and Exhibition,Vienna,Austria, 20-22 March. SPE-152333-MS. http://dx.doi.org/10.2118/152533-MS
25. Simonson, E. R., Abou-Sayed, A. S., and Clifton, R. J.,1978. Containment of Massive Hydraulic Fractures. Society of Petroleum Engineers Journal,Vol. 18,Issue 01,Feb 1978. SPE-6089-PA. http://dx.doi.org/10.2118/6089-PA
26. Sneddon, I. N., 1946. The opening of Griffith Crack Under Internal Pressure. Quart. Appl. Math **4**: 262-267.
27. U.S. Energy Information Administration, A. E. O. E. R. (2015). U.S. Energy Information Administration, Annual Energy Outlook 2015 Early Release. http://www.eia.gov/todayinenergy/detail.cfm?id=21072 Date of accessed: 2/25/2016
28. Warnock, W. E., Jr., Harris, P. C., et al., 1985. Successful Field Applications of CO_2-Foam Fracturing Fluids in the Arkansas-Louisiana-Texas Region. Journal of Petroleum Technology, 37 (01), January. SPE-11932-PA. http://dx.doi.org/10.2118/11932-PA
29. Warpinski, N. R., Lorenz, J. C., Branagan, P. T., et al.,1991. Examination of a Cored Hydraulic Fracture in a Deep Gas Well (includes associated papers 26302 and 26946). SPE Production & Facilities, 8 (03),August. http://dx.doi.org/10.2118/22876-PA
30. Xiang, J., 2011. A PKN Hydraulic Fracture Model Study and Formation Permeability Determination, MS Thesis,Petroleum Engineering, Texaa A&M University, Dec 2011.
31. Xiao, C., Balasubramanian, S. N. and Clapp, L. W., 2017. Rheology of Viscous CO2 Foams Stabilized by Nanoparticles under High Pressure, Industrial & Engineering Chemistry Research 2017 56 (29), 8340-8348 DOI: 10.1021/acs.iecr.7b01404.

SCREENING IMPROVED RECOVERY METHODS IN TIGHT-OIL FORMATIONS BY INJECTING AND PRODUCING THROUGH FRACTURES

Harpreet Singh[1]
[1]National Energy Technology Laboratory, Morgantown, WV, USA

Abstract

Tight-oil reservoirs exhibit two characteristic behaviors that limit potential for improved recovery: i) limited fluid movement from the unfractured matrix limits the potential to sweep additional oil towards the production wells with injection of a displacing fluid, ii) the wettability of these reservoirs tend to be oil-wet which holds oil in relatively smaller pores, as compared to gas or water, making it difficult to mobilize that trapped oil. Primary oil production from these formations can decline to half of initial rates in the first year due to low permeability in the unfractured reservoir matrix, and a large amount of unrecovered oil remains in smaller pores that is not able to move out. Declining oil rates and low ultimate recoveries from oil-rich shale formations could benefit from application of secondary fluid injection schemes.

To address these challenges, this study investigates the potential to improve oil recovery from tight-oil reservoirs by a method that includes injecting a fluid into the fractured reservoir to produce hydrocarbons from adjacent fractures intersecting the same wellbore. The injection and production through fractures has a potential economic advantage over huff-n-puff scheme such that there is no lag-period between injection and production. We compare the estimates of incremental oil recovery obtained by injecting water, immiscible CO_2, and surfactant over the oil recovered by primary depletion (with no injection). The results suggest that injection of fluids may not always improve recovery from tight-oil reservoirs, especially in a heterogeneous reservoir (having different rock types) whose average matrix permeability is lower than 0.01 mD. In cases where recovery is improved over primary depletion, water flood and surfactant flood perform almost equally well with no noticeable difference between the two. Although, surfactant results in favorable fluid properties to mobilize oil, the reason surfactant did not perform well in tight-oil formation studied here is because the unswept oil is inaccessible due to small permeability, and not because of high capillary pressure.

Keywords: tight-oil; shale; EOR; water; surfactant; immiscible CO_2;

1. Introduction

Oil and gas production from unconventional tight-oil resources has become an important source of energy security in the United States (US) and it is emerging to be as important in other countries, especially Russia, Argentina, and Canada will see significant tight-oil production by 2040 (Aloulou, 2016). In the U.S., the contribution of oil produced from unconventional shale resources has jumped from 0.5 million barrels per day (BPD) in 2005 to ~5 million BPD in 2015, i.e. ~52% of the total US crude oil production (US EIA, 2016). This increase in production from unconventional shale resources has been mainly due to advancements in drilling efficiency and well configuration (Rystad Energy, 2014). However, similar improvements have not been achieved for well performance; it has been observed (Rystad Energy, 2014) that year after year wells in mature plays (e.g. Eagle Ford and Bakken) have shown similar initial rates and decline trends. Production from tight-oil resources has two drawbacks with a potential for improvement: i) low ultimate recovery factor (Gaswirth and Marra, 2015; Gaswirth et al., 2013), and ii) decline in daily production rate by more than 50% within the first year of production from an average well in the Bakken (Lund, 2014; Wachtmeister et al., 2017) and Eagle Ford (Delaihdem, 2013; Gong, 2013) plays. For instance, estimated oil resources in the Bakken formation is estimated to be 300 billion barrels (Flannery, 2006), out of which 7.38 billion barrels is estimated to be technically recoverable using current best-available technology (Gaswirth and Marra, 2015; Gaswirth et al., 2013), i.e. just ~2% of the original oil in place (OOIP). In addition to low recovery factor, the production rate of a new well from tight-oil reservoirs have been reported to decline by more than half within a year (Delaihdem, 2013; Gong, 2013; Lund, 2014; Wachtmeister et al., 2017), primarily because of fast depletion from fractures and slow fluid movement from low matrix permeability.

To arrest the declining production rates and recover additional oil from tight-oil resources, traditional methods of improved oil recovery (IOR) can be considered (e.g. water/gas/surfactant flooding etc.). However, unlike IOR in conventional reservoirs with relatively high permeability, unconventional tight-oil resources present two major challenges to IOR: i) very limited fluid movement outside the fractures limits the possibility of an efficient sweep between the wells, ii) the wettability of these reservoirs tend to be oil-wet that keeps the oil in the smallest pores compared to gas or water, making it difficult to drive oil out.

We investigate the potential of IOR from tight-oil reservoirs by a method that includes numerical reservoir simulation-based evaluation of the performance of fluid injection through a set of fractures to stimulate additional hydrocarbon recovery from a different set of fractures intersecting the same wellbore (Dombrowski et al., 2015) as shown in Figure 1. This injection-production scheme is less expensive compared to IOR using several injection and production wells in tight-oil reservoir (e.g. Ghaderi et al., 2017; Hustak et al., 2017), and unlike huff-n-puff, does not require shutting-in the well.

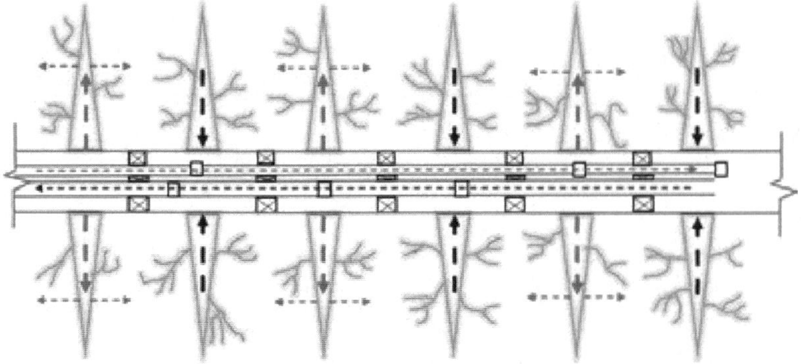

Figure 1: Schematic for fluid injection and production through hydraulic fractures in tight-oil reservoir. Dashed red and black colored arrows represent the direction of the injected and produced fluids, respectively.

To investigate the potential of improving recovery in tight-oil reservoirs, we study the injection of three different fluids (water, immiscible CO_2, and surfactant) in a single wellbore that is intersected by three injection-fractures and three recovery-fractures. This representation used for simulation of what is very likely a network of engineered fractures and pre-existing natural fractures. We compare the estimates of oil recovery obtained by injecting water, immiscible CO_2, and surfactant with the oil recovered by primary depletion (from all six fractures) with no injection. The impact of fluid injection on oil recovery is assessed for three different average permeabilities (very low, low, and relatively high), three different injection rates (low, high, and very high), and when injection is constrained by constant bottom-hole pressure in a homogeneous and heterogeneous medium with rock characteristics representative of a typical tight-oil reservoir.

2. Method

Below, we discuss the development of the reservoir model and the fluid model used in simulations. The reservoir model is composed of discrete hydraulic fractures that are connected with the matrix grid by a nearest neighborhood connection approach (Lie, 2016). The flow simulations are performed using Matlab Reservoir Simulation Toolbox (MRST), a full-physics reservoir simulator (Lie, 2016), with fully implicit solutions of blackoil model.

2.1. Reservoir Model

2.1.1. Injection and Production in Tight-Oil Reservoirs

A schematic of a single horizontal well (henceforth referred to as *lateral*) intersecting six hydraulic fractures in a tight-oil reservoir is depicted in Figure 2. The conventional approach of recovering hydrocarbons from tight reservoirs would lead to low injection rates and very limited fluid movement due to extremely low permeability, therefore this drawback can be addressed by either having extremely long fractures between injection and production wells (which is not technically possible at present) or drilling additional infill wells that are closely spaced. The former option is difficult to achieve with current state of fracturing technology where fracture half-length is on the order of a few hundred feet (Zuloaga-Molero et al., 2016), while the option of drilling closely spaced additional wells may prove uneconomical for reservoirs with low rates of production. Dombrowski et al. (2015) proposed an economical approach that addresses these concerns by a method that includes injecting a fluid through a set of fractures and producing hydrocarbons from a different set of fractures within the same wellbore. As the gravity forces have relatively small impact on light oil in tight reservoirs compared to viscous forces, the hydrocarbons can be recovered from both above and below the fractures.

In this study, numerical simulation of a single horizontal wellbore intersecting six hydraulic fractures is used to investigate the injection of fluids as a measure to improve recovery after production rate drops significantly. Initially, all six fractures are used for production by primary depletion (without any injection) for ~1800 days when the production rate is significantly lower (< 50 %) than its initial value.

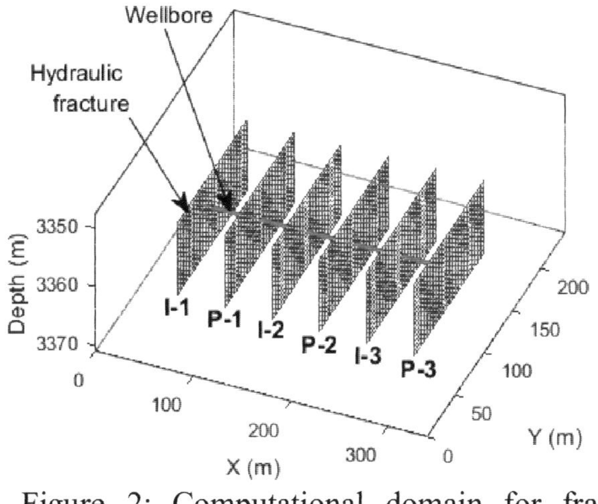

Figure 2: Computational domain for fractured tight-oil reservoir with six hydraulic fractures intersecting the lateral. After 1800 days (~ 5 years) of production by primary depletion, three existing hydraulic fractures are used to inject fluid (I-1, I-2, I-3) and the remaining three fractures are used for production (P-1, P-2, P-3).

Although complex fracture geometries (including interaction of hydraulic and natural fractures) are common in tight-oil reservoirs, such geometries are not considered here in order to control the number of variables and to avoid the complexity beyond necessary.

2.1.2. Reservoir Properties

The reservoir and fracture properties are taken from other studies (Sanchez-Rivera et al., 2015; Zhu et al., 2017; Zuloaga-Molero et al., 2016) on Bakken tight-oil. The parameters for reservoir including well, and hydraulic fractures (shown in Figure 2) are given in Table I. The hydraulic fractures in Figure 2 are modeled using embedded discrete fracture and are linked to the matrix grid through nearest neighborhood connection approach (Lie, 2016).

Table I: Parameters for (a) reservoir including well, and (b) hydraulic fractures.

(a)

Reservoir parameters	Value	Unit
Initial oil saturation (S_{oi})	80	%
Initial reservoir pressure	8000	psi
Production pressure (BHP)	1800	psi
Reservoir temperature	240	F
Reservoir depth	11000	ft
Formation compressibility	5e-06	psi^{-1}
Time when injection begins	1800	$days$
Time when injection ends	20	$years$
Mean permeability (matrix)	0.1, 0.01, 0.005	mD
Mean porosity (matrix)	8	%

(b)

Hydraulic Fracture Parameters	Value	Unit
Aperture	5e-03	ft
Permeability	10	D
Porosity	50	%
Spacing	160	ft
Half-length	200	ft
Height	40	ft

2.2. Fluid Model

2.2.1. Fluid Properties

In-situ fluid properties

The PVT analysis conducted on in-situ fluids from different tight-oil reservoirs (Ghaderi et al., 2017; Hawthorne et al., 2017a; Li and Luo, 2017; Mavor, 2013) reveal the crude oil from these reservoirs to be light, with low density and viscosity. The density, viscosity, compressibility, and formation volume factor of light oil and brine in a typical tight-oil formation like Bakken at initial reservoir conditions (Zhu et al., 2017; Zuloaga-Molero et al., 2016) are shown in Table II. Only brine and oil are considered as the in-situ fluid present at initial conditions in the reservoir. The bubble point pressure of Bakken oil is ~1480 psi (Ghaderi et al., 2017), therefore, it is safe to assume that no free gas exists at initial reservoir condition (8000 psi, 240 ^oF), with all the gas dissolved in oil. Another reason to assume no initial free gas in the reservoir is due to limited capability of surfactant flooding module in MRST that does not provide an option for the gas phase; therefore, having no initial gas in the reservoir would allow for consistent comparison from all four types of recovery processes, i.e. primary depletion, water flooding, gas flooding, and surfactant flooding.

Table II: Properties of formation (a) light oil, and (b) brine, with fluid properties reported at in situ conditions.

(a)

Oil Properties	Value	Unit
Density ($\rho_{o,i}$)	40.47	lb/ft^3
Viscosity ($\mu_{o,i}$)	0.142	cp
Compressibility (c_o)	6.65e-05	bar^{-1}
Formation volume factor	$0.94e^{c_o(p-p_{ini})}$	−

(b)

Brine Properties	Value	Unit
Density ($\rho_{w,i}$)	1000	kg/m^3
Viscosity ($\mu_{w,i}$)	0.61	cp
Compressibility (c_w)	4.28e-05	bar^{-1}
Formation volume factor	$0.99e^{c_w(p-p_{ini})}$	−

Injection fluid properties

Three different fluids are injected to study their impact on improving recovery from tight-oil formations, which are water, immiscible CO_2, and surfactant. The properties of CO_2 (Andersen, 2017) at initial reservoir conditions (8000 psi, 240 F) are given in Table III. We assume CO_2 is immiscible with reservoir oil, i.e. both CO_2 and oil exist as separate phases.

Table III: Properties of CO_2.

Gas Properties	Value	Unit
Density ($\rho_{g,i}$)	803.5	kg/m^3
Viscosity ($\mu_{g,i}$)	0.075	cp
Compressibility (c_g)	6.25e-04	bar^{-1}
Formation volume factor	$e^{c_g(p-p_{ini})}$	−

Although surfactant flooding has been studied extensively in conventional reservoir rocks (Sheng, 2015) to improve recovery, only few studies (He et al., 2014; Mirchi et al., 2015; Shuler et al., 2016; Wang et al., 2011, 2012, 2016; Xu et al., 2011) exist about their application in tight reservoir rocks like shale, especially on real reservoir rock. Surfactant's ability to reduce the interfacial tension between oil and water (σ_{ow}) results in changing the relative permeability and decreasing the residual oil saturation. The decrease in interfacial tension (IFT) by adding surfactant results in increasing the capillary number (N_c), which is a measure of viscous force relative to the capillary force, i.e.:

$$N_c = \frac{|k\nabla p|}{\sigma_{ow}} \quad (1)$$

Therefore, the consequential effect of an increased capillary number is the reduction in residual saturation of oil. The reduction in residual saturation is a function of miscibility condition, which is defined in terms of miscibility factor ($m \in [0, 1]$) that is calculated by interpolating the logarithmic value of the capillary number (Jørgensen, 2013).

$$m(N_c) = \frac{\log_{10}(N_c) - \log_{10}(N_c^{misc})}{\log_{10}(N_c^{immisc}) - \log_{10}(N_c^{misc})} \qquad (2)$$

The surfactant adsorption is assumed to be negligible such that it does not result in any change in porosity or permeability of the rock. Relevant details on the surfactant model implementation can be found in the Appendix or in Jørgensen (2013) for specific details. The parameters used for surfactant flooding are taken from Xu et al. (2011) who conducted surfactant flooding experiments for a tight-oil formation (Yanchang) in China with various concentrations of surfactant. The IFT and viscosity as a function of surfactant concentration, and the magnitude of miscibility for different capillary numbers (function of σ_{ow}) as reported in Xu et al. (2011) are given in Table IV:

Table IV: (a) interfacial tension and viscosity as functions of surfactant concentration, and (b) miscibility factor as a function of capillary number.

(a)

C_s (kg/m^3)	σ_{ow} (N/m)	μ_w (cp)
0	0.05	0.61
0.1	5e-04	-
0.5	1e-05	-
1	1e-06	-
30	1e-06	0.8
100	1e-06	1

(b)

$log_{10}(N_c)$	$m(N_c)$
-10	0
-5.5	0
-4	0.5
-3	1
2	1

2.2.2. Relative Permeabilities

Relative permeability data govern the flow movement and relative ease of flow for multi-fluids in tight-oil reservoirs. Unfortunately, there are not many experimental data for relative permeability in tight-oil reservoirs, however, there is one notable study by Honarpour et al. (2012) that published relative permeability data for tight-oil Eagle Ford samples categorized as calcite-rich, organic-rich, and fractured. Besides this study, Liu et al. (2010) published the relative permeability data for a tight-oil reservoir, Yanchang oilfield, in China. The data from these two studies suggest that end-point relative permeability values in tight-oil reservoirs are lower than generally higher values in conventional reservoir rocks (Moghadasi et al., 2015) and the range of saturation within which fluids are mobile is tighter than conventional rocks. To the authors' knowledge there are no experimental data available tight-oil reservoir in Bakken, but many simulation studies for Bakken (Sanchez-Rivera et al., 2015; Zhu et al., 2017; Zuloaga-Molero et al., 2016) have used data from Yu et al. (2014) who assumed Brooks-Corey model for relative permeability and estimated its parameters by history-matching pressure and production data from the Middle Bakken. However, the relative permeability parameters estimated by Yu et al. (2014) for the Middle Bakken show two contradictions:

i) their estimated end-point relative permeability values are highly optimistic ($k_{rw}^{o} = k_{row}^{o} = k_{rog}^{o} = 1; k_{rg}^{o} = 0.85$) when compared to experimental observations (Honarpour et al., 2012; Liu et al., 2010) from other tight-oil formations, e.g. Eagle Ford in Texas (Honarpour et al., 2012) and Yanchang oilfield in China (Liu et al., 2010; Xu et al., 2011).

ii) their estimated parameters (end-point and residual saturations) suggest that Middle Bakken is water-wet ($S_{wr} > S_{or};\ k_{rw}^{o} = k_{row}^{o}$), however, experimental survey for Bakken (Wang et al., 2012) reveals that it is slightly oil-wet. A review on wettability of various shale formations around the world by Singh (2016) revealed that most shales tend to be oil-wet.

Therefore, we use the relative permeability data that are representative of a slightly oil-wet medium ($k_{row}^{o} \leq k_{rw}^{o};\ S_{wr} = S_{or}$) and a tight-oil formation ($k_{row}^{o}, k_{rw}^{o} < 1$). The relative permeability data for oil-water flow are taken from an experimental surfactant flooding study (Xu et al., 2011) on tight-oil cores from Yanchang oilfield in China, while for oil-gas flow the relative permeability parameters from Yu et al. (2014) are modified slightly to account for oil-wet and tight-oil medium properties as discussed in Zhang et al. (2016). The Brooks-Corey relative permeability model for oil-water and gas-liquid flow can be defined as:

$$k_{rw}(S_w) = k_{rw}^{0}.(S_{wn})^{n_w} \quad (3)$$

$$k_{row}(S_w) = k_{row}^{0}.(1 - S_{wn})^{n_{ow}} \quad (4)$$

$$k_{rg}(S_g) = k_{rw}^{0}.(S_{gn})^{n_g} \quad (5)$$

$$k_{rog}(S_g) = k_{rog}^{0}.(1 - S_{gn})^{n_{og}} \quad (6)$$

Where, k_{ri} is the relative permeability of phase $i(= o, w, g)$, k_{ri}^{o} is the end-point relative permeability of phase i, n_i is the Corey curve exponent for k_{ri}, and S_{in} is the normalized saturation of phase i. The parameters in above relative permeability model for different fluids are defined in Table V. The

Table V: Relative permeability model parameters.

Water-Oil					
S_{wr}	S_{or}	k_{rw}^{0}	k_{row}^{0}	n_w	n_{ow}
0.2	0.2	0.6	0.5	3	3
Gas-Oil					
S_{gr}	S_{or}	k_{rg}^{0}	k_{rog}^{0}	n_g	n_{og}
0.0	0.2	0.85	0.7	3.15	5.75
Water-Surfactant-Oil					
S_{wr}	S_{or}	k_{rw}^{0}	k_{row}^{0}	n_w	n_{ow}
0.05	0.05	1	1	1.5	1.5

2.3. HETEROGENEITY

Based on laboratory (Dong et al., 2017; Hawthorne et al., 2017b; Tuero et al., 2017) and field (Kenomore et al., 2017; Steiner et al., 2016; Witkowsky et al., 2012) observations of shale rocks, it can be broadly divided into three rock types, i.e. inorganic (poor quality), organic (good quality), and intermediate (fair quality) that lies between the inorganic and organic. A spatially heterogeneous reservoir model based on this concept is developed using sequential indicator simulation algorithm (Deutsch and Journel, 1997) by considering marginal probabilities of 0.6, 0.2, and 0.2 for the inorganic, intermediate, and organic rock types, respectively. These marginal probabilities are chosen to reflect the volumetric percentages of organic pores found in many shales, which is about 20% by volume (Witkowsky et al., 2012) based on 1.1 g/cc kerogen grain density. The spatial correlation of each rock type is defined using three nested spherical variograms along 60 degrees to 120 degrees.

The rock properties of inorganic, organic, and intermediate type are considered based on laboratory-scale and field-scale observations of shale rock (Hawthorne et al., 2017b; Kenomore et al., 2017; Tuero et al., 2017; Witkowsky et al., 2012) that show organic rock type, which are also associated with micro-fractures (Hawthorne et al., 2017b), have larger pore sizes (Dong et al., 2017) in comparison to intermediate and inorganic rock types that occupy larger pore volume of the rock (Tuero et al., 2017) with significant saturation of hydrocarbons (Tuero et al., 2017). Therefore, these qualitative observations are used to assign rock properties (porosity, permeability, and initial oil saturation) to each rock type as shown in Figure 3 .

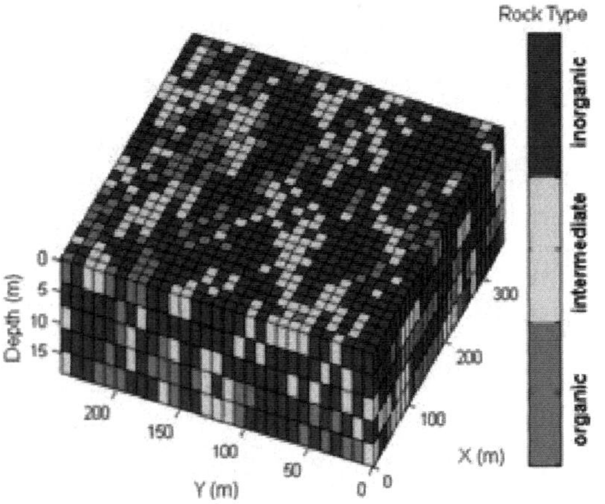

Geology		Φ (%)	k (mD): Scenario I	II	III	S_{oi} (%)
Heterogeneous	organic	3	0.4	0.04	0.02	60
	intermediate	6	0.1	0.01	0.005	70
	inorganic	10	0.01	0.001	0.0005	90
	Mean	8	0.1	0.01	0.005	80
Homogeneous		8	0.1	0.01	0.001	80

Figure 3: Reservoir model is composed of three different rock types (shown by colors) for a spatially heterogeneous model, where each rock type has a distinct porosity, permeability, and initial oil saturation value. The average values of porosity, permeability, and initial oil saturation for the heterogeneous model are equal to their corresponding values in a homogeneous model. Three matrix permeability scenarios considered are very low (mean 0.005 mD), low (0.01 mD), and relatively high (0.1 mD).

2.4 INJECTION RATES

Three injection rates are chosen and denoted as low (~7 rb/d), high (32 rb/d), and very high (48 rb/d). The magnitude of each injection rate is denoted as low, high, or very high based on range of injection rates adopted in other studies (Sanchez-Rivera et al., 2015; Zhu et al., 2017) for tight-oil reservoirs with different permeability and fracture spacing.

3. Results

The reservoir is depleted naturally without any fluid injection for ~1800 days when a fluid (water, surfactant, or immiscible CO_2) is injected for the remaining life of the reservoir (~20 years) to arrest the declining production rates and recover additional oil.

The impact of injected fluid on oil recovery is assessed for three different average permeabilities (very low: 0.005 md, low: 0.01 mD, and relatively high: 0.1 mD) and different injection rates (~ low: 7 rb/d, high: 32 rb/d, and very high: 48 rb/d) in reservoir barrels (rb/d) for both homogeneous and heterogeneous scenarios. A similar set of results is obtained to see the impact of injecting fluids constrained by a constant bottom-hole pressure (BHP ~ 8100 psi) instead of a constant injection rate.

3.1. OIL RATES

Figure 4 shows oil rates for four types of recovery processes, i.e. by primary depletion (no injection), by water injection, by surfactant injection, and by immiscible CO_2 injection. Each plot on left column of Figure 4 is for different reservoir matrix permeability in a homogeneous medium, whereas a corresponding plot on the right side of this figure is for heterogeneous medium with an average permeability equal to its corresponding homogeneous value. The results presented in Figure 4 are for an injection rate of ~7 rb/d for all three injection fluids. Figure 5 and Figure 6 show similar results as presented in Figure 4, but for higher injection rates of ~32 rb/d and ~48 rb/d, respectively. Figure 7 shows similar set of results as presented in Figure 4, but these results are obtained when the fluid injection is constrained by a constant BHP (~8100 psi) instead of a constant injection rate.

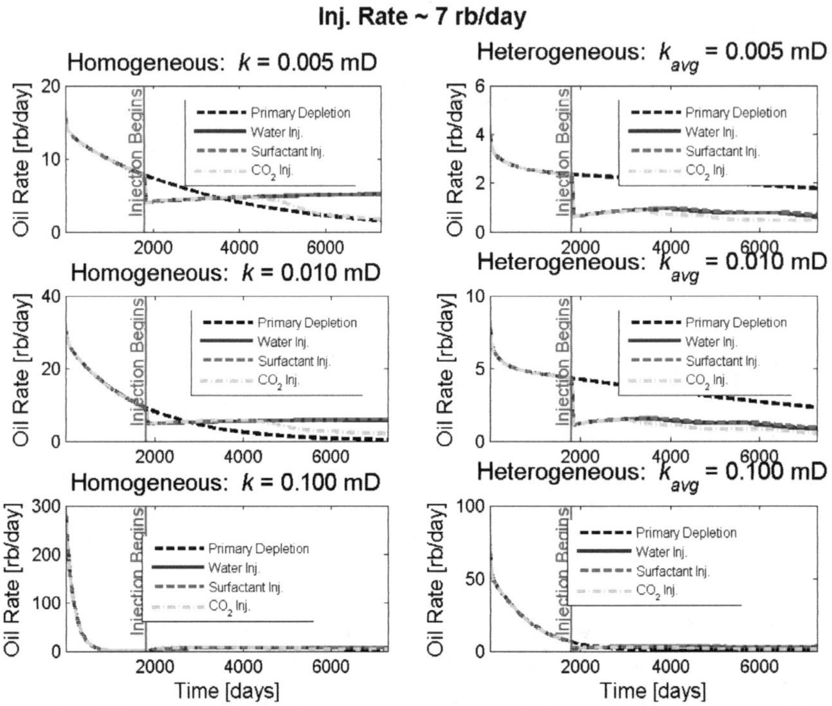

Figure 4: Oil rates from homogeneous and heterogeneous media with three improved recovery methods (injection rate ~ 7 rb/day). Primary depletion represents the base case where oil is produced without any fluid injection.

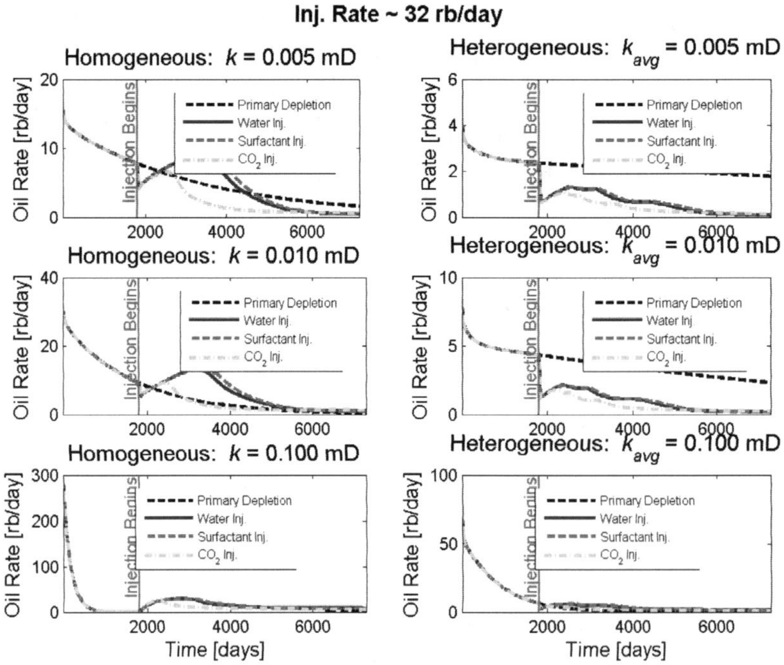

Figure 5: Oil rates from homogeneous and heterogeneous media with three improved recovery methods (injection rate ~ 32 rb/day). Primary depletion represents the base case where oil is produced without any fluid injection.

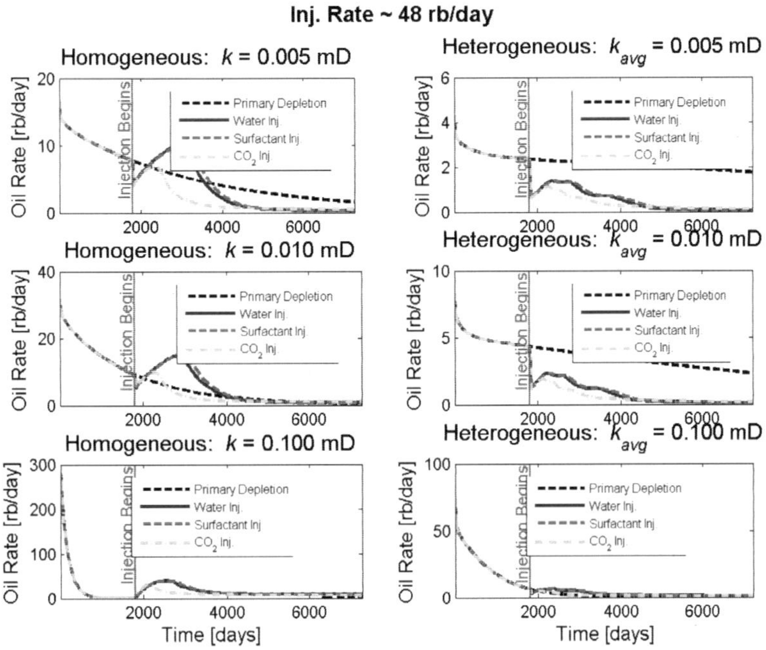

Figure 6: Oil rates from homogeneous and heterogeneous media with three improved recovery methods (injection rate ~ 48 rb/day). Primary depletion represents the base case where oil is produced without any fluid injection.

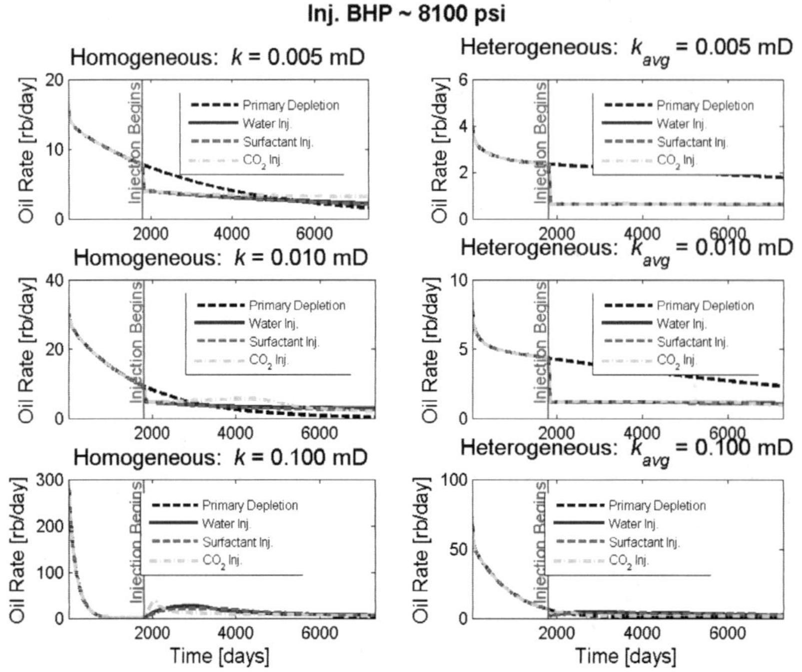

Figure 7: Oil rates from homogeneous and heterogeneous geological models with three improved recovery methods (injection BHP ~ 8100 psi). Primary depletion represents the base case where oil is produced without any fluid injection.

3.2. Oil Recovery

The oil rates plotted in Figure 4 through Figure 7 are used to find oil recovered as percent of the original oil in place (OOIP).

Figure 8, Figure 9, Figure 10, and Figure 11 show oil recovered as percent of the original oil in place (OOIP) for the corresponding oil rates plotted in Figure 4 through Figure 7, respectively. Among the recovery values from all scenarios, maximum oil is recovered (~1.2 % of OOIP; bottom left plot in Figure 10) by injecting surfactant (or water) at the largest rate of ~48 rb/d in a homogeneous medium with 0.1 mD permeability. The maximum oil recovered is ~65% more than the primary depletion (bottom left plot in Figure 14) by injecting surfactant (or water, with no noticeable difference compared to surfactant), whereas the oil recovered by immiscible CO_2 is about 25% less than that by water or surfactant, but it is 40% more than the primary depletion. This is because, as shown in Figure 6, the oil rates by water (or surfactant) get a larger bump in rates than by immiscible CO_2 initially, and the rate bump provided by water (or surfactant) is able to be sustained for a longer period than by immiscible CO_2. The reason for the higher bump provided by water (or surfactant) as opposed to immiscible CO_2 is that earlier CO_2 breakthrough (~2,200 days) is predicted in comparison to water (or surfactant) breakthrough (~2800 days). Minimum oil is recovered (~0.04 % of OOIP; top right plot in Figure 10) with the largest injection rate of ~48 rb/d in a heterogeneous medium with 0.005 mD average permeability. In this case, there is no substantial difference in oil recovered between the three injected fluids (top right plot in Figure 14), whereas the oil recovered by primary depletion (no injection) is almost 50% greater than the oil recovered by injection. The corresponding oil rates for the minimum oil recovered are shown by Figure 6, where it can be observed that the oil rates by water (or surfactant) get a larger bump than by immiscible CO_2 initially, however, the magnitude of oil rates ($< 2\ rb/d$) with all three fluids is not substantial enough to make any difference in the percent of oil recovered.

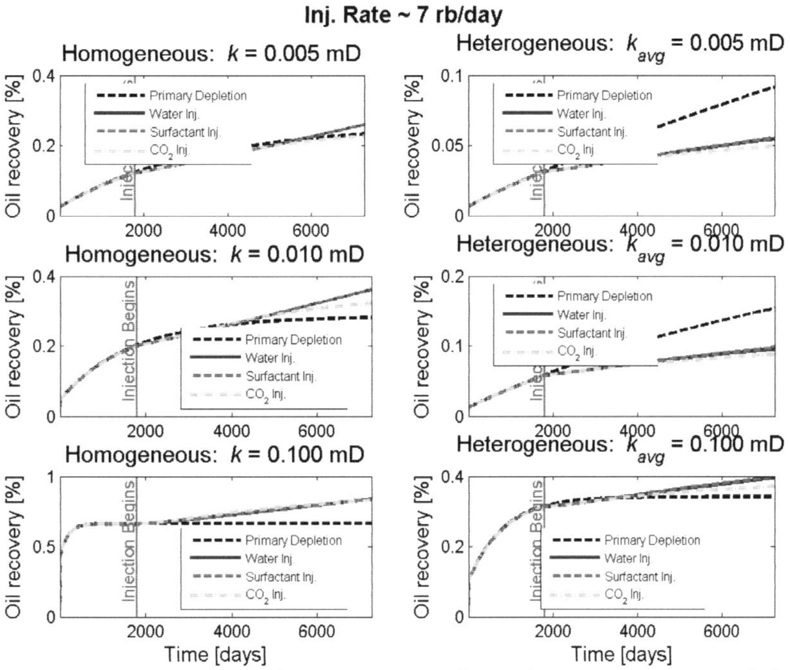

Figure 8: Percent cumulative recovery from homogeneous and heterogeneous geological models with three improved recovery methods (injection rate ~ $7\ rb/day$). Primary depletion represents the base case where oil is produced without any fluid injection.

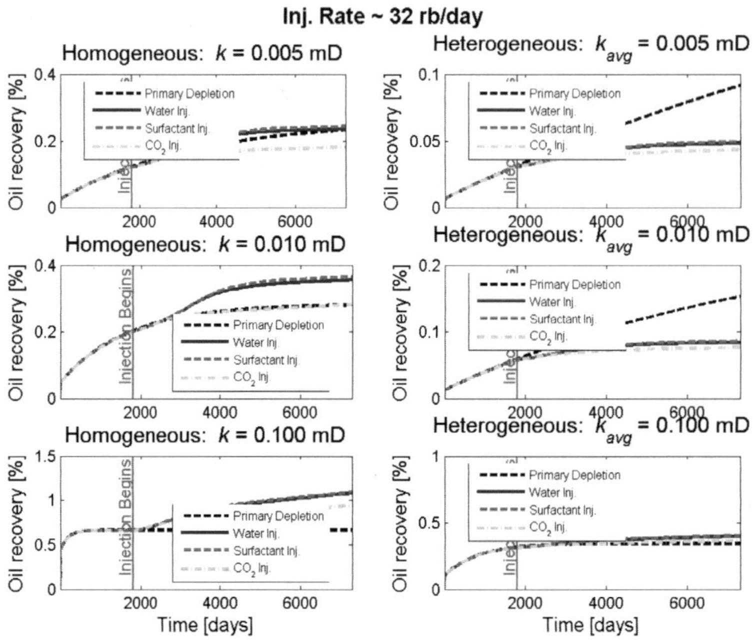

Figure 9: Percent cumulative recovery from homogeneous and heterogeneous geological models with three improved recovery methods (injection rate ~ 32 rb/day). Primary depletion represents the base case where oil is produced without any fluid injection.

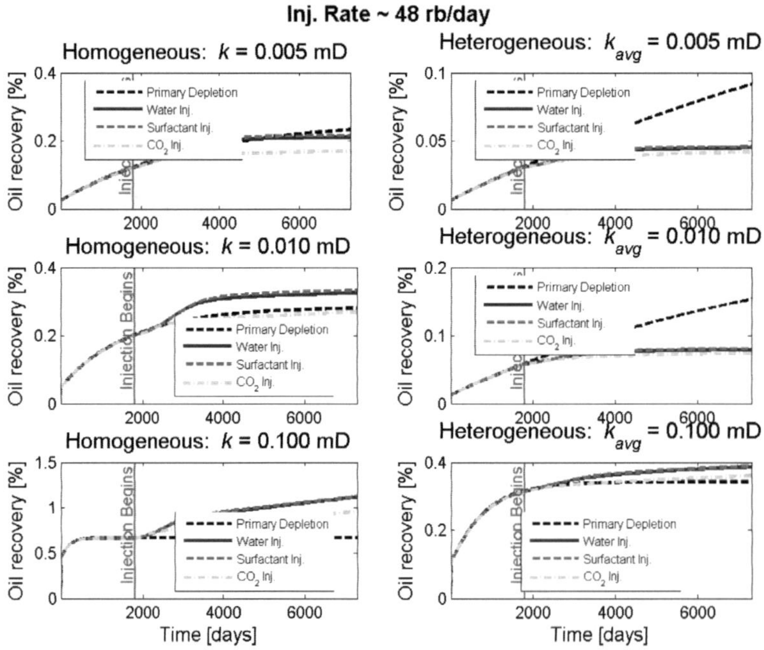

Figure 10: Percent cumulative recovery from homogeneous and heterogeneous geological models with three improved recovery methods (injection rate ~ 48 rb/day).

Primary depletion represents the base case where oil is produced without any fluid injection.

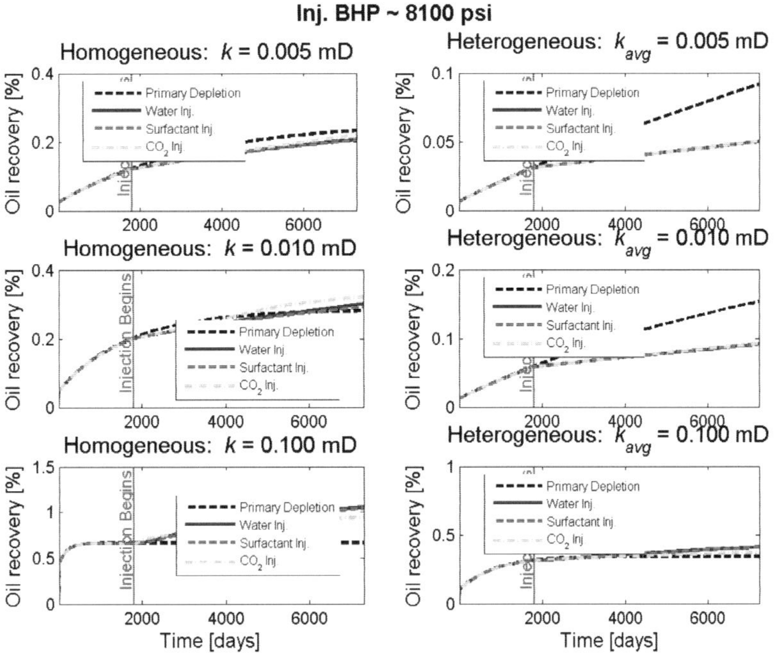

Figure 11: Percent cumulative recovery from homogeneous and heterogeneous geological models with three improved recovery methods (injection BHP ~ 8100 *psi*). Primary depletion represents the base case where oil is produced without any fluid injection.

3.3. Incremental Oil Recovery

To show the impact of injecting different fluids on recovery, incremental recovery (IR) with respect to primary depletion is calculated as shown:

$$Incremental\ recovery\ [\%] = \left(\frac{recovery_{inj} - recovery_{PD}}{recovery_{PD}}\right) \times 100 \qquad (7)$$

Where subscript inj represents the method of recovery by fluid injection (water, surfactant, CO_2), and PD is the method of recovery by primary depletion with no injection (base case). Figure 12, Figure 13, Figure 14, and Figure 15 show relative change in oil recovered as percent with respect to primary depletion (PD) for the corresponding oil recoveries plotted in Figure 8, Figure 9, Figure 10, and Figure 11, respectively.

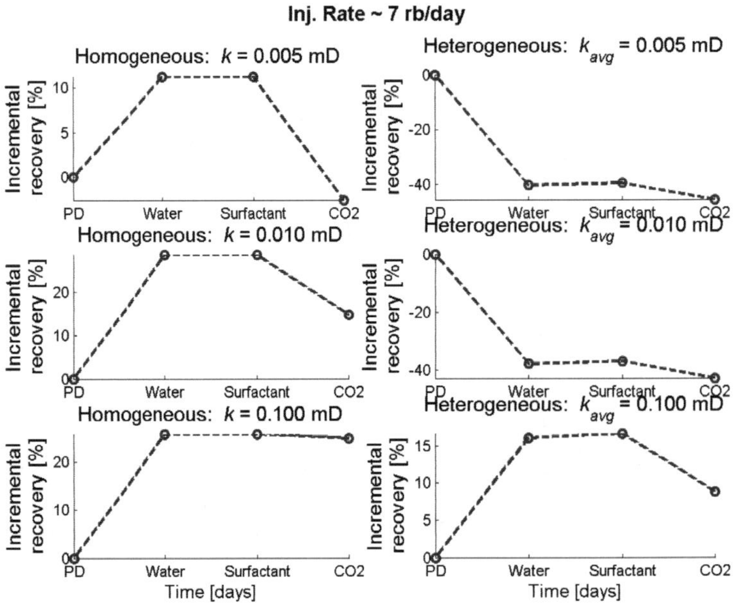

Figure 12: Incremental recovery (IR %) with respect to primary depletion (base case) for the recovery values plotted in Figure 8.

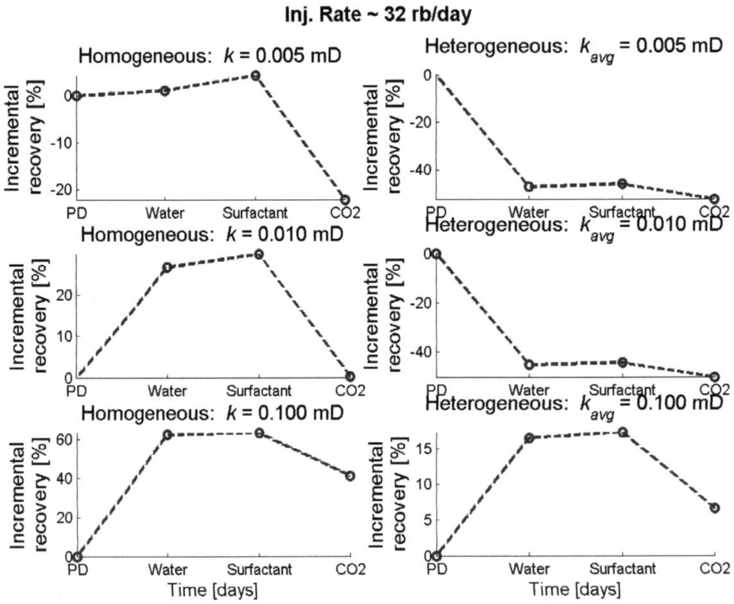

Figure 13: Incremental recovery (IR %) with respect to primary depletion (base case) for the recovery values plotted in Figure 9.

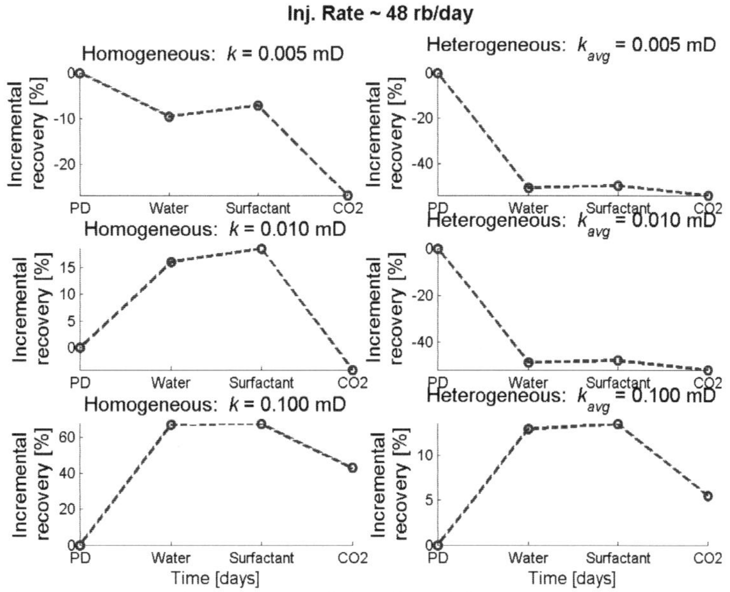

Figure 14: Incremental recovery (IR %) with respect to primary depletion (base case) for the recovery values plotted in Figure 10.

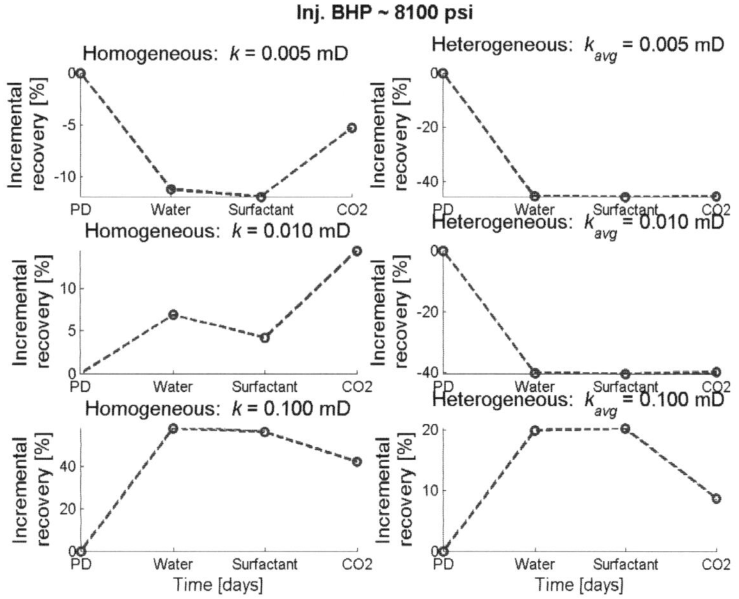

Figure 15: Incremental recovery (IR %) with respect to primary depletion (base case) for the recovery values plotted in Figure 11.

Impact of an Additional Lateral

To study the impact of number of laterals and hydraulic fractures on the incremental recovery, we replace a single lateral in the middle with two similar laterals (each intersecting six hydraulic fractures) at equal distance from the boundaries. For detailed results with two laterals, readers may refer Singh and Cai (2018). The results presented in Singh and Cai (2018) show that the percent of incremental recovery due to fluid injection is not affected by having an additional lateral in a homogeneous medium. Although the percent of IR is affected in heterogeneous medium due to relative change in spatial heterogeneity with two laterals as compared to a single lateral in the middle (Figure 2), the behavior of incremental recovery due to injection of each fluid (water, surfactant, or immiscible CO_2) is still the same; that is, the relative performance exhibited by each fluid in increasing fluid recovery with respect to the base case still shows the same behavior despite the relative change in heterogeneity.

3.4. Effect of Surfactant Concentration

The surfactant concentration used for this study is 50 kg/m^3 (or 5% by mass of water), which provides minimum surface tension and maximum miscibility factor of 1 as shown by Table IV. The minimum surface tension and maximum miscibility is obtained with surfactant concentration of 30 kg/m^3 (or 3% by mass of water). Any surfactant concentration above this value did not change the recovery values.

3.5. Effect of Capillary Pressure

Capillary pressure curve for water-oil and gas-liquid was obtained by using the laboratory-scale capillary pressure data from middle Bakken cores (Karimi and Kazemi, 2015) and extrapolating them by fitting a curve over the data. The expression obtained after extrapolating the data beyond the water saturation range of ~62-81% is shown as following:

$$p_c(psi) = \begin{cases} (-9.34 \times 10^{-42}).e^{143.3.S_w}, & water\ displacing\ oil \\ (7.78 \times 10^{3}).e^{-4.91.S_o}, & gas\ displacing\ oil \end{cases} \quad (8)$$

The capillary pressure at in-situ oil saturation of 80% as given by this curve is negligible in comparison to the initial reservoir pressure of ~8000 psi. Even after producing for 20 years, the pressure did not drop enough to be comparable to the highest possible value of capillary pressure by the above expression. Therefore, using the above curve as capillary pressure for the reservoir fluids in a model (mimicking properties of Bakken) did not affect the recovery estimates. The reason why capillary pressure did not play a role in affecting recovery is because the pore space that is inaccessible is due to its small permeability, and not because of low pressure support. The average pressure support in the reservoir even by the end of 20 years of production is much higher that can overcome capillary forces.

It should be noted that, although, there are some studies (e.g. Zhang et al., 2017) that show numerically that capillary pressure can have significant impact on oil recovery (upto 20 % higher), their methodology uses Young-Laplace equation for the estimates of capillary pressure, which usually gives impractical results for shales as discussed in a review by Singh (2016).

4. Discussions

Results suggest that fluid injection can improve recovery over the base case (with no injection), but, the recovery is not improved universally across all the studied scenarios. Fluid injection did not improve recovery (negative IR %) in 10 out of 24 cases studied, where 8 out of those 10 unsuccessful cases were from heterogeneous media with very low (0.005 mD) to low (0.01 mD) average matrix permeability. The remaining 2 unsuccessful cases are for very low (0.005 mD) matrix permeability in homogeneous media with very high injection rate (~48 rb/d), or when the production is constrained by a constant BHP, respectively. In the 14 successful or improved recovery (positive IR %) cases where injecting fluids improved the recovery, surfactant injection performed the best in 3 cases while immiscible CO_2 injection performed the best in 1 case. In the remaining 10 out of the 14 improved recovery cases, there was no clear winner among the three fluids; in most cases, surfactant and water performed almost equally, but better than immiscible CO_2.

4.1. Effect of Matrix Permeability

An increase in matrix permeability results in increasing oil recovery for all the recovery methods, including primary depletion, and for the three methods of recovery by fluid injection. This impact is consistent irrespective of injection rate or well constraint, and the relative change in oil recovery between different recovery methods increases with the increase in matrix permeability as shown in Figure 12, Figure 13, Figure 14, and Figure 15.

4.2. Effect of Injection Rate

An increase in injection rate leads to decrease in oil recovery matrix with very low (0.005 mD) to low (0.01 mD) matrix permeability, but it leads to increase in oil recovery for matrix with relatively higher permeability (0.1 mD). This can be seen in in Figure 12, Figure 13, Figure 14, and Figure 15, which show that at small injection rate (~ 7 rb/d) the recovery by injection for reservoir with very low (0.005 mD) to low (0.01 mD) matrix permeability is greater than recovery by primary depletion; but, with increase in injection rates (~ 32 rb/d and 48 rb/d) there is a relative drop in recovery for reservoir with very low (0.005 mD) to low (0.01 mD) matrix permeability. However, for reservoir with relatively high (0.1 mD) matrix permeability, the recovery increases relatively with increase in injection rate from 7 rb/d to 48 rb/d. This is possibly due to the fact that high injection rate (equivalent to relatively high convective flow and low diffusion) is useful only when the rock is sufficiently permeable that it allows the fluid flow; whereas low injection rate (equivalent to relatively low convective flow and high diffusion) is useful when the rock is extremely tight that the primary mode of flow is through diffusion.

4.3. Effect of Injection Constrained by Constant BHP

An injector constrained by a constant BHP is relatively more effective in improving recovery in a reservoir that has a relatively high matrix permeability (0.1 mD) than in a reservoir having a very low (0.005 mD) to low (0.01 mD) matrix permeability. Injector constrained by a constant BHP is most effective for immiscible CO_2 injection, which tends to perform better with constant BHP than with constant injection rate. An injector constrained by a constant BHP performs better with the injection of immiscible CO_2 than with water or surfactant when the reservoir matrix permeability is very low (0.005 mD) to low (0.01 mD); whereas, effectiveness of immiscible CO_2 in case of reservoir with relatively high (0.1 mD) matrix permeability is not as good as water or surfactant, but it's still relatively better than its performance in the cases where injection is constrained by a constant rate.

4.4. Effect of Heterogeneity

Injecting fluids in reservoirs with very low (0.005 mD) to low (0.01 mD) average matrix permeability does not lead to a better recovery. In fact, switching three fractures from producers to injectors for the purpose of injecting fluids in these very low (0.005 mD) to low (0.01 mD) permeability reservoirs lead to a net decrease in recovery relative to primary depletion; that is, producing through all six fractures with no injection leads to a higher recovery than by injecting fluids through half of the fractures. However, injecting fluids in reservoirs with relatively larger average matrix permeability of 0.1 mD does lead to a better recovery compared to primary depletion; lower injection rates are relatively more effective than a higher injection rate in increasing the recovery, whereas production constrained by a constant BHP is even better than low injection rates to increase recovery rate.

4.5. Effect of Type of Fluid

Results suggest that fluid injection improved recovery (positive IR %) in 14 out of 24 cases studied, where surfactant injection performed the best in 3 cases while immiscible CO_2 injection performed the best in 1 case. In the remaining 10 IR cases, there was no clear winner among the three fluids; in 9 cases out of 10, surfactant and water performed almost equally, but better than immiscible CO_2, while only in one case all the three injection fluids (water, surfactant, and immiscible CO_2) performed equally.

The 10 cases where injecting fluids did not improve recovery relative to primary depletion are:

1) Heterogeneous reservoirs with very low (0.005 mD) to low (0.01 mD) average matrix permeability irrespective of well constraint or injection rate [8 cases]
2) Very high injection rates (~48 rb/d) in reservoir with very low (0.005 mD) matrix permeability [1 cases], and
3) Production constrained by a constant BHP in reservoir with very low (0.005 mD) matrix permeability [1 case].

The 14 cases where injecting fluids improved the recovery relative to primary depletion are:

1) Surfactant injection performed the best in reservoir with very low (0.005 mD) matrix permeability with high injection rate (~32 rb/d) [1 case], and in low (0.01 mD) matrix permeability with high to very high injection rates (~32 rb/d and 48 rb/d) [2 cases]. The improvement in recovery by surfactant ranged between 1-3% higher than water, relative to primary depletion.
2) Immiscible CO_2 injection performed the best in reservoir with low (0.01 mD) matrix permeability when the production is constrained by a constant BHP [1 case].
3) In the remaining 10 IR cases, there was no clear winner among the three fluids; in 9 cases out of 10, surfactant and water performed almost equally, but better than immiscible CO_2, while only in one case all the three injection fluids (water, surfactant, and immiscible CO_2) performed equally.

In general, the oil rates by water (or surfactant) get a larger bump in rates initially than by immiscible CO_2 (as shown by Figure 4 through Figure 7) because of earlier breakthrough of immiscible CO_2 in comparison to water (or surfactant) by few years in most cases. This is because gas being a lower viscosity fluid than the water (or surfactant) will have a higher velocity (by Darcy's law) and that results in its earlier breakthrough.

A brief summary of above results is presented in form of matrix (Table VI and Table VII) to compare the performance of methods used for oil recovery in a homogeneous and heterogeneous medium, respectively. Scenarios marked as X is shown in red colored box to indicate that injection of fluid did not help improve recovery for that particular scenario. Also, scenarios marked with 'Water or Surf' (or 'Water or Surf or Immiscible CO_2') indicate that there was no single injection fluid that performed the best in that particular scenario, and that any one of the injection fluids mentioned in that scenario will result in a similar recovery as obtained by the rest of the fluids mentioned for that scenario.

Table VI: Recovery technique for each scenario that recovered the most oil in a homogeneous medium. Scenarios marked as X is shown in red colored box to highlight that injection of fluid did not improve recovery.

HOMOGENEOUS	*Low Injection Rate*	*High Injection Rate*	*Very High Injection Rate*	*Constrained by BHP*
Very Low Permeability	Water or Surf	Surfactant	X	X
Low Permeability	Water or Surf	Surfactant	Surfactant	Immiscible CO_2
High Permeability	Water or Surf or Immiscible CO_2	Water or Surf	Water or Surf	Water or Surf

Table VII: Recovery technique for each scenario that recovered the most oil in a heterogeneous medium. Scenarios marked as X is shown in red colored box to highlight that injection of fluid did not improve recovery.

HETEROGENEOUS	*Low Injection Rate*	*High Injection Rate*	*Very High Injection Rate*	*Constrained by BHP*
Very Low Permeability	X	X	X	X
Low Permeability	X	X	X	X
High Permeability	Water or Surf	Water or Surf	Water or Surf	Water or Surf

4.6. RELATING RECOVERY WITH MECHANISM OF INJECTED FLUID

Recovery with Water and Immiscible CO_2

In general, the oil rates by water (or surfactant) get a larger bump in rates initially than by immiscible CO_2 (as shown by Figure 4 through Figure 7) because of earlier breakthrough of immiscible CO_2 in comparison to water (or surfactant) by few years in most cases. This is because gas being a lower viscosity fluid than the water (or surfactant) will have a higher velocity (by Darcy's law) and that results in its earlier breakthrough. Both water flooding and immiscible CO_2 flooding work by increasing the pressure that drives the oil out by immiscible displacement. Although, the relative permeability of gas with oil is different than the relative permeability of water with oil, it does not unilaterally dictate the oil recovery; this is because, relative permeability of oil with gas favors higher mobility of both the gas and the oil, which is good for driving the oil out faster, however, it also simultaneously results in faster breakthrough of gas due to its higher mobility. This is especially important in a tight-oil reservoir where the tight pore space dominates the reservoir, and even a slightly higher permeable pathways can lead to preferential pathway for the gas that may result in its breakthrough. In the current injection-production scheme, the gas is injected through the fractures, and it tends to breakthrough from the other fractures that are working as production zones. However, water being a heavier fluid with about ten times higher viscosity than the CO_2, is not likely to breakthrough as fast as gas because of its much lower velocity compared to CO_2 ($\propto \mu_{gas}/\mu_{water}$~0.1). Therefore, without early breakthrough, water is able to sustain higher pressure than CO_2 for a longer period of time.

Recovery with Surfactant

Unlike water and immiscible CO_2 flooding, surfactant flooding works by mixing with the oil to form a miscible solution that reduces the surface tension of the oil and helps bring down the capillary forces; therefore, reduced capillary forces push the oil out from the small pores that hold the oil due to high capillary pressure. Therefore, the success of a surfactant flooding in extracting oil depends not just on its chemical properties that can reduce the surface tension of oil (also referred as wettability alteration), but also on the amount of oil trapped by capillary forces. If the average reservoir pressure is sufficiently high, such as in the studied scenario (and also the scenario with most tight-oil formations, which are over pressured) the amount of oil that can be displaced by surfactant depends on the capability of the surfactant to first access the tight rock to mobilize additional oil for recovery. In conventional reservoirs, surfactant flooding generally performs better over water flooding because the surfactant is able to access the pores that are otherwise inaccessible due to high capillary pressure with water. Although, surfactant results in favorable fluid properties to mobilize oil, the reason surfactant did not perform well in tight-oil formation studied here is because the unswept oil is inaccessible due to small permeability, and not because of high capillary pressure. The average pressure support in the reservoir even by the end of 20 years of production is much higher that can overcome capillary forces. Even though the surfactant performs better in 2 cases out of 24 cases, the increase in recovery (1-3% higher than water) is not sufficiently higher to warrant its supremacy over water in tight-oil formations, primarily because it cannot access the tight pore space.

5. Summary and Conclusions

Production rates in tight-oil reservoirs decline rapidly within a year after the well is put to production. To arrest the declining production rates and recover additional oil from tight-oil resources, traditional methods of IOR (water flooding, immiscible CO_2 flooding, surfactant flooding) were investigated. To overcome the challenge of negligible fluid movement and sweep outside the fractures, the fluid was injected through a set of fractures and produced through another set of fractures intersecting the same lateral. We compare the estimates of oil recovery obtained by injecting water, surfactant, and immiscible CO_2, with the oil recovered by primary depletion (with no injection).

Results suggest that fluid injection did not improve recovery in 10 out of 24 cases studied, where 8 out of those 10 unsuccessful cases were from heterogeneous medium with very low (0.005 mD) to low (0.01 mD) average matrix permeability. The remaining 2 unsuccessful cases are for very low (0.005 mD) matrix permeability in homogeneous media with very high injection rate (~48 rb/d) and when the injection is constrained by a constant BHP, respectively. In the 14 successful or improved recovery cases, surfactant injection performed the best in 3 cases while immiscible CO_2 injection performed the best in 1 case. In the remaining 10 improved recovery cases, there was no clear winner among the three fluids; in 9 cases out of 10, surfactant and water performed almost equally, but better than immiscible CO_2, while only in one case all the three injection fluids (water, surfactant, and immiscible CO_2) performed similarly.

In conclusion, the results summarized in Table VI and Table VII suggest that an enhanced recovery method (e.g. water, immiscible CO_2, or surfactant) may not always improve recovery from tight-oil reservoirs, especially in a heterogeneous reservoir whose average matrix permeability is less than a threshold of 0.01 mD. Although, surfactant results in favorable fluid properties to mobilize oil, the reason surfactant did not perform well in tight-oil formation studied here is because the unswept oil is inaccessible due to small permeability, and not because of high capillary pressure. Therefore, it can be said that reservoir heterogeneity coupled with the average matrix permeability of a tight-oil reservoir seems to play the most significant role in exploring the potential of enhanced recovery.

There have been some lab-scale EOR studies (Li and Sheng, 2016, 2017; Li et al., 2017) on tight cores, but we should be cautious in directly adapting the results from lab-scale studies; that is because, unlike EOR from conventional reservoirs where reservoir heterogeneity is the primary variable that decides the performance of the recovery, there are multiple factors that may decide the performance of an EOR in tight-oil formations. One of those factors, besides other factors (e.g. fractures), is reservoir heterogeneity coupled with the average matrix permeability, which seems to play the most important role in the feasibility of an EOR method as discussed above. Most of the lab-scale studies look at performance of a certain EOR method on a core that is fairly homogeneous (with no permeability variation) and do not account for other important parameters (e.g. fractures) that may impact the performance of some EOR methods, specifically the EOR by gas injection. Therefore, we must be careful in adapting the results of those studies to support field scale response.

6. Acknowledgment

This research was supported in part by an appointment to the National Energy Technology Laboratory Research Participation Program, sponsored by the U.S. Department of Energy and administered by the Oak Ridge Institute for Science and Education.

7. References

Aloulou, F. (2016). World tight oil production to more than double from 2015 to 2040.

Andersen, O. (2017). Simplified models for numerical simulation of geological CO2 storage. The University of Bergen.

Delaihdem, D.K. (2013). Decline curve analysis and enhanced shale oil recovery based on Eagle Ford Shale data. Thesis. University of Alaska Fairbanks.

Deutsch, C.V., and Journel, A.G. (1997). GSLIB: Geostatistical Software Library and User's Guide (Oxford University Press, USA).

Dombrowski, Fonseca, Karanikas, Moya, and Lamantia (2015). Fluid injection in light tight oil reservoirs. US Patent US20130228337.

Dong, T., Harris, N.B., Ayranci, K., Twemlow, C.E., and Nassichuk, B.R. (2017). The impact of composition on pore throat size and permeability in high maturity shales: Middle and Upper Devonian Horn River Group, northeastern British Columbia, Canada. Marine and Petroleum Geology *81*, 220–236.

Flannery, J. (2006). Integrated Analysis of the Bakken Petroleum System, U.S. Williston Basin. pp. 138–145.

Gaswirth, S.B., and Marra, K.R. (2015). U.S. Geological Survey 2013 assessment of undiscovered resources in the Bakken and Three Forks Formations of the U.S. Williston Basin Province. AAPG Bulletin *99*, 639–660.

Gaswirth, S.B., Marra, K.R., Cook, T.A., Charpentier, R.R., Gautier, D.L., Higley, D.K., Klett, T.R., Lewan, M.D., Lillis, P.G., Schenk, C.J., et al. (2013). Assessment of undiscovered oil resources in the Bakken and Three Forks Formations, Williston Basin Province, Montana, North Dakota, and South Dakota, 2013 (Reston, VA: U.S. Geological Survey).

Ghaderi, S.M., Clarkson, C.R., Ghanizadeh, A., Barry, K., and Fiorentino, R. (2017). Improved Oil Recovery in Tight Oil Formations: Results of Water Injection Operations and Gas Injection Sensitivities in the Bakken Formation of Southeast Saskatchewan. (Society of Petroleum Engineers), p. SPE-185030-MS.

Gong, X. (2013). Assessment of Eagle Ford Shale Oil and Gas Resources. Thesis. Texas A&M University.

Hawthorne, S.B., Miller, D.J., Grabanski, C.B., Sorensen, J.A., Pekot, L.J., Kurz, B.A., Gorecki, C.D., Steadman, E.N., Harju, J.A., and Melzer, S. (2017a). Measured Crude Oil MMPs with Pure and Mixed CO2, Methane, and Ethane, and Their Relevance to Enhanced Oil Recovery from Middle Bakken and Bakken Shales. (Society of Petroleum Engineers), p.

Hawthorne, S.B., Jin, L., Kurz, B.A., Miller, D.J., Grabanski, C.B., Sorensen, J.A., Pekot, L.J., Bosshart, N.W., Smith, S.A., Burton-Kelly, M.E., et al. (2017b). Integrating Petrographic and Petrophysical Analyses with CO2 Permeation and Oil Extraction and Recovery in the Bakken Tight Oil Formation. (Society of Petroleum Engineers), p.

He, K., Xu, L., Gao, Y., Neeves, K.B., Yin, X., Bai, B., Ma, Y., and Smith, J. (2014). Validating Surfactant Performance in the Eagle Ford Shale: A Correlation between the Reservoir-on-a-Chip Approach and Enhanced Well Productivity. (Society of Petroleum Engineers), p.

Honarpour, M.M., Nagarajan, N.R., Orangi, A., Arasteh, F., and Yao, Z. (2012). Characterization of Critical Fluid PVT, Rock, and Rock-Fluid Properties - Impact on Reservoir Performance of Liquid Rich Shales. (Society of Petroleum Engineers), p.

Hustak, C., Dieva, R., Baker, R., MacIsaac, B., Frankiw, K., and Clark, B. (2017). Waterflooding a Multi-layered Tight Oil Reservoir Developed with Hydraulically Fractured Horizontal Wells. (Society of Petroleum Engineers), p. SPE-185021-MS.

Jørgensen, K. (2013). Implementation of a Surfactant Model in MRST With Basis in Schlumberger's Eclipse. Norwegian University of Science and Technology.

Karimi, S., and Kazemi, H. (2015). Capillary Pressure Measurement using Reservoir Fluids in a Middle Bakken Core. (Society of Petroleum Engineers), p.

Kenomore, M., Hassan, M., Dhakal, H., and Shah, A. (2017). Total organic carbon evaluation of the Bowland Shale Formation in the Upper Bowland of the Widmerpool Gulf. Journal of Petroleum Science and Engineering *150*, 137–145.

Li, L., and Sheng, J.J. (2016). Experimental study of core size effect on CH4 huff-n-puff enhanced oil recovery in liquid-rich shale reservoirs. Journal of Natural Gas Science and Engineering *34*, 1392–1402.

Li, L., and Sheng, J.J. (2017). Numerical analysis of cyclic CH4 injection in liquid-rich shale reservoirs based on the experiments using different-diameter shale cores and crude oil. Journal of Natural Gas Science and Engineering *39*, 1–14.

Li, S., and Luo, P. (2017). Experimental and simulation determination of minimum miscibility pressure for a Bakken tight oil and different injection gases. Petroleum *3*, 79–86.

Li, L., Sheng, J.J., Xu, J., and others (2017). Gas Selection for Huff-n-Puff EOR in Shale Oil Reservoirs Based upon Experimental and Numerical Study. In SPE Unconventional Resources Conference, (Society of Petroleum Engineers), p. SPE-185066-MS.

Lie, K.-A. (2016). An introduction to reservoir simulation using MATLAB: User guide for the Matlab Reservoir Simulation Toolbox (MRST). SINTEF ICT. http://www.sintef.no/Projectweb/MRST/Publications.

Liu, R., Liu, H., Li, X., Wang, J., and Pang, C. (2010). Calculation of Oil and Water Relative Permeability for Extra Low Permeability Reservoir. (Society of Petroleum Engineers), p.

Lund, L. (2014). Decline curve analysis of shale oil production: The case of Eagle Ford. Uppsala University.

Mavor, M. (2013). Reservoir Fluid Properties Required for Low Permeability Oil Reservoir Analysis, by Matt Mavor; #90186 (2013). In Hydrocarbon Charge Considerations in Liquid-Rich Unconventional Petroleum Systems, (Canada: AAPG), p.

Mirchi, V., Saraji, S., Goual, L., and Piri, M. (2015). Dynamic interfacial tension and wettability of shale in the presence of surfactants at reservoir conditions. Fuel *148*, 127–138.

Moghadasi, L., Guadagnini, A., Inzoli, F., and Bartosek, M. (2015). Interpretation of two-phase relative permeability curves through multiple formulations and Model Quality criteria. Journal of Petroleum Science and Engineering *135*, 738–749.

Rystad Energy (2014). SHALE COMPANIES HAVE PROVEN MORE EFFICIENT BUT NOT MORE PRODUCTIVE (Rystad Energy).

Sanchez-Rivera, D., Mohanty, K., and Balhoff, M. (2015). Reservoir simulation and optimization of Huff-and-Puff operations in the Bakken Shale. Fuel *147*, 82–94.

Sheng, J.J. (2015). Status of surfactant EOR technology. Petroleum *1*, 97–105.

Shuler, P.J., Lu, Z., Ma, Q., and Tang, Y. (2016). Surfactant Huff-n-Puff Application Potentials for Unconventional Reservoirs. (Society of Petroleum Engineers), p.

Singh, H. (2016). A Critical Review of Water Uptake by Shales. Journal of Natural Gas Science and Engineering *34*, 751–766.

Singh, H., and Cai, J. (2018). Screening improved recovery methods in tight-oil formations by injecting and producing through fractures. International Journal of Heat and Mass Transfer *116*, 977–993.

Steiner, S., Ahsan, S.A., Raina, I., Dasgupta, S., and Lis, G.P. (2016). Interpreting Total Organic Carbon TOC in Source Rock Oil Plays. (Society of Petroleum Engineers), p.

Tuero, F., Crotti M, M., and Labayen, I. (2017). Water Imbibition EOR Proposal for Shale Oil Scenarios. (Society of Petroleum Engineers), p.

US EIA (2016). Annual Energy Outlook 2016 (Government Printing Office).

Wachtmeister, H., Lund, L., Aleklett, K., and Höök, M. (2017). Production Decline Curves of Tight Oil Wells in Eagle Ford Shale. Nat Resour Res *26*, 365–377.

Wang, D., Butler, R., Liu, H., and Ahmed, S. (2011). Surfactant Formulation Study For Bakken Shale Imbibition. (Society of Petroleum Engineers), p.

Wang, D., Butler, R., Zhang, J., and Seright, R. (2012). Wettability Survey in Bakken Shale With Surfactant-Formulation Imbibition. SPE Reservoir Evaluation & Engineering *15*, 695–705.

Wang, D., Zhang, J., Butler, R., and Olatunji, K. (2016). Scaling Laboratory-Data Surfactant-Imbibition Rates to the Field in Fractured-Shale Formations. SPE Reservoir Evaluation & Engineering *19*, 440–449.

Witkowsky, J.M., Galford, J.E., Quirein, J.A., and Truax, J.A. (2012). Predicting Pyrite and Total Organic Carbon from Well Logs for Enhancing Shale Reservoir Interpretation. (Society of Petroleum Engineers), p.

Xu, F., Guo, X., Wang, W., Zhang, N., Jia, S., and Wang, X. (2011). Case Study: Numerical Simulation Of Surfactant Flooding In Low Permeability Oil Filed. (Society of Petroleum Engineers), p.

Yu, W., Lashgari, H., and Sepehrnoori, K. (2014). Simulation Study of CO2 Huff-n-Puff Process in Bakken Tight Oil Reservoirs. (Society of Petroleum Engineers), p.

Zhang, K., Seetahal, S., Alexander, D., Lv, J., Hu, Y., Lu, X., Zhang, D., Wu, K., and Chen, Z. (2016). Effect of Confinement on Gas and Oil Relative Permeability During CO2 Flooding in Tight Oil Reservoirs. (Society of Petroleum Engineers), p.

Zhang, Y., Yu, W., Sepehrnoori, K., and Di, Y. (2017). A Comprehensive Numerical Model for Simulating Fluid Transport in Nanopores. Sci Rep *7*.

Zhu, P., Balhoff, M.T., and Mohanty, K.K. (2017). Compositional modeling of fracture-to-fracture miscible gas injection in an oil-rich shale. Journal of Petroleum Science and Engineering *152*, 628–638.

Zuloaga-Molero, P., Yu, W., Xu, Y., Sepehrnoori, K., and Li, B. (2016). Simulation Study of CO2-EOR in Tight Oil Reservoirs with Complex Fracture Geometries. Sci Rep *6*.

Appendix

SURFACTANT MODEL (FROM JØRGENSEN, 2013)

$$f(N_c) = \begin{cases} N_c^{misc};\ k_\alpha^{misc}(s_\alpha);\ s_{\alpha,res}^{misc}, & max(N_c) \\ N_c^{immisc};\ k_\alpha^{immisc}(s_\alpha);\ s_{\alpha,res}^{immisc}, & min(N_c) \end{cases} \quad \text{(A-1)}$$

Where, $\alpha \in \{w, o\}$

The residual saturation values for the relative permeability are obtained by weighted average of residual saturation values at immiscible (without surfactant) and fully miscible conditions.

$$s_{\alpha,res}^{N_c} = m.\, s_{\alpha,res}^{misc} + (1-m).\, s_{\alpha,res}^{immisc} \tag{A-2}$$

In addition to reducing the IFT and residual oil saturation, surfactant also changes the relative permeability due to miscible flow such that the effective relative permeability after adding surfactant is the weighted average of the relative permeabilities with immiscible (without surfactant) and with fully miscible conditions.

$$k_{r\alpha}(s_\alpha) = m.\, k_{r\alpha}^{misc}\left(\bar{s}_\alpha^{misc}, N_c\right) + (1-m).\, k_{r\alpha}^{immisc}\left(\bar{s}_\alpha^{immisc}, N_c\right) \tag{A-3}$$

The discrete relative permeability values cannot be directly scaled (a continuous function) to match the new end-points, therefore, it require modified saturation values that are calculated as following.

$$\bar{s}_\alpha^{misc}(N_c) = \left(\frac{s_\alpha - s_{\alpha,res}^{N_c}}{1 - s_{\alpha,res}^{N_c} - s_{\alpha',res}^{N_c}}\right).\left(1 - s_{\alpha,res}^{misc} - s_{\alpha',res}^{misc}\right) + s_{\alpha,res}^{misc} \tag{A-4}$$

$$\bar{s}_\alpha^{immisc}(N_c) = \left(\frac{s_\alpha - s_{\alpha,res}^{N_c}}{1 - s_{\alpha,res}^{N_c} - s_{\alpha',res}^{N_c}}\right).\left(1 - s_{\alpha,res}^{immisc} - s_{\alpha',res}^{immisc}\right) + s_{\alpha,res}^{immisc} \tag{A-5}$$

Also, addition of surfactant to the water increases its viscosity in a way that the effective viscosity of water after adding surfactant is given as:

$$\mu_w^{surf}(c_s, p) = \frac{\mu_w(p)}{\mu_w(p_{ref})} \cdot \mu_s(c_s) \tag{A-6}$$

487521 Maximizing Uniformity of Hydraulic Fracture Stimulation of Horizontal Wells through Stress Shadow Balancing and Limited Entry Methods

Abstract:

The combined practice of horizontal drilling and hydraulic fracturing has made production of petroleum resources from ulta-low permeability shale formations both technically feasible and economically viable. However, a major concern about ongoing development of shale reservoirs is a trend towards increasing development intensity in order to try to increase gas/oil production. That is, there can be a tendency towards solutions that entail more well pads, more water, more chemicals, more sand, and therefore more potential for negative environmental and community impacts. Reversing a trend towards increasing development intensity requires a focus on developing approaches that make the most of every well that is drilled, striving towards an industry that is more efficient in terms of gas/oil recovery from a given level of development activity. And, in fact, there ought to be substantial gains that can be made through efforts to make the most of every foot of every well because one thing is clear about the current technology of horizontal drilling and hydraulic fracturing: It is still achieving results that are far from optimal. Of particular concern is the fact that data from producing wells show a tendency for 1/3 or more of the drilled length of horizontal wells to be completely unproductive. A likely cause is that fluid that is injected with the intention of generating hydraulic fractures from all perforation clusters â€“ the clusters of holes through the well casing that allow fluid exchange between the well and the reservoir - is instead diverted to only a portion of the clusters leaving many sections of the well unstimulated.

Thus motivated, here we present a numerical parametric study showing two methods for optimizing uniformity of stimulation through design of the clusters of perforation holes that are used as the entry points for fluid from the wellbore into the formation. We show that non-uniform perforation cluster locations combined with designing the entry friction associated with the perforation clusters (â€œlimited entryâ€) can be optimized to produce the maximum surface area for an array of hydraulic fractures. Optimization is carried out using a reduced order, approximate simulator capable of carrying out an entire coupled simulation, which would normally take hours or days, in seconds to minutes on a desktop computer. Candidate designs are then tested using a fully-coupled, Planar 3D, multi-hydraulic fracture simulator that uses an Implicit Level Set Algorithm (ILSA). We find that for stages completed with 5 perforation clusters, optimizing the cluster spacing leads to over 70% increase in fracture surface area generated by the same injected volume. For uniform cluster spacing, optimal limited entry design alone is able to improve the fracture surface area by approximately 10%. When combined and optimized, improvements can approach a doubling of the fracture surface area leading to a potential for a similar order of increase in hydrocarbon recovery.

Rowan University

Novel Bio-Based Polyesters & Polycarbonates from Xylochemicals

2017 AIChE Annual Meeting
Minneapolis, MN October 29 – November 3 2017
"Bioplastics, Biocomposites and Value-Added Uses of Biofuel Coproducts for Sustainable Manufacturing"

Joseph F. Stanzione, III, Ph.D.
Department of Chemical Engineering
Sustainable Materials Research Laboratory

 1

Plastics: The Facts

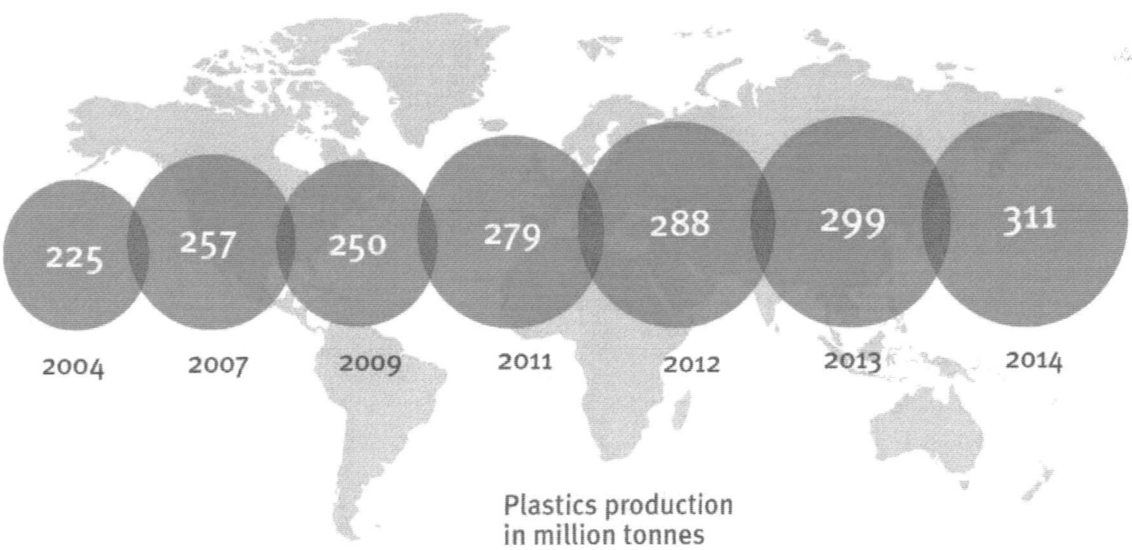

- **Plastic production is growing globally[1]**
- **Currently around 330 Mtonnes**

[1] PlasticsEurope, *Plastics - the Facts* (2015)

 2

Plastics: The Facts

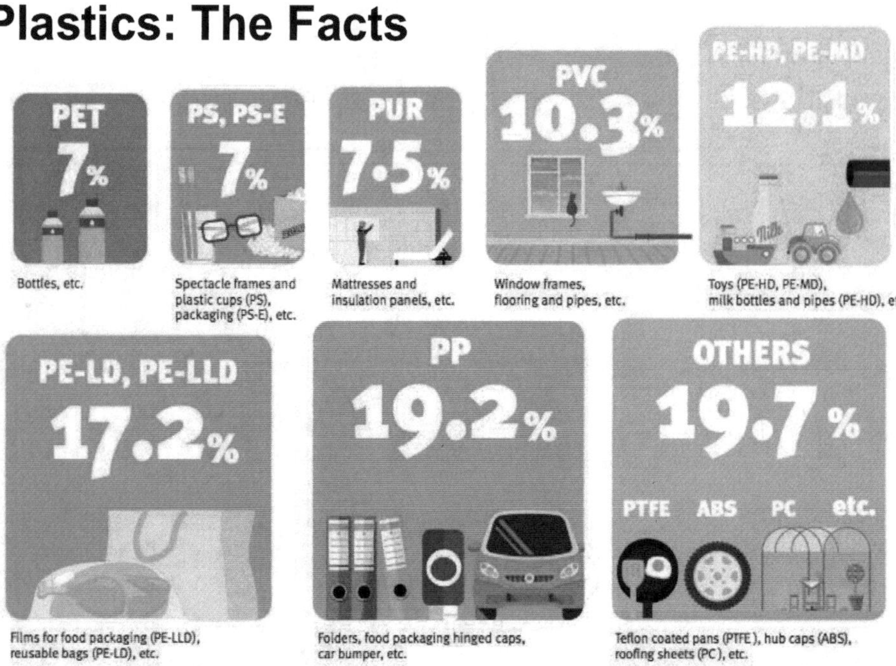

Less than 1.5% are renewably sourced[2]

[2] European Bioplastics, *Bioplastics facts and figures* (2017)

Plastics: The Future

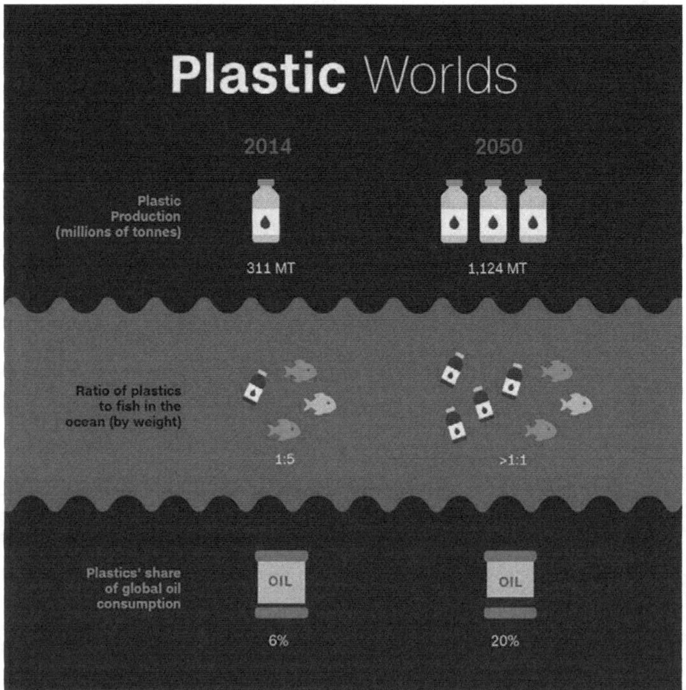

By 2050:

- **Expected to triple**
- **More plastic than fish in the ocean**
- **20% of global annual oil consumption**

[3] World Economic Forum (2016)

Trying to Replace PET

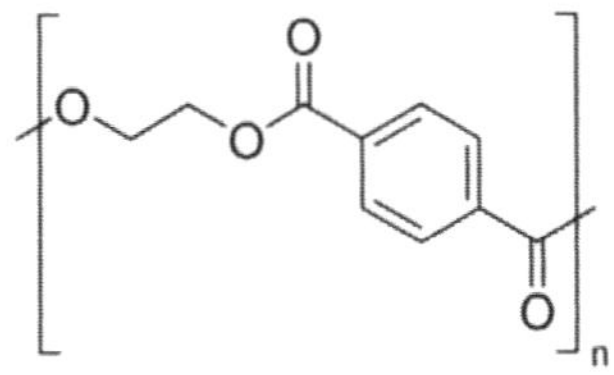

- **Very good mechanical properties**
- **T_g = (67-81) °C, T_m ≈ 267 °C**
- Petroleum-derived (green routes in progress)
- Not easily degraded

High Tg Bio-Based Aliphatic Polyesters from Bicyclic D-Mannitol
Cristina Lavilla, Abdelilah Alla, Antxon Martínez de Ilarduya, and Sebastián Muñoz-Guerra*
Departament d'Enginyeria Química, Universitat Politècnica de Catalunya, ETSEIB, Diagonal 647, 08028 Barcelona, Spain

- Sugar derived
- T_gs range from -40 to 70 °C

Renewable (semi)aromatic polyesters from symmetrical vanillin-based dimers†
Audrey Llevot,[a,b] Etienne Grau,[a,b] Stéphane Carlotti,[a,b] Stéphane Grelier[a,b] and Henri Cramail*[a,b]

- Lignin derived
- T_gs up to 140 °C
- Insoluble in any solvent

Bio-based alternatives in the synthesis of aliphatic–aromatic polyesters dedicated to biodegradable film applications
Nicolas Jacquel [c], René Saint-Loup [c], Jean-Pierre Pascault [a, b], Alain Rousseau [a, b], Françoise Fenouillot [a, b, *]
[a] Université de Lyon, F-69003, France
[b] INSA-Lyon, Ingénierie des Matériaux Polymères, IMP, UMR5223, F-69621 Villeurbanne, France
[c] Roquette Frères, Division Recherche, F-62080 Lestrem, France

- Sugar derived (isosorbide)
- Negative T_gs, high T_ms

etc…!

Additional Concerns: Toxicity

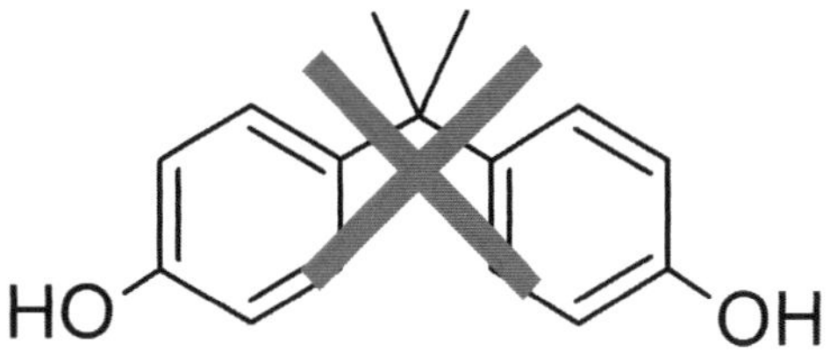

Bisphenol A (BPA)

Bisguaiacol (BG): A Lignin-Derived Bisphenol

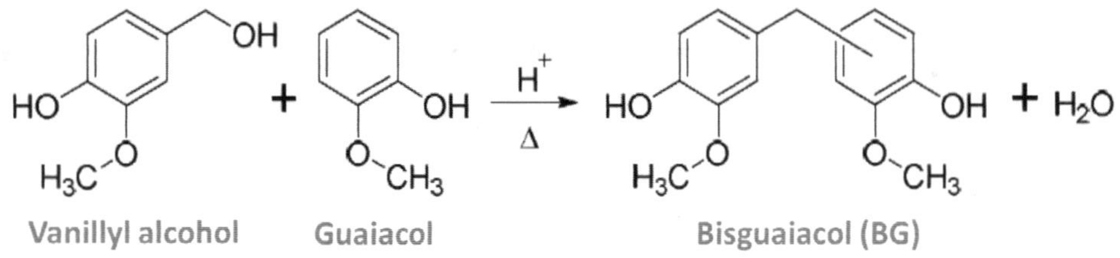

- **Similar structure to BPA**[4]
- **Fully renewable**
- **Reduced toxicity**[5]
- **Isomeric distribution**
  - **82%** ***p,p'***
  - **15%** ***m,p'***
  - **3%** ***o,p'***

[4] Hernandez *et al.*, *ACS Sustain Chem Eng* (2016)
[5] Hong *et al.*, *Int. J. Environ. Res. Publ. Health(* 2016)

 7

Q: Can we use BG to synthesize high-performance renewable thermoplastics?

 8

BG-based Polycarbonates

Joseph Mauck,
M.S. CHE

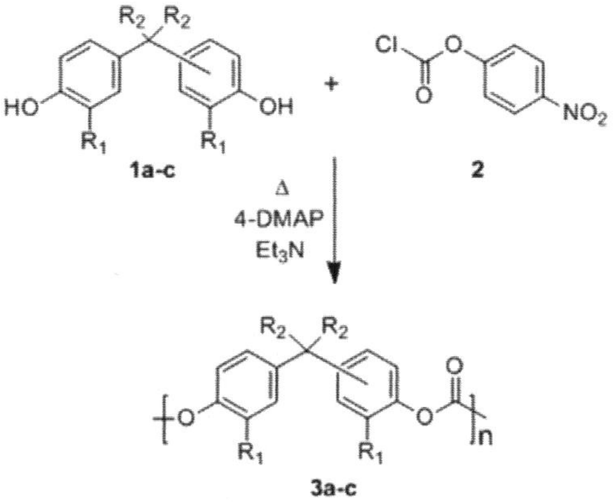

1a,3a: R_1=H, R_2=CH_3
1b,3b: R_1=H, R_2=H
1c,3c: R_1=OCH_3, R_2=H

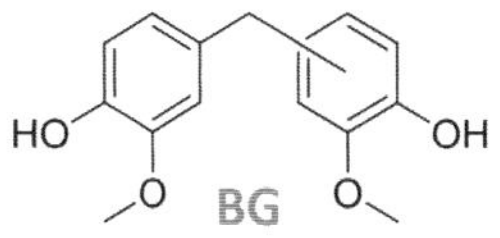

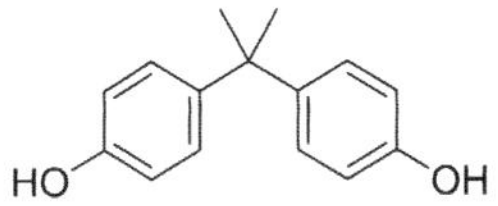

BPA

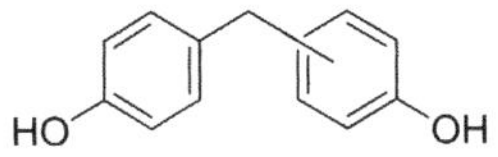

BPF

[6] J. Mauck *et al.*, *J Biobased Materials & Bioenergy* (2017)

BG-based Polycarbonates: Thermal Properties

Sample	*IDT* (°C)	$T_{50\%}$ (°C)	T_{max} (°C)	Char Content (%)	T_g (°C)
BG-PC	375	455	450	24.2	100
BPA-PC	440	543	545	23.2	150
BPF-PC	369	546	501	40.5	119

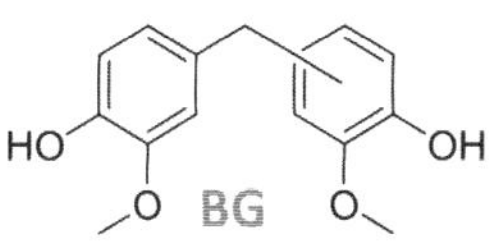

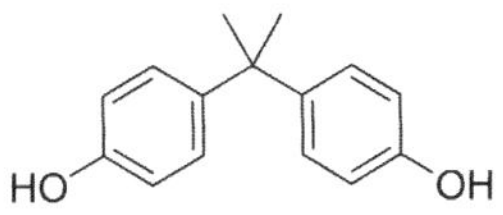

BPA

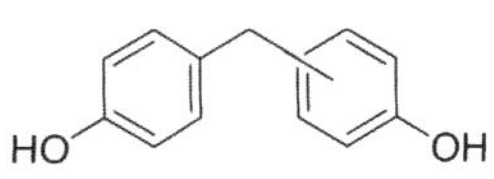

BPF

[6] J. Mauck *et al.*, *J Biobased Materials & Bioenergy* (2017)

Q: Can we use BG to synthesize polyesters?

 11

Preparation of BG-based Polyesters

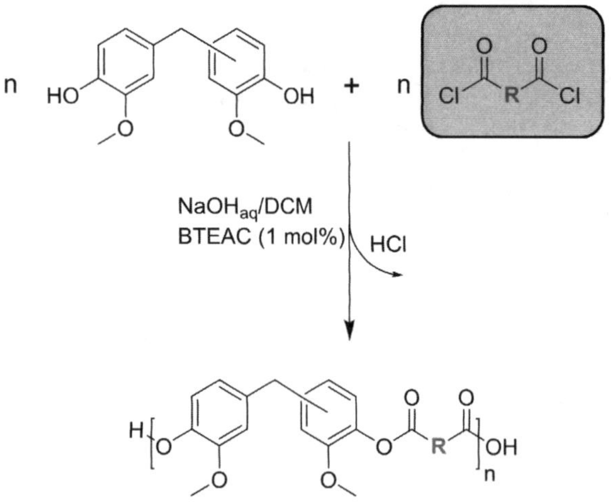

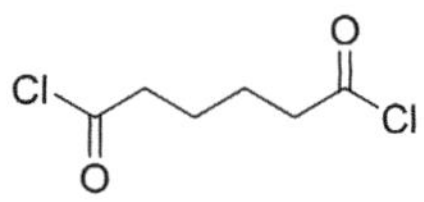

Adipoyl Chloride

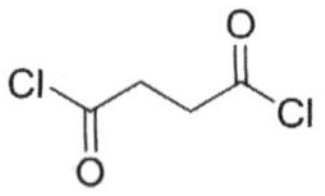

Succinyl Chloride

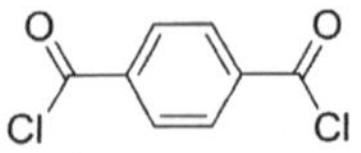

Terephthaloyl Chloride

[7] S. Curia *et al.*, *In preparation* (2017)

 12

BG-based polyesters

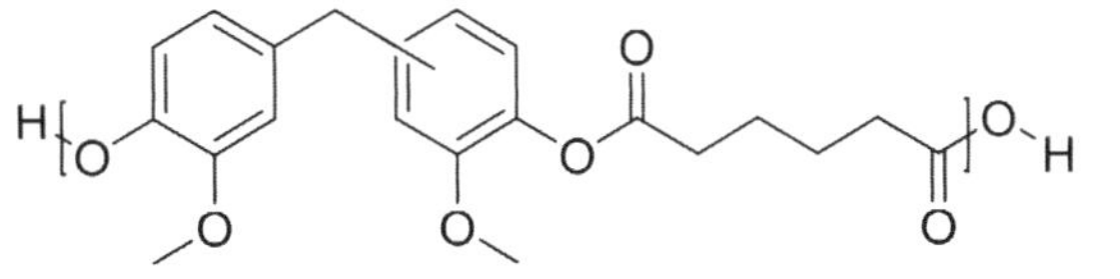

Poly(BG adip)
M_n=6.6 kg mol^{-1} (Đ=1.82)

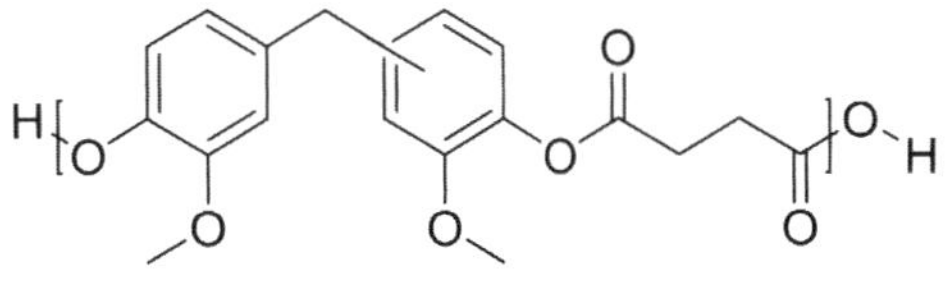

Poly(BG succ)
M_n=7.2 kg mol^{-1} (Đ=2.16)

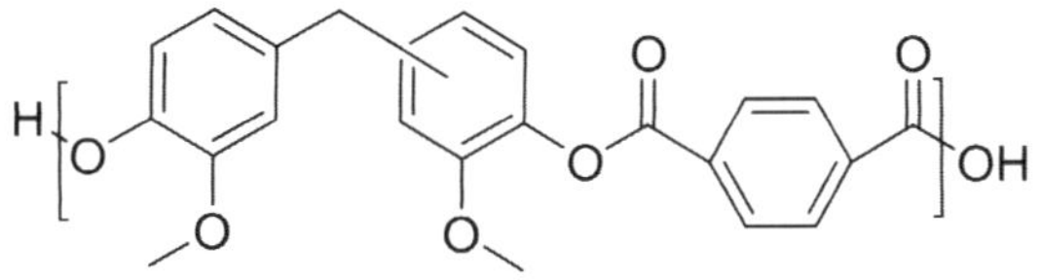

Poly(BG tere)
M_n=26.9 kg mol^{-1} (Đ=2.29)

[7] S. Curia *et al.*, *In preparation* (2017)

IR Analysis

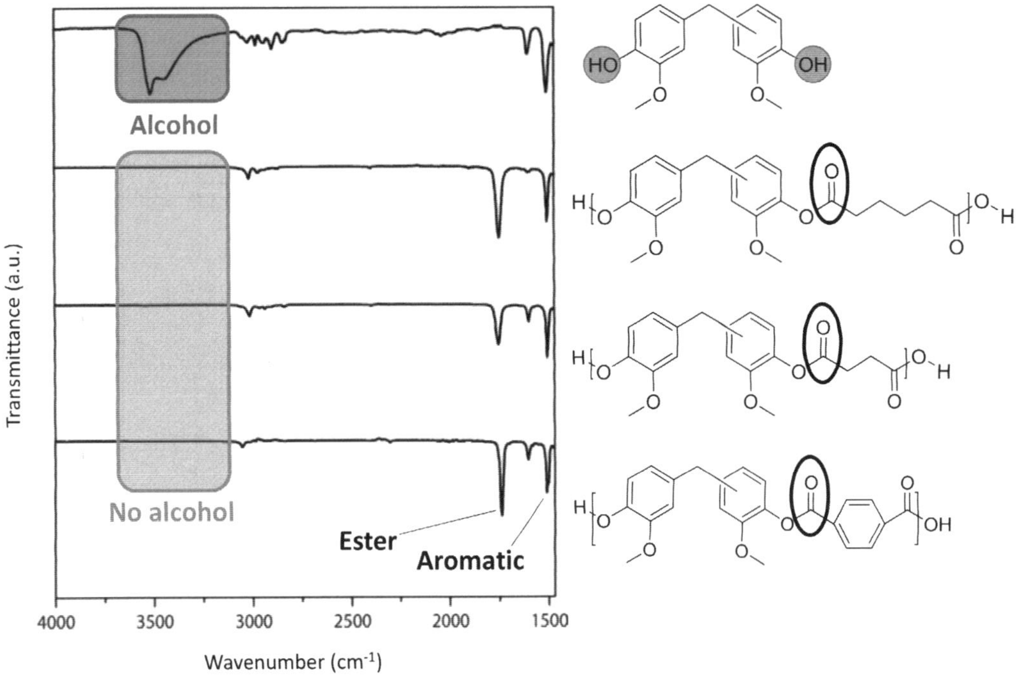

^{1}H-NMR Analysis (in $CDCl_3$)

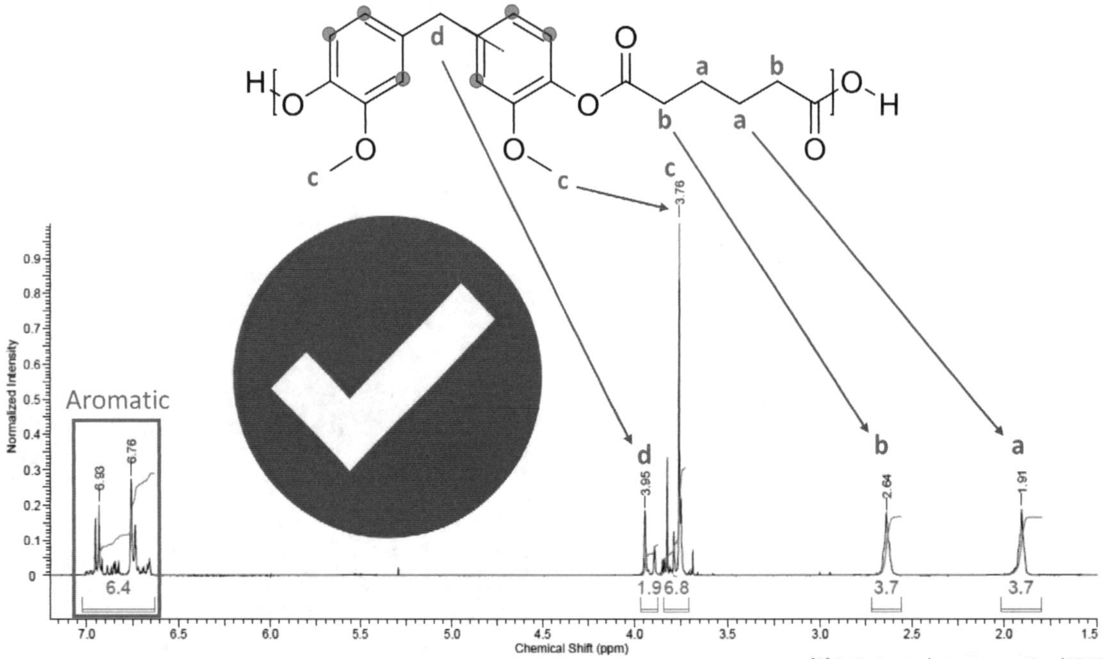

[7] S. Curia *et al.*, *In preparation* (2017)

 15

BG-based Polyesters: DSC Analysis

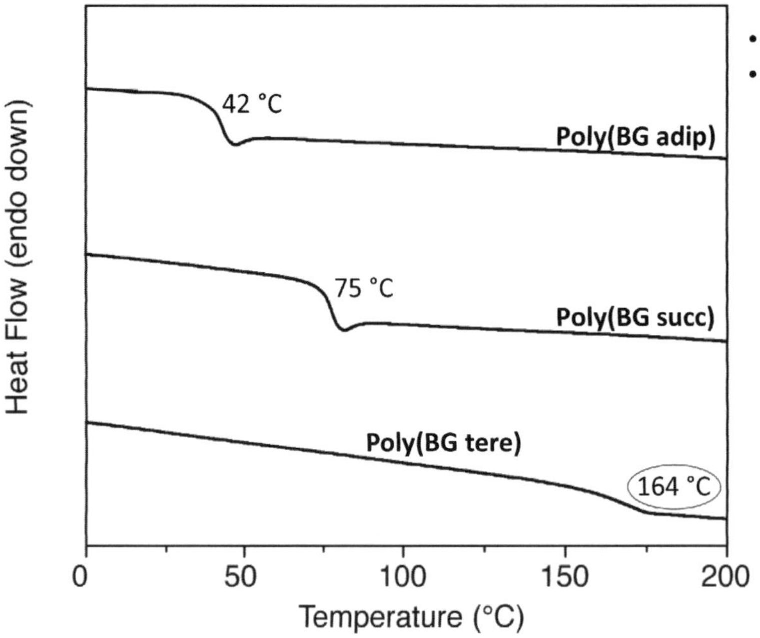

- **10 °C min^{-1}**
- **N_2 50 mL min^{-1}**

[7] S. Curia *et al.*, *In preparation* (2017)

 16

BG-based Polyesters: Thermal Stability (N_2)

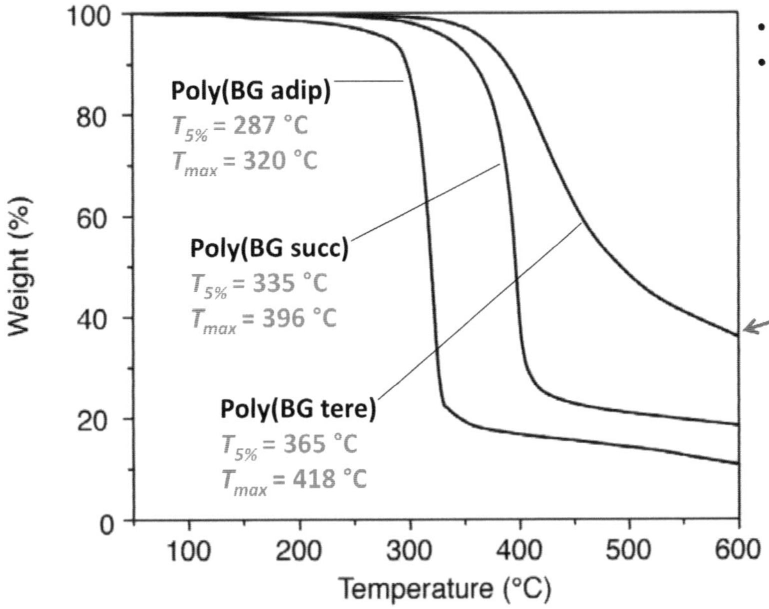

- 10 °C min^{-1}
- N_2 25 mL min^{-1}

36% char @ 600 °C

[7] S. Curia *et al.*, *In preparation* (2017)

 17

BG-based Polyesters: Thermal Stability (Air)

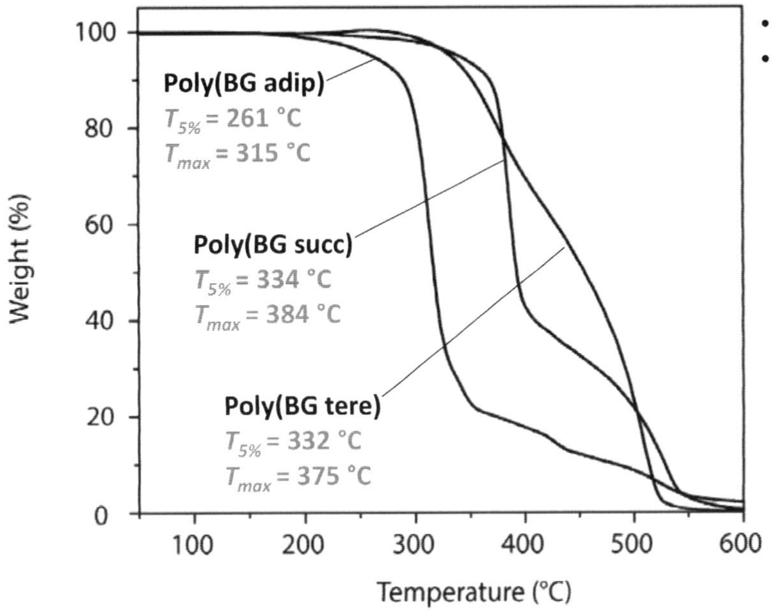

- 10 °C min^{-1}
- Air 25 mL min^{-1}

T_{max} still above 300 °C

[7] S. Curia *et al.*, *In preparation* (2017)

 18

BG+FDCA: a Fully Renewable Polyarylate

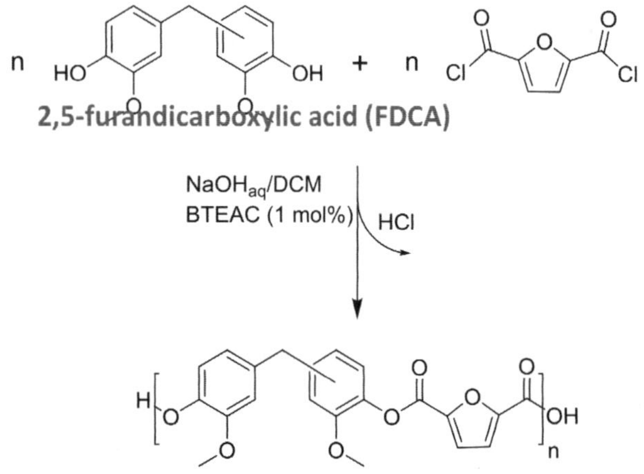

Poly(BG FDCA)
M_n=6.3 kg mol^{-1} (Đ=1.57)

[7] S. Curia *et al.*, *In preparation* (2017)

Poly(BG-FDCA): DSC

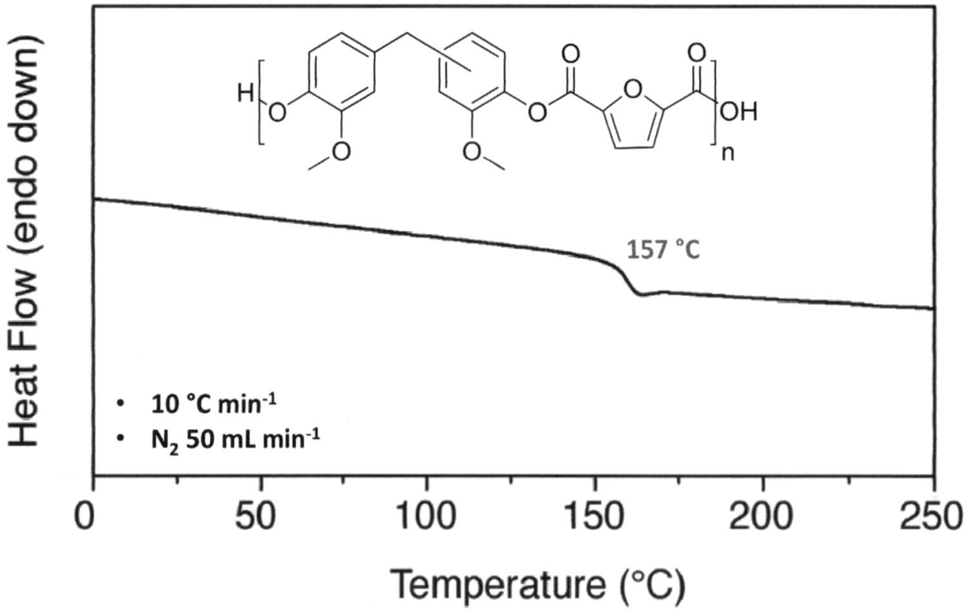

[7] S. Curia *et al.*, *In preparation* (2017)

Poly(BG-FDCA): TGA N_2 *vs* air

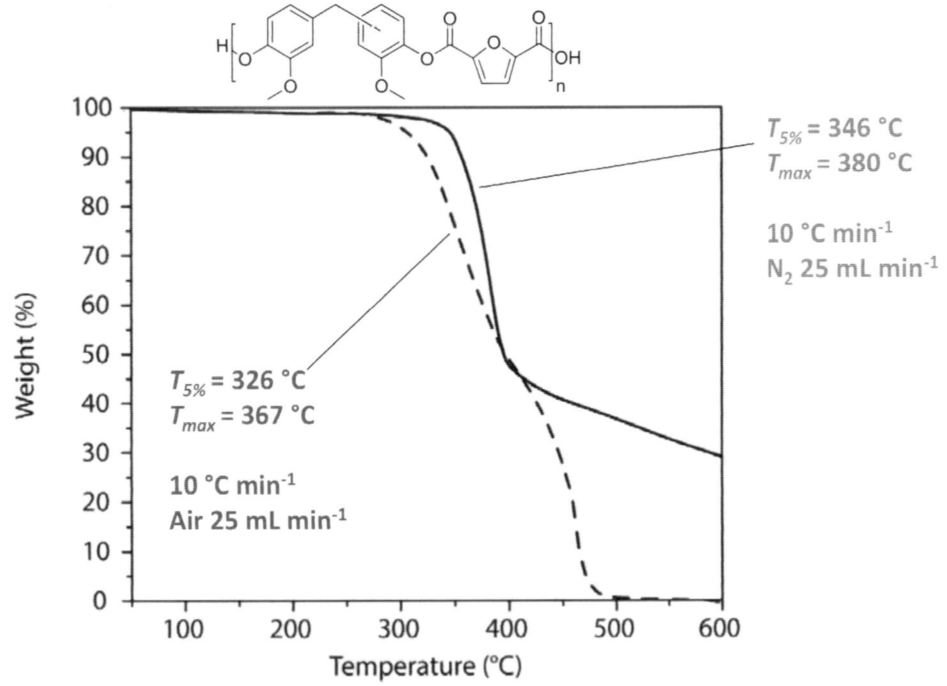

[7] S. Curia *et al.*, *In preparation* (2017)

Conclusions

- **BG** can be used to synthesise **useful polycarbonates & polyesters**
- **BG-based polyesters** show **high thermal resistance** and **tuneable T_g**
- **Poly(BG FDCA)** has a remarkably **high T_g (157 °C)**

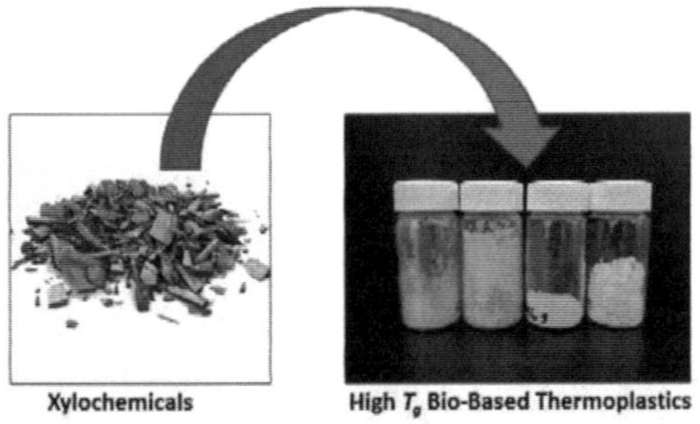

Future Investigations

- Hydrolytic degradation & biodegradation tests
- End-cap with functional molecules
- Chain extension, melt polymerization (achieve higher MW)
- Gas permeability/diffusivity
- Mechanical tests

Xylochemicals — High T_g Bio-Based Thermoplastics

Acknowledgements

- Sustainable Materials Research Laboratory
- Joseph R. Mauck
- Silvio Curia, Ph.D.
- John J. La Scala, Ph.D. (ARL)
- Rowan University
    - ❖ Department of Chemical Engineering
    - ❖ Department of Chemistry & Biochemistry
- U.S. ARL for Financial Support via Cooperative Agreement W911NF-16-2-0225

Silvio Curia, Ph.D.

Rowan University

ARL U.S. ARMY RDECOM

South Jersey Technology Park at Rowan University

Thank you!
Any questions?

Crater Lake National Park

Rocky Mountain National Park

503981 Light Weight Biocomposite from Toughened Polyolefin and Biocarbon

Abstract:

In converting biomass to bioenergy through thermochemical processes, a considerable amount of solid residue generates in each conversion cycle (up to 35 wt.%) which mainly consists of renewable carbon (biocarbon). However, the biocarbon is considered as a byproduct of the pyrolysis units. Unless a high-volume application exists for the biocarbon, the significant amount of this byproduct can challenge the long-term sustainability of pyrolysis units. In this project, biocarbon is intended to be used as filler in plastic industry targeting toughened polypropylene (PP) based automotive parts. In order to achieve a suitable balance of mechanical properties, biocomposites were compatibilized using a maleated polypropylene. Furthermore, the PP phase was nucleated using a Î² nucleating agent to increase the impact toughness without significant loss in the stiffness. Tensile properties and notched Izod impact strength were measured as response variable. Atomic force microscopy (AFM) was utilized to analyze the PP-biocarbon interface in presence and absence of the compatibilizer. The effect of the nucleating agent was investigated by differential scanning calorimetry (DSC) and correlated to the improvement of biocomposites impact toughness. The results suggest that the morphological changes caused by the compatibilizer and toughening effect of Î² crystals can generate high impact toughness in composites with 20 wt.% loaded biocarbon. Therefore, the biocarbon can be utilized in high loading levels similar to current petroleum-based composites with comparable mechanical properties. At the end, substituting conventional mineral filled composites with biocarbon-based biocomposites can provide the required mechanical properties with added benefits of lower density (â‰ˆ15%), 30% higher bio-based content and cheaper price. These benefits would translate into improved fuel economy of cars and reduction in greenhouse gas production. Considering the high-volume usage of plastic parts especially in automotive industry, this approach would make the biomass to bioenergy conversion more sustainable.

Keywords: Biocarbon, Toughened biocomposites, Polypropylene.

Acknowledgement: We gratefully acknowledge the financial support of the Ontario Ministry of Agriculture, Food and Rural Affairs (OMAFRA)/University of Guelph - Bioeconomy for Industrial Uses Research Program Theme Project # 030055; Ontario Research Fund, Research Excellence Program; Round-7 (ORF-RE07) from the Ontario Ministry of Research, Innovation and Science (MRIS) (Project #052644 and 052665), and the Natural Sciences and Engineering Research Council (NSERC), Canada Discovery Grant Project # 401111 to carry out this research work

497443 Secondary Fermentation of Corn Ethanol Co-Products for Improved Amino Acid Qualities

Abstract:

Secondary Fermentation of Corn Ethanol Co-Products for improved Amino Acid Qualities

In 2016 5.28 billion bushels of corn were used to produce about 14.79 billion gallons of ethanol in the United States. As a result, about 36 million tons of Dried Distillers Grains with Solubles (DDGS) were manufactured and fed to livestock (U.S. Dept. of Energy, 2017). DDGS are a common feed supplement in cattle rations as it is inexpensive and has positive feeding characteristics. One of the drawbacks of DDGS is that it is lacking in in key amino acids such as tryptophan, arginine, and lysine. Historically these amino acids have been supplemented by external addition of feed grade amino acids to rations. The research carried out attempts to fortify DDGS with higher amounts of tryptophan through secondary fermentation of Whole Stillage (WS) by bacteria. When cultured on WS the bacteria secretes extracellular tryptophan, which serves to close the amino acid gap in corn ethanol co-products. The bacteria used is Generally Regarded as Safe (GRAS) and has been used to produce feed ingredients historically. Because of the research, profitability of corn ethanol operations will be increased due to the improved feeding value of the co-products produced.

References

United State Department of Energy, â€œEthanol Fuel Basics,â€ Energy Efficiency & Renewable Energy http://www.afdc.energy.gov/fuels/ethanol_fuel_basics.html 03/16/2017

500583 Extraction High-Value Chemicals from Ethanol Co-Products: A Feasibility Assay on Phytate Extraction with Life Cycle and Techno Economic Assessment

Abstract:

Background

The well-established dry-grind process today accounts for also over 80% of the operating ethanol facilities in the United States, producing not only ethanol, but also distiller's grains. Distiller's grains, commonly commercialized as dried distiller's grains with solubles (DDGS) accounts for most of the heavy fraction from the ethanol distillation and rectification operations, and is processed through a complex downstream processing after the fermentation is unloaded. Despite being a nutritious animal feed, inclusion of DDGS as feed ration is still limited due to compositional properties, such as the high presence of non-starch polysaccharides, which are non-digestible to monogastric animals, the fatty acid profile, and the overdose of phosphorus, which is largely non-available to some animals. Phosphorus behavior in ethanol downstream processing is of particular interest, not only because it can lead to environmental issues, but also because it can potentially be an additional economic source to ethanol producers. Over 40% of the phosphorus present in ethanol co-products is present as phytate, the main form of organic phosphorus in cereals and grains, which is not completely hydrolyzed during the whole ethanol process. Phytate has a high market value, and today is a platform chemical utilized in a wide array of industries, and it is mostly exploited because of its high chelating potential (He et al. 2017). Phytate is also the raw material to produce *myo*-inositol, a cyclic polyol with a variety of uses by pharmaceutical and nutraceutical companies.

Phytate is found to be dissolved on the ethanol co-products(He et al. 2017). During the dry-grind downstream processing, thin stillage, i.e., the liquid-rich fraction derived from the centrifugation of whole stillage, accounts for the highest concentration of phytate among all the co-products. Recent research (He et al. 2017) has proposed an extraction method using ion exchange separation, which can efficiently separate phytate from thin stillage, with little to no effect on the residual nutritional profile of the final DDGS. Even though the concentration of phytate being relatively low on ethanol co-products, at around 1 g L^{-1} in thin stillage, due to the enormous volumetric production by dry-grind plants, the production leveraging costs in extracting phytate might become an industrial alternative.

The high volumes of DDGS produced by ethanol plants, alongside with its limited dietary inclusion, pose a challenge for higher revenues to ethanol plants. The second-most important co-product from dry-grind ethanol facilities, distiller's corn oil, represent today a mature product suitable for biodiesel production and other uses of low-grade vegetable oil(Reis et al. 2017). Removal of distiller's corn oil, on the other hand, has been criticized by animal nutritionists claiming a loss in energy value from DDGS. Despite still representing about 20% of the total revenue stream from dry-grind ethanol plants, DDGS are still ranking in low positions when compared to higher end animal feeds, such as soybean meal. The extraction of high-value chemicals through mature sound technologies may represent, thus, an alternative for higher economic security of ethanol plants. This current research describes a techno-economic analysis and a life-cycle assessment of a phytate extraction process on a large-scale dry-grind corn-to-ethanol plant in the United States, and its posterior *in-loco* conversion to *myo*-inositol.

Methods

The scope of this work analyzed the utilization of corn-to-ethanol thin stillage produced through the dry-grind process as raw material to extract phytate using an ion exchange process, followed by hydrolysis to generate *myo*-inositol. The process was modelled according to He et al. (2017) , and by additional empirical analysis, when required. Thin stillage was obtained from a dry-grind ethanol plant in the state of Iowa, USA. The process of converting thin stillage to a phytate brine solution is divided into the major steps: (1) filtration, (2) flow through ion exchange column, (3) desorption of phytate using brine solution, (4) reconditioning of column using sodium hydroxide, and (5) column washing using water. The conversion of phytate brine solution to calcium phytate is done by (6) addition of calcium chloride and sodium hydroxide, followed by product filtration, and production of *myo*-inositol is achieved by (7) hydrolysis of phytate brine solution, (8) precipitation of phosphate with calcium chloride and sodium hydroxide, and (9) evaporation and drying of *myo*-inositol.

The life cycle assessment of calcium phytate extracted from corn-to-ethanol thin stillage was estimated using GREET model. Calcium phytate extraction, and posterior conversion to *myo*-inositol were implemented on two stationary models on GREET – Dry mill ethanol production w/ corn oil extraction, and Dry mill ethanol production w/o corn oil extraction. Techno-economic assessment for ethanol plants was considered using the BuGal model, with added market and technical data for the modified phytate extraction and conversion units. Since the basis of the BuGal model for techno-economic analysis has been modified to analyze both conditions (with and without distiller's corn oil extraction), the two stationary models on GREET to describe corn-to-ethanol process were utilized.

Results and discussion

LCA results indicate that the proposed phytate extraction does not impact significantly most of the environmental impact factors of ethanol. As a gallon of ethanol was considered as functional unit, the results indicate that for a system with phytate extraction and oil extraction, CO_2 emissions will increase from 3.21 kg to 3.28 kg per gallon of ethanol, whilst for a system without distiller's corn oil extraction, 3.23 kg of CO_2 will be increased to 3.30 kg of CO_2 per gallon of ethanol produced. NOx, SOx, N2O, and other gases also increased with phytate extraction by a factor of about 1%. Analyses were carried out for techno economic assessment in 4 different cases, (1) with distiller's corn oil extraction and phytate extraction, (2) with distiller's corn oil extraction and without phytate extraction, (3) without oil extraction and without phytate extraction extraction, (4) without oil extraction and with phytate extraction, and were consistently all across different cases.

Unlike major recent upgrades to the dry-grind ethanol industry, such as distiller's corn oil removal, extraction of phytate and production of *myo*-inositol are related to products found in dilute concentrations. The low concentration of phytate in ethanol co-products requires low energy input on an ethanol basis, while still providing a high-value co-product. BuGal model calculates the operating expenses per gallon of denatured ethanol sold as USD 0.2673 for the control conditions, i.e., without phytate extraction, and USD 0.2686 for a system considering 20% of thin stillage being redirected to phytate extraction, due to increase in energy consumption for additional processing. Regarding chemical costs, phytate extraction and conversion to inositol likely increased the value from USD 0.1201 to USD 0.1226 per gallon of denatured ethanol, mainly due to the requirement of NaCl for phytate desorption from the ion exchange process, additional process water, and calcium chloride addition. These costs are balanced by the calcium phytate, *myo*-inositol, and calcium phosphate as fertilizer (derived from the phytate hydrolysis) sales, at prices estimated at USD 12, USD 10, and USD 0.20 per kg, respectively. BuGal estimates an average large-scale dry-grind operation, with factor of equity and debt being 0.7 and 0.3, respectively, with an interest rate charged on debt as 0.07, and the rate of return required by investors on the plant's equity as 0.12. Even though these numbers are variable, the economic considerations on such assumptions are beyond the scope of this work, and remain constant through all assumptions. Following similar numbers as Kwiatkowski et al. (2006), 2.795 gallons of ethanol are estimated to be produced from each bushel of corn, and 17.5 lb of

DDGS, and CO_2 are produced on a bushel basis. Denatured ethanol sales are estimated to be at USD 1.70 per gallon, an estimate within current market prices , and that phytate extraction would not affect the price of DDGS, thus, being marketed at a price of USD 161 per ton, also within current market estimates.

Conclusion

The removal of phytate from thin stillage may represent a novel process in the dry-grind ethanol industry. The removal of phytate via ion exchange does not increase significantly (around 1% of increase) the emissions involved on the production of one gallon of ethanol, and that it can be a source of profit if operated within a certain range of thin stillage processed within a plant. The results also presented herein describe the potential of converting the extracted phytate to myo-inositol. Myo-inositol has a more flexible market when compared to phytate, and could also represent another source of revenue for ethanol plants.

References

Kwiatkowski JR, McAloon AJ, Taylor F, Johnston DB (2006) Modeling the process and costs of fuel ethanol production by the corn dry-grind process Industrial crops and products 23:288-296

He Q, Rodrigues Reis, CE, Wang F, Hu B (2017) Phytate extraction from coproducts of the dry-grind corn ethanol process. RSC Advances. doi:10.1039/C6RA27409A

Reis CER, Rajendran A, Hu B (2017) New technologies in value addition to the thin stillage from corn-to-ethanol process Reviews in Environmental Science and Bio/Technology 16:175-206 doi:10.1007/s11157-017-9421-6

495895 Effect of Carbonaceous Susceptors on Selective Production of Phenolics and Syngas from Lignin Via Microwave Pyrolysis

Abstract:

At present, global energy supply and production of chemicals are largely dependent on fossil fuels such as coal, petroleum and natural gas. However, the utilization of these resources is limited owing to their non-renewable nature and its environmental adversity. Lignocellulosic biomass is projected as the only source of renewable carbon on earth. The primary structural building blocks of biomass are cellulose, hemicellulose, and lignin; of these components, contributes to 30% and 40% by weight and energy of the total biomass, respectively. Energy, syngas, fuels and chemicals can be obtained from lignin through thermochemical processing techniques like combustion, gasification and pyrolysis. Fast pyrolysis is a promising process to produce bio-oil and chemicals in high yields. High yields of aromatic hydrocarbons and organics can be obtained due to the aromatic functionality in the lignin structure. In order to extract resources from lignin on a large scale by fast pyrolysis technology, challenges such as high yield of condensable products, product selectivity and energy efficiency are to be improved. Achieving fast pyrolysis conditions with improved efficiency is possible by microwave assisted heating. Microwave assisted fast pyrolysis technology can be used to overcome the techno-economic barriers by reducing the processing time, improving the heat distribution, enhancing the product selectivity, and aiding in better heating efficiency.

The intention of the current study is to enhance the yield of bio-oil and its quality from lignin by microwave assisted pyrolysis. The effects of (i) lignin (L) to activated carbon (AC) ratio , (ii) particle size of activated carbon (0.45 mm to 3 mm), and (iii) carbonaceous susceptors (AC, graphite and charcoal) on heating rate, product yields (oil, gas and char fractions), product composition and process efficiency were investigated at a constant microwave power of 480 W. It was found that, owing to the high amount of physically bound moisture in the mesoporous capillaries of AC (1mm), high bio-oil yield of 66 wt. % at optimum conditions of 10: 90 g/g of L: AC was observed. In L:AC blend of 100 g, increase in mass of AC from 10 g to 90 g decreased the heating rate significantly from 128 ^{o}C/min to 32 ^{o}C/min, correspondingly a decrease in energy efficacy of the system from 95% to 34% was observed. From the oil composition analyses, it was found that bio-oil comprises of phenolics (simple phenols, guaiacols and syringols), aromatic hydrocarbons, naphthalenes, indene derivatives and furans. The highest yield of total phenolics (64.2%) was obtained at 10: 90 g/g of L: AC (1mm). Initial lignin sample mass of 9 to 40 wt. % was converted into simple phenolics and its composition increased with increase in L: AC ratio.

The carbonaceous materials, such as activated carbon, graphite, charcoal and bio-char are found to be the efficient microwave absorbers because of their high microwave coupling nature even at ambient temperatures **[1,2]**. For this reason, we have chosen activated carbon, graphite and charcoal as the susceptors to investigate their effect on lignin pyrolysis. The effect of susceptors on the heating rate, product mass and energy yields were investigated**.** The yield (wt. %) of bio-oil is in line with the heating rates and varies in the following order: charcoal (36 wt. %) > graphite (34.5 wt. %) > AC (0.45mm) (29.2 wt. %). Phenolic derivatives are dominant in oil fraction, 97% of oil obtained with AC (0.45mm) contains phenolics followed by charcoal (79%) and graphite (72%). Selectivity (%) of simple phenolics

is varying in the following order: 60.1 (AC (0.45mm)) > 55.8 (charcoal) > 40.9 (graphite). A significant enhancement in heating value was noticed both in char (28-30 MJ/kg) and oil (28-33 MJ/kg) fractions compared to raw lignin (25 MJ/kg). Tiny micro plasma spots generated during the pyrolysis process increased tiny cylindrical and spherical micro-pores within the residual char. Yield of CO_2 decreased and that of H_2 increased with increase in pyrolysis temperature and its evolution profiles were influenced by the nature of susceptor and lignin to susceptor mass ratio.

References:

1. T. Ohra-aho, J. Linnekoski. Catalytic pyrolysis of lignin using analytical pyrolysis-GC-MS. J. Anal. Appl. Pyrol. 113 (2015) 186-92.
2. S. Adhikari, V. Srinivasan, O. Fasina. Catalytic pyrolysis of raw and thermally treated lignin using different acidic zeolites. Energy & Fuels 28 (2014) 4532-38.

497990 Catalytic Conversion of Lignin in Ionic Liquids *Via* Catalysis and Biocatalysis

Abstract:

Lignin makes up 20-30% of the plant biomass and is the most abundant aromatic polymers in nature. Converting lignin into high-value chemicals adds revenues for a biorefinery thus improves the economic viability of biofuel production. Ionic liquids (ILs) have received increasing interest because of their potentials in fractionating and pretreating lignocellulosic biomass. Given the unique properties of aqueous ILs that provide for lignin solubility and biocompatibility to enzymes and microbes, we foresee a great opportunity to develop a new strategy for lignin extraction and depolymerization *via* catalysis and biocatalysis in aqueous ILs. The stability and activity of commercial fungal and plant laccases and a recombinant bacterial laccase were evaluated in difference concentration of ILs. Results suggest that certain ILs were fully compatible with laccases; while some appeared as non-competitive inhibitors. Immobilization of laccase on solid matrix, i.e. membrane surface helped to retain the enzyme activity and overcome the inhibition effects. We also explored catalytic oxidation route in ILs by screening a range of metal catalysts and oxidizing agents. Molecular weight distribution of lignin fraction was followed using gel permeation chromatography while the oxidation products such as vanillic acid, syringaldehyde, and syringic acid were identified and quantified using a GC-MS. Cleavage of the lignin structures was further investigated by NMR to understand the plausible oxidation mechanisms. Results from this study provide insights into the oxidative degradation pathway of lignin in the presence of IL and the selectivity of catalyst towards product formation. In situ lignin conversion in IL will also help to tackle the challenges associated with IL recycle and recovery.

498243 Depolymerization of Lignin Using Peracetic Acid Under Mild Conditions

Abstract:

Lignin is the largest non-carbohydrate component in lignocellulosic biomass, so utilization of lignin can significantly improve total biomass utilization. Depolymerization of lignin is an important step for effective utilization of lignin. Effective and selective depolymerization of lignin into mono-aromatics and/or dicarboxylic acids does not only convert lignin into value-added products, but they also enhance the total biomass utilization by valorizing the co-products from biomass. Diverse lignin depolymerization strategies have been discussed; however, there are few commercially feasible lignin conversion processes because of low conversion yield, poor product selectivity, expensive solvents and catalysts, and severe reaction conditions. PAA removed and/or decomposed lignin. In this study, Peracetic acid (PAA) was applied to decompose lignin under mild reaction temperature with minimal side reactions. Structural changes on lignin were evaluated by diverse analytical methods such as GPC and NMR. In addition, quantitative analysis with depolymerization products was conducted by GC/MS and HPLC.

498934 Low-Energy Catalytic Electrolysis for Simultaneous Hydrogen Evolution and Lignin Depolymerization

Abstract:

Here, a new proton-exchange-membrane electrolysis is presented, in which lignin was used as the hydrogen source at the anode for hydrogen production. Either polyoxometalate (POM) or $FeCl_3$ was used as the catalyst and charge-transfer agent at the anode. Over 90% Faraday efficiency was achieved. In a thermal-insulation reactor, the heat energy could be maintained at a very low level for continuous operation. Compared to the best alkaline--water electrolysis reported in literature, the electrical-energy consumption could be 40% lower with lignin electrolysis. At the anode, the Kraft lignin (KL) was oxidized to aromatic chemicals by POM or $FeCl_3$, and reduced POM or Fe ions were regenerated during the electrolysis. Structure analysis of the residual KL indicated a reduction of the amount of hydroxyl groups and the cleavage of ether bonds. The results suggest that POM- or $FeCl_3$-mediated electrolysis can significantly reduce the electrolysis energy consumption in hydrogen production and, simultaneously, depolymerize lignin to low-molecular-weight value-added aromatic chemicals.

Green Chemistry

PAPER

View Article Online
View Journal | View Issue

Check for updates

Cite this: Green Chem., 2017, 19, 4939

Synergistic maximization of the carbohydrate output and lignin processability by combinatorial pretreatment†

Zhi-Hua Liu,[a,b,c] Michelle L. Olson,[d] Somnath Shinde,[e] Xin Wang,[a,b,c] Naijia Hao,[e] Chang Geun Yoo,[f] Samarthya Bhagia,[e] John R. Dunlap,[g] Yunqiao Pu,[f] Katy C. Kao,[d] Arthur J. Ragauskas,[e,f,h] Mingjie Jin[i,j] and Joshua S. Yuan *[a,b,c]

Lignocellulosic biorefineries have gained much attention worldwide as a potential solution to the challenges of energy demand and global climate change. However, the industrial implementation of biorefineries has been hindered by low fermentable sugar yields and low lignin processability. Combinatorial pretreatments with a low holding temperature were investigated in an effort to synergistically improve the carbohydrate output and lignin processability from corn stover. Upon combinatorial pretreatment with 1% H_2SO_4 for 30 min followed by 1% NaOH for 60 min at 120 °C, glucan and xylan conversion increased by 11.2% and 8.3% respectively relative to single pretreatment. This combinational pretreatment removed the amorphous portion, disrupted the rigid structure, and increased the water holding capacity of corn stover, thus increasing the hydrolysis performance. With whole fractionation by combinatorial pretreatment, glucose and xylose yields were 88.4% and 72.6%, respectively, representing increases of 10.0% and 8.1%. The lignin yield was 19.7% in the solid residue and 77.6% in the liquid stream, which increased by 33.4%. When grown in fed-batch fermentation mode, a record level of polyhydroxyalkanoate (PHA) concentration (1.0 g l^{-1}) was obtained using lignin as a carbon source by *Pseudomonas putida* KT2440. Lignin characterization results showed that combinatorial pretreatment increased the G- and H-lignin content, reduced the β–β and β-O-4 groups, and fractionated more aromatic monomers, thus facilitating lignin processability into PHA. These results highlighted the use of combinational pretreatment at a low holding temperature as a means to synergistically maximize the carbohydrate output and lignin processability, which provides a unique set of features to improve the biorefining performance.

Received 7th July 2017,
Accepted 6th September 2017

DOI: 10.1039/c7gc02057k

rsc.li/greenchem

1 Introduction

Lignocellulosic biorefineries offer a potential solution to the current energy and environmental challenges facing our generation.[1–3] Despite their potential, using biorefineries to produce biofuels and other bio-based products from lignocellulosic biomass (LCB) has yet to be fully commercialized.[4,5] Several major issues hinder the industrial implementation of biorefineries, including unsatisfactory pretreatment performance, low fermentable sugar yield, and lack of lignin utilization for fungible products.

LCB biorefining involves several technologies and processes to fractionate LCB into its primary building blocks: cellulose, hemicellulose, and lignin, which can be further converted to biofuels, chemicals, and materials. The critical challenges in biorefineries are full utilization of these three main components of the plant cell wall.[6] Generally, cellulose and hemicellulose are converted into fermentable sugars, which are

[a] *Synthetic and Systems Biology Innovation Hub (SSBiH), Texas A&M University, College Station, TX, 77843, USA. E-mail: syuan@tamu.edu*
[b] *Department of Plant Pathology and Microbiology, Texas A&M University, College Station, TX 77843, USA*
[c] *Institute for Plant Genomics and Biotechnology, Texas A&M University, College Station, TX 77843, USA*
[d] *Department of Chemical Engineering, Texas A&M University, College Station, TX 77843, USA*
[e] *Department of Chemical & Biomolecular Engineering, University of Tennessee, Knoxville, Tennessee 37996, USA*
[f] *Biosciences Division, Oak Ridge National Laboratory, Oak Ridge, Tennessee 37831, USA*
[g] *Advanced Microscopy and Imaging Center, University of Tennessee, Knoxville, TN 37996, USA*
[h] *Center for Renewable Carbon, Department of Forestry, Wildlife, and Fisheries, University of Tennessee Institute of Agriculture, Knoxville, TN 37996, USA*
[i] *School of Environmental and Biological Engineering, Nanjing University of Science and Technology, Nanjing 210094, China*
[j] *Guangdong Cleamol LTD., Foshan 528225, China*

† Electronic supplementary information (ESI) available. See DOI: 10.1039/c7gc02057k

then used to produce ethanol in cellulosic ethanol projects. In contrast, lignin is viewed as residual waste, or it is burned to produce energy with low-value utilization. Lignin (15–30%, dry weight) is the second most abundant polymer after cellulose, yet it is significantly underutilized in biorefineries and in the pulp and paper industry.[7–10] Recently, these industries have recognized that valorization of the lignin-containing residue is essential in enhancing the overall biorefinery competitiveness.[11–15] However, to achieve an appreciable product yield, all current lignin bioconversion platforms require biological or chemical fractionations of the lignin-containing biorefinery residue, which are costly and time-consuming.[11,14,16,17] The development of pretreatment technologies to both achieve higher carbohydrate output and derive more processable lignin stream thus has the potential to deliver more cost-effective and sustainable biorefineries.

Generally, most of the integrated biorefineries consist of three major unit operations: pretreatment, hydrolysis, and fermentation.[5,18] The goal of the pretreatment is to remove the structural and compositional impediments that hinder the hydrolysis efficiency and thus to increase the sugar productivity and yield lignin-rich streams. Hydrolysis is used to convert cellulose to glucose and hemicellulose to xylose, arabinose, *etc.*[7,19] During fermentation, monosaccharides and fractionated lignin can be converted into bioproducts *via* microorganisms.

Various pretreatments have been developed to deconstruct LCB and increase the hydrolysis efficiency. The pretreatment conditions mainly include catalyst types, holding temperature, and residence time.[20–24] Pretreatments vary in severity, which is ranked using a pretreatment severity (PS) score. For example, dilute acid pretreatment dissolves most of the hemicelluloses into monosaccharides, which in turn enhances the hydrolysis of cellulose. However, dilute acid pretreatment with a high PS causes irreversible degradation of hemicellulose and formation of inhibitors.[25] Acidic pretreatment may also form "pseudo-lignin" at a high holding temperature, which can hinder the hydrolysis performance and high-value utilization of LCB.[26] Liquid hot water uses only water to deconstruct LCB by the auto-hydrolysis effect at a high holding temperature. However, it usually results in incomplete LCB matrix opening and low hydrolysis efficiency. Alkaline pretreatment alters the lignin fraction to increase the hydrolysis efficiency.[27,28] Unfortunately, alkaline pretreatment degrades hemicellulose and releases monomeric lignin compounds, generating enzyme inhibitors. When one considers all these advantages and disadvantages, it is unlikely that a single pretreatment can become the universal method to maximize the yield of fermentable sugars and lignin as well as the lignin processability at the same time.

To overcome these challenges, we aimed to develop a novel pretreatment technology that would not only achieve a high carbohydrate and lignin output, but also enhance the processability of lignin containing the biorefinery residue. We hypothesized that the application of combinatorial pretreatment may enhance both carbohydrate and lignin yields and improve the lignin processability. To test this hypothesis, we exploited several combinatorial pretreatment technologies to deconstruct corn stover (CS). Pretreatment efficiency was assessed by measuring the enzymatic hydrolysis of pretreated CS. We then characterized the pretreated CS and lignin fractions to determine the effects of each pretreatment on the properties of different solids. Lignin processability was assessed by measuring the direct bioconversion of the biorefinery residue to polyhydroxyalkanoate (PHA). These analyses helped us to get a deep understanding of how combinatorial pretreatment improves the fermentable sugar output as well as lignin processability.

2 Materials and methods

2.1 Combinatorial pretreatment strategies of corn stover

The composition analysis of corn stover (CS) was conducted following the Laboratory Analysis Protocol (LAP) of the National Renewable Energy Laboratory (NREL), Golden, CO, USA. To study the effects of pre-washing on the composition transformation, CS was washed 20 times with distilled water based on its dry weight (dw). The compositions of CS are given in ESI 1.†

Seven pretreatment strategies were used, alone or in combination: dilute sulfuric acid, liquid hot water, sodium hydroxide, and ethanol pretreatment (Table 1). All pretreatments were conducted in a 1.0 L screw bottle heated by using an Amsco® LG 250 Laboratory Steam Sterilizer (Steris, USA). The

Table 1 Combinatorial pretreatment strategies of corn stover

	Step 1			PS	Step 2			PS
Case	Chemicals	Conditions	SL (w/w)	$\log R''_0$	Chemicals	Conditions	SL (w/w)	$\log R''_0$
1	1% NaOH	120 °C, 60 min	10%	8.4				
2	50% Ethanol	120 °C, 60 min	10%	3.2				
3	50% Ethanol + 1% NaOH	120 °C, 60 min	10%	8.2				
4	Liquid hot water	120 °C, 30 min	10%	2.8	1% NaOH	120 °C, 60 min	10%	8.4
5	Liquid hot water	120 °C, 30 min	10%	2.8	50% Ethanol + 1% NaOH	120 °C, 60 min	10%	8.2
6	1% H_2SO_4	120 °C, 30 min	10%	7.3	1% NaOH	120 °C, 60 min	10%	8.4
7	1% H_2SO_4	120 °C, 30 min	10%	7.3	50% Ethanol + 1% NaOH	120 °C, 60 min	10%	8.2

SL, solids loading; S1 stands for Step 1; S2 stands for Step 2; PS, pretreatment severity; % was calculated based on the weight percent, w/w.

holding temperature and pressure profiles of each pretreatment strategy are shown in ESI 2.† Fig. 1 provides a flow diagram of the fractionation process for single and combinatorial pretreatments, describing the points where fermentable sugars and lignin were obtained.

During single pretreatment (Cases 1, 2 and 3), 50 g CS (dw) were loaded into the screw bottle at 10% (w/w) solid loading for pretreatment. After pretreatment, the pretreated slurry was filtered by vacuum filtration to separate pretreated solids from the liquid stream. The pretreated solids were post-washed 10 times with distilled water. The liquid stream and washing stream were collected for further analysis. For combinatorial pretreatments (Cases 4, 5, 6, and 7), CS (loaded as described above) was pretreated first with liquid hot water or dilute sulfuric acid in Step 1, followed by EtOH + NaOH in Step 2. The morphological properties of the pretreated solids exposed to each of the pretreatments are shown in ESI 3.†

Pretreatment severity ($\log R_0$) is calculated as follows:[29]

$$\log R_0 = t \exp[(T - T_b)/\omega] \quad (1)$$

where t is the residence time, min; T is the pretreatment temperature, °C; T_b is the base temperature, 100 °C; ω is the fitted value based on the activation energy, 14.75.

The severity factor ($\log R''_0$) was used to compare the severity of combinatorial pretreatment, where the effects of pH value are taken into account (Table 1).[20]

$$\log R''_0 = \log R_0 + |\text{pH-7}| \quad (2)$$

2.2 Enzymatic hydrolysis of pretreated corn stover

The pretreated CS was hydrolyzed using the cellulase preparations Cellic CTec2 and HTec2. Filter paper activity (FPU) of Cellic CTec2 is 96 FPU ml^{-1}, while the cellobiase activity of β-glucosidase is 1270 CBU ml^{-1}. The protein content of Cellic CTec2 and HTec 2 determined by the Bradford assay is 178 ± 19.9 mg ml^{-1} and 103 ± 9.6 mg ml^{-1}, respectively.

Enzymatic hydrolysis assays were conducted at 3% (w/v) solid loading in citrate buffer solution (50 mM, pH 4.8) with an enzyme loading of 10 FPU g^{-1} solid in a 250 mL

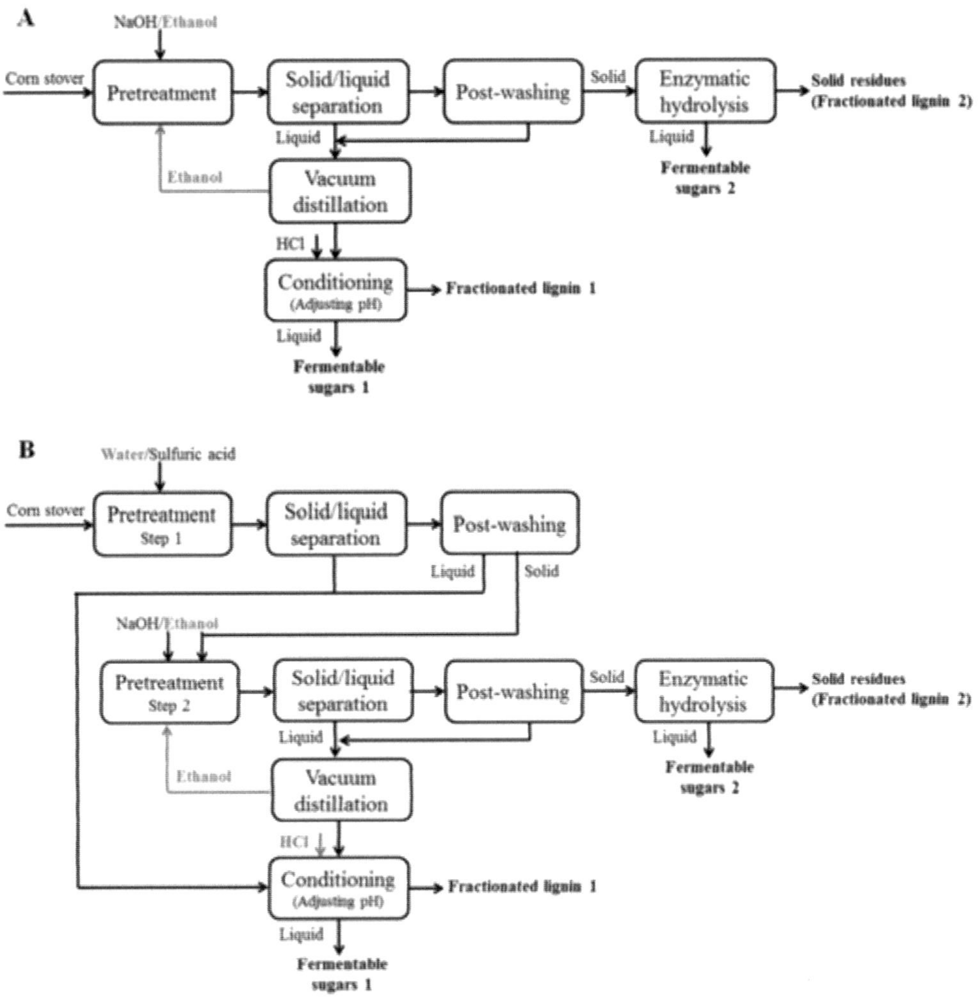

Fig. 1 The fractionation process flow diagram of fermentable sugars and lignin from corn stover by (A) single pretreatments (Cases 1–3) and (B) combinatorial pretreatments (Cases 4–7).

Erlenmeyer flask. The volumetric ratio of CTec2 : HTec2 was 10 : 1 used in hydrolysis assays. The mixture was stirred with a shaker at 50 °C and 200 rpm for 168 h. A 1.0 ml sampling of supernatant was collected at 6, 12, 24, 48, 96 and 168 h for sugar analysis. Sugar conversions are calculated as follows:

$$\text{Glucan conversion}\ (\%) = (\text{Glucose in hydrolysis} \times 162/180)/\text{Glucan in pretreated corn stover} \times 100\% \quad (3)$$

$$\text{Xylan conversion}\ (\%) = (\text{Xylose in hydrolysis} \times 132/150)/\text{Xylan in pretreated corn stover} \times 100\% \quad (4)$$

2.3 Characterization of untreated and pretreated corn stover

The biomass crystallinity of the untreated corn stover (UCS) and pretreated CS was measured using a D8 Fucos X-ray diffractometer (Bruker AXS Co., Germany). The samples with a particle size less than 125 μm were placed in a quartz sample holder using a Rigaku Miniflex diffractometer in conjunction with a CuKα radiation source (k = 0.154 nm), and then scanned at a speed of 1° min^{-1} ranging from 2θ = 5–40°, and with a step size of 0.02°. The crystallinity index (CrI) is calculated as follows:

$$\text{CrI} = [(I_{002} - I_{\text{amorphous}})/I_{002}] \times 100 \quad (5)$$

where I_{002} is the intensity for the crystalline portion of LCB (*i.e.*, cellulose) at about 2θ = 22.1°, and $I_{\text{amorphous}}$ is the peak for the amorphous portion (*i.e.*, cellulose, hemicellulose, and lignin) at about 2θ = 18.2°.

The water holding capacity (WHC) of CS solids before and after pretreatment was determined by the centrifuge separation method.[30] 1.0 g sample (dw) was mixed well with 10 ml water in a 50 mL beaker for 1 h, and then centrifuged for 5 min at a certain separation factor (F = 1000). The separation factor (F) was defined as the ratio of rotational acceleration and gravity acceleration:

$$F = \omega^2 R/g = (2\pi Rn/60)^2/Rg = Rn^2/895 \quad (6)$$

where ω is angular velocity, rad s^{-1}; R is the distance of the rotator and the axis of rotation, m; n is rotating speed, rpm; g is gravity acceleration, 9.8 m s^{-2}. WHC was defined as the adsorbed water weight per gram of dry sample.

2.4 PHA fermentation of lignin stream

The liquid stream from pretreatment, which contains mainly lignin, was used for polyhydroxyalkanoate (PHA) fermentation. The fermentation process and PHA analysis were conducted following a previous report.[31] The lignin stream was adjusted using 1.0 M HCl to pH 7.0, and then sterilized by using an autoclave at 121 °C for 20 min. For medium preparation, the lignin stream was used as a carbon source, dissolved to a certain soluble substrate concentration (SSC) with ddH_2O, and then transferred into 250 mL Erlenmeyer flasks in a working volume of 100 ml. *Pseudomonas putida* KT2440 cell pellets for inoculation were obtained by centrifuging the seed culture at 5000 rpm for 10 min. Fermentation was carried out at pH 7.0, 28 °C, and 200 rpm for 18 h.

2.5 Characterization of fractionated lignin from different pretreatments

2.5.1 2D HSQC NMR analysis. For lignin recovery from different pretreatments, the liquid stream was filtered with a 0.22 μm filter membrane and then acidified to a pH less than 2.0 with concentrated HCl. The lignin precipitates were collected in tared centrifuge tubes by centrifugation at 10 000 for 10 min. The lignin solids after acidification were washed with ddH_2O and then freeze-dried by using a lyophilizer at −55 °C for 24 h.

Native lignin from corn stover (cellulolytic enzyme lignin, CEL) was isolated according to procedures described in the previous study.[32] In brief, the extractive-free corn stover was air-dried and ball-milled using a planetary ball mill (Retsch PM 100) at 600 rpm with zirconium dioxide vessels (50 ml) containing ZrO_2 ball bearings (10 mm × 10) for 2 h (5 min grinding and 5 min break). The ball-milled CS was then hydrolyzed using Cellic CTec2 (0.1 ml g^{-1} biomass) and HTec2 (0.1 ml g^{-1} biomass) at 50 °C for 24 h. The supernatants were removed by centrifugation, and the solid residues were hydrolyzed again under the same conditions with fresh enzyme mixtures. The residue was then extracted with 96% dioxane at ambient temperature for 48 h. The CS native lignin was recovered using a rotary evaporator, freeze-dried, and vacuum dried for further analysis.

For the 2D HSQC NMR analysis, the lignin samples (30–50 mg) were dissolved in DMSO-d_6 (0.6 ml). Two-dimensional (2D) ^1H–^{13}C heteronuclear single quantum coherence (HSQC) nuclear magnetic resonance (NMR) experiment was performed using a Bruker Avance-III 400 MHz spectrometer equipped with a 5 mm Broadband Observe probe (5 mm BBO 400 MHz W1 with Z-gradient probe, Bruker) and a Bruker standard pulse sequence ('hsqcetgpsi2'). The spectra were obtained with a spectral width of 11 ppm in F2 (^1H, 2048 data points) and 190 ppm in F1 (^{13}C, 256 data points). Sixty-four scans with a 1 s delay were used for each sample.

2.5.2 ^{31}P NMR analysis. ^{31}P NMR spectra were acquired according to published methods with some modifications.[33] In detail, the lignin sample was weighed into a vial with a PTFE cap. The stock solution of pyridine/$CDCl_3$ (0.7 ml, v/v = 1.6/1), including 1.25 mg ml^{-1} $Cr(acac)_3$ and 2.5 mg ml^{-1} internal standard *endo-N*-hydroxy-5-norbene-2,3-dicarboxylic acid imide (NHND) was then added to the vial. The vial was shaken until the lignin was dissolved completely. Prior to the ^{31}P NMR experiment, 70 μl of the phosphorylating reagent 2-chloro-4,4,5,5-tetramethyl-1,3,2-dioxaphospholane (TMDP) was added to the vial. Quantitative ^{31}P NMR spectra were acquired on a Varian 500 MHz spectrometer using an inverse-gated decoupling pulse sequence, 90° pulse angle, 1.2 s acquisition time, 25 s pulse delay, and 64 scans. The phase was adjusted automatically and the baseline was corrected using a Bernstein polynomial fit with a parameter of 6. The quantitative calculation of the hydroxyl groups was based on the amount of the internal standard.

2.5.3 Gel-permeation chromatography (GPC). The lignin samples (dried under vacuum at 40 °C overnight) were acetylated with acetic anhydride/pyridine (1/1, v/v) at ambient temperature for 24 h in a sealed flask under an inert atmosphere. The lignin concentration in the solution was approximately 2 mg ml^{-1}. After 24 h, the solution was diluted with approximately 20 ml of ethanol and stirred for an additional 30 min, after which the solvents were removed with a rotary evaporator, followed by drying in a vacuum oven at 40 °C. Prior to GPC analysis, the acetylated lignin samples were dissolved in tetrahydrofuran (1.0 mg ml^{-1}), filtered through a 0.45 μm filter, and placed into a 1 ml auto-sampler vial. The molecular weight distributions of the acetylated lignin samples were analyzed on an Agilent GPC SECurity 1200 system equipped with four Waters Styragel columns (HR1, HR2, HR4, HR6), an Agilent refractive index (RI) detector, and an Agilent UV detector (270 nm), using tetrahydrofuran (THF) as the mobile phase (1.0 ml min^{-1}), with an injection volume of 20.0 μl. A standard polystyrene sample was used for calibration. The number-average molecular weight (M_n) and weight-average molecular weight (M_w) were determined by GPC.

2.5.4 Fourier transform infrared (FTIR) analysis. FTIR spectroscopy experiments of lignin samples were carried out using a Spectrum One FTIR Spectrometer from PerkinElmer (USA). The resolution was set at 4 cm^{-1}, 16 scans were recorded for each analysis and the scanning range was from 600 cm^{-1} to 4000 cm^{-1}. Five analyses were performed at five locations per sample. For calculating ratios, absorption intensities were considered and normalization was performed on the band at 1511 cm^{-1} (C=C stretching vibrations of the aromatic skeletal of lignin).

2.6 Scanning electron microscopy (SEM) of untreated and pretreated corn stover

The samples for SEM analysis were mounted onto stubs with carbon tape and sputter-coated with gold. SEM was then carried out on a Zeiss EVO MA15 at an accelerating voltage of 20 kV with a back scatter detector at 100 to 1000 times magnification. Raw images were adjusted for brightness and contrast in ImageJ software.[34] Images were merged using Adobe Photoshop CC v. 2017.

2.7 Derivatives analysis by gas chromatography-mass spectrometry (GC-MS)

The liquid streams produced from different pretreatments were filtered with a 0.22 μm filter membrane and then acidified to pH 1–2 with concentrated HCl. 3.0 ml of acidified supernatant were mixed well with 1.0 ml butanedioic acid-d_6 (0.04 mg ml^{-1}) as an internal standard, and then extracted with three volumes of methyl *tert*-butyl ether (MTBE) at 4500 rpm for 30 min. The organic layer was collected and dried under a stream of nitrogen gas. Prior to GC-MS analysis, 1.0 ml MTBE was added to dissolve the sample.

GC-MS was performed on a GCMS-QP2010SE (Shimadzu Scientific Instruments, Inc.). The analytical column used was a Shimadzu SH-Rxi-5Sil column (30 m × 250 μm × 0.25 μm). The eluted sample was analyzed using helium as carrier gas at a flow rate of 1.0 ml min^{-1}. The temperature profile of the GC-method was 3 min at 50 °C, and then it was increased to 290 °C at 8 °C min^{-1}. Ions were generated with a 70 eV electron beam at an ionization current of 40 μA. Mass spectral peak quantification was performed using GCMSsolution software Ver. 2.6. The acquired peak height was normalized against that of an internal standard.

2.8 Analysis methods and mass balance

The sugars were analyzed using HPLC in an HPLC 1260 Infinity (Agilent Technologies, CA) equipped with an Aminex HPX-87P carbohydrate analysis column (Bio-Rad Laboratories, CA) and a refractive index detector at 85 °C with HPLC grade water as the mobile phase at a flow rate of 0.6 ml min^{-1}. The sugar yield in the whole fractionation process (pretreatment + hydrolysis) was calculated as follows:

$$\text{Glucose yield}\ (\%) = \left[\text{Glucose}_{\text{Liquid of pretreatment}} + \text{GO}_{\text{Liquid of pretreatment}} \times (180/162) + \text{Glucose}_{\text{Hydrolysis}}\right] / \left[\text{Glucan}_{\text{Feedstock}} \times (180/162)\right] \quad (7)$$

$$\text{Xylose yield}\ (\%) = \left[\text{Xylose}_{\text{Liquid of pretreatment}} + \text{XO}_{\text{Liquid of pretreatment}} \times (150/132) + \text{Xylose}_{\text{Hydrolysis}}\right] / \left[\text{Xylan}_{\text{Feedstock}} \times (150/132)\right] \quad (8)$$

where GO represents the glucose oligomers and XO represents the xylose oligomers. Error bars in the figures represent the standard deviation of the replicates. Significance was determined using the Student *t*-test. A *p*-value less than 0.05 was considered significant.

3 Results and discussion

3.1 Effects of pre-washing on the composition transformation of untreated corn stover

Untreated corn stover (UCS) was analyzed to establish the baseline composition. The total sugar content in UCS was 56.4%, and the total lignin content was 22.0% (Table 2). For the biorefinery purpose, feedstock with a high sugar content has high potential to increase the biorefining efficiency.[35] Interestingly, water and ethanol extractives in UCS were 10.1% and 3.2%, respectively, and the free glucose in the water extractives was only 0.7% (ESI 1†).

Pre-washing was evaluated for an impact on the components of UCS (ESI 1†). The results showed that pre-washing increased the glucan, xylan, and lignin content. The total sugars in washed UCS were 58.8%, which increased by 4.3% compared to those in UCS. The water extractives were decreased by 50% after pre-washing. Notably, free glucose was removed almost completely, and the ash content decreased more than 36% after pre-washing, as confirmed by the

Table 2 Components of untreated corn stover and pretreated corn stover with each pretreatment

Components/%	Glucan	Xylan	Araban	Galactan	Lignin (AIL)	Lignin (ASL)	Ash	Total
UCS	33.1 (1.3)	17.5 (0.6)	0.9 (0.3)	4.4 (0.3)	19.7 (0.1)	2.3 (0.2)	3.6 (0.2)	81.5
Case 1	44.9 (1.3)	21.1 (0.5)	0.8 (0.2)	4.9 (0.5)	10.8 (1.6)	1.2 (0.1)	2.4 (0.2)	86.1
Case 2	34.4 (0.7)	19.1 (1.6)	1.4 (0.4)	4.9 (0.4)	19.4 (1.9)	1.9 (0.1)	2.6 (0.2)	83.8
Case 3	44.5 (2.1)	21.5 (1.3)	1.4 (0.5)	4.8 (0.1)	11.0 (1.3)	1.4 (0.2)	3.0 (0.1)	87.6
Case 4 S1	35.8 (1.3)	19.3 (1.1)	1.0 (0.3)	5.0 (0.1)	21.0 (0.6)	2.0 (0.1)	2.0 (0.2)	86.1
Case 4 S2	49.5 (0.5)	22.1 (0.1)	0.9 (0.1)	4.5 (0.0)	10.4 (0.7)	1.4 (0.1)	1.8 (0.3)	90.7
Case 5 S1	35.8 (1.3)	19.3 (1.1)	1.0 (0.1)	5.0 (0.1)	21.0 (0.6)	2.0 (0.2)	2.0 (0.2)	86.1
Case 5 S2	45.6 (2.3)	23.8 (0.1)	1.4 (0.6)	4.5 (0.2)	10.3 (1.4)	1.4 (0.1)	1.9 (0.3)	88.9
Case 6 S1	43.9 (0.1)	9.0 (0.0)	—	4.2 (0.2)	28.8 (1.3)	1.3 (0.0)	2.2 (0.4)	89.3
Case 6 S2	74.4 (1.9)	8.1 (0.5)	—	3.8 (0.5)	11.0 (2.4)	0.9 (0.3)	1.6 (0.3)	99.9
Case 7 S1	43.9 (0.1)	9.0 (0.0)	—	4.2 (0.2)	28.8 (1.3)	1.3 (0.1)	2.2 (0.4)	89.3
Case 7 S2	65.9 (0.9)	8.8 (0.3)	—	3.6 (0.3)	14.3 (1.6)	0.7 (0.1)	1.9 (0.3)	95.3

UCS, untreated corn stover; S1 stands for Step 1; S2 stands for Step 2; AIL, acid-insoluble lignin; ASL, acid-soluble lignin. Water and ethanol extractives are not included in the total content of UCS presented in this table. All data in the table are mean values of duplicate experiments; standard deviations are shown in parentheses.

washing stream analysis. Previous studies have confirmed that free sugars are easily degraded in the pretreatment to generate inhibitors of enzymes and microbes.[36,37] The high ash content may impede the performance of hydrolysis and fermentation.[38,39] Thus, lower free sugar and ash content should improve the bioconversion of LCB. The results indicate that pre-washing removes the non-structural components and increases the content of structural components, both of which may improve the biorefinery performance.

3.2 Combinatorial pretreatment increases the carbohydrate content

Considering that non-structural components exist in UCS, pretreatment conditions should be designed to remove them to avoid the degradation of non-structural components and improve the carbohydrate output. The hypothesis of this study was that combinatorial pretreatment with mild severity may reduce inhibitor generation by removing non-structural components first, maximize the carbohydrate release, and improve lignin processability for more cost-effective and sustainable biorefineries. In order to validate the hypothesis, four types of combinatorial pretreatments were designed and evaluated (Table 1), where liquid hot water (Cases 4 and 5) and dilute sulfuric acid (Cases 6 and 7) in Step 1 were used to remove non-structural components with a low holding temperature of 120 °C. Following Step 1, NaOH (Cases 4 and 6) or EtOH + NaOH (Cases 5 and 7) in Step 2 was used to further deconstruct CS and fractionate carbohydrate and lignin. Single pretreatment was used as a control with NaOH (Case 1), EtOH (Case 2), and EtOH + NaOH (Case 3).

The results showed that the composition of pretreated CS varies under different pretreatments (Table 2). Single pretreatment of 50% EtOH (Case 2) and liquid hot water (Cases 4S1 and 5S1) led to the lowest glucan content (about 35%). Combinatorial pretreatments with 1% H_2SO_4 followed by 1% NaOH (Case 6S2) and 50% EtOH + 1% NaOH (Case 7S2) resulted in the highest glucan content at 74.4% and 65.9%, respectively. These glucan contents were 2.3 and 2.0 times as that of UCS. Interestingly, pretreatments with 1% NaOH (Cases 1, 4S2 and 6S2) had a slightly higher glucan content than those with 50% EtOH + 1% NaOH (Cases 3, 5S2, and 7S2). Importantly, combinational pretreatments (Cases 5–7S2) resulted in a higher glucan content than single pretreatments (Cases 1–3). The increased glucan content generally indicates the removal of extractives, ash, hemicellulose, and/or lignin.[36,40] These results highlight the effectiveness of combinational pretreatments to increase the glucan content.

Pretreatments in Cases 1–5 resulted in the xylan content ranging from 19.1–23.8% while pretreatments in Cases 6–7 produced the xylan content at lower than 9.0%. The reason was that pretreatments in Cases 1–5 were more effective at dissolving the soluble components, such as extractives and ash, and thus increased the xylan content. However, pretreatment of 1% H_2SO_4 (Cases 6S1 and 7S1) dissolved most of the hemicellulose (mainly xylan) into the liquid stream, leading to the decreased xylan content.[21,41] Pretreatments with NaOH (Cases 1, 4S2, and 6S2) produced less xylan content than those with EtOH + NaOH (Cases 3, 5S2 and 7S2). These results, taken together, suggest an inverse relationship between the xylan and glucan contents. Araban and galactan contents were also evaluated for each pretreatment. Araban in pretreated CS produced from Cases 6 and 7 was not detected because pretreatment with H_2SO_4 dissolves araban. The galactan content in each pretreatment was increased by 5–20% over that in UCS.

In support of the hypothesis, the total sugars in pretreated CS reached 77.1% for Case 4, 75.3% for Case 5, 86.3% for Case 6 and 78.3% for Case 7, which were greater than those for Cases 1–3. The results suggest that combinatorial pretreatment increases the total sugar contents of UCS. Previous studies have confirmed that the high-value utilization of LCB is hindered by the high ash content.[38,39,42,43] Pretreatment step 1 in combinatorial pretreatments (Cases 4–7S1) removed most of the ash, and thus could facilitate the high-value utilization of three main components.

Different combinatorial pretreatments resulted in significant differences of sugar recovery (Fig. 2). Glucan recovery was more than 90% for pretreatment Cases 1, 2, 4–7S1, and 4–7S2.

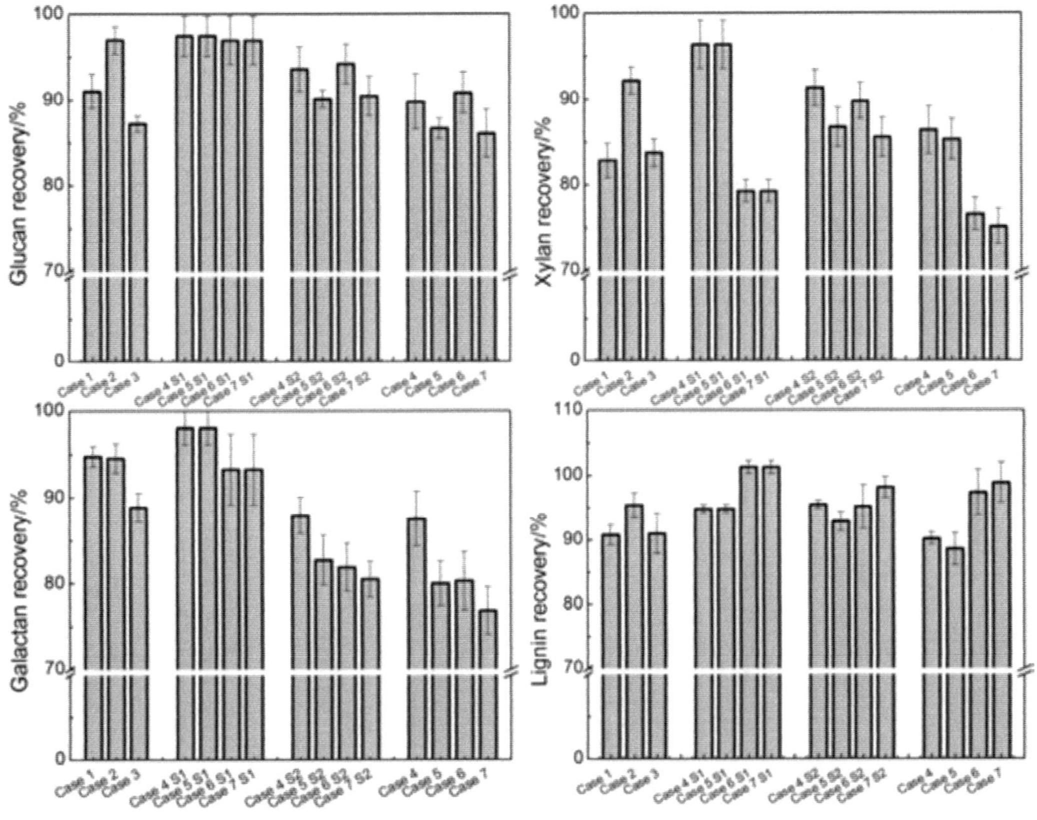

Fig. 2 The recovery of glucan, xylan, galactan, and lignin by each pretreatment of corn stover. S1 stands for Step 1; S2 stands for Step 2. The conditions of each pretreatment strategy are shown in Table 1.

Xylan recovery was more than 80% for pretreatment Cases 1–5 and 6–7S2, while it was 75–77% for Cases 6–7S1 and 6–7. This result was due to the degraded xylan in Cases 6–7S1, which was supported by the xylan content analysis. Pretreatment Cases 1, 4S2, 4, 6S2, and 6 (NaOH) obtained 3.6–4.3% more glucan, 1.2–5.2% more xylan, and 1.1–9.4% more galactan than pretreatment Cases 3, 5S2, 5, 7S2 and 7 (EtOH + NaOH). Overall, combinatorial pretreatment Cases 4–7 resulted in comparable glucan recovery compared to single pretreatments. Xylan recovery results were mixed: in Cases 6 and 7, combinatorial pretreatments recovered less xylan than single pretreatments, whereas in Cases 4 and 5, single and combinatorial pretreatments had comparable xylan recovery.

3.3 Combinatorial pretreatment dissolves more of the lignin fraction from corn stover

In biorefinery design, the pretreatment efficiency should also be a factor in lignin transformation and processability.[11,12,14] As shown in Table 2, pretreatments with EtOH (Case 2), liquid hot water (Cases 4–5S1) and H_2SO_4 (Cases 6–7S1) resulted in 19.4%, 21% and 28.8% lignin content, respectively, which was slightly higher than that in UCS. However, pretreatment in Cases 1, 3, and 4–7S2 produced the lignin content at lower than 14.3%, suggesting that the combinatorial pretreatment with NaOH was more effective at lignin fractionation into the liquid stream. The dissolution of lignin also contributed to the increased sugar contents, consisting with the above results of sugar analysis. In addition, the efficient lignin dissolution by combinatorial pretreatment will enable the bioconversion of lignin-containing waste to by-pass the additional fractionation step.

Lignin recovery was greater than 90% with all pretreatments and depended on the combinatorial pretreatments (Fig. 2). Pretreatment in Cases 6–7S1 (which included 1% H_2SO_4) led to the highest lignin recovery, while pretreatment in Cases 1 and 3 led to the lowest one. Pretreatment Cases 4S2 and 4 led to higher lignin recovery than Cases 5S2 and 5, respectively. Pretreatment Cases 6S2 and 6 resulted in lower lignin recovery than Cases 7S2 and 7. Lignin recovery by combinatorial pretreatments (Cases 4–7) was similar to or higher than that by single pretreatments (Cases 1–3). These results from combinatorial pretreatments help to recover the lignin fraction from corn stover.

3.4 Combinatorial pretreatment improves carbohydrate conversion

Pretreated CS produced from different pretreatments exhibited a range of enzymatic hydrolysis (Fig. 3). The initial glucan conversion, measured before 12 h, was the greatest in Case 7S2 (74.2%), followed by Case 5S2 (69.1%), Case 1 (65.4%), Case 3

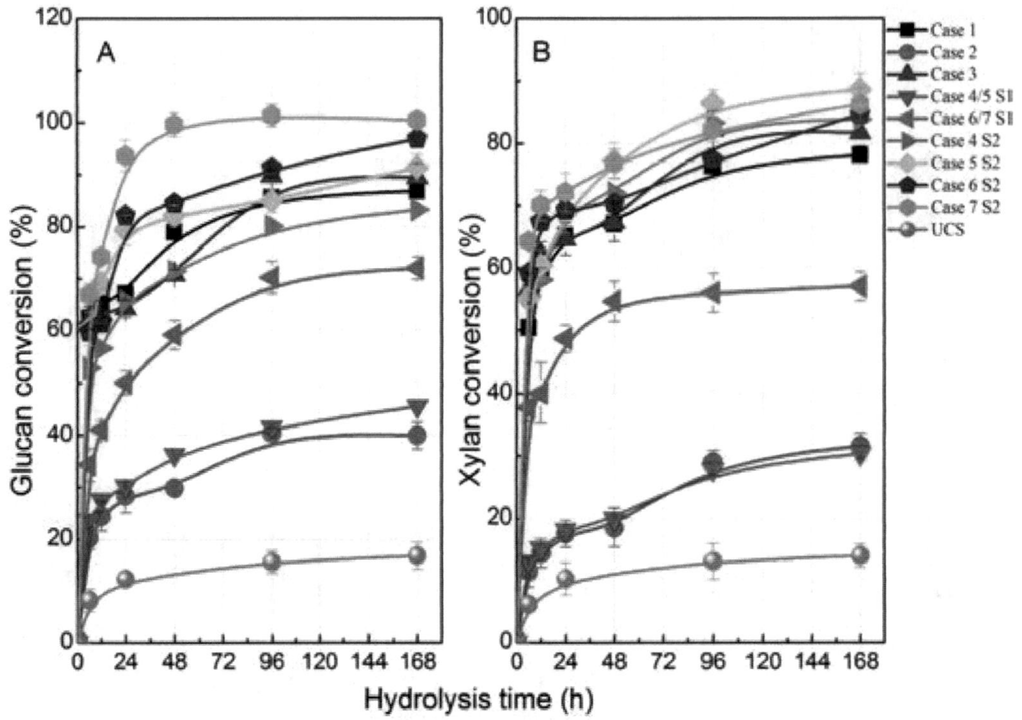

Fig. 3 Enzymatic hydrolysis of untreated corn stover (UCS) and pretreated corn stover by different pretreatments. The initial sugar conversion is calculated before 12 h hydrolysis. S1 stands for Step 1; S2 stands for Step 2. The conditions of each pretreatment strategy are shown in Table 1.

(63.2%), Case 6S2 (61.3%), Case 4S2 (54.0%), Cases 6/7S1 (41.1%), Cases 4/5S1 (27.5%), and Case 2S (24.5%). Most of the pretreatments with NaOH (Cases 1 and 6) or EtOH + NaOH (Cases 3, 5, and 7) led to more than 60% initial glucan conversion. The final glucan conversion, measured at 168 h, was the greatest in Case 7S2 (100.4%), followed by Case 6S2 (96.9%), Case 5S2 (91.4%), Case 3 (89.6%), Case 1 (86.8%), Case 4S2 (78.1%), Cases 6/7S1 (72.0%), Cases 4/5S1 (45.6%), and Case 2 (40.0%). Glucan conversion was 4.3–6.0 times higher in pretreated CS from Cases 1, 3, 6–7S1 and 4–7S2 than that in UCS. Most of the pretreatments of NaOH (Case 1 and 4) or EtOH + NaOH (Case 3, 5, and 7) resulted in more than 90% glucan conversion. Furthermore, pretreatment Cases 3, 5S2, and 7S2 had higher glucan conversion than those in Cases 1, 4S2, and 6S2, respectively. Combinational pretreatments in Cases 5–7S2 resulted in more glucan conversion than single pretreatments (Cases 1–3). Xylan conversion data had shown a similar trend. In all cases, combinational pretreatments (Cases 4–7S2) produced more than 81% xylan conversion, a value higher than that of any single pretreatments (Cases 1–3). These results highlight that combinatorial pretreatments (Cases 4–7) outperform single pretreatment on the carbohydrate conversion and improve the hydrolysis efficiency of CS.

The enzymatic hydrolysis of a LCB substrate depends on its structural property.[44,45] The biomass crystallinity result showed that the CrI value of pretreated CS increased over UCS by 7.6–18.8% for single pretreatments (Cases 1–3 and 4–7S1) and by 21.1–33.8% for combinatorial pretreatments (Cases 4–7S2) (ESI 4†). Combinational pretreatments (Cases 5–7S2) increased the CrI values by 2.4–15.2% over single pretreatments (Cases 1–3). Pretreatments in Cases 3, 5S2 and 7S2 resulted in higher CrI values than those in Cases 1, 4S2 and 6S2, respectively. The same trends were observed between CrI and glucan conversion. An increase in CrI value generally indicates the removal of the amorphous portion (*i.e.*, cellulose, hemicellulose, and lignin). The biomass crystallinity results correlate well with the composition analysis, in that combinatorial pretreatment can better remove the amorphous portion and increase the cellulose crystallinity.

Combinatorial pretreatments might have also altered the physicochemical properties of CS to increase the accessible surface area of carbohydrates to enzymes. Water holding capacity (WHC) is a measure of this property and represents the adsorption of water on/in the carbohydrates.[30,46] The WHC of the pretreated CS increased by 14.7–56.9% with single (Cases 1–3 and 4–7S1) and by 53.3–105.5% with combinatorial pretreatments (Cases 4–7S2) (ESI 5†). Interestingly, combinational pretreatments (Cases 5–7S2) showed a 14.3–46.3% higher WHC than single pretreatments (Cases 1–3). Pretreatments in Cases 3, 5S2, and 7S2 led to higher WHC than those in Cases 1, 4S2, and 6S2, correlating with the crystallinity data. The WHC trends were consistent with glucan conversion.

The ultrastructure of UCS and pretreated CS as observed through SEM imaging indicated a large variation in biomass morphology (ESI 6†). The variations divided into roughly three shapes: long fiber-like strands, long or wide cuboid-like bulky structures with near 90° edges, and ellipsoid-like bulk structures with curved edges (EIS 6C1). The results showed that there were discernible changes in the ultrastructure of the pretreated CS compared to those of UCS. All pretreatments resulted in an overall decrease in the size of CS. Ellipsoid shaped particles had the largest changes in morphology with pretreatment. These particles suggested the distorted structure of CS. The cuboid-like bulky structures indicated that the pretreatments led to delamination, seen as appearance of fibrillary surfaces. The thin fiber strands indicated disentangling of the fiber bundle at the ends. These results suggested that all pretreatment led to the deconstruction of CS and increased the accessible surface area of CS to enzymes. One interesting feature was the absence of lignin droplets that are often produced after liquid hot water and dilute acid pretreatments (Cases 4–7) due to the movement of lignin from the inner regions to the outer surface.[26,47] This was likely due to the use of alkali and/or organosolv pretreatments in Step 2 that solubilized the lignin molecules on the surface of these particles into liquid streams. This result should be helpful to increase enzymatic hydrolysis and bioconversion performance of the lignin stream.

Overall, these results suggested that combinational pretreatments removed the amorphous portion, altered the structural and compositional properties of CS, increased the accessible surface area of cellulose, and thus improved the hydrolysis performance.

3.5 Combinatorial pretreatment improves the sugar and lignin yields significantly

Techno-economic analyses have emphasized the importance of the sugar yield on the biorefinery optimization.[41,48] As shown in Fig. 4, the glucose yield from the whole fractionation process (pretreatment + hydrolysis) was the highest for Case 7 (91.4%), followed by Case 6S2 (91.3%), Case 7S2 (90.9%), Case 6 (88.4%), Case 5S2 (82.9%), Case 5 (82.8%), and Case 1 (80.5%). For all other cases, the glucose yield was less than 80%. Pretreatments with 50% EtOH + 1% NaOH (Cases 5S2, 5, 7S2, and 7) led to higher glucose yields than those with 1% NaOH (Cases 4S2, 4, 6S2, and 6), which was consistent with glucan conversion. The combinatorial pretreatment of Cases 5–7S2 and 5–7 increased glucose yields by 3.0–16.7% over single pretreatment of Cases 1 and 3. Similar trends were observed for xylose yields. The xylose yield increased by 7.2–20.1% in pretreatment Cases 4, 4S2, 6, and 6S2 over that for Case 1 with NaOH, and it increased by 4.0–11.4% for Cases

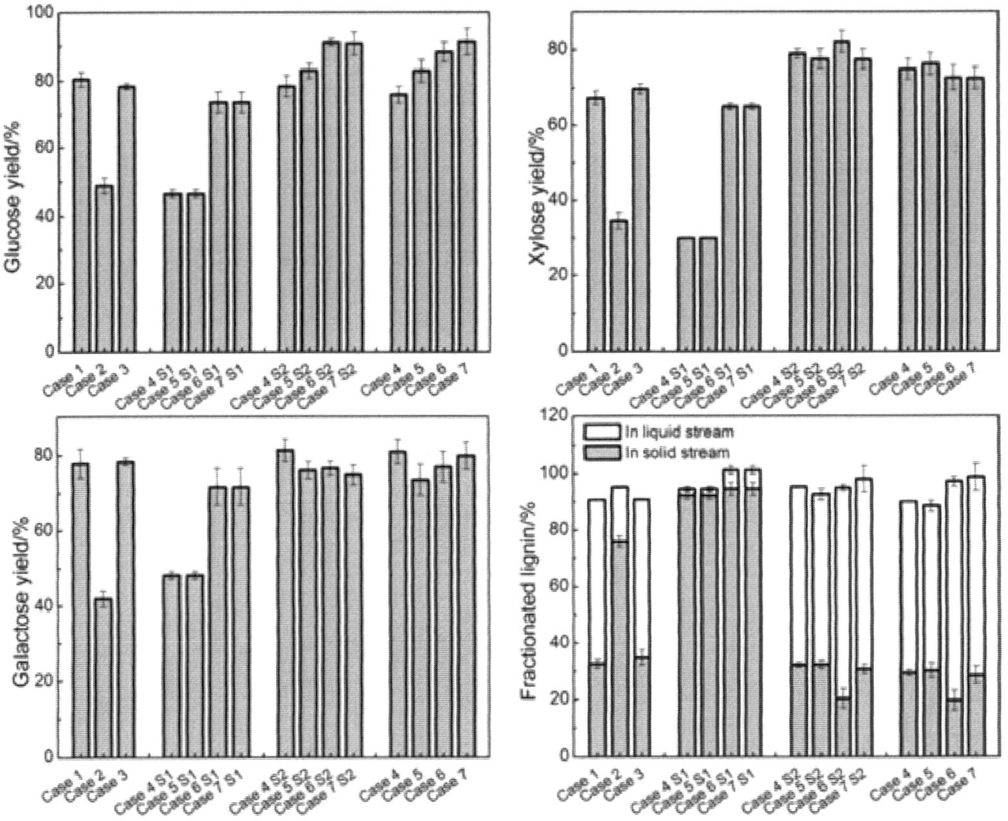

Fig. 4 Yields of fermentable sugar and fractionated lignin in the whole fractionation process (pretreatment + hydrolysis) of corn stover by different pretreatment strategies. S1 stands for Step 1; S2 stands for Step 2. The conditions of each pretreatment strategy are shown in Table 1.

5, 5S2, 7, and 7S2 over that for Case 3 with EtOH + NaOH. The total xylose yields from CS exposed to combinatorial pretreatments were 77.6–80.5% for Cases 4–7S2 and 72.1–76.3% for Cases 4–7. Thus, these results show that combinatorial pretreatments improved the fermentable sugar yields.

Lignin was measured in the solid fraction and liquid stream produced after each pretreatment. More than 90% of the lignin was retained in the residue solids after hydrolysis with pretreatment Cases 4–7S1, and 75% of the lignin was retained with pretreatment Case 2 (Fig. 4). However, only 20% lignin was retained in the residue solids after pretreatment Cases 6S2 and 6, and about 30% lignin was retained by other pretreatments. Correspondingly, more than 60% of the lignin was fractionated and dissolved into the liquid stream by combinatorial pretreatments (Cases 4–7S2 and 4–7). The highest lignin yield in the liquid stream was 77.8% produced from combinatorial pretreatment Case 6. Pretreatment in Cases 1, 4S2, 4, 6S2, and 6 (NaOH) led to 4.0–11.1% more lignin in the liquid stream than that in Cases 3, 5S2, 5, 7S2, and 7 (EtOH + NaOH). These results paralleled the trend observed for sugar recovery, but were opposite to those trends seen in the glucose yield. Compared with single pretreatment Cases 1 and 3, combinatorial pretreatments in Cases 4S2, 4, 5S2, and 5 increased the lignin yield in the liquid stream by 4.2–8.8%, while Cases 6S2, 6, 7S2, and 7 increased this yield by 20.4–33.5%. Overall, these results indicate that combinatorial pretreatments, especially Cases 6, improve the lignin output from CS, thus enhancing the biorefining performance.

3.6 Combinatorial pretreatment improves lignin valorization to polyhydroxyalkanoate (PHA) without additional fractionation

Besides the carbohydrate and lignin output, an important yet unexplored aspect of pretreatment is the processability of the lignin stream. Recent studies indicated that lignin can be further processed into high value products like PHA *via* bioconversion. Despite the progress, the product yield is still low and an additional fractionation step or enzymatic depolymerization is often needed to promote lignin degradation and conversion. Combinatorial pretreatment may improve the lignin processability and bioconversion due to two reasons. First, the combination of different conditions will maximize the lignin fractionation and solubilization, which further provides suitable carbon sources for bioconversion. Second, the low holding temperature will reduce the inhibitors to improve the fermentation performance. The hypothesis was validated by both fermentation and structure characterization of lignin.

The lignin fractionated by different pretreatments (Lignins 1–7 as labeled by their corresponding pretreatment Cases 1–7) was used to generate PHA through fermentation with *P. putida* KT2440. Working with a soluble substrate concentration (SSC) of 20 g l^{-1} and an initial OD of 0.5, the highest cell dry weight was 2.53 g l^{-1}, fermented from lignin 6 (pretreatment with 1% H_2SO_4 followed by 1% NaOH) using engineered *P. putida* KT2440 (Fig. 5A). The PHA concentration from lignins 4, 5, and 6 produced by combinatorial pretreatments was 0.66 to 1.38 times higher than that from lignin 1 (a single pretreatment) (Fig. 5B). The highest PHA concentration was 0.48 g l^{-1}, which was also fermented from lignin 6 using engineered *P. putida* KT2440. The trends for PHA content were similar to those for PHA yield (Fig. 5C and D). PHA contents from lignins 4, 5, and 6 were 0.56–1.0 times higher than that from lignin 1. The highest PHA content was 0.19 g g^{-1} dried cell, produced by fermenting lignin 6 using engineered *P. putida* KT2440. The PHA yields from lignin 4, 5, and 6 were 0.21–0.58 times higher than from lignin 1 using engineered *P. putida* KT2440. The highest PHA yield was 14.1% from lignin 6 by engineered *P. putida* KT2440. As a result, combinatorial pretreatments improved the lignin processability, as suggested by the increase in PHA fermentation performance with engineered *P. putida* KT2440.

To produce a high PHA concentration from the lignin stream, fed-batch fermentation was conducted using engineered *P. putida* KT2440 with the addition of 20 g l^{-1} SCC at 0 h plus 20 g l^{-1} SSC at 18 h (Fig. 6). The highest cell dry weight was 5.3 g l^{-1} from lignin 6 produced from the pretreatment of 1% H_2SO_4 followed by 1% NaOH. The PHA concentration pro-

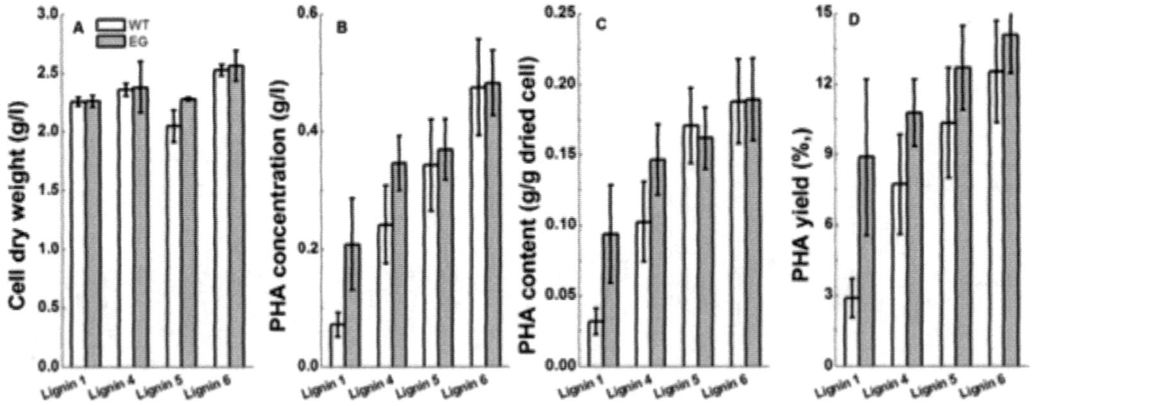

Fig. 5 PHA fermentation from the lignin stream produced by different CS pretreatments using wild type (WT) and engineered (EG) *Pseudomonas putida* KT2440. The fermentation conditions were: 20 g l^{-1} soluble substrate concentration (SSC), OD 0.5, pH 7.0, 28 °C, 200 rpm, and 18 h. "Lignin 1" represents the fractionated lignin produced by pretreatment Case 1 described in Table 1.

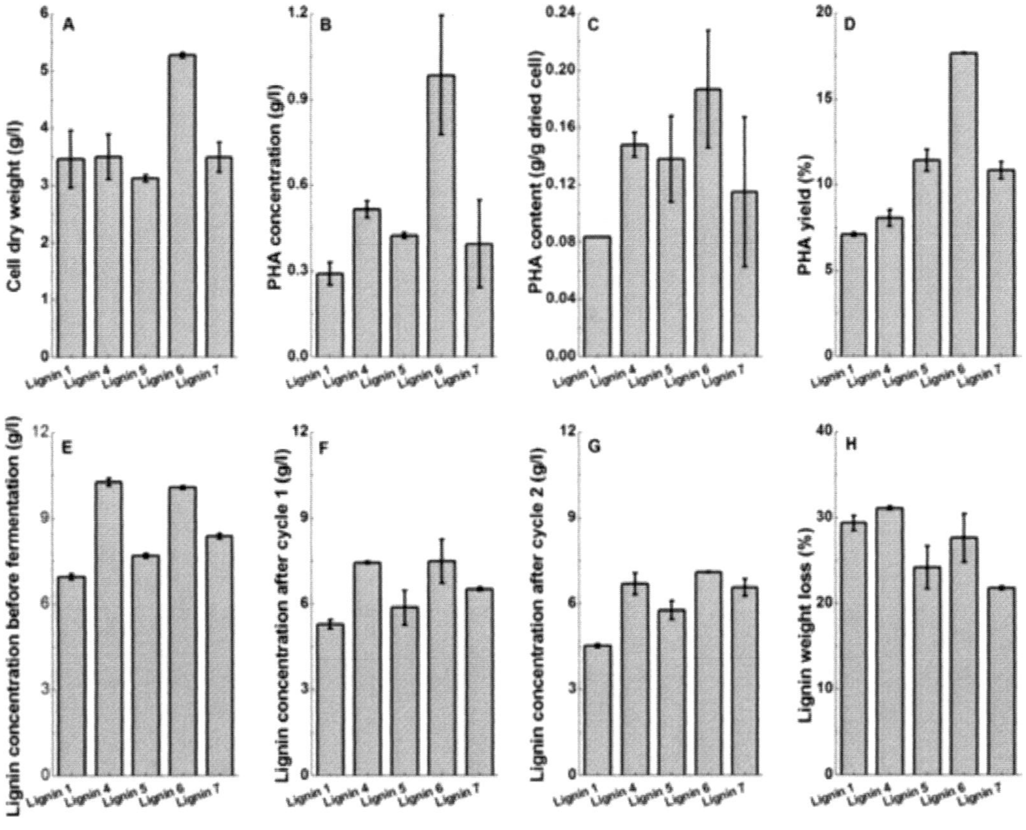

Fig. 6 PHA fermentation from the lignin streams produced by different pretreatments using engineered (EG) *Pseudomonas putida* KT2440 grown in fed-batch fermentation mode. The fermentation conditions were: 20 g l^{-1} soluble substrate concentration (SSC) in cycle 1 and 20 g l^{-1} SSC in cycle 2, OD 2.0, pH 7.0, 28 °C, 200 rpm, and 18 h for cycle 1 and 18 h for cycle 2. "Lignin 1" represents the fractionated lignin produced by pretreatment Case 1 described in Table 1.

duced from lignins 4, 5, 6, and 7 was 0.77, 0.46, 2.39, and 0.36 times higher than that from lignin 1, respectively. The highest PHA concentration was 1.0 g l^{-1}, generated from lignin 6. This result represents the highest PHA concentration reported using lignin as a carbon source for fermentation. From this sample, the PHA content was 0.19 g g^{-1} dried cell and the PHA yield was 17.6%. Notably, the PHA content and the yield from lignins 4–7 were 0.36–1.22 and 0.61–1.48 times higher than that from lignin 1, respectively. The lignin weight loss was around 30% for lignins 1, 4, and 6, which supports the results described above (Fig. 6E–H). These results highlight that combinatorial pretreatments lead to a higher cell dry weight as well as increased PHA concentration, content, and yield, thus improving the lignin bioconversion performance.

3.7 Lignin characterization using 2D- and ^{31}P-NMR revealed the mechanisms for improved lignin processability

The lignin fractionated from each pretreatment was characterized to reveal the molecular mechanisms for improved lignin processability by HSQC-NMR, quantitative ^{31}P NMR, and gel-permeation chromatography (GPC), and FTIR. Fig. 7 presents the aromatic regions of the HSQC NMR spectra, including lignin subunits and hydroxycinnamates.[33] Previous studies have confirmed that H- and G-lignin are more readily degraded and utilized by a lignin-degradable microbe than S-lignin.[11,49–51]Fig. 8 shows that fractionated lignin by a single pretreatment (Case 1) had the highest lignin S/G ratio (1.22), implying the least readily degradable lignin composition. Combinatorial pretreatments in Cases 4–7 increased the (G + H)/(G + H + S) ratios by 9.0–35% over single pretreatment in Case 1. The increases in G- and H-lignin may be the underlying cause for the improved bioconversion of lignin to PHA.

The selection of pretreatments also affected the lignin composition. Compared with the lignin fractionated using a NaOH pretreatment in Step 2 (*i.e.*, lignin 4 and 6), the lignin fractionated using EtOH + NaOH pretreatment in Step 2 (*i.e.*, lignin 5 and 7) had more hydroxycinnamates such as FA and *p*CA. Most importantly, the lignin fractioned from combinatorial pretreatments in Cases 5, 6, and 7 had more *p*CA than that from single pretreatment in Case 1. Previous studies have confirmed that FA and pCA are primary aromatic compounds produced from alkaline pretreatment, and are readily consumed by *P. putida* KT2440 to produce cell biomass and PHA.[51,52] These results demonstrate that combinatorial pretreatments increase the hydroxycinnamate content, and thus facilitate the bioconversion of lignin to PHA.

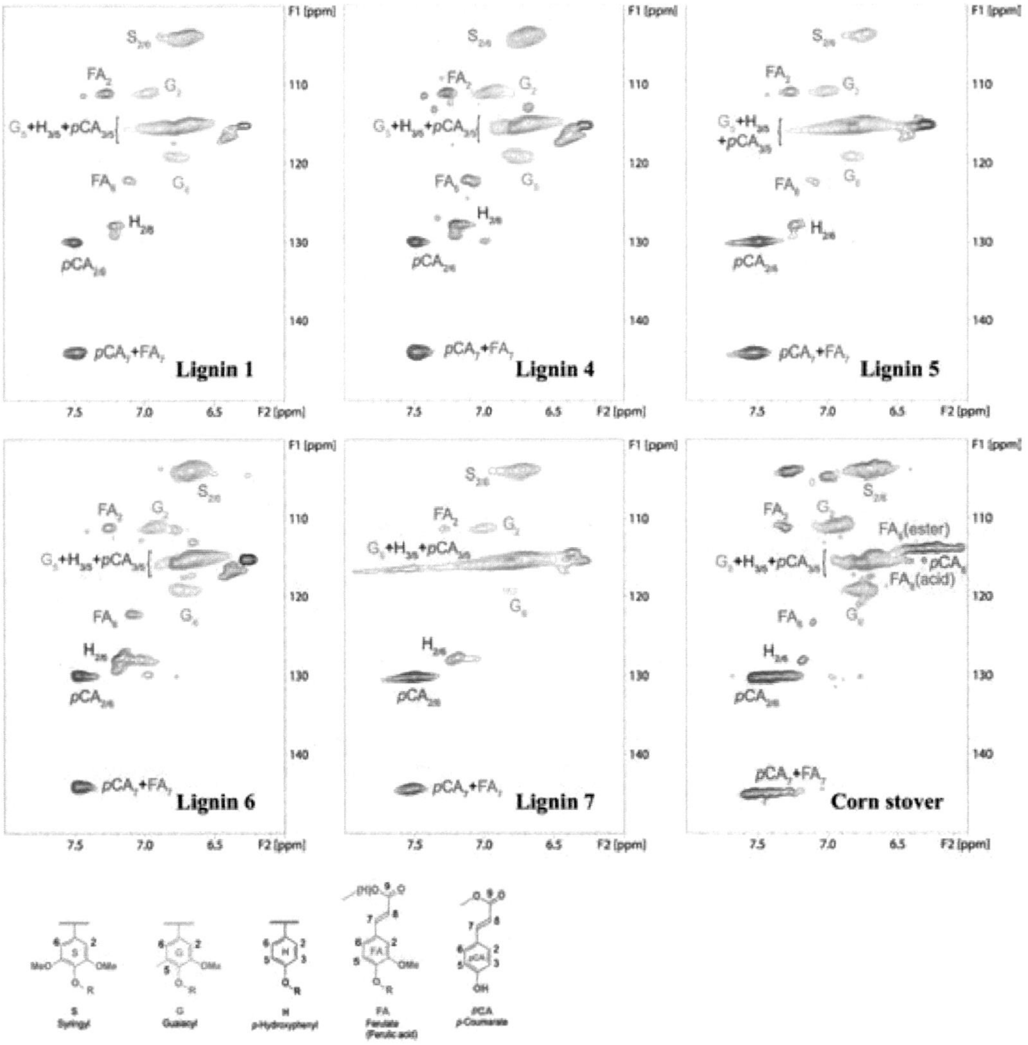

Fig. 7 Aromatic regions of 2D HSQC NMR spectra from corn stover native lignin and the fractionated lignin produced after each pretreatment. "Lignin 1" represents the fractionated lignin produced by pretreatment Case 1 described in Table 1.

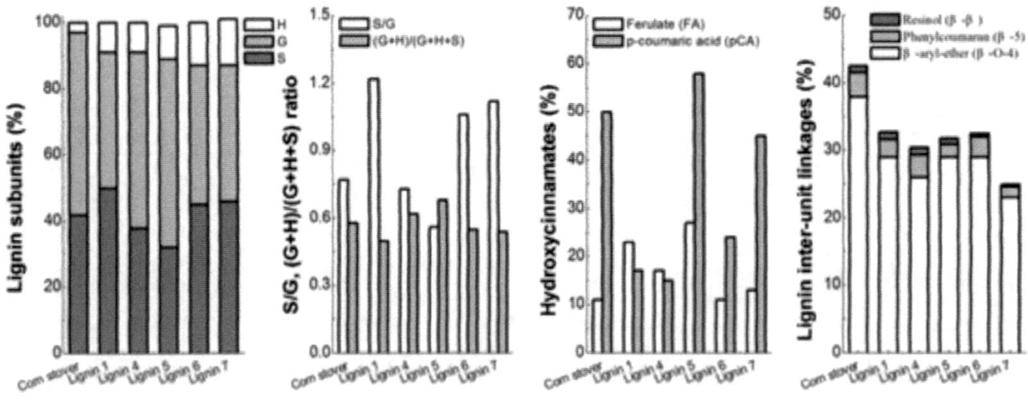

Fig. 8 Semi-quantitative information of fractionated lignin produced after each pretreatment as detected using 2D HSQC NMR. Contents were expressed as a percentage relative to the total lignin subunits (G+H+S). "Lignin 1" represents the fractionated lignin produced by pretreatment Case 1 as described in Table 1.

The occurrence of three major lignin interunit linkages (β-aryl ether, phenylcoumaran, and resinols) was estimated by measuring the correlation of the α position of β-aryl ether (β-O-4), phenylcoumaran (β-5), and resinols (β–β), respectively (Fig. 8). All pretreatments significantly decreased the amount of β-O-4 in fractionated lignin compared to that in CS native lignin. Combinatorial pretreatments led to similar amounts of β-O-4 to the single pretreatment in Case 1, or a bit less. The amounts of β-5 in the lignin produced by pretreatments in Cases 5 and 7 were less than those from pretreatments in Cases 1, 4, and 6. These results indicate that combinatorial pretreatment with EtOH + NaOH is more effective at cleaving the β-5 linkage than NaOH alone. The selection of pretreatments also affected the amount of β–β. For instance, the lignin produced by combinatorial pretreatments (Cases 4–7) decreased the amount of β–β by 18–73% compared to single pretreatment (Case 1). The lignin produced by pretreatments in Cases 6 and 7 showed less β–β than Cases 4 and 5. These results suggest that combinatorial pretreatment reduces the amounts of β–β and β-O-4 present in the lignin, indicating more degradation of the lignin polymer to facilitate the bioconversion of lignin to PHA.

The hydroxyl groups including the aliphatic, C5 substituted phenolic, guaiacyl, *p*-hydroxy phenyl, and acid hydroxyl groups in fractionated lignin and native lignin from CS were determined using ^{31}P NMR. As shown in Fig. 9, all pretreatment strategies significantly reduced the aliphatic hydroxyl groups in fractionated lignin compared with those in CS native lignin. Combinatorial pretreatments in Cases 5, 6, and 7 decreased aliphatic hydroxyl groups by 29.5–38.1% over single pretreatment in Case 1. The lignin produced by pretreatment in Cases 6 and 7 showed higher C5 substituted phenolic hydroxyl group abundance (0.52 mmol g^{-1}). It is possible that this result is caused by the condensation of G units because both syringyl and C5 condensed hydroxyl groups are counted as the C5 substituted phenolic hydroxyl groups. The guaiacyl and *p*-hydroxy phenyl phenolic hydroxyl groups from combinatorial pretreatments in Cases 5, 6, and 7 were 1.4–1.7 and 1.9–2.9 times as that from single pretreatment in Case 1, respectively. The lignin fractionated by pretreatments in Cases 6 and 7 contained 2.2 and 2.3 times more phenolic hydroxyl groups than those fractionated by pretreatment in Case 1, respectively.

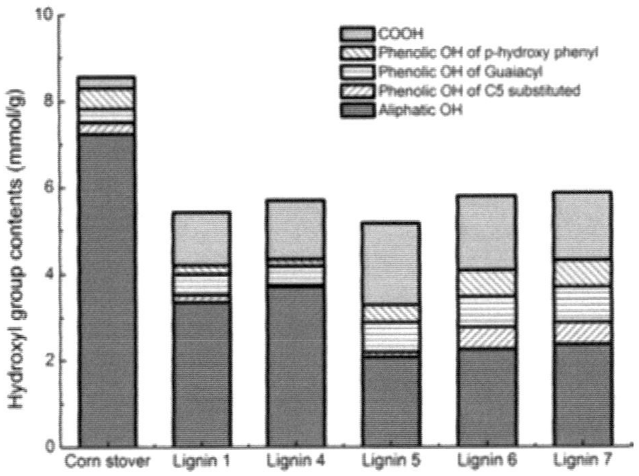

Fig. 9 Contents of hydroxyl groups in fractionated lignin produced after each pretreatment as detected using ^{31}P NMR. "Lignin 1" represents the fractionated lignin produced by pretreatment Case 1 described in Table 1.

All the pretreatments significantly increased the abundance of carboxyl groups in the fractionated lignin compared to that in the CS native lignin. The lignin fractionated by combinatorial pretreatments in Cases 4–7 contained 10.1–54.5% more carboxyl groups than that fractionated by pretreatment in Case 1. The lignin fractionated by combinatorial pretreatments in Cases 5 and 6 produced most carboxyl groups: 1.87 and 1.70 mmol g^{-1}, respectively. During biodegradation and bioconversion of lignin, fewer or shorter side chain groups containing aromatic rings could help to increase the lignin utilization efficiency by lignin-degradable microbes.[50,51] Combinatorial pretreatments led to fewer aliphatic hydroxyl groups, more phenolic hydroxyl groups and more carboxyl groups, resulting in more breakage of the lignin linkage, and thus facilitating the lignin bioconversion to PHA.

The relative intensities of the infrared spectra of the lignin from different pretreatments showed the decrease in functional groups corresponding to the removal of glucan and xylan, which was consistent with the composition analysis results (ESI 7†). Lignins 3 and 5 showed similar functionalities, while lignins 2 and 4 showed similar functionalities in the FTIR spectra. The ester bond signal at 1730 cm^{-1} was weaker in the lignin samples from all the pretreatments, suggesting that ester linkages between lignin and carbohydrates were cleaved. The band at 1640 cm^{-1} assigned to conjugated carbonyl groups of lignin showed slightly less relative adsorption in the lignin fractioned by pretreatment than that in CS native lignin. The absorption of unconjugated carbonyl groups at 1708 cm^{-1} was increased for all the fractioned lignin compared to CS native lignin. The results indicated the presence of unconjugated ketone, aldehyde and/or acid groups in all the fractioned lignin. These results were consistent with the above composition analysis and NMR results.

The molecular weight distribution of lignin fractions produced from different pretreatments was analyzed. The different pretreatment also showed an effect on the fragmentation of lignin. As shown in Fig. 10, the number-average molecular weight (M_n) of fractionated lignin ranged from 700 to 900 g mol^{-1} and the weight-average molecular weight (M_w) ranged from 1350 to 1981 g mol^{-1}, while CS native lignin showed a M_n of 1497 g mol^{-1} and a M_w of 2962 g mol^{-1}. All pretreatments significantly decreased the molecular weight of lignin. Combinatorial pretreatment in Cases 4 and 6 exhibited lower M_n and M_w of lignin than Cases 5 and 7. Generally, low molecular weight lignin is easy to convert by lignin-degradable microbes.[53,54] The lignin produced from pretreatment in Cases 4 and 6 had higher polydispersity index (PDI) values than those in Cases 5 and 7. A higher polydispersity index (PDI) value may also facilitate lignin utilization by lignin-degradable

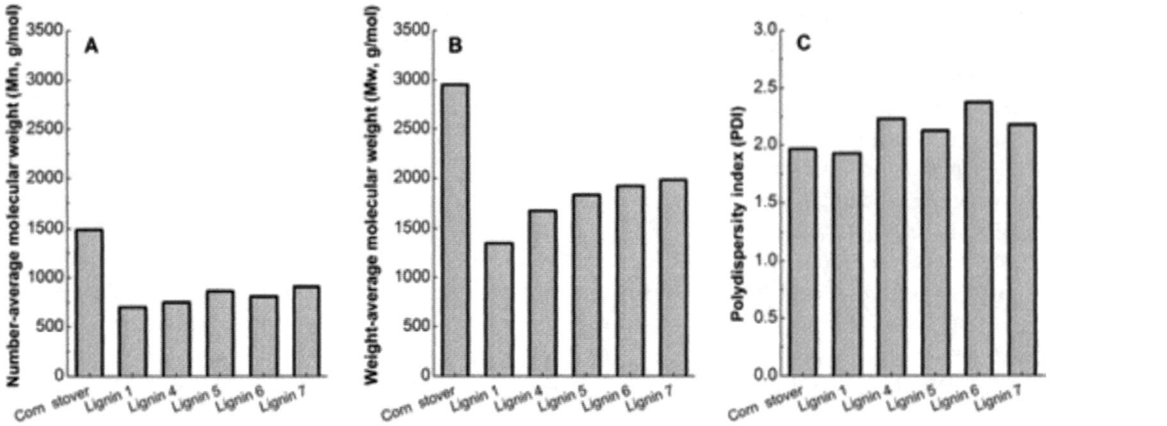

Fig. 10 Molecular weight distributions of corn stover native lignin and fractionated lignin produced after each pretreatment. "Lignin 1" represents the fractionated lignin produced by pretreatment Case 1 described in Table 1.

microbes. These results highlighted that combinatorial pretreatment reduced the molecular weight of lignin and increased the PDI value, facilitating its bioconversion to PHA. Overall, the study suggested that the bioconversion performance of lignin could be impacted by the molecular weight, PDI, the functional groups and the type of lignin.

3.8 Derivatives of carbohydrates and lignin with different pretreatments

The pretreatment of LCB normally generates derivatives, such as weak acids, furans, and phenolic compounds.[37,55,56] In particular, aromatic derivatives can be more readily processed by lignin-degraded microbes, and the amount of aromatic derivatives will impact the processability and fermentation performance of the lignin stream. The results showed that the relative abundance and the type of derivative varied with the pretreatment option (ESI 8†). Malic acid, hydroxybutyric acid and valeric acid were the major non-volatile carboxylic acid derivatives formed, accounting for 60–95% of the quantifiable carboxylic acid detected. The pretreatments at a low holding temperature produced few furan aldehydes, which are usually not stable under alkaline conditions.[55] Combinatorial pretreatments in Cases 4–7S2 led to higher concentrations of carboxylic acids and furan aldehydes than those in Cases 1 and 3, likely due to their higher pretreatment severity. However, the concentrations of weak acids and furans produced after combinatorial pretreatments were obviously lower compared to those in previous reports.[36]

In total, 35 different aromatic monomers were quantified in the liquid stream from the different pretreatments. *p*-Coumaric acid, ferulic acid, 4-hydroxy-3-methyl-acetophenone, 4-vinylguaiacol, coumarone, acetoxycinnamic acid, ethyl ferulate, vanillin, guaiacol, and syringaldehyde were the major derivatives that formed from the lignin polymer, which accounted for about 67–81% of the quantifiable derivatives detected. The aromatic monomers produced from pretreatments in Cases 3, 5S2, and 7S2 were 6.3, 1.6, and 1.7 times higher than those in Cases 1, 4S2, and 6S2, respectively. Notably, the evaporation and concentration effects caused by ethanol distillation after pretreatment (Cases 3, 5, and 7) should contribute to this result (Fig. 1). Combinatorial pretreatments in Cases 4S2 and 6S2 produced 4.2 and 3.1 times more aromatic monomers than single pretreatment in Case 1. Aromatic monomers are more easily converted by *P. putida* KT2440 compared to the high molecular weight lignin polymer.[31,51,57] Therefore, combinatorial pretreatment produced more aromatic monomers, and thus facilitated lignin bioconversion to PHA.

3.9 Combinatorial pretreatment holds promise for improving biorefining efficiency and cost-effectiveness

Overall, combinatorial pretreatments offered several significant advantages over single pretreatment. Firstly, combinatorial pretreatment increased the carbohydrate output. The mass balance of the whole fractionation process (pretreatment + hydrolysis) showed that combinatorial pretreatment in Case 6 led to 96.6% glucan, 84.6% xylan, and 89.2% galactan conversions, which increased by 11.2%, 8.3%, and 21.0%, respectively, over single pretreatment in Case 1 (Fig. 11). In the whole fractionation process, glucose and xylose yields were 88.4% and 72.6%, respectively, representing increases of 10.0% and 8.1% over single pretreatment in Case 1. On one hand, pretreatment Step 1 removed degradable components, water extractives, ash, and other impurities, which reduced the formation of potential inhibitors. On the other hand, pretreatment Step 2 further deconstructed CS and improved hydrolysis performance as indicated by the high WHC and the SEM image.

Secondly, combinatorial pretreatment enhanced lignin processability and bioconversion performance for PHA. The lignin yield from combinatorial pretreatment in Case 6 was 77.6% in the liquid stream, which represents an increase of 33.4% over single pretreatment. The lignin stream was used by *P. putida* KT2440 to produce PHA. In the fed-batch fermentation mode,

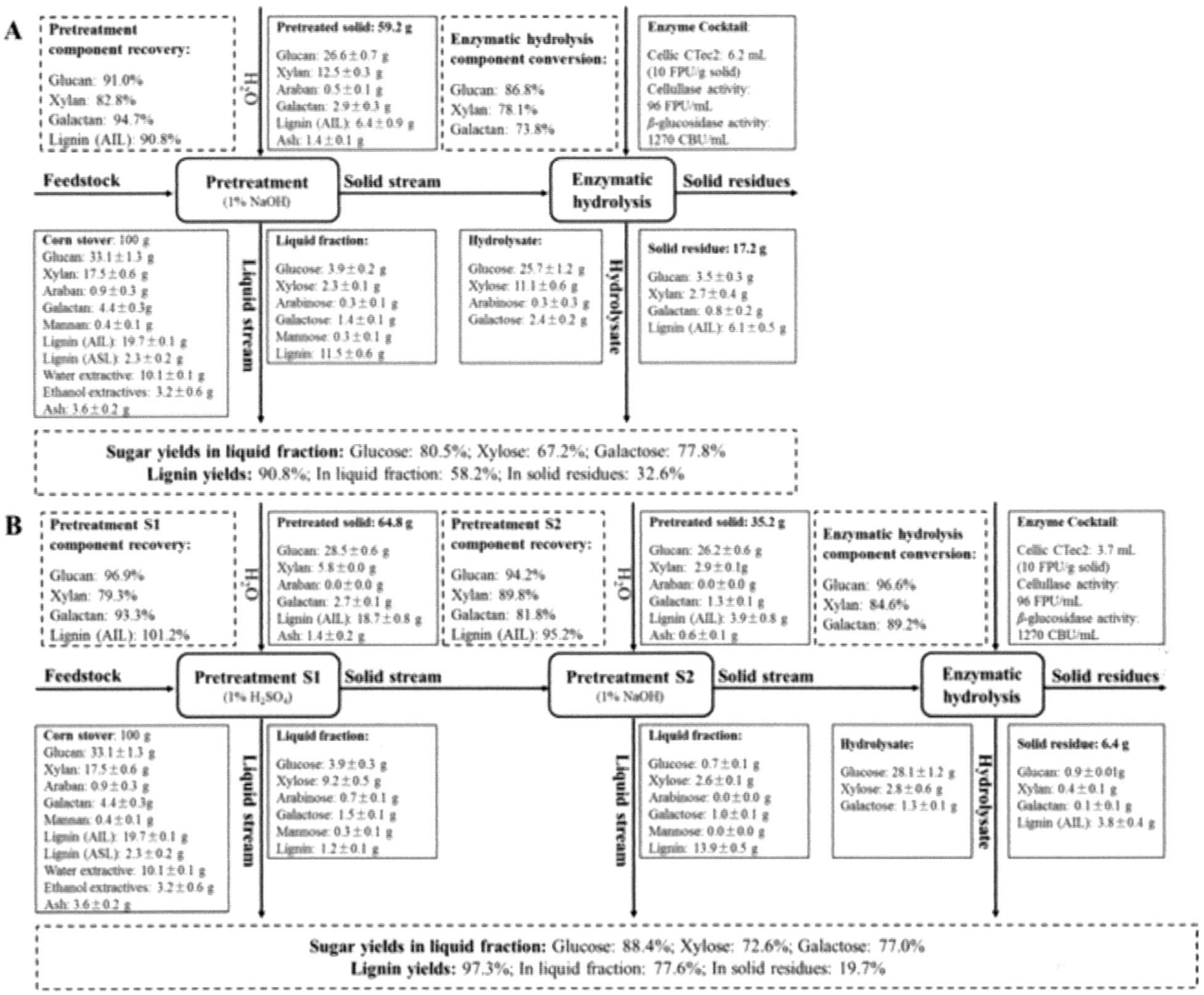

Fig. 11 Mass balance around the whole fractionation process of corn stover by single (A, Case 1) and combinatorial pretreatment (B, Case 6) strategies. S1 stands for Step 1; S2 stands for Step 2.

a record PHA concentration (1.0 g l^{-1}) was produced. Such an increase of lignin processability could be due to the high G- and H-type lignin content, the improved cleavage of β–β and β-O-4 linkages, the broader distribution of the molecular weight, the reduced aliphatic hydroxyl groups, and the increased phenolic hydroxyl and carboxyl groups in the lignin. Additionally, the reduced inhibitor and non-structural components in the lignin stream further improved the lignin processability compared to traditional pretreatment, contributing the record PHA yield for biorefinery waste utilization.

Thirdly, the low holding temperature of combinatorial pretreatment offers a range of advantages. Conventional hydrothermal pretreatment is carried out between 160–240 °C holding temperature.[5,15,18] Such a high holding temperature leads to increased pretreatment severity and improved hydrolysis performance, but it also leads to sugar loss, high-energy consumption and inhibitor formation.[15,18] Combinatorial pretreatment here was conducted at a low holding temperature (120 °C), which reduced sugar degradation, inhibitor generation, and the need for energy consumption (Fig. 1 and ESI 2†).

Fourth, conventional acidic or alkaline pretreatments require conditioning of the liquid stream before fermentation. Combinatorial pretreatments using 1% H_2SO_4 in Step 1 and 1% NaOH in Step 2 helped to eliminate the need for conditioning. Instead, the liquid streams from Step 1 and Step 2 were mixed, which balanced the pH (Fig. 1). In addition, the use of acid pretreatment followed by base pretreatment allowed the maximum dissolving of lignin and avoided the lignin condensation and precipitation. Together with the low holding temperature, the combination of acid and base pretreatments also avoids the formation of pseudo-lignin, a major cause of low hydrolysis efficiency in acid pretreatment.[26,36] Fifth, the ethanol can be recycled by distillation and reused, reducing water consumption.[58] The use of ethanol as a model organic solvent proves the concept that other solvents more readily for recovery can also be used in the process. Altogether, these advantages highlighted that combinatorial pretreatment synergistically increased the carbohydrate output and enhanced the lignin processability, which would open new avenues for optimizing biorefining performance, cost-effectiveness, and sustainability. The research thus proved an important concept for synergistic maximization of carbohydrate output and lignin processability *via* combinatorial pretreatment at a low holding temperature. Further optimization can be carried out by evaluating the combination of different chemical concentrations, holding temperatures, and residence time.

4 Conclusions

Combinatorial pretreatments with a low holding temperature were developed to synergistically increase both the carbohydrate output and the lignin processability. Using combinatorial pretreatments with 1% H_2SO_4 followed by 1% NaOH, glucose and xylose yields were 88.4% and 72.6%, respectively, representing an increase of 10.0% and 8.1% over single NaOH pretreatment. The lignin yield from this combinatorial pretreatment was 77.6% in the liquid stream, representing an increase of 33.4%. Using *P. putida* KT2440, the highest observed PHA concentration was 1.0 g l^{-1}, with a 17.6% yield. Combinational pretreatments maximized the output of both carbohydrate and lignin and enhanced the lignin processability with a record level of PHA concentration. In summary, combinational pretreatments may improve energy production from CS and maximize its carbon utilization in biorefinery applications.

Author contributions

ZHL and JSY designed the study. ZHL carried out the experiments, performed the statistical analysis, and drafted the manuscript. MLO and KK determined the sugar samples. XW determined the derivative samples. SS, NH, CY, YP, and AJR determined the lignin characterization. SB and JRD conducted SEM analysis. MJ and JSY revised the manuscript. All authors provided critical inputs to the manuscript and read and approved the final draft.

Conflicts of interest

There are no conflicts to declare.

Acknowledgements

The work was supported financially by the U.S. DOE (Department of Energy) EERE (Energy Efficiency and Renewable Energy) BETO (Bioenergy Technology Office) (DE-EE0007104 and DE-EE0006112).

References

1 M. E. Himmel, S.-Y. Ding, D. K. Johnson, W. S. Adney, M. R. Nimlos, J. W. Brady and T. D. Foust, *Science*, 2007, **315**, 804–807.
2 S. Chu and A. Majumdar, *Nature*, 2012, **488**, 294–303.
3 D. M. Alonso, S. G. Wettstein, M. A. Mellmer, E. I. Gurbuz and J. A. Dumesic, *Energy Environ. Sci.*, 2013, **6**, 76–80.
4 M. J. Jin, C. Gunawan, N. Uppugundla, V. Balan and B. E. Dale, *Energy Environ. Sci.*, 2012, **5**, 7168–7175.
5 J. J. Bozell and G. R. Petersen, *Green Chem.*, 2010, **12**, 539–554.
6 P. C. A. Bruijnincx, R. Rinaldi and B. M. Weckhuysen, *Green Chem.*, 2015, **17**, 4860–4861.
7 P. Azadi, O. R. Inderwildi, R. Farnood and D. A. King, *Renewable Sustainable Energy Rev.*, 2013, **21**, 506–523.
8 D. S. Argyropoulos and C. Crestini, *ACS Sustainable Chem. Eng.*, 2016, **4**, 5089–5089.
9 L. D. Sousa, M. J. Jin, S. P. S. Chundawat, V. Bokade, X. Y. Tang, A. Azarpira, F. C. Lu, U. Avci, J. Humpula, N. Uppugundla, C. Gunawan, S. Pattathil, A. M. Cheh, N. Kothari, R. Kumar, J. Ralph, M. G. Hahn, C. E. Wyman, S. Singh, B. A. Simmons, B. E. Dale and V. Balan, *Energy Environ. Sci.*, 2016, **9**, 1215–1223.
10 K. H. Kim, B. A. Simmons and S. Singh, *Green Chem.*, 2017, **19**, 215–224.
11 G. T. Beckham, C. W. Johnson, E. M. Karp, D. Salvachúa and D. R. Vardon, *Curr. Opin. Biotechnol.*, 2016, **42**, 40–53.
12 A. J. Ragauskas, G. T. Beckham, M. J. Biddy, R. Chandra, F. Chen, M. F. Davis, B. H. Davison, R. A. Dixon, P. Gilna and M. Keller, *Science*, 2014, **344**, 1246843.
13 S. Xie, A. J. Ragauskas and J. S. Yuan, *Ind. Biotechnol.*, 2016, **12**, 161–167.
14 M. De Bruyn, J. Fan, V. L. Budarin, D. J. Macquarrie, L. D. Gomez, R. Simister, T. J. Farmer, W. D. Raverty, S. J. McQueen-Mason and J. H. Clark, *Energy Environ. Sci.*, 2016, **9**, 2571–2574.
15 D. Salvachua, E. M. Karp, C. T. Nimlos, D. R. Vardon and G. T. Beckham, *Green Chem.*, 2015, **17**, 4951–4967.
16 Z. Chen and C. Wan, *Renewable Sustainable Energy Rev.*, 2017, **73**, 610–621.
17 S. X. Xie, Q. Li, P. Karki, F. J. Zhou and J. S. Yuan, *ACS Sustainable Chem. Eng.*, 2017, **5**, 2817–2823.
18 Z. Y. Zhang, M. D. Harrison, D. W. Rackemann, W. O. S. Doherty and I. M. O'Hara, *Green Chem.*, 2016, **18**, 360–381.
19 S. Roth and A. C. Spiess, *Bioprocess Biosyst. Eng.*, 2015, **38**, 2285–2313.
20 M. Pedersen and A. S. Meyer, *New Biotechnol.*, 2010, **27**, 739–750.
21 N. Mosier, C. Wyman, B. Dale, R. Elander, Y. Y. Lee, M. Holtzapple and M. Ladisch, *Bioresour. Technol.*, 2005, **96**, 673–686.
22 Z. H. Liu, L. Qin, B. Z. Li and Y. J. Yuan, *ACS Sustainable Chem. Eng.*, 2015, **3**, 140–146.

23 X. Z. Meng, T. Wells, Q. N. Sun, F. Huang and A. Ragauskas, *Green Chem.*, 2015, **17**, 4239–4246.
24 A. Mittal, T. B. Vinzant, R. Brunecky, S. K. Black, H. M. Pilath, M. E. Himmel and D. K. Johnson, *Green Chem.*, 2015, **17**, 1546–1558.
25 Y. Q. He, J. Zhang and J. Bao, *Biotechnol. Biofuels*, 2016, **9**, 19.
26 P. Sannigrahi, D. H. Kim, S. Jung and A. Ragauskas, *Energy Environ. Sci.*, 2011, **4**, 1306–1310.
27 Y. Chen, M. A. Stevens, Y. M. Zhu, J. Holmes and H. Xu, *Biotechnol. Biofuels*, 2013, **6**, 8.
28 R. J. Stoklosa, A. del Pilar Orjuela, L. da Costa Sousa, N. Uppugundla, D. L. Williams, B. E. Dale, D. B. Hodge and V. Balan, *Bioresour. Technol.*, 2017, **226**, 9–17.
29 R. P. Overend and E. Chornet, *Philos. Trans. R. Soc. London, Ser. A*, 1987, **321**, 523–536.
30 Z. H. Liu and H. Z. Chen, *Bioresour. Technol.*, 2015, **193**, 345–356.
31 L. Lin, Y. Cheng, Y. Pu, S. Sun, X. Li, M. Jin, E. A. Pierson, D. C. Gross, B. E. Dale, S. Y. Dai, A. J. Ragauskas and J. S. Yuan, *Green Chem.*, 2016, **18**, 5536–5547.
32 C. G. Yoo, M. Li, X. Z. Meng, Y. Q. Pu and A. J. Ragauskas, *Green Chem.*, 2017, **19**, 2006–2016.
33 P. Yunqiao, C. Shilin and R. J. Arthur, *Energy Environ. Sci.*, 2011, **4**, 3154–3166.
34 C. A. Schneider, W. S. Rasband and K. W. Eliceiri, *Nat. Methods*, 2012, **9**, 671–675.
35 C. Krishnan, L. D. Sousa, M. J. Jin, L. P. Chang, B. E. Dale and V. Balan, *Biotechnol. Bioeng.*, 2010, **107**, 441–450.
36 Z. H. Liu, L. Qin, M. J. Jin, F. Pang, B. Z. Li, Y. Kang, B. E. Dale and Y. J. Yuan, *Bioresour. Technol.*, 2013, **132**, 5–15.
37 L. J. Jonsson, B. Alriksson and N. O. Nilvebrant, *Biotechnol. Biofuels*, 2013, **6**, 16.
38 C. A. Mullen, A. A. Boateng, R. B. Dadson and F. M. Hashem, *Energy Fuels*, 2014, **28**, 7014–7024.
39 B. Yu and H. Z. Chen, *Bioresour. Technol.*, 2010, **101**, 9114–9119.
40 D. Szczerbowski, A. P. Pitarelo, A. Zandoná Filho and L. P. Ramos, *Carbohydr. Polym.*, 2014, **114**, 95–101.
41 M. Kapoor, S. Soam, R. Agrawal, R. P. Gupta, D. K. Tuli and R. Kumar, *Bioresour. Technol.*, 2017, **224**, 688–693.
42 H. Z. Chen and Z. H. Liu, *Biotechnol. Biofuels*, 2014, **7**, 53.
43 S. V. Vassilev, D. Baxter, L. K. Andersen and C. G. Vassileva, *Fuel*, 2013, **105**, 40–76.
44 S. Park, J. O. Baker, M. E. Himmel, P. A. Parilla and D. K. Johnson, *Biotechnol. Biofuels*, 2010, **3**, 10.
45 H. Z. Chen and Z. H. Liu, *Biotechnol. J.*, 2015, **10**, 866–885.
46 Z. H. Liu and H. Z. Chen, *ACS Sustainable Chem. Eng.*, 2016, **4**, 1274–1285.
47 M. J. Selig, S. Viamajala, S. R. Decker, M. P. Tucker, M. E. Himmel and T. B. Vinzant, *Biotechnol. Prog.*, 2007, **23**, 1333–1339.
48 Z. H. Liu, L. Qin, J. Q. Zhu, B. Z. Li and Y. J. Yuan, *Biotechnol. Biofuels*, 2014, **7**, 167.
49 R. K. Le, T. Wells, P. Das, X. Z. Meng, R. J. Stoklosa, A. Bhalla, D. B. Hodge, J. S. Yuan and A. J. Ragauskas, *RSC Adv.*, 2017, **7**, 4108–4115.
50 T. Wells and A. J. Ragauskas, *Trends Biotechnol.*, 2012, **30**, 627–637.
51 J. G. Linger, D. R. Vardon, M. T. Guarnieri, E. M. Karp, G. B. Hunsinger, M. A. Franden, C. W. Johnson, G. Chupka, T. J. Strathmann and P. T. Pienkos, *Proc. Natl. Acad. Sci. U. S. A.*, 2014, **111**, 12013–12018.
52 A. Rodriguez, D. Salvachúa, R. Katahira, B. A. Black, N. S. Cleveland, M. Reed, H. Smith, E. E. K. Baidoo, J. D. Keasling, B. A. Simmons, G. T. Beckham and J. M. Gladden, *ACS Sustainable Chem. Eng.*, 2017, **5**, 8171–8180.
53 C. Zhao, S. Xie, Y. Pu, R. Zhang, F. Huang, A. J. Ragauskas and J. S. Yuan, *Green Chem.*, 2016, **18**, 1306–1312.
54 Z. Wei, G. Zeng, F. Huang, M. Kosa, D. Huang and A. J. Ragauskas, *Green Chem.*, 2015, **17**, 2784–2789.
55 H. B. Klinke, B. K. Ahring, A. S. Schmidt and A. B. Thomsen, *Bioresour. Technol.*, 2002, **82**, 15–26.
56 E. Palmqvist and B. Hahn-Hagerdal, *Bioresour. Technol.*, 2000, **74**, 17–24.
57 M. E. Brown and M. C. Chang, *Curr. Opin. Chem. Biol.*, 2014, **19**, 1–7.
58 X. B. Zhao, S. M. Li, R. C. Wu and D. H. Liu, *Biofuels, Bioprod. Biorefin.*, 2017, **11**, 567–590.

501845 Cleavage of β-*O*-4 Ether Bonds in Acidic Lithium Bromide Trihydrate for Lignin Depolymerization

Abstract:

Depolymerization of lignin is essential in pulping and biorefining processes. It was observed in our previous studies that lignin could be extensively depolymerized in acidic lithium bromide trihydrate media. In this study, the mechanism of lignin depolymerization in acidic lithium bromide trihydrate was investigated. First, it was demonstrated that the native lignin in lignocellulosic biomass was easily depolymerized under mild conditions (e.g. in 60 % LiBr with 40 mM HCl at 110 °C). Six types of biomass, including hardwood, softwood, and herbaceous biomass, were treated in acidic lithium bromide trihydrate, and the isolated lignin fractions were characterized with GPC and NMR. The result indicated that the β-*O*-4 ether linkages of the native lignin were selectively cleaved, leading to lignin depolymerization (M*w*: 2000-4000). Second, lignin model compound (guaiacylglycerol-β-guaiacyl ether) was used to elucidate the cleavage mechanism of the β-*O*-4 ether bonds. It was found that the ether bonds were cleaved via the enol ether pathway initiated by the a-C catioinzation, and both LiBr and HCl catalyzed the cleavage. In addition, new C-C bonds were formed, resulting from the nucleophilic attack of the benzyl carbocations by the electron-rich lignin aromatic rings. Third, this lignin depolymerization strategy was applied to technical lignins isolated from pulping and biorefining processes. It was found that the residual β-*O*-4 ether linkages of the lignin could be selectively cleaved as well.

501979 Oxidative Functionalization of Lignins for the Synthesis of Polyamides and Polyesters

Abstract:

Lignin, a renewable resource available from lignocellulosic biomass, has the potential to replace petrochemical feedstock in the production of synthetic polymers. Valorization of lignin coproduct from enzymatic conversion offers opportunity to improve the economic feasibility of lignocellulosic sugar production. Applications of lignin as a polymer material, however, is hindered by brittleness and thermal instability of the material. Unless blended with other polymers (e.g. PLA, PP, PVA, PET) or added as a filler in composites, technical lignins alone are not suited for material applications. The complexity in lignin structure and condensation chemistry also complicates the use of lignin as a source of polymer building blocks.

We here present a new process that converts a wide variety of lignins to high molecular weight (>10,000 Da) polymer that contains over 80% lignin by weight. In this process, enzymatic lignin and organosolv lignin are functionalized via Cu-catalyzed oxidation to form oligomers with acyl chloride groups. Using these oligomers and difunctional linkers, we have successfully synthesized block copolymers of lignin that contain amide and ester linkages.

Characterization of the synthesized polymer demonstrated the preservation of native beta-aryl ether linkages in lignin, while FTIR data confirms the formation of amide and ester linkages. Our studies on lignin polymerization chemistry revealed the association between polymerization parameters and mechanical properties. This new polymer material has potential in coating, adhesives, and drug delivery applications.

497117 Lignin Aerobic Oxidation Catalyzed By Sn ( ) Using Dioxygen in Biphasic System

Abstract:

Lignin aerobic oxidation catalyzed by Sn ( ) using dioxygen in biphasic system

Chao Liu, Fei Lin, Huiyan Zhang, Rui Xiao*

Key Laboratory of Energy Thermal Conversion and Control of Ministry of Education, Southeast University, Nanjing 210096, PR China

*Corresponding author. Tel.: +86-25-83795726. Fax: +86-25-57714489. E-mail: ruixiao@seu.edu.cn.

Abstract: In recent years, with the shortage of fossil energy, reasonable exploitation and utilization of renewable biomass resource have attracted wide attention in the field of energy and chemical industry. Lignin, as the second most abundant natural polymer on the earth, is an important bioresource. It accounts for 15-35% content and 40% energy content of lignocellulose biomass, and is the only renewable aromatic resource. However, so huge potential valued organic polymer was usually directly subjected to combustion as a fuel conventionally, which means the high value-added utilization of lignin has not been completely achieved. Therefore, it is necessary to develop the rational and valuable utilization of lignin in the field of biomass conversion. Herein, thermochemical conversion is regarded as the effective methods compared with biochemical conversion. However, the shortcomings of pyrolysis and hydrogenolysis, such as the complex products and harsh reaction conditions, limit their development. Thus, attentions turn to be paid on lignin oxidative degradation. In this study, a biorefinery lignin obtained from the hydrothermal treatment before cooking was selected as the rare material to investigate the aerobic oxidation characteristics of lignin using Sn ( ) as the catalyst and dioxygen as the oxidant in biphasic system. The oxidative degradation was carried out in a bath reactor. After reaction, the gases were collected and off-line determined by gas chromatography. For further analysis, the mixture was filtrated, rotary evaporated, and then the internal standard substance (n-dodecane) was added. Results showed that the char yield was less than 5%, which was much lower than that from the monophasic system. The phenomenon suggested that the biphasic system could inhibit the repolymerization very well. Oxidative degradation products were aromatic monomers, including phenolic aldehydes, phenolic ketones, and phenolic acids. These products benefited from the creaking of Cα-Cβ and Cβ-O bonds within lignin, as well as the oxidation of the hydroxyl group at α-position. Furthermore, according to the different aromatic units, these products could be divided in p-hydroxylphenyl, guaiacyl, and syringyl compounds, which were consistent with the initial substructures within lignin. Meanwhile, demethoxylation reaction was recognized, which could be proofed by the reduction of syringyl compounds and the existence of CH_4 in gaseous products. After reaction, the molecular weight of liquid products dropped to 300 Da, which suggested the formation of dimeric and trimeric compounds. Generally, the catalytic activity of Sn ( ) combined with the biphasic system was high than those reported V catalysts in monophasic organic systems.

Keywords: Lignin; aerobic oxidation; aromatic monomers; Sn ( ) catalyst; biphasic system

499168 A Novel Transport-Reaction Model for the Estimation of Topochemical Changes during the Pretreatment of Plant Biomass Using Raman Spectroscopy

Abstract:

Porous materials have a complex network of pores and fibers which affect their structural and transport properties. In case of plant cell walls, the internal structure consisting of cellulose fibers enclosed in a lignocellulose matrix and pore spaces, has a significant influence on the efficiency of the biomass conversion process, the properties of the materials and their end-use applications. Since plant biomass is a renewable source and serves as a starting material for a wide range of bioproducts including biofuels, bioplastics, pulp and paper, etc., it is extremely important that we understand the effect of structure on conversion processes. The first step in the conversion process is the pretreatment step, which achieves the disruption of the recalcitrant structure of lignocellulose and increases the efficiency of the subsequent hydrolysis. One or more of the cell wall components are dissolved during this process which causes significant changes in the structure and topochemistry. These changes have to be studied in order to determine the efficiency of pretreatment strategies. In this regard, several 2-D and 3-D characterization techniques have been used. However, in order to get a true sense of the structure of plant cell walls, 3D characterization techniques such as X-ray computed tomography (X-ray CT) and TEM computed tomography (TEM-CT) have been very effective. The X-ray CT offers an insight into the micron scale structure of plant cell walls whereas the TEM-CT is a more powerful technique and visualizes the nanoscale architecture of these cell walls and hence, provides a complete understanding of the 3D structure. The changes in structural characteristics such as porosity, specific surface area and 3D pore size distribution during the pretreatment process can be determined from the processed TEM-CT images.

The structural information obtained from TEM-CT can be correlated with topochemical information from Raman spectroscopy to fully understand the physiochemical structure of plant cell walls. The pretreatment of plant biomass is treated as a diffusion-reaction process. A transport-reaction model based on a hybrid random walk and reaction is proposed to estimate the topochemical changes in the cell walls. A fixed number of particles of the reagent used for the pretreatment process, which is representative of its concentration, diffuse through the lumen and pore spaces of the cell wall and follow a random walk path until they encounter the cell wall interface. At this interface, they either react with one more cell wall components or diffuse through the cell wall or bounce back into pore space. This is decided by the reaction probability as well ratio of diffusivities of the pore and cell walls. The random path followed by these particles and the corresponding changes in spatial concentration of the cell wall components such as lignin, cellulose and hemicellulose are monitored throughout the pretreatment process. Thus, such a stochastic dynamic approach keeps track of both the structural changes and spatial concentration changes of the reactants. The transport-reaction model was tested initially for a model system consisting of a periodic array of spheres. This was extended to biomass images obtained from Raman Spectroscopy. A 3D stack of Raman images was obtained by duplicating a single Raman image giving rise to a cylindrical network of pores and cell wall fibers on which the model was tested. The model can also be applied to TEM-CT images after making changes to account for nanoscale transport. Thus, such an approach gives a complete understanding of the conversion process and is very useful in determining the efficiency of such processes.

489500 Engineering Innovative Polyelectrolyte Complex Membranes with Enhanced Pervaporation Performance in Ethanol Dehydration

Abstract:

Pervaporation has been a significant technology in recent years with wide industrial applications in organic dehydration especially for separating azeotropic, close-boiling, isomeric and heat sensitive liquid mixtures. Among these organic-water systems, ethanol dehydration is one of the most challenging problems as an azeotrope can be formed when traditional distillation is employed. Complete separation is hardly to achieve. In comparison, utilizing membranes for pervaporation can greatly reduce the energy consumption and realize desired separation effect. Compared with inorganic materials, organic membranes have already been applied in the water-ethanol separation field due to their prominent superiorities such as extensive availability of raw materials, facile processability, and suitability for large-scale production.

Conventional hydrophilic polymers used for ethanol dehydration mainly include poly(vinyl alcohol) (PVA), chitosan (CS), polyacrylonitrile (PAN), polyacrylic acid (PAA), and sodium carboxymethyl cellulose (CMCNa) [1, 2] etc. Among them, PVA, applied for commercial areas already, was the most attractive and economical material, since its high hydrophilicity, facile film-forming ability, thermal and chemical-resistance, and low manufacturing cost. However tiny free volume, low selectivity, and poor stability in aqueous solution made its pervaporation performance not as satisfactory as expected.

To obtain an ideal performance of ethanol dehydration, the difference of physical and chemical properties in ethanol and water must be fully exploited, the most distinct of which is polarity. Meanwhile, in order to ensure a higher flux, a novel organic membrane should possess large free volume, attracting water molecules and allowing it to pass quickly while rejecting ethanol molecules. Considering above factors, polymers with higher charge densities are likely to have the potential to meet these requirements. Focusing on this, the polyelectrolyte complexes (PECs), which have been paid close attention in recent years, are an promising candidate to be selected to prepare novel pervaporation membranes for ethanol dehydration.

PECs are a sort of multicomponent polymers with many ionizable groups in backbones, which could be applied in many fields such as gene/DNA delivery, drug release, protein adsorption, microencapsulation, pervaporation, and fuel cells [3, 4]. Because of their both good hydrophilicity and ionic cross-linking structure, the former can ensure a larger flux, while the latter favors a higher selectivity by limiting the excessive swelling and the counterions loss. Therefore, they are a kind of excellent material and very suitable for ethanol dehydration. Nevertheless, traditional strategies to prepare polyelectrolyte complex membranes (PECMs), i.e., layer-by-layer assembly and excessive acid blending method [5], could not realize notable pervaporation performance owing to the underutilization of charges of ionized groups. Thus, new methods to synthesize innovative PECM have been constantly explored in scientific community.

For the past few years, Zhao et al. [6, 7] proposed a new route, so-called protection-deprotection, to fabricate soluble and processable PECs, and homogeneous polyelectrolyte complex membranes (HPECMs). As it worked out the key question of PECs, many distinguished properties, i.e., no glass transition, high hydrophilicity, tunable surface charge, and structural stability could be further developed to design novel functional materials. Along this way, several kinds of needle-like PECs, sizes of which ranged from 300 to 700 nm, and HPECMs based on CMCNa for polyanions were synthesized [8-12]. Preferable performances were already obtained in ethanol, isopropanol, acetone, and dioxane dehydration compared with these of membranes prepared by conventional methods.

However, in view of particle characteristics, the larger size and the anisotropy of needle-like PECs cause the nonuniform distribution of long-range arrangement as well as charge densities, which will weaken dehydration performance to a certain extent. In contrast, we speculate that smaller sphere-like PEC particles can avoid these problems ingeniously. It is believed that tiny spherical PECs should possess higher charge density, spatial isotropy, and long-range homogeneity. All of these are beneficial to enhance pervaporation performance of HPECMs. By controlling operating conditions, such as temperature and stirring speed, the conformation of polyelectrolytes in solution will change and PECs with different morphologies can be further generated [13, 14].

Fig. 1. FESEM image of HPECM0.31 surface with a magnified top-view insert.

The effect of conformation and morphology of HPECMs on pervaporation performance in ethanol dehydration was researched. Based on sodium carboxymethyl cellulose (CMCNa) and poly(diallyldimethylammonium chloride) (PDDA), a series of sphere-like PEC particles with a diameter between 30-50 nm, and HPECMs with tunable compositions or ionic complexation degrees (ICD) were prepared, as shown in Fig. 1. The properties of HPECMs were varied obviously with ICD ranging from 0.23 to 0.57. With the increase of ICD, water contact angles decreased while the tendency of Zeta potential was just opposite, indicating that both the hydrophilicity and charge densities were strengthened. This was exactly contrary to the characteristics of those needle-like PECs and HPECMs. When subjected to ethanol dehydration, sphere-like morphology HPECMs revealed more superior pervaporation performance and desired anti-trade-off phenomenon, as seen from Fig. 2. The flux increased rapidly, while the water content in permeate remained basically unchanged and were always kept above 99.30 wt% in the range of 17-70 °C. For HPECM0.31 and HPECM0.57, the permeation fluxes were 2.25 $kg \cdot m^{-2} \cdot h^{-1}$ and 2.93 $kg \cdot m^{-2} \cdot h^{-1}$, and the water concentration in permeate reached up to 99.35 wt% and 99.38 wt%, respectively, with 10 wt% water in feed at 70 °C. Under the premise of maintaining similar separation effect, the flux of spherical morphology HPECMs in this work was nearly as twice as that of needle-like HPECMs and about 10-15 times higher than that of traditional PVA and CS membranes. Interestingly, the HPECMs flux of this study was even much bigger than part of inorganic membranes and hybrid membranes, almost 1-2.5 times higher, and the selectivity was also outstanding to the majority of them. Moreover, excellent separation effect, good repeatability and stable performance could be continuously retained in practical operation. All these results implied that sphere-like PEC particles and isotropic HPECMs were indeed in favor of enhancing pervaporation performance in ethanol dehydration.

Fig. 2. Effect of operating temperature on flux (solid) and water in permeate (hollow) of HPECM0.31 (triangle) and HPECM0.57 (circle) for ethanol dehydration with 10 wt% water in feed.

Acknowledgements

This work was supported by the National Natural Science Foundation of China (Grant No. 21136008).

References

[1] Bolto B., Hoang M., Xie Z. L. A review of membrane selection for the dehydration of aqueous ethanol by pervaporation. Chem. Eng. Process. 50 (2011) 227-235.

[2] Chapman P. D., Oliveira T., Livingston A. G., Li K. Membranes for the dehydration of solvents by pervaporation. J. Membr. Sci. 318 (2008) 5-37.

[3] Tsai H., Chen W., Kuo C., Lee K., Lai J. Study on the pervaporation performance and long-term stability of aqueous iso-propanol solution through chitosan/polyacrylonitrile hollow fiber membrane. J. Membr. Sci. 309 (2008) 146-155.

[4] Kishimura A., Koide A., Osada K., Yamasaki Y., Kataoka K. Encapsulation of myoglobin in PEGylated polyion complex vesicles made from a pair of oppositely charged block lonomers: A physiologically available oxygen carrier. Angew. Chem. Int. Ed. 46 (2007) 6085-6088.

[5] Zhao Q., An Q. F., Ji Y. L., Qian J. W., Gao C. J. Polyelectrolyte complex membranes for pervaporation, nanofiltration and fuel cell applications. J. Membr. Sci. 379 (2011) 19-45.

[6] Zhao Q., Qian J. W., An Q. F., Yang Q., Zhang P. A facile route for fabricating novel polyelectrolyte complex membrane with high pervaporation performance in isopropanol dehydration. J. Membr. Sci. 320 (2008) 8-12.

[7] Zhao Q., Qian J. W., An Q. F., Gui Z. L., Jin H. T., Yin M. J. Pervaporation dehydration of isopropanol using homogeneous polyelectrolyte complex membranes of poly(diallyldimethylammonium chloride)/sodium carboxymethyl cellulose. J. Membr. Sci. 329 (2009) 175-182.

[8] Zhao Q., Qian J. W., An Q. F., Gao C. J., Gui Z. L., Jin H. T. Synthesis and characterization of soluble chitosan/sodium carboxymethyl cellulose polyelectrolyte complexes and the pervaporation dehydration of their homogeneous membranes. J. Membr. Sci. 333 (2009) 68-78.

[9] Wang X. S., An Q. F., Zhao Q., Lee K., Qian J. W., Gao C. J. Homogenous polyelectrolyte complex membranes

incorporated with strong ion-pairs with high pervaporation performance for dehydration of ethanol. J. Membr. Sci. 435 (2013) 71-79.

[10] Wang X. S., An Q. F., Liu T., Zhao Q., Hung W., Lee K., Gao C. J. Novel polyelectrolyte complex membranes containing free sulfate groups with improved pervaporation dehydration of ethanol. J. Membr. Sci. 452 (2014) 73-81.

[11] Jin H. T., An Q. F., Zhao Q., Qian J. W., Zhu M. H. Pervaporation dehydration of ethanol by using polyelectrolyte complex membranes based on poly (N-ethyl-4-vinylpyridinium bromide) and sodium carboxymethyl cellulose. J. Membr. Sci. 347 (2010) 183-192.

[12] Liu T., An Q. F., Zhao Q., Lee K., Zhu B. K., Qian J. W., Gao C. J. Preparation and characterization of polyelectrolyte complex membranes bearing alkyl side chains for the pervaporation dehydration of alcohols. J. Membr. Sci. 429 (2013) 181-189.

[13] Starchenko V., Muller M., Lebovka N. Growth of Polyelectrolyte Complex Nanoparticles: Computer Simulations and Experiments. J. Phys. Chem. C. 112 (2008) 8863-8869.

[14] Muller M., Reihs T., Ouyang W. Needlelike and Spherical Polyelectrolyte Complex Nanoparticles of Poly(L-lysine) and Copolymers of Maleic Acid. Langmuir. 21 (2005) 465-469.

498212 Biphasic Tandem Catalytic Process for Renewable Biofuel Production

Abstract:

The market demand for renewable liquid fuels has surged in the transportation sector in the past decades. However, today the cost of liquid hydrocarbon fuels from biomass is still too high to compete with petroleum derived transportation fuels. One of the grand challenges in biofuels production is the low carbon atom efficiency to the fuel products. Herein we have successfully developed a novel â€œone-potâ€ biphasic tandem catalytic process (biTCP) that was used to convert terpenoids and lipids to hydrocarbons at an extraordinarily high carbon efficiency.Â For demonstration, the mono- and diterpenoids, e.g., the biocrudes of eucalyptus and grindelia squarrosa, have been converted into cycloalkanes, which are high-density jet fuel components, with the homogeneous acid and the hydrophobic supported metal catalysts in aqueous and organic solvents, respectively, in the biTCP. This novel process was also able to highly efficiently convert fatty acids and triglycerides to renewable diesel. The reaction mechanism was investigated to gain a fundamental understanding of the effects of catalyst properties and process conditions on the conversion of the two different types of biomass feedstock.

504164 Nano Carbon Structures from Cellulosic Biomass for use as Functional Materials

Abstract:

We report herein the production of hierarchical nanocarbon materials from cellulose fibers. The carbonized cellulosic fibers with branching out CNTs (CC-CNT) offer interesting surface and electrochemical properties that are of interests for chemical processing and biosensing applications. In our tests for chemical adsorption tests, CC-CNT offered highly selective adsorption of hydrocarbons from gas mixtures, especially chemicals with C=C bond and aromatic structures. More interestingly, the chemicals absorbed and accumulated can be quickly released from the surface of the materials upon application of modest electrical current. Such an electrically switchable adsorption behavior may promise applications in a broad areas including air quality control, chemical sensing and fuel storage.

498421 Highly Selective FT Synthesis for Production of JP-8 Jet Fuel from Coal and Coal/Biomass Mixtures

Abstract:

An indirect liquefaction process employing advanced process intensification approaches is being developed to reduce costs and accelerate commercialization of a coal to liquid (CTL) process for converting coal and coal-biomass mixtures to JP-8 jet fuel. In this Air Force and DOE/NETL sponsored cooperative agreement, Southern Research is leading a team consisting of IntraMicron, Inc., Southwest Research Institute, and Nexant, Inc., along with Chevron as selective Fischer-Tropsch (FT) catalyst supplier and the National Carbon Capture Center (NCCC) as the host site. The process reduces the number of parasitic syngas cleaning steps and eliminates costly upgrading of wax produced in conventional indirect liquefaction processes, resulting in improved economics and lower lifecycle greenhouse gas (GHG) emissions for CTL-based jet fuel production. These improvements directly reduce capital costs of indirect CTL processes, resulting in significantly improved economics and cost competitiveness of CTL when compared to petroleum refining. Specific objectives for FT experiments are to demonstrate a selective and active second generation cobalt-zeolite hybrid Fischer-Tropsch (FT) catalyst from Chevron that predominantly produces liquid range C5-C20 hydrocarbons with high productivity (5X conventional FT catalysts) and jet fuel selectivity while eliminating the production and associated handling of solid C21+ wax. In fall of 2015, Southern operated a bench-scale FT reactor with a 2" diameter heat exchange reactor tube using IntraMicron technology in a nominal 5 gallons per day skid-mounted fixed-bed installed at NCCC. In those experiments, both productivity and selectivity goals were achieved during a 300 hour experiment. Another experiment has just been completed in spring 2017 with the reactor diameter increased to 4". This work will compare the results from the two long-term tests of the hybrid FT jet-selective catalyst using the skid-mounted fixed-bed FT reactor system at NCCC.

501942 Microfibrous Entrapped Catalyst Structure for Highly Exothermic Reaction: Fischer-Tropsch Synthesis

Abstract:

An Abstract Submission to 2017 AICHE Meeting

Session: Alternative Fuels

Topical Area: Catalysis and Reaction Engineering Division

Microfibrous Entrapped Catalyst Structure for Highly Exothermic Reaction: Fischer-Tropsch Synthesis

Xinquan Cheng and Bruce J. Tatarchuk*

Center for Microfibrous Materials Manufacturing, Department of Chemical Engineering, 212 Ross Hall, Auburn University, AL 36849

**Corresponding author. Tel.: +1 334 844 2023; fax: +1 334 844 2065; E-mail address: Tatarbj@auburn.edu*

Abstract

Fischer-Tropsch synthesis (FTS) is a well-known technique for converting gas to liquid fuels. However, FTS is a highly exothermic reaction, which producing extremely amount of heat during the reaction process. More important, the product distribution of FTS is strongly dependent on the reaction temperature. High temperatures favor the formation of undesirable light products like methane. Therefore, a reactor system with an excellent heat transfer property is necessary for highly exothermic reaction like FTS. Our group developed a novel catalyst structure called Microfibrous Entrapped Catalyst (MFEC) structure. MFEC is a catalyst network supported by micron-sized metal fibers with high thermal conductivity. Previous efforts on this unique catalyst structure have shown significant enhancement in intra-bed heat transfer and maintains a stable reaction temperature.

In this investigation, both iron-based and cobalt-based FTS catalysts have been tested for FTS reactions in a large tubular reactor (34.0 mm I.D.) packed with Cu MFEC. Typically, the MFEC reactor can be prepared in two steps: 1) the prepared catalyst particles were entrapped into a sintered copper microfiber media to form a Cu MFEC structure; 2) circular disks with a diameter of 36.0 mm were punched out from the Cu MFEC structure to pack into the tubular reactor. The disk size (36.0 mm) were designed 6 % larger than the inside diameter of the tubular reactor (34.0 mm) in order to increase the contact points between the copper fiber media and the reactor wall. By doing this, the inside wall heat transfer coefficient increased greatly when compared with the traditional packed bed. A four-point thermocouple was sticked in from the side when the reactor was loaded half way, and then loading the other half. In that way, the radial temperature distribution profiles can be measured precisely. Take iron-based catalyst as an example, the reaction demonstrated a radial temperature gradient around 5 ºC. While with the same test conditions, the comparative packed bed reached a radial temperature gradient around 139 ºC. It was also found that the Cu MFEC based FTS reaction required negligible amount of time to reach its steady state compared with the traditional packed bed. Considering these attributes and advantages, Cu MFEC structure is a promising reactor system for highly exothermic reactions such as the FTS reaction.

488458 Hydrogen Production from Thermal and Catalytic Gasification of Biomass in a Fluidized CREC-Riser Simulator: Thermodynamic Modeling and Experimental Results

Abstract:

Biomass gasification is an alternative path of critical future value. It produces synthesis gas, a mixture of hydrogen and carbon monoxide, the first one being an energetic vector. (1, 2)

Gasification of several biomasses was performed in a CREC Riser Simulator reactor. Thermal experiments were performed at 550 °C and 600 °C with broza, bark and CANMET Pellets, under helium or CO_2. On the other hand, catalytic experiments were developed using glucose and metoxy-phenol. Temperature, steam/biomass ratio, catalyst loading and reaction time were changed in a wide range of operating conditions. Two fluidizable catalysts were used: 20%Ni/5%La_2O_3/g-Al_2O_3 and 15%Ni-5%CeO_2/γ-Al_2O_3. They were synthesized via incipient wetness impregnation and characterized using PSD, DA, TPD, TPR and BET. (2)

Thermal conversion showed that hydrogen increases with temperature, with this being consistent for the three feedstocks employed (Broza, Canmet Pellet and Bark). Furthermore, the gasifying agent influences hydrogen production, with maximum yields obtained for hydrogen at 600 C, under inert atmospheres (helium). As well, higher temperatures led to a decrease in the H_2 /CO ratio for "Broza". For the "Canmet Pellet", it showed a slight H_2 /CO increase, while for "bark" remained almost constant. The maximum value of the ratio H_2 / CO was obtained at 550 °C, being 5.88.

Regarding the catalytic experiments with fluidizable methoxy-phenol (representing lignin) were used. It was shown hydrogen production increased with temperature. The change with temperature was more notorious from 600 °C to 650 °C than from 650 °C a 700 °C. Hydrogen production augmented as well, with reaction time, with a major change observed in the 10 and 20 second range.

Concerning reaction equilibrium eight chemical reactions were first considered(1). The Gauss- Jordan procedure was applied to identify five independent equations. On this basis a thermodynamic model, with five algebraic nonlinear equations and five unknowns was considered. The equations were solved with Newton-Raphson method. Experimental hydrogen production was systematically compared with chemical equilibrium values and this for H_2, CO, CO_2 and CH_4. An experimental maximum hydrogen molar fraction of 0.54 for S/B = 2.5 was observed. A similar situation was noticed with CO_2 produced with however, CO and methane being reduced at those conditions.

(1) Salaices E., Serrano B. *de Lasa H.* "Steam Gasification of a Cellulose Surrogate Over Fluidizable Ni/alpha alumina: A Kinetics Model" *AIChEJ* Vol.58, Issue 5, 1588–1599 (2012), *IF: 2.03*

(2) Mazumder, A., *de Lasa H.* Fluidizable La_2O_3 Promoted Ni/ γ-Al_2O_3 Catalyst for Steam Gasification of Biomass: Effect of Catalyst Preparation Conditions. *Applied Catalysis B,* Vol.169-169, 250-265 (2015). *IF:3.42*

500597 A Generalized Kinetic Model for Transesterification and Saponification

Abstract:

Owing to the depletion of fossil fuel resources and increasing environmental concerns, the search for alternative fuels such as biodiesel has gained enormous impetus in the past two decades. Plant and vegetable oils are so far the most prominent feedstocks for these alternative fuels. Rudolf Diesel was the first to test peanut oil as an alternative fuel for his engine about a century ago (Shay, 1993). During 1930s and 1940s, vegetable oils were used as alternative fuels in case of emergency (Singh and Singh, 2010). However, vegetable oils cannot directly substitute diesel due to their high viscosity, low volatility, and poor cold flow properties. Hence, they are processed to obtain biodiesels with properties similar to those of the petroleum-based diesel. This has attracted much attention in recent years (Verma and Sharma, 2016). Biodiesel is a non-toxic, renewable, and biodegradable fuel composed mainly of fatty acid esters. It is industrially produced via transesterification of vegetable oils with alcohol in the presence of a catalyst.

Though biodiesel holds promise as an alternative fuel, its production offers a significant challenge to researchers due to higher production costs compared to the production of petroleum based diesel. Hence, it becomes crucial to model and optimize the biodiesel production process with the aim to achieve lower costs along with meeting the standard property specifications. There are numerous research articles and literature review in modeling, simulation and optimization of biodiesel production process (Aransiola et al., 2014). However, in all of the studies, modeling the reaction kinetics is the foremost important step. The studies regarding kinetics of biodiesel production can be divided into two categories. In the first category (Vicente et al., 2006; Eze et al., 2014), the kinetic models are oversimplified and strongly dependent on the parent oil used for that study. Hence, there are a lot of inconsistencies in the reported values of the kinetic parameters. In the second category (Likozar and Levec, 2014a,b), the effect of the saponification reactions is not accounted for in the model which restricts their application to the case of water-free chemicals only. Therefore, in this article, we address these issues by proposing a kinetic parameter ranking and estimation framework that has the following key components,

1. A generalized kinetic model that considers both both transesterification and saponification reactions.
2. The model is suitable for any biodiesel feed containing the glycerides formed using the five most common fatty acids *viz.*, Oleic (O), Linoleic (L), Linolenic (Ln), Palmitic (P) and Stearic (S).
3. A detailed sensitivity analysis of all model parameters.
4. Application of a Bayesian approach to estimate the most important six parameters.

The purpose of this work is to develop a generalized kinetic model for both transesterification and saponification reactions. It consists of a differential mass balance for the process of transesterification and saponification of each glyceride and ester formed by the major fatty acids present in oil. Eq. (1) shows the general form of differential equations in the model.

$dC_X/dt = f_X(C,\Theta)$ (1)

where x represents the different species such as glycerides, esters, methanol, glycerol, soaps, and free fatty acids; C_x represents the concentration of species x; and Θ represents the set of kinetic rate constants. Here, heterogeneity in the reaction system is neglected and the model is assumed to be in pseudo-homogeneous phase. This is a sound assumption for the industrial-scale process as they employ high mixing intensity. Our model consists of 71 species and 257 reactions. User-defined operational inputs for the model are temperature of the reactor, molar ratio of alcohol to oil, catalyst concentration (wt.% based on oil), and the composition of oil. The model response/output is the yield of fatty acid methyl ester (biodiesel) at different time points. The model parameters include the pre-exponential (Arrhenius) factors and the activation energies with respect to all the reactions. This sums up to 2 x 257 = 514 parameters. However, it is not practically feasible to estimate all the parameters of a reaction mechanism simultaneously. Thus, we employ finite difference approach based sensitivity analysis to identify the most significant parameters for estimation. Typically, the number of parameters estimated is limited by the available experimental observations. Here, we estimate the 6 most sensitive parameters using Bayesian approach of parameter estimation and 32 experimental observations. As the direct application of the Bayesian method is challenging over wide ranges in parameter space, it is prudent to employ a preliminary step of optimization to narrow down the ranges using sum of squared errors as an objective function. This is performed in 2 stages *viz.* a quasi-random global search as the first stage followed by an optimization using the Hooke-Jeeves algorithm. We then employ a Bayesian approach described in (Mosbach et al., 2012; Beers, 2006) over the optimized parameter ranges to study the effects of uncertainties in parameters on the model response. Due to the large number of evaluations of the model required for sampling from the distributions, we employed a surrogate model instead of directly evaluating the original model. The surrogate model was developed using the High Dimensional Model Representation (HDMR) technique.

Resulting distributions showed that we obtain a single optimal value obtained for each of the parameters respectively. We validate our model by conducting parametric studies over varied ranges of temperature, molar ratio of oil to alcohol and catalyst concentration. Our analysis indicates that the model could predict the cetane number and higher heating value of biodiesel formed using various parent oils with an average accuracy of around 3%. Finally, we performed mechanism reduction in order to find the minimum set of reactions that can represent the kinetics of 4 major oils *viz.*, Cottonseed oil, Palm oil, Rapeseed oil and Soybean oil without significant loss of accuracy. We observe that the deviation in the dynamic concentration profiles increased from less than 1% to around 10% as we vary the size of the mechanism from 211 reactions to 175 reactions.

References

1. Aransiola, E., Ojumu, T., Oyekola, O., Madzimbamuto, T., Ikhu-Omoregbe, D., 2014. A review of current technology for biodiesel production: State of the art. Biomass and bioenergy 61, 276–297.
2. Beers, K. J., 2006. Numerical methods for chemical engineering: applications in Matlab. Cambridge Univer-sity Press.
3. Eze, V. C., Harvey, A. P., Phan, A. N., 2015. Determination of the kinetics of biodiesel saponification in alcoholic hydroxide solutions. Fuel 140, 724–730.
4. Eze, V. C., Phan, A. N., Harvey, A. P., 2014. A more robust model of the biodiesel reaction, allowing identifica-tion of process conditions for significantly enhanced rate and water tolerance. Bioresource Technology 156, 222–231.
5. Keera, S., El Sabagh, S., Taman, A., 2011. Transesterification of vegetable oil to biodiesel fuel using alkaline catalyst. Fuel 90 (1), 42–47.
6. Likozar, B., Levec, J., 2014a. Effect of process conditions on equilibrium, reaction kinetics and mass transfer for triglyceride transesterification to biodiesel: experimental and modeling based on fatty acid compo-sition. Fuel Processing Technology 122, 30–41.
7. Likozar, B., Levec, J., 2014b. Transesterification of canola, palm, peanut, soybean and sunflower oil with methanol, ethanol, isopropanol, butanol and tert-butanol to biodiesel: Modelling of chemical equilib-rium, reaction kinetics and mass transfer based on fatty acid composition. Applied Energy 123, 108–120.
8. Marjanovic,´ A. V., Stamenkovic,´ O. S., Todorovic,´ Z. B., Lazic,´ M. L., Veljkovic,´ V. B., 2010. Kinetics of the base-catalyzed sunflower oil ethanolysis. Fuel 89 (3), 665–671.
9. Mendow, G., Veizaga, N., Querini, C., 2011. Ethyl ester production by homogeneous alkaline

transesterifica-tion: influence of the catalyst. Bioresource technology 102 (11), 6385–6391.
10. Mosbach, S., Braumann, A., Man, P. L., Kastner, C. A., Brownbridge, G. P., Kraft, M., 2012. Iterative improvement of bayesian parameter estimates for an engine model by means of experimental design. Combustion and Flame 159 (3), 1303–1313.
11. Richard, R., Thiebaud-Roux, S., Prat, L., 2013. Modelling the kinetics of transesterification reaction of sun-flower oil with ethanol in microreactors. Chemical engineering science 87, 258–269.
12. Shay, E. G., 1993. Diesel fuel from vegetable oils: status and opportunities. Biomass and bioenergy 4 (4), 227–242.
13. Singh, S., Singh, D., 2010. Biodiesel production through the use of different sources and characterization of oils and their esters as the substitute of diesel: a review. Renewable and Sustainable Energy Reviews 14 (1), 200–216.
14. Verma, P., Sharma, M., 2016. Review of process parameters for biodiesel production from different feedstocks. Renewable and Sustainable Energy Reviews 62, 1063–1071.
15. Vicente, G., Martinez, M., Aracil, J., 2006. Kinetics of brassica c arinata oil methanolysis. Energy & fuels 20 (4), 1722–1726.
16. Vicente, G., Martínez, M., Aracil, J., Esteban, A., 2005. Kinetics of sunflower oil methanolysis. Industrial & Engineering Chemistry Research 44 (15), 5447–5454.

495814 Microwave Assisted Co-Pyrolysis of High Ash Indian Coal and Rice Husk and Detailed Product Characterization

Abstract:

With a share of over 60% in the commercial energy sector, coal has become Indiaâ€™s energy security. The relative abundance of coal in India compared to other fossil fuels makes it a natural choice as the primary source of fuel. The reserves to production ratio at the current rate of production for coal is around 230 years as against 15 and 23 years, respectively, for indigenous oil and natural gas. This makes domestic coal a very secure energy source for the country. India has got 192 billion tons of coal reserves but the majority contain high amount of ash [1]. Coal is a complex heterogeneous material which consists of a number of distinct organic (maceral) and inorganic (mineral) compounds. Although regained interest in the production of transportation fuels from coal is positive sign, it is not sufficient as liquefaction of coal yields relatively low liquid yield compared with biomass liquefaction.

Coal and biomass to liquid hydrocarbon fuels offer great hope in commercialization than other alternatives as it not only reduces the carbon footprint in the environment but also facilitates the utilization of locally available low-rank coals [2]. To improve the utilization of coal and to increase the carbon and energy density, co-pyrolysis, co-gasification and co-combustion processes have been developed. Among these, co-pyrolysis of coal and biomass is a promising strategy to produce liquid bio-oil that can be either combusted for power/heat or can be further catalytically upgraded to liquid transportation fuels [3]. Co-processing of coal and lignocellulosic biomass can provide a solution for the use of biomass on a commercial scale. Conventional pyrolysis involves heat transfer in the form of conduction, convection and radiation, while in microwave heating electromagnetic energy is converted to thermal energy from inside of the sample to the surface. Due to its uniform, volumetric, rapid, selective and efficient heating mechanism, microwave heating reduces the processing time and enables the selective production of chemicals [4]. Moreover, secondary cracking of primary pyrolysates can be controlled in microwave heating using different microwave absorbing materials/susceptors.

In this study, microwave assisted co-pyrolysis of high ash Indian coals and rice husk is investigated to evaluate the liquid fuel potential and to understand the interaction effect in the liquid yields and its composition. Microwave co-pyrolysis experiments were conducted in a bench scale unit, and the effects of various parameters such as microwave power, coal: rice husk ratio, feed: susceptor ratio on product yields and oil composition were thoroughly evaluated. The bio-oil composition was semi-quantitatively analyzed using gas chromatograph-mass spectrometry. Gas composition was analyzed using an online gas analyzer. The oil yield was not significantly affected by the microwave power and feed: susceptor ratio for a particular coal: rice husk ratio. Additive effect was observed with increase in rice husk ratio. The bio-oil yields were 11±2 wt.%, 15± 2 wt.%, 18 ± 2 wt.%, 21 ± 2 wt.% and 26 ± 2 wt.% for 100:0, 75:25, 50:50, 25:75 and 0:100 mass ratios of coal to rice husk. The use of graphite as susceptor improved the heating rate significantly owing to its high dielectric tangent value. High microwave powers increased the second cracking of condensables, which resulted in reduction of oil yield as compared to oil yield from low microwave power. Coal oil is a mixture of aromatics, aliphatic HC and phenolic compounds, while pure rice husk oil was a mixture of furans,

aromatics, phenolic compounds, guaiacols, syringols and cyclooxygenates. The addition of coal to rice husk demoted the yield of phenolic compounds and promoted the aromatic content. HHV of bio-oil was 32-35 MJ kg^{-1}, while HHV of char was 11-15 MJ kg^{-1}. Although char was highly porous, its HHV was low due to the high amount of ash that was present in both feedstocks (20% ash in rice husk and 33% ash in Indian coal). Hydrogen gas constituted a major fraction in the pyrolysis gases. Addition of rice husk promoted CO_2 evolution.

References

[1] Haque E. Indian Coal: Production and Ways to Increase Coal Supplies. Intl J Sci Res Publ. 2(2013), 1-3.

[2] Arvelakis S, Folkedahl B, Dam-Johansen K, Hurley J. Studying the melting behavior of coal, biomass, and coal/biomass ash using viscosity and heated stage XRD data. Energy Fuel 20(2006), 1329-1340.

[3] Guan Y, Ma Y, Zhang K, Chen H, Xu G, Liu W, Yang Y. Co-pyrolysis behaviors of energy grass and lignite. Energy Convers Manage. 93(2015), 132-140.

[4] Shraâ€™ah A A, Helleur R. Microwave pyrolysis of cellulose at low temperature. J Anal Appl Pyrol. 105(2014), 91-99.

496717 Highly Selective Catalytic Conversion of Furfural to Î³-Butyrolactone

Abstract:

Highly selective catalytic conversion of furfural to ¦Ã-butyrolactone

Tiefeng Wang* and Xiaodan Li, Yafei Li

Beijing Key Laboratory of Green Reaction Engineering and Technology

Department of Chemical Engineering, Tsinghua University, Beijing 100084, China

**Corresponding author: wangtf@tsinghua.edu.cn*

The depletion of fossil fuel reserves and growing demand for energy are drawing efforts to find alternative sources of energy and organic carbon. Biomass-derived molecules can serve as renewable substitutes for petroleum-based building blocks to produce fuels and chemicals.[1] Furfural, a renewable platform compound produced by hydrolysis and dehydration of xylan contained in lignocellulose, has already been used to produce fine chemicals such as furfuryl alcohol, tetrahydrofurfuryl alcohol, furan, and furfural resin in industry.[2-4] During the past decades, novel technologies and processes have been developed for furfural production. This will necessarily promote the further utilization of furfural to produce more value-added chemicals.

One of the most promising products is ¦Ã-butyrolactone (GBL), which has wide applications in the fields of solvents, medicine, spice and fine organic synthesis.[5] Currently, GBL is mainly produced by dehydrocyclization of 1,4-butanediol (BDO) and hydrogenation of maleic anhydride, but they usually have problems of harsh operating conditions, complicated process, and environmental pollution.[6] Therefore, it is attractive to sustainably produce GBL from a renewable biomass-derived molecule with high atom economy. This work reports a two-step process to produce GBL from furfural with 2(5*H*)-furanone as the intermediate. In the first step, furfural was oxidized to 2(5*H*)-furanone with a promising yield over 60%. In the second step, the purified 2(5*H*)-furanone was further hydrogenated to GBL using SiO_2 supported metal catalysts.

The catalytic oxidation of furfural to 2(5*H*)-furanone was studied using hydrogen peroxide as the oxidant and formic acid as the catalyst. A 60-62% yield of 2(5*H*)-furanone, with a total yield of diacids (*i.e.*, maleic acid and succinic acid) in the range of 15-20% as by-products, was obtained in a bi-phasic system using 1,2-dichloroethane or ethyl acetate as the solvent. The good solubility in the organic phase and the strong oxidizing nature of performic acid, which was the real active oxidant generated from formic acid and hydrogen peroxide, contributed to the promising yield of 2(5*H*)-furanone.[7]

Having obtained 2(5*H*)-furanone by furfural oxidation, we next turn to the hydrogenation of 2(5*H*)-furanone to GBL. With only one more additional C=C bond than GBL, 2(5*H*)-furanone is an ideal intermediate for converting furfural to GBL. The selective hydrogenation of the C=C bond in 2(5*H*)-furanone was first studied in a liquid phase over a series of SiO_2 supported monometallic catalysts (M = Pt, Pd, Rh, Ru, Ni, Co and Cu) prepared by incipient wetness impregnation. The catalyst activity followed the trend of $Pd0.5/SiO_2 \sim Rh0.5/SiO_2 > Ru0.5/SiO_2 > Pt0.5/SiO_2$, and $Ni4.5/SiO_2 > Co4.5/SiO_2 > Cu4.5/SiO_2$. Furthermore, the initial TOF for 2(5*H*)-furanone hydrogenation over these monometallic catalysts showed a volcano-type dependence on their metal *d*-band center values, with Pd having the highest activity.[8] This correlation further guided our design of non-noble metal-based bimetallic catalysts to reduce cost.

Ni was shown to be active and selective for the hydrogenation of the C=C bond in 2(5*H*)-furanone. The activity and selectivity of a metal catalyst can be significantly altered by incorporating a second metal due to electronic and geometric effects.[9] Thus, a series of Ni-based bimetallic catalysts, Ni-M/SiO_2 (M= Fe, Co, Cu and Zn), were prepared and evaluated for the selective hydrogenation of 2(5*H*)-furanone to GBL. Among the prepared bimetallic catalysts, Ni-Fe/SiO_2 showed the best performance. The effect of addition of Fe on the catalyst performance was further studied with different Ni/Fe ratios. It was found that the introduction of a suitable amount of Fe significantly enhanced the activity while maintaining the selectivity of the Ni/SiO_2 catalyst. Characterization measurements by XRD, BET, TEM and chemisorption were used to investigate the promotional effect of Fe, showing that the formation of NiFe alloy, optimized reducibility and metal particle size distribution contributed to the enhanced catalytic performance of the Ni-Fe/SiO_2 bimetallic catalyst.

In summary, a two-step process was developed for producing GBL from furfural, providing a new strategy for converting biomass-derived platform compound to high value-added fine chemicals. In the oxidation step, the highest yield of the intermediate, 2(5*H*)-furanone, was obtained in a bi-phasic system with formic acid as the catalyst, ethyl acetate as the solvent, and hydrogen peroxide as the oxidant. In the subsequent hydrogenation step, a series of Group VIII monometallic catalysts were first investigated and their catalytic performances were correlated with intrinsic properties of the metals, based on which an efficient non-noble bimetallic catalyst, Ni-Fe/SiO_2, was further designed.

References

[1] P. Gallezot, *Chem. Soc. Rev.*, **2012**, 41, 1538.

[2] D. M. Alonso, J. Q. Bond and J. A. Dumesic, *Green Chem.*, **2010**, 12, 1493.

[3] J. P. Lange, E. van der Heide, J. van Buijtenen and R. Price, *ChemSusChem*, **2012**, 5, 150.

[4] S. Dutta, S. De, B. Saha and Md. I. Alam, *Catal. Sci. Technol.*, **2012**, 2, 2025.

[5] X. Li, Y. Y. Cui, X. L. Yang, W. L. Dai and K. N. Fan, *Appl. Catal. A*, **2013**, 458, 63.

[6] J. R. Ebner and V. M. Francetti, *Chem. Rev.*, **1988**, 88, 55.

[7] X. D. Li, X. C. Lan and T. F. Wang, *Catal. Today*, **2016**, 276, 97.

[8] X. D. Li, X. C. Lan and T. F. Wang, *Green Chem.*, **2016**, 18, 638.

[9] S. Zafeiratos, S. Piccinin and D. Teschner, *Catal. Sci. Technol.*, **2012**, 1787.

497585 In-Situ Drifts Studies over Cobalt Catalyst for Hydrogen Production from Ethanol

Abstract:

Ethanol dehydrogenation mechanism was investigated over cobalt catalyst prepared by novel cellulose assisted combustion synthesis techniques [1-4]. Nanoparticles synthesis mechanism was studied using TGA/DTA analysis. The effect of synthesis precursor ratio was also studied and correlated with microstructural properties of the product. The amount of fuel content in the combustion solution was found to greatly affect the phase and microstructure of the synthesized catalysts. Along with the microstructure, changes in surface cobalt oxidation states as a function of fuel ratio was also studied for various fuel to oxidizer ratios using XPS. The cobalt catalyst was found to be highly active for ethanol dehydrogenation reaction over a temperature of 50°C– 400°C. In situ DRIFTS studies were conducted to understand the ethanol-metal interaction and deduce a reaction pathway by monitoring the gas phase and surface adsorbed species. The reaction mechanism was found to proceed via ethoxy formation and subsequent generation of aldehyde species at 50°C – 200°C and formation of carbon dioxide at 400°C. Carbon formation was also observed as shown in the TEM image (Fig. 1) of the spent catalyst, which supposedly comes from sequential removal of H from adsorbed $*CH_3$ to produce hydrogen and carbon.

Figure 1: TEM image of spent Co catalyst after ethanol dehydrogenation reaction

-

References

[1] A. S. Mukasyan, P. Dinka, Advanced engineering materials 9 (2007) 653-657.

[2] A. Kumar, A.S. Mukasyan and E.E. Wolf, Applied catalysis.A, General 372 (2010) 175-183.

[3] A. Kumar A.S. Mukasyan, E.E. Wolf, Industrial engineering chemistry research 49 (2010) 11001-11008.

[4] A. Ashok, A. Kumar, R.R. Bhosale, M.A.H. Saleh and Leo J.P. van den Broeke, RSC Advances 5 (2015) 28703-28712.

500197 Towards the Design of Sulfur-Tolerant CO_2-Reforming Catalysts

Abstract:

Mixed metal oxide catalysts, combining late transition metals (TMs) doped into earth oxides (REOs), offer hydrocarbon conversion (e.g., tar reforming) activities and selectivities that meet or exceed supported metal systems,[1] while also offering continued catalytic performance in the presence of sulfur poisons and sintering agents such as water.[2] These materials also have potential as sulfur tolerant CO_2-reforming or tri-reforming catalysis. The objectives of the proposed work are to acquire a "proof of principle" data set towards the development of sulfur- and water-tolerant (especially the former) TM-doped rare-earth oxysulfide catalysts for high temperature reforming, and 2) develop a mechanistic rationale for how surface sulfidation alters reforming and coke formation kinetics. Integrated experimental and computational efforts will provide a mechanistic rationale for designing sulfur- and water-tolerant catalysts. Density functional theory (DFT) methods examine C-H and C-C bond activation and coke formation over partially sulfided metal oxide surfaces, identifying materials promising for retaining, or possibly enhancing, conversion in the presence of sulfur. Synthesis, testing, and characterization evaluate the computationally predicted catalyst performance and provide feedback and further hypothesis development to aid the computational efforts.

To date, we have found that Ce-Zr supports at the 1:1 or higher elemental ratio are superior to the Ce-La supports at similar elemental ratios. The latter are relatively inactive. We have also found that coking rates in the absence of water are strongly dependent upon stoichiometric equivalence between CO_2 and CH_4. Structurally, both the Ce-Zr and Ce-La families of supports show no post-mortem evidence of separate TM (Ni, Mn, Fe) or TM oxide phases, at least for TM:REO ratios kept below thresholds determined in past work on tar reforming. Characterizations of the used catalysts have included XANES, XAFS, Raman, XRD and chemisorption techniques.

Extended coking rate measurements by DSC-TGA show that for the TM-doped REO catalysts a coking plateau is attained, while for simpler Co- and Ni-supported Al_2O_3 catalysts the coking rates are nonzero and almost constant. This suggests that regardless of their short-term performance, the latter family of catalysts supported on Al_2O_3 will ultimately deactivate more rapidly.

1. R. Li, A. Roy, J. Bridges and K.M. Dooley, "Tar Reforming in Model Gasifier Effluents: Transition Metal/Rare Earth Oxide Catalysts," Ind. Eng. Chem. Res. 53, 7999–8011 (2014)
2. M. Krcha, K. Dooley, and M. Janik, "Alkane Reforming on Partially Sulfided CeO_2 (1 1 1) Surfaces," J. Catal., 330, 167-176 (2015)

500717 Development of Catalytic Process for CO_2 Utilization

Abstract:

The utilization of CO2 as a raw material in the synthesis of chemicals and liquid energy carriers offers a direct way to mitigate the increasing CO2 build-up in the atmosphere. However, high-energy reagents and harsh reaction conditions are typically required for transforming CO2 into other chemicals. Renewable hydrogen is a high-energy material that can be stored by hydrogenating CO2 to valuable secondary energy carriers, such as methane, methanol, and formic acid. Indeed, formic acid/formate has been widely used in agriculture and industry and is the potential hydrogen storage. Herein, we developed a process of hydrogenating liquid amine captured CO2, the intermediates in the Bioenergy with Carbon Capture and Storage (BECCS) process, over supported palladium nano-catalysts, to produce formates. We further developed the rechargeable â€œhydrogen batteryâ€ based on the redox equilibrium between amine captured CO2 / formate amine adducts. The stored hydrogen can be catalytically released under mild conditions as the feed to fuel cells for regenerating electricity. The strategies of the process development has been focused on the efficiency (rate of CO2 hydrogenation and dehydrogenation reactions) and energy use (the amount of energy required to utilize CO2 in existing products). This development also includes gaining insight into the relationships among the various structures of amines, solvent effect, distribution of intermediates, catalytic performance, and reaction mechanism.

487340 Catalytic Methanation of Carbon Dioxide with Ni/MgO Catalyst in Dual-Step Reductive Calcination

Abstract:

The heterogeneously catalyzed gas-phase conversion of carbon dioxide with hydrogen to produce methane has gained major interest in recent years [1]. One of the reasons for this increased interest is the quest for hydrogen storage systems. Renewably produced hydrogen is stored via methanation of the greenhouse gas carbon dioxide. Methane can partially substitute natural gas in an established gas grid. Our research focuses on process integration in this context and addresses potential carbon dioxide sources. Calcination is a widely used preparatory step in mineral carbonate beneficiation. During calcination carbon dioxide is released in high concentration into the off-gas. If calcination is performed in hydrogen atmosphere (= reductive calcination), a highly concentrated (CO_2/H_2) process gas stream suitable for direct gas phase methanation can be prepared. Our research [2,3] revealed that reductive calcination is a low temperature process compared to calcination in oxidizing atmosphere, and carbon dioxide is partially converted into methane and/or carbon monoxide in a single-step process. The yield of methane and carbon monoxide can be partially controlled by reaction temperature and pressure. Selective methanation, however, is not achieved in the calcining step. Therefore, methanation of the process gas from reductive calcination by catalytic methanation at ambient pressure with a Ni/MgO catalyst was investigated [4]. Although methanation of CO_2 according to the Sabatier reaction has been investigated since decades, little information about process performance at ambient pressure is available. At moderate temperature (534 K) and ambient pressure CO_2-conversion close to equilibrium conversion beyond 90% is possible. Adsorption of reactants and products does not affect kinteics. At moderate temperature conversion does draw advantages from interaction of carbon dioxide with MgO. Latter effect was confirmed by direct methanation of $MgCO_3$ with hydrogen without catalytic support.

1. Bailera, M., Lisbona, P., Romeo, L.M., and Espatolero, S. (2017) Power to Gas projects review: Lab, pilot and demo plants for storing renewable energy and CO2. Renew. Sustain. Energy Rev., 69 (January 2016), 292–312.

2. Baldauf-Sommerbauer, G., Lux, S., Aniser, W., and Siebenhofer, M. (2017) Synthesis of carbon monoxide from hydrogen and magnesite/dolomite. Chemie Ing. Tech., 89 (1–2), 172–179.

3. Baldauf-Sommerbauer, G., Lux, S., Aniser, W., and Siebenhofer, M. (2016) Reductive calcination of mineral magnesite: hydrogenation of carbon dioxide without catalysts. Chem. Eng. Technol., 39 (11), 2035–2041.

4. Baldauf-Sommerbauer, G., Lux, S., and Siebenhofer, M. (2016) Methanation of carbon dioxide with a nickel/magnesium oxide catalyst. in: AIChE Annual Meeting 2016, San Francisco, USA.

501005 Glycerol Transfer Hydrogenation of CO_2 Using Ir and Ru Carbene Organometallics Immobilized on Active Hydrotalcites in a Packed Bed Flow Reactor

Abstract:

Utilization of CO_2 has become a focal point for reducing carbon emissions. A long-standing method involves direct hydrogenation and transfer hydrogenation (TH) using isopropanol as a hydrogen donor to convert CO_2 to value-added commodity chemicals such as formic acid. Less explored is the use of glycerol, a low value byproduct of biodiesel synthesis made from trans-esterification of triglycerides, as a substitute for isopropanol due to thermodynamic and kinetic challenges. Using CO_2 as a substrate and glycerol as a hydrogen donor for TH yields formate and lactate under basic condition following a Cannizzaro reaction. Herein we explored the use of ruthenium and iridium N-heterocyclic carbenes (NHC) homogeneous catalysts immobilized on a tunable hydrotalcite (HT) supports for CO_2 TH, converting CO_2 to potassium formate and glycerol to potassium lactate in a packed bed flow reactor. HT was synthesized using 3:1 ratio of Al:Mg in flow, and composition was varied by incorporating Cu or Fe in place of Al up to a total of 15 wt. %. Comparison of the heterogenous catalysts to their respective homogenous catalysts indicated that the HT is non-innocent during the reaction. The use of a flow reactor allowed for continuous data collection of kinetic parameters as well as catalytic efficiencies within a single reaction setup and demonstrated that immobilized catalysts are stable for over 300 hours on stream with no sign of deactivation in the presence of 0.25 - 2.0 M potassium hydroxide (KOH) at 150-225ËšC. Precise control over the gas-to-liquid ratio within the reactor provided insight into the efficiency of catalysts at utilizing CO_2, giving scale-up details on required gas stream purity and operating pressure requirements. Reactions were initially carried out in 45 bars of CO_2 in a batch reactor and then scaled to a flow reactor using less than 35 bar CO_2. Upon depletion of accessible carbonate in the solution, the formate yield stabilizes while lactate/lactic acid yield continues to form under acid conditions at a reduced rate. Due to similar turn-over-frequencies (TOF) in 0.25-2.0 M KOH exceeding 4.4 K and 1.5 K per hour for lactate and formate, respectively.

489777 Combining CO_2 Reduction with Propane Oxidative Dehydrogenation over Bimetallic Catalysts

Abstract:

Combining CO_2 reduction with propane oxidative dehydrogenation over bimetallic catalysts

Elaine Gomez and Jingguang G. Chen*

Department of Chemical Engineering, Columbia University, New York City, NY 10027, USA

*jgchen@columbia.edu

Abstract: Propylene is one of the most diverse petrochemical building blocks used for the production of many chemicals (e.g. polypropylene, propylene oxide, and acrylonitrile). The inherent variability and insufficiencies in the co-production of propylene from steam crackers has raised concerns regarding the global propylene production gap and has directed industry to more on-purpose propylene technologies[1]. The oxidative dehydrogenation of propane by CO_2 (CO_2-ODHP) can potentially fill this gap while consuming a greenhouse gas. Utilizing propane and CO_2 presents several advantages over the conventional processes of direct or O_2 assisted propane dehydrogenation, including lower operating temperatures, less coke formation, and higher equilibrium conversions.[2] The reactions of CO_2 with propane may occur through two distinct pathways, oxidative dehydrogenation ($CO_2 + C_3H_8 \rightarrow C_3H_6 + CO + H_2O$) and dry reforming ($3CO_2 + C_3H_8 \rightarrow 6CO + 4H_2$). The two reactions should occur simultaneously at temperatures around 823 K and above with considerable conversions, allowing the formation of both dehydrogenation products (propylene) and reforming products (synthesis gas). Thus, it is of great interest to identify catalytic systems that can selectively break either the C-H bond to produce propylene or the C-C bonds to produce synthesis gas ($CO + H_2$).

Non-precious Fe_3Ni and precious Ni_3Pt supported on CeO_2 have been identified as promising catalysts for CO_2-ODHP and dry reforming, respectively, in flow reactor studies conducted at 823 K. For oxidative dehydrogenation, nonprecious metal based Fe_3Ni had higher stability and propylene production rates. Fe_3Ni showed distinct C_3H_6 yield compared to its parent metals and the Ni_3Pt catalyst showed both greater activity and stability than the corresponding monometallics. In-situ X-ray absorption spectroscopy measurements revealed the oxidation states of metals under reaction conditions and density functional theory was utilized to identify the most favorable reaction pathways over the two types of catalysts. Additional flow reactor studies have been performed on the Fe_3Ni and Ni_3Pt bimetallic systems with other oxide supports (TiO_2, ZrO_2, SiO_2, γ-Al_2O_3, and V_2O_5) to examine the effects of strong support metal interactions, surface defects, inert surfaces, irreducibility, as well as alkane activation on both the CO_2-ODHP and DRP pathways. Overall, two types of bimetallic catalysts have been identified for the $CO_2 + C_3H_8$ system; one is selective toward reforming (C-C bond cleavage) and the other favors dehydrogenation (C-H bond scission).

References

(1) Plotkin, J. S. *ACS News Industrial Chemistry & Engineering*. 2016, pp 1–4.

(2) Wang, S.; Zhu, Z. H. *Energy and Fuels* **2004**, *18* (4), 1126–1139.

494828 CO_2 Splitting Using Miec Membranes

Abstract:

In this presentation, we discuss fuel production using membrane-supported CO_2 splitting. A mixed ionic-electronic conductive (MIEC) $La_{0.9}Ca_{0.1}FeO_{3-\delta}$ (LCF-91) membrane is utilized. At intermediate temperatures (850 – 990°C), CO_2 undergoes thermochemical reduction on the feed-side producing CO, while oxygen incorporates into the perovskite as lattice oxygen. Oxygen diffuses to the sweep side, driven by the chemical potential gradient across the membrane generated in this case by fuel (either H_2 or CO in this study) oxidation on that side.

LCF-91 membranes have been examined in our laboratory before for air separation [1-3] and water splitting [4, 5]. In this work, we report a detailed study on CO_2 reduction on the same membrane. We examine the dependence of CO_2 splitting rate on various operating conditions such as CO_2 concentration on the feed-side, fuel (H_2 or CO) concentration on the sweep side and temperatures. CO_2 splitting rate is proportional to the oxygen flux across, and the measured maximum flux is 0.19 μmol/cm^2•s on a 1.3 mm membrane at 990°C, when 9.5% H_2 is added to the sweep side (balance with argon gas). Long term study shows almost a constant flux, i.e., stable performances for 106 hours. Following the experiment, morphology changes, carbonate formation and calcium/iron enrichment are observed on the feed-side surface, while no obvious changes are found on the sweep-side surface as revealed by SEM, EDS and XRD. A resistance-network model is developed to describe the oxygen flux, in which CO_2 direct-incorporation mechanism is used on the feed side, and the Mars-van Krevelen (MvK) mechanism for the oxidation reaction on the sweep side (for both H_2 and CO) [6]. Using the flux measurements, surface kinetics parameters are derived, which shows that H_2 has larger oxidation rate constant than CO. The model also shows that in CO_2 splitting, the rate limiting step changes from the CO_2 splitting reaction on the feed-side to the fuel (H_2 or CO) oxidation reaction on the sweep-side at higher temperatures. This is different from the H_2O splitting cases, where fuel oxidation is always the limiting step within the same temperature range tested. This is supported by the change in the slope of the Arrhenius plot of the oxygen flux. Moreover, the transition temperature for H_2 is higher than that of CO (both used as fuels).

Porous layers made from the same materials are added on both side of the membrane to enhance the oxygen flux. Results shows that the flux is raised by about 0.02 μmol/cm^2•s when CO is used as a fuel. And the same transition temperature is observed in the Arrhenius plot of oxygen flux with or without porous layers.

Reference

[1] Hunt, A., Dimitrakopoulos, G., and Ghoniem, A. F., 2015, "Surface oxygen vacancy and oxygen permeation flux limits of perovskite ion transport membranes," J. Membr. Sci., 489, pp. 248-257.

[2] Dimitrakopoulos, G., and Ghoniem, A. F., 2016, "A two-step surface exchange mechanism and detailed defect transport to model oxygen permeation through the $La_{0.9}Ca_{0.1}FeO_{3-\delta}$ mixed-conductor," J. Membr. Sci., 510, pp. 209-219.

[3] Dimitrakopoulos, G., and Ghoniem, A. F., 2017, "Developing a multistep surface reaction mechanism to model the impact of H_2 and CO on the performance and defect chemistry of mixed-conductors," J. Membr. Sci., 529, pp. 114-132.

[4] Wu, X. Y., Chang, L., Uddi, M., Kirchen, P., and Ghoniem, A. F., 2015, "Toward enhanced hydrogen generation from water using oxygen permeating LCF membranes," PCCP, 17(15), pp. 10093-10107.

[5] Wu, X. Y., Ghoniem, A. F., and Uddi, M., 2016, "Enhancing co-production of H_2 and syngas via water splitting and POM on surface-modified oxygen permeable membranes," AIChE J., 62(12), pp. 4427-4435.

[6] McFarland, E. W., and Metiu, H., 2013, "Catalysis by Doped Oxides," Chem. Rev., 113(6), pp. 4391-4427.

494005 Molten Metal Catalysts for Methane Pyrolysis

Abstract:

Molten metals were investigated as catalysts for the thermochemical pyrolysis of methane. Conversion of methane to hydrogen and solid carbon is demonstrated in bubble column reactors where the solid carbon can be readily separated and removed enabling continuous processes. Twenty-one molten metal systems consisting of elemental metals and binary alloys are compared for methane pyrolysis activity. Nickel-bismuth was found to be the most active catalyst, and 95% methane conversion was observed at 1060 °C in a 1.1 meter bubble column. Initial characterization of the carbon product indicated predominately sp^2 graphitic carbon as the solid product. Although the hydrogen produced from partial oxidation of methane by pyrolysis retains only approximately half the combustion energy potential as complete methane combustion, pyrolysis produces CO_2-free hydrogen, and catalytic molten metals may provide a means for performing reactive separation of the hydrogen from the potentially valuable carbon co-product.

500486 Increasing Activity, Reducing Coking, and Promoting Unexpected Reaction Pathways during Methane Steam Reforming By Applying Uniform Electric Fields in a Scalable Reactor

Abstract:

It has been shown both computationally and in small-scale studies using STM tips for single-molecule analysis that very high electric fields can change both the kinetics and thermodynamics of chemical reactions. This could be an exciting and unexplored new area for reaction engineering, but generating uniform electric fields in a larger scale continuous flow reactor has proven to be a significant challenge. We have designed a system capable of generating uniform electric fields over Ni catalysts for methane steam reforming, which has been compared to density functional theory computations by our collaborating group in the past. This talk focuses on the experimental observations, with special focus on the properties of the catalyst itself after methane steam reforming in the presence of electric fields. We report both expected findings and a new mystery â€“ and our attempts to unravel said mystery. On the expected side, our DFT calculations correctly predicted that, in a positive electric field, reaction rates would increase consistent with the lowering of the energy barrier of the first methane C-H bond scission; H_2O would bind more readily to the surface; H_2O would have a more difficult time decomposing; and that coke would have a more difficult time forming. Unexpectedly, the negative field apparently promotes alternate reaction pathways which may lead to C-C coupling reactions to form larger unsaturated hydrocarbons without coking. Both post-reaction catalysts were scrutinized using visual observations including macroscopic photographs and SEM imaging, x-ray photoelectron spectra, temperature-programmed oxidation, and x-ray diffraction patterns.

495752 Developing Multi-Scale Models of Bimetallic Catalysts for the Hydrodeoxygenation of Bio-Oil Compounds

Abstract:

With the ever-increasing need to find a sustainable source of renewable energy, bio-oil produced via the fast pyrolysis of biomass is a promising source of liquid fuel; however, the resulting bio-oil contains oxygenated products that contribute to poor fuel quality [1]. Hydrodeoxygenation (HDO) is used to refine the bio-oil by reducing the oxygen content, ideally using a minimal amount of H_2 in order to form H_2O. Bimetallic catalysts such as Pd on Fe have demonstrated synergistic behavior that contributes to a more cost-effective and longer-lasting catalyst [2-4]. Similarly, Pt/Sn catalysts have also exhibited traits favorable for the HDO of phenolic compounds [5], namely a propensity to promote lower temperature desorption of aromatic compounds like benzene [6]. Identifying the cause of the different synergetic effects of these systems can provide insight into the synthesis of a superior HDO bimetallic catalyst. However, the behavior of bimetallic catalysts is complex and necessitates a thorough understanding of the nanoscale behavior of these systems in order to develop a truly predictive model on relevant time scales. In particular, we present a DFT study that examines the importance of including coverage effects into theoretical models in order to accurately predict experimental behavior.

Quantifying the lateral interactions between adspecies is necessary in the development of truly predictive, theoretical models for bimetallic catalysts, as they will significantly affect the coverage and diffusion of adspecies and therefore the heterogeneous catalytic environment. However, such a characterization of the lateral interactions between adspecies for the highly complex HDO reaction has not yet been attempted. To that end, we have characterized the benzene-benzene lateral interactions on a Pt (111) surface and a Pt_3Sn (111) surface alloy. This coverage-dependent adsorption behavior was incorporated into a kinetic model for each surface and used to simulate the temperature-programmed desorption (TPD) spectra. We benchmarked our simulated spectra by comparing it to experimental TPD [6]. We found that it is imperative to incorporate the effect of surface coverage in these theoretical models if the desorption behavior is to be accurately predicted (see Figure 1). Furthermore, we found that a mean-field model is sufficient in describing the benzene-benzene lateral interactions. Using our models, we were able to interpret the desorption behavior at an atomistic level and provide a deeper level of insight into these systems. While it was previously speculated that the broad desorption peak for benzene on Pt (111) was the result of desorption from two adsorption sites of different binding strengths [6], our results indicate that the behavior is due to benzene desorbing from a single site but experiencing strong lateral interactions on the surface. We also found that the adsorption of benzene was significantly weaker on the Pt_3Sn (111) than on Pt (111), and as such, we were able to assign the monolayer desorption of benzene to a TPD peak previously attributed entirely to multilayer benzene desorption [6]. In this regard, we were also able to attribute the higher temperature desorption peaks observed on Pt_3Sn (111) to benzene desorbing from defects in the alloy surface, where Pt atoms were not substituted for Sn atoms in the alloying process.

Our investigation into modeling the lateral interactions of an aromatic molecule on both the Pt (111) and the Pt_3Sn

(111) surfaces warrants the application of our approach onto systems less experimentally characterized, such as Fe surfaces doped with different noble metals.

Figure 1. TPD spectra for benzene on Pt (111) simulated with coverage effects (blue) and without coverage effects (light green) compared to experiment (black) [6].

References

[1] H. Wang, J. Male, Y. Wang, ACS Catal. 3 (2013) 1047-1070.

[2] Y. Hong, H. Zhang, J. Sun, K.M. Ayman, A.J.R. Hensley, M. Gu, M.H. Engelhard, J.-S. McEwen, Y. Wang, ACS Catal. 4 (2014) 3335-3345.

[3] J. Sun, A.M. Karim, H. Zhang, L. Kovarik, X.S. Li, A.J. Hensley, J.-S. McEwen, Y. Wang, J. Catal. 306 (2013) 47-57.

[4] A.J.R. Hensley, Y. Hong, R. Zhang, H. Zhang, J. Sun, Y. Wang, J.-S. McEwen, ACS Catal. 4 (2014) 3381-3392.

[6] C. Xu, Y.L. Tsai, B.E. Koel, J. Phys. Chem. 98 (1994) 585-593.

496761 Mechanistic Insights into Hydroeoxygenation of Phenol on Bimetallic Phosphide Catalyst

Abstract:

Bimetallic transition metal catalysts provide immense opportunity to tailor catalytic function via composition modulation to promote desired chemical transformation. Addition of non-metal components such as oxygen, nitrogen, and phosphorous can provide for better charge density distribution to enhance Lewis acidic character. Recent work by Rensel et al.[1] has shown that FeMoP catalyst is highly selective for Câ€"O bond cleavage for aryl ethers. Furthermore, the catalytic function of FeMoP can be tuned by altering the Fe/Mo ratio to improve benzene selectivity upon hydrodeoxygenation of phenol.[2] In this presentation, we will share density functional theory calculations that provide mechanistic insights into hydrodeoxygenation of phenol on $Fe_XMo_{2-X}P$ catalyst. We find that optimal Fe/Mo composition has smallest activation energy barrier for Câ€"O bond cleavage which correlates well with experimental TOFs. Bader charges and density of states analysis provide pointers to the change in the electronic structure of the catalyst upon changing Fe/Mo composition.

495987 Catalytic Conversion of Furfural to Methylfuran: Investigating Reaction Mechanisms on Ni and the Effect of Boron Doping on the Activity and Selectivity of the Catalyst

Abstract:

Fast pyrolysis of lignocellulosic biomass is emerging as a promising route for the production of liquid transportation fuels. However, the bio-oil produced from pyrolysis has a high oxygen content (~45%) which results in a poor heating value, thus limiting its direct usability as a fuel. The catalytic hydrodeoxygenation (HDO) of bio-oil in the presence of hydrogen is an effective technique for the removal of oxygen groups from bio-oil. Furfural is one of the major bio-oil components and is a potential starting material for the production of compounds like 2-methylfuran (MF), which can be blended into transportation fuels. The challenge is to develop catalysts which can selectively break C – O bonds, while preserving the C – C bonds. In this context, hydrogenolysis of furfural (C_{formyl} – O cleavage) to produce MF is desirable, as the high octane number of MF allows easy blending into gasoline/diesel pools. Vapor phase HDO of furfural over Ni catalysts leads to furfuryl alcohol (via hydrogenation of formyl group) as the dominant product at low temperatures, while furans (via decarbonylation by C – C cleavage) and C4 products (via furan ring opening) dominate at higher temperatures[1-2]. We investigated the potential competitive pathways for the catalytic conversion of furfural on model Ni(111) surfaces using Density Functional Theory. Our calculations show that decarbonylation (C – C cleavage) and ring opening (C_{ring} – O cleavage) of furfural to form furan and ring opened C4 products are kinetically favored pathways on Ni(111) surface, in agreement with experimental studies. Recent studies have suggested that the activity and selectivity of catalyst can be improved by doping boron[3-6]. On Ni(111) surfaces boron binds strongly at the subsurface positions, resulting in a corrugated Ni surface, similar to a stepped surface. The presence of the subsurface, interstitial boron expands the lattice of Ni and also alters its electronic properties by donating electrons. The incorporation of B in the Ni surface was found to increase the C – C cleavage and ring opening barriers and drastically lower the C_{formyl} – O cleavage barriers, relative to the Ni(111) surfaces. We propose the hydrogenolysis of furfuryl alcohol to form MF to be kinetically favorable over C – C cleavage (furan formation) and ring opening (C4 products), leading to significantly enhanced MF selectivity in the vapor phase hydrogenation of furfural on boron doped Ni surfaces.

References

1. S. Sitthisa, D. E. Resasco, Catalysis Letters 2011, 141, 784-791.
2. S. Sitthisa, W. An, D. E. Resasco, Journal of Catalysis 2011, 284, 90-101.
3. J. Xu, M. Saeys, Journal of Physical Chemistry C 2009, 113, 4099-4106.
4. T. Q. Thang, A. V. Nguyen, D. C. Huynh, T. H. Pham, S. H. Mushrif, Catalysis Science and Technology 2016, 6, 5871-5883.
5. B. Yang, R. Burch, C. Hardacre, P. Hu and P. Hughes, Journal of Physical Chemistry C, 2014, 118, 3664-3671.

6. J. Xu, L. Chen, K. F. Tan, A. Borgna and M. Saeys, Journal of Catalysis, 2009, 261, 158-165.

494725 First-Principles Insights into the Mechanisms and Sites for Base Catalyzed Aldol Condensation and Esterification over Copper

Abstract:

Base-catalyzed aldol condensation and esterification are important reactions in the synthesis of large organic molecules from biomass-derived precursors.[1] These molecules can then subsequently be converted to fuels and useful chemicals. Such reactions are typically catalyzed on homogenous catalysts in the presence of a basic promoter or on basic metal oxides, that catalyze C-C and C-O bond formation.[1-3] Recent studies have shown that non-basic materials such as Cu supported on SiO_2 can actively catalyze both condensation and esterification, when equilibrated mixtures of propanol, propanal and H_2 are reacted.[4] These supported Cu systems result in the formation of 47% 3-pentanone (C-C formation via condensation) as well as 34% propyl propionate (C-O formation via esterification).[4] The results suggest that mixtures of propanol, propanal and H_2 readily equilibrated via the formation of surface propoxide intermediates, that can act as a base, thus, catalyzing both condensation and esterification. Ab-initio density functional theoretical calculations were carried along with detailed kinetic analyses to elucidate elementary steps and sites involved in propanal conversion to propanol and the subsequent C-C and C-O bond formation steps. Detailed charge analyses show that alkoxides which readily form on Cu as well as other group 11 metals readily extract electron density from the metal surface and behave as basic intermediates. For the aldol condensation path, the surface bound propoxide abstracts the weakly acidic hydrogen from a bound propanal intermediate to generate an enolate species in a rate-determining step having a Gibbs free energy barrier of 69 kJ/mol. The enolate subsequently undergoes rapid C-C coupling with a second vicinally-bound propanal to generate a Î²-alkoxide intemediate. The Î²-alkoxide can either eliminate hydrogen to the Cu or react with a surface propanal to form a 2-formyl-3-pentanone intermediate that subsequently decarbonylate to form 3-pentanone. The same alkoxide intermediates can also catalyze esterification. The calculations show that the bound propoxide can readily carry out a nucleophilic attack on the carbonyl group of a surface bound propanal in a rate-determining step to form a hemiacetalate intermediate with a Gibbs free energy barrier of 66 kJ/mol. The hemiacetalate further reacts on Cu to form the propyl propionate. The theoretical results indicate that the rate limiting steps for both C-C and C-O formation essentially involve the same propoxide and propanal surface intermediates, which is fully consistent with experimental findings. We predict similar activation barriers for both C-C and C-O formation with C-O formation being more favorable by 3 kJ/mol, while experiments report that C-C formation is slightly more favorable than C-O formation. In addition, we show that esterification and condensation preferentially proceed over coordinatively-saturated Cu 111 terrace sites. The alkoxide as well as other critical intermediates bind too strongly to the edge and corner sites and as such are blocked and much less active than the terrace sites. The active alkoxide intermediates bind much more weakly to the terrace sites and thus catalyze condensation and esterification. This is is fully consistent with the experimental results which show that larger particles are more active than the smaller Cu particles.

References

[1]Chheda, J.N., and Dumesic, J. A. Catal. Today (2007), 123, 59.

[2]Hamilton, C. A., Jackson, S.D., and Kelly, G.J. Appl. Catal. A (2004), 263, 63.

[3]West, R.M., Liu, Z.Y., Peter, M., Gartner, C.A., and Dumesic, J.A. J. Mol. Catal. A: Chem. (2008), 296, 18.

[4]Sad, M.E., Neurock, M., and Iglesia, E. J.Am.Chem.Soc. (2011), 133, 20384.

500107 Advancing the Selective Oxidation of Ethylene Glycol Via Combining Novel Catalyst Design and Density Functional Theory (DFT) Calculations

Abstract:

Ethylene glycol (EG) is the simplest diol that can be produced from hydration of ethylene oxide or hydrogenolysis of biomass (cellulose) [1]. In recent work, significant research work is focused on EG electroxidation for fuel cells [2]. This status is attributed to the fact that EG has higher boiling point (198 °C) making it a safer liquid fuel and has higher energy density compared to ethanol and methanol. However, in order to harvest the total amount of energy during this process, EG is fully converted into CO_2 that aggravates global warming. In this context, conversion of EG into value-added products chemically could be more attractive. A number of papers are available on EG oxidation in aqueous alkali medium using a non-hazardous oxidizing agent – oxygen gas [3-6]. However, the data reported on this reaction system are mainly on EG conversion and little is known on the reaction mechanism or the viability of applying different catalysts. Moreover, glycolic acid is usually the main product, while formic acid and CO_2 are believed as the side products of over oxidation rather than desirable oxalic acid (dicarboxylic acid) [6]. At the same time, current oxalic acid production suffers from the problems of using corrosive and stoichiometric reagents (nitric acid and sulfuric acid), with low yield and selectivity [7,8]. Suitable catalysts for a single-step conversion of EG to oxalic acid are highly desirable, that would be important to both industrial application and fundamental understanding of alcoholic group oxidation in polyols.

In this study, a series of monometallic catalysts (including Ag, Pd, Ru, Rd, Fe, Pt) supported on CeO_2 were synthesized by using a novel solvothermal method. The oxidation tests were conducted at a mild condition (70 °C, ambient pressure O_2) in aqueous alkali solution. The catalysts were characterized by H_2 chemisorption, inductively coupled plasma (ICP), transmission electron microscopy (TEM) for the metal content, dispersion and morphology information. Volcano plots of normalized TOF values vs. density functional theory (DFT) calculated Gibbs Free energy was established for EG oxidation. Furthermore, a bimetallic Pt-Fe catalyst showed a significant rate enhancement and selectivity in one-step conversion of EG-to-oxalic acid. The reaction mechanism for EG catalytic oxidation by O_2 on (100) and (111) surfaces was studied using DFT calculations which rationalizes the experimental observations that the bimetallic Pt-Fe catalyst has a significantly higher activity (TOF) compared to the monometallic catalysts (Pt and Fe). Fundamental concepts in the computational chemistry and its relation to experimentally observed performance of bimetallic catalysts will be presented in this work.

References

1. Yue, H., Zhao, Y., Ma, X., & Gong, J. (2012). Ethylene glycol: properties, synthesis, and applications. *Chemical Society Reviews*, *41*(11), 4218-4244.

2. Serov, A., & Kwak, C. (2009). Review of non-platinum anode catalysts for DMFC and PEMFC application. *Applied Catalysis B: Environmental*, *90*(3), 313-320.

3. Sakharov, A. M., Sakharov, P. A., & Zaikov, G. E. (2012). Catalytic Oxidation of Ethylene Glycol by Dioxygen in Alkaline Medium. The New Example of One-Stage Oxidative Cleavage of C–C Bond. *Molecular Crystals and Liquid Crystals*, *555*(1), 168-176.

4. Berndt, H., Pitsch, I., Evert, S., Struve, K., Pohl, M. M., Radnik, J., & Martin, A. (2003). Oxygen adsorption on Au/Al 2 O 3 catalysts and relation to the catalytic oxidation of ethylene glycol to glycolic acid. *Applied Catalysis A: General*, *244*(1), 169-179.

5. Case, J. M. (2009). *Gold catalysts prepared by ion exchange for use in ethylene glycol oxidation: An exploratory study* (Doctoral dissertation, University of Cape Town).

6. Kopeykina, L. Y., Vodyankina, O. V., Kurina, L. N., & Golovko, A. K. Selective ethylene glycol oxidation on the oxide catalysts. In *Science and Technology, 2001. KORUS'01. Proceedings. The Fifth Russian-Korean International Symposium on* (Vol. 2, pp. 177-179). IEEE.

7. Isshiki, T., Suzuki, T., Yashima, Y., & Yonemitsu, E. (1972). *U.S. Patent No. 3,678,107.* Washington, DC: U.S. Patent and Trademark Office.

8. Knepper, W., & Rohl, H. (1975). *U.S. Patent No. 3,864,393.* Washington, DC: U.S. Patent and Trademark Office.

488083 Quantum Chemical Characterization of Catalytic Ester Decarbonylation: Olefins from Biomass

Abstract:

Quantum Chemical Characterization of Catalytic Ester Decarbonylation: Olefins from Biomass

Synthetic transformations of carboxylic acids into olefins are of paramount importance since bio-mass derived carboxylic acids could then be a renewable source for polyolefin syntheses. An important goal, then, is to develop catalytic transformations that sustainably transform these structurally diverse acid educts into olefin products with high selectivity. Production of olefins via transition-metal-catalyzed decarbonylation of bio-derived carboxylic acids is one approach that has been explored and demonstrated to be possible with low toxicity, and tolerant of diversity.[1-2]. As a subclass, linear a-olefins (LAOs) are important commodity chemicals that find utility in a wide range of industrial applications. LAOs are usually generated from petroleum, which is a nonrenewable and depleting resource. A green and sustainable approach would be to produce LAOs by decarbonylative dehydration of fatty acids, which are inexpensive and renewable.

Direct routes for the conversion of carboxylic acids to olefins typically proceed with high activation energies. However, carboxylic acid derivatives such as esters are more reactive and can serve as key precursors in carboxylic acid transformations, and in particular transition-metal-based catalysts have been shown capable of transforming them to high-value olefins. However, controlling selectivity for terminal alkene formation is a major challenge. A highly selective, palladium-catalyzed decarbonylation of *p*-nitrophenyl esters of fatty acids to their corresponding LAOs has been very recently developed.[3] A cooperative ligand system including phosphines and N-heterocyclic carbenes is reported to exhibit enhanced selectivity to terminal olefins. Herein, we investigate the decarbonylation mechanism of *p*-nitrophenylbutyrate to rationalize the switching reactivity and selectivity trends based on either individual or cooperative roles of phosphine and heterocyclic carbene ligands. The primary goal of the project is to elucidate the origin of terminal alkene **A** selectivity vs the thermodynamically more stable internal alkene **B** which can be generated as an undesirable alternative product (Scheme 1).

Scheme 1. Pd-catalyzed decarbonylation of *p*-nitrophenylester to a-olefins. XantPhos and IPr ligands comprise the dual ligand system.

To gain mechanistic insight at the atomic level of detail, we performed density functional theory (DFT) modeling. We focused on the Pd-catalyzed decarbonylation of *p*-nitrophenylbutyrate in order to fully characterize reaction coordinates for the formation of terminal **A** and internal **B** alkene products and to explain the roles of the ligands in promoting a-selectivity. To assess selectivity, we have thoroughly compared the competing reaction mechanisms using three ligand systems: IPr, XantPhos, and a mixture of IPr and XantPhos. Calculations suggest that the bulkiness of IPr combined with the bidentate nature of XantPhos prompt a rapid decoordination of the terminal alkene, thus precluding the alkene isomerization pathway.

References

1. Rodriguez, N.; Goossen, L. J., Decarboxylative coupling reactions: a modern strategy for C-C-bond formation. *Chemical Society Reviews* 2011, *40* (10), 5030-5048.
2. Le Nôtre, J.; Scott, E. L.; Franssen, M. C. R.; Sanders, J. P. M., Selective preparation of terminal alkenes from aliphatic carboxylic acids by a palladium-catalyzed decarbonylation–elimination reaction. *Tetrahedron Letters* 2010, *51* (29), 3712-3715.
3. Alex John, Levi T. Hogan, Marc A. Hillmyer and William B. Tolman, Olefins from biomass feedstocks: catalytic ester decarbonylation and tandem Heck-type coupling, Chem. Commun., 2015, 51, 2731.

500166 Multi-Scale Models of Oxygen on Iron-Based Hydrodeoxygenation Catalysts: Elucidating the Effect of External Electric Fields and Surface Dopants

Abstract:

Fe catalysts are of great interest as they have been shown to be highly selective for the catalytic upgrading, via hydrodeoxygenation (HDO), of bio-oil compounds. Unfortunately, these catalysts alone are unacceptably prone to oxidative deactivation. One method of improving the oxidation resistance of Fe catalysts has been to dope the Fe with a second metal (e.g. Pd, Pt, etc.). However, the addition of the second metal increases the catalyst cost and complexity, thereby making it harder to experimentally optimize such systems to both increase activity and prevent oxidative deactivation. Previous work has shown that externally applied electric fields can dramatically improve the coke resistance of a Ni catalyst used in methane steam reforming (MSR).[6-9] These results suggest that external electric fields are another adjustable parameter that can be used to increase activity, tune selectivity, or prevent catalyst deactivation. Due to the highly polar nature of oxygen-containing compounds, it is expected that electric fields will have a large effect on HDO and could potentially be used in conjunction with doping to create a highly stable and selective Fe-based HDO catalyst. To this end, we must develop multi-scale models that can guide the selection of dopants in a rational manner—with both surface coverage and electric fields explicitly accounted for.

Here, we develop density functional theory (DFT) parameterized lattice gas models of oxygen on both undoped and doped Fe(100) surfaces to show the direct effect of an applied electric field on oxygen adsorption. Typical DFT models consider a uniform coverage distribution of adsorbates on the surface, when in reality, this is not the case.[2-4] Our cluster expansions of oxygen on Fe(100) surfaces both with and without an externally applied electric field accounts for the more realistic non-uniform adsorbate distributions and characterizes the lateral interactions between oxygen adatoms.[5] The apparent sensitivity of the lateral interactions on an applied external electric field reveals their influence on the oxidative deactivation of Fe catalysts.

Our preliminary results for O/Fe(100) indicate that an applied electric field has a significant effect on the attractive/repulsive interactions between adsorbed oxygen atoms. As seen in Figure 1, the clusters that are most important to the predictiveness of the model also change after the application of an electric field. These changes suggests that electric fields will have major effects on the Fe oxidation process as the electric field has a significant effect on the dynamic surface behavior of oxygen. Using such a lattice gas model for oxygen on both undoped and doped Fe(100), we will characterize the synergy between surface dopants and external electric fields in order to minimize the oxidation potential of the Fe surface. Our preliminary results suggest that one can expect significant deviations in catalytic activity after the application of external electric fields, but also, they suggest that mean-field models will not be sufficiently predictive for the work needed moving forward. Overall, the work presented here provides the first ingredients necessary for the selection of potential dopants for optimizing Fe-based HDO catalysts in the presence of external electric fields.

Figure 1: Effect of an electric field on oxygen cluster lateral interactions on the Fe(100) surface.

References:

1. Wang, H., Male, J. & Wang, Y. Recent Advances in Hydrotreating of Pyrolysis Bio-Oil and Its Oxygen-Containing Model Compounds. *ACS Catal.* **3,** 1047–1070 (2013).

2. Govender, A., Curulla Ferré, D. & Niemantsverdriet, J. W. The Surface Chemistry of Water on Fe(100): A Density Functional Theory Study. *ChemPhysChem* **13,** 1583–1590 (2012).

3. Blonski, P., Kiejna, A. & Hafner, J. Theoretical study of oxygen adsorption at the Fe (110) and (100) surfaces. *Surface Science* **590,** 88–100 (2005).

4. Freitas, R. R. Q., Rivelino, R., de Brito Mota, F. & de Castilho, C. M. C. Dissociative Adsorption and Aggregation of Water on the Fe(100) Surface: A DFT Study. *J. Phys. Chem. C* **116,** 20306–20314 (2012).

5. Frey, K., Schmidt, D. J. & Wolverton, C. Implications of coverage-dependent O adsorption for catalytic NO oxidation on the late transition metals. *Catalysis Science & ...* (2014). doi:10.1039/C4CY00763H

6. Che, F., Gray, J. T., Ha, S. & McEwen, J.-S. Improving Ni Catalysts Using Electric Fields: A DFT and Experimental Study of the Methane Steam Reforming Reaction. *ACS Catal.* (2016). doi:10.1021/acscatal.6b02318

7. Che, F., Ha, S., McEwen, J. S., Hensley, A. J. & Zhang, R. Elucidating the Role of the Electric Field at the Ni/YSZ Electrode: A DFT Study. *J. Phys. Chem. C* (2016). doi:10.1021/acs.jpcc.6b01292

8. Che, F., Ha, S. & McEwen, J. S. Hydrogen Oxidation and Water Dissociation over an Oxygen-Enriched Ni/YSZ Electrode in the Presence of an Electric Field: A First Principles-Based Microkinetic .... *Industrial & Engineering Chemistry ...* (2017). doi:10.1021/acs.iecr.6b04028

9. Che, F., Gray, J. T., Ha, S. & McEwen, J. S. Catalytic water dehydrogenation and formation on nickel: Dual path mechanism in high electric fields. *Journal of Catalysis* (2015). doi:10.1016/j.jcat.2015.09.010

495994 Techno-Economic Comparison of an Industrial Scale Pyrolysis of Seaweed for Liquid Fuel Production: H_2 Production vs. H_2 Purchase Scenarios

Abstract:

In this study, process design, simulation and technoeconomical assessment is performed using Aspen plus software and experimental data for indirectly heated pyrolysis of *Saccharina japonica* seaweed. The scale of the process was set to 350,000 tons per year of dry feed. For the upgrading and stabilization of the liquid pyrolysis product, hydrogen is required. Experimentally determined molar ratio of hydrogen to feed is 5.5 which at this scale translates to 220 kmol/hr. The base case process design includes on site hydrogen production through natural gas reforming and pressure swing adsorption, which presents a significant capital cost investment of estimated 55% compared to other parts of the process. This study compares the base case with hydrogen purchase and provides economical insight in terms of expected lower capital cost investment and lower Minimum Product Selling Price.

498430 Techno-Economic and Life Cycle Analysis for the Production of Renewable Acrylonitrile from Non-Food Biomass

Abstract:

The industrial sector anticipates an 11-18% annual increase in the market for carbon fiber, specifically driven by the motivation to reduce weight for vehicles. Currently, the highest performance carbon fibers are produced by heat treatment or controlled pyrolysis of polyacrylonitrile (PAN). The precursor acrylonitrile (ACN) monomer is currently produced via ammoxidation of propylene, a high volume commodity chemical subject to the volatility of the petrochemical market. Propylene accounts for approximately 70% of the total ACN production cost. The goal of this work is to develop a cost-effective, low environmental impact process for production of ACN from biomass-derived sugars, such that the ACN meets all chemical specifications of and is a direct replacement for conventional propylene-based ACN.

Southern Research (SR) has developed the biomass to ACN (B2ACN) process under a cooperative agreement with the Department of Energy during Phase I. The process is currently being designed for Phase II pilot studies at a scale of kg/hr. Briefly, B2ACN is a multi-step catalytic process for the conversion of sugars from non-food biomass to ACN at mild conditions. The process utilizes known pretreatment methods for recovery of sugars from any type of biomass. In the first reaction step, sugars are converted to glycols/glycerol using a novel multi-functional catalyst. After desired separation, glycerol is then converted to a gas phase intermediate acrolein in the second reaction step, which is followed by ammoxidation of acrolein to acrylonitrile in the final reaction step. Critical performance attributes of the acrylonitrile product are compared to a propylene-derived standard for production readiness in the PAN and carbon fiber processes. The process has been optimized for heat recovery and separations. Phase I experimental results and initial estimates show a significant reduction in greenhouse gas (GHG) emissions of up to ~37% and cost reduction of up to ~15-22% compared to propylene-derived ACN. This paper will discuss the techno-economic and lifecycle impact on the production of ACN with respect to key variables, including feed materials, impurity/by-product separations, catalysts, utilities, water utilization, and green field capital expenditures tailored to downstream applications. The motivation of this study is to provide a guide to end consumers and brand owners to realize the potentials and pitfalls of the process for further improvements and adoption.

499367 Relative Sustainability of Natural Gas Assisted High-Octane Gasoline Blendstock Production from Biomass

Abstract:

Biomass-derived hydrocarbon fuel technologies are being developed and pursued for better economy, environment, and society benefits underpinning the sustainability of transportation energy. Increasing availability and affordability of natural gas (NG) in the US can play an important role in assisting renewable fuel technology development, primarily in terms of economic feasibility. When a biorefinery is co-processing NG with biomass, the current low cost of NG coupled with the higher NG carbon conversion efficiency potentially allow for cost competitiveness of the fuel while achieving a minimum GHG emission reduction of 50% or higher compared to petroleum fuel. This study evaluates the relative sustainability of the production of high-octane gasoline blendstock via indirect liquefaction (IDL) of biomass (and with NG co-feed) through methanol/dimethyl ether intermediates. The sustainability metrics considered in this study include minimum fuel selling price (MFSP), carbon conversion efficiency, life cycle GHG emissions, life cycle water consumption, fossil energy return on investment (EROI), GHG emission avoidance cost, and job creation. Co-processing NG can evidently improve the MFSP. Evaluation of the relative sustainability can shed light on the biomass-NG synergistic impacts and sustainability trade-offs associated with the IDLâ€™s high-octane gasoline blendstock production.

495797 *In situ* Catalytic Fast Pyrolysis of *Nannochloropsis sp.* using Co-Mo Catalysts for Value Added Chemicals

Abstract:

Growing global energy need demands an urge for exploitation of renewable and environmentally friendly energy resources. In order to reduce our dependency on fossil fuels, carbon neutral sources such as biomass are seen as potential source of energy. Besides lignocellulosic biomass, fast growing plants such as microalgae are looked upon as a prospective source of energy. Microalgae do not require arable land for its cultivation, and so they do not compete with food crops. Thermochemical conversion of microalgal biomass is a promising strategy to convert them into valuable chemicals and fuel molecules. Fast pyrolysis of microalgae is recognized as a promising technique to produce bio-oil. However, high oxygen and nitrogen content of bio-oil limit its direct application in engines or transportation fuels. The various oxygenates like aldehyde, ketone, carboxylic acid and nitrogen containing nitriles, pyrroles, indoles result in low heating value, poor storage stability and low thermal stability. The problem of complex composition of bio-oil can be addressed by deploying catalysts for enhanced selectivity with an aim of promoting deoxygenation, denitrogenation and dehydration resulting in a stable product.

In this study *Nannochloropsis oculata* (*N. oculata*), a high lipid and protein containing microalgae, was fast pyrolyzed in the absence and presence of catalysts. The effect of temperature on catalytic and non-catalytic fast pyrolysis and the effect of catalyst to algae ratio on product distribution was studied. The algae was characterized by proximate, elemental, heating value and thermal stability analyses using various analytical techniques such as thermogravimetric analyzer (TGA), CHNS analyzer and bomb calorimeter. Fast pyrolysis experiments were performed in an analytical micropyrolyzer and the products were analyzed in an online gas chromatograph-mass spectrometer (Py-GC/MS) at different temperatures from 400-600°C. Bifunctional Co-Mo supported on γ-Al_2O_3 was prepared and characterized before using it as a catalyst for Py-GC/MS experiments. Thermal stability, pore size distribution and catalyst composition were studied using various characterization techniques like TGA, porosimetery and ICP-OES (Infrared coupled plasma – optical emission spectroscopy), respectively. This is the first study to investigate the effect of Co-Mo catalysts on *N. oculata* under the fast pyrolysis conditions.

Aromatic and aliphatic hydrocarbons were the major products (c.a. 50%) produced during the non-catalytic fast pyrolysis at 500°C. The major products were benzene, styrene and alkenes (C_{14} – C_{19}), nitriles (C_4 – C_6). The yields of oxygenated and nitrogen containing compounds were low in the pyrolysates. Catalyst: algae ratios of 1:9, 1:3, 1:1 and 2:1 were tested for catalytic fast pyrolysis. The use of catalyst resulted in the formation of 1-isocyanobutane and dimethylketene (combined yield of 25-40%) as the major compounds. Long chain nitriles, long chain alkanes and alkenes, polyaromatic and monoaromatic hydrocarbons were the major product groups. Pyrolysates obtained from catalytic fast pyrolysis had lower amount of oxygen containing organic compounds (< 3%) as compared with non-catalytic fast pyrolysis (c.a. 21%). This was evident from the increase in the approximate HHV (33-40 MJ kg^{-1}) of organics in the pyrolysates from 18 MJ kg^{-1} (raw dried *N. oculata*). Highest deoxygenation was observed with CoMo

catalyst to algae ratio 1:3. Mo and support (γ-Al_2O_3) as catalysts resulted in long chain nitriles. The plausible reaction mechanism for the conversion of microalgae to important chemicals was proposed in this study. More interesting results on (i) effect of metal loading, (ii) effect of temperature and (iii) effect of catalyst: algae ratio will be discussed during the presentation.

501045 Wood Adhesive As a Coproduct of Cellulosic Alcohol: Crosslinking and Bond Performance

Abstract:

The conversion of woody biomass to fermentable sugars through enzymatic saccharification (hydrolysis) often leaves behind solid residues that are rich in lignin. Our laboratory had demonstrated that these residues can be converted into a thermosetting wood adhesive without pre-purifying for lignin. This presentation reports the evolution of crosslinking during thermal curing of the adhesive. Aspen wood flour was enzymatically hydrolyzed, milled, and blended with a biobased crosslinker before application onto wood sticks to form lap shear specimens. The lap shear assemblies were hot-pressed for different durations of time and then mechanically tested to infer bond strength development. Variously cured adhesives were also examined for characteristic functional groups of the crosslink using infrared spectroscopy and conductometric titration. Results showed that the amount of functional groups was closely correlated to the lap shear bond strength of the cured and partially cured adhesive. This finding suggests that the bond performance of the adhesive can be potentially optimized by manipulating crosslinking conditions during thermal curing. This outcome signifies an improved understanding of saccharification residues for adhesive coproducts aiming at increasing the economic viability of the bioenergy system.

497553 Design, Fabrication, Testing, and Operation of a Continuous Reactor for Hydrothermal Carbonization

Abstract:

Hydrothermal carbonization (HTC, thermal hydrolysis, or wet torrefaction) is a treatment process which converts moist feedstocks into homogenized, carbon rich, and energy dense solid fuel, called hydrochar. One of the main advantages of HTC compared to other thermochemical treatment processes is the use of moisture as reaction medium and reactant, thereby precluding the need for drying prior to treatment. Thermodynamic properties of water change greatly in the subcritical region from 180-280 °C, and as a result, subcritical water behaves as a non-polar solvent and mild acid and base catalyst simultaneously. When subjected to HTC biomass undergoes rapid hydrolysis and other reactions, releasing oxygen-containing volatiles and producing a highly hydrophobic hydrochar.

Although HTC offers a relatively simple and straightforward solution to process diverse biomass feedstocks, the requirements of high pressure and high temperature make the process complex and potentially costly to design and operate. The lab-scale batch process has already been demonstrated in various laboratories around the world, but batch process is not cost-effective for industrial-scale deployment. A continuous process offers a relatively small footprint, energy recovery and efficiency, and notably, economies of scale.

In this study, a bench-scale continuous HTC reactor system has been designed, constructed, commissioned, and operated with cellulose and with whole biomass feedstocks. The maximum temperature and pressure of the reactor are 250 °C and 40 bar. The throughput of the reactor system was maintained at 0.3 L/min with solids loading of ~5%, while the reaction time was maintained at 5 min. The basic reactor configuration is a section of straight pipe, and numerous engineering challenges were encountered during design and commissioning. The primary challenges are related to instrumentation durability, temperature control, and flow control.

Results are presented for hydrothermal conversion of microcrystalline cellulose. Both solid and liquid products were tested for their physico-chemical properties and compared with the corresponding products from batch process generated in a 2L batch Parr reactor. It was found that temperature and pressure were stable during operation and products are relatively similar to that of batch process. Similar results were found for HTC of cow manure, when comparing solid yield, char fuel value and total organic content in the aqueous phase, when corrected for water volume.

495138 Fate of Nitrogen during Hydrothermal Treatment of Septage

Abstract:

Septic tanks are a popular method in rural areas to treat human waste. Waste enters the septic tank and settles so that solids collect in the bottom of the tank. Here organic matter is aerobically digested. Digestion only occurs partially, and is variable with each design of tank and the conditions of the tank (such as temperature and extent of cleaning). Septic tanks offer many advantages, such as the containment of the digestate and the containment of human waste. However, the resulting solid and liquid products must be pumped and disposed of, or further treated. One concern with the use of septic tank is the long term effects of untreated pollutants that are only present in the microgram/liter range, which have been shown to have serious impacts on aquatic ecology of surrounding water systems [1],[2]. Failure of septic tanks can also result in contamination of surrounding water sources with the various macronutrients present in the digestate and effluent of the tank.

Hydrothermal treatment has been investigated as a possible method for the treatment of municipal solid waste and sewage sludge [3]. Septage, or the combination of both septic sludge and effluent, can also be treated by hydrothermal means. Hydrothermal treat is an attractive option for additional septage treatment due to its high water content; there is no need for any drying or separation for solid treatment. More so, the high nitrogen content of septage can be recovered and reused as fertilizer. In this study septage was collected from local health department and hydrothermally treated. A full nitrogen balance was performed between soluble nitrogen species, as well as solid nitrogen content determined by CHNS. Experimental trials for this treatment used a batch reactor with a reaction time of 30 minutes. It was found that hydrothermal treatment at temperatures between 180-260 °C results in increased nitrogen in the process liquid compared to the untreated septage. In addition, treatment results in increased ammonia concentration in the liquid phase and a pH between 6 and 8. This generation of soluble nitrogen species is a result of the solid phase decomposing as it is heated.

References

[1]. Schwarzenbach, René P., et al. "The challenge of micropollutants in aquatic systems." *Science* 313.5790 (2006): 1072-1077.

[2]. Withers, Paul JA, et al. "Do septic tank systems pose a hidden threat to water quality?." *Frontiers in Ecology and the Environment* 12.2 (2014): 123-130.

[3]. Zhao, Peitao, et al. "Clean solid biofuel production from high moisture content waste biomass employing hydrothermal treatment." *Applied Energy* 131 (2014): 345-367.

491306 Hydrothermal Treatment of Paper Mill Sludge

Abstract:

Each year the pulp and paper industry generates large amounts of effluent treatment sludge. The sludge is made up of three different fractions namely, primary, secondary, and tertiary sludge. Depending on the layout of the sludge handling in the individual mill, these three sludges can either be extracted separately or mixed into a stream called mixed sludge. Usually secondary and tertiary sludge are mixed together. The primary sludge, often rich in fiber material and easy to dewater, can easily be used as fuel source. However, water is difficult to remove from mixed sludge.[1-2] Moreover, a low fuel heating value of mixed sludge results against economical combustion. Therefore, a new technology that converts mixed sludge to viable solid fuel may be required for mixed sludge valorization.

Hydrothermal carbonization (HTC) has received considerable attention during the last several years as a prominent alternative for upgrading biomass feedstock ranging from wood and algae to agricultural and forest residues and industrial or municipal wastes. One of the main advantages of HTC is the use of aqueous solutions under high temperature (180-260 °C) and autogenously generated pressure (5-50 bar).[3] As a result, moist and wet feedstocks are favorable for HTC.[4]

This goal of this study was to determine the effects of HTC conditions on paper mill sludge dewaterbility and fuel value. A 100 ml Parr batch reactor was be used for HTC. Reaction temperature was varied from 180-260 °C, while reaction time was varied from 30 min-1h. The solid hydrochars were analyzed for energy content, ultimate analysis, FTIR, and NMR. Meanwhile, the process liquid is characterized with GCMS, HPLC, UV-VIS, NMR, and pH. The energy content of the hydrochar was increased with reaction temperature and longer reaction time as a result of the decrease in oxygen content. The resulted hydrochar is more hydrophobic than the feedstock. Therefore, dewaterbility is enhanced with the hydrothermal treatment of paper mill sludge.

References

1. Makela, M.; Benavente, V.; Fullana, A., Hydrothermal carbonization of industrial mixed sludge from a pulp and paper mill. *Bioresource Technol* **2016,** *200*, 444-450.

2. Makela, M.; Yoshikawa, K., Simulating hydrothermal treatment of sludge within a pulp and paper mill. *Appl Energ* **2016,** *173*, 177-183.

3. Reza, M. T.; Andert, J.; Wirth, B.; Busch, D.; Pielert, J.; Lynam, J. G.; Mumme, J., Hydrothermal Carbonization of Biomass for Energy and Crop Production. *Applied Bioenergy* **2014,** *1* (1).

4. Reza, M. T.; Coronella, C.; Holtman, K. M.; Franqui-Villanueva, D.; Poulson, S. R., Hydrothermal Carbonization of Autoclaved Municipal Solid Waste Pulp and Anaerobically Treated Pulp Digestate. *Acs Sustain Chem Eng* **2016,** *4* (7), 3649-3658.

486662 Structural Analysis of Humins Formed in the BrÃ¸nsted-Catalyzed Dehydration of Fructose

Abstract:

Humins are carbonaceous and polymeric by-products formed during acid-catalyzed hydrothermal processing of sugars to 5-hydroxymethylfurfural (HMF). These are formed via cross-polymerization of HMF with sugars in acidic solutions, and cause significant carbon loss and unfavorable process economics of biorefineries. The deposition of these solid particles on the reactors and processing equipment also increases the maintenance costs. Currently humins are combusted to supply heat in biorefineries. Minimization of humins formation or their valorization is essential for improving the process economics of the biorefineries. In this work, we developed methods to understand microstructure of humins through solubility experiments. First, we have conducted dissolution experiments in various solvents and correlated the solubility data to molecular properties of solvents to develop suitable descriptors. By doing this, we have identified solvents for high solubility of humins at room temperature and provided insights into the mechanism of humins dissolution. Next, Liquid Chromatography- Mass Spectrometry (LC-MS) was used to determine and compare the masses of components in the humins solubilized in different solvents. Finally, a six stage dissolution experiment was done to investigate the structural inhomogeneity of humins. Both Fourier Transform Infrared Spectroscopy (FT-IR) and LC-MS were used to characterize the solubilized and insoluble humins obtained at each stage. Based on our results, an inhomogeneous structure where macromolecular and molecular components are connected through weak forces, was proposed.

Super Oleophilic and Water Repellent Graphene Oxide Sponge for Catalytic Coupling of Furans and Oil Absorption

Saikat Dutta

Advisers: Basu Saha, Dion G. Vlachos

Catalysis Center for Energy Innovation

University of Delaware

Newark, Delaware

Session: Hydrothermal Carbonization

November 2, 2017

AIChE Annual Meeting 2017, Minneapolis, MN

U.S. DEPARTMENT OF ENERGY | Office of Science

CCEI is an Energy Frontier Research Center funded by the U.S. Department of Energy, Office of Basic Energy Sciences

Catalysis Center for Energy Innovation

University of Delaware

Why Graphene Sponge

- Graphene is generally hydrophobic, can be tuned to hydrophilic

Hydrophobic properties

Graphene@SiC CA 92°

SiC CA 69°

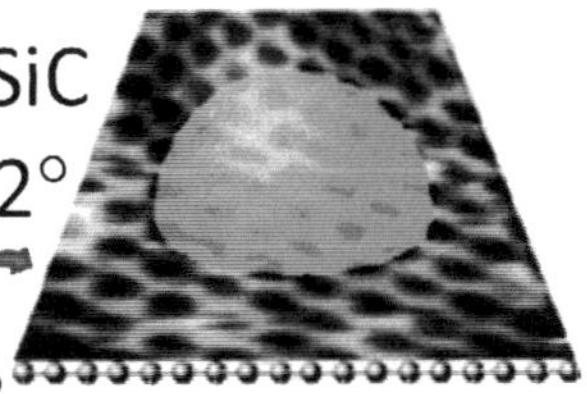

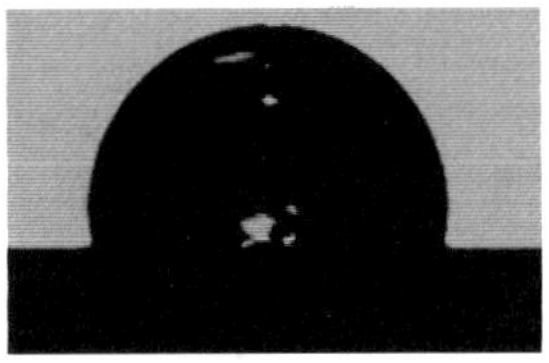

O_2 plasma

- Thickness Independent Hydrophobicity

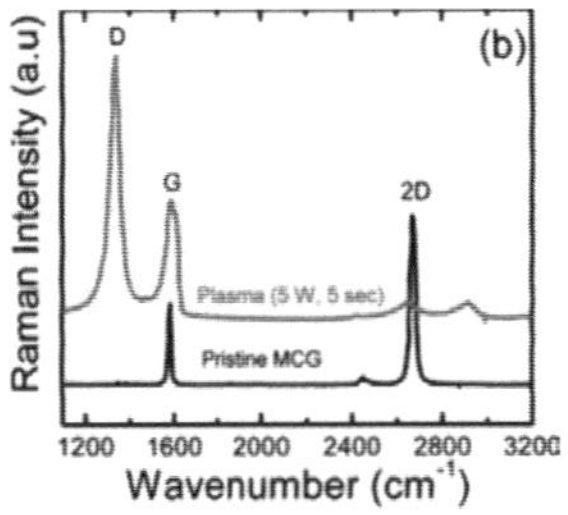

Energy Environ. Sci. **2012**, *5*, 7908

Langmuir **2010,** *26*, 3798

- More defects, more hydrophilic
- GO sponges are not explored extensively

2

Sponge Materials: Why Is It Important

Marine Oil Spill

Threats to the planet

- High frequency
- Long-lasting adverse environmental impacts
- Significant economic impact

Current Methods

- Skimming on calm water
- Chemical dispersants spread

Sponge Materials

carbon-based aerogels

Adv. Mater. **2013**, *25*, 2554

porous boron-nitride

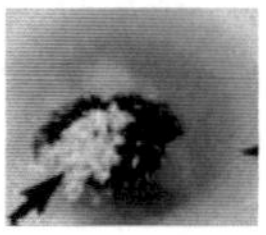

Nat. Commun. **2013**, *4*, 1777

modified commercial sponges

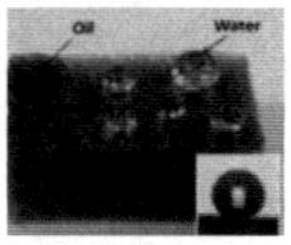

Energy Environ. Sci. **2012**, *5*, 7908

GO-wrapped sponge

Nat. Nanotechnol. **2017**, *12*, 434

3

Graphene Oxide Sponge (GOS)

- Explore hydrophobic and hydrophilic behaviors of GOS

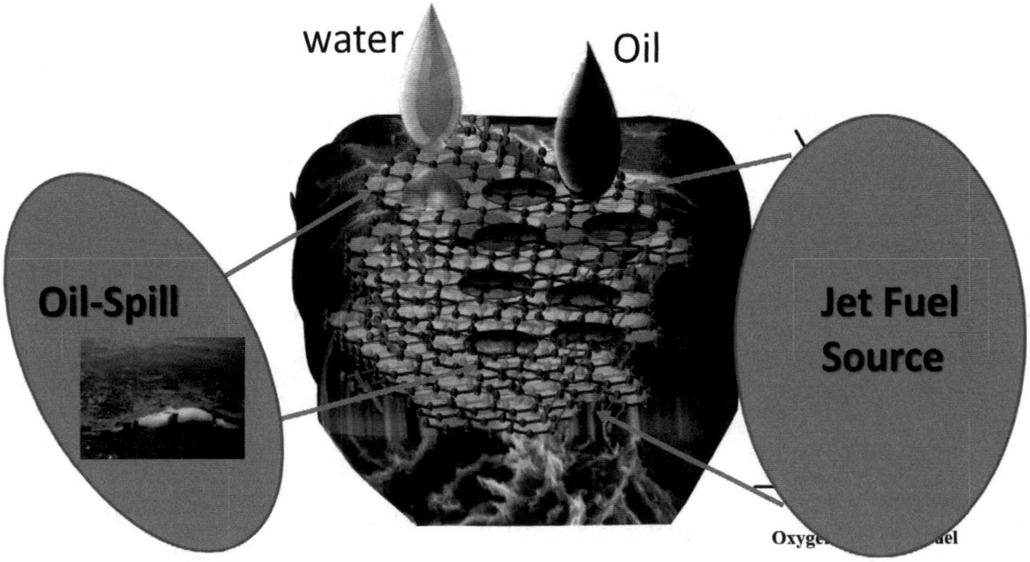

4

Hydrophilic IGO to GOS

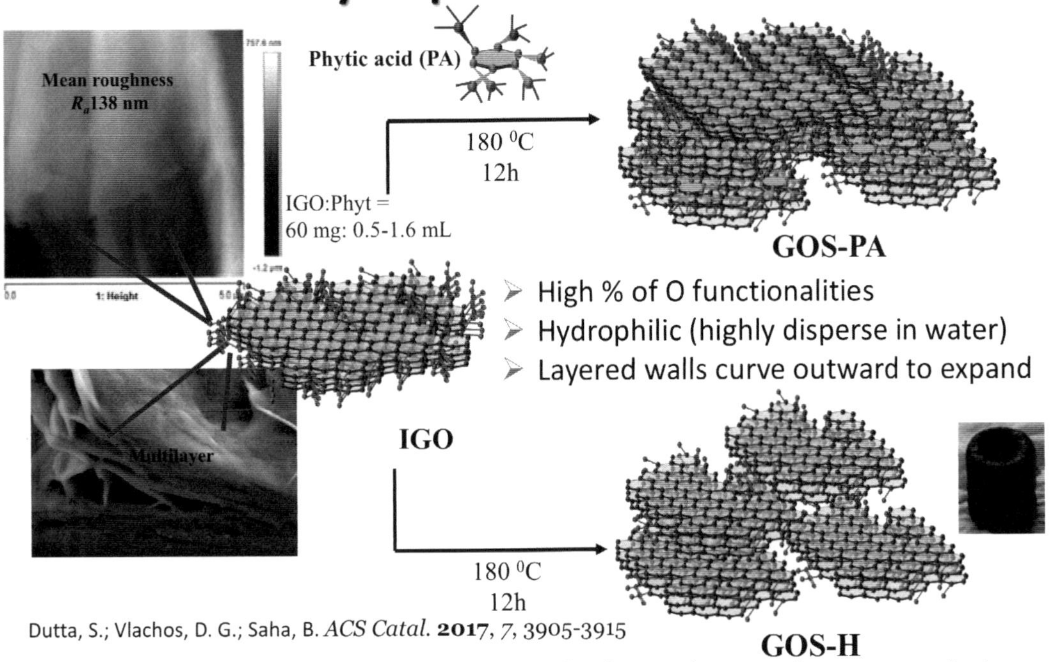

- High % of O functionalities
- Hydrophilic (highly disperse in water)
- Layered walls curve outward to expand

Dutta, S.; Vlachos, D. G.; Saha, B. *ACS Catal.* **2017**, *7*, 3905-3915

- Hydrothermal treatment is simplest method to reduce surface oxygen (gain hydrophobicity) compared to cross-linking/freeze-drying

5

Hydrophobicity of GOS from XPS C1s Peak

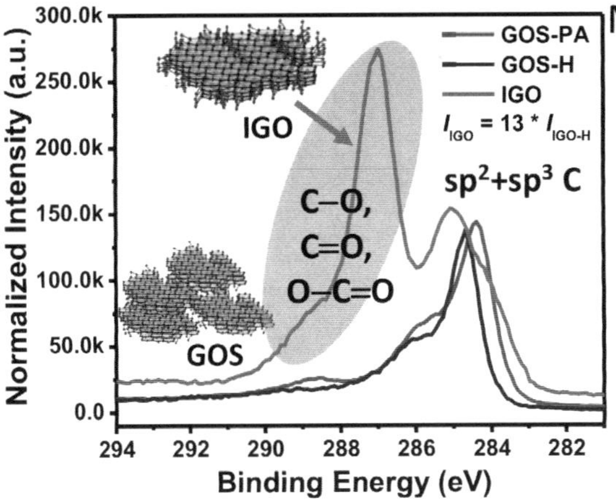

More surface oxygen (hydrophilic)

Less surface oxygen (hydrophobic)

- Increase in hydrophobicity: selective for oil/organic absorption

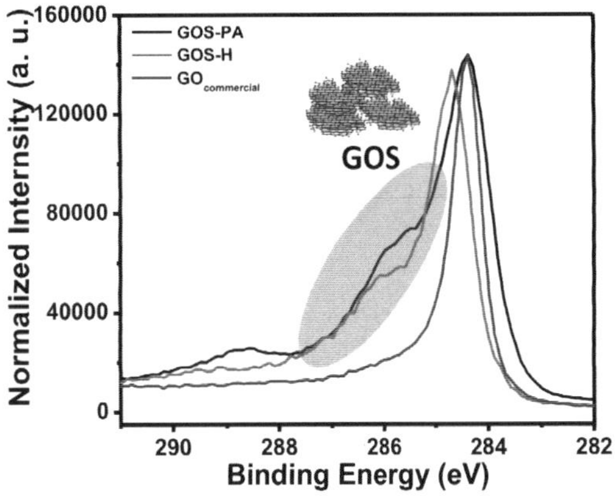

- Surface oxygen GOS > $GO_{commercial}$
- GOS more reactive

6

Defect Analysis: Raman Spectra

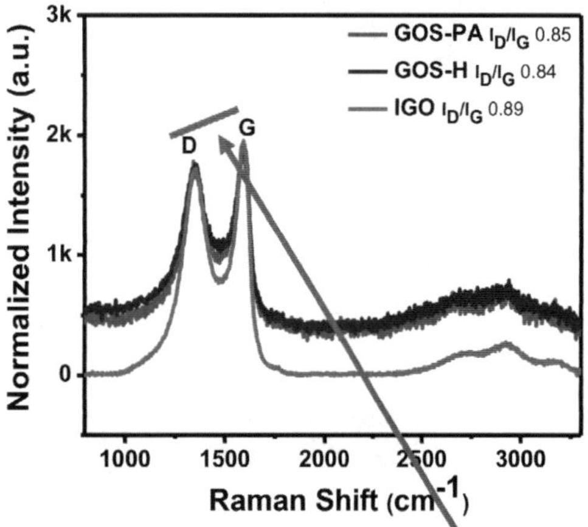

O_2 plasma (introduction of defect)
G@SiC

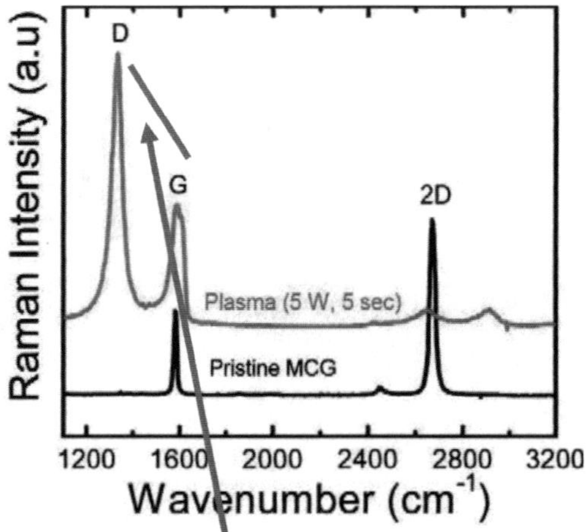

Langmuir **2010,** *26*, 3798

- GOS and IGO comparable crystalline defect
- G contains less crystalline defect before plasma treatment

7

Catalysis Center for Energy Innovation
UNIVERSITY OF DELAWARE

Raman Imaging of D and G Bands

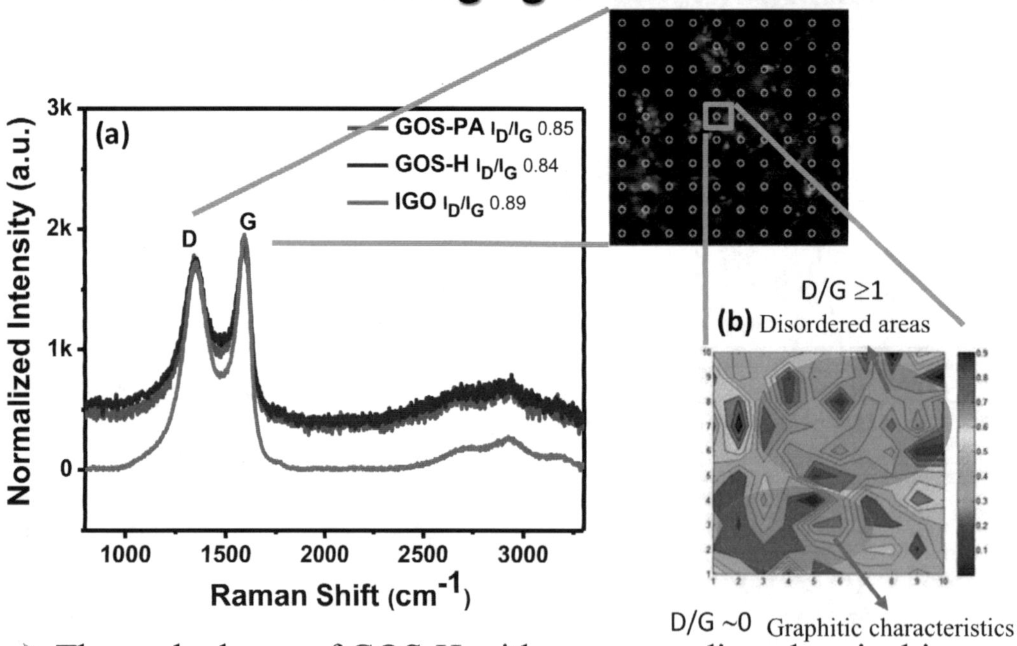

- The probed area of GOS-H with corresponding chemical image displaying the ratio of D-band and G-band intensity (b)

8

SEM Characterization of GOS-PA and GOS-H

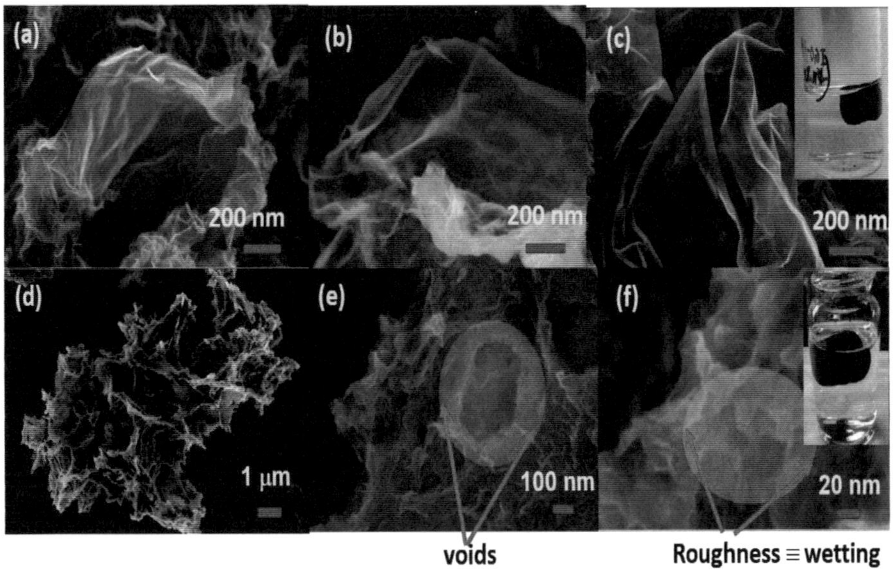

- GOS-H contains more rough surface as compared to GOS-PA

9

Catalysis Center for Energy Innovation

Water-Contact Angle of GOS

Sample	Contact Angle (Degrees)	Error (%)
HOPG (std)	67.3	3.3%
GOS-H	86.5	0.7%
GOS-PA	84.2	1.8%

HOPG

GOS-H

GOS-PA

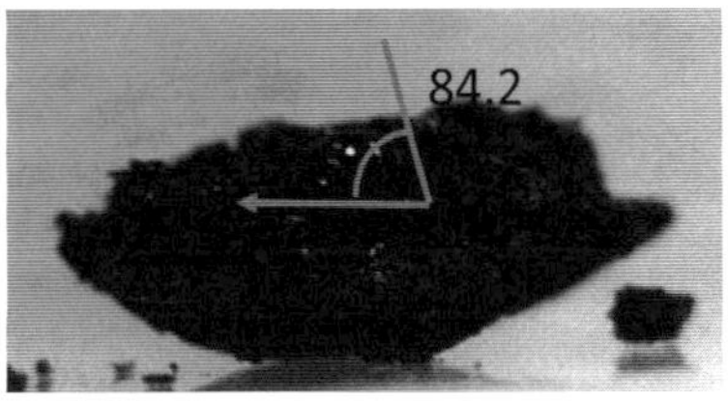

- GOS hydrophobic with contact angle ~90°
- Enhancement of contact angle is countered by surface oxygen

10

Oil Absorption by GOS-H

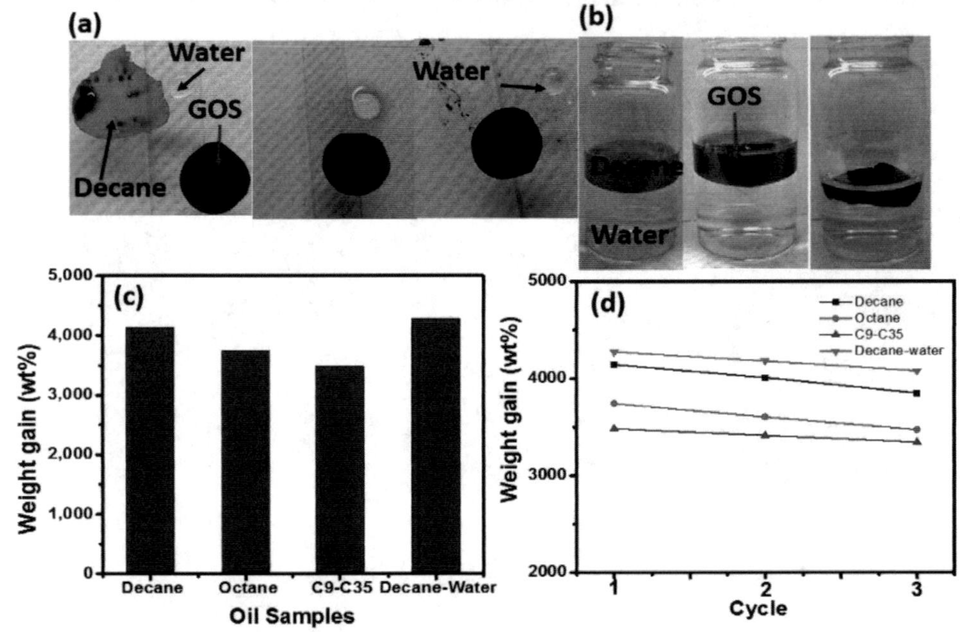

- Selective oil absorption by GOS in 15 min
- Exhibit recyclability for oil absorption

Origin of Hydroxyalkylation-Alkylation (HAA) Catalysis

- Surface-oxygen active sites of IGO enables tailor the C-numbers of lignocellulosic fuel precursors

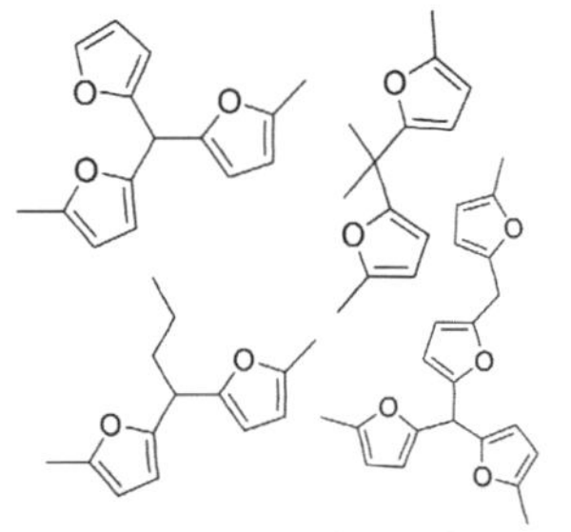

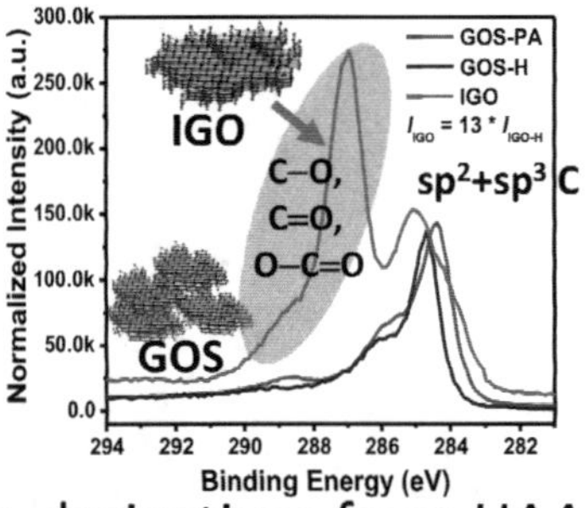

Entry	C (%)	O(%)	P(%)	S(%)
IGO-H	73.4	25.1		1.4
IGO-PA	72.5	24.2	2.1	1.2
IGO	59.2	36.6		3.1
HOPG	98.2	0.41		

- C13-C21 furan-derivatives from HAA with IGO

Dutta, S.; Saha, B. *ACS Catal.* **2017**, *7*, 5491-5499

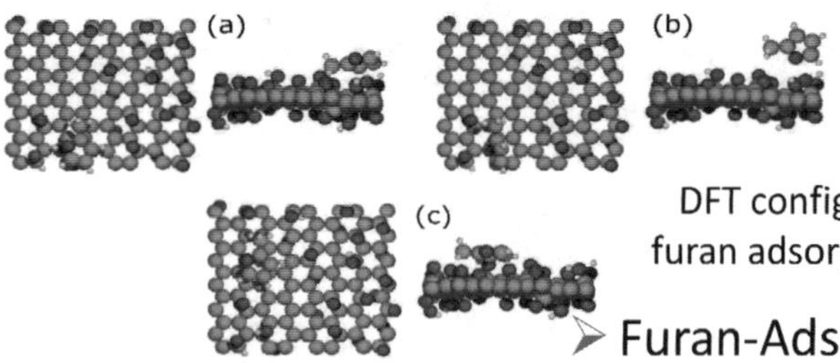

DFT configuration of furan adsorption on IGO

- Furan-Adsorption on a Oxygenated Surface

Dutta, S.; Vlachos, D. G.; Saha, B. *ACS Catal.* **2017**, *7*, 3905-3915

Catalysis Center for Energy Innovation

HydroxyAlkylation-Alkylation (HAA) of Furans

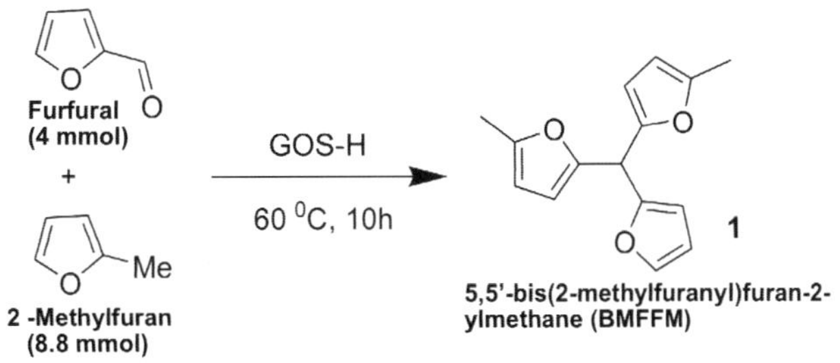

GOS-H recycle

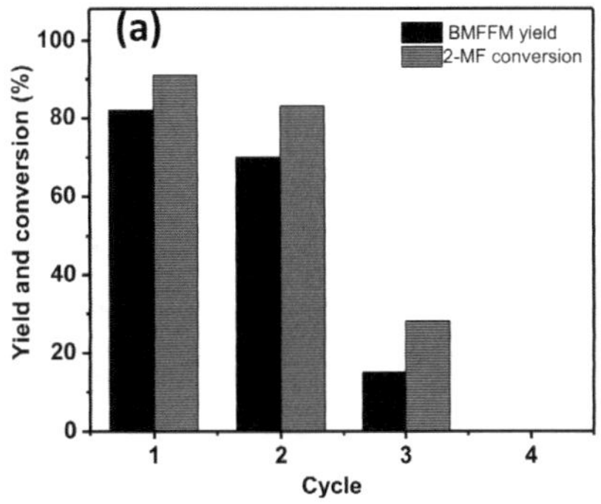

- Carbon extension of furans to desired fuel precursors
- Intermediate surface oxygen % with hydrophilic behavior GOS-H for HAA
- Loss of activity due to further reduction of surface oxygen (XPS)

13

Catalysis Center for Energy Innovation

Summary

- GOS derived from aqueous dispersion of IGO.
- GOS contains surface roughness responsible for wettability
- Raman imaging, XPS analysis, and SEM were used for characterization
- Percentage of surface oxygen in GOS is higher than graphene or rGO sponge
- GOS-H is effective for HAA reactions for furans
- GOS-H offers selective oil absorption from oil-water mixture

Moving Forward

- Technique for cross-linked GOS for mechanical stability

14

Acknowledgements

Joseph P. Smith (Raman Spectroscopy)
Prof. Karl S. Booksh (Raman Spectroscopy)
Dr. Weiqing Zheng (SEM,TEM imaging)
Rachel Pupillo (XPS)
Dion G. Vlachos (insightful discussions)
Jeff Everhart (GC analysis)

U.S. Department of Energy, Office of Science, Office of Basic Energy Sciences, under Award No. DE-SC0001004

U.S. DEPARTMENT OF ENERGY | Office of Science

502391 Simultaneous Isomerization and Reactive Extraction Followed By Back Extraction of Sugars from Biomass Hydrolysate for High Purity and Yield of Ketose Sugars

Abstract:

Biomass hydrolysates are often low in sugar concentration and contain contaminants associated with the pretreatment method and sugar and lignin degradation. Cost-effective and high-yield downstream processing requires sugar streams at higher concentration, free of contaminants generated during the pretreatment process. Instead of removing contaminants from the hydrolysate, we have taken the approach of removing sugars from the hydrolysate, leaving contaminants behind. In the process of removing sugars, we enzymatically isomerize them from their aldose (glucose, xylose) to their ketose (fructose, xylulose) forms. These ketose sugars are ideally suited for production of furans â€“ HMF from fructose and furfural from xylulose.

Unfortunately, the isomerization reactions do not have favorable equilibria with respect to ketose formation. To overcome this limitation, we have previously developed a method of simultaneous-isomerization-and-reactive-extraction (SIRE) to produce the ketose isomer in high yield. We have demonstrated that with a xylose/xylulose mixture, SIRE followed by 2-stage back-extraction (BE) allows high yield recovery of xylulose in nearly pure form.

For mixtures of xylose and glucose found in biomass hydrolysate, separation and purification of the ketose sugars (xylulose and fructose) from the hydrolysate will be more complex. The extraction affinities of glucose and fructose are lower than xylulose, but higher than xylose. SIRE-BE has not been previously evaluated for biomass hydrolysates with high sugar concentrations (up to 90 g/L) for recovery of a high purity, mixed C5 and C6 ketose product stream. We have developed strategies for high purity recovery of both xylulose and fructose from mixed sugar streams using SIRE-BE. We will present these strategies as well as xylulose and fructose yields and purity obtained under different experimental configurations.

494869 Development CO Sensor for Syngas Fermentation for Production of Fuels and Chemicals

Abstract:

Syngas fermentation is an emerging technology to produce fuels and chemicals from gasified biomass, coal, petcoke and solid wastes. This technology can also be used to make fuels and chemicals from gas streams that contain syngas components (CO, CO_2 and H_2). Some acetogens can convert these gases to organic acids and alcohols via the acetyl-CoA pathway. Challenges for commercialization of syngas fermentation technology include mass transfer limitations and low alcohol productivity associated with low cell mass concentration in the bioreactor and inhibition effects of high CO concentrations on key enzymes such as hydrogenase. Our recent results showed that the control of syngas feed rate to match the kinetics requirements of cells resulted in a stable operation of syngas fermentation (US Patent Application Publication, US 2016/0281114 A1 and International Patent Application No. PCT/US2015/60720). However, this required the use of gas chromatography and mathematical models to estimate the dissolved CO concentration in the medium, which was time consuming. Direct monitoring of dissolved CO and H_2 concentrations in the fermentation medium is a key to improve productivity and process stability. A real-time monitoring system for the dissolved CO and H_2 concentrations is being developed, which includes a built in-house dissolved CO sensor, a sample handling unit, a central control unit and an off-the-shelf dissolved H_2 sensor. The dissolved CO sensor was developed based on the spectral responses of CO in the mid infrared range (MIR). The CO sensor was housed in a designed cage with a careful alignment between a MIR LED as a light source and a photodetector as a signal receiver. A preamplifier and a lock-in-amplifier were used to preprocess the raw signal before being collected by the central control unit. The CO sensor was tested with gas samples containing CO concentration range generally found in bioreactors at typical operating pressures and achieved a sensitivity of over 10 mV/ppm CO. The central control unit was also developed for data acquisition and process control. The dissolved H_2 sensor was calibrated to measure 0 to 3 ppm of dissolved H_2 in the fermentation medium. Both CO and H_2 sensors were integrated in the bioreactor control platform to allow efficient delivery of these gases to the biocatalysts to enhance conversion efficiency, productivity and operation stability of syngas fermentation. Preliminary results showed the feasibility of the use of continuous mentoring of CO and H_2 concentrations in the bioreactor. Further work is focused on improving the sensorsâ€™ performance and its integration with the bioreactor control for efficient continuous syngas fermentation.

501393 Techno-Economic Analysis (TEA) of a Novel Hybrid Enzyme- and Chemo-Catalytic Process for Producing Furans from Biomass Hydrolysate

Abstract:

Lignocellulosic biomass is a sustainable feedstock for production of furans – 5-Hydroxymethy furfural (HMF) and furfural – versatile platform molecules that can be converted to drop-in fuels and precursors for synthetic materials. However, the lack of efficient pathways from to furans from lignocellulosic sugars constitutes a major barrier to their industrial production. We have recently achieved high yields of furans from biomass hydrolysates using a novel hybrid enzyme- and chemo-catalytic technology (*Green Chem.*, **2017**, 19, 1782). The present work assesses the techno-economic feasibility of this novel technology. The process was simulated for steady-state operation of an nth plant at receiving hydrolysate containing glucose and xylose in a 3:1 ratio. The feed rate was 1000 metric tons of mixed sugars (dry-basis) per day. The equipment sizing was done based on the process-unit hold-up volumes and experimentally-determined residence times to estimate total capital costs. Using discounted cash flow analysis in CAPCOST®, a Minimum Furan Selling Price (MFSP) of $2.48/kg was estimated for a "base case" hydrolysate containing a sugar concentration of 40 g/L (Case I)). The model estimates that by processing a more concentrated mixed-sugars stream (160 g/L), from upstream high-solids processing, MFSP can be lowered to $2.33/kg due to reduction in total capital costs (Case II). Further, MFSP reduction to $2.18 can be realized by solvent optimization (Case III), and to $ 1.48/kg with on-site chemical regeneration (Case IV). Sensitivity analysis revealed that the key factors affecting MFSP are overall yield of furan (from sugar) and the hydrolysate costs. TEA performed for furan production by other state-of-the-art technologies predict a furan selling price which is 75% higher than what is possible with our recently-developed method.

496508 Electrocatalytic Hydrogenation of Lignin-Derived Bio-Oil Model Compounds Using Ruthenium on Activated Carbon Cloth to Produce Liquid Fuel Intermediates and Value-Added Products

Abstract:

Biomass fast pyrolysis (BFP) uses heat in the absence of oxygen to convert biomass to bio-oil, biochar and combustible gas. The biochar co-product can be used for the removal of pollutants in the environment and has potential for use in soil amendment and carbon sequestration. The combustible gas co-product is typically burned for process heat needed by the pyrolysis system. The major product, bio-oil, has the potential to displace liquid hydrocarbon fuels; however, bio-oil is highly oxygenated, corrosive, low in energy content and unstable during storage. Electrocatalytic hydrogenation (ECH) is employed to reduce and deoxygenate reactive compounds and improve bio-oil properties. The present study particularly focuses on the electrocatalytic stabilization of compounds derived from the pyrolytic depolymerization of lignin. As lignin is a component of biomass comprising up to 30% of the mass and 40% of the energy stored in biomass, it offers great potential as feedstock for BFP. Lignin model compounds representative of bio-oil components were subjected to ECH under mild conditions using ruthenium on activated carbon. To date, several model lignin monomers and dimers have been reduced to simpler compounds such as cyclohexanol and phenol. The effect of different electrolytes, substrate concentrations, current density, and catalyst preparation methods on conversion and current efficiency was studied. The coupling of pyrolysis and electrocatalysis for depolymerization and upgrading of lignins, that are byproducts of the biomass pretreatment and extraction methods, will enable the maximization of yields from biomass conversion to fuels and value-added products.

499405 Thermal Deconstruction Opens Biomass for Acid Hydrolysis to Sugars

Abstract:

We have explored thermal deconstruction, which involves rapidly heating then cooling biomass in an inert environment before complete pyrolysis can occur. While we produce some bio-oil and light gases, solid or liquid phase reactions greatly modify our biomass. As shown through our microscopy and chemical analysis, thermal deconstruction cracks the biopolymers and loosens their lignocellulosic structure. This change allows molecules to diffuse more readily into and out of the biomass. Additionally, we have proven that as the biopolymers fragment into oligomers they become more susceptible to further depolymerization reactions. In particular, when cellulose is thermally deconstructed it rapidly forms anhydro-oligosaccharidesâ€"oligosaccharides where the reducing end has dehydrated. We can hydrolyze these oligomers to glucose under more mild conditions than required for untreated cellulose. Dilute sulfuric acid catalyzes hydrolysisâ€"primarily in cellulose and hemicellulose oligomersâ€"producing glucose and xylose, which can be fermented to ethanol or other products. Our process allows for higher monosaccharide yields than those obtainable from fast pyrolysis.

IOWA STATE UNIVERSITY
Bioeconomy Institute

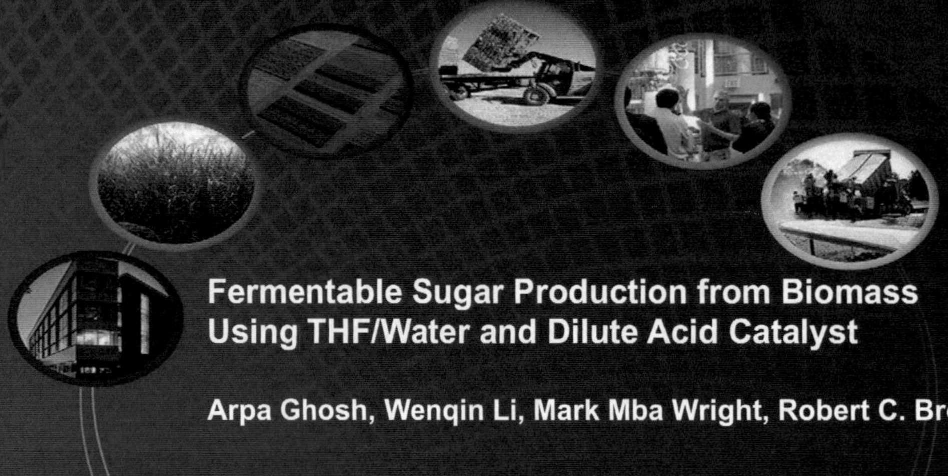

Fermentable Sugar Production from Biomass Using THF/Water and Dilute Acid Catalyst

Arpa Ghosh, Wenqin Li, Mark Mba Wright, Robert C. Brown

AIChE Annual Meeting 2017, Integrated Thermochemical and Biochemical Processing for Renewable Fuels and Chemicals Session, November 2nd, 2017

Challenges in producing cellulosic sugars via conventional enzymatic pathway

- **Cost of enzyme production**[1,2]
  - Covers 20% of total production cost of fermentable carbohydrates
  - Estimated cost contribution up to $1.47 per gallon of ethanol
- **Slow rates of bioconversion**[3]
  - It takes days or even weeks to reach glucose yields >70%
- **Achieving high concentration of sugar products**[4]
  - Sugar substrate required for fermentation is high
- **Substrate specific enzyme production could be uneconomical**

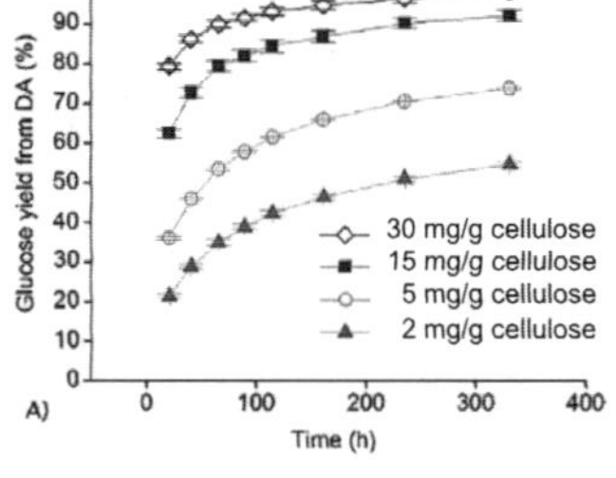

DA: Dilute acid pretreatment (common in biochemical processing of cellulosic biomass)[3]

Solubilized carbohydrates production using polar aprotic solvents

- Accelerated rate of monosaccharide production from biomass in polar aprotic solvents[5,6]
- Dilution prevents secondary reactions
- Mild temperature (150-350 °C)
- Carbohydrates are solubilized
- Low consumption of acid

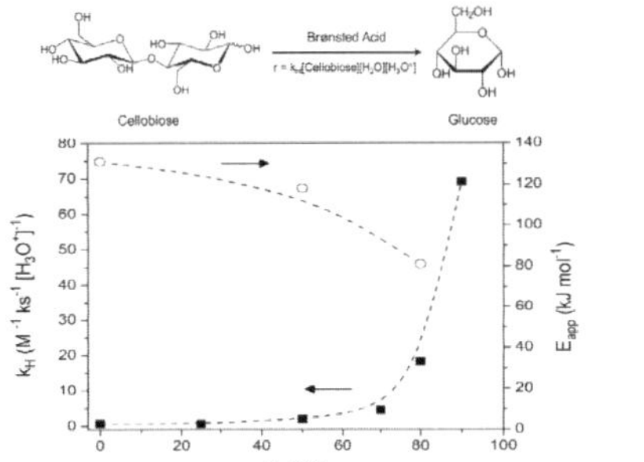

(Mellmer, 2014)

Recent progress in the field:

- Production of 68% of theoretical soluble carbohydrates from biomass at 157-217 °C using γ-valerolactone (GVL) mediated process[7]
- Other effective polar aprotic solvents: Hot, pressurized 1,4-dioxane, acetone, sulfolane, acetonitrile

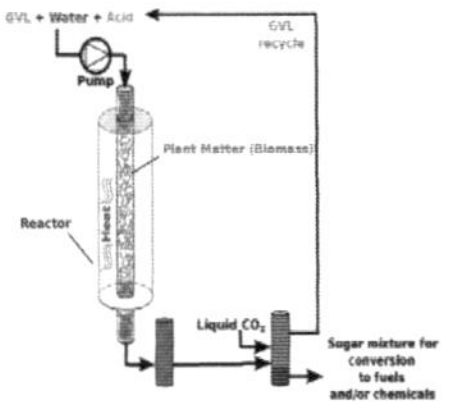

(Luterbacher, 2014)

IOWA STATE UNIVERSITY
Bioeconomy Institute

Challenges in developing cellulosic sugar production process using polar aprotic solvents

Which solvent to choose for cellulosic sugar production?

- Inadequate information on the effects of different polar aprotic solvents on solubilized carbohydrate production

Challenges with high polarity solvents

- Often expensive and thus demand quantitative recovery
- Recovery of solvent is often complicated such as the use of liquid CO_2 extraction for GVL recovery[8]

Challenges with low polarity solvents

- High temperature and high pressure processing associated with these solvents increases capital and operating costs

IOWA STATE UNIVERSITY
Bioeconomy Institute

Properties and costs of polar aprotic solvents tested

Polar aprotic solvent	Normal boiling point (°C)	Polar solubility parameter [a] (MPa$^{1/2}$)	Cost [b] ($/Kg)
1,4-Dioxane	101	2.1	105
Tetrahydrofuran (THF)	66	7.0	121
Acetone	56	13.1	80
Acetonitrile	82	18.7	153
Gamma-valerolactone (GVL)	207-208	22.1	524

[a] At reaction condition of 350 °C, 5-19 MPa; [b] From Sigma Aldrich for ACS Reagent solvents

IOWA STATE UNIVERSITY
Bioeconomy Institute

5

Use of acid catalyst removes much of the difference among solvents

No catalyst:

The higher the polar solubility parameter, the higher levoglucosan yield as a result of higher solubilization of cellulose[9]

20 mg cellulose at 350 °C

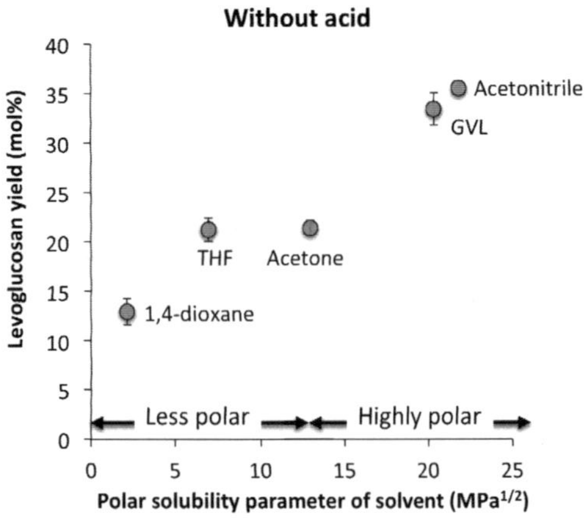

Mean yields reported based on at least duplicates.
STD error is shown in error bars.

Bioeconomy Institute

6

Use of acid catalyst removes much of the difference among solvents

No catalyst:

The higher the polar solubility parameter, the higher levoglucosan yield as a result of higher solubilization of cellulose[9]

With acid catalyst:

- Dramatically increased in low polarity solvents compared to high polarity solvents
- Levoglucosan yields equalized among solvents

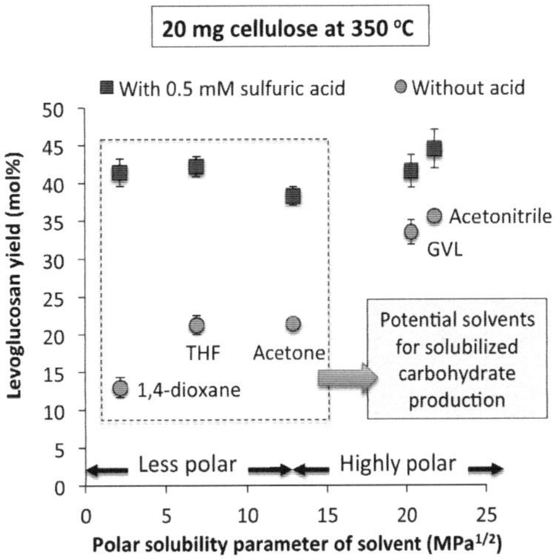

Mean yields reported based on at least duplicates.
STD error is shown in error bars.

Developing solvent liquefaction process to overcome challenges of current technologies

Goals

- Significant increase of sugar production rates with respect to enzymatic hydrolysis (>> 70% of theoretical glucose yield within 72 hours)
- Competitive solubilized carbohydrates yields with current solvent liquefaction methods (approaching 70% of theoretical yield)
- Using relatively low-cost and easily recoverable solvent

THF-Water with dilute sulfuric acid could serve as an effective solvent for sugar production

Tetrahydrofuran (THF)

Approach

Process design:

Stage-1. Solubilize lignin and hemicellulose in a mild pretreatment

- Partial removal of lignin and hemicellulose should render cellulose more susceptible to deconstruction[3]
- THF is a good solvent for dissolving lignin and decomposing hemicellulose in presence of acid catalyst[10]

Stage-2. Solubilize cellulose-rich solids in solvent liquefaction

THF is an effective solvent for acid-catalyzed cellulose depolymerization

Process economics:

THF is relatively inexpensive and has lower boiling point than water (66 °C)

9

Biomass two-step solvent process using THF/water with sulfuric acid catalyst

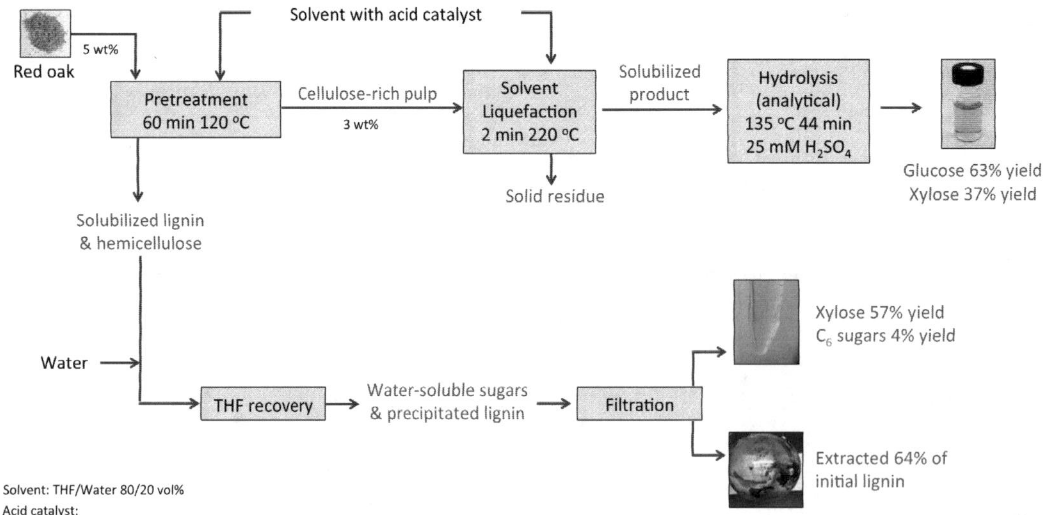

Solvent: THF/Water 80/20 vol%
Acid catalyst:
Pretreatment- 2.5-10 mM H_2SO_4 (0.5-2 wt% of red oak), Solvent liquefaction- 2.5 mM H_2SO_4

10

Prolonged pretreatment at high acid loading gives delignified biomass and solubilized hemicellulose keeping cellulose intact

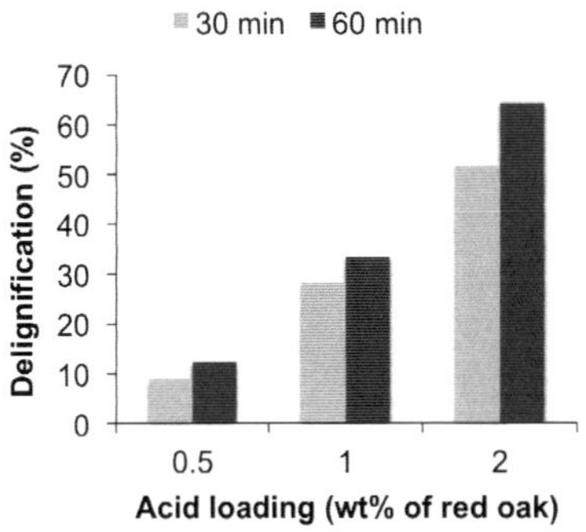

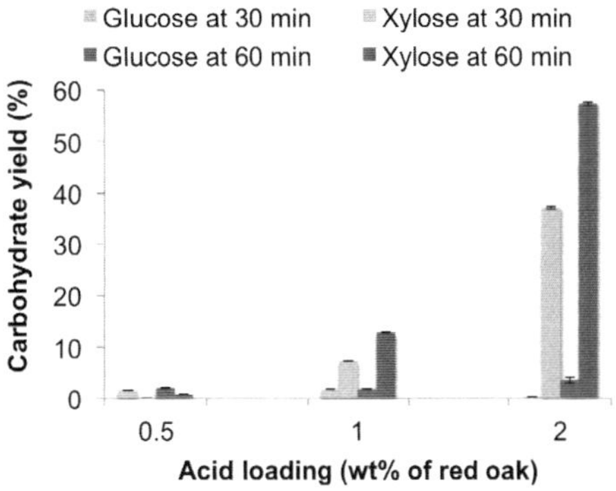

- **64% lignin extracted from water-insoluble fraction of pretreatment liquid**
- **57% xylose and 4% glucose recovered in water-soluble fraction of pretreatment liquid**
- **Resulting pulp yield at 60 min pretreatment at 2 wt% acid loading: 56 wt%**

11

Effect of pretreatment on enhancing yields of solubilized carbohydrates from red oak

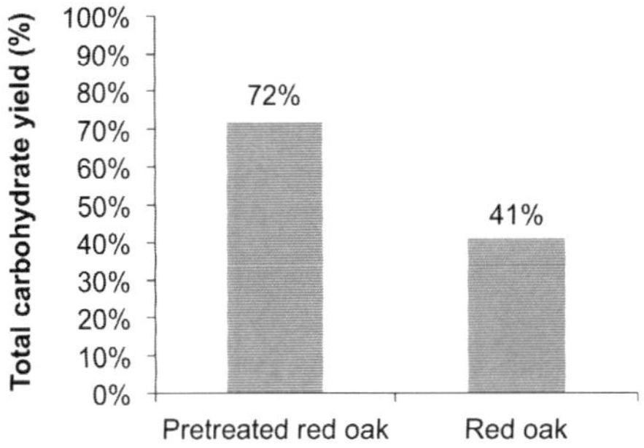

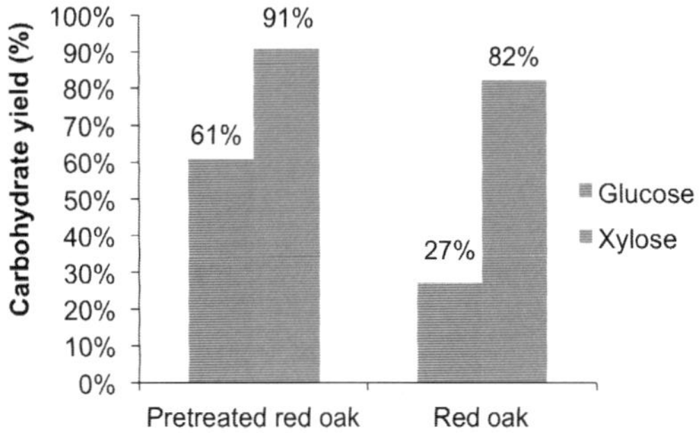

Carbohydrate yields from combined pretreatment and solvent liquefaction of red oak was significantly higher than solvent liquefaction without pretreatment

IOWA STATE UNIVERSITY
Bioeconomy Institute

Mean yields reported based on at least duplicates. STD error is within 10% of mean yield.

12

Pretreated red oak is enriched with cellulose making it easier to deconstruct by solvent liquefaction

- Lignin and xylan are reduced by 53% and 67% via pretreatment
- Glucan only reduces by 14%
- Cellulose-rich pretreated red oak could be more susceptible to deconstruction by solvent and acid

(Analysis by: Celignis Ltd.)

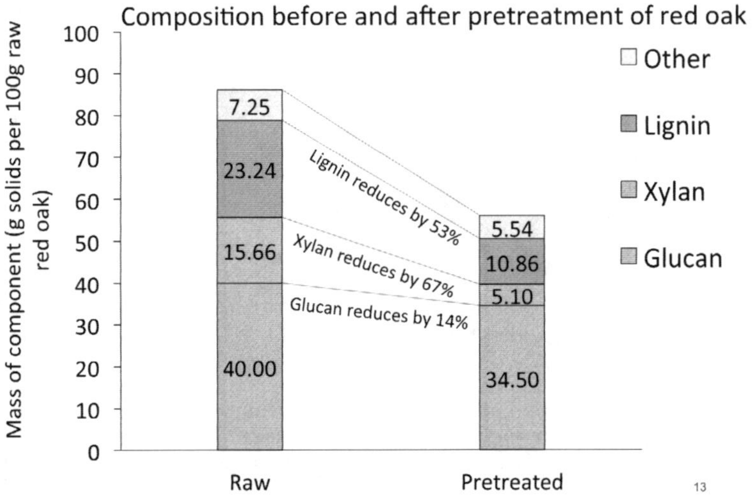

SEM images reveal significantly modified structure of lignocellulosics by THF/Water pretreatment

- Polymeric matrix of biomass was loosened up
- More penetrable structure with large pores and microgrooves
- Wrinkles around edges imply loss of structural strength; attributed to delignification

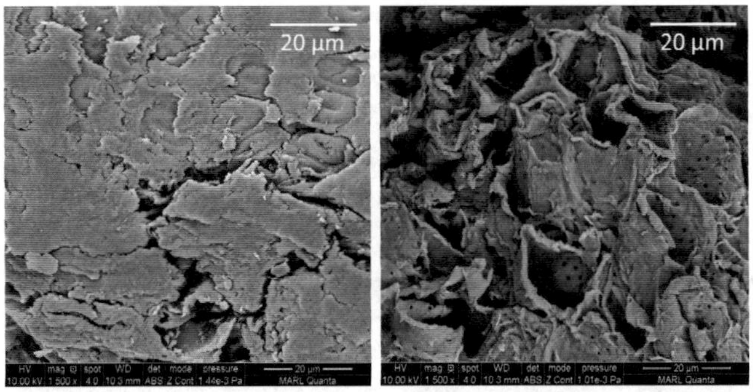

Untreated red oak **THF/Water pretreated red oak**

SEM images at 1500X magnification as taken originally

IOWA STATE UNIVERSITY
Bioeconomy Institute

One-step simple distillation to recover THF and sugars from solubilized products

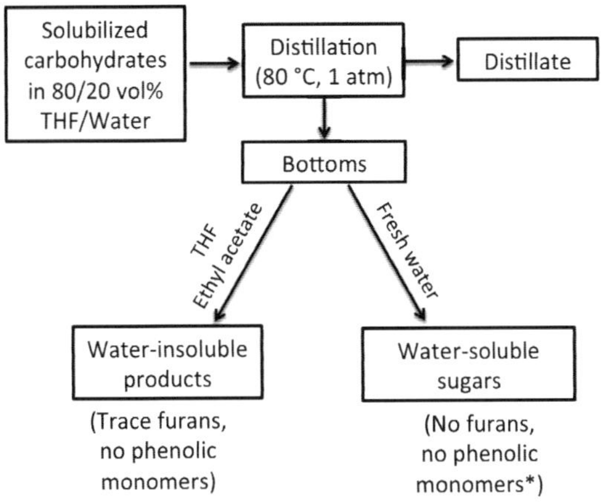

*possibly contains water-insoluble non-carbohydrate oligomers

Bioeconomy Institute

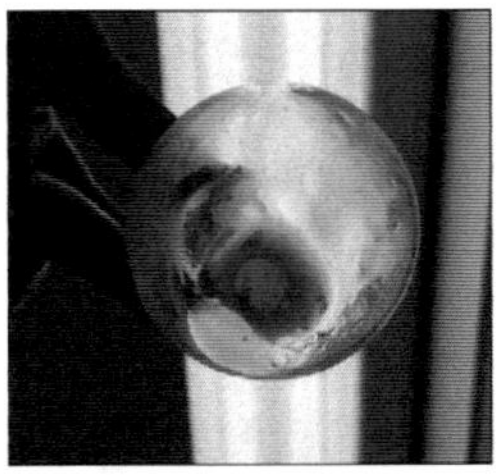

Highly concentrated sugar solution after distillation

Distillation Fraction	THF recovery (%)
Distillate (Almost pure THF, very low level of contaminants)	106.3
Bottoms (recovered products of solvent liquefaction)	0.3

Most of the sugars recovered in bottoms with no migration into distillate

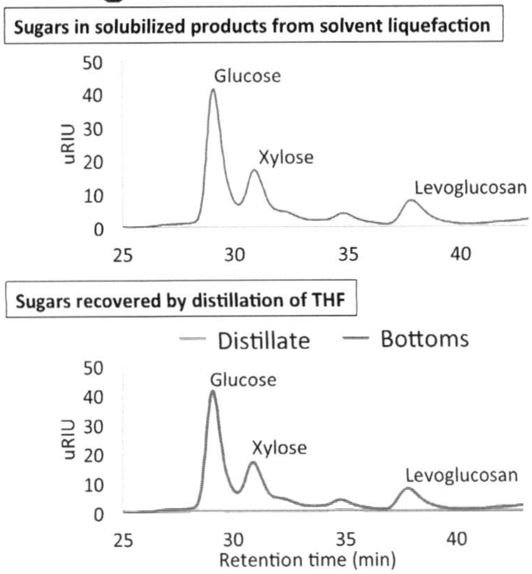

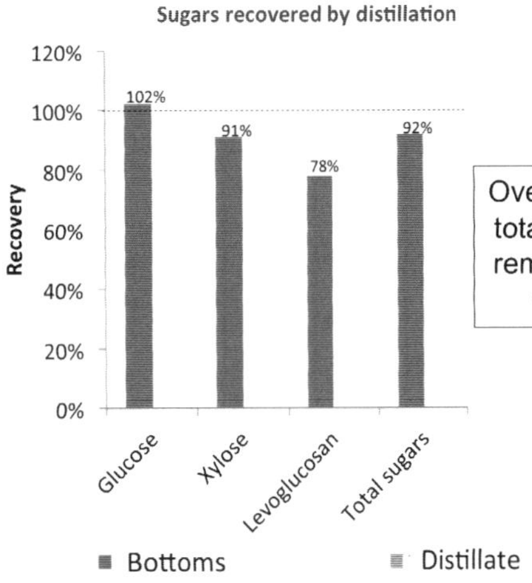

Over 90% of total sugars remained in water

16

Fermentation substrate: high concentration sugar solution as biofuel precursor at rapid rate

Process	Pretreated Biomass loading (wt%)	Glucose concentration (g/L)	Estimated glucose production (g/L/h)
Enzymatic hydrolysis[12]	**15**	**114**	0.7
GVL flow through[7]	**7**	**127**	**64**
THF batch *(this work)*	**3**	**90**	2715*
THF flow through *(predicted)*	7	215	6463

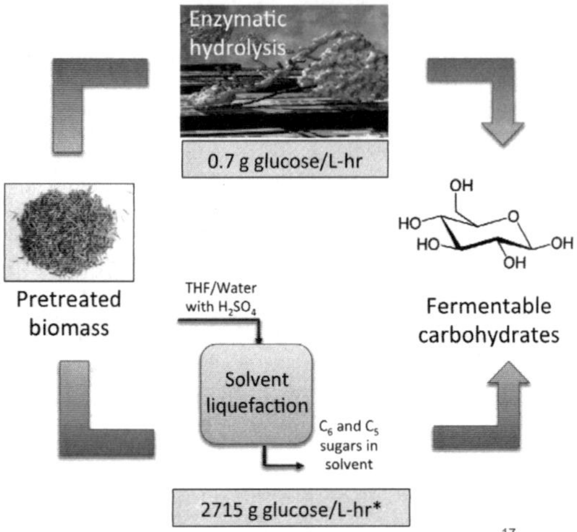

Enzymatic hydrolysis figure source: (Himmel et al., 2007)

*excluding analytical hydrolysis step of 44 min

17

Comparable or better economics of cellulosic ethanol production than current processes

- MFSP for GVL and THF processes were \$3.33/gal and \$2.80/gal ethanol, respectively
- Comparable/better than over both enzymatic hydrolysis and GVL processes
- Key assumptions: same feedstock cost and solvent to biomass ratio at large scale production

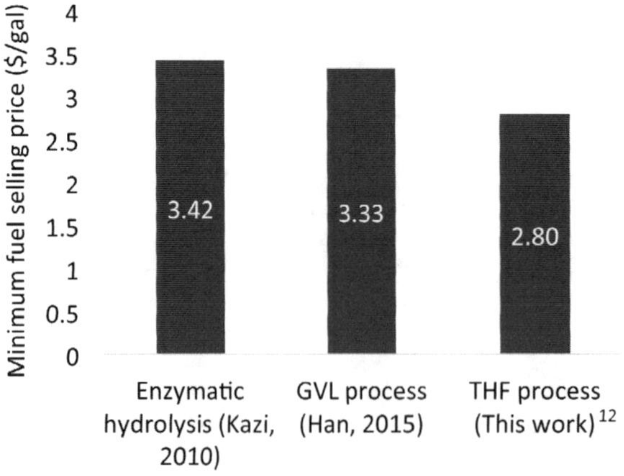

IOWA STATE UNIVERSITY
Bioeconomy Institute

18

Summary

- THF was identified as a low boiling point polar aprotic solvent that could effectively and rapidly depolymerize cellulose into solubilized carbohydrates with high yields of monomeric sugars
- A novel non-enzymatic strategy for cellulosic sugar production using THF/ water mixture shows competitive sugar yields with current solvent-mediated processes
- Preliminary techno-economic analysis indicates that THF-water process could be more economical than current processes for cellulosic ethanol production

IOWA STATE UNIVERSITY
Bioeconomy Institute

19

References

1. Lynd, L. R., Laser, M. S., Bransby, D., Dale, B. E., Davison, B., Hamilton, R., ... & Wyman, C. E. (2008). How biotech can transform biofuels. *Nature biotechnology*, *26*(2), 169.
2. Klein-Marcuschamer, D., Oleskowicz-Popiel, P., Simmons, B. A., & Blanch, H. W. (2012). The challenge of enzyme cost in the production of lignocellulosic biofuels. *Biotechnology and bioengineering*, *109*(4), 1083-1087.
3. Nguyen, T. Y., Cai, C. M., Kumar, R., & Wyman, C. E. (2015). Co-solvent Pretreatment Reduces Costly Enzyme Requirements for High Sugar and Ethanol Yields from Lignocellulosic Biomass. *ChemSusChem*, *8*(10), 1716-1725.
4. Dowe, N., & McMILLAN, J. (2001). SSF experimental protocols: lignocellulosic biomass hydrolysis and fermentation. *National Renewable Energy Laboratory (NREL) Analytical Procedures.*
5. Mellmer, M. A., Alonso, D. M., Luterbacher, J. S., Gallo, J. M. R., & Dumesic, J. A. (2014). Effects of γ-valerolactone in hydrolysis of lignocellulosic biomass to monosaccharides. *Green Chemistry*, *16*(11), 4659-4662.
6. Weingarten, R., Tompsett, G. A., Conner, W. C., & Huber, G. W. (2011). Design of solid acid catalysts for aqueous-phase dehydration of carbohydrates: The role of Lewis and Brønsted acid sites. *Journal of Catalysis*, *279*(1), 174-182.
7. Luterbacher, J. S., Rand, J. M., Alonso, D. M., Han, J., Youngquist, J. T., Maravelias, C. T., Pfleger B. F. , Dumesic, J. A. (2014). *Science, 343*, 277-280
8. Han, J., Luterbacher, J. S., Alonso, D. M., Dumesic, J. A., & Maravelias, C. T. (2015). A lignocellulosic ethanol strategy via nonenzymatic sugar production: Process synthesis and analysis. *Bioresource technology*, *182*, 258-266.
9. Ghosh, A., Brown, R. C., & Bai, X. (2016). Production of solubilized carbohydrate from cellulose using non-catalytic, supercritical depolymerization in polar aprotic solvents. *Green Chemistry*, *18*(4), 1023-1031.
10. Cai, C. M., Zhang, T., Kumar, R., & Wyman, C. E. (2013). THF co-solvent enhances hydrocarbon fuel precursor yields from lignocellulosic biomass. *Green Chemistry*, *15*(11), 3140-3145.
11. Hodge, D.B., Karim, M.N., Schell, D.J., McMillan, J.D. (2008) *Bioresour. Technol, 99*, 8940-8948
12. Li, W., Ghosh, A., Bbosa, D., Brown, R.C., Wright M.M. (2017). Techno-economic analysis of ethanol production from solvent liquefaction of lignocellulosic biomass. *ACS Sustainable. Chem. Eng.* (To be submitted)

20

Acknowledgement

- Major Professor: Prof. Robert C. Brown
- Undergraduate Research Assistants
- Bioeconomy Institute Staff, Scientists and Graduate Students
- Iowa Energy Center, Iowa EPSCoR

IOWA EPSCoR

IOWA STATE UNIVERSITY
Bioeconomy Institute

21

Thank you..

Questions?

IOWA STATE UNIVERSITY
Bioeconomy Institute

22

Batch mode solvent liquefaction process

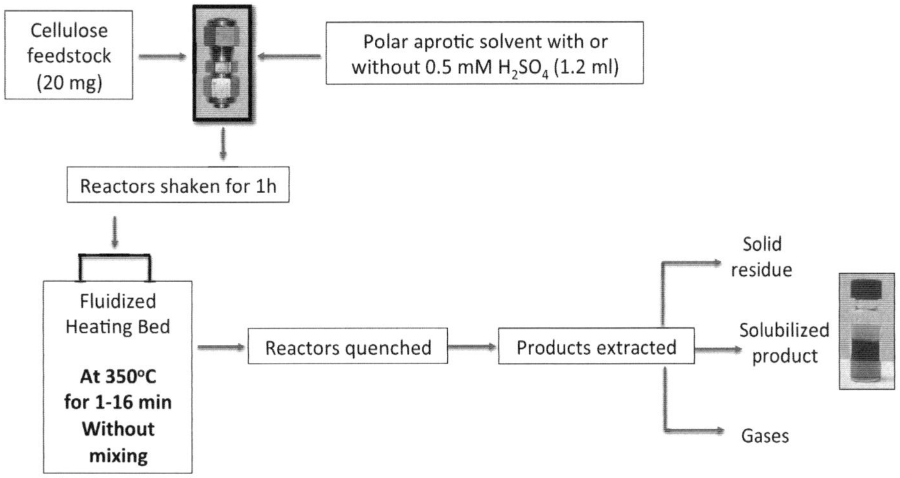

Red oak composition

Composition of lignocellulose in red oak[a]

Component	Composition (wt%)	Total C_5 polysaccharides (wt%)	Total C_6 polysaccharides (wt%)	Total lignin (wt%)
Glucan	40.00	16.00	42.22	23.24
Xylan	15.66			
Mannan	1.30			
Arabinan	0.34			
Galactan	0.92			
Klason lignin	20.29			
Acid-soluble lignin	2.95			
Extractives	6.85			
Ash	0.40			

[a] Determined according to NREL procedure by Celignis, Inc.

Solubilized carbohydrates quantified based on theoretical yields

$$Glucose\ yield\ (\%) = \frac{\text{concentration of glucose in solution} * \text{volume of solution}}{1.111 * \text{glucan content in red oak} * \text{initial mass of red oak}} \times 100\%$$

$$Xylose\ yield\ (\%) = \frac{\text{concentration of xylose in solution} * \text{volume of solution}}{1.136 * \text{xylan content in red oak} * \text{initial mass of red oak}} \times 100\%$$

$$Total\ carbohydrate\ yield\ (\%) = \frac{\text{glucose mass} + \text{xylose mass}}{(f_{C5} * \text{total pentosan content in red oak}) + (f_{C6} * \text{total hexosan content in red oak}) * \text{initial mass of red oak}} \times 100\%$$

Conversion factors:
f_{C5} = 1.36 for pentosan to pentose formation
f_{C6} = 1.11 for hexosan to hexose formation

Dehydration product yield and delignification

$$\text{Dehydration product yield (wt\%)} = \frac{\text{Mass of dehydration product}}{\text{Initial mass of biomass}} \text{ X } 100\%$$

$$\text{Delignification (\%)} = \frac{\text{Mass of lignin extracted from pretreatment}}{\text{Initial lignin in biomass}} \text{ X } 100\%$$

Dehydration products: products from dehydration of carbohydrate monomers such as 5-HMF, furfural

Mass balance and conversion of polysaccharides in two-step solvent liquefaction

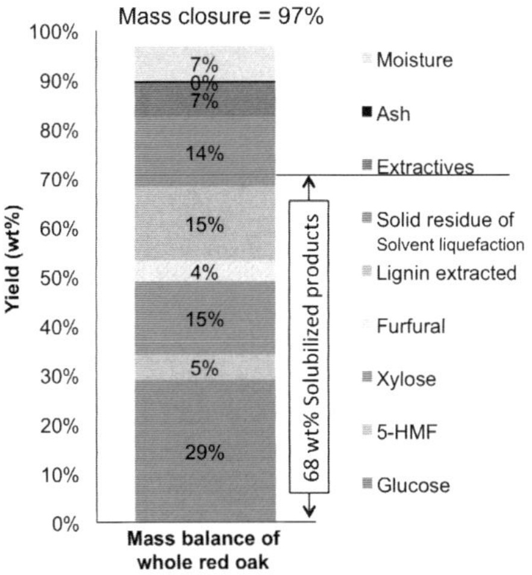

Process step	Yield (wt%)
Pretreatment	
Glucose	2
Xylose	10
Solvent liquefaction	
Glucose	28
5-HMF	5
Xylose	7
Furfural	4
Total carbohydrate derived products	56
Total Initial polysaccharides	58
Polysaccharide conversion*	97%

*Carbon-based molar conversion could not be done as source of furfural may both be pentosan/hexosan

27

Effect of delignification on recovery of xylose during pretreatment and glucose during subsequent solvent liquefaction

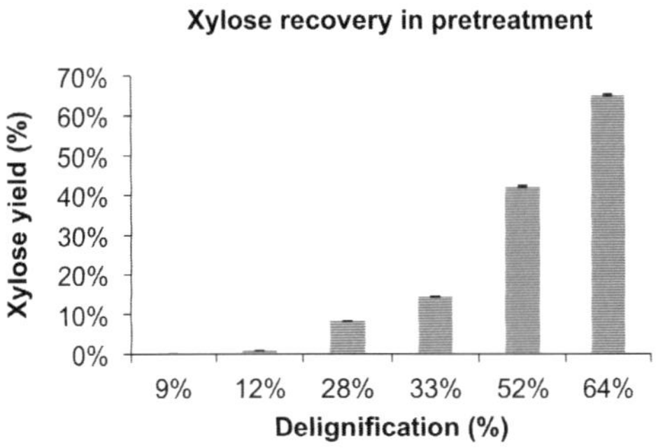

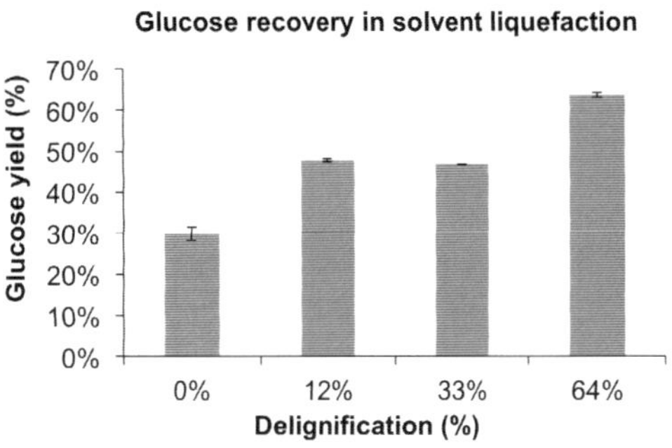

- **Delignification increases xylose recovery from hemicellulose during pretreatment**
- **Delignification increases glucose recovery from cellulose during subsequent solvent liquefaction**

IOWA STATE UNIVERSITY
Bioeconomy Institute

Mean yields reported based on at least duplicates. STD error is within 10% of mean yield.

28

Proposed block diagram for cellulosic ethanol production

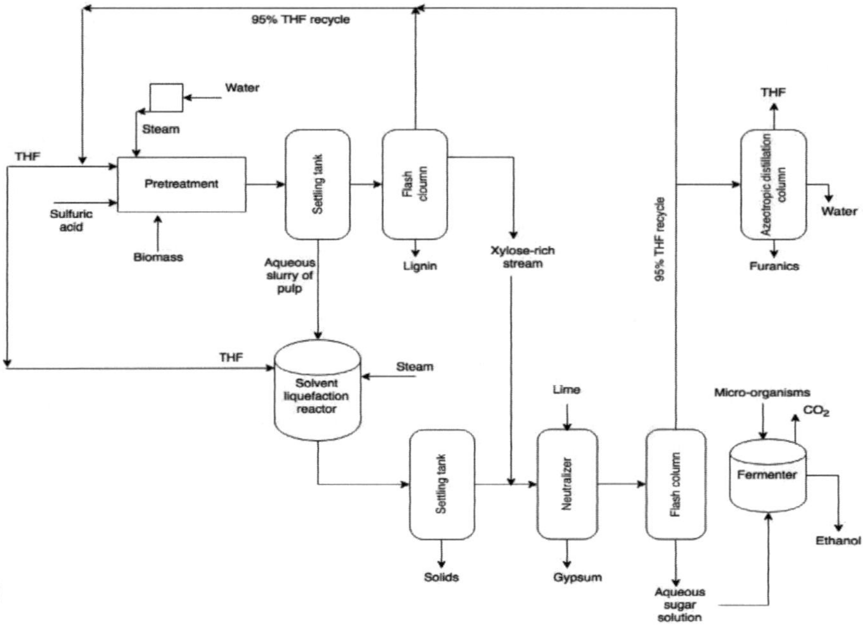

IOWA STA
Bioeconomy In

29

488549 Optimization and Performance Improvement of Bioethanol Biorefinery through the Integration of Thermochemical-Utilized Lignin Residue

Abstract:

In global economy, energy crisis has become a main issue since the global energy demand is increasing exceptionally, whilst natural reserves of crude oil are depleting as the time passes. Therefore, it is necessary to find alternative renewable energy sources. These sources include wind, solar, geothermal, hydro- and bioenergy. Bioenergy (e.g. biofuels) is the most promising renewable energy source and bioethanol is one of the main biofuels that is mostly used to be as a transportation fuel. During the production of bioethanol, a huge amount of lignin is produced as a byproduct. The global market of lignin is expected to exceed USD 6 billion by 2022, growing at 4.9% from 2015 to 2022. Due to that, strong industrial R&D has been exploring the commercialization of lignin downstream applications to produce valuable products to increase its revenue.

In this research, a biorefinery was designed to produce bioethanol by simultaneous saccharification and fermentation (SSF) of softwood (spruce) biomass after the extraction of lignin. The performance of this process is enhanced and optimized using Aspen Plus (V.2006) simulation software by investigating the potential of integrating extracted lignin back into the biorefinery. The simulation results are generally not 100% accurate but it provides a reflective prospect to process design and economic analysis of suggested processes in a limited period of time.

Firstly, the recovery of lignin from the biomass was maximised to ensure maximum production rate of bioethanol. The optimal removal rate of lignin was 20.100 tonne per hour (tph), corresponded to bioethanol production rate of 19.117 tph from a total feed charge of about 160tph.

The primary aim of this research is either to generate combined heat and power (CHP) or activated biochar. The first case was generating CHP in steam turbines that can be integrated back into the biorefinery to meet its energy requirements to minimise operating costs. The optimal turbine power of about 470 MW was produced from steam generated by the gasification of lignin. The flow rate of the steam was 2516.138 tph with turbine mechanical and isentropic efficiencies of 97% and 80% respectively. The optimal power generation was achieved by recycling steam through the gasifier to generate as much as 180 times more power than the unrecycled steam. This increase was achieved even that there was a trade-off with the gasifier power requirement, where more energy was required by the gasifier to raise the LP (low pressure) steam to superheated steam temperature. Also, the continuous recycling of steam had minimised wastewater treatment.

Heat integration was carried out to recover the waste heat from the entire system at a power duty of about 54 MW. Heat integration provided a part of the biorefinery energy requirements; reducing further its operating costs. For example, the power recovered and integrated from cooling the steam leaving the last turbine to the boiler feed water (BFW) temperature of about 105°C resulted in annual cost savings of about 37 million US dollar.

The second case was producing activated biochar at a flow rate of 4.703 tph from the pyrolysis of lignin. Biochar was firstly produced at a flow rate of 7.236 tph and then thermally treated to produce activated biochar with about 35% weight loss. The activation of biochar was achieved in the presence of CO_2 at a flow rate of 1.266 tph and activation temperature of 850°C. Using steam for biochar activation showed no benefit since it reduces power generation and has more GHG emissions when compared to CO_2 activating agent. The product streams were cooled to atmospheric conditions for heat recovery. It was concluded that utilising steam generated from the pyrolysis reactor to produce activited biochar was not benefical in terms of power, cost and environment as a whole. Producing activated biochar had reduced the power generation of the whole process and minimised the potential of heat integration, which resulted in an increase in total production cost. Also the unused carbon dioxide from lignin pyrolysis was discharged to the atmostphere and more CO was formed, making the pyrolysis process harmful to the enviroment.

497373 CO_2-Philic Polymer Membranes for High Flux CO_2 Separation

Abstract:

The vast majority of the worldâ€™s energy is presently derived from the burning of fossil fuels, which releases vast quantities of carbon dioxide (CO_2) into the environment and results in undesirable climate change. Practical and cost efficient methods of CO_2 separation and capture would thus solve one of the most challenging problems today. This presentation summarizes our effort on the development of novel polymer membranes functionalized with CO_2-philic groups for high flux CO_2 separation. Our strategy focuses on tuning solubility selectivity in addition to diffusivity selectivity for achieving high permeability membranes combined with good selectivity. Various synthetic techniques including ROMP and post functionalization were used and the careful design permits to prepare well-defined novel high permeable polymers containing CO_2-philic groups. This study demonstrated the addition of CO_2-philic groups (e.g. amidoxime and PEO) significantly increased the solubility selectivity of CO_2 over N_2. The membrane performance is also highly dependent on the balance of gas/functional group interaction, intra/inter- molecular interaction (H-bonding etc.) of membrane, packing, and polymer dynamics. Tuning the balance of the interaction enables to achieve the CO_2 separation performance over the Robeson upper bound, e.g. CO_2 permeability 3000-8000 Barrer and CO_2/N_2 selectivity 18-20. The structure-property relationships especially on CO_2 uptake, CO_2 and N_2 permeability, CO_2/N_2 selectivity to the polymer structure will be discussed.

496766 Highly Cross-Linked Polymers for Membrane H_2/CO_2 Separation at Elevated Temperatures

Abstract:

Highly Cross-linked Polymers for Membrane H_2/CO_2 Separation at Elevated Temperatures

Maryam Omidvar, Mark Swihart, and Haiqing Lin

Department of Chemical and Biological Engineering

University at Buffalo, The State University of New York)

Membrane technology has attracted significant interest for pre-combustion H_2 purification and CO_2 capture. The key to the success of this technology is membrane materials with high H_2 permeability and H_2/CO_2 selectivity at elevated temperatures of syngas processing (>100 °C). Polyimides such as P84 have been explored for H_2/CO_2 separation due to their strong size-sieving ability and resulting high H_2/CO_2 diffusivity selectivity. In this study, we demonstrate that these polymers can be cross-linked chemically to further improve H_2/CO_2 selectivity. We systematically evaluate the effect of cross-linking P84 using amines such as propane diamines, butane diamines and p-xylene diamine on the H_2/CO_2 separation properties. The cross-linking can significantly increase H_2/CO_2 selectivity while retaining moderate H_2 permeability. For example, cross-linking of P84 with butane diamine with 42% conversion increases H_2/CO_2 selectivity from 2.4 to 8.0, while decreasing H_2 permeability from 40 to 24 Barrers at 100 °C. The combination of 24 Barrers for H_2 permeability and 8.0 for H_2/CO_2 selectivity is very close to the Robesonâ€™s upper bound for H_2/CO_2 separation, demonstrating the potential of this approach for membrane development. These cross-linked polymers are thoroughly evaluated using FTIR, DSC, TGA, WAXD, and XPS. The effect of cross-linking degree on the physical properties will be discussed, including glass transition temperature, fractional free volume and pure- and mixed-gas H_2/CO_2 separation properties at various temperatures. This presentation will also discuss the fundamental structure/property relationship in these highly cross-linked polymers with promising H_2/CO_2 separation properties.

499468 High-Throughput Direct Synthesis of Cubdc MOF Nanosheets and Their Application in CO_2 Separation

Abstract:

Membranes based on metal organic frameworks (MOFs) have attracted increased research attention especially for their application in separation of carbon dioxide[1,2]. High aspect ratio MOF nanosheets, having thickness on the order of few nanometers and large lateral dimensions are ideal for fabrication of high performing membranes for gas separations.

CuBDC MOF nanosheets first introduced by the Gascon group were synthesized via a gradient synthesis strategy[3]. They show promising performance in the separation of carbon dioxide from natural gas streams.

In this work, the bottom-up synthesis strategy for CuBDC nanosheets was modified to eliminate the need for gradient synthesis. Instead CuBDC nanosheets were obtained by a direct mixing scheme consisting of homogeneous mixing of the metal and linker solutions. Nanosheets with an average thickness of 30 nm and average lateral dimensions of 2.5 micron were obtained. The aspect ratio of CuBDC nanosheets was optimized by varying parameters such as synthesis time, temperature and concentration of the synthesis solutions. It was found that lower temperature favors higher aspect ratios. The direct synthesis scheme developed herein is advantageous as it is scalable and high-throughput.

Mixed matrix membranes were fabricated by incorporating the CuBDC nanosheets in a polymer matrix and tested for CO_2/N_2 separation performance. A significant increase in the ideal selectivity of CO_2/N_2 was obtained reiterating the promising separation performance of CuBDC nanosheets.

[1] J. E. Bachman, J. R. Long, *Energy Environ. Sci* **2016**, *9*, 2031–2036.

[2] Y. Peng, Y. Li, Y. Ban, H. Jin, W. Jiao, X. Liu, W. Yang, *Science* **2014**, *346*, 1356–1359.

[3] T. Rodenas, I. Luz, G. Prieto, B. Seoane, H. Miro, A. Corma, F. Kapteijn, F. X. Llabrés i Xamena, J. Gascon, *Nat. Mater.* **2015**, *14*, 48–55.

483115 New Negative Emissions Technology: Indirect Ocean Capture Separating CO_2 from Air

Abstract:

Negative emissions technologies (NETs), which result in the removal of CO_2 from the atmosphere, will likely prove crucial to limiting global warming to 2 °C. Few viable NET options exist, and detailed, transparent cost assessments are scarce, hindering policy development and strategic investment.

We propose a new platform of membrane-based negative emissions technologies, which we term Indirect Ocean Capture, and provide detailed experimental results demonstrating real-world feasibility. Our research presents the design, pilot-scale construction, testing, and fully integrated techno-economic analysis of a new platform of negative emissions technologies for removing CO_2 from the atmosphere. The technological platform has two unique implementations but both rely on bi-polar membrane electrodialysis in series with hydrophobic membranes used for water-gas separations. The fully automated system extracts CO_2 in the following manner: A custom build bi-polar membrane electrodialysis system is used to efficiently produce HCl and NaOH in low concentrations. In the first implementation the continuous production of HCl at a stable concentration is pumped into degassed seawater. The seawater is acidifed to pH 4.5 and pumped across hydrophobic membrane contactors under vacuum pressures. High purity CO_2 is evolved from the seawater and extracted across the membrane. The continuous production of NaOH at a stable concenrtaion is pumped into the acidified seawater to return the seawater to its original alkalinity, and then returned to the oceans. In the second implementation, NaOH is added to seawater such that the seawater achieves a $9.3 < pH < 9.6$. $CaCO_3$ is precipitated from solution while basic seawater is removed. HCl is pumped into the basic seawater to return the seawater to its original pH, and then returned to the oceans.

We built a pilot-scale platform of this technology and quantified the performance of this previously untested negative emissions platform. Four different membranes from two different companies were analyzed to optimize acid and base production during electrodialysis. Further membranes for CO_2 separation from water were evaulated for efficiency. Finally, we used our experimental measurements as input into a fully integrated techno-economic quantitative model that calculates the cost of indirect ocean capture, including the effects of co-location with other industrial processes that use seawater, and the resultant costs of equivalent emissions in dollars per ton of captured CO_2. Our analysis takes the important step of determining the optimized design in pilot-scale, as well as quantifying the economics for a real-world implementation verified through industrial partner. This represents the first and only detailed process design and cost assessment for indirect ocean capture.

495079 Study of Carrier Saturation Phenomenon in Facilitated Transport Membrane for CO_2 Capture from Low CO_2 Concentration Sources

Abstract:

Amine-containing facilitated transport membranes exhibit superior CO_2 separation performance because the CO_2 transport is enhanced by the reversible reaction between CO_2 and the active carrier, amine, incorporated in the membrane. Continuously increasing the CO_2 partial pressure leads to the constant CO_2 flux, which does not increase with CO_2 partial pressure, i.e., the carrier saturation phenomenon. At the carrier saturation, all available carrier molecules have been reacted with the high pressure CO_2. However, this carrier saturation phenomenon can be beneficial for CO_2 capture from low CO_2 concentration (~1% CO_2) sources, e.g., the residual flue gas after the primary CO_2 capture system (with 90% CO_2 removal) and coal mine gas streams. For CO_2 separation from low concentration sources, an improved membrane separation performance can be obtained. With reducing the CO_2 concentration in the feed gas, both CO_2 permeance and CO_2/N_2 selectivity increased. These were mainly due to more available amine carriers for CO_2 molecular transport at lower CO_2 concentration conditions. In this study of CO_2 separation from low concentration feed, the membrane used consisted of polyvinylamine serving as the fixed-site carrier and piperazine glycinate acting as the mobile carrier. A CO_2 permeance of 785 GPU (1 GPU = 10^{-6} cm^3 (STP)·cm^{-2}·s^{-1}·$cmHg^{-1}$) and a CO_2/N_2 selectivity of 150 were obtained with a feed of 20% CO_2 and 80% N_2 on a dry basis (16.6% CO_2 and 66.2% N_2 along with the saturation water vapor at 17.2%) at the typical flue gas temperature of 57°C. By reducing the CO_2 concentration in the feed, a significant CO_2 permeance increase was observed when the CO_2 concentration was below 5%. For instance, the membrane showed a CO_2 permeance of 950 GPU and a CO_2/N_2 selectivity of 190 with a feed of 1% CO_2 and 99% N_2 on a dry basis. The observed performance increase was due to the carrier saturation phenomenon as more carrier molecules were available for CO_2 transport at a lower CO_2 concentration. Additionally, it was found that the effect of SO_2 was less significant when the CO_2 concentration was low. The CO_2 permeance dropped by approximately 2.5% with an introduction of 3 ppm SO_2 in the feed gas of 20% CO_2 and 80% N_2 on a dry basis at 57°C. However, no performance reduction was observed after exposing the membrane to 3 ppm SO_2 with the feed gas of 1% CO_2 and 99% N_2 on a dry basis at the same temperature. Therefore, the studied membrane can be applied for CO_2 separation from low concentration sources with an exceptional separation performance.

502281 Recent Applications of Separex Membranes Beyond CO_2 Removal

Abstract:

Gas separation membranes have been used commercially for over 35 years to process natural gas, most commonly with the goal of removing CO_2. This presentation will explore two recent applications of Honeywell UOP's Separex membranes: (1) Condensing mode operation of the membrane and (2) Modular design for fuel gas conditioning.

The operation of membranes in condensing mode has commonly included carbon dioxide enhanced oil recovery (CO_2-EOR), natural gas liquids (NGL) recovery and operations targeting reduction of pretreatment costs. We have developed a new membrane element, "NGL–Extract", for these condensing mode applications. NGL-Extract elements replace older generation membrane elements in commercial plants that produce NGL from a feed with a high CO_2 content. The results showed up to 70% increase in gas production while utilizing less membrane area than the older elements. In addition, results from field and laboratory tests confirmed that NGL–Extract elements are able to operate robustly under condensing conditions even with the presence of common contaminants.

The other application under discussion is fuel gas conditioning. Here the target is to use raw field gas as a substitute for purchased fuel. Raw field gas typically includes contaminants such as CO2 and H2S. Gas engines need a consistent supply of gas with a high methane number and low contaminants and heavy hydrocarbons to have higher efficiencies and prevent engine knocking. Honeywell UOP's Fuel Gas Conditioning unit with Separex membranes recovers C2+ to provide a lean engine fuel for power applications. The rich gas stream as the byproduct from fuel gas conditioning is sent back to the main gas pipeline for NGL recovery. In the approach presented here a fit for purpose fuel-gas conditioning system was developed. It utilizes Separex membranes to perform the separation and delivers high pressure treated gas. The units are pre-engineered and consist of a catalog of pre-treatment options, membrane and project-specific value blocks that can be arranged to deliver specific solutions. This flexible design approach covers a wide range of operating pressures, feed and product flowrates.

493661 a New Experimental Technique for the Study of Gas Permeation of Binary Mixture

Abstract:

It is well known that the gas permeation follows the mechanisms of 'sorption-diffusion' in microporous inorganic membranes and 'solution-diffusion' in polymeric membranes. [1,2] However, the traditional experimental technologies for investigating those diffusion/permeation processes, such as diffusion cell, are costly, difficult to operate (e.g. using GC or MS for the composition analysis of permeate), and prone to experimental errors (e.g. the use of helium as the sweeping gas as down-stream, which may cause back-diffusion to upstream).

In this study, we proposed and validated a new experimental technique for effective study of such processes. Binary gas permeation (Helium and CO_2) were measured on a Carbon Molecular Sieve Membrane (CMSM) derived from Kapton polyimide at 700. The transient permeation time lag and steady state permeation fluxes were measured for the binary mixture as well as for each pure gas component.

It was found that the proposed method, which can readily condense the CO_2 component into solid phase via a physical/nondestructive method, offers the good advantages over the traditional methods, by revealing the 'true' permeation behavior of each component in the binary mixture. The method can provide us with deep insight into the gas permeation processes together with the Maxwell-Stefan theory. The figure below shows the permeation data of one set of He-CO_2 mixture.

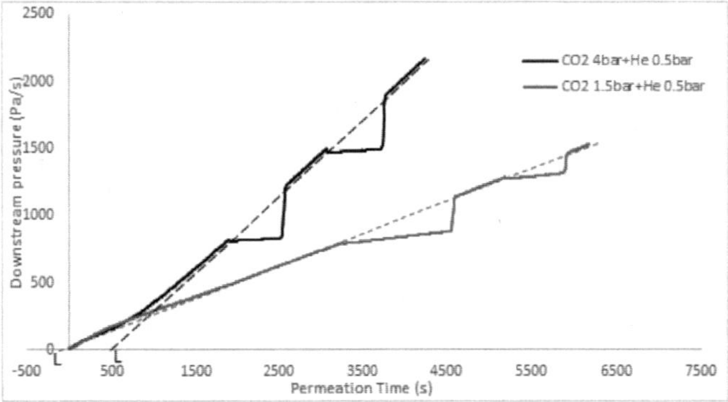

Figure 1. The pressure of binary permeated side against time by using a new method.

Key words: Time lag; adsorption; diffusion; permeation; gas mixture

1. Koresh, J. E.; Soffer, A., Mechanism of permeation through molecular-sieve carbon membrane. Part 1.- The effect of adsorption and the dependence on pressure. *Journal of the Chemical Society, Faraday Transactions 1: Physical Chemistry in Condensed Phases* **1986,** *82* (7), 2057-2063.

2. W.J. Koros, et al, Polymeric membrane materials for solution-diffusion based permeation separations, Progress in Polymer Science, 13, 1988, 339-401

501943 A Comparative Assessment of Sol-Gel and Solid State Syntheses Derived Redox Materials for H_2 Production *Via* Thermochemical Water-Splitting Process

Abstract:

Redox materials of the type $M_{1x}M_{2y}Fe_2O_4$ (M_1 or M_2 = Zn, Mn, Co, Ni) were prepared by the solid state and sol-gel syntheses routes. In the solid state synthesis route, metal oxide precursors were mixed in the required amounts in water and ball milled for 36 hours. Resultant powered material was calcined in air at higher temperatures to achieve the desired phase formation. In the sol-gel synthesis approach, metal salt precursors were sonicated in ethanol and gel formation was achieved using propylene oxide. The gel was aged, dried and finally calcined in air. The solid state diffusion properties of these redox ferrites were studied using the variable temperature impedance spectroscopy whereas the morphological properties were investigated using scanning/transmission electron microscopy (SEM/TEM), powdered X-ray diffraction (XRD), and Brunner-Emmett-Teller (BET) specific surface area analyzer. A solid state or sol-gel derived redox material was loaded inside a continuous Inconel tubular reactor and thermochemical water-splitting reaction was performed at isothermal conditions. Materials syntheses, characterization and H_2 volume generation over multiple thermochemical cycles will be presented.

501733 Overview of Impact of Electrode Catalyst Loadings on Thin Flexible Fuel Cell (TFFC) Performance

Abstract:

Thin Flexible Fuel Cell (TFFC) is a flat and flexible device without any rigid support capable of being integrated into portable electronics. TFFC uses hydrogen and oxygen as fuel to provide high volumetric and gravimetric energy densities capable of competing with battery technology as a power source. The group is currently investigating a TFFC with an air-breathing design that maximizes the cathode interface with ambient air. The current challenges of making TFFC commercialized are the cost and performance. Enhanced cell performance is critical for a competitive cost of the device. The group investigated the impact of catalyst loading on the cell performance to increase the overall performance of the device. The catalyst loading of the cathode electrode was varied while keeping the anode catalyst loading constant and vice versa. The effects of the various catalyst loadings were characterized using various analytical techniques. Polarization curves and Electrochemical Impedance Spectroscopy (EIS) were used to separate and quantify the sources of losses for the TFFC. Power curves provided the maximum power density of the device. Cyclic voltammetry (CV) was implemented to characterize the electrochemically active surface area (ECSA) and utilization of the various catalyst layers. Rotating Disc Electrodes (RDE) tests were performed for a direct assessment of the catalyst. We will report the results of these assessments and summarize the resulting guidance for optimizing catalyst loading on similar fuel cell structures.

497849 NiMo-Ceria-Zirconia Catalytic Reforming Layer for Solid Oxide Fuel Cells Running on Isooctane

Abstract:

Solid oxide fuel cells (SOFCs) are a promising energy conversion device because they can efficiently and directly convert the chemical energy of fuels into electrical energy. Due to the high operating temperatures (typically 600-1000 °C), SOFCs hold particular promise because of their ability to use a variety of complex liquid fuels, such as conventional liquid transportation fuels (e.g., gasoline, diesel-like fuels and jet fuels) and next generation liquid bio-fuels (e.g., biodiesel), either by using external reforming systems or directly feeding these fuels into their inexpensive transition metal-based anodes

Nickel-based anodes are commonly used in SOFCs owing to their low cost, good chemical stability, and excellent catalytic activity toward hydrogen oxidation and reforming of small hydrocarbon molecules. However, the Ni-based anodes are well known for promoting severe surface carbon deposition. Excessive formation of carbon species on the anode leads to a rapid deterioration of the cell performance by physically blocking access of the reactants to the active catalyst sites. To operate the Ni-based SOFCs by feeding liquid isooctane fuel, a catalytic layer can be applied over the Ni-based anode to act as an internal micro-reformer. The micro-reforming layer is designed to convert the mixture of isooctane and air to synthesis gas (hydrogen and carbon monoxide), which is believe to suppress the carbon deposition over the Ni-based anode.

In this work, a NiMo-Ceria-Zirconia (NiMo-CZ) catalyst was used as a micro-reforming layer for SOFCs running on isooctane. The catalyst layer was applied on top of the conventional anode supported single cell with a configuration of Ni-yttira-stabilized zirconia (YSZ) anode, YSZ/$Ce_{0.8}Sm_{0.2}O_{1.9}$ bi-layer electrolyte and $La_{0.6}Sr_{0.4}Co_{0.2}Fe_{0.8}O_{3\text{-Î´}}$ cathode. Our results showed that application of the novel catalyst layer was an effective way to enhance the electrochemical oxidation of complex hydrocarbons at the anode. At 750 Â°C, the single cell exhibited a low polarization resistance of 1.36 Î© cm^2 and a good maximum power density of 405 mW cm^{-2} in isooctane/air. At the current density of 500 mA cm^{-2}, the cell voltage presented a fairly low degradation rate of 3.0 mV h^{-1} during over 12 h stability test. The excellent electrochemical performance suggests the high catalytic activity of the NiMo-CZ catalyst for reforming isooctane and suppressing performance degradation of the single cell in the isooctane fuel.

500402 Homogeneous Reaction Kinetics of Carbohydrates with Viologen Catalysts in Biofuel Cell Applications

Abstract:

The production of renewable energy from biomass has been demonstrated using processes such as Fischer-Tropsch synthesis, direct fermentation of biomass sugars, biomass conversion to syngas followed by fermentation, and utilization of biomass carbohydrates in biofuel cells.Â Processes have been used to produce either electrical power, heat, liquid biofuels, or hydrogen. However, limitations and challenges for commercial viability still exist with all of these processes. Challenges inhibiting biofuel cell processes include low power output, low biological catalyst durability, and low carbohydrate conversion. Recently, a potentially transformative breakthrough has occurred with the discovery of a homogeneous non-biological viologen catalyst that facilitates carbohydrate conversion to electricity.

The viologen catalyst serves as a mediator or â€œelectron shuttleâ€ to extract electrons from the carbohydrate to form a reduced viologen. The reduced viologen is then oxidized at the anode of a biofuel cell to generate power. The combination of homogeneous viologen reduction by the fuel coupled with oxidation of the viologen at the electrode provides a continuous electron shuttle, enabling energy to be extracted from carbohydrates. For a biofuel cell process, the viologen catalyst provides greater durability relative to that achievable with biological catalysts.

Our initial studies have shown that viologens are effective in shuttling electrons from a wide variety of carbohydrates (C1 to C6 compounds). The homogeneous reaction rates and the efficiency of electron extraction were found to vary with reaction conditions such as pH, temperature, and reactant concentrations. Optimization of the reaction rates and efficiency are critical towards developing a viable biofuel cell using biomass carbohydrates as the feedstock.

Studies were conducted to determine the homogeneous rate law that describes the reaction between the carbohydrate and viologen. Specifically, studies were performed using NMR and spectroscopic techniques to develop a mechanistic description of the reaction rate applicable for a wide range of conditions. Studies of viologen stability were also performed to quantify potential viologen decomposition conditions that could compromise carbohydrate conversion efficiency. This work will present the mechanistic model of the carbohydrate-viologen reaction and provide insights towards optimizing reactor conditions that enhance reaction rates and carbohydrate conversion efficiency.

500124 Hydrogen Fueled Micro-Fuel Cell with Microfluidic Channels on a PFSA Plane

Abstract:

Microfluidic devices have shown considerable promise in a wide range of applications from medical screening to portable energy sources. Micro fuel cells (Î¼FCs) have been attracting much attention as a leading candidate for prospective portable power sources and battery replacements [1-4]. Their ability to create an efficient and clean source of energy combined with the ease of portability makesÂ Î¼FCs available to meet the needs of various portable electronic applications in the future [4, 5]. To this purpose, making more efficient devices with cost effective processes and materials is crucial.

Many efforts on fuel cell miniaturization are focused on silicon-based techniques as silicon is the most common substrate in MEMS technology [6, 7]. However, combining silicon devices with polymeric fuel cells at mm or sub-mm scale presents many challenges, none of which have been solved in a completely satisfactory manner [8]. Furthermore, in recent years due to material property issues associated with PDMS such as bulk absorption of small molecules and evaporation through the device, there has been a tendency towards the employment of thermoplastics for microfluidic systems [9].

Distinct from previous attempts on fabricating in-planeÂ Î¼FCs in other studies, in this work we present a novel On-Membrane Micro Fuel Cell design where the micro-flow channels for fuel and oxidant input are fabricated by using a combination of laser micromachining and hot-bonding process. In preparation of the device, a laser patterned acrylic sheet was pressed onto the Nafion substrate by a gas-cushion hot-bonding and the inlet tubes were mounted on the acrylic sheet. The micro device performance was characterized by I-V polarization curves and impedance spectroscopy under dry and humidified conditions. The combined in-plane and through-plane flux of protons in the membrane can be compared to literature data.

With this fabrication technology high aspect ratio structures can be fabricated over large surface areas, which prompts a commercially successful manufacturing of polymer-based micro-components.

Refrences:

1. Sundarrajan, S., Allakhverdiev, S. I., & Ramakrishna, S. (2012). Progress and perspectives in micro direct methanol fuel cell. *International Journal of Hydrogen Energy*, *37*(10), 8765-8786.

2. Taylor, A. D., Lucas, B. D., Guo, L. J., & Thompson, L. T. (2007). Nanoimprinted electrodes for micro-fuel cell applications. *Journal of Power Sources*, *171*(1), 218-223.

3. Z. Yuan, J. Yang, Y. Zhang and X. Zhang, "The optimization of air-breathing micro direct methanol fuel cell using response surface method," *Energy*, vol. 80, pp. 340-349, 2015.

4. Kamarudin, S. K., Daud, W. R. W., Ho, S. L., & Hasran, U. A. (2007). Overview on the challenges and developments of micro-direct methanol fuel cells (DMFC).*Journal of Power Sources*, *163*(2), 743-754.

5. Cao, J., Xu, J., Chen, Z., Wang, W., Huang, Q., & Zou, Z. (2013). A Silicon-based micro direct methanol fuel cell stack with a serial flow path design.*International Journal of Energy Research*, *37*(4), 370-376.

6. Shah, K., Shin, W. C., & Besser, R. S. (2004). A PDMS micro proton exchange membrane fuel cell by conventional and non-conventional microfabrication techniques. *Sensors and Actuators B: Chemical*, *97*(2), 157-167.

7. Lu, G. Q., Wang, C. Y., Yen, T. J., & Zhang, X. (2004). Development and characterization of a silicon-based micro direct methanol fuel cell.*Electrochimica Acta*, *49*(5), 821-828.

8. Omosebi, A., & Besser, R. S. (2013). Fabrication and performance evaluation of an in membrane micro-fuel cell. *Journal of Power Sources*, *242*, 672-676.

9. Sackmann, Eric K., Anna L. Fulton, and David J. Beebe. "The present and future role of microfluidics in biomedical research." *Nature* 507.7491 (2014): 181-189.

496142 Design and Analysis of Tri-Generation System Powered By Seaweed Biogas

Abstract:

Over the last two decades, fuel cells have been intensely researched. Biogas-powered solid oxide fuel cells (SOFCs) have become prospective devices for stationary power generation. In this work, simulation and economic analysis were performed for a 3 MW biogas-fueled SOFC system. The composition of the anode outlet of the proposed SOFC system contains CO, H_2, H_2O, and CO_2 which is comparable to typical high temperature water-gas-shift (WGS) feed to co-produce hydrogen. The conversion of CO in water-gas shift reaction was 88% with the hydrogen concentration of 35 mol %. Pressure swing adsorption is used to purify the hydrogen. In order to utilize the heat from the WGS reaction, organic rankine cycle (ORC) was used. To our knowledge, this has been the first work on design and analysis of an SOFC integrated into a combined heat, hydrogen and power (CHHP) system.

498740 Overcoming the Challenges of Using Corn Stover As Feedstock in Autothermal Pyrolysis

Abstract:

Corn stover biomass is abundant across the Midwestern United States and is considered a potentially promising feedstock from which biorenewable fuels and chemicals could be produced. However, significant challenges to economically and efficiently convert the lignocellulose fractions into valuable products remain. Fast pyrolysis is an endothermic process capable of thermally deconstructing biomass into bio-oil vapors, non-condensable gases (NCG's), and solid biochar at an elevated temperature (500°C). Autothermal operation differs from traditional pyrolysis with the addition of a small amount of air introduced in the reactor to generate sufficient exothermic energy to drive pyrolytic reactions and achieve a neutral energy balance. Pilot scale testing in a bubbling fluidized bed reactor resulted in low equivalence ratios (ER) during autothermal operation that ranged from 8-11% of the theoretical air requirements for stoichiometric combustion; these ER's are almost half of those compared to autothermal gasifiers. Significant process intensification was reached when air-blown autothermal operation was achieved using 100% air as the fluidizing gas and a maximum biomass feed of 20 kg hr^{-1} was obtained. Bio-oil yields greater than 50 wt. % were achieved along with biochar and NCG's yields around 20 and 30 wt. % respectively. Six bio-oil product streams were collected using a fractional recovery system consisting of three alternating pairs of condensers and electrostatic precipitators; this allowed bio-oil compounds across a range of molecular weights to be collected based on the compound's dew point. Early technical challenges of processing corn stover involved the accumulation of biochar in the reactor and degradation of bio-oil heavy ends in the first stage fraction (SF1). These were overcome by adjusting the fluidization gas velocity to promote better char particle attrition and elutriation from the reactor as well implementing a new SF1 quench tower to facilitate the condensation and collection of viscous bio-oil products. Autothermal process intensification represents a significant advancement in pyrolysis technology and removes the heat transfer bottleneck traditionally associated with thermochemical conversion. Under these conditions, the reactor's footprint was reduced by 40% while simultaneously increasing biomass throughput by a factor of four. This advancement is expected to significantly improve the scalability and process economics of corn stover pyrolysis systems.

495587 Development of a Kinetics Model for Autothermal Pyrolysis in a Fluidized Bed Reactor

Abstract:

Autothermal pyrolysis is achieved by reacting a small amount of oxygen (equivalence ratios of 0.06-0.10) with biomass in a fluidized bed reactor, which partially combusts the products of pyrolysis to overcome the endothermic enthalpy of pyrolysis. Experiments performed at Iowa State have shown that autothermal pyrolysis produces no significant loss in bio-oil yield or quality as compared to traditional non-oxidative pyrolysis. Current experimental data shows that half the heat comes from oxidation of non-condensable gases (NCGs) such as carbon monoxide, methane, ethane, and ethylene and an additional one-quarter comes from oxidation of char. However, currently available kinetic mechanisms for gas-phase combustion of NCGs do not predict significant oxidation at pyrolysis temperatures (400 - 600ËšC). On the other hand, many of the volatile organic compounds that subsequently condense to bio-oil are predicted to readily oxidize at these temperatures, thus prompting a closer examination of the gas-phase kinetics of autothermal pyrolysis.

The role of gas-phase oxidative reactions in providing the enthalpy of pyrolysis during autothermal pyrolysis are explored in this study using both experimental and computational methods. A fluidized bed operated at temperatures between 400-600ËšC was used to estimate the oxidation rates of bio-oil model compounds at equivalence ratios of 0.06. Gas analysis allowed product distributions to be determined. These results were used for modeling oxidative reactions during autothermal pyrolysis.

492975 Coupling Effects of Mass Transfer and Chemical Kinetics during the Co-Pyrolysis of Cellulose and High Density Polyethylene

Abstract:

The increasing amount of municipal solid waste (MSW) causes growing environmental concern for the humanity. Growing global population needs to secure food, energy and water resources, and with decreasing amount of available land, landfills of the waste do not constitute a long-term solution. Biomass and plastics are two of the most dominant components of MSW, and a method that can concurrently convert both waste streams into energy or useful products will be beneficial for humanity. Pyrolysis is one of the simplest ways to convert waste into high-value products. When biomass is pyrolyzed alone, however, it forms pyrolysis oil with high oxygen content, which causes low heating value, thermal instability and corrosiveness. Plastic wastes, which are rich in hydrogen and low in oxygen, could be helpful for enhancing the quality of pyrolytic oil. Many studies have shown that co-pyrolysis of biomass and plastics provides oil with higher yields and caloric value than the one produced from biomass pyrolysis alone. Despite this promise, mixed results regarding their interactions are reported in the literature, and the fundamental interactions between biomass and plastics during co-pyrolysis are still not well understood. As a result, there is a need to unravel the interactions between biomass and plastics at the molecular level. In our work, the coupling effects of the mass transfer and chemical kinetics during cellulose/plastics co-pyrolysis is studied to explain the contradictory results in the literature. An experimental apparatus was constructed to allow fast pyrolysis of cellulose, high density polyethylene (HDPE), and their mixtures. The samples with different mixing patterns but same compositions were reacted. The differences of the product yields were observed, suggesting a competition between the intraphase diffusion and interphase diffusion of the biomass-derived molecules.

495855 Fast Pyrolysis of Four Different Microalgae: Apparent Kinetic Parameters Evaluation and Product Analysis

Abstract:

Fast Pyrolysis of Four Different Microalgae: Apparent Kinetic Parameters Evaluation and Product Analysis

Ribhu Gautam*, R. Vinu

Department of Chemical Engineering and National Center for Combustion Research and Development, Indian Institute of Technology Madras, Chennai, Tamil Nadu, India

Economic growth and development is resulting in fast depletion of fossil fuel reserves. Presently, the mitigation of environmental impacts caused due to the excessive use of fossil fuels has gained importance. Over the past few years, research for replacement of conventional fossil fuels is gaining importance with focus on biomass as it has got high potential and is an abundant energy source. Compared to terrestrial biomass, marine biomass has advantages such as high growth rate, high per hectare yield and high photosynthetic efficiency. Marine biomasses have flexibility to grow in sea water, fresh water ponds, industrial effluents and waste water. Microalgae, one of the most prominent marine biomass, are abundantly distributed water plants. They consist of lipids, proteins and carbohydrates and these lipids are converted into biodiesel. Some microalgae have lipid content as high as 80 wt.%, and are hence considered as potential feedstocks for biodiesel and biofuel production.

Fast pyrolysis is a promising thermochemical processing technique to convert microalgae into liquid, gas and solid products in the absence of oxygen at moderate temperatures. High heating rates (>1000°C s^{-1}) involved in fast pyrolysis leads to faster completion of reactions within seconds. This results in high yields of liquid bio-oil as compared to slow pyrolysis. The existing kinetic models of algae pyrolysis are all based on thermogravimetric mass loss data obtained at slow heating conditions. The use of these data limits the application of these models for fast pyrolysis. Therefore there is an immense need to obtain kinetic parameters under fast pyrolysis conditions and validate them with the existing rate parameters.

This study reports a new experimental technique to evaluate the mass loss profiles of microalgae under fast pyrolysis conditions using an analytical pyrolyzer. Fast pyrolysis of four different microalgae (*Chlorella vulgaris (C. vulgaris)*, *Nannochlorpsis oculata (N. oculata)*, *Schizochytrium limacinum (S. limacinum)* and *Arthrospira platensis (A. platensis)*) was studied in the temperature range of 400-700°C and sample residence times in the range of 2-60 s. The apparent kinetic parameters of fast pyrolysis of the algae were evaluated using first order, one dimensional (1D), 2D and 3D diffusion models. Prior to the fast pyrolysis experiments the dried microalgae samples were characterized by TGA, proximate analyser, CHNS analyser, bomb calorimeter for thermal stability, proximate analysis, ultimate analysis and heating value respectively. Fast pyrolysis experiments were carried out in an analytical pyrolyzer with 4.5±0.2 mg of sample in a quartz tube. The mass loss data of the microalgae samples was generated in isothermal conditions via

gravimetry. The most conventional rate loss equation was considered for the estimation of rate constants.

Where *f(a)* is the dependence of conversion on the reaction model, *k*(T) is dependence of rate on temperature. 'a' termed as normalized conversion can be calculated with the formula a = $(m_0-m_t)/(m_0-m_8)$, where m_0, m_t and m_8 are initial sample mass, sample mass at time *t* and final mass obtained respectively. As the reaction proceeds from initial to completion, a increases from 0 to 1. This cannot be linked to specific reactions during the process, so a is the degree of overall completion of reaction. Now,

The function of *g(a)* has been approximated as $-ln(1-a)$, a^2, $[(1-a)ln(1-a)+a]$ and $[1-(1-a)^{1/3}]^2$ for first order, 1D, 2D and 3D diffusion respectively for solid state reactions. The rate constants for each reaction temperature was obtained from the slope of the *g(a)* v/s t graph. The estimation of apparent activation energy was done by fitting temperature and rate constant data in Arrhenius Equation represented below

First order kinetic model fitted the experimental data well with very high R^2 for all the four microalgae samples. The apparent activation energies (E_a) and pre-exponential factors (A) for fast pyrolysis of the algae varied in the range of 4-24 $kJmol^{-1}$ and 0.03–1.7 s^{-1} respectively. The obtained rate parameters were validated by constructing a kinetic compensation plot, which showed that $Ln(A)=0.19E_a-0.43$ (R^2=0.97) justifying the kinetic compensation effect. The pyrolysates obtained from fast pyrolysis in an analytical pyrolyzer at 500°C were characterized using gas chromatograph/mass spectrometry for all the four microalgae samples. The pyrolysates were characterized in a GC/MS coupled with pyrolyzer. Fast pyrolysis *S. limacinum* yielded highest aliphatic and aromatic hydrocarbons owing to its high lipid content. A detailed kinetic analysis comprising all the kinetic models and the presence of various compounds observed in the pyrolysates will be discussed during the presentation.

498168 Overcoming Lignin Agglomeration during Pyrolytic Sugar Production in a Fluidized Bed

Abstract:

Fast pyrolysis of acid infused biomass has enabled over 300% increase in sugars derived from the biomass carbohydrate fraction. Our research goal was to investigate the problem of lignin and carbohydrate interaction after acid infusion: promoting agglomeration and ultimately inhibiting volatilization of both the sugars and the phenolics. Increasing the production of sugars during fast pyrolysis is important because the sugars can be used for fermentation to biofuels and important chemicals currently derived from petroleum. Understanding the interactions is crucial to the advancement of thermochemical conversion technology utilizing acid pretreatment for enhanced sugar production. Alkali and alkaline earth metals (AAEM) found in biomass have negative impact on sugar yields during fast pyrolysis, catalyzing the formation of low molecular weight species. Inductively coupled plasma (ICP) determination of alkali and alkaline earth metals was performed on the biomass to determine the amount of acid infusion needed to neutralize the negative effects of the AAEM. The acid infused biomass samples were pyrolyzed using a 2 kg/h fluidized bed reactor. The biomass, the equivalency ratio which influences reactor operability, the amount of acid infusion, and neutralization steps, were varied followed by evaluation using various instrumentation. Thermogravimetric Analysis (TGA) was performed on each of the biomass samples for comparison to the starting biomass. Fourier transform infrared spectroscopy (FTIR) of the biomass was utilized for each of the pretreatment approaches. Specific wavelengths were selected that are attributed to either lignin or cellulose. Net integration of the intensities of the same absorption bands were compared between experimental parameters. Both hydrolysable and water soluble sugars quantities were established between the different pretreatments by high performance liquid chromatography (HPLC), while the relative molecular weight distribution of the phenolics was determined by gel permeation chromatography (GPC). Mass balances were obtained for each reactor run. With this approach we have accomplished uninterrupted fast pyrolysis of acid infused biomass with dramatically increased sugar production.

489438 Thermochemical Methylation of Lignin to Produce High Value Aromatic Compounds

Abstract:

Thermochemical methylation of lignin to produce high value aromatic compounds

Patrick A. Johnston[a] and Robert C. Brown[a,b]

[a]Bioeconomy Institute, Iowa State University, Ames, IA 50011

[b]Department of Mechanical Engineering, Iowa State University, Ames, IA 50011

Technical lignin, the byproduct from cellulosic ethanol production, is a potential feedstock for producing coumaric and ferulic acids. These biorenewable aromatic hydroxycinnamic acids can be upgraded to high value chemicals for the manufacturing of cosmetics, flavoring agents, and sunscreens. We hypothesize that during the fast pyrolysis of lignin aromatic acid esters can be recovered from coumaric, ferulic and other hydroxycinnamic acids using a methylation agent. In this study, technical lignin from cellulosic ethanol production was pyrolyzed in a micropyrolyzer with a methylation agent to produce gas chromatography (GC) detectable aromatic acid esters. Corn stover lignin from the enzymatic hydrolysis process and the organosolv process was combined with different methylation agents to produce, identify and quantify the hydroxycinnamic acid esters. High performance liquid chromatography (HPLC) with a diode array detector (DAD) was used to separate and quantify the extracted hydroxycinnamic acids from lignin for determination of yield recoveries.

To determine the actual total theoretical yield of coumaric and ferulic acid in the lignin an alkaline extraction method was used. The theoretical yields of coumaric and ferulic in the enzymatic hydrolysis lignin was 5.6% and 1.9%, respectively. The theoretical yields of coumaric and ferulic in the organosolv lignin was 7.7% and 2.9%, respectively.

The recoveries of the hydroxycinnamic acids from enzymatic acid hydrolysis lignin with different loading of methylation agent were compared using a micropyrolyzer at 500°C. Using tetramethylammonium hydroxide (TMAH) at 1μL produced recoveries of 40% coumaric and 37% ferulic compared to the theoretical extracted amount from the lignin. The recoveries at 5μL produced coumaric and ferulic acids concentrations of 68% and 55%, respectively. The recoveries at 10μL produced coumaric and ferulic acids concentrations of 84% and 69%, respectively.

The recoveries of the hydroxycinnamic acids from organosolv lignin with different loading of methylation agent were compared using a micropyrolyzer at 500°C. Using tetramethylammonium hydroxide (TMAH) at 1μL produced recoveries of 35% coumaric and 27% ferulic compared to the theoretical extracted amount from the lignin. The recoveries at 5μL produced coumaric and ferulic acids concentrations of 69% and 52%, respectively. The recoveries at 10μL produced coumaric and ferulic acids concentrations of 89% and 64%, respectively.

496975 Nano-Engineered Ni Catalyst for Tri-Reforming of Methane

Abstract:

Tri-reforming of methane has been proposed to utilize the components from a direct flue gas without CO_2 separation step with desirable H_2/CO product ratio of 2. The activity of a tri-reforming catalyst increases with an active site depending on metal content and/or size. However, the catalyst can lead to sintering of metal particles during the highly endothermic reaction such as dry or steam reforming of methane reactions. In this study, we prepared the core-shell structure of Ni@CeO_2 catalyst, and applied to the try-reforming of methane. CeO_2 exhibits high oxygen vacancies with reversible valence change property (Ce^{4+}/Ce^{3+}), which is an active site in reaction accompanying CO_2. The prepared catalysts were characterized with transmission electron microscopy (TEM), X-ray diffraction (XRD), Fourier transform infrared spectroscopy (FTIR), and H_2-temperature-programmed reduction (TPR). Preliminary results showed that the core-shell structure prevents the activation of methane onto the Ni surface. Therefore, diffusion of methane through the CeO_2 shell should be increased to obtain higher activity in tri-reforming.

498248 Density Functional Theory Study of Dry Reforming of Methane on Pure Nickel As Well As Transition Metal Overlayer Deposited Nickel Surfaces

Abstract:

Dry reforming of methane (DRM) reaction utilizes CO_2, a major greenhouse gas to convert natural gas (mainly composed of methane) to synthesis gas, an important feedstock which could further be converted into valuable chemicals and cleaner fuel. This reaction presents a clear incentive in terms of its positive impact on the global environment and thus it has gained increasing attention from the scientific community lately. The superior catalytic activity of Nickel based catalysts and also their relatively lower costs make them the most promising catalyst for this reaction. However, these catalysts also deactivate rapidly owing to their high susceptibility to coke formation and filamentous carbon deposition. This severe catalyst deactivation is one of the major drawbacks that is obstructing the widespread commercialization of DRM. Several alternative catalysts have been explored for this reaction, including noble metals such as Rh and Ru. Even though these metals are found to be more reactive, as well as more resistant to carbon deposition, their high cost generally hinders their use [1]. One of the techniques that has been proposed to suppress the coke formation on the nickel surface is the substitution of single foreign transition metal atom which could modify the electronic structure [2].

In the current work, we present our exhaustive work on the solid state density functional theory (DFT) model results to study theoretically the wide network of elementary reactions comprising the DRM reaction on various facets of pure nickel catalyst, such as Ni(111), Ni(100) and Ni(110) surfaces. Calculations were performed using rev-PBE as exchange-correlation functional within the generalized gradient approximation (GGA) as implemented in the software VASP. Long term interactions such as the van der Waals forces are quite critical for the accurate description of the adsorption energies. Thus the DFT-D3 method of Grimme is used to account for dispersion corrections. Adsorption energies were calculated for all the DRM reaction intermediate species and then subsequently the activation barriers were calculated for all the elementary reactions in the DRM cycle. The catalytic activity of these pure nickel surfaces in terms of DRM reaction rate are then compared to the rates obtained on the single overlayer deposited nickel (X/Ni) surfaces where X is a transition metal such as Cu, Rh and Pd. Electronic structure analysis of various pure as well as overlayer modified surfaces is performed in terms of the d-band theory of catalysis. Our results indicate that the catalyst stability is greatly improved by a transition metal overlayer deposition on nickel surface.

References:

1. Pakhare, D. and J. Spivey, *A review of dry (CO2) reforming of methane over noble metal catalysts.* Chemical Society Reviews, 2014. **43**(22): p. 7813-7837

2. Fan, C.; Zhu, Y.; Xu, Y.; Zhou, Y.; Zhou, X. and Chen, D; *Origin of synergistic effect over Ni-based bimetallic surfaces: A density functional theory study.* The Journal of Chemical Physics, 2012. 137: 014703

500455 A Feasibility Study of Biogas Reforming to Improve Energy Efficiency and to Reduce Nitrogen Oxide Emissions

Abstract:

The feasibility of employing the catalytic steam reforming of biogas to increase its energy content was determined by converting the methane content of it into a hydrogen-rich syngas mixture; and using this reformate product intermixed with raw biogas in an Internal Combustion (IC) engine to enhance combustion stability and to reduce NO_X emissions. The field-testing component of the project involved catalytically reforming a side stream of biomethane from a landfill gas collection system at a California landfill via a waste-energy chemical recuperation process, in which waste heat from a gas engine was used to promote the reforming reaction of biogas. In the study, the total flow of raw biogas diverted to the engine remained constant. A fraction of that biogas was, however, separated and directed to a catalytic reactor packed with a Pt/Al_2O_3 catalyst. The reformer exit stream was then dehydrated (dried), blended with the remaining portion of biogas, and burned in the engine to produce electricity. When operating on the blended biogas mixture, combustion stability was enhanced and the engine ran smoothly at the full speed of 3600 rpm with a 60 Hz output frequency. On the other hand, when burning raw biogas without any reformate gas in the fuel mixture, the test engine ran poorly, sputtering and never reaching 3600 rpm. It was determined that reforming increased the energy value of the biogas by nominally 24% and estimated that a 50/50 blended reformed gas with biogas would have nominally 12% higher energy content. It was also demonstrated that the engine operating on mixtures of reformer gas had significantly lower NO_x emissions than when operating on pure propane under the same load conditions. Similar comparative testing could not be performed on raw landfill gas alone because the engine would not operate at full speed, as noted above. The conducted research has validated the technical and economic feasibility of using reforming products during biogas combustion in an internal combustion engine to reduce NO_X emissions and to enhance combustion stability. This, in our opinion is an important contribution to the scientific/technical literature and a significant advance for the NO_x emissions reduction and the field of biogas and engine waste heat utilization.

498372 Confined Nickel Nanoparticles Supported on Silica with and without Ceria Promoter for the Partial Oxidation of Isooctane

Abstract:

A confined Ni-based catalyst was synthesized using ceria promoters over silica supports via a wet impregnation method in ethylene glycol solvent, tested for its catalytic performance toward a partial oxidation reaction (POX) of isooctane at high space velocities, and compared with Rh- based catalyst that is commonly used for the same reaction. The catalytic performance of the catalysts was investigated at 750 ^{o}C, atmospheric pressure, GHSV of 43,681 h^{-1} and O_2/C ratio of 0.5. Various characterizations of prepared catalytic samples (including XRD and TEM analysis) demonstrated that the addition of ceria promoter and utilization of ethylene glycol solvent lead to the reduced Ni particle size and the improved Ni metal dispersion. Our reforming test results showed that isooctane conversion, syngas yields, and long-term stability of Ni supported catalyst samples significantly improve as the Ni metal dispersion increases. Thus, the Ni supported catalyst sample that is prepared with ethylene glycol solvent and ceria promoter showed the best reforming performances toward the POX of isooctane, which even outperformed the supported Rh catalyst.

501690 Aqueous Phase Reforming of Ethanol over Cobalt Doped Bismuth Vanadate

Abstract:

Cation doped bismuth vanadate ($BiMeVO_x$) has generated interest as a heterogeneous catalyst support due to its high oxygen conductivity. This characteristic is due to the unusual morphology of the material, in which crystalline perovskite sheets of $(Bi_2O_2)^{2+}$ alternate with $(VO_4)^{2-}$. This anisotropy is responsible for the material's high ionic conductivity, while limiting diffusion to a plane axis. In order to stabilize the desirable tetragonal phase, which ordinarily occurs only above 570 °C, ~15% of the vanadium is replaced with a dopant (Fe^{3+}, Cu^{2+}, Ti^{4+}, Co^{2+}). This creates defects in the structure which increases the number of oxygen vacancies, further improving the ionic conductivity.

In this study, BiCoVOx has been investigated as a support for nickel and platinum catalysts for the aqueous phase reforming of ethanol. This process is of interest because it allows dilute fermentation broths to be directly and inexpensively converted into syngas, an important precursor for synthetic natural gas and specialty chemicals. Testing has shown BiCoVOx supported catalysts perform favorably against an alumina support. In order to best exploit the material's unique properties, supports were prepared both by conventional combustion synthesis, yielding a highly textured, porous substrate; and by molten salt synthesis, producing micro-platelets of tunable dimensions. This highly ordered product has the highest oxygen conductivity along its axis. Thus, the best activity can be achieved by selective deposition of the catalyst on the platelet edges, which has been accomplished via strong electrostatic adsorption (SEA). To cheaply investigate this, a protocol for selective fluorescent dye uptake has been developed. By using this, a proposed SEA study can be quickly analyzed by fluorescent microscopy, to determine spatial deposition on materials that exhibit heterogeneous surface charging.

491253 Nickel-Based Catalysts for Aqueous-Phase Reforming of Short-Chain Alcohols

Abstract:

Aqueous-phase reforming (APR) is a technology to convert organic compounds in aqueous medium primarily to hydrogen and carbon dioxide. Because of the energetic efficiency of the process, APR has been considered as a promising route to upgrade the organic residue found in biorefinery water fractions to value added products. APR of oxygenated hydrocarbons to produce hydrogen requires catalysts that promote C-C bond cleavage and water-gas-shift (WGS) reaction. The highest activity and hydrogen selectivity in reforming reactions have typically been achieved over Pt-based catalysts. However, comparable activities were achieved over cheaper Ni catalysts[1]. However, the most tested nickel-based catalysts have been observed to deactivate rapidly[2].

The target of this work was to prepare and test suitable nickel modified catalysts using zirconium and cerium mixed oxides as supports. These supports have been chosen because of their good hydrothermal stability which is the main issue to be considered in APR studies.

Nickel was introduced to CeO_2-ZrO_2 supports by incipient wetness impregnation method. Reduction of the catalyst was done in-situ in APR reactor before the reaction. Successful modification of all tested oxide supports by nickel was proven by various characterization methods (AAS, N_2 physisorption, TEM and XRD).

APR experiments were performed in a continuously operated stainless steel tubular reactor using downward flow. Reaction conditions of experiments were 230 °C, 32 bars and 5 wt% aq. solutions of short chain alcohols, namely methanol, ethanol, propan-1-ol and butan-1-ol, were used as a feedstock.

Deep characterization of prepared catalysts revealed differences in pore size, surface areas, and metal particle size.

The prepared catalysts showed very high activities in APR of short-chain alcohols and conversions over 50 % were achieved. In the case of methanol reforming, hydrogen, methane, CO and CO_2 were detected as the main gaseous products. In case of other alcohols APR, also significant formation of light alkanes and alkenes was detected in gaseous phase and of aldehydes in liquid phase products. Comparable conversions of longer alcohols were determined but lower hydrogen efficiency was achieved due to the significant formation of side products.

Furthermore, the possibility of the catalyst reuse (after their re-reduction) in the APR of methanol was tested and proven. However, some loss of activity, which was addressed to the Ni particles sintering, was observed.

References:

[1] Wikberg, H.; Grönberg, V.; Jermakka, J.; Kemppainen, K.; Kleen, M.; Laine, C.; et al. Tappi J. **2015**, 14, 195-207.

[2] Coronado, I.; Stekrova, M.; Reinikainen, M.; Simell, P.; Lefferts, L.; Lehtonen, J. Int J Hydrogen Energy **2016**, 41, 11003-11032.

489614 Production of Hydrogen-Rich Syngas from Steam Reforming of Acetone over Ni-Co-Mg-Al Hydrotalcite Catalysts

Abstract:

Production of Hydrogen-rich SynGas from Steam Reforming of Acetone over Ni-Co-Mg-Al Hydrotalcite Catalysts

Sanchari Basu and Narayan C. Pradhan

Department of Chemical Engineering, Indian Institute of Technology,

Kharagpur - 721302, India.

E-mail:88sanchari@gmail.com, ncp@che.iitkgp.ernet.in

Abstract

In present time, the world is witnessing gradual exhaustion of conventional fossil fuels.This is driving us to look for alternate sources of energy. Hydrogen is a potential source of green energy. Today the most common process for hydrogen generation is the catalytic steam reforming of smaller hydrocarbons and alcohols such as methane, ethane, methanol, ethanol, glycerol, etc. The prominent catalysts for steam reforming are nickel supported on various oxides, hydrotalcites, etc. There are very few work reported on steam reforming of acetone, which is a low-cost by product of the cumene route for phenol. Also, acetone is an important intermediate in ethanol and acetic acid steam reforming. So, acetone steam reforming process can provide a better understanding of ethanol and acetic acid steam reforming processes.

In the present work, the steam reforming of acetone has been investigated over Ni-Co-Mg-Al hydrotalcite catalysts. The catalysts were made by varying the Ni:Co mole ratio (0.33, 1 and 3). Co-precipitation method was used to prepare the catalysts. The synthesized catalysts were characterized by XRD, N_2-BET, FESEM-EDS, FTIR and H_2-TPR techniques. CHNS analysis was performed to estimate the extent of carbon deposition on the spent catalyst. The steam reforming activity was evaluated in a continuous packed bed reactor at atmospheric pressure. The operating parameters were temperature, 450 ^{o}C - 550 ^{o}C; water/acetone molar ratio, 6 and space-time, 12.10 kg cat h/kmol acetone. The reactant and product compositions were determined by gas chromatography. Initially, the experiments were conducted at 550°C to compare the activities of different catalysts as shown in Figure 1. All the catalysts exhibited around 99% conversion. The methane selectivity was very less for all the catalysts. However, a maximum hydrogen selectivity of ~ 80% was obtained over Ni-Co-Mg-Al (0.33). Therefore, all further experiments were conducted over Ni-Co-Mg-Al (0.33) catalyst. The effect of temperature is shown in Figure 2. The conversion of acetone was >90% throughout the temperature range. The selectivity of hydrogen increased with temperature up to 500 ^{o}C and remained constant beyond this temperature. The selectivity of CO_2 and CO was high initially and decreased at high temperature. The catalyst did not show any considerable selectivity for CH_4. The stability test was carried out for 4 h at 500 ^oC and water/acetone mole ratio of 6. The time-on-stream behavior of the catalyst is shown in Figure 3. As can be seen from Figure 3, there is very little change in conversion over a period of 4h. Therefore, the catalyst is quite stable at the reforming conditions.

Figure 1: Activity of different catalysts at 550 ^oC.

Water:Acetone = 6:1 (molar), Space-time = 12.1 kg cat h/kmol acetone;

A: Ni-Co-Mg-Al (0.33), B: Ni-Co-Mg-Al(1) &C: Ni-Co-Mg-Al(3)

Figure 2: Effect of temperature on acetone conversion and product selectivity.

Water:Acetone = 6:1 (molar), Space-time = 12.1 kg cat h/kmol acetone;

Catalyst: Ni-Co-Mg-Al (0.33)

Figure 3: Time-on-stream behavior of the Catalyst A

References

1. Braga A. H., Sodre E. R., Santos J. B. O., Marques C. M. D. P, Bueno J. M. C., Steam reforming of acetone over Ni- and Co-based catalysts: Effect of the composition of reactants and catalysts on reaction pathways. *Applied Catalysis B: Environmental*, 195 (2016)16-28.

2. Guil-Lopez R., Navarro R.M., Ismail A.A., Al-Sayari S.A., Fierro J.L.G., Influence of Ni environment on the reactivity of Ni-Mg-Al catalysts for the acetone steam reforming reaction. *International Journal of Hydrogen energy*, 40 (2015) 5289-5296.

3. Hu X., Zhang L., Lu G., Pruning of the surface species on Ni/Al_2O_3 catalyst to selective production of hydrogen via acetone and acetic acid steam reforming. *Applied Catalysis A: General*, 427(2012) 49-57.

4. Navarro R. M., Guil-Lopez R., Ismail A. A., Al-Sayari S. A., Fierro J. L. G., Ni- and PtNi-catalysts supported on Al_2O_3 for acetone steam reforming: Effect of the modification of support with Ce, La and Mg. *Catalysis Today*, 242 (2015) 60-70.

5. Sun J., Mei D., Karim A.M., Datye A.K., Wang Y., Minimizing the formation of coke and methane on Co nanoparticles in steam reforming of biomass-derived oxygenates. *ChemCatChem*, 5 (2013) 1-5.

497658 Advancing the Production of Olefins and Aromatics from Natural Gas Via Methanol: Chemical Looping for Syngas Generation

Abstract:

Due to technological advances in the shale gas industry, the supply of natural gas in the United States has greatly increased (37% since 2008[1]) and its price has significantly dropped ($2.52/MMBtu in 2016[2]). The shale gas boom has not only revitalized the U.S. economy, but also rejuvenated interests in C1 catalysis toward the production of chemicals typically derived from petroleum refining[3]. Olefins (ethylene, propylene, butane isomers, butadiene) and aromatics (benzene, toluene, xylenes) are examples of valuable chemicals in high demand.

As diesel continues to command a larger portion of the crude oil usage, less naphtha will be available for petrochemical production. Given its availability, low cost, and hydrogen-rich composition, natural gas is an attractive feedstock for the on-purpose production of chemicals[4]. Conversion of natural gas to a methanol intermediate for further upgrading to chemicals is a potential avenue for chemicals production and is a promising commercial candidate for natural gas based refineries. Currently, methanol is almost entirely produced from syngas.

However, a judicious utilization of natural gas resources necessitates a systematic analysis of all available processing routes. To this aim, a gas to olefins and aromatics (GTOA) superstructure has been developed[5,6] and includes several olefins and aromatics production technologies. Previous works investigated reforming as the primary method to convert natural gas to syngas and demonstrated the profitability of GTOA processes. Recently, the benefits of chemical looping[7,8] for syngas generation have been observed in lowering the breakeven oil prices for a gas to liquids (GTL) process[9]. In this work, we incorporate chemical looping into the GTOA superstructure as an alternative to reforming and study its impact on chemicals production.

Process synthesis of the GTOA superstructure forms a large-scale nonconvex mixed-integer nonlinear model (MINLP). A tailored deterministic global optimization branch-and-bound algorithm[10] is used to solve the MINLP and determine optimal natural gas refineries with the highest profit. Several case studies are investigated to compare chemical looping against reforming for natural gas conversion. The effect of different plant capacities and product ratios are explored as well. Key topological selection of process technologies will be discussed.

1. U.S. Energy Information Administration. "Monthly Energy Review – March 2017". Available at: http://www.eia.gov/totalenergy/data/monthly/pdf/mer.pdf. Accessed April 2017.

2. U.S. Energy Information Administration. "Henry Hub Natural Gas Spot Price". Available at:

http://www.eia.gov/dnav/ng/hist/rngwhhdm.htm. Accessed April 2017.

3. Mokrani T., Scurrel M. Gas conversion to liquid fuels and chemicals: the methanol route-catalysis and processes development. Catalysis Reviews 2009, 51, 1-145

4. Floudas, C. A.; Niziolek, A. M.; Onel, O.; Matthews, L. R. Multi-scale systems engineering for energy and the environment: Challenges and opportunities. AIChE Journal 2016, 62 (3), 602-623.

5. Onel, O.; Niziolek, A. M; Floudas, C. A Optimal Production of Light Olefins from Natural Gas via the Methanol Intermediate. Industrial & Engineering Chemistry Research 2016, 55 (11), 3043-3063.

6. Niziolek, A.M.; Onel, O.; Floudas, C.A. Production of Benzene, Toulene, and Xylenes from Natural Gas via Methanol: Process Synthesis and Global Optimization. AIChE Journal 2016, 62 (5), 1531-1556.

7. Luo, S.; Zeng, L.; Xu, D.; Kathe, M.; Chung, E.; Deshpande, N.; Qin, L.; Majumder, A.; Hsieh, T.-L.; Tong, A.; Sun, Z.; Fan, L.-S. Shale gas-to-syngas chemical looping process for stable shale gas conversion to high purity syngas with a H2:CO ratio of 2:1. Energy & Environmental Science 2014, 7 (12), 4104–4117.

8. de Diego, L. F.; Ortiz, M.; García-Labiano, F.; Adánez, J.; Abad, A.; Gayán, P. Hydrogen production by chemical-looping reforming in a circulating fluidized bed reactor using Ni-based oxygen carriers. Journal of Power Sources 2009, 192 (1), 27–34.

9. Tso, W.W.; Niziolek, A.M.; Onel, O.; Demirhan, C.D.; Floudas, C.A.; Pistikopoulos, E.N.; Enhancing Natural Gas-to-Liquids (GTL) Processes Through Chemical Looping for Syngas Production: Process Synthesis and Global Optimization. *In preparation*.

10. Baliban, R. C; Elia, J. A; Misener, R.; Floudas, C. A Global optimization of a MINLP process synthesis model for thermochemical based conversion of hybrid coal, biomass, and natural gas to liquid fuels. Computers & Chemical Engineering 2012, 42, 64-86.

497121 Minimizing the Economic Impact of Amine Scrubbing Using High Fidelity Modeling and Optimization

Abstract:

Minimizing the economic impact of amine scrubbing using high fidelity modeling and optimization

R.C. Pattison, C. Tsay, Y. Zhang, P.T. Frailie, G.T. Rochelle, and M. Baldea

McKetta Department of Chemical Engineering, The University of Texas at Austin, Austin, TX 78712

In an effort to minimize carbon emissions from existing power generation units, amine scrubbing technologies have garnered significant attention owing to the energetically favorable set of reactions between carbon dioxide and amine solvents. A large body of research has been dedicated to this field, with a primary focus on characterizing the reaction kinetics, mass transfer properties, and thermodynamic properties for various solvents [1-2]. Process modeling and simulation studies have also been carried out, with a focus on predicting the performance and cost of the absorption and stripping units [3]. However, relatively little research has been focused on the holistic economic analysis and optimization of the entire process using detailed unit models.

Owing to the complexity of these systems and the numerous design decisions that impact process performance, identifying the most economically compelling designs is extremely challenging, and calls for the use of modern optimization tools. These must be used to consider all the design decisions and constraints simultaneously, such that capital cost and energy use are minimized while simultaneously guaranteeing that the process meets required removal specifications and operational constraints.. While detailed rate-based mass transfer separation models are available in various software programs, the corresponding equations are typically highly nonlinear, coupled and stiff; they are difficult to solve (particularly when embedded within process flowsheets), and thus, not suitable for process-wide optimization.

In this presentation, we introduce a novel framework for plant-wide design optimization of amine-based carbon capture processes. We describe a set of experimentally-validated first-principles and empirical rate-based models of the absorption and stripping columns, along with empirical but detailed capital cost estimates for the process units. The numerical underpinnings of our approach are derived from the pseudo-transient continuation methods and the relaxation-based flowsheet optimization algorithms introduced in our previous work [4-5].

We demonstrate our framework with two extensive case studies, considering amine-based carbon capture plants having different carbon dioxide removal requirements (90% and 99%) from flue gas. The design optimization objective function is designed to minimize the present value (based on both capital and operating expenses) of the plant in the two cases while ensuring that the removal requirements are met and operational constraints are satisfied. The results show significant cost reduction when compared to a heuristics-guided design.

References

[1] Frailie, P.T.; Plaza, J.; Van Wagener, D.; Rochelle, G.T. Modeling piperazine thermodynamics. *Energy Procedia*, **2011**, 4, 35-42.

[2] Sherman, B.; Frailie, P.T.; Li, L.; Salta, N.; Rochelle, G.T. Thermodynamic and Kinetic Modeling of Piperazine/ 2-Methylpiperazine. *Energy Procedia*, **2014,** 63, 1243-1255.

[3] Frailie, P.T. *Modeling of carbon dioxide absorption/stripping by aqueous methyldiethanolamine/ piperazine*. Doctoral Dissertation, 2014.

[4] Pattison, R.C.; Baldea, M. Equation-oriented flowsheet simulation and optimization using pseudo-transient models. *AIChE J.*, **2014**, 60, 4104-4123.

[5] Pattison, R.C.; Tsay, C.; Baldea, M. Pseudo-transient models for multiscale, multiresolution simulation and optimization of intensified reaction/separation/recycle processes: Framework and a dimethyl ether production case study. *Comp. & Chem. Eng.*, **2017.**

IDAES
Institute for the Design of Advanced Energy Systems

Optimal model synthesis for solvent-based CO_2 capture systems by simultaneously using multi-scale data

Paul T. Akula[a], John Eslick[b], Debangsu Bhattacharyya[a], David Miller[b]

[a] *Department of Chemical and Biomedical Engineering, West Virginia University, Morgantown, WV*
[b] *National Energy Technology Laboratory, Pittsburgh, PA*

AIChE Annual Meeting, Minneapolis, MN
November 2nd , 2017

Outline

- **Background**
- **Model Development**
  - **Packed Column**
  - **Wetted- Wall Column (WWC)**
- **Parameter Estimation**
  - **Sequential**
  - **Simultaneous**
- **Challenges**
- **IDAES Platform**
- **Results**
- **Conclusion**

IDAES Institute for the Design of Advanced Energy Systems | NATIONAL ENERGY TECHNOLOGY LABORATORY | Berkeley Lab | Sandia National Laboratories | Carnegie Mellon | West Virginia University | U.S. DEPARTMENT OF ENERGY

Solvent-Based CO_2 Capture System

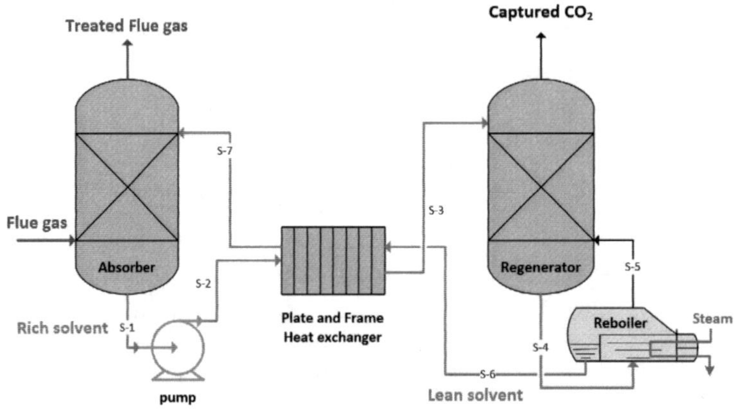

IDAES NATIONAL ENERGY TECHNOLOGY LABORATORY Sandia National Laboratories Carnegie Mellon WestVirginiaUniversity U.S. DEPARTMENT OF ENERGY

Packed Column

Mass Balance

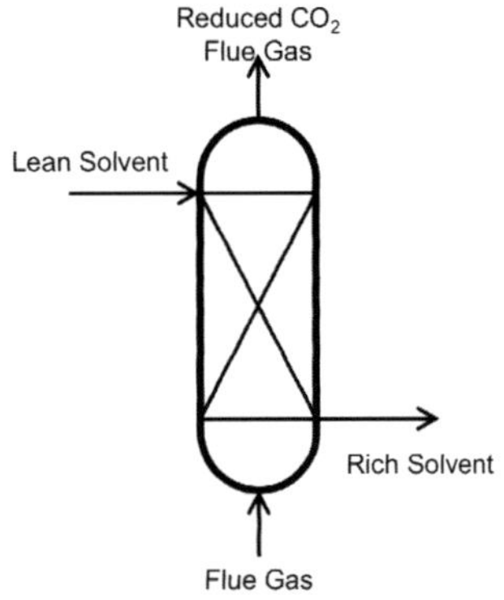

Vapor Phase

$$\varepsilon_V \frac{\partial C_{i,V}}{\partial t} = -u_V \frac{\partial C_{i,V}}{\partial z} - N_{i,V} \qquad i = 1.....n_{c^V}$$

Liquid Phase

$$\varepsilon_L \frac{\partial C_{i,L}}{\partial t} = u_L \frac{\partial C_{i,L}}{\partial z} + N_{i,L} \qquad i = 1.....n_{c^V}$$

Mass transfer rate

$$N_{i,V} = K_{V,i} a_e C_{tot,V} \left(y_i - y_i^{*,eq} \right) + N_{tot,V}\left(y_i \right) \qquad i = 1.....n_{c^V}$$

$$N_{i,V} = N_{i,L}$$

- 1D PDE-based model with plug flow
- Rate-based
- Enhancement factor approach
- NRTL model for liquid phase nonideality

IDAES NATIONAL ENERGY TECHNOLOGY LABORATORY Sandia National Laboratories Carnegie Mellon WestVirginiaUniversity U.S. DEPARTMENT OF ENERGY

Packed Column

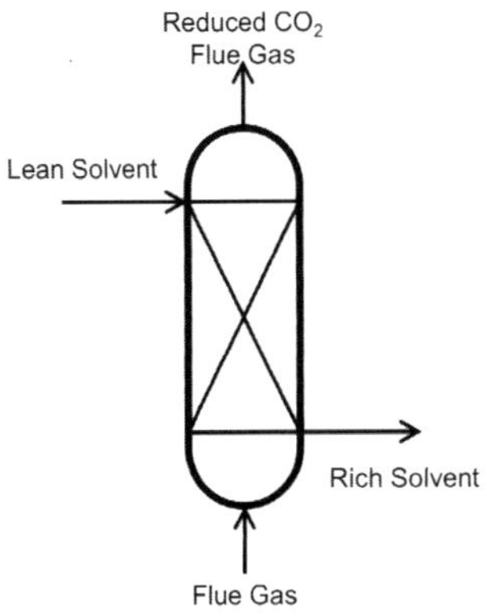

Energy Balance

Vapor Phase

$$\varepsilon_V C_{tot,V} C_{p,V} \frac{\partial T_V}{\partial t} = -u_V C_{tot,V} C_{p,V} \frac{\partial T_V}{\partial z} + Q_{h,V}$$

Liquid Phase

$$\varepsilon_L C_{tot,L} C_{p,L} \frac{\partial T_L}{\partial t} = u_L C_{tot,L} C_{p,L} \frac{\partial T_L}{\partial z} - Q_{h,L}$$

Heat transfer rate

$$Q_{h,V} = a_s h_{VL} (T_L - T_V)$$

$$Q_{h,L} = a_s h_{VL} (T_L - T_V) + N_{CO_2,V} \Delta H_{CO_2} - N_{H_2O,g} \Delta H_{H_2O}$$

Packed Column

Vapor Liquid Equilibrium

For MEA and H2O

$$y_i^* P = \gamma_i x_i^* p_i^v; \quad i = \mathrm{MEA, H2O}$$

For CO2

$$y_i^* P = \gamma_i x_i^* \left(\frac{H_i}{\gamma_i^\infty}\right); \quad where\ \gamma_i^\infty = \lim_{x_i \to 0} \gamma_i\ for\ i = \mathrm{CO2};$$

Henry's constant

$$\ln\left(\frac{H_i}{\gamma_i^\infty}\right) = \sum_s \ln\left(\frac{H_{is}}{\gamma_{is}^\infty}\right) \frac{x_s (V_{cs})^{2/3}}{\sum_{s'} x_{s'} (V_{cs'})^{2/3}} \qquad s = s' = \mathrm{MEA, H2O}$$

Kinetics

For CO_2

$$\frac{1}{K_{V,CO_2}} = \frac{1}{k_{V,CO_2}} + \frac{m_{He_{CO_2}}}{k_{L,CO_2} E}$$

- Physical equilibrium
- Chemical equilibrium
- diffusivity

Enhancement factor

$$E_{abs} = \frac{\sqrt{k_{rx,CO_2} D_{L,CO_2}}}{k_{L,CO_2}}$$

$$E_{desop} = 1 + \frac{\left(D_{L,MEACOO^-} / D_{L,CO_2}\right) \sqrt{K} C^*_{MEA,L}}{\left[1 + 2 D_{L,MEACOO^-} \sqrt{\frac{K C^*_{CO_2}}{D_{L,MEA}}}\right] \left(\sqrt{C^*_{CO_2,L}} + \sqrt{C^{int}_{CO_2,L}}\right)}$$

Wetted-Wall Column (WWC)

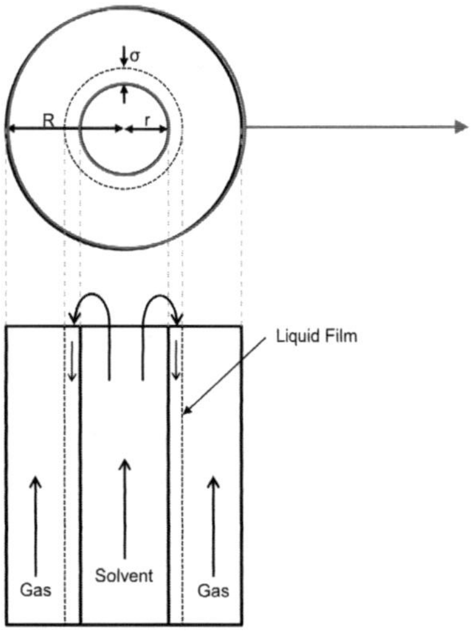

Equivalent Column diameter

$$D = 2\sqrt{(R+r)(R-r)}$$

Specific Interfacial Area

$$a_p = a_e = \frac{\text{interfacial area}}{\text{volume of annulus}} = \frac{2(r+\delta)}{(R^2-r^2)}$$

Void fraction and hydraulic diameter

$$\varepsilon_P = 1, \quad d_h = \frac{4A_c}{P} = 2(R-r)$$

Liquid hold up

$$\varepsilon_L = \frac{\text{volume of liquid film}}{\text{volume of annulus}} = \frac{(r+\delta)^2 - r^2}{R^2 - r^2}$$

Average liquid velocity

$$u_L = \frac{\rho g \delta^2}{3\mu}$$

Sub-Models Containing Estimated Parameters

Reaction Equilibrium Model
(WWC & Packed Column)

$$K_{eq,i} = \exp\left(a_i + \frac{b_i}{T} + c_i \ln T\right)$$

Equilibrium constant

$$K_1 = \frac{C_{RNHCOO^-} C_{RNH_3^+}}{C^*_{CO_2,L} C^2_{RNH_2}}$$

$$K_2 = \frac{C_{RNH_3^+} C_{HCO_3^-}}{C^*_{CO_2,L} C_{RNH_2}}$$

Mass transfer Model
(WWC & Packed Column)

$$k_{v,CO_2} = \frac{C_v}{\sqrt{\varepsilon_p - \varepsilon_L}} \frac{\sqrt{a}}{d_h} D_{v,CO_2} \left(\frac{u_v}{a\nu_v}\right)^{3/4} \left(\frac{\mu_v}{\rho_v D_{v,CO_2}}\right)^{1/3}$$

$$k_{L,CO_2} = 12^{1/6} C_L \left(\frac{u_L}{\varepsilon_L}\right)^{1/2} \left(\frac{D_{v,CO_2}}{d_h}\right)^{1/2}$$

Interfacial area Model
(Packed Column Only)

$$\left(\frac{a_e}{\beta_1}\right)^{\beta_2} = \frac{\rho_L}{\sigma_L} u_L^{\beta_3}$$

Typical Approach To Parameter Estimation: Sequential

1st Step Optimization
(smaller scale apparatus or eqp.)

$$\min_{\theta_1} (y - \eta)^T \Sigma_1^{-1} (y - \eta)$$

s.t.

$$\dot{\eta} = f(\eta, u, \theta_1, \theta_2)$$
$$g(\eta, u, \theta_1, \theta_2) \leq 0$$
$$u_{L1} \leq u \leq u_{U1}$$
$$y_{L1} \leq y \leq y_{U1}$$

Small scale (Wetted wall Column)

- Mass transfer
- Diffusivity
- Reaction kinetic Parameters

→

2nd Step Optimization
(larger scale apparatus or eqp.)

$$\min_{\theta_2} (y - \varphi)^T \Sigma_2^{-1} (y - \varphi)$$

s.t.

$$\dot{\varphi} = f(\varphi, u, \theta_1, \theta_2)$$
$$h(\varphi, u, \theta_1, \theta_2) \leq 0$$
$$u_{L2} \leq u \leq u_{U2}$$
$$y_{L2} \leq y \leq y_{U2}$$

Large scale (Packed Column)

- Interfacial area
- Holdup
- Mass transfer
- Diffusivity
- Reaction kinetic Parameters

Limitations of the Sequential Approach

- Mass Transfer and chemical reactions occur simultaneously.
- Experiment must isolate chemical reactions & mass transfer.
- Small scale data may not be representative of large scale equipment.

Our Approach: Simultaneous

$$\min_{\theta_1, \theta_2} (y-\eta)^T \Sigma_1^{-1} (y-\eta) + (y-\varphi)^T \Sigma_2^{-1} (y-\varphi)$$

s.t.

$$\dot{\eta} = f(\eta, u, \theta_1, \theta_2)$$
$$\dot{\varphi} = f(\varphi, u, \theta_1, \theta_2)$$
$$h(\varphi, u, \theta_1, \theta_2) \le 0$$
$$g(\eta, u, \theta_1, \theta_2) \le 0$$
$$min(u_{L1}, u_{L2}) \le u \le max\,(u_{U2}, u_{U2})$$
$$min(y_{L1}, y_{L2}) \le y \le max(y_{U2}, y_{U2})$$

Data:
Small scale (Wetted wall Column)
Large scale (Packed Column)

Simultaneous Parameter Estimation

$$\min_{\boldsymbol{\theta} \in \{\theta_1, \theta_2 \dots\}} \sum_{i=1}^{NG} w_i \left(\sum_{j=1}^{ND} (\varepsilon_{1,i,j}^T \sigma_{1,i,j}^{-2} \varepsilon_{1,i,j} + \varepsilon_{2,i,j}^T \sigma_{2,i,j}^{-2} \varepsilon_{2,i,j}) \right)$$

subject to:

WWC model $\forall i, \forall j$

packed Column $\forall i, \forall j$

$$\boldsymbol{\theta}^L \le \boldsymbol{\theta} \le \boldsymbol{\theta}^U$$

where:

$\boldsymbol{\theta}$ = vector of estimated parameters

σ = standard deviation of data points

NG = number of data set

ND = number of data points in data set

w_i = weight for data set i

$$\varepsilon_{1,i}^T \sigma_{1,i}^{-2} \varepsilon_{1,i} = \frac{(CO_2 flux_{wwc,i}^{exp} - CO_2 flux_{wwc,i}^{cal})^2}{\sigma_{CO_2 flux,wwc,i}^2}$$

$$\varepsilon_{2,i}^T \sigma_{2,i}^{-2} \varepsilon_{2,i} = \frac{(Loading_{abs,i}^{exp} - Loading_{abs,i}^{cal})^2}{\sigma_{loading,abs,i}^2}$$

Data

Bench scale -WWC
Pilot scale - NCCC

Wide range of operating conditions

- Liquid flowrate
- Gas flowrate
- Solvent loading
- Temperature
- Flue gas concentration

Data Sources

- National Carbon Capture Center (NCCC) Pilot Plant
- Bench Scale WWC[1]

[1]Experimental data retrieved from Dugas, R.E . (2009). *Carbon Dioxide Absorption, Desorption, and Diffusion in Aqueous Piperazine and Monoethanolamine.* (UT Austin Phd Thesis)

Challenges

Size

- Number of equations (absorber ~15,000, WWC~6,000)
- Number of data points (absorber data from NCCC ~30 set, WWC data ~200 set)
- Total number of equation (simultaneously considered):~1,650,000

Nature

- Highly nonlinear equations
- Different scales (WWC height~9 cm and diameter~ 1.5 cm and NCCC absorber height ~20 m and diameter~ 0.64 m)

Capabilities Required

- Commercial simulators: primarily sequential
- Process and property sub-models
- Large # of equations
- Large amount of experimental data

Core Pyomo Capabilities Enabling IDAES

- **Hierarchical modeling**
  - Blocks
  - External functions

PYOMO → **Property block & Unit models**

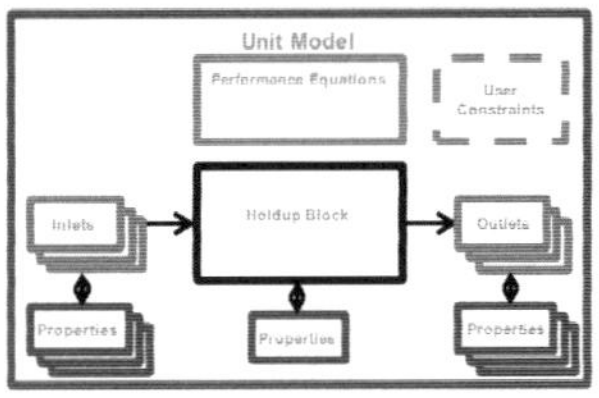

- **(Partial) differential equations** PYOMO DAE → **Dynamic process models, optimal control**

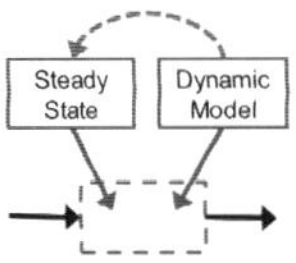

- **Stochastic programming** PySP → **Uncertainty quantification**

- **Generalized Disjunctive Programming [Pyomo.GDP]** → **Superstructure optimization**

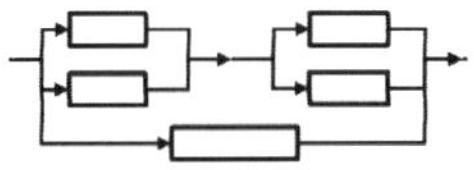

Pyomo: Extensible Object-oriented Algebraic Modeling in Python

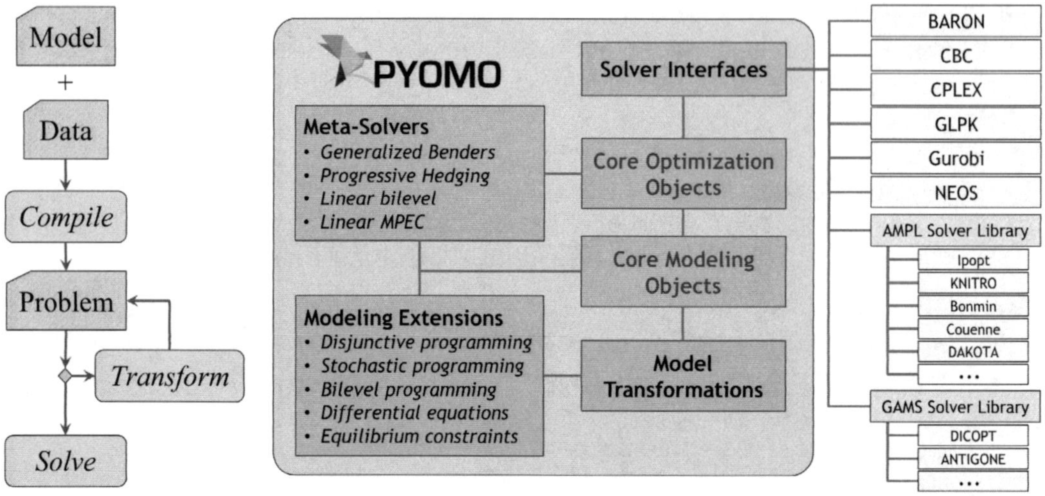

IDAES NATIONAL ENERGY TECHNOLOGY LABORATORY Sandia National Laboratories Carnegie Mellon West Virginia University U.S. DEPARTMENT OF ENERGY

Results

WWC & Packed Column

	Before opt	After opt
OBJECTIVE	**14.35**	**0.2058**

Model	Parameters	Before opt	After opt
Equilibrium reaction	a_1	198.4	200.0
	b_1	-1986	-1803
	c_1	-32.71	-32.98
	a_2	166.6	164.0
	b_2	-2102	-2200
	c_2	-26.51	-25.95
Mass transfer	Cv	0.357	0.1495
	CL	0.5	0.4903
Interfacial area	β_1	157.0	39.87
	β_2	8.621	23.95
	β_3	1.333	0.9962

IDAES NATIONAL ENERGY TECHNOLOGY LABORATORY Sandia National Laboratories Carnegie Mellon West Virginia University U.S. DEPARTMENT OF ENERGY

Results

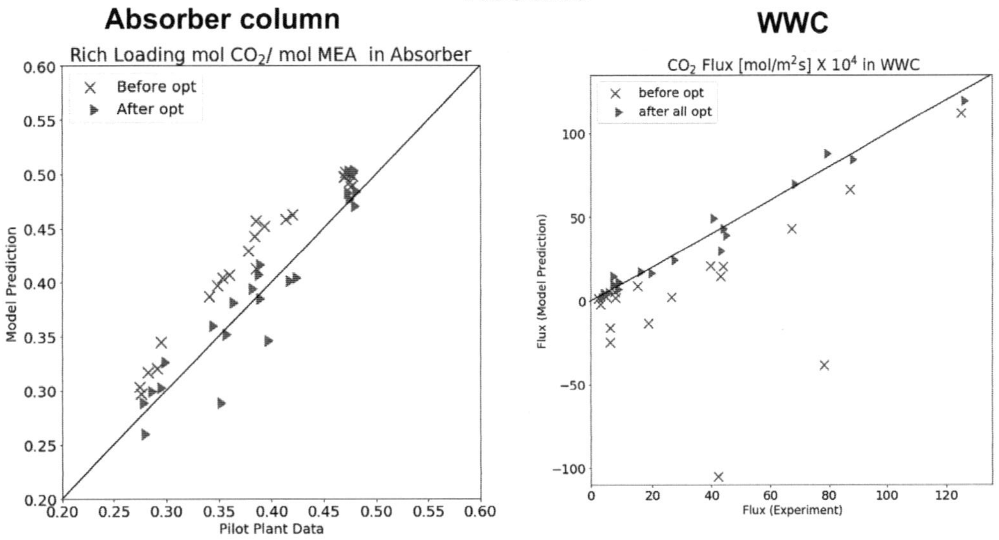

Conclusion

- Process models for packed bed and wetted wall columns
- IDAES framework demonstrated for parameter estimation
  - Complex process models
  - Multi-scale data
- Superior fitting

Continuing work:

- Larger subsets of model parameters with larger data sets
- Pyomo decomposition
  - Enable larger data sets and more complex models

ACKNOWLEDGEMENT & DISCLAIMER

Acknowledgement

This research was funded by the Lawrence Berkeley National Laboratory through Subcontract NO.7268587 as part of the U.S. DOE's Institute for the Design of Advanced Energy Systems (IDAES).

Disclaimer

This presentation was prepared as an account of work sponsored by an agency of the United States Government. Neither the United States Government nor any agency thereof, nor any of their employees, makes any warranty, express or implied, or assumes any legal liability or responsibility for the accuracy, completeness, or usefulness of any information, apparatus, product, or process disclosed, or represents that its use would not infringe privately owned rights. Reference herein to any specific commercial product, process, or service by trade name, trademark, manufacturer, or otherwise does not necessarily constitute or imply its endorsement, recommendation, or favoring by the United States Government or any agency thereof. The views and opinions of authors expressed herein do not necessarily state or reflect those of the United States Government or any agency thereof.

THANK YOU FOR YOUR ATTENTION

494599 The Value of Chemical Looping Combustion in Future Electricity Systems

Abstract:

Power generation via Chemical Looping Combustion (CLC) is inherently low in carbon dioxide emissions while maintaining the advantages of conventional thermal power plants such as dispatchability, flexibility, or firmness of capacity. The energy penalty incurred by the carbon dioxide removal process is performed at a fraction of other carbon capture techniques [1], such as post-combustion capture using amine solvents.

In this work, we quantify the role and value of three different CLC technologies and their integration into the future electricity system; CLC with natural gas coupled to a combined cycle (NGCC-CLC), integrated gasification combined cycle with CLC (IGCC-CLC) and CLC with in situ gasification and combustion of solid fuel (iG-CLC). We develop a national-scale capacity expansion and unit commitment model based on a mixed-integer linear program (MILP), minimising total system cost while limiting carbon emission, and maintaining security, and operability requirements [2]. Detailed operational constraints (e.g., start-up, ramp rates, minimum stable generation) on an hourly time discretisation enable the analysis of individual power plant performance characteristics and an assessment of their system-level impact.

We investigate the effects of the integration of CLC power plants into a 2030 and 2050 electricity system of the United Kingdom on the optimal capacity mix, the operational and system cost, and dispatch patterns. All three CLC power generation technologies prove valuable to the electricity system, as their deployment reduces system-wide investment and operational cost. Power generation from intermittent renewables becomes less economical as CLC technologies become available. Consequently, the requirements for carbon-intense reserve capacity, energy storage and overseas electric interconnection reduces. NGCC-CLC operating at virtually zero-carbon dioxide emission and moderate operation cost compares most favourably to IGCC-CLC and iG-CLC. The most cost-effective operational strategy for CLC power plants is in a steady manner, however, their ability for flexible operation can complement intermittent renewable power generation.

Additionally, we present the sensitivities to capital and operating expenditure and efficiency. A reduction in the efficiency of NGCC-CLC power plants from 46 % to 42 % reduces their annual average utilisation rate by only 1 %. For iG-CLC, the sensitivity to the cost of the oxygen carrier material is small; a variation of -50 % to +70 % causes a deviation in the average utilisation factor of -2 % to 3 %. For iG-CLC and IGCC-CLC, a 20-40 % increase in capital cost can reduce optimal capacity deployment levels considerably. NGCC-CLC deployment is more robust against investment cost increase. The results emphasises the high value of CLC power generation in a carbon-constrained system and can indicate and guide a prioritisation of future research efforts.

[1] C. Ekstrom, F. Schwendig, O. Biede, F. Franco, and G. Haupt. Technoeconomic evaluations and benchmarking of pre-combustion CO_2 capture and oxy-fuel processes developed in the European ENCAP project. Energy Procedia, 1(1):4233-4240, 2009.

[2] C. F. Heuberger, I. Staffell, N. Shah, and N. Mac Dowell. Levelised Value of Electricity - A Systemic Approach to Technology Valuation. In 26th European Symposium on Computer Aided Process Engineering, volume 38, pages 721-726, 2016.

498958 Superstructure-Based Optimization of Membrane-Based Carbon Capture Systems

Abstract:

This work aims to optimize the plant design, process configuration and operating conditions of a post-combustion membrane-based CO_2 capture plant. A new superstructure-based mathematical model is presented, modeling discrete and continuous decision variables to minimize the cost of electricity (COE). The proposed approach tackles a major issue in carbon capture and storage (CCS) for the power generation industry, involving the simultaneous optimization of process configurations and operating conditions (gas temperature, pressure ratio, sweep feed, etc.).

The development of efficient and cost-effective CCS systems is important to mitigate the CO_2 emission in fossil fuel-based power generation. Available technologies for post-combustion carbon capture include liquid-solvents, gas-permeation, and solid sorbent systems. Solvent and Sorbent â€" based systems can considered mature technologies for CCS, while membrane systems are generally less well-studied.

This work focuses on membrane-based systems for CCS applied to gas permeation, where the membranes with high CO_2 selectivity separate the CO_2 from other components in the flue gas. The main driving force of the process is the partial pressure between the feed and the permeate sides. Membrane systems often consist of multiple stages of membranes and compressors with intercooling. Most of the applications consider membrane models discretizing the membrane area and solving the separation problem for each membrane-section. This often results in a large-scale model typically analyzed through process simulation-based frameworks (ASPEN, gPROMs, etc.)[1] developed primarily for accurately modeling performance as opposed to optimization. Therefore, the number of configurations that can be optimized by conventional simulation-based frameworks is limited.

Significant efforts have been made for developing multi-stage systems, where previous studies focused on optimizing the number of stages and their operation[2,3], instead of analyzing advanced process configurations[4]. Advanced process configurations can lead to important savings in operating and capital costs. This work analyzes advanced process configurations while considering rigorous process models in a mathematical optimization framework.

A set of streams, and multiple processes (membrane stages, mixers, splitters, pumps and compression units with intercooling) is considered. Discrete design decisions determine the plant layout by selecting the optimal number of stages (where each stage consists of a membrane, compressor, pump, and heat exchanger units) and the process configuration. At the operating level, continuous decision variables involve the optimal flows, temperatures, concentrations, pressure ratio, and sweep feed. The proposed model is formulated as a mixed integer nonlinear programming model that minimized the cost of electricity to derive an optimal design and operation of a post-combustion carbon capture plant. The optimization of several process configurations led to lower CCS costs.

[1] Eslick J. C., Tong C., Lee A., Dowling A. W., Mebane D. S. (2015). Optimization under uncertainty with rigorous process models. Proceedings AIChE Annual Meeting 2015.

[2] Hasan M. M., Baliban R. C., Elia J. A., Floudas C. A. (2012). Modeling, Simulation, and Optimization of Postcombustion CO_2 Capture for Variable Feed Concentration and Flow Rate. 1. Chemical Absorption and Membrane Processes. Industrial and Engineering Chemistry Research. 51, 15642-15664.

[3] Arias A. M., Mussati M. C., Mores P. L., Scenna N. J. (2016). Optimization of multi-stage membrane systems for CO_2 capture from flue gas. International Journal of Greenhouse Gas Control. 53, 371-390.

[4] Merkel T. C., Lin H., Wei X., Baker R. (2010). Power plant post-combustion carbon dioxide capture: An opportunity for membranes. Journal of Membrane Science. 359, 126-139.

Imperial College London

Bio-energy with carbon capture and storage (BECCS): Opportunities for efficiency improvement

Mai Bui[a,b], Mathilde Fajardy[a,b], Niall Mac Dowell[a,b]

[a]Centre for Environmental Policy, Imperial College London, London SW7 1NA UK
[b]Centre for Process Systems Engineering, Imperial College London, London SW7 2AZ UK

cep

Centre for Process Systems Engineering

2nd November 2017

EPSRC
Engineering and Physical Sciences Research Council

Imperial College London

Meeting the below 2°C and 1.5°C COP21 targets

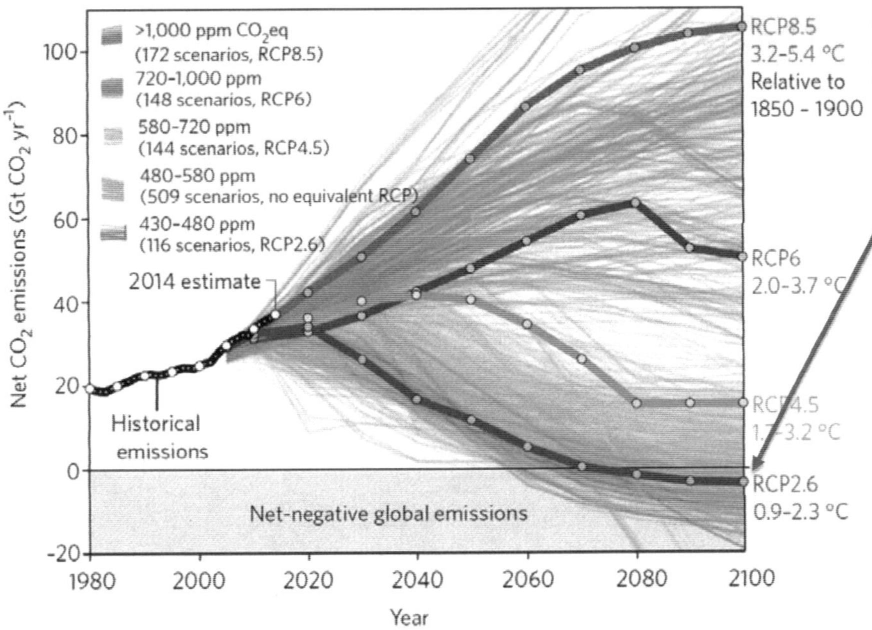

- Mitigation costs 138% more without CCS
- Without CCS, <50% of IAM models solve the problem
- Most feasible solutions use BioEnergy + CCS substantially

→ BECCS is a key negative emissions technology

IPCC Fifth Assessment Synthesis Report November 2014, Fuss, S., et al. (2014). Nature Climate Change, 4(10), 850–853

2

Imperial College London

Negative emissions technology: BECCS

Atmospheric CO_2 is captured and permanently stored → Negative CO_2 balance

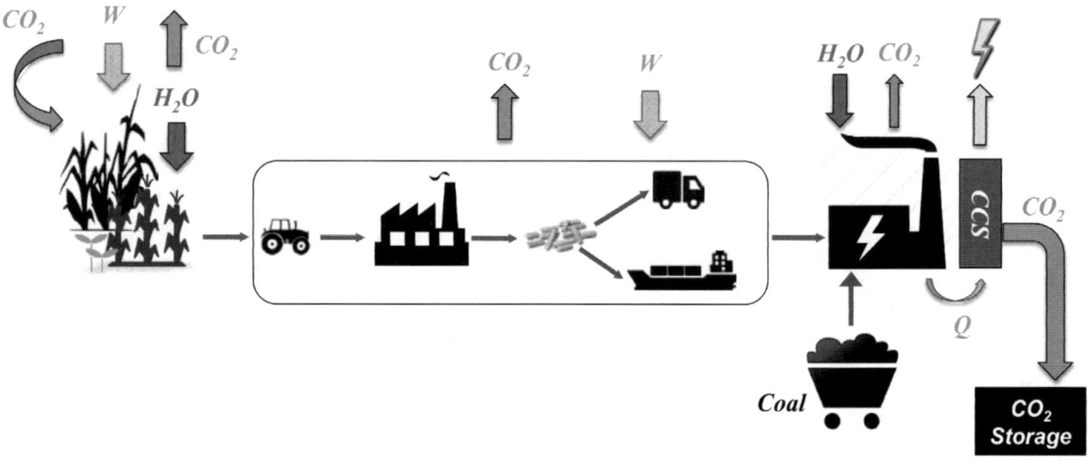

Reference: Fajardy, M. & Mac Dowell, N. (2017). Can BECCS deliver sustainable and resource efficient negative emissions? Energy and Environmental Science, 10 (6), 1389-1426.

3

Imperial College London

Fuel properties and composition

Property	Coal	Wood	Herbaceous
Ash content (wt% dry)	8.5 – 10.9	0.4 – 4.0	5 – 7.5
Moisture content (wt% wet)	6 – 10	5 – 60	15 – 25
Lower heating value, LHV (MJ/kg)	26.0 – 28.3	18.4 – 20.0	17.1 – 17.5
C (wt% dry)	76 – 87	47 – 52	45 – 47
H (wt% dry)	3.5 – 5.5	5.8 – 6.7	5.7 – 6.0
O (wt% dry)	2.8 – 11.3	38 – 46	40 – 46
Cl (wt% dry)	< 0.1	0.01 – 0.05	0.09 – 0.97
S (wt% dry)	0.5 – 3.1	0.02 – 0.10	0.05 – 0.2
N (wt% dry)	0.8 – 1.5	0.1 – 0.8	0.4 – 1.04

Efficiency penalty due to: Biomass (moisture & low heating value) + CO_2 capture process requires heat

4

Imperial College
London

Fuel properties and composition

Property	Coal	Wood	Herbaceous
Ash content (wt% dry)	8.5 – 10.9	0.4 – 4.0	5 – 7.5
Moisture content (wt% wet)	6 – 10	5 – 60	15 – 25
Lower heating value, LHV (MJ/kg)	26.0 – 28.3	18.4 – 20.0	17.1 – 17.5
C (wt% dry)	76 – 87	47 – 52	45 – 47
H (wt% dry)	[illegible]	[illegible]	5.7 – 6.0
O (wt% dry)	[illegible]	[illegible]	40 – 46
Cl (wt% dry)	< 0.1	0.01 – 0.05	0.09 – 0.97
S (wt% dry)	0.5 – 3.1	0.02 – 0.10	0.05 – 0.2
N (wt% dry)	0.8 – 1.5	0.1 – 0.8	0.4 – 1.04

SO_X and NO_X emissions

Reference: Bui, M., Fajardy, M. & Mac Dowell, N. (2017). Bio-Energy with CCS (BECCS) performance evaluation: Efficiency enhancement and emissions reduction. Applied Energy, 195, 289–302.

5

Imperial College
London

Heating and condensation of gas phase

Solids and liquids formed at various stages

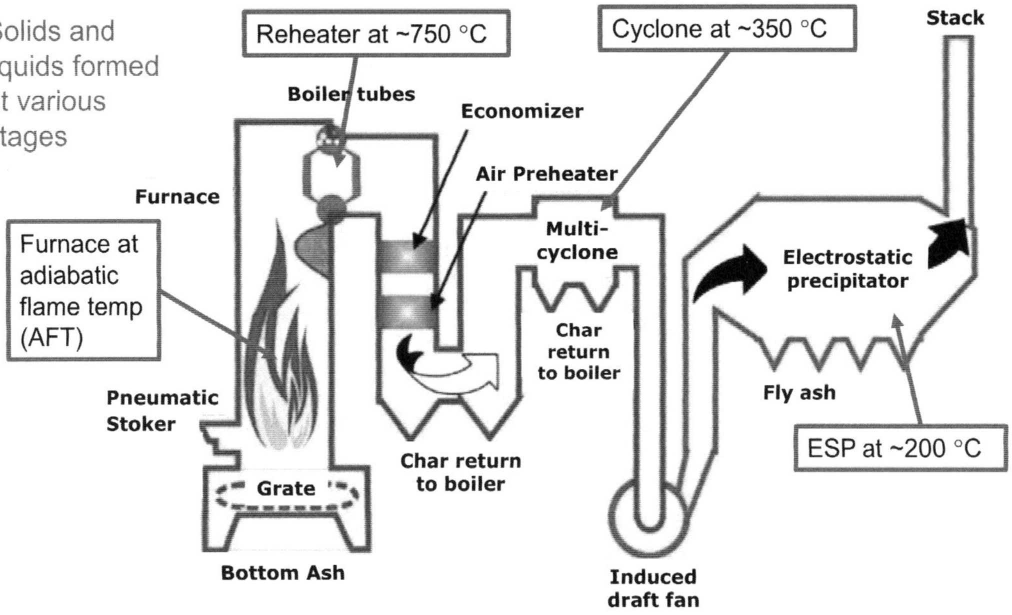

Image: Overend, Ralph P. Direct combustion of biomass, Encyclopedia of Life Support Systems

6

Imperial College
London

Flue gas condensation in a power plant

Unit	Temperature (°C)	Solid/liquid removed (%)	Solid/liquid carried to next step (%)
Furnace[1]	AFT	80	20
Reheater[2]	750	15 (fouling)	85
Multi-cyclone[3]	350	80	20
ESP[4]	200	99	1

[1] Biomass combustion ash is typically 80% bottom ash, 20% fly ash (Jensen et al., 2013; Hansen et al., 2014)

[2] Biomass ash fouling at ~750 °C based on T_{15} (varies with Cl & K content) (Mansikkasalo et al., 2014)

[3] Multi-cyclones: separation efficiency of 80-95% (US EPA, 2016)

[4] ESP: separation efficiency of 99-99.9% (US EPA, 2016)

References: Jensen, P. A., *et al.* (2013). Characterization and quantification of deposits build up and removal in straw suspension fired boilers. Energinet.dk project no. 7217
Hansen, S. B., *et al.* (2014). Deposit probe measurements in large biomass-fired grate boilers and pulverized-fuel boilers. Energy & Fuels, 28 (6), 3539–3555.
Mansikkasalo, J., *et al.* (2014). High power generation from recovery boilers, what are the limits? [Online]. 2014 International Chemical Recovery Conference, Finland.
USEPA (2016). Air Pollution Control Technology Fact Sheets. United States Environmental Protection Agency (USEPA) Technology Transfer Network.

7

Imperial College
London

Automation of FactSage: Macros

FactSage 7.0 is a thermo-chemical modelling tool – accesses thermodynamic data from the FACT and SGTE databases

Macros can be used to automate a large number of simulations:

- Incremental temp increase
- Removal of a particular phases from the reaction

Designed macros used to simulate the following:

- Combustion Process: (200°C to AFT)
- Flue gas condensation divided into:
  - i. Furnace to Reheater (AFT to 750°C)
  - ii. Reheater to Multi-cyclone (750°C to 400°C)
  - iii. Multi-cyclone to ESP (400°C to 200°C)
  - iv. ESP to Stack (200°C to 120°C)

Can observe flue gas composition changes during heating and cooling of the flue gas

8

Imperial College London

Flue gas combustion-condensation graph

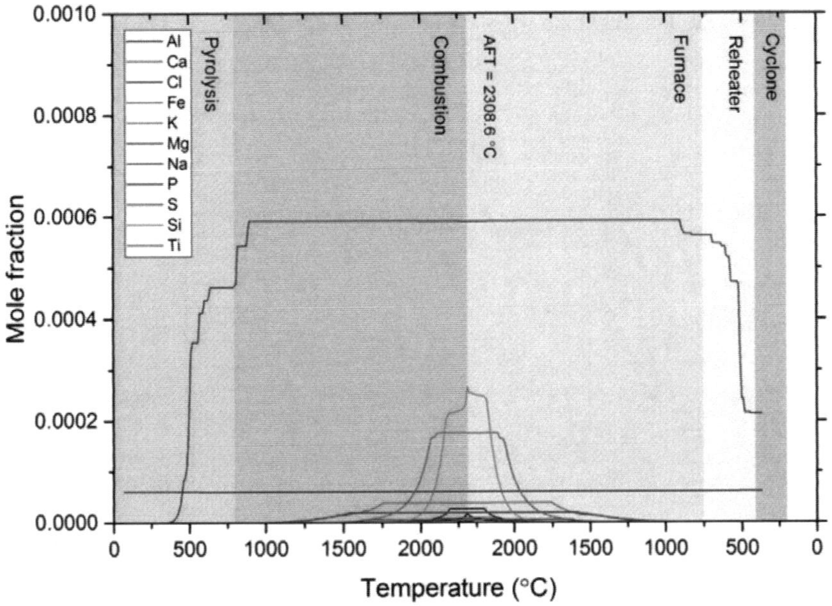

9

Imperial College London

Energy flow in a thermal power plant

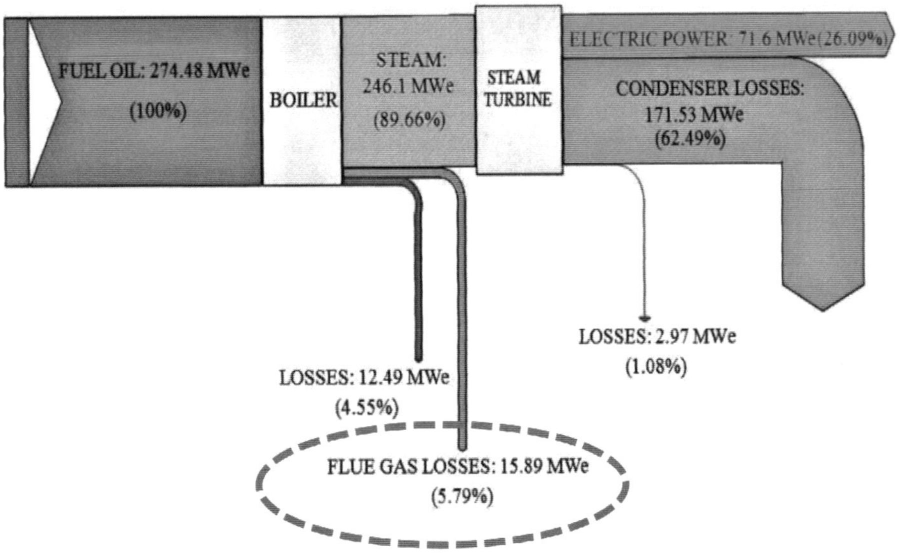

Irrazabal Bohorquez *et al.*: Repowering: An option for refurbishment of old thermal power plants in Latin-American countries. Proceedings of ASME Turbo Expo 2010: Power for Land, Sea and Air GT2010 (June 14-18, 2010, Glasgow, UK). DOI: 10.1115/GT2010-23058
http://www.sankey-diagrams.com/tag/power-plant/

10

Imperial College
London

Flue gas heat recovery to improve power plant efficiency

Power Plant **CO_2 capture plant**

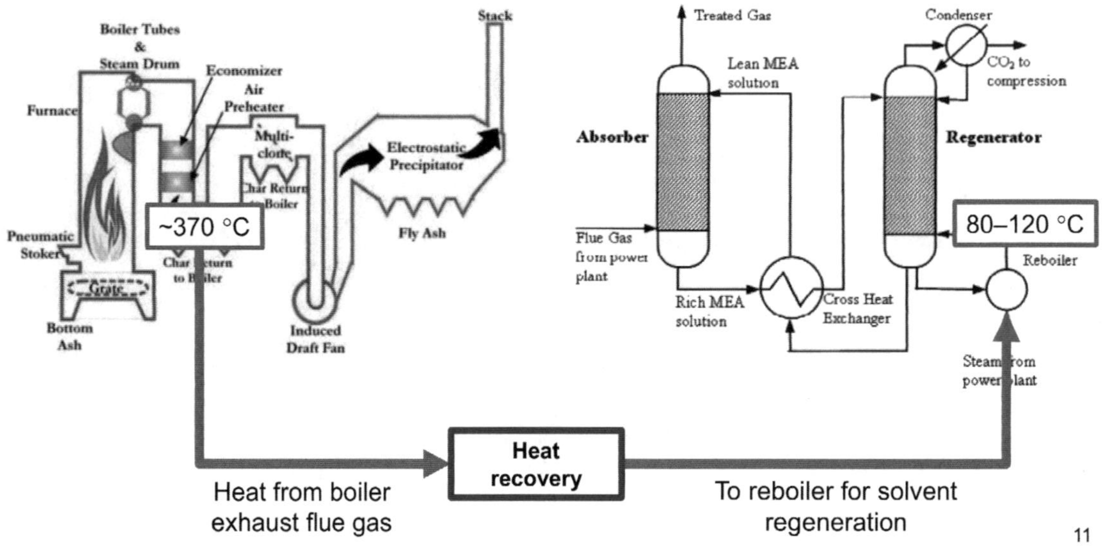

11

Imperial College
London

BECCS performance evaluation study: Objectives

- Assess the performance of a BECCS system using a 500 MW ultra-supercritical pulverized fuel power plant with 90% CO_2 capture.
- The effects of the following factors were evaluated:
  i. Coal types: high sulphur 2.51 wt% and medium sulphur 0.86 wt%
  ii. Biomass type (wheat straw and clean wood chips)
  iii. Variable moisture content of the biomass (5%, 16%, 50%)
  iv. Biomass co-firing proportion %
  v. Solvent selection (MEA, Cansolv, biphasic solvents)
- Key performance indicators include:
  - Combustion adiabatic flame temperature (AFT)
  - Pollutant emissions e.g. SO_X and NO_X
  - Recoverable heat from the flue gas
  - Energy efficiency
  - Carbon intensity

12

Imperial College London

Co-firing scenarios modelled in FactSage

Scenario	Coal type	Biomass type
A	Medium sulphur coal	Wheat straw 5% moisture
B	Medium sulphur coal	Wheat straw 16% moisture
C	Medium sulphur coal	Wood chip 5% moisture
D	Medium sulphur coal	Wood chip 50% moisture
E	High sulphur coal	Wheat straw 5% moisture
F	High sulphur coal	Wheat straw 16% moisture
G	High sulphur coal	Wood chip 5% moisture
H	High sulphur coal	Wood chip 50% moisture

13

Imperial College London

Biomass Co-firing: Adiabatic Flame Temperature (AFT)

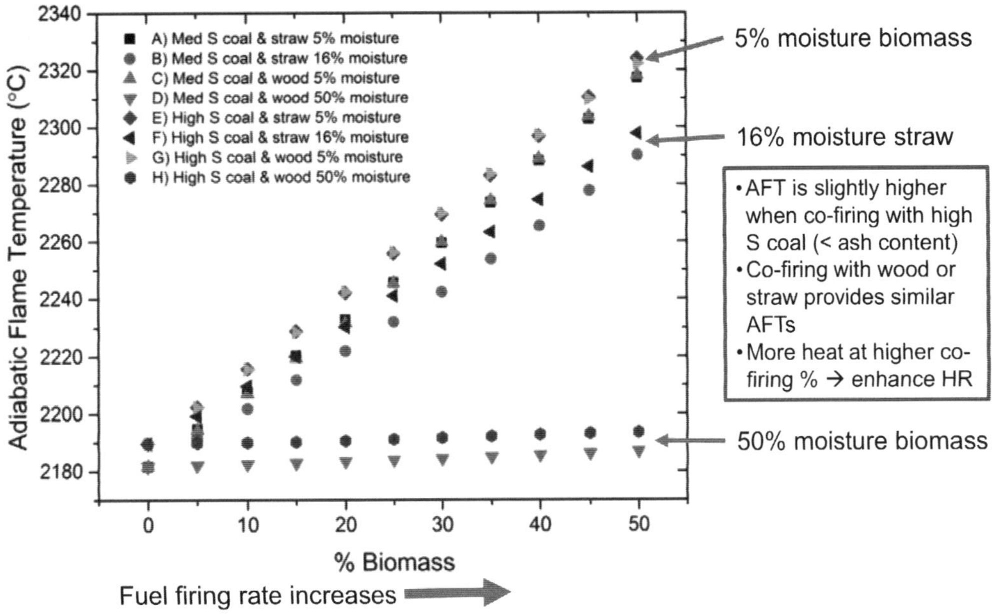

- AFT is slightly higher when co-firing with high S coal (< ash content)
- Co-firing with wood or straw provides similar AFTs
- More heat at higher co-firing % → enhance HR

Reference: Bui, M., Fajardy, M. & Mac Dowell, N. (2017). Applied Energy, 195, 289–302.

14

Imperial College
London

Coal co-firing with biomass : SO_X emissions

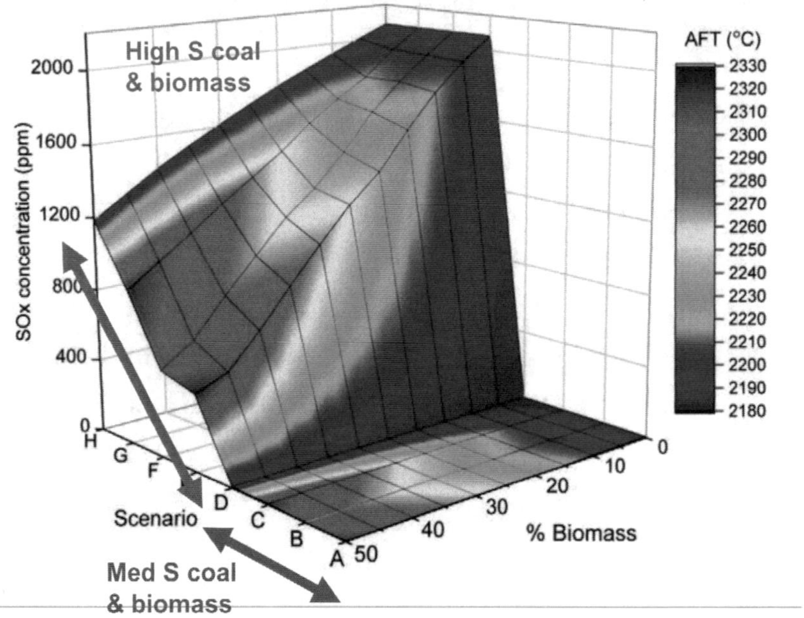

Reference: Bui, M., Fajardy, M. & Mac Dowell, N. (2017). Applied Energy, 195, 289–302.

15

Imperial College
London

Coal co-firing with biomass : SO_X emissions

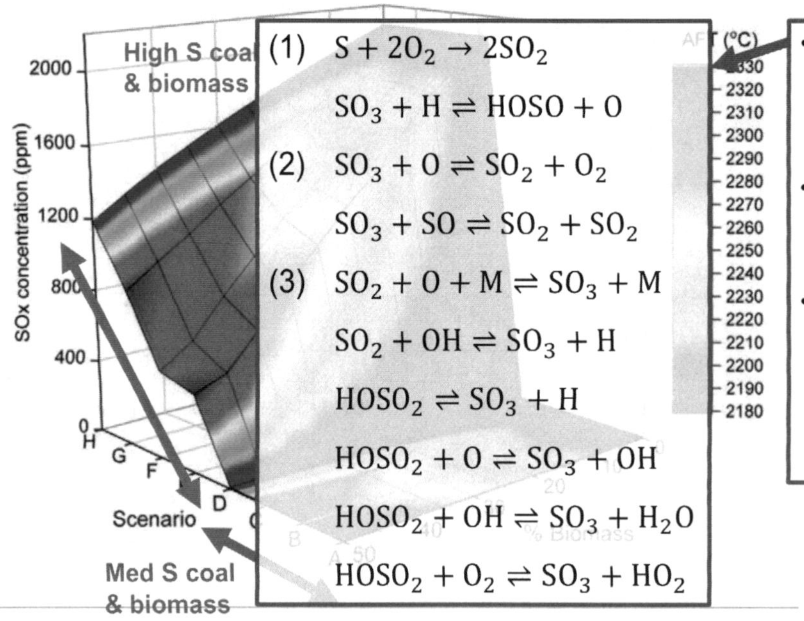

(1) $S + 2O_2 \rightarrow 2SO_2$

$SO_3 + H \rightleftharpoons HOSO + O$

(2) $SO_3 + O \rightleftharpoons SO_2 + O_2$

$SO_3 + SO \rightleftharpoons SO_2 + SO_2$

(3) $SO_2 + O + M \rightleftharpoons SO_3 + M$

$SO_2 + OH \rightleftharpoons SO_3 + H$

$HOSO_2 \rightleftharpoons SO_3 + H$

$HOSO_2 + O \rightleftharpoons SO_3 + OH$

$HOSO_2 + OH \rightleftharpoons SO_3 + H_2O$

$HOSO_2 + O_2 \rightleftharpoons SO_3 + HO_2$

- SO_X formation mechanism involves equilibrium reactions
- SO_2 formation thermodynamically favoured
- Temperature ↓ : Equilibrium shifts towards SO_3 formation (iron oxide can catalyse reaction 2)

Reference: Bui, M., Fajardy, M. & Mac Dowell, N. (2017). Applied Energy, 195, 289–302.

16

Imperial College London

Coal co-firing with biomass : SO_X emissions

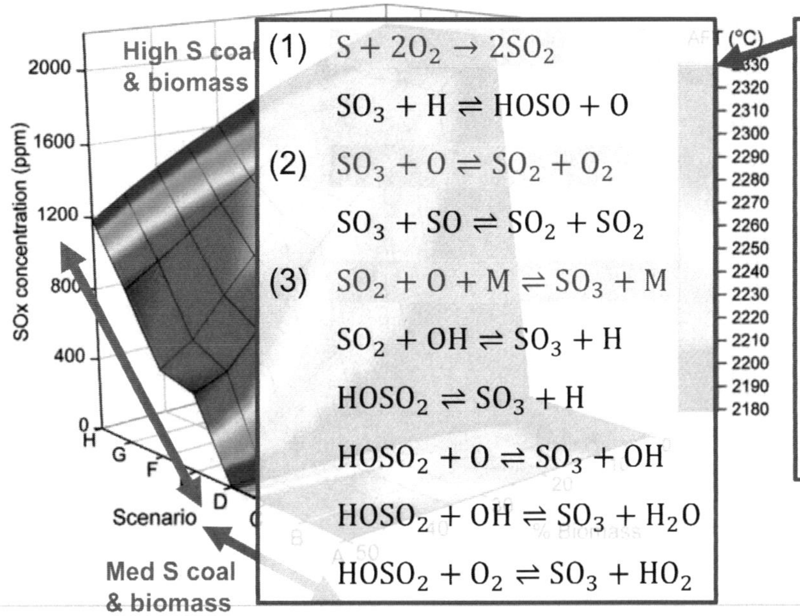

(1) $S + 2O_2 \rightarrow 2SO_2$

$SO_3 + H \rightleftharpoons HOSO + O$

(2) $SO_3 + O \rightleftharpoons SO_2 + O_2$

$SO_3 + SO \rightleftharpoons SO_2 + SO_2$

(3) $SO_2 + O + M \rightleftharpoons SO_3 + M$

$SO_2 + OH \rightleftharpoons SO_3 + H$

$HOSO_2 \rightleftharpoons SO_3 + H$

$HOSO_2 + O \rightleftharpoons SO_3 + OH$

$HOSO_2 + OH \rightleftharpoons SO_3 + H_2O$

$HOSO_2 + O_2 \rightleftharpoons SO_3 + HO_2$

- SO_X formation mechanism involves equilibrium reactions
- SO_2 formation thermodynamically favoured
- Temperature ↓ : Equilibrium shifts towards SO_3 formation (iron oxide can catalyse reaction 2)

Reference: Bui, M., Fajardy, M. & Mac Dowell, N. (2017). Applied Energy, 195, 289–302.

17

Imperial College London

Coal co-firing with biomass : SO_X emissions

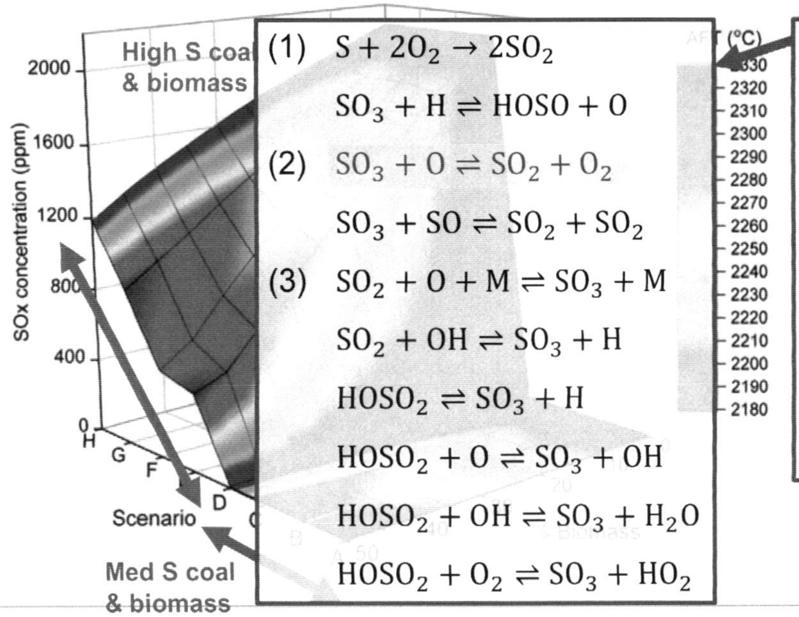

(1) $S + 2O_2 \rightarrow 2SO_2$

$SO_3 + H \rightleftharpoons HOSO + O$

(2) $SO_3 + O \rightleftharpoons SO_2 + O_2$

$SO_3 + SO \rightleftharpoons SO_2 + SO_2$

(3) $SO_2 + O + M \rightleftharpoons SO_3 + M$

$SO_2 + OH \rightleftharpoons SO_3 + H$

$HOSO_2 \rightleftharpoons SO_3 + H$

$HOSO_2 + O \rightleftharpoons SO_3 + OH$

$HOSO_2 + OH \rightleftharpoons SO_3 + H_2O$

$HOSO_2 + O_2 \rightleftharpoons SO_3 + HO_2$

- SO_X formation mechanism involves equilibrium reactions
- SO_2 formation thermodynamically favoured
- Temperature ↓ : Equilibrium shifts towards SO_3 formation (iron oxide can catalyse reaction 2)

Reference: Bui, M., Fajardy, M. & Mac Dowell, N. (2017). Applied Energy, 195, 289–302.

18

Imperial College London

Coal co-firing with biomass : SO_x emissions

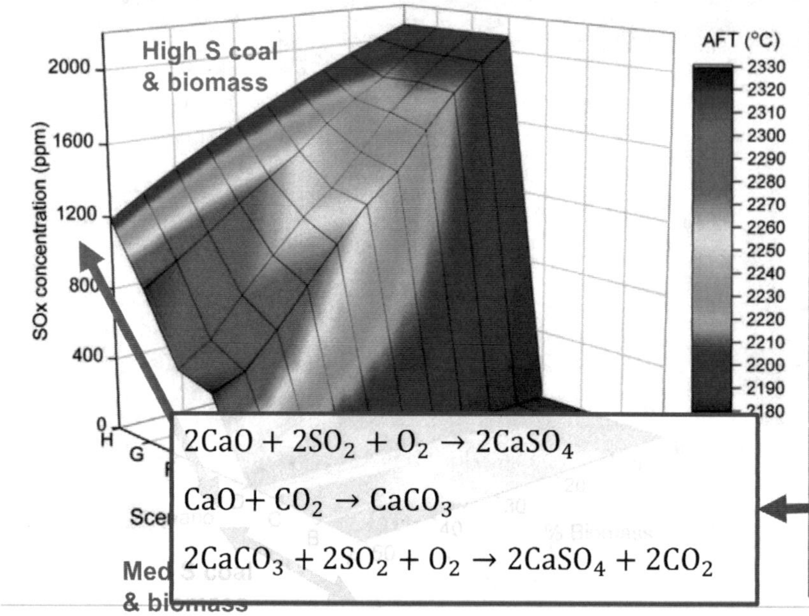

$$2CaO + 2SO_2 + O_2 \rightarrow 2CaSO_4$$

$$CaO + CO_2 \rightarrow CaCO_3$$

$$2CaCO_3 + 2SO_2 + O_2 \rightarrow 2CaSO_4 + 2CO_2$$

- SO_X formation mechanism involves equilibrium reactions
- SO_2 formation thermodynamically favoured
- Temperature ↓ : Equilibrium shifts towards SO_3 formation (iron oxide can catalyse reaction 2)
- Alkali oxides in ash capture SO_2 and SO_3

Reference: Bui, M., Fajardy, M. & Mac Dowell, N. (2017). Applied Energy, 195, 289–302.

19

Imperial College London

High sulphur coal & biomass: SO_x emissions

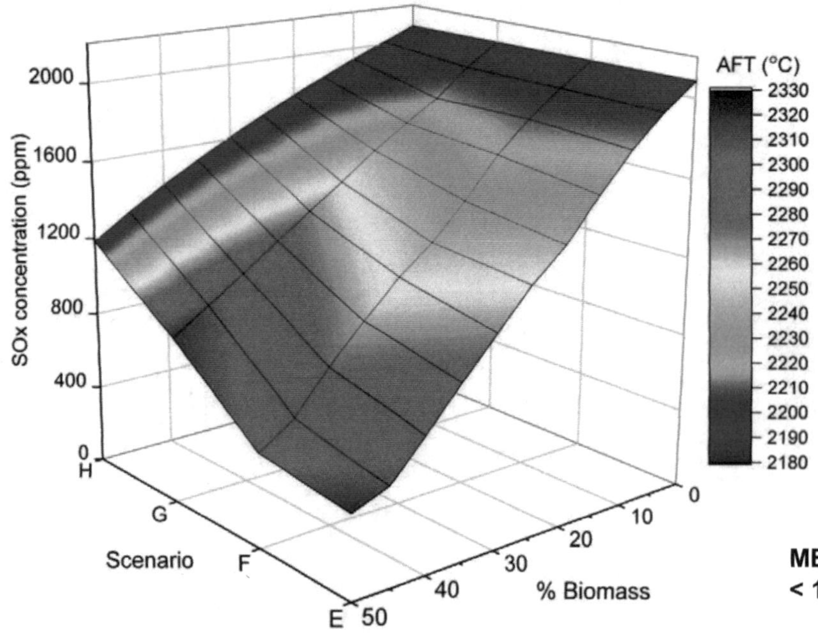

- SO_X decreases linearly with % biomass due to decreased S content

MEA tolerance to SO_X < 10 ppm

Reference: Bui, M., Fajardy, M. & Mac Dowell, N. (2017). Applied Energy, 195, 289–302.

20

Imperial College London

Medium sulphur coal & biomass: SO_X emissions

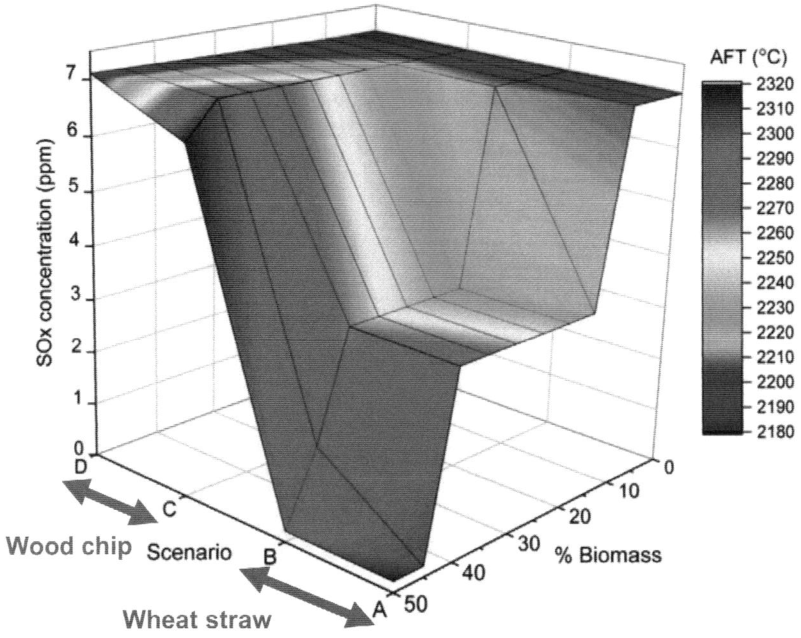

- FactSage results show that CaO, MgO, Na_2O, K_2O captures SO_X
- In FactSage, Al_2O_3, Fe_2O_3 & MnO_2 also involved in SO_X ↓
- Wheat straw has higher sulphur content than the wood chip (0.1 wt% Vs 0.05 wt%)
- Wheat straw has more ash than wood chip (5.5 dry wt% Vs 2 dry wt%)
- Alkali oxides in ash → reduce SO_X emissions

MEA tolerance to SO_X < 10 ppm

Reference: Bui, M., Fajardy, M. & Mac Dowell, N. (2017). Applied Energy, 195, 289–302.

21

Imperial College London

BECCS: Recoverable heat from the flue gas

Analysis of scenario F (high S coal and 16% moisture wheat straw)

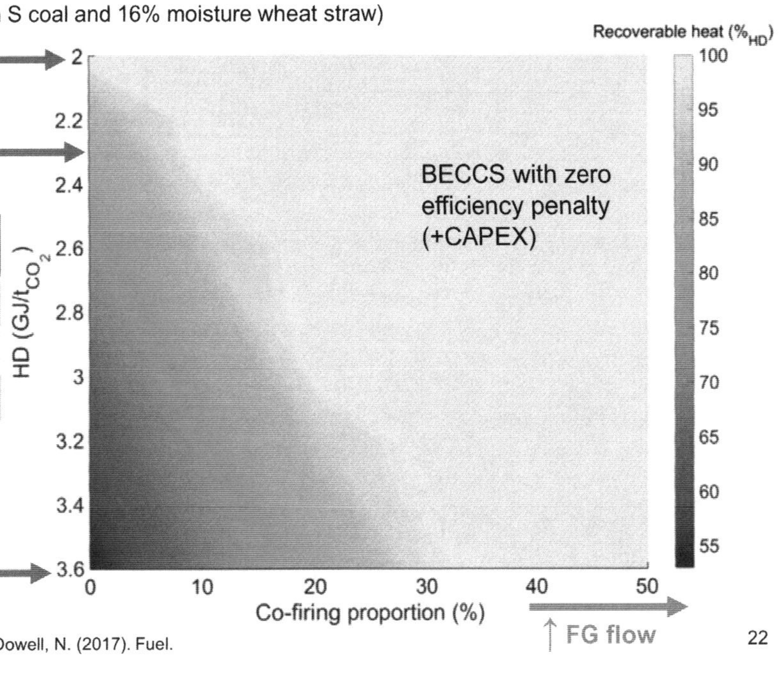

Solvent	HD (MJ/t_{CO_2})	Reg T (°C)
MEA	3600	120
Cansolv	2300	120
Biphasic	2000	80

Data for biphasic solvent systems published by Ye *et al.* (2015) and Raynal *et al.* (2011)

Reference: Bui, M., Fajardy, M. & Mac Dowell, N. (2017). Fuel.

22

Imperial College
London

BECCS power plant efficiency improvement

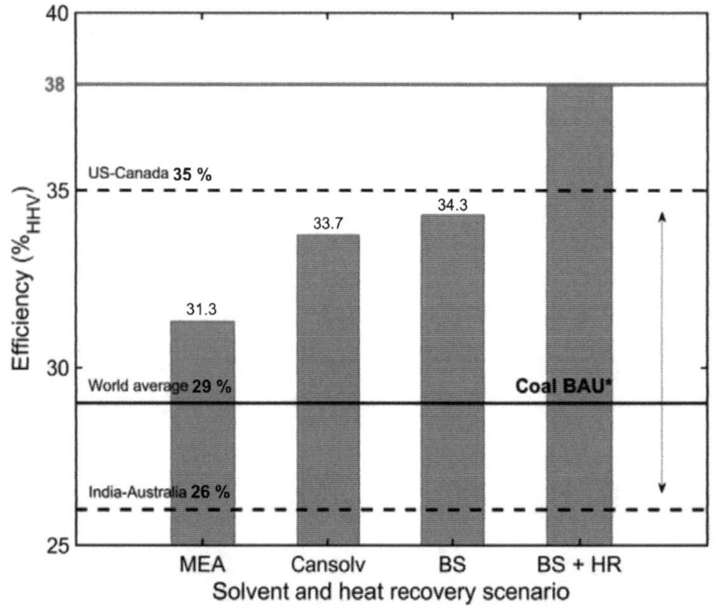

- Efficiency for a 500 MW BECCS system **co-firing 50% biomass**
- Efficiency can be significantly improved by using (i) high performance solvents, (ii) heat recovery

BS = biphasic solvent

HR = heat recovery

BAU = Business As Usual (average efficiencies around the world)

Reference: Bui, M., Fajardy, M. & Mac Dowell, N. (2017). Bio-Energy with carbon capture and storage (BECCS): Opportunities for performance improvement. Fuel.

23

Imperial College
London

BECCS carbon intensity

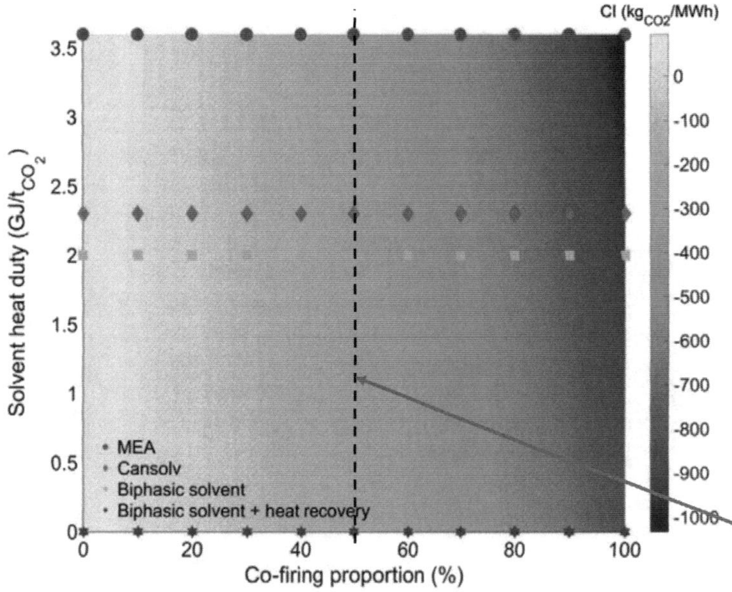

As system efficiency increases, less fuel is burned per MWh produced → less CO_2 is captured per MWh of electricity generated

At 50% biomass co-firing and 90% CO_2 capture:

System	Efficiency %HHV	Carbon intensity kg CO_2/MWh
MEA	31.3	-296
Cansolv	33.7	-275
BS	34.3	-271
BS + HR	38.0	-244

Reference: Bui, M., Fajardy, M. & Mac Dowell, N. (2017). Fuel.

24

Imperial College London

CO_2 mitigation potential of a 500 MW BECCS system

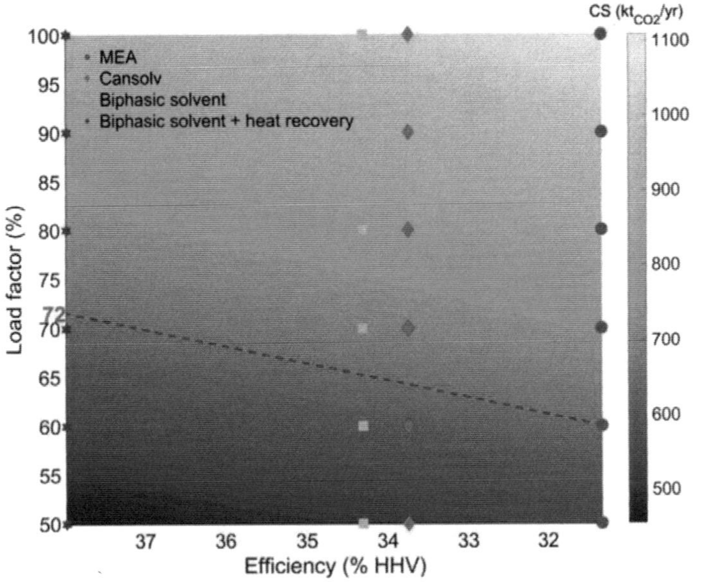

- Power plant load factor is a function of power plant efficiency.
- Improving efficiency reduces marginal costs of electricity generation, enabling operation at higher load factor (Mac Dowell & Shah, 2015; Mac Dowell & Staffell, 2016)
- The efficiency of a BECCS system will have an impact on its dispatch rate within an electricity market
  → Influences CO_2 mitigation potential

Objective: (i) more efficient at electricity generation, or (ii) more carbon negative & capture CO_2?

Reference: Bui, M., Fajardy, M. & Mac Dowell, N. (2017). Applied Energy, 195, 289–302.

25

Imperial College London

Effect of biomass carbon footprint (BCF) and co-firing proportion on the system carbon intensity (CI)

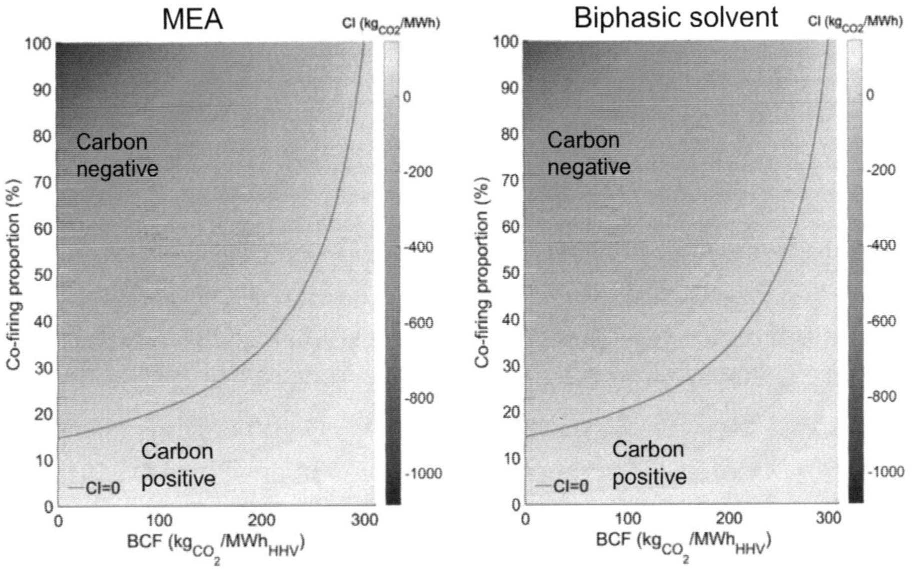

Bui, M., Fajardy, M. & Mac Dowell, N. (2017) Fuel.

Low efficiency MEA system is more carbon positive above the CI = 0 limit & more carbon negative below this limit. To ensure the BECCS system is actually carbon negative, the impact of biomass carbon footprint (BCF) must be considered.

Imperial College London

Conclusion: BECSS Performance Matrix

Performance parameter	Ash content	Sulphur content	Moisture content	HHV	Biomass co-firing %	Solvent heat duty
NO_X ↓	dependent on combustion conditions					
SO_X ↓	↑	↓	↓		↑	
AFT ↑	↓		↓		↑	
Heat recovery ↑	↓		↓	↓*	↑	↓
Efficiency ↑	↓		↓	↑↓	↓	↓
CO_2 negativity ↑	↑		↑	↑	↑	↑
Exhaust gas flow rate ↑				↓	↑	

Enhance efficiency: high performance solvents (low heat duty) and higher heat recovery (higher AFT and flue gas flow rate)
Greater carbon negativity: lower efficiency systems (burn more biomass fuel, thus more CO_2 is captured and permanently stored), low biomass carbon footprint.
Minimise pollutant emissions: Increasing biomass co-firing proportion significantly reduces SO_X emissions, fuel selection (low S, metal oxides), combustion control is required to regulate NO_X emissions.

27

Imperial College London

Conclusion

- BECCS has the potential provide negative emissions, whilst providing reliable firm power generation capacity.
- Opportunities to improve energy efficiency of BECCS include enhancing heat recovery and using high performance solvents for CO_2 capture (*e.g.*, biphasic).
- Power generation efficiency increased from $31\%_{HHV}$ (MEA system) to $38\%_{HHV}$ (biphasic system with heat recovery).
- Important to consider the complex trade-offs between carbon intensity, power plant efficiency and biomass carbon footprint.
- Inefficient BECCS → removes more CO_2 from the atmosphere.
- However, inefficient systems might have lower dispatch rate compared to high efficiency systems.

Bui, M., Fajardy, M. & Mac Dowell, N. (2017). Applied Energy, 195, 289–302.
Bui, M., Fajardy, M. & Mac Dowell, N. (2017). Energy Procedia, 114 (Supplement C), 6010-6020.
Bui, M., Fajardy, M. & Mac Dowell, N. (2017). Bio-Energy with carbon capture and storage (BECCS): Opportunities for performance improvement. Fuel.

28

493767 A Process Synthesis Approach to Natural Gas Liquefaction

Abstract:

A new approach to the design of natural gas liquefaction processes, which could significantly reduce their energy consumption, is proposed. A graphical technique based on exergy and enthalpy is used to determine optimal paths for refrigeration systems. The analysis reveals that keeping refrigeration cycles in the two-phase region, including compression and heat removal, is the most efficient way of supplying the work required to remove heat from sub-ambient processes. The approach is validated using an Aspen Plus simulation of a natural gas liquefaction process with pure refrigerant in cascade; it is shown that work efficiency higher than 90% can be achieved when the process fluid as well as the working fluids in the cascade cycles are kept in the two phase region at every stage.

494507 Insights of Biomass Recalcitrance in *Populus* for Biomass Valorization

Abstract:

For the effective utilization of biomass, understanding of its properties, in particular, biomass recalcitrance-related properties is essential. Diverse approaches are applicable to reveal the secret of biomass recalcitrance. Among many, physicochemical characteristics of biomass feedstock do not only elucidate the biomass recalcitrance, but also provide clues to overcome its physical and chemical barriers for biomass conversion. Transcriptome analysis also gives important overview about transcriptional regulation network of biomass synthesis related gene families. *Populus* has been investigated as a promising biomass feedstock for alternative fuels and chemicals. However, biomass recalcitrance-related properties and their causable variation of gene expression are not sufficiently understood. Here, recalcitrance of *Populus* natural variants was elucidated in three aspects: (1) fermentable sugar release, (2) physicochemical properties of biomass, and (3) relative variation of gene expression. Sugar release and physicochemical characteristics of *Populus* natural variants including chemical composition, cellulose and ligninâ€™s molecular weights/degree of polymerization, cellulose crystallinity, lignin composition, and cellulose accessibility were measured by diverse analytical methods such as HPLC, GPC, 2D ^{13}C-^{1}H HSQC NMR, solid-state NMR, and others. Differential Gene expression of each variant was also investigated by RNA-seq analysis.

499309 Evaluation of Industrial Hemp Varieties As Potential Biomass Feedstock for Biofuels and Bioproducts

Abstract:

Industrial hemp is an annual herbaceous crop historically grown for fiber in the U.S. and worldwide. Despite existing applications of industrial hemp based fiber and oil products, hemp-based biofuels and bioproducts represent a new potential application area. Several key questions still remain for the technical and economic feasibility of using hemp as a bioenergy crop. Furthermore, it is necessary to screen industrial hemp species and process conditions while aiming at maximized production of biofuel and bioproducts. Hence, this study takes combined agronomy, laboratory, and economic analysis approaches to evaluate industrial hemp in comparison with other biomass feedstocks e.g. kenaf, switchgrass and biomass sorghum. Several industrial hemp varieties, currently widely cultivated in North America, Europe and Asia, were evaluated. Experimental results show an ethanol yield of 85-100 gallons /dry ton hemp stems using dilute alkali or acid pretreatment, enzymatic hydrolysis and fermentation, which is comparable to the other three tested feedstocks. Agronomy data suggest that the per hectare yield of industrial hemp stem alone was at a similar level to the other energy crops such as switchgrass and sorghum; while the hemp plants may require reduced inputs. Field trial also showed that ~1200 kg/ha hemp grain can be harvested in addition to stems. A preliminary cost analysis indicates that industrial hemp could generate higher per hectare gross profit than the other crops if both hemp grains and biofuels from hemp stem were counted. These combined evaluation results demonstrate that industrial hemp has great potential to become a promising regional commodity crop for producing both biofuels and value-added products.

503939 Thermochemical Conversion of Blended Herbaceous and Woody Biomass Feedstocks

Abstract:

The United States has a vast potential supply of renewable lignocellulosic biomass, which can be utilized as feedstocks for the production of renewable fuels and chemicals via biomass conversion technologies. In order to support a large-scale bioeconomy, it will be required that multiple types of biomass be utilized such as dedicated bioenergy crops, forestry and agricultural residues, and municipal wastes. Blending several types of biomass to produce a uniform feedstock has the potential to control feedstock quality specifications. In order to exploit the potential benefits of biomass blending, we must gain a comprehensive understanding of its effects on thermochemical conversion. In this study, we target pyrolysis as the thermochemical conversion technology of interest as it is well suited to produce liquid fuels and chemicals through thermal decomposition and condensation of pyrolysis vapors into bio-oil. The chemical composition of the feedstock used has a significant impact of the yield and quality of the bio-oil produced. In this work, we first utilized pyrolysis gas chromatography mass spectroscopy (py-GC/MS) to measure the vapor-phase pyrolysis products produced from switchgrass, two pine residue feedstocks, and binary blends of these. Next, we used an intermediate-scale pyrolysis reactor to test the effect of blending on bio-oil yield and quality. Results indicate that the composition and content of alkali and alkaline earth metals in the feedstocks have a significant effect on the pyrolysis products, and that blending is a potential strategy to produce high quality and low variability feedstocks. Further results and discussion regarding the effects of blending on the yield, quality, and chemical composition of vapor-phase and liquid-phase pyrolysis products will be presented.

491952 Feedstock Preprocessing, Fractionation, and Blending to Improve Overall Cost, Supply, and Quality Considerations for Catalyzed and Uncatalyzed Fast Pyrolysis

Abstract:

The success of lignocellulosic biofuels and biochemical industries depends on an economical and reliable supply of high-quality biomass. As such, biomass cost, supply, and quality are critical parameters to consider when choosing feedstocks and locations for biorefineries. Biomass cost is dependent upon feedstock type, location, availability, transportation costs, and the quality specifications required by the conversion technology. Biomass quality depends on inherent feedstock properties, field and storage conditions, and any preprocessing methods used to improve quality. Biomass quantity depends on location, field conditions, annual weather, and collection and storage conditions. Because of these variables, it is necessary to develop strategies that enable supply chain improvements in biomass quality and consistency to improve lower overall conversion costs and improve yields, while also securing a less risky feedstock supply.

The presented work examines the interdependencies of these variables and how they affect the biomass blends required by biomass depots and/or biorefineries to achieve the lowest cost feedstock with sufficient quality at the quantities needed for biorefinery operation. Four biomass depots were proposed in South Carolina to each produce 200,000 tons of feedstock per year. These depots supply a centrally located 800,000 ton biorefinery that converts the feedstocks to bio-oil using either catalyzed or uncatalyzed fast pyrolysis. The four depots utilize biomass based upon availability, but the feedstock or feedstock blend still met the minimum quality requirements for the biorefinery. Costs were minimized by using waste biomass resources such as construction and demolition waste, logging residues, and forest residuals. Preprocessing methods such as air classification and water washing or acid leaching were used to upgrade biomass quality. Air classification was found to be an effective method for separating low- and high-ash fractions of woody residues. For example, a high-ash fraction of logging residues was generated that contained approximately 50 wt% of the total original ash content, but less than 8 wt% of the original organic content. Dilute-acid leaching was then effectively used to remove between 85% (25°C, 0.5 wt% acid) and 98% (90°C, 1.0 wt% acid) of the alkali and alkaline earth metals from the high-ash fraction, making it suitable for recombination with the low-ash fraction. Various combinations of such preprocessing methods were tested to produce an array of blends that were available for recombination to produce least-cost blendstocks that met various conversion specifications. For both uncatalyzed and catalyzed fast pyrolysis, all four depots produced blends that met quality and quantity specifications at a cost lower than using a single feedstock.

499592 New Approach to Ammonia Pretreatment Integrates Better Feedstock Logistics with Improved Sugar Conversion

Abstract:

The implementation of a large scale bioeconomy, where lignocellulosic residues can be used for production of commodity products (fuels, chemicals, materials, food additives, animal feed, etc.) highly depends on the ability of moving large amounts of biomass across long distances. To be able to effectively transport low density materials, such as agricultural residues, it is necessary to apply densification methods (e.g., pelletization). Though densification is expensive, it opens a range of new possibilities for feedstock logistics (e.g., ability to use various transportation methods, safer feedstock storage, etc.) and the ability to implement very large scale lignocellulosic biorefineries, as these require economic transportation of lignocellulosic feedstocks within a wide feedstock production radius. By increasing the size of the biorefinery, it is possible to take advantage of economies of scale and thus, reduce the capital cost per unit of processed biomass. To extract further value from using densified biomass, it is important to explore biomass conversion methods that benefit from having high density feedstock, which can generate higher sugar yields with lower energy inputs.

In this work, we have developed a new ammonia pretreatment strategy called Compacted Biomass with Reduced Ammonia (COBRA) pretreatment. In COBRA pretreatment, liquid ammonia is added to biomass pellets in ratios lower than 1:1, allowing native cellulose I to be converted to cellulose III at relatively low temperatures (~70 °C) and residence times of 3 to 4 hours. As pellets occupy less volume per unit mass relative loose biomass, lower ammonia volumes are required to completely submerge the solid biomass. Thus, as pellets contain low moisture content, the presence of liquid ammonia around the cellulose fibers will allow the formation of a highly digestible cellulose III. As a result, COBRA pretreated biomass can be effectively digested with 60% less enzyme relative to the traditional AFEX pretreatment. Furthermore, as COBRA pretreatment uses densified biomass, it also allows the reduction of reactor size compared to AFEX, even though it uses longer residence times. By using low ammonia loadings, small sized reactors, relatively low temperatures and pressures, low enzyme loading and taking advantage of economies of scale, COBRA pretreatment has the potential to reduce overall processing costs in the biorefinery. Here, we will present a complete optimization of COBRA pretreatment for corn stover, an enzyme cocktail optimization to maximize sugar yields and determination of sugar fermentability using *Saccharomyces cerevisiae* 424A (LNH-ST). In addition, a complete mass balance and process economics will be presented.

Danmarks Tekniske Universitet DTU

Screening method for selection of biomass feedstock and location in biorefneries

A. Jaksland, M. Bertran, J.D. Woodley, R. Gani

2/11 AIChE Annual Meeting 2017

DTU Chemical Engineering
Department of Chemical and Biochemical Engineering

Danmarks Tekniske Universitet DTU

Outline

- Motivation and Background
- Biomass selection
- One parameter screening method
- Case study
  - Biomass selection
  - Superstructure generation and solution
- Results
- Evaluation in three parameters
- Conclusion

DTU Chemical Engineering
Department of Chemical and Biochemical Engineering

Danmarks Tekniske Universitet DTU

Motivation

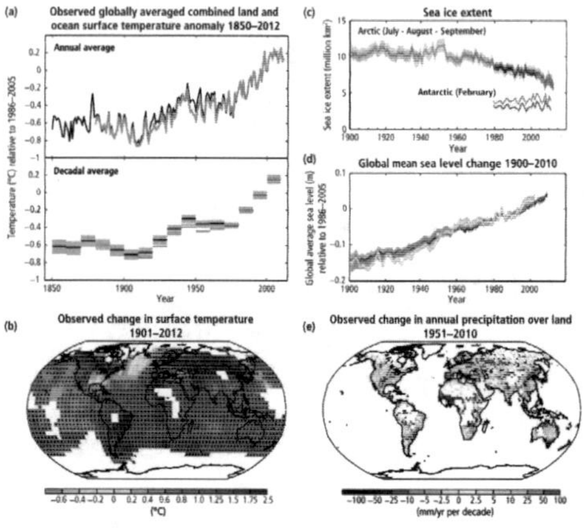

Background

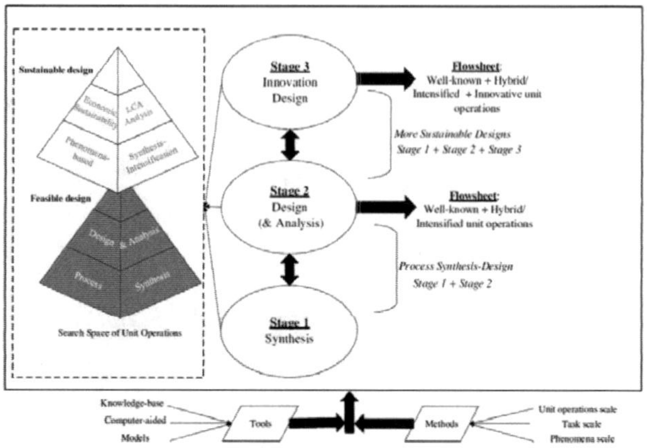

DTU Chemical Engineering
Department of Chemical and Biochemical Engineering

1) IPCC. Climate Change 2014: Synthesis Report. Contribution of Working Groups I, II and III to the Fifth Assessment Report of the intergovernmental Panel on Climate Change. 2014.
2) Deenesh K. Babi, Johannes Holtbruegge, Philip Lutze, Andrzej Gorak, John M. Woodley, and Raqul Gani. Sustainable process synthesis-intensication. Computers and Chemical Engineering, 81:218{244, 2015.

Danmarks Tekniske Universitet DTU

Superstructure Optimisation Framework

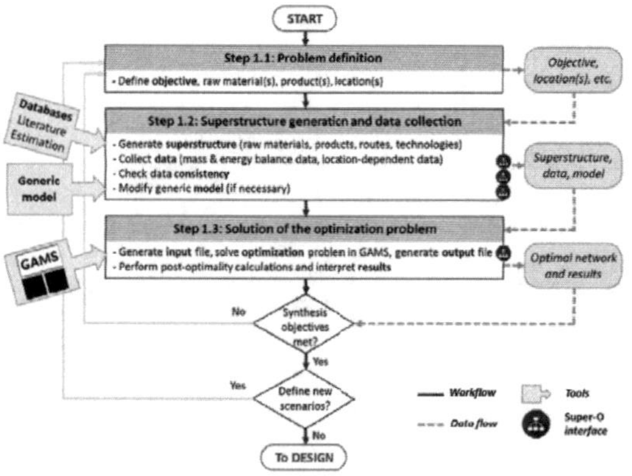

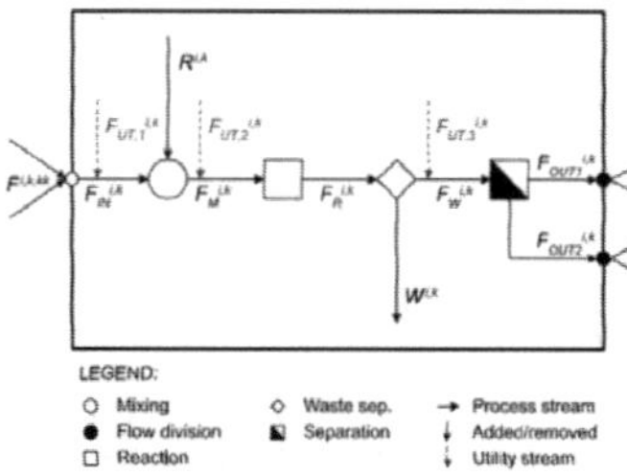

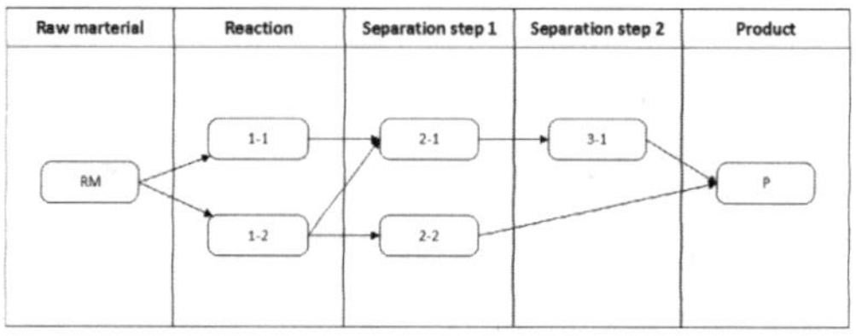

DTU Chemical Engineering
Department of Chemical and Biochemical Engineering

3) Maria Ona Bertran, Rebecca Frauzem, Lei Zhang, and Raqul Gani. A Generic Methodology for Superstructure Optimization of Different Processing Networks. Computer Aided Chemical Engineering, 38:685{690, 2016.

Danmarks Tekniske Universitet DTU

Biomass as Feedstock

Category	Potential biomass supply in 2050 (EJ/y)
(I) Biomass production on surplus agricultural land	0-988
(II) Biomass production on degraded lands	8-110
(III) Agricultural residues	10-32
(IV) Forest residues	10-16
(V) Animal Manure	9-25
(VI) Organic waste	1-3
(VII) Bio-materials	Minus 83-116
Total	33-1130

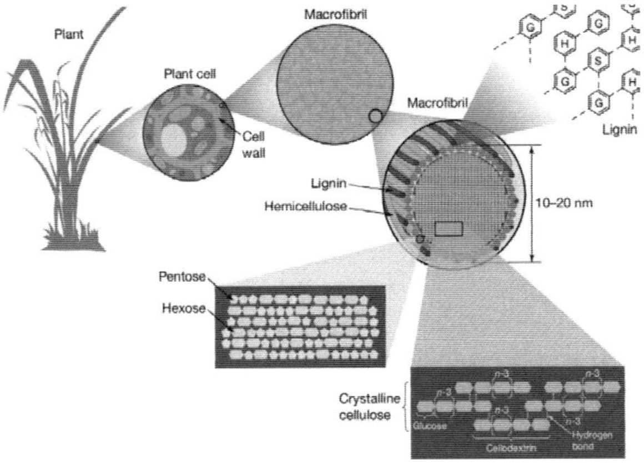

DTU Chemical Engineering
Department of Chemical and Biochemical Engineering

4) Moniqu Hoogwijk, Andre Faaij, Richard van den Broek, G?oran Berndes, Dolf Gielen, and Wim Turkenburg. Exploration of the ranges of the global potential of biomass for energy. Biomass and Bioenergy, 25:119(133, 2003.

Danmarks Tekniske Universitet DTU

Biomass Selection

Location and specie dependent variables: Composition, density, harvest field yield, transportation cost, at-field price

Three parameters: Harvest field yield, at-field price and transportation cost

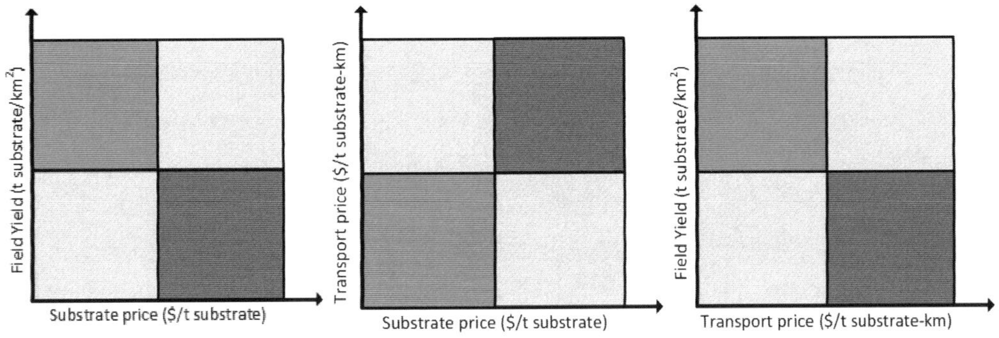

DTU Chemical Engineering
Department of Chemical and Biochemical Engineering

Biomass Selection

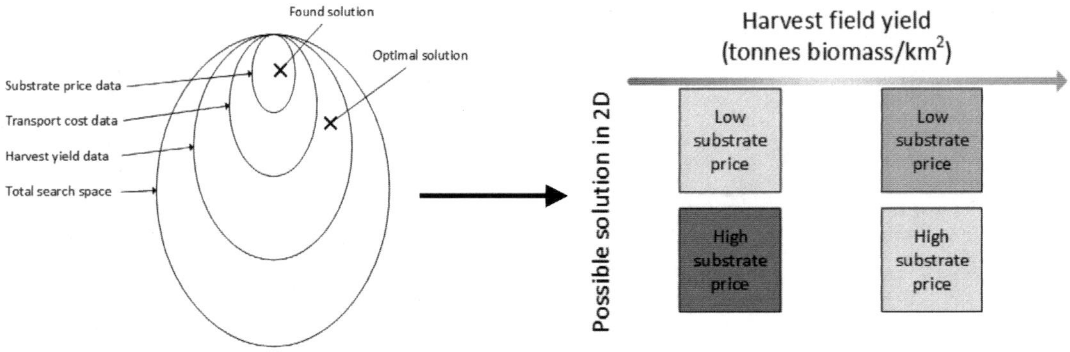

Biomass Selection

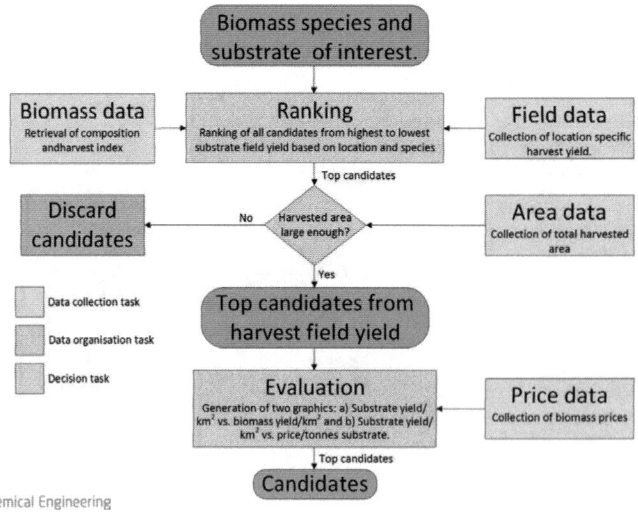

Case study – Step 1

- 750.000 tonnes biomass/year
- Biomass specie and location: Selected with one parameter screening
- Intermediate: glucose
- Biomass to glucose: Milling, steam explosion and enzymatic hydrolysis
- Products: 1,3-propanediol, 2,3-butanediol, succinic acid, itaconic acid
- Ideal reactions based on redox potential
- Transport and storage cost of biomass not included

Case study – Step 1

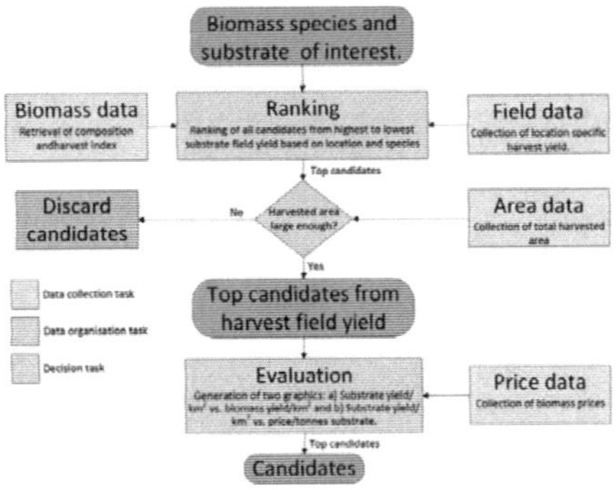

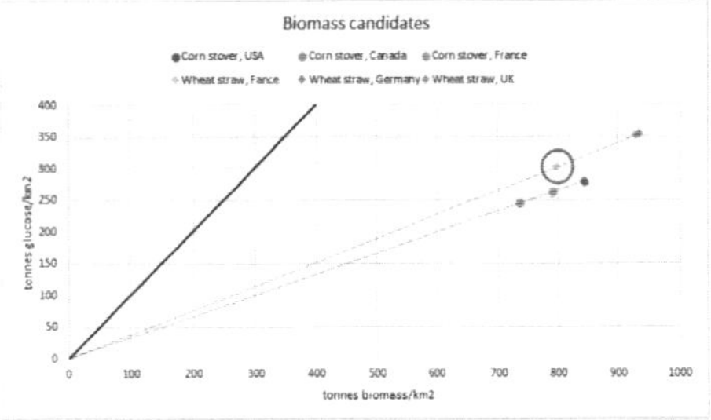

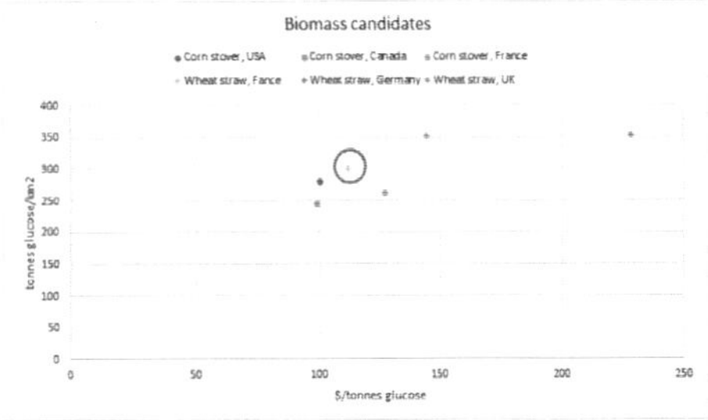

Case study Step 2 & 3

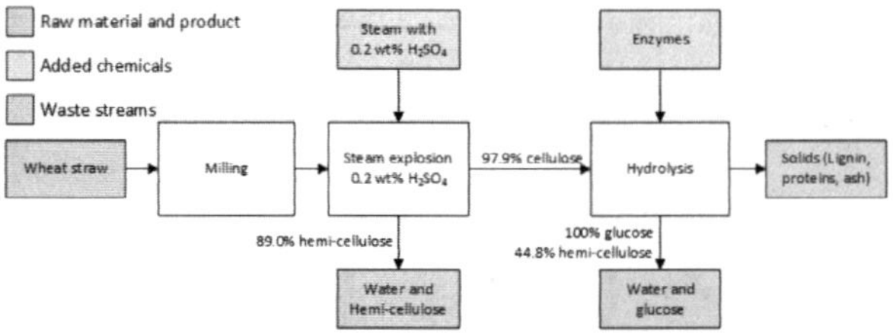

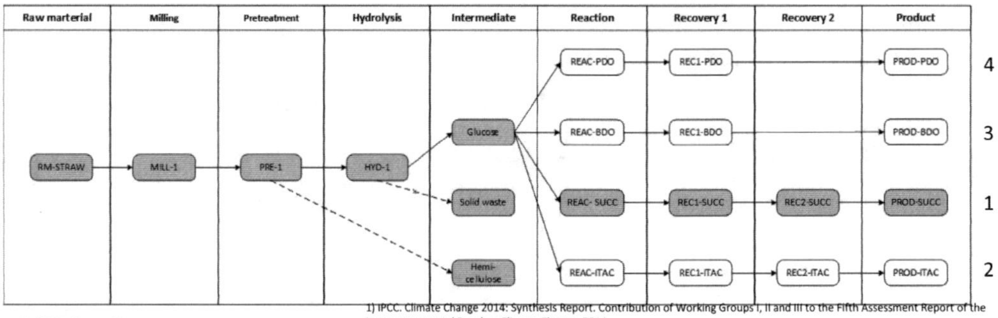

1) IPCC. Climate Change 2014: Synthesis Report. Contribution of Working Groups I, II and III to the Fifth Assessment Report of the intergovernmental Panel on Climate Change. 2014.

2) Deenesh K. Babi, Johannes Holtbruegge, Philip Lutze, Andrzej Gorak, John M. Woodley, and Raqul Gani. Sustainable process synthesis-intensication. Computers and Chemical Engineering, 81:218{244, 2015.

Results

Succinic acid		Price	Amount	Value
Raw materials	Wheat straw	0.045$/kg	85616.4	3852.7 $/hr
Added chemicals	Cellulase	15.67 $/kg	684,6	10728.1 $/hr
Products	Succinic acid	3 $/kg	29560.7	88682.1 $/hr
Utilities	Steam	0.00983 $/MJ	3680214.4 MJ/hr	36176.5 $/hr
	Cold water	0.00035 $/MJ	45830.5 MJ/hr	16.0 $/hr
	Electricity	0.0295 $/MJ	3532.6 MJ/hr	104.1 $/hr
	Total	-	3729577.5 MJ/hr	36296.7 $/hr
Operating profit				37804.6$/hr
Capital investment				59000635.0 $
Payback time				1561 hr = 65 days
Hourly profit over 20 years				37467.8 $/hr

Results – Comparison

	Product	Theo. yield	Prac. yield	Operating profit	Capital investment	Payback time
Scenario 1	1,3-propanediol	0.84 g/g	0.57 g/g	12186.8 \$/hr	38.6m \$	132 days
	2,3-butanediol	0.75 g/g	0.48 g/g	26856.5 \$/hr	35.5m \$	55 days
	Itaconic acid	0.96 g/g	0.81 g/g	29053.1 \$/hr	47.0m \$	67 days
	Succinic acid	1.12 g/g	0.89 g/g	37804.6 \$/hr	59.0m \$	65 days

Sensitivity Analysis

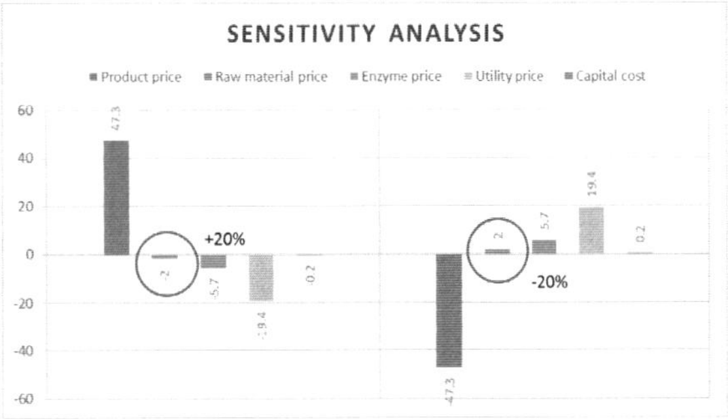

Biomass Selection in Three Parameters

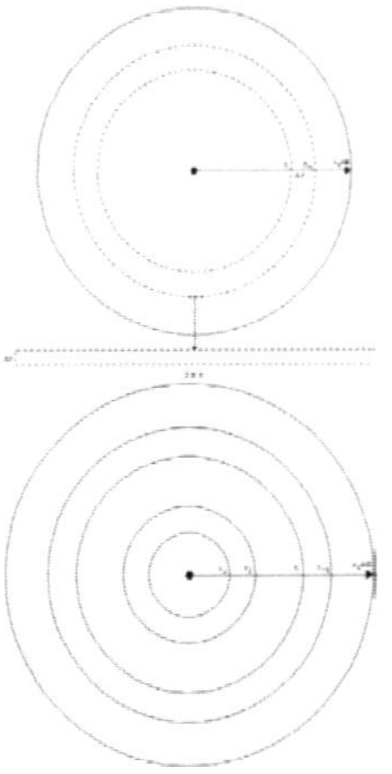

Radius as a function of capacity and yield:

$$R = \sqrt{\frac{C}{\pi Y}}$$

Transportation cost normalised by capacity as a function of radius, truck load and fuel cost:

$$\frac{T}{C} = \frac{\frac{4R^3\pi Y}{3L}F}{C} = \frac{4R^3\pi Y}{3LR^2\pi Y}F = \frac{4R}{3L}F$$

Transportation normalised by capacity as a function of capacity, yield, truck load and fuel cost

$$\frac{T}{C} = \frac{4\sqrt{\frac{C}{\pi Y}}}{3L}F = \frac{4\sqrt{C}}{3L\sqrt{Y\pi}}F$$

Symbol	Description	Unit
Y	Biomass harvest field yield	kg biomass/km^2
L	Truck load	kg biomass
R	Outer diameter	km
F	Fuel cost	\$/km
C	Biomass plant capacity	kg biomass
T	Total transport cost	\$
T_i	Transport cost of annulus	\$
B	At-field biomass price	\$/kg biomass
r	Radius (axis)	km
Δr	Differential radius piece	km

Biomass Selection in Three Parameters

- Transportation estimate
- France vs. Brazil
  - 796.58 t/km^2 vs. 239.27 t/km^2
  - Estimation gives: 42.3 \$/t vs. 39.35 \$/t
  - 941.5 km^2 vs. 3134.3 km^2

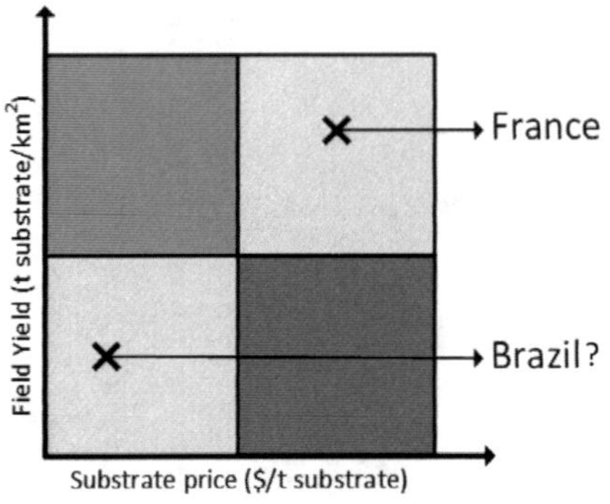

$$\frac{4\sqrt{C}}{3L\sqrt{Y_1\pi}}F + B_1 = \frac{4\sqrt{C}}{3L\sqrt{Y_2\pi}}F + B_2$$

Conclusion

- Location and specie dependent biomass variables can be reduced to three parameters: Harvest field yield, at-field cost and transportation cost, all based on desired substrate.
- A one-parameter based on harvest field yield has been developed.
- In a case-study a biorefinery network superstructure was developed with feedstock found from the screening.
- An operational profit of 37804.6 \$/hr was found for succininc acid. The payback time was 65.
- A sensitivity analysis (+/- 20%) showed that the feedstock price only impacted the operating profit by +/- 2%.
- Expanding the potential selection of biomass to the original three parameters showed that inclusion of transport cost is small compared to at-field price.

Danmarks Tekniske Universitet

Thank you for listening
Questions?

DTU Chemical Engineering
Department of Chemical and Biochemical Engineering

501689 Visualization Study of Dynamic Shrinkage during Biomass Pyrolysis

Abstract:

Biomass pyrolysis produces both renewable fuels and feedstock for chemical production. Yields and selectivity for pyrolysis is a function of numerous factors including rates of heating and quenching (residence times for heterogeneous and homogeneous events) and condensed phase properties including the dynamic shrinkage of the biomass. The shrinkage of the condensed matter affects many physical properties including the porosity, permeability, overall density, the volume occupied by volatiles (gas and tar), the volume occupied by the solid (wood and char) and consequently the total volume of the particle, all of which change continuously as biomass is thermally decomposed. In this study two different types biomass microspheres (tall fescue and switchgrass) and microcrystalline cellulose were produced and pyrolyzed in a micro-reactor at three different heating rates (10, 20 and 50ËšC/min). The dynamic shrinkage was captured in real time using a CMOS camera and side-by-side weight loss experiments were performed using thermal gravimetric analysis (TGA). The results show that for pyrolysis of single biomass microspheres, the normalized volume-based shrinkage and normalized weight loss are identical for both cellulose and switchgrass. However, for tall fescue at temperatures above 360 ËšC the shrinkage is higher than the weight loss indicating that shrinkage is not directly proportional to the amount of decomposed matter. Although, shrinkage is independent of particle size for these biomass types, yet crystalline cellulose shrinks more than switchgrass and tall fescue. These results will be used to calibrate microstructural pyrolysis models currently being developed by the same team.

492599 Effect of Feedstock and Pyrolysis MethodÂ on the Yield of Hydrotreated Refinery Intermediate

Abstract:

Biomass can be converted into a hydrocarbon mixture with properties similar to refinery intermediates such as naphtha or gas-oil *via* fast pyrolysis followed by hydrotreating of the pyrolysis liquid. However the feedstock affects the yield and properties of the pyrolysis liquid in ways that are not well understood. The properties of the pyrolysis liquid in turn affect the operability of the hydrotreating and the number of stages required to accomplish the hydroprocessing. Catalytic upgrading of the fast pyrolysis vapors produces a higher quality pyrolysis oil but in lower yield than the non-catalytic method. Thus the most carbon efficient, energy efficient, and cost-effective path for a given feedstock is still in question.

Therefore Idaho National Laboratory (INL), the National Renewable Energy Laboratory (NREL), and Pacific Northwest National Laboratory (PNNL) have been collaborating to study the effect of feedstock properties and vapor-phase upgrading (VPU) on the yield and quality of the hydrotreated products. INL has provided feedstocks, NREL has performed pyrolysis/VPU, and PNNL has hydrotreated the liquids. Preliminary results suggest that air classified forest residue is a promising low-cost waste stream and miscanthus is a promising energy crop for these processes. Modeling has identified lignin, volatile matter, and alkali content of the biomass as key predictors of pyrolysis yield.

Ongoing work is exploring the effect of the type of lignin and modeling the effect of oil properties on the hydrotreating.

ZSM-5 Coating on SiC Foam Support as a Composite Catalyst for Microwave-assisted Pyrolysis of Biomass

Nan Zhou

Shiyu Liu, Yaning Zhang, Liangliang Fan, Yanling Cheng, Paul Chen, Roger Ruan

11/2/2017

UNIVERSITY OF MINNESOTA

Research Background

❑ ZSM-5

- A zeolite, $Na_nAl_nSi_{96-n}O_{192}$
- Shape selectivity: pore size matches kinetic diameter of benzene [1]
- Al substitution of Si→ Brønsted Acid Sites[1]
- Increase the aromatic content, lower the oxygen content[2]

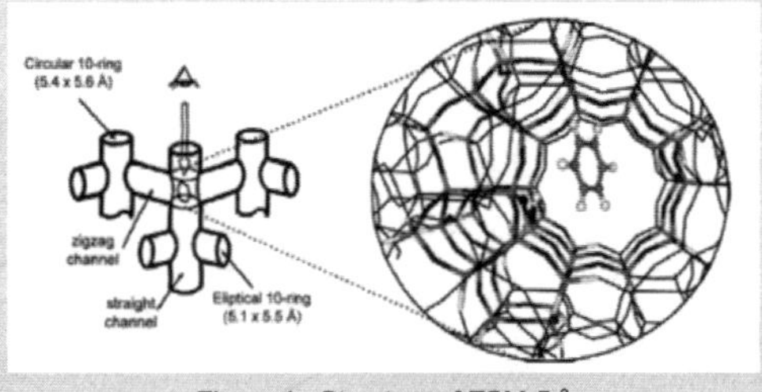

Figure 1. Structure of ZSM-5 [3]

❑ *In situ* vs. *ex situ* catalytic pyrolysis

- In-situ catalytic pyrolysis: difficult separation with char, easy coking, high catalyst/biomass ratio needed, difficult independent temperature control
- *Ex-situ* catalytic pyrolysis: external heating required
- ZSM-5 packed bed: temperature gradient, pressure drop, channeling effect

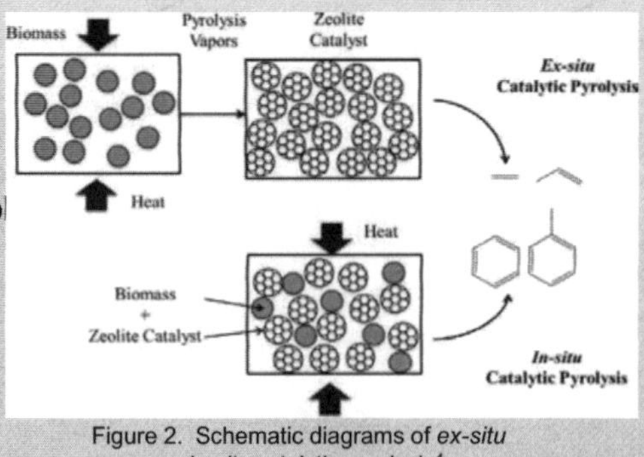

Figure 2. Schematic diagrams of *ex-situ* vs. *in-situ* catalytic pyrolysis[4]

UNIVERSITY OF MINNESOTA

SiC Foam Support

❑ Foam support

- Uniform temperature
- Enhanced mass transfer of the reactant vapor with catalysts
- Controllable diffusion path of the product molecules

❑ SiC

- High thermal conductivity (200 W/m° C)
- Low thermal expansion coefficient (4.8 ppm/° C)
- High chemical inertness at elevated temperatures
- Cheap SiC foams available as industrial filters
- High microwave absorption >> ZSM-5 or other ceramics

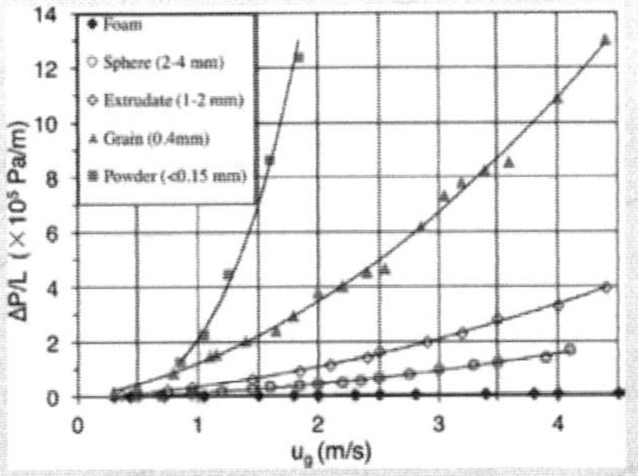

Figure 1. Pressure drop measurements on the various states of silicon carbide materials[5]

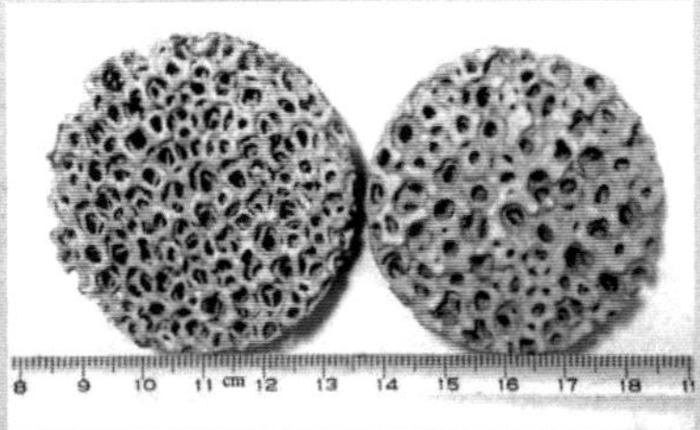

Figure 2. SiC foams of different pore sizes

UNIVERSITY OF MINNESOTA

Composite Catalyst Preparation

❑ SiC foams of different pore sizes purchased (10, 20, 30, 60 pores per inch).

❑ ZSM-5 coatings prepared by a hydrothermal synthesis method. [6]

- Precursor gel: tetraethyl orthosilicate (TEOS), tetrapropylammonium hydroxide (TPAOH), sodium chloride, and sodium aluminate ($NaAlO_2$), water
- Molar ratio: 2.16:5.62: 3.43: 0.13:1000
- Temperature: 166 °C
- Synthesis time 48 h
- Sonication for 5 mins to remove loosely attached powders
- Repeat the synthesis procedure→ control the loadings of the ZSM-5
- Calcination at 500 °C to remove organic template
- Ion exchange with 1M NH_4Cl solution at 85 °C to replace the Na^+ with NH_4Cl
- Calcination at 500 °C to activate ZSM-5 to H-ZSM-5

Figure 1. SiC foams with (right) and without (left) synthesized ZSM-5

❑ Final catalyst loading: ~ 13.4% (w.) ZSM-5 after 3*48h.

UNIVERSITY OF MINNESOTA

Characterization of the Catalysts

❑SEM

Figure 1. SEM micrographs of ZSM-5/SiC foam composites of different magnifications: 20 (a), 500 (b), 2500 (c), and 10000 (d).

A almost complete coverage of the support surface by ZSM-5 crystals with the typical shape of hexagonal prisms and a size of ~0.4 μm

UNIVERSITY OF MINNESOTA

Characterization of the Catalysts

❑XRD

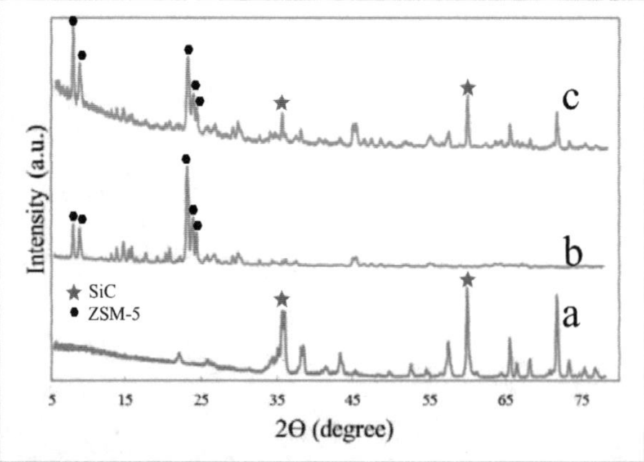

Figure. XRD patterns of (a) SiC foam, (b) ZSM-5 powder, and (c) ZSM-5/SiC foam composite

Both powder and composite catalyst exhibit the MFI structure with major peaks located at 7.9° and 8.9° and the characteristic triplet at 23.5°.

UNIVERSITY OF MINNESOTA

Pyrolysis Experiment Set-up

- Corn stover (20 -35 mesh, 6.7% moisture, 73.2% volatile, 18.1% fixed carbon, 2.0% ash)
- 550 °C pyrolysis, 450 °C catalytic bed

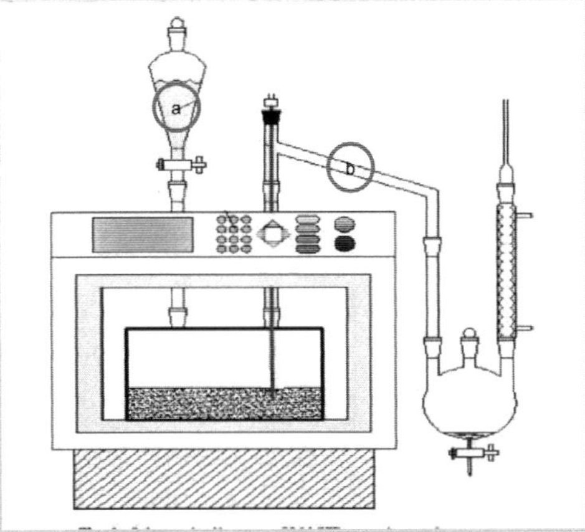

Figure 1. Schematic diagram of a microwave-assisted pyrolysis system[7]

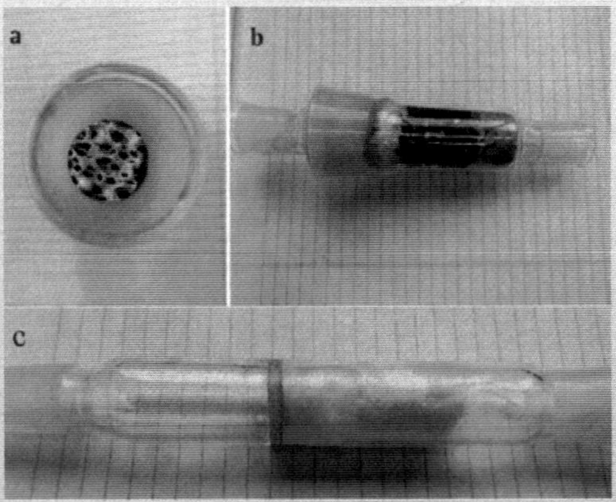

Figure 2. Axial (a) and radial (b) views of the secondary composite catalyst bed and (c) catalyst packed bed.

UNIVERSITY OF MINNESOTA

Test of Catalytic Performance

Table. Product yields (wt. %) under different MAP conditions *

Experiment Runs	Catalyst Bed	Catalyst to Biomass Ratio	Char	Bio-oil	Gas
1	No catalyst bed	0	29.0±0.9	37.5±1.4	33.6±2.3
2	Bare SiC support	0	27.4±1.4	37.8±3.3	34.8±4.7
3	In-situ	1/4	33.9±0.5	23.4±1.6	42.8±2.1
4	Ex-situ, packed bed	1/4	26.4±2.4	32.5±1.3	41.1±1.6
5	Composite catalyst bed	1/4	27.9±0.6	34.3±1.4	37.8±2.0

* Note: Pyrolysis temperature kept at 550 °C, and catalyst bed temperature kept at 450 °C except for Experiment 1 and 3 with no catalyst beds

UNIVERSITY OF MINNESOTA

Test of Catalytic Performance

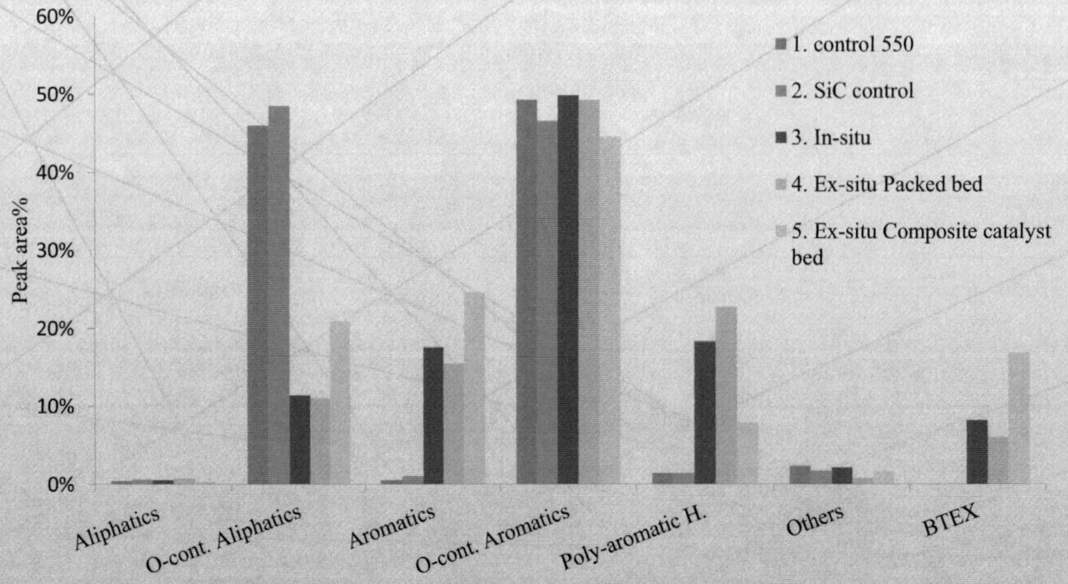

Figure. Bio-oil compositions under different MAP conditions

UNIVERSITY OF MINNESOTA

Summary

- ZSM-5 coatings was successfully synthesized on the SiC foam with a catalyst loading of 15% wt. after 3 synthesis.
- Compared with non-catalytic control experiments, the bio-oil yields with composite catalyst bed slightly decreased, but the bio-oil quality was significantly improved with 27% aromatics produced.
- Compared with other catalytic pyrolysis experiments, the composite catalyst bed produced significantly more BTEX hydrocarbons and less PAHs.

UNIVERSITY OF MINNESOTA

Acknowledgement

- Dr. Roger Ruan's group
- This project is supported in part by the Minnesota Environment and Natural Resources Trust Fund as recommended by the Legislative-Citizen Commission on Minnesota Resources (LCCMR), MCES, Xcel energy, and UMN MNDrives and Center for Biorefining.
- Parts of this work were carried out in the Characterization Facility, University of Minnesota, which receives partial support from NSF through the MRSEC program.

UNIVERSITY OF MINNESOTA

References

1. Hoff, T. C., Gardner, D. W., Thilakaratne, R., Wang, K., Hansen, T. W., Brown, R. C., & Tessonnier, J. P. (2016). Tailoring ZSM-5 Zeolites for the Fast Pyrolysis of Biomass to Aromatic Hydrocarbons. ChemSusChem, 9(12), 1473-1482.
2. Carlson, T. R., Tompsett, G. A., Conner, W. C., & Huber, G. W. (2009). Aromatic production from catalytic fast pyrolysis of biomass-derived feedstocks. *Topics in Catalysis, 52*(3), 241.
3. Raksakoon, C., & Limtrakul, J. (2003). Adsorption of aromatic hydrocarbon onto H-ZSM-5 zeolite investigated by ONIOM study. Journal of Molecular Structure: THEOCHEM, 631(1), 147-156.
4. Wang, K., Johnston, P. A., & Brown, R. C. (2014). Comparison of in-situ and ex-situ catalytic pyrolysis in a micro-reactor system. Bioresource technology, 173, 124-131.
5. Ba, H., Liu, Y., Mu, X., Doh, W. H., Nhut, J. M., Granger, P., & Pham-Huu, C. (2015). Macroscopic nanodiamonds/β-SiC composite as metal-free catalysts for steam-free dehydrogenation of ethylbenzene to styrene. Applied Catalysis A: General, 499, 217-226.
6. Ivanova, S., et al. "ZSM-5 coatings on β-SiC monoliths: possible new structured catalyst for the methanol-to-olefins process." The Journal of Physical Chemistry C 111.11 (2007): 4368-4374.
7. Liu, S., Zhang, Y., Fan, L., Zhou, N., Tian, G., Zhu, X., ... & Ruan, R. (2017). Bio-oil production from sequential two-step catalytic fast microwave-assisted biomass pyrolysis. *Fuel, 196*, 261-268.

UNIVERSITY OF MINNESOTA

Questions?

Thank you!

UNIVERSITY OF MINNESOTA

493017 Bio-Fuel Production from Sequential Two-Step Catalytic Fast Microwave-Assisted Biomass Pyrolysis

Abstract:

Currently most research focused on the one-step catalytic fast pyrolysis where the catalysts are mixed with the feedstock in the pyrolysis reactor (Xie et al., 2015; Zhang et al., 2017). However, the one-step process has several disadvantages, including fixed cracking and catalytic upgrading temperature, and high coke formation on the catalyst causing rapid catalyst deactivation, etc. In the present study, a new two-step approach combining a fast microwave-assisted pyrolysis (fMAP) step for bio-crude production and a separate packed-bed catalysis step for bio-oil upgrading in one system was developed. In this two-step system, the catalyst bed is placed between the pyrolysis and condensation processes outside of the microwave cavity. The cracking and catalytic upgrading operations can be flexibly and independently controlled, thus enabling the optimum catalyst performances targeted at desired products under optimized reaction temperature and residence time. This also prevents the unreacted char and mineral components from contacting the upgrading catalyst, and can extend the life of the catalyst and reduce the catalyst loading and therefore the cost. In this study, the effects of fMAP temperature, catalyst bed temperature and catalyst-to-biomass ratio on product fractional yield and selectivity were investigated. Eight major components in bio-oils were identified and quantified with an external standard method. The comparison between catalytic and non-catalytic pyrolytic bio-oils provided insights into the possible reaction pathway of the sequential two-step catalytic fMAP. The results showed that the optimum fMAP temperature and catalyst bed temperature were 550 and 425 , respectively. The XRD results showed that the crystal structure and crystallinity of the HZSM-5 catalyst were intact before and after the pyrolysis and regeneration process, indicating the HZSM-5 had good stability in the catalytic upgrading process. In addition, the two-step process was superior to the one-step process in terms of the amount of catalyst needed to achieve similar bio-oil quality. Thus, the newly developed two-step fast microwave-assisted pyrolysis process allowed more flexibility for the high quality bio-oil production that meets the requirements for downstream processing and applications.

References

Xie, Q.L., Addy, M., Liu, S.Y., Zhang, B., Cheng, Y.L., Wan, Y.Q., Li, Y., Liu, Y.H., Lin, X.Y., Chen, P., Ruan, R. 2015. Fast microwave-assisted catalytic co-pyrolysis of microalgae and scum for bio-oil production. *Fuel*, **160**, 577-582.

Zhang, Y., Chen, P., Liu, S., Peng, P., Min, M., Cheng, Y., Anderson, E., Zhou, N., Fan, L., Liu, C., Chen, G., Liu, Y., Lei, H., Li, B., Ruan, R. 2017. Effects of feedstock characteristics on microwave-assisted pyrolysis - A review. *Bioresour Technol*, **230**, 143-151.

504078 Microwave Assisted Fast Catalytic Pyrolysis and Gasification for Solid Wastes Conversion and Utilization

Abstract:

Various solid wastes, e.g., as crop residues, processing byproducts, food wastes, municipal solid wastes, and waste plastics and used tires are potential feedstock for production of renewable energy, chemicals and materials. Gasification and pyrolysis are efficient ways for conversion and utilization of these solid wastes. Fluidized bed processes are the most common methods employed in fast pyrolysis and gasification. Major issues with the fluidized bed processes include complex and expensive system and process, high ash particles in products due to violent agitation, low energy density of syngas due to the dilution by carrying gases, etc. In this presentation, we will present our improvement and new development of a fast microwave assisted downdraft catalytic pyrolysis and gasification processes using novel microwave heating mechanism in which microwave absorbents and catalysts are used to significantly improve the heating characteristics, and the yield and quality of the desirable products. Solid feedstocks are directly dropped on to the hot microwave absorbents efficiently maintained at desirable temperatures, resulting in higher temperature rise of the feedstock and therefore much more efficient absorption of the microwave energy also, and in turn instant gasification and pyrolysis. A separated packed-bed catalysis for the volatiles was also developed to improve the quality and yield of the pyrolytic and gasification products. Results and discussion on the effects of key process variables such as microwave absorbent type, particle size, and loading, microwave power input and control, feedstock loading method, raw material and catalyst temperatures, and ratio of raw material and catalyst loading on product yields and quality, and energy consumption will be presented.

503964 Synthesis of Anisotropic Calcined Coke from Fast Pyrolysis Bio-Oils

Abstract:

Aluminum smelters utilize calcined petroleum coke to make anodes for large-scale conversion of alumina into aluminum. The industry faces ongoing coke quality deterioration, such as increasing impurity content (sulfur and heavy metals), decreasing bulk density, and increasing concentrations of undesired textures. Furthermore, it is desirable to reduce SO_2 and CO_2 emissions.

Renewable alternatives to petroleum coke have been explored by several groups. Nevertheless, up to now it is not possible to produce coke with the required quality for aluminum smelting. Biomass-derived cokes usually have the desired low concentrations of sulfur and metals. However, bulk densities are much too low, and the cokes do not have the required anisotropic texture.

This talk will illustrate a process for production of renewable calcined coke from various fast pyrolysis bio-oils. Biomass is inherently low in undesired impurities. Coke manufacture by coking of biomass oil (as opposed to biomass carbonization) allows production of high-density coke. Finally, coke texture depends on the oxygen content of the coker feed. The desired anisotropic texture is favored by a low oxygen content. A modified pyrolysis process and bio-oil distillation procedure were used. This considerably reduced the oxygen content of the coker feed and yielded coke with the desired texture. Nickel and vanadium concentrations fell below 10 ppm. Sulfur was below 500 ppm. Other metals concentrations can vary, depending on the biomass source.

503091 Thermal Conversion of Biomass into Clean Energy and Resource Using Gasification Integrated with Power Generation and Bottom Ash Reutilization

Abstract:

Gasification is a thermal-based process that slowly converts biomass waste into syngas (a mixture of CO and H_2) by chemical reaction at high temperatures (> 700 °C) with very tiny amount of oxygen to avoid complete combustion. Therefore, gasification not only provides environmental friendly way of discharging biomass wastes, but also produces syngas as a clean energy fuel, offering an alternative clean process for recovering energy from biomass waste. In this work, a small-scale waste-to-energy system combined gasification and a set of internal combustion engine power generation was investigated by using redwood waste pellets was used as feedstock. It was found that the redwood waste pellets have a higher heating value of 21 MJ/kg and it can produce the maximum syngas yield of 47.9 Nm^3/h by gasification at a power load of 10 kW. Subsequently, total energy and exergy efficiencies of the integrated system can harvest 10.7% and 10.1%, respectively. In addition to the syngas, gasification also produces two main solid residues as unavoidable by-products, i.e. bottom ash â€“ primarily minerals and metals with minimal carbon, and char â€“ unreformed carbon. Although bottom ash is not classified as hazardous wastes, it may create potential risks to human health and environment. Therefore, the conversion of those by-products into beneficial materials could provide an environmental friendly and economical way to re-utilize the solid residues from gasification process, and hence creating a clean waste-to-energy system. In this work, bottom ash produced from gasification of redwood pellets was found to have significant amount of silica (>30 wt. %) and it can be successfully converted into zeolite materials (aluminosilicate) by using alkali hydrothermal method. The batch adsorption experiments showed that the original bottom ash itself has low adsorption capability, while the adsorption capability is significantly increased after the alkali hydrothermal reaction of bottom ash, which is mainly due to the formation of zeolite (which is a mixture of Na-A and hydroxylsodalite phases).

499787 Techno-Economic Feasibility of Liquid Fuel Production Using Supercritical Water Gasification of Algal Biomass

Abstract:

The potential of microalgae as a renewable energy source through synthetic fuel production has been actively researched in the recent years. However, most economic viability studies have discredited algal biomass due to the prohibitively expensive cost of algae production. In this work, a techno-economic sensitivity analysis of liquid fuels production via solar-driven non-catalytic supercritical water gasification (SCWG) of microalgae is presented. The system consists of four major parts â€" heliostat field for capturing the solar energy, the gasification reactor, post-processing of syngas in water-gas shift and methane reforming reactors, and production of liquid fuels using Fischer-Tropsch process. SCWG of wet algal biomass is carried out in a tubular continuous flow reactor, modelled as a series of parallel tubes inside a cavity receiver. The base case scenario has been standardized using 50 MWth input from the concentrated solar thermal plant. The plant design is optimized using exergy accounting in each unit operation and extensive heat exchange network optimization. Three scenarios for syngas refining have been considered - using steam methane reforming, partial oxidation and dry reforming, and autothermal reforming. The capital investment, operating costs, and total product values are calculated considering an operating duration of 30 years for the plant, and the data are reported based on the 2016 cost year.

The findings of this study have been compared with a recently published non-solar catalytic SCWG process, wherein a significant fraction of the product synthetics natural gas is used to provide the process heat for the gasification reactor. This substantially reduces the yield of product gas per unit mass of the expensive algae feedstock. The merits of using a renewable source (such as the sun) for driving the endothermic process are evaluated here. The current work highlights that from an economic standpoint the focus of process design should be on three aspects â€" maximizing the overall carbon efficiency, minimizing exergy losses and effective utilization of the captured solar energy.

490714 Pyrolysis and CO_2 Gasification Kinetics of Manure-Derived Hydrochar

Abstract:

Availability of biomass has made bioenergy a potential source of energy for future. But biomass usually contains high moisture content with low calorific value and bulk density which make the conversion process difficult. Hydrothermal carbonization (HTC) is a thermochemical pre-treatment process that can convert a high moisture containing biomass into a solid carbon rich fuel. Gasification is also considered as a promising technology for biomass energy conversion. It is a process that can convert biomass into gaseous fuel through partial oxidation in presence of an oxidizing agent. This study aims to examine the reaction kinetics of pyrolysis and gasification process of hydrothermally treated manure derived hydrochar by determining the associated reaction parameters and factors affecting those parameters. In this study, hydrochar was first pyrolyzed in presence of N_2 and then gasified with CO_2 at 800°C, 900°C and 1000°C with a residence time of 30 min, 60 min and 90 min at each temperature. Results from thermogravimetric analysis illustrated that with the increase of temperature, residence time became less important factor in pyrolysis zone. Modified form of Kissingerâ€™s method was applied for peak temperature of each residence time to determine the kinetic parameters. Results showed that reaction parameters were more dependent on temperature than on residence time in pyrolysis zone. The activation energy for pyrolysis zone was found to be within (172.6-276.6) KJ/mole in (800-1000)Â°C temperature range with three different residence time while for gasification zone the value was (210Â ±19) MJ/mole.

503075 Syngas Production from Sewage Sludge By Solar-Driven CO_2 Gasification

Abstract:

Conventional CO_2 gasification of sewage sludge requires a certain portion of feedstock to be combusted to supply necessary reaction heat, which releases a large amount of gaseous byproducts to the environment. The high-temperature energy, harvested from high-flux concentrated solar radiation, has been proved to supply the gasification process heat. In this work, the solar thermochemical gasification of CO_2-based sewage sludge was investigated theoretically by coupling concentrated solar energy as the heat source of high-temperature process heat. A dynamic model using CO_2 as a gasifying agent was developed to couple radiative transfer with fluid flow, heat and mass transfer, and thermochemical reaction kinetics. A fluidized gasifier with electric furnace was applied to simulate the thermal atmosphere of solar gasifier and to validate the theoretical model. Sensitivity of parameters on solar gasifier performance was assessed in terms of operating conditions, material properties, solar shape, and geometric structure. Thermodynamic-based analysis revealed that solar driven CO_2 gasification could potentially improve the quality of product syngas. Simulations indicate that the solar-to-syngas energy conversion increases with increasing flux density of solar irradiation. Furthermore, theoretical upgraded ratio of syngas using solar gasification has been investigated as well under various solar irradiances.

Keywords: CO_2 gasification; fluidized gasifier; sewage sludge; syngas; concentrated solar energy.

499369 Photocatalytic Hydrogen Generation Onto Mesoporous Titania Doped with Pt

Abstract:

In this research mesoporous TiO_2 photocatlysts were synthesized using the sol-gel technique. A Pluronic F127 triblock-copolymer, a structure-directing agent, was incorporated as a soft template into the sol-gel. In addition, and during a separate synthesis, the sol-gel was doped with a Pt precursor. Photocatalysts were prepared with 1.00 wt. %, 2.50 wt. %, 5.00 wt. % Pt nominal loadings, respectively. They were calcined at 500°C and 550°C following synthesis. Morphological and structural properties were studied by: a) X-ray diffraction, b) UV-Vis spectrophotometry, c) N_2 adsorption-desorption (BET, BJH), and d) X-ray photoelectron spectroscopy (XPS). Optical band gap values for mesoporous TiO_2 and modification with Pt were calculated by Kubelka-Munk (K-M) function coupled with Tauc plot methodology. It was observed that the prepared semiconductors displayed pore sizes in the 10 to 40 nm range with bimodal distributions. Their photocatalytic activity for hydrogen production via water splitting was established in a Photo-CREC Water-II reactor under near-UV light irradiation. The aqueous solution contained 2% v/v ethanol, employed as a renewable organic scavenger. The prepared materials exhibited that the mesoporous 2.50 wt. % Pt-TiO_2 has the highest photoactivity for hydrogen generation. This suggests the key role played by the loading of platinum as a TiO_2 dopant, reducing the optical band gap, increasing electron storage and diminishing, as a result, electron-hole recombination. The measured Quantum Yield (QY), obtained using a rigorous approach, was established for the mesoporous 2.50 wt. % Pt-TiO_2 at a promising level of 22.6 %.

495916 TiO_2 Nanoparticles By Flame Synthesis for Photocatalytic Hydrogen Evolution

Abstract:

Energy shortage and environment deterioration have been increasingly urgent and severe over the past decades. The limited non-renewable fossil fuels can hardly meet the growing energy demand. Thus, the exploitation of new green energy resources has become a global concern in recent years. Sunlight and water are abundant natural resources, which can be utilized to produce renewable energy in the form of hydrogen through water splitting. Thus, our work has been focused on this promising area. Despite the great progress achieved by now, the research in this field is still in an early age. The efficiencies of photocatalytic water splitting for the production of hydrogen are still relatively low and remain to be improved. Our flame-made TiO_2 has many advantages over the traditional TiO_2 which is prepared by wet chemistry. The small uniform particle size (~15nm) and the large specific surface area (~150 m^2/g) make our TiO_2 a good candidate for this reaction and it can improve the hydrogen generation rate largely. The simple procedure and stable mass production (~100mg/h) can largely save the cost of time and money, compared to the wet chemistry method. Besides, the excellent durability and environmental-friendly property can make our TiO_2 nanoparticles as a promising substitution for the toxic heavy metal catalysts, like CdS, CdO, MoP. The highest hydrogen generation rate that we have achieved by now is 24178 Î¼mol $g^{-1}h^{-1}$. So far, this performance has been proved to be better than most results in other works.

UCLA

$Ni(OH)_2$ Decorated Jagged Pt Nanowires Have High Activities for Hydrogen Evolution Reaction

Mufan Li, Kaining Duanmu, Zipeng Zhao, Peng Li, Huilong Fei, Wenxin Chen, Yuanming Zhu, Rong Yu, Jun Luo, Ketao Zang, Zhaoyang Lin, Mengning Ding, Chengzhang Wan, Jin Huang, Hongtao Sun, Yu Huang, Philippe Sautet, Xiangfeng Duan

Nov. 2, 2017

Synthesis of J-PtNWs/$Ni(OH)_2$

UCLA

NWs = Nanowires
CV = Cyclic Voltammetry

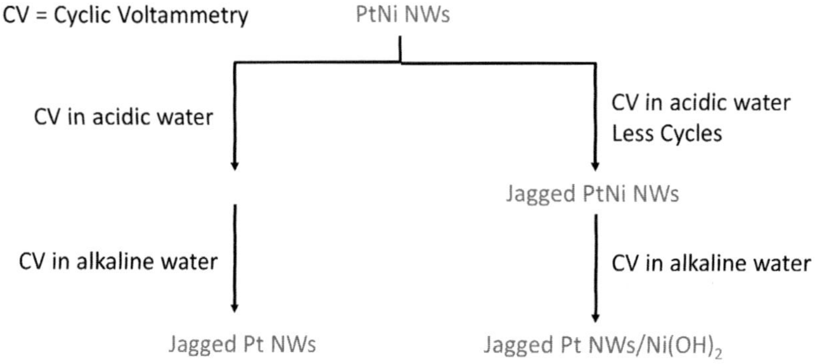

Characterization of J-PtNWs/Ni$(OH)_2$

UCLA

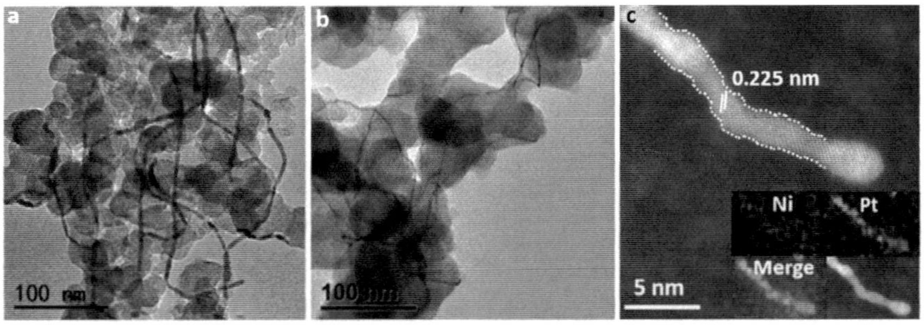

TEM of PtNi NWs TEM of J-PtNWs/Ni$(OH)_2$ STEM of J-PtNWs/Ni$(OH)_2$

Electrocatalytic activity for HER

UCLA

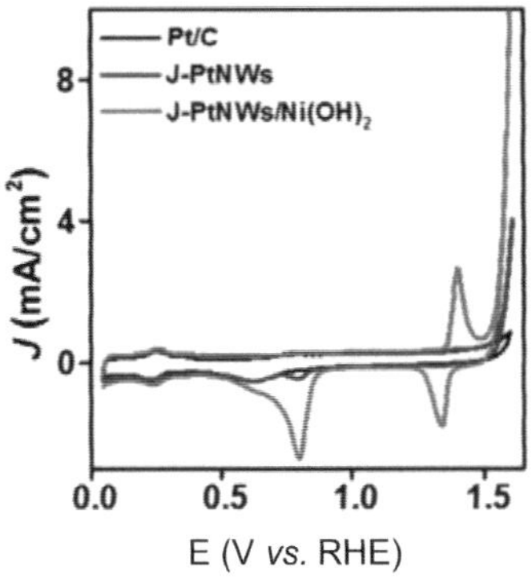

Cyclic voltammetry (CV) diagram

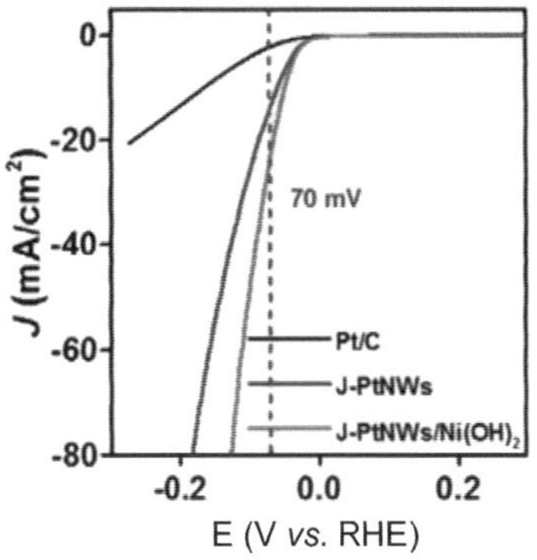

Linear sweep voltammetry diagram

Electrocatalytic activity for HER

UCLA

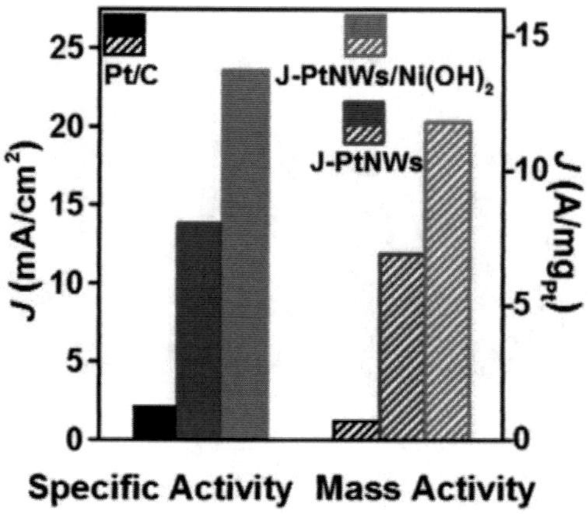

Volcano curve

UCLA

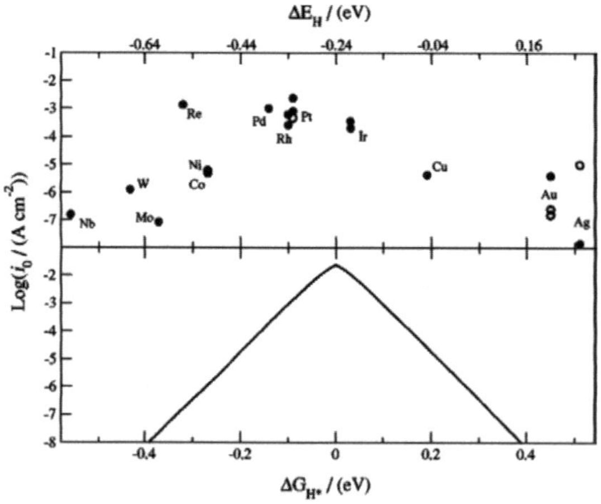

Nørskov, Bligaard, Logadottir, Kitchin, Chen, Pandelov, Stimming *J. Electrochem. Soc.* **2005**, *152*, J23.

Cavity model 1

UCLA

$\Delta\Delta G_H(\text{site}) = \Delta G_H(\text{site}) - \Delta G_H(\text{Pt111})$

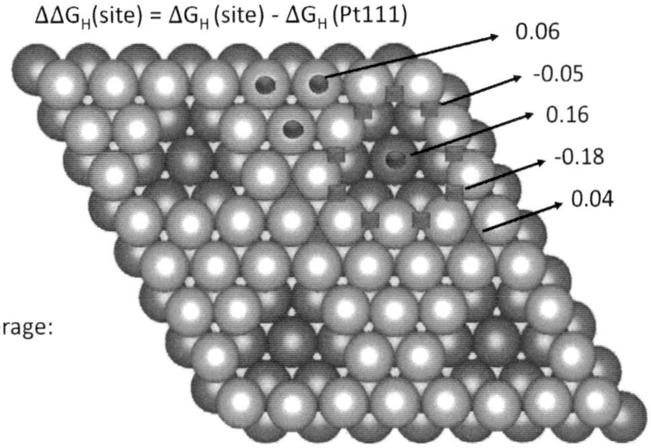

Full coverage:
16/16

Cavity model 2

UCLA

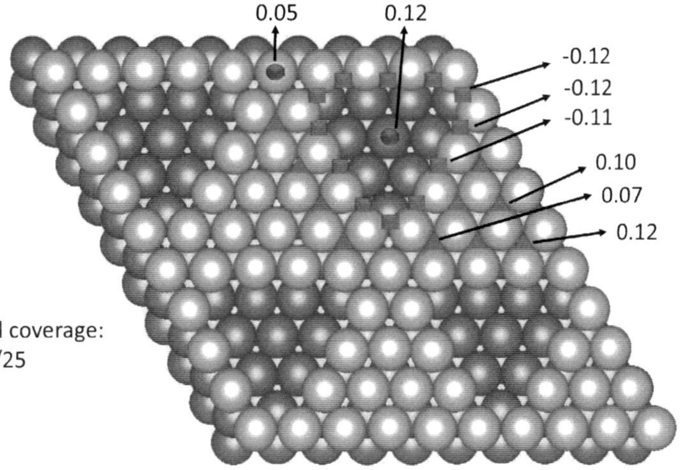

Full coverage:
24/25

Step model 3

UCLA

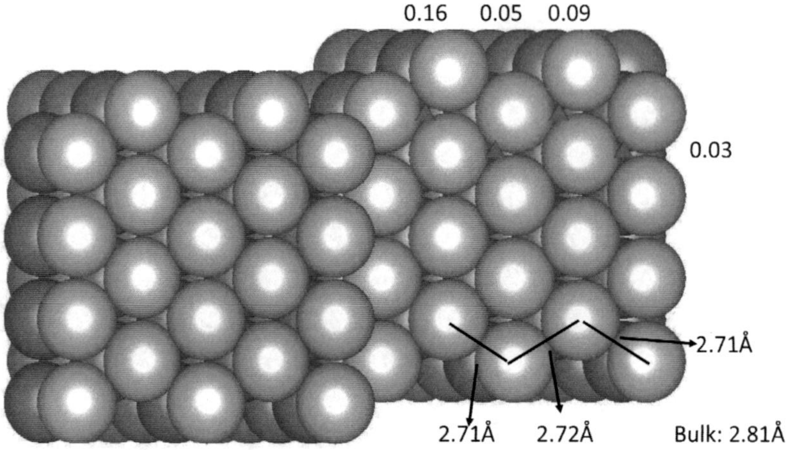

HER for J-PtNWs

UCLA

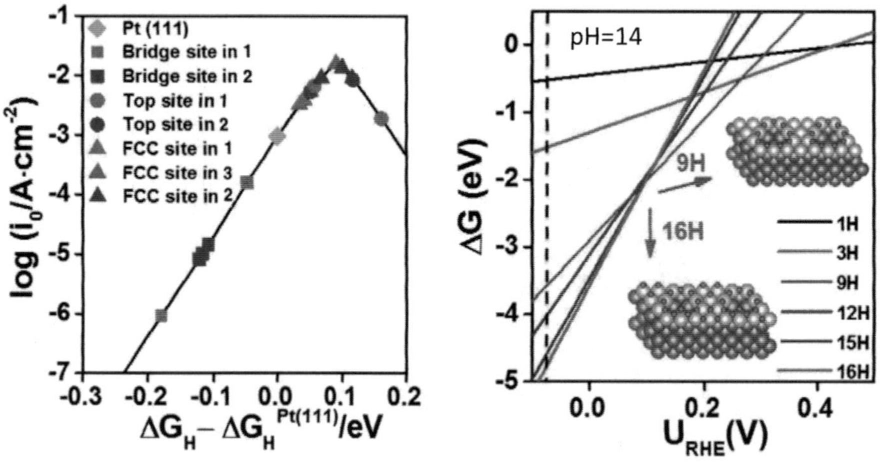

Model J-PtNWs/$Ni(OH)_2$ – low coverage

UCLA

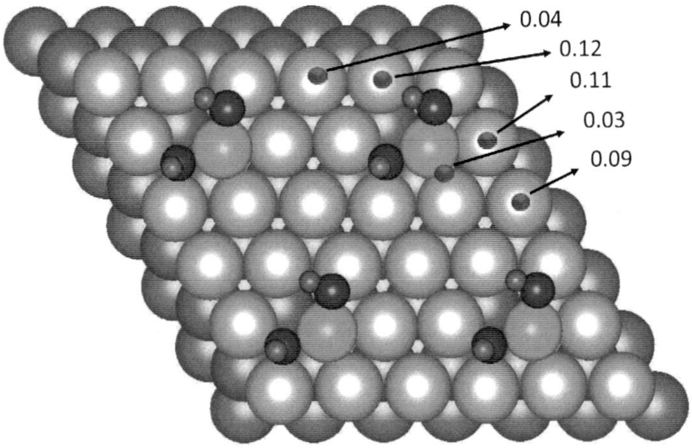

Model J-PtNWs/$Ni(OH)_2$ – high coverage

UCLA

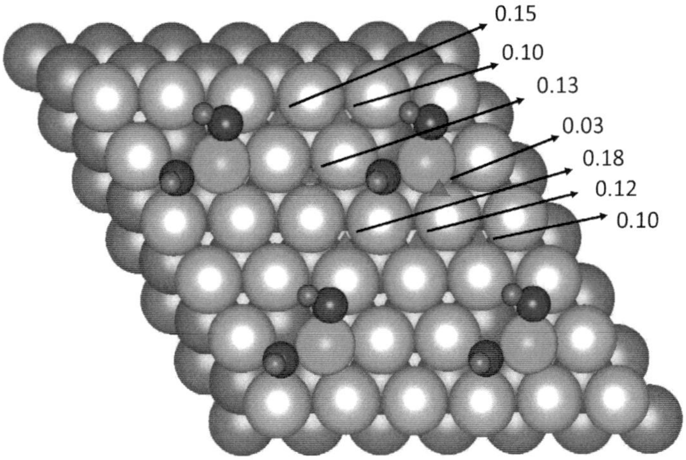

HER for model J-PtNWs/Ni(OH)$_2$

UCLA

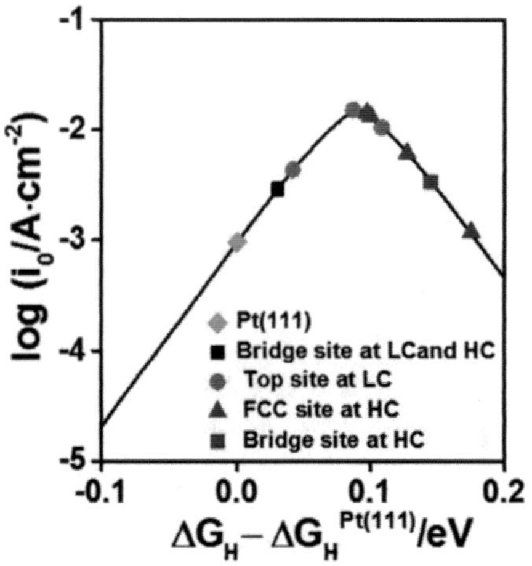

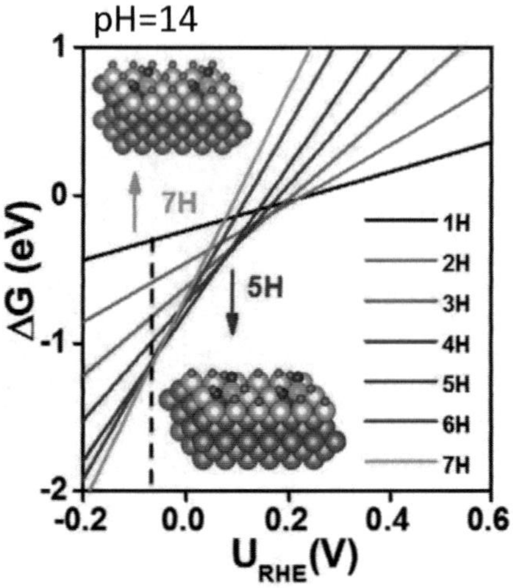

Conclusions

UCLA

- Synthesis of J-PtNWs and J-PtNWs/Ni(OH)$_2$
- HER activities for Pt/C, J-PtNWs and J-PtNWs/Ni(OH)$_2$ are 1: 10: 17
- The defective surface of J-PtNWs creates both active sites and non-active sites. In the experimental condition, hydrogen can adsorb on both of them.
- The decoration of Ni(OH)$_2$ can increase the activity of Pt, and all the sites around Ni(OH)$_2$ have higher activities.

Acknowledgments

UCLA

Sautet Group

Prof. Xiangfeng Duan

Mufan Li

XSEDE
Extreme Science and Engineering Discovery Environment

NSF

UCLA

Thank You

500531 Selective and Highly Stable H_2 Production from Formic Acid Decomposition over Graphite-Supported Moc

Abstract:

Formic acid (FA) is a good â€œgreenâ€ energy carrier candidate since it can be produced from CO_2, stored as a liquid, and converted to H_2 and CO_2 on demand to power H_2 fuel cells. Several interesting catalytic approaches have been investigated on both ends of this cycle, but the final reforming reaction still needs more work to become a viable commercial fuel source. Unlike the synthesis reaction, which will be conducted at a production plant and can use any materials or processes within the cost and safety limits, the on-demand reforming reaction must meet a number of criteria. First, FA reforming cannot be conducted over expensive, difficult-to-prepare, or unsafe catalysts â€“ for both economic and safety reasons. Any catalyst chosen for a commercial FA reformer must also maintain very good long-term stability, which oftentimes means a loss in activity. We have synthesized MoC nanoparticles (~2 nm) supported on graphite which demonstrate higher FA reforming activity, better CO_2/CO selectivity, and greater long-term stability without H_2 pretreatment than unsupported Mo_2C. We also propose a simple reaction mechanism for both reaction pathways based on previous observations and DFT studies on other materials as well as our own observations, and thoughts on how to further improve the reaction without using expensive or difficult-to-prepare catalysts, including by additive co-feed and with electric fields.

494871 Catalytic Trends for Nanoscale Catalysts from First-Principles Modeling and Tuning of Catalytic Reactivity for Fuel Production

Abstract:

It is still a rather open area regarding our fundamental understanding for effective manipulation of the activity and selectivity of catalyst for chemicals and fuel production. In this study, a first-principles-based kinetic model has been developed in order to understand the catalytic trends of reactivity and selectivity for hydrogen production on nanoscale single crystal nickel particles. The earth abundant nickel is a widely utilized industrial catalyst for hydrogen production through reforming, which is often accompanied by high temperature water-gas shift reaction (WGSR) that further boosts the hydrogen yield. However, nickel also promotes a number of side reactions, such as methanation and hydrogenolysis, both of which would lower hydrogen selectivity, and thus making the effort to tune the activity and selectivity of nickel catalyst an urgent fundamental research topic.

The mechanistic guidelines were first established using periodic density functional theory (DFT) calculations on the Ni(111), Ni(211) and Ni(100) single crystal facets, followed by the establishment of the most thermodynamically and kinetically relevant hydrogen production pathways. Furthermore, the reaction routes involving CO methanation consuming hydrogen will also be considered. In this investigation, the adsorbate-adsorbate interactions - between the key WGSR and methanationÂ reaction intermediate pairs, such as CO-CO, CO-H_2O, CO-OH, CO-O, and CO-CH â €“ have been extensively considered and shown significant impact on the kinetic trends on nickel single crystal surfaces. It has also been noted that, although most of the lateral interactions are repulsive in nature, catalyst lattice structures and intermolecular hydrogen bonding will create exceptions and complicate the trend analysis. Mean-field microkinetic modelings based on the molecular-level mechanism showed that the catalyst reactivity varies at different temperature regimes. At temperature below 570 K, the H_2 production rate is higher on the open flat surface, such as Ni(100), than on the stepped surface, e.g., Ni(211). Above 570 K, this trend has been reversed. Detailed analysis of the hydrogen production route also revealed that the contribution from the redox and carboxyl pathways also vary at different facets. This study will enrich our understanding of the hydrogen chemistry at the molecular level, and would also provide theoretical insights into tuning the properties of nanoscale transition metal catalysts by manipulating the surface structure and morphology.

495800 Sorption-Enhanced Water Gas Shift (SE-WGS) Reaction Processes for the Production of High-Purity Hydrogen

Abstract:

The interest of hydrogen, regarded as an alternative energy carrier, has been increasing with expectation of the upcoming hydrogen economy. As one of methods for the bulk hydrogen production, water gas shift (WGS) reaction processes using synthesis gas feeds produced by the gasification of carbonaceous feedstocks has been highlighted. The performance of WGS reaction is significantly increased by applying the sorption-enhanced reaction (SER) concept. In the reaction of sorption-enhanced WGS (SE-WGS) reactions, the WGS reaction and CO_2 removal by sorption are carried out simultaneously in a single reactor. For the successful fulfillment of SE-WGS reaction, adopting the novel materials that can capture high-temperature CO_2, and managing homogenous packing of catalyst and sorbent are crucial. In this study, the SE-WGS reactions were experimentally demonstrated using Cu/ZnO/Al_2O_3 catalyst and double salt-based sorbent pellets. High-purity hydrogen could be directly from the SE-WGS reaction as a result of applying double salt-based CO_2 sorbent, and the effect of the degree of homogeneity for the catalyst and sorbent in a column was further investigated by applying divided section packing concept. The productivity of high-purity hydrogen could be enhanced when more catalyst was packed in the former sections and more sorbent in the later sections.

487615 Aldehyde Water Shift Reaction: Comparative Investigation of Supported Metal Catalysts

Abstract:

There has been significant interest in the use of H_2O as a reactant for biomass conversion given its high concentration in typical feed streams. Steam reforming and 5-hydroxymethylfurfural oxidation are two important examples.[1], [2] The aim of research described in this paper was to investigate the use of H_2O as a terminal oxidant for the partial oxidation of aldehydes. The aldehyde water shift (AWS) reaction oxidizes the aldehyde and produces H_2 (Equation 1).

$CH_3CHO + H_2O \rightarrow CH_3COOH + H_2$ (Eq. 1)

Prior investigations of the AWS reaction have focused on homogeneous catalysts.[3] The commercial implementation of these types of catalysts can be cost-prohibitive. To our knowledge, no heterogeneous catalyst has been reported for this reaction. We explored the use of Mo_2C-, CeO_2-, and Al_2O_3-supported metals (Cu, Pt, Au), the design of which was inspired by that for highly active water gas shift (WGS) catalysts. In particular, the most active heterogeneous WGS catalysts are bifunctional and possess distinct sites for H_2O dissociation and CO oxidation.[4] The target metal loadings were equivalent to 0.1 monolayer. The catalysts were loaded into a quartz flow-through reactor and the reactions were then carried out at 200-280^oC.

The AWS activities (Figure. 1) and product distributions (Figure. 2) were strong functions of the support. The observed AWS activities and activation energies suggest that the rate-limiting step, which is believed to be H_2O dissociation, occurred on the support surface. Among all the catalysts studied, the Mo_2C-supported catalysts possessed the highest AWS activities, with activities at least 5 times higher than those for the other materials. For the Mo_2C-supported catalysts, the admetal did not significantly affect the activity. The Cu/CeO_2 showed the highest AWS activity among the CeO_2-supported materials. Like Mo_2C, CeO_2 is active for H_2O dissociation. The high activity of the Cu/CeO_2 catalyst clearly illustrates the importance of Cu particles for this material. In addition to the AWS reaction, the major side reactions were the Cannizzaro reaction, which produces ethanol and acetic acid, and aldol condensation, which forms crotonaldehyde. The Mo_2C-supported catalysts and Cu/CeO_2 were more than 60% selective to the AWS reaction. Our results suggest that the cooperation between the sites for H_2O dissociation and aldehyde oxidation may have contributed to the high AWS activities and selectivities. In contrast, the other CeO_2- and Al_2O_3-supported catalysts produced primarily crotonaldehyde via the aldol condensation reaction, which appeared to be catalyzed by weak acid sites. These and other results will be presented in the paper.

References

[1] Davda, R. R.; Shabaker, J. W.; Huber, G. W.; Cortright, R. D.; Dumesic, J. A. *Appl. Catal. B Environ.* **2005**, *56*, 171-186

[2] Zhang, Z.; Deng, K. *ACS Catal.* **2015**, *5*, 6529-6544

[3] Brewster, T. P.; Goldberg, J. M.; Tran, J. C.; Heinekey, D. M.; Goldberg, K. I. *ACS Catal.* **2016**, *6*, 6302-6305

[4] Schweitzer, N. M.; Schaidle, J. A. ; Ezekoye, O. K.; Pan, X. ; Linic, S.; Thompson, L. T. *J. Am. Chem. Soc.* **2011**, *133*, 2378-2381

499441 The Selection and Optimization of Heat Carrier Materials to Maximize Liquid Yields and Heat Recovery in Auger Pyrolyzers

Abstract:

Auger pyrolysis reactors are a promising technology in the future of commercial pyrolysis processes. Advantages of auger pyrolyzers over traditional fluidized bed pyrolyzers include the minimal requirement of an inert gas and the lack of sensitive hydrodynamic bed conditions. Additionally, auger pyrolyzers have the flexibility of different modes of heat transfer such as conventional indirect heating or directing heating via heat carrier materials to provide higher heat transfer rates. To mitigate the risk upon scale-up, a better understanding and optimization is required of auger pyrolyzer heat carrier materials. This work focuses on the selection of heat carrier materials and their ability to be recycled within the auger pyrolyzer for maximum product yields and energy recovery.

A twin-screw, lab-scale auger reactor was used to pyrolyze up to 1 kg/hr of red oak at 515°C for all trials. The effect of heat transfer properties on pyrolysis product yields was evaluated for several heat carrier materials. Stainless steel shot, silicon carbide, and sand were investigated. High liquid yields of ~65 wt. % were achieved. Sand was then chosen to further evaluate and optimize heat carrier parameters such as mass flow ratios, particle size, and long-term performance. Two sizes of sand deemed fine (250-600 μm) and coarse sand (600-1000 μm) were evaluated at heat carrier to biomass ratios of up to 20:1. The sand was then recycled at up to 5 recycles producing consistent yields of bio-oil (~60 wt. %), char (~17 wt. %) and gas (~18wt %). Regeneration of the heat carriers was found to be required trial to both remove any carbon residue and maximize heat recovery. The results from this work suggest that the selection and optimization of heat carrier for auger pyrolyzers is important in reaching high liquid yields, while minimizing processing costs attributed to attrition rates and heat recovery.

*Funding for this project was provided by Phillips 66 Company.

494870 A Kinetic Study of the Fast Micro-Pyrolysis of Hybrid Poplar

Abstract:

Hybrid poplar from the clone DN34 was studied to determine the rate of production of bio-oil vapor species as well as gas and char production with respect to time during the fast pyrolysis process. 300-660 ug samples were pyrolyzed using a micropyrolzer at 500ºC at very high heating rates and negligible vapor residence times. Individual poplar samples were run in triplicate at discrete time points ranging from 1-20 seconds in the micropyrolzer. Several bio-oil compounds from each individual sample were analyzed using GC/MS. Select bio-oil compounds, derived from the hemicellulose, cellulose, and lignin fractions of the wood, were used to determine their rates of production from thermal degradation of the biomass. These rates were quantified using standards to determine the weight percent relative to the original raw biomass with respect to time. Pyrolysis kinetic reaction models were fit to this experimental data in order to determine suitability of each model. A first order exponential decay model for degradation of the solid biomass was fit against the data along with a six-step consecutive degradation model previously developed by our group (Klinger et al. 2015). The experimental data suggests that vapor compounds derived from the hemicellulose fraction of poplar are produced at faster rates than that of the lignin or cellulose fractions, consistent with prior data from thermogravimetric analysis (TGA). When applying the first order exponential decay model, the reaction rates calculated did prove that holocellulose compounds reaction rates were approximately twice that of the lignin compounds reaction rate but did not provide the best fit. The six-step degradation model provided a better fit to all of the biooil compounds and the char data, and the stoichiometric parameters derived from the model fit showed that the first three reactions involved mostly the hemicellulose derived compounds while reaction steps 4-6 released cellulose- and lignin-derived compounds.

Reference:

Klinger, J., Bar-Ziv, E., & Shonnard, D. (2015). Unified kinetic model for torrefactionâ€"pyrolysis. Fuel Processing Technology, 138, 175-183.

497410 Experimental Investigation of Tar Recycling in Biomass Gasification

Abstract:

Tar formation is one of the major problems that hinders the development of biomass gasification process as it lowers the utilization efficiency of the syngas products and has negative effect on downstream process equipment. Extensive efforts have been made on tar control and reduction, while limited investigation have been conducted on tar recycling and reusing in biomass gasification. In this work, biomass and tar mixture at different ratios was fed to gasifiers of varied sizes to understand the impact of tar recycling on gasification. The syngas composition, tar production, ash agglomeration and reactor shut down time were studied. The experiment results show that tar recycling upgrades the syngas because of the decomposition of hydrocarbons, while lowers the syngas quality due to the extra moisture involvement in the gasification process. It is also found syngas compositions depend on the biomass/tar ratios and temperatures. The ash agglomeration problem becomes severe after the feeding of tar and makes it harder to control the reactor.

499239 A Convolutional Neural Network Model for Biomass Gasification in Fluidized Bed

Abstract:

Biomass is a renewable resource and its utilization has received great attention due to its life cycle carbon-neutrality and the potential to substitute fossil fuel to produce a variety of energy-related products. Thermochemical gasification is an important route for conversion of biomass that results in a product gas consisting H2, CO, CO2, CH4 and other light hydrocarbons that can be used as fuel gas to generate power or as well as raw material to produce variety of chemicals. Among the existing gasifiers, fluidized beds offer many advantages such as high conversion efficiency and great flexibility regarding feedstock type and size.

Modeling and simulation of FB gasifier is a problem of considerable theoretical and computational difficulty mainly due to presence of several chemical and physical processes in the reactor and the lack of proper characterization of solid fuel. The largest challenge of the existing modeling works is lack of generality and predictive capability. Here, we present a neural based (black-box) model that can predict the composition of product gas, amount of tar and solid conversion under a wide range of operating conditions and different types of feedstock.

Initially in this study, elemental composition (C, H, N, O, S), proximate analysis (Volatile matter, Fixed carbon, Ash), structural component (Cellulose, Hemicellulose, Lignin) and ash composition of 35 types of biomass were analyzed using principal component analysis (PCA), Factor analysis (FA), and a variety of machine learning approaches. A new method of biomass characterization is proposed based on composite statistical ensembles to improve understanding of nature of biomass.

Subsequently, over 200 data sets of biomass gasification in fluidized bed were collected, structured and analyzed. Multiple measures of associations between product gas characteristics, biomass constituents, and operating condition were analyzed to detect underlying features of the system that were most useful in developing the predictive model. The architecture of the neural network was founded based on the subjective knowledge of the process of biomass gasification in fluidized bed and statistical inferences of the data. The model was trained using the Bayesian regularization technique that is known for yielding more general models. It should be noted that the outputs of BBM are designed to only be used as constitutive equations and in the last step the equilibrium formulation is used as the physical framework of the model to explicitly close the mass and energy balance.

501642 Upgrading Biomass Using Low Temperature Plasma

Abstract:

Linear hydrocarbons are not easily obtained from biomass, even after thermochemical deconstruction and upgrading. The established thermochemical methods of deconstructing biomass include pyrolysis and gasification, which provide possible routes to hydrocarbons. The resulting products of biomass thermochemical deconstruction can then be upgraded by catalytic reactions such as hydrodeoxygenation of pyrolysis-oil or Fischerâ€“Tropsch reaction of biomass-derived syngas.Â A major challenge with these routes is that they produce highly heterogeneous product mixtures with low selectively. Catalyst fouling and inactivation is also problematic. A single-step process, which does not require metal catalysts, to produce oxygen-free hydrocarbons with high selectivity would be attractive. In this talk, we will report on our recent work to upgrade lignocellulosic biomass into deoxygenated hydrocarbons in a single step using a catalyst-free, low temperature plasma process. Effects of process parameters such as plasma power, pressure, gas composition (e.g. H_2 fraction), and biomass feedstock composition (e.g. lignin vs. cellulose) will be discussed. The evidence demonstrates that the plasma is capable of rapidly generating volatile deoxygenated hydrocarbons with high yield and selectivity without the need for catalysts. A preliminary technoeconomic assessment will be presented.

500694 Cellulose Pretreatment and Dissolution: Selection of Solvent and Processing Conditions

Abstract:

Efficient utilization of biomass is hindered by the recalcitrance to dissolution of semicrystalline cellulose. Pretreatment is often used to alter the structure of cellulosic biomass in order to make cellulose more accessible to solvents and enzymes. The pretreatment involves physical and/or chemical processing which affects the degree of crystallinity and size of biomass particles. We examine here the effects of (i) solvent properties, pretreatment steps and temperature, and (ii) fiber diameter and degree of crystallinity, on the kinetics of cellulose swelling and dissolution. To this end we have combined (a) experimental results on cotton fiber swelling, change in crystallinity and dissolved amount when treated under different solvent conditions, with (b) a phenomenological model that accounts for the phenomena governing the dissolution of solid cellulose, e.g., solvent penetration, transformation from crystalline to amorphous domains, specimen swelling, and polymer chain untangling. [Ghasemi, M.; Singapati, A. Y.; Tsianou, M.; Alexandridis, P., Dissolution of nanostructured polymer fibers: Numerical modeling and parametric analysis. AIChE Journal 2017, 63 (4), 1368-1383. DOI: 10.1002/aic.15615] The insights obtained from this analysis would facilitate the rational selection of solvents and the design of pretreatment processes to reduce the size and degree of crystallinity of cellulosic biomass particles, leading to enhanced biomass utilization. [Ghasemi, M.; Tsianou, M.; Alexandridis, P., Assessment of solvents for cellulose dissolution. Bioresource Technol. 2017, 228, 330-338. DOI: 10.1016/j.biortech.2016.12.049]

496185 Ionic Liquids: Solvent Design for Diverse Sustainable Biomass Valorisation

Abstract:

The processing of residual lignocellulosic biomass within the biorefinery frames impelled the use of green innovative technologies aiming at sustainable valorisation of this feedstock. The main constrain is related to the complex and recalcitrant macromolecular three-dimensional structure of biomass. The efficient and selective pre-treatment of biomass into main constituents i.e. cellulose, hemicellulose and lignin opens a possibility to produce diverse biofuels, biochemicals and biomaterials as it happens in fossil refinery. Ionic liquids (ILs) are one of the most promising solvents which offers greener solutions in biomass conversion technologies into a great variety of products.[1] The present work discloses two different scenarios of lignocellulosic biomass (wheat straw) processing with ILs, where i) alkaline or ii) acidic character of ILs remark biomass fractionation and subsequent product formation.

ILs with alkaline character demonstrated ability to dissolve the entire biomass allowing fractionation of each single fraction of biomass. Such approach produced high-purity fractions of cellulose, hemicellulose and lignin and IL was recycled and reused with success confirming the economic and environmental feasibility of the proposed method.[2] Besides, a fraction rich in phenolic compounds was also produced by employing polymeric resins and supercritical CO_2 ($scCO_2$) extraction. Silica C18 resin showed the highest extraction efficiency of phenolic compounds, while Amberlite XAD-7 was the most selective one. Nevertheless, a residual quantity of IL was found in the extracted samples and further purification was achieved by applying $scCO_2$ extraction. An extract containing vanillin, tricin, syringaldehyde, *p*-coumaric and vanillic acids, was produced with no IL presence.[3] The discussed approach is presented in Figure 1.

Figure 1. Biomass fractionation after pre-treatment with [emim][OAc] and further extraction and purification of phenolic compounds.

Another strategy of biomass valorisation was based on selective removal of the hemicellulose fraction of biomass solely using acidic ILs, i.e. 1-butyl-3-methylimidazolium hydrogen sulphate ([bmim][HSO_4]) and 1-ethyl-3-methylimidazolium hydrogen sulphate ([emim][HSO_4]). The acidity of [HSO_4]-based ILs hydrolyses hemicellulose and produce pentoses or furfural leaving a solid fraction constituted by cellulose and lignin for further valorisation in the perspective of the biorefinery concept.[4] The capacity of these ILs to hydrolyse and/or convert biomass into pentoses and furfural can be attained by controlling the reaction conditions, such as temperature, residence time and water content. A maximum pentose yield of 81.6 mol% was obtained after processing of wheat straw at 141 °C during 1.5 hours with [emim][HSO_4]/water mixture containing 58 wt. % of water. The obtained result is for now, far the best result reported in literature for biomass processing with ILs.

The present work demonstrates the great versatility of ILs to assist biomass processing for the manufacture of a wide range of products. Particularly, the ILs chemical character governing pathways in the fractionation of biomass allowing designing specific strategies towards the production of target chemicals.

Acknowledgments

IF/00471/2015 grants and by BBRI - Biomass and Bioenergy Research Infrastructure (ROTEIRO/0189/2013) and the Associated Laboratory for Sustainable Chemistry-Clean Processes and Technologies-LAQV which is financed by national funds from FCT/MEC (UID/QUI/50006/2013) and co-financed by the ERDF under the PT2020 Partnership Agreement (POCI-01-0145-FEDER-007265). This work was done in the frame of COST EUBIS TD1203 action.

References

1. R. Bogel-Lukasik, *Ionic Liquids in the Biorefinery Concept*, RSC, Cambridge, UK, 2015.

2. S. P. Magalhães da Silva, A. M. da Costa Lopes, L. B. Roseiro and R. Bogel-Lukasik, *RSC Adv.*, 2013, **3**, 16040-16050.

3. A. M. da Costa Lopes, M. Brenner, P. Fale, L. B. Roseiro and R. Bogel-Lukasik, *ACS Sustain. Chem. Eng.*, 2016, **4**, 3357-3367.

4. A. V. Carvalho, A. M. da Costa Lopes and R. Bogel-Lukasik, *RSC Adv.*, 2015, **5**, 47153-47164.

490695 Pretreatment of Biomass and Coal with Ionic Liquids for Advanced Fuel and Chemical ProductionÂ

Abstract:

Title: Pretreatment of Biomass and Coal with Ionic Liquids for Advanced Fuel and Chemical Production

Authors: C. Luke Williams[1], Chenlin Li[2], Jared C. Allen[1], and Karen Delezene-Briggs[3],
(1) Biofuels and Renewable Energy Technology, Idaho National Laboratory, Idaho Falls, ID
(2) Chemical and Radiation Measurement, Idaho National Laboratory, Idaho Falls, ID
(3) Biological and Chemical Processing, Idaho National Laboratory, Idaho Falls, ID

The United States currently faces major energy security challenges due to increasing energy demand. To reduce imports of foreign energy and fuels the development of domestic energy and materials sources, from a variety of carbon-based feedstocks, will be required. However, to efficiently transform nationally available resources, like biomass and coal, to energy and fuels will require both advanced processing techniques and catalytic upgrading. Ionic liquids are a class of solvent capable of altering the structure of many carbonaceous materials at low temperatures and atmospheric pressures. Due to the relatively mild conditions required ionic liquid pretreatment can be integrated with many advanced fuel/chemical processing techniques. In this work multiple types of biomass and several grades of coal are treated with ionic liquids. After treatment, the physical and chemical alterations in the materials are investigated using: scanning electron microscopy, X-ray diffraction, compositional analysis, enzymatic hydrolysis, and thermogravimetric analysis. It can be seen that ionic liquid pretreatment has a drastic effect on the catalytic activity of enzymatic hydrolysis for the production of biomass derived sugars, as well as altering the combustion properties of coal. Overall, ionic liquid pretreatment has the potential to improve the catalytic production of energy and materials from both biomass and fossil feedstocks by altering material structures.

499207 *In situ* Raman Microscopy to Monitor Changes in Cellulose Crystallinity During Acid Pre-Treatment

Abstract:

Dilute acid pre-treatment is a common technique for decreasing lignocellulosic biomass recalcitrance. During this process hemicellulose is removed, providing structural access for enzymes to hydrolyze cellulose. The effect of acid pre-treatment processes on the cellulose structure is well understood, in particular the change of crystallinity, partly due to uncertainty from sample preparation techniques carried out prior to *ex situ* analysis. However, cellulose crystallinity is the primary determinant of its recalcitrance to hydrolysis. Raman microscopy has been used extensively to study cellulose structural changes. Furthermore, it is a convenient and versatile tool allowing *in situ* analysis under hydrothermal environments, as water is mostly Raman transparent. Thus, understanding how cellulose crystallinity changes *in situ* would allow for improvement of pre-treatment processes by optimizing temperature and acid concentration of treatment.

In this work, we designed and built a custom view-cell reactor capable of operating at conditions of up to 220 °C and 30 bar to investigate cellulose structural changes under pre-treatment conditions. The crystallinity of ball-milled cellulose increased during the first 60 min of dilute acid pretreatment at 150 °C. After 60 min, the increasing crystallinity appeared to cease, coinciding with the appearance of a new band at 1600 cm^{-1}. To determine whether cellulose decrystallizes under pre-treatment conditions, highly crystalline cellulose was thermally treated in dilute acid. The crystallinity appeared to increase up to a temperature of 160 °C, after which it appeared to decrease. These observations could be attributed to the preferential hydrolysis of amorphous regions in the cellulose structure resulting in an increase in the relative amount of crystalline regions.

To verify the observed increase in crystallinity during dilute acid hydrolysis we carried out control tests in hydrothermal conditions in the absence of acid. At such conditions the hydrolysis rate should be rather low compared to dilute acid. Partially decrystallized cellulose started to fluoresce relatively early, within 20 minutes of reaction time at 150 °C preventing crystallinity analysis. However, similarly to reactions under acidic conditions, a new band at 1584 cm^{-1} appeared within 20 minutes of reaction time. Cool-down and washing of the cellulose with water did not completely eliminate the fluorescence and the new band at 1584 cm^{-1}, suggesting partial water solubility and surface degradation. In the absence of acid, the crystallinity of highly crystalline cellulose appeared to be relatively stable, which is consistent with low hydrolysis rate in the absence of acid.

To understand the nature of the new bands formed during acidic and hydrothermal hydrolysis, we employed IR spectroscopy. We heated cellulose in nitrogen environment and new bands appeared consistent with formation of C=O and C=C at temperatures above 230°C. Further, we compared a Raman spectrum of hydrothermal char, produced by degradation of glucose to the Raman spectra of cellulose under dilute acid and hydrothermal conditions. The G band of the hydrochar appeared at 1582 cm^{-1} similar to our observations at 1584 cm^{-1} as observed in hydrothermal treatment. These results suggest formation of degradation compounds such as HMF and humins during dilute acid and

hydrothermal hydrolysis at temperatures as low as 150 °C.

In situ Ramans studies of the ball-milled and highly crystalline cellulose further understanding of the changes of cellulose structure that occur during dilute acid pre-treatment.

(748e) Hydrochloric Acid Pretreated-Digested Carica Papaya Petioles, Towards Holistic Biogas Potential Assessment Frameworks

This presentation describes research that explores the hypothesis that simple feedstock pretreatment-digestion methods can be applied to rapidly evaluate biochemical biogas potential of anaerobic digestion in resource constrained development. Biomass feedstock, *i.e. Carica papaya* petioles were obtained from Edo State, Nigeria. Samples were analyzed with Fourier transform infrared spectroscopy after hydrochloric acid pre-treatment towards characterization of spectra with characteristic absorbance ratios indicative of physicochemical changes in biomass that include cross-linked lignin (CLL) ratio, the lateral order index (LOI), and hydrogen bond intensity (HBI) (Tables 1 & 2).

Wavenumber cm^{-1}	Raw	1%	3%	5%
897	36.1	55.4	61.5	65.5
1323	45.3	64.1	71.2	70.8
1424	43.9	60.0	66.9	67.5
1508	39.7	55.0	62.1	63.7
1599	48.0	58.8	66.3	67.6
3401	51.6	72.5	78.7	74.3

	Raw	1%	3%	5%
LOI	1.2	1.1	1.1	1.0
HBI	1.1	1.1	1.1	1.0
CLL	0.8	0.9	0.9	0.9

Figure 1 (left): Absorbance of wavenumbers associated with biomass indices indicative chemical nature of biomass for raw Carica papaya petiole, 1%, 3% and 5% hydrochloric acid pretreated samples. Wavenumber assignments: 897 cm^{-1} absorbance, i.e. glycosidic bond β-(1,4) in cellulose, 1420 -1430 cm^{-1} associated with crystalline structure of cellulose, 3400 cm^{-1}, i.e. O–H stretching H-bonds between molecules, 1320 cm^{-1}, i.e. CH rocking vibration of glucose rings, 1600cm^{-1} i.e. condensed lignin guaiacyl residues and 1508cm^{-1}, i.e. cross-linked structures of lignin guaiacyl residues.

Figure 2 (right): Summary of chemical bond ratios of Carica papaya petioles at respective pretreatment acid concentrations

Cumulative biogas yields from samples were fit to the modified Gompertz form to derive mathematical parameters related to biological reactor performance. To date, optimum sample treatment suggests that a 150% increase in biogas yield compared to non-pretreated *Carica papaya* petioles is possible. Overall, results suggest that rapid biogas potential assessment can be performed in a few days-time and provides for a visual means to characterize individual digestions in constrained research setting and collaborations (Figure 3).

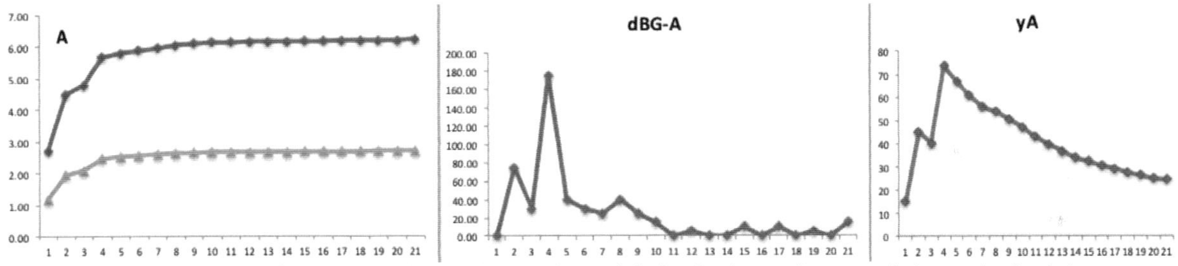

Figure 3: After selecting the Gompertz form, gas production data can be modelled either by graphing cumulative production in log or natural log forms (top and bottom lines of LEFT column), by change in the daily production (MIDDLE column), and cumulative production averaged by day of production (right column). Visual discerning maximum production (LEFT column), days of maximum production (MIDDLE column) and periods of maximum productivity (RIGHT column) are apparent.

The research reported is collaborative, in support of larger sustainability goals, and the transfer of knowledge from National lab settings such as the Department of Energy's National Renewable Energy Laboratory to biofuels research and development efforts. Overall, a goal is to assess biofuels production in more complex or integrated socio-economic models (Figure 4)

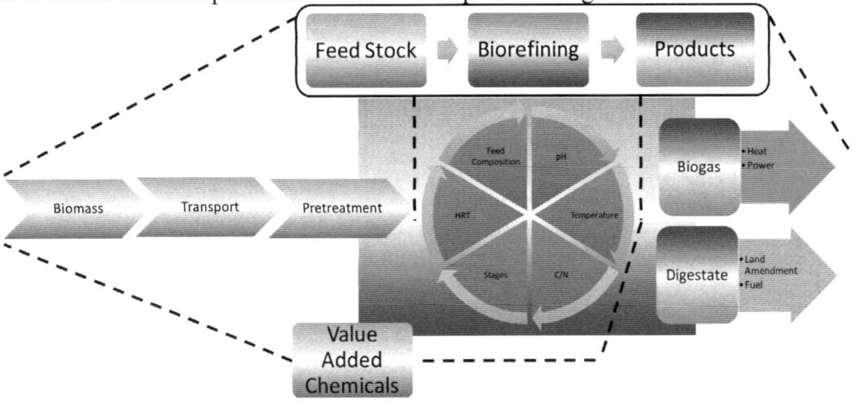

Figure 4: A framework that considers the complexity of biogas production. We propose that there are additional considerations such as: 1) natural value added chemicals, 2) socio-economic considerations, and 3) community appropriate, rural, or smaller scales to consider in addition to the traditional process specific considerations.

500004 Supercritical CO_2 Pretreatment of Wheat Straw: Hydrolysis Performance, Enzymatic Yields and Comprehensive Mass Balances

Abstract:

Wheat straw, with an estimated annual production over 170 million tons,[1, 2] has been considered the most promising agricultural residue for the production of bio-products and 2nd generation bioethanol in Europe. The processing of wheat straw into fermentable sugars represents a great challenge, mainly due to its complex and highly recalcitrant architecture. In this work, the conversion of wheat straw into pentoses using supercritical CO_2 followed by an enzymatic saccharification of cellulosic fraction is for the first time investigated. This technology benefits from the *in-situ* formation of carbonic acid. Without an additional chemical, this technology outperformed liquid hot water pretreatment allowing to produce total sugars yield as high as 84 % in comparison to 67 % with liquid hot water (and under more severe reaction conditions). Both chemical and physical effects of supercritical CO_2 resulted in a glucan to glucose conversion yield of 82 mol% representing a 26 % improvement over those achieved in liquid hot water.[3] Firstly, the effect of various process conditions, such as temperature, residence time and initial CO_2 pressure on conversion of hemicelluloses into pentoses (in both oligomeric and monomeric form) were scrutinised in an effort to define the optimal operational conditions. Then, several material characterisation techniques were used to investigate thoroughly the physical influence of supercritical CO_2on the processed solids.

Acknowledgment

IF/00471/2015 grants and by BBRI - Biomass and Bioenergy Research Infrastructure (ROTEIRO/0189/2013) and the Associated Laboratory for Sustainable Chemistry–Clean Processes and Technologies–LAQV which is financed by national funds from FCT/MEC (UID/QUI/50006/2013) and co-financed by the ERDF under the PT2020 Partnership Agreement (POCI-01-0145-FEDER-007265).

References

1. S. Kim and B. E. Dale, *Biomass Bioenerg.*, 2004, **26**, 361-375.

2. D. Montane, X. Farriol, J. Salvado, P. Jollez and E. Chornet, *J. Wood Chem. Technol.*, 1998, **18**, 171-191.

3. A. R. C. Morais, A. C. Mata and R. Bogel-Lukasik, *Green Chem.*, 2014, **16**, 4312-4322.

494899 Dilute Alkali Flowthrough Pretreatment of Softwood

Abstract:

Alkali pretreatment effectively disrupts biomass plant cell wall. In this study, dilute alkali flowthrough pretreatment of softwood was investigated at 200-270 °C for 2-10 min at a flow rate of 25 mL/min, under pH of, 8, 9, 11, and 12, respectively, comparing with water flowthrough pretreatment at pH 6.4. The composition analysis of pretreated solid residues and the recovered liquid streams showed the removal profile of biomass components and the recovery profile of sugars/chemicals as a function of pretreatment severity. 2D ^{13}C-^{1}H NMR analysis revealed that complete solubilization of biomass under high pH conditions resulted in pretreated liquid stream rich in various compounds, including monomeric and oligomeric aromatics, and carbohydrates degradation products. Results indicated that initial pH greatly affected the biomass degradation pathways and efficiency. The plausible mechanism of biomass degradation by dilute alkaline under various pretreatment severities was proposed.

498634 Production of Renewable Hydrogen from Wastewater Using Microbial Electrochemical Systems

Abstract:

Hydrogen is one of the key starting materials used in the chemical industry. Large amounts of hydrogen are also required in biorefinery schemes due to the high O/C and low H/C ratio of biomass. The H2 industry in the USA is large, but virtually all of the H2 today comes from reforming fossil fuels. Capturing renewable H2 stored in organic-rich wastewater would provide a new source of H2 that would also accomplish the wastewater treatment. Dark fermentation, a process that uses fermentative bacteria to convert organic matter to H2 without the presence of light, can achieve relatively high H2 production rates. However, the H2 yield is very low due to the formation of large quantities of side products. Microbial electrolysis cell (MEC) overcomes the fermentation barrier and provides a new approach for generating H2 from biomass. However, the key challenges for realizing the practical applications of MECs, especially for wastewater treatment include: 1) low H2 production rate; and 2) H2 uptake by H2 consumers, such as methanogens and homoacetogens. In this presentation, we will present our recent development of a hybrid microbial electrochemical system that combines the strengths of fermentation and MEC processes in one reactor for efficient H2 production. We will also present some low-cost strategies we developed against methanogens and homoacetogens for enhancing H2 recovery in the hybrid system. Future works, including scaling up and techno-economic analysis, will also be discussed.

500058 Scale-up and Process Intensification of Agricultural and Municipal Solid Waste Conversion Using Ionic Liquid Based Process

Abstract:

The sustainable production of both biofuels and chemicals depends largely on access to efficient and cost-effective biomass conversion technologies. Among the various pretreatment technologies being developed, ionic liquid (IL) pretreatment has been receiving significant recent attention as a potential "greener" process that enables fractionation of a wide range of feedstocks due to its high biomass processing efficiency. This presentation will describe recent advances in IL process development that we have demonstrated in the conversion of different types of municipal solid waste blends to fermentable sugars. We will highlight two integrated IL technologies: acidolysis with chloride based IL pretreatment and one-pot biomass conversion using a "bionic liquid" (amino acid based IL). For the acidolysis process, an acid catalyst was used to hydrolyze polysaccharides to monosaccharides following IL pretreatment, which provides an enzyme-free process with significant reductions in processing times and material costs. The one-pot biomass conversion combines the pretreatment and enzymatic saccharification into one vessel, eliminating solid-liquid separation and extensive water-washing steps with the potential to offer better economics, scalability and environmental sustainability. Process scalability and biomass conversion efficiency along with the feedstock cost and quality will be presented and discussed, as well as identifying and overcoming the remaining technical and economic challenges of these approaches to provide perspectives and opportunities for future process development and optimization.

497641 Map of Our Scrap: A GIS-Based Analysis on the Potential of Food Wastes for Bioenergy Production

Abstract:

The growth of our cities has been accompanied by the inevitable production of wastes. With the continuous increase in the amount of our wastes, their utilization has been initiated all over the world. One of the attractive utilization feedstock is food waste (FW) due to the capability of communities to segregate and direct this waste vector to established conversion technologies such as anaerobic digestion. Anaerobic digestion produces biogas consists of methane (CH_4), hydrogen (H_2), and carbon dioxide (CO_2), which are high energy-density chemicals. The digestion of food wastes typically produce 0.198 kWh/kg food waste [1]. As of now, the best sites for the installation of biogas-to-power turbine-generators are wastewater treatment plants (WWTPs) [2]. The transport of FW to WWTPs, however, consumes fuel, which may be counter-productive to a waste-to-energy initiative.

This work evaluated the energy balance associated with the digestion of food wastes at a nearest WWTP transported from the following key sources: supermarkets, hospitals, restaurants, colleges, and schools, using geographical information system (GIS) platform. The study area was Lafayette, Louisiana. The analysis was done in QGIS (Version 2.18, a.k.a. Las Palmas).

The preliminary results support the following inferences.

- This GIS-based analysis for the potential of FW as feedstock for bioenergy production showed the need for additional organic wastes to be combined with FW to possibly increase the power generation of WWTP-based biogas-to-energy turbine generators.
- Wastewater sludges produced in significant amount during water treatment in WWTPs are viable candidate organic wastes that may significantly increase biogas production. These sludges do not require transportation, hence, omitting the energy expenditure through transportation fuel.
- FWs from restaurants may significantly increase the potential bioenergy production, which may supply a utility-scale (1 MW) power plant in the city.

This city case study may be adopted to a state-wide analysis to evaluate the bioenergy potential of FW via anaerobic digestion for the state of Louisiana, USA.

Citations:

[1] Moriarty, K. (2013) Feasibility study of anaerobic digestion of food waste in St. Bernard, Louisiana. NREL, Doc. TP-7A30-57082

[2] Report: Massachusetts Department of Environmental Protection (2002) Identification, Characterization, and Mapping of Food Waste and Food Waste Generators In Massachusetts

497451 Biochar from Excelsior Residues for Plantation Production of Hybrid Poplar

Abstract:

Excelsior is a long fiber material made from wood that was used in many packing applications until the rise in the use of expanded plastics. Today, a common use of excelsior is in natural erosion control materials on construction sites. Residues from excelsior production include bark, excelsior fines, and small bits of other materials used with wood in the erosion control blankets and rolls. In this project, we investigate the potential of a three-way material pathway: making excelsior from plantation-grown hybrid poplar (instead of forest-harvested aspen), producing biochar and process heat through slow pyrolysis of excelsior residues, and using the biochar to amend soils within the hybrid poplar plantation to improve productivity. In the Southwestern U.S., many soils have elevated pH, so there is an opportunity to tailor biochar production for properties that would be advantageous for these soils, such as biochars with lower pH and lower ash contents. Results are presented from the first hybrid poplar growth study on soil amended with excelsior residue-derived biochar, as well as our initial findings related to the feasibility of the hybrid poplar-excelsior-residue biochar material pathway.

498725 Co-Pyrolysis of Plastics and Biomass Waste

Abstract:

Fast pyrolysis has been demonstrated previously for conversion of mixed waste constituents such as pine, sugarcane bagasse, corn stover, various plastics, and mixtures thereof into value-added liquid bio-oil fuel, biochar, and combustible gases. This process reduces the waste stream volume by up to 98% with low emissions and minimal impact on the environment. In an effort to understand the interactions between biomass and fossil-derived plastics, several pine/polystyrene(PS) binary mixtures were pyrolyzed in a bench-scale (1 kg/h), continuous, bubbling fluidized bed reactor at temperatures between 500 °C and 650 °C. Bio-oil properties from pyrolysis of pine/PS mixtures of 70%/30%, 80%/20%, and 90%/10% by mass were measured to determine the effect of plastic concentration on heating value, moisture content, and viscosity. Fractional condensation of heavier and lighter bio-oil fractions was demonstrated with two condensers in series, which allowed for the condensing temperature to be varied for collecting different oil fractions. The fluidizing gas temperature, residence time, reactor bed temperature, and condensing temperatures were varied to determine the effects on collection efficiency and bio-oil product characteristics such as heating value, viscosity, and water content. Samples of the bio-oil product were collected at each condenser and analyzed using gas chromatography coupled with mass spectrometry (GC/MS). Addition of plastics to the feedstock appears to have prompted a shift in bio-oil composition that indicates some interaction between the PS and pine pyrolysis intermediates. Significantly more aromatics were formed with the mixtures than when pine was pyrolyzed alone. The lower temperature fractions contained an abundance of the styrene monomer while the higher temperature fractions contained significantly heavier aromatics, including tolylacetic acid, bibenzyl, Î±-methylbibenzyl, and 6-phenyl-2-hexenal. Future work is focusing on catalytic upgrading of the heavier and lighter bio-oil fractions.

499711 Recovery of Rare Earth Elements from Coal Ash with a Recycling Acid Leach Process

Abstract:

Recovery of Rare Earth Elements from Coal Ash with a Recycling Acid Leach Process

Rick Peterson1, Mike Heinrichs1, Rachid Taha1, Darwin Argumedo1, Justin Glier1, Slawek Winecki1

1Battelle Memorial Institute, Energy Resource Group, 505 King Avenue, Columbus, OH 43201

KEYWORDS: rare earth elements, coal ash, coal liquefaction

ABSTRACT

Rare earth elements (REE) have a wide range of uses in catalysis, alloying, magnets, and optics, and have become critical for the batteries, motors, and generators that enable sustainable technologies and infrastructure. Despite these critical applications, the United States has developed very little in the way of REE resources, with China producing more than 80% of the global supply of the elements1. A technology that will allow for the economic recovery of REE from coal will reduce dependence on foreign supplies and ensure US capability for renewable and defense products.

1 U.S. Geological Survey, "Mineral Commodity Summaries 2017," 19 January 2017. [Online]. Available: https://minerals.usgs.gov/minerals/pubs/mcs/2017/mcs2017.pdf.

Battelle has investigated the application of its Acid Digestion Process (ADP) to recovery of REE from coal ash produced by pulverized coal combustors (PCC), fluidized bed combustors (FBC), and Battelle's bio-based direct coal liquefaction process. The ADP was shown to recover over 90% of the nitric acid used in the process for recycle, greatly reducing chemical consumption costs. This presentation will cover results of a sampling and characterization effort for varying coal ash feedstocks, as well as a feasibility study that assessed the economics of the ADP process when considering available sources.

The sampling and characterization portion of the project analyzed coal byproducts to allow selection of a suitable feedstock for Battelle's process, and selection of key target materials for recovery to drive the process economics. Within the REE in coal, dysprosium (Dy) represents the highest value, and combined with neodymium (Nd), terbium (Tb), and praseodymium (Pr), these four elements account for over 75% of the REE value in the ash. Accordingly, Battelle's process will be tailored to focus on these REE components. Scandium (Sc) represents the highest overall value within the ash, and will be a primary byproduct. Vanadium (V), yttrium (Y), lithium (Li), and cobalt (Co) also represent a significant value, and will be evaluated as potential byproducts going forward.

Based on the analytical results, the feasibility study focused on a specific PCC fly ash as a feedstock for the ADP. The PCC plant fly ash had a higher Total REE+Y+Sc concentration than

all other operating plant ashes sampled, at 545 ppm +/- 13.4 ppm, as well as a higher Heavy REE/Light REE ratio at 0.37 +/- 0.01. Additionally, it contains significant amounts of scandium (36 ppm +/- 1.4 ppm), vanadium (279 ppm +/- 12 ppm), yttrium (104 ppm +/- 5.3 ppm), cobalt (44 ppm +/- 2.5 ppm), and lithium (~166 ppm), which can be valuable process byproducts with robust market outlets. Although the coal liquefaction ash had a greater concentration of total REE, there is some risk for implementation since Battelle's liquefaction process is not yet commercialized. The low temperature ash appeared to be more leachable, but it also demonstrated lower selectivity due to the high calcium content common with FBC systems. Furthermore, the overall capacity of FBCs is much lower than PCCs. Therefore, there would be more risk in feedstock sourcing.

Feasibility testing compared leaching efficiency and selectivity of PCC ash with samples of liquefaction ash and FBC ash to ensure and validate assumptions regarding leaching kinetics, acid loading limits, and product purities. This information was used to refine a process model built in CHEMCAD modeling software and fed into overall process economics and unit sizing used in the technoeconomic assessment.

The feasibility study focused on construction of a REE recovery plant sized to process up to 30 tonnes of ash per hour, which is approximately the ash production rate of a 2,000 MW power plant operating at full capacity.

Operating costs for both a First-of-a-Kind and an Nth-of-a-Kind plant were determined. These costs to recover rare earths from coal ash were then compared to the value of rare earths present in a variety of coal ashes around the country. Values for coal ash were obtained through determination of the concentrations of rare earth elements in coal samples adjusted to account for changes in concentration associated with the combustion process. Current market pricing information for REEs was used to determine the value associated with each of the coal sources.

The results of the feasibility study indicate that when a First-of-a-Kind plant is considered, approximately 5% of U.S. coal sources contain sufficient rare earth material to cover the costs of the recovery plant. When an Nth-of-a-Kind plant is considered, over 20% of all coal sources reviewed contain sufficient recoverable rare earths to cover recovery costs at current REE market prices. When historical high REE prices are considered, more than 25% of coal sources can be treated with a First-of-a-Kind plant, and more than 47% of sources with an Nth-of-a-Kind plant. These results are based on the current understanding and developmental status of the recovery process. It should be noted that further downstream purification costs are not included in this assessment. A probabilistic sensitivity analysis was conducted which examines how the cost to recover rare earth materials may change due to technology developments, uncertainties in capital requirements for the plant, or prices of REEs due to shifting market dynamics. Under certain sensitivity scenarios, the recovery plant can be economically viable processing an even greater range of coal sources.

This project is ongoing, and the presentation will also cover pending results of experiments designed to improve REE leaching efficiency, produce additional valuable byproducts, and upgrade the REE concentration in process products. Additionally, it will briefly address the design of a continuous bench scale ADP for REE recovery from coal ash.

This project was performed with funding from the Department of Energy's National Energy Technology Laboratory (DE-FE0027012) and the Ohio Coal Development Office (OER-CDO-D-15-10)

498633 Brief Overview of Rare Earth Research at NETL R&ic

Abstract:

Everything that is in the Earth's crust is also present in coal to some extent, and the challenge is always to use abundant domestic coal in clean and environmentally friendly ways. Trace elements present within the coal such as mercury, arsenic, selenium, cadmium, antimony, phosphorus, and chlorine, as well as sulfur and nitrogen, have long been of interest due to their roles in producing pollution emissions from combustion or gasification plants. Critical elements, such as some of the rare earths present within the coal, have recently been of great interest due to their importance in the world's economy.

Coal is an important resource, both in the United States and around the world. The United States generates approximately 30 percent of its electricity through coal combustion and, at the current overall rate of consumption, has more than a 250-year supply of coal. According to the US Energy Information Administration, approximately 1 Gt of coal has been mined annually in the United States over the past twenty-five years (1990-2014), although the 2015 total was 896 million tons, and 2016 was 739 million tons. The recent decline in domestic coal production is due to the availability of inexpensive natural gas; regulatory uncertainty with regards to carbon dioxide emissions; and the retirement of older coal-fired power plants.

In the United States, most coal is burned for power generation, but substantial quantities are also used in the manufacture of steel, chemicals, and activated carbons. Numerous industries, such as mining, power, rail transportation, manufacturing, chemical, steel, activated carbon, and fuels, are involved in the production, transportation, and use of coal.

Rare earth elements (REEs), which comprise the lanthanide elements, as well as scandium and yttrium, are present in the abundant coal and coal by-products produced domestically and worldwide. Widely used in high-technology products such as catalysts, cell phones, hard drives, hybrid engines, lasers, fluorescent lamps, batteries, magnets, medical devices, and televisions, REEs are of significant value to US national security, energy independence, economic growth, and the country's environmental future.

Most of the common inorganic lanthanide compounds, such as the phosphates found in coal, have very high melting, boiling, and thermal decomposition temperatures, allowing them to concentrate in combustion and gasification by-products. Rare earths are commercially produced from ores containing monazite (rare earth phosphate mineral) or bastnasite (rare earth carbonate-fluoride mineral), as well as from ion exchangeable clays. The lanthanides are found in coal as well as combustion by-products such as ash, coal preparation residues, gasification slags, and mining by-products. Rare earths have also been found in interesting concentrations in the strata above and below some coal seams, thereby making every process in the mining and utilization of coal a potential source of rare earth elements.

Also of interest, some coal and coal by-products have elevated concentrations of heavy rare earths (HREEs) —which are lowest in supply, rank high in criticality and price, and are projected to increase in demand— making them potentially attractive targets for REE recovery despite their overall lower concentration.

The National Energy Technology Laboratory's Research and Innovation Center (NETL-RIC) recently initiated research to support the measurement of concentration, identification of rare earth compounds, and recovery of rare earths from abundant domestic coal by-products. The NETL Rare Earth EDX Database (https://edx.netl.doe.gov/ree/) is a resource for rare earth information related to coal and by-products. Many other research organizations have also initiated efforts for the analytical characterization and recovery of rare earths from unconventional sources such as coal by-products.

Much of the recent research on coal use in the United States has focused on the capture of pollutants such as acid gases, particulates, mercury, and carbon dioxide. The possible recovery of rare earth and other critical elements from abundant coal and by-products is an exciting new research area and represents a dramatic paradigm shift for coal. Some of the early exciting results on the occurrence and recovery of lanthanides from abundant coal by-products will be discussed.

497530 Extraction of Rare Earth Elements from Fly Ash Using NaOH Hydrothermal and Ultrasound Pretreatment

Abstract:

Extraction of valuable and critical trace elements from coal and coal byproducts such as fly ash has been studied for over 100 years. Recently the volatility of rare earth element (REE) prices, and the closing of the only rare earth mine in the U.S., has renewed interest in trace element and REE recovery from abundant domestic coal and fly ash. Acid leaching, using 2M HCl to extract REE from a fly ash was studied with less than 10% total REE extraction. However, when a sodium hydroxide hydrothermal pretreatment was employed, a dramatic increase in REE extraction was observed. The sodium hydroxide pretreatment greatly increased the REE extraction to approximately 90% in the HCl acid leach. Sonication experiments using NaOH solution in combination with hydrothermal pretreatment were also examined and will be discussed.

490447 Recovery of Rare Earth Elements from North Dakota Lignite and Lignite-Related Feedstocks

Abstract:

According to the U.S. Department of Energy National Energy Technology Laboratory, rare earth elements (REE) provide significant value to our national security, energy independence, environmental future, and economic growth. REEs are utilized in a suite of high importance end-uses, such as cell phones, hybrid vehicles, magnets, computer components, catalysts and many others. The unique magnetic properties of REEs, in particular, make them crucial materials for the growing renewable energy and electric automotive markets. REEs are used in the strongest permanent magnets currently known, and are used in generators for wind turbines and in motors for hybrid/electric vehicles. These same types of magnets, as well as other critical REE-based products are used in a host of military defense applications, as well.

China, in part due to its deposits of a unique REE resource (ion-adsorbed clays) that combines high quantities of heavy and critical REE, as well as simple and low cost extraction, dominates the global supply market. In 2010, China established new quotas on exports of REEs, which resulted in huge increases in REE prices peaking in 2011 due to an expected supply shortage for critical applications. As a result, production at the California-based Mountain Pass Mine was re-started after several years of dormancy. However, after peaking in price in 2011, prices have dropped substantially to slightly above 2010 levels, challenging the profitability of non-China based production which consists mainly of hard rock carbonatite deposits, that are deficient in critical REE and heavy REE. Some researchers have noted that mining of the Mountain Pass and similar resources will neither mitigate the crisis in REE resources nor eliminate the shortage of the most critical REE, but will only result in overproduction of excessive cerium.

China accounted for about 83% of the total global REE supply in 2016, down from about 95% prior to 2010. Meanwhile, the US production was zero, with the Mountain Pass mine having declared bankruptcy and closing operations in the last quarter of 2015. The U.S. is currently 100% import reliant for REEs. Although still dominating global supply of the heavy REE, some researchers have estimated that the Chinese ion-adsorbed clays resource will only last another 15-20 years. The Chinese clays represent essentially the entire global supply of heavy REE and most of the critical REE. Further, the bulk of Chinese reserves and production is from a carbonatite-type deposit (Bayan Obo) that contains only trace amounts of heavy REE (98.7% light REE), and supplies roughly 80% of the global light REE demand. Due to its limited supply, and because the Chinese clay resource is rich in heavy and critical REE, while most other traditional resources are deficient in these less common and more valuable elements, it is imperative that new domestic sources of REEs, especially the heavy and critical REE, be identified and processes be developed to produce them. Coal and coal byproducts have recently been identified as one of these potential new resources for REEs.

As part of the US Department of Energy effort to identify alternative domestic sources of Rare Earth Elements (REE), the University of North Dakota was awarded a project to determine the feasibility of recovery of REE from North Dakota lignite coal and related feedstocks. The project team includes the University of North Dakota, Barr Engineering, and Pacific Northwest National Laboratory (PNNL) with the support of the cost share partners North American Coal Company, Great River Energy, and the North Dakota Industrial Commission/Lignite Energy Council as well as technical support from the North Dakota Geological Survey. The overall goal of this Phase I project was to develop an economically viable and environmentally benign process to concentrate REE to a goal of 2 weight percent to form a synthetic mixed REE ore. As part of the project, an extensive sampling and analysis campaign was completed to identify promising feedstocks, which was followed by laboratory-scale concentration testing and a technical and economic feasibility study. Finally, the design of a bench-scale test system was developed, which is to be constructed and tested in a subsequent Phase II project, if awarded by DOE. This paper will provide the results of the Phase I project and the plans for a subsequent Phase II project.

The originally proposed concept involved recovery of REE from the reject stream of the Great River Energy lignite drying system located at the Coal Creek Station power plant in North Dakota. However, upon further investigation in the project, it was discovered that on an ash-basis, the REE are more concentrated in certain locations within the coal seams than in the associated roof/floor sediments. Therefore, the focus for the remainder of the project was shifted towards development of methods to concentrate the REEs from these selected coal seams.

Laboratory testing of a novel concentrating method has shown very high recovery of about 80-90% of the REEs from two potential feedstocks, ranging from 580 to 2300 ppm REE content on an ash basis, with excellent selectivity/recovery of the critical and heavy REE. Details of a processing scheme to economically concentrate the REE to the goal of 2 weight percent using a simple and environmentally benign method have been established. The project team has also developed the concepts for integration of the REE recovery process within an existing coal conversion facility that will provide cost-saving synergies, multiple saleable product streams, and infrastructure to reduce costs. Based on the results of the Phase I work, the project team believes North Dakota lignite-related feedstocks to be a highly promising alternative domestic source of REE that will limit dependence on foreign supply.

489276 Froth Flotation Enrichment of Rare Earth Elements (REEs) from Different Feedstocks

Abstract:

Flotation is an aqueous, efficient, inexpensive, and well-developed commercial technology that has been widely applied in a variety of industrial processes, including mineral processing. The flotation process first generates gas bubbles within a solid-liquid or liquid-liquid suspension. The gas bubbles then attach to solid particles or immiscible liquid droplets and rise to top of the liquid surface. The floated solid particles or immiscible liquid droplets finally are skimmed off to achieve the separation. As particles/immiscible droplets are hydrophobic, they attach easily to gas bubbles and can be separated by the flotation technology.

The effectiveness of surfactants, i.e., frother, depressant, and collector, can strongly influence the flotation process, including the enrichment of rare earth elements (REEs) from coal, coal by-products or fly ashes. The properties of particle size, temperature, and pH will also affect REE enrichment. Here, we report the performance of different chemicals/process parameters for the froth flotation of different feeds in the mechanic agitated flotation cell. Different types of frothers and collectors were studied at varying flotation conditions for REE enrichment. The discussion will focus on the performance of these surfactants and the REE enrichment efficiency at these different conditions.

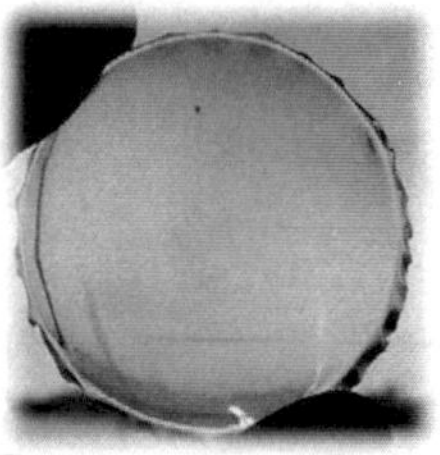

Rowan University

Strategic Assemblies of Wood-Derived Building Blocks for the Sustainable Redesign of BPA-Based Polymers

Joseph F. Stanzione, III, Ph.D.

Department of Chemical Engineering

Sustainable Materials Research Laboratory

2017 AIChE Annual Meeting
Minneapolis, MN October 29 – November 3 2017
"Bio-Based Polymers"

 1

About Rowan University (Glassboro, NJ)

 2

About Rowan University (Glassboro, NJ)

- Comprehensive public research university
- Three campuses:
  - Glassboro
  - Camden
  - Stratford
- ~ 17,500 students
- Schools/Colleges:
  - Business
  - Biomedical Science & Health Professions
  - Communication & Creative Arts
  - Cooper Medical School
  - Education
  - Engineering
  - Global Learning & Partnerships
  - Humanities & Social Sciences
  - Performing Arts
  - School of Osteopathic Medicine
  - Science & Mathematics

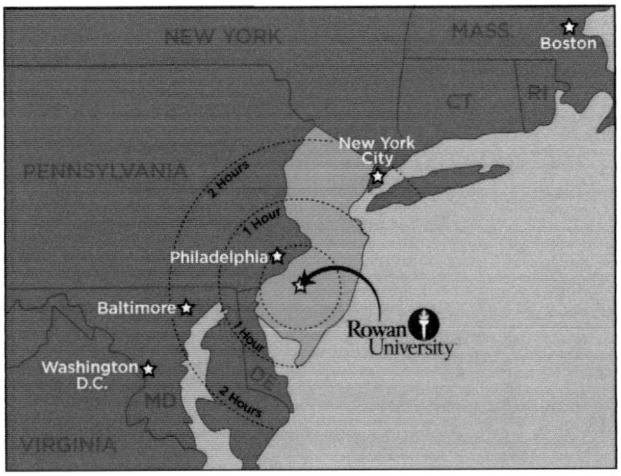

RowanUniversity
HENRY M. ROWAN
COLLEGE OF ENGINEERING

Plastics: The Facts

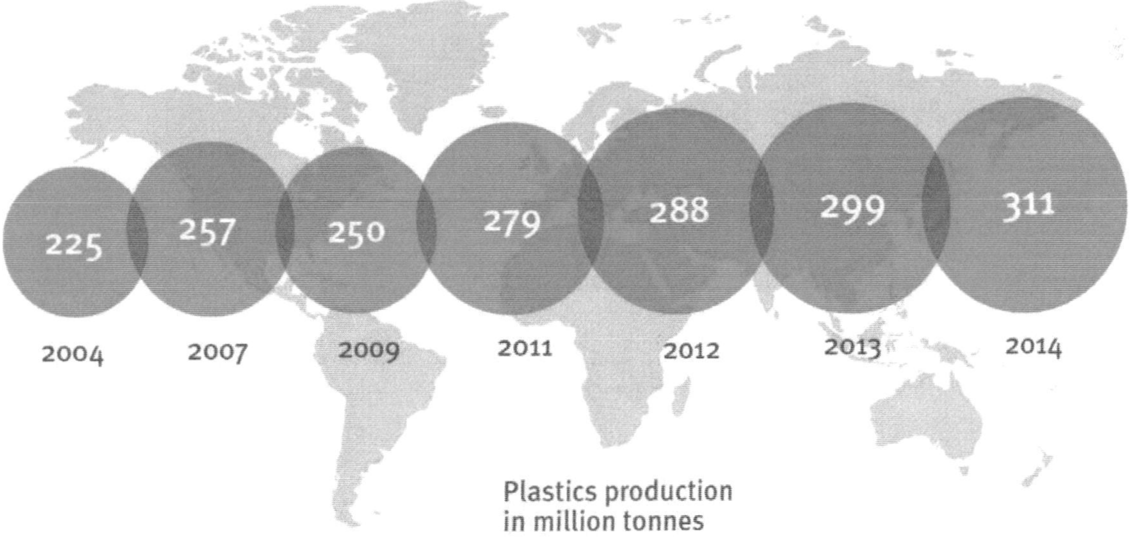

- **Plastic production is growing globally[1]**
- **Currently around 330 Mtonnes**

[1] PlasticsEurope, *Plastics - the Facts* (2015)

RowanUniversity
HENRY M. ROWAN
COLLEGE OF ENGINEERING

Plastics: The Future

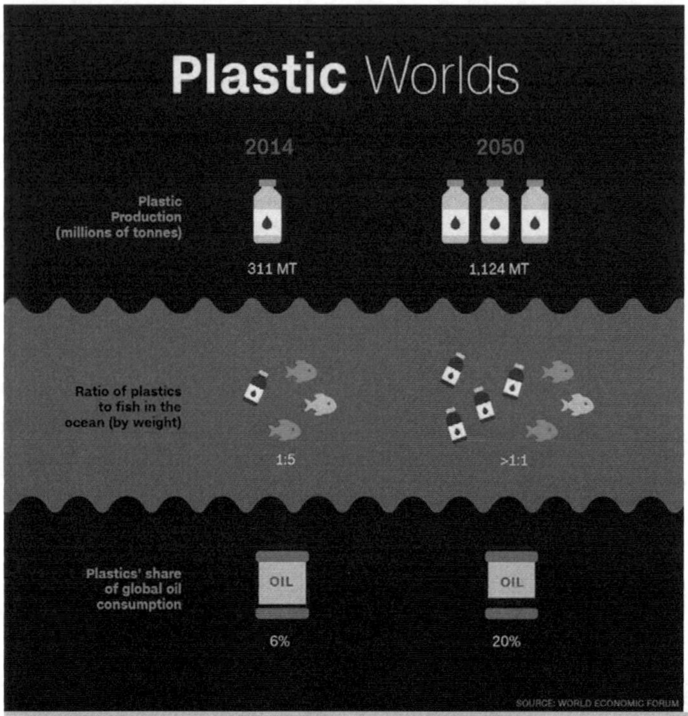

By 2050:

- Expected to triple
- More plastic than fish in the ocean
- 20% of global annual oil consumption

RowanUniversity
HENRY M. ROWAN
COLLEGE OF ENGINEERING

Plastics: The Facts

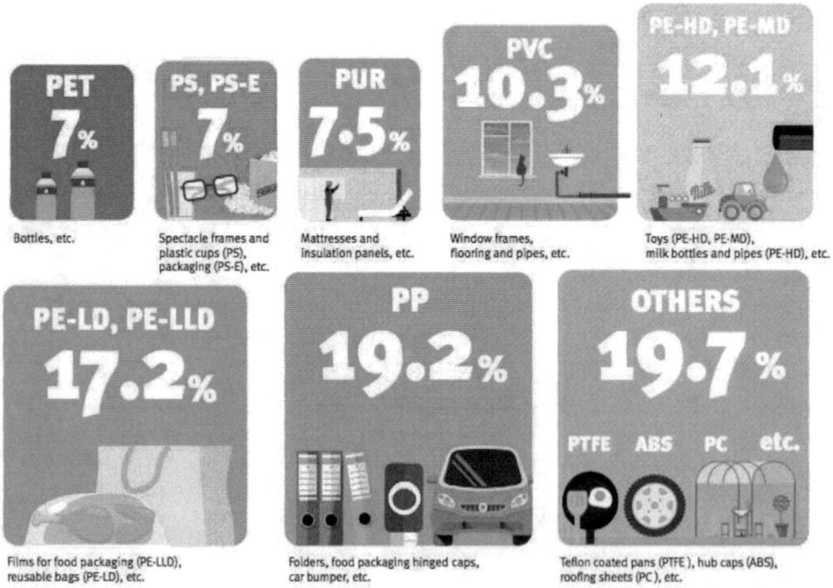

Less than 1.5% of plastics is renewably sourced[1]

[1] European Bioplastics, *Bioplastics facts and figures* (2017).

RowanUniversity
HENRY M. ROWAN
COLLEGE OF ENGINEERING

Paradigm Change in Polymer Manufacturing Occurring

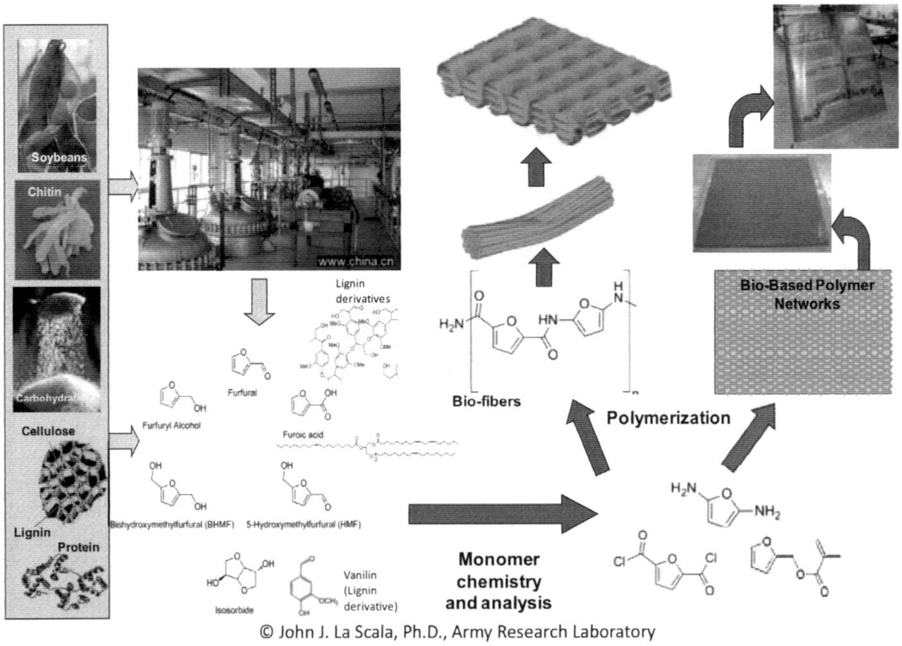

© John J. La Scala, Ph.D., Army Research Laboratory

 7

RowanUniversity
HENRY M. ROWAN
COLLEGE OF ENGINEERING

SMRL's Biomass Utilization to Date

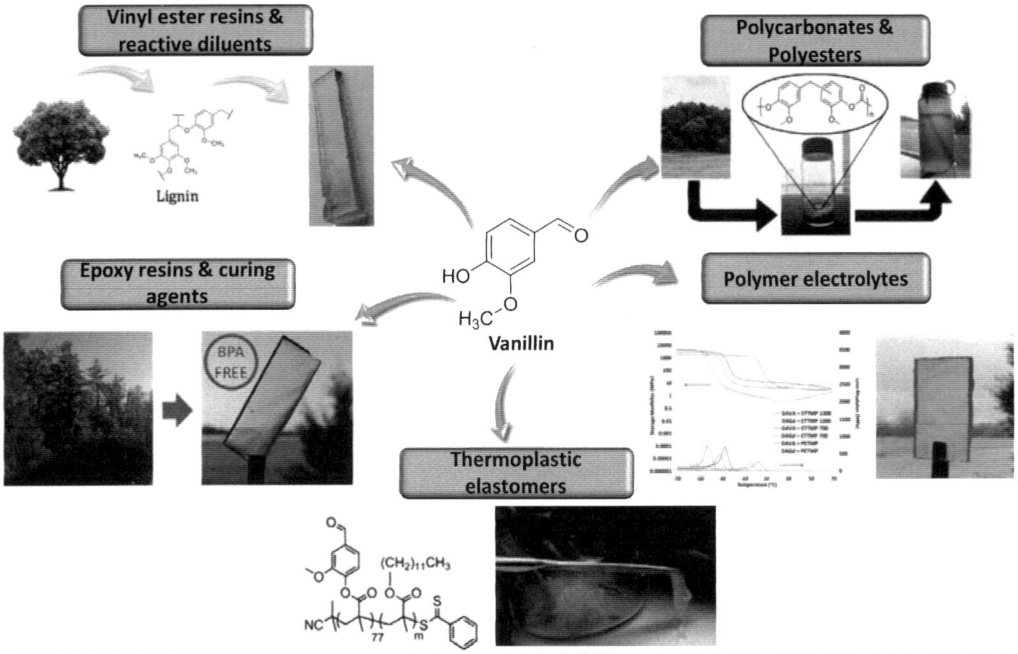

 8

RowanUniversity
HENRY M. ROWAN
COLLEGE OF ENGINEERING

Vanillin-based resin created[1]

Reaction 1

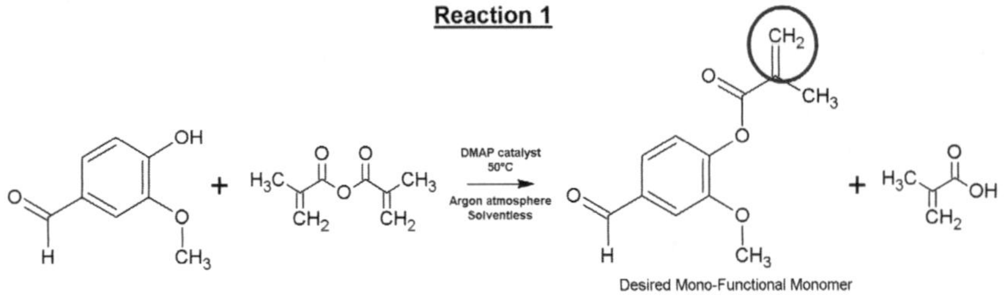

Reaction 2

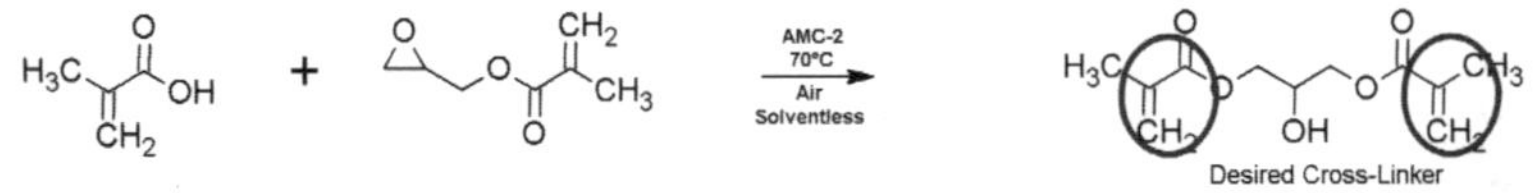

- Green Attributes:
  - All reactants can be bio-based
  - No by-products
  - Solventless reactions
  - Catalyzed reactions
  - Low reaction temperatures
  - **Atom economy/efficiency = 100 %**

Resin viscosity (at 25°C) = 87.25 ± 0.5 cP

Glass transition temperature = 160°C

Glassy storage modulus = 3.6 ± 0.3 GPa

Bio-based Content = 34.8 %

1. Stanzione III, J.F.; et al. *Green Chemistry* 14 (2012) 2346-2352.

 9

RowanUniversity
HENRY M. ROWAN
COLLEGE OF ENGINEERING

Diglycidyl ether of bisphenol-A (DGEBA) alternatives synthesized[1-2]

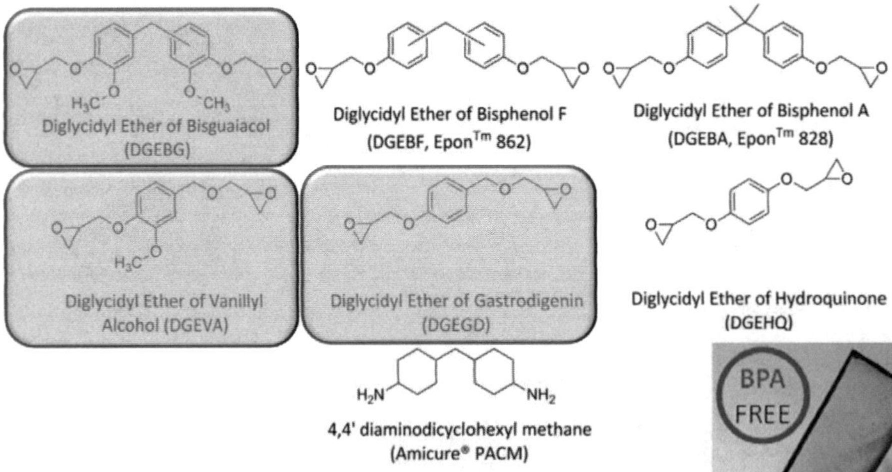

1. Hernandez, E.D.; et al. *ACS Sustainable Chemistry & Engineering* **2016** 4 (8) 4328-4339.
2. U.S. Patent Application Serial # 15/032.583.

 10

RowanUniversity
HENRY M. ROWAN
COLLEGE OF ENGINEERING

Thermogravimetric properties comparable to commercial epoxy resins[1-2]

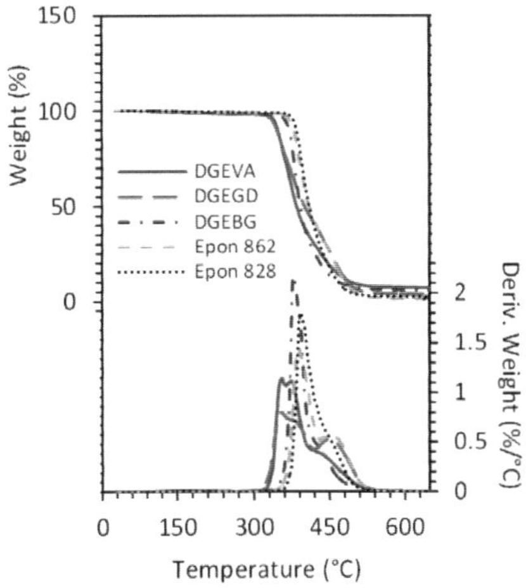

Epoxy Resins	IDT^a (°C)	$T_{50\%}$ (°C)	T_{Max} (°C)	Char Content (%)
DGEVA	341	384	352	6.9
DGEGD	337	400	354	3.6
DGEBG	363	392	376	5.6
Epon 862	371	410	388	1.2
Epon 828	379	410	392	2.3

[a] *IDT* measured as the temperature at 5 % weight loss.

Resins cured with stoichiometric equivalents of PACM

1. Hernandez, E.D.; et al. *ACS Sustainable Chemistry & Engineering* **2016** 4 (8) 4328-4339.
2. U.S. Patent Application Serial # 15/032.583.

RowanUniversity
HENRY M. ROWAN
COLLEGE OF ENGINEERING

Thermomechanical properties comparable to commercial epoxy resins[1-2]

Epoxy Resins	T_g (°C)	E' (GPa)
DGEVA	100	3.02
DGEGD	123	2.29
DGEBG	104	3.35
Epon 862	131	2.57
Epon 828	149	2.37

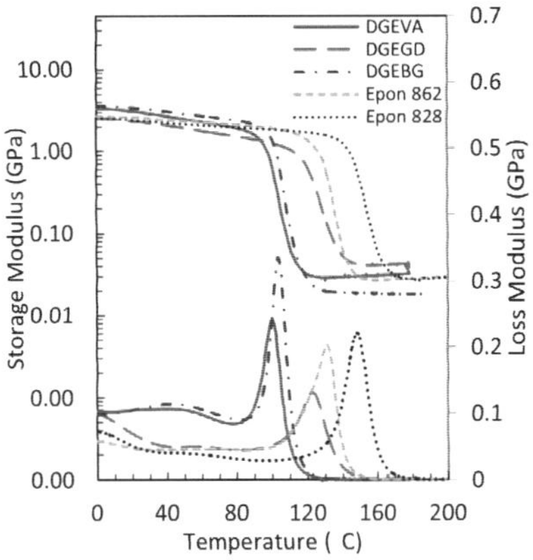

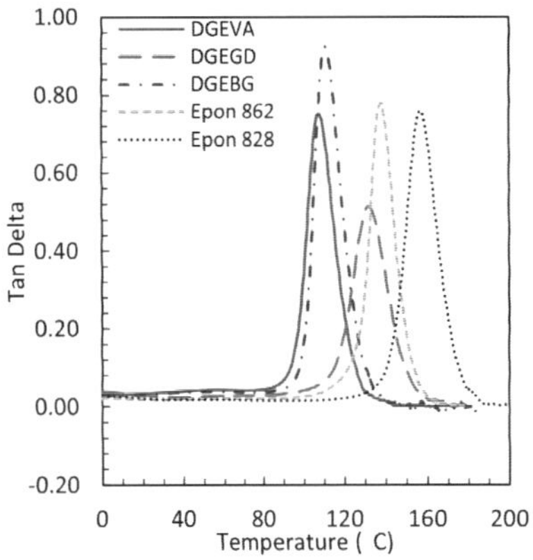

Resins cured with stoichiometric equivalents of PACM
E' measured at 25 °C

1. Hernandez, E.D.; et al. *ACS Sustainable Chemistry & Engineering* **2016** 4 (8) 4328-4339.
2. U.S. Patent Application Serial # 15/032.583.

RowanUniversity
HENRY M. ROWAN
COLLEGE OF ENGINEERING

High bio-based content cured epoxy-amine resins produced[1-4]

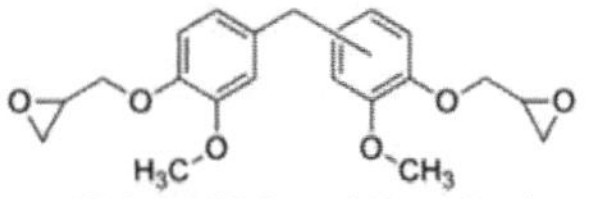

Diglycidyl Ether of Bisguaiacol (DGEBG)

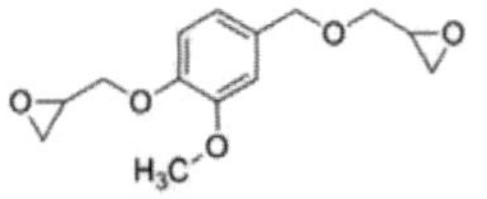

Diglycidyl Ether of Vanillyl Alcohol (DGEVA)

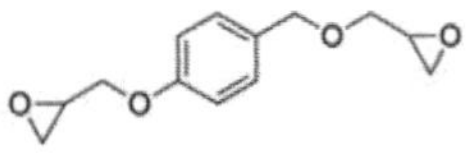

Diglycidyl Ether of Gastrodigenin (DGEGD)

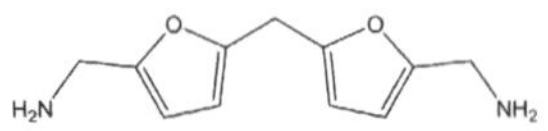

5,5′-methylenedifurfurylamine (DFDA)

> 96.5 % bio-derived carbon content

1. Mauck, J.R; et al. *Macromolecular Chemistry & Physics* **2017** DOI:10.1002/macp.201700013 .
2. U.S. Patent Application Serial # 15/032.583.
3. Hu, F.S.; et al. *Macromolecules* **2014** 47 (10) 3332-3342.
4. Hu, F. S.; et al. *Macromolecular Chemistry & Physics* **2015** 216 (13) 1441-1446.

Drexel UNIVERSITY

 13

RowanUniversity
HENRY M. ROWAN
COLLEGE OF ENGINEERING

Thermomechanical properties comparable to commercial epoxy resins[1-4]

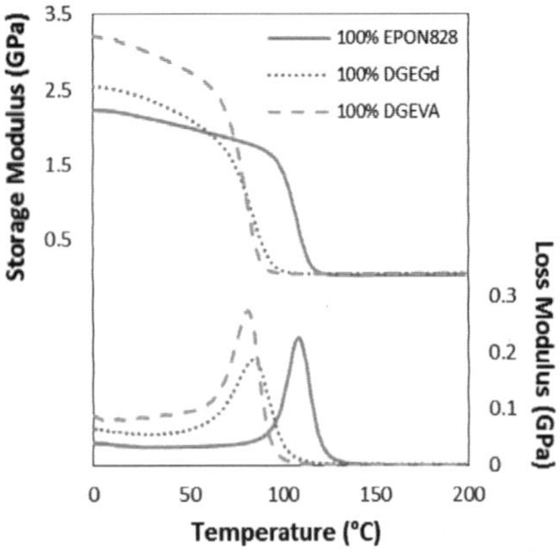

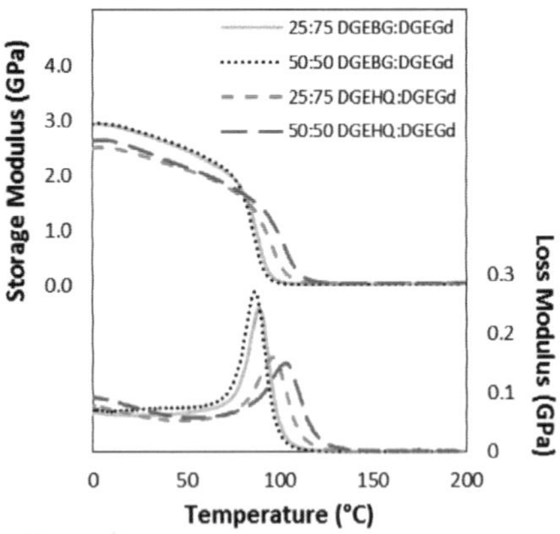

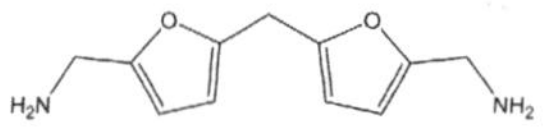

5,5′-methylenedifurfurylamine (DFDA)

1. Mauck, J.R; et al. *Macromolecular Chemistry & Physics* **2017** DOI:10.1002/macp.201700013 .
2. U.S. Patent Application Serial # 15/032.583.
3. Hu, F.S.; et al. *Macromolecules* **2014** 47 (10) 3332-3342.
4. Hu, F. S.; et al. *Macromolecular Chemistry & Physics* **2015** 216 (13) 1441-1446.

Drexel UNIVERSITY

 14

RowanUniversity
HENRY M. ROWAN
COLLEGE OF ENGINEERING

Bisguaiacol (BG)-based Polycarbonate Synthesized[1]

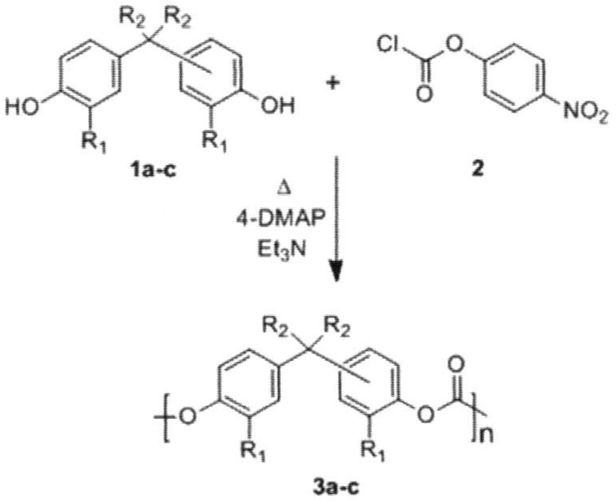

1a,3a: R_1=H, R_2=CH_3
1b,3b: R_1=H, R_2=H
1c,3c: R_1=OCH_3, R_2=H

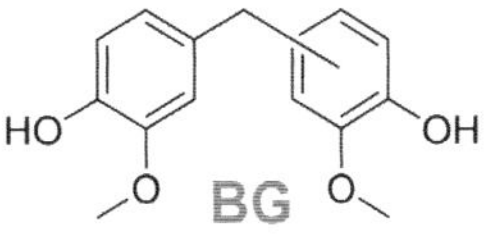

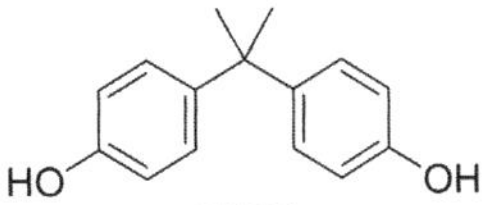

BPA

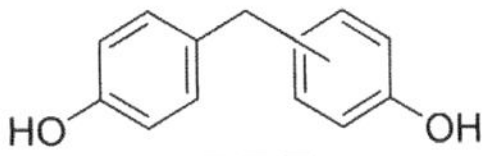

BPF

[1] J. Mauck *et al.*, *J Biobased Materials & Bioenergy* (2017) In Press.

RowanUniversity
HENRY M. ROWAN
COLLEGE OF ENGINEERING

BG-based Polycarbonate Possesses Comparable Thermal Properties[1]

Sample	*IDT* (°C)	$T_{50\%}$ (°C)	T_{max} (°C)	Char Content (%)	T_g (°C)
BG-PC	375	455	450	24.2	100
BPA-PC	440	543	545	23.2	150
BPF-PC	369	546	501	40.5	119

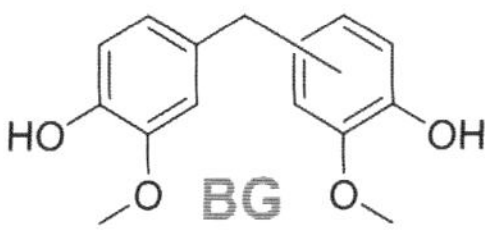

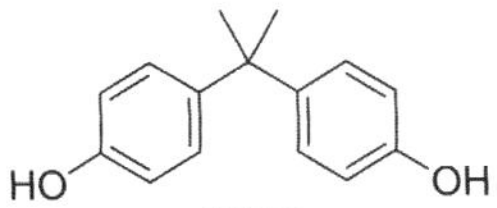

BPA

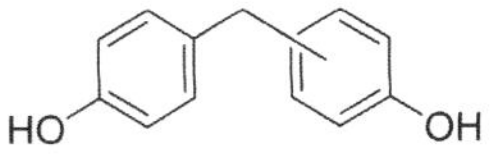

BPF

[1] J. Mauck *et al.*, *J Biobased Materials & Bioenergy* (2017) In Press.

RowanUniversity
HENRY M. ROWAN
COLLEGE OF ENGINEERING

BG-based Polyesters Synthesized[1]

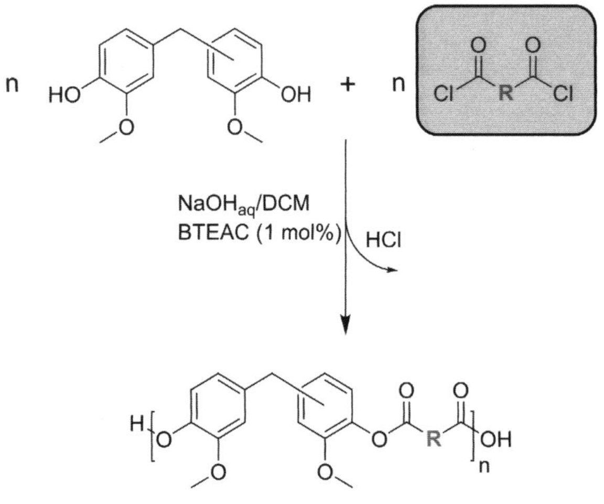

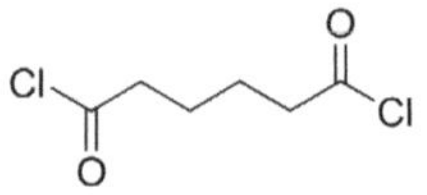

Adipoyl Chloride

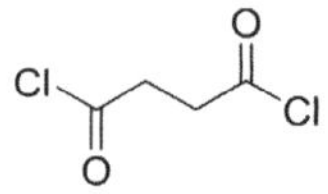

Succinyl Chloride

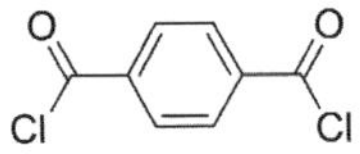

Terephthaloyl Chloride

[1] S. Curia *et al.*, *In preparation* (2017)

 17

RowanUniversity
HENRY M. ROWAN
COLLEGE OF ENGINEERING

BG-based Polyesters Synthesized[1]

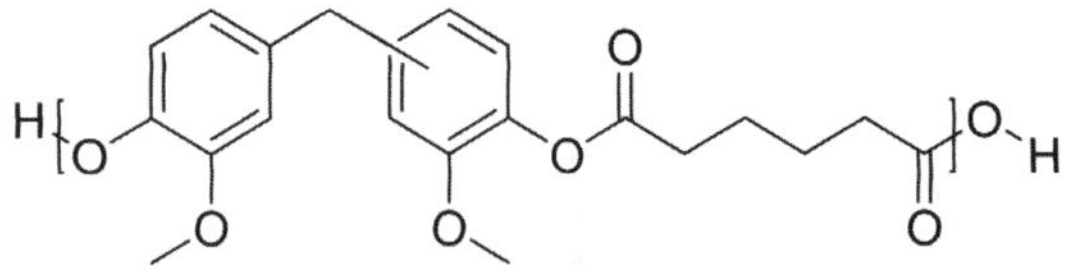

Poly(BG adip)
M_n=6.6 kg mol^{-1} (Đ=1.82)

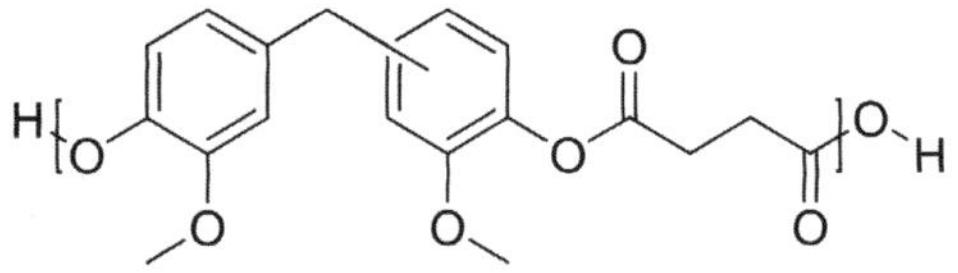

Poly(BG succ)
M_n=7.2 kg mol^{-1} (Đ=2.16)

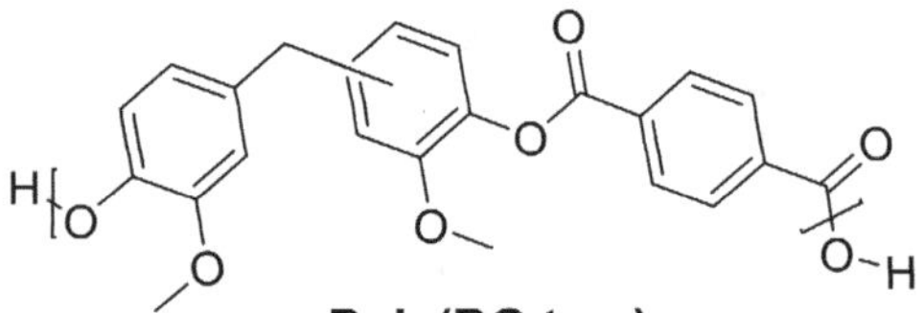

Poly(BG tere)
M_n=26.9 kg mol^{-1} (Đ=2.29)

[1] S. Curia *et al.*, *In preparation* (2017)

 18

RowanUniversity
HENRY M. ROWAN
COLLEGE OF ENGINEERING

BG-based Polyesters Synthesized[1]

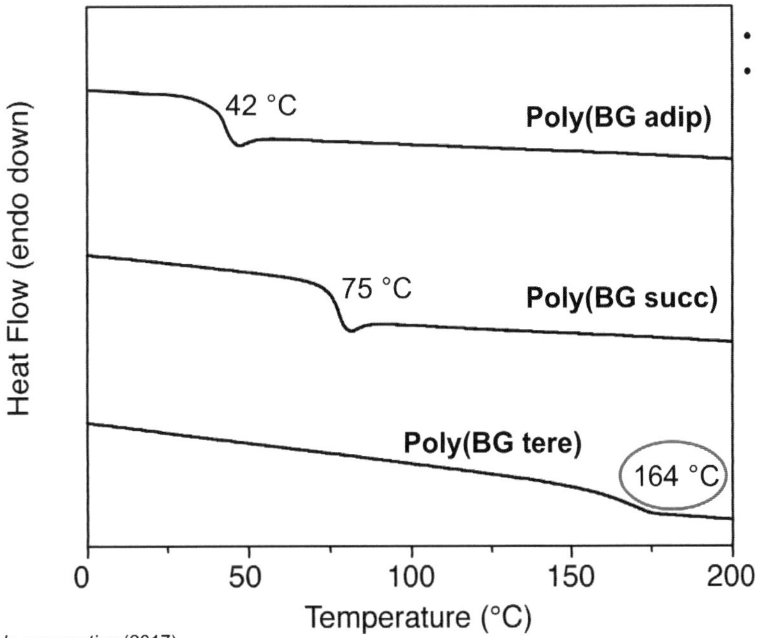

- 10°C/min
- N_2 50 mL/min

[1] S. Curia *et al.*, *In preparation* (2017)

 19

RowanUniversity
HENRY M. ROWAN
COLLEGE OF ENGINEERING

Thermal stability a strong function of comonomer type (in N_2)[1]

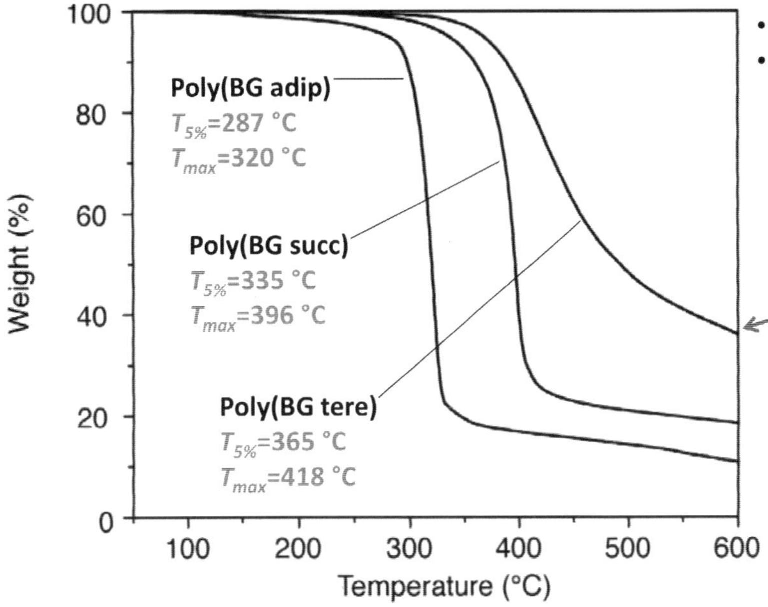

- 10 °C min^{-1}
- N_2 25 mL min^{-1}

36% char
@ 600 °C

[1] S. Curia *et al.*, *In preparation* (2017)

 20

RowanUniversity
HENRY M. ROWAN
COLLEGE OF ENGINEERING

Thermal stability a strong function of comonomer type (in Air)[1]

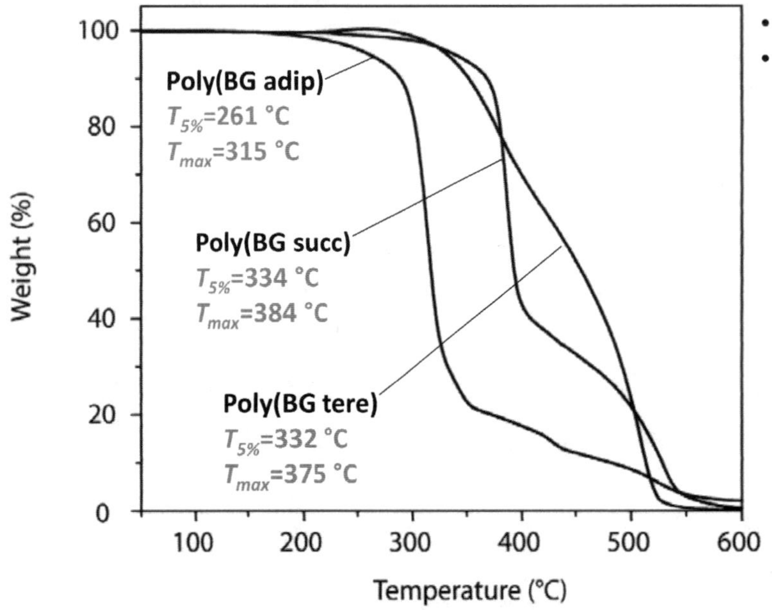

- 10 °C min^{-1}
- Air 25 mL min^{-1}

T_{max} still above 300 °C

[1] S. Curia *et al.*, *In preparation* (2017)

 21

RowanUniversity
HENRY M. ROWAN
COLLEGE OF ENGINEERING

Fully Bio-based Polyaryate Synthesized[1]

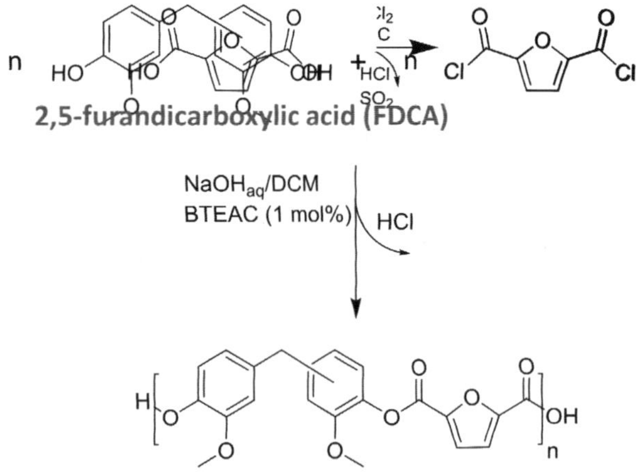

Poly(BG FDCA)
M_n=6.3 kg mol^{-1} (Đ=1.57)

[1] S. Curia *et al.*, *In preparation* (2017)

 22

RowanUniversity
HENRY M. ROWAN
COLLEGE OF ENGINEERING

Promising Thermal Stability Demonstrated[1]

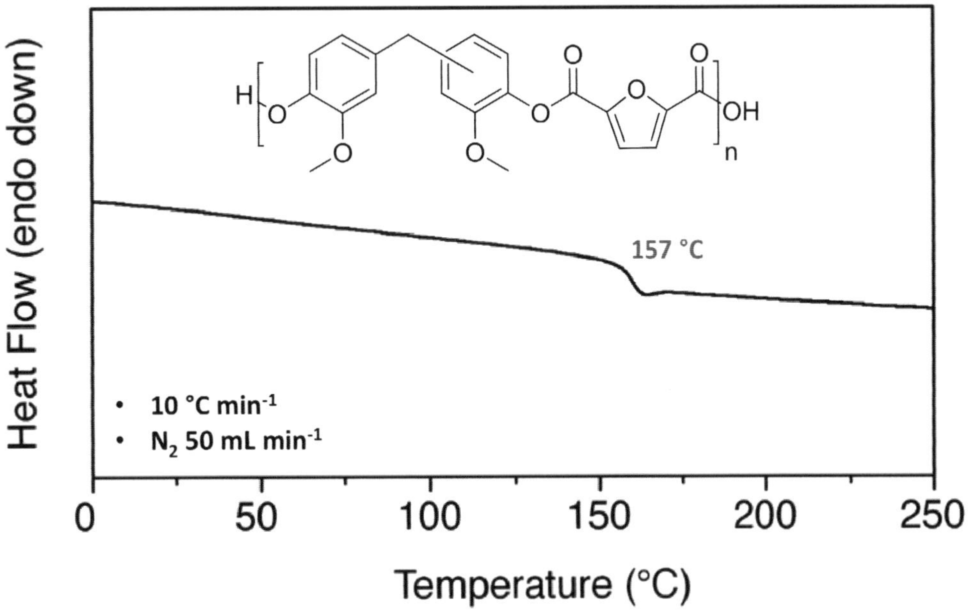

[1] S. Curia *et al.*, *In preparation* (2017)

RowanUniversity
HENRY M. ROWAN
COLLEGE OF ENGINEERING

Promising Thermal Stability Demonstrated[1]

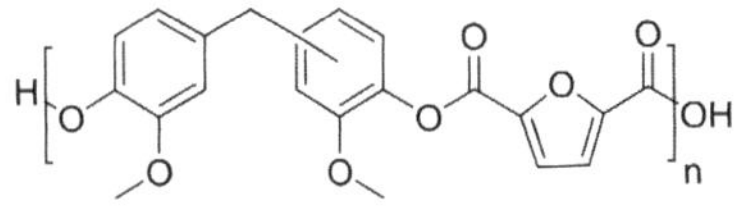

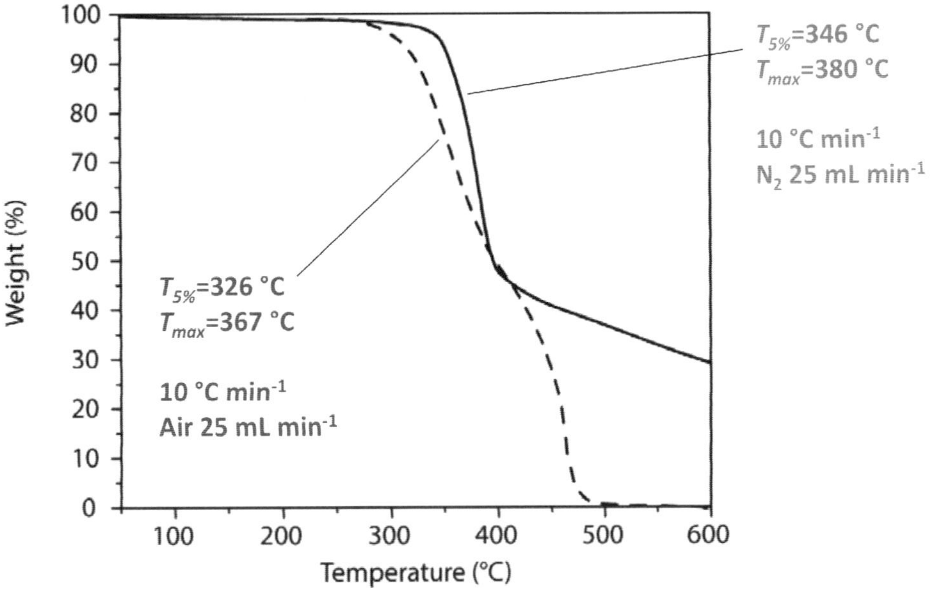

[1] S. Curia *et al.*, *In preparation* (2017)

Rowan University
HENRY M. ROWAN
COLLEGE OF ENGINEERING

Conclusions & Acknowledgements

- Sustainable Materials Research Laboratory
  - Eric D. Hernandez
  - Joseph R. Mauck
  - Silvio Curia, Ph.D.
- U.S. ARL for Financial Support via Cooperative Agreements
  - W911NF-06-2-001
  - W911NF-14-2-0086
  - W911NF-16-2-0225

ARL
U.S. ARMY RDECOM
DREXEL UNIVERSITY Chemical and Biological Engineering
Rowan University
HENRY M. ROWAN
COLLEGE OF ENGINEERING

 25

Rowan University
HENRY M. ROWAN
COLLEGE OF ENGINEERING

Questions?

Crater Lake National Park

Rocky Mountain National Park

 26

500886 Preparation and Characterization of Cardanol Based Vinyl Ester Resins As Cross-Linker Units

Abstract:

Abstract

Cardanol which is the main component of the thermally threated cashew nutshell liquid is a versatile monomer which combines a phenyl ring with a C15 alkyd chain with different degrees of unsaturation. Because of its unique structure and reactive functionalities, cardanol based building blocks has a growing potential to supply aromatic building blocks for many applications such as corrosion prevention coatings. In this study, cardanol glycidyl ether (LITE 2513HP, CardoliteÂ®) was chemically modified from the unsaturation sites of the side chain by means of epoxidation in the presence of hydrogen peroxide and formic acid. The structure of the modified cardanol resin (SCECGE) was confirmed via proton nuclear magnetic resonance, Fourier Transform infrared spectroscopy, and epoxy equivalent weight titrations. This modified resin was cured with petroleum based epoxy hardener 4-4â€™-methylene biscyclohexanamine (PACM). Also for comparison, another cardanol based di-phenol di-epoxy resin NC514 (Cardolite®) and petroleum based epoxy resin diglycidyl ether of bisphenol-A (DGEBA) were thermally cured with listed epoxy hardeners at stoichiometric ratios. The extent of curing reaction and curing mechanism of different epoxy/amine systems were explored via mid-IR techniques and differential scanning calorimetry (DSC) to determine that the secondary epoxides are not as reactive as the primary glycidyl epoxides. The glass transition, modulus, and strength of the cured SCECGE resins were lower than that of the NC514 and DGEBA because of the lower extent of reaction of the secondary epoxides. In addition, the water barrier properties of this cardanol based SCECGE resin were also determined via contact angle and, water and vapor sorption studies to explore the coating performance of the material.

499647 Ionic Liquids: Green Solvents for Dry Native Cellulose and Chitosan

Abstract:

Cellulose and chitosan (the deacetylated form of chitin) are the most abundant renewable polymeric natural resources on our planet. Native cellulose and chitosan can dissolve in certain ionic liquids. Having nearly zero vapour pressure and being almost completely recyclable make these liquids strong alternatives for the traditional solvents with quite large environmental impacts and derivative effects on the polymers. It is crucial to gain better understanding of the physical state of cellulose and chitosan in ionic liquids. We report solution rheology of six native cellulose samples of different molecular weight in two different ionic liquids. Based on the concentration dependences of viscosity and longest relaxation time, 1-ethyl-3-methylimidazolium acetate (EMImAc) and 1-butyl-3-methyl imidazolium chloride (BMImCl]) are Î¸-solvents for cellulose. The strange failure of the Cox-Merz rule (shear viscosity larger than linear complex vsicosity in the shear-thinning region) for cellulose solutions in ionic liquids suggests that cellulose is not a simple flexible polymer in solution. This was hypothesized to be related to intra-chain hydrogen bonding at shear rates that align and stretch the longest chains. Trace amounts of water compete for hydrogen bonds and can impart a yield stress to the solution rheology, which grows in magnitude as more water is incorporated. However, simply heating the solution to 80°C drives off the water and reverts the solutions to viscoselastic liquids with no yield stress. Rheology of five different chitosan samples (with deacetylation ratio ~75% and of different molecular weight) was also studied in EMImAc. Unlike those of dry cellulose, chitosan solutions contained non-crystalline aggregates, identified using linear viscoelastic tests. This was related to the chemical crosslinks present in the samples due to reaction between amine and hydroxyethyl groups on the chitosan backbone while being extracted from crustacean shells at elevated temperatures in strong base. The chitosan solutions were much less sensitive than the cellulose ones to the presence of water in the system. Upon adding water (<3 wt.%) to the system. the solutions of dry cellulose transition from having terminal behaviour to weak gels, whereas the chitosan solutions already show non-terminal characteristic and do not tend to change at small water content (<20 wt.%).

490787 Environmentally Friendly Flame Retardants Based on Adhesive Catecholamine

Abstract:

There has been much effort expended over the last few decades to develop a wide variety of flame retardant (FR) additives for the purposes of preventing fire spread and reducing the risk of fire-related injuries from everyday-use materials such as polymers and textiles. Some of the most frequently used FR chemicals are halogenated and organophosphorous materials due to their cost-effectiveness and extensive industrial availability. However, there have been serious concerns raised regarding their impact to human health and the environment. This has spawned efforts in industry and academia to development replacement FR materials. In this presentation, we will describe a novel surface coating FR system based on a catecholamine derivative called polydopamine (pDA) that was applied to polyurethane (PU) foam in a single step [1, 2]. PU foam is a particularly attractive test material because it is highly flammable without FRs and it is one of the most widely-used polymers in our daily life. The dopamine precursor to pDA is a naturally occurring chemical found in our body and other living organisms, and this material has an ability to undergo self-polymerization into pDA under mild basic pH conditions. The conformal pDA coating was applied to PU during a simple aqueous dip coating procedure. The catechol functional group in pDA inherently gives this material a nearly universal surface adhesion capability and imparts radical scavenging capability, one of many important features of additives that serve as FRs. pDA coated PU foam exhibited a remarkably reduced flammability compared to pure PU foam and also self-extinguished in some cases. Data regarding the mechanism of flame retardancy of pDA, including morphological observation using scanning electron microscopy (SEM) and quantitative thermal analysis using micro combustion calorimetry, cone calorimetry and thermogravimetric analysis, will also be discussed. We believe this study introduces new opportunities for utilizing naturally derived materials for future green FR applications.

References:

[1] Bioinspired Catecholic Flame Retardant Nanocoating for Flexible Polyurethane Foams, J.H. Cho, V. Vasager, K. Shanmuganathan, S. Nazarenko, and C.J. Ellison, *Chemistry of Materials,* 27, 6784â€"6790, 2015.

http://pubs.acs.org/doi/10.1021/acs.chemmater.5b03013

[2] Development of Kinetic Parameters for Polyurethane Thermal Degradation Modeling Featuring a Bioinspired Catecholic Flame Retardant, B.C. Roberts, A.R. Jones, O.A. Ezekoye, C.J. Ellison, and M.E. Webber, *Combustion and Flame*, 177, 184-192, 2017.

http://www.sciencedirect.com/science/article/pii/S0010218016303777

490792 Soybean Oil Based Thermoset Materials with High Biorenewable Content

Abstract:

In recent years, there has been growing interest in substituting petroleum-based materials with bio-renewable feedstock due to concerns of environmental pollution and the depletion of fossil fuel resources. Thus, the conversion of various bio-sourced monomer precursors into useful polymeric materials has emerged as an attractive strategy. Vegetable oils are a prime example onto which many useful chemical functionalities have been introduced to make them suitable alternatives to petroleum-based monomers. While a large number of vegetable oil derivatives have been blended/incorporated with petroleum-based materials to prepare thermosetting polymers, existing examples normally incorporate low total biorenewable content into the final product. With the goal of generating thermosets with high biorenewable content, recent work [1] from our group will be described where two different soybean oil derivatives with different functional groups (e.g., one as a multifunctional thiol and the other as a multifunctional ene) were photocured to prepare cross-linked thermosets. For example, thiol functional groups were introduced to produce mercaptanized soybean oil by reaction of commercially available epoxidized soybean oil with cysteine functionalities. Then, as-prepared mercaptanized soybean oil was mixed with the commercially available acrylated counterpart, diluent, and photoinitiator, followed by UV irradiation under ambient conditions to produce crosslinked films and fibers. Not only do the resulting materials possess high biorenewable content (e.g., as high as 90 % biosourced carbon) but the fact that they are thermosets makes them potentially suitable for high performance applications.

[1] Soybean Oil Based Fibers Made Without Solvent or Heat, D.W. Janes, K. Shanmuganathan, D.Y. Chou and C.J. Ellison, *ACS Macro Letters*, 1, 1138-1142 (Sept. 2012).

http://pubs.acs.org/doi/abs/10.1021/mz300325g

492684 Self Assembly of Coil-Hyperbranched Poly(styrene-block-acrylated epoxidized soybean oil) Block Copolymers

Abstract:

Biorenewable thermoplastic elastomers have shown its economic potential to substitute their petroleum counterparts. In this study we preset a series of coil-hyperbranched poly(styrene-b-acrylated epoxidized soybean oil) (PS-PAESO) diblock copolymers through reversible addition-fragmentation chain transfer (RAFT) polymerization, which demonstrate its competitiveness toward poly(styrene-b-butadiene-b-styrene) (SBS) in applications such as pressure sensitive adhesives and bitumen modification. Even with significant dispersity in both molecular weight and molecular architecture, triglyceride-based diblock copolymers undergo microphase separation. Through tuning the degree of acrylic functionality in AESO monomer, the influence of degree of branching of PAESO block can be shown on the phase diagram by constituting an additional parameter which has never been discussed before. By thoroughly understanding the self assembly behavior of PS-PAESO, the ability to adjusting the performance in applications will be enhanced. We have constructed a library of PS-PAESO polymers sampling the parameter space defined by the molecular weight, composition, and functionality of AESO. The self assembly behavior is demonstrated through dynamic shear rheology, transmission electron microscopy and the corresponding small angle X-ray scattering.

482356 Physicochemical Properties of Polylactide/Delta-Valerolactone/Organosolv Lignin Atactic Terpolymers

Abstract:

Compostable graft terpolymers of stereotactic lactide, delta-valerolactone, and switchgrass organosolv lignin were synthesized via ring-opening polymerization to improve on polylactide homopolymer properties for commercial applications. Organosolv lignin from swtichgrass has been shown to have properties that improve some of the deficiencies of polylactide, including polylactideâ€™s limited use in food, beverage and medical applications due to its ease to degrade hydrolytically. Delta-valerolactone has been shown to add flexibility to copolymers. It was found that controlling the ratios of the stereoisomers of lactide had significant effects the tensile properties of the resultant terpolymers. Terpolymers containing various ratios of L(-) lactide to D(+) lactide showed greatly increased elasticity and lower Youngâ€™s modulus compared to terpolymers containing only L(-) lactide.

جامعة الملك فهد للبترول والمعادن
King Fahd University of Petroleum & Minerals

Oil Field Chemicals from Macromolecular Renewable Resources: A Case Study of Date Pit Nanoparticles for Drilling Fluid Formulation

Jimoh K. Adewole, Musa O. Najimu

College of Petroleum Engineering & Geosciences
King Fahd University of Petroleum & Minerals,
Dhahran 31261, Saudi Arabia

Presentation Outlines

- Innovative Drilling Fluid Technologies
- Drilling Fluid and its Composition
- Date Pit as a Multicomponent Polymer
- Nanoparticle Based Drilling Fluids/Additives
- Preliminary Results and Discussions
- Conclusion & Recommendations

KFUPM

Smart Drilling Fluid Technologies

Time

Costs

Environmental

Improved Drilling Operations:

- ✓ **Physically small**
- ✓ **Chemical/Thermally stable**
- ✓ **Biologically degradable**
- ✓ **Environmentally benign**

Drilling Fluid and its Composition

- Drilling fluids are mixture natural and synthetic chemical compounds
- Used as circulating fluid in drilling operations
- Successful completion of wells depends to a large extent on the properties of the drilling fluid used during the operation
- Types
  - ✓ Water-base mud
  - ✓ Oil-base mud
  - ✓ Gas

Drilling Fluid and its Composition

Functions of Drilling Fluids

- ✓ Transportation of cuttings
- ✓ Cooling and Cleaning of bit
- ✓ Friction Reducer

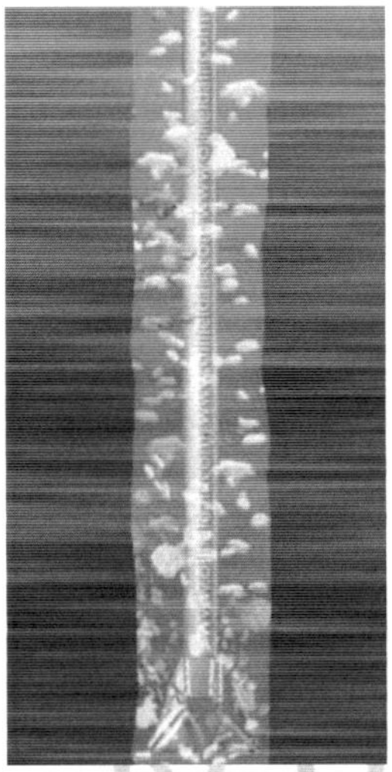

Drilling Fluid and its Composition

Functions of Drilling Fluids

- ✓ Maintaining borehole stability
- ✓ Prevent inflow of fluids from permeable rocks that have been penetrated
- ✓ Formation of thin filter cake to seal pores/other openings

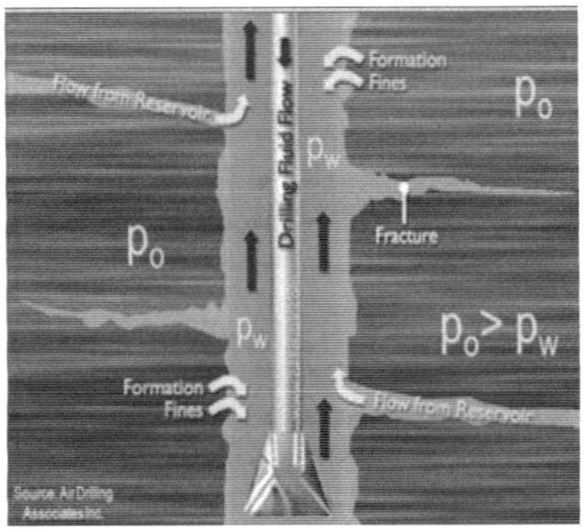

Drilling Fluid and its Composition

Performance Requirements of Drilling Fluids

- Non-damaging/offensive
- No expensive completion methods
- No interfere with productivity of the fluid bearing formation
- No corrosion/excessive wear of drilling equipment

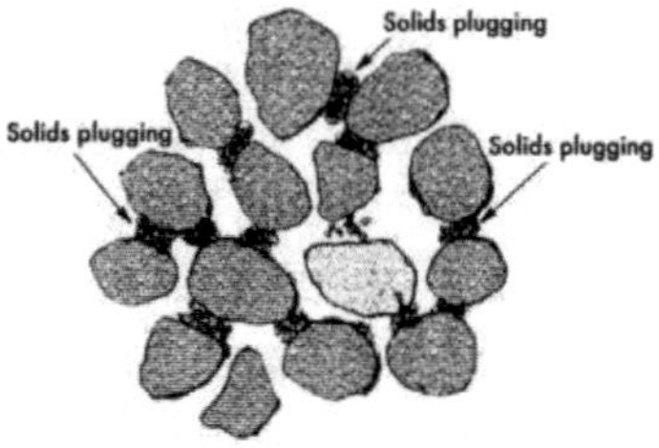

Formation damage caused by solids plugging (petrowiki.org)

Drilling Fluid and its Composition

Composition of drilling mud

Drilling Fluid Additives	Function	Possible Source
Water	Base fluid	-
Bentonite	Base solid	Clay
Barite	Weighting agent	barium sulfate
Calcium carbonate	Weighting agent	Inorganic
Xanthan Gum	Thermal stability/Viscosity control	Modified bio materials
Starch/cellulose and its derivatives	Viscosifiers/clay stabilizer	Natural polymers
Lignin and its derivatives	Viscosity and gel strength control	Modified bio materials
Quebracho	dispersant	heartwood
Polyol	shale inhibitor	Bio-materials

KFUPM

© King Fahd University of Petroleum & Minerals | Dhahran, 31261 | Saudi Arabia | +966 (3) 860 0000

Date Pit as Multicomponent Polymer

Date Fruit Production in KSA

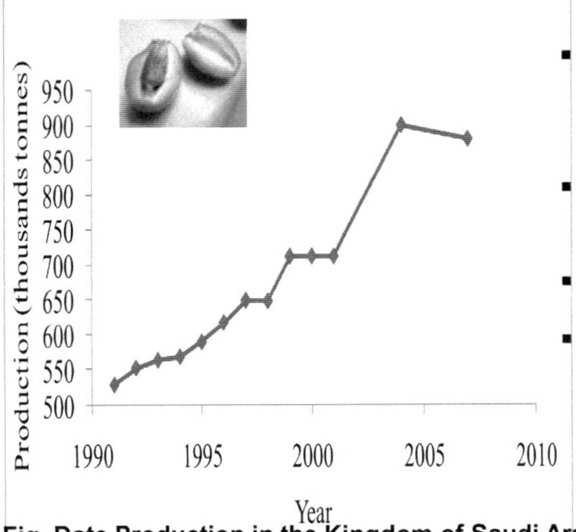

Fig. Date Production in the Kingdom of Saudi Arabia between 1990 and 2007 (Adewole et al 2013)

- Saudi Arabia ranked among the top ten date producing countries in the world
- 4th largest producer in 2001 (13.3% of global production)
- 3rd largest producer in 2007
- Improved Agriculture Tech can increase production

Date Pit as Multicomponent Polymer

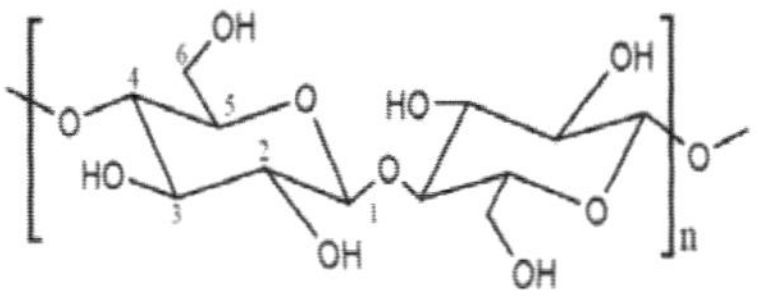

Fig. Molecular Structure of Cellulose

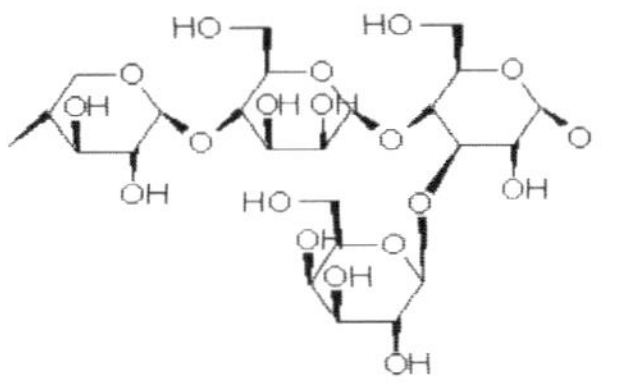

Fig. Molecular Structure of Hemicellulose

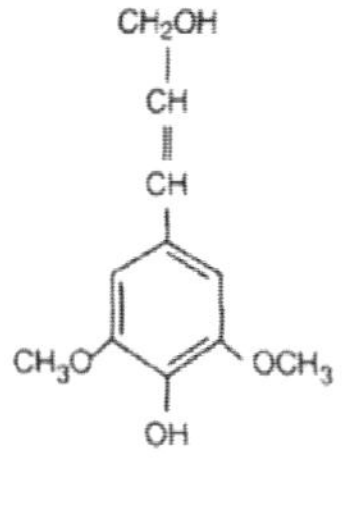

Fig. Molecular Structure of Lignin

Table. Chemical Composition of Date Pit Polysaccharides Fraction (Adewole & Sultan (2014)

Component	Molecular Formula	Composition (Wt%)
Hemicellulose	(C5H8O4)n and (C6H10O5)n	23.3 - 55.5
Lignin	$C_9H_{10}O_2$, $C_{10}H_{12}O_3$, $C_{11}H_{14}O_4$	14.7 - 23.2
Cellulose	$C_6H_{10}O_5$	20.18 – 56.7
Ash	-	1.1 – 5.3

Date Pit as Multicomponent Polymer

Cellulose in Drilling Fluid

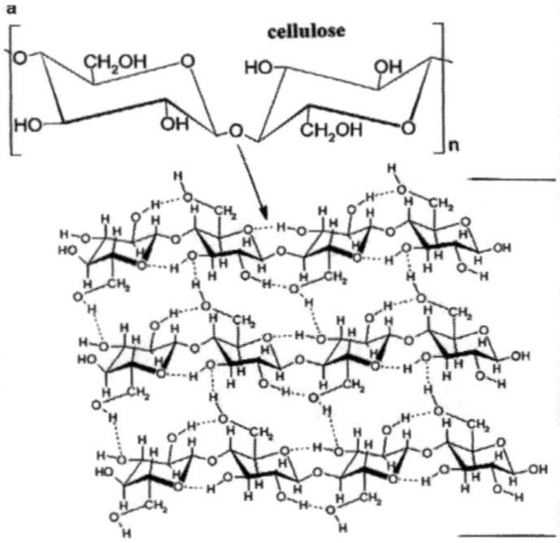

Fig. Structure of Cellulose

Kamel, 2007

Cell-MPs & NPs as:

Natural morphology of fiber
Cellulose whiskers:
15nm in diameter & ~1um long
Microfibrils (vegetable):
2-10nm diameter & several um in length
Cellulose Crystallites:
Length 300 - 2000 nm
Diameter 5 - 20 nm
Aspect ratio 20 - 60
Tensile strength 10 000MPa
E-modulus 150 GPa
Used for Viscosity enhancement and Filtration Control

Date Pit as Multicomponent Polymer

Lignin in Drilling Fluid

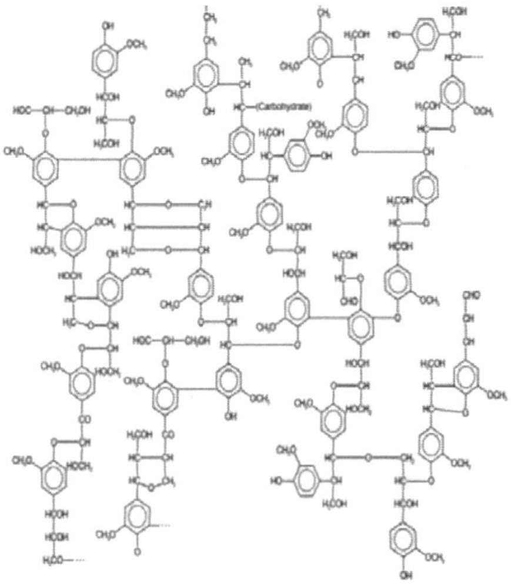

Fig. Structure of Lignin

Lignin

High contents of aromatics moieties
Could be spliced into monoaromatic monomers
Serves as stiffening materials for cellulose and hemicellulose
Its derivative:
Lignosulfonate
Polyanionic lignin
Uses:
Thinner
Deflocculants

Date Pit as Multicomponent Polymer

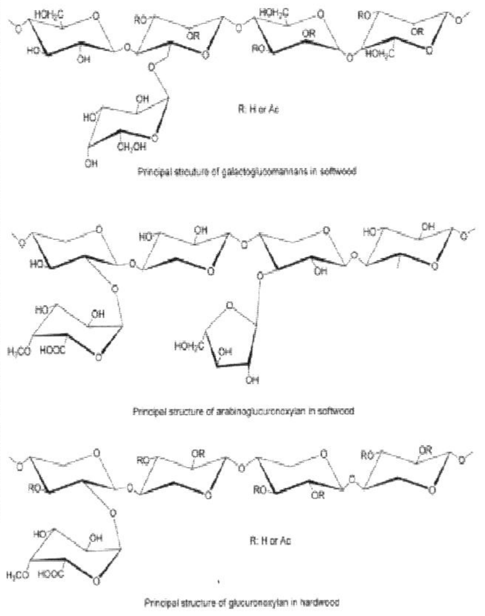

Hemicellulose in Drilling Fluid

Hemicellulose

Different composition & Structures
Combination of : pentose, arabinose, galactose, hexoses
Used:
Polyelectrolyte
rheology modifiers

Fig. Structure of Hemicellulose

Date Pit as Multicomponent Polymer

Processing

Functionalization

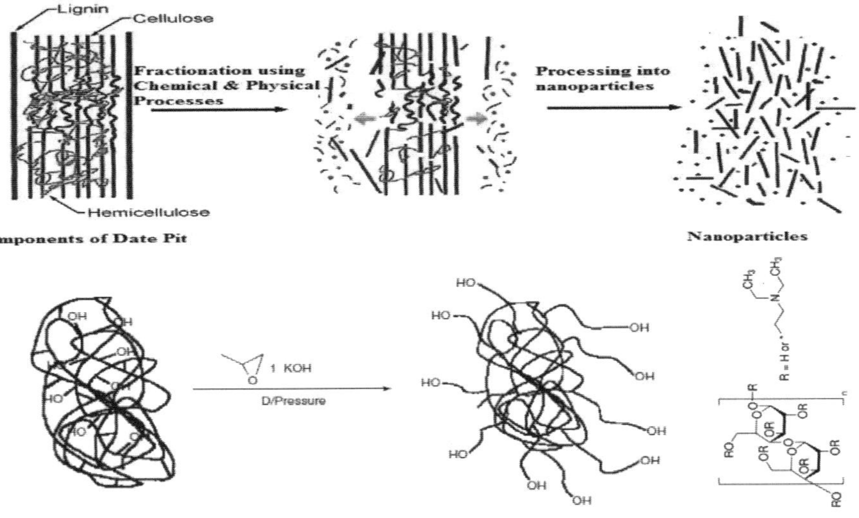

Lignocelluose processing into Nanoparticles (Gandini & Belgacem, 2008; Lejeune & Deprez, 2010)

Polymeric Nanoparticles for Drilling

Why Nanoparticles?

- Competitive performance properties
  - ✓ Generally nano-based drilling fluids will form compact and less permeable mud cake that will reduce fluid invasion/spurt losses
  - ✓ Penetrant particles during spurt loss will be small enough to prevent throat blockage
  - ✓ It will also contribute to the plastering capability
  - ✓ Thin mud cake

Polymeric Nanoparticles for Drilling

Why Bio-based Multicomponent Polymeric Nanoparticles?

- Naturally occurring composite
- Renewable nature
- Availability and Sustainability of sources
- Nonfood agricultural based
- Low cost
- Low density
- High specific strength and modulus
- Relatively easily functionalized
- Non-toxic and Biodegradable

KFUPM

© King Fahd University of Petroleum & Minerals | Dhahran, 31261 | Saudi Arabia | +966 (3) 860 0000

Nano-materials for Drilling Fluids: Literature Review

- Studies on Nano-based drilling fluids using commercial nanoparticles (Amanullah et al, 2011 & Zakaria et al, 2012)
  - ✓ Significant improvement in performance using Low concentration
  - ✓ Reduce the tendency of the formation of internal mud cake (facilitate effective cleaning of borehole before completing the well)
  - ✓ Deposition of thin mud cake may lead to decrease in differential pipe sticking in highly permeable formations
  - ✓ Formation of low solid content of unweighted nana-based drilling fluids will enhance the rate of penetration in hard rock formations

Nano-materials for Drilling Fluids: Literature Review

- Cellulose Nanoparticles investigated as modifiers for Rheology and Fluid loss in bentonite water-based fluids (Li et al, 2015)
- Cellulose nanocrystals (CNC) was prepared (by hydrolysis) from microfibrillated cellulose (MFC)
  - ✓ CNC has better surface interaction with bentonite layers
  - ✓ Superior rheological properties
  - ✓ Higher thermal stability
  - ✓ Less fluid loss volume
  - ✓ Thinner filter cake

Project Objectives

- Main objective is to investigate the potential of locally available multicomponent (non-petrochemical) raw materials for applications as oil field chemicals.
  - ✓ Drilling Fluid Additives

Date Pit/Seed

- Minimize the cost of production
  - ✓ Particle size (conserve the composition)
  - ✓ Weight percent date pit
  - ✓ Date pit oil content

Drilling Fluids

- Maximize project efficiency
  - ✓ Filtration properties
  - ✓ Borehole stability/Density
  - ✓ Rheology/ Flow properties
  - ✓ Thermal properties

Date Pit Processing Method

Fig. Date pit overall processing sequence

Preliminary Results and Discussions

Drilling Fluid Density

- $P_{static\ mud\ column}\ f(density, depth)$
- Density: the weight per unit volume
- Units: lb/gal or ppg, $^{psi}/_{ft}$ or $^{kg/cm^2}/_{m}$
- Balanced drilling: $p_w - p_o \geq 200psi$
- Prevent inflow of formation fluids
- Lay down a thin, low permeability filter cake on the walls
- Maintain borehole stability
- Buoyant effect of mud $\propto$ density

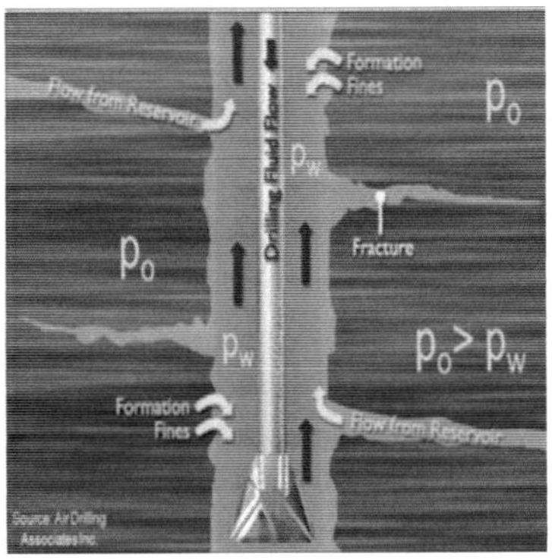

Preliminary Results and Discussions

Effect Date Pit Concentration on Density

Densities of Common Mud Components (lb/gal)	
Oil	6.66
Water	8.33
Salt	18.3
Clay	20.8
Barite	34.2

Change observed at loading greater than 10%

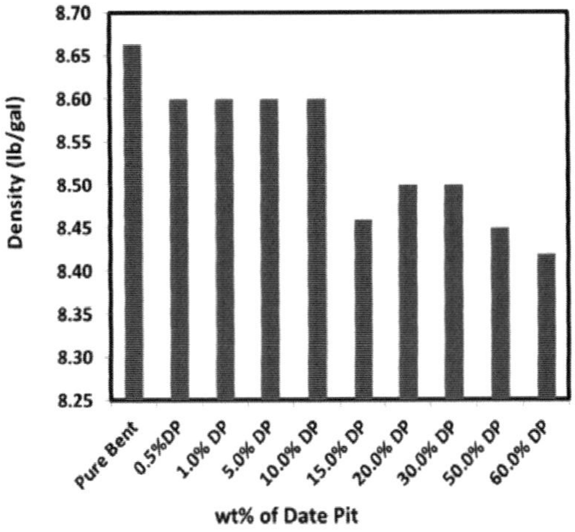

Fig. Change in density with date pit loading

Preliminary Results and Discussions

Rheology & Flow Properties

- **Flow properties is responsible for removal of drill cuttings**
- Unsatisfactory flow properties can lead to:
  - ✓ Filling bottom of hole with drill cuttings
  - ✓ Reduced penetration rate
  - ✓ Hole enlargement
  - ✓ Stuck pipe
  - ✓ Loss of circulation
  - ✓ Blow out

A Blowout Scene

Preliminary Results and Discussions

Rheology & Flow Properties

- The shear stress/ shear rate relationship of drilling fluids depends on its compositions
- Behaviour close to Bingham plastics
- Yield point and Plastic viscosity

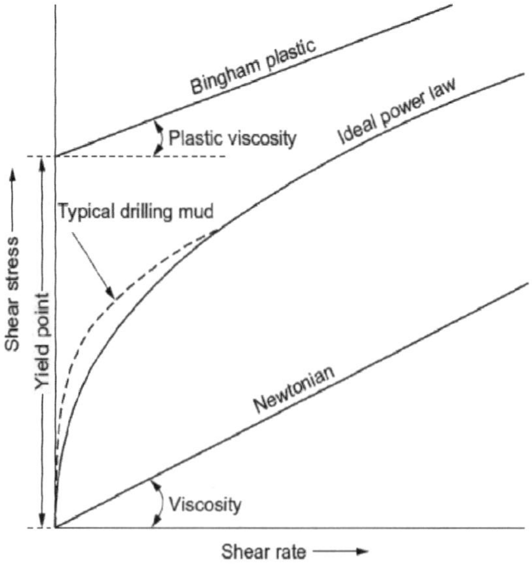

Fig. Consistency curves for common flow model

Preliminary Results and Discussions

Rheology & Flow Properties

- Decrease in viscosity due to increase in shear rate (shear thinning)
- high shear rate in the drill pipe reduces pumping pressures
- high viscosity in the annulus increases cuttings carrying capacity

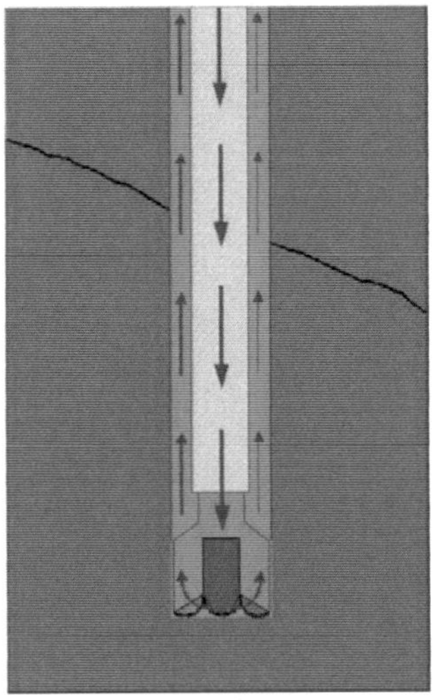

Fig. Typical flow during drilling operation

Preliminary Results and Discussions

Effect Date Pit Loading & Particle Size

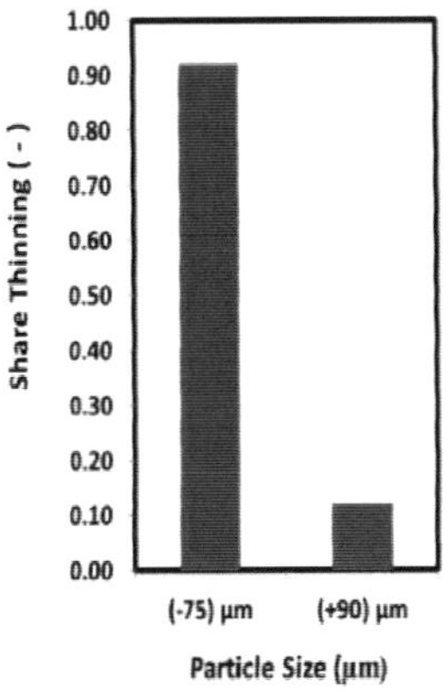

Fig. Effect of Particle Size

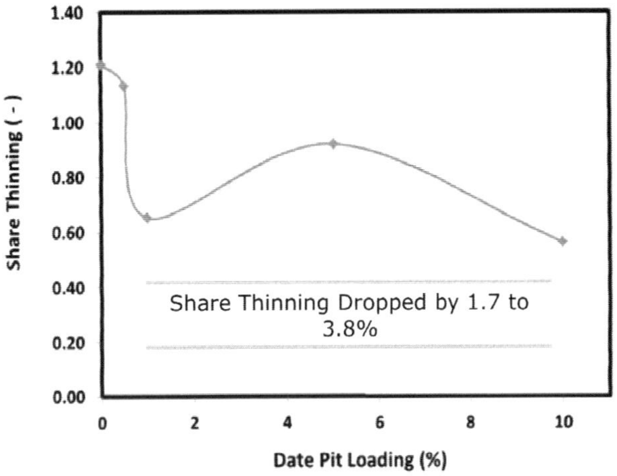

Fig. Effect of wt% Date Pit Loading

Preliminary Results and Discussions

Effect Date Pit Particle Size Distribution

Date pit has the potential to be used as Rheological modifier if used as smaller particles (such as nanoparticles)

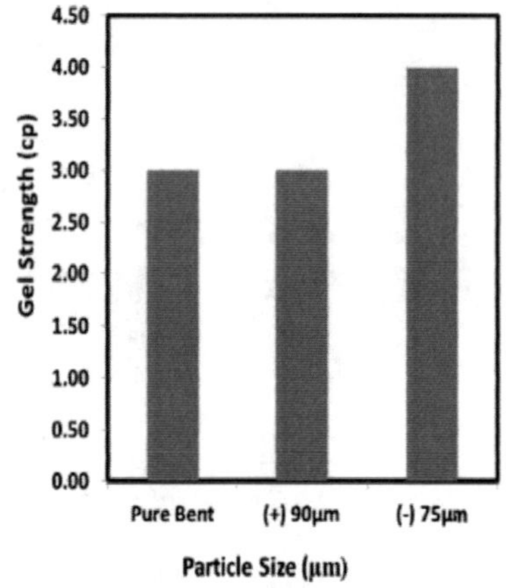

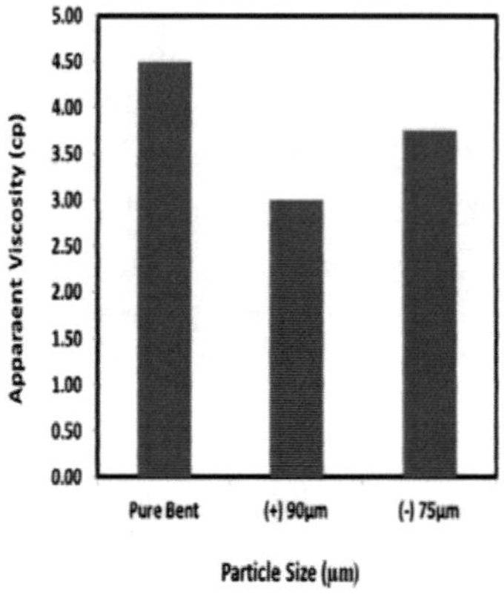

Fig. Effect of particle size on Gel Strength and Viscosity

Preliminary Results and Discussions

Rheology & Flow Properties: Effect Date Pit Loading

At 60wt% date pit loading:

- Gel strength dropped by 50%
- Apparent viscosity dropped by 27% drop

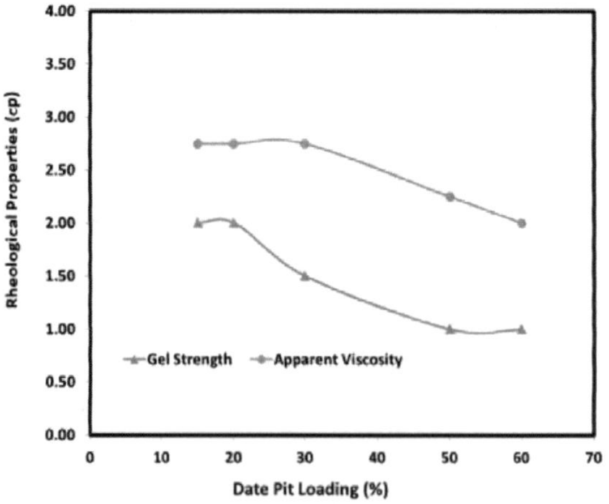

Fig. Effect of date pit loading

Adewole and Najimu, (2017) Journal of Energy Resources Technolog

© King Fahd University of Petroleum & Minerals | Dhahran, 31261 | Saudi Arabia | +966 (3) 860 0000

Preliminary Results and Discussions

Filtration Properties

- Another major requirement for successful hole completion is the ability of the drilling mud to seal permeable formations exposed by the bit
- Done by using low permeability filter cake
- Permeability of filter cake depends on:
  - ✓ Particle size distribution in the mud
  - ✓ Electrochemical conditions (Caenn et al 2011)
- Filtration properties depends on the nature of formation to be drilled
- In permeable formations; control are required to prevent thick filter cake:
  - ✓ borehole gauge reduction
  - ✓ differential sticking
- In unconsolidated sands; to prevent slumping into the hole by rapid formation of filter cake

Preliminary Results and Discussions

Filtration Properties: Effect Date pit oil content

For fluid loss prevention, de-oiling process can be eliminated to improve date pit performance and reduce cost of production

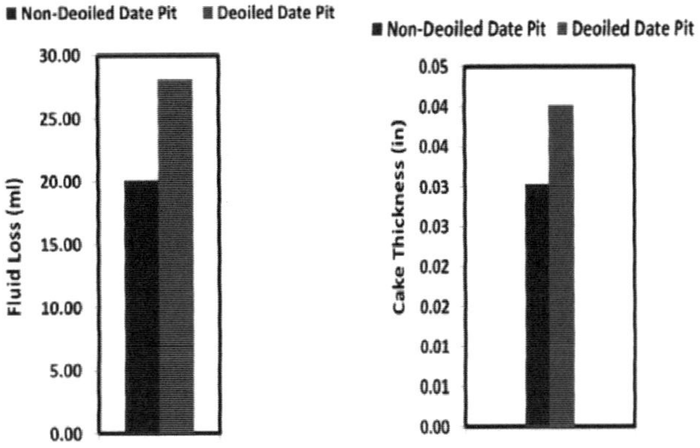

Fig. 15 Effect of Oil Content

Adewole and Najimu, (2017) Journal of Energy Resources Technology

Preliminary Results and Discussions

Filtration Properties: Effect Date Pit Oil Content

The effect of oil content increased with time

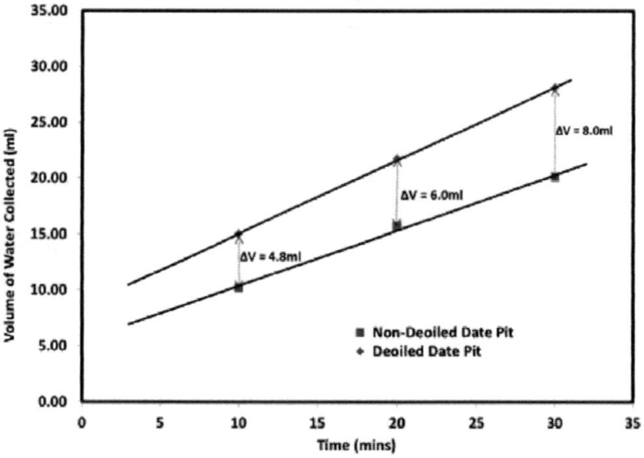

Fig. 16 Effect of Oil Content with time

Adewole and Najimu, (2017) Journal of Energy Resources Technology

Preliminary Results and Discussions

Filtration Properties: Effect Date Pit Particle Size Distribution

Table. Effect of size on Filtration properties (Adewole and Najimu, 2017)

Sample	Fluid Loss after 30 mins (ml)	Residue Thickness (in)
Pure Bentonite	38	0.034
(-) 75μm	24.8	0.035
(+) 180μm	28.1	0.040
(+) 710μm	30.2	0.044

Adewole and Najimu, (2017) Journal of Energy Resources Technology

Preliminary Results and Discussions

Effect of Date Pit Loading (Low Concentration)

- Low wt% date pit loading
- Continued drop in Fluid Loss and residue thickness
- Agreement with published results (Amanullah et al 2016), J. Petrol Sc & Eng

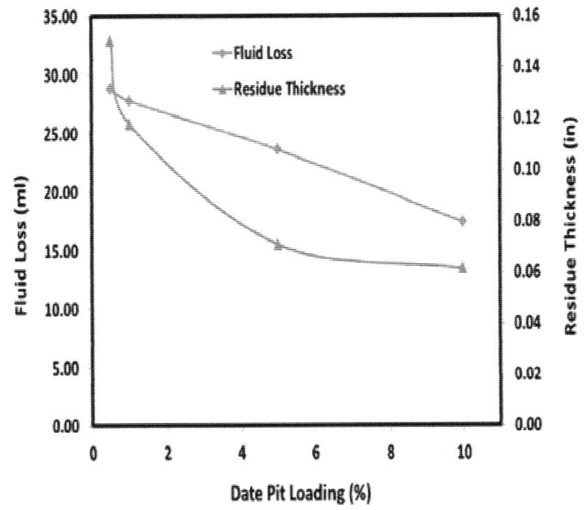

Fig. Effect of date pit concentration on filtration properties

Preliminary Results and Discussions

Effect Date Pit Loading (High Concentration)

- High wt% date pit loading may lead to formation of soft and thick mud cake which increase the potential of differential sticking.
- Reduce the ease of recovery of stuck pipe

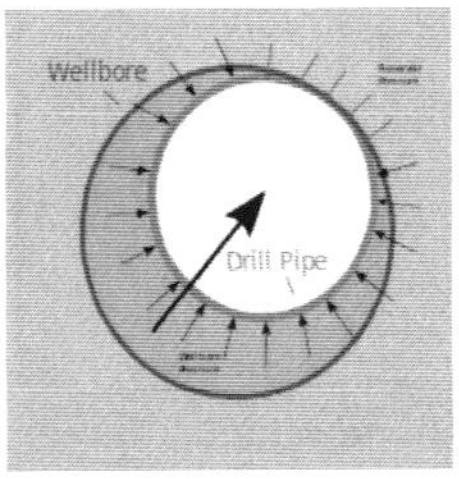

Differential sticking (wikipedia.org)

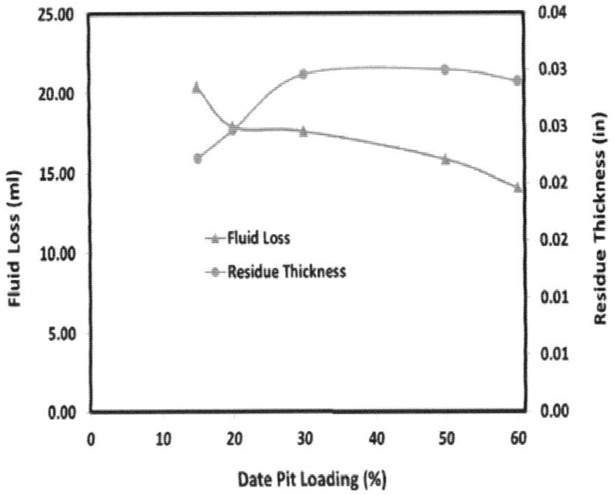

Fig. Effect of date pit concentration on filtration properties

Preliminary Results and Discussions

Thermal Properties: Effect Date Pit Loading

Temperature and rate of degradation are used to specify thermal stability

- Critical Temperature
- ✓ Starch 107°C
- ✓ Cellulosic polymers 135°C
- ✓ Polyacrylates 204°C
- ✓ Certain acrylic polymers 260°C

Onset Degradation Temperature	
Date Pit	**250**
Bentonite	**455**

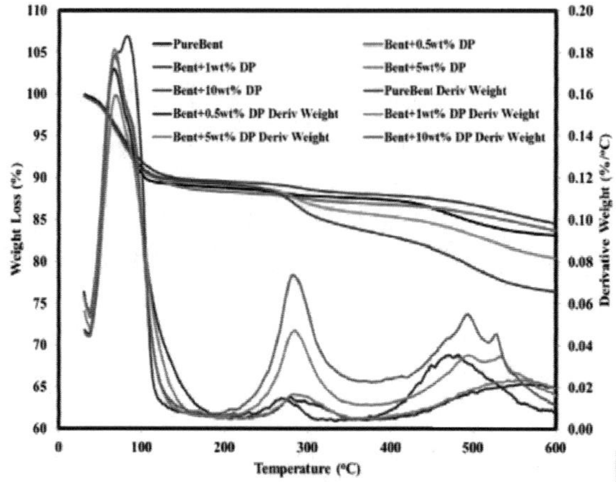

Fig. Thermogravimetric curves for date pit composite

Adewole and Najimu, (2017) Journal of Energy Resources Technology

Preliminary Results and Discussions

Thermal Properties: Effect Date Pit Oil Content

- Weight loss difference due to oil content
- Smoke Points (oxidative stability) of vegetable oils range between 107 - 271°C

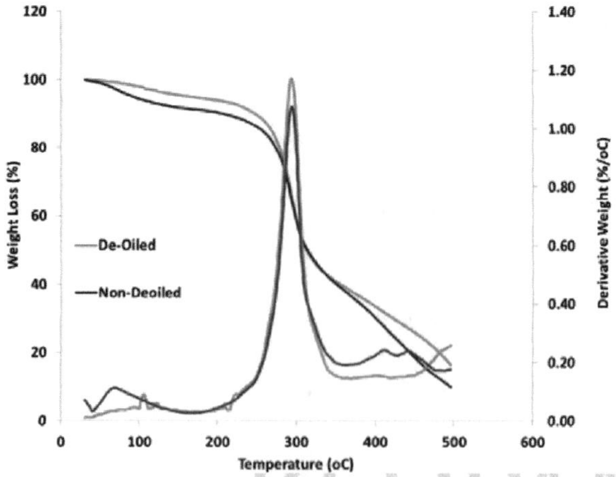

Fig. Thermogravimetric curves for de-oiled and non de-oiled date pit

Preliminary Results and Discussions

Smart Behaviour of Natural Composite

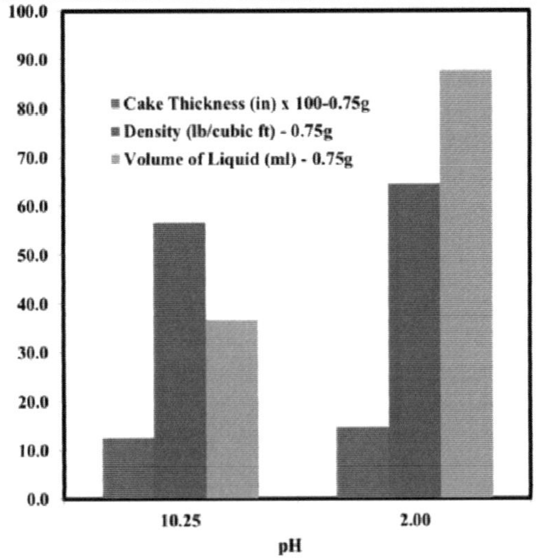

Fig. pH Responsiveness of date pit based drilling fluid

Conclusion and Recommendations

- De-oiling processes could be eliminated to reduce cost
  - ✓ reduction in energy requirement for processing
  - ✓ reduced time of heating
- Multifunctional at lower particle size
  - ✓ fluid loss prevention
  - ✓ rheological property modifier
- Better at low concentration (overbalance drilling)
  - ✓ date pit loading of 1-10wt%
  - ✓ higher wt% above 15% will affect density
  - ✓ increase mud thickness
- Demonstrated potential for use in high temperature environment

Research Collaboration Needs

- Compositional Analysis of naturally occurring multicomponent materials (Lignocellulose)
- Functionalization of these materials to enhance its relevant properties

KFUPM

Acknowledgement

- **College of Petroleum Engineering and Geosciences, King Fahd University Petroleum Minerals, Saudi Arabia**
- **Dr Ahmad A. Adewunmi, CIPR/CPG, KFUPM**
- **Dr Shahzad M. Kamal, CIPR/CPG, KFUPM**
- **Dr Hakeem Abbas, CENT, KFUPM**
- **Mr Musa O. Najimu, ChE, KFUPM**
- **Mr Mobeen Murtaza, CPG, KFUPM**
- **Mr Idrees Bakare, CENT, KFUPM**

KFUPM

This presentation should be cited as:

Adewole and Najimu (2017)

Referenced as:

Jimoh K. Adewole and Musa O. Najimu (2017), Oil Field Chemicals from Macromolecular Renewable Resources: A Case Study of Date Pit Nanoparticles for Drilling Fluid Formulation, AIChE Annual Meeting in Minneapolis, MN, USA, Oct 29 - Nov 3, 2017

KFUPM

EERC

GEOLOGIC STORAGE OF CARBON DIOXIDE IN THE CENTRAL PLAINS OF NORTH AMERICA

2017 AIChE Annual Meeting
Minneapolis, Minnesota
November 3, 2017

Neil Wildgust
Principal CCS Scientist

Critical Challenges. **Practical Solutions.**

PCOR PARTNERSHIP

- Region includes:
  - Nine states.
  - Four Canadian provinces.
  - Over 3.6 million km^2.
- Several completed field projects.
- Over 4 million tons of CO_2 stored and monitored in association with CO_2 enhanced oil recovery (EOR).
- More than 120 partners!

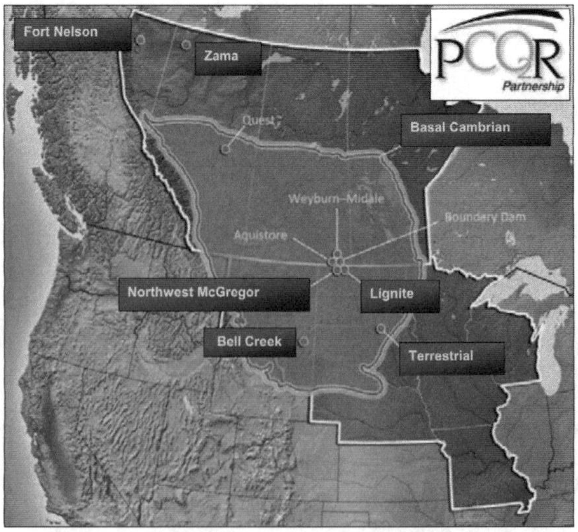

EERC

Critical Challenges. **Practical Solutions.**

WHERE WE BEGAN

Phase I (2003–2005)

- Source, sink, seal characterization
- Regional vision
- 21 topical reports, multiple fact sheets
- Decision Support System (DSS)

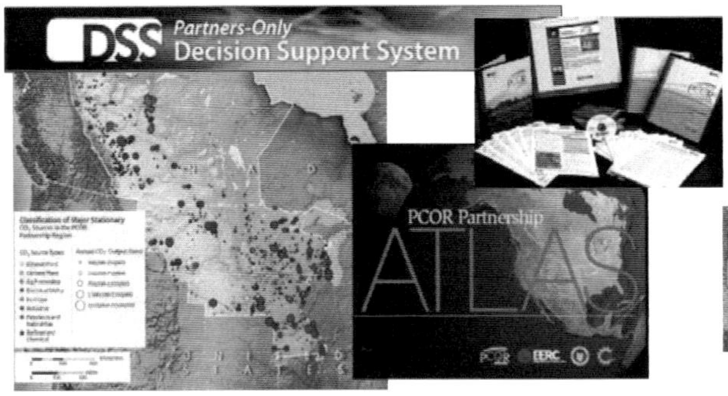

Phase II (2005–2009)

- Field validation

WHERE WE HAVE BEEN: COMPLETED PHASE III EFFORTS

Fort Nelson Carbon Capture and Storage (CCS) Feasibility Study

- CCS in a deep saline formation.

Zama Acid Gas EOR and Storage

- Operated by Apache Canada Ltd.
- Static and dynamic modeling activities completed.

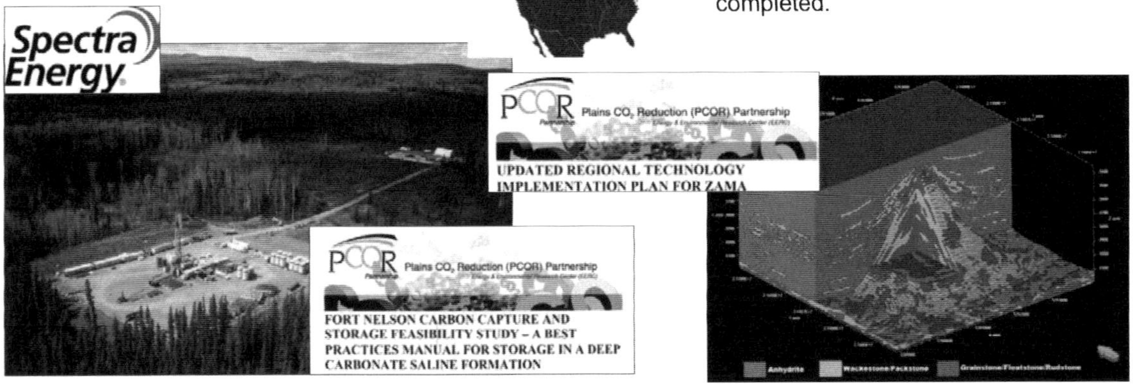

WHERE WE ARE NOW: AQUISTORE

- Injection of CO_2 from the Boundary Dam power plant began in April 2015.
- Serves as buffer storage for CO_2 that is not used for EOR.

WHERE WE ARE NOW: BELL CREEK

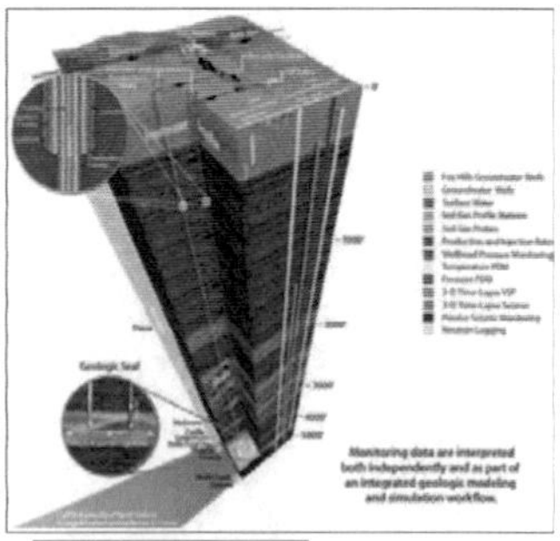

- CO_2 storage associated with a commercial CO_2 EOR project.
- Operated by Denbury Onshore LLC.
- CO_2 is sourced from ConocoPhillips' Lost Cabin natural gas-processing plant and Exxon's Shute Creek gas-processing plant.

Adaptive Management Approach

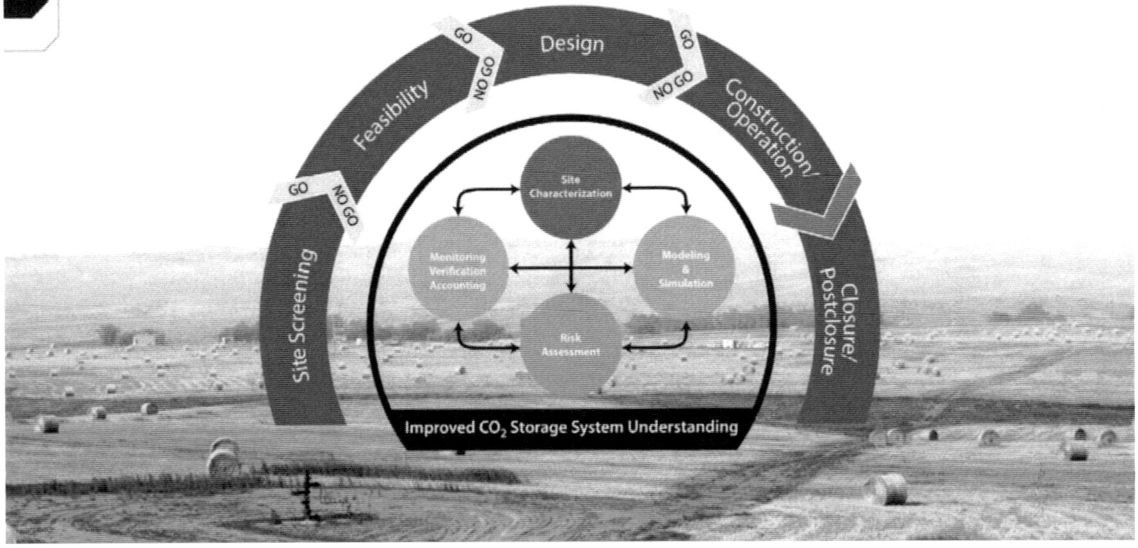

SIMULATION-GUIDED MONITORING, VERIFICATION, AND ACCOUNTING (MVA)

- Simulation was used to predict location and saturation of CO_2.
- 2-D seismic line used to confirm ability of seismic to detect CO_2 in the reservoir.
- Results supported decision to conduct large 3-D survey.

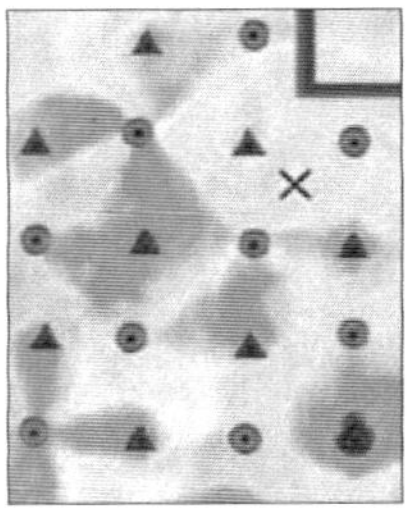

Predictive Simulation Results (CO_2 plumes)

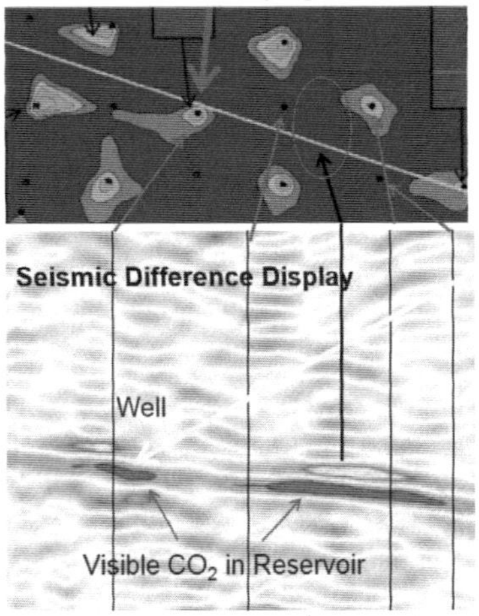

EERC

Critical Challenges. **Practical Solutions.**

First Repeat 4-D Difference Interpretation

1.Utilization of Resources
2.Technical Approach
3.Rate of Progress
4.Potential Risks Considered
5.Plan to Complete

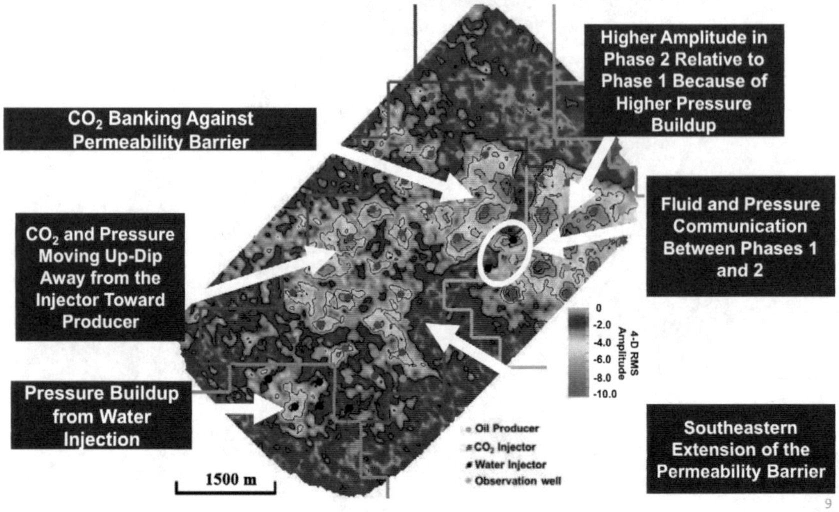

9

MVA-GUIDED SIMULATION

- Pressure and fluid communication revealed by 4-D seismic helped explain Phase 1 model history-matching issues.
- A combined Phase 1 and 2 was developed in response.

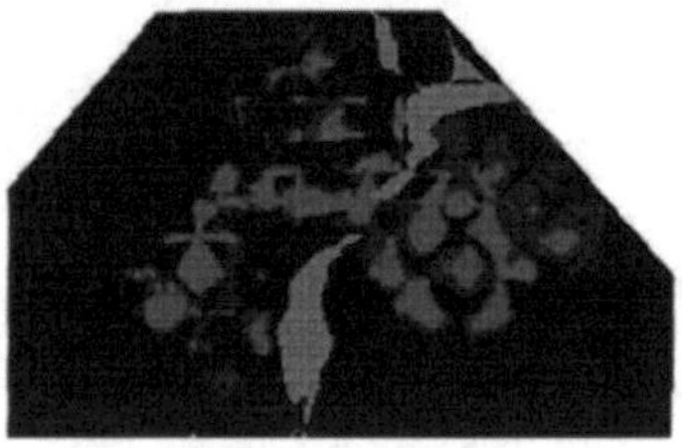

WAG, 1 HCPVI

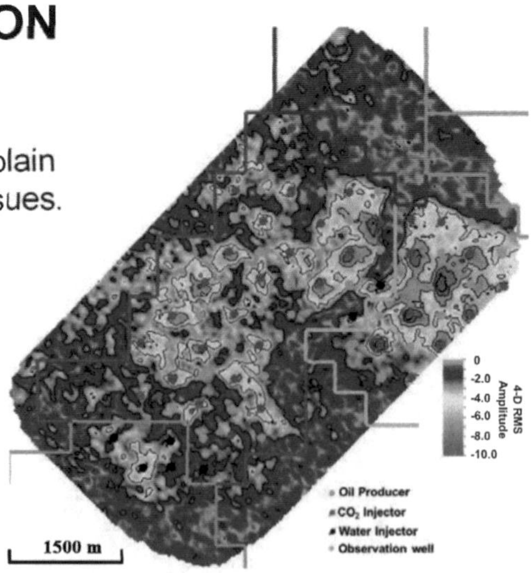

EERC

Critical Challenges. **Practical Solutions.**

LESSONS LEARNED/KEY FINDINGS

- MVA techniques are applicable to both EOR and saline storage in any region.
- The region has outstanding potential for storage opportunities.
- Adaptive management approach is readily applicable to future geologic CO_2 storage projects in this region and others.
- PCOR Partnership provided platform to test various techniques and technologies. A significant time gap can exist between data collection and analysis and integration.

11

MVA EVOLUTION

- Faster processing for quicker integration
  - Improve performance predictions
  - Inform operational decisions with actionable results
- Intelligent monitoring
- Low environmental impact
- No impact on operations
- Semiautonomous and scalable
- Viable and cost-effective long term

SASSA (SCALABLE, AUTOMATED, SEMIPERMANENT SEISMIC ARRAY)

KRAUKLIS WAVE (K-WAVE)

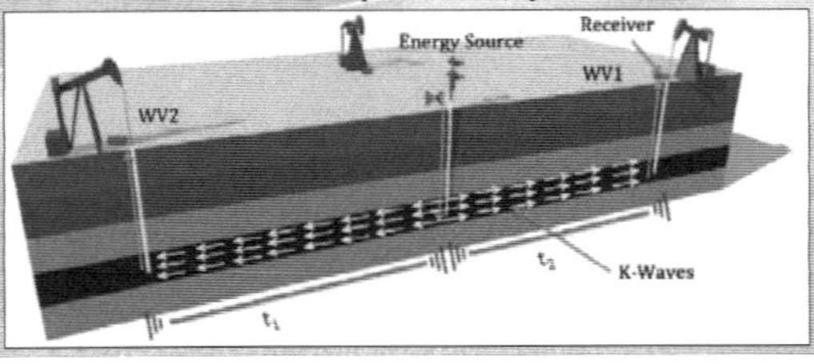

INTELLIGENT MONITORING SYSTEM (IMS)

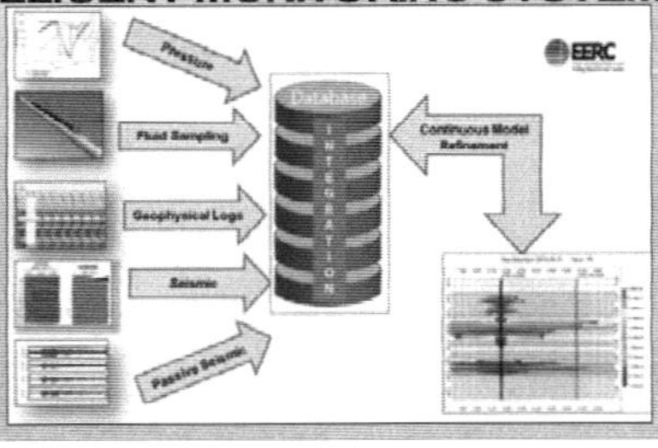

12

BRINE EXTRACTION AND STORAGE TEST (BEST)

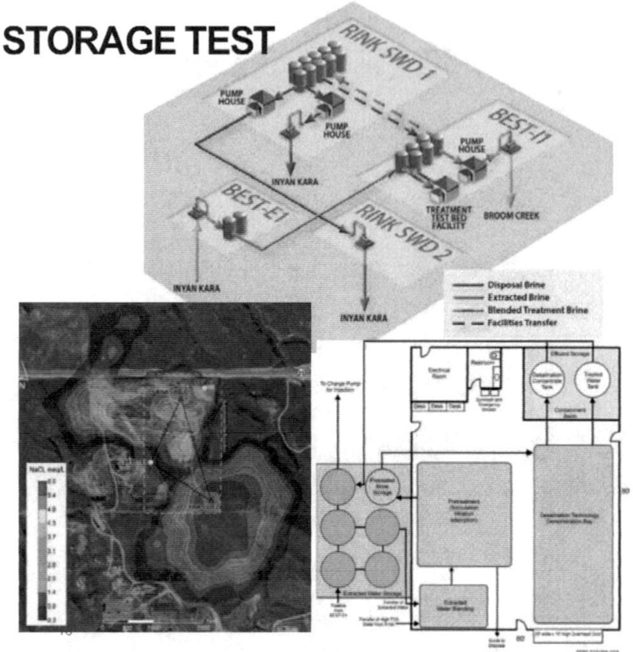

- Proxy for managing and measuring pressure and CO_2 plume movement associated with future saline CO_2 storage projects.
- **Phase I:** Development of a field implementation plan for a BEST in North Dakota (complete).
- **Phase II:** Field demonstration and validation of active reservoir management (ARM) strategies and extracted brine treatment technologies (under way).

EERC

ND INTEGRATED CARBON STORAGE COMPLEX FEASIBILITY STUDY (CARBONSAFE PHASE II)

Objective and Project Partners

- Determine the feasibility of developing a commercial-scale CO_2 storage complex able to store 50+ million metric tons of CO_2 in central North Dakota safely, permanently, and economically.

U.S. DEPARTMENT OF ENERGY | NATIONAL ENERGY TECHNOLOGY LABORATORY

ALLETE clean energy

INDUSTRIAL COMMISSION OF NORTH DAKOTA
LIGNITE RESEARCH COUNCIL

BNI energy
AN ALLETE COMPANY

BASIN ELECTRIC POWER COOPERATIVE
A Touchstone Energy® Cooperative

Prairie Public

Minnkota Power COOPERATIVE, INC.
Your Touchstone Energy® Partner

Schlumberger Carbon Services

CMG

NORTH AMERICAN COAL CORPORATION

EERC

Critical Challenges. Practical Solutions.

ND CARBONSAFE PROJECT SUMMARY

- DOE's CarbonSAFE* program aims to develop integrated CO_2 storage complexes, constructed and permitted for operation in the 2025 time frame over a series of development phases.

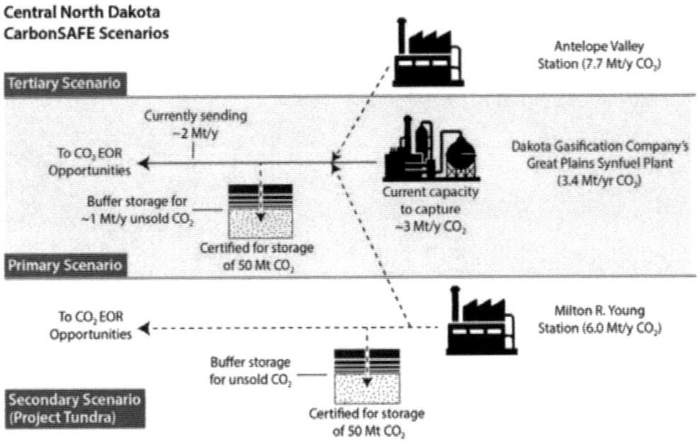

*CarbonSAFE = Carbon Storage Assurance and Facility Enterprise

EERC

Critical Challenges. **Practical Solutions.**

NEBRASKA INTEGRATED CARBON STORAGE CAPTURE PRE-FEASIBILITY STUDY (CARBONSAFE PHASE I)

- **Goal**: Determine conceptual feasibility of CO_2 capture from Nebraska Public Power District's Gerald Gentlemen Station (GGS) with geologic storage.
- **Objectives**
  - Establish CCS coordination team
  - Identify challenges to commercial-scale CCS in western Nebraska, and develop mitigation plan
  - Conduct conceptual evaluations:
    - ♦ Western Nebraska subsurface for geologic storage
    - ♦ CO_2 emissions from GGS for potential capture

U.S. DEPARTMENT OF ENERGY | NATIONAL ENERGY TECHNOLOGY LABORATORY

Nebraska Public Power District
Always there when you need us

Schlumberger Carbon Services

CMG COMPUTER MODELLING GROUP LTD

EERC

Critical Challenges. **Practical Solutions.**

CONTACT INFORMATION

Energy & Environmental Research Center
University of North Dakota
15 North 23rd Street, Stop 9018
Grand Forks, ND 58202-9018

www.undeerc.org
701.777.5193 (phone)
701.777.5181 (fax)

Neil Wildgust
Principal CCS Scientist
nwildgust@undeerc.org

EERC 17 Critical Challenges. **Practical Solutions.**

497196 Defining CO_2 Storage Options in the Upper Ohio River Valley – Advanced Characterization of Geologic Reservoirs and Caprocks Authors: Caitlin Mcneil (Presenter

Abstract:

Development of Carbon Dioxide Capture, Utilization, and Storage (CCUS) technologies represents one of the most important components in the portfolio of options for greenhouse gas reduction. The emerging international consensus under mechanisms such as the Paris Accord and the possibility of future federal or state regulations for limiting CO_2 emissions essentially rule out continued use of coal as a power source in the future without implementation of CCUS technology. The upper Ohio River Valley region including eastern Ohio has a strong history of coal production and a dependence on coal for its industrial base, which includes power generation and steel manufacturing. However, the potential reservoirs and containment zones for CO_2 storage in the area are very poorly understood due to lack of deep well data. Thus, knowledge of capacity, containment security, and injectivity is rudimentary at best and a sustained effort is needed to map the deep geology and quantify the storage potential in the region. This project attempts to start developing a CO_2 storage framework using pre-existing and new data on potential storage reservoirs and containment intervals to estimate CO_2 storage resources in eastern Ohio. The geologic data were compiled from oil and gas well logs and cores available at the Ohio Division of Geological Survey, purchase of seismic data, and through collaboration with brine-injection well operators in eastern Ohio to collect advanced well logs and operational data. The available information was used to construct cross-sections and maps of formations depths, thickness, and porosity across the study area. Based on these maps, the prospective storage resources were estimated. The porosity and limited available permeability data were used to develop preliminary numerical reservoir models in selected areas to evaluate the feasibility of sustained injection. Finally, caprocks were evaluated using geomechanical and reservoir modeling methods to determine their effectiveness for containing injected CO_2. The project findings were disseminated to the stakeholders at various venues.

The overall findings of the work show that there is significant CO_2 storage potential in eastern Ohio based on porosity estimates. This is also proven by the presence of a number of high-performance injection wells in the region. The storage resource is vertically diversified into a number of geologic layers, both carbonates and sandstones, and is laterally heterogeneous. This heterogeneous nature means that advanced characterization and sustained geologic characterization using advanced tools is needed to improve confidence in storage estimates and long-term injectivity. The assessment also indicates that the primary caprock for the Cambrian-Ordovician reservoirs in the area has appropriate characteristics to retain the injected CO_2. In this study, the focus of the caprock analysis was limited to assessing the sealing effectiveness and mechanical integrity of the two geologic units immediately overlying the reservoirs [i.e., immediately overlying the Knox Unconformity], the Black River Group and the Wells Creek Formation, even though there are hundreds to thousands of feet of additional low permeability limestone and shale (Trenton Formation, Point Pleasant/Utica) above these units. Thus, the caprock analysis is conservative because it doesn't take into consideration the entire caprock system.

This project has been conducted under grant agreement D-13-22 funded by Ohio Development Services Agency's Ohio Coal Development Office and the Midwestern Regional Carbon Sequestration Partnership (MRCSP) led by Battelle with US Department of Energy funding. The project also benefited significantly through collaboration with a number brine injection well operators, who allowed access to their wells for collection of critical geologic characterization and injection operation data. The present project complements a number of past and ongoing efforts to prove CO_2 storage and utilization options in Ohio and surrounding regions.

499289 Impact of Brine/CO_2 Exposure on the Transport and Mechanical Properties of Mt Simon Rock Samples

Abstract:

When sandstone rocks are exposed to CO_2-saturated brine their transport and mechanical properties can change due to brine/CO_2-induced chemical reactions. The present study investigates the change in the flow-through characteristics, porosity and the deformation of Mt Simon rock samples caused by exposure to brine/CO_2. The subsamples, extracted from the Mt. Simon sandstone, were first characterized for their mechanical and transport properties, and then aged in CO_2-saturated brine at pressures of 2500 psi and temperature of 50â°C for one week. Following that, tests were carried out to measure the change in transport and mechanical properties of the rock. The changes in porosity were measured by the helium porosimetry method. The alteration in pore size distribution and flow-through characteristics were determined by the flow permporometry method. The changes in the brine composition were measured by ion chromatography. In addition, a brine-saturated Mt. Simon sandstone core was aged in a specialized core holder while the deformation of the core during CO_2 invasion was detected in situ via an array of strain gauges.

498620 Assignment and Calibration of Relative Permeability By Hydrostratigraphic Units: A Novel Approach for Multiphase Flow Analyses; Case Study Example: CO_2-EOR Operations at the Farnsworth Unit, Texas

Abstract:

Among the most critical factors for geological CO_2 storage site screening, selection and operation is effective simulation of multiphase flow and transport. Relative permeability is probably the greatest source of potential uncertainty in multiphase flow simulation, second only to intrinsic permeability heterogeneity. The specific relative permeability relationship assigned greatly impacts forecasts of CO_2 trapping mechanisms, phase behavior, and long-term plume movement. A primary goal of this study is to evaluate the impacts and implications on CO_2-EOR model forecasts of different methods of assigning relative permeability relationships.

We applied a novel approach to assigning relative permeability relationships in numerical models. In previous research, most models assign relative permeability functions on the basis of geologic formation or rock type. In this study, we assign and calibrate relative permeability by hydrostratigraphic units, extending the seminal work of Maxey (1964) to multiphase flow. Ongoing work by the SWP has identified distinct regions that are believed to exhibit similar flow characteristics using the Improved Hydraulic Flow Units (IHFU) method. Core from each of these flow regions are being scanned by micro-computer-tomography (micro-CT) and relative permeability testing will be completed when the scans are finished. Specific, calibrated relative permeability relationships will be assigned to each hydrostratigraphic unit. Results of forward simulations with the newly-calibrated models will be compared to those of existing models that utilize a single relative permeability relationship for the whole model domain.

The study site is the Farnsworth Unit (FWU) in the northeast Texas Panhandle, an active CO_2-EOR operation. The target formation is the Morrow â€˜Bâ€™ Sandstone, a clastic formation composed of medium to course sands. The study was undertaken as part of the Southwest Regional Partnership on Carbon Sequestration (SWP) under the phase III project.

Development and Evaluation of an Iridium Oxide Based Chemical Sensor for Downhole CO_2 Monitoring—Part II

Sai Wang, University of North Dakota--Department of Petroleum Engineering;

Kegang Ling, University of North Dakota--Department of Petroleum Engineering;

Hongsheng Wang, University of North Dakota--Department of Petroleum Engineering;

Ning Liu, University of Louisiana at Lafayette—Department of Petroleum Engineering;

Yun Ji, University of North Dakota--Department of Chemical Engineering

Abstract

Geological carbon sequestration represents a long-term storage of CO_2, in which large-scale CO_2 is injected into the subsurface geologic formations. The in-situ CO_2 monitoring approach is highly desirable during the storage process. Compare with other monitoring technologies, a preferred method for in-situ underground CO_2 migration monitoring is to install down-hole CO_2 sensors directly in the underground. This paper presents the design and construction process of the iridium oxide based chemical sensor, including the IrO_2 working electrode electrodeposition with iridium oxide nanoparticles, Ag/AgCl reference electrode fabrication and the Severinghaus-type CO_2 sensor preparation. Then the sensor was tested under high pressure environment. Under this condition, the sensor displayed a very good CO_2 sensing performance and a linear relationship was obtained between the sensor's response potential and the Log[C_{CO2}]. In order to evaluate the CO_2 sensor performance in the carbon sequestration process, the sensor was settled in the simulated CO_2 storage condition. The results indicated that this chemical sensor could provide the CO_2 movement monitoring data continuously, timely and accurately.

Introduction

In recent decades, the CO_2 geologic storage has been considered as a feasible means for reducing anthropogenic CO_2[1, 2]. The fundamental logic of the geological storage of CO_2 is to inject the supercritical CO_2 at the storage depths, which need the CO_2 to keep on either supercritical or in solution in the water/brine present in the formation. The injected CO_2 is expected to stay in the subsurface formations for a long period. However, there are aspects of geologic storage which need to take further study, particularly in regard to the public concerns of CO_2 migration and possible leakage to surface. Since then, an in-situ CO_2 monitoring approach is highly desirable for this purpose. At present, a broad range of technologies based on geophysical, microbiological, geoelectrical and geochemical principle were exploited [3]. While due to the various disadvantages of those technologies, such as the largely time consuming or costly requirement, they were not identified as the potential methods for the application.

Because the chemical reaction between CO_2 and H_2O are pH dependent, which is transforming the CO_2 detection to the pH measurement, a preferred method of monitoring in-situ CO_2 movement and migration is to directly install the CO_2 sensors at storage depth[4]. The sensor must hold the ability to withstand harsh environmental conditions especially under high pressure circumstance. This paper describes the design and construction process of the iridium oxide based chemical sensor which fulfill the harsh environmental condition requirements. Differed from the traditional sensor, this type of the CO_2 sensor used an iridium-oxide electrode instead of a glass electrode to improve the resistant ability for the harsh downhole environments encountered. The developed CO_2 sensor includes a couple of iridium-oxide electrodes (To meet this purpose, iridium oxide nanoparticles was electrodeposited on the surface of gold substrate), a thin gas-permeable silicone membrane, a porous metal supporting material, and the bicarbonate-based electrolyte solution. With conducting several tests in different scenarios, the sensor was proved as a potential technology which can provide the CO_2 movement monitoring data continuously, timely and accurately.

Experimental

1. Sensor Preparation

1.1 Electrode Fabrication

To prepare electrodeposition solution, alkaline hydrolysis of iridium(IV) salt method was used to synthesize iridium oxide nanoparticles [5, 6]. 2.0 mM aqueous K_2IrCl_6 solution was adjusted to pH 13 with 10 wt% of aqueous NaOH to obtain a yellow solution, which then was heated at 90 °C for 20 minutes. The resulting solution was kept in an ice bath to obtain a blue solution. The cold solution was then adjusted to pH 1 by rapidly adding 3 M HNO_3 and stirred continuously for 80 minutes until the solution became deep blue. The solution was then stored in a refrigerator at 2 °C and could be adjusted to pH 1-13 by addition of 10 wt% of aqueous NaOH solution.

Then the as-prepared working electrode is electrochemically deposited with iridium oxide nanoparticles as the deposition solutions. The electrodeposition protocol was carried with cyclic voltammetry[7]. The experiments were carried out with Solartron Analytical 1470E CellTest System with Solartron Analytical 1242A frequency response analyzer. In this work, a three-electrode cell system was used where an iridium oxide layer was deposited from the deposition solution on the metal substrate working electrode by using an Ag|AgCl and Pt as reference and counter electrode, respectively. A triangular wave cyclic voltammetry method between -0.5 to 0.65 V (vs. Ag|AgCl) was applied at 200 mV/s of scan rate for 500 cycles.

Next, an electrochemical method was applied for the reference electrode preparation[8]. The bare silver wire was electroplated in 0.1-M HCl electrolyte solution to form a thin coating of silver chloride on the silver wire surface. Along this route the silver wire was used as the anode while the bare copper wire was treated as the cathode. Finally, this silver/silver chloride electrode was insulated by the plastic heat shrank tubing.

1.2 Sensor construction

For the sensor applied under high pressure conditions and simulated reservoir conditions, the construction procedure was described as follows. Firstly, the working electrode and reference electrode were fixed through a stainless steel cap and then stabilized by silicone sealant. The stabilized electrodes were dried for two days under air condition at room temperature (25 ± 2°C). Then a stain steel porous filter element was settled for the function of supporting the whole frame, which means that it can enable the sensor to work under harsh downhole environment, especially under high pressure condition. What followed is the above stabilized electrodes were inserted into the support unit where inside being full of bicarbonate-based internal electrolyte solution ($NaHCO_3$-NaCl electrolyte solution). Finally, as the above mentioned sealing method, both outlets of the gas-permeable membrane were blocked off by the silicone sealant. Meanwhile, some other tiny materials were applied in the sensor manufacture for the purpose of support and potential transfer.

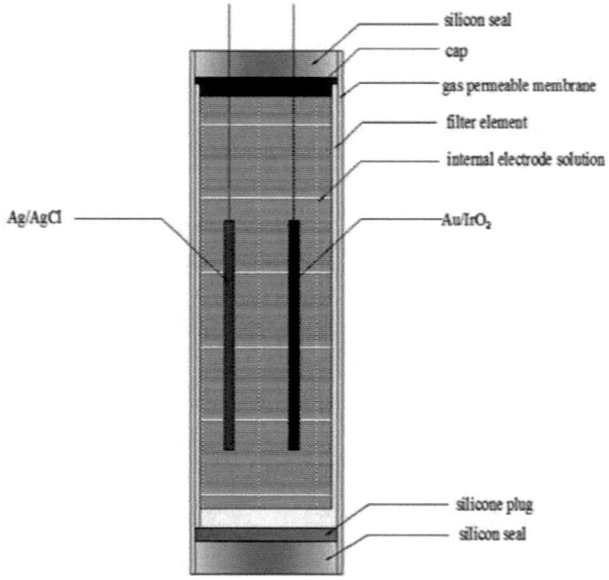

Fig. 1 Schematic of the CO_2 sensor[9]

2. *Sensor test under high pressure condition(2000psi)*

The following steps were designed to investigate the CO_2 sensor's performance. All potentiometric measurements were carried out at room temperature of 20°C±2°C. Firstly, the CO_2 sensor was settled in the stainless steel tank. Then the tank was injected with 0.2M citric acid and tightly sealed. The stainless steel tubing and the connecting wires from the sensor came out from the top of the tank. The ISCO pump was used for controlling the pump injection and the tank pressure. The connecting wires of the sensor were connected with a 34401A 61/2 digit multi-meter. Before carrying out the tests, the base potential of the CO_2 sensor was measured by maintaining the CO_2 sensor into the citric acid solution under a certain pressure. After the potential reached a stable value, the certain amount of the $NaHCO_3$ solution (0.1M) was injected in the tank and then the tank was sealed immediately. Finally, the ESCO pump began to increase the tank pressure to 2000 psi. The sensor output signal was measured by the digit multi-meter.

3. *Sensor performance under simulated reservoir conditions*

Lab-scale core flooding test was conducted to evaluate the performance of the CO_2 chemical sensor in CO_2 geological storage process. A sandstone core was applied to simulate the storage process of CO_2 in a sandstone saline aquifer. After the sensor was installed in the core, 2.0% brine was pumped into the core with flow rate of 100 ml/h until 1 pore volume was achieved. The injection pressure was around 2000 psi. Then CO_2/brine was injected into the core with the phase ratio of CO_2/brine = 1:1 and total flowrate = 100 ml/h. The response of the CO_2 sensor during this process was recorded.

Results and Discussion

1. Working Principle of the CO_2 sensor

The pH is a critical parameter need to be evaluated in the CO_2 sequestration process. Generally, the pH is defined as the degree of acidity or basicity of the solution. In its more common explanation, the pH was defined as the minus logarithm of the concentration of H^+ [10], which expressed as Eq.1.

$$\mathbf{pH = -log[H^+]} \tag{1}$$

During the CO_2 injection process, some CO_2 molecules could dissolve into the brine aquifer and change the pH from neutral to acid dramatically. Eq.2 describes the hydrolysis of CO_2. The CO_2 molecules react with H_2O molecules to generate carbonic acid. The carbonic acid is not stable and then dissociates into bicarbonate and H^+ subsequently. As a result, the produced H^+ will decrease the pH of the surrounding circumstance. From Eq. 2, the concentration of CO_2 is proportional to the concentration of hydrogen ion.

$$\mathbf{CO_{2\,(aq)} + H_2O \leftrightarrow H_2CO_3 \leftrightarrow H^+ + HCO_3^-} \tag{2}$$

From Eq. 2, the concentration of CO_2 is proportional to the concentration of hydrogen ion. Where the transformed equation was generated as Eq.3,

$$\mathbf{[CO_2] = k\,[H^+]} \tag{3}$$

The pH measurement uses the electrodes to transfer the chemical activity of the H^+ into the electronic signal, which is the potential differential. Inevitably the well-known Nernstian function [11] then was substituted into the pH measurement. This equation could be described as:

$$\mathbf{E = E^0 + (2.303RT/F)\log[H^+] = E^0 + 59.15\log[H^+]} \tag{4}$$

Where,

E^0 =Standard electrode potential	R = Universal gas constant, 8.314 472 J /K/mol
T = Absolute temperature	F = Faraday constant

By combining Eq. 3 and Eq.4, the potential differential between the electrodes can be described as Eq. 5. This generated equation reveals that there is a linear relationship between ΔE and the logarithm of the CO_2 concentration [12].

$$\Delta E = E - E^0 = 2.303RT/F * \log[CO_2] \qquad [5]$$

2. ***Potential response characteristics of the CO_2 sensor at 2,000 psi***

Fig. 2 shows the potential differential curves for the CO_2 sensors to changes of the CO_2 concentration from 0.001 M to 0.006 M at room temperature and 2000 psi. At the pressure of 2000 psi, the CO_2 sensor showed very similar sensing behavior except for the initial stabilized potential response. It can be found that the potential differential is directly proportional to the CO_2 concentration. The response time of the CO_2 sensor was found to be dependent on the CO_2 concentration. The response time was 6295 seconds for the highest CO_2 concentration of 0.006 M, while a little faster response was observed at low concentration ranged from 0.001 M to 0.004 M.

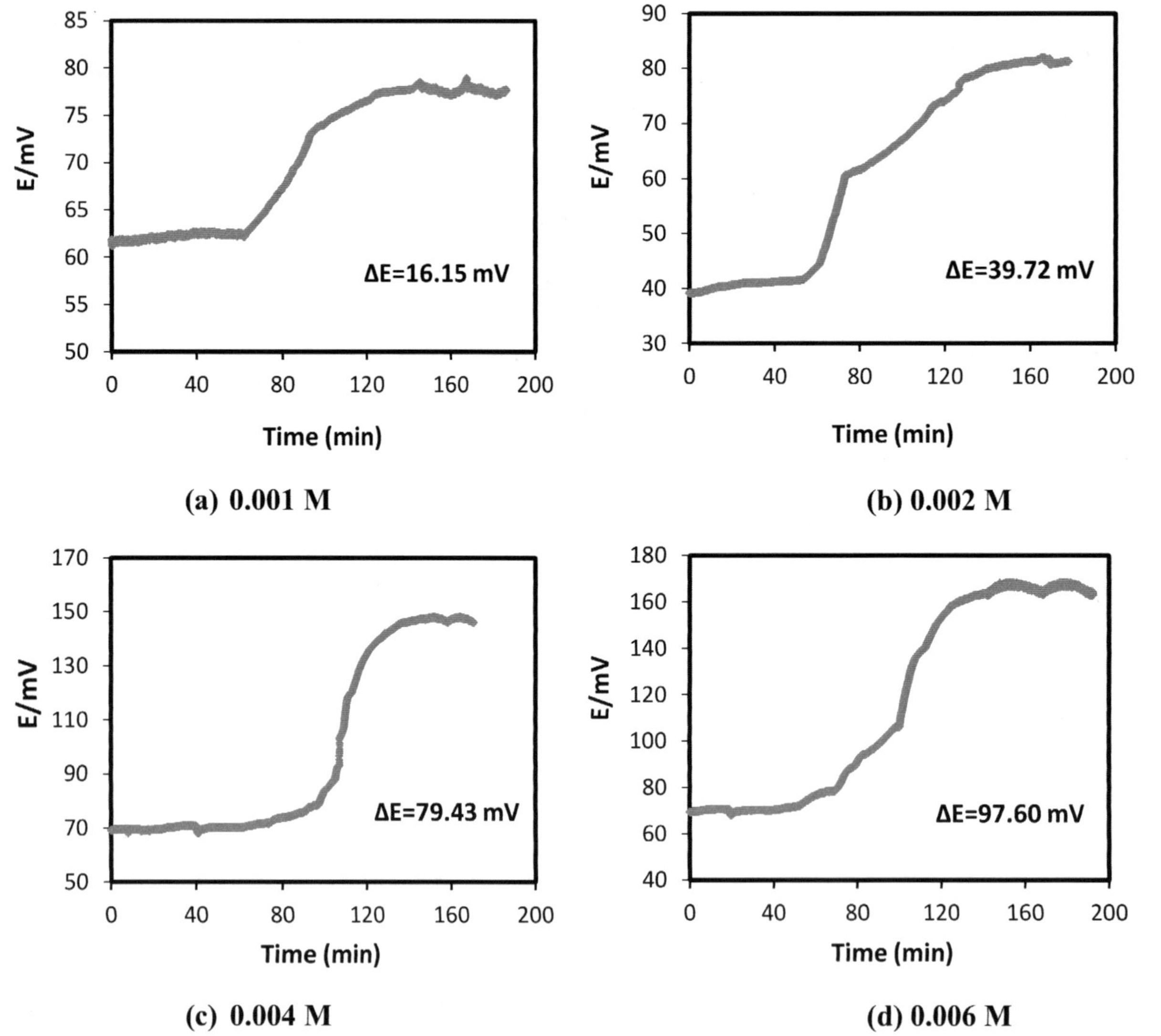

Fig. 2 Potential response of CO_2 sensor with the addition of $NaHCO_3$ in the 0.2 M citric acid-HCl solution at 2,000 psi, [$NaHCO_3$] is (a) 0.001M, (b) 0.002 M, (c) 0.004 M and (d) 0.006 M.

Fig. 3 illustrates the sensitivity curves of the CO_2 sensor at room temperature and 2000 psi. The linear relationship between the sensor response potential differential and the logarithm of the CO_2 concentration are quite good in the tested range of 0.001M to 0.006M. The slope of the CO_2 sensor versus Log [CO_2] is 107.5 mV/Log [CO_2], and the correction coefficient is 0.9889.

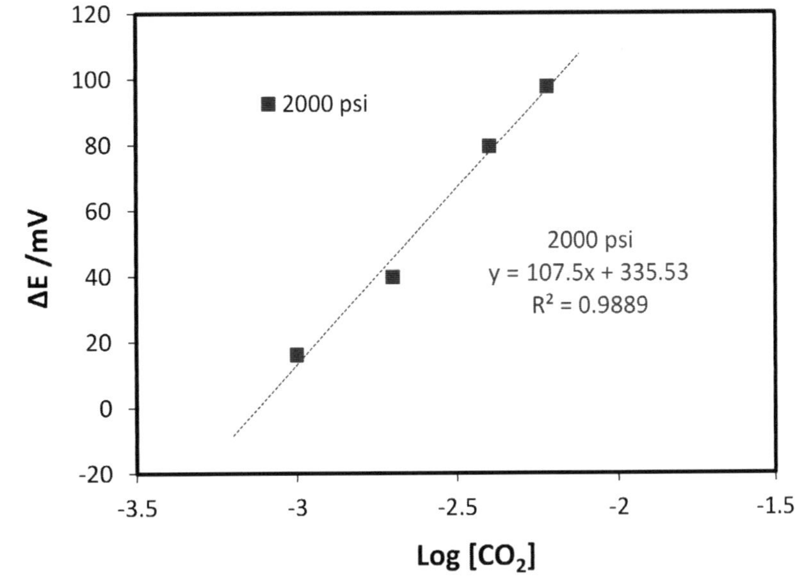

Fig. 3 Calibration curves of the CO_2 sensor tested at room temperature and 2,000 psi.

3. CO_2 sensor performance under simulated reservoir conditions

The results demonstrate that, with one pore volume of 2.0% brine was pumped into the flooding system, the output signal of the sensor was keep constantly around 42 mV. Then, when CO_2 started to inject into the core, the output potential was increased to 241 mV. At the beginning, the output potential of the sensor kept roughly constant before the CO_2 plume arrived. When the injected CO_2 arrived the sensor's location, due to the CO_2 molecule diffusion, the potential started to increase until reach the equilibrium[9, 13]. The output potential change of the sensor is 200 mV.

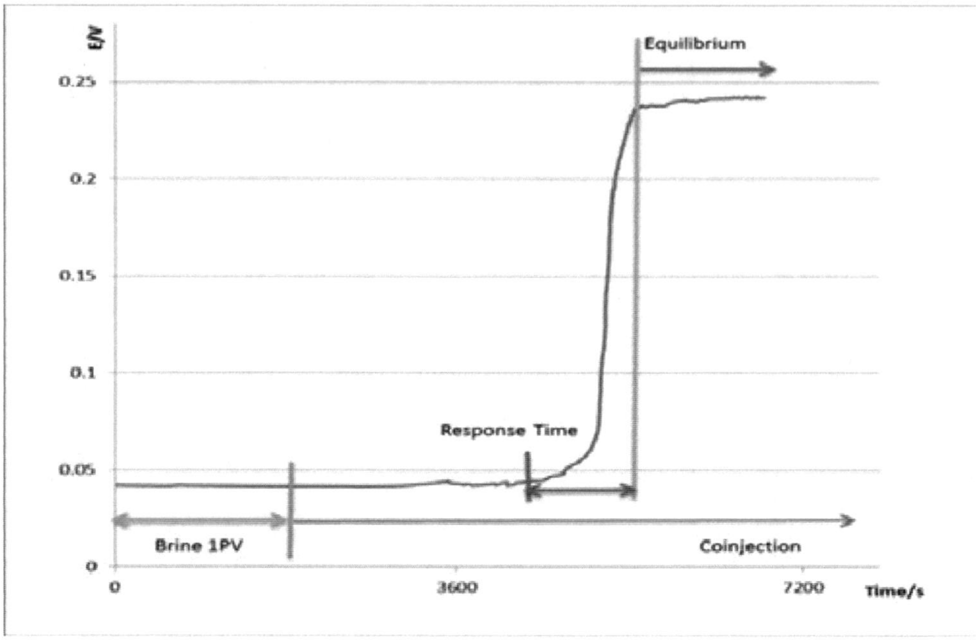

Fig. 4 Potential response of CO_2 sensor vs. Time. Brine injection rate: 100mL/h; Co-injection rate: 100 mL/h

Conclusions

An iridium oxide-based electrode was prepared by electrodeposition of the iridium oxide nanoparticles on a metal substrate with cyclic voltammetry. Then the prepared electrode was combined to develop a downhole CO_2 sensor which can provided the sensing performance under high pressure condition. With displaying Nernst performance, the sensor performed a good linear relationship between the sensor response potential and the logarithm of the CO_2 concentration. Lab-scale core flooding test demonstrated that the CO_2 sensor can measure the CO_2 movement under the reservoir condition, especially for the high pressure conditions. Also, the results indicated that the Severinghaus-type CO_2 chemical sensor may reserve the potential application in the geological carbon storage.

Acknowledgement

The authors are grateful to The Petroleum Engineering Department of University of North Dakota. This research is supported in part by the funding from Vice President for Research & Economic Development program.

Reference

[1] Allen, D.E., Strazisar, B.R., Soong, Y., Hedges, S.W., Modeling carbon dioxide sequestration in saline aquifers: significance of elevated pressures and salinities. Fuel Process. Technol. 86, 1569–1580. 2005.

[2] Wollenweber, J., Alles, S., Busch, A., Krooss, B.M., Stanjek, H., Littke, R., Experimental investigation of the CO_2 sealing efficiency of caprocks. Int. J. Green h. Gas Control 4, 231–241.2010.

[3] Giese R., Henniges J., Lüth S., Morozova D., Schmidt-Hattenberger C., Würdemann H., Zimmer M., Cosma C., Juhlin C., Monitoring at the CO_2 Sink Site: A Concept Integrating Geophysics, Geochemistry and Microbiology. Energy Procedia, 1 (1): 2251–2259, 2009.

[4] Ning L, Yang. L., Li L.X., Grigg R., McPherson B., Robert. L., Development of A CO_2 Chemical Sensor and Sensor Strategy for CO_2 Monitoring in Carbon Sequestration. Seventh Annual Carbon Capture and Sequestration Conference, 2008, Pittsburgh, PA.

[5] Zhao, Y., Vargas-Basbosa, N.M., Hernandez-Pagan, E.A., Mallouk, T.E., Anodic deposition of colloidal iridium oxide thin films from hexahydroxyiridate(IV) solutions. Small, 14, 2087-2093. 2011a.

[6] Zhao, Y., Hernandez-Pagan, E.A., Vargas-Barbosa, N.M., Dysart, J.L., Mallouk, T.E., A high yield synthesis of ligand-free iridium oxide nanoparticles with high electrocatalytic activity. Journal of Physical Chemistry Letters, 2, 402-406. 2011b.

[7] Munawar Khalil, Sai Wang, Jianjia Yu, Robert L. Lee, and Ning Liu, Electrodeposition of Iridium Oxide Nanoparticles for pH Sensing Electrodes. Journal of the Electrochemical Society, 163 (9) B485-B490 (2016).

[8] Wen-Ding H., Hung C., Sanchali D., Mu C., A Flexible pH Sensor Based on the Iridium Oxide Sensing Film. Sensors and Actuators A, 169: 1– 11, 2011.

[9] Sai W., Kegang L., Ning L., Hongsheng W., Development and Evaluation of an Iridium Oxide Based Chemical Sensor for Downhole CO_2 Monitoring--Part I, Carbon Management Technology Conference, 17-20 July 2017. Houston, Texas, USA,.

[10] Skoog D.A., West D.M., Holler F.J., Fundamentals of Analytical Chemistry, 7th ed., Philadelphia, PA: Saunders College Publishing, 1996.

[11] Orna M. V., Stock J., Electrochemistry, Past and Present. Columbus, Ohio: American Chemical Society, 1989. ISBN: 0-8412-1572-3.

[12] Sai Wang, Jingshan San, Ning Liu, A Chemical CO_2 Sensor Monitoring CO_2 Movement Under Reservoir Conditions, Carbon Management Technology Conference, 17-19 Nov,2015, Sugarland, Texas, USA

[13] Sai Wang, Jingshan San, Jianjia Yu, Robert Lee, Ning Liu, A Downhole CO_2 Sensor to Monitor CO_2 Movement In-situ for Geologic Carbon Storage, Internal Journal of Greenhouse Gas Control. 2016.

498673 Monitoring Carbon Sequestration Using Charged Wellbore Controlled Source Electromagnetics and Integrated Reservoir Models

Abstract:

Carbon capture and sequestration (CCS) can provide an effective means by which to reduce global CO2 levels. To ensure effectiveness and safety, regulatory frameworks mandate 99% storage permanence over the lifetime of a carbon sequestration project. As such, robust monitoring, verification, and accounting (MVA) technologies are needed in order for large scale CCS to be realized. Many existing MVA measurements require the costly drilling of boreholes and installation of downhole sensors. Borehole data provide minimal spatial information and their expense is prohibitive against the installation of dense monitoring grids. Surface based measurements can detect massive leaks but do not respond to changing reservoir conditions

Geophysical imaging techniques could theoretically provide this needed monitoring information if sufficient physical property change can be attributed to CO2 injection. The density of supercritical CO2 is a function of reservoir conditions, but is often similar to brine. As such gravity and 4D seismic monitoring will often fail to detect appreciable signal variation with injection activities. Seismic attributes sensitive to subsurface pressure changes can be utilized, but the expense of repeated seismic data acquisition is prohibitive for long-term monitoring. In most cases, CO2 is injected into deep saline aquifers or depleted oil and gas reservoirs. The supercritical CO2 will be much less electrically conductive than the brine which it displaces and mixes with. Therefore, electrical or electromagnetic methods, which are sensitive to electrical conductivity, are a promising techniques for monitoring CCS activities.

Many options exist for electrical survey configurations. The desired characteristics for this application are low cost—requiring minimal new infrastructure—and sufficient depth of investigation. In offshore exploration controlled source electromagnetics (CSEM) are increasingly relied upon as a de-risking tool. In this case the hydrocarbons are the resistive target within the conductive brine. We leverage this experience by applying CSEM to the monitoring of an active onshore enhanced oil recovery carbon capture and sequestration project (EOR-CCS) project by utilizing well casings as deep-reaching transmitting electrodes. Electrical current can be injected into the formation of interest by utilizing legacy wellbore casings that penetrate the injection reservoir as long deep electrodes. By monitoring electrical and magnetic fields on the surface, information about saturation changes in the subsurface can be determined through inversion and interpretation of the data. Two major advantages of the method is the ease of access on the surface and the ability to inject current to the depth of reservoir formation.

In order to improve the resolution and reliability of the quantitative interpretation, we develop a methodology for tightly integrated multiphase reservoir flow simulations with the geophysical data inversion. Electrical conductivity can be related to phase saturation using relations such as Archie's Equation, which will be calibrated and verified using auxiliary data including production and injection records. This multiphysical approach is designed to integrate structural, physical property, and reservoir simulations systematically. Structural information is retained from the reservoir models and associated characterization efforts, and incorporated into the inversion of the CSEM data as geometrical constraints. Furthermore, physical property information derived from petrophysical data, reservoir porosity, and reservoir simulation-produced fluid saturations will be incorporated into the inversion as direction constraints.

In this paper, we will present the preliminary study on the site selection, numerical simulations consisting of reservoir simulations, generation of electrical conductivity models, and electromagnetic-modeling based EM survey design, and preliminary analyses of data acquired in the baseline field survey.

488989 Development of LIBS Sensor for Sub-Surface CO_2 Leak Detection in Carbon Sequestration

Abstract:

To ensure the success of USDOEâ€™s CCS program it is important that the carbon dioxide that is injected underground for storage remains there permanently (99% permanence over 1000 years). In fact, the integrity of the storage can be ensured by monitoring for CO_2 leakage, and injection sites are the logical places to perform such monitoring. The sub-surface environment is notoriously harsh, with large potential mechanical, thermal, and chemical stresses, making long-term stability and survival a challenge to any potential in situ monitoring method. The laser induced breakdown spectroscopy (LIBS) has been demonstrated as a promising technology for chemical monitoring of harsh environments and hard to reach places. LIBS has a real-time monitoring capability and can be used for the elemental and isotopic analysis of solid, liquid, and gas samples. The flexibility of the probe design and the use of fiber-optics has made LIBS particularly suited for remote measurements. Our experiments using a stainless steel high-pressure cell demonstrated that the LIBS is a suitable technique to perform measurements in high-temperature and high-pressure (HTHP) environment, which makes it attractive for downhole measurements. The discussion will include the progress in fabricating a compact, robust, and simple LIBS sensor and distribution of many laser spark sources across a wide area for widespread subsurface leak detection.

487678 Study on Mechanism of Clinker Formation in Bamboo Combustion Process

Abstract:

In this study, we burn the bamboo powder packed layer placed on stainless steel mesh to know how the clinker is made and also we examined the possibility of clinker reduction by adding bark powder to bamboo powder. We carried out the mass change of the mesh after burning, mesh surface observation and also measured the adsorption amount and size of clinker. The following results were obtained.As a result, In the process of clinker formation, two stages were confirmed. At stage 1, about 10 Î¼m of fine white particles adsorbed onto the stainless steel mesh at the volatile component combustion of about 1000 K. In step 2, it was suggested that during combustion of char having a surface temperature of about 1200 K, these small grains melted and turned into large chunks of about 200 Î¼m.Furthermore, it was found that clinker clogging on the surface of the stainless steel mesh is reduced by mixing bamboo powder with bark and burning it. This proved that bark contains CaO with high melting point.

493264 Modeling Pyrolysis-Induced Microstructural Changes in Biomass: A Cellular Automata Approach

Abstract:

Lignocellulosic biomass is a chemically and morphologically heterogeneous material. Furthermore, this heterogeneity is in-part responsible for the vast number of thermal decomposition products seen in pyrolysis events. While modeling of biomass pyrolysis has been a subject of much research in past years at length-scales from the macro to molecular, the majority of these works have focused on a range of continuum-based approaches. Though effective at capturing global outcomes, these approaches are less tractable as frameworks for capturing microstructural effects and upscaling molecular information. This work demonstrates the use of kinetic cellular automata (kCA) as an alternative platform for the modeling and simulation of biomass pyrolysis. Asides from being effective at capturing transport and chemical processes in highly heterogeneous system, kCA is capable of modeling microstructural changes that occur as a result of chemical and physical transformations. A number of earlier benchmark trials demonstrated the convergence of the kCA to global continuum outcomes. Application of the kCA to actual biomass microstructures show promise for this platform as an intermediate length-scale tool capable of predicting char morphologies that mimic experimental outcomes at meso-scale resolution.

487672 Shrinking and Heat and Mass Transfer during Pyrolysis of Woody Biomass

Abstract:

Heat and mass transfer during pyrolysis of major components bed have been investigated numerically. For Cellulose powder, the calculation temperature and velocity vector distributions had a quasi-steady state during pyrolysis because the gas volume increased near the side wall of the packed bed and then the effective thermal conductivity decreased. The calculation temperature with the effect of volume shrinking of the packed bed agreed qualitatively with the experimental ones for all kinds of powder when the temperature plateau appeared.

493123 Plasma Gasification of Wood Pellets Using a Laboratory Scale Plasma Reactor to Produce Synthesis Gas: Effect of Increasing Temperature and Addition of Oxygen to the Process

Abstract:

A laboratory scale plasma reactor (max 15Kw power) has been used to convert wood pellets (biomass) into synthesis gas that can be used for downstream processes. A series of experiments have been conducted to see the effects of increasing temperature and adding oxygen to the process on conversion of hydrogen and carbon monoxide. The results show that increase in temperature from 400°C to 1000°C results in overall increase in biomass conversion (mass %) from 30% to 82%. In the case where no additional oxygen is supplied, the carbon and hydrogen conversion increases with increase in temperature up to 82% and 98% respectively but the molar ratio of carbon monoxide and hydrogen is unchanged at 1:1 for the temperatures studied. Addition of controlled oxygen increases the carbon monoxide yield and also results in the CO/H2 ratio of 3:2 for temperature range 800°C -1200°C. Simultaneous increase in reactor temperature and oxygen amounts to the process resulted in increased biomass conversion and a reduction of residues in the reactor.

497756 Multistage Fluidized Bed Reactor for Gasification

Abstract:

An ideal gasification process for a solid organic feedstock entails direct conversion of the feedstock into a synthesis gas, primarily CO and H_2, without any volatile or char byproducts. These byproducts coat the internal surfaces of the gasifier, rapidly degrading the performance of the gasifier and decreasing the yield of synthesis gas.

A three stage fluidized bed reactor has been developed to produce a clean synthesis gas in high yield. High quality synthesis gas can be used as a fuel for a combustion turbine or as feedstock for a chemical refinery. The first stage is the fluidized bed reactor where both the feedstock and heat transfer particles are fluidized in order to quickly heat the feedstock to a high temperature to achieve rapid pyrolysis and gasification. The second stage has a larger cross-sectional area than the first stage. This increase in area minimizes entrainment of solids in the second stage and increases the residence time to gasify char and decompose volatiles into low molecular weight gases. Typically the second stage operates at a high temperature in order to minimize condensation. The diameter of the third stage is greater than the diameter of the second stage in order to permit more char gasification and volatiles decomposition. If most of the volatiles have decomposed before entering stage three then the gases can be partially cooled in stage three. Following stage three the product stream passes through a cyclone to remove some of the entrained solids. Then the gases pass through a condenser to recover residual volatiles. Following the condenser gas samples are captured periodically for chemical analysis.

In the initial trials using this reactor, pine powder was used as the feedstock and alumina particles were used as the heat transfer medium. The temperature profile was set at 600 ^{o}C for stage 1, 800 ^{o}C for stage 2 and 300 ^{o}C in stage 3. Steady state operation was not achieved due to tar deposits on the alumina particles and the reactor walls. A close approach to steady state has been achieved by switching from alumina to zirconia particles and setting each stage at 800 ^{o}C. CO_2 was used as the fluidizing gas. Steady state was maintained for about 100 minutes and then the reactor was shut down. At this high temperature less fouling was apparent and the corresponding carbon yield in the gas fraction was about 74 %.

495901 Overall View of Lignocellulosic Biomass Decomposition in Supercritical Water

Abstract:

Supercritical water gasification is a promising technology for gasifying wet biomass. However, detail reaction kinetics of each component, cellulose, hemicellulose, and lignin have not been elucidated, and the interaction between the components has to be elucidated. In this study, mixture of cellulose, hemicellulose, and/or lignin was delivered into a continous reactor for supercritical water gasification and its reaction characteristics were determined. Reaction temperature was changed around critical point of water so that both subcritical and supercritical conditions were considered while the reaction pressure was kept at 25 MPa. Intermediate compounds were analyzed by HPLC, and comparison between avaraged single copound result and mixture was made. The interaction was clearly observed, especially in terms of hydrogen consumption. Reaction network was develped, and reaction rate constants were deteremined for each of the reaction by least square method to fit the experimental data with the calculation by the model. The Arrhenisc plot for these reactions classified ionic reaction and radical reaction. For the case of radical reacitons, Arrhenius plot showed a straight line as usual, but for the case of ionic reactions, reaction rate kinetics deviated from the straight line at high temperature, indicating ions get unstable under supercritical condition. The overall reaction kinetics was also determined, and discussed.

AUTHOR INDEX

AUTHOR INDEX

AUTHOR INDEX

AUTHOR INDEX

AUTHOR INDEX

AUTHOR INDEX

AUTHOR INDEX

AUTHOR INDEX

AUTHOR INDEX

AUTHOR INDEX

AUTHOR INDEX

AUTHOR INDEX

AUTHOR INDEX

AUTHOR INDEX

AUTHOR INDEX

AUTHOR INDEX

AUTHOR INDEX

AUTHOR INDEX